KU-538-973

H. Zollner

Handbook of Enzyme Inhibitors

WILEY-VCH

Helmward Zollner

Handbook of Enzyme Inhibitors

3rd, revised and enlarged edition

Part B

J–Z

WILEY-VCH

Weinheim · New York · Chichester · Brisbane · Singapore · Toronto

Prof. Dr. Helmward Zollner
Institut für Biochemie
Karl-Franzens-Universität Graz
Schubertstrasse 1
A-8010 Graz
Austria

This book was carefully produced. Nevertheless, author and publisher do not warrant the information contained therein to be free of errors. Readers are advised to keep in mind that statement, data, illustrations, procedural details or other items may inadvertenly inacurate.

Deutsche Bibliothek – CIP-Einheitsaufnahme

Zollner, Helmward:
Handbook of enzyme inhibitors / Helmward Zollner. –
Weinheim ; New York; Chichester ; Brisbane ; Singapore; Toronto : WILEY-VCH.
ISBN 3-527-30103-8

Pt. A. – 3., rev. and enl. ed. – 1999

Printed on acid-free and chlorine-free paper

Composition: Kühn & Weyh, D-79111 Freiburg
Printing: Strauss Offsetdruck GmbH, D-69509 Mörlenbach
Bookbinding: J. Schäffer, D-67269 Grünstadt
Printed in the Federal Republic of Germany.

To my sons Gernot and Johannes

Preface to the Third Edition

Ten years ago the First Edition was published comprising one volume with 1,000 enzymes and 2,000 inhibitors. The Third Edition has grown to four volumes with more than 19,000 inhibitors and 5,000 enzymes or isoenzymes emphasizing the importance of this field of research. Inhibitors have substantially contributed to our understanding of metabolic processes and of enzyme mechanisms and have made possible the labeling and mapping of active sites of enzymes. Moreover, inhibitors gained great interest as pharmaceuticals, pesticides, and herbicides.

This book is intended for biochemists, pharmacists, biologists, toxicologists, enzymologists, and other scientists working in related fields. Its aim is to provide rapid information on enzyme inhibitors and help to avoid mistakes that could be made due to limited knowledge about the side effects of an inhibitor, both in planning and interpreting experimental results.
A lot of effort was put into the standardization of enzyme names and inhibitor names as well as into the elimination of errors. The huge amount of data made this a tedious task. Particularly with inhibitor names and enzymes not yet classified and named by the nomenclature committee, and due to the fact that scientists most frequently use trivial names, standardization may not always have been perfect.
I would like to thank all persons who supported me and this work. I would also like to thank the publisher for their understanding and support as well as all those readers who made suggestions and proposals for improvement.

Graz, March 1999 H. Zollner

Preface to the Second Edition

The great demand for the *Handbook of Enzyme Inhibitors* – the first edition and a reprint are out of stock – and an enormous amount of new information were the impetus for the publisher and myself to edit a completely revised and updated version of this book. The organizational structure of the first edition was maintained, but the additional data have necessitated a division into two volumes, which, however, makes the text more manageable. The first volume contains an enzyme → inhibitor list, the second an inhibitor → enzyme list, glossary and EC numbers list.

A great deal of effort was put into the standardization of enzyme names and inhibitor names as well as into the elimination of errors. The huge amount of data made this a tedious task. Particularly with inhibitor names and enzymes not yet classified and named by the nomenclature committee, standardization may not always have been perfect.

I would like to thank the many colleagues who readily provided me with reprints of their papers and thus supported my work. I would like to mention Dr. G. Ranner who was a great help to me in compiling the data. I would also like to thank the publisher for their understanding and support as well as all those readers who made suggestions and proposals for improvement.

Graz, in June 1992 H. Zollner

Preface to First Edition

Inhibitors have substantially contributed to our understanding of metabolic processes and of enzyme mechanisms and have made possible the labeling and mapping of active sites of enzymes. Moreover, inhibitors gained great interest as pharmaceuticals and pesticides. In this handbook more than 5,000 inhibitors for about one thousand enzymes are listed. Two extensive lists are provided: (i) enzymes listed alphabetically, with their respective inhibitors, (ii) inhibitors listed alphabetically, with enzymes inhibited. Thus, it is made possible to search for an inhibitor of a particular enzyme *and* for the enzymes which are inhibited by a particular inhibitor. If available, the following data are given for all entries: Ki value or effective concentration, the type of inhibition, the source of the enzyme and the substrate for which these parameters have been determined. This book is intended for biochemists, biologists, pharmacists, toxicologists, enzymologists and other scientists working in related fields. Its aim is to provide rapid information about enzyme inhibitors and to help avoid mistakes that could be made due to limited knowledge about the side effects of an inhibitor, both in planning of investigations and in interpreting the experimental results. The author would like to encourage the reader to make comments and critical suggestions concerning the presentation of the data and to bring possible errors to his attention. Because of the rapidly growing literature as well as the vast amount of data, the author is aware that the list of inhibitors is not complete. It would be greatly appreciated if the users would provide information about important inhibitors missing from the text (The author's address is given on p. IV). I am grateful to Prof. H. Esterbauer for stimulating discussions and wish to acknowledge the help of Dr. G. Ranner in correcting the manuscript and proof reading.

Graz, 1988 H. Zollner

Contents

Part A

Part B

User's Guide

This handbook consists of four listings: In the **Enzyme → Inhibitor List,** the enzymes are ordered alphabetically. Each entry is headed by the enzyme name and the EC number. For each enzyme the inhibitors (column 1), the type of inhibition (column 2), effective inhibitor concentrations or *K*i values (column 3), some comments (column 4), and the reference numbers (column 5) are given. The references are listed at the end of the enzyme entry. In the **Inhibitor → Enzyme List,** the inhibitors are listed alphabetically together with the enzymes they affect. For easy comparison of the effectiveness of the inhibitors, the type of inhibition and effective concentration or *K*i value is included. In the **Trivial Name → Synonyma or Systematic Names List**, common names of inhibitors as used in this handbook (see below) are listed alphabetically together with the respective systematic names. Finally, EC numbers of enzymes covered in this handbook (in ascending order) together with the recommended names (see below) are given in the **EC Numbers List**.

Enzyme names and EC numbers. The names recommended by the International Union of Biochemistry are used throughout (Enzyme Nomenclature 1992; Academic Press, San Diego, 1992). Names other than the recommended ones are generally not listed. If required, the reader should consult the EC Numbers List to identify recommended names. For enzymes which could not be assigned with certainty or which are not yet classified the name used in the paper cited was adopted; in such cases no EC numbers are listed.

Inhibitor names. Common names (synonyms) are frequently used instead of systematic names since the latter are often rather long and cumbersome. 15-Hydroxy-[[2-*O*-(3-methyl-1-oxobutyl)-3,4-di-*O*--sulpho-β-D-glucopyranosyl]oxy]-19-norkaur-16-en-18-oic acid dipotassium salt is really not a catchy name compared to its common name, atractyloside. The synonyms used are the common ones, as recommended by the Merck Index (Merck & Co., Inc. Rahaway 1983). The systematic names (chemical abstracts names) can be found in the Common → Systematic Inhibitor Names List.

Type of inhibition. In this column the type of inhibition is reported. C = competitive, NC = non-competitive, UC = un-competitive, MI = mixed type, pC = partial-competitive, IR = irreversible inhibition; SS = suicide substrate.

Effective concentration. This column contains either the percent inhibition or the *K*i or the *I*50 values. For the percent inhibition at a given inhibitor concentration, the first figure signifies the percent inhibition and the second figure signifies the concentration. In a few cases the concentration is also a percent value. *I*50 is the concentration needed for half-maximum inhibition. The *K*i values are inhibition constants as reported in the reference. No effort has been made to distinguish between the different modes of calculation. When both *K*i value and effective concentration are reported in the reference, the *K*i-value is given.

Comments. In the comment column the substrate for which the kinetic parameters were determined, the organism or tissue from which the enzyme was isolated, and some self-explanatory comments are given.

References. The conventional system of abbreviations has been adopted except for a few journals for which short hand notation is used. J.B.C. = J. Biol. Chem., B.B.A. = Biophys. Biochim. Acta, B.B.R.C. = Biochim. Biophys. Res. Commun., E.J.B. = Eur. J. Biochem., P.N.A.S. = Proc. Natl. Acad. Sci. U.S.A. When more than one paper was found dealing with the same subject, the latest is cited in which the reader can easily find earlier references.

Column heads. The information presented in both the Enzyme → Inhibitor and the Inhibitor → Enzyme List are largely self-explanatory. Hopefully, all the details required for the use of this handbook are given above. Thus, in order to save space column heads are omitted throughout the lists. However, a book-mark is provided on which the column heads for both lists are printed. By holding the appropriate side of the book-mark on top of the list referred to, the columns can be conveniently identified.

J

J 104118
FARNESYL-DIPHOSPHATE FARNESYLTRANSFERASE — $I50 = 520$ pM

J 104123
FARNESYL-DIPHOSPHATE FARNESYLTRANSFERASE

J 104871
FARNESYL PROTEINTRANSFERASE — $I50 = 3.9$ nM

J852
CAROTENE DESATURASE — $I50 = 1.8$ nM
PHYTOENE DESATURASE — $I50 = 4.8$ μM

JATRORRHIZINE
RETICULINE OXIDASE
(S)-TETRAHYDROPROTOBERBERINE OXIDASE — $I50 = 540$ μM

JATRORRHIZINE CHLORIDE
RNA-DIRECTED DNA POLYMERASE — $I50 = 107$ μM
RNA-DIRECTED DNA POLYMERASE — $I50 = 113$ μM

JAVANICIN
GLUTATHIONE REDUCTASE (NADPH) — U — $Ki = 70$ μM
TRYPANOTHIONE REDUCTASE — U — $Ki = 50$ μM

JCR89C
CHOLINE KINASE — $I50 = 1$ μM

JG 365
PROTEINASE HIV-2 — $Ki = 660$ pM

JG-365
PROTEINASE HIV-1 — $Ki = 240$ pM

(S)-JG-365
PROTEINASE HIV-1 — $I50 = 3.4$ nM

JM 1583
RNA-DIRECTED DNA POLYMERASE — $I50 = 1.9$ μM

JM 1590
RNA-DIRECTED DNA POLYMERASE — $I50 = 1.3$ μM

JM 1591
RNA-DIRECTED DNA POLYMERASE — $I50 = 900$ nM

JM 1596
RNA-DIRECTED DNA POLYMERASE — $I50 = 3.6$ μM

JM 1809
RNA-DIRECTED DNA POLYMERASE — $I50 = 80$ nM

JM 2766
RNA-DIRECTED DNA POLYMERASE — $I50 = 190$ nM

JM 2815
RNA-DIRECTED DNA POLYMERASE — $I50 = 1.7$ μM

JMV 390-1
LEUCYL AMINOPEPTIDASE — $Ki = 52$ nM
MEMBRANE ALANINE AMINOPEPTIDASE — $Ki = 52$ nM
NEPRILYSIN — $Ki = 40$ nM
NEUROLYSIN — $Ki = 58$ nM
NEUROLYSIN — $Ki = 58$ nM
PEPTIDYL-DIPEPTIDASE A — $Ki = 70$ μM
THIMET OLIGOPEPTIDASE — $Ki = 31$ nM

JOBS TEAR α-AMYLASE INHIBITOR

α-AMYLASE

JODIDE

CAMPHOR 1,2-MONOOXYGENASE

JS-X 112

PROTEIN KINASE C — $I50 = 17\ \mu M$

JTE 522

CYCLOOXYGENASE 2 — $I50 = 85$ nM

JTP 2724

RENIN — $I50 = 680$ pM

JTP 4819

PROLYL OLIGOPEPTIDASE — $I50 = 5.4$ nM
PROLYL OLIGOPEPTIDASE — $I50 = 830$ pM

JUGLONE

Ca^{2+}-TRANSPORTING ATPASE — $I50 = 12\ \mu M$
CHOLINE O-ACETYLTRANSFERASE
NAD ADP-RIBOSYLTRANSFERASE
PEPTIDYLPROLYL ISOMERASE PARVULIN LIKE — IR

JUSTICIDIN E

ARACHIDONATE 5-LIPOXYGENASE — $I50 = 500$ nM

K

K^+

ACETYL-CoA CARBOXYLASE
N-ACETYLGALACTOSAMINE-6-SULFATASE — 46% 17 mM
ADENOSINE KINASE
ADENYLATE CYCLASE
(R)AMINOPROPANOL DEHYDROGENASE — > 200 mM
ARGINIIE KINASE
ASPARTATE-SEMIALDEHYDE DEHYDROGENASE
κ-CARRAGEENASE
CITRATE (si)-SYNTHASE
CLOSTRIPAIN
CYTOCHROME-b5 REDUCTASE
17-O-DEACETYLVINDOLINE O-ACETYLTRANSFERASE
DIHYDROFOLATE REDUCTASE
DIPEPTIDYL-PEPTIDASE 2
FRUCTOSE-BISPHOSPHATASE — $Ki = 68$ mM
GLUTAMATE FORMIMINOTRANSFERASE
IMP DEHYDROGENASE — > 100 mM
ISOLEUCINE-tRNA LIGASE
NAD(P) TRANSHYDROGENASE (AB-SPECIFIC)
NEOLACTOSYLCERAMIDE α-2,3-SIALYLTRANSFERASE
NUCLEOSIDE-TRIPHOSPHATASE — 50% 100 mM
OXOGLUTARATE DEHYDROGENASE (LIPOAMIDE) — > 50 mM
POLYDEOXYRIBONUCLEOTIDE SYNTHASE (ATP)
PROLINE-tRNA LIGASE
PROTEINASE I ACHROMOBACTER — $Ki = 5$ mM
PROTEIN KINASE C — 60% 150 mM
RIBONUCLEASE III — > 140 mM
RIBONUCLEOSIDE-TRIPHOSPHATE REDUCTASE
RNA-DIRECTED RNA POLYMERASE — 90% 50 mM
tRNA(GUANINE-N^2-) METHYLTRANSFERASE — > 100 mM
SUCROSE 1^F-FRUCTOSYLTRANSFERASE — $Ki = 122$ mM
THYMIDINE KINASE
dTMP KINASE

TRITRIRACHIUM ALKALINE PROTEINASE
UREASE

K^++

tRNA(GUANINE-N^1-) METHYLTRANSFERASE		> 100 mM

K 13

AMINOPEPTIDASE B		13% 230 μg/ml
PEPTIDYL-DIPEPTIDASE A		*I*50 = 350 nM
PEPTIDYL-DIPEPTIDASE A-I		

K 76

STEROL ESTERASE		*I*50 = 200 μM

K 150

DIHYDROFOLATE REDUCTASE		*I*50 = 39 nM
DIHYDROFOLATE REDUCTASE		*I*50 = 19 μM
DIHYDROFOLATE REDUCTASE		*I*50 = 420 nM

K 252a

Ca^{2+} DEPENDENT PROTEIN KINASE α		*I*50 = 800 nM
Ca^{2+} DEPENDENT PROTEIN KINASE β		*I*50 = 800 nM
Ca^{2+} DEPENDENT PROTEIN KINASE γ		*I*50 = 300 nM
CAM KINASE		*I*50 = 300 nM
MYOSIN-LIGHT-CHAIN KINASE	C	*K*i = 20 nM
MYOSIN-LIGHT-CHAIN KINASE		*K*i = 20 nM
MYOSIN-LIGHT-CHAIN KINASE		*I*50 = 190 nM
PHOSPHORYLASE KINASE		*I*50 = 1.7 nM
PROTEIN KINASE A		*I*50 = 200 nM
PROTEIN KINASE A1	C	*K*i = 18 nM
PROTEIN KINASE C	C	*K*i = 25 nM
PROTEIN KINASE C		*I*50 = 50 nM
PROTEIN KINASE (Ca^{2+} DEPENDENT)		*I*50 = 50 nM
PROTEIN KINASE ECTO		
PROTEIN KINASE G	C	*K*i = 20 nM
PROTEIN-2 KINASE MICROTUBULE-ASSOCIATED		
PROTEIN TYROSINE KINASE EGFR		*I*50 = 9.5 μM
PROTEIN TYROSINE KINASE NGFR		*K*i = 1.4 nM
PROTEIN TYROSINE KINASE PROTEIN TYROSINE KINASE NGFR gp140trk		*I*50 = 3 nM
PROTEIN TYROSINE KINASE pp60c-Src		*I*50 = 600 ng/ml

K 252b

MYOSIN-LIGHT-CHAIN KINASE		*K*i = 147 nM
PROTEIN KINASE A1	C	*K*i = 90 nM
PROTEIN KINASE C	C	*K*i = 20 nM
PROTEIN KINASE ECTO		
PROTEIN KINASE G	C	*K*i = 100 nM
PROTEIN TYROSINE KINASE NGFR		*K*i = 2.5 nM
PROTEIN TYROSINE KINASE pp60c-Src		*I*50 = 160 ng/ml

K 6803

3′,5′-CYCLIC-NUCLEOTIDE PHOSPHODIESTERASE (Ca^{2+}/CALMODULIN)		*I*50 > 100 μM

K 6807

3′,5′-CYCLIC-NUCLEOTIDE PHOSPHODIESTERASE (Ca^{2+}/CALMODULIN)		*I*50 = 43 μM

K 6917

3′,5′-CYCLIC-NUCLEOTIDE PHOSPHODIESTERASE (Ca^{2+}/CALMODULIN)		*I*50 = 25 μM

K 6920

3′,5′-CYCLIC-NUCLEOTIDE PHOSPHODIESTERASE (Ca^{2+}/CALMODULIN)		*I*50 = 46 μM

K 7027

3′,5′-CYCLIC-NUCLEOTIDE PHOSPHODIESTERASE (Ca^{2+}/CALMODULIN)		*I*50 = 23 μM

K 252A

PROTEIN TYROSINE KINASE NGFR

K 76COOH
COMPLEMENT FACTOR I

K 76 MONOCARBOXYLIC ACID
myo-INOSITOL-1(or 4)-MONOPHOSPHATASE NC $Ki = 500\ \mu M$

K-13
PEPTIDYL-DIPEPTIDASE A $Ki = 350$ nM

KAEMPFERIDE
11β-HYDROXYSTEROID DEHYDROGENASE 1 73% 1.2 μM
PROTEIN TYROSINE KINASE p56lck $I50 = 30\ \mu g/ml$

KAEMPFEROL
ADENOSINE DEAMINASE $I50 = 10\ \mu g/ml$
ADENOSINE DEAMINASE $I50 = 32\ \mu M$
ARACHIDONATE 5-LIPOXYGENASE $I50 = 20\ \mu M$
ARACHIDONATE 5-LIPOXYGENASE $I50 = 1\ \mu M$
AROMATASE $Ki = 27\ \mu M$
ARYL SULFOTRANSFERASE P $I50 = 390$ nM
CHALCONE ISOMERASE $Ki = 4.2\ \mu M$
CYCLOOXYGENASE $I50 = 20\ \mu M$
CYTOCHROME P450 1A1 $I50 = 110$ nM
GLUTATHIONE REDUCTASE (NADPH) 28% 100 μM
GLUTATHIONE TRANSFERASE 76% 100 μM
GLUTATHIONE TRANSFERASE $I50 = 76\ \mu M$
HYALURONIDASE PH20 38% 50 μM
HYALURONOGLUCOSAMINIDASE
11β-HYDROXYSTEROID DEHYDROGENASE $I50 = 322\ \mu M$
11β-HYDROXYSTEROID DEHYDROGENASE 1 42% 1.2 μM
INTEGRASE $I50 = 98\ \mu M$
IODIDE PEROXIDASE $I50 = 1.2\ \mu M$
MYOSIN-LIGHT-CHAIN KINASE $I50 = 13\ \mu M$
NADH DEHYDROGENASE (EXOGENOUS) $I50 = 200\ \mu M$
PHENYLALANINE AMMONIA-LYASE
PROLYL OLIGOPEPTIDASE $I50 = 1.7$ ppm
PROTEINASE HIV-1 $I50 = 250\ \mu M$
PROTEIN KINASE C $I50 = 2\ \mu M$
PROTEIN KINASE CASEIN KINASE G 50% 2.5 μM
PROTEIN TYROSINE KINASE
PROTEIN TYROSINE KINASE EGFR $I50 = 3.2\ \mu g/ml$
PROTEIN TYROSINE KINASE p56lck $I50 = 8\ \mu g/ml$
RNA-DIRECTED DNA POLYMERASE $I50 = 320\ \mu M$
RNA-DIRECTED DNA POLYMERASE $I50 = 233\ \mu M$
TRYPSIN $Ki = 30\ \mu M$
XANTHINE OXIDASE $I50 = 1.1\ \mu M$

KAEMPFEROL-O^3-α-ARABINOPYRANOSIDE
PROTEIN TYROSINE KINASE p56lck $I50 = 200\ \mu g/ml$

KAEMPFEROL-3-O-β-GLUCOSIDE
α-GLUCOSIDASE

KAEMPFEROL-O^3-α-RHAMNOPYRANOSIDE
PROTEIN TYROSINE KINASE p56lck $I50 = 400\ \mu g/ml$

KAINIC ACID
3-HYDROXYANTHRANILATE 3,4-DIOXYGENASE

KALAFUNGIN
PROTEINASE 3C $I50 = 3.3\ \mu M$

KALI-DY
COAGULATION FACTOR XIa $Ki = 8.2$ nM
KALLIKREIN $Ki = 15$ pM

KALIURETIC PEPTIDE
Na^+/K^+-EXCHANGING ATPASE

KALLISTATIN
KALLIKREIN
KALLIKREIN TISSUE

KAN 104141
PROTEIN TYROSINE KINASE (erbB2) — *I*50 = 68 μM
PROTEIN TYROSINE KINASE p56lck — *I*50 = 3.6 μM
PROTEIN TYROSINE KINASE PDGFR — *I*50 = 50 μM

KAN 108308
PROTEIN KINASE A — *I*50 = 100 μM
PROTEIN TYROSINE KINASE (erbB2) — *I*50 = 8 μM
PROTEIN TYROSINE KINASE p56lck — *I*50 = 20 μM
PROTEIN TYROSINE KINASE PDGFR — *I*50 = 29 μM

KAN 108329
PROTEIN TYROSINE KINASE (erbB2) — *I*50 = 21 μM
PROTEIN TYROSINE KINASE p56lck — *I*50 = 4 μM
PROTEIN TYROSINE KINASE PDGFR — *I*50 = 19 μM

KANAMINE
GENTAMICIN 3′-N-ACETYLTRANSFERASE

KANAMYCIN A
GENTAMICIN 3′-N-ACETYLTRANSFERASE
RIBOZYME HAMMERHEAD — 32% 500 μM

KANAMYCIN B
PHOSPHOLIPASE A1 — *I*50 = 110 μM

KANCHANOMYCIN
RNA POLYMERASE

KATIONS MONOVALENT
BENZON NUCLEASE

KBR
CREATINE KINASE

KB-R 7785
MATRIX METALLOPROTEINASES

KBT 3022
CYCLOOXYGENASE

KC 11404
ARACHIDONATE 5-LIPOXYGENASE — *I*50 = 900 nM

KC 1300
DIHYDROFOLATE REDUCTASE — *I*50 = 705 nM
DIHYDROFOLATE REDUCTASE — *I*50 = 926 nM
DIHYDROFOLATE REDUCTASE — *I*50 = 101 nM

KCL
ACYL[ACYL-CARRIER-PROTEIN]-UDP-N-ACETYLGLUCOSAMINE O-ACYLTRANSFERASE
D-ALANINE-POLY(PHOSPHORIBITOL)LIGASE
ALANOPINE DEHYDROGENASE
ASPARAGINE-tRNA LIGASE
ASPARTATE 1-DECARBOXYLASE — 63% 200 mM
CATHEPSIN D — 52% 300 mM
CHORISMATE MUTASE
CREATINE KINASE
3-DEOXY-MANNO-OCTULOSONATE-8-PHOSPHATASE — *I*50 = 100 mM
DEOXYRIBONUCLEASE IV (PHAGE T4-INDUCED) — 80% 50 mM
DNA (CYTOSINE-5-)-METHYLTRANSFERASE

DYNEIN ATPASE		
EXODEOXYRIBONUCLEASE (LAMDA INDUCED)		99% 200 mM
GLUTAMATE-tRNA LIGASE		
GLYCEROL 2-DEHYDROGENASE		> 100 mM
GLYCEROL-3-PHOSPHATE DEHYDROGENASE (NAD)		
GLYCINE DEHYDROGENASE (CYTOCHROME)		> 75 mM
GLYCINE DEHYDROGENASE (DECARBOXYLATING)		
GLYCINE HYDROXYMETHYLTRANSFERASE		
GTP CYCLOHYDROLASE I		87% 1 M
HEPAROSAN-N-SULFATE-GLUCURONATE 5-EPIMERASE		
HISTONE ACYLTANSFERASE		*I*50 = 175 mM
HYDROXYLAMINE REDUCTASE (NADH)		
LEUCINE-tRNA LIGASE		> 50 mM
LYSINE-tRNA LIGASE		
LYSOZYME		
LYSYL ENDOPEPTIDASE		*K*i = 45 mM
MANNOSE ISOMERASE		
METHIONINE-tRNA LIGASE		
NAD-DINITROGEN-REDUCTASE ADP-D-RIBOSYLTRANSFERASE		
NITROUS OXIDE REDUCTASE		> 50 mM
OXALACETATE TAUTOMERASE		20% 40 mM
PHOSPHOPANTOTHENATE-CYSTEINE LIGASE		50% 50 mM
PROTEIN TYROSINE-PHOSPHATASE (Mg^{2+} DEPENDENT)		52% 100 mM
PYRUVATE DECARBOXYLASE		
[PYRUVATE DEHYDROGENASE (LIPOAMIDE)]-PHOSPHATASE		
SERINE-tRNA LIGASE		85% 500 mM
STEROL ESTERASE		
TRYPTOPHAN-tRNA LIGASE		85% 500 mM
VALINE-tRNA LIGASE		
KCNO		
MALATE DEHYDROGENASE (DECARBOXYLATING)		
KDTF5		
COAGULATION FACTOR VIIa		*K*i = 235 pM
KELATORPHAN		
DIPEPTIDYL AMINOPEPTIDASE		*I*50 = 900 pM
LEUKOTRIENE A4 HYDROLASE		*I*50 = 10 µM
LEUKOTRIENE A4 HYDROLASE		*I*50 = 5 nM
MEMBRANE ALANINE AMINOPEPTIDASE		*I*50 = 380 nM
NEPRILYSIN		*I*50 = 1.4 nM
KELLETIN I		
DNA POLYMERASE α		52% 10 µM
DNA POLYMERASE β		51% 100 µM
RNA POLYMERASE		50% 50 µM
KEMPTIDE-AP_4		
PROTEIN KINASE A		*I*50 = 68 µM
KERAMAMIDE A1		
Ca^{2+}-TRANSPORTING ATPASE		*I*50 = 300 µM
KERATAN SULPHATE		
N-ACETYLGALACTOSAMINE-6-SULFATASE	C	*K*i = 40 µM
CHONDROITIN AC LYASE		80% 0.06%
KERATO SULPHATE		
HYALURONOGLUCOSAMINIDASE		
6-KESTOSE		
SUCROSE-PHOSPHATASE		
KETAMINE		
ACETYLCHOLINESTERASE	REV	*K*i = 520 µM
MORPHINE 6-DEHYDROGENASE		
NITRIC-OXIDE SYNTHASE		

KETANSERIN
CYTOCHROME P450 2D6 — $Ki = 220\ \mu M$

KETENE
SOLANIN

KETOACE
PEPTIDYL-DIPEPTIDASE A — C — $Ki = 600$ pM

KETOACYL-CoA COMPOUNDS LONG CHAIN
ACETYL-CoA C-ACYLTRANSFERASE

7-KETO-8-AMINOPELARGONIC ACID
ADENOSYLMETHIONINE-8-AMINO-7-OXONONANOATE TRANSAMINASE

11-KETO-β-BOSWELLIC ACID
ARACHIDONATE 5-LIPOXYGENASE

2-KETO-(3R,S)-3-BROMOPYRUVIC ACID
2-DEHYDRO-3-DEOXYPHOSPHOGLUCONATE ALDOLASE — IR

KETOCARBAPHOSPHONATE
3-DEHYDROQUINATE SYNTHASE — IR

6-KETOCHOLESTANOL
CHOLESTEROL EPOXIDE HYDROXYLASE

7-KETOCHOLESTANOL
CHOLESTEROL EPOXIDE HYDROXYLASE

7-KETOCHOLESTEROL
CHOLESTEROL EPOXIDE HYDROXYLASE

KETOCONAZOLE

AROMATASE		$I50 = 7.3\ \mu M$
AROMATASE		$I50 = 4.7\ \mu M$
AROMATASE		$I50 = 103\ \mu M$
CHOLESTEROL MONOOXYGENASE (SIDE-CHAIN-CLEAVING)		$I50 = 3\ \mu M$
CHOLESTEROL MONOOXYGENASE (SIDE-CHAIN-CLEAVING)		$I50 = 1.7\ \mu M$
CHOLESTEROL MONOOXYGENASE (SIDE-CHAIN-CLEAVING)		$I50 = 23\ \mu M$
CYTOCHROME P450 1A2		$Ki = 32\ \mu M$
CYTOCHROME P450 2A6	C	$Ki = 24\ \mu M$
CYTOCHROME P450 3A		$I50 = 200$ nM
CYTOCHROME P450 3A4	MIX	$Ki = 15$ nM
CYTOCHROME P450 2B6		$I50 = 4.1\ \mu M$
CYTOCHROME P450 2C8		$I50 = 13\ \mu M$
CYTOCHROME P450 2C19		$I50 = 28\ \mu M$
CYTOCHROME P450 2D6		$I50 = 17\ \mu M$
CYTOCHROME P450 2E1		$I50 = 90\ \mu M$
7-ETHOXYCOUMARIN O-DEACETYLASE		
7-ETHOXYRESORUFIN O-DEETHYLASE		
FLAVONOID 3′-MONOOXYGENASE		
1,3-β-GLUCAN SYNTHASE		18% 10 μg/ml
3β-HYDROXY-5-ENE-STEROID OXIDOREDUCTASE/ISOMERASE	C	$Ki = 2.9\ \mu M$
17α-HYDROXYPROGESTERONE ALDOLASE		
17α-HYDROXYPROGESTERONE ALDOLASE		$Ki = 3.6\ \mu M$
20α-HYDROYSTEROID OXIDOREDUCTASE (NADH)		$Ki = 900$ nM
LANOSTEROL C14 DEMETHYLASE		$Ki = 66$ nM
LANOSTEROL C14 DEMETHYLASE		$I50 = 50$ ng/ml
METHYLTETRAHYDROPROTOBERBERINE 14-MONOOXYGENASE		
NITRIC-OXIDE SYNTHASE (CALMODULIN DEPENDENT)	C	$I50 = 20\ \mu M$
PROGESTERONE 21 16α-HYDROXYLASE		$I50$ 100 μM
PROGESTERONE 21 6β-HYDROXYLASE		$I50$ 100 μM
STEROID 17α-HYDROXYLASE/17,20 LYASE		
STEROID 11β-MONOOXYGENASE		$I50 = 700$ nM
STEROID 17α-MONOOXYGENASE		
STEROID 17α-MONOOXYGENASE		$Ki = 40\ \mu M$

STEROID 21-MONOOXYGENASE
STEROID 17α-MONOOXYGENASE/17,20 LYASE *I*50 = 260 nM
STEROID 5α-REDUCTASE *I*50 = 100 nM
UNSPECIFIC MONOOXYGENASE

3-KETODECANOYL-CoA
ACYL-CoA DEHYDROGENASE (LONG CHAIN)
ACYL-CoA DEHYDROGENASE (MEDIUM CHAIN)

3-KETO-GLYCYRRHETIC ACID
11β-HYDROXYSTEROID DEHYDROGENASE *I*50 = 400 nM
3α-HYDROXYSTEROID DEHYDROGENASE (B-SPECIFIC) *I*50 = 6 μM

α-KETOISOCAPROIC ACID
OXOGLUTARATE DEHYDROGENASE (LIPOAMIDE)

α-KETOISOVALERIC ACID
3-ISOPROPYLMALATE DEHYDROGENASE
OXOGLUTARATE DEHYDROGENASE (LIPOAMIDE)
2-OXOISOVALERATE DEHYDROGENASE (ACYLATING) C

α-KETO-β-METHYLVALERIC ACID
OXOGLUTARATE DEHYDROGENASE (LIPOAMIDE)
2-OXOISOVALERATE DEHYDROGENASE (ACYLATING) C

α-KETOOCTANOIC ACID
PYRUVATE DEHYDROGENASE (LIPOAMIDE)

2′-KETOPANTETHEINE
PANTOTHENATE KINASE

S(2-KETOPENTADECYL)-CoA
GLYCYLPEPTIDE N-TETRADECANOYLTRANSFERASE C *K*i = 110 nM

KETOPROFEN
CARBONYL REDUCTASE (NADPH) 80% 1 mM
CYCLOOXYGENASE *I*50 = 1.5 μM
CYCLOOXYGENASE 1 *I*50 = 7.5 μM
CYCLOOXYGENASE 2 *I*50 = 7.6 μM
PROSTAGLANDIN-ENDOPEROXIDE SYNTHASE *I*50 = 75 nM
PROSTAGLANDIN-ENDOPEROXIDE SYNTHASE 1 *I*50 = 500 nM
PROSTAGLANDIN-ENDOPEROXIDE SYNTHASE 2 *I*50 = 2.3 μM
PROSTAGLANDIN F SYNTHASE

dexKETOPROFEN
CYCLOOXYGENASE *I*50 = 3.5 μM

KETOTIFEN
ARACHIDONATE LIPOXYGENASE

γ-KETOTRIAZOLE
OBTUSIFOLIOL 14-DEMETHYLASE

α-KETOVALERIC ACID
[3-METHYL-2-OXOBUTANOATE DEHYDROGENASE (LIPOAMIDE)] KINASE

KETOVOLAC TROMETHAMINE
CYCLOOXYGENASE 1 *I*50 = 32 μM
CYCLOOXYGENASE 2 *I*50 = 61 μM

KEXSTATIN
PROTEINASE KEX-2 *I*50 = 1.4 μg
SUBTILISIN

KF 13218
THROMBOXANE A SYNTHASE *I*50 = 27 nM

KF 17625
3′,5′-CYCLIC-NUCLEOTIDE PHOSPHODIESTERASE I *I*50 = 6.2 μM

3′,5′-CYCLIC-NUCLEOTIDE PHOSPHODIESTERASE II — $I50 = 32\ \mu M$
3′,5′-CYCLIC-NUCLEOTIDE PHOSPHODIESTERASE IV — $I50 = 53\ \mu M$

KF 17828
STEROL O-ACYLTRANSFERASE — $I50 = 23$ nM

KF 18678
STEROID 5α-REDUCTASE — $I50 = 3.3$ nM

KF 19514
3′,5′-CYCLIC-NUCLEOTIDE PHOSPHODIESTERASE I — $I50 = 270$ nM
3′,5′-CYCLIC-NUCLEOTIDE PHOSPHODIESTERASE IV — $I50 = 400$ nM

KF 20405
STEROID 5α-REDUCTASE 1 — $I50 = 480$ pM
STEROID 5α-REDUCTASE 2 — $I50 = 280$ pM

KF 8940
ARACHIDONATE 5-LIPOXYGENASE — NC — $Ki = 70\ \mu M$

$K[FE(CN)_6]$
β-FRUCTOFURANOSIDASE

Kg^+
EXOPOLYPHOSPHATASE

KHAFREFUNGIN
INOSITOLPHOSPHOCERAMIDE SYNTHASE — $I50 = 600$ poM

KI 6783
PROTEIN TYROSINE KINASE PDGFR — $I50 = 100$ nM

K_2IrCl_6
β-AMYLASE

KIDNEY BEAN α-AMYLASE INHIBITOR
α-AMYLASE

KIEVITONE
CHALCONE ISOMERASE — $Ki = 9.2\ \mu M$
NARINGENIN-CHALCONE SYNHTASE

KIFUNENSINE
MANNOSYL-OLIGOSACCHARIDE 1,3-1,6-α-MANNOSIDASE — $I50 = 50$ nM

KIH 200
PHOSPHATE CARRIER MITOCHONDRIA — $I50 = 7\ \mu M$

KIH 201
PHOSPHATE CARRIER MITOCHONDRIA — $I50 = 7\ \mu M$

KIH 202
PHOSPHATE CARRIER MITOCHONDRIA — $I50 = 7\ \mu M$

KIH 2023
ACETOLACTATE SYNTHASE

KIH 6127
ACETOLACTATE SYNTHASE

(+)-KINEFUNESINE
MANNOSIDASE I

KINETIN
5-HYDROXYFURANOCOUMARIN 5-O-METHYLTRANSFERASE
8-HYDROXYFURANOCOUMARIN 8-O-METHYLTRANSFERASE
INDOLE-3-ACETATE β-GLUCOSYLTRANSFERASE

H-KININOGEN
CATHEPSIN B — C — $Ki = 400$ nM

CATHEPSIN H	C	Ki = 1.1 nM
CATHEPSIN L	C	Ki = 19 pM

KININOGEN HIGH MW

CATHEPSIN H		Kd = 25 nM
CATHEPSIN L		Kd = 160 pM
LEGUMAIN		32% 5 μM
PAPAIN		Kd = 14 pM

KININOGEN HK

CALPAIN	MIX	Ki = 2.3 nM

KININOGEN HMW (BOVINE)

CATHEPSIN B		Ki = 12 nM
CATHEPSIN H		Ki = 15 nM
CATHEPSIN L		Ki = 700 pM
DIPEPTIDYL-PEPTIDASE 1		Ki = 3.5 nM
PAPAIN		Ki = 21 nM

KININOGEN HUMAN DOMAIN 3

ACTINIDAIN		Ki = 6 nM
CATHEPSIN B		Ki = 63 nM
CATHEPSIN L		Ki = 1 pM
PAPAIN		Ki = 1.8 pM

L-KININOGEN (HUMAN)

ACTINIDAIN		Ki = 23 nM
CALPAIN II		Ki = 1 nM
CATHEPSIN B		Ki = 600 nM
CATHEPSIN H		Ki = 1.2 nM
CATHEPSIN L	C	Ki = 17 pM
DIPEPTIDYL-PEPTIDASE		Ki > 130 nM
PAPAIN		Ki = 3.2 pM

L-KININOGEN

CALPAIN	C	Ki = 1 nM
CRUZIPAIN		Kd = 41 pM

KININOGEN LK

CALPAIN	NC	Ki = 2.7 nM

KININOGEN LMW (BOVINE)

CATHEPSIN B		Ki = 23 nM
CATHEPSIN H		Ki = 44 nM
CATHEPSIN L		Ki = 1.8 nM
PAPAIN		Ki < 5 pM

KININOGENS

CATHEPSIN H
CATHEPSIN L
FICAIN
PAPAIN

KININOGEN T (RAT)

CATHEPSIN B		Ki = 170 nM
CATHEPSIN H		Ki = 470 nM
CATHEPSIN L		Ki = 2.8 nM
FICAIN		Ki = 4.2 nM
PAPAIN		Ki = 70 pM

p57^{KIP2}

CYCLIN D2 DEPENDENT KINASE 4
CYCLIN DEPENDENT KINASE 2

KIWI PECTIN METHYLESTERASE INHIBITOR

PECTINESTERASE

KK 505
THROMBOXANE A SYNTHASE 87% 1 μM

KKKKKRFAFKKAFKLAGFAKKNKK
PROTEIN KINASE C NC $I50 = 200$ nM

K3L
PROTEIN KINASE R

KLAAENHCLRIKILGDCYYC
ADENYLATE CYCLASE V $I50 = 20$ μM

KM 043
DNA POLYMERASE α $I50 = 250$ nM
DNA POLYMERASE β $I50 = 3.6$ μM
RNA-DIRECTED DNA POLYMERASE $I50 = 11$ μM

KME 4
ARACHIDONATE 5-LIPOXYGENASE $I50 = 600$ nM
CYCLOOXYGENASE $I50 = 5$ μM

$KMnO_4$
CATHEPSIN D

$KMnO_4$
CELLULASE 100% 100 μM
ENDO-1,3(4)-β-GLUCANASE
ENDO-1,4-β-XYLANASE
GLUCAN 1,4-α-GLUCOSIDASE
GLUCAN 1,6-α-ISOMALTOSIDASE

KN 62
CAM KINASE
CAM KINASE II $I50 = 400$ nM
CAM KINASE III

KN 93
CAM KINASE II 90% 5 μM
3′,5′-CYCLIC-NUCLEOTIDE PHOSPHODIESTERASE (Ca^{2+}/CALMODULIN) 26% 5 μM

KNI 93
PROTEINASE HIV $I50 = 5$ nM
PROTEINASE HIV-1 $I50 = 5$ nM

KNI 102
PROTEINASE HIV $I50 = 89$ nM
PROTEINASE HIV-1 $I50 = 89$ nM

KNI 122
PROTEINASE HIV-1 $I50 = 100$ nM

KNI 174
PROTEINASE HIV-1 $I50 = 2.8$ nM

KNI 227
PROTEINASE HIV-1 $I50 = 2.3$ nM

KNI 272
PROTEINASE HIV
PROTEINASE HIV-1 $Ki = 5.5$ pM

KNI 577
PROTEINASE HIV-1 88% 50 nM

KOJIBIOSE
GLUCOSIDASE I $I50 = 500$ μM
MANNOSYL-OLIGOSACCHARIDE GLUCOSIDASE

KOJIC ACID

D-AMINO-ACID OXIDASE		
CATECHOL OXIDASE		*I*50 = 200 nM
LACCASE		*I*50 = 130 nM
MONOPHENOL MONOOXYGENASE	C	*K*i = 12 μM
POLYPHENOL OXIDASE		*K*i = 30 μM
POLYPHENOL OXIDASE		*K*i = 20 μM
POLYPHENOL OXIDASE		*K*i = 90 μM
POLYPHENOL OXIDASE		*K*i = 70 μM
POLYPHENOL OXIDASE		*K*i = 50 μM
POLYPHENOL OXIDASE		*K*i = 60 μM

N-KOJIC-L-PHENYLALANYL-KOJIATE

TYROSINASE

KOJISTATIN

FICAIN		*I*50 = 80 ng/ml
PAPAIN		*I*50 = 23 ng/ml
STEM BROMELAIN		*I*50 = 331 ng/ml

KONBU′ACIDIN

CYCLIN DEPENDENT KINASE 4		*I*50 = 20 μg/ml

KONINGIC ACID

GLYCERALDEHYDE-3-PHOSPHATE DEHYDROGENASE (NADP) (PHOSPHORYLATING)		*I*50 = 5.1 μM
GLYCERALDEHYDE-3-PHOSPHATE DEHYDROGENASE (PHOSPHORYLATING)		*I*50 = 200 nM
GLYCERALDEHYDE-3-PHOSPHATE DEHYDROGENASE (PHOSPHORYLATING)		*I*50 = 170 nM
GLYCERALDEHYDE-3-PHOSPHATE DEHYDROGENASE (PHOSPHORYLATING)		*I*50 = 200 nM
GLYCERALDEHYDE-3-PHOSPHATE DEHYDROGENASE (PHOSPHORYLATING)		*I*50 = 160 nM
GLYCERALDEHYDE-3-PHOSPHATE DEHYDROGENASE (PHOSPHORYLATING)		*I*50 = 240 nM
GLYCERALDEHYDE-3-PHOSPHATE DEHYDROGENASE (PHOSPHORYLATING)		*I*50 = 560 μM
GLYCERALDEHYDE-3-PHOSPHATE DEHYDROGENASE (PHOSPHORYLATING)		*I*50 = 28 μM
GLYCERALDEHYDE-3-PHOSPHATE DEHYDROGENASE (PHOSPHORYLATING)		*I*50 = 140 nM
GLYCERALDEHYDE-3-PHOSPHATE DEHYDROGENASE (PHOSPHORYLATING)		*I*50 = 13 μM
GLYCERALDEHYDE-3-PHOSPHATE DEHYDROGENASE (PHOSPHORYLATING)		*I*50 = 300 nM
GLYCERALDEHYDE-3-PHOSPHATE DEHYDROGENASE (PHOSPHORYLATING)		*I*50 = 270 nM
GLYCERALDEHYDE-3-PHOSPHATE DEHYDROGENASE (PHOSPHORYLATING)	IR	*K*i = 1.6 μM

K_2PtCl_4

β-AMYLASE
β-FRUCTOFURANOSIDASE

K_2PtCl_6

β-AMYLASE

KRI 1230

RENIN		*I*50 = 25 nM

KRI 1314

RENIN		*I*50 = 2.4 nM
RENIN	C	*K*i = 100 pM

KS 501

3′,5′-CYCLIC-NUCLEOTIDE PHOSPHODIESTERASE (Ca^{2+}/CALMODULIN)		
MYOSIN-LIGHT-CHAIN KINASE		*I*50 = 33 μM

KS 502

CAM KINASE I		
CAM KINASE II		
3′,5′-CYCLIC-NUCLEOTIDE PHOSPHODIESTERASE (Ca^{2+}/CALMODULIN)		
MYOSIN-LIGHT-CHAIN KINASE		*I*50 = 78 μM

KS 504a

3′,5′-CYCLIC-NUCLEOTIDE PHOSPHODIESTERASE (Ca^{2+}/CALMODULIN)		*I*50 = 122 μM

KS 504b

3′,5′-CYCLIC-NUCLEOTIDE PHOSPHODIESTERASE (Ca^{2+}/CALMODULIN)		*I*50 = 109 μM

Inhibitor / Enzyme		
KS 505a		
3′,5′-CYCLIC-NUCLEOTIDE PHOSPHODIESTERASE (Ca^{2+} CALMODULIN)		*K*i = 60 nM
3′,5′-CYCLIC-NUCLEOTIDE PHOSPHODIESTERASE (Ca^{2+}/CALMODULIN)		*I*50 = 65 nM
KS 619-1		
3′,5′-CYCLIC-NUCLEOTIDE PHOSPHODIESTERASE (Ca^{2+}/CALMODULIN)		
K_2SO_4		
CATHEPSIN D		70% 100 mM
KT-1		
3′,5′-CYCLIC-NUCLEOTIDE PHOSPHODIESTERASE		*I*50 = 1.5 µM
KT 5720		
MAP KINASE		
PROTEIN KINASE A1	C	*K*i = 60 nM
KT 5822		
PROTEIN KINASE A1	C	*K*i = 37 nM
PROTEIN KINASE C	C	*K*i = 79 nM
PROTEIN KINASE G	C	*K*i = 2.4 nM
KT 5823		
PROTEIN KINASE C		*K*i = 4 µM
PROTEIN KINASE G		*K*i = 234 nM
KT 5926		
MYOSIN-LIGHT-CHAIN KINASE	C	*K*i = 18 nM
PROTEIN KINASE A	C	*K*i = 1.2 µM
PROTEIN KINASE C	C	*K*i = 723 nM
PROTEIN KINASE G	C	*K*i = 158 nM
PROTEIN TYROSINE KINASE NGFR		*K*i = 137 nM
KT 8108		
CASPASE-1		*I*50 = 3.3 µM
KT 8109		
CASPASE-1		*I*50 = 240 nM
KT 8110		
CASPASE-1		*I*50 = 70 nM
KT 8111		
CASPASE-1		*I*50 = 1.7 µM
KT 8112		
CASPASE-1		*I*50 = 1 µM
KUANONIAMINE D		
DNA TOPOISOMERASE (ATP-HYDROLYSING)		*I*90 = 127 µM
KUDINGOSIDE A		
STEROL O-ACYLTRANSFERASE		*I*50 = 2.7 mM
KUDINGOSIDE B		
STEROL O-ACYLTRANSFERASE		*I*50 = 2.9 mM
KUEHNEROMYCIN A		
RNA-DIRECTED DNA POLYMERASE	NC	*K*i = 200 µM
RNA-DIRECTED DNA POLYMERASE	NC	*K*i = 40 µM
KUKOAMINE A		
TRYPANOTHIONE REDUCTASE		*K*i = 1.8 µM
KUMARA TRYPSIN INHIBITOR		
PROTEINASE FIBROBLAST CELL SURFACE		
TRYPSIN		

KURASOIN A
FARNESYL PROTEINTRANSFERASE

KURASOIN B
FARNESYL PROTEINTRANSFERASE

KW 3033

STEROL O-ACYLTRANSFERASE		*I*50 = 8 nM

KW 3635

ARACHIDONATE 5-LIPOXYGENASE		*I*50 = 71 μM

KW 5092

ACETYLCHOLINESTERASE		*I*50 = 68 nM
CHOLINESTERASE		*I*50 = 24 μM

KY 234

THROMBOXANE A SYNTHASE		*I*50 = 44 nM

KYNURENIC ACID

2-AMINOADIPATE AMINOTRANSFERASE		30% 1 mM
NAD ADP-RIBOSYLTRANSFERASE		*I*50 = 670 μM

KYNURENINE

2-AMINOADIPATE AMINOTRANSFERASE		50% 1 mM
CYSTEINE-CONJUGATE β-LYASE		57% 1 mM

L-KYNURENINE

KYNURENINE-OXOGLUTARATE AMINOTRANSFERASE		> 6 mM
TRYPTOPHANASE	C	*K*i = 710 μM

L

L 86-8275

CYCLIN DEPENDENT KINASE 2	C	*K*i = 41 nM
PROTEIN KINASE A		*I*50 = 145 μM
PROTEIN KINASE C		*I*50 = 6 μM
PROTEIN TYROSINE KINASE EGFR		*I*50 = 25 μM

L 155212

PEPTIDYL-DIPEPTIDASE A		*I*50 = 2.7 nM
XAA-PRO AMINOPEPTIDASE		100% 1mM
XAA-PRO AMINOPEPTIDASE		100% 1 mM
XAA-PRO AMINOPEPTIDASE		*I*50 = 300 nM

L 243799

CARBONATE DEHYDRATASE II		*I*50 = 4 nM

L 245151

CARBONATE DEHYDRATASE II		*I*50 = 4 nM

L 363563

ENDOTHIAPEPSIN		*K*i = 40 nM

L 363564

CATHEPSIN D		*K*i = 70 nM
CATHEPSIN E		*K*i = 8 nM
GASTRICSIN		*K*i = 440 nM
PEPSIN		*K*i = 1 μM
PROTEINASE ENDOTHIA		*K*i = 40 nM
RENIN		*K*i = 2.3 nM

L 364099

CATHEPSIN D		*K*i = 730 nM
CATHEPSIN E		*K*i = 60 nM

ENDOTHIAPEPSIN		Ki = 420 nM
GASTRICSIN		Ki = 860 nM
PEPSIN		Ki = 2.2 μM
PROTEINASE ENDOTHIA		Ki = 420 nM
RENIN		Ki = 160 pM
L 365505		
PROTEINASE HIV-1		I50 = 1 nM
RENIN		I50 = 73 nM
L 370518		
THROMBIN		Ki = 90 pM
L 371912		
THROMBIN		Ki = 5 nM
L 372460		
THROMBIN		Ki = 1.5 nM
L 373890		
THROMBIN		Ki = 500 pM
TRYPSIN		Ki = 570 nM
L 374087		
THROMBIN		Ki = 500 pM
TRYPSIN		Ki = 3.2 μM
L 375052		
THROMBIN		
L 612710		
5α-REDUCTASE	IR	
L 619323		
INTEGRASE		I50 = 1.7 μM
L 634549		
FUMARATE REDUCTASE (NADH)		I50 = 115 μM
L 636619		
THROMBIN		
L 645066		
5α-REDUCTASE		
L 645151		
ARACHIDONATE 5-LIPOXYGENASE		I50 = 400 nM
L 647947		
ELASTASE PANCREATIC		
L 651392		
ARACHIDONATE 5-LIPOXYGENASE		I50 = 300 nM
L 651580		
5α-REDUCTASE		
STEROID 5α-REDUCTASE 1		I50 = 40 nM
STEROID 5α-REDUCTASE 2		I50 = 960 nM
L 651896		
ARACHIDONATE 5-LIPOXYGENASE		I50 = 250 nM
ARACHIDONATE 12-LIPOXYGENASE		I50 = 5.9 μM
CYCLOOXYGENASE		I50 = 57 μM
L 652117		
CATHEPSIN G		
ELASTASE LEUKOCYTE		

L 652243
ARACHIDONATE 5-LIPOXYGENASE — $I50 = 10\ \mu M$
CYCLOOXYGENASE — $I50 = 1\ \mu M$

L 653180
THYMIDINE KINASE

L 654066
Δ4 3-KETOSTEROID 5α REDUCTASE
STEROID 5α-REDUCTASE

L 656224
ARACHIDONATE 5-LIPOXYGENASE — $I50 = 400$ nM
ARACHIDONATE 5-LIPOXYGENASE — $I50 = 400$ nM
ARACHIDONATE 12-LIPOXYGENASE — $I50 = 40\ \mu M$
CYCLOOXYGENASE — $I50 = 20\ \mu M$

L 658758
ELASTASE LEUKOCYTE — IR
ELASTASE PANCREATIC
ELASTASE SPUTUM — $Ki = 1.7\ \mu M$

L 659286
ELASTASE LEUKOCYTE — $Ki = 400$ nM
ELASTASE SPUTUM — $Ki = 350$ nM

L 659699
HYDROXYMETHYLGLUTARYL-CoA REDUCTASE (NADPH) — IR — $I50 = 100$ nM
HYDROXYMETHYLGLUTARYL-CoA SYNTHASE — $I50 = 100$ nM

L 660281
HYDROXYMETHYLGLUTARYL-CoA SYNTHASE — $I50 = 400$ nM

L 660347
HYDROXYMETHYLGLUTARYL-CoA SYNTHASE — $I50 = 100$ nM

L 660631
ACETYL-CoA C-ACETYLTRANSFERASE

L 662583
CARBONATE DEHYDRATASE I — $Ki = 10\ \mu M$
CARBONATE DEHYDRATASE III TOPICAL
CARBONATE DEHYDRATASE IV — $Ki = 10$ nM

L 669262
HYDROXYMETHYLGLUTARYL-CoA REDUCTASE (NADPH) — $I50 = 100$ pg/ml

L 670630
ARACHIDONATE 5-LIPOXYGENASE — $I50 = 23$ nM
ARACHIDONATE 5-LIPOXYGENASE — $I50 = 500$ nM

L 671329
1,3-β-GLUCAN SYNTHASE — $I50 = 640$ nM

L 671776
myo-INOSITOL-1(or 4)-MONOPHOSPHATASE — NC — $Ki = 450$ nM
1-D-myo-INOSITOL-TRISPHOSPHATE 3-KINASE — $I50 = 3$ mM

L 679336
HYDROXYMETHYLGLUTARYL-CoA REDUCTASE (NADPH)

L 680831
ELASTASE LEUKOCYTE

L 680833
ELASTASE LEUKOCYTE — IR — $I50 = 9\ \mu M$

L 680845
ELASTASE LEUKOCYTE

L 681110
Na^+/K^+-EXCHANGING ATPASE (P TYPE) — *K*i = 11 µM

L 681176
PEPTIDYL-DIPEPTIDASE A

L 682679
PROTEINASE HIV — *I*50 = 600 pM
PROTEINASE HIV-1 — *I*50 = 420 pM

L 683453
STEROID 5α-REDUCTASE

L 683845
ELASTASE LEUKOCYTE — IR

L 685434
PROTEINASE HIV-1 — *I*50 = 300 pM

L 685502
PROTEINASE HIV-1 — *I*50 = 450 pM

L 686398
3′,5′-CYCLIC-NUCLEOTIDE PHOSPHODIESTERASE III — *K*i = 27 µM
3′,5′-CYCLIC-NUCLEOTIDE PHOSPHODIESTERASE IV-D — *K*i = 5 µM

L 687630
PROTEINASE HIV-1 — *I*50 = 37 nM

L 687781
1,3-β-GLUCAN SYNTHASE — *I*50 = 160 nM

L 687908
PROTEINASE HIV — *I*50 = 30 pM

L 689065
ARACHIDONATE 5-LIPOXYGENASE — *I*50 = 400 nM

L 689502
PROTEINASE HIV-1 — *K*i = 300 pM

L 690330
myo-INOSITOL-1(or 4)-MONOPHOSPHATASE — C — *K*i = 420 nM
myo-INOSITOL-1(or 4)-MONOPHOSPHATASE — C — *K*i = 190 nM
myo-INOSITOL-1(or 4)-MONOPHOSPHATASE — C — *K*i = 300 nM
myo-INOSITOL-1(or 4)-MONOPHOSPHATASE — C — *K*i = 600 nM
myo-INOSITOL-1(or 4)-MONOPHOSPHATASE — C — *K*i = 1.9 µM
myo-INOSITOL-1(or 4)-MONOPHOSPHATASE — C — *K*i = 2.1 µM
myo-INOSITOL-1(or 4)-MONOPHOSPHATASE — C — *K*i = 270 nM

L 691816
ARACHIDONATE 5-LIPOXYGENASE — *I*50 = 17 nM
ARACHIDONATE 5-LIPOXYGENASE — *I*50 = 8 nM

L 693549
PROTEINASE HIV-1 — *I*50 = 100 pM

L 693612
CARBONATE DEHYDRATASE
CARBONATE DEHYDRATASE II

L 694458
ELASTASE
ELASTASE LEUKOCYTE

L 694599
FARNESYL-DIPHOSPHATE FARNESYLTRANSFERASE — *I*50 = 230 pM

L 694746
PROTEINASE HIV-1 — $I50 = 1nM$

L 696229
RIBONUCLEASE H
RNA-DIRECTED DNA POLYMERASE — $Ki = 23$ nM
RNA-DIRECTED DNA POLYMERASE — C — $I50 = 18$ nM

L 696418
INTERSTITIAL COLLAGENASE — $Ki = 750$ nM
STROMELYSIN 1 — $Ki = 310$ nM

L 696474
PEPSIN — C — $I50 = 52$ μM
PROTEINASE HIV-1 — C — $Ki = 3$ μM
RNA-DIRECTED DNA POLYMERASE — $I50 = 3$ μM

L 697198
ARACHIDONATE 5-LIPOXYGENASE

L 697350
FARNESYL-DIPHOSPHATE FARNESYLTRANSFERASE — $I50 = 370$ pM

L 697639
PROTEINASE HIV-1 — $I50 = 19$ nM
RNA-DIRECTED DNA POLYMERASE — $I50 = 20$ nM

L 697661
RNA-DIRECTED DNA POLYMERASE — $I50 = 2.7$ μM

L 697693
RNA-DIRECTED DNA POLYMERASE — $I50 = 27$ nM

L 697695
RNA-DIRECTED DNA POLYMERASE — $I50 = 17$ nM

L 697908
PROTEINASE HIV-1 — $I50 = 30$ pM

L 699333
ARACHIDONATE 5-LIPOXYGENASE — $I50 = 22$ nM
LEUKOTRIENE-C4 SYNTHASE — NC — $Ki = 5.8$ μM

L 700417
PROTEINASE HIV-1 — $I50 = 670$ pM

L 700497
PROTEINASE HIV-1 — $I50 = 1.8$ nM

L 702019
RNA-DIRECTED DNA POLYMERASE — $I50 = 30$ nM

L 702539
ARACHIDONATE 5-LIPOXYGENASE

L 704486
PROTEINASE HIV-1 — $I50 = 7.8$ nM

L 708714
ARACHIDONATE 5-LIPOXYGENASE — $I50 = 300$ nM

L 708780
ARACHIDONATE 5-LIPOXYGENASE — $I50 = 190$ nM

L 724180
INTEGRASE — $I50 = 9$ μM

L 731078
FARNESYL-DIPHOSPHATE FARNESYLTRANSFERASE — $I50 = 30$ nM

L 731120
FARNESYL-DIPHOSPHATE FARNESYLTRANSFERASE $I50 = 260$ nM

L 731280
FARNESYL-DIPHOSPHATE FARNESYLTRANSFERASE $I50 = 767$ nM

L 731735
FARNESYL PROTEINTRANSFERASE $I50 = 18$ nM
FARNESYL PROTEINTRANSFERASE $I50 = 18$ nM

L 735021
FARNESYL-DIPHOSPHATE FARNESYLTRANSFERASE $I50 = 140$ pM

L 735524
PROTEINASE HIV $I50 = 400$ pM
PROTEINASE HIV-1 C $Ki = 340$ pM
PROTEINASE HIV-2 C $Ki = 3.3$ nM

L 735525
PROTEINASE HIV-1 $Ki = 520$ pM
PROTEINASE HIV-2 $Ki = 3.3$ nM

L 735882
ENDONUCLEASE $I50 = 1.8$ μM

L 737126
RNA-DIRECTED DNA POLYMERASE $I50 = 300$ pM

L 738372
PROTEINASE HIV-1 $I50 = 23$ nM
RNA-DIRECTED DNA POLYMERASE C $Ki = 140$ nM

L 739010
ARACHIDONATE 5-LIPOXYGENASE $I50 = 20$ nM

L 739633
INTEGRASE $I50 = 7$ μM

L 739749
FARNESYL PROTEINTRANSFERASE $I50 = 240$ nM

L 739750
FARNESYL PROTEINTRANSFERASE $I50 = 400$ pM
FARNESYL PROTEINTRANSFERASE $I50 = 1.8$ nM
GERANYLGERANYL PROTEINTRANSFERASE $I50 = 3$ μM

L 739787
FARNESYL PROTEINTRANSFERASE $I50 = 350$ nM

L 743726
RNA-DIRECTED DNA POLYMERASE $Ki = 2.9$ nM

L 744832
FARNESYL PROTEINTRANSFERASE $I50 = 300$ pM

L 745337
CYCLOOXYGENASE 2 $I50 = 23$ nM

L 746530
ARACHIDONATE 5-LIPOXYGENASE $I50 = 27$ nM

L 748496
PROTEINASE HIV-1 $I50 = 120$ pM

L 754394
CYTOCHROME P450 3A SSI
PROTEINASE HIV-1

L 758354

GELATINASE A		$I50 = 17$ nM
GELATINASE A1		$Ki = 17$ nM
STROMELYSIN 1		$Ki = 10$ nM

L 761000

CYCLOOXYGENASE 2		$I50 = 2$ nM

L 761065

CYCLOOXYGENASE 2		$I50 = 20$ nM

L 766112

CYCLOOXYGENASE 2		$I50 = 16$ nM

L 768277

CYCLOOXYGENASE 1		$I50 = 43$ μM
CYCLOOXYGENASE 2		$I50 = 10$ nM

L 784512

CYCLOOXYGENASE 2

L 8027

THROMBOXANE A SYNTHASE

L 868276

CYCLIN DEPENDENT KINASE 2

L 662583AMIDE

CARBONATE DEHYDRATASE II		$Ki = 400$ pM

(–)L 689065

ARACHIDONATE 5-LIPOXYGENASE		$I50 = 40$ nM

L-754394

CYTOCHROME P450 3A

La^{3+}

ALCOHOL DEHYDROGENASE		40% 1 mM
D-AMINO-ACID OXIDASE		
ARYLESTERASE	C	$Ki = 210$ μM
Ca^{2+}/Mg^{2+}-TRANSPORTING ATPASE		
Ca^{2+}-TRANSPORTING ATPASE		
GLUCOSE-6-PHOSPHATE 1-DEHYDROGENASE		91% 830 μM
GLUTAMATE DEHYDROGENASE		83% 830 μM
GLUTAMATE FORMIMINOTRANSFERASE		
ISOCITRATE DEHYDROGENASE		67% 830 μM
L-LACTATE DEHYDROGENASE		83% 830 μM
NAD(P) TRANSHYDROGENASE (AB-SPECIFIC)		
NAD PYROPHOSPHATASE		53% 400 μM
PARAOXON HYDROLASE	NC	$Ki = 280$ μM
PHOSPHATASE ACID II		16% 30 mM
1-PHOSPHATIDYLINOSITOL 4-KINASE		
3-PHOSPHOSHIKIMATE 1-CARBOXYVINYLTRANSFERASE		
PROTEIN-GLUTAMINE γ-GLUTAMYLTRANSFERASE		
PROTEIN KINASE (Ca^{2+} DEPENDENT)		97% 2 mM

LAB 158241F

OBTUSIFOLIOL 14α-DEMETHYLASE		$I50 = 300$ nM

LAB 170250F

LANOSTEROL C14 DEMETHYLASE		
OBTUSIFOLIOL 14-DEMETHYLASE		
OBTUSIFOLIOL 14α-DEMETHYLASE		$I50 = 50$ nM
OBTUSIFOLIOL 14α-DEMETHYLASE		

LACETENOCIN

DEMETHYLMACROCIN O-METHYLTRANSFERASE

LACTACYSTIN

CATHEPSIN A LIKE ENZYME		74% 1 μM
MULTICATALYTIC ENDOPEPTIDASE COMPLEX (CHYMOTRYPSIN LIKE ACTIVITY		
MULTICATALYTIC ENDOPEPTIDASE COMPLEX (PEPTIDYL GLUTAMYL PEPTIDE		
MULTICATALYTIC ENDOPEPTIDASE COMPLEX 6S		
MULTICATALYTIC ENDOPEPTIDASE COMPLEX 20S (CASINOLYTIC ACTIVITY)		*I*50 ≈ 200 μM
MULTICATALYTIC ENDOPEPTIDASE COMPLEX (TRYPSIN LIKE ACTIVITY)		

α-LACTALBUMIN

N-ACETYLLACTOSAMINE SYNTHASE

β-LACTAM ANTIBIOTICS

D-ALANYL-D-ALANINE CARBOXYPEPTIDASE
PROCOLLAGEN-PROLINE, 2-OXOGLUTARATE-4-DIOXYGENASE

β-LACTAMASE INHIBITORY PROTEIN

β-LACTAMASE TEM-1		*K*i ≈ 100 pM

β-LACTAME ANTIBIOTICS

DNA POLYMERASE α

LACTATE

β-GALACTOSIDASE
PYRUVATE KINASE
TRIMETHYLYSINE DIOXYGENASE
UREIDOGLYCOLLATE DEHYDROGENASE

LACTATE 2-PHOSPHATE

PHOSPHOGLYCERATE MUTASE

LACTIC ACID

NICOTINATE-NUCLEOTIDE PYROPHOSPHORYLASE (CARBOXYLATING)

D-LACTIC ACID

ALANOPINE DEHYDROGENASE		
3-HYDROXYBUTYRATE DEHYDROGENASE	C	*K*i = 80 μM
3-HYDROXYBUTYRATE DEHYDROGENASE		
LACTATE 2-MONOOXYGENASE		
D-MALATE DEHYDROGENASE (DECARBOXYLATING)		
PHOSPHOENOLPYRUVATE CARBOXYKINASE (GTP)	C	*K*i = 5 mM
PROLINE DEHYDROGENASE	C	*K*i = 2.1 mM

L-LACTIC ACID

ALANOPINE DEHYDROGENASE		
3-HYDROXYBUTYRATE DEHYDROGENASE		
PHOSPHOENOLPYRUVATE DECARBOXYLASE	C	*K*i = 7.6 mM
PROLINE DEHYDROGENASE	C	*K*i = 1.4 mM
TAUROPINE DEHYDROGENASE	C	*K*i = 8.5 mM
TAUROPINE DEHYDROGENASE		

LACTIMIDE

AMIDASE

LACTITOL

GALACTOSIDE 3(4)-L-FUCOSYLTRANSFERASE	C	*K*i = 17 mM

closter-LACTOCYSTIN β-LACTONE

MULTICATALYTIC ENDOPEPDTIDASE COMPLEX (CHYMOTRYPSIN LIKE ACTIVIT

LACTONE

PEPTIDOGLYCAN β-N-ACETYLMURAMIDASE

β-LACTONE 1233A

HYDROXYMETHYLGLUTARYL-CoA REDUCTASE (NADPH)

LACTORFERRIN

TRYPTASE		*K*i = 24 nM
TRYPTASE		*K*i = 546 μM

LACTOSE

ARABINOGALACTAN ENDO-1,4-β-GALACTOSIDASE		
CELLULOSE 1,4-β-CELLOBIOSIDASE		
α-GALACTOSIDASE		
β-GALACTOSIDASE		*K*i = 1.2 mM

LACTOSYL CERAMIDE

GALACTOSYLGALACTOSYLGLUCOSYLCERAMIDASE

S-D-LACTOYLGLUTATHIONE

LACTOYLGLUTATHIONE LYASE

α-LACTYLAMINO-β-HYDROXY-βAMINOPIMELIC ACID

GLUTAMATE DECARBOXYLASE	C	*K*i = 2.4 μM

LACTYLPEPSTATIN

CATHEPSIN D	*K*i = 100 nM
CHYMOSIN	*K*i = 1.9 μM
MUCOROPEPSIN	*K*i = 700 nM
PEPSIN	*K*i = 50 nM
PEPSIN	*K*i = 400 pM
PEPSIN	*K*i = 1 nM
PEPSIN L	*K*i = 70 nM
PEPSIN L	*K*i = 5.8 μM
PROTEINASES ASPARTIC	*K*i = 4 μM

LAGENARIA LEUCANTHA TRYPSIN INHIBITOR I

TRYPSIN	*K*i = 360 pM

LAGENARIA LEUCANTHA TRYPSIN INHIBITOR II

TRYPSIN	*K*i = 65 pM

LAGENARIA LEUCANTHA TRYPSIN INHIBITOR-II

COAGULATION FACTOR Xa	*K*i = 41 μM
COAGULATION FACTOR XIIa	*K*i = 1.4 μM
KALLIKREIN PLASMA	*K*i = 27 μM

LAGENARIA LEUCANTHA TRYPSIN INHIBITOR III

TRYPSIN	*K*i = 300 pM

LAGENARIA LEUCANTHA TRYPSIN INHIBITOR-III

COAGULATION FACTOR Xa	*K*i = 19 μM
COAGULATION FACTOR XIIa	*K*i = 4.2 μM
KALLIKREIN PLASMA	*K*i = 200 μM

LAMINARAN

ENDONUCLEASES

LAMINARIBOSE

β-GLUCOSIDASE

LAMINARIN

GLUCAN ENDO-1,3-β-GLUCOSIDASE

LANNATE

ACETYLCHOLINESTERASE	3.1.1.7

5α-LANOSTA-8,24-DIEN-3β-OL

CHOLESTEROL OXIDASE

LANOSTEROL

11β-HYDROXYSTEROID DEHYDROGENASE	*I*50 = 650 μM
STEROL O-ACYLTRANSFERASE	

LANOSTEROL ANALOGUE 7

Δ(24)-STEROL C-METHYLTRANSFERASE	NC	*K*i = 20 μM

LANOSTEROL ANALOGUE 8

Δ(24)-STEROL C-METHYLTRANSFERASE	NC	*K*i = 40 μM

Inhibitor / Enzyme	Type	Value
LANSOPRAZOLE		
CYTOCHROME P450 3A	NC	Ki = 170 μM
CYTOCHROME P450 2C9	C	Ki = 52 μM
CYTOCHROME P450 2C19	C	Ki = 3.2 μM
CYTOCHROME P450 2D6		Ki = 22 μM
DEXTROMETHROPHAN O-DEMETHYLASE		
H^+/K^+-EXCHANGING ATPASE		I50 = 2.1 μM
UREASE		I50 = 9.3 μM
LANTHIONINE		
DIAMINOPIMELATE EPIMERASE	C	Ki = 180 μM
2S,6S-LANTHIONINE		
DIAMINOPIMELATE EPIMERASE	C	Ki = 420 μM
L-LANTHIONINE		
DIAMINOPIMELATE DECARBOXYLASE		
meso-LANTHIONINE		
DIAMINOPIMELATE EPIMERASE	MIX	Ki = 180 μM
LAPACHOL		
DIHYDROOROTATE DEHYDROGENASE	UC	Ki = 2.1 μM
LACTOYLGLUTATHIONE LYASE	C	Ki = 38 μM
LACTOYLGLUTATHIONE LYASE	C	Ki = 8 μM
NAD(P)H DEHYDROGENASE (QUINONE)		I50 = 300 nM
VITAMIN K EPOXIDE REDUCTASE		I50 = 3 μM
VITAMIN K-EPOXIDE REDUCTASE (WARFARIN SENSITIVE)		
VITAMIN K QUINONE REDUCTASE		
β-LAPACHONE		
DNA TOPOISOMERASE		
LARGE INHIBITOR OF METALLOPROTEINASES		
GELATINASE A		
INTERSTITIAL COLLAGENASE		
STROMELYSIN		
LASALOCID		
Ca^{2+}-TRANSPORTING ATPASE		
LATERITIN		
STEROL O-ACYLTRANSFERASE		I50 = 5.7 μM
LATOSYL CERAMIDE		
GALACTOSYLCERAMIDASE	C	
LAUDANOSINE		
BUFURALOL-1′-HYDROXYLASE (CYT P-450dbl)	C	I50 = 33 μM
LAURIC ACID		
GLUCOKINASE		I50 = 500 μM
HEXOKINASE		I50 = 1 mM
NAD(P)-ARGININE ADP-RIBOSYLTRANSFERASE		I50 = 1.1 mM
PEPTIDYL-DIPEPTIDASE A	NC	Ki = 2.5 mM
PYRUVATE KINASE		
LAURIC ACID SUCROSE ESTER		
LIPOXYGENASE 1		I50 = 2.3 mM
n-LAUROYLTYRAMINEPHOSPHATE		
ESTRONE SULFATASE		Ki = 3.6 μM
LAUROYL VANILLYLAMIDE		
NADH DEHYDROGENASE (UBIQUINONE)		I50 = 2.5 μM
LAURYLCARNITINE		
PROLYL OLIGOPEPTIDASE	NC	Ki = 250 μM

LAURYL-CoA

ACETYL-CoA CARBOXYLASE	C	Ki = 74 μM
GLUCOSE-6-PHOSPHATASE		33% 50 μM
GLUCOSE-6-PHOSPHATE 1-DEHYDROGENASE	C	Ki = 870 nM

LAURYL GALLATE

NADH DEHYDROGENASE		I50 = 9.3 μM
NADH DEHYDROGENASE		I50 = 9.5 μM

LAURYLGALLATE

PROTEIN KINASE A		I50 = 1.5 μM

N-LAURYLSARCOSINE

α-GLUCOSAMIDINE-N-ACETYLTRANSFERASE		100% 1%
HEPARAN-α-GLUCOSAMINIDINE N-ACETYLTRANSFERASE		

LAURYLSULPHATE

ASPERGILLUS ALKALINE PROTEINASE
β-FRUCTOFURANOSIDASE
GLUCAN ENDO-1,3-β-GLUCOSIDASE
GLUCONATE 5-DEHYDROGENASE
PEPTIDYL-GLUTAMINASE
PROTEIN-GLUTAMINE GLUTAMINASE
XYLAN 1,4-β-XYLOSIDASE

LAURYLSULPHONIC ACID

β-GALACTOSIDASE		100% 10 mM

LAURYLSULPHONYLFLUORIDE

N-(LONG-CHAIN-ACYL)ETHANOLAMINE DEACYLASE		I50 = 2 nM
N-(LONG-CHAIN-ACYL)ETHANOLAMINE DEACYLASE		I50 = 3 nM

LAVENDUSTIN A

PHOSPHATIDYLINOSITOL KINASE		I50 = 6.4 μg/ml
PROTEIN KINASE C		I50 = 14 μM
PROTEIN TYROSINE KINASE EGFR		I50 = 4.4 ng/ml
PROTEIN TYROSINE KINASE EGFR		I50 = 60 nM
PROTEIN TYROSINE KINASE pp60F527		I50 = 18 μM

LAVENDUSTIN B

PROTEIN TYROSINE KINASE EGFR		I50 = 490 ng/ml

LAVENDUSTIN C6

PROTEIN TYROSINE KINASE pp60F527		I50 = 5 μM

LAWSONE

LACTOYLGLUTATHIONE LYASE	C	Ki = 214 μM
NAD ADP-RIBOSYLTRANSFERASE		

LAZABEMIDE

AMINE OXIDASE MAO-B

LB 71350

PROTEINASE HIV-1

LDL OXIDIZED

PHOSPHATIDYLCHOLINE-STEROL O-ACYLTRANSFERASE

LEAD

RIBOSYLPYRIMIDINE NUCLEOSIDASE

LEAD ACETATE

GLUTATHIONE TRANSFERASE

LECITHIN

ASPULVINONE DIMETHYLALLYLTRANSFERASE		
BUTYRATE-CoA LIGASE		60% 10 mM
SPHINGOMYELIN PHOSPHODIESTERASE		
STEROL ESTERASE		

LECTIN HELIX POMATIA
POLYPEPTIDE N-ACETYLGALACTOSAMINYLTRANSFERASE

LECTIN SOY BEAN
POLYPEPTIDE N-ACETYLGALACTOSAMINYLTRANSFERASE

LEECH TRYPTASE INHIBITOR
CHYMOTRYPSIN — $Ki = 20$ nM
TRYPSIN — $Ki = 900$ pM
TRYPTASE I — $Ki \approx 1.4$ nM

LEFLUNOMIDE
PROTEIN TYROSINE KINASE p56lck — $I50 = 160$ µM
PROTEIN TYROSINE KINASE p59fyn — $I50 \approx 175$ µM

LEKOTRIEN A4
ARACHIDONATE 5-LIPOXYGENASE — IR — $I50 = 22$ µM

LEMINOPRAZOLE
H^+/K^+-EXCHANGING ATPASE

LENTARON
AROMATASE — C

LENTIGINOSINE
GLUCAN 1,4-α-GLUCOSIDASE — $I50 = 32$ µM

LEOKUTRIEN C4
15-HYDROXYPROSTAGLANDIN DEHYDROGENASE (NADP) — C — $I50 = 2.3$ µM

LEPTOPHOS
ACETYLCHOLINESTERASE

LESPEDEZA CUNEATA CELLULASE INHIBITOR
GLUCAN ENDO-1,3-β-GLUCOSIDASE

LETD PROTEIN
DNA TOPOISOMERASE (ATP-HYDROLYSING)

LETHODIM
ACETYL-CoA CARBOXYLASE

LEU
TRIPEPTIDE AMINOPEPTIDASE — $I50 = 500$ µM

LEU-ALA
UBIQUITIN-PROTEIN LIGASE

LEU-ALA-GLU-GLY-SER-ALA-ALA-2,3,5,6-TETRAFLUORO-D-TYR-GLU-GLU-OH
PROTEIN TYROSINE KINASE EGFR — $Ki = 9.6$ µM

LEU-ARG
DIPEPTIDYL-PEPTIDASE 3 — $Ki = 9.8$ µM

LEU-ARG-ARG-ALA-D,L-(2-AMINO-4-DIETHYLPHOSPHONOBUTYRIC ACID)LEU-GLY A
PROTEIN KINASE A — $Ki = 9$ µM

LEU-ARG-PRO
PEPTIDYL-DIPEPTIDASE A — $I50 = 270$ nM

LEUCAENA LEUCOCEPHALA SERINE PROTEINASE INHIBITOR
COAGULATION FACTOR XIIa
KALLIKREIN PLASMA — $Ki = 6.3$ nM
PLASMIN — $Ki = 320$ pM
TRYPSIN — $Ki = 25$ nM

LEUCINAL
LEUCYL AMINOPEPTIDASE — $Ki = 60$ nM
MEMBRANE ALANINE AMINOPEPTIDASE — $Ki = 760$ nM
TRIPEPTIDE AMINOPEPTIDASE

LEUCINE		
ACETOLACTATE SYNTHASE		
ACETOLACTATE SYNTHASE III		35% 25 mM
ALANINE AMINOPEPTIDASE	NC	$I50 = 1.4$ mM
D-ALANINE γ-GLUTAMYLTRANSFERASE		
ALANOPINE DEHYDROGENASE		
D-AMINO-ACID OXIDASE		
2-AMINOHEXANOATE TRANSAMINASE		
ARGINASE	C	$Ki = 14$ mM
2-ETHYLMALATE SYNTHASE		$I50 = 450$ μM
GLUTAMATE SYNTHASE (FERREDOXIN)		
GLUTAMATE SYNTHASE (NADH)		
2-ISOPROPYLMALATE SYNTHASE		
LEUCYL AMINOPEPTIDASE	C	$Ki = 10$ mM
MEMBRANE ALANINE AMINOPEPTIDASE		
ORNITHINE-OXO-ACID TRANSAMINASE		
PHENYLALANINE DEHYDROGENASE		
PHOSPHATASE ALKALINE	U	$Ki = 9.2$ mM
1-PYRROLINE-5-CARBOXYLATE DEHYDROGENASE		
SACCHAROPINE DEHYDROGENASE (NAD, L-LYSINE FORMING)		
SACCHAROPINE DEHYDROGENASE (NADP, L-GLUTAMATE FORMING)		
SACCHAROPINE DEHYDROGENASE (NADP, L-LYSINE FORMING)		
VALINE-PYRUVATE TRANSAMINASE	C	$Ki = 3$ mM
L-LEUCINEAMIDE		
ASPARAGINASE		
D-LEUCINE		
PHENYLALANINE DEHYDROGENASE		
PHOSPHATASE ALKALINE	U	$Ki = 53$ mM
LEUCINEHYDROXAMATE		
AMINOPEPTIDASE		50% 20 μM
L-LEUCINEHYDROXAMATE		
S-BENZYL-L-CYSTEINE-p-NITROANILIDE HYDROLASE		
L-LEUCINE		
S-BENZYL-L-CYSTEINE-p-NITROANILIDE HYDROLASE		
DIPEPTIDASE	C	$Ki = 65$ mM
(S)-2-HYDROXY-ACID OXIDASE		
2-ISOPROPYLMALATE SYNTHASE		$I50 = 6$ μM
LEUCINE DEHYDROGENASE		
LEUCYL AMINOPEPTIDASE	C	
ORNITHINE CARBAMOYLTRANSFERASE	C	$Ki = 1.2$ mM
PHOSPHATASE ALKALINE	U	
LEUCINETHIOL		
LEUCYL AMINOPEPTIDASE	C	
MEMBRANE ALANINE AMINOPEPTIDASE	C	$Ki = 22$ nM
LEUCINOL		
LEUCYL AMINOPEPTIDASE		
LEUCINOSTATIN		
H^+-TRANSPORTING ATPASE		
LEUCINYL-ALANINE		
PEPTIDYL-DIPEPTIDASE A		$Ki = 400$ μM
LEUCOANTHOCYANIDIN		
o-DIPHENOL OXIDASE		
PHENOL OXIDASE	NC	
LEUCOBENZAURIN		
GLUTATHIONE TRANSFERASE		

LEUCOCYANIDOL
L-ASCORBATE OXIDASE

LEUCOPTERIN
XANTHINE DEHYDROGENASE

LEUCOSIDE
ARACHIDONATE 5-LIPOXYGENASE — $I50 = 290\ \mu M$

N-(LEUCYL)-o-AMINOBENZENESULPHONIC ACID
LEUCYL AMINOPEPTIDASE

LEUCYLCHLOROMETHYLKETONE
CATHEPSIN L — 50% 30 μM
TYROSYL AMINOPEPTIDASE — 93% 1 mM

LEUCYLETHYLESTER
LEUCYL AMINOPEPTIDASE — C — $Ki = 80$ mM

LEU-ENKEPHALIN
CARBOXYPEPTIDASE H — C — $Ki = 6.5$ mM

LEU-GLY
UBIQUITIN-PROTEIN LIGASE

D-LEU-GLY
DIPEPTIDASE — $Ki = 350\ \mu M$

LEU-GLY-GLY
TRIPEPTIDE AMINOPEPTIDASE — $I50 = 975\ \mu M$

LEUHISTIN
AMINOPEPTIDASE B — $I50 = 13\ \mu g/ml$
GLUTAMYL AMINOPEPTIDASE — $I50 = 10\ \mu g/ml$
MEMBRANE ALANINE AMINOPEPTIDASE — $I50 = 74\ \mu M$
MEMBRANE ALANINE AMINOPEPTIDASE — C — $Ki = 230$ nM

LEUKOCYTE ELASTASE INHIBITOR
ELASTASE LEUKOCYTE

LEUKOTRIENE A3
LEUKOTRIENE A4 HYDROLASE — IR

LEUKOTRIENE A4
LEUKOTRIENE A4 HYDROLASE — IR

LEUKOTRIENE A5
LEUKOTRIENE A4 HYDROLASE — IR

7,11-trans-9-cis-LEUKOTRIENE A4
LEUKOTRIENE A4 HYDROLASE — IR

7-trans-9,11-cis-LEUKOTRIENE A4
LEUKOTRIENE A4 HYDROLASE — IR

14,15-LEUKOTRIENE A4
LEUKOTRIENE A4 HYDROLASE — IR

LEUKOTRIENE A4 METHYLESTER
LEUKOTRIENE A4 HYDROLASE — IR

LEUKOTRIENE C4
N-ACETYLGALACTOSAMINE-4-SULFATASE — C

LEUKOTRIENE D4
N-ACETYLGALACTOSAMINE-4-SULFATASE — C

LEUKOTRIENE E4
N-ACETYLGALACTOSAMINE-4-SULFATASE — C

LEU-NHOH		
MEMBRANE ALANINE AMINOPEPTIDASE		$Ki = 14\ \mu M$
D-LEU-NHOH		
BACTERIAL LEUCYL AMINOPEPTIDASE		$Ki = 2$ nM
LEUCYL AMINOPEPTIDASE		$Ki = 1.3\ \mu M$
LEUPEPTIN		
ACROSIN		$Ki = 210$ nM
CALPAIN		$Ki = 10$ nM
CALPAIN I		$Ki = 320$ nM
CALPAIN II		$Ki = 430$ nM
CANCER PROCOAGULANT		71% 100 μM
CANCER PROCOAGULANT		71% 100 μM
CATHEPSIN B		97% 1 μM
CATHEPSIN B	IR	$Ki = 4.1$ nM
CATHEPSIN H		
CATHEPSIN L		99% 100 μM
CATHEPSIN L		50% 100 nM
CATHEPSIN R		
CHYMASE	NC	$I50 = 258\ \mu M$
CHYMOTRYPSIN		$I50 = 1.1$ mM
CLOSTRIPAIN		
COAGULATION FACTOR Xa		$I50 = 7.9\ \mu M$
COMPLEMENT SUBCOMPONENT C1r	C	
COMPLEMENT SUBCOMPONENT C1s		
DNA POLYMERASE β		17% 1 mM
ELASTASE		68% 100 μM
ELASTASE PANCREATIC II		93% 1 μM
HISTOLYSAIN		$Ki = 43$ nM
HUMICOLIN		$I50 = 100\ \mu M$
KALLIKREIN PLASMA		
KALLIKREIN TISSUE	C	$Ki = 700$ nM
KALLIKREIN URINE		$Ki = 19\ \mu M$
LACTOSYLCERAMIDE α-2,3-SIALYLTRANSFERASE		72% 1 μM
LEGUMAIN		62% 1 mM
LYS-GINGIVAIN		100% 100 μM
MULTICATALYTIC ENDOPEPTIDASE COMPLEX		
MULTICATALYTIC ENDOPEPTIDASE COMPLEX (TRYPSIN LIKE ACTIVITY)		
MULTICATALYTIC ENDOPEPTIDASE COMPLEX (TRYPSIN LIKE ACTIVITY)		
PAPAIN		$Ki = 2.2$ nM
PLASMIN		$Ki = 3.4\ \mu M$
PRO-OPIOMELANOCORTIN CONVERTING ENZYME		58% 1 mM
PROTEINASE ASPARTIC PLASMODIUM		
PROTEINASE BOAR EPIDIDYMAL SPERM		99% 10 μM
PROTEINASE CATHEPSIN S LIKE		93% 100 nM
PROTEINASE Ca^{2+} THIOL DROSOPHILA MELANOGASTER		
PROTEINASE CYSTEINE LEISHMANIA MEXICANA		100% 5 μg/ml
PROTEINASE DF		$Ki = 17$ nM
PROTEINASE II MYXOBACTER AL-1		
PROTEINASE In		$I50 = 84\ \mu M$
PROTEINASE I PACIFIC WHITING		100% 10 μM
PROTEINASE METALLO ASCIDIAN HEMOCYTE I		27% 100 μM
PROTEINASE METALLO ASCIDIAN HEMOCYTE II		34% 100 μM
PROTEINASE SERINE PANCREATIC MICROSOMAL		94% 2 μM
THIMET OLIGOPEPTIDASE		32% 120 μM
THROMBIN		$Ki = 45\ \mu M$
TONIN	C	$Ki = 140\ \mu M$
TRYPSIN		99% 100 μM
TRYPSIN		$Ki = 35$ nM
TRYPSIN		
TRYPSIN LIKE ACTIVITY		13% 1.3 μg/ml
TRYPTASE	C	$I50 = 1.4\ \mu M$
TRYPTASE TL-2		87% 10 μM

UCA PUGILATOR COLLAGENOLYTIC PROTEINASE
YEAST PROTEINASE B — 33% 50 μM

LEUPEPTIN ACETYL
KALLIKREIN — $I50 = 163$ μM
PLASMIN — $I50 = 234$ μM
THROMBIN — $I50 = 28$ mM
TRYPSIN — $I50 = 218$ μM

LEUPEPTIN PROLYL
KALLIKREIN — $I50 = 170$ μM
THROMBIN — $I50 = 22$ mM
TRYPSIN — $I50 = 211$ μM

LEU-PHE-$CF_2CH_2CH_2CO$-leu-ARG-OME
CHYMOTRYPSIN — $Ki = 3$ nM

H-D-LEU-PHE-CH_2Ph
CHYMOTRYPSIN — $Ki = 3{,}6$ μM

H-D-LEU-PHE-$NHCH_2C_6H_4Fp$
CHYMOTRYPSIN — $Ki = 610$ nM

LEU-POLYOXIN B
CHITIN SYNTHASE — $I50 = 20$ ng/ml

LEU-PRO-PRO
PEPTIDYL-DIPEPTIDASE A — 100% 10 μM

LEU-THIOL
GLUTAMYL AMINOPEPTIDASE — $I50 = 51$ nM
LEUKOTRIENE A4 HYDROLASE — 75% 2.5 μM

LEU-TRP
ENDOPEPTIDASE TREPONEMA DENTICOLA ARCC35405 — $Ki = 3.2$ mM
THERMOLYSIN — C — $Ki = 85$ μM

LEU-TYR
PEPTIDYL-DIPEPTIDASE A — C — $I50 = 39$ μM

LEU-VAL
UBIQUITIN-PROTEIN LIGASE

LEU-VAL-PHE-{$CHOHCH_2N$}PHE-ILE-VAL-NH_2
PROTEINASE HIV-1 — $I50 = 5$ nM

LEVAFIX E-5BNA
CHOLINE-PHOSPHATE CYTIDYLYLTRANSFERASE

LEVALLORPHAN
GUANOSINE-3′,5′-BIS(DIPHOSPHATE) 3′-PYROPHOSPHATASE

LEVAMISOLE
Na^+/K^+-EXCHANGING ATPASE
Na^+/K^+-EXCHANGING ATPASE
4-NITROPHENYLPHOSPHATASE
5′-NUCLEOTIDASE
PHOSPHATASE ALKALINE — $I50 = 900$ μM
PHOSPHATASE ALKALINE — $I50 = 900$ μM
PHOSPHATASE ALKALINE — $I50 = 50$ μM
PHOSPHATASE ALKALINE — $I50 = 1.4$ mM
PHOSPHATASE ALKALINE — $I50 = 33$ μM
PHOSPHATASE ALKALINE — U — $Ki = 45$ μM
PHOSPHATASE ALKALINE — $I50 = 1.2$ mM
PHOSPHATASE ALKALINE (PHOSPHODIESTERASE ACTIVITY) — U — $Ki = 500$ μM

LEVODOPA
PYRIDOXAL KINASE — 16% 100 μM

Inhibitor / Enzyme	Type	Value
LEVOFLOXACIN		
DNA GYRASE		$I50$ = 31 μg/ml
DNA TOPOISOMERASE (ATP-HYDROLYSING)		$I50$ = 650 ng/ml
DNA TOPOISOMERASE IV		$I50$ = 2.3 μg/ml
LEVULINATE		
PORPHOBILINOGEN SYNTHASE	C	
LEVULINIC ACID		
GLUTAMATE DECARBOXYLASE		
PORPHOBILINOGEN SYNTHASE		29% 33 mM
LEX 032		
PROTEINASE 3		Ki = 12 nM
LEXITROPSIN		
INTEGRASE		
LGC 40863		
ACETOLACTATE SYNTHASE		
Li^+		
ACETOIN DEHYDROGENASE		
ADENYLATE CYCLASE		
ADENYLATE KINASE		
ALDEHYDE REDUCTASE		> 200 mM
L-ASCORBATE PEROXIDASE		
ASPARAGINE SYNTHASE (GLUTAMINE-HYDROLYSING)		
CARBONYL REDUCTASE (NADPH)		
CARBOSINE SYNTHASE		
CARNITINE 3-DEHYDROGENASE		
CITRATE (si)-SYNTHASE		
DIAMINOPIMELATE DEHYDROGENASE		
DIPEPTIDYL-PEPTIDASE 2		
EXOPOLYPHOSPHATASE		
FRUCTOSE-BISPHOSPHATASE		
FRUCTOSE-BISPHOSPHATASE	NC	
FRUCTOSE-BISPHOSPHATASE		Ki = 80 μM
GLUCAN 1,3-β-GLUCOSIDASE		
GLUCOSE-1,6-BISPHOSPHATE SYNTHASE		
GLUTAMATE FORMIMINOTRANSFERASE		
GLYCOGEN SYNTHASE KINASE 3		
GLYCOGEN SYNTHASE KINASE 3β		$I50$ = 2 mM
GUANINE DEAMINASE A		
GUANYLATE KINASE		
INORGANIC PYROPHOSPHATASE		
INOSITOL-BISPHOSPHATE PHOSPHATASE		Ki = 2.5 mM
INOSITOL-1,4-BISPHOSPHATE 1-PHOSPHATASE		Ki = 300 μM
INOSITOL-1,4-BISPHOSPHATE 1-PHOSPHATASE	U	Ki = 600 μM
INOSITOL-1,4-BISPHOSPHATE 1-PHOSPHATASE	UC	Ki = 6 mM
INOSITOL-1,4-BISPHOSPHATE 4-PHOSPHOHYDROLASE	UC	Ki = 600 μM
myo-INOSITOL-1(or 4)-MONOPHOSPHATASE	C	Ki = 1.5 mM
myo-INOSITOL-1(or 4)-MONOPHOSPHATASE	C	Ki = 1.1 mM
myo-INOSITOL-1(or 4)-MONOPHOSPHATASE	C	Ki = 110 μM
myo-INOSITOL-1(or 4)-MONOPHOSPHATASE	UC	Ki = 600 μM
myo-INOSITOL-1(or 4)-MONOPHOSPHATASE		Ki = 1 mM
3-ISOPROPYLMALATE DEHYDROGENASE		
α-MANNOSIDASE		
6-PHOSPHO-β-GALACTOSIDASE		18% 2 mM
PHOSPHOPYRUVATE HYDRATASE		
POLYDEOXYRIBONUCLEOTIDE SYNTHASE (ATP)		
PROLINE-tRNA LIGASE		
PROPANEDIOL DEHYDRATASE		
PROTEINASE I ACHROMOBACTER		Ki = 24 mM
RHAMNULOSE-1-PHOSPHATE ALDOLASE	C	Ki = 6 mM
SELENEPHOSPHATE SYNTHASE		

Inhibitor / Enzyme	Type	Value
STARCH SYNTHASE		> 20 mM
SUCROSE α-GLUCOSIDASE		
THYMIDINE KINASE		
UDP-N-ACETYLGLUCOSAMINE 2-EPIMERASE		47% 1 mM
UREA CARBOXYLASE		> 100 mM
LIAROZOLE		
AROMATASE		*I*50 = 3 nM
AROMATASE		
17α-HYDROXYPROGESTERONE ALDOLASE		
RETINOIC ACID 4-HYDROXYLASE		
RETINOIC ACID 4-HYDROXYLASE		
STEROID 17α-HYDROXYLASE/17,20 LYASE		
STEROID 17α-HYDROXYLASE/17,20 LYASE		
STEROID 11β-MONOOXYGENASE		*I*50 = 2.5 μM
STEROID 11β-MONOOXYGENASE		*I*50 = 2.5 μM
STEROID 17α-MONOOXYGENASE		
LIBR		
GLYCOGEN (STARCH) SYNTHASE		
LiCl		
ACYL[ACYL-CARRIER-PROTEIN]-UDP-N-ACETYLGLUCOSAMINE O-ACYLTRANSFERASE		
ASPARAGINE-tRNA LIGASE		
CREATINE KINASE		
GLYCINE HYDROXYMETHYLTRANSFERASE		
LYSYL ENDOPEPTIDASE		*K*i = 90 mM
OXALACETATE TAUTOMERASE		32% 40 mM
LICOARYLCOUMARIN		
3′,5′-CYCLIC-NUCLEOTIDE PHOSPHODIESTERASE (cycAMP)		*I*50 = 10 μM
LICOCHALCONE B		
XANTHINE OXIDASE		*I*50 = 30 μM
LICOCOUMARONE		
XANTHINE OXIDASE		*I*50 = 13 μM
LICORICIDIN		
3′,5′-CYCLIC-NUCLEOTIDE PHOSPHODIESTERASE (cycAMP)		*I*50 = 49 μM
LICORICONE		
3′,5′-CYCLIC-NUCLEOTIDE PHOSPHODIESTERASE (cycAMP)		*I*50 = 23 μM
LIDOCAINE		
ADENYLATE CYCLASE	NC	
CYTOCHROME-C OXIDASE		*K*i = 40 mM
STEROL O-ACYLTRANSFERASE		*I*50 = 800 μM
LIDOFLAZINE		
ADENOSINE DEAMINASE		*K*i = 30 μM
NUCLEOSIDE TRANSPORT		*I*50 = 5 μM
LIGHT GREEN SF YELLOWISH		
PROTEINASE HIV-1		*I*50 = 8 μM
LIGNAN GLUCOSIDES		
3′,5′-CYCLIC-NUCLEOTIDE PHOSPHODIESTERASE (cycAMP)		
LIGNANS		
Na^+/K^+-EXCHANGING ATPASE		
LIGNIN		
ADP-RIBOSE GLYCOHYDROLASE	C	*K*i = 18 μg/ml
LIGNIN ALKALI		
PROTEINASE HIV-1		*I*50 = 50 μg/ml

LIGNOCERIC ACID	
PROTEIN KINASE (Ca^{2+} DEPENDENT)	*I*50 = 4 μM
LILLY COMPOUND 18947	
METHANE MONOOXYGENASE	100% 1 mM
LILLY COMPOUND 53375	
METHANE MONOOXYGENASE	100% 1 mM
LIMA BEAN INHIBITOR	
COAGULATION FACTOR XIIa	
LIMA BEAN TRYPSIN INHIBITOR	
ACROSIN	
BOAR SPERM ACIDIC ARGININE AMIDASE 1	*I*50 = 47 μg
BOAR SPERM ACIDIC ARGININE AMIDASE 2	*I*50 = 680 μg
CATHEPSIN G	
CHYMASE	*I*50 = 18 nM
LEUKOCYTE-MEMBRANE NEUTRAL ENDOPEPTIDASE	
PLASMIN	
PROTEINASE ALKALINE	100% 100 μg/ml
TRYPTASE	*I*50 = 2.6 μM
LIMETTIN	
CYTOCHROME P450 2B1	*I*50 = 86 μM
(+)-LIMONENE	
GERANIOL DEHYDROGENASE	
LIMONENE	
CYTOCHROME P450 6D1	*I*50 = 66 μM
LIMULUS CYSTATIN	
CATHEPSIN L	*K*i = 170 pM
FICAIN	*K*i = 520 pM
PAPAIN	*K*i = 80 pM
LIMULUS INTRACELLULAR COAGULATION INHIBITOR 1	
FACTOR C	
KALLIKREIN TISSUE	
KALLIKREIN TISSUE	
LIMULUS LIPOPOLYSACCHARIDE SENSITIVE PROTEINASE	
PLASMIN	
PLASMIN	
t-PLASMINOGEN ACTIVATOR	
α-THROMBIN	
α-THROMBIN	
TRYPSIN	
LIMULUS INTRACELLULAR COAGULATION INHIBITOR 2	
KALLIKREIN TISSUE	
PLASMIN	
α-THROMBIN	
LIMULUS POLYPHEMUS TRYPSIN INHIBITOR	
CHYMOTRYPSIN	97% 8 μM
TRYPSIN	*K*i = 3.3 nM
LINALOOL	
GERANIOL DEHYDROGENASE	
LINALOYLPHOSPHATE	
ISOPENTENYL-DIPHOSPHATE Δ-ISOMERASE	51% 5 mM
LINAQUINONE	
NAD(P)-ARGININE ADP-RIBOSYLTRANSFERASE	*I*50 = 30 μM

LINATASTINE		
ARACHIDONATE 5-LIPOXYGENASE		*I*50 = 320 nM

LINCOMYCIN		
PEPTIDYLTRANSFERASE		

LINOLEIC ACID		
ARACHIDONATE-CoA LIGASE	C	
CHITIN SYNTHASE		
3′,5′-CYCLIC-NUCLEOTIDE PHOSPHODIESTERASE (Ca^{2+}/CALMODULIN)		*I*50 = 85 µM
3′,5′-CYCLIC-NUCLEOTIDE PHOSPHODIESTERASE (cGMP INHIBITABLE)		*I*50 = 97 µM
3′,5′-CYCLIC-NUCLEOTIDE PHOSPHODIESTERASE (cycGMP STIMULATED)		*I*50 = 46 µM
3′,5′-CYCLIC-NUCLEOTIDE PHOSPHODIESTERASE (ROLIPRAM SENSITIVE)		*I*50 = 106 µM
CYCLOOXYGENASE		*I*50 = 500 µM
DNA METHYLTRANSFERASE ECO RI.		*I*50 = %= µM
DNA POLYMERASE α		*I*50 = 38 µM
DNA POLYMERASE α		95% 25 µg/ml
DNA POLYMERASE β		*I*50 = 45 µM
DNA POLYMERASE I		*I*50 = 88 µM
DNA POLYMERASE II		44% 25 µg/ml
GLUTATHIONE TRANSFERASE		*I*50 = 75 µM
GLUTATHIONE TRANSFERASE		*I*50 = 55 µM
GLUTATHIONE TRANSFERASE 11-11		*I*50 = 120 µM
HYDROPEROXIDE ISOMERASE		
15-HYDROXYPROSTAGLANDIN-D DEHYDROGENASE		
9-HYDROXYPROSTAGLANDIN DEHYDROGENASE	NC	
MULTICATALYTIC ENDOPEPTIDASE COMPLEX (CHYMOTRYPSIN LIKE ACTIVITY		
MULTICATALYTIC ENDOPEPTIDASE COMPLEX (PEPTIDYLGLUTAMYLPEPTIDE LI		
MYOSIN-LIGHT-CHAIN KINASE		
NAD ADP-RIBOSYLTRANSFERASE		*I*50 = 48 µM
NAD(P)-ARGININE ADP-RIBOSYLTRANSFERASE		*I*50 = 90 µM
PROSTAGLANDIN-ENDOPEROXIDE SYNTHASE		*I*50 = 2 mM
PROTEIN KINASE A		*I*50 = 20 µM
STEROID 5α-REDUCTASE		

LINOLEIC ACID DERIVATIVES		
LINOLEATE ISOMERASE		

LINOLENIC ACID		
CYCLOOXYGENASE		*I*50 = 23 mM
DNA METHYLTRANSFERASE ECO RI.		*I*50 = %= µM
GLYCEROL-3-PHOSPHATE DEHYDROGENASE (NAD)		
NAD ADP-RIBOSYLTRANSFERASE		*I*50 = 110 µM
NAD(P)-ARGININE ADP-RIBOSYLTRANSFERASE		*I*50 = 110 µM
PROSTAGLANDIN-ENDOPEROXIDE SYNTHASE		*I*50 = 1 mM
PROTEIN KINASE A		*I*50 = 60 µM

α-LINOLENIC ACID		
DNA POLYMERASE α		*I*50 = 56 µM
STEROID 5α-REDUCTASE		

γ-LINOLENIC ACID		
DNA POLYMERASE α		*I*50 = 69 µM
NAD ADP-RIBOSYLTRANSFERASE		*I*50 = 120 µM
NAD(P)-ARGININE ADP-RIBOSYLTRANSFERASE		*I*50 = 180 µM
STEROID 5α-REDUCTASE		

trans-LINOLENIC ACID		
LIPOXYGENASE	C	

LINOLENOYL-CoA		
ACETYL-CoA CARBOXYLASE		*K*i = 27 nM
[PYRUVATE DEHYDROGENASE (LIPOAMIDE)] KINASE		

N-LINOLENOYLDOPAMINE		
ARACHIDONATE 5-LIPOXYGENASE		*I*50 = 3.2 nM

LINOLEOYLAMIDE
CONODIPINE-M (PLA2)

N-LINOLEOYLDOPAMINE
ARACHIDONATE 5-LIPOXYGENASE — *I*50 = 2.7 nM

LINOLEYL HYDROXAMATE
ARACHIDONATE 5-LIPOXYGENASE
ARACHIDONATE 15-LIPOXYGENASE

LINOLIC ACID
1,3-β-GLUCAN SYNTHASE — *I*50 = 30 μM

LINURON
3-PHYTASE

(S)-LIPEDOSIDE B-III
STEROL O-ACYLTRANSFERASE — *I*50 = 269 μM

LIPID A
PHOSPHOLIPASE D (GLYCOSYLPHOSPHATIDYLINOSITOL SPECIFIC) — *I*50 = 1.5 μM

LIPID ASSOCIATED COAGULATION INHIBITOR
COAGULATION FACTOR Xa

LIPIDHYDROPEROXYDES
PROSTACYCLIN SYNTHASE

LIPIDS
Na^+/K^+-EXCHANGING ATPASE

LIPOCALIN
PAPAIN

LIPOCORTIN I
1-PHOSPHATIDYLINOSITOL PHOSPHODIESTERASE — 100% 670 nM

LIPOCORTIN II
1-PHOSPHATIDYLINOSITOL PHOSPHODIESTERASE — 100% 670 nM
PHOSPHOLIPASE A2 — 90% 200 μM

LIPOCORTIN LIKE PROTEINS
PHOSPHOLIPASE A2

LIPOCORTINS
PHOSPHOLIPASE A2

LIPOHEXIN
PAPAIN — *I*50 = 34 μM
PROLYL OLIGOPEPTIDASE — C — *K*i = 3.5 μM

LIPOIC ACID
(S)-2-HYDROXY-ACID OXIDASE — NC
PHOSPHORYLASE b
PROLINE RACEMASE

α-LIPOIC ACID
NADPH-FERRIHEMOPROTEIN REDUCTASE — IR

D,L-LIPOIC ACID
ACETYL-CoA HYDROLASE — NC — *K*i = 5 μM

LIPOMODULIN
PHOSPHOLIPASE A2

LIPOPHOSPHOGLYCAN
PROTEIN KINASE C — C — *K*i < 1 μM

LIPOPHOSPHOGLYCAN FRAGMENTS
PROTEIN KINASE C

LIPOPOLYSACCHERIDE
LYSOZYME

LIPOSIDOMYCIN
PHOSPHO-N-ACETYLMURAMOYL-PENTAPEPTIDE-TRANSFERASE — *I*50 = 38 ng/ml

LIPOTEICHONIC ACID
N-ACETYLMURAMOYL-L-ALANINE AMIDASE — *K*i = 19 mg/ml
MANNOSYL-GLYCOPROTEIN ENDO-β-N-ACETYLGLUCOSAMINIDASE

LIPOXAMYCIN
SERINE C-PALMITOYLTRANSFERASE — *I*50 = 21 nM

LIPSTATIN
TRIACYLGLYCEROL LIPASE

LIQUIRITIGENIN
3′,5′-CYCLIC-NUCLEOTIDE PHOSPHODIESTERASE (cycAMP) — *I*50 = 1.8 mM
XANTHINE OXIDASE — *I*50 = 11 nM

LIQUIRITIN
PROLYL OLIGOPEPTIDASE — *I*50 = 120 ppm

LIRIODENINE
PROTEIN TYROSINE-PHOSPHATASE CD45 — *I*50 = 128 μM

LISINOPRIL
LYSINE (ARGININE) CARBOXYPEPTIDASE
PEPTIDYL-DIPEPTIDASE A — *I*50 = 630 pM
PEPTIDYL-DIPEPTIDASE A — *I*50 = 1 nM

LITHIUMFLUORIDE
H^+-TRANSPORTING ATPASE

LITHOCHOLATE 3-SULPHATE
GLUTATHIONE TRANSFERASE

LITHOCHOLIC ACID
CHOLESTEROL 7α-MONOOXYGENASE
trans-1,2-DIHYDROBENZENE-1,2-DIOL DEHYDROGENASE 1 — *K*i = 2.7 μM
trans-1,2-DIHYDROBENZENE-1,2-DIOL DEHYDROGENASE 2 — *K*i = 20 nM
DIHYDRODIOL DEHYDROGENASE 2 — 98% 100 μM
DIHYDRODIOL DEHYDROGENASE 4 — 55% 100 μM
GLUTATHIONE TRANSFERASE — *K*i = 5 μM
GLUTATHIONE TRANSFERASE — C — *K*i = 41 μM
11β-HYDROXYSTEROID DEHYDROGENASE — *I*50 = 300 nM
3α-HYDROXYSTEROID DEHYDROGENASE (B-SPECIFIC) — *I*50 = 8 μM
3α-HYDROXYSTEROID/DIHYDRODIOL DEHYDROGENASE 2 — U — *K*i = 40 nM
MORPHINE 6-DEHYDROGENASE
PROTEINASE HIV-1 — *I*50 = 10 μM
PROTEIN KINASE A — *I*50 = 4.2 μM
PROTEIN KINASE C — *I*50 = 57 μM
UDP-GLUCURONOSYLTRANSFERASE — C — *K*i = 30 μM

LITHOSPERMATE Mg^{2+}
LYSYL HYDROXYLASE
PROCOLLAGEN-PROLINE, 2-OXOGLUTARATE-4-DIOXYGENASE

LITHOSPERMIC ACID
ADENYLATE CYCLASE

LITOXETIN
CYTOCHROME P450 1A2 — *I*50 = 60 μM

LIVER ACAT INHIBITOR
STEROL O-ACYLTRANSFERASE — *I*50 = 20 μM

LIVIDOMYCIN B		
GENTAMICIN 3′-N-ACETYLTRANSFERASE		
LIXAZINONE		
3′,5′-CYCLIC-NUCLEOTIDE PHOSPHODIESTERASE III		*I*50 = 10 μM
LL-Z1272β		
FARNESYL PROTEINTRANSFERASE		*I*50 = 700 nM
LM 1228		
PHOSPHOLIPASE A2		
LND 623		
Na^+/K^+-EXCHANGING ATPASE		
LND 796		
Na^+/K^+-EXCHANGING ATPASE		*I*50 = 600 nM
Na^+/K^+-EXCHANGING ATPASE		*I*50 = 1.6 nM
Na^+/K^+-EXCHANGING ATPASE		*I*50 = 6 nM
Na^+/K^+-EXCHANGING ATPASE		*I*50 = 1.8 μM
Na^+/K^+-EXCHANGING ATPASE		*I*50 = 8 μM
Na^+/K^+-EXCHANGING ATPASE		*I*50 = 4 nM
Na^+/K^+-EXCHANGING ATPASE		*I*50 = 20 nM
Na^+/K^+-EXCHANGING ATPASE		*I*50 = 5 μM
LOBELINE		
CARBONATE DEHYDRATASE		
α-LOBELINE		
BUFURALOL-1′-HYDROXYLASE (CYT P-450dbl)	C	*I*50 = 113 nM
LOCUSTA MIGRATORIA PROTEINASE INHIBITOR I		
α-CHYMOTRYPSIN		Kd = 120 pM
ELASTASE LEUKOCYTE		Kd = 18 nM
LOCUSTA MIGRATORIA PROTEINASE INHIBITOR II		
α-CHYMOTRYPSIN		Kd = 250 pM
ELASTASE PANCREATIC		Kd = 40 nM
LOMEFLOXACIN		
CYTOCHROME P450 LA2		23% 500 μM
LOMETREXOL		
PHOSPHORIBOSYLGLYCINAMIDE FORMYLTRANSFERASE		*K*i = 25 nM
PHOSPHORIBOSYLGLYCINAMIDE FORMYLTRANSFERASE		*K*i = 60 nM
LOMUSTINE		
CYTOCHROME P450 2D6	C	*K*i = 7.7 μM
LONAPALENE		
ARACHIDONATE 5-LIPOXYGENASE		*I*50 = 300 nM
ARACHIDONATE 5-LIPOXYGENASE		*I*50 = 700 nM
CYCLOOXYGENASE		*I*50 = 1.3 μM
LONETREXOL		
PHOSPHORIBOSYLGLYCINAMIDE FORMYLTRANSFERASE		*K*i = 60 nM
LONG CHAIN SUBSTRATE		
ENOYL-CoA HYDRATASE (LONG CHAIN)		
LOOPHYTUM CRISTAGALII PROTEIN FARNESYL TRANSFERASE INHIBITOR		
FARNESYL PROTEINTRANSFERASE		*I*50 = 150 nM
GERANYLGERANYL PROTEINTRANSFERASE		*I*50 = 5.3 μM
LORAZEPAM		
UDP-GLUCURONOSYLTRANSFERASE		

LOSARTAN

3′,5′-CYCLIC-NUCLEOTIDE PHOSPHODIESTERASE III		$I50 = 13\ \mu M$
3′,5′-CYCLIC-NUCLEOTIDE PHOSPHODIESTERASE IV		$I50 = 26\ \mu M$

LOVASTATIN

CYTOCHROME P450 2C9		$I50 = 140\ \mu M$
HYDROXYMETHYLGLUTARYL-CoA REDUCTASE (NADPH)		$I50 = 2.6$ nM
HYDROXYMETHYLGLUTARYL-CoA REDUCTASE (NADPH)		$I50 = 45$ nM
HYDROXYMETHYLGLUTARYL-CoA REDUCTASE (NADPH)		$I50 = 6.6$ nM
HYDROXYMETHYLGLUTARYL-CoA REDUCTASE (NADPH)		$I50 = 13$ nM
HYDROXYMETHYLGLUTARYL-CoA REDUCTASE (NADPH)		$I50 = 4.4$ nM
HYDROXYMETHYLGLUTARYL-CoA REDUCTASE (NADPH)		$I50 = 3.1$ nM
STEROL O-ACYLTRANSFERASE	C	$Ki = 36\ \mu M$
STEROL O-ACYLTRANSFERASE		$I50 = 5\ \mu g/ml$

LOVIRIDE

RNA-DIRECTED DNA POLYMERASE		$I50 = 500$ nM

LOXISTATIN

PROTEINASES CYSTEINE

LOXOPROFEN

CARBONYL REDUCTASE (NADPH)		77% 1 mM

LP 149

PROTEINASE FELINE IMMUNODEFICIENCY VIRUS		$I50 = 260$ nM
PROTEINASE HIV		$I50 = 1.7$ nM

.L-PRO.L-BOROPO

DIPEPTIDYL-PEPTIDASE 4		$Ki = 16$ pM

LQ VENOM

H^+-TRANSPORTING ATP SYNTHASE		$Ki = 1.6\ \mu g/ml$

LR-D 009

PLASMIN		$I50 = 1\ \mu M$
THROMBIN		$I50 = 18$ nM
THROMBIN		$I50 = 18$ nM
TRYPSIN		$I50 = 100$ nM

LS 82034

PROTOPORPHYRINOGEN OXIDASE		$Ki = 4$ nM
PROTOPORPHYRINOGEN OXIDASE		$Ki = 12$ nM

LS 82556

PROTOPORPHYRINOGEN OXIDASE		$I50 = 300$ nM

LSD

ARYL-ACYLAMIDASE

LSD 894

CYTOCHROME P450 3A1

LSL 60101

AMINE OXIDASE MAO-A		$I50 = 58\ \mu M$
AMINE OXIDASE MAO-B		$I50 = 16\ \mu M$

LSL 61122

AMINE OXIDASE MAO-A		$I50 = 100\ \mu M$
AMINE OXIDASE MAO-B		$I50 = 32\ \mu M$

Lu^{3+}

GLUCOSE-6-PHOSPHATE 1-DEHYDROGENASE		100% 830 μM
ISOCITRATE DEHYDROGENASE		67% 830 μM
L-LACTATE DEHYDROGENASE		100% 830 μM

LUBORL WX

PHOSPHATIDATE PHOSPHATASE		30% 3 mg/ml

LUBROL
EPOXIDE HYDROLASE
NITRIC-OXIDE REDUCTASE

LUBROL 17A-10
GLYCEROL-3-PHOSPHATE O-ACYLTRANSFERASE

LUBROL PX
ADENYLATE CYCLASE
1-PHOSPHATIDYLINOSITOL 4-KINASE

LUBROL WX
1-ACYLGLYCEROPHOSPHOCHOLINE O-ACYLTRANSFERASE
CHOLESTENONE 5α-REDUCTASE

LUCENSOMYCIN
Na^+/K^+-EXCHANGING ATPASE

LUCIFERYL SULPHATE
RENILLA-LUCIFERIN 2-MONOOXYGENASE

LUDARTIN
AROMATASE — $Ki = 23\ \mu M$

LUFFA ACUTANGULA TRYPSIN INHIBITOR 1
TRYPSIN

LUFFA ACUTANGULA TRYPSIN INHIBITOR 2
TRYPSIN

LUFFA CYLINDRICA TRYPSIN INHIBITOR I
TRYPSIN

LUFFA CYLINDRICA TRYPSIN INHIBITOR II
TRYPSIN

LUFFA CYLINDRICA TRYPSIN INHIBITOR-II
COAGULATION FACTOR Xa — $Ki = 780\ \mu M$
COAGULATION FACTOR XIIa — $Ki = 75$ nM
KALLIKREIN PLASMA — $Ki = 20\ \mu M$

LUFFA CYLINDRICA TRYPSIN INHIBITOR III
COAGULATION FACTOR Xa — $Ki = 100\ \mu M$
COAGULATION FACTOR XIIa — $Ki = 3.8$ nM
KALLIKREIN PLASMA — $Ki = 38\ \mu M$
TRYPSIN

LUFFARIELLOLIDE
PHOSPHOLIPASE A2 — $I50 = 9.3\ \mu M$

LULIBERIN
PYROGLUTAMYL-PEPTIDASE I — $Ki = 4.5$ mM

LUMBRICIUS TERRESTRIS TRYPSIN INHIBITOR
TRYPSIN

LUMICHROME
ALKANAL MONOOXYGENASE (FMN -LINKED)
NAD(P)H DEHYDROGENASE (FMN)
RIBOFLAVIN KINASE — C — $Ki = 48\ \mu M$
STEROID 5α-REDUCTASE — $I50 = 9.7\ \mu M$

LUMIFLAVIN
ALKANAL MONOOXYGENASE (FMN -LINKED)
RIBOFLAVIN KINASE — C — $Ki = 31\ \mu M$
STEROID 5α-REDUCTASE — $I50 = 700$ nM

LUMINOL
NAD ADP-RIBOSYLTRANSFERASE

LUNG NICOTINATE-NUCLEOTIDE PYROPHOSPHORYLASE INHIBITOR
NICOTINATE-NUCLEOTIDE PYROPHOSPHORYLASE (CARBOXYLATING)

LUNG SURFACTANT PROTEIN A
PHOSPHOLIPASE A2 — *K*i = 5 μg/ml

LUPEOL
FARNESYL PROTEINTRANSFERASE — *I*50 = 65 μg/ml
11β-HYDROXYSTEROID DEHYDROGENASE — 52% 100 μM

LUTENSIN A
AMINO OXIDASE MAO — *I*50 = 6.6 μM

LUTENSIN B
AMINO OXIDASE MAO — *I*50 = 6.6 μM

LUTEOLIN
ARACHIDONATE 5-LIPOXYGENASE — *I*50 = 102 nM
ARACHIDONATE 12-LIPOXYGENASE — *I*50 = 20 nM
AROMATASE — *K*i = 4.8 μM
CAFFEATE O-METHYLTRANSFERASE
CHOLINE O-ACETYLTRANSFERASE — *I*50 = 780 μM
CYTOCHROME P450 1A1 — *I*50 = 210 nM
GLUTATHIONE TRANSFERASE — *I*50 = 42 μM
HYALURONOGLUCOSAMINIDASE
INTEGRASE — *I*50 = 33 μM
LUTEOLIN-7O-DIGLUCURONIDE 4′-O-GLUCURONOSYLTRANSFERASE
LUTEOLIN-7O-GLUCURONIDE 7-O-GLUCURONOSYLTRANSFERASE — *I*50 = 100 μM
LYSO PLATELET ACTIVATING FACTOR ACETYLTRANSFERASE — *I*50 = 45 μM
NADH DEHYDROGENASE (EXOGENOUS) — *I*50 = 400 μM
NADH OXIDASE — *I*50 = 48 nmole
NARINGENIN-CHALCONE SYNHTASE
1-PHOSPHATIDYLINOSITOL 3-KINASE — *I*50 = 8 μM
PROLYL OLIGOPEPTIDASE — *I*50 = 0.2 ppm
PROTEIN KINASE C — *I*50 = 1 μM
PROTEIN TYROSINE KINASE p56lck — *I*50 = 175 μM
SUCCIN OXIDASE — *I*50 = 32 nmole
TRYPSIN — *K*i = 9.8 μM
UDP-GLUCURONOSYLTRANSFERASE — *K*i = 446 μM
XANTHINE OXIDASE — *I*50 = 590 nM

LUTEOLIN-7-R1
XANTHINE OXIDASE — *I*50 = 76 μM

LUTEOSKYRIN
RNA POLYMERASE

LUZOPEPTIN A
RNA-DIRECTED DNA POLYMERASE — *I*50 = 7 nM
RNA-DIRECTED DNA POLYMERASE — *I*50 = 68 nM

LY 046601
PROTEINASE 3C — *I*50 = 56 μM

LY 113174
AROMATASE — *I*50 = 24 nM

LY 121019
1,3-β-GLUCAN SYNTHASE — NC — *K*i = 16 μM

LY 134046
PHENYLETHANOLAMINE N-METHYLTRANSFERASE — *K*i = 264 nM

LY 178550
THROMBIN

LY 181984
NADH OXIDASE — 90% 1 μM

NADH OXIDASE		$I50 = 50$ nM
PROTEIN DISULFIDE ISOMERASE		
LY 186126		
3′,5′-CYCLIC-NUCLEOTIDE PHOSPHODIESTERASE		
LY 191704		
5α-REDUCTASE		
5α-REDUCTASE 1		$I50 = 8$ nM
5α-REDUCTASE 2		$I50 = 10$ μM
STEROID 5α-REDUCTASE 1	NC	$Ki = 3.2$ nM
STEROID 5α-REDUCTASE 2		$Ki = 20$ nM
STEROID 5α-REDUCTASE 2	NC	$Ki = 17$ μM
LY 207320		
17α-HYDROXYPROGESTERONE ALDOLASE		$I50 = 56$ nM
STEROID 17α-MONOOXYGENASE		$I50 = 76$ nM
STEROID 5α-REDUCTASE		$I50 = 60$ nM
LY 2079		
1-PHOSPHATIDYLINOSITOL 3-KINASE		$I50 = 6.2$ μg/ml
LY 219612		
PROTEINASE HIV-1		$I50 = 2$ nM
LY 222306		
PHOSPHORIBOSYLGLYCINAMIDE FORMYLTRANSFERASE		$Ki = 770$ pM
LY 231514		
DIHYDROFOLATE REDUCTASE		$Ki = 7.0$ nM
THYMIDYLATE SYNTHASE		$Ki = 340$ nM
LY 233569		
ARACHIDONATE 5-LIPOXYGENASE		$I50 = 100$ nM
LY 245769		
LEUKOTRIENE-B4 20-MONOOXYGENASE		
LEUKOTRIENE-E4 20 MONOOXYGENASE		
LY 254155		
PHOSPHORIBOSYLGLYCINAMIDE FORMYLTRANSFERASE		$Ki = 2.1$ nM
LY 266111		
5α-REDUCTASE		
LY 287045		
THROMBIN		
LY 289598		
PROTEINASE HIV-1		$I50 = 3.4$ nM
LY 289612		
PROTEINASE HIV		$Ki = 1.6$ nM
PROTEINASE HIV-1		$I50 = 1.5$ nM
LY 294002		
1-PHOSPHATIDYLINOSITOL 3-KINASE		$I50 = 1.4$ μM
LY 294291		
THROMBIN		
LY 298207		
DIHYDROFOLATE REDUCTASE		$Ki = 5$ pM
PHOSPHORIBOSYLGLYCINAMIDE FORMYLTRANSFERASE		$Ki = 62$ μM
LY 300046		
RNA-DIRECTED DNA POLYMERASE		$I50 = 17$ nM

LY 300217
PROTEINASE HIV-1 — $I50 = 2.2$ nM

LY 300502
STEROID 5α-REDUCTASE 1

LY 302146
1,3-β-GLUCAN SYNTHASE

LY 307987
IMP DEHYDROGENASE

LY 309887
PHOSPHORIBOSYLGLYCINAMIDE FORMYLTRANSFERASE — $Ki = 6.5$ nM

LY 311727
PHOSPHOLIPASE A2 (14KD SECRETORY) — $I50 = 23$ nM
PHOSPHOLIPASE A2 (SECRETORY)

LY 314613
PROTEINASE HIV-1

LY 315920
PHOSPHOLIPASE A2 — $I50 = 48$ nM
PHOSPHOLIPASE A2 — $I50 = 228$ nM
PHOSPHOLIPASE A2 (NON PANCREATIC SECRETED) — $I50 = 9$ nM

LY 316340
PROTEINASE HIV-1

LY 316373
DIHYDROFOLATE REDUCTASE — $Ki = 120$ pM
PHOSPHORIBOSYLGLYCINAMIDE FORMYLTRANSFERASE — $Ki = 54$ μM

LY 317644
PROTEIN KINASE C — $Ki = 6.4$ μM
PROTEIN KINASE Cβ2 — $I50 = 270$ nM

LY 326188
PROTEINASE HIV-1 — $I50 = 9$ nM

LY 326449
PROTEIN KINASE C — $Ki = 550$ nM
PROTEIN KINASE Cβ2 — $I50 = 32$ nM

LY 333531
PROTEIN KINASE Cβ — $I50 = 600$ nM
PROTEIN KINASE Cα — $I50 = 360$ nM
PROTEIN KINASE Cβ1 — $I50 = 4.7$ nM
PROTEIN KINASE Cβ2 — $I50 = 5.9$ nM
PROTEIN KINASE Cδ — $I50 = 250$ nM
PROTEIN KINASE Cγ — $I50 = 300$ nM
PROTEIN KINASE Cn — $I50 = 52$ nM

LY 335230
PROTEINASE 3C — $Ki = 490$ nM

LY 338387
PROTEINASE 3C — $Ki = 470$ nM

LY 343814
PROTEINASE 3C — $I50 = 20$ μM

LY 51641
AMINE OXIDASE MAO-A

LY 54761
AMINE OXIDASE MAO-B

LY 64139
PHENYLETHANOLAMINE N-METHYLTRANSFERASE

LY 802132
1-PHOSPHATIDYLINOSITOL 3-KINASE | | $I50 = 8.4\ \mu g/ml$

LY 805921
1-PHOSPHATIDYLINOSITOL 3-KINASE | | $I50 = 1.7\ \mu g/ml$

LY 806303
COAGULATION FACTOR Xa | | $I50 = 100$ nM
KALLIKREIN | | $I50 = 7.2\ \mu M$
PLASMIN | | $I50 = 240\ \mu M$
PLASMINOGEN ACTIVATOR nt | | $I50 = 160\ \mu M$
THROMBIN | | $I50 = 30$ nM
TRYPSIN | | $I50 = 29\ \mu M$

LY 83583
GUANYLATE CYCLASE
NITRIC-OXIDE SYNTHASE | | $Ki = 16\ \mu M$

LYCORINE
GALACTONOLACTONE DEHYDROGENASE | | 89% 10 μM

LYMPHOSTIN
PROTEIN TYROSINE KINASE (Lck) | | $I50 = 50$ nM

LYNALYLE CATION
GERANYL-DIPHOSPHATE CYCLASE

LY 231514 PENTAGLUTAMATE
DIHYDROFOLATE REDUCTASE | | $Ki = 7.2$ nM
PHOSPHORIBOSYLGLYCINAMIDE FORMYLTRANSFERASE | | $Ki = 65$ nM
THYMIDYLATE SYNTHASE | | $Ki = 1.3$ nM
THYMIDYLATE SYNTHASE | | $Ki = 3.4$ nM

$(LYS)_3$
PROTEINASE ALFA ALFA | | $Ki = 6$ nM

LYS-ALA-CH_2
DIPEPTIDYL-PEPTIDASE 2

LYS-ARG-TRP-LYS-LYS-ASN-PHE-ILE-PHE-VAL
MYOSIN-LIGHT-CHAIN KINASE | C | $I50 = 1\ \mu M$

cyclo(LYS-m-aB[$CH_2S^+(C_6H_5)CH_3$]GLY_4)
PLASMIN | IR |
u-PLASMINOGEN ACTIVATOR | IR | $Ki = 500\ \mu M$
TRYPSIN | | $Ki = 8.7\ \mu M$

D-LYS-8-CYCLOSPORIN
PEPTIDYLPROLYL ISOMERASE | | $I50 = 50$ nM

LYS-CYS-(N-ME)VAL-TIC-METH
FARNESYL PROTEINTRANSFERASE | | $I50 = 5$ nM

.-l-LYSIMETHYLESTER
LYSINE DECARBOXYLASE | C | $Ki = 8.6$ mM

LYSINE
ALANINE AMINOPEPTIDASE | NC | $I50 = 10$ mM
ALANOPINE DEHYDROGENASE
L-AMINOADIPATE-SEMIALDEHYDE DEHYDROGENASE
L-AMINOADIPATE-SEMIALDEHYDE DEHYDROGENASE | | $Ki = 50\ \mu M$
AMINOPEPTIDASE B | NC | $I50 = 10$ mM
5-AMINOVALERATE TRANSAMINASE
ARGINASE | C | $Ki = 1.2$ mM
ARGINASE | C | $Ki = 930\ \mu M$

D-ARGINASE		
ARGININE DECARBOXYLASE	C	Ki = 3.8 mM
CARBOXYPEPTIDASE B	C	Ki = 13 mM
CATHEPSIN H	C	Ki = 7.6 mM
COAGULATION FACTOR VIIa		I50 = 1 mM
DIAMINOPIMELATE DECARBOXYLASE		
DIHYDRODIPICOLINATE SYNTHASE	C	Ki = 210 μM
DIHYDRODIPICOLINATE SYNTHASE		
DIHYDRODIPICOLINATE SYNTHASE		Ki = 300 μM
GLUTAMATE DEHYDROGENASE (NAD(P))		
GLUTAMATE SYNTHASE (FERREDOXIN)		
HOMOCITRATE SYNTHASE	C	Ki = 8 μM
HOMOSERINE KINASE		
LEUCYL AMINOPEPTIDASE	NC	I50 = 25 mM
LEUCYL AMINOPEPTIDASE	C	Ki = 46 mM
LYSINE (ARGININE) CARBOXYPEPTIDASE		
LYSYL ENDOPEPTIDASE		Ki = 10 mM
D-ORNITHINE 4,5-AMINOMUTASE		
ORNITHINE CARBAMOYLTRANSFERASE		
ORNITHINE DECARBOXYLASE		Ki = 12 mM
PEPTIDYL-LYS METALLOENDOPEPTIDASE		Ki = 1 mM
PLASMINOGEN ACTIVATOR		50% 500 μM
SACCHAROPINE DEHYDROGENASE (NAD, L-LYSINE FORMING)		
TRYPSIN		Ki = 460 μM
XAA-ARG DIPEPTIDASE		

β-LYSINE

D-ORNITHINE 4,5-AMINOMUTASE

LYSINE ESTERS

AGKISTRODON SERINE PROTEINASE

L-LYSINE ETHYLESTER

ASPARTATE KINASE

L-LYSINE HOMOPOLYMERS

PROTEINASE ALFA ALFA LEAF

LYSINEHYDROXAMATE

LYSINE-tRNA LIGASE

L-LYSINE HYDROXAMIC ACID

PROTEINASE ASTACUS	I50 = 22 mM

L-β-LYSINE

LYSINE 2,3-AMINOMUTASE
D-LYSINE 5,6-AMINOMUTASE

LYSINE-p-NITROANILIDE

SACCHAROPINE DEHYDROGENASE (NADP, L-LYSINE FORMING)

LYSINETHIOL

AMINOPEPTIDASE B	Ki = 910 pM

LYSINYLADENYLATE

NUCLEOSIDE PHOSPHOACYLHYDROLASE	Ki = 340 μM

cyclo[LYS(9)-LYS(3)-LYS(9)-GLU]GLY-OH

CHYMOTRYPSIN	I50 = 51 μM

LYSOLECITHIN

1-ACYLGLYCEROPHOSPHOCHOLINE O-ACYLTRANSFERASE	Ki = 930 μM
PHOSPHATIDATE CYTIDILYLTRANSFERASE	
PHOSPHATIDYLCHOLINE-STEROL O-ACYLTRANSFERASE	
PROTEIN-DISULFIDE REDUCTASE (GLUTATHIONE)	
STEROL ESTERASE	

LYSOPHOSPHATIDIC ACID		
CARBOXYLESTERASE (LONG CHAIN)		*K*i = 7.5 μM
DNA POLYMERASE β		
DOLICHYL-PHOSPHATASE		81% 100 μM
GLYCEROL-3-PHOSPHATE O-ACYLTRANSFERASE		
1-PHOSPHATIDYLINOSITOL 3-KINASE		
LYSOPHOSPHATIDIC ACIDS		
Na^+/K^+-EXCHANGING ATPASE		
LYSOPHOSPHATIDYLCHOLINE		
CDP: PHOSPHOCHOLINE CYTIDYLTRANSFERASE		*I*50 = 25 μM
Na^+/K^+-EXCHANGING ATPASE		
PALMITOYL-CoA HYDROLASE		*K*i = 16 nM
PHOSPHATIDYLCHOLINE-STEROL O-ACYLTRANSFERASE		
1-PHOSPHATIDYLINOSITOL PHOSPHODIESTERASE		
PROTEIN-DISULFIDE REDUCTASE (GLUTATHIONE)		
3-sn-LYSOPHOSPHATIDYLCHOLINE		
CHOLINE-PHOSPHATE CYTIDYLYLTRANSFERASE		
L-α-LYSOPHOSPHATIDYLCHOLINE		
CHOLESTENONE 5α-REDUCTASE		
LYSOPHOSPHATIDYLGLYCEROL		
β-GALACTOSIDE α-2,3-SIALYLTRANSFERASE		
LYSOPHOSPHATIDYLGLYCEROL (EGG)		
PHOSPHOLIPASE D (GLYCOSYLPHOSPHATIDYLINOSITOL SPECIFIC)		*I*50 = 20 μM
LYSOPHOSPHATIDYLINOSITOL		
DNA POLYMERASE β		
PALMITOYL-CoA HYDROLASE		*K*i = 30 nM
LYSOPHOSPHATIDYLSERINE		
ETHANOLAMINEPHOSPHOTRANSFERASE		
β-GALACTOSIDE α-2,3-SIALYLTRANSFERASE		
PHOSPHOLIPASE D1a		*I*50 = 3.5 μM
PHOSPHOLIPASE D1b		*I*50 = 2 μM
PHOSPHOLIPASE D2		*I*50 = 1.9 μM
LYSOPHOSPHOLIPIDS		
α-N-ACETYLNEURAMINATE α-2,8-SIALYLTRANSFERASE		
1,3-β-GLUCAN SYNTHASE		*I*50 = 325 μM
LYSOSPHINGOLIPIDS		
PROTEIN KINASE C		
LYSOSPHINGOMYELINOLINE		
PHOSPHOLIPASE Cδ		*I*50 = 15 μM
LYSOSPHOSPHATIDYLCHOLINE		
PHOSPHOLIPASE Cδ		*I*50 = 7 μM
LYSOSPHOSPHATIDYLSERINE		
PHOSPHOLIPASE Cδ		*I*50 = 10 μM
LYSPHOSPHATIDIC ACID		
PHOSPHOLIPASE D (GLYCOSYLPHOSPHATIDYLINOSITOL SPECIFIC)		
LYS-TRP		
PEPTIDYL-DIPEPTIDASE A	C	*I*50 = 1.6 μM
LYSYL-ALANINE		
PEPTIDYL-DIPEPTIDASE A		*K*i = 50 μM
L-LYSYLGLYCINE		
SACCHAROPINE DEHYDROGENASE (NADP, L-LYSINE FORMING)		

LYS[Z(NO_2)]-PYRROLIDIDE
DIPEPTIDYL-PEPTIDASE 4 — $I50 = 2\ \mu M$

α-LYTIC PROTEINASE PROPART
α-LYTIC ENDOPEPTIDASE — $Ki = 10\ pM$

D-LYXOSE
GLUCOKINASE — C — $Ki = 83\ mM$
XYLOSE ISOMERASE — C — $Ki = 70\ \mu M$

L-LYXOSE
α-L-RHAMNOSIDASE — C
D-XYLOSE 1-DEHYDROGENASE

M

M 2
3′,5′-CYCLIC-NUCLEOTIDE PHOSPHODIESTERASE III — $I50 = 610\ nM$

M 37
THROMBIN — $Ki = 2\ nM$

M 16209
ALDEHYDE REDUCTASE — $I50 = 240\ nM$

M 16287
ALDEHYDE REDUCTASE — $I50 = 80\ nM$

M 79175
ALCOHOL DEHYDROGENASE (NADP)
ALDEHYDE REDUCTASE

M 13 PROCOAT LEADER PEPTIDE
SIGNAL PEPTIDASE I

4-MA
5α-REDUCTASE 1 — $Ki = 5\ nM$
5α-REDUCTASE 2 — $I50 = 230\ pM$
STEROID 5α-REDUCTASE — C — $Ki = 5\ nM$
STEROID 5α-REDUCTASE 1 — $Ki = 8\ nM$
STEROID 5α-REDUCTASE 2 — $Ki = 4\ nM$
STEROID 5β-REDUCTASE — C

MACROCARPAL A
ALDEHYDE REDUCTASE — $I50 = 2\ \mu M$

MACROCARPAL B
ALDEHYDE REDUCTASE — $I50 = 2.8\ \mu M$

MACROCARPAL D
ALDEHYDE REDUCTASE — $I50 = 2\ \mu M$

MACROCARPAL G
ALDEHYDE REDUCTASE — $I50 = 2.7\ \mu M$

MACROCIN
MACROCIN O-METHYLTRANSFERASE — $> 120\ \mu M$

α-MACROGLOBULIN
BROMELAIN
FICAIN

α1-MACROGLOBULIN
β-GALACTOSIDE α-2,6-SIALYLTRANSFERASE

α2 MACROGLOBULIN
PROTEINASE I PACIFIC WHITING 32% 10 μM

α2-MACROGLOBULIN
AGKISTRODON SERINE PROTEINASE
ASTACIN
ATROLYSIN C
ATROYLSIN A
CALPAIN
CALPAIN II
CATHEPSIN B
CATHEPSIN G
CHYMOSIN
CHYMOTRYPSIN
CLOSTRIPAIN
ELASTASE LEUKOCYTE
ELASTASE PANCREATIC
ENDOPROTEINASEN
ENTEROPEPTIDASE
INTERSTITIAL COLLAGENASE
JARARHAGIN
KALLIKREIN PLASMA
LACTOSYLCERAMIDE α-2,3-SIALYLTRANSFERASE 67% 1 U
MINIPLASMIN
MYELOBLASTIN
PAPAIN
PLASMIN
PROTEINASE 3 IR
PROTEINASE II MYXOBACTER AL-1
PROTEINASE NEUTRAL BACILLUS SUBTILIS
PROTEINASES ALL CLASSES
PROTEINASE SERINE CROTALUS ADAMANTEUS
PROTEINASES LYSOSOMAL
SERRALYSIN
STAPHYLOCOCCAL CYSTEINE PROTEINASE
SUBMANDIBULAR PROTEINASE A
SUBTILISIN
TEYCHOPHYTON MENTAGRAPHYTES KERATINASE
THERMOLYSIN
TRICHOPHYTON MENTAGROPHYTES KERATINASE
TRYPSIN

α-MACROGLUBULIN
CHYMOTRYPSIN
PAPAIN
TRYPSIN

α2-MACROGLUBULIN
CANCER PROCOAGULANT 19% 20 μg/ml

MACROPHAGE 23-KD STRESS PROTEIN
PROTEIN TYROSINE KINASE c-ABL

MAESANIN
ARACHIDONATE 5-LIPOXYGENASE 80% 10 μM

MAGNOLIANIN
ARACHIDONATE 5-LIPOXYGENASE $I50$ = 450 nM

MAGNOLOL
1-ALKYLGLYCEROPHOSPHOCHOLINE O-ACETYLTRANSFERASE $I50$ = 150 μM
11β-HYDROXYSTEROID DEHYDROGENASE 44% 100 μM
STEROL O-ACYLTRANSFERASE $I50$ = 86 μM

MAIZE ENDOSPERM β-FRUCTOFURANOSIDASE INHIBITOR
β-FRUCTOFURANOSIDASE

MAIZE ENDOSPERM INHIBITOR
α-1,4-GLUCAN-PROTEIN SYNTHASE (UDP-FORMING)

MAJOR HISTOCOMPATIBILITY COMPLEX CLASS II ASSOCIATED p41 INVARIANT CHAIN FRAGMENT

CATHEPSIN L		Ki = 1.7 pM
CRUZIPAIN		Ki = 580 pM

MAKALUVAMINE A
DNA TOPOISOMERASE (ATP-HYDROLYSING)

MAKALUVAMINE B
DNA TOPOISOMERASE (ATP-HYDROLYSING)

MAKALUVAMINE C
DNA TOPOISOMERASE (ATP-HYDROLYSING)

MAKALUVAMINE D
DNA TOPOISOMERASE (ATP-HYDROLYSING)
DNA TOPOISOMERASE (ATP-HYDROLYSING)

MAKALUVAMINE E
DNA TOPOISOMERASE (ATP-HYDROLYSING)

MAKALUVAMINE F
DNA TOPOISOMERASE (ATP-HYDROLYSING)

MAKALUVAMINE G

DNA TOPOISOMERASE		I50 = 3 μM

(R)-MALAOXON
ACETYLCHOLINESTERASE

MALATE
ASPARTATE CARBAMOYLTRANSFERASE
FUMARATE HYDRATASE
GLUTAMATE DEHYDROGENASE (NAD(P))
GLUTAMATE SYNTHASE (NADPH)
OXALOACETATE DECARBOXYLASE
PHOSPHOENOLPYRUVATE CARBOXYLASE
PHOSPHORIBULOKINASE

D-MALATE
D-LACTATE DEHYDROGENASE (CYTOCHROME)

L-MALATE
D-LACTATE DEHYDROGENASE (CYTOCHROME)

MALATHION
ACETYLCHOLINESTERASE

MALEIC ACID

ACONITATE HYDRATASE		
ADENOSINE-PHOSPHATE DEAMINASE		Ki = 22 mM
ADENYLOSUCCINATE SYNTHASE	C	Ki = 900 μM
ALANINE TRANSAMINASE		
D-ALANINE TRANSAMINASE		
D-AMINO-ACID TRANSAMINASE	C	Ki = 1.5 mM
ASPARTATE 4-DECARBOXYLASE		41% 5.6 mM
ASPARTATE TRANSAMINASE	C	Ki = 2.3 mM
ATP DEAMINASE		
2,3-DIKETO-L-GULONATE DECARBOXYLASE		90% 40 mM
DIPEPTIDASE		80% 1 mM
β-FRUCTOFURANOSIDASE		Ki = 17 mM
FUMARATE HYDRATASE	C	Ki = 11 mM
GLUTAMATE FORMIMINOTRANSFERASE		
γ-GLUTAMYLTRANSFERASE		
GLUTATHIONE PEROXIDASE		
GLYOXYLATE-L-ASPARTATE AMINOTRANSFERASE		

erythro-3-HYDROXYASPARTATE DEHYDRATASE	C	*K*i = 32 mM
4-HYDROXYGLUTAMATE TRANSAMINASE		
[HYDROXYMETHYLGLUTARYL-CoA REDUCTASE (NADPH)]-PHOSPHATASE		
ISOCITRATE DEHYDROGENASE (NADP)		
ISOCITRATE LYASE	C	*K*i = 360 μM
METHYLITACONATE Δ-ISOMERASE	C	*K*i = 7 mM
NICOTINATE-NUCLEOTIDE PYROPHOSPHORYLASE (CARBOXYLATING)		
NITROETANE REDUCTASE		
PHOSPHATIDYLSERINE DECARBOXYLASE		
PHOSPHOENOLPYRUVATE DECARBOXYLASE		*I*50 = 1.3 mM
PROLINE RACEMASE		
SUCCINATE DEHYDROGENASE		
SUCCINATE DEHYDROGENASE (UBIQUINONE)		
SUCROSE-PHOSPHATE SYNTHASE		
XAA-PRO DIPEPTIDASE		*K*i = 410 μM
MALEIC ANHYDRIDE		
INORGANIC PYROPHOSPHATASE		25% 25 mM
LACTOSE SYNTHASE		
PHOSPHOLIPASE A2		
MALEIC HYDRAZIDE		
PORPHOBILINOGEN SYNTHASE		*I*50 = 2.8 mM
SUPEROXIDE DISMUTASE		*I*50 = 40 μM
MALEIMIDE		
ASPARTATE-AMMONIA LIGASE		70% 3 mM
2-(4′-MALEIMIDYLAMINO)NAPHTHALENE-6-SULPHONIC ACID		
GLUCOSE CARRIER		
MALHAMENSILIPIN A		
PROTEIN TYROSINE KINASE pp60v-Src		*I*50 = 35 μM
MALIC ACID		
CITRATE (pro-3s)-LYASE		
β-FRUCTOFURANOSIDASE		*K*i = 19 mM
GLUTAMATE DEHYDROGENASE (NADP)		
NUCLEOTIDASE		
OXALOACETATE DECARBOXYLASE	C	*K*i = 4.5 mM
PHOSPHATIDYLSERINE DECARBOXYLASE		*K*i = 1.3 mM
PHOSPHOENOLPYRUVATE CARBOXYKINASE (PYROPHOSPHATE)		
PHOSPHOENOLPYRUVATE CARBOXYLASE		
PROCOLLAGEN-LYSINE 5-DIOXYGENASE		
PROCOLLAGEN-PROLINE, 2-OXOGLUTARATE-4-DIOXYGENASE		
PYRUVATE, WATER DIKINASE		
THIOSULFATE SULFURTRANSFERASE		
D-MALIC ACID		
D-ASPARTATE OXIDASE	C	*K*i = 10 μM
FUMARATE HYDRATASE	C	*K*i = 6.3 mM
(S)-2-METHYLMALATE DEHYDRATASE		*K*i = 2.6 mM
SINAPOYLGLUCOSE-MALATE O-SINAPOYLTRANSFERASE		
L(+)-TARTRATE DEHYDRATASE		78% 2 mM
L-MALIC ACID		
CYANATE HYDROLASE		*I*50 = 600 μM
GLUTAMATE DEHYDROGENASE		
INORGANIC PYROPHOSPHATASE		40% 500 μM
ISOCITRATE LYASE	NC	*K*i = 5.9 mM
L-LACTATE DEHYDROGENASE (CYTOCHROME)		
MALATE SYNTHASE	C	*K*i = 246 μM
NICOTINATE-NUCLEOTIDE PYROPHOSPHORYLASE (CARBOXYLATING)		
PHOSPHOENOLPYRUVATE CARBOXYKINASE (ATP)		
L(+)-TARTRATE DEHYDRATASE		62% 2 mM

MALLOTOCHROMENE		
RNA-DIRECTED DNA POLYMERASE		
MALLOTOJAPONIN		
RNA-DIRECTED DNA POLYMERASE	C	Ki = 6.1 µM
MALONATE		
OXALOACETATE DECARBOXYLASE		
SUCROSE-PHOSPHATE SYNTHASE		
MALONDIALDEHYDE		
ALDEHYDE DEHYDROGENASE (NAD)		
ALDEHYDE DEHYDROGENASE (NAD(P))		
XANTHINE OXIDASE		
MALONIC ACID		
ACETOACETATE DECARBOXYLASE		
D-AMINO-ACID TRANSAMINASE	C	Ki = 1.3 mM
D-ASPARTATE OXIDASE	C	Ki = 60 µM
CREATINE KINASE		
CYANATE HYDROLASE		I50 = 15 µM
DOPAMINE β-MONOOXYGENASE		
FUMARATE HYDRATASE	C	Ki = 40 mM
FUMARATE REDUCTASE (NADH)		
FUMARYLACETOACETASE		
3-HYDROXYBUTYRATE DEHYDROGENASE		
[HYDROXYMETHYLGLUTARYL-CoA REDUCTASE (NADPH)]-PHOSPHATASE		
ISOCITRATE LYASE	C	Ki = 850 µM
ISOCITRATE LYASE		
LACTATE 2-MONOOXYGENASE		
MALONATE-SEMIALDEHYDE DEHYDROGENASE	C	Ki = 570 µM
OXOGLUTARATE DEHYDROGENASE (LIPOAMIDE)		Ki = 600 nM
PHOSPHOGLYCERATE KINASE		Ki = 38 mM
PREPHENATE DEHYDROGENASE		
PROCOLLAGEN-PROLINE, 2-OXOGLUTARATE-4-DIOXYGENASE		
SUCCINATE-CoA LIGASE (ADP-FORMING)	C	
SUCCINATE DEHYDROGENASE		
SUCCINATE DEHYDROGENASE	C	10 mM
SUCCINATE DEHYDROGENASE (UBIQUINONE)		
SUCCINATE-HYDROXYMETHYLGLUTARATE CoA-TRANSFERASE	C	Ki = 2.1 mM
TRIMETHYLYSINE DIOXYGENASE		
TRIOSE-PHOSPHATE ISOMERASE		Ki = 180 µM
XAA-PRO DIPEPTIDASE		Ki = 400 µM
XANTHINE OXIDASE		20% 600 µM
D-MALONIC ACID		
RIBOSE-5-PHOSPHATE ISOMERASE	C	Ki = 1.3 mM
MALONIC ACID SEMIALDEHYDE		
SUCCINATE-SEMIALDEHYDE DEHYDROGENASE (NAD(P))	C	Ki = 100 µM
MALONIC SEMIALDEHYDE		
2-HYDROXY-3-OXOPROPIONATE REDUCTASE		
SUCCINATE-SEMIALDEHYDE DEHYDROGENASE (NAD(P))		
MALONIDALDEHYDE		
GLUCOSE-6-PHOSPHATASE		
PAPAIN		
RIBONUCLEASE		
TRIACYLGLYCEROL LIPASE		
MALONOBEN		
OXIDATIVE PHOSPHORYLATION UNCOUPLER		
MALONYL-CoA		
ACETOACETATE-CoA LIGASE	NC	Ki = 145 µM
ACYL-CoA DEHYDROGENASE (NADP)		

CARNITINE O-OCTANOYLTRANSFERASE	C	Ki = 6.4 µm
CARNITINE O-OCTANOYLTRANSFERASE		87% 17 µM
CARNITINE O-PALMITOYLTRANSFERASE		I50 = 1.6 µM
CARNITINE O-PALMITOYLTRANSFERASE		I50 = 160 nM
CARNITINE O-PALMITOYLTRANSFERASE		I50 = 25 nM
CARNITINE O-PALMITOYLTRANSFERASE		I50 = 2.7 µM
CARNITINE O-PALMITOYLTRANSFERASE		I50 = 36 nM
CARNITINE O-PALMITOYLTRANSFERASE		I50 = 17 nM
CARNITINE O-PALMITOYLTRANSFERASE		I50 = 100 nM
CARNITINE O-PALMITOYLTRANSFERASE 1		I50 = 1.2 µM
CARNITINE O-PALMITOYLTRANSFERASE 1		I50 = 19 µM
FATTY ACID SYNTHASE		
METHYLMALONYL-CoA DECARBOXYLASE		Ki = 100 µM
METHYLMALONYL-CoA EPIMERASE		78% 2 mM
METHYLMALONYL-CoA MUTASE	C	Ki = 120 µM
[3-METHYL-2-OXOBUTANOATE DEHYDROGENASE (LIPOAMIDE)] KINASE		
NARINGENIN-CHALCONE SYNHTASE		
PYRUVATE CARBOXYLASE		

MALOTILATE

ARACHIDONATE 5-LIPOXYGENASE		I50 = 4.7 µM

MALOXAN

CHOLINESTERASE

MALTITOL

CYCLOMALTODEXTRIN GLUCANOTRANSFERASE		
GLUCAN 1,4-α-GLUCOSIDASE	C	Ki = 3.3 mM

D-MALTOBIONOLACTONE

α-AMYLASE

MALTOHEPTAOSE

1,4-α-GLUCAN BRANCHING ENZYME

MALTOHEXAOSE

1,4-α-GLUCAN BRANCHING ENZYME

MALTOL

LACTOYLGLUTATHIONE LYASE		
LIPOXYGENASE		I50 = 8 mM
THROMBOXANE A SYNTHASE		I50 = 8 mM

MALTOOCTAOSE

1,4-α-GLUCAN BRANCHING ENZYME

MALTOSE

α-AMYLASE	NC	Ki = 8.8 mM
β-AMYLASE		
CYCLOMALTODEXTRIN GLUCANOTRANSFERASE		
DEXTRANSUCRASE		
β-FRUCTOFURANOSIDASE		
β-D-FUCOSIDASE		
GDP-MANNOSE 6-DEHYDROGENASE		
D-GLUCANSUCRASE		Ki = 800 µM
α-GLUCOSIDASE	C	Ki = 35 mM
β-GLUCOSIDASE		45% 15 mM
MYCODEXTRANASE		Ki = 70 mM
SUCROSE α-GLUCOSIDASE		
α,α-TREHALASE		

D-MALTOSE

GLUCOKINASE

MALTOTRIITOL

α-AMYLASE		Ki = 340 µM

Inhibitor / Enzyme	Type	Value
MALTOTRIOSE		
α-AMYLASE	MIX	$K\text{i} = 3.8$ mM
MYCODEXTRANASE		$K\text{i} = 152$ mM
(2R)-MALY-CoA		
MALYL-CoA LYASE		
MAMORDIA CHARANTIA TRYPSIN INHIBITOR A		
TRYPSIN		
MANDELIC ACID		
D-LACTATE DEHYDROGENASE (CYTOCHROME)		
MANDELONITRILE LYASE		19% 50 μM
MANDELONITRILE LYASE		11% 20 μM
MANDELONITRILE LYASE		$K\text{i} = 230$ μM
α-MANGOSTAIN		
3′,5′-CYCLIC-NUCLEOTIDE PHOSPHODIESTERASE (cycAMP)		$I50 = 24$ μM
γ-MANGOSTAIN		
3′,5′-CYCLIC-NUCLEOTIDE PHOSPHODIESTERASE (cycAMP)		$I50 = 50$ μM
MANGOSTANOL		
3′,5′-CYCLIC-NUCLEOTIDE PHOSPHODIESTERASE (cycAMP)		$I50 = 47$ μM
MANGOSTIN		
MYOSIN-LIGHT-CHAIN KINASE		$I50 = 120$ μM
PROTEIN KINASE A		$I50 = 13$ μM
PROTEIN KINASE (Ca^{2+} DEPENDENT)		$I50 = 33$ μM
α-MANGOSTIN		
Ca^{2+}-TRANSPORTING ATPASE		$I50 = 5$ μM
γ-MANGOSTIN		
MYOSIN-LIGHT-CHAIN KINASE		$I50 = 110$ μM
PROTEIN KINASE A		$I50 = 2$ μM
PROTEIN KINASE (Ca^{2+} DEPENDENT)		$I50 = 6$ μM
MANNAN		
MANNAN ENDO-1,4-β-MANNOSIDASE		
MANNAN YEAST		
PROTEINASE SERINE ARTHROBACTER		
MANNITOL		
α,α-TREHALASE		
D-MANNITOL		
MANNONATE DEHYDRATASE	C	$K\text{i} = 30$ mM
XYLOSE ISOMERASE		
MANNITOL-1-PHOSPHATE		
MANNITOL 2-DEHYDROGENASE		
MANNOSE-6-PHOSPHATE ISOMERASE		$K\text{i} = 800$ μM
MANNITOL-6-PHOSPHATE		
GLUCOSE-6-PHOSPHATE ISOMERASE		$K\text{i} = 130$ μM
D-MANNITOL-6-PHOSPHATE		
myo-INOSITOL-1-PHOSPHATE SYNTHASE	C	$K\text{i} = 120$ μM
MANN-MAN-GLCN-PI		
DOL-P-MAN: MANα1-6-MANα1-GLCNα1-GPIα1-2 MANNOSYLTRANSFERASE		
D-MANNOAMIDRAZONE		
β-GLUCOSIDASE		$K\text{i} = 570$ μM
β-GLUCOSIDASE		$K\text{i} = 200$ μM
MANNOSIDASE		$K\text{i} = 170$ nM
α-MANNOSIDASE		$K\text{i} = 170$ nM

MANNOHEPTULOSE
KETOHEXOKINASE

α 1- > 6 MANNOHEXAOSE
MANNAN ENDO-1,4-β-MANNOSIDASE

MANNOHEXOSE(REDUCED α1- > 6)

MANNAN ENDO-1,6-β-MANNOSIDASE	C	54% 100 μg/ml

D-MANNO-1,4-LACTONE

β-MANNOSIDASE		42% 40 mM

D-MANNO-1-4-LACTONE

α-MANNOSIDASE		*K*i = 23.8 mM

MANNO(1- > 4)LACTONES
α-MANNOSIDASE

MANNO(1- > 5)LACTONES
α-MANNOSIDASE

MANNONATE
FRUCTURONATE REDUCTASE
TAGATURONATE REDUCTASE

D-MANNONHYDROXIMOLACTAM

α-MANNOSIDASE		*K*i = 150 nM

D-MANNONIC ACID AMIDE

MANNONATE DEHYDRATASE	NC	*K*i = 20 μM

D-MANNONIC AMIDE

MANNONATE DEHYDRATASE	NC	*K*i = 20 μM

D-MANNONOHYDROXYMOLACTONE N-PHENYLCARBAMATE

α-MANNOSIDASE	C	*K*i = 80 μM
β-MANNOSIDASE	C	*K*i = 25 nM

MANNONOJIRITETRAZOLE

α-MANNOSIDASE		*K*i = 180 μM
α-MANNOSIDASE		*K*i = 700 μM
β-MANNOSIDASE		*K*i = 160 μM

D.MANNONOLACTAM AMIDRAZONE

ARYL α-MANNOSIDASE		*I*50 = 400 nM
α-MANNOSIDASE		*I*50 = 1 μM
β-MANNOSIDASE		*I*50 = 150 μM
MANNOSIDASE I		*I*50 = 4 μM
MANNOSYL-OLIGOSACCHARIDE 1,3-1,6-α-MANNOSIDASE		*I*50 = 100 nM

MANNONO-(1,4)-LACTONE

α-MANNOSIDASE	C	*K*i = 10 mM

MANNONO-(1,5)-LACTONE

α-MANNOSIDASE	C	*K*i = 120 μM

α-D-MANNOPYRANOSIDE

OLIGO-1,6-GLUCOSIDASE		56% 10 mM

MANNOPYRANOSYLAMINE

β-MANNOSIDASE		60% 40 mM

MANNOSAMINE

α-MANNOSIDASE		
MANNOSIDASE I		*I*50 = 3 mM
PROTEIN-GLUCOSYLGALACTOSYLHYDROXYLYSINE GLUCOSIDASE		*K*i = 24 μM
PROTEIN-GLUCOSYLGALACTOSYLHYDROXYLYSINE GLUCOSIDASE	C	*K*i = 5.6 μM

D-MANNOSAMINE

β-N-ACETYLHEXOSAMINIDASE		

d-MANNOSAMINE

MANNOKINASE		

MANNOSE

β-N-ACETYLHEXOSAMINIDASE		75% 10 mM
GLUCAN ENDO-1,3-β-GLUCOSIDASE		
GLUCOKINASE	C	*K*i = 140 μM
GLUCOSE OXIDASE		
β-GLUCURONIDASE		37% 100 mM
LEVANSUCRASE		
MANNOKINASE	C	*K*i = 400 μM
α-MANNOSIDASE		*K*i = 85 mM
MANNOSYL-GLYCOPROTEIN ENDO-β-N-ACETYLGLUCOSAMINIDASE		95% 500 mM
MANNOSYL-GLYCOPROTEIN ENDO-β-N-ACETYLGLUCOSAMINIDASE		70% 50 mM
MANNOSYL-OLIGOSACCHERIDE GLUCOSIDASE		25% 5 mM
THIOGLUCOSIDASE		
α,α-TREHALASE		

D-MANNOSE

ACYL-PHOSPHATE PHOSPHOTRANSFERASE		
exo-β-MANNOSIDASE		

MANNOSE-1-PHOSPHATE

ALDOSE-1-PHOSPHATE NUCLEOTIDYLTRANSFERASE		40% 4.2 mM
HEXOSE-1-PHOSPHATE GUANYLYLTRANSFERASE	C	*K*i = 47 μM
NUCLEOSIDE-TRIPHOSPHATE-HEXOSE 1-PHOSPHATE NUCLEOTIDYLTRANSFERASE		*K*i = 47 μM
SUGAR PHOSPHATASE		

MANNOSE-6-PHOSPHATE

N-ACETYLGLUCOSAMINE-1-PHOSPHODIESTER N-ACETYLGLUCOSAMINIDASE		38% 10 mM
GLUCOSE-6-PHOSPHATE ISOMERASE		*K*i = 450 μM

D-MANNOSE-6-PHOSPHATE

myo-INOSITOL-1-PHOSPHATE SYNTHASE	C	

α-MANNOSIDES

MANNOSYL-GLYCOPROTEIN ENDO-β-N-ACETYLGLUCOSAMINIDASE		

MANNOSTATIN

α-MANNOSIDASE		
α-MANNOSIDASE	C	*I*50 = 160 μM
MANNOSYL-OLIGOSACCHARIDE 1,3-1,6-α-MANNOSIDASE	C	*I*50 = 15 μM

MANNOSTATIN A

α-GLUCOSIDASE		*I*50 = 40 ng/ml
α-MANNOSIDASE	C	*K*i = 48 nM
α-MANNOSIDASE		*I*50 = 20 ng/ml

MANNOSTATIN B

α-MANNOSIDASE		*K*i = 48 nM
α-MANNOSIDASE		*I*50 = 20 ng/ml

MANNOSTATIN SULPHONE

α-MANNOSIDASE		*K*i = 126 nM

MANNOSYLAMINE

α-MANNOSIDASE		*K*i = 7 μM
β-MANNOSIDASE		

α 1- > 6 MANNOTRIOSE REDUCED

MANNAN ENDO-1,4-β-MANNOSIDASE		92% 10 mM

MANNOTRIOSE(REDUCED α1- > 6)

MANNAN ENDO-1,6-β-MANNOSIDASE	C	92% 100 μg/ML

MANNO-VALIDAMINE

α-MANNOSIDASE		*I*50 = 8 μM
α-MANNOSIDASE		*I*50 = 56 μM
α-MANNOSIDASE II		*I*50 = 34 μM

MANNUHEPTULOSE

GLUCOKINASE		

MANOALIDE

ARACHIDONATE 5-LIPOXYGENASE		*I*50 = 300 nM
1-PHOSPHATIDYLINOSITOL PHOSPHODIESTERASE		*I*50 = 3 μM
PHOSPHOLIPASE A2		*I*50 = 4.7 μM
PHOSPHOLIPASE A2		*I*50 = 16 μM
PHOSPHOLIPASE A2	IR	*I*50 = 1.8 μM
PHOSPHOLIPASE A2 (HIGH MW)		*I*50 = 30 μM
PHOSPHOLIPASE A2 (LOW MW)		*I*50 = 80 nM
TRANSACYLASE (COA INDEPENDENT)		*I*50 = 5 μM

MANOALIDE2

PHOSPHOLIPASE A2 (14KD)		*I*50 = 6 μM

MANOALOGUE

PHOSPHOLIPASE A2		*I*50 = 26 μM

MANUMYCIN

FARNESYL PROTEINTRANSFERASE		*I*50 = 5 μM
FARNESYL PROTEINTRANSFERASE		*K*i = 1.2 μM
GERANYLGERANYL PROTEINTRANSFERASE I		*I*50 = 180 μM

MANUMYCIN A

CASPASE-1		*I*50 = 11 μM

MANUMYCIN B

CASPASE-1		*I*50 = 650 nM

MANUMYCIN G

CASPASE-1		*I*50 = 200 nM

MAP 30

INTEGRASE		*I*50 = 1 μM

α-MAPI

PROTEINASE HIV-1		*I*50 = 1.3 μM

β-MAPI

PROTEINASE HIV-1		*I*50 = 32 μM

MAPROUNIC ACID

RNA-DIRECTED DNA POLYMERASE		*I*50 = 79 μM
RNA-DIRECTED DNA POLYMERASE		*I*50 = 25 μM

MAPROUNIC ACID ACETATE

RNA-DIRECTED DNA POLYMERASE		*I*50 = 136 μM

MARCARBOMYCIN

PEPTIDOGLYCAN GLUCOSYLTRANSFERASE		

MARGINOYL-CoA

ACETYL-CoA CARBOXYLASE	C	*K*i = 4.3 μM

MARIMASTAT

ACE SECRETASE		*I*50 = 8.3 μM
COLLAGENASE FIBROBLAST		*I*50 = 5 nM
COLLAGENASE NEUTROPHIL		*I*50 = 2 nM
GELATINASE A		*I*50 = 6 nM
GELATINASE B		*I*50 = 3 nM
INTERSTITIAL COLLAGENASE		*I*50 = 5 nM
MATRILYSIN		*I*50 = 20 nM

NEPRILYSIN		$I50 = 4\ \mu M$
PEPTIDYL-DIPEPTIDASE A		30% 100 µM
α-SECRETASE		$I50 = 1.2\ \mu M$
STROMELYSIN 1		$I50 = 200$ nM
TNFα CONVERTING ENZYME		$I50 = 3.8\ \mu M$

MARINE TURTLE EGG WHITE BASIC TRYPSIN/SUBTILISIN INHIBITOR
CHYMOTRYPSIN
TRYPSIN

MARINE WORM CYTOTOXIN A-IV

MYOSIN-LIGHT-CHAIN KINASE		$I50 = 35\ \mu M$
PROTEIN KINASE C		$I50 = 1.3\ \mu M$

MARINE WORM NEUROTOXIN B-IV

MYOSIN-LIGHT-CHAIN KINASE		$I50 = 13\ \mu M$
PROTEIN KINASE C		$I50 = 125\ \mu M$

MARINOBUFAGENIN
Na^+/K^+-EXCHANGING ATPASE

MARINOSIN
Na^+/K^+-EXCHANGING ATPASE

MARINOSTATINS
CHYMOTRYPSIN
ELASTASE
SUBTILISIN

MARSALYL
NADPH-FERRIHEMOPROTEIN REDUCTASE

MASTICADIENOIC ACID
PHOSPHOLIPASE A2

MATLYSTATIN B

GELATINASE A		$I50 = 1.7\ \mu M$
GELATINASE B		$I50 = 570$ nM
MEMBRANE ALANINE AMINOPEPTIDASE		$I50 = 3.1\ \mu M$
STROMELYSIN		$I50 = 350$ nM
STROMELYSIN 1		$I50 = 350$ nM

MATLYSTATINS
COLLAGENASE

MATTEUORIENATE A

ALDEHYDE REDUCTASE		$I50 = 1\ \mu M$

MATTEUORIENATE B

ALDEHYDE REDUCTASE		$I50 = 1\ \mu M$

MATTEUORIENATE C

ALDEHYDE REDUCTASE		$I50 = 2.3\ \mu M$

MAZ 525
H^+/K^+-EXCHANGING ATPASE

Mb^{2+}
OLIGO-1,6-GLUCOSIDASE

MB 22948

3′,5′-CYCLIC-NUCLEOTIDE PHOSPHODIESTERASE (cycAMP)		$I50 = 93\ \mu M$
3′,5′-CYCLIC-NUCLEOTIDE PHOSPHODIESTERASE (cycAMP)		$I50 = 100\ \mu M$
3′,5′-CYCLIC-NUCLEOTIDE PHOSPHODIESTERASE (cycAMP)	C	$Ki = 250\ \mu M$
3′,5′-CYCLIC-NUCLEOTIDE PHOSPHODIESTERASE (cycAMP)		$I50 = 58\ \mu M$
3′,5′-CYCLIC-NUCLEOTIDE PHOSPHODIESTERASE (Ca^{2+}/CALMODULIN)	C	$Ki = 9.9\ \mu M$
3′,5′-CYCLIC-NUCLEOTIDE PHOSPHODIESTERASE (cycGMP)		$I50 = 1\ \mu M$
3′,5′-CYCLIC-NUCLEOTIDE PHOSPHODIESTERASE (cycGMP)		$I50 = 56\ \mu M$

- 3′,5′-CYCLIC-NUCLEOTIDE PHOSPHODIESTERASE (cycGMP) — C — Ki = 160 nM
- 3′,5′-CYCLIC-NUCLEOTIDE PHOSPHODIESTERASE (cycGMP) (Ca^{2+}/CALMODULIN) — I50 = 15 µM
- 3′,5′-CYCLIC-NUCLEOTIDE PHOSPHODIESTERASE (cycGMP STIMULATED) — I50 = 708 µM
- 3′,5′-CYCLIC-NUCLEOTIDE PHOSPHODIESTERASE (cycGMP STIMULATED) — I50 = 50 µM
- 3′,5′-CYCLIC-NUCLEOTIDE PHOSPHODIESTERASE I (Ca^{2+}/CALMODULIN) — I50 = 16 µM

MB 39279

- PROTOPORPHYRINOGEN OXIDASE — I50 = 500 nM

MBP1

- MULTICATALYTIC ENDOPEPTIDASE COMPLEX 26 S

MCEI-I

- ELASTASE — Ki = 300 nM

MCI 154

- 3′,5′-CYCLIC-NUCLEOTIDE PHOSPHODIESTERASE I — I50 = 220 µM
- 3′,5′-CYCLIC-NUCLEOTIDE PHOSPHODIESTERASE III — I50 = 2.5 µM

MCTI-I

- TRYPSIN — Ki = 67 pM

MCTI-II

- TRYPSIN — Ki = 25 pM

MDL

- NEPRILYSIN — Ki = 5 nM

MDL 100173

- NEPRILYSIN — Ki = 80 pM
- PEPTIDYL-DIPEPTIDASE A — Ki = 111 pM

MDL 100407

- NEPRILYSIN — Ki = 400 pM

MDL 101028

- ELASTASE LEUKOCYTE — I50 = 10 µM

MDL 101146

- EALASTASE LEUKOCYTE — Ki = 20 nM
- ELASTASE LEUKOCYTE — Ki = 25 nM

MDL 101232

- α-CHYMOTRYPSIN — Ki = 1.2 µM
- ELASTASE LEUKOCYTE — Ki = 4 nM

MDL 101628

- NEPRILYSIN — Ki = 70 pM

MDL 101680

- CATHEPSIN G — Ki = 2.2 µM
- ELASTASE LEUKOCYTE — Ki = 16 nM

MDL 101731

- RIBONUCLEOSIDE-DIPHOSPHATE REDUCTASE — IR

MDL 102111

- ELASTASE LEUKOCYTE — Ki = 170 nM

MDL 102823

- ELASTASE LEUKOCYTE — Ki = 59 nM

MDL 104168

- PROTEINASE HIV — Ki = 20 nM

MDL 17043

- 3′,5′-CYCLIC-NUCLEOTIDE PHOSPHODIESTERASE II — I50 = 450 µM
- 3′,5′-CYCLIC-NUCLEOTIDE PHOSPHODIESTERASE II — I50 = 860 µM
- 3′,5′-CYCLIC-NUCLEOTIDE PHOSPHODIESTERASE III — I50 = 14 µM

MDL 18962
AROMATASE SSI $K\mathrm{i} = 4.5$ nM

MDL 19987
PROTEIN KINASE C $I50 = 65$ μM

MDL 201000
CATHEPSIN B

MDL 201003
CATHEPSIN B

MDL 25637
α,α-TREHALASE $I50 = 140$ nM
α,α-TREHALASE $I50 = 20$ nM

MDL 27032
3′,5′-CYCLIC-NUCLEOTIDE PHOSPHODIESTERASE II $I50 = 158$ μM
3′,5′-CYCLIC-NUCLEOTIDE PHOSPHODIESTERASE IVa $I50 = 172$ μM
3′,5′-CYCLIC-NUCLEOTIDE PHOSPHODIESTERASE IVb $I50 = 120$ μM
PROTEIN KINASE A $I50 = 33$ μM
PROTEIN KINASE C $I50 = 37$ μM
PROTEIN KINASE C $I50 = 42$ μM

MDL 27088
PEPTIDYL-DIPEPTIDASE A

MDL 27428
PROTEIN KINASE C $I50 = 22$ μM

MDL 27855
NEPRILYSIN $K\mathrm{i} = 5$ nM

MDL 27986
PROTEIN KINASE C $I50 = 631$ μM

MDL 28170
CALPAIN $K\mathrm{i} = 10$ nM
CALPAIN I $I50 = 11$ nM
CATHEPSIN B $K\mathrm{i} = 25$ nM
α-CHYMOTRYPSIN $K\mathrm{i} = 54$ μM
PLASMIN $K\mathrm{i} = 270$ μM

MDL 28574
α-GLUCOSIDASE I $I50 = 1.3$ μM

MDL 28815
C8- > C7 STEROL ISOMERASE
STEROL C8-C7 ISOMERASE

MDL 28842
ADENOSYLHOMOCYSTEINASE SSI

MDL 72145
AMINE OXIDASE MAO-A SSI $K\mathrm{i} = 130$ μM
AMINE OXIDASE MAO-A $I50 = 4.2$ μM
AMINE OXIDASE MAO-B SSI $K\mathrm{i} = 40$ μM
AMINE OXIDASE MAO-B $I50 = 100$ nM
AMINE OXIDASE (SEMICARBAZIDE SENSITIVE) $I50 = 6$ nM

MDL 72161
AMINE OXIDASE (SEMICARBAZIDE SENSITIVE) $I50 = 2.5$ nM

MDL 72274
AMINE OXIDASE (SEMICARBAZIDE SENSITIVE) $I50 = 8$ nM

MDL 72392
AMINE OXIDASE MAO-A

MDL 72521
AMINE OXIDASE (POLYAMINE) IR

MDL 72527
AMINE OXIDASE (POLYAMINE)

MDL 72638
AMINE OXIDASE MAO-B

MDL 72887
AMINE OXIDASE MAO-B SSI $I50 = 6$ nM

MDL 72974
AMINE OXIDASE MAO-A $I50 = 470$ nM
AMINE OXIDASE MAO-B SSI $I50 = 4$ nM
AMINE OXIDASE MAO-B $I50 = 270$ pM
AMINE OXIDASE (SEMICARBAZIDE SENSITIVE) IR $I50 = 20$ nM
AMINE OXIDASE (SEMICARBAZIDE SENSITIVE) IR $I50 = 2$ nM
AMINE OXIDASE (SEMICARBAZIDE SENSITIVE) IR $I50 = 5$ nM
AMINE OXIDASE (SEMICARBAZIDE SENSITIVE) IR $I50 = 80$ nM

MDL 73669
PROTEINASE HIV $Ki = 5$ nM

MDL 73745
ACETYLCHOLINESTERASE $Ki = 47$ pM

MDL 73811
ADENOSYLMETHIONINE DECARBOXYLASE SSI $Ki = 1.5$ μM
ADENOSYLMETHIONINE DECARBOXYLASE SSI $Ki = 560$ nM

MDL 73945
GLUCAN 1,4-α-GLUCOSIDASE $I50 = 50$ μM
α-GLUCOSIDASE $I50 = 500$ μM
LACTASE $I50 = 800$ μM
OLIGO-1,6-GLUCOSIDASE $I50 = 1$ mM
SUCROSE α-GLUCOSIDASE $I50 = 6$ μM

MDL 74147
CATHEPSIN D $Ki = 4$ μM
CHYMOTRYPSIN $Ki = 2.5$ μM
PEPSIN $Ki = 40$ μM
RENIN $I50 = 2$ μM
RENIN $I50 = 16$ nM
RENIN $I50 = 100$ μM
RENIN $I50 = 22$ nM

MDL 74428
PURINE-NUCLEOSIDE PHOSPHORYLASE

MDL 74695
PROTEINASE HIV-1

MDL 100948 A
ELASTASE LEUKOCYTE $Ki = 60$ nM

MDL 72274A
AMINE OXIDASE MAO-B

MDL 72974A
AMINE OXIDASE MAO-A $I50 = 340$ nM
AMINE OXIDASE MAO-B $I50 = 600$ pM
AMINE OXIDASE (SEMICARBAZIDE SENSITIVE)
SSAO

MEARNSITRIN
ALDEHYDE REDUCTASE $I50 = 1.4$ μM

MEARSALIC ACID
- CHOLESTEROL 7α-MONOOXYGENASE

MEBENDAZOLE
- FUMARATE REDUCTASE (NADH) — *I*50 = 23 μM

MECARBAM
- ACETYLCHOLINESTERASE

MECLOFENAMATE HYDROXAMIC ACID
- ARACHIDONATE 5-LIPOXYGENASE — *I*50 = 3.9 μM
- CYCLOOXYGENASE — *I*50 = 1.1 μM

MECLOFENAMIC ACID
- ARACHIDONATE 5-LIPOXYGENASE — *I*50 = 24 μM
- CYCLOOXYGENASE — *I*50 = 100 nM
- CYCLOOXYGENASE 1 — *I*50 = 1.5 μM
- CYCLOOXYGENASE 2 — *I*50 = 9.7 μM
- GLUTATHIONE TRANSFERASE
- 3α-HYDROXYSTEROID DEHYDROGENASE (A SPECIFIC)
- 3α-HYDROXYSTEROID DEHYDROGENASE (B-SPECIFIC) — *I*50 = 2.9 μM
- PROSTAGLANDIN-ENDOPEROXIDE SYNTHASE
- PROSTAGLANDIN-ENDOPEROXIDE SYNTHASE 1 — *I*50 = 2.5 μM
- PROSTAGLANDIN-ENDOPEROXIDE SYNTHASE 1 — *I*50 = 10 μM
- PROSTAGLANDIN-ENDOPEROXIDE SYNTHASE 2 — *I*50 = 300 nM
- PROSTAGLANDIN-ENDOPEROXIDE SYNTHASE 2 — *I*50 = 18 μM

MECLOFENAMIC ACID METHYLHYDROXAMATE
- ARACHIDONATE 5-LIPOXYGENASE — *I*50 = 16 μM

MECLOFENAMIC ACID N-METHYLHYDROXAMATE
- ARACHIDONATE 5-LIPOXYGENASE — *I*50 = 1.5 μM
- CYCLOOXYGENASE — *I*50 = 15 μM

(R)-MEDALATE
- BENZOYLFORMATE DECARBOXYLASE — *K*i = 1.4 mM

(S)-MEDALATE
- BENZOYLFORMATE DECARBOXYLASE — *K*i = 6.7 mM

MEDICA 16
- ATP CITRATE (pro-S)-LYASE — *I*50 = 29 μM
- ATP CITRATE (pro-S)-LYASE — *I*50 = 54 μM

MEDIPINE
- BUFURALOL-1′-HYDROXYLASE (CYT P-450dbl) — C — *I*50 = 4 μM

MEDROXYPROGESTERONE
- CARBONYL REDUCTASE (NADPH) 1 — 21% 50 μM
- CARBONYL REDUCTASE (NADPH) 3 — 16% 50 μM
- 3α-HYDROXYSTEROID DEHYDROGENASE (B-SPECIFIC) — *I*50 = 200 nM
- 3α-HYDROXYSTEROID DEHYDROGENASE (B-SPECIFIC) — C — *K*i = 31 nM

MEDROXYPROGESTERONE ACETATE
- 3α(or 17β)-HYDROXYSTEROID DEHYDROGENASE
- 3α-HYDROXYSTEROID DEHYDROGENASE (B-SPECIFIC) — C — *K*i = 5 nM
- 3α-HYDROXYSTEROID DEHYDROGENASE (B-SPECIFIC) — 95% 100 nM
- 3α-HYDROXYSTEROID DEHYDROGENASE (B-SPECIFIC)
- INDANOL DEHYDROGENASE — C — *K*i = 100 nM

MEFENAMIC ACID
- D-AMINO-ACID OXIDASE
- CYCLOOXYGENASE 1 — *I*50 = 40 nM
- CYCLOOXYGENASE 2 — *I*50 = 3 μM
- 3α-HYDROXYSTEROID DEHYDROGENASE (B-SPECIFIC) — *I*50 = 12 μM
- PHOSPHOLIPASE A1 — *I*50 = 1 μM
- PHOSPHOLIPASE A2 — *I*50 = 22 μM
- PROSTAGLANDIN F SYNTHASE

MEFLOQUINE

AMINOPYRINE-N-DEMETHYLASE	NC	Ki = 54 μM
AROMATASE	C	Ki = 72 μM
MYOSIN-LIGHT-CHAIN KINASE		I50 = 33 μM
p-NITROANISOLE-O-DEMETHYLASE (CYT P-450)	NC	Ki = 317 μM
PROTEIN KINASE A		47% 400 μM
PROTEIN KINASE A		I50 = 268 μM
PROTEIN KINASE C	C	Ki = 60 μM

MEFSALYL

H^+-TRANSPORTING PYROPHOSPHATASE		100% 100 μM

MEGA 10

GLYCERYL-ETHER MONOOXYGENASE		Ki = 1.7 mM

MEGAZOL

FUMARATE REDUCTASE (NADH)		I50 = 2.1 μM

MEGESTROL ACETATE

3α-HYDROXYSTEROID DEHYDROGENASE (B-SPECIFIC)		53% 100 μM
Na^+/K^+-EXCHANGING ATPASE		39% 10 μM

D-MEHIONINE

PHENYLALANINE DEHYDROGENASE

MEHYL-4-NITROBENZENESULPHONATE

1-D-myo-INOSITOL-TRISPHOSPHATE 6-KINASE

MEI3(+)

PROTEIN KINASE STE11(+)

MELANOGENIC INHIBITOR

DOPACHROME TAUTOMERASE
MONOPHENOL MONOOXYGENASE

MELANOIDIN

TRYPSIN		I50 < 1 μg/ml
TRYPSIN		Ki = 5.8 %

MELANOXAZAL

TYROSINASE		I50 = 30 μg/ml

MELARSEN

PHOSPHOENOLPYRUVATE CARBOXYKINASE (ATP)

MELARSEN OXIDE

GLUTATHIONE REDUCTASE (NADPH)		Ki = 250 μM

MELARSOPROL

GLYCEROL-3-PHOSPHATE DEHYDROGENASE (NAD(P))		I50 = 1.4 mM
MALATE DEHYDROGENASE		I50 = 1.7 mM

MELATONIN

ARALKYLAMINE N-ACETYLTRANSFERASE		I50 = 10 μM
CAM KINASE II		30% 1 μM
NITRIC-OXIDE SYNTHASE		I50 = 100 μM
TRYPTOPHAN 2,3-DIOXYGENASE	C	Ki = 2.7 μM

MELEMELEONE B

PROTEIN TYROSINE KINASE pp60ySrc		I50 = 28 μM

MELEZITOSE

INULINASE
SUCROSE-PHOSPHATASE

MELIBIOSE

GALACTINOL-RAFFINOSE GALACTOSYLTRANSFERASE
α-GALACTOSIDASE
β-GLUCOSIDASE

MELINAMIDE		
STEROL O-ACYLTRANSFERASE	C	*K*i = 180 μM
STEROL O-ACYLTRANSFERASE		*I*50 = 3.2 μM
D,L-MELINAMIDE		
STEROL O-ACYLTRANSFERASE		*I*50 = 35 nM
STEROL O-ACYLTRANSFERASE		*I*50 = 130 nM
DL-MELINAMIDE		
STEROL O-ACYLTRANSFERASE		*I*50 = 500 nM
MELITTIN		
CALMODULIN-LYSINE N-METHYLTRANSFERASE		*I*50 = 3 μM
CALPAIN I		*I*50 = 2.6 μM
Ca^{2+}-TRANSPORTING ATPASE		
H^+/K^+-EXCHANGING ATPASE		*K*i = 500 nM
H^+-TRANSPORTING ATPASE		*I*50 = 1 μM
MYOSIN-LIGHT-CHAIN KINASE	NC	*K*i = 1.4 μM
Na^+/K^+-EXCHANGING ATPASE		
PHOSPHOLIPASE A2 (SECRETORY)	NC	
PHOSPHORYLASE KINASE	C	*K*i = 1 μM
PROTEIN KINASE C	NC	*K*i = 1.3 μM
MEMBRANACIN		
NADH DEHYDROGENASE (UBIQUINONE)		*I*50 = 580 pM
MEMNONIELLA ECHINATA FACTOR A		
myo-INOSITOL-1(or 4)-MONOPHOSPHATASE		*I*50 = 70 μM
MEMNONIELLA ECHINATA FACTOR B		
myo-INOSITOL-1(or 4)-MONOPHOSPHATASE		*I*50 = 460 μM
MEMNONIELLA ECHINATA FACTOR C		
myo-INOSITOL-1(or 4)-MONOPHOSPHATASE		*I*50 = 200 μM
MEN 10979		
RNA-DIRECTED DNA POLYMERASE		*I*50 = 180 nM
MENADIONE		
ALDEHYDE OXIDASE		85% 200 nM
ALDEHYDE REDUCTASE	C	*K*i = 43 mM
Ca^{2+}-TRANSPORTING ATPASE		*K*i= 34 μM
trans-CINNAMATE 4-MONOOXYGENASE		
myo-INOSITOL OXYGENASE		
NAD ADP-RIBOSYLTRANSFERASE		*I*50 = 420 μM
NAD(P)-ARGININE ADP-RIBOSYLTRANSFERASE		*I*50 = 120 μM
NAD(P)-ARGININE ADP-RIBOSYLTRANSFERASE		*I*50 = 200 μM
NITRIC-OXIDE SYNTHASE I	NC	*K*i = 27 μM
PHOSPHOPROTEIN PHOSPHATASE cdc25		
PLASMANYLETHANOYLAMINE DESATURASE		
THIOREDOXIN REDUCTASE (NADPH)		23% 100 μM
MENADIONE BISULPHITE		
NITROGENASE		
MENADIONE SODIUM BISULPHITE		
NAD ADP-RIBOSYLTRANSFERASE		*I*50 = 720 μM
NAD(P)-ARGININE ADP-RIBOSYLTRANSFERASE		*I*50 = 440 μM
XYLOSE REDUCTASE	NC	*K*i = 64 mM
MENAZON		
ACETYLCHOLINESTERASE		
MENOGARIL		
DNA TOPOISOMERASE (ATP-HYDROLYSING)		*I*50 = 10 μM
MEOCO-PHE-GLY-CN		
PAPAIN		*K*i = 40 nM

MEO-GLY-GLY-LEU-(2S,3S)-tEPS-LEU-PRO-OH
CATHEPSIN B
PAPAIN

2-ME-9-OH ET
DNA TOPOISOMERASE (ATP-HYDROLYSING)

MEO-SUCC-ALA-ALA-PRO-VAL-CH_2Cl
MYELOBLASTIN | IR

MEO-SUCC-ALA-ALA-PRO-VAL CHOLROMETHYL KETONE
ELASTASE PANCREATIC | IR

MEO-SUCC-ALA-ALA-PRO-VAL CHOLROMETHYLKETONE
ELASTASE LEUKOCYTE | IR

MEO-SUC-PHE-ALA-PRO-PHE-CHLOROMETHYLKETONE
PROTINASE CERCARIAL SHISTOSOMA MANSONI | *K*i = 1 μM

ME-O-SUC-VAL-PRO-PHE-CO_2CH_2
CATHEPSIN G | *K*i = 1.1 μM

MEPACRINAMYTAL
CYTOCHROME-b5 REDUCTASE

MEPACRINE
ACETYLCHOLINESTERASE | 100% 250 μM
NITRATE REDUCTASE
NITRATE REDUCTASE (CYTOCHROME)
1-PHOSPHATIDYLINOSITOL-4,5-BISPHOSPHATE PHOSPHODIESTERASE | 60% 2 μM
1-PHOSPHATIDYLINOSITOL PHOSPHODIESTERASE | *I*50 = 80 μM
PHOSPHOLIPASE A2 | 80% 250 μM
PHOSPHOLIPASE A2 | *I*50 = 2.2MM
PHOSPHOLIPASE A2 | REV | *I*50 > 1 mM
PHOSPHOLIPASE C | 20% 250 μM
SULFITE REDUCTASE (NADPH)
TRIACYLGLYCEROL LIPASE | 10% 250 μM

D-MEPHE-PRO-ARG-H
COAGULATION FACTOR Xa | *I*50 = 7.6 μM
PLASMIN | *I*50 = 670 nM
PROTEIN Ca | *I*50 = 8.4 μM
THROMBIN | *I*50 = 8.9 nM
TRYPSIN | *I*50 = 13 nM

D-MEPHG-PRO-ARG-H
COAGULATION FACTOR Xa | *I*50 = 2.2 μM
KALLIKREIN | *I*50 = 12 μM
PLASMIN | *I*50 = 500 nM
PLASMINOGEN ACTIVATOR nt | *I*50 = 43 μM
THROMBIN | *I*50 = 20 nM
TRYPSIN | *I*50 = 20 nM

MEPHOSFOLAN
ACETYLCHOLINESTERASE

MEPROBAMATE
CARBONATE DEHYDRATASE I | *K*i = 14 mM

MERBARONE
DNA TOPOISOMERASE (ATP-HYDROLYSING) | *I*50 = 20 μM

4-MERCAPTOACETANILIDE
PROTEINASE ASTACUS | *I*50 = 900 μM

MERCAPTOACETIC ACID
BETAINE-HOMOCYSTEINE S-METHYLTRANSFERASE
THERMOLYSIN | *K*i = 8 μM
THETIN-HOMOCYSTEINE S-METHYLTRRANSFERASE

2-MERCAPTOACETIC ACID

GLUTAMATE DECARBOXYLASE	C	Ki = 330 µM

8-MERCAPTOACYCLOVIR

PURINE-NUCLEOSIDE PHOSPHORYLASE		Ki = 4.8 µM

(R)-4-[N-(3-MERCAPTO-2-AMINOPROPYL)]AMINO-3′-3CARBOXYBIPHENYL

FARNESYL PROTEINTRANSFERASE		I50 = 150 nM
GERANYLGERANYL PROTEINTRANSFERASE		I50 = 100 µM

19-MERCAPTOANDROST-4-EN-3,17-DIONE

AROMATASE	SSI	Ki = 34 nM

3-MERCAPTOANILINE

PROTEINASE ASTACUS		I50 = 750 µM

4-MERCAPTOANILINE

PROTEINASE ASTACUS		I50 = 500 µM

2-MERCAPTOBENZIMIDAZOLE

LACTOPEROXIDASE	SSI	

2-MERCAPTOBENZTHIAZOLE

CATECHOL OXIDASE		
LACCASE		

(3-MERCAPTO-2-BENZYLPROPANOYL)-ALA-GLY-NH_2

ELASTASE PSEUDOMONAS AERUGINOSA		Ki = 64 nM
THERMOLYSIN		Ki = 750 nM

4-MERCAPTOBUTYRIC ACID

GLUTAMATE DECARBOXYLASE	C	

(trans-2-MERCAPTOCYCLOHEXYLCARBONYL)-ALA-PRO

PEPTIDYL-DIPEPTIDASE A		I50 = 3 nM

1-MERCAPTODECANE

ALCOHOL DEHYDROGENASE		Ki = 550 pM

5-MERCAPTO-DEOXYURIDINE TRIPHOSPHATE

RNA-DIRECTED DNA POLYMERASE		

MERCAPTODICARBANOBORATE

ADENOSINETRIPHOSPHATASE		I50 = 250 µM
CYTOCHROME-C OXIDASE		I50 = 1.2 mM
GLYCEROL-3-PHOSPHATE DEHYDROGENASE		I50 = 20 µM

8-MERCAPTO-9-[(1,3-DIHYDROXYPROPYL-2)AMINO]ETHYLGUANINE

PURINE-NUCLEOSIDE PHOSPHORYLASE		Ki = 2.2 µM
PURINE-NUCLEOSIDE PHOSPHORYLASE		Ki = 600 nM

2-MERCAPTOETHANESULPHONIC ACID

CATECHOL OXIDASE		I50 = 140 µM
GLUTAMATE DECARBOXYLASE		

MERCAPTOETHANOL

N-ACETYLMURAMOYL-L-ALANINE AMIDASE		
ALCOHOL DEHYDROGENASE (NAD(P))		
ALDEHYDE DEHYDROGENASE (NADP)		
(R)AMINOPROPANOL DEHYDROGENASE		
ARACHIDONATE 8-LIPOXYGENASE		
AROMATIC-L-AMINO-ACID DECARBOXYLASE		
CATECHOL 1,2-DIOXYGENASE		
CHOLINE SULFOTRANSFERASE		
CMP-N-ACYLNEURAMINATE PHOSPHODIESTERASE		100% 10%
CYSTATHIONINE γ-LYASE		
DEOXYCYTIDINE DEAMINASE		
DIHYDROOROTATE OXIDASE		

ETHANOLAMINE OXIDASE	
FORMATE DEHYDROGENASE (NADP)	
GALACTOSE OXIDASE	
GLUCOSAMINE-6-PHOSPHATE ISOMERASE	79% 1 mM
GLUCOSE 1-DEHYDROGENASE	
GLUCURONATE ISOMERASE	90% 1 mM
GLUTAMYL AMINOPEPTIDASE	
INDOLE-3-ACETALDEHYDE OXIDASE	
INDOLE 2,3-DIOXYGENASE	
D-LACTALDEHYDE DEHYDROGENASE	
LIPOXYGENASE	
NITRITE REDUCTASE	
2-NITROPROPNAE DIOXYGENASE	
3′-NUCLEOTIDASE	
OXALATE OXIDASE	
3-PHYTASE	
QUERCETIN 2,3-DIOXYGENASE	
RHODOTORULAPEPSIN	63% 1 mM
RIBITOL 2-DEHYDROGENASE	
RNA-DIRECTED DNA POLYMERASE	
SIALIDASE	
SORGHUM ASPARTIC PROTEINASE	79% 5 mM
SUCCINATE DEHYDROGENASE	
THIAMIN PYRIDINYLASE	
THIOL OXIDASE	
peptide-TRYPTOPHAN 2,3-DIOXYGENASE	
XAA-TRP AMINOPEPTIDASE	*I*50 = 600 μM

2-MERCAPTOETHANOL

β-N-ACETYLHEXOSAMINIDASE	80% 2 mM
ACETYLSPERMIDINE DEACETYLASE	
ACYL[ACYL-CARRIER-PROTEIN] DESATURASE	
AEROMONOLYSIN	
ALCOHOL DEHYDROGENASE	*I*50 = 2.2 mM
ALCOHOL SULFOTRANSFERASE	
ALDEHYDE REDUCTASE	
AMIDOPHOSPHORIBOSYLTRANSFERASE	
AMINE OXIDASE (COPPER-CONTAINING)	
AMINOACYLASE	
4-AMINOBUTYRATE TRANSAMINASE	
5-AMINOLEVULINATE SYNTHASE	
AMINOPEPTIDASE B	
L-ASCORBATE PEROXIDASE	
ASPARAGINASE	
ASPERGILLUS NUCLEASE S1	
BRANCHED-CHAIN-AMINO-ACID TRANSAMINASE	
CARBAMOYL-PHOSPHATE SYNTHASE (GLUTAMINE-HYDROLYSING)	
CARBONYL REDUCTASE (NADPH)	
CATECHOL OXIDASE	
trans-CINNAMATE 4-MONOOXYGENASE	
CMP-N-ACYLNEURAMINATE PHOSPHODIESTERASE	46% 25 mM
COLLAGENASE CLOSTRIDIUM HISTOLYTICUM	
COMPLEMENT FACTOR I	
DIHYDROOROTASE	
2,3-DIHYDROXYBENZOATE 2,3-DIOXYGENASE	
2,5-DIKETO-D-GLUCONATE REDUCTASE	
DIPEPTIDASE	
DIPEPTIDYL-PEPTIDASE 4	
FAD PYROPHOSPHATASE	
GALACTOLIPASE	
β-GALACTOSIDASE	
β-GLUCOGALLIN O-ACETYLTRANSFERASE	
GLUCONATE DEHYDRATASE	56% 10 mM
GLUCURONATE REDUCTASE	
GLUTAMATE DECARBOXYLASE	59% 15 mM

γ-GLUTAMYL HYDROLASE		$I50 = 10$ μM
γ-GLUTAMYLTRANSFERASE		
GLUTATHIONE THIOLESTERASE		
GLYCEROL 2-DEHYDROGENASE		
GLYCEROL DEHYDROGENASE (NADP)		
GLYCEROL-3-PHOSPHATE O-ACYLTRANSFERASE		
HEME OXYGENASE (DECYCLIZING)		81% 100 μM
HEPARAN-α-GLUCOSAMINIDINE N-ACETYLTRANSFERASE		
HISTIDINE AMMONIA-LYASE		
4-HYDROXYMANDELATE OXIDASE		100% 1 mM
7α-HYDROXYSTEROID DEHYDROGENASE		
INDOLEACETALDOXIME DEHYDRATASE		75% 1 mM
INDOLEACETYLGLUCOSE-INOSITOL O-ACYLTRANSFERASE		
myo-INOSITOL-1(or 4)-MONOPHOSPHATASE		
ISOCITRATE DEHYDROGENASE (NADP)		
ISOCITRATE LYASE		
ISOPENTENYL-DIPHOSPHATE Δ-ISOMERASE		
3-ISOPROPYLMALATE DEHYDRATASE		
LACCASE		
LEUCOLYSIN		
LEUCYL AMINOPEPTIDASE		
LINOLEOYL-CoA DESATURASE		
LYSOPEPTASE		
MALATE DEHYDROGENASE (OXALACETATE-DECARBOXYLATING) (NADP)		
MANDELONITRILE LYASE		
MANDELONITRILE LYASE		21% 1 mM
MANNITOL 2-DEHYDROGENASE (NADP)		
METHANE MONOOXYGENASE		85% 100 μM
MICROCOCCUS CASEOLYTICUS NEUTRAL PROTEINASE		
MONOPHENOL MONOOXYGENASE		$Ki = 3.5$ μM
MONOPHENOL MONOOXYGENASE		$I50 = 25$ μM
NARINGENIN-CHALCONE SYNHTASE		
NEPRILYSIN		38% 2 mM
NUCLEOTIDE PYROPHOSPHATASE		96% 16 mM
PEPTIDYL-DIPEPTIDASE A		
PHENYLALANINE 4-MONOOXYGENASE		
PHOSPHODIESTERASE		
PHOSPHODIESTERASE I		100% 10 mM
POLYPHENOL OXIDASE		40% 33 μM
PREGASTRIC ESTERASE		
PROCOLLAGEN N-ENDOPEPTIDASE		78% 1 mM
PROTEINASE ASTACUS		$I50 = 5.4$ mM
PROTOPORPHYRINOGEN OXIDASE		25% 5 mM
RIBULOSE-BISPHOSPHATE CARBOXYLASE		
SARCOSINE OXIDASE		
SINAPOYLGLUCOSE-SINAPOYLGLUCOSE O-SINAPOYLTRANSFERASE		100% 1 mM
STEAROYL-CoA DESATURASE		
THIOPURINE S-METHYLTRANSFERASE		
TRIMETHYLSULFONIUM-TETRAHYDROFOLATE N-METHYLTRANSFERASE		
UREASE	C	$Ki = 550$ μM
UROCANATE HYDRATASE		
XAA-HIS DIPEPTIDASE		95% 1 mM
XAA-METHYL-HIS DIPEPTIDASE		
XAA-PRO DIPEPTIDASE		

MERCAPTOETHYLAMINE

β-GALACTOSIDASE		
HISTIDINE AMMONIA-LYASE	C	$Ki = 4.6$ mM
THERMOLYSIN		$Ki = 800$ nM

2-MERCAPTOETHYLAMINE

AMINOPEPTIDASE		94% 1 mM
CYSTATHIONINE γ-LYASE		
SPERMIDINE SYNTHASE	C	67% 100 μM
SPERMINE SYNTHASE	C	12% 1 mM

S-(2-MERCAPTOETHYL)(2-MERCAPTOMETHYL-3-PHENYLPROPANOIC ACID
SERINE-TYPE CARBOXYPEPTIDASE NCΔ *K*i = 4.2 μM

6-MERCAPTOGUANINE
HYPOXANTHINE PHOSPHORIBOSYLTRANSFERASE C *K*i = 29 μM

6-MERCAPTOHEXANOYL-L-THREONINEPHOSPHATE
COENZYME-M-7-MERCAPTOHEPTANOYLTHREONINE-PHOSPHATE-HETEROSULFIDE HYDROGENASE

2-MERCAPTO-4,5-IMIDAZOLEDICARBOXYLIC ACID
FATTY-ACID PEROXIDASE

2-MERCAPTOINOSINE
IMP CYCLOHYDROLASE C *K*i = 94 nM
IMP CYCLOHYDROLASE C *K*i = 94 nM

cis-6-MERCAPTOMETHYLDIHYDROOROTIC ACID
DIHYDROOROTASE C *K*i = 140 nM

2-MERCAPTOMETHYLGLUTAMIC ACID
γ-GLUTAMYL HYDROLASE

2-MERCAPTOMETHYL-3-GUANIDINOETHYL THIOPROPIONIC ACID
LYSINE (ARGININE) CARBOXYPEPTIDASE

DL-2-(MERCAPTOMETHYL) 3-(GUANIDINOETHYLTHIO)PROPIONIC ACID
CARBOXYPEPTIDASE H *K*i = 44 nM

2-MERCAPTOMETHYL-5-GUANIDINOPENTANOIC ACID
CARBOXYPEPTIDASE A C *K*i = 12 μM
CARBOXYPEPTIDASE B C *K*i = 400 pM

D,L-2-(MERCAPTOMETHYL) 5-GUANIDINOPENTANOIC ACID
CARBOXYPEPTIDASE B *K*i = 400 pM

MERCAPTOMETHYLINIDAZOLE
PEROXIDASE *I*50 = 100 nM

(S,S)-3-MERCAPTO-6-METHYL-4-[[[1(S)-[METHYLAMINO)CARBONYL]2(3-INDOLYL)ETHYL]AMINO]CARBONYL]HEPTANOIC ACID METHYL-ESTER
COLLAGENASE *I*50 = 2.5 nM

2-[(R,S)MERCAPTOMETHYL]-4-METHYLPENTANOIC ACID
INTERSTITIAL COLLAGENASE *I*50 = 320 μM

2-[(R,S)MERCAPTOMETHYL]-4-METHYLPENTANOIC ACID ALA-GLY-GLN
INTERSTITIAL COLLAGENASE *I*50 = 14 μM

3-(R)-(MERCAPTOMETHYL)-2-OXO-1-AZACYCLODECANE-10(S)CARBOXYLIC ACID
NEPRILYSIN *I*50 = 3 nM

N-[[1-[(2(S)MERCAPTO-3-METHYL-1-OXOBUTYL)AMINO-1-CYCLOPENTYL]CARBONYL]-L-TYROSINE
NEPRILYSIN *I*50 = 1.5 nM
PEPTIDYL-DIPEPTIDASE A *I*50 = 7 nM

N[2(S)-(MERCAPTOMETHYL)-1-OXOPENTYL]-S-METHYL-L-CYSTEINE
NEPRILYSIN *I*50 = 1.5 nM
PEPTIDYL-DIPEPTIDASE A *I*50 = 77 nM

N[8R,S)-2-MERCAPTOMETHYL-1-OXO-3-PHENYLPROPYL]PHENYLALANINE
NEPRILYSIN *I*50 = 2 nM

[R-(R*,S*)]-3-(3-MERCAPTO-2-METHYL-1-OXOPROPYL)-4-THIAZOLIDINECARBOXYLIC ACID
PEPTIDYL-DIPEPTIDASE A

(3R)-6-(MERCAPTOMETHYL)-5-OXO-1-THIA-4-AZACYCLOTRIDECANE-3-CARBOXYLIC ACID
NEPRILYSIN *I*50 = 18 nM
PEPTIDYL-DIPEPTIDASE A *I*50 = 12 nM

N[2-(MERCAPTOMETHYL)-3-PHENYLPROPANOYL]-L-TYROSINE

NEPRILYSIN		$I50 = 2$ nM
PEPTIDYL-DIPEPTIDASE A		$I50 = 25$ nM

3-MERCAPTO-2-D-METHYLPROPANOYLPROLINE

PENICILLOPEPSIN		

D-3-MERCAPTO-2-METHYLPROPANOYL-L-PROLINE

PEPTIDYL-DIPEPTIDASE A	C	$Ki = 17$ nM

1-[(2S)-3-MERCAPTO-2-METHYLPROPANOYL]-PYROGLUTAMIC ACID

PEPTIDYL-DIPEPTIDASE A		$I50 = 9$ nM

2-MERCAPTONICOTINIC ACID

XAA-PRO DIPEPTIDASE		$Ki = 560$ µM

1-MERCAPTO-9-OXO-11,15-DIHYDROXYPROSTA-5,13-DIENE

PROSTAGLANDIN-ENDOPEROXIDE SYNTHASE		

3-MERCAPTO-2-OXOGLUTARIC ACID

ISOCITRATE DEHYDROGENASE (NAD)	C	$Ki = 520$ nM
ISOCITRATE DEHYDROGENASE (NADP)	C	$Ki = 5$ nM

(2-MERCAPTOPHENYLACETYL)-2-AMINONAPHTHYLPROPIONYL-PROLINE

PEPTIDYL-DIPEPTIDASE A		$I50 = 14$ nM

(1S,2R,5S)-(–)-[(2-MERCAPTO-5-PHENYLCYCLOPENTANECARBONYL)AMINO]ACETIC ACID

NEPRILYSIN		$I50 = 430$ nM
THERMOLYSIN		$I50 = 29$ µM

2-MERCAPTOPHENYLPHOSPHONIC ACID

PHOSPHATASE ALKALINE		$Ki = 210$ nM

MERCAPTOPICOLINIC ACID

PHOSPHOENOLPYRUVATE CARBOXYKINASE (GTP)		

2-MERCAPTOPICOLINIC ACID

GLUCOSE-6-PHOSPHATASE		16% 4 mM

3-MERCAPTOPICOLINIC ACID

GLUCOSE-6-PHOSPHATASE		22% 4 mM
GLUTAMATE DEHYDROGENASE (NADP)		30% 300 µM
PHOSPHOENOLPYRUVATE CARBOXYKINASE (ATP)	NC	$Ki = 46$ µM
PHOSPHOENOLPYRUVATE CARBOXYKINASE (GTP)		97% 4 mM
6-PHOSPHOFRUCTOKINASE		30% 300 µM

3-MERCAPTOPROPANOYL-L-PROLINE

PEPTIDYL-DIPEPTIDASE A	C	$Ki = 12$ nM

2-MERCAPTOPROPIONIC ACID

GLUTAMATE DECARBOXYLASE	C	$Ki = 53$ µM
PROTEINASE ASTACUS		$I50 = 37$ mM

3-MERCAPTOPROPIONIC ACID

ALCOHOL DEHYDROGENASE		$I50 = 45$ mM
FATTY ACID OXYDATION		
GLUTAMATE DECARBOXYLASE		$I50 = 15$ µM
GLUTAMATE DECARBOXYLASE	C	$Ki = 1.8$ µM
GLUTAMATE DECARBOXYLASE	C	$Ki = 2,7$ µM
GLUTAMATE DECARBOXYLASE	C	$Ki = 130$ µM
O-SUCCINYLHOMOSERINE (THIOL)-LYASE	NC	$Ki = 1$ mM
UROCANATE HYDRATASE		
XAA-PRO DIPEPTIDASE		$Ki = 630$ µM

3-MERCAPTOPROPIONYL CoA

ACYL-CoA DEHYDROGENASE		

α-MERCAPTOPROPIONYLGLYCINE

GLUTATHIONE PEROXIDASE		

N-(2-MERCAPTOPROPIONYL)GLYCINE
CATECHOL OXIDASE — $I50 = 58\ \mu M$
PROTEINASE ASTACUS — $I50 = 8.7\ mM$

MERCAPTOPROPIONYL-L-PROLYL-L-ALANINE
ACHROMOBACTER IOPHAGUS COLLAGENASE — C — $Ki = 2.1\ \mu M$

MERCAPTOPROPIONYL-L-PROLYL-L-ALANINE METHYLESTER
ACHROMOBACTER IOPHAGUS COLLAGENASE — C — $Ki = 8\ \mu M$

MERCAPTOPROPIONYL-L-PROLYL-L-ARGININE
ACHROMOBACTER IOPHAGUS COLLAGENASE — C — $Ki = 500\ nM$

MERCAPTOPROPIONYL-L-PROLYL-L-ARGININE METHYLESTER
ACHROMOBACTER IOPHAGUS COLLAGENASE — C — $Ki = 600\ nM$

MERCAPTOPROPIONYL-L-PROLYL-L-HYDROXYPROLINE
ACHROMOBACTER IOPHAGUS COLLAGENASE — C — $Ki = 1\ \mu M$

MERCAPTOPROPIONYL-L-PROLYL-L-PHENYLALANINE
ACHROMOBACTER IOPHAGUS COLLAGENASE — C — $Ki = 2.8\ \mu M$

MERCAPTOPROPIONYL-L-PROLYL-L-PROLINE
ACHROMOBACTER IOPHAGUS COLLAGENASE — C — $Ki = 800\ nM$

MERCAPTOPROPYLGUANIDINE
NITRIC-OXIDE SYNTHASE I — $I50 = 80\ \mu M$
NITRIC-OXIDE SYNTHASE II — $I50 = 7\ \mu M$
NITRIC-OXIDE SYNTHASE III — $I50 = 4\ \mu M$

6-MERCAPTOPURINE
HYPOXANTHINE PHOSPHORIBOSYLTRANSFERASE — C — $Ki = 12\ \mu M$
PURINE-NUCLEOSIDE PHOSPHORYLASE — $Ki = 73\ \mu M$
XANTHINE DEHYDROGENASE — C — $Ki = 8\ \mu M$

6-MERCAPTOPURINE RIBONUCLEOTIDE
IMP DEHYDROGENASE

6-MERCAPTOPURINERIBOSIDE
INOSINE NUCLEOSIDASE
PROTEIN KINASE N — $Ki = 5\ nM$

6-MERCAPTOPURINERIBOSIDE-5′-MONOPHOSPHATE
IMP CYCLOHYDROLASE — C — $Ki = 200\ nM$

6-MERCAPTOPURINE RIBOSIDE 5′-PHOSPHATE
ADENYLOSUCCINATE SYNTHASE
IMP CYCLOHYDROLASE — C — $Ki = 200\ nM$

2-MERCAPTOPYRIDINE
CATECHOL OXIDASE — $I50 = 100\ \mu M$

MERCAPTOPYRUVIC ACID
4-HYDROXY-2-OXOGLUTARATE ALDOLASE — C — $Ki = 5\ \mu M$

3-MERCAPTOPYRUVIC ACID
GLUTAMINE-PYRUVATE TRANSAMINASE
2-HYDROXY-3-OXOPROPIONATE REDUCTASE

2-MERCAPTO-4(3H)-QUINAZOLINONE
NAD ADP-RIBOSYLTRANSFERASE — $I50 = 44\ \mu M$

8-MERCAPTOQUINOILINE
ENDOTHELIN CONVERTING ENZYME — $I50 = 5.1\ \mu M$

2-MERCAPTOSUCCINATE
ADENYLOSUCCINATE SYNTHASE — C — $Ki = 3.2\ mM$

MERCAPTOSUCCINIC ACID
GLUTAMATE DECARBOXYLASE 30% 1 mM
GLUTAMATE DECARBOXYLASE 100% 1 mM
GLUTATHIONE PEROXIDASE

1-MERCAPTO-9,11,15-TRIHYDROXYPROSTA-5,13-DIENE
PROSTAGLANDIN-ENDOPEROXIDE SYNTHASE

MERCAPTOTRIPEPTIDES
COLLAGENASE CLOSTRIDIUM HISTOLYTICUM

MERCAPTO UNDECAHYDRO DODECABORON
H^+-TRANSPORTING ATP SYNTHASE

8-MERCAPTOXANTHINE
XANTHINE PHOSPHORIBOSYLTRANSFERASE

6-MERCAPTUPURINE
ADENINE PHOSPHORIBOSYLTRANSFERASE

MERCPTOETHANOL
HYDROGEN DEHYDROGENASE

p-MERCUIRPHENYLSULPHONIC ACID
D-LACTATE DEHYDROGENASE (CYTOCHROME) 60% 500 nM

MERCURIACETIC ACID
NAD KINASE

MERCURIALS
CYSTATHIONINE γ-LYASE
FERREDOXIN-NADP REDUCTASE
SERINE PROTEINASE STREPTOMYCES THERMOPHILIC

MERCURIBENZOIC ACID
ALANINE TRANSAMINASE
AROMATIC-L-AMINO-ACID DECARBOXYLASE
CATHEPSIN L
TRYPTOPHAN SYNTHASE

p-MERCURIBENZOIC ACID
CLOSTRIDIUM HISTOLYTICUM AMINOPEPTIDASE
α-L-FUCOSIDASE
4-NITROPHENYLPHOSPHATASE ALKALINE NC $Ki = 1.2$ mM
PHOSPHATE CARRIER MITOCHONDRIA NC
SUGAR PHOSPHATASE

MERCURIC ACETATE
GLUTATHIONE TRANSFERASE
LYSOPEPTASE

p-MERCURICBENZOIC ACID
5-FORMYLTETRAHYDROFOLATE CYCLO-LIGASE
LEUCINE-tRNA LIGASE

MERCURIC CHLORIDE
GLUTATHIONE TRANSFERASE

5-MERCURI-CTP
N-ACETYLNEURAMINATE CYTIDYLYLTRANSFERASE

MERCURY CONTAINIG THIOL REAGENTS
YEAST CYSTEINE PROTEINASE D
YEAST CYSTEINE PROTEINASE E
YEAST CYSTEINE PROTEINASE F

MERCURY THIOLATE
PROLYL OLIGOPEPTIDASE (THIOL DEPENDENT)

MERLASEN OXIDE
GLUTATHIONE REDUCTASE (NADPH)

MERLASOPROL
GLUCOSE-6-PHOSPHATE 1-DEHYDROGENASE — *I*50 = 820 μM
HEXOKINASE — *I*50 = 5.4 mM
TRYPANOTHIONE REDUCTASE — 45% 50 μM

MER-N 5075A
PROTEINASE HIV-1 — *I*50 = 18 μM

MER-NF5003 B
PROTEINASE MYELOBLASTOSIS VIRUS — *I*50 = 17 μM

MER-NF5003 E
PROTEINASE MYELOBLASTOSIS VIRUS — *I*50 = 13 μM

MER-NF5003 F
PROTEINASE MYELOBLASTOSIS VIRUS — *I*50 = 8 μM

MERSALYL

Enzyme	Type	Value
D-AMINO-ACID OXIDASE		
AQUACOBALAMIN REDUCTASE (NADPH)		
ARGININE DEIMINASE		
CATECHOL 1,2-DIOXYGENASE		
CATHEPSIN L		
DICARBOXYLATE CARRIER MITOCHONDRIA	C	
2-ENOATE REDUCTASE		100% 1 mM
FERREDOXIN-NITRITE REDUCTASE		
GLUTATHIONE DEHYDROGENASE (ASCORBATE)		
9-HYDROXYPROSTAGLANDIN DEHYDROGENASE		70% 1 mM
11-HYDROXYTHROMBOXANE B2 DEHYDROGENASE		86% 1 mM
LEVANSUCRASE		
LYSINE 2-MONOOXYGENASE		
METHYLENETETRAHYDROFOLATE REDUCTASE (NADPH)		
MONODEHYDROASCORBATE REDUCTASE (NADH)		
NADH DEHYDROGENASE (UBIQUINONE)		
NADPH-CYTOCHROME-c2 REDUCTASE	C	*K*i = 55 μM
PHOSPHATE CARRIER MITOCHONDRIA		*I*50 = 30 μM
PYRUVATE CARRIER MITOCHONDRIA		
tRNA ADENYLYLTRANSFERASE		
tRNA CYTIDYLTRANSFERASE		100% 1 mM
RUBREDOXIN-NAD REDUCTASE		
SACCHAROPINE DEHYDROGENASE (NAD, L-LYSINE FORMING)		
SERINE-TYPE CARBOXYPEPTIDASE		85% 100 μM
STEROID 21-MONOOXYGENASE		
THREONINE DEHYDRATASE		
UDP-GALACTURONYLTRANSFERASE		

MERSDALYL
HYDROGENASE II

MERTHIOLATE

1-ACYLGLYCEROL-3-PHOSPHATE O-ACYLTRANSFERASE	48% 50 μM
1-ACYLGLYCEROPHOSPHOCHOLINE O-ACYLTRANSFERASE	97% 50 μM
GLUCAN ENDO-1,3-α-GLUCOSIDASE	100% 1 mM
LYSOPHOSPHATIDYLETHANOLAMINE ACYL-CoA ACYLTRANSFERASE	68% 50 μM
LYSOPHOSPHATIDYLINOSITOL ACYL-CoA ACYLTRANSFERASE	77% 50 μM
LYSOPHOSPHATIDYLSERINE ACYL-CoA ACYLTRANSFERASE	77% 50 μM
SIALIDASE	

(4R,5R,1′′S)1-MERTHYL-(3′-CARBOXYPROPYL)4,5-BIS[1′′[(FLUORENYLMETHOXYCARBONYL)AMINO]2′′-PHENYLETHYL]-1,3-DIOXANE
PROTEINASE HIV-1

MESACONIC ACID
FUMARATE HYDRATASE — C — *K*i = 25 mM
FUMARATE REDUCTASE (NADH)

2-METHYLENEGLUTARATE MUTASE	C	*K*i = 1.2 mM
METHYLITACONATE Δ-ISOMERASE	C	*K*i = 1.2 mM
MESOHEM		
GLUTATHIONE TRANSFERASE		
MESOPORPHYRIN		
5-AMINOLEVULINATE SYNTHASE		
TRYPTOPHAN 2,3-DIOXYGENASE	C	*K*i = 500 nM
MESOPORPHYRIN IRON		
5-AMINOLEVULINATE SYNTHASE		
MESOPORPHYRIN IX		
UROPORPHYRINOGEN DECARBOXYLASE		32% 8 μM
MESOPORPHYRIN TIN		
HEME OXYGENASE (DECYCLIZING)		100% 3.3 μM
MESO-TARTRATE		
ASPARTATE CARBAMOYLTRANSFERASE		
L-ASPARTATE OXIDASE		
FUMARATE HYDRATASE		10 μM
L(+)-TARTRATE DEHYDRATASE	C	*K*i = 8 mM
L(+)-TARTRATE DEHYDRATASE	C	*K*i = 8 mM
MESOXALIC ACID		
3-HYDROXYBUTYRATE DEHYDROGENASE		
MALATE DEHYDROGENASE (OXALACETATE-DECARBOXYLATING) (NADP)	C	*K*i = 100 μM
METAARSENITE		
ARYLFORMAMIDASE		100% 10 μM
METABISULPHITE		
POLYPHENOL OXIDASE		47% 16 μM
METAL CHELATORS		
LEUCYL AMINOPEPTIDASE		
NEUTRAL PROTEINASE ASPERGILLUS ORYZAE		
METAL-CHELATORS		
CYTOSOL NON-SPECIFIC DIPEPTIDASE		
OCTADECANAL DECARBOXYLASE		
XAA-HIS DIPEPTIDASE		
METAL IONS		
GLUCOSE 1-DEHYDROGENASE		
GTP CYCLOHYDROLASE I		
3α-HYDROXYSTEROID DEHYDROGENASE (B-SPECIFIC)		
METAL IONS HEAVY		
5-AMINOPENTANAMIDASE		
FRUCTOSE-BISPHOSPHATE ALDOLASE		
HYALURONOGLUCOSAMINIDASE		
4-HYDROXYPROLINE EPIMERASE	C	*K*i = 16 mM
MYXOBACTER AL-1 PROTEINASE		
PROTEINASE II SPOROTRICHUM		
PROTEINASE I SPOROTRICHUM		
URATE OXIDASE		
METAPHOSPHATES		
AMP DEAMINASE		
METAPYRONE		
CYTOCHROME P450		
LEUKOTRIENE-B4 20-MONOOXYGENASE		

METASILICATE
β-FRUCTOFURANOSIDASE

METAVANADATE
PHOSPHATASE ALKALINE (PHOSPHODIESTERASE ACTIVITY) C $Ki = 20\ \mu M$

MET-ENKEPHALIN
CARBOXYPEPTIDASE H C $Ki = 12\ mM$
TRIPEPTIDE AMINOPEPTIDASE $I50 = 520\ \mu M$

MET-ENKEPHALIN-ARG^{6}GLU^{7}LEU^8
CARBOXYPEPTIDASE H C $Ki = 7\ mM$

MET-ENKEPHALIN-ARG^{6}PHE^7
CARBOXYPEPTIDASE H C $Ki = 5.5\ mM$

METHADONE
ALDEHYDE OXIDASE $I50 = 310\ nM$
CYTOCHROME P450 (O-DEMETHYLATION) $Ki = 300\ nM$

METHAMIDOPHOS
ACETYLCHOLINESTERASE
CHOLINESTERASE $I50 = 1.8\ \mu M$
PORPHOBILINOGEN SYNTHASE $I50 = 7.1\ mM$
SUPEROXIDE DISMUTASE $I50 = 420\ \mu M$

METHANE
AMMONIA MONOOXYGENASE C $Ki = 3.2\ mM$

S-(METHANETHIO)CYSTEINE
METHIONINE ADENOSYLTRANSFERASE C $Ki = 600\ \mu M$

S-(METHANETHIO)HOMOCYSTEINE
METHIONINE ADENOSYLTRANSFERASE C $Ki = 600\ \mu M$

L-3-METHANOHOMOSERINEPHOSPHATE
THREONINE SYNTHASE C $Ki = 10\ \mu M$

METHANOL
1-ACYLGLYCEROPHOSPHOCHOLINE O-ACYLTRANSFERASE
ALDEHYDE OXIDASE
AMINOPEPTIDASE (HUMAN LIVER)
CATALASE
DIACYLGLYCEROL CHOLINEPHOSPHOTRANSFERASE
ECDYSONE OXIDASE 31% 5%
ENDO-1,4-β-XYLANASE
EPOXIDE HYDROLASE
ETHANOLAMINEPHOSPHOTRANSFERASE
FORMATE DEHYDROGENASE
STARCH SYNTHASE
XANTHINE DEHYDROGENASE
XANTHINE OXIDASE
XANTHINE OXYDASE FERROXYDASE ACTIVITY

5,6-METHANO-LEUKOTRIENE A4
ARACHIDONATE 5-LIPOXYGENASE $I50 = 44\ \mu M$
CYCLOOXYGENASE $I50 > 440\ \mu M$

METHAPYRILENE
ALDEHYDE OXIDASE $I50 = 116\ \mu M$

METHAZOLAMIDE
CARBONATE DEHYDRATASE $I50 = 9.8\ nM$
CARBONATE DEHYDRATASE I $Ki = 10\ \mu M$
CARBONATE DEHYDRATASE II $Ki = 8\ nM$
CARBONATE DEHYDRATASE III $Ki = 100\ \mu M$
CARBONATE DEHYDRATASE IV $Ki = 56\ nM$

METHICILLIN

D-ALANYL-D-ALANINE CARBOXYPEPTIDASE		
AMINOPEPTIDASE (HUMAN LIVER)	NC	Ki = 2 mM
β-LACTAMASE	C	Ki = 930 nM
β-LACTAMASE RTEM	IR	

METHIDATHION

ACETYLCHOLINESTERASE

METHIMAZOLE

IODIDE PEROXIDASE	SSI	
LACTOPEROXIDASE		
MYELOPEROXIDASE		40% 10 μM

METHIOCARB

ACETYLCHOLINESTERASE

METHIOLATE

3-OXO-5β-STEROID Δ^4-DEHYDROGENASE

METHIONINE

O-ACETYLHOMOSERINE (THIOL)-LYASE		50% 10 mM
O-ACETYLHOMOSERINE (THIOL)-LYASE		14% 10 mM
ADENOSYLHOMOCYSTEINE NUCLEOSIDASE		
ALANINE AMINOPEPTIDASE	NC	I50 = 6.1 mM
D-AMINO-ACID OXIDASE		
ASPARAGINE-OXO-ACID TRANSAMINASE		
ASPARTATE-SEMIALDEHYDE DEHYDROGENASE		
CHOLINE KINASE		
L-3-CYANOALANINE SYNTHASE		I50 = 300 μM
CYSTATHIONINE β-LYASE		
CYSTATHIONINE γ-LYASE		14% 5 mM
CYSTINYL AMINOPEPTIDASE		
[GLUTAMATE-AMMONIA-LIGASE] ADENYLYLTRANSFERASE		
GLUTAMATE SYNTHASE (FERREDOXIN)		
GLUTAMATE SYNTHASE (NADH)		
GLUTAMATE SYNTHASE (NADPH)		
GLYCINE HYDROXYMETHYLTRANSFERASE		
HOMOCYSTEINE S-METHYLTRANSFERASE		
HOMOSERINE DEHYDROGENASE		
MEMBRANE ALANINE AMINOPEPTIDASE		
5-METHYLTETRAHYDROPTEROYLTRIGLUTAMATE-HOMOCYSTEINE METHYLTRANSFERASE		
MYELOPEROXIDASE		38% 50 μM
PHENYLALANINE (HISTIDINE) TRANSAMINASE		
O-SUCCINYLHOMOSERINE (THIOL)-LYASE		61% 2 mM

METHIONINE AMIDE

METHIONINE-tRNA LIGASE	NC	Ki = 97 mM

D-METHIONINE

D-ALANINE TRANSAMINASE		
METHIONINE ADENOSYLTRANSFERASE		Ki = 400 μM
METHIONINE DECARBOXYLASE		
METHIONINE-tRNA LIGASE	U	Ki = 640 μM

METHIONINE ENKEPHALINAMIDE D-ALA2

ADENYLATE CYCLASE

METHIONINE ETHYLESTER

HOMOCYSTEINE S-METHYLTRANSFERASE

DL-METHIONINE HYDROXAMIC ACID

PROTEINASE ASTACUS		I50 = 345 μM

L-METHIONINE

O-ACETYLSERINE (THIOL)-LYASE	NC	Ki = 18 mM
ASPARTATE KINASE		15% 20 mM

HOMOSERINE O-ACETYLTRANSFERASE
HOMOSERINE KINASE 13% 20 mM

METHIONINE METHYLESTER
HOMOCYSTEINE S-METHYLTRANSFERASE

L-METHIONINE SULFONE
GLUTAMATE-AMMONIA LIGASE 80% 10 mM
GLUTAMATE SYNTHASE (NADPH)

L-METHIONINE-S-SULFOXIMINE PHOSPHATE
GLUTAMATE-AMMONIA LIGASE Ki < 1 nM

METHIONINE SULPHONE
METHIONINE-tRNA LIGASE C Ki = 15 mM

L-METHIONINE SULPHONE
GLUTAMATE SYNTHASE (NADH)

METHIONINE SULPHOXIDE
PROTEIN-METHIONINE-S-OXIDE REDUCTASE 20% 100 μM

L-METHIONINESULPHOXIDE
ASPARAGINE SYNTHASE (GLUTAMINE-HYDROLYSING) 24% 1 mM

L-METHIONINE-S,R-SULPHOXIDE
GLUTAMATE-AMMONIA LIGASE 12% 25 mM
GLUTAMATE SYNTHASE (NADH)
GLUTAMATE SYNTHASE (NADPH)

L-METHIONINE SULPHOXIME
GLUTAMATE-CYSTEINE LIGASE
GLUTAMATE KINASE
GLUTAMIN-(ASPARAGIN-)ASE
METHIONINE-tRNA LIGASE C Ki = 17 mM

L-METHIONINE-(R,S) SULPHOXIME
GLUTAMATE-AMMONIA LIGASE C Ki = 2 μM

L-METHIONINE-S-SULPHOXIME
GLUTAMATE-AMMONIA LIGASE 81% 1 mM

L-METHIONINE-S,R-SULPHOXIME
ASPARAGINE-OXO-ACID TRANSAMINASE
GLUTAMATE-AMMONIA LIGASE 75% 2 mM
GLUTAMATE-CYSTEINE LIGASE
GLUTAMATE SYNTHASE (NADH)
GLUTAMATE SYNTHASE (NADPH)

L-METHIONINOL-AMP
METHIONINE-tRNA LIGASE C Ki = 8.6 nM

D,L-METHIONINOL
METHIONINE-tRNA LIGASE C Ki = 17 μM

METHIXENE
INOSITOL-1,4,5-TRISPHOSPHATE 5-PHOSPHATASE 73% 1 mM

METHOCARBAMOL
CARBONATE DEHYDRATASE I Ki = 70 μM

METHOMYL
ACETYLCHOLINESTERASE

METHOTHREXATE
GLUTATHIONE SYNTHASE Ki = 100 μM

METHOTHREXATE γ-PENTGLUTAMATE
PROTEIN KINASE CASEIN KINASE II C Ki = 90 μM

Inhibitor / Enzyme	Type	Value
METHOTREXATE		
ACETYLCHOLINESTERASE		$I50 = 730\ \mu M$
ARYLAMINE N-ACETYLTRANSFERASE		
DEOXYCYTIDYLATE HYDROXYMETHYLTRANSFERASE		
DIHYDROFOLATE REDUCTASE	C	$Ki = 52$ pM
DIHYDROFOLATE REDUCTASE		$I50 = 24$ nM
DIHYDROFOLATE REDUCTASE		$I50 = 1$ nM
DIHYDROPTERIDINE REDUCTASE	C	$Ki = 38\ \mu M$
GLUTAMATE FORMIMINOTRANSFERASE		
GLYCINE HYDROXYMETHYLTRANSFERASE		
GLYCINE METHYLTRANSFERASE		
MANDELATE 4-MONOOXYGENASE		63% 250 μM
METHIONYL-tRNA FORMYLTRANSFERASE		
METHYLENETETRAHYDROFOLATE CYCLOHYDROLASE		
NADPH OXIDASE		
NITRIC-OXIDE SYNTHASE II		
PHENYLALANINE 4-MONOOXYGENASE	NC	$Ki = 38\ \mu M$
PHOSPHORIBOSYLAMINOIMIDAZOLECARBOXAMIDE FORMYLTRANSFERASE		$Ki = 140\ \mu M$
PHOSPHORIBOSYLAMINOIMIDAZOLECARBOXAMIDE FORMYLTRANSFERASE	NC	$Ki = 72\ \mu M$
THYMIDYLATE SYNTHASE	NC	$Ki = 52\ \mu M$
THYMIDYLATE SYNTHASE		$Ki = 13\ \mu M$
THYMIDYLATE SYNTHASE	NC	$Ki = 410\ \mu M$
METHOTREXATE γ-(m-BROMOANILIDE)		
DIHYDROFOLATE REDUCTASE		$I50 = 30$ nM
METHOTREXATE γ-(m-CARBOXYANILIDE)		
DIHYDROFOLATE REDUCTASE		$I50 = 35$ nM
METHOTREXATE POLYGLUTAMATE		
THYMIDYLATE SYNTHASE		
METHOTREXATE TETRAGLUTAMATE		
PHOSPHORIBOSYLAMINOIMIDAZOLECARBOXAMIDE FORMYLTRANSFERASE		$Ki = 57$ nM
THYMIDYLATE SYNTHASE		$Ki = 47$ nM
METHOXSALEN		
CYTOCHROME P450 1A		
CYTOCHROME P450 1A2		
CYTOCHROME P450 2A		
CYTOCHROME P450 2A6		
CYTOCHROME P450 3A		
CYTOCHROME P450 3A4		
CYTOCHROME P450 3A5		
CYTOCHROME P450 2B		
CYTOCHROME P450 2B6		
METHOXYACETATE		
DIMETHYLGLYCINE DEHYDROGENASE		
SARCOSINE DEHYDROGENASE	C	$Ki = 1.1$ mM
SARCOSINE OXIDASE	C	$Ki = 1.8$ mM
p-METHOXYACETOPHENONE		
ALDEHYDE DEHYDROGENASE (NAD)		$Ki = 7\ \mu M$
E-β-METHOXYACRYLATE STILBENE		
CYTOCHROME bf		
METHOXYAMINE		
DEOXYRIBONUCLEASE (PYRIMIDINE DIMER)		
HYDROXYMETHYLBILANE SYNTHASE		
2-METHOXY-4[N[2(,r)AMINO-3-MERCAPTOPROPYL]AMINO]-3′-CARBOXYBIPHENYL		
FARNESYL PROTEINTRANSFERASE		$I50 = 40$ nM
GERANYLGERANYL PROTEINTRANSFERASE		$I50 = 44\ \mu M$

4-METHOXY-4-ANDROSTENE-3,17-DIONE
AROMATASE C $Ki = 1.1\ \mu M$

N^{G}-METHOXYARGININE
NITRIC-OXIDE SYNTHASE II $Ki = 21\ \mu M$

2-METHOXYBENZAMIDE
NAD ADP-RIBOSYLTRANSFERASE $I50 = 20\ \mu M$

3-METHOXYBENZAMIDE
NAD ADP-RIBOSYLTRANSFERASE $I50 = 1\ \mu M$
NAD ADP-RIBOSYLTRANSFERASE 98% 50 μM
NAD ADP-RIBOSYLTRANSFERASE $I50 = 17\ \mu M$

4-METHOXYBENZAMIDE
NAD ADP-RIBOSYLTRANSFERASE $I50 = 1.1$ mM

3-METHOXYBENZAMIDINE
NAD(P)-ARGININE ADP-RIBOSYLTRANSFERASE

(S)-1-[(S)-2-(4-METHOXYBENZAMIDO)-3-METHYLBUTYRYL)-N[(S)2-METHYL-1-(TRIFLUOROACETYL)PROPYL]PYRROLIDINE-2-CARBOXAMIDE
CARBOXYPEPTIDASE A $Ki = 170\ \mu M$
ELASTASE PANCREATIC $Ki = 200$ nM
CHYMOTRYPSIN $Ki = 200\ \mu M$
ELASTASE LEUKOCYTE $Ki = 6.7$ nM
CATHEPSIN G $Ki = 250\ \mu M$

3-METHOXYBENZOIC ACID
4-METHOXYBENZOATE MONOOXYGENASE (O-DEMETHYLATING)

2-(4-METHOXYBENZOYL)-1-NAPHTOIC ACID
ELASTASE CERCARIAL $Ki = 3\ \mu M$

N-(p-METHOXYBENZOYL)-N-PHENYLMETHYLOXY-β-ALALNINE
ALDEHYDE REDUCTASE $I50 = 3.5\ \mu M$

p-METHOXYBENZYLAMINODECAMETHYLENE GUANIDINE SULPHATE
NAD(P)-ARGININE ADP-RIBOSYLTRANSFERASE

[(4-METHOXYBENZYL)OXY]ACETIC ACID
PEPTIDYLGLYCINE MONOOXYGENASE C $Ki = 480\ \mu M$

2[3(4-METHOXYBENZYL)2,4,8-TRIOXO-6-PHENYL-1,2,3,4,7,8-HEXAHYDROPYRIDO[3,4-d]PYRIMIDIN-7-YL]N-[3,3,3-TRIFLUORO-1-ISOPROPYL-2-OXOPROPYL)ACETAMIDE
ELASTASE LEUKOCYTE $Ki = 5.7$ nM

8-METHOXYCAFFEINE
DNA TOPOISOMERASE (ATP-HYDROLYSING)

(3R,S)[[4-METHOXYCARBONYL)PHENYL]CARBONYL]-VAL-N-[3-1,1,1-TRIFLUORO-4-METHYL-2-OXOPENTYL)]-L-PROLINAMIDE
ELASTASE LEUKOCYTE $I50 = 1.8$ nM

3(E)-METHOXYCARBONYL-2,4,6-TRIENAL
PHOSPHOLIPASE A2 IR

4-METHOXYCHALCONE
GLUTATHIONE TRANSFERASE

4-METHOXYCOUMARIN
NAD(P)H DEHYDROGENASE (QUINONE)

6′-METHOXYDEHYDROLUCIFERIN
PHOTINUS-LUCIFERIN 4-MONOOXYGENASE (ATP-HYDROLYSING)

N3-4-METHOXY-L-2,3-DIAMINOPROPIONIC ACID
GLUCOSAMINE SYNTHASE C $Ki = 7.5\ \mu M$

3-METHOXY-2,3-DIHYDROLUTEOLIN
PROTEIN KINASE C $I50 = 18\ \mu M$

2-METHOXY-2,4-DIPHENYL-3-DIHYDROFURANONE
H$^+$/K$^+$-EXCHANGING ATPASE

METHOXY E 3810
H$^+$/K$^+$-EXCHANGING ATPASE — $I50 = 160$ nM

METHOXYETHANOL
CAFFEATE O-METHYLTRANSFERASE

5-[1-[2-(2-METHOXYETHOXY)ETHYL](2-METHOXYETHYL)AMINO]ETHYL]-THIENO[2,3-b]THIPHENE-2-SULPHONAMIDE HCl
CARBONATE DEHYDRATASE II — $I50 = 500$ pM

O-(2-METHOXYETHYL) O-(p-NITROPHENYL) n-UNDECYLPHOSPHONATE
TRIACYLGLYCEROL LIPASE

3-METHOXYETHYL-4′-PHENYLCHALCONE EPOXIDE
EPOXIDE HYDROLASE — $I50 = 500$ nM
EPOXIDE HYDROLASE — 19% 100 nM

7-METHOXYFLAVONE
PROTEIN TYROSINE KINASE ptabl50 — C — $Ki = 58$ μM

N4-METHOXY-5-FLUORO-dCMP
THYMIDYLATE SYNTHASE — $Ki = 22$ μM

METHOXYFLURANE
Ca^{2+}-TRANSPORTING ATPASE

N4-(METHOXYFUMAROYL)-L-2,4-DIAMINOBUTANOIC ACID
GLUTAMINE-FRUCTOSE-6-PHOSPHATE AMINOTRANSFERASE (ISOMERIZING) — SSI — $Ki = 270$ nM

N3-(4-METHOXYFUMAROYL)-L-2,3-DIAMINOPROPANOYL-L-LEUCINE
GLUTAMINE-FRUCTOSE-6-PHOSPHATE AMINOTRANSFERASE (ISOMERIZING) — $I50 = 20$ μM

N3-(4-METHOXYFUMAROYL)-L-2,3-DIAMINOPROPANOYL-L-TYROSINE
GLUTAMINE-FRUCTOSE-6-PHOSPHATE AMINOTRANSFERASE (ISOMERIZING) — $I50 = 70$ μM

N3-(4-METHOXYFUMAROYL)-L-2,3-DIAMINOPROPIONIC ACID METHYLESTER
GLUTAMINE-FRUCTOSE-6-PHOSPHATE AMINOTRANSFERASE (ISOMERIZING) — $I50 = 60$ μM

N3-(4-METHOXYFUMAROYL)-L-2,3-DIAMINOPROPIONIC ACID
GLUCOSAMINE-FRUCTOSE-6-PHOSPHATE AMINOTRANSFERASE (ISOMERIZING) — IR
GLUCOSAMINE SYNTHASE — C — $Ki = 100$ μM
GLUTAMINE-FRUCTOSE-6-PHOSPHATE AMINOTRANSFERASE (ISOMERIZING) — C — $Ki = 350$ nM
GLUTAMINE-FRUCTOSE-6-PHOSPHATE AMINOTRANSFERASE (ISOMERIZING) — $I50 = 4$ μM

METHOXYFURANE
ALKANAL MONOOXYGENASE (FMN -LINKED)

3-METHOXY-4-(β-D-GLUCOPYRANOSYLOXY)DIFLUOROMETHYLBENZENE
β-GLUCOSIDASE — SI — $Ki = 60$ μM

14-METHOXYHALENAQUINONE
PROTEIN TYROSINE KINASE pp60 v-arc — $I50 = 5$ μM

6-METHOXY HARMALAN
AMINE OXIDASE MAO-A — $I50 = 2$ μM

6-METHOXY HARMANE
AMINE OXIDASE MAO-A — $I50 = 500$ nM

2-METHOXY-4H-BENZO-1,3,2-DIOXAPHOSPHORIN 2-SULPHIDE
ACETYLCHOLINESTERASE

2-[α-(METHOXY-1H-IMIDZOLYL)-2′,4′-DICHLOROBENZYL]BENZIMIDAZOLE
AROMATASE — $Ki = 26$ μM

3-METHOXY-4-HYDROXYBENZOIC ACID
PROTOCATECHUATE 4,5-DIOXYGENASE

3-METHOXY-4-HYDROXYMANDELIC ACID
ALCOHOL DEHYDROGENASE (NADP)

3-METHOXY-4-HYDROXYPHENYLACETIC ACID
4-HYDROXYPHENYLPYRUVATE DIOXYGENASE

3-METHOXY-4-HYDROXYPHENYLPYRUVIC ACID
4-HYDROXYPHENYLPYRUVATE DIOXYGENASE

7-METHOXY-2-(IMIDAZOL-4-YLMETHYLENE)-1-TETRALONE
AROMATASE — *I*50 = 41 nM
THROMBOXANE A SYNTHASE — 15% 50 μM

6-METHOXY-2-(IMIDAZOL-4-YLMETHYL)-1-TETRALONE
AROMATASE — *I*50 = 580 nM
STEROID 17α-MONOOXYGENASE/17,20 LYASE — *I*50 = 1.2 μM
THROMBOXANE A SYNTHASE — 44% 50 μM

5-METHOXYINDOLE-2-CARBOXYLIC ACID
DIHYDROLIPOAMIDE DEHYDROGENASE — *K*i = 3 mM

4′-METHOXYKAEMPFEROL
PROTEIN KINASE C — *I*50 = 9 μM

8-METHOXYKAEMPFEROL
8-HYDROXYQUERCETINE 8-O-METHYLTRANSFERASE

O-METHOXYLAMINE
NITRATE REDUCTASE (NADH)

6′-METHOXYLUCIFERIN
PHOTINUS-LUCIFERIN 4-MONOOXYGENASE (ATP-HYDROLYSING)

6-METHOXYMELLEIN
6-HYDROXYMELLEIN-O-METHYLTRANSFERASE

N4-METHOXY-5-METHYL-dCMP
THYMIDYLATE SYNTHASE — *K*i = 380 μM

6-METHOXY-1-METHYL-3,4-DIHYDRONAPHTHALENE
MYELOPEROXIDASE — *I*50 = 3 μM

17β-METHOXY-17-METHYL-(5α)-1H′-ANDROSTANE[3,2c]PYRAZOLE
STEROID 5α-REDUCTASE — *I*50 = 130 nM

8-METHOXYMETHYL-1-ISOBUTYL-3-METHYLXANTHINE
3′,5′-CYCLIC-NUCLEOTIDE PHOSPHODIESTERASE II — *I*50 = 19 μM
3′,5′-CYCLIC-NUCLEOTIDE PHOSPHODIESTERASE IV — *I*50 = 72 μM

8-METHOXYMETHYL-1-METHYL-3-ISOBUTYLXANTHINE
3′,5′-CYCLIC-NUCLEOTIDE PHOSPHODIESTERASE (cycAMP) — *I*50 = 212 μM
3′,5′-CYCLIC-NUCLEOTIDE PHOSPHODIESTERASE (cycGMP) — *I*50 = 5 μM

7-(METHOXY)-α-METHYLNAPHTHALENE-2-ACETIC ACID
CYCLOOXYGENASE — *I*50 = 500 μM

2[[(4-METHOXY-3-METHYL-2-PYRIDYL)METHYL]SULPHINYL]CYCLOPENTAO[1,2-f]BENZIMIDAZOLE
UREASE — 65% 10 μM

12-METHOXY-Nb-METHYLVOACHALOTINE
RNA-DIRECTED DNA POLYMERASE

(E)-6-[2-(2-METHOXY-1-NAPHTHALENEYL)ETHENYL]3,4,5,6-TETRAHYDRO-4-HYDROXY-2H-PYRAN-2-ONE
HYDROXYMETHYLGLUTARYL-CoA REDUCTASE (NADPH) — C — *K*i = 800 μM

6-METHOXYNAPHTHYL ACETIC ACID
PROSTAGLANDIN-ENDOPEROXIDE SYNTHASE 1 — *I*50 = 70 μM
PROSTAGLANDIN-ENDOPEROXIDE SYNTHASE 2 — *I*50 = 20 μM

6-METHOXY-2-NAPHTHYLACETIC ACID
CYCLOOXYGENASE
PROSTAGLANDIN-ENDOPEROXIDE SYNTHASE 1 — *I*50 = 280 μM
PROSTAGLANDIN-ENDOPEROXIDE SYNTHASE 2 — *I*50 = 55 μM

(+)1-METHOXY-6-(NAPHTH-2-YL-METHOXY)-1-(THIAZOL-2-YL)INDANE
ARACHIDONATE 5-LIPOXYGENASE — *I*50 = 130 nM

2-METHOXY-5-NITROPONE
THIOGLUCOSIDASE

1-METHOXYOXALYL-3,5-DICAFFEOYLQUINIC ACID
INTEGRASE — *I*50 = 270 ng/ml

2-(7-METHOXY-9-OXATHIOXANTH-3-YL)-5-METHYL-1,3,4-OXADIAZOLE-10,10-DIOXIDE
AMINE OXIDASE MAO-A — *I*50 = 1 nM
AMINE OXIDASE MAO-B — *I*50 = 100 nM

2-METHOXYPHENOL
CATECHOL 2,3-DIOXYGENASE — *K*i = 230 μM

p-METHOXYPHENOL
RIBONUCLEOTIDE REDUCTASE R2 — *I*50 = 11 μM

9-(4-METHOXYPHENYL)GUANINE
GUANINE DEAMINASE — *I*50 = 380 nM

2-(4-METHOXYPHENYL)-4-HYDROXY-5-PHENYLTHIAZOLE
ARACHIDONATE 5-LIPOXYGENASE — *I*50 = 370 nM

5-[[4-METHOXY-3-(PHENYLMETHOXY)PHENYL]METHYL]-2,4-PYRIMIDINEDIAMINE
DIHYDROFOLATE REDUCTASE — *K*i = 100 pM

3-(4-METHOXY]PHENYL]-5-[(METHYLAMINO)METHYL]-2-OXAZOLIDINONE
AMINE OXIDASE MAO-B — SSI — *K*i = 62 mM

N-(2,4,6-METHOXYPHENYL)-N′-(1-METHYLHEPTADECANSULPHONYL)UREA
STEROL O-ACYLTRANSFERASE — *I*50 = 85 nM

6-(4-METHOXYPHENYL)PYRAZOLO[2,3-d]-1,2,4-TRIAZIN-3-YL ACETIC ACID
ALDEHYDE REDUCTASE — *I*50 = 2 μM

5-METHOXY-2-[N-(2-PROPYNYL)AMINOMETHYL]-1-METHYLINDOLE
AMINE OXIDASE MAO-A — IR — *I*50 = 30 nM
AMINE OXIDASE MAO-B — IR — *I*50 = 3 μM

5-METHOXYPSORALEN
CYTOCHROME P450 6D1 — *I*50 = 540 nM
PROTEIN KINASE A — *I*50 = 240 μM

8-METHOXYPSORALEN
CYTOCHROME P450 2A6 — NC — *K*i = 260 nM

6-METHOXYPURINE ARABINOSIDE
THYMIDINE KINASE — C — *K*i = 200 μM

5-METHOXY-2-(4-PYRIDYLMETHYL)-1-TETRALONE
STEROID 17α-MONOOXYGENASE/17,20 LYASE — *I*50 = 13 μM

(+)-ex0-6-METHOXY-1-(4-PYRIDYL)-1a-2,3,7b-TETRAHYDRO-1H-CYCLOPROPA[α]NAPHTHALENE
AROMATASE — *I*50 = 30 nM

6-METHOXYQUINALIDINE
AMINE OXIDASE MAO-A — NC — *K*i = 33 μM

6-METHOXYQUINOLINE
AMINE OXIDASE MAO-A — C — *K*i = 103 μM

METHOXYSUCCINYL-AAPVCMK
ELASTASE SPUTUM — *K*i = 29 μM

METHOXYSUCCINYL-ALA-ALA-PRO-ALA-CH_2Cl
PROTEINASE K

METHOXYSUCCINYL-ALA-ALA-PRO-(l)BORO-PHE-OH
CHYMASE — $Ki = 58$ nM
CHYMASE ATYPICAL — $Ki = 2.3$ μM

METHOXYSUCCINYL-ALA-ALA-PRO-(l)BORO-PINACOL
CHYMASE — $Ki = 79$ nM
CHYMASE ATYPICAL — $Ki = 6.4$ μM

METHOXYSUCCINYL-ALA-ALA-PRO-VAL-CHLOROMETHYLKETONE
ELASTASE LEUKOCYTE
α-1-PROTEINASE

7-METHOXYTACRINE
ACETYLCHOLINESTERASE — $I50 = 9.3$ μM

2[(4-METHOXY-6,7,8,9-TETRAHYDRO-5H-CYCLOHEPTA[b]PYRIDIN-9-YL)SULPHINYL]BENZIMIDAZOLE
H^+/K^+-EXCHANGING ATPASE — $I50 = 3.3$ μM

(R+s,9S*)(±)-2-(4-METHOXY-6,7,8,9-TETRAHYDRO-5H-CYCLOHEPTA[b]PYRIDIN-9-YLSULPHINYL)-1H-BENZIMIDAZOLE
H^+/K^+-EXCHANGING ATPASE — $I50 = 5.8$ μM

7[[3(4-METHOXYTETRAHYDRO-2H-PYRAN-4-YL)PHENYL]METHOXY]4-PHENYLNAPHTHO[2,3-c]FURAN-1(3H)ONE
ARACHIDONATE 5-LIPOXYGENASE — $I50 = 14$ nM

6-METHOXY-TETRAHYDRONORHARMANE
AMINE OXIDASE MAO-A — $I50 = 8$ μM

6-METHOXY-1,2,3,4-TETRAHYDRO-9H-PYRIDO[3,4-b]INDOLE
AMINE OXIDASE (FLAVIN-CONTAINING)

4-METHOXYTRANYLCYPROMINE
AMINE OXIDASE MAO

4′-METHOXY-3′,5,7-TRIHYDROXYFLAVONE
XANTHINE OXIDASE — Kei = 10 nM

8-METHOXY-5,7,2′-TRIHYDROXYFLAVONE
3′,5′-CYCLIC-NUCLEOTIDE PHOSPHODIESTERASE (cycAMP) — $I50 = 6$ μM

4-METHOXY-2,3,6-TRIMETHYL-PHENYLSULPHONYL-ASPARTYL-D-4-AMIDINOPHENYLALANINE PIPERIDINE
THROMBIN — $Ki = 2.5$ μM
TRYPSIN — $Ki = 312$ μM

3-METHOXYTYROSINE
TYROSINE TRANSAMINASE

3-METHOXY-L-TYROSINE
MONOPHENOL MONOOXYGENASE — $I50 = 420$ μM

METHOXYVINYLGLYCINE
1-AMINOCYCLOPROPANE-1-CARBOXYLATE SYNTHASE — C — 44% 10 μM

METHYLACETAMIDATE
TYPE II SITE SPECIFIC DEOXYRIBONUCLEASE

METHYLACETIMIDATE
ASPARTATE TRANSAMINASE
GLUTAMATE DEHYDROGENASE (NAD(P))
1-D-myo-INOSITOL-TRISPHOSPHATE 5-KINASE
1-D-myo-INOSITOL-TRISPHOSPHATE 6-KINASE
RESTRICTION ENDONUCLEASE EcoRI

METHYLACETONYLPHOSPHONATE
ACETOACETATE DECARBOXYLASE

METHYL-2-ACETONYLPHOSPHONATE
3-HYDROXYBUTYRATE DEHYDROGENASE

(R*,R*)-(±)-METHYL 3-ACETYL-4-[3-(CYCLOPENTYLOXY)-4-METHOXYPHENYL]-3-METHYL-1-PYRROLIDINE CARBOXYLATE
3′,5′-CYCLIC-NUCLEOTIDE PHOSPHODIESTERASE IV 2B2

D-METHYL ACETYLENIC PUTRESCINE
ORNITHINE DECARBOXYLASE

(R,R)-δ-METHYL-α-ACETYLENICPUTRESCINE
ORNITHINE DECARBOXYLASE IR

β-METHYLACETYLGLUCOPYRANOSYLAMINE
PHOSPHORYLASE b $Ki = 32\ \mu M$

3-O-METHYL-N-ACETYLGLUCOSAMINE
N-ACETYLGLUCOSAMINE KINASE C $Ki = 17\ \mu M$
N-ACETYLMANNOSAMINE KINASE NC $Ki = 80\ \mu M$

α-METHYL-N-ACETYLMURAMYL-ALANYL-D-GLUTAMATE
UDP-N-ACETYLMURAMOYL-D-GLUTAMATE-2,6-DIAMINOPIMELATE LIGASE 50% 10 mM

N-METHYLACRYLOHYDROXAMIC ACID
ARACHIDONATE 5-LIPOXYGENASE $I50 = 100\ \mu M$

N^6-METHYLADENINE(6-METHYLAMINOPURINE)
ADENOSINE NUCLEOSIDASE

2′-O-METHYLADENOSINE
ADENYLATE CYCLASE

5′-S-METHYLADENOSINE
PROTEIN-2 KINASE MICROTUBULE-ASSOCIATED $Ki = 50\ \mu M$

2′-C-METHYLADENOSINEDIPHOSPHATE
RIBONUCLEOSIDE-DIPHOSPHATE REDUCTASE IR

2′-C-METHYLADENOSINE-5′-DIPHOSPHATE
RIBONUCLEOSIDE-DIPHOSPHATE REDUCTASE (ADENOSYLCOBALAMIN DEPENDENT) C $Ki = 3.7\ \mu M$

2′-O-METHYLADENOSINE-3′-MONOPHOSPHATE
3′-NUCLEOTIDASE $Ki = 1.3$ mM

N6-METHYLADENOSINE
ADENOSYLHOMOCYSTEINASE C $Ki = 190\ \mu M$
AMP NUCLEOSIDASE

2′-O-METHYLADENOSINE-5′-TRIPHOSPHATE
THREONINE-tRNA LIGASE

3′-O-METHYLADENOSINE-5′-TRIPHOSPHATE
THREONINE-tRNA LIGASE

METHYLALLOSAMIDIN
CHITINASE $I50 = 8.8\ \mu g/ml$

METHYL 2-ALLYL-3-BENZENEPROPANOATE
α-CHYMOTRYPSIN

METHYLALLYL THIOSULPHINATE
FARNESYL PROTEINTRANSFERASE $I50 = 400\ \mu M$
GERANYLGERANYL PROTEINTRANSFERASE $I50 = 53\ \mu M$

METHYLAMINE
ω-AMIDASE
ANTHRANILATE SYNTHASE
GLYCINE-tRNA LIGASE C $Ki = 6.3$ mM
LYSYL ENDOPEPTIDASE $Ki = 4.6$ mM
METHANEDIOL OXIDASE
PROTEINASE I ACHROMOBACTER $Ki = 4.6$ mM

N-METHYLAMINO ACIDS
SARCOSINE OXIDASE

2-METHYL-3-AMINO-4,5-BIS(HYDROXYMETHYL)PYRIDINE
PYRIDOXAL KINASE

METHYL 3-AMINO-3-DEOXY-β-D-GALACTOPYRANOSYL(1- > 4)-2-ACETAMIDO-2-DEOXY-β-D-GLUCOPYRANOSE
N-ACETYLLACTOSAMINIDINE α-1,3-GALACTOSYLTRANSFERASE *K*i = 104 μM

METHYL-6-AMINO-DEOXY-α-D-GLUCOPYRANOSIDE
DEXTRANSUCRASE 100% 12 mM

METHYL-6-AMINO-DEOXY-β-D-GLUCOPYRANOSIDE
DEXTRANSUCRASE 80% 12 mM

METHYL-6-AMINO-DEOXY-α-D-MANNOPYRANOSIDE
DEXTRANSUCRASE 95% 12 mM

METHYL-4-AMINO-4-DEOXY-4-N(5′-THIO-α-D-GLUCOPYRANOSYL)-α-D-GLUCOPYRANOSIDE
GLUCAN 1,4-α-GLUCOSIDASE G2 *K*i = 4 μM

2-(METHYLAMINO)ETHANOL
ETHANOLAMIN AMMONIA-LYASE C *K*i = 39 mM
ETHANOLAMINE KINASE

N-[2-(METHYLAMINO)ETHYL]-5-ISOQUINOLINESULPHONAMIDE DIHYDROCHLORIDE
MYOSIN-LIGHT-CHAIN KINASE *K*i = 68 μM
PROTEIN KINASE A *K*i = 1.2 μM
PROTEIN KINASE C *K*i = 15 μM
PROTEIN KINASE G *K*i = 480 nM

4-METHYL-5-AMINO-1-FORMYLISOQUINOLINETHIOSEMICARBAZONE
RIBONUCLEOSIDE-DIPHOSPHATE REDUCTASE 95% 5 μM

1-METHYL-AMINOGLUTETHIMIDE
AROMATASE *I*50 = 32 μM
CHOLESTEROL MONOOXYGENASE (SIDE-CHAIN-CLEAVING) *I*50 = 40 μM

(3αR,4R,5R,6S,6αS)2(METHYLAMINO)-4-HYDROXY-6-(HYDROXYMETHYL)-3α,5,6,6α-TETRAHYDRO-4H-CYCLOPENTOXAZOL-5-YL 2-ACETAMIDO-4-O-(2-ACETAMIDO 2-DEOXY-β-
CHITINASE *I*50 = 500 ng/ml
CHITINASE *I*50 = 57 μg/ml

(±)-METHYL 4-(AMINOIMINOETHYL)-β-[3-(AMINOIMINOETHYL)PHENYL]BENZENE
COAGULATION FACTOR Xa *I*50 = 9 nM

2-METHYLAMINOMALONIC ACID
ASPARTATE 4-DECARBOXYLASE

METHYL 4-O-(5-AMINO-9-α-MALTOTOSYLOXY-7-THIANONYL)-α-D-GLUCOPYRANOSIDE
α-AMYLASE *K*i = 420 μM

5′-[[3-(METHYLAMINO)PROPYL]AMINO]-5′-DEOXYADENOSINE
ADENOSYLMETHIONINE DECARBOXYLASE 40% 1 mM
SPERMIDINE SYNTHASE 43% 0.1 mM
SPERMINE SYNTHASE 85% 0.1 mM

3-METHYLAMINOPROPYLSULPHONATE
γ-BUTYROBETAINE DIOXYGENASE

N-METHYL-AMINOPROPYLTHUIOUREA
NITRIC-OXIDE SYNTHASE I *I*50 = 15 μM
NITRIC-OXIDE SYNTHASE II *I*50 = 7 μM
NITRIC-OXIDE SYNTHASE III *I*50 = 3 μM

6-METHYLAMINOPURINE
ADENINE DEAMINASE C *K*i = 50 μM
ADENINE DEAMINASE

N-METHYL-9-AMINO-1,2,3,4-TETRAHYDROACRIDINE
ACETYLCHOLINESTERASE — Ki = 80 nM

3-METHYLAMINYLPHENYLARSENOXIDE
FUMARATE HYDRATASE

6-METHYLANDROSTA-1,4-DIENE-3,17-DIONE
AROMATASE — IR — I50 = 43 nM

1-METHYL-1,4-ANDROSTENE-3,17-DIONE
AROMATASE — SSI

3-METHYLANTHRANILIC ACID
INDOLE-3-GLYCEROL-PHOSPHATE SYNTHASE

4-METHYLANTHRANILIC ACID
INDOLE-3-GLYCEROL-PHOSPHATE SYNTHASE

5-METHYLANTHRANILIC ACID
INDOLE-3-GLYCEROL-PHOSPHATE SYNTHASE

6-METHYLANTHRANILIC ACID
6-METHYLSALYCILATE DECARBOXYLASE — C — Ki = 40 μM

5′-METHYLANTHRANILOYL ADENOSINE
ADENOSINE KINASE

2-METHYLANTHRAQUINONE
7-ETHOXYCOUMARIN O-DEETHYLASE — I50 = 52 μM

METHYL ARACHIDONYLFLUOROHOSPHATE
PHOSPHOLIPASE A2 (Ca^{2+} INDEPENDENT) — I50 = 500 nM

METHYLARACHIDONYLFLUOROPHOSPHONATE
N-(LONG-CHAIN-ACYL)ETHANOLAMINE DEACYLASE — I50 = 2.5 nM

N^G-METHYLARGININE
ARGINASE — C — Ki = 75 mM
NITRIC-OXIDE SYNTHASE I — Ki = 230 nM
NITRIC-OXIDE SYNTHASE I — Ki = 700 nM
NITRIC-OXIDE SYNTHASE II — Ki = 860 nM
NITRIC-OXIDE SYNTHASE II — I50 = 17 μM
NITRIC-OXIDE SYNTHASE II — Ki = 2.5 μM
NITRIC-OXIDE SYNTHASE III — Ki = 410 nM
NITRIC-OXIDE SYNTHASE III — I50 = 5 μM
NITRIC-OXIDE SYNTHASE III — I50 = 5 μM

N^G-METHYL-L-ARGININE
NITRIC-OXIDE SYNTHASE — SSI — Ki = 2.7 μM

3-METHYLARGININOSUCCINATE
ARGININOSUCCINATE LYASE — C — Ki = 300 μM

α-METHYLASPARTATE
ARGININOSUCCINATE SYNTHASE

2-METHYL-D,L-ASPARTIC ACID
ARGININOSUCCINATE SYNTHASE — C — Ki = 1.8 mM

N-METHYL-D,L-ASPARTIC ACID
ASPARTATE-AMMONIA LIGASE — 65% 10 mM

METHYLATED RECEPTOR PROTEIN
PROTEIN-GLUTAMATE O-METHYLTRANSFERASE

(24R,S)-24-METHYL-25-AZACYCLOARTANOL
Δ(24)-STEROL C-METHYLTRANSFERASE — NC — Ki = 20 nM

22-METHYL-4-AZA-21-NOR-5α-CHOl-1-ENE-3,20-DIONE		
STEROID 5α-REDUCTASE		*I*50 = 25 nM
STEROID 5α-REDUCTASE		*I*50 = 10 nM
STEROID 5α-REDUCTASE		*I*50 = 100 nM
4-METHYL-4-AZA-3-OXO-5α-PREGNANE-20(S)-CARBOXYLATE		
5α-REDUCTASE		
METHYL-2-AZOACETAMIDOHEXANOATE		
THYROID ASPARTIC PROTEINASE		
2-METHYLBENZALDEHYDE		
ARYL-ALCOHOL DEHYDROGENASE		
3-METHYLBENZALDEHYDE		
BENZALDEHYDE DEHYDROGENASE (NAD) II		75% 10 μM
2-METHYLBENZAMIDE		
NAD ADP-RIBOSYLTRANSFERASE		*I*50 = 1.5 mM
3-METHYLBENZAMIDE		
ALCOHOL DEHYDROGENASE		*K*i = 33 μM
NAD ADP-RIBOSYLTRANSFERASE		*I*50 = 19 μM
4-METHYLBENZAMIDE		
NAD ADP-RIBOSYLTRANSFERASE		*I*50 = 1.8 mM
1-METHYLBENZIMIDAZOLIDINE-2-THIONE		
LACTOPEROXIDASE		
1-METHYL-5-[3-(exo-BENZOBICYCLO[2.2.1]HEPT-2-YLOXY)-4-METHOXYPHENYL]-2-IMIDAZOLIDINONE		
3′,5′-CYCLIC-NUCLEOTIDE PHOSPHODIESTERASE (cycAMP) (Ca^{2+} INDEPENDENT)		*I*50 = 290 μM
2-METHYLBENZOIC ACID		
6-METHYLSALYCILATE DECARBOXYLASE	NC	
3-METHYLBENZOIC ACID		
D-AMINO-ACID OXIDASE		
4-METHYLBENZOIC ACID		
BENZOATE 4-MONOOXYGENASE		
METHYLBENZOPRIM		
DIHYDROFOLATE REDUCTASE		
2-METHYLBENZOTHIAZOLINONE HYDRAZIDE		
AMINE OXIDASE (COPPER-CONTAINING)	C	*K*i = 200 μM
3-METHYL-2-BENZOTHIAZOLINONE HYDRAZONE		
CYSTATHIONINE γ-LYASE		
3-METHYLBENZOTHIAZOLONE		
KYNURENINE-OXOGLUTARATE AMINOTRANSFERASE		
3-METHYL-2-BENZOTHIAZOLONE HYDRAZONE		
4-AMINOBUTYRATE TRANSAMINASE		
ORNITHINE-OXO-ACID TRANSAMINASE		
2-METHYLBENZOXAZINONE		
CHYMOTRYPSIN		*K*i = 1.5 μM
ELASTASE LEUKOCYTE		*K*i = 5 μM
THROMBIN		*K*i = 370 μM
2-METHYLBENZOXAZOLE-4-CAROXAMIDE		
NAD ADP-RIBOSYLTRANSFERASE		*I*50 = 9.5 μM
4-(3-METHYLBENZOYL)THIOPHENE-2-SULPHONYLCHLORIDE		
CARBONATE DEHYDRATASE II		*I*50 = 2 nM

N-α-METHYLBENZYLAMINOBENZOTRIAZOLE
CYTOCHROME P450 2B4

p-METHYLBENZYL HYDROPEROXIDE
ALCOHOL DEHYDROGENASE | | *K*i = 4 nM

METHYL-3-(N-BENZYLOXYCARBONYL-L-PHENYLALANYL)AMINO-2-OXOPROPIONIC ACID
CATHEPSIN B | | *K*i = 5 μM
PAPAIN | | *K*i = 1 μM

1-METHYL-5-[3-(exo-BICYCLO[2.2.1]HEPT-5-EN-2-YLOXY)-4-METHOXYPHENYL]-2-IMIDAZOLIDINONE
3′,5′-CYCLIC-NUCLEOTIDE PHOSPHODIESTERASE (cycAMP) (Ca^{2+} INDEPENDENT) | | *I*50 = 130 μM

1-METHYL-5-[3-(exo-BICYCLO[2.2.1]HEPT-2-YLOXY)-4-METHOXYPHENYL]ETHYLAMINE
3′,5′-CYCLIC-NUCLEOTIDE PHOSPHODIESTERASE (cycAMP) (Ca^{2+} INDEPENDENT) | | *I*50 = 220 μM

1-METHYL-5-[3-(endo-BICYCLO[2.2.1]HEPT-2-YLOXY)-4-METHOXYPHENYL]-2-IMIDAZOLIDINONE
3′,5′-CYCLIC-NUCLEOTIDE PHOSPHODIESTERASE (cycAMP) (Ca^{2+} INDEPENDENT) | | *I*50 = 460 μM

METHYL(BIS-3-CHLOROETHYL)AMINE
BETAINE-ALDEHYDE DEHYDROGENASE | C |

1-METHYL-2-(BROMOMETHYL)-4,7-DIMETHOXYBENZIMIDAZOLE
XANTHINE OXIDASE | NC | *K*i = 34 μM

N′-METHYL-N2(BUTA-2,3-DIENYL)BUTANAL, 4-DIAMINE
AMINE OXIDASE (POLYAMINE) | IR |

3-METHYL-1-BUTANOL
ACETYLESTERASE | NC | *K*i = 1 mM

3-METHYL-2-BUTANONE
3-METHYL-2-OXOBUTANOATE HYDROXYMETHYLTRANSFERASE | | 27% 5 mM

2-METHYLBUTENOIC ACID
2-ENOATE REDUCTASE | | > 25 mM

4-METHYL-3-BUTENYLGUANIDINE
NADH DEHYDROGENASE (UBIQUINONE) | | 20 nMOLE/g

METHYLBUTHIONINESULPHOXIME
GLUTAMATE-CYSTEINE LIGASE

1-METHYLBUTYLAMINE
SPERMIDINE SYNTHASE | | *I*50 = 450 μM

2-METHYLBUTYLAMINE
SPERMIDINE SYNTHASE | | *I*50 = 20 μM

3-METHYLBUTYLAMINE
SPERMIDINE SYNTHASE | | *I*50 = 7.8 μM

S-2-METHYL-n-BUTYLHOMOCYSTEINE SULPHOXIMINE
GLUTAMATE-CYSTEINE LIGASE

S-METHYL N-BUTYLTHIOCARBAMATE
ALDEHYDE DEHYDROGENASE 2 | | *I*50 = 8.7 μM

METHYLBUTYRIC ACID
SUBTILISIN

N-METHYL CALYSTEGINE B2
α-GALACTOSIDASE

S-(METHYLCARBAMOYL)CYSTEINE
GMP SYNTHASE (GLUTAMINE-HYDROLYSING)

S-(N-METHYLCARBAMOYL)CYSTEINE
GLUTATHIONE REDUCTASE (NADPH) | IR |

Inhibitor / Enzyme	Type	Value
S-(N-METHYLCARBAMOYL)GLUTATHIONE		
GLUTATHIONE REDUCTASE (NADPH)	IR	
1-[3-(N-METHYLCARBAMOYL)PHENYL]-3-BENZYLQUINAZOLINE-2,4-DIONE		
3′,5′-CYCLIC-NUCLEOTIDE PHOSPHODIESTERASE (Ca^{2+} INDEPENDENT)		*I*50 = 21 μM
3-N-METHYLCARBAMOYL)THIOXANTHEN-9-ONE 10,10-DIOXIDE		
AMINE OXIDASE MAO-A		*I*50 = 60 nM
METHYL-3-CARBOXYPYRIDINE-2-CARBOXYLATE		
NICOTINATE-NUCLEOTIDE PYROPHOSPHORYLASE (CARBOXYLATING)		
METHYLCATECHOL		
TYROSINE 3-MONOOXYGENASE		
3-METHYLCATECHOL		
CATECHOL 1,2-DIOXYGENASE		
PROTOCATECHUATE 4,5-DIOXYGENASE	C	60% 330 μM
4-METHYLCATECHOL		
CATECHOL 1,2-DIOXYGENASE		
MONOPHENOL MONOOXYGENASE		
PROTOCATECHUATE 3,4-DIOXYGENASE		
PROTOCATECHUATE 4,5-DIOXYGENASE	C	37% 330 μM
METHYL 4-S-β-CELLOBIOSYL-4-THIO-β-CELLOTETRAOSIDE		
CELLULASE		*K*i = 270 μM
METHYLCELLULOSE		
GLUCAN 1,4-β-GLUCOSIDASE	C	
METHYLCELLUSOLVE		
AMINOPEPTIDASE (HUMAN LIVER)		
1-METHYL-1-[(4-CHLOROPHENYL)SULPHONYL]-3-n-PROPYLUREA		
ALDEHYDE DEHYDROGENASE (NAD)		100% 1 mM
20-METHYLCHOLESTEROL		
STEROL O-ACYLTRANSFERASE		
2-METHYLCHROMENE		
NAD(P)-ARGININE ADP-RIBOSYLTRANSFERASE		*I*50 = 6.3 mM
3-METHYLCIANIDANOL		
CYTOCHROME P450		*I*50 = 240 μM
m-METHYLCINNAMIC ACID		
PHENYLALANINE AMMONIA-LYASE		77% 1 mM
PHENYLALANINE AMMONIA-LYASE		31% 1 mM
o-METHYLCINNAMIC ACID		
PHENYLALANINE AMMONIA-LYASE		73% 1 mM
PHENYLALANINE AMMONIA-LYASE		77% 1 mM
3-METHYLCINNAMOYLHYDRAZONO-2-PROPIONIC ACID		
FATTY ACID TRANSFER MITOCHONDRIA		
N-METHYL-E-CINNAMYLAMINE		
AMINE OXIDASE MAO-A	IR	
2-(3-METHYLCINNAMYLHYDRAZONO)PROPIONIC ACID		
CARNITINE ACYLCARNITINE TRANSLOCASE		
1-METHYL-1,2-CIS-CYCLOPROPANEDICARBOXYLIC ACID		
2-METHYLENEGLUTARATE MUTASE	NC	*K*i = 38 mM
METHYLCOBALAMIN		
ETHANOLAMIN AMMONIA-LYASE	IR	
INTEGRASE		*I*50 = 17 μM

4-METHYLCOUMARIN 7-O-SULPHAMATE
ESTRONE SULFATASE — *I*50 = 380 nM

2-METHYLCROTONYL-CoA
ACETOACETYL-CoA REDUCTASE
ACYL-CoA DEHYDROGENASE (NADP)

3-METHYLCROTONYL-CoA
ACYL-CoA DEHYDROGENASE (NADP)
ISOVALERYL-CoA DEHYDROGENASE — C — *K*i = 100 µM
OXOGLUTARATE DEHYDROGENASE (LIPOAMIDE) — C — *K*i = 70 µM

N-METHYL-C18-SPHINGOSINE
PROTEIN KINASE C — *I*50 = 4 mol%

(1S,2R)-3-METHYLCYCLOHEXA-3,5-DIENE-1,2-DIOL
TOLUENE DIOXYGENASE

3-METHYL-1,2-CYCLOHEXANEDIONE
CARBONYL REDUCTASE (NADPH) — 84% 1 mM

3-METHYL-2-CYCLOHEXEN-1-OL
ACETYLCHOLINESTERASE — C — *K*i = 640 µM

3-METHYL-2-CYCLOHEXEN-1-ONE
ACETYLCHOLINESTERASE — C — *K*i = 380 µM

3-METHYLCYCLOHEXYLAMINE
SPERMIDINE SYNTHASE — *I*50 = 300 µM

cis-4-METHYLCYCLOHEXYLAMINE
SPERMIDINE SYNTHASE — *I*50 = 430 µM

N-METHYLCYCLOHEXYLAMINE
SPERMIDINE SYNTHASE — *I*50 = 103 µM

trans METHYLCYCLOHEXYLAMINE
SPERMIDINE SYNTHASE

trans-4-METHYLCYCLOHEXYLAMINE
SPERMIDINE SYNTHASE — *I*50 = 1.7 µM

METHYL CYCLO[(2S)2[[(1R)1(N(L-N(3-METHYLBUTANOYL)VALYLASPARTYL)AMINO)3-METHYLBUTYL]HYDROXYPHOSPHINYLOXY]3-(3AMINOMETHYL)PHENYLPROPANOATE
PENICILLOPEPSIN — *K*i = 100 pM

4-METHYLCYCLOPENTYLAMINE
SPERMIDINE SYNTHASE — *I*50 = 15 µM

1.METHYL-1,2-trans-CYCLOPROPANEDICARBOXYLIC ACID
METHYLITACONATE Δ-ISOMERASE — C — *K*i = 3.3 mM

S-(S-METHYL)CYSTEAMINE
GLUTAMATE-CYSTEINE LIGASE

α-METHYL-DL-CYSTEINE
CYSTEINE TRANSAMINASE

S-METHYLCYSTEINE
ASPARAGINE-OXO-ACID TRANSAMINASE
CYSTEINE DIOXYGENASE
CYSTEINE-tRNA LIGASE
PHENYLALANINE 4-MONOOXYGENASE

S-METHYL-L-CYSTEINE
ALLIIN LYASE — *K*i = 770 µM
PHENYLALANINE 4-MONOOXYGENASE
TRYPTOPHANASE — C — *K*i = 8.8 mM

5′-METHYL-7-DEAZA-ADENOSINE
SPERMIDINE SYNTHASE 64% 100 μM

9-METHYL-5-DEAZAISOFOLIC ACID
DIHYDROFOLATE REDUCTASE *I*50 = 250 nM
PHOSPHORIBOSYLAMINOIMIDAZOLECARBOXAMIDE FORMYLTRANSFERASE *I*50 = 141 μM
THYMIDYLATE SYNTHASE *I*50 = 480 nM

9-METHYL-5-DEAZA-5,6,7,8-TETRAHYDROISOFOLIC ACID
PHOSPHORIBOSYLGLYCINAMIDE FORMYLTRANSFERASE *I*50 = 26 μM

6-METHYL-5-DEAZOTETRAHYDROPTERINE
PHENYLALANINE 4-MONOOXYGENASE

4-METHYL-N-DEMETHYLALLOSAMIDIN
CHITINASE *I*50 = 600 ng/ml

5-METHYLDEOXYCYTIDINE
DIHYDROOROTASE C *K*i = 100 μM

β-METHYLDEOXYFUCONOJIRIMYCIN
β-GALACTOSIDASE 27% 1 mM

N-METHYLDEOXYFUCONOJIRIMYCIN
β-N-ACETYLHEXOSAMINIDASE 59% 1 mM
α-L-FUCOSIDASE C *K*i = 50 nM

METHYL O-(6-DEOXY-β-D-GALACTOPYRANOSYL)-81- > 4)-2-ACETAMIDO-2-DEOXY-β-D-GLUCOPYRANOSIDE
β-GALACTOSIDE α-2,6-SIALYLTRANSFERASE *K*i = 760 μM

β-METHYLDEOXYMANNONOJIRIMYCIN
α-L-FUCOSIDASE C *K*i = 10 nM

METHYLDEOXYNOJIRIMYCIN
GLUCAN 1,3-β-GLUCOSIDASE C *K*i = 12 μM
PHOSPHATIDYLCHOLINE-STEROL O-ACYLTRANSFERASE

N-METHYL-1-DEOXYNOJIRIMYCIN
GLUCOSIDASE I
MANNOSYL-OLIGOSACCHARIDE GLUCOSIDASE

METHYL 2-DEOXY-2-{(1S,5R,6S,7R,8S)-6,7,8-TRIHYDROXY-1-(HYDROXYMETHYL)-2-OXA-4-AZABICYCLO[3.3.0]OCTAN-3-YLIDENEAMINO}α-D-GLYCOPYRANOSIDE
α-GLUCOSIDASE *I*50 = 12 nM

METHYL-3,6-DI-O-(2-ACETAMIDO-2-DEOXY-β-D-GLUCOPYRANOSYLOXYETHYL)α-D-MANNOPYRANOSIDE
β-.-ACETYLGLUCOSAMINYL-GLYCOPEPTIDE β-1,4-GALACTOSYLTRANSFERASE IR

N-METHYL-1,3-DIAMINOPROPANE
SPERMINE SYNTHASE C 34% 100 μM

METHYL-2-DIAZOACETAMIDOHEXANOIC ACID
PEPSIN A
PHYSAROPEPSIN
RHODOTORULAPEPSIN
THYROID ASPARTIC PROTEINASE

METHYL-2-DIAZOAZETAMIDOHEXANOIC ACID
CATHEPSIN D
RHIZOPUSPEPSIN

METHYL 3′,3′′-DICHLORO-4′,4′′-DIMETHOXY-5′,5′′-BIS(METHOXYCARBONYL)-6,6-DIPHENYL-5-HEXANPAOTEW
RNA-DIRECTED DNA POLYMERASE *I*50 = 300 nM

9-METHYL-5,8-DIDEAZAISOAMINOPTERIN
DIHYDROFOLATE REDUCTASE *I*50 = 4.6 nM
THYMIDYLATE SYNTHASE *I*50 = 33 nM

2-METHYL-3,4-DIDEHYDROGLUTAMIC ACID
GLUTAMATE DECARBOXYLASE SSI

METHYL-6,7-DIDEOXY-7-[(3R,4R,5R)-3,4-DIHYDROXY-5-HYDROXYMETHYLPIPERININYL]α-GLUCOHEPTOPYRANOSIDE
α-GLUCOSIDASE *K*i = 59 μM
β-GLUCOSIDASE *K*i = 59 μM
GLUCAN 1,4-α-GLUCOSIDASE *K*i = 63 nM
α-MANNOSIDASE *K*i = 1.2 mM
OLIGO-1,6-GLUCOSIDASE *K*i = 110 μM

METHYL-4,6-DIDEOXY-4-[(1S)-(1,4/5,6)-4,5,6-TRIHYDROXY-3-(HYDROXYMETHYL)-2-CYCLOHEXEN-1-YLAMINO]-α-D-MANNOPYRANOSIDE
α-GLUCOSIDASE 74% 100 μG/ML
β-GLUCOSIDASE 8% 100 μg/ml
α-MANNOSIDASE 41% 100 μg/ml

S-METHYL-N,N-DIETHYLDITHIOCARBAMATE SULPHONE
ALDEHYDE DEHYDROGENASE 100% 300 μM

S-METHYL-N,N-DIETHYLDITHIOCARBAMATE SULPHOXIDE
ALDEHYDE DEHYDROGENASE 100% 500 μM

S-METHYL-N,N-DIETHYLTHIOCARBAMATE SULPHONE
ALDEHYDE DEHYDROGENASE (NAD) (LOW Km) *I*50 = 420 nM

METHYLDIETHYLTHIOCARBAMOYLSULPHOXIDE
ALDEHYDE DEHYDROGENASE

S-METHYL-N,N-DIETHYLTHIOLCARBAMATE SULPHONE
ALDEHYDE DEHYDROGENASE (low Km) *I*50 = 3.8 μM

S-METHYL-N,N-DIETHYLTHIOLCARBAMATE SULPHOXIDE
ALDEHYDE DEHYDROGENASE (NAD)

1-L-METHYLDIFARNESYL DIPHOSPHATE
ARISTOLOCHENE SYNTHASE SSI

METHYL-2,5-DIHYDROCINNAMIC ACID
DNA TOPOISOMERASE (ATP-HYDROLYSING)

METHYL-2-[(3,4-DIHYDRO-3,4-DIOXO-1-NAPHTHALENYL)AMINO]BENZOIC ACID
ARACHIDONATE 5-LIPOXYGENASE *I*50 = 100 nM

cis-5-METHYLDIHYDROOROTIC ACID
DIHYDROOROTATE DEHYDROGENASE *K*i = 212 μM

6-METHYLDIHYDROPTERIN
DIHYDRONEOPTERIN ALDOLASE

(±) erythro 9-(2-METHYL-3,4-DIHYDROXYBUTYL)GUANINE
PURINE-NUCLEOSIDE PHOSPHORYLASE *K*i = 2.5 μM

9-(2-METHYL-3,4-DIHYDROXYBUTYL)GUANINE
PURINE-NUCLEOSIDE PHOSPHORYLASE *K*i = 1.3 μM
PURINE-NUCLEOSIDE PHOSPHORYLASE *K*i = 1.5 μM
PURINE-NUCLEOSIDE PHOSPHORYLASE *K*i = 1.8 μM

2-METHYL-2,5-DIHYDROXYCINNAMIC ACID
CAM KINASE II *I*50 = 150 ng/ml

METHYL-2,5-DIHYDROXYCINNAMIC ACID
PROTEIN KINASE C *I*50 = 52 μM
PROTEIN TYROSINE KINASE
PROTEIN TYROSINE KINASE EGFR *I*50 = 1.2 μM

METHYL 3,4-DIHYDROXYCINNAMIC ACID
Δ5 DESATURASE 90% 100 μM
LINOLEOYL-CoA DESATURASE 98% 100 μM

METHYL-(13S)-ent-11β,16-DIHYDROXY-8α,13-EPOXILABD-14-EN-18-OATE
ADENYLATE CYCLASE

METHYL-7,8-DIHYDROXYISOQUINOLINE-3-CARBOXYLATE
PROTEIN TYROSINE KINASE p56lck *I*50 = 200 nM

α-METHYL-3,4-DIHYDROXYPHENYLALANINE
PHENYLALANINE DECARBOXYLASE

2-METHYL-3,4-DIHYDROXYPHENYLALANINE
AROMATIC-L-AMINO-ACID DECARBOXYLASE NC

METHYL-3(R,S),5(R,S)-DIHYDROXY-7[1-PHENYL-2-ISOPROPYL-4(4-FLUOROPHENYL-5-METHYL-1H-PYRROL-3-YL]HEPTANOATE
HYDROXYMETHYLGLUTARYL-CoA REDUCTASE (NADPH) *I*50 = 5 nM

N-METHYL-4-(3′,4′-DIHYDROXYPHENYL)-1,2,3,6-TETRAHYDROPYRIDINE
AMINE OXIDASE C *K*i = 4 μM

N-METHYL-N,N-DIISOPROPYLETHANOLAMINE
CHOLINE KINASE

1-(METHYL-3-(N,N-DIMETHYLCARBAMOYLOXY)-2-PYRIDYLMETHYLENE)-4-(2,6-DIFLUOROPHENYL)DIAZINECARBOPXAMIDE IODIE
ACETYLCHOLINESTERASE IR *K*i = 39μM

4-(E,8E)-2-[N-METHYL-N-DIMETHYL-3(E),7-NONADIENYL)AMMONIO]5,9,13-TRIMETHYL-4,8,12-TETRADECATRIEN-1-YL-DIPHOSPHATE
FARNESYL-DIPHOSPHATE FARNESYLTRANSFERASE *I*50 = 25 μM

(±)-METHYL 1,2-DI-O-OCTYLGLYCER-5-YL PHOSPHATE
PHOSPHOLIPASE C *I*50= 12 μM

6-METHYL-2,2′-DIPHENYLENEIODONIUM
PROTOPORPHYRINOGEN OXIDASE *K*i = 6.4 nM

METHYL DIPHOSPHATE
ISOPENTENYL-DIPHOSPHATE Δ-ISOMERASE *K*i = 100 μM

5-METHYL-4,7-DITHIADECA-1,9-DIENE
LIPOXYGENASE *I*50 = 319 μM

N-METHYL-N-DODECYLHYDROXYLAMINE
ARACHIDONATE 5-LIPOXYGENASE *I*50 = 800 nM
CYCLOOXYGENASE *I*50 = 20 μM

O-METHYL-N-DODECYLHYDROXYLAMINE
ARACHIDONATE 5-LIPOXYGENASE *I*50 = 100 μM
CYCLOOXYGENASE *I*50 = 1 μM

N-METHYL-2n-DODECYL-3-METHYLQUINOLINIUM
NADH DEHYDROGENASE (UBIQUINONE) *I*50 = 730 nM

METHYLDOPA
11β-HYDROXYSTEROID DEHYDROGENASE

2-METHYLDOPA
CARBOSINE SYNTHASE

N-METHYLDOPAMINE
DIHYDROPTERIDINE REDUCTASE

1-METHYL EMODIN
PROTEIN TYROSINE KINASE (v-ABL) *I*50 = 7.4 μM
PROTEIN TYROSINE KINASE EGFR *I*50 = 16 μM
PROTEIN TYROSINE KINASE c-Src *I*50 = 230 μM

8-METHYL EMODIN
PROTEIN TYROSINE KINASE (v-ABL) *I*50 = 15 μM
PROTEIN TYROSINE KINASE EGFR *I*50 = 14 μM
PROTEIN TYROSINE KINASE c-Src *I*50 = 380 μM

3-METHYLENANDROST-4-EN-17-ONE		
AROMATASE	C	$Ki = 4.7\ \mu M$
METHYLENCYCLOPROPYLACETYL-CoA		
METHYLMALONYL-CoA MUTASE		$Kic = 470\ \mu M$
4-METHYLENEADENOSINE TRIPHOSPHATE		
GLUCOKINASE	C	$Ki = 2$ mM
METHYLENEALAPRILAT		
PEPTIDYL-DIPEPTIDASE A		
γ-METHYLENEAMINOPTERIN		
DIHYDROFOLATE REDUCTASE		$I50 = 88$ nM
3-METHYLENE-1,4-ANDROSTADIEN-17-ONE		
AROMATASE	IR	$Ki = 80$ nM
β-METHYLENE-D,L-APSARTATE		
SULFINOALANINE DECARBOXYLASE	IR	
β-METHYLENE-D,L-ASPARTATE		
SULFINOALANINE DECARBOXYLASE	SSI	
β-METHYLENE-ASPARTIC ACID		
GLUTAMATE DECARBOXYLASE		$I50 = 100\ \mu M$
β,γ-METHYLENE ATP		
NUCLEOSIDE-TRIPHOSPHATASE		
2,3-α-METHYLENEBENZYLPENICILLIN		
D-ALANYL-D-ALANINE CARBOXYPEPTIDASE		
2,2′-METHYLENEBIS[3,4,6-TRICHLOROPHENOL]		
CHITIN SYNTHASE		
METHYLENE BLUE		
ALDEHYDE DEHYDROGENASE (low Km)	C	$Ki = 8.4\ \mu M$
CAMPHOR 1,2-MONOOXYGENASE		
CYCLOHEXYLAMINE OXIDASE		
GLUTATHIONE REDUCTASE (NADPH)	C	$Ki = 6.4\ \mu M$
GLUTATHIONE REDUCTASE (NADPH)	C	$Ki = 16\ \mu M$
GUANYLATE CYCLASE		
GUANYLATE CYCLASE		$I50 = 60\ \mu M$
(R)-6-HYDROXYNICOTINE OXIDASE		
(S)-6-HYDROXYNICOTINE OXIDASE		
NADPH DIAPHORASE OF NOS		
NITRIC-OXIDE SYNTHASE		$I50 = 5.3\ \mu M$
NITRIC-OXIDE SYNTHASE		$Ki = 2.7\ \mu M$
PROGESTERONE MONOOXYGENASE		
STEAROYL-CoA DESATURASE		
TAURINE DEHYDROGENASE		
24(28)-METHYLENE CYCLOARTANOL		
Δ(24)-STEROL C-METHYLTRANSFERASE	C	$Ki = 14\ \mu M$
METHYLENECYCLOPROPANE ACETIC ACID		
ACYL-CoA DEHYDROGENASE		
3-METHYLENECYCLOPROPANE-trans-1,2-DICARBOXYLIC ACID		
XAA-PRO DIPEPTIDASE		$Ki = 72\ \mu M$
METHYLENECYCLOPROPIONIC ACID		
BUTYRYL-CoA DEHYDROGENASE		
METHYLENECYCLOPROPIONINATE		
ISOVALERYL-CoA DEHYDROGENASE		

(METHYLENECYCLOPROPYL)ACETYL-CoA		
ACYL-CoA DEHYDROGENASE	SSI	
ACYL-CoA DEHYDROGENASE (MEDIUM CHAIN)	SI	
ACYL-CoA DEHYDROGENASE (SHORT CHAIN)	SI	
BUTYRYL-CoA DEHYDROGENASE	SSI	
GLUTARYL-CoA DEHYDROGENASE	SSI	
ISOVALERYL-CoA DEHYDROGENASE		100% 10 μM
(R)-(–)-(METHYLENECYCLOPROPYL)ACETYL-CoA		
ACYL-CoA DEHYDROGENASE (SHORT CHAIN)		
(S)-(–)-(METHYLENECYCLOPROPYL)ACETYL-CoA		
ACYL-CoA DEHYDROGENASE		
(METHYLENECYCLOPROPYL)FORMYL-CoA		
ENOYL-CoA HYDRATASE		
METHYLENECYCLOPROPYLGLYCINE		
ACETYL-CoA C-ACETYLTRANSFERASE		
ACETYL-CoA C-ACYLTRANSFERASE		
(METHYLENECYCLOPROPYL)PYRUVIC ACID		
BUTYRYL-CoA DEHYDROGENASE		
5,10-METHYLENE-5-DEAZATETRAHYDROFOLIC ACID		
DIHYDROFOLATE REDUCTASE		*I*50 = 5 μM
4-METHYLENE DIAMINOPIMELIC ACID		
DIAMINOPIMELATE EPIMERASE	NC	*K*i = 950 μM
1-[3,4-(METHYLENEDIOXY)BENZOYL-3-[2-(1-BENZYL-4-PIERIDINYL)ETHYL]THIOUREA		
ACETYLCHOLINESTERASE		
4-[[3,4(METHYLENEDIOXY)BENZYL]AMINO]-6-CHLOROQUINAZOLINE		
3′,5′-CYCLIC-NUCLEOTIDE PHOSPHODIESTERASE V		*I*50 = 19 nM
4-((3,4-(METHYLENEDIOXY)BENZYL)AMINO)-6,7,8-TRIMETHOXYQUINAZOLINE		
3′,5′-CYCLIC-NUCLEOTIDE PHOSPHODIESTERASE		*I*50 = 360 nM
6,7-METHYLENEDIOXY-4′-METHOXYISOFLAVAN		
ARACHIDONATE 5-LIPOXYGENASE		*K*i = 600 nM
METHYLENEDIPHOSPHATE		
PHOSPHOENOLPYRUVATE CARBOXYKINASE (PYROPHOSPHATE)	C	*K*i = 150 μM
METHYLENEDIPHOSPHONATE		
GERANYL-DIPHOSPHATE CYCLASE	C	*K*i = 100 μM
H^+-TRANSPORTING PYROPHOSPHATASE		*K*i = 41 μM
H^+-TRANSPORTING PYROPHOSPHATASE		*I*50 = 100 μM
H^+-TRANSPORTING PYROPHOSPHATASE		*K*i = 190 μM
H^+-TRANSPORTING PYROPHOSPHATASE		*K*i = 68 μM
H^+-TRANSPORTING PYROPHOSPHATASE VACUOLAR		*K*i = 1.8 μM
α-PINENE CYCLASE	C	*K*i = 140 μM
XAA-PRO DIPEPTIDASE		*K*i = 18 μM
16-METHYLENE ESTRADIOL-17β		
ESTRADIOL 17β-DEHYDROGENASE	SSI	
16-METHYLENE ESTRONE		
ESTRADIOL 17β-DEHYDROGENASE	IR	
γ-METHYLENE-D-GLUTAMATE		
GLUTAMATE-CYSTEINE LIGASE		
10,11-METHYLENEHAXADEC-10-ENOIC ACID		
Δ9 DESATURASE		
Δ-11 DESATURASE		

11,12-METHYLENEHAXADEC-11-ENOIC ACID
Δ9 DESATURASE
Δ-11 DESATURASE

12,13-METHYLENEHAXADEC-12-ENOIC ACID
Δ-11 DESATURASE

cis-9,10-METHYLENEHEXADECANOIC ACID
GLYCEROL-3-PHOSPHATE O-ACYLTRANSFERASE

10,11-METHYLENEHEXA-10-DECENOIC ACID
Z11 DESATURASE

11,12-METHYLENEHEXA-11-DECENOIC ACID
Z11 DESATURASE

12,13-METHYLENEHEXA-12-DECENOIC ACID
Z11 DESATURASE

12,13-METHYLENEHEXADEC-12-ENOIC ACID
Δ11 DESATURASE

6-METHYLENE-17α-HYDROXYPROGESTERONE
STEROID 5α-REDUCTASE IR

4-(METHYLENE-1-IMIDAZOYLMETHYL)BENZONITRILE
AROMATASE *I*50 = 700 pM

γ-METHYLENEMETHOTREXATE
DIHYDROFOLATE REDUCTASE *I*50 = 44 nM

β-METHYLENE MYCOPHENOLIC ADENINEDINUCLEOTIDE
IMP DEHYDROGENASE 2 *K*i = 300 nM

γ-METHYLENENGLUTAMINE
GLUTAMINE-tRNA LIGASE

3-METHYLENEOCTANOYL-CoA
ACYL-CoA DEHYDROGENASE (MEDIUM CHAIN) SSI

3-METHYLENE-19-OXOANDROST-4-EN-17-ONE
AROMATASE C *K*i = 24 μM

METHYLENEPENEMS
β-LACTAMASE

2,3-α-METHYLENEPENICILLANIC ACID
β-LACTAMASE *I*50 = 12 μM

3,4-METHYLENEPENTANOIC ACID
ACYL-CoA DEHYDROGENASE

4,5-METHYLENEPENTANOIC ACID
ACYL-CoA DEHYDROGENASE

β,γ-METHYLENE PHOSPHONATE ANALOGUE OF TTP
THIAMIN-TRIPHOSPHATASE *K*i = 3 mM

4-METHYLENE-L-PROLINE
PROLINE DEHYDROGENASE SSI

6-METHYLENE STEROID
STEROID 5α-REDUCTASE SSI

METHYLENESUCCINIC ACID
SUCCINATE DEHYDROGENASE

5,10-METHYLENETETRAHYDROFOLIC ACID
METHIONYL-tRNA FORMYLTRANSFERASE

β-METHYLENE-THIAZOLE-4-CARBOXAMIDE ADENINE DINUCLEOTIDE		
IMP DEHYDROGENASE 1	U	*K*i = 95 nM
IMP DEHYDROGENASE 2	U	*K*i = 145 nM
3,4-METHYLENETRANYLCYPROMINE		
AMINE OXIDASE MAO		
METHYLENPENICILLANIC ACID		
β-LACTAMASE	SSI	*I*5O = 1.4 nM
N-METHYL-psi-EPHEDRINE		
ACETYLCHOLINESTERASE	NC	*K*i = 630 μM
METHYL 1′-EPICARVIOSIN		
α-GLUCOSIDASE		41% 100 μg/ml
β-GLUCOSIDASE		33% 1 g/ml
α-MANNOSIDASE		71% 100 μg/ml
3-METHYL-3,4-EPOXYBUTYLDIPHOSPHATE		
ISOPENTENYL-DIPHOSPHATE Δ-ISOMERASE	IR	*K*i = 370 nM
1,1′-[(METHYLETHANEDIYLIDINE)DINITRILO]BIS(3-AMINOGUANIDINE)		
ADENOSYLMETHIONINE DECARBOXYLASE	SSI	
ADENOSYLMETHIONINE DECARBOXYLASE	IR	
N-METHYLETHANOLAMINE		
GLYCEROPHOSPHOCHOLINE PHOSPHODIESTERASE		
N-METHYL -1-ETHYL-3(4-AZONIA-4,4-DIMETHYLPENTYL)-CARBODIIMIDE		
CELLULASE		
2-METHYL-2-ETHYLSUCCINIC ACID		
CARBOXYPEPTIDASE A		*K*i = 100 nM
CARBOXYPEPTIDASE B		*K*i = 4.5 μM
METHYLETHYLSULPHOXIDE		
METHIONINE-S-OXIDE REDUCTASE		
N-METHYL 4-(o-FLUOROBENZYL)PYRIDINIUM		
AMINE OXIDASE MAO-A		*K*i = 29 μM
AMINE OXIDASE MAO-B		*K*i = 2 mM
2-METHYL-6-FLUOROSPIROCHROMAN-4,5′-IMIDAZOLIDINE-2′,4′-DIONE		
ALDEHYDE REDUCTASE		*I*50 = 28 nM
GLUCURONATE REDUCTASE		*I*50 = 160 nM
N(6)METHYL FORMYCIN A		
PURINE-NUCLEOSIDE PHOSPHORYLASE	C	*K*i = 300 nM
4-METHYL-2-FORMYLPYRIDINE THIOSEMICARBAZONE		
RIBONUCLEOSIDE-DIPHOSPHATE REDUCTASE		*I*50 = 180 nM
9-(5-METHYL-2-FURYL)ADENINE		
ADENYLATE CYCLASE		*I*50 = 3.4 mM
METHYL-β-D-GALACTOPYRANOSIDE		
β-GALACTOSIDASE		
METHYL GANODERIC ACID A		
FARNESYL PROTEINTRANSFERASE		*I*50 = 38 μM
METHYL GANODERIC ACID C		
FARNESYL PROTEINTRANSFERASE		*I*50 = 317 μM
GERANYLGERANYL PROTEINTRANSFERASE		*I*50 = 3.4 mM
3-METHYL-GIBBERELLIN C		
GIBBERELLIN 3β-DIOXYGENASE		

N-METHYL-α-D-GLUCAMINE
DEXTRANSUCRASE

METHYL-α-D-GLUCOPYRANOSIDE
α-GLUCOSIDASE

3-O-METHYL-α-D-GLUCOPYRANOSIDE
URIDINE PHOSPHORYLASE

1-O-METHYL-α-D-GLUCOSE
OLIGO-1,6-GLUCOSIDASE 42% 62 mM

6R,6C-METHYLGLUCOSE
GLUCOKINASE
GLUCOSE-6-PHOSPHATASE
PHOSPHOGLUCOMUTASE C $Ki = 500\ \mu M$

6S,6C-METHYLGLUCOSE
GLUCOKINASE

N-METHYL-β-GLUCOSE-C-CARBOXAMIDE
PHOSPHORYLASE b $Ki = 160\ \mu M$

O-METHYLGLUCOSE
CYCLOMALTODEXTRIN GLUCANOTRANSFERASE

6-O-METHYLGLUCOSEPOLYSACCHARIDE
PALMITOYL-CoA HYDROLASE

α-METHYLGLUCOSIDE
GLUCAN 1,4-α-GLUCOSIDASE
THIOGLUCOSIDASE

METHYL-α-GLUCOSIDE
β-AMYLASE $Ki = 40$ mM

METHYL-α-D-GLUCOSIDE
ALDOSE 1-EPIMERASE
CYCLOMALTODEXTRIN GLUCANOTRANSFERASE
DEXTRANSUCRASE
GLUCAN 1,6-α-GLUCOSIDASE
4-α-GLUCANOTRANSFERASE C $Ki = 8$ mM

β-METHYLGLUCOSIDE
THIOGLUCOSIDASE

METHYL-β-GLUCOSIDE
α,α-TREHALASE
XYLAN 1,4-β-XYLOSIDASE

METHYL-β-D-GLUCOSIDE
4-α-GLUCANOTRANSFERASE C $Ki = 10$ mM

3-METHYLGLUTAMIC ACID
GLUTAMATE-CYSTEINE LIGASE

4-METHYLGLUTAMIC ACID
[GLUTAMATE-AMMONIA-LIGASE] ADENYLYLTRANSFERASE
GLUTAMATE-CYSTEINE LIGASE

DL-METHYLGLUTAMIC ACID
GLUTAMATE DECARBOXYLASE C

N-METHYLGLUTAMIC ACID
GLUTAMATE DEHYDROGENASE

METHYLGLUTAMINE
GLYCINE DEHYDROGENASE (DECARBOXYLATING)

METHYL-γ-GLUTAMINE		
ASPARAGINE SYNTHASE (GLUTAMINE-HYDROLYSING)		25% 1 mM
2-METHYLGLUTARIC ACID		
HYDROXYGLUTAMATE DECARBOXYLASE	C	*K*i = 8.4 μM
S-METHYLGLUTATHIONE		
γ-GLUTAMYLTRANSFERASE		
GLUTATHIONE TRANSFERASE		
LACTOYLGLUTATHIONE LYASE		
2-METHYL-3-GLUTATHIONYL-1,4-BENZOQUINONE		
15-HYDROXYPROSTAGLANDIN DEHYDROGENASE (NADP)	MIX	*I*50 = 170 nM
2-METHYL-3-GLUTATHIONYL-1,4-NAPHTHOQUINONE		
15-HYDROXYPROSTAGLANDIN DEHYDROGENASE (NADP)	MIX	*I*50 = 70 nM
METHYLGLYOXAL		
AMINOLEVULINATE TRANSAMINASE		
FORMALDEHYDE DEHYDROGENASE (GLUTATHIONE)		
GLUTATHIONE REDUCTASE (NADPH)	IR	
HYDROXYACYLGLUTATHIONE HYDROLASE		
MALATE SYNTHASE		
L-THREONINE 3-DEHYDROGENASE		
METHYLGLYOXAL BIS(3-AMINOPROPYLAMIDINOHYDRAZONE)		
ADENOSYLMETHIONINE DECARBOXYLASE	C	*K*i = 26 μM
ORNITHINE DECARBOXYLASE	C	*K*i = 38 μM
METHYLGLYOXAL BIS(CYCLOHEXYLAMIDINOHYDRAZONE)		
ADENOSYLMETHIONINE DECARBOXYLASE	C	*K*i = 19 μM
SPERMIDINE SYNTHASE	C	*K*i = 28 μM
METHYLGLYOXAL BIS(GUANYLHYDRAZONE)		
ADENOSYLMETHIONINE DECARBOXYLASE	C	*K*i = 5 μM
ADENOSYLMETHIONINE DECARBOXYLASE		*K*i = 32 μM
ADENOSYLMETHIONINE DECARBOXYLASE		*K*i = 13 μM
ADENOSYLMETHIONINE DECARBOXYLASE		*I*50 = 1.5 μM
ADENOSYLMETHIONINE DECARBOXYLASE		*K*i = 600 nM
ADENOSYLMETHIONINE DECARBOXYLASE		*K*i = 604 μM
ADENOSYLMETHIONINE DECARBOXYLASE I		*K*i = 130 nM
ADENOSYLMETHIONINE DECARBOXYLASE II		*K*i = 230 nM
AMINE OXIDASE (COPPER-CONTAINING)	NC	*K*i = 700 nM
ARGININE DECARBOXYLASE		
NADH DEHYDROGENASE (UBIQUINONE)		*I*50 > 15 mM
PUTRESCINE OXIDASE		
METHYLGUANIDINE HYDROGENCARBONATE		
PUTRESCINE OXIDASE		
7-METHYLGUANINE		
QUEUINE tRNA-RIBOSYLTRANSFERASE		
1-METHYLGUANOSINE		
1-METHYLADENOSINE NUCLEOSIDASE		
7-METHYLGUANOSINE		
PYROPHOSPHATASE(m7G(5′)pppN)		*I*50 = 25 μM
2-METHYL HARMINIUM		
NADH DEHYDROGENASE (UBIQUINONE)		*I*50 = 185 μM
SUCCINATE DEHYDROGENASE (UBIQUINONE)		*I*50 = 225 μM
2-METHYL HARMOLIUM		
NADH DEHYDROGENASE (UBIQUINONE)		*I*50 = 208 μM
SUCCINATE DEHYDROGENASE (UBIQUINONE)		*I*50 = 225 μM

1-METHYL-4-(4′-HEPTYLPHENYL)PYRIDINIUM+
NADH DEHYDROGENASE (UBIQUINONE) *I*50 = 510 nM

METHYL-β-D-HEXOSIDE
ARYL β HEXOSIDASE

2-(1-METHYLHEXYL)-2,4-DINITROPHENOL
NADH DEHYDROGENASE (UBIQUINONE) *I*50 = 8 μM
SUCCINATE DEHYDROGENASE *I*50 = 4.3 μM
UBIQUINOL-CYTOCHROME-C REDUCTASE *I*50 = 4 μM

N-METHYL-N-(2-HEXYL)PROPARGYLAMINE
AMINE OXIDASE MAO-B

4-[4-(2-METHYL-1H-IMIDAZOL-1-YL)BUTYL]PHENYLACETYL-SER-LYS-N-(2-CYCLOHEXYLETHYL)AMIDE
GLYCYLPEPTIDE N-TETRADECANOYLTRANSFERASE *I*50 = 56 nM
GLYCYLPEPTIDE N-TETRADECANOYLTRANSFERASE *I*50 = 14 μM

(R)-N.[2-[4-[4-(2-METHYL-1H-IMIDAZOL-1-YL)BUTYL]PHENYL]-1-OXOPROPYL-L-SER-N-(2-CYCLOHEXYLETHYL)L-LYSYLAMIDE
GLYCYLPEPTIDE N-TETRADECANOYLTRANSFERASE *I*50 = 20 nM
GLYCYLPEPTIDE N-TETRADECANOYLTRANSFERASE *I*50 = 8.2 μM

(3α,5α,1,20R)-23-(1-METHYL-1H-IMIDAZOL-2-YL)-4,4,14-TRIMETHYL-24-NORCHOL-8-EN-3-OL
Δ(24)-STEROL C-METHYLTRANSFERASE *K*i = 5 nM
Δ24 STEROL REDUCTASE *K*i = 48 nM

METHYLHISTAMINE
HISTAMINE N-METHYLTRANSFERASE

1-METHYLHISTAMINE
HISTAMINE N-METHYLTRANSFERASE

METHYLHISTIDINE
DIMETHYLHISTIDINE N-METHYLTRANSFERASE

α-METHYLHISTIDINE
HISTIDINE DECARBOXYLASE 93% 1 mM

4-METHYL-1-HOMOPIPERAZINYLDITHIOCARBONIC ACID
DOPAMINE β-MONOOXYGENASE

2-METHYL-4[3H]QUINAZOLINONE
NAD ADP-RIBOSYLTRANSFERASE C *I*50 = 1.1 μM

METHYLHYDRAZINE
MANGANESE PEROXIDASE

(2R,4S,5S)[[2-METHYL-4-HYDROXY-5[[[BENZYLOXYCARBONYL]ALA]AMINO]6PHENYLHEXANOYL]VAL]VAL-METHYL ESTER
PROTEINASE HIV-1 *K*i = 600 pM

METHYL 4-HYDROXYCINNAMIC ACID
LINOLEOYL-CoA DESATURASE 98% 100 μM

2-METHYL-9-HYDROXYELLIPTICINE
DNA TOPOISOMERASE (ATP-HYDROLYSING)

N-METHYLHYDROXYLAMINE
ANTHRANILATE SYNTHASE
HYDROXYLAMINE REDUCTASE (NADH)

N-METHYL-2-(5-HYDROXY-1-METHYLINDOLYL)METHYLAMINE
AMINE OXIDASE MAO-A C *K*i = 110 μM
AMINE OXIDASE MAO-B C *I*50 = 140 μM

α-METHYL-5-HYDROXYPHENYLALANINE
PHENYLALANINE DECARBOXYLASE

2-METHYL-p-HYDROXYPHENYLPYRUVIC ACID
4-HYDROXYPHENYLPYRUVATE DIOXYGENASE

N-METHYL-4-(4′-HYDROXYPHENYL)-1,2,3,6-TETRAHYDROPYRIDINE		
AMINE OXIDASE	C	Ki = 150 μM
2-METHYL-3-(1′-HYDROXYPHYTYL)-1,4-NAPHTHOQUINONE 2,3-EPOXIDE		
VITAMIN K EPOXIDE REDUCTASE	C	Ki = 43 μM
6-METHYL-7-HYDROXY-8-RIBITYLLUMAZINE		
RIBOFLAVIN SYNTHASE		
2-METHYL-5-HYDROXYTRYPTOPHAN		
AROMATIC-L-AMINO-ACID DECARBOXYLASE	NC	
4-METHYL-2-HYDROXYVALERATE		
D-2-HYDROXY-4-METHYLVALERATE DEHYDROGENASE	C	Ki = 40 mM
(3S)-29-METHYLIDENE-2,3-OXIDOSQUALENE		
LANOSTEROL SYNTHASE		Ki = 2.5 μM
29-METHYLIDENE-2,3-OXIDOSQUALENE		
LANOSTEROL SYNTHASE	SSI	Ki = 4.4 μM
OXIDOSQUALENE SYNTHASE		I50 = 500 nM
1-METHYLIMIDAZOLE		
CARBONATE DEHYDRATASE		
HISTIDINOL DEHYDROGENASE	C	Ki = 20 mM
THROMBOXANE A SYNTHASE		
2-METHYLIMIDAZOLE		
HISTIDINOL DEHYDROGENASE	C	Ki = 14 mM
4-METHYLIMIDAZOLE		
HISTIDINOL DEHYDROGENASE	C	Ki = 1.3 mM
N-METHYLIMIDAZOLE		
HISTIDINE DECARBOXYLASE	C	Ki = 7.2 mM
5-METHYLIMIDAZOLO (2,3-b]ISOQUINOLINIUM HYDROGENSULPHATE		
RNA-DIRECTED DNA POLYMERASE		I50 = 45 μM
(R,S,S)2-{4-[4-(2-METHYLIMIDAZOL-3-YL)BUTYL]PHENYL}PROPIONYL-SERYL-LYSINE 2-PHENYLETHYLAMIDE		
PROTEIN N-MYRISTOYLTRANSFERASE		I50 = 20 nM
1-METHYLIMIDZOLE		
FATTY-ACID PEROXIDASE		
2-METHYLIMIDZOLONEPROPIONIC ACID		
UROCANATE HYDRATASE		
METHYLINDOLE-3-ACETIC ACID		
AMINE OXIDASE MAO-A		I50 = 88 μM
1-METHYLINOSINE		
1-METHYLADENOSINE NUCLEOSIDASE		
METHYLIODIDE		
CARBON-MONOXIDE DEHYDROGENASE		
2-METHYL-3-IODOTYROSINE		
TYROSINE 3-MONOOXYGENASE		Ki = 500 nM
N-METHYLISATIN		
AMINE OXIDASE MAO-A		I50 = 95 μM
AMINE OXIDASE MAO-B		I50 = 14 μM
N-METHYLISATIN-β-THIOSEMICARBAZONE		
RNA-DIRECTED DNA POLYMERASE		
METHYLISOCITRIC ACID		
ISOCITRATE DEHYDROGENASE (NADP)		

D-threo-α-METHYLISOCITRIC ACID
ISOCITRATE DEHYDROGENASE (NADP)

METHYL ISOCYANATE
Mg^{2+}-TRANSPORTING ATPASE
Na^+/K^+-EXCHANGING ATPASE

N-METHYL-N-ISOPROPYLETHANOLAMINE
CHOLINE KINASE

1-METHYLISOQUINOLINE

AMINE OXIDASE MAO-A	C	*K*i = 86 µM

METHYL-N-(2-ISOQUINOLINE-5-SULPHONAMIDOETHYL)-3-AMINOPROPIONATE

PROTEIN KINASE A	C	*I*50 = 12 µM
PROTEIN KINASE C	C	*I*50 = 22 µM

METHYL N-(2-ISOQUINOLINESULPHONAMIDOETHYL)-3-AMINOPRPIONIC ACID

PROTEIN KINASE A		*I*50 = 2 µM
PROTEIN KINASE C		*I*50 = 20 µM

N-METHYLISOQUINOLINIUM

NADH DEHYDROGENASE (UBIQUINONE)	U	*I*50 = 650 µM

N-METHYL-ISOQUINOLINIUM$^+$

AMINE OXIDASE MAO-A	C	*K*i = 41 µM
AMINE OXIDASE MAO-B	NC	*K*i = 285 µM

S-METHYLISOTHIOCITRULLINE

NITRIC-OXIDE SYNTHASE I		Kd = 1.2 nM
NITRIC-OXIDE SYNTHASE II		*K*i = 40 nM
NITRIC-OXIDE SYNTHASE III		*K*i = 11 nM

S-METHYLISOTHIOUREA

NITRIC-OXIDE SYNTHASE I		*K*i = 160 nM
NITRIC-OXIDE SYNTHASE II		*K*i = 120 nM
NITRIC-OXIDE SYNTHASE II		*I*50 = 6 µM
NITRIC-OXIDE SYNTHASE III		*K*i = 200 nM
NITRIC-OXIDE SYNTHASE III		*I*50 = 2 µM

δ-(S-METHYLISOTHIOUREIDO)-L-NORVALINE

NITRIC-OXIDE SYNTHASE I		*I*50 = 5.7 nM
NITRIC-OXIDE SYNTHASE II		*I*50 = 248 nM
NITRIC-OXIDE SYNTHASE III		*I*50 = 238 nM

2-METHYLLACTIC ACID

KETOL-ACID REDUCTOISOMERASE		*I*50 = 1 mM

32-METHYLLANOST-7-ENE-3β,32 DIOL

LANOSTEROL C14 DEMETHYLASE		*I*50 = 330 nM

METHYLLIDOCAINE
1-ACYLGLYCEROPHOSPHOCHOLINE O-ACYLTRANSFERASE

5-cis-METHYLLUCIFERIN
PHOTINUS-LUCIFERIN 4-MONOOXYGENASE (ATP-HYDROLYSING)

4-METHYLLYSINE
LYSINE 2-MONOOXYGENASE

N^β-METHYLLYSINE
LYSINE 2-MONOOXYGENASE

N^δ-METHYLLYSINE
LYSINE 2-MONOOXYGENASE

N-METHYLMALEIMIDE

2-METHYLACYL-CoA DEHYDROGENASE		70% 2 mM
PHOSPHATIDATE PHOSPHATASE		

METHYLMALONATE
3-HYDROXYBUTYRATE DEHYDROGENASE

METHYLMALONIC ACID

3-HYDROXYBUTYRATE DEHYDROGENASE	C	$Ki = 275\ \mu M$
3-HYDROXYBUTYRATE DEHYDROGENASE	C	$Ki = 15\ \mu M$
ISOCITRATE LYASE		$I50 = 15$ mM
PANTOATE-β-ALANINE LIGASE		38% 1 mM
SUCCINATE DEHYDROGENASE	C	$Ki = 2.3$ mM
SUCCINATE DEHYDROGENASE	C	$Ki = 4.5$ mM

METHYLMALONYL-CoA

ACETOACETYL-CoA REDUCTASE		
ACYL-CoA DEHYDROGENASE (NADP)		
AMINO-ACID N-ACETYLTRANSFERASE		
CARNITINE O-PALMITOYLTRANSFERASE 1		
FATTY ACID SYNTHASE		$Ki = 8.4\ \mu M$
MALONYL-CoA DECARBOXYLASE		$Ki = 50\ \mu M$
MALONYL-CoA DECARBOXYLASE		49% 1 mM
METHYLMALONYL-CoA CARBOXYLTRANSFERASE		$Ki = 4\ \mu M$
[3-METHYL-2-OXOBUTANOATE DEHYDROGENASE (LIPOAMIDE)] KINASE		
OXOGLUTARATE DEHYDROGENASE (LIPOAMIDE)	C	$Ki = 75\ \mu M$
PYRUVATE DEHYDROGENASE (LIPOAMIDE)	MIX	$Ki = 350\ \mu M$

D-METHYLMALONYL-CoA
METHYLMALONYL-CoA MUTASE

D,L-METHYLMALONYL-CoA

METHYLMALONYL-CoA EPIMERASE		89% 2 mM

(S)-METHYLMALONYL-CoA
METHYLMALONYL-CoA CARBOXYLTRANSFERASE

METHYL-β-MALTOSIDE

4-α-GLUCANOTRANSFERASE	C	$Ki = 2.5$ mM

α-METHYL.-d-MANNOSAMINE
MANNOKINASE

α-METHYLMANNOSE
UDP-N-ACETYLGLUCOSAMINE-LYSOSOMAL-ENZYME N-ACETYLGLUCOSAMINEPHOSPHOTRANSFERASE

3-O-METHYLMANNOSEPOLYSACCHARIDE
PALMITOYL-CoA HYDROLASE

METHYL-α-MANNOSIDE

α-MANNOSIDASE		
MANNOSYL-GLYCOPROTEIN ENDO-β-N-ACETYLGLUCOSAMINIDASE		55% 50 mM
MANNOSYL-GLYCOPROTEIN ENDO-β-N-ACETYLGLUCOSAMINIDASE		54% 50 mM

METHYL-α-D-MANNOSIDE

α-L-RHAMNOSIDASE	C	

METHYL-D-MANNOSIDE

α-MANNOSIDASE		63% 100 μM

S-METHYLMERCAPTOETHYLGUANIDINE

NITRIC-OXIDE SYNTHASE I		$I50 = 8\ \mu M$
NITRIC-OXIDE SYNTHASE II		$I50 = 1.4\ \mu M$
NITRIC-OXIDE SYNTHASE III		$I50 = 43\ \mu M$

6-METHYLMERCAPTOGUANINE
QUEUINE tRNA-RIBOSYLTRANSFERASE

1-METHYL-2-MERCAPTOIMIDAZOLE

IODIDE PEROXIDASE		
LACTOPEROXIDASE	SSI	
MONOPHENOL MONOOXYGENASE	MIX	$Ki = 4.6\ \mu M$

6-METHYLMERCAPTOPURINE
PROTEIN KINASE (MYELIN BASIC) — *I*50 = 10 μM
PROTEIN KINASE N — *K*i = 10 nM
THIOPURINE S-METHYLTRANSFERASE

6-METHYLMERCAPTOPURINE RIBOSIDE
ADENOSINE KINASE — C

6-METHYLMERCAPTOPURINERIBOSIDE
ADENOSINE KINASE — 100% 100 μM

METHYLMERCAPTOPURINE RIBOSIDE 5′-MONOPHOSPHATE
AMIDOPHOSPHORIBOSYLTRANSFERASE — *I*50 = 90 μM

6-METHYLMERCAPTOPURINE RIBOSIDE-5′-PHOSPHATE
IMP DEHYDROGENASE 1 — C — *K*i = 520 μM
IMP DEHYDROGENASE 2 — C — *K*i = 1 mM

METHYLMERCURIC BROMIDE
NAD KINASE

METHYLMERCURIC CHLORIDE
ACYL-CoA DEHYDROGENASE
BUTYRYL-CoA DEHYDROGENASE
FERROCHELATASE — 95% 50 μM
GLUTATHIONE TRANSFERASE
LONG-CHAIN-ACYL-CoA DEHYDROGENASE

METHYLMERCURIC IODIDE
2-METHYLACYL-CoA DEHYDROGENASE — 50% 100 μM

N-METHYLMESOPORPHYRIN
FERROCHELATASE

METHYLMETHANETHIOSULPHONIC ACID
DIPHOSPHOMEVALONATE DECARBOXYLASE — IR
IMP DEHYDROGENASE — IR
NAD(P) TRANSHYDROGENASE (AB-SPECIFIC)
PATCHOULOL SYNTHASE — *I*50 = 250 μM
PYRUVATE DEHYDROGENASE (CYTOCHROME)
L-THREONINE 3-DEHYDROGENASE
XYLAN 1,4-β-XYLOSIDASE

3-METHYLMETHANETHIOSULPHONIC ACID
HYDROXYMETHYLGLUTARYL-CoA REDUCTASE

2-METHYLMETHIONINE
METHIONINE-tRNA LIGASE — C — *K*i = 8.6 mM

N-METHYL-6-METHOXYISOQUINOLINIUM
AMINE OXIDASE MAO-A — *I*50 = 810 nM

N-METHYL-N-(7-METHOXYNAPHTHALENE-2-PROPYL)HYDROXYLAMINE
ARACHIDONATE 5-LIPOXYGENASE — *I*50 = 800 nM

N-METHYL-4-(4′-METHOXYPHENYL)-1,2,3,6-TETRAHYDROPYRIDINE
AMINE OXIDASE — C — *K*i = 20 μM

METHYL-2-METHOXYPHOSPHINYLACETIC ACID
3-HYDROXYBUTYRATE DEHYDROGENASE

N-METHYL-6-METHOXY-1,2,3,4-TETRAHYDROISOQUINOLINE
NADH DEHYDROGENASE (UBIQUINONE) — *I*50 = 360 μM

METHYL-4-METHYL-3-OXO-4-AZA-5α-ANDROST-1-ENE-17β-CARBOXYLATE
STEROID 5α-REDUCTASE — *I*50 = 2.9 μM
STEROID 5α-REDUCTASE — *I*50 = 200 nM

1-METHYL-4-(4′-METHYLPHENYL)PYRIDINIUM+
NADH DEHYDROGENASE (UBIQUINONE) — *I*50 = 24 μM

(E)-2-METHYL-3-[1-METHYL-5-(3-PYRRIDINYLMETHYL)-1H-PYRROL-2-YL]-2-PROPIONIC ACID
THROMBOXANE A SYNTHASE — *I*50 = 50 nM

(2-METHYL-5-(METHYLSULPHONYL)BENZOYL)GUANIDINE
Na^+/H^+-EXCHANGER

(2S,4S)-4-METHYL-1-[N2-[(3-METHYL-1,2,3,4-TETRAHYDRO-8-QUINOLINYL)SULPHONYL]ARGINYL PIPERIDINE CARBOXYLATE
THROMBIN — *K*i = 280 μM

(2R,4R)-4-METHYL-1-[N2-[(3-METHYL-1,2,3,4-TETRAHYDRO-8-QUINOLINYL)SULPHONYL]ARGINYL]-2-PIPERIDINE CARBOXYLIC ACID
THROMBIN — *K*i = 19 nM
PLASMIN — *K*i = 800 μM
TRYPSIN — *K*i = 5 μM
COAGULATION FACTOR Xa — *K*i = 210 μM
KALLIKREIN — *K*i = 1.5 mM

(2R,4S)-4-METHYL-1-[N2-[(3-METHYL-1,2,3,4-TETRAHYDRO-8-QUINOLINYL)SULPHONYL]ARGINYL]-2-PIPERIDINE CARBOXYLIC ACID
THROMBIN — *K*i = 240 nM
TRYPSIN — *K*i = 30 μM

(2S,4R)-4-METHYL-1-[N2-[(3-METHYL-1,2,3,4-TETRAHYDRO-8-QUINOLINYL)SULPHONYL]ARGINYL]-2-PIPERIDINE CARBOXYLIC ACID
THROMBIN — *K*i = 1.9 μM

7-METHYL-8-METHYLTHIO-PYRROLO[1,2-a]PYRAZINE DISULPHIDE
GLUTATHIONE REDUCTASE (NADPH)

[(S)(1R*,2S*,3R*)]-2-[[3-METHYL-2-[(MORPHOLINOSULPHONYL)AMINO]BUTYRYL]AMINO]PENT-4-ENOIC ACID
RENIN — *I*50 = 1.4 nM

2-METHYL-1,4-NAPHTHOQUINONE
15-HYDROXYPROSTAGLANDIN DEHYDROGENASE (NAD) — IR

(Z)-5-METHYL-2-[2-(1-NAPHTHYL)ETHENYL]-4-PIPERIDINOPYRIDINE
ACYLPHOSPHATASE — NC — *I*50 = 90 μM

(Z)-5-METHYL-2-[2-(1-NAPHTHYL)ETHENYL]-4-PIPERIDONOPYRIDINE
H^+/K^+-EXCHANGING ATPASE
Na^+/K^+-EXCHANGING ATPASE — *I*50 = 50 μM

N-METHYL-4-(1-NAPHTHYLVINYL)PYRIDINE
CHOLINE O-ACETYLTRANSFERASE

5-METHYL-1,6-NAPHTHYRIDIN-2(1H)-ONE
3′,5′-CYCLIC-NUCLEOTIDE PHOSPHODIESTERASE (cycAMP) LOW Km — *I*50 = 580 nM
3′,5′-CYCLIC-NUCLEOTIDE PHOSPHODIESTERASE (cycGMP) — *I*50 = 485 μM

5-METHYLNICOTINAMIDE
NAD ADP-RIBOSYLTRANSFERASE — *I*50 = 350 μM

1-METHYLNICOTINAMIDE CHLORIDE
NAD ADP-RIBOSYLTRANSFERASE — *I*50 = 3.8 mM

N′-METHYLNICOTINAMIDE
ALDEHYDE DEHYDROGENASE (NAD(P))

N^1-METHYLNICOTINAMIDE
NICOTINAMIDE N-METHYLTRRANSFERASE

METHYLNICOTINIC ACID
CARBONATE DEHYDRATASE I — 58% 10 μM

6-METHYLNICOTINIC ACID
3-HYDROXY-2-METHYLPYRIDINE CARBOXYLATE DIOXYGENASE

N-METHYLNICOTINIC ACID
NICOTINATE N-METHYLTRRANSFERASE

METHYL p-NITROBENZENE SULPHONIC ACID
BROMELAIN

METHYL-p-NITROBENZENESULPHONIC ACID
BROMELAIN IR
CHYMOTRYPSIN
1-D-myo-INOSITOL-TRISPHOSPHATE 5-KINASE

[(2S,4S,5R)2[1-METHYL-1(2-NITRO-4-METHYLPHENOXY)ETHYL]4(3-PYRIDYL)-1,3-DIOXAN-5-YL]HEXAOIC ACID
PROSTACYCLIN I2 SYNTHASE *I*50 = 28 μM
THROMBOXANE A SYNTHASE *I*50 = 19 nM

N-METHYL-N′-NITRO-N-NITROSOGUANIDINE
NICOTINAMIDE-NUCLEOTIDE ADENYLYLTRANSFERASE IR 20% 1.3 mM

N$^{\omega}$-(5-METHYL-2-NITROPHENYL)ORNITHINE
NITRIC-OXIDE SYNTHASE I *I*50 = 2.5 μM
NITRIC-OXIDE SYNTHASE II *I*50 = 11 μM
NITRIC-OXIDE SYNTHASE III *I*50 = 5 μM

O-METHYL-O-(p-NITROPHENYL)-n-UNDECANEPHOSPHONATE
TRIACYLGLYCEROL LIPASE

N-METHYL-N-NITROSOANILINE
PROTEIN TYROSINE-PHOSPHATASE IR

2-METHYL-2-NITROSOPROPANE
ALDEHYDE DEHYDROGENASE (NAD) *I*50 = 150 -μM

6-METHYL-5-NITROURACIL
URIDINE PHOSPHORYLASE C *K*i = 10 μM

4,23-METHYL-4-21-NOR-5α-CHOLANE-3,20-DIONE
STEROID 5α-REDUCTASE *I*50 > 1 nM
STEROID 5α-REDUCTASE *I*50 = 400 nM
STEROID 5α-REDUCTASE *I*50 = 10 nM

N-METHYLNORSALSOLINOL
AMINE OXIDASE MAO-A C *K*i = 71 μM
AMINE OXIDASE MAO-A C *K*i = 44 μM
AMINE OXIDASE MAO-B NC *K*i = 289 μM
TYROSINE 3-MONOOXYGENASE NC *I*50 = 10 μM

1-METHYL-3-OCTADECANOYLINDOLE-2-CARBOXYLIC ACID
PHOSPHOLIPASE A2 *I*50 = 8 μM

N-METHYLOCTADECYLAMINE
CAM KINASE *I*50 = 40 μM
MYOSIN-LIGHT-CHAIN KINASE *I*50 = 10 μM

3-METHYL-trans-OCTENOYL-CoA
ACYL-CoA DEHYDROGENASE (MEDIUM CHAIN) SSI

METHYLOCTOPINE
D-OCTOPINE DEHYDROGENASE

35-METHYL OKADAIC ACID
PHOSPHOPROTEIN PHOSPHATASE 1 *K*i = 165 nM
PHOSPHOPROTEIN PHOSPHATASE 2A *K*i = 19 pM

2-METHYLOPSOPYRROLEDICARBOXYLATE
HYDROXYMETHYLBILANE SYNTHASE 50% 30 nM

α-METHYLORNITHINE
GLUTAMATE N-ACETYLTRANSFERASE
TYROSINE-ARGININE LIGASE *I*50 = 2.5 mM

2-METHYLORNITHINE
ORNITHINE DECARBOXYLASE

5-METHYLOROTIC ACID
DIHYDROOROTASE $Ki = 166\ \mu M$
DIHYDROOROTASE
DIHYDROOROTATE DEHYDROGENASE $Ki = 9.6\ \mu M$
DIHYDROOROTATE DEHYDROGENASE C $Ki = 18\ \mu M$
OROTATE REDUCTASE (NADH)

β-METHYLOXALOACETIC ACID
METHYLMALONYL-CoA CARBOXYLTRANSFERASE

METHYL OXLUCIFERIN
RENILLA-LUCIFERIN 2-MONOOXYGENASE

1-[4-METHYL-3-OXO-4-AZA-5α-ANDROSTANE-17β-CARBONYL]-1,3-DICYLCOHEXYLUREA
STEROID 5α-REDUCTASE $I50 = 41\ nM$
STEROID 5α-REDUCTASE $I50 = 83\ nM$

1-[4-METHYL-3-OXO-4-AZA-5α-ANDROSTANE-17β-CARBONYL]-1,3-DIISOPROPYLUREA
STEROID 5α-REDUCTASE $I50 = 55\ nM$
STEROID 5α-REDUCTASE $I50 = 53\ nM$

3-METHYL-2-OXOBUTYRIC ACID
3-METHYL-2-OXOBUTANOATE DEHYDROGENASE (LIPOAMIDE)
[3-METHYL-2-OXOBUTANOATE DEHYDROGENASE (LIPOAMIDE)] KINASE

2-METHYL-1-OXO-1H-PHENALENE-3-ACETIC ACID
ALDEHYDE REDUCTASE $I50 = 240\ nM$

4-METHYL-2-OXOPENTANOIC ACID
[3-METHYL-2-OXOBUTANOATE DEHYDROGENASE (LIPOAMIDE)] KINASE

D,L-3-METHYL-2-OXOPENTANOIC ACID
3-METHYL-2-OXOBUTANOATE DEHYDROGENASE (LIPOAMIDE)

METHYL 1-OXO-3-PHENYL-1H-INDENE-2-CARBOXYLIC ESTER
PROTEIN TYROSINE KINASE FGFR $I50 = 5.1\ \mu M$

D,L-3-METHYL 5-OXOPROLINE
5-OXOPROLINASE (ATP-HYDROLYSING) C

1-METHYL-7-(2′-OXOPROPYL)-3-n-PROPYLXANTHINE
3′,5′-CYCLIC-NUCLEOTIDE PHOSPHODIESTERASE I $I50 = 120\ \mu M$
3′,5′-CYCLIC-NUCLEOTIDE PHOSPHODIESTERASE IV $I50 = 3.1\ \mu M$

3-METHYL-2-OXOVALERIC ACID
PYRUVATE CARRIER MITOCHONDRIA

4-METHYL-2-OXOVALERIC ACID
PYRUVATE CARRIER MITOCHONDRIA

METHYLPARATHION
ACETYLCHOLINESTERASE $I50 = 9\ \mu M$
Ca^{2+}/Mg^{2+}-TRANSPORTING ATPASE $Ki = 166\ \mu M$
CHOLINESTERASE $I50 = 90\ \mu M$
L-LACTATE DEHYDROGENASE
MALATE DEHYDROGENASE
Na^{+}/K^{+}-EXCHANGING ATPASE $Ki = 110\ \mu M$

2-METHYL-PENEMS
D-ALANYL-D-ALANINE CARBOXYPEPTIDASE

8-METHYLPENTADECANOIC ACID
DNA METHYLTRANSFERASE ECO RI. $I50 = 26\ \mu M$

3-METHYL-1-PENTAFLUOROPHENYLMETHYL-6-SULPHOOXY-2-(4-SULPHOOXYPHENYL)4TRIFLUOROMETHYLINDOLE
STERYL-SULFATASE $I50 = 80\ \mu M$

3-METHYL-n-PENTANOIC ACID
BRANCHED-CHAIN-AMINO-ACID TRANSAMINASE

7-METHYL-4,5,6,9,10-PENTATHIATRIDECA-1,12-DIENE
LIPOXYGENASE | | $I50 = 20\ \mu M$

N-METHYL-N-(7-PENTOXYNAPHTHALENE-2-METHYL)HYDROXYLAMINE
ARACHIDONATE 5-LIPOXYGENASE | | $I50 = 1\ \mu M$

1-METHYLPENTYLAMINE
SPERMIDINE SYNTHASE | | $I50 = 150\ \mu M$

1-METHYL-4′-PENTYL-4-PHENYLPYRIDINIUM+
AMINE OXIDASE MAO-A | C | $Ki = 130$ nM
AMINE OXIDASE MAO-B | C | $Ki = 43\ \mu M$

1-METHYL-4-(4′-PENTYLPHENYL)PYRIDINIUM+
NADH DEHYDROGENASE (UBIQUINONE) | | $I50 = 1.3\ \mu M$

N-METHYL-N-(2-PENTYL)PROPARGYLAMINE
AMINE OXIDASE MAO-B

2′-O-METHYLPERLATOLIC ACID
AMINE OXIDASE MAO-B | | $I50 = 81\ \mu M$

α-METHYLPHENETHYLHYDRAZINE
PYRIDOXAL KINASE

METHYLPHENIDATE
CYTOCHROME P450 IID1 (DEBRISOQUINE) | | $Ki = 15\ \mu M$

METHYLPHENYLAZOFORMATE
CATHEPSIN L | | 74% 1 mM

p-METHYLPHENYLBORONIC ACID
β-LACTAMASE | IR | $Ki = 6\ \mu M$

N-(2-METHYLPHENYL)-3,3-DIFLUOROAZETIDIN-2-ONE
ELASTASE LEUKOCYTE | IR | 3.4.21.37
ELASTASE PANCREATIC | IR |

7-METHYL-4-PHENYL-3,4-DIHYDROCOUMARIN
ALDEHYDE DEHYDROGENASE (NAD) 1 | C | $Ki = 1.5\ \mu M$
ALDEHYDE DEHYDROGENASE (NAD) 2 | C | $Ki = 250$ nM

1-METHYL-4-PHENYL-2,3-DIHYDROPYRIDINIUM+
AMINE OXIDASE MAO-A | SS |
AMINE OXIDASE MAO-B | SS |

2(2-METHYLPHENYL)-4,6-DIMETHYL-1(2SULPHONAMIDO-1,3,4-THIADIAZOL-5-YL)PYRIDINIUMPERCHLORATE
CARBONATE DEHYDRATASE II | | $I50 = 100$ nM

4,5-METHYL-o-PHENYLENEDIAMINE
LACCASE

METHYL 4-(2-PHENYLHYDRAZONO-2H-1,4-BENZOTHIAZIN-3-YL)BUTANOATE
ARACHIDONATE 5-LIPOXYGENASE | | $I50 = 700$ nM
ARACHIDONATE 15-LIPOXYGENASE | | $I50 = 440$ nM

METHYLPHENYLMETHANETHIOSULPHONATE
EPOXIDE HYDROLASE | | 89% 1 mM
EPOXIDE HYDROLASE | | 89% 1 mM

2-METHYL-8-(PHENYLMETHOXY)IMIDAZO[1,2a]PYRIDINE 3-ACETONITRILE
H^+/K^+-EXCHANGING ATPASE | C |

N-(METHYL PHENYLPHOSPHONYL)GLUTAMIC ACID
γ-GLUTAMYL HYDROLASE

(+)-(R)-2-METHYL-1-PHENYL-1-PROPANOL
TYROSINE-ESTER SULFOTRANSFERASE IV | C |

1-METHYL-4-PHENYLPYRIDINIUM+

AMINE OXIDASE MAO-A	C	*K*i = 3 µM
AMINE OXIDASE MAO-B	C	*K*i = 230 µM
NADH DEHYDROGENASE (UBIQUINONE)		*I*50 = 110 µM
OXOGLUTARATE DEHYDROGENASE (LIPOAMIDE)		*I*50 = 19 mM
TYROSINE 3-MONOOXYGENASE		

N-METHYL-4-PHENYLPYRIDINIUM+

NADH DEHYDROGENASE (UBIQUINONE)

2′-(4-METHYLPHENYL)PYRIMIDINE-5-SPIROCYCLOPROPANE-2,4,6-(1H, 3H, 5H)-TRIONE

OROTATE REDUCTASE (NADH)

1-METHYL-4-PHENYL-1,2,3,6-TETRAHYDROPYRIDINE

AMINE OXIDASE MAO-A	SS	
AMINE OXIDASE MAO-B	SS	
NADH DEHYDROGENASE (UBIQUINONE)		
OXOGLUTARATE DEHYDROGENASE (LIPOAMIDE)		
TYROSINE 3-MONOOXYGENASE		*I*50 = 1 µM

N-METHYL-4-PHENYL-1,2,3,6-TETRAHYDROPYRIDINE

AMINE OXIDASE	C	*K*i = 680 µM

METHYLPHOSPHATE

ACYLPHOSPHATASE	C	*K*i = 3.5 mM
PHOSPHOENOLPYRUVATE DECARBOXYLASE	C	*K*i = 570 µM

α-METHYLPHOSPHINOTHRICIN

GLUTAMATE-AMMONIA LIGASE	C	*K*i = 125 µM

γ-METHYLPHOSPHINOTHRICIN

GLUTAMATE-AMMONIA LIGASE	C	*K*i = 407 µM

D,L-α-METHYLPHOSPHINOTHRICIN

GLUTAMATE-AMMONIA LIGASE	C	*K*i = 5.7 µM
GLUTAMATE-AMMONIA LIGASE	C	*K*i = 137 µM
GLUTAMATE-AMMONIA LIGASE 1		*K*i = 12 µM
GLUTAMATE-AMMONIA LIGASE 2		*K*i = 8 µM

D,L-γ-METHYLPHOSPHINOTHRICIN

GLUTAMATE-AMMONIA LIGASE	C	*K*i = 25 µM
GLUTAMATE-AMMONIA LIGASE	C	*K*i = 347 µM
GLUTAMATE-AMMONIA LIGASE 1		*K*i = 17 µM
GLUTAMATE-AMMONIA LIGASE 2		*K*i = 123 µM

3-METHYLPHOSPHOENOLPYRUVIC ACID

2-DEHYDRO-3-DEOXYPHOSPHOHEPTONATE ALDOLASE

(Z)-3-METHYLPHOSPHOENOLPYRUVIC ACID

XAA-PRO DIPEPTIDASE		*K*i = 6.2 nM

METHYLPHOSPHONIC ACID

(2-AMINOETHYL)PHOSPHONATE-PYRUVATE TRANSAMINASE		
XAA-PRO DIPEPTIDASE		*K*i = 160 µM

5-METHYLPHOSPHONO-D-ARABINO HYDROXIMOLACTONE

GLUTAMINE-FRUCTOSE-6-PHOSPHATE AMINOTRANSFERASE (ISOMERIZING)		*K*i = 400 µM
PHOSPHOGLUCOSE ISOMERASE		*K*i = 32 µM

METHYLPHOSPHONOFLUORIDIC ACID 1,2,2-TRIMETHYLPROPYLESTER

CHOLINESTERASE	IR	

2-METHYL-1,4-PHTHALAZINEDIONE

NAD ADP-RIBOSYLTRANSFERASE		*I*50 = 45 µM

7[(4-METHYLPIPERAZINO)METHYL]-10,11-(METHYLENEDIOXY)(20S)CAMTOTHECIN TRIFLUOROACETATE

DNA TOPOISOMERASE		*I*50 = 300 nM

N,N′-(3-(2-(N-(3-(N-METHYLPIPERAZINYL)AMINOPROPYLOXY)AMINO-4-BROMO)PHENYLTHIO)PHENYL)-1,5-PENTYLENEDIAMIDE		
TRYPANOTHIONE REDUCTASE		*I*50 = 550 μM
METHYLPREDNISOLONE		
ACYL-CoA DEHYDROGENASE (MEDIUM CHAIN)		55% 20 μM
CALPAIN		*I*50 = 3.2 mM
6-METHYLPREDNISOLONE		
DIHYDRODIOL DEHYDROGENASE 2		29% 100 μM
DIHYDRODIOL DEHYDROGENASE 4		97% 100 μM
6α-METHYLPREDNISOLONE		
GLUTATHIONE DEHYDROGENASE (ASCORBATE)		90% 5 μM
3α-HYDROXYSTEROID DEHYDROGENASE (A SPECIFIC)		
3α-HYDROXYSTEROID DEHYDROGENASE (B-SPECIFIC)		65% 10 μM
3α-HYDROXYSTEROID DEHYDROGENASE (B-SPECIFIC)	C	*K*i = 7.5 μM
METHYLPREDNISOLONE 21-SUCCINATE		
ACYL-CoA DEHYDROGENASE (MEDIUM CHAIN)		57% 20 μM
N-10-METHYLPROPARGYL-5,8-DIDEAZAFOLIC ACID		
DIHYDROFOLATE REDUCTASE		*I*50 = 250 μM
THYMIDYLATE SYNTHASE		*I*50 = 4.4 μM
N-METHYL-N-(8-PROPOXYNAPHTHALENE-1-ETHYLENE)HYDROXYLAMINE		
ARACHIDONATE 5-LIPOXYGENASE		*I*50 = 5 μM
N-METHYL-N-(8-PROPOXYNAPHTHALENE-2-ETHYLENE)HYDROXYLAMINE		
ARACHIDONATE 5-LIPOXYGENASE		*I*50 = 600 nM
N-METHYL-N-(7-PROPOXYNAPHTHALENE-2-ETHYL)HYDROXYLAMINE		
ARACHIDONATE 5-LIPOXYGENASE		*I*50 = 400 nM
N-METHYL-N-(7-PROPOXYNAPHTHALENE-2-METHYL)HYDROXYLAMINE		
ARACHIDONATE 5-LIPOXYGENASE		*I*50 = 900 nM
2-METHYLPROPYLAMINE		
SPERMIDINE SYNTHASE		*I*50 = 45 μM
17β[N-(METHYL-2-PROPYL)CARBAMOYL]ANDROST-3,5-DIENE-3-CARBOXYLIC ACID		
STEROID 5α-REDUCTASE	U	*K*i = 5 nM
1-METHYLPROPYL 2-IMIDAZOYL DISULPHIDE		
THIOREDOXIN REDUCTASE (NADPH)	C	*K*i = 13 μM
N-(2-METHYL-2-PROPYL)-3-OXO-4-AZA-5α-ANDROST-1-ENE-17β-CARBOXAMIDE		
STEROID 5α-REDUCTASE	C	
1-METHYL-4′-PROPYL-4-PHENYLPYRIDINIUM+		
AMINE OXIDASE MAO-A	C	*K*i = 200 nM
AMINE OXIDASE MAO-B	C	*K*i = 100 μM
1-METHYL-4-(4′-PROPYLPHENYL)PYRIDINIUM+		
NADH DEHYDROGENASE (UBIQUINONE)		*I*50 = 3.8 μM
3-METHYL-1-PROPYL-5-SULPHOOXY-2-(4-SULPHOOXYPHENYL)INDOLE		
STERYL-SULFATASE		*I*50 = 7 mM
3-METHYL-1-PROPYL-6-SULPHOOXY-2-(3-SULPHOOXYPHENYL)INDOLE		
STERYL-SULFATASE		*I*50 = 850 μM
N-METHYL-N-(2-PROPYNYL)-2-(5-BENZYLOXY-1-METHYLINDOLYL)METHYLAMINE		
AMINE OXIDASE MAO-A	IR	*K*i = 70 nM
AMINE OXIDASE MAO-B	IR	*K*i = 70 nM
N-METHYL-N-(2-PROPYNYL)-2-(5-HYDROXY-1-METHYLINDOLYL)METHYLAMINE		
AMINE OXIDASE MAO-A	IR	*K*i = 90 nM
AMINE OXIDASE MAO-B	IR	*K*i = 100 μM

α-METHYLPROTHIONINE SULPHOXIME
GLUTAMATE-CYSTEINE LIGASE

METHYLPROTODISCOIN
XANTHINE OXIDASE 43% 100 μM

N-METHYLPROTOPORPHYRIN IX
NITRIC-OXIDE SYNTHASE I *I*50 = 6 μM
NITRIC-OXIDE SYNTHASE II *I*50 = 5 μM
NITRIC-OXIDE SYNTHASE III *I*50 = 8 μM

N-METHYLPROTOPORPHYRIN
FERROCHELATASE C *K*i = 7 μM

O-METHYLPSYCHOTRINE SULPHATE HEPTAHYDRATE
RNA-DIRECTED DNA POLYMERASE *I*50 = 31 μM
RNA-DIRECTED DNA POLYMERASE *I*50 = 9 μM

O-METHYLPSYCHOTRINE SULPHATE
RNA-DIRECTED DNA POLYMERASE NC *K*i = 10 μM

N[1]0-METHYLPTEROYLGLUTAMIC ACID
DIHYDROFOLATE REDUCTASE

10-METHYLPTEROYLORNITHINE
FOLYLPOLYGLUTAMATE SYNTHASE *K*i = 94 μM

2-METHYLPUTRESCINE
ORNITHINE DECARBOXYLASE C *K*i = 1 mM

4-METHYLPYRAZOLE
ALCOHOL DEHYDROGENASE *K*i = 13 μM
CINNAMYL-ALCOHOL DEHYDROGENASE *I*50 = 4 mM
CYTOCHROME P450 2A6 C *K*i = 78 μM
FATTY-ACYL-CoA SYNTHASE C
RETINOL DEHYDROGENASE

5-METHYLPYRAZOLE-3-CARBOXYLATE
D-AMINO-ACID OXIDASE

ω-METHYLPYRIDOXAL
ALANINE RACEMASE C *K*i = 530 μM

N-METHYLPYRIDOXAL
PYRIDOXAMINE-PYRUVATE TRANSAMINASE

ω-METHYLPYRIDOXAMINE
ALANINE RACEMASE C *K*i = 2.7 mM

3-METHYL-1-(4-PYRIDYLMETHYL)-6-SULPHOOXY-2-(4-SULPHOOXYPHENYL)INDOLE
STERYL-SULFATASE *I*50 = 450 μM

1-METHYLPYRROLE ACETIC ACID
3α-HYDROXYSTEROID DEHYDROGENASE (B-SPECIFIC)

3-METHYL-QUERCETIN
XANTHINE OXIDASE *I*50 = 32 μM

2-METHYL-3(3H)QUINAZOLINE
NAD(P)-ARGININE ADP-RIBOSYLTRANSFERASE *I*50 = 1 mM

2-METHYL-4(3H)-QUINAZOLINONE
NAD ADP-RIBOSYLTRANSFERASE *I*50 = 5.6 μM

4-METHYLQUINOLINE
AMINE OXIDASE MAO-A C *K*i = 31 μM
AMINE OXIDASE MAO-A C *K*i = 27 μM
AMINE OXIDASE MAO-B C *K*i = 205 μM

6-METHYLQUINOLINE

AMINE OXIDASE MAO-A	C	Ki = 11 μM
AMINE OXIDASE MAO-A	C	Ki = 23 μM
AMINE OXIDASE MAO-B	C	Ki = 112 μM
AMINE OXIDASE MAO-B	NC	Ki = 930 μM

7-METHYLQUINOLINE

AMINE OXIDASE MAO-A	NC	Ki = 50 μM
AMINE OXIDASE MAO-A	C	Ki = 27 μM
AMINE OXIDASE MAO-B	NC	Ki = 238 μM
AMINE OXIDASE MAO-B	NC	Ki = 72 μM

8-METHYLQUINOLINE

AMINE OXIDASE MAO-A	NC	Ki = 25 μM
AMINE OXIDASE MAO-A	NC	Ki = 58 μM
AMINE OXIDASE MAO-B	NC	Ki = 538 μM

2-METHYLRESORCINOL

ORCINOL 2-MONOOXYGENASE

3-METHYLSALICYLIC ACID

6-METHYLSALYCILATE DECARBOXYLASE	NC	

4-METHYLSALICYLIC ACID

6-METHYLSALYCILATE DECARBOXYLASE	NC	

5-O-METHYLSALICYLIC ACID

GENTISATE 1,2-DIOXYGENASE		Ki = 1.5 mM

N-METHYL-SALSOLINIUM$^+$

AMINE OXIDASE MAO-A		Ki = 9.2 μM
AMINE OXIDASE MAO-B	NC	Ki = 78 μM

N-METHYL-R-SALSOLINOL

AMINE OXIDASE MAO-A	C	Ki = 36 μM
AMINE OXIDASE MAO-A	C	Ki = 86 μM
AMINE OXIDASE MAO-B	NC	Ki = 433 μM

N-METHYL-S-SALSOLINOL

AMINE OXIDASE MAO-A	C	Ki = 81 μM

1-METHYL-2-SELENOIMIDAZOLE

THYROXINE DEIODINASE		28% 300 μM

6-METHYL-2-SELENOURACIL

THYROXINE DEIODINASE		I50 = 500 nM

2-METHYLSERINE

GLYCINE HYDROXYMETHYLTRANSFERASE	C	Ki= 7 mM

D,L-α-METHYLSERINE

GLYCINE HYDROXYMETHYLTRANSFERASE

D,L-O-METHYLSERINE

GLYCINE HYDROXYMETHYLTRANSFERASE

O-METHYL-D,L-SERINE

D-SERINE DEHYDRATASE	C	Ki = 1.3 mM

METHYLSPINAZARIN

CATECHOL O-METHYLTRANSFERASE

7-O-METHYLSPINOCHROME B

CATECHOL O-METHYLTRANSFERASE

METHYLSPIRO-(4-AZA-5α-ADROSTAN-17(R),2′-THIIRAN)-3-ONE

STEROID 5α-REDUCTASE		I50 = 5 nM
STEROID 5α-REDUCTASE		I50 = 1 μM

3′-METHYLSPIRO-[IMIDAZOLIDINE-4,4′-(1′H)QUINOZOLNE]2,2′,5(3′H)TRIONE
ALDEHYDE REDUCTASE

METHYLSUCCINIC ACID
SUCCINATE DEHYDROGENASE (UBIQUINONE)
XAA-PRO DIPEPTIDASE — *K*i = 220 μM

D-METHYLSUCCINIC ACID
SUCCINATE DEHYDROGENASE

METHYL-O-SUCCINYL-ALA-ALA-PRO-PHE-COOCH3
CATHEPSIN G — *K*i = 7 μM

4-METHYLSULPHANYL-2-OXOBUTANOATE
GLUTAMINE-PYRUVATE TRANSAMINASE

5′-[4-(METHYLSULPHONYL)BENZOYL]ADENOSINE
PROTEIN TYROSINE KINASE p60v-abl — *I*50 = 63 μM

N-METHYLSULPHONYL-12,12-DIBROMODODECA-11-ENAMIDE
CYTOCHROME P4A

3-METHYL-5-SULPHOOXY-2-(4-SULPHOOXYPHENYL)INDOLE
STERYL-SULFATASE — *I*50 = 500 μM

3-METHYL-6-SULPHOOXY-2-(3-SULPHOOXYPHENYL)INDOLE
STERYL-SULFATASE — *I*50 = 1.6 mM

5-METHYL-3,4,5,6-TERTAHYDRO-2′-DEOXYURIDINE-MONOPHOSPHATE
DEOXYCYTIDINE DEAMINASE — 67% 400 nM

17α-METHYLTESTOSTERONE
3α-HYDROXYSTEROID DEHYDROGENASE (B-SPECIFIC) — 53% 100 μM

17β-METHYLTESTOSTERONE
TESTOSTERONE 17β-DEHYDROGENASE

5-METHYLTETRAHYDRODEOXYURIDINE 5′-MONOPHOSPHATE
dCMP DEAMINASE — 70% 400 nM

5-METHYLTETRAHYDROFOLIC ACID
GLYCINE METHYLTRANSFERASE
METHIONYL-tRNA FORMYLTRANSFERASE
PHOSPHORYLASE b — *I*50 = 1.9 mM

10-METHYLTETRAHYDROFOLIC ACID
FORMIMINOTETRAHYDROFOLATE CYCLODEAMINASE — C — *K*i = 24 μM
FORMIMINOTETRAHYDROFOLATE CYCLODEAMINASE — C — *K*i = 24 μM

N5-METHYLTETRAHYDROFOLIC ACID
GLYCINE HYDROXYMETHYLTRANSFERASE — C — *K*i = 92 μM

1-METHYLTETRAHYDROISOQUINOLINE
AMINE OXIDASE MAO-A — C — *K*i = 30 μM
AMINE OXIDASE MAO-B — C — *K*i = 116 μM

2-METHYL-1,2,3,4-TETRAHYDROISOQUINOLINE
AMINE OXIDASE MAO-A — C — *K*i = 27 μM
AMINE OXIDASE MAO-B — C — *K*i = 1 μM

3-METHYL-1,2,3,4-TETRAHYDROISOQUINOLINE
PHENYLETHANOLAMINE N-METHYLTRANSFERASE — *K*i = 3 μM

N-METHYLTETRAHYDROISOQUINOLINE
AMINE OXIDASE MAO-A — C — *K*i = 41 μM
AMINE OXIDASE MAO-B — NC — *K*i = 25 μM

N-METHYL-1,2,3,4-TETRAHYDROISOQUINOLINE
NADH DEHYDROGENASE (UBIQUINONE) — *I*50 = 6.5 mM

N-METHYL1,2,3,4-TETRAHYDROISOQUINOLINE		
OXOGLUTARATE DEHYDROGENASE (LIPOAMIDE)		*I*50 = 2 mM
2-METHYL-2[p-(1,2,3,4-TETRAHYDRO-1-NAPHTHYL)PHENOXY]PROPIONIC ACID		
ACETYL-CoA CARBOXYLASE	NC	*K*i = 230 μM
2-METHYL-2-[p(1,2,3,4-TETRAHYDRO-1-NAPHTYL)-PHENOXY]PROPIONIC ACID		
ACETYL-CoA CARBOXYLASE	SSI	*K*i = 90 μM
D,L-6-METHYL-5,6,7,8-TETRAHYDROPTERIN		
GTP CYCLOHYDROLASE I		*I*50 = 29 μM
PURINE-NUCLEOSIDE PHOSPHORYLASE		
5-METHYL-TETRAHYDROPTEROYL-α-GLUTAMIC ACID		
5-METHYLTETRAHYDROPTEROYLTRIGLUTAMATE-HOMOCYSTEINE METHYLTRANSFERASE	C	
5-METHYLTETRAHYDROPTEROYLHEXAGLUTAMATE		
GLYCINE METHYLTRANSFERASE		
10-METHYLTETRAHYDROPTEROYLORNITHINE		
FOLYLPOLYGLUTAMATE SYNTHASE		*K*i = 1.9 μM
5-METHYLTETRAHYDROPTEROYLPENTAGLUTAMATE		
GLYCINE METHYLTRANSFERASE		82% 50 μM
5-METHYL-5,6,7,8-TETRAHYDROPTEROYLTRIGLUTAMATE		
5-METHYLTETRAHYDROFOLATE-HOMOCYSTEINE S-METHYLTRANSFERASE		
6-METHYL-4,5,8,9-TETRATHIADODECA-1,11-DIENE		
LIPOXYGENASE		*I*50 = 28 μM
2-METHYL-3-[4-(THIAZOL-5-YLMETHYL)PHENYL]-2-PROPENOIC ACID		
THROMBOXANE A SYNTHASE		*I*50 = 220 nM
9-(4-METHYLTHIBUTYL)ADENINE		
METHYLTHIOADENOSINE NUCLEOSIDASE		*K*i = 790 nM
5′-METHYLTHIOADENOSIE		
tRNA(CYTOSINE-5-) METHYLTRANSFERASE		
METHYLTHIOADENOSINE		
ADENOSYLMETHIONINE CYCLOTRANSFERASE		
1-AMINOCYCLOPROPANE-1-CARBOXYLATE SYNTHASE		47% 100 μM
HISTAMINE N-METHYLTRANSFERASE		
tRNA(GUANINE-N^7-) METHYLTRANSFERASE		
2-METHYLTHIOADENOSINE		
SPERMIDINE SYNTHASE	C	77% 100 μM
SPERMINE SYNTHASE	C	57% 10 μM
5′-METHYLTHIOADENOSINE		
ADENOSYLHOMOCYSTEINE NUCLEOSIDASE		*K*i = 27 nM
ADENOSYLMETHIONINE CYCLOTRANSFERASE		
DNA (CYTOSINE-5-)-METHYLTRANSFERASE		
SPERMIDINE SYNTHASE		77% 100 μM
SPERMINE SYNTHASE		*K*i = 300 nM
SPERMINE SYNTHASE		92% 100 μM
METHYLTHIOADENOSINE SULPHOXIDE		
GLUCOSE-6-PHOSPHATASE	IR	
17β-(METHYLTHIO)ANDROST-4-EN-3-ONE		
STEROID 5α-REDUCTASE		*I*50 = 1 μM
STEROID 5α-REDUCTASE		*I*50 = 1 nM
4-METHYLTHIOBENZOIC ACID		
MONOPHENOL MONOOXYGENASE	C	*K*i = 8.3 μM

S-(N-METHYLTHIOCARBAMOYL)-D,L-CYSTEINE		
AMINO-ACID RACEMASE		
3-(METHYLTHIO)CATECHOL		
CATECHOL 1,2-DIOXYGENASE	C	*K*i = 600 nM
S-METHYLTHIOCITRULLINE		
NITRIC-OXIDE SYNTHASE I		*K*i = 1.2 nM
NITRIC-OXIDE SYNTHASE II		*K*i = 11 nM
NITRIC-OXIDE SYNTHASE III		*K*i = 40 nM
S-METHYL-L-THIOCITRULLINE		
NITRIC-OXIDE SYNTHASE		73% 1 μM
NITRIC-OXIDE SYNTHASE I		*I*50 = 1.1 μM
NITRIC-OXIDE SYNTHASE I		*K*i = 1.2 nM
NITRIC-OXIDE SYNTHASE II		*I*50 = 2.2 μM
NITRIC-OXIDE SYNTHASE II		*K*i = 40 nM
NITRIC-OXIDE SYNTHASE III		*K*i = 11 nM
5′-METHYLTHIO-9-DEAZAADENOSINE		
5′-METHYLTHIOADENOSINE PHOSPHORYLASE		*K*i = 200 nM
PURINE-NUCLEOSIDE PHOSPHORYLASE		*K*i = 200 nM
METHYLTHIO-7-DEAZAADENOSINE		
SPERMIDINE SYNTHASE		
SPERMINE SYNTHASE		98% 100 μM
5′-METHYLTHIO-5′-DEOXYTUBERCIDIN		
5′-METHYLTHIOADENOSINE PHOSPHORYLASE	C	*K*i = 600 μM
5′-METHYLTHIOFORMYCIN		
ADENOSYLHOMOCYSTEINE NUCLEOSIDASE		*K*i = 27 nM
METHYLTHIOINOSINE		
RIBOSE-5-PHOSPHATE-AMMONIA LIGASE		
5′-METHYLTHIOINOSINE		
ADENOSYLHOMOCYSTEINE NUCLEOSIDASE		19% 3.2 μM
3-[(METHYLTHIO)METHYL]CATECHOL		
CATECHOL 2,3-DIOXYGENASE		
4-METHYLTHIO-2-OXOBUTANOIC ACID		
D-AMINO-ACID OXIDASE	C	*K*i = 1.9 mM
4-(METHYLTHIO)PHENYLDIPROPYL PHOSPHATE		
ACETYLCHOLINESTERASE		
METHYLTHIOPROPYLAMINE		
SPERMIDINE SYNTHASE		
6-METHYLTHIOPURINE		
THIOPURINE S-METHYLTRANSFERASE		
4-METHYLTHIO-1-βD-RIBOFURANOSYLPYRAZOLO-(3,4-d)PYRIMIDINE		
RIBOSYLPYRIMIDINE NUCLEOSIDASE		*K*i = 65 μM
5′-METHYL-THIORIBOSE		
METHYLTHIOADENOSINE NUCLEOSIDASE	C	*K*i = 1 mM
5′-METHYLTHIOTUBERCIDIN		
ADENOSYLHOMOCYSTEINE NUCLEOSIDASE		*K*i = 4.1 μM
SPERMIDINE SYNTHASE		
SPERMINE SYNTHASE		
6-METHYL-2-THIOURACIL		
THYROXINE DEIODINASE		*I*50 = 1 μM

S-METHYLTHIOUREA
NITRIC-OXIDE SYNTHASE I | | *I*50 = 1.3 μM
NITRIC-OXIDE SYNTHASE II | | *I*50 = 300 nM

5-(METHYLTHIO)VARACIN
PROTEIN KINASE C

METHYL 1-THIO-β-D-XYLOPYRANOSIDE
XYLAN 1,4-β-XYLOSIDASE | | *K*i = 2.8 mM

1-METHYL-5-[3-exo-TRICYCLO[5.2.1.0^2,6]DEC-8-YLOXY)-4-METHOXYPHENYL]-2-IMIDAZOLIDINONE
3′,5′-CYCLIC-NUCLEOTIDE PHOSPHODIESTERASE (cycAMP) (Ca^{2+} INDEPENDENT) | | *I*50 = 490 μM

1S,2R,3S,4R,5R-METHYL[2,3,4-TRIHYDROXY-5-(HYDROXYMETHYL)CYCLOPENTYL]AMINE
α-MANNOSIDASE | | *I*50 = 62 nM
MANNOSYL-OLIGOSACCHARIDE 1,3-1,6-α-MANNOSIDASE | | *I*50 = 1 μM

5-METHYL-6-[[(3,4,5-TRIMETHOXYPHENYL)AMINO]METHYL-2,4-QUINAZOLINEDIAMINE
DIHYDROFOLATE REDUCTASE

(24R,S)-24-METHYL-25,26,27-TRISNOR-24-DIMETHYLAMINO-N-OXIDE CYCLOARTANOL
Δ(24)-STEROL C-METHYLTRANSFERASE | NC | *K*i = 15 nM

(24R,S)-24-METHYL-25,26,27-TRISNOR-24-DIMETHYLARSONIUM CYCLOARTANOL IODIDE
Δ(24)-STEROL C-METHYLTRANSFERASE | NC | *K*i = 30 nM

(24R,S)-24-METHYL-25,26,27-TRISNOR-24-DIMETHYLSULPHONIUM CYCLOARTANOL IODIDE
Δ(24)-STEROL C-METHYLTRANSFERASE | NC | *K*i = 25 nM

(24R)-24-METHYL-25,26,26-TRISNOR-24-TRIMETHYLAMMONIUM CYCLOARTANOL IODIDE
Δ(24)-STEROL C-METHYLTRANSFERASE | NC | *K*i = 45 nM

6-METHYL-4,5,8-TRITHIAUNDECA-1,10-DIENE
LIPOXYGENASE | | *I*50 = 31 μM

α-METHYLTRYPTOPHAN
PHENYLALANINE DECARBOXYLASE

1-METHYLTRYPTOPHAN
TRYPTOPHAN 2-C-METHYLTRANSFERASE

1-METHYL-D,L-TRYPTOPHAN
INDOLEAMINE-PYRROLE 2,3-DIOXYGENASE | C | *K*i = 6.6 μM

2-METHYLTRYPTOPHAN
TRYPTOPHAN-tRNA LIGASE | C

3-METHYLTRYPTOPHAN
TRYPTOPHAN TRANSAMINASE

4-METHYLTRYPTOPHAN
PREPHENATE DEHYDRATASE

5-METHYLTRYPTOPHAN
PREPHENATE DEHYDRATASE
TRYPTOPHAN 2,3-DIOXYGENASE
TRYPTOPHAN-tRNA LIGASE | C

6-METHYLTRYPTOPHAN
TRYPTOPHAN-tRNA LIGASE | C

D,L-4-METHYLTRYPTOPHAN
ANTHRANILATE SYNTHASE | | 75% 20 μM

N-METHYL TYRAMINE
TYRAMINE N-METHYLTRANSFERASE | C | *K*i = 30-60 μM

α-METHYL-D,L-TYROSINE
PREPHENATE DEHYDROGENASE

α-METHYL-m-TYROSINE
PHENYLALANINE DECARBOXYLASE

α-METHYL-p-TYROSINE
TYROSINE 3-MONOOXYGENASE

2-METHYL-D,L-TYROSINE
TYROSINE 3-MONOOXYGENASE

N-METHYLTYROSINE DIKETOPIPERAZINE
CALPAIN

METHYL-m-TYROSINE
AROMATIC-L-AMINO-ACID DECARBOXYLASE C *K*i = 320 μM

N-METHYLTYROSINE-N-METHYLTYROSYL-LEUCYL-ALANINE
CALPAIN

7-O-METHYL-UCN 01
PROTEIN KINASE A *I*50 = 240 nM
PROTEIN KINASE C *I*50 = 1.4 nM

7-O-METHYL-UCN 02
PROTEIN KINASE A *I*50 = 380 nM
PROTEIN KINASE C *I*50 = 28 nM

4-METHYLUMBELLIFERONE
β-GLUCOSIDASE C *K*i = 973 μM

4-METHYLUMBELLIFERONE NONANOATE
PHOSPHATASE ACID

N-METHYL-N-UNDECYLHYDROXYLAMINE
ARACHIDONATE 5-LIPOXYGENASE *I*50 = 800 nM

O-METHYL-N-UNDECYLHYDROXYLAMINE
ARACHIDONATE 5-LIPOXYGENASE *I*50 = 50 μM
CYCLOOXYGENASE *I*50 = 3 μM

6-METHYLURACIL
THYMIDINE PHOSPHORYLASE
URIDINE PHOSPHORYLASE

3-METHYLURIC ACID
URATE OXIDASE

7-METHYLURIC ACID
URATE OXIDASE

9-METHYLURIC ACID
URATE OXIDASE

5-METHYLURIDINE
CYTIDINE DEAMINASE *K*i = 440 μM

2′-C-METHYLURIDINE-5′-DIPHOSPHATE
RIBONUCLEOSIDE-DIPHOSPHATE REDUCTASE (ADENOSYLCOBALAMIN DEPENDENT) C *K*i = 84 μM

β-D-5-METHYLURIDINE-5′-TRIPHOSPHATE
RNA-DIRECTED DNA POLYMERASE

β-L-5-METHYLURIDINE-5′-TRIPHOSPHATE
RNA-DIRECTED DNA POLYMERASE

2-METHYLUROCANIC ACID
UROCANATE HYDRATASE

METHYLVALYL-4-CYCLOSPORIN
PEPTIDYLPROLYL ISOMERASE *I*50 = 11 nM

N-METHYLVALYL-PHENYL(1-AMINO-2-(4-HYDROXYPHENYL)ETHYLPHOSPHONIC ACID)
PEPTIDYL-DIPEPTIDASE A — *I*50 = 340 nM

N-METHYLVALYL-TYROSYL(1-AMINO-2-(4-HYDROXYPHENYL)ETHYLPHOSPHONIC ACID)
PEPTIDYL-DIPEPTIDASE A — *I*50 = 91 nM

METHYL VIOLOGEN
NITROGENASE

METHYLVIOLOGEN REDUCED
2-ENOATE REDUCTASE

METHYLXANTHINE
3′,5′-CYCLIC-NUCLEOTIDE PHOSPHODIESTERASE

METHYL-β-D-XYLOPYRANOSIDE
XYLAN 1,4-β-XYLOSIDASE — *K*i = 47 mM

METHYL-β-D-XYLOSIDE
ARYL β HEXOSIDASE

METHYMETHANETHIOSULPHONATE
GLYCINE HYDROXYMETHYLTRANSFERASE

5-METHYOXY-2-METHYL-1-(PHENYLMETHYL)-1H-INDOLE-3-ACETIC ACID
PHOSPHOLIPASE A2 (NON PANCREAITC SECRETED) — *I*50 = 14 µM

METHYRAPONE
CYTOCHROME P450 2A6 — C — *K*i = 28 µM
EPOXIDE HYDROLASE

METIAMIDE
NADH DEHYDROGENASE (UBIQUINONE) — *I*50 = 600 µM

METIPAMID
11β-HYDROXYSTEROID DEHYDROGENASE

MET-LEU-TYR
TRIPEPTIDE AMINOPEPTIDASE — *I*50 = 13 µM

METOCLOPRAMIDE
BUFURALOL-1′-HYDROXYLASE (CYT P-450dbl) — C — *I*50 = 50 µM
CARBOXYLESTERASE — *I*50 = 331 µM
CARBOXYLESTERASE — *I*50 = 4.4 µM
CARBOXYLESTERASE — *I*50 = 309 µM
CARBOXYLESTERASE — *I*50 = 331 µM
CARBOXYLESTERASE — *I*50 = 316 µM
CHOLINESTERASE
HISTAMINE N-METHYLTRANSFERASE — *I*50 = 657 µM

METOCURINE
HISTAMINE N-METHYLTRANSFERASE — C — *K*i = 15 µM

METOPRINE
HISTAMINE N-METHYLTRANSFERASE — *I*50 = 56 nM
HISTAMINE N-METHYLTRANSFERASE — C — *K*i = 58 nM
HISTAMINE N-METHYLTRANSFERASE

METOTHREXATE
PHOSPHORIBOSYLAMINOIMIDAZOLECARBOXAMIDE FORMYLTRANSFERASE — C — *K*i = 596 µM
PHOSPHORIBOSYLGLYCINAMIDE FORMYLTRANSFERASE — C — *K*i = 9.3 µM

METOTHREXATE PENTAGLUTAMATE
5-FORMYLTETRAHYDROFOLATE CYCLO-LIGASE — C — *K*i = 16 µM

METOXY E 3810
H^+/K^+-EXCHANGING ATPASE — *I*50 = 160 nM

MET-PHE
PEPTIDYL-DIPEPTIDASE A | C | *I*50 = 45 μM

METRIFONATE
ACETYLCHOLINESTERASE | | *I*50 = 1.8 μM

METRIZAMIDE
GUANYLATE CYCLASE

METRONIDAZOLE
ALCOHOL DEHYDROGENASE | | 55% 8 mM
AMINE OXIDASE DIAMINE | | *K*i = 100 μM
AMINE OXIDASE DIAMINE | | *K*i = 250 μM
NITROGENASE

METROPOLOL
LIPOPROTEIN LIPASE | | *I*25 = 20 mM

METSULFURON
ACETOLACTATE SYNTHASE

MET-THIOL
GLUTAMYL AMINOPEPTIDASE | | *I*50 = 11 nM

MET-TYR
PEPTIDYL-DIPEPTIDASE A | NC | *I*50 = 193 μM

METYRAPOL
STEROID 11β-MONOOXYGENASE | | 41% 400 μM

METYRAPONE
ECDYSONE 20-MONOOXYGENASE
ESTRADIOL 2-HYDROXYLASE
ESTRADIOL 16α-HYDROXYLASE
3(or 11)β-HYDROSTEROID DEHYDROGENASE 1 | C | *K*i = 30 μM
(−)-LIMONENE 3-MONOOXYGENASE
(−)-LIMONENE 6-MONOOXYGENASE
(−)-LIMONENE 7-MONOOXYGENASE
MENTHOL MONOOXYGENASE
METHYLTETRAHYDROPROTOBERBERINE 14-MONOOXYGENASE
PROGESTERONE 11α-MONOOXYGENASE
STEROID 11β-MONOOXYGENASE | | 40% 400 μM
UNSPECIFIC MONOOXYGENASE

MEVALONATE-5-PYROPHOSPHATE
[HYDROXYMETHYLGLUTARYL-CoA REDUCTASE (NADPH)]-PHOSPHATASE | | 70% 50 μM

MEVASTIN
HYDROXYMETHYLGLUTARYL-CoA REDUCTASE (NADPH)

MEVINOLIN
HYDROXYMETHYLGLUTARYL-CoA REDUCTASE (NADPH) | | *I*50 = 8 nM
TRICHODIENE SYNTHASE | | *I*50 = 100 μM

MEVINOLININIC ACID
HYDROXYMETHYLGLUTARYL-CoA REDUCTASE (NADPH) | | *K*i = 600 pM

MEVINPHOS
ACETYLCHOLINESTERASE

cis-MEVINPHOS
CARBOXYLESTERASE | | *K*ic = 3.2 μM

trans-MEVINPHOS
CARBOXYLESTERASE | | *K*ic = 5.6 μM

MEZLOCILLIN
NEPRILYSIN | | *K*i = 473 μM

MF 268
CHOLINESTERASE

MFT 279
AROMATASE $I50 = 2.4$ nM

MF TRICYCLIC
CYCLOOXYGENASE 2 $I50 = 16$ nM

Mg^{2+}
ACETATE-CoA LIGASE
ACETOIN DEHYDROGENASE
ACETOLACTATE SYNTHASE
ACETYL-CoA C-ACETYLTRANSFERASE
ACETYL-CoA C-ACYLTRANSFERASE 20% 25 mM
ACETYL-CoA CARBOXYLASE
N-ACETYLGALACTOSAMINE-4-SULFATASE 40% 10 mM
N-ACETYLGALACTOSAMINE-6-SULFATASE 40% 20 mM
N-ACETYLGALACTOSAMINE-6-SULFATASE 55% 17 mM
N-ACETYL-β-GLUCOSAMINIDASE
N^4(β-N-ACETYLGLUCOSAMINYL)-L-ASPARAGINASE 20% 1 mM
ACETYLSEROTONIN O-METHYLTRANSFERASE
ACTINOMYCIN LACTONASE
2-ACYLGLYCEROL O-ACYLTRANSFERASE
ACYLGLYCERONE-PHOSPHATE REDUCTASE
ADENOSYLMETHIONINE DECARBOXYLASE 100% 500 μM
ADENYLATE CYCLASE
ADENYLOSUCCINATE LYASE
ADENYLYLSULFATASE
ADP DEAMINASE
ADP-RIBOSE PYROPHOSPHATASE
AGMATINE DEIMINASE
D-ALANYL-D-ALANINE CARBOXYPEPTIDASE
ALDEHYDE DEHYDROGENASE (PYRROLOQUINOLINE-QUINONE) 15% 500 μM
ALKENYLGLYCEROPHOSPHOETHANOLAMINE HYDROLASE
1-ALKYLGLYCEROPHOSPHOCHOLINE O-ACETYLTRANSFERASE 98% 1 mM
AMINO-ACID N-ACETYLTRANSFERASE
AMINOCARBOXYMUCONATE-SEMIALDEHYDE DEHYDROGENASE
5-AMINOPENTANAMIDASE 100% 5 mM
AMINOPEPTIDASE 23% 1 mM
o-AMINOPHENOL OXIDASE
(R)AMINOPROPANOL DEHYDROGENASE > 3 mM
AMMONIA KINASE
α-AMYLASE
β-AMYLASE
APYRASE
ARGININE DECARBOXYLASE
ARGININE-tRNA LIGASE
ARYL-ALDEHYDE DEHYDROGENASE
ARYLFORMAMIDASE 100% 1 mM
ARYL SULFOTRANSFERASE 10 mM
L-ASCORBATE PEROXIDASE
ASPERGILLUS NUCLEASE S1
BENZOATE 4-MONOOXYGENASE
BUTYRATE-CoA LIGASE
CARBON MONOXIDE OXIDASE
CARBOXYPEPTIDASE 41% 1 mM
κ-CARRAGEENASE
CATECHOL 2,3-DIOXYGENASE
CATECHOL O-METHYLTRANSFERASE > 2 mM
CATHEPSIN B 55% 500 μM
CHITINASE
CHODROTIN 6-SULFOTRANSFERASE
CHOLINESUFATASE
CHONDROITIN AC LYASE 51% 10 mM

CITRATE (pro-3s)-LYASE	C	Ki = 25 mM
CITRATE (si)-SYNTHASE		
CREATINASE		35% 1 mM
CREATININASE		
CYCLOMALTODEXTRIN GLUCANOTRANSFERASE		
17-O-DEACETYLVINDOLINE O-ACETYLTRANSFERASE		
DEXTRANASE		44% 15 mM
DIHYDROFOLATE REDUCTASE		
DIHYDROPYRIMIDINASE		
2,3-DIHYDROXYBENZOATE 2,3-DIOXYGENASE		100% 1 mM
DIPEPTIDASE		
DIPEPTIDYL-PEPTIDASE 3		
DIPEPTIDYL-PEPTIDASE 4		
DNA α-GLUCOSYLTRANSFERASE		
DNA POLYMERASE α		
DOLICHOL KINASE		94% 100 μM
DOLICHYL-PHOSPHATE-MANNOSE-PROTEIN MANNOSYLTRANSFERASE		> 5 mM
ECDYSONE 20-MONOOXYGENASE		
ESTRADIOL 17β-DEHYDROGENASE		
FLAVONE APIOSYLTRANSFERASE		
FLAVONOL 3-O-GLUCOSYLTRANSERASE		> 5 mM
FMN ADENYLYLTRANSFERASE		
FRUCTAN β-FRUCTOSIDASE		20% 1 mM
FRUCTOSE-2,6-BISPHOSPHATASE		
GALACTOLIPASE		53% 47%
α-GALACTOSIDASE		
β-GALACTOSIDASE		
GALACTOSIDE 2-L-FUCOSYLTRANSFERASE		
GLIOCLADIUM PROTEINASE		
1,4-α-GLUCAN BRANCHING ENZYME		
GLUCOSE 1-DEHYDROGENASE		
GLUTAMATE-AMMONIA LIGASE		
GLUTAMATE DEHYDROGENASE (NADP)		
GLUTAMATE FORMIMINOTRANSFERASE		
GLUTAMATE SYNTHASE (NADH)		
γ-GLUTAMYLTRANSFERASE		100% 1 mM
GLUTATHIONE PEROXIDASE		19% 5 mM
GLYCERALDEHYDE-3-PHOSPHATE DEHYDROGENASE (NADP)		
GLYCEROL-3-PHOSPHATE O-ACYLTRANSFERASE		
GLYCEROL-3-PHOSPHATE O-ACYLTRANSFERASE		
GLYCEROL-3-PHOSPHATE DEHYDROGENASE		
GLYCEROPHOSPHOINOSITOL INOSITOLPHOSPHODIESTERASE		
GLYCERYL-ETHER MONOOXYGENASE		
GLYCINE ACYLTRANSFERASE		
GLYCINE BENZOYLTRANSFERASE		
GLYCOLALDEHYDE DEHYDROGENASE		
GLYCOPROTEIN O-FATTY-ACYLTRANSFERASE		
GMP REDUCTASE		
GTP CYCLOHYDROLASE A2		I50 = 2.5 mM
GTP CYCLOHYDROLASE I		I50 = 5 mM
GUANINE DEAMINASE		
HEXOKINASE	C	Ki = 9 μM
HIGH-MANNOSE-OLIGOSACCHARIDE β-1,4-N-ACETYLGLUCOSAMINYLTRANSFERASE		
HISTIDINOL DEHYDROGENASE		46% 500 μM
H^+-TRANSPORTING ATP SYNTHASE	C	Ki = 2.8 mM
3-HYDROXYANTHRANILATE 3,4-DIOXYGENASE		20% 1 mM
HYDROXYMETHYLBILANE SYNTHASE		
HYDROXYMETHYLGLUTARYL-CoA SYNTHASE		
HYDROXYPYRUVATE REDUCTASE		100% 1 mM
7α-HYDROXYSTEROID DEHYDROGENASE		
HYOSCYAMINE (6S)-DIOXYGENASE		
INORGANIC PYROPHOSPHATASE		Ki = 35 μM
myo-INOSITOL-1-PHOSPHATE SYNTHASE		I50 = 40 mM
INULINASE		

ISOLEUCINE-tRNA LIGASE		
L-LACTATE DEHYDROGENASE (CYTOCHROME)		
1,4-LACTONASE		
LACTOSE SYNTHASE		> 4mM
LEUCINE N-ACETYLTRANSFERASE		
LEUCYL AMINOPEPTIDASE		
LEUCYLTRANSFERASE		
LYSINE 2-MONOOXYGENASE		
LYSINE-tRNA LIGASE		66% 2.5 mM
LYSOPHOSPHOLIPASE		
LYSOZYME		
MALATE DEHYDROGENASE		
MANNAN 1,2-(1,3)-α-MANNOSIDASE		
α-MANNOSIDASE		26% 10 mM
METHYLENETETRAHYDROFOLATE CYCLOHYDROLASE		
NAD(P) TRANSHYDROGENASE (AB-SPECIFIC)		
NAD SYNTHASE		
NICOTINATE-NUCLEOTIDE PYROPHOSPHORYLASE (CARBOXYLATING)		> 500 μM
NITRATE REDUCTASE (NADH)		
3′-NUCLEOTIDASE		
OLIGO-1,6-GLUCOSIDASE		23% 2 mM
OXALACETATE TAUTOMERASE		40% 40 μM
PALMITOYL-CoA HYDROLASE		40% 2.4 mM
PALMITOYL-CoA HYDROLASE		29% 10 mM
PARAOXON HYDROLASE		Ki = 860 μM
PEPTIDE α-N-ACETYLTRANSFERASE		
PHENOL O-METHYLTRANSFERASE		
PHOSPHATIDATE PHOSPHATASE		
1-PHOSPHATIDYLINOSITOL-4,5-BISPHOSPHATE PHOSPHODIESTERASE		
1-PHOSPHATIDYLINOSITOL PHOSPHODIESTERASE		
PHOSPHOACETYLGLUCOSAMINE MUTASE		
PHOSPHOADENYLYLSULFATASE		23% 200 μM
PHOSPHOENOLPYRUVATE CARBOXYKINASE (PYROPHOSPHATE)		Ki = 2.1 mM
6-PHOSPHOFRUCTOKINASE		
6-PHOSPHO-β-GALACTOSIDASE		25% 2 mM
PHOSPHOGLUCONATE 2-DEHYDROGENASE		
PHOSPHOGLUCONATE DEHYDROGENASE (DECARBOXYLATING)		> 50 mM
PHOSPHOGLYCERATE PHOSPHATASE		
6-PHOSPHO-3-KETOHEXULOSE ISOMERASE		25% 1 mM
3-PHYTASE		
POLYGALACTURONASE		
POLYNUCLEOTIDE ADENYLYLTRANSFERASE		
POLYNUCLEOTIDE ADENYLYLTRANSFERASE		
PREPHENATE DEHYDROGENASE		
PROCOLLAGEN N-ENDOPEPTIDASE		
PROLINE DEHYDROGENASE		
PROLINE RACEMASE		25% 37 mM
PROPANEDIOL DEHYDRATASE		
PROTEIN-ARGININE N-METHYLTRANSFERASE		
PROTEIN-DISULFIDE REDUCTASE (GLUTATHIONE)		
PROTEIN KINASE C		
PROTEIN TYROSINE-PHOSPHATASE		99% 10 mM
PROTEIN TYROSINE SULFOTRANSFERASE		
PYROPHOSPHATE-PROTEIN PHOSPHOTRANSFERASE		
RHODOTORULAPEPSIN		22% 5 mM
RIBONUCLEASE T1		
RIBONUCLEASE U2		
RIBONUCLEOSIDE-DIPHOSPHATE REDUCTASE		
RIBONUCLEOSIDE-TRIPHOSPHATE REDUCTASE		
RIBOSE-5-PHOSPHATE ISOMERASE	NC	Ki = 90 mM
tRNA(ADENINE-N^1-) METHYLTRANSFERASE		> 2 mM
tRNA(ADENINE-N^6)-METHYLTRANSFERASE		
tRNA(CYTOSINE-5-) METHYLTRANSFERASE		
tRNA(GUANINE-N^1-) METHYLTRANSFERASE		> 2 mM

Enzyme	Type	Value
tRNA (5-METHYLAMINOMETHYL-2-THIOURIDYLATE) METHYLTRANSFERASE		> 5 mM
SERINE CARBOXYPEPTIDASE		
SERRATIA MARCESCENS NUCLEASE		
SIALIDASE		
SIGNAL PEPTIDASE I		> 1 mM
SINAPOYLGLUCOSE-SINAPOYLGLUCOSE O-SINAPOYLTRANSFERASE		35% 10 mM
SPLEEN EXONUCLEASE		
STARCH SYNTHASE		
STIZOLOBATE SYNTHASE		
STIZOLOBINATE SYNTHASE		
α,α-TREHALASE		13% 1 mM
TREHALOSE O-MYCOLYLTRANSFERASE		91% 16 mM
TRIACYLGLYCEROL LIPASE		*I*50 = 20 mM
TRIMETHYLYSINE DIOXYGENASE		
TRITRIRACHIUM ALKALINE PROTEINASE		
TYROSINE-ESTER SULFOTRANSFERASE		
TYROSYL PROTEIN SULFOTRANSFERASE		
UDP-N-ACETYLMURAMATE DEHYDROGENASE		
UDP-GLUCOSE-HEXOSE-1-PHOSPHATE URIDYLYLTRANSFERASE		
XYLAN 1,4-β-XYLOSIDASE		
YEAST RIBONUCLEASE		

MG 14

Enzyme	Type	Value
CONODIPINE-M (PLA2)		
PHOSPHOLIPASE A2		

Mg^{2+}/ADP

Enzyme	Type	Value
KETOHEXOKINASE		
MANNOKINASE	C	
NUCLEOSIDE-DIPHOSPHATE KINASE		

Mg^{2+}/ATP

Enzyme	Type	Value
ALKYLGLYCEROPHOSPHATE 2-O-ACETYLTRANSFERASE		
HEXOKINASE	C	*K*i = 25 μM
H^+-TRANSPORTING ATP SYNTHASE	C	*K*i = 2.8 mM
6-PHOSPHOFRUCTOKINASE		

$MgCl_2$

Enzyme	Type	Value
AMINOACYL-tRNA HYDROLASE		> 100 mM
CYTOCHROME-b5 REDUCTASE		*I*50 = 18 mM
DEOXYRIBONUCLEASE (APURINIC OR APYRIMIDIC)		> 20 mM
GLYCEROL 2-DEHYDROGENASE		
PLASMIN		
PROPIONATE-CoA LIGASE		

Mg^{2+}/EDTA COMPLEX

Enzyme	Type	Value
GLUCOSE-6-PHOSPHATASE		

Mg^{2+}/OXALACETATE COMPLEX

Enzyme	Type	Value
GLUCOSE-6-PHOSPHATASE		

Mg^{2+}/α-OXOGLUTARATE COMPLEX

Enzyme	Type	Value
GLUCOSE-6-PHOSPHATASE		

Mg^{2+} PROTOPORPHYRIN IX MONOMETHYL ESTER

Enzyme	Type	Value
MAGNESIUM PROTOPORPHYRIN O-METHYLTRANSFERASE		

MG-PYROPHOSPHATE

Enzyme	Type	Value
UDP-GLUCOSE-1-PHOSPHATE URIDYLYLTRANSFERASE		

$MgSO_4$

Enzyme	Type	Value
CYCLAMATE SULFOHYDROLASE		285 10 mM
STEROL ESTERASE		

MHP 133

Enzyme	Type	Value
ACETYLCHOLINESTERASE		*K*i =

MIANSERINE

SPHINGOMYELIN PHOSPHODIESTERASE		52% 10 μM

MICHELLAMINE B

RNA-DIRECTED DNA POLYMERASE		*I*50 = 122 μM
RNA-DIRECTED DNA POLYMERASE		*I*50 = 167 μM

MICONAZOLE

ADENOSINETRIPHOSPHATASE		
ARACHIDONATE EPOXYGENASE		
AROMATASE		*I*50 = 470 nM
AROMATASE		*I*50 = 2 μM
AROMATASE		*K*i = 35 nM
CYTOCHROME P450 2A6	C	*K*i = 220 nM
7-ETHOXYCOUMARIN O-DEACETYLASE		
7-ETHOXYRESORUFIN O-DEETHYLASE		
H^+-TRANSPORTING ATPASE		
(–)-LIMONENE 3-MONOOXYGENASE		
(–)-LIMONENE 6-MONOOXYGENASE		
(–)-LIMONENE 7-MONOOXYGENASE		
NITRIC-OXIDE SYNTHASE		44% 100 μM
NITRIC-OXIDE SYNTHASE (CALMODULIN DEPENDENT)	C	*I*50 = 8 μM
STEROID 21-MONOOXYGENASE		
STEROID 17α-MONOOXYGENASE/17,20 LYASE		*I*50 = 2.6 μM
UNSPECIFIC MONOOXYGENASE		

α-MICROBIAL ALKALINE PROTEINASE INHIBITOR

PROTEINASE HIV		*I*50 = 2.7 μM

β-MICROBIAL ALKALINE PROTEINASE INHIBITOR

PROTEINASE HIV		*I*50 = 45 μM

MICROCIN B17

DNA GYRASE		
DNA TOPOISOMERASE (ATP-HYDROLYSING)		

MICROCOCCUS LYSODEIKTICUS INHIBITOR

tRNA(ADENINE-N^6)-METHYLTRANSFERASE		

MICROCYSTIN LR

MYOSIN-LIGHT-CHAIN PHOSPHATASE		*K*i = 50 nM
PHOSPHOPROTEIN PHOSPHATASE		*I*50 = 1.6 nM
PHOSPHOPROTEIN PHOSPHATASE 1		*I*50 = 100 pM
PHOSPHOPROTEIN PHOSPHATASE 1		*K*i = 50 pM
PHOSPHOPROTEIN PHOSPHATASE 1		*I*50 = 6 nM
PHOSPHOPROTEIN PHOSPHATASE 2A		*I*50 = 800 pM
PHOSPHOPROTEIN PHOSPHATASE 2A		*I*50 = 100 pM
PHOSPHOPROTEIN PHOSPHATASE 2A		*K*i = 13 pM
PHOSPHOPROTEIN PHOSPHATASE 2A		*I*50 = 40 pM

MICROCYSTIN RR

PHOSPHOPROTEIN PHOSPHATASE		*I*50 = 3.4 nM
PHOSPHOPROTEIN PHOSPHATASE 1		*I*50 = 3 nM
PHOSPHOPROTEIN PHOSPHATASE 2A		*I*50 = 1 nM
PHOSPHOPROTEIN PHOSPHATASE 2A		*I*50 = 1.4 nM

MICROCYSTIN YR

PHOSPHOPROTEIN PHOSPHATASE		*I*50 = 1.4 nM
PHOSPHOPROTEIN PHOSPHATASE 2A		*I*50 = 1.3 nM

MICROGININ

PEPTIDYL-DIPEPTIDASE A		*I*50 = 7 μg/ml

MICROGININ 299A

LEUCYL AMINOPEPTIDASE		*I*50 = 4.6 μg/ml

MICROGININ 299B
LEUCYL AMINOPEPTIDASE *I*50 = 6.5 μg/ml

MICROPEPTIN 90
PLASMIN *I*50 = 100 ng/ml
TRYPSIN *I*50 = 2 μg/ml

MICROPEPTIN 103
CHYMOTRYPSIN *I*50 = 1 μg/ml

MICROPEPTIN A
PLASMIN *I*50 = 26 ng/ml
TRYPSIN *I*50 = 71 ng/ml

MICROPEPTIN 88A
CHYMOTRYPSIN *I*50 = 400 ng/ml
ELASTASE *I*50 = 3.5 μg/ml

MICROPEPTIN 478 A
PLASMIN *I*50 = 100 ng/ml

MICROPEPTIN B
PLASMIN *I*50 = 35 ng/ml
TRYPSIN *I*50 = 250 ng/ml

MICROPEPTIN 478 B
PLASMIN *I*50 = 400 ng/ml

MICROPEPTIN 88C
CHYMOTRYPSIN *I*50 = 5 μg/ml

MICROPEPTIN 88D
CHYMOTRYPSIN *I*50 = 10 μg/ml

MICROPEPTIN 88E
CHYMOTRYPSIN *I*50 = 5.2 μg/ml

MICROPEPTIN 88F
CHYMOTRYPSIN *I*50 = 3.4 μg/ml

β-MICROSEMINOPROTEIN
Na^+/K^+-EXCHANGING ATPASE *I*50 = 90 μM

MICROVIRIDIN
TYROSINASE

MICROVIRIDIN B
CHYMOTRYPSIN *I*50 = 2.5 μg/ml
ELASTASE *I*50 = 44 ng/ml
TRYPSIN *I*50 = 32 μg/ml
TRYPSIN *I*50 = 58 μg/ml

MICROVIRIDIN C
CHYMOTRYPSIN *I*50 = 4.9 μg/ml
ELASTASE *I*50 = 84 ng/ml

MICROVIRIDIN D
CHYMOTRYPSIN *I*50 = 1.2 μg/ml
ELASTASE *I*50 = 700 ng/ml

MICROVIRIDIN E
CHYMOTRYPSIN *I*50 = 1.1 μg/ml
ELASTASE *I*50 = 600 ng/ml

MICROVIRIDIN F
ELASTASE *I*50 = 5.8 μg/ml

MICROVIRIDIN G
ELASTASE *I*50 = 19 ng/ml

MICROVIRIDIN H		
ELASTASE		$I50 = 31$ ng/ml
MIFENTIDINE		
CYTOCHROME P450 3A1		
CYTOCHROME P450 2B1		
CYTOCHROME P450 2E1		
MILACEMIDE		
AMINE OXIDASE MAO-A	IR	$Ki = 115$ µM
AMINE OXIDASE MAO-A		
AMINE OXIDASE MAO-B		
MILRINONE		
3′,5′-CYCLIC-NUCLEOTIDE PHOSPHODIESTERASE		$I50 = 1$ µM
3′,5′-CYCLIC-NUCLEOTIDE PHOSPHODIESTERASE (cycAMP)		$I50 = 15$ µM
3′,5′-CYCLIC-NUCLEOTIDE PHOSPHODIESTERASE (cycAMP)		$I50 = 200$ nM
3′,5′-CYCLIC-NUCLEOTIDE PHOSPHODIESTERASE (cycAMP) cycGMP INHIBITED		$I50 = 400$ nM
3′,5′-CYCLIC-NUCLEOTIDE PHOSPHODIESTERASE (cycAMP) IV		$I50 = 450$ µM
3′,5′-CYCLIC-NUCLEOTIDE PHOSPHODIESTERASE (cycAMP) IV		$I50 = 18$ µM
3′,5′-CYCLIC-NUCLEOTIDE PHOSPHODIESTERASE (cycAMP) LOW Km		
3′,5′-CYCLIC-NUCLEOTIDE PHOSPHODIESTERASE (cycAMP) LOW Km		$I50 = 680$ nM
3′,5′-CYCLIC-NUCLEOTIDE PHOSPHODIESTERASE (cycGMP)		$I50 = 100$ µM
3′,5′-CYCLIC-NUCLEOTIDE PHOSPHODIESTERASE I		$I50 = 114$ µM
3′,5′-CYCLIC-NUCLEOTIDE PHOSPHODIESTERASE I		$I50 = 452$ µM
3′,5′-CYCLIC-NUCLEOTIDE PHOSPHODIESTERASE Ib		$I50 = 270$ µM
3′,5′-CYCLIC-NUCLEOTIDE PHOSPHODIESTERASE IB		$I50 = 269$ µM
3′,5′-CYCLIC-NUCLEOTIDE PHOSPHODIESTERASE II		$I50 = 263$ µM
3′,5′-CYCLIC-NUCLEOTIDE PHOSPHODIESTERASE II		$I50 = 98$ µM
3′,5′-CYCLIC-NUCLEOTIDE PHOSPHODIESTERASE II		$I50 = 1$ mM
3′,5′-CYCLIC-NUCLEOTIDE PHOSPHODIESTERASE II (cycAMP)		$I50 = 1.1$ µM
3′,5′-CYCLIC-NUCLEOTIDE PHOSPHODIESTERASE II (cycGMP STIMULATED)		$I50 = 263$ µM
3′,5′-CYCLIC-NUCLEOTIDE PHOSPHODIESTERASE III		$I50 = 1$ µM
3′,5′-CYCLIC-NUCLEOTIDE PHOSPHODIESTERASE III		$I50 = 2$ µM
3′,5′-CYCLIC-NUCLEOTIDE PHOSPHODIESTERASE III		$I50 = 1.1$ µM
3′,5′-CYCLIC-NUCLEOTIDE PHOSPHODIESTERASE III		$I50 = 400$ nM
3′,5′-CYCLIC-NUCLEOTIDE PHOSPHODIESTERASE III		$I50 = 1.7$ µM
3′,5′-CYCLIC-NUCLEOTIDE PHOSPHODIESTERASE III		$Ki = 450$ nM
3′,5′-CYCLIC-NUCLEOTIDE PHOSPHODIESTERASE III		$I50 = 710$ nM
3′,5′-CYCLIC-NUCLEOTIDE PHOSPHODIESTERASE III		$I50 = 600$ nM
3′,5′-CYCLIC-NUCLEOTIDE PHOSPHODIESTERASE III		$I50 = 2.7$ µM
3′,5′-CYCLIC-NUCLEOTIDE PHOSPHODIESTERASE III		$I50 = 724$ nM
3′,5′-CYCLIC-NUCLEOTIDE PHOSPHODIESTERASE III		$Ki = 260$ nM
3′,5′-CYCLIC-NUCLEOTIDE PHOSPHODIESTERASE III		$I50 = 1.5$ µM
3′,5′-CYCLIC-NUCLEOTIDE PHOSPHODIESTERASE III		$I50 = 2.2$ µM
3′,5′-CYCLIC-NUCLEOTIDE PHOSPHODIESTERASE III (cycAMP)		$I50 = 1.5$ µM
3′,5′-CYCLIC-NUCLEOTIDE PHOSPHODIESTERASE III (cycGMP INHIBITED)		$I50 = 890$ nM
3′,5′-CYCLIC-NUCLEOTIDE PHOSPHODIESTERASE IV		$I50 = 40$ µM
3′,5′-CYCLIC-NUCLEOTIDE PHOSPHODIESTERASE IV		$I50 = 37$ µM
3′,5′-CYCLIC-NUCLEOTIDE PHOSPHODIESTERASE IV		$I50 = 20$ µM
3′,5′-CYCLIC-NUCLEOTIDE PHOSPHODIESTERASE IV		$I50 = 55$ µM
3′,5′-CYCLIC-NUCLEOTIDE PHOSPHODIESTERASE IV		$I50 = 12$ µM
3′,5′-CYCLIC-NUCLEOTIDE PHOSPHODIESTERASE IV		$I50 = 18$ µM
3′,5′-CYCLIC-NUCLEOTIDE PHOSPHODIESTERASE LOW Km		$I50 = 630$ nM
3′,5′-CYCLIC-NUCLEOTIDE PHOSPHODIESTERASE V		$I50 = 145$ µM
3′,5′-CYCLIC-NUCLEOTIDE PHOSPHODIESTERASE V (cycGMP)		$I50 = 145$ µM
MILTEFOSINE		
PROTEIN KINASE C (CYTOSOLIC)		$I50 = 577$ µM
PROTEIN KINASE C (MEMBRANE)		$I50 = 184$ µM
MIMOSINE		
DEOXYHYPUSINE MONOOXYGENASE		
MONOPHENOL MONOOXYGENASE	C	$Ki = 13$ µM
RIBONUCLEOSIDE-DIPHOSPHATE REDUCTASE		

L-MIMOSINE
MONOPHENOL MONOOXYGENASE — Ki = 45 μM

MINOCYCLINE
COLLAGENASE
GELATINASE B — I50 = 300 μM
PHOSPHOLIPASE A2 — I50 = 8 μM

MINOIODOACETIC ACID
ACYL-LYSINE DEACYLASE

MINOXIDIL
LYSYL HYDROXYLASE
STEROID 5α-REDUCTASE — I50 = 1 μM

MIOFLAZINE
ADENOSINE TRANSPORT — I50 = 72 nM
NUCLEOSIDE TRANSPORT — I50 = 2.3 μM

MIPAFOX
CARBOXYLESTERASE — 50% 50 μM
NEUROPATHY TARGET ESTERASE
ORGANOPHOSPHATE ACID ANHYDRASE

MIS
PROTEIN TYROSINE KINASE EGFR

MISONIDAZOLE
FUMARATE REDUCTASE (NADH) — I50 = 119 μM

MITHRAMYCIN
RNA POLYMERASE

MITINDOMIDE
DNA TOPOISOMERASE (ATP-HYDROLYSING)

MITOCHONDRIAL CALMITINE PROTEINASE INHIBITOR
CALMITINE PROTEINASE

MITOCHONDRIAL INHIBITOR
L-THREONINE 3-DEHYDROGENASE

MITOMYCIN
THIOREDOXIN REDUCTASE (NADPH) — 13% 100 μM

MITOMYCIN C
UBIQUITIN-PROTEIN LIGASE (E1)

MITOXANTHRONE
INTEGRASE — I50 = 3.8 μM
PROTEIN KINASE A — I50 = 60 μM
PROTEIN KINASE (Ca^{2+} DEPENDENT) — 76% 160 μM

MITOXANTRONE
DNA TOPOISOMERASE (ATP-HYDROLYSING)
MYOSIN-LIGHT-CHAIN KINASE — I50 = 2 μM
PROTEIN KINASE C — C — Ki = 6.3 μM
PROTEIN KINASE C — I50 = 4 μM

N^G-MITROARGININE METHYL ESTER
NADPH DEHYDROGENASE — 23% 3 mM

MIXANPRIL
NEPRILYSIN — Ki = 1.7 nM
PEPTIDYL-DIPEPTIDASE A — Ki = 4.2 nM

MIXANTRONE
DNA TOPOISOMERASE (ATP-HYDROLYSING)

MIXED DISULPHIDES

CATHEPSIN B	IR	

MIZORIBINE 5′-MONOPHOSPHATE

IMP DEHYDROGENASE 1	C	Ki = 8 nM
IMP DEHYDROGENASE 2	C	Ki = 4 nM

MJ 33

PHOSPHOLIPASE A2
PHOSPHOLIPASE A2 (ACID)

MJ 72

PHOSPHOLIPASE A2

MJ 133

PHOSPHOLIPASE A2

(S)(+) MJ33

PHOSPHOLIPASE A2

MK 386

5α-REDUCTASE 1		I50 = 20 nM
STEROID 5α-REDUCTASE 1		I50 = 20 nM
STEROID 5α-REDUCTASE 2		I50 = 154 nM
STEROID 5α-REDUCTASE 2		I50 = 31 μM

MK 401

PHOSPHOGLYCERATE KINASE	NC	Ki = 6 mM

MK 417

CARBONATE DEHYDRATASE I		Ki > 5 mM
CARBONATE DEHYDRATASE II		Ki = 2 nM
CARBONATE DEHYDRATASE II		Ki = 610 pM
CARBONATE DEHYDRATASE III		Ki = 80 μM
CARBONATE DEHYDRATASE IV		Ki = 33 nM

MK 421

PEPTIDYL-DIPEPTIDASE A		I50 = 1.5 nM

MK 0434

STEROID 5α-REDUCTASE

MK 434

STEROID 5α-REDUCTASE		I50 = 2.3 nM
STEROID 5α-REDUCTASE		I50 = 1.3 μM

MK 0507

CARBONATE DEHYDRATASE

MK 507

CARBONATE DEHYDRATASE		Ki = 280 pM

MK 0521

PEPTIDYL-DIPEPTIDASE A		I50 = 3 nM

MK 591

ARACHIDONATE 5-LIPOXYGENASE		I50 = 3.4 nM

MK 886

ARACHIDONATE 5-LIPOXYGENASE		
ARACHIDONATE 12-LIPOXYGENASE		61% 500 μM
ARACHIDONATE 15-LIPOXYGENASE		31% 500 μM
3′,5′-CYCLIC-NUCLEOTIDE PHOSPHODIESTERASE III		I50 = 29 μM
3′,5′-CYCLIC-NUCLEOTIDE PHOSPHODIESTERASE IV		I50 = 8.4 μM
PHOSPHOLIPASE A2		73% 50 μM

MK 906

5α-REDUCTASE		I50 = 10 nM
5α-REDUCTASE 1		Ki = 230 nM

5α-REDUCTASE 2		$Ki = 5$ nM
STEROID 5α-REDUCTASE	C	$Ki = 6$ nM

MK 927

CARBONATE DEHYDRATASE		$I50 = 1.2$ nM
CARBONATE DEHYDRATASE II		$Ki = 700$ pM

MK 0963

STEROID 5α-REDUCTASE

MK 963

STEROID 5α-REDUCTASE		$I50 = 2.5$ μM
STEROID 5α-REDUCTASE		$I50 = 13$ μM

MK 4174

CARBONATE DEHYDRATASE II		$Ki = 2$ nM
CARBONATE DEHYDRATASE IV		$Ki = 30$ nM

ML 7

MYOSIN-LIGHT-CHAIN KINASE
TWITCH KINASES (Ca^{2+}/S100 DEPENDENT)

ML 9

MYOSIN-LIGHT-CHAIN KINASE		$I50 = 40$ μM
MYOSIN-LIGHT-CHAIN KINASE	C	$Ki = 3.8$ μM
PROTEIN KINASE A		$Ki = 32$ μM
PROTEIN KINASE C		$Ki = 6$ μM
TWITCH KINASES (Ca^{2+}/S100 DEPENDENT)		

ML 3000

ARACHIDONATE 5-LIPOXYGENASE		$I50 = 370$ nM
ARACHIDONATE 5-LIPOXYGENASE		$I50 = 230$ nM
CYCLOOXYGENASE		$I50 = 160$ nM
CYCLOOXYGENASE 1		$I50 = 220$ nM

ML 236A

HYDROXYMETHYLGLUTARYL-CoA REDUCTASE (NADPH)

MLR 52

PROTEIN KINASE C

Mn^{2+}

ACETOIN DEHYDROGENASE		
ACETOLACTATE DECARBOXYLASE		25% 400 μM
N-ACETYLGALACTOSAMINE-4-SULFATASE		16% 10 mM
N-ACETYLGALACTOSAMINE-6-SULFATASE		60% 17 mM
N-ACETYLGLUCOSAMINE-6-PHOSPHATE DEACETYLASE		
N^4(β-N-ACETYLGLUCOSAMINYL)-L-ASPARAGINASE		
N-ACETYL-γ-GLUTAMYL-PHOSPHATE REDUCTASE		
2-ACYLGLYCEROL O-ACYLTRANSFERASE		
ACYLGLYCERONE-PHOSPHATE REDUCTASE		
N-ACYLHEXOSAMINE OXIDASE		
ADENINE DEAMINASE		
ADENOSINE KINASE		> 1 mM
ADENOSINE NUCLEOSIDASE		33% 10 mM
ADENOSINE-PHOSPHATE DEAMINASE		$I50 = 40$ μM
ADENOSINETRIPHOSPHATASE ECTO		42% 500 μM
ADENOSYLMETHIONINE CYCLOTRANSFERASE		
ADENOSYLMETHIONINE DECARBOXYLASE		100% 500μM
ADENYLATE CYCLASE		
ADENYLOSUCCINATE SYNTHASE		
ADP-RIBOSE PYROPHOSPHATASE		59% 4 mM
AGARASE		
AGMATINE DEIMINASE		
ALAKN-1-OL DEHYDROGENASE (ACCEPTOR)		
ALANINE DEHYDROGENASE		
D-ALANYL-D-ALANINE CARBOXYPEPTIDASE		$I50 = 200$ μM

Enzyme	Inhibition
ALCOHOL DEHYDROGENASE (ACCEPTOR)	
ALDEHYDE DEHYDROGENASE (PYRROLOQUINOLINE-QUINONE)	10% 500 μM
1-ALKYLGLYCEROPHOSPHOCHOLINE O-ACETYLTRANSFERASE	*I*50 = 5 mM
ALKYLGLYCEROPHOSPHOETHANOLAMINE PHOSPHODIESTERASE	> 10 mM
ALLANTOIN RACEMASE	10% 100 μM
AMINO-ACID N-ACETYLTRANSFERASE	
D-AMINO-ACID OXIDASE	
AMINOPEPTIDASE	48% 1 mM
AMMONIA KINASE	
AMP DEAMINASE	
AMP NUCLEOSIDASE	
AMP-THYMIDINE KINASE	
β-AMYLASE	
ANHYDROTETRACYCLINE MONOOXYGENASE	
ARABINOSE-5-PHOSPHATE ISOMERASE	
ARGININE DECARBOXYLASE	
ARYL-ALDEHYDE DEHYDROGENASE	
ARYLAMINE N-ACETYLTRANSFERASE	
L-ASCORBATE PEROXIDASE	
ASPARTATE-AMMONIA LIGASE	
ASPARTATE CARBAMOYLTRANSFERASE	
ASPARTATE RACEMASE	37% 22 mM
β-ASPARTYLDIPEPTIDASE	
ASPERGILLUS ALKALINE PROTEINASE	
ATP DEAMINASE	
AUREOLYSIN	
BENZOATE 4-MONOOXYGENASE	
BIS(5′-ADENOSYL)-TRIPHOSPHATASE	
BUTYRATE-ACETOACETATE CoA-TRANSFERASE	> 5 mM
CAFFEOYL-CoA O-METHYLTRANSFERASE	
CALCIDOL 1-MONOOXYGENASE	
CARBOSINE SYNTHASE	
CARBOXYNORSPERMIDINE DECARBOXYLASE	50% 1 mM
CARNITINE O-ACETYLTRANSFERASE	33% 10 mM
CARNITINE 3-DEHYDROGENASE	
κ-CARRAGEENASE	
CATHEPSIN D	
CATHEPSIN L	60% 1 mM
CELLULASE	
CERAMIDASE	
CHITINASE	
CHITIN DEACYTYLASE	32% 1 mM
CHODROTIN 6-SULFOTRANSFERASE	
CHOLINE-PHOSPHATE CYTIDYLYLTRANSFERASE	
CHONDROITIN ABC LYASE	
CHONDROITIN AC LYASE	65% 10 mM
CHONDRO-4-SULFATASE	
CMP-N-ACYLNEURAMINATE PHOSPHODIESTERASE	
COLLAGENASE CLOSTRIDIUM HISTOLYTICUM	
COPROPORPHYRINOGEN OXIDASE	> 5mM
CORTICOSTERONE 18-MONOOXYGENASE	
2′,3′-CYCLIC-NUCLEOTIDE 2′-PHOSPHODIESTERASE	14% 10 mM
3′,5′-CYCLIC-NUCLEOTIDE PHOSPHODIESTERASE	
β-CYCLOPIAZONATE DEHYDROGENASE	
CYSTINYL AMINOPEPTIDASE	89% 10 mM
CYTOSINE DEAMINASE	
17-O-DEACETYLVINDOLINE O-ACETYLTRANSFERASE	
2-DEHYDRO-3-DEOXYPHOSPHOHEPTONATE ALDOLASE	
2-DEHYDRO-3-DEOXYPHOSPHOOCTONATE ALDOLASE	40% 1 mM
DEOXYHYPUSINE MONOOXYGENASE	36% 5 μM
DEOXYRIBONUCLEASE (APURINIC OR APYRIMIDIC)	
DEOXYRIBONUCLEASE II	92% 5 mM
DESULFOGLUCINOLATE SULFOTRANSFERASE	> 10 mM
α-DEXTRIN ENDO-1,6-α-GLUCOSIDASE	

DIACYLGLYCEROL O-ACYLTRANSFERASE		
DIPEPTIDYL-PEPTIDASE 3		
DISCADENINE SYNTHASE		
DNA (CYTOSINE-5-)-METHYLTRANSFERASE		
DNA TOPOISOMERASE		
DOLICHOL O-ACYLTRANSFERASE		60% 20 mM
DOLICHYL-PHOSPHATE β-D-MANNOSYLTRANSFERASE		
ENDO-1,4-β-XYLANASE		
trans-2-ENOYL-CoA REDUCTASE (NAD)		
FERROCHELATASE	C	*K*i = 15 μM
FLAVONE APIOSYLTRANSFERASE		
FLAVONOL 3-O-GLUCOSYLTRANSERASE		
FLAVONOL-3O-GLYCOSIDE XYLOSYLTRANSFERASE		
FORMALDEHYDE DEHYDROGENASE		
FORMALDEHYDE TRANSKETOLASE		
β-FRUCTOFURANOSIDASE		
α-L-FUCOSIDASE		
GALACTINOL GALACTOSYLTRANSFERASE		
GALACTINOL-RAFFINOSE GALACTOSYLTRANSFERASE		
α-GALACTOSIDASE		
GALACTOSYLCERAMIDE SULFOTRANSFERASE		> 20 mM
GALACTURAN 1,4-α-GALACTURONIDASE		
GDP-MANNOSE 4,6-DEHYDRATASE		*I*50 = 16 mM
GERANYLGERANYLPYROPHOSPHATASE		
GIBBERELLIN 3β-DIOXYGENASE		
GLIOCLADIUM PROTEINASE		
1,4-α-GLUCAN BRANCHING ENZYME		
GLUCAN ENDO-1,3-α-GLUCOSIDASE		100% 1 mM
GLUCAN ENDO-1,3-β-GLUCOSIDASE		
GLUCAN 1,3-β-GLUCOSIDASE		
GLUCAN 1,4-α-GLUCOSIDASE		12% 1 mM
GLUCOSAMINE-PHOSPHATE N-ACETYLTRANSFERASE		
GLUTAMATE-AMMONIA LIGASE		> 2mM
GLUTAMATE-CYSTEINE LIGASE		
GLUTAMATE DEHYDROGENASE (NADP)		
GLUTAMATE FORMIMINOTRANSFERASE		
GLUTAMATE SYNTHASE (FERREDOXIN)		
GLUTAMATE SYNTHASE (NADPH)		
GLUTAMYL AMINOPEPTIDASE		
γ-GLUTAMYL HYDROLASE		
GLUTATHIONE-CYSTINE TRANSHYDROGENASE		
GLYCEROL-1-PHOSPHATASE		28% 5 mM
GLYCEROL-2-PHOSPHATASE		
GLYCEROPHOSPHOCHOLINE PHOSPHODIESTERASE		
GLYCERYL-ETHER MONOOXYGENASE		
GLYCINE C-ACETYLTRANSFERASE		
GLYCOGEN SYNTHASE (CASEIN) KINASE	C	*I*50 = 55 μM
GLYCOLIPID 2-α-MANNOSYLTRANSFERASE		
GLYCOPEPTIDE α-N-ACETYLGALACTOSAMINIDASE		44% 2.5 mM
GLYCOPROTEIN O-FATTY-ACYLTRANSFERASE		
GMP REDUCTASE		
GTP CYCLOHYDROLASE I		
2-HALOACID DEHALOGENASE		54% 1 mM
HEPAROSAN-N-SULFATE-GLUCURONATE 5-EPIMERASE		
HIPPURATE HYDROLASE		
HISTONE ACYLTANSFERASE		
HOMOCITRATE SYNTHASE		
HYALURONATE LYASE		
3-HYDROXYBUTYRATE DEHYDROGENASE		
6β-HYDROXYHYOSCYAMINE EPOXIDASE		
HYDROXYLAMINE OXIDASE		
6-HYDROXYMELLEIN-O-METHYLTRANSFERASE		
HYDROXYMETHYLBILANE SYNTHASE		36% 500 μM
HYOSCYAMINE (6S)-DIOXYGENASE		

HYOSCYAMINE (6S)-DIOXYGENASE		
INDOLE 3-ACETALDEHYDE REDUCTASE (NADPH)		
INDOLYLACETYL-MYO-INOSITOL GALACTOSYLTRANSFERASE		
INORGANIC PYROPHOSPHATASE		39% 100 μM
INORGANIC PYROPHOSPHATASE		Ki = 60 μM
INOSITOL-1,4-BISPHOSPHATE 1-PHOSPHATASE		I50 = 6 μM
myo-INOSITOL-1(or 4)-MONOPHOSPHATASE	C	Ki = 2 μM
INOSITOL POLYPHOSPHATE 4-PHOSPHATASE	C	I50 = 1.5 mM
ISOFLAVONE 4′-O-METHYLTRANSFERASE		
3-ISOPROPYLMALATE DEHYDROGENASE		
KALLIKREIN TISSUE		
KERATAN-SULFATE ENDO-1,4-β-GALACTOSIDASE		100% 10 mM
LACTALDEHYDE DEHYDROGENASE		
D-LACTALDEHYDE DEHYDROGENASE 1		
D-LACTALDEHYDE DEHYDROGENASE 2		
1,4-LACTONASE		
LEUCINE N-ACETYLTRANSFERASE		
LEUCINE DEHYDROGENASE		
LEUCYL AMINOPEPTIDASE		I50 = 20 mM
LEUCYLTRANSFERASE		
LEVANASE		
LYSINE 2-MONOOXYGENASE		
LYSOPHOSPHOLIPASE		69% 10 mM
LYSYL AMINOPEPTIDASE		
MALATE DEHYDROGENASE		100% 333 μM
MANNAN 1,4-β-MANNOBIOSIDASE		29% 2 mM
MANNOSE-6-PHOSPHATE ISOMERASE		
α-MANNOSIDASE		42% 10 mM
MEMBRANE ALANINE AMINOPEPTIDASE		
METHIONINE DECARBOXYLASE		
METHIONINE-tRNA LIGASE		
4-METHYLGLUTAMINE SYNTHETASE		
N^6-METHYL-LYSINE OXIDASE		
(S)-2-METHYLMALATE DEHYDRATASE		
MEVALONATE KINASE		> 2.5 mM
MONODEHYDROASCORBATE REDUCTASE (NADH)		
MUCINAMINYLSERINE MUCINAMINIDASE		44% 2.5 mM
NAD ADP-RIBOSYLTRANSFERASE		
NAD(P) TRANSHYDROGENASE (AB-SPECIFIC)		
NAD SYNTHASE		
NICOTINAMIDASE		
3′-NUCLEOTIDASE		
5′-NUCLEOTIDASE		
NUCLEOTIDE PYROPHOSPHATASE		9% 10 mM
OLIGO-1,6-GLUCOSIDASE		
ORNITHINE CARBAMOYLTRANSFERASE		> 10 mM
ORNITHINE DECARBOXYLASE		
OXALATE OXIDASE		
2-OXOALDEHYDE DEHYDROGENASE (NAD)		
2-OXOALDEHYDE DEHYDROGENASE (NADP) I		I50 = 5 mM
2-OXOALDEHYDE DEHYDROGENASE (NADP) II		11% 5 mM
3-OXOSTEROID Δ^1-DEHYDROGENASE		
PALMITOYL-CoA HYDROLASE		40% 2.4 mM
PARAOXON HYDROLASE		Ki = 110 μM
PECTATE LYASE		
PEPTIDE α-N-ACETYLTRANSFERASE		
PEPTIDOGLYCAN GLUCOSYLTRANSFERASE		
PEPTIDYL-DIPEPTIDASE A		39% 30 mM
PEROXIDASE		I50 = 52 μM
PEROXIDASE (MANGANESE INHIBITED)		
PHENOL β-GLUCOSYLTRANSFERASE		
PHENYLETHANOLAMINE N-METHYLTRANSFERASE		
PHOSPHATASE ALKALINE		
PHOSPHATE ACETYLTRANSFERASE		

PHOSPHATIDATE PHOSPHATASE	53% 47%
PHOSPHATIDYLGLYCEROPHOSPHATASE	80% 100 μM
PHOSPHATIDYLINOSITOL 3-KINASE (LIPID KINASE ACTIVITY)	
PHOSPHOACETYLGLUCOSAMINE MUTASE	100% 1 mM
PHOSPHOADENYLYLSULFATASE	71% 200 μM
PHOSPHODIESTERASE	
PHOSPHOENOLPYRUVATE CARBOXYKINASE (PYROPHOSPHATE)	*K*i = 140 μM
6-PHOSPHO-β-GALACTOSIDASE	50% 10 mM
PHOSPHOGLYCERATE PHOSPHATASE	
PHOSPHOKETOLASE	
PHOSPHOLIPASE C	
PHOSPHOLIPASE D	
PHOSPHOMANNOMUTASE	
PHOSPHOPYRUVATE HYDRATASE	*K*i = 31 μM
PHOSPHORYLASE PHOSPHATASE	
Δ1PIPERIDINE-2-CARBOXYLATE RDUCTASE	
POLYGALACTURONASE	
PRENYL-PYROPHOSPHATEASE	
PROCOLLAGEN N-ENDOPEPTIDASE	49% 5 mM
PROLYL OLIGOPEPTIDASE	65% 100 μM
PROPANEDIOL DEHYDRATASE	
PROTEIN-ARGININE N-METHYLTRANSFERASE	
PROTEINASE ASPARTIC ASPERGILLUS NIGER VAR.MARCOSPORUS	
PROTEINASE In	76% 1 mM
PROTEIN TYROSINE-PHOSPHATASE	75% 1 mM
PROTEIN TYROSINE-PHOSPHATASE CD45	37% 1 mM
PROTEIN TYROSINE-PHOSPHATASE (Mg^{2+} DEPENDENT)	71% 250 μM
PROTEIN TYROSINE SULFOTRANSFERASE	> 30 mM
PYRIDOXINE 4-DEHYDROGENASE	
PYRROLINE-5-CARBOXYLATE REDUCTASE	
PYRUVATE DEHYDROGENASE (LIPOAMIDE)	
QUEUINE tRNA-RIBOSYLTRANSFERASE	
RENILLA-LUCIFERIN 2-MONOOXYGENASE	
RHODOTORULAPEPSIN	13% 5 mM
RIBONUCLEASE H	*I*50 = 50 μM
RIBONUCLEASE III	
RIBONUCLEASE T1	
RIBONUCLEASE T2	
RIBOSYLPYRIMIDINE NUCLEOSIDASE	
SACCHAROPINE DEHYDROGENASE (NAD, L-LYSINE FORMING)	
SACCHAROPINE DEHYDROGENASE (NADP, L-LYSINE FORMING)	
SALICYL ALCOHOL β-D-GLUCOSYLTRANSFERASE	20% 2 mM
SARCOSINE OXIDASE	37% 2 mM
SECONDARY-ALCOHOL OXIDASE	
SERINE CARBOXYPEPTIDASE	
SERRALYSIN	
SIALIDASE	*I*50 = 2 mM
SINAPOYLGLUCOSE-CHOLINE O-SINAPOYLTRANSFERASE	> 5 mM
SPLEEN EXONUCLEASE	
STAPHYLOCOCCAL CYSTEINE PROTEINASE	
STEROL 3β-GLUCOSYLTRANSFERASE	
STIZOLOBATE SYNTHASE	
STIZOLOBINATE SYNTHASE	
SUCROSE 1^F-FRUCTOSYLTRANSFERASE	44% 10 mM
SULFATE ADENYLYLTRANSFERASE	
THETIN-HOMOCYSTEINE S-METHYLTRRANSFERASE	
THIAMIN PYRIDINYLASE	
L-THREONINE 3-DEHYDROGENASE	
THYMIDINE KINASE	
TRIACYLGLYCEROL LIPASE	
TRIMETHYLAMINE-N-OXIDE REDUCTASE	
TRIMETHYLYSINE DIOXYGENASE	
TRIPEPTIDE AMINOPEPTIDASE	48% 1 mM
TRIPEPTIDE AMINOPEPTIDASE	54% 1 mM

TRITRIRACHIUM ALKALINE PROTEINASE
TYROSINE TRANSAMINASE
UDP-N-ACETYLGLUCOSAMINE 2-EPIMERASE
UDP-N-ACETYLMURAMATE DEHYDROGENASE
UDP-GLUCOSE-HEXOSE-1-PHOSPHATE URIDYLYLTRANSFERASE
URATE OXIDASE
UREASE
URIDINE NUCLEOSIDASE
UROPORPHYRINOGEN-III SYNTHASE 94% 10 μM
VALINE DEHYDROGENASE (NADP)
XAA-HIS DIPEPTIDASE *I*50 = 100 μM
XYLAN 1,4-β-XYLOSIDASE
XYLOSE ISOMERASE
D-XYLULOSE REDUCTASE

$MNCL_2$
CYCLAMATE SULFOHYDROLASE 59% 10 mM
GLYCEROL 2-DEHYDROGENASE
THYMIDYLATE 5′-PHOSPHATASE

MN(II)TETRAKIS(1-METHYL-4-PYRIDYL)PORPHYRIN
GUANYLATE CYCLASE *I*50 = 800 nM
NITRIC-OXIDE SYNTHASE I *I*50 = 5.5 μM
NITRIC-OXIDE SYNTHASE II *I*50 = 9 μM
NITRIC-OXIDE SYNTHASE III *I*50 = 23 μM

MNIOPETAL A
RNA-DIRECTED DNA POLYMERASE *I*50 = 41 μM
RNA-DIRECTED DNA POLYMERASE *I*50 = 4 μM

MNIOPETAL B
RNA-DIRECTED DNA POLYMERASE *I*50 = 42 μM
RNA-DIRECTED DNA POLYMERASE *I*50 = 1.7 μM

MNIOPETAL C
RNA-DIRECTED DNA POLYMERASE *I*50 = 7 μM
RNA-DIRECTED DNA POLYMERASE *I*50 = 93 μM

MNIOPETAL D
RNA-DIRECTED DNA POLYMERASE *I*50 = 77 μM
RNA-DIRECTED DNA POLYMERASE *I*50 = 6 μM

MNIOPETAL E
RNA-DIRECTED DNA POLYMERASE *I*50 = 59 μM

MNIOPETAL F
RNA-DIRECTED DNA POLYMERASE *I*50 = 30 μM

MNNADATE
HYDROXYMETHYLBILANE SYNTHASE

Mo^{2+}
BENZOATE 4-MONOOXYGENASE
CARBOSINE SYNTHASE

Mo^{5+}
INDOLEACETALDOXIME DEHYDRATASE 100% 1 mM

Mo^{6+}
ALCOHOL OXIDASE
CYCLOMALTODEXTRIN GLUCANOTRANSFERASE
METHANOL OXIDASE
STIZOLOBATE SYNTHASE
STIZOLOBINATE SYNTHASE

Mo^{2-}
GLYCOLALDEHYDE DEHYDROGENASE

MOCLOBEMIDE

AMINE OXIDASE MAO		
AMINE OXIDASE MAO-A		*I*50 = 6 μM

MOENOMYCIN

PEPTIDOGLYCAN GLUCOSYLTRANSFERASE		
α,α-TREHALOSE-PHOSPHATE SYNTHASE		*I*50 = 10 μM

MOENOMYCIN A

TRANSGLYCOSYLASE		78% 100ng/ml

MOLINATE SULPHOXIDE

ALDEHYDE DEHYDROGENASE (NAD) (LOW Km)		*I*50 = 1 μM

MOLVIZARIN

NADH DEHYDROGENASE (UBIQUINONE)		*K*i = 1 nM

MOLYBDATE

AMINOPEPTIDASE		
ARYLSULFATASE	C	*K*i = 130 μM
ASPARAGINE-AMMONIA LIGASE (ADP-FORMING)		80% 2 mM
FAD PYROPHOSPHATASE		
GLUTATHIONE-CYSTINE TRANSHYDROGENASE		
myo-INOSITOL-1(or 4)-MONOPHOSPHATASE		
ISOCITRATE DEHYDROGENASE (NAD)		
MONOTERPENYL-PYROPHOSPHATASE		
NITRATE REDUCTASE (NADH)		
PHOSPHATASE ACID	C	*K*i = 37 μM
PHOSPHATASE ACID	C	*K*i = 41 nM
PHOSPHATASE ACID	C	*K*i = 180 nM
PHOSPHATASE ACID		100% 10 mM
PHOSPHATASE ACID	C	*K*i = 900 nM
PHOSPHATASE ACID PURPLE		
PHOSPHATASE ALKALINE		39% 100 μM
PHOSPHATASE ALKALINE	C	*K*i = 440 μM
PHOSPHOAMIDASE		
6-PHOSPHOFRUCTO-2-KINASE		
PHOSPHOGLUCONATE DEHYDROGENASE (DECARBOXYLATING)		
3-PHOSPHOSHIKIMATE 1-CARBOXYVINYLTRANSFERASE		
PROTEIN TYROSINE-PHOSPHATASE	NC	95% 10 μM
PROTEIN TYROSINE-PHOSPHATASE CDC25B		88% 100 μM
PROTEIN TYROSINE-PHOSPHATASE (Mg^{2+} DEPENDENT)		66% 1 mM
QUINATE DEHYDROGENASE		
TRIMETHYLAMINE-N-OXIDE REDUCTASE		
TRIPEPTIDE AMINOPEPTIDASE		

MOMORDICA CHARANTIA ELASTASE INHIBITOR I

ELASTASE PANCREATIC		*K*i = 970 μM

MOMORDICA CHARANTIA ELASTASE INHIBITOR II

ELASTASE PANCREATIC		*K*i = 9.4 μM

MOMORDICA CHARANTIA ELASTASE INHIBITOR III

ELASTASE PANCREATIC		*K*i = 4 μM

MOMORDICA CHARANTIA ELASTASE INHIBITOR IV

ELASTASE PANCREATIC		*K*i = 4.7 μM

MOMORDICA CHARANTIA TRYPSIN INHIBITOR I

TRYPSIN		*K*i = 12 nM

MOMORDICA CHARANTIA TRYPSIN INHIBITOR-I

COAGULATION FACTOR Xa		*K*i = 100 μM
COAGULATION FACTOR XIa		*K*i = 18 μM
COAGULATION FACTOR XIIa		*K*i = 13 nM
KALLIKREIN PLASMA		*K*i = 110 μM

MOMORDICA CHARANTIA TRYPSIN INHIBITOR II
TRYPSIN $Ki = 820$ pM

MOMORDICA CHARANTIA TRYPSIN INHIBITOR-II
COAGULATION FACTOR Xa $Ki = 1.4$ μM
COAGULATION FACTOR XIIa $Ki = 56$ nM
KALLIKREIN PLASMA $Ki = 100$ μM

MOMORDICA CHARANTIA TRYPSIN INHIBITOR III
TRYPSIN $Ki = 190$ μM

MOMORDICA CHARANTIA TRYPSIN INHIBITOR-III
COAGULATION FACTOR Xa $Ki = 59$ μM
COAGULATION FACTOR XIIa $Ki = 1.6$ μM
KALLIKREIN PLASMA $Ki = 140$ μM

MONANKARIN A
AMINE OXIDASE MAO $I50 = 16$ μM

MONANKARIN C
AMINE OXIDASE MAO $I50 = 11$ μM

MONILIFORMIN
ACETOLACTATE SYNTHASE IR $Ki = 170$ μM
OXOGLUTARATE DEHYDROGENASE (LIPOAMIDE) IR $Ki = 40$ μM
PYRUVATE DECARBOXYLASE IR $Ki = 1$ mM
PYRUVATE DEHYDROGENASE (LIPOAMIDE)

N^G-MONIMETHYLARGININE ACETATE
NADPH DEHYDROGENASE 14% 1 mM

MONIODOACETIC ACID
ADENINE DEAMINASE

MONKEY BRAIN MONOAMINE OXIDASE INHIBITOR
AMINE OXIDASE MAO

MONKEY KIDNEY SERINE PROTEINASE INHIBITOR
COAGULATION FACTOR Xa
u-PLASMINOGEN ACTIVATOR
THROMBIN
TRYPSIN

MONOACETYLENIC ARACHIDONIC ACID
CYCLOOXYGENASE
LIPOXYGENASE

MONOACYLAMINO-GLYCERO-PHOSPHOCHOLINES
PHOSPHOLIPASE A2

(R)-MONO(2-AMINOETHYL)MONO[2-[(HEPTYLHYDROXYPHOSPHINYL)OXY]-3-(OCTYLOXY)PROPYL]PHOSPHATE
PHOSPHOLIPASE A2 $I50 = 1{,}7$ μM

MONOAMINOGUANIDINE
ALDEHYDE REDUCTASE NC $Ki = 25$ mM

N^2-MONOBUTYRYL-cGMP
3′,5′-CYCLIC-GMP PHOSPHODIESTERASE $I50 = 550$ μM

MONOCHLORAMINE
CATALASE

MONOCHLORODICARBONONARBORON
H^+-TRANSPORTING ATP SYNTHASE

MONOCHLOROTRIFLUORO-p-BENZOQUINONE
THIOGLUCOSIDASE

MONOCROTOPHOS
ACETYLCHOLINESTERASE — $I50 = 90\ \mu M$

MONOCYTE/NEUTROPHIL ELASTASE INHIBITOR
CATHEPSIN G
ELASTASE
PROTEINASE 3

MONODANSYLCYSTEAMINE
GLUTAMATE-CYSTEINE LIGASE

8′-MONOENE ANACARDIC ACID
LIPOXYGENASE 1 — $I50 = 50\ \mu M$

10′-MONOENE ANACARDIC ACID
ARACHIDONATE 5-LIPOXYGENASE — $I50 = 6\ \mu M$
PROSTAGLANDIN-ENDOPEROXIDE SYNTHASE

MONOETHYLFUMARATE
PEPTIDYLGLYCINE MONOOXYGENASE — SSI — $Ki = 1.3\ \mu M$

MONO(2-ETHYLHEXYL)PHTHALATE
GLUTATHIONE TRANSFERASE — $I50 = 320\ \mu M$
PHOSPHOLIPASE A2 — $Ki = 370\ \mu M$

N-MONOFLUOROACETYLNEURAMINIC ACID
N-ACETYLNEURAMINATE LYASE — $Ki = 69\ \mu M$

2-MONOFLUOROAGMATINE
ARGININE DECARBOXYLASE — IRR — $Ki = 9\ \mu M$

MONOFLUOROFUMARATE
ARGININOSUCCINATE LYASE — IR

α-MONOFLUOROGLUTARIC ACID
GLUTAMATE DEHYDROGENASE (NAD(P))

2-MONOFLUORO HISTAMINE
HISTIDINE DECARBOXYLASE — SSI

2-MONOFLUORO 5-HYDROXYTRYPTOPHAN
AROMATIC-L-AMINO-ACID DECARBOXYLASE — SSI

2-MONOFLUOROMETHYL AGMATINE
ARGININE DECARBOXYLASE — SSI

2-MONOFLUOROMETHYL β-ALANINE
4-AMINOBUTYRATE TRANSAMINASE — SSI

2-MONOFLUOROMETHYL 4-AMINOBUTYRIC ACID
4-AMINOBUTYRATE TRANSAMINASE — SSI

2-MONOFLUOROMETHYL 5-AMINOPENTANOIC ACID
4-AMINOBUTYRATE TRANSAMINASE — SSI

2-MONOFLUOROMETHYLARGININE
ARGININE DECARBOXYLASE — IRR — $Ki = 65\ \mu M$

(E)-2-MONOFLUOROMETHYLDEHYDROARGININE
ARGININE DECARBOXYLASE — IRR — $Ki = 34\ \mu M$

Δ-MONOFLUOROMETHYLDEHYDROORNITHINE
ORNITHINE DECARBOXYLASE — IR — $Ki = 13\ \mu M$

2-MONOFLUOROMETHYL-trans-DEHYDROORNITHINE
ORNITHINE DECARBOXYLASE — SSI

Δ-MONOFLUOROMETHYLDEHYDROORNITHINE METHYLESTER
ORNITHINE DECARBOXYLASE — IR — $Ki = 4.3\ \mu M$

2-MONOFLUOROMETHYL DEHYDROPUTRESCINE
ORNITHINE DECARBOXYLASE SSI

2-MONOFLUOROMETHYL GLUTAMIC ACID
GLUTAMATE DECARBOXYLASE SSI

(S)-2-MONOFLUOROMETHYL HISTIDINE
HISTIDINE DECARBOXYLASE SSI

MONOFLUOROMETHYLORNITHINE
ORNITHINE DECARBOXYLASE

2-MONOFLUOROMETHYL PUTRESCINE
ORNITHINE DECARBOXYLASE SSI

6-MONOFLUOROMEVALONIC ACID
DIPHOSPHOMEVALONATE DECARBOXYLASE $Ki = 10$ nM

MONOFLUOROPUTRESCINE
ORNITHINE DECARBOXYLASE

7-MONOFLUOROSPERMIDINE
SPERMINE SYNTHASE

MONOGYLCERIDES
TRIACYLGLYCEROL LIPASE C

MONOHYDROXYETHYLMONOVINYLDEUTEROPORPHYRIN
GLUTATHIONE TRANSFERASE

MONOIODOACETIC ACID
N-ACETYL-β-ALANINE DEACYLASE 73% 1 mM
β-N-ACETYLHEXOSAMINIDASE
ALDEHYDE DEHYDROGENASE (PYRROLOQUINOLINE-QUINONE) 15% 500 μM
β-GALACTOSIDASE
α-MANNOSIDASE 20% 100 μM
PROTEIN-ARGININE DEIMINASE

MONOLYSOCARDIOLIPIN
PHOSPHOLIPASE A2

MONOMETHYL-4-ACETAMIDO-2,4-DIDEOXY-D-glycero-D-galacto-β-OCTOPYRANOSONO-1-1PHOSPHONATE
SIALIDASE $I50 = 200$ μM

MONOMETHYLACETYLPHOSPHONIC ACID
PYRUVATE DEHYDROGENASE (LIPOAMIDE) C $Ki = 50$ nM

MONOMETHYLAMINOPROPIONITRILE
PROTEIN-LYSINE 6-OXIDASE $I50 = 10$ mM

L-N^G-MONOMETHYLARGININE
NITRIC-OXIDE SYNTHASE I $I50 = 338$ nM
NITRIC-OXIDE SYNTHASE II $I50 = 3.5$ μM
NITRIC-OXIDE SYNTHASE III $I50 = 950$ nM

L-N^g-MONOMETHYL ARGININE
NITRIC-OXIDE SYNTHASE $I50 = 2.4$ μM

N^G-MONOMETHYLARGININE
NITRIC-OXIDE SYNTHASE $I50 = 13$ μM
NITRIC-OXIDE SYNTHASE IR $I50 = 30$ μM
NITRIC-OXIDE SYNTHASE $I50 = 3$ μM
NITRIC-OXIDE SYNTHASE I $Ki = 650$ nM
NITRIC-OXIDE SYNTHASE I $I50 = 8.3$ μM
NITRIC-OXIDE SYNTHASE I $Ki = 700$ nM
NITRIC-OXIDE SYNTHASE II $I50 = 42$ μM
NITRIC-OXIDE SYNTHASE II $Ki = 3.9$ μM
NITRIC-OXIDE SYNTHASE II $I50 = 18$ μM

MONOMETHYLETHANOLAMINE
ETHANOLAMINE KINASE

MONOMETHYLPHOSPHATE
PEPTIDYL DIPEPTIDASE I 50% 50 mM

MONONUCLEOTIDES
L-LACTATE DEHYDROGENASE

MONOOELIN
PHOSPHATIDYLCHOLINE-DOLICHOL O-ACYLTRANSFERASE

MONOOLEIN
PHOSPHOLIPASE A2

2-MONOOLEINGLYCEROL
LIPOPROTEIN LIPASE 31% 80 μM

1-MONOOLEOYLGLYCEROL
GLYCEROL-3-PHOSPHATE O-ACYLTRANSFERASE

2-MONOOLEOYLGLYCEROL
GLYCEROL-3-PHOSPHATE O-ACYLTRANSFERASE

1-MONOOLEOYLGLYCEROLAMIDE
GLYCEROL-3-PHOSPHATE O-ACYLTRANSFERASE

2-MONOOLEOYLGLYCEROLAMIDE
GLYCEROL-3-PHOSPHATE O-ACYLTRANSFERASE

1-MONOOLEOYLGLYCEROL-3-PHOSPHATE
GLYCEROL-3-PHOSPHATE O-ACYLTRANSFERASE

1-MONOOLEYLGLYCEROL
DIACYLGLYCEROL KINASE C *K*i = 91 μM

2-MONOPALMITOYLGLYCEROL
LIPASE (HORMONE SENSITIVE) NC *I*50 = 500 μM

MONOSTERIN
PHOSPHATIDYLCHOLINE-DOLICHOL O-ACYLTRANSFERASE

MONOSULPHATE DISSACHARIDE
KERATAN-SULFATE ENDO-1,4-β-GALACTOSIDASE

MONOVALENT IONS
ACETOACETATE DECARBOXYLASE

MONOVANADATE
6-PHOSPHOFRUCTO-2-KINASE *I*50 = 260 μM

MOPS
CHODROTIN 4-SULFOTRANSFERASE

MOPS BUFFER
CREATINE KINASE

MORANTEL
FUMARATE REDUCTASE (NADH) C

MORDANT BLUE 9
PROTEINASE HIV-1 *I*50 = 68 μM

MORDANT BROWN 1
PROTEINASE HIV-1 *I*50 = 4 μM

MORDANT ORANGE 6
PROTEINASE HIV-1 *I*50 = 34 μM

MORDANT ORANGE 10

PROTEINASE HIV-1		*I*50 = 25 μM

MORE

GLUCAN ENDO-1,3-β-GLUCOSIDASE		

MORELLOFLAVONE

PHOSPHOLIPASE A2 (SECRETORY)		*I*50 = 600 nM
PHOSPHOLIPASE A2 (SECRETORY)		*I*50 = 900 nM

MORIN

ADENOSINE DEAMINASE		*I*50 = 200 μM
ALCOHOL DEHYDROGENASE (NADP)		
ARACHIDONATE 5-LIPOXYGENASE		*I*50 = 160 μM
3′,5′-CYCLIC-NUCLEOTIDE PHOSPHODIESTERASE		51% 1 mM
CYCLOOXYGENASE		*I*50 = 180 μM
DNA LIGASE I		*I*50 = 236 μM
2-ENOATE REDUCTASE		
ETHOXY RESORUFIN DEALKYLASE	C	
GLUTATHIONE REDUCTASE (NADPH)		30% 100 μM
GLUTATHIONE TRANSFERASE		*I*50 = 41 μM
GLUTATHIONE TRANSFERASE		*I*50 = 14 μM
GLUTATHIONE TRANSFERASE		48% 100 μM
INTEGRASE		*I*50 = 76 μM
IODIDE PEROXIDASE		*I*50 = 2.1 μM
LACTOYLGLUTATHIONE LYASE		*K*i = 30 μM
LIPOXYGENASE		89% 30 mM
NADH OXIDASE		*I*50 = 430 nmole
NAD(P)-ARGININE ADP-RIBOSYLTRANSFERASE		*I*50 = 300 μM
1-PHOSPHATIDYLINOSITOL-4-PHOSPHATE 5-KINASE	U	*K*i = 5.2 μM
1-PHOSPHATIDYLINOSITOL-4-PHOSPHATE 5-KINASE	C	*K*i = 9.4 μM
PROTEINASE HIV-1		*I*50 = 24 μM
PROTEIN KINASE C		*I*50 = 10 μM
PROTEIN TYROSINE KINASE p56lck		*I*50 = 50 μg/ml
RNA-DIRECTED DNA POLYMERASE		
SUCCIN OXIDASE		*I*50 = 730 nmole
TETRAHYDROBERBERINE OXIDASE		
(S)-TETRAHYDROPROTOBERBERINE OXIDASE		*I*50 = 180 μM
TRYPSIN		*K*i = 31 μM
XANTHINE OXIDASE		*I*50 = 10 μM

MORPHICETIPTIN

MEMBRANE ALANINE AMINOPEPTIDASE		*K*i = 173 μM

MORPHOLINE

CYTOCHROME P450 2E1	U	*K*i = 8.7 mM

5-[(4-MORPHOLINOPHENYL)THIO]-2,4-DIAMINOQUINAZOLINE

DIHYDROFOLATE REDUCTASE		*I*50 = 130 nM
DIHYDROFOLATE REDUCTASE		*I*50 = 70 μM

4-MORPHOLINOSULPHONIC ACID

β-LACTAMASE		*K*i = 23 mM

N-(4-MORPHOLINOSULPHONYL)-4-PHE-N-[1(S)-(CYCLOHEXYLMETHYL)-3,3-DIFLUORO-2(R)HYDROXY-4-OXO4[[2(2PYRIDINYL)ETHYL]AMINO-BUTYL]4,5DIDEHYDRO-NORVALAMIDE

RENIN		*I*50 = 210 pM

MORPHOLINOUNDECYLCARBAMOYLOXYESEROLINE

ACETYLCHOLINESTERASE		*I*50 = 40 nM

N-[4(4-MORPHOLINYLSULPHONYL)BENZOYL]-L-VAL-N-[3,3,4,4,4-PENTAFLUORO-1(1-METHYLETHYL)2-OXOBUTYL]-L-PROLINAMIDE

ELASTASE LEUKOCYTE		*K*i = 2 nM

N-(4-MORPHOLINYLSUPHONYL)-PHE-LYS(C=S(NHCH$_3$))-ACHPA-AEM

RENIN		*I*50 = 3.2 nM

MORUSIN

ARACHIDONATE 5-LIPOXYGENASE		*I*50 = 2.9 µM
ARACHIDONATE 12-LIPOXYGENASE		*I*50 = 3.4 µM
ARACHIDONATE 15-LIPOXYGENASE		*I*50 = 3.3 µM
CYCLOOXYGENASE		*I*50 = 1.6 µM

(18Z)-29-MOS: (6Z,10E,14E,18E)-22,23-EPOXY-2,10,15,19,23-PENTAMETHYL-6-VINYL-2,6,10,14,18-TETRACOSAPENTANE

LANOSTEROL SYNTHASE		*I*50 = 400 nM
LANOSTEROL SYNTHASE		*I*50 = 1 µM

MOTAPIZONE

3′,5′-CYCLIC-NUCLEOTIDE PHOSPHODIESTERASE Ib		*I*50 = 400 µM
3′,5′-CYCLIC-NUCLEOTIDE PHOSPHODIESTERASE IB		*I*50 = 398 µM
3′,5′-CYCLIC-NUCLEOTIDE PHOSPHODIESTERASE II		*I*50 = 617 µM
3′,5′-CYCLIC-NUCLEOTIDE PHOSPHODIESTERASE II (cycGMP STIMULATED)		*I*50 = 630 µM
3′,5′-CYCLIC-NUCLEOTIDE PHOSPHODIESTERASE III		*I*50 = 759 nM
3′,5′-CYCLIC-NUCLEOTIDE PHOSPHODIESTERASE III (cycGMP INHIBITED)		*I*50 = 800 nM
3′,5′-CYCLIC-NUCLEOTIDE PHOSPHODIESTERASE III (cycGMP INHIBITED)		*I*50 = 34 nM
3′,5′-CYCLIC-NUCLEOTIDE PHOSPHODIESTERASE IV		*I*50 = 138 µM
3′,5′-CYCLIC-NUCLEOTIDE PHOSPHODIESTERASE IV		*I*50 = 54 µM
3′,5′-CYCLIC-NUCLEOTIDE PHOSPHODIESTERASE IV		*I*50 = 54 µM
3′,5′-CYCLIC-NUCLEOTIDE PHOSPHODIESTERASE V		*I*50 = 79 µM
3′,5′-CYCLIC-NUCLEOTIDE PHOSPHODIESTERASE V (cycGMP)		*I*50 = 79 µM

MOTUPORIN

PHOSPHOPROTEIN PHOSPHATASE 1		> 1 nM
PHOSPHOPROTEIN PHOSPHATASE 1c		*I*50 = 600 pM
PHOSPHOPROTEIN PHOSPHATASE 2Ac		*I*50 = 600 pM

MOUSE LIVER DNASE INHIBITOR

DEOXYRIBONUCLEASE II		

MOUSE SEMINALSECRETORY TRYPSIN INHIBITOR

TRYPSIN	C	*K*i = 150 pM

MOUSE SEMINAL VESICLE TRYPSIN INHIBITOR

TRYPSIN		

MOXAVERINE

3′,5′-CYCLIC-NUCLEOTIDE PHOSPHODIESTERASE (cycAMP) (Ca^{2+}/CALMODULIN)		*I*50 = 18 µM

MOXIPRIL

PEPTIDYL-DIPEPTIDASE A		*I*50 = 2.1 nM

MP 134

PROTEINASE HIV		*K*i = 4.8 nM

MP 167

PROTEINASE HIV		*K*i = 700 pM

$MpDp^+$

BUFURALOL-1′-HYDROXYLASE (CYT P-450dbl)	C	*I*50 = 40 µM

MPP^+

AMINE OXIDASE MAO-A	C	*K*i = 3 µM
AMINE OXIDASE MAO-B	C	*K*i = 230 µM
BUFURALOL-1′-HYDROXYLASE (CYT P-450dbl)	C	*I*50 = 124 µM

MPP^+(4PHPY-C1)

NADH OXIDASE		*I*50 = 4 µM

MPTP

BUFURALOL-1′-HYDROXYLASE (CYT P-450dbl)	C	*I*50 = 37 µM

MPTP OXIDATION PRODUCTS

AMINE OXIDASE MAO		

MR 889
ELASTASE LEUKOCYTE C Ki = 1.4 µM
PROTEINASES SERINE

MR 20492
AROMATASE Ki = 10 nM

MR 387A
MEMBRANE ALANINE AMINOPEPTIDASE I50 = 198 nM

MR 387B
AMINOPEPTIDASE B I50 = 260 nM
MEMBRANE ALANINE AMINOPEPTIDASE I50 = 164 nM

MS 282a
3′,5′-CYCLIC-NUCLEOTIDE PHOSPHODIESTERASE (Ca^{2+}/CALMODULIN) I50 = 4.2 µM
MYOSIN-LIGHT-CHAIN KINASE I50 = 3.8 µM

MS 282b
MYOSIN-LIGHT-CHAIN KINASE I50 = 5.2 µM

MS 347a
MYOSIN-LIGHT-CHAIN KINASE I50 = 9.2 µM
PROTEIN KINASE C I50 = 16 µM

MS 444
MYOSIN-LIGHT-CHAIN KINASE I50 = 10 µM

MS 681a
MYOSIN-LIGHT-CHAIN KINASE I50 = 110 nM

MS 681b
MYOSIN-LIGHT-CHAIN KINASE I50 = 290 nM

MS 681c
MYOSIN-LIGHT-CHAIN KINASE I50 = 95 nM

MS 681d
MYOSIN-LIGHT-CHAIN KINASE I50 = 260 nM

MS 857
3′,5′-CYCLIC-NUCLEOTIDE PHOSPHODIESTERASE (Ca^{2+}/CALMODULIN) I50 = 16 µM
3′,5′-CYCLIC-NUCLEOTIDE PHOSPHODIESTERASE (cGMP INHIBITABLE) I50 = 1.5 µM

MS 888
RNA-DIRECTED DNA POLYMERASE I50 = 200 µM

MS 1060
RNA-DIRECTED DNA POLYMERASE

MS 1126
RNA-DIRECTED DNA POLYMERASE

MS 347A
MYOSIN-LIGHT-CHAIN KINASE I50 = 9.2 µM

MST 16
DNA TOPOISOMERASE (ATP-HYDROLYSING) I50 = 300 µM

MTR-ASN(PEG5000-=me)-D-ADF-PIP
THROMBIN Ki = 600 nM
TRYPSIN Ki = 1.5 mM

MUCIDIN
NADH DEHYDROGENASE (UBIQUINONE) 16% 3.3 µM
SUCCINATE DEHYDROGENASE (UBIQUINONE)

MUCIN
PEPTIDYL-DIPEPTIDASE A

MUCOCHLORIC ACID
ASPARTATE-AMMONIA LIGASE 65% 5 mM

trans,trans-MUCONALDEHYDE
CYTOCHROME-C REDUCTASE (NADPH) 25% 1.5 mM

cis,cis-MUCONIC ACID
4-OXALOCROTONATE TAUTOMERASE

trans,trans-MUCONIC ACID
FUMARATE REDUCTASE (NADH)

MUGINEIC ACID
GELATINASE A $I50 = 410$ nM
GELATINASE B $I50 = 1$ μM

MU-LEU-HPH-PSI(CH=CH)SO_2PH)
CATHEPSIN K $Ki = 110$ nM
CATHEPSIN L $Ki = 120$ nM
CATHEPSIN S $Ki = 13$ nM

MU-LEU-HPH-VSPH
PROTEINASE CYSTEINE PLASMODIUM FALCIPARUM $I50 = 3$ nM
PROTEINASE CYSTEINE PLASMODIUM VINCKEI $I50 = 200$ nM

MU-PHE-ARG-VSPH
PROTEINASE CYSTEINE PLASMODIUM FALCIPARUM $I50 = 50$ nM
PROTEINASE CYSTEINE PLASMODIUM VINCKEI $I50 = 50$ nM

MU-PHE-HOMOPHE-FLUOROMETHYLKETONE
CATHEPSIN B

MURAMIC ACID
N-ACETYLMURAMOYL-L-ALANINE AMIDASE

MUREIDOMYCIN A
PHOSPHO-N-ACETYLMURAMOYL-PENTAPEPTIDE-TRANSFERASE $I50 = 50$ ng/ml

MUREIDOMYCIN C
PHOSPHO-N-ACETYLMURAMOYL-PENTAPEPTIDE-TRANSFERASE

MURINE HEPATOME MEMBRANE ASSOCIATED CYSTEINE PROTEINASE INHIBITOR
PROTEINASES CYSTEINE

MURIN FIBROBLAST PROTEIN KINASE R INHIBITOR
PROTEIN KINASE R

MUSCIMOL
4-AMINOBUTYRATE TRANSAMINASE

MUSCUS PROTEINASE INHIBITOR
CHYMOTRYPSIN A $Ki = 71$ pM
ELASTASE PANCREATIC $Ki = 67$ nM
TRYPSIN I $Ki = 500$ nM
TRYPSIN II $Ki = 18$ nM

MUSCEL PROTEINPHOSPHATASE INHIBITORS
PHOSPHOPROTEIN PHOSPHATASE

MUSTARD TRYPSIN INHIBITOR-2
α-CHYMOTRYPSIN Kd = 500 nM
β-TRYPSIN Kd = 160 pM

MUTASTEIN
GLUCAN SYNTHASE INSOLUBLE

MUZOLIMINE
PYRIDOXAL KINASE 27% 100 μM

MVT 101
PROTEINASE HIV-1 — $Ki = 780$ nM

MY 5445
3′,5′-CYCLIC-NUCLEOTIDE PHOSPHODIESTERASE (cycAMP) — C — $Ki = 915$ µM
3′,5′-CYCLIC-NUCLEOTIDE PHOSPHODIESTERASE (cycGMP) — C — $Ki = 1.3$ µM

MYCALOLIDE B
MYOSIN ATPASE Mg^{2+} — $I50 = 4$ µM

O-MYCAMINOSYLTYLONOLIDE
DEMETHYLMACROCIN O-METHYLTRANSFERASE

MYCELIANAMIDE
ACETYLCHOLINESTERASE

MYCENON
ISOCITRATE LYASE — C — $Ki = 7.4$ µM

MYCOBACTERIUM TUBERCULOSIS TREHALOSE PHOSPHATE SYNTHASE INHIBITOR
α,α-TREHALOSE-PHOSPHATE SYNTHASE

MYCOPHENOLIC ACID
IMP DEHYDROGENASE — UC — $Ki = 14$ µM
IMP DEHYDROGENASE — MIX — $Ki = 8$ nM
IMP DEHYDROGENASE 1 — $Ki = 37$ nM
IMP DEHYDROGENASE 1 — U — $Ki = 11$ nM
IMP DEHYDROGENASE 2 — $I50 = 14$ nM
IMP DEHYDROGENASE 2 — U — $Ki = 6$ nM
IMP DEHYDROGENASE 2 — $Ki = 9.5$ nM

MYCOPHENOLIC ACID GLUCURONIDE
IMP DEHYDROGENASE 2 — $I50 = 14$ µM

MYCORBIN
α,α-TREHALOSE-PHOSPHATE SYNTHASE

MYCOTOXIN
CALF THYMUS RIBONUCLEASE H

MYCRICETRIN
INTEGRASE — $I50 = 40$ µM

MYCROCYSTIN
PHOSPHOPROTEIN PHOSPHATASE 1 — $I50 = 300$ pM

MYCROCYSTIN LR
PHOSPHOPROTEIN PHOSPHATASE 2A — $I50 = 60$ nM

MYOCRISIN
INTERSTITIAL COLLAGENASE — NC — $Ki = 62$ nM

MYOGLOBIN
ALANYL AMINOPEPTIDASE
5-AMINOLEVULINATE SYNTHASE
AMINOPEPTIDASE B
CALPAIN I
CATHEPSIN B
CATHEPSIN H

MYOINOSITOL
GALACTINOL-RAFFINOSE GALACTOSYLTRANSFERASE
α-GALACTOSIDASE — NC — $Ki = 750$ mM

MYOINOSITOLPENTAKIS(DIHYDROGENPHOSPHATE)
myo-INOSITOL-1(or 4)-MONOPHOSPHATASE — 11% 500 µM

MYOINOSITOL-2-PHOSPHATE
1,2-CYCLIC-INOSITOL-PHOSPHATE PHOSPHODIESTERASE — $I50 = 4$ µM

MYOINOSITOLTETRAKIS(DIHYDROGENPHOSPHATE)

myo-INOSITOL-1(or 4)-MONOPHOSPHATASE		74% 500 μM

MYOINOSITOL-1,4,5-TRISPHOSPHORTHIOATE

INOSITOL-1,3,4,5-TETRAKISPHOSPHATE 3-PHOSPHATASE		

MYOSIN LIGHT CHAIN KINASE PEPTIDE (480-501)

MYOSIN-LIGHT-CHAIN KINASE		*I*50 = 46 nM

MYOTOXIN A

Ca^{2+}-TRANSPORTING ATPASE		

MYRICETIN

ADENOSINE DEAMINASE		*I*50 = 283 μM
ADENOSINE DEAMINASE		*I*50 = 32 μM
ALDEHYDE REDUCTASE		
ARACHIDONATE 5-LIPOXYGENASE		*I*50 = 13 μM
ARACHIDONATE 12-LIPOXYGENASE		*I*50 = 2 μM
ARYL SULFOTRANSFERASE P		*I*50 = 340 nM
CYCLOOXYGENASE		*I*50 = 56 μM
CYTOCHROME P450 1A1		*I*50 = 200 nM
DNA LIGASE I		*I*50 = 91 μM
GLUTATHIONE REDUCTASE (NADPH)		45% 100 μM
GLUTATHIONE TRANSFERASE		*I*50 = 25 μM
H^+-TRANSPORTING ATP SYNTHASE		
HYALURONOGLUCOSAMINIDASE		
INTEGRASE		*I*50 = 7.6 μM
IODIDE PEROXIDASE		*I*50 = 600 nM
LACTOYLGLUTATHIONE LYASE		*K*i = 5 μM
LIPOXYGENASE		91% 30 mM
MYOSIN-LIGHT-CHAIN KINASE	C	*K*i = 1.7 μM
NADH OXIDASE		*I*50 = 35 nmole
1-PHOSPHATIDYLINOSITOL 3-KINASE		*I*50 = 1.8 μM
PHOSPHOGLYCERATE KINASE		*I*50 = 1.1 μM
PROTEINASE HIV-1		*I*50 = 22 μM
PROTEIN KINASE A	C	*K*i = 28 μM
PROTEIN KINASE C	C	*K*i = 12 μM
PROTEIN KINASE C		*I*50 = 1 μM
PROTEIN KINASE CASEIN KINASE I	C	*K*i = 9 μM
PROTEIN KINASE CASEIN KINASE II	C	*K*i = 600 nM
PROTEIN TYROSINE KINASE (INSULIN RECEPTOR)		*K*i = 2.6 μM
PROTEIN TYROSINE KINASE p40		*I*50 = 31 μM
PROTEIN TYROSINE KINASE p56lck		*I*50 = 10 μg/ml
PROTEIN TYROSINE KINASE pp130fps		*K*i = 1.8 μM
RNA-DIRECTED DNA POLYMERASE		*K*i = 80 nM
SUCCIN OXIDASE		*I*50 = 45 nmole
TRYPSIN		*K*i = 2.8 μM
XANTHINE OXIDASE	MIX	Kei = 33 nM
XANTHINE OXIDASE		*I*50 = 2.4 μM

MYRICETRIN

ADENOSINE DEAMINASE		*I*50 = 70 μg/ml
ALDEHYDE REDUCTASE		*I*50 = 3.8 μM
β-FRUCTOFURANOSIDASE		*I*50 = 490 μM
GLUTATHIONE TRANSFERASE		*I*50 = 41 μM
INTEGRASE		*I*50 = 40 μM
MALTASE		*I*50 = 420 μM

MYRICIACETIN II

ALDEHYDE REDUCTASE		*I*50 = 15 μM

MYRICIACITRIN

ALDEHYDE REDUCTASE		*I*50 = 3.2 μM

MYRICIACITRIN I

MALTASE		*I*50 = 600 μM
SUCROSE α-GLUCOSIDASE		*I*50 = 700 μM

MYRICIACITRIN II
ALDEHYDE REDUCTASE | $I50 = 15\ \mu M$

MYRICIAPHENONE B
ALDEHYDE REDUCTASE | $I50 = 29\ \mu M$
MALTASE | $I50 = 440\ \mu M$
SUCROSE α-GLUCOSIDASE | $I50 = 310\ \mu M$

MYRIOCIN
SERINE C-PALMITOYLTRANSFERASE

MYRISTATE
PHOSPHOLIPASE D (GLYCOSYLPHOSPHATIDYLINOSITOL SPECIFIC) | $I50 = 340\ \mu M$

MYRISTIC ACID
ETHANOLAMINEPHOSPHOTRANSFERASE
GLUCOKINASE | $I50 = 140\ \mu M$
HEXOKINASE | $I50 = 2.4$ mM
15-HYDROXYPROSTAGLANDIN DEHYDROGENASE (NAD) | C
NAD(P)-ARGININE ADP-RIBOSYLTRANSFERASE | $I50 = 860\ \mu M$
PYRUVATE KINASE

MYRISTOLEIC ACID
PROTEIN KINASE A | $I50 = 200\ \mu M$
STEROID 5α-REDUCTASE

MYRISTOYL-ACP
ACYL[ACYL-CARRIER-PROTEIN]-UDP-N-ACETYLGLUCOSAMINE O-ACYLTRANSFERASE

(1S,2R)-D-erythro-2-(N-MYRISTOYLAMINO)-1-PHENYL-1-PROPANOL
CERAMIDASE ALAKLINE | $Ki \approx 13\ \mu M$

N-MYRISTOYL-ARG-LYS-ARG-THR-LEU-ARG-ARG-LEU
PROTEIN KINASE C | $I50 = 5\ \mu M$

N-MYRISTOYL-O-CAPROYL-L-SERINE
PROTEIN N-MYRISTOYLTRANSFERASE | 74% 2 mM

MYRISTOYL-CARBA(DETHIO)-CoA
GLYCYLPEPTIDE N-TETRADECANOYLTRANSFERASE | C | $Ki = 300$ nM

MYRISTOYL-CARBA(DETHIO)COA
GLYCYLPEPTIDE N-TETRADECANOYLTRANSFERASE

MYRISTOYL-CoA
ACETYL-CoA CARBOXYLASE | $Ki = 680$ nM
ACETYL-CoA HYDROLASE | 48% 5 mM
DYNEIN ATPASE | $> 20\ \mu M$
GLUCOSE-6-PHOSPHATASE | U | $Ki = 1.2\ \mu M$
[PYRUVATE DEHYDROGENASE (LIPOAMIDE)] KINASE

N-MYRISTYL-LYS-ARG-THR-LEU-ARG
PROTEIN KINASE C | $I50 = 75\ \mu M$

MYRORIDIN K
PHOSPHOLIPASE C γ1 (PHOSPHATIDYLINOSITOL SPECIFIC) | $I50 = 8.3\ \mu M$

MYR-psi-PKC
PROTEIN KINASE C | $I50 = 8\ \mu M$

MYRYSTIC ACID
PEPTIDYL-DIPEPTIDASE A

MYRYSTOYL-CoA
ARYLALKYLAMINE N-ACETYLTRANSFERASE | $I50 = 2.4\ \mu M$

N-MYRYSTOYLGLYCOPEPTIDES
GLYCYLPEPTIDE N-TETRADECANOYLTRANSFERASE

MYXALAMID B
NADH DEHYDROGENASE (UBIQUINONE)

MYXALAMID PI
NADH DEHYDROGENASE (UBIQUINONE)

MYXOMA VIRUS SERINE PROTEINASE INHIBITOR
COMPLEMENT SUBCOMPONENT C1s
PLASMIN
t-PLASMINOGEN ACTIVATOR
u-PLASMINOGEN ACTIVATOR

MYXOPYRONIN A
RNA POLYMERASE

MYXOPYRONIN B
RNA POLYMERASE

MYXOTHIAZOL
CYTOCHROME bc1
MENAQUINOL OXIDASE 55% 10 μM
NADH DEHYDROGENASE (UBIQUINONE) 84% 3 μM
NITRIC-OXIDE REDUCTASE
UBIQUINOL-CYTOCHROME-C REDUCTASE

N

N 614
THROMBOXANE A SYNTHASE

N 751
PROTEIN KINASE C $I50 = 14$ μM

Na^+
ACETYL-CoA CARBOXYLASE
ACTINOMYCIN LACTONASE
ADENINE PHOSPHORIBOSYLTRANSFERASE
ADENOSINE KINASE
ADENYLATE CYCLASE
(R)AMINOPROPANOL DEHYDROGENASE > 200 mM
ARGINIIE KINASE
ASPARTATE-SEMIALDEHYDE DEHYDROGENASE
CHITINASE
CHITIN DEACYTYLASE 30% 40 mM
CITRATE (si)-SYNTHASE
CLOSTRIPAIN
DIHYDROFOLATE REDUCTASE
DIPEPTIDYL-PEPTIDASE 2
EXODEOXYRIBONUCLEASE (LAMDA INDUCED)
FRUCTOSE-BISPHOSPHATASE
GALACTOLIPASE 94% 50 mM
1,3-β-GLUCAN SYNTHASE
GLUTAMATE FORMIMINOTRANSFERASE
HYDROGEN DEHYDROGENASE
HYDROXYMETHYLBILANE SYNTHASE
INORGANIC PYROPHOSPHATASE
myo-INOSITOL-1-PHOSPHATE SYNTHASE 71% 100 mM
ISOLEUCINE-tRNA LIGASE
KERATAN-SULFATE ENDO-1,4-β-GALACTOSIDASE 82% 10 mM
β-LYSINE 5,6-AMINOMUTASE
LYSINE-tRNA LIGASE
NAD(P) TRANSHYDROGENASE (AB-SPECIFIC)
4-NITROPHENYLPHOSPHATASE
NUCLEOSIDE-TRIPHOSPHATASE 50% 100 mM

PECTATE LYASE	
PHOSPHOLIPASE A1	20% 100 mM
PHOSPHOPYRUVATE HYDRATASE	
PHOSPHORIBOSYLFORMYLGLYCINAMIDINE SYNTHASE	
POLYDEOXYRIBONUCLEOTIDE SYNTHASE (ATP)	
PROLINE-tRNA LIGASE	
PROPANEDIOL DEHYDRATASE	
PROTEINASE I ACHROMOBACTER	*K*i = 15 mM
PROTEIN KINASE C	60% 150 mM
RHODOPSIN KINASE	90% 100 mM
RIBONUCLEASE III	> 140 mM
RIBONUCLEOSIDE-TRIPHOSPHATE REDUCTASE	
tRNA(GUANINE-N^1-) METHYLTRANSFERASE	> 100 mM
tRNA(GUANINE-N^2-) METHYLTRANSFERASE	> 100 mM
SELENEPHOSPHATE SYNTHASE	
STARCH SYNTHASE	> 20 mM
Δ(24)-STEROL C-METHYLTRANSFERASE	
SUCROSE α-GLUCOSIDASE	
THIAMIN-TRIPHOSPHATASE	
THYMIDINE KINASE	
dTMP KINASE	
α,α-TREHALOSE PHOSPHORYLASE	
TYROSYL PROTEIN SULFOTRANSFERASE	
UREA CARBOXYLASE	> 100 mM
UREASE	

NA 0344

PROTEIN KINASE C	*I*50 = 260 nM
PROTEIN KINASE C	*I*50 = 114 nM

NA 0345

PROTEIN KINASE C	*I*50 = 110 nM
PROTEIN KINASE C	*I*50 = 68 nM

NA 0346

PROTEIN KINASE C	*I*50 = 90 nM

NA 0346

PROTEIN KINASE C	*I*50 = 62 nM

NA 382

PROTEIN KINASE C

$Na_2B_4O_7$

CYCLAMATE SULFOHYDROLASE	94% 10 mM

NaBr

GLUTAMINE-tRNA LIGASE
NAD-DINITROGEN-REDUCTASE ADP-D-RIBOSYLTRANSFERASE

NaCl

ACYL[ACYL-CARRIER-PROTEIN]-UDP-N-ACETYLGLUCOSAMINE O-ACYLTRANSFERASE	
ACYLGLYCEROL LIPASE	47% 1 M
ADENYLATE CYCLASE	
D-ALANINE-POLY(PHOSPHORIBITOL)LIGASE	
ALANOPINE DEHYDROGENASE	
AMINE OXIDASE (COPPER-CONTAINING)	
AMINOPEPTIDASE (HUMAN LIVER)	
ARGINASE	31% 1 M
ASPARAGINE-tRNA LIGASE	
ASPERGILLUS ORYZAE NEUTRAL PROTEINASE	> 4%
BETAINE-ALDEHYDE DEHYDROGENASE	> 50 mM
κ-CARRAGEENASE	
CATHEPSIN D	10%
CHORISMATE MUTASE	
CREATINE KINASE	

DEOXYRIBONUCLEASE (APURINIC OR APYRIMIDIC)		> 10 mM
DEOXYRIBONUCLEASE I		62% 100 mM
DEOXYRIBONUCLEASE IV (PHAGE T4-INDUCED)		85% 50 mM
trans-1,2-DIHYDROBENZENE-1,2-DIOL DEHYDROGENASE		
DIPEPTIDYL-PEPTIDASE 2	C	*K*i = 1.8 mM
DNA (CYTOSINE-5-)-METHYLTRANSFERASE		
DNA-DEOXYINOSINE GLYCOSIDASE		70% 100 mM
EXODEOXYRIBONUCLEASE Y		57% 100 mM
FERREDOXIN-NADP REDUCTASE		
GALACTOSYLCERAMIDASE		75% 200 mM
GALACTOSYLGALACTOSYLGLUCOSYLCERAMIDASE		
GALACTURAN 1,4-α-GALACTURONIDASE		
GLUTAMATE DECARBOXYLASE		*I*50 = 18 mM
GLUTAMINE-tRNA LIGASE		
GLYCEROL-3-PHOSPHATE DEHYDROGENASE (NAD)		
GLYCINE DEHYDROGENASE (CYTOCHROME)		> 75 mM
GLYCINE HYDROXYMETHYLTRANSFERASE		
GTP CYCLOHYDROLASE I		95% 1 M
HEPAROSAN-N-SULFATE-GLUCURONATE 5-EPIMERASE		
HISTONE ACYLTANSFERASE		*I*50 = 160 mM
13-HYDROXYDOCOSANOATE 13-β-GLUCOSYLTRANSFERASE		
HYDROXYLAMINE REDUCTASE (NADH)		
L-IDURONIDASE		
LEUCYL AMINOPEPTIDASE		> 18%
LIPOPROTEIN LIPASE		
LYSOZYME		90% 130 mM
LYSYL ENDOPEPTIDASE		*K*i = 86 mM
MANNOSE ISOMERASE		
NAD-DINITROGEN-REDUCTASE ADP-D-RIBOSYLTRANSFERASE		
NITRATE REDUCTASE (NADH)		
NITROUS OXIDE REDUCTASE		> 50 mM
OXALACETATE TAUTOMERASE		62% 40 mM
PALMITOYL-CoA HYDROLASE		
PHENYLETHANOLAMINE N-METHYLTRANSFERASE		
PHOSPHOPANTOTHENATE-CYSTEINE LIGASE		50% 50 mM
PLASMIN		
POLY(A)-SPECIFIC RIBONUCLEASE		> 150 mM
POLYGALACTURONASE		> 150 mM
POLYRIBONUCLEOTIDE SYNTHASE (ATP)		94% 400 μM
PROTEIN-ARGININE N-METHYLTRANSFERASE		> 100 mM
PROTEIN-GLUCOSYLGALACTOSYLHYDROXYLYSINE GLUCOSIDASE		42% 800 mM
PROTEIN-HISTIDINE PROS-KINASE		
PROTEIN KINASE (Ca^{2+} DEPENDENT)		*I*50 = 100 μM
PROTEIN TYROSINE SULFOTRANSFERASE		> 200 mM
PYRROLINE-5-CARBOXYLATE REDUCTASE		
REC A PEPTIDASE		
RIBONUCLEASE (POLY-(U)-SPECIFIC)		23% 150 mM
RIBONUCLEASE T2		
RIBONUCLEASE U2		
tRNA(CYTOSINE-5-) METHYLTRANSFERASE		
mRNA GUANYLYLTRANSFERASE		> 100 mM
tRNA (5-METHYLAMINOMETHYL-2-THIOURIDYLATE) METHYLTRANSFERASE		
mRNA(NUCLEOSIDE-2′O-)-METHYLTRANSFERASE		
SIGNAL PEPTIDASE I		> 160 mM
STEROL ESTERASE		
SULFITE REDUCTASE (FERREDOXIN)		
N-SULFOGLUCOSAMINE-6-SULFATASE		50% 160 mM
THIOGLUCOSIDASE		
TRYPTASE		
D-XYLULOSE REDUCTASE		

NACN

CELLULASE		

$NACO_3$

CYCLAMATE SULFOHYDROLASE		18% 10 mM

NAD

ACETOIN DEHYDROGENASE	C	
ACETYL-CoA C-ACYLTRANSFERASE		
ADENYLATE CYCLASE		
ADENYLATE KINASE		
ALANINE DEHYDROGENASE		
ALANOPINE DEHYDROGENASE	C	*K*i = 130 μM
ALCOHOL DEHYDROGENASE	C	*K*i = 128 μM
ALCOHOL DEHYDROGENASE	C	*K*i = 610 μM
L-ASPARTATE OXIDASE	C	
BILIVERDIN REDUCTASE		
CDP-DIACYLGLYCEROLPYROPHOSPHATASE		79% 3.3 mM
CREATINE KINASE		
3′,5′-CYCLIC-NUCLEOTIDE PHOSPHODIESTERASE		
CYTOCHROME-b5 REDUCTASE		
(R)-DEHYDROPANTOATE DEHYDROGENASE		> 300 μM
DIHYDROPICOLINATE REDUCTASE		
DIHYDROPTERIDINE REDUCTASE		
ENDOPEPTIDASE TREPONEMA DENTICOLA ARCC35405		*K*i = 9.6 mM
FAD PYROPHOSPHATASE		
FRUCTURONATE REDUCTASE		
GLUTAMATE SYNTHASE (NADH)	C	*K*i = 120 μM
GLUTATHIONE PEROXIDASE		30% 1.5 mM
GLYCEROL-3-PHOSPHATE DEHYDROGENASE (NAD)		
3-HYDROXY-2-METHYLPYRIDINE CARBOXYLATE DIOXYGENASE		
20α-HYDROXYSTEROID DEHYDROGENASE		
21-HYDROXYSTEROID DEHYDROGENASE (NADP)		
L-IDITOL 2-DEHYDROGENASE		
myo-INOSITOL OXYGENASE		
LEUCINE DEHYDROGENASE		
MALATE DEHYDROGENASE (OXALACETATE-DECARBOXYLATING) (NADP)		
MALATE OXIDASE	C	*K*i = 180 μM
MALATE OXIDASE		
METHYLELETETRAHYDROFOLATE DEHYDROGENASE (NADP)		
METHYLENETETRAHYDROFOLATE CYCLOHYDROLASE		*K*i = 4.4 mM
NADH DEHYDROGENASE		
NADH DEHYDROGENASE (UBIQUINONE, ROTENONE INSENSITIVE)		
NADH PEROXIDASE		29% 3 mM
NAD NUCLEOSIDASE		70% 5 mM
NAD(P)H DEHYDROGENASE (FMN)	C	
NADPH-FERRIHEMOPROTEIN REDUCTASE		
NAD(P) TRANSHYDROGENASE (B-SPECIFIC)	U	
NICOTINAMIDASE	NC	*K*i = 530 μM
NICOTINAMIDE PHOSPHORIBOSYLTRANSFERASE		40% 100 μM
NITRATE REDUCTASE (NADH)	C	*K*i = 2.2 mM
NITRATE REDUCTASE (NADPH)		
NITRITE REDUCTASE (NAD(P)H)		
5′-NUCLEOTIDASE	C	38% 100 μM
PHOSPHODIESTERASE		14% 200 μM
PROSTAGLANDIN E2 9-KETOREDUCTASE		
PYRROLINE-5-CARBOXYLATE REDUCTASE		
RETINAL DEHYDROGENASE	NC	
RIBOSE-5-PHOSPHATE ISOMERASE		10 mM
RIBOSE-PHOSPHATE PYROPHOSPHOKINASE		
RUBREDOXIN-NAD REDUCTASE		
Δ(24)-STEROL C-METHYLTRANSFERASE		
TAUROPINE DEHYDROGENASE	C	*K*i = 400 μM
UDP-GLUCOSE 1-EPIMERASE		
D-XYLOSE 1-DEHYDROGENASE		

α-NAD

FORMALDEHYDE DEHYDROGENASE (GLUTATHIONE)		

NAD ADP-RIBOSYLTRANSFERASE		40% 500 μM
NAD NUCLEOSIDASE		70% 5 mM
NICOTINAMIDE PHOSPHORIBOSYLTRANSFERASE		

NAD ANALOGUES

GLYCEROL-3-PHOSPHATE DEHYDROGENASE (NAD)		

NADH

ACETALDEHYDE DEHYDROGENASE (ACETYLATING)		
ACETOIN DEHYDROGENASE		> 300 μM
ACETYL-CoA C-ACETYLTRANSFERASE		
ACETYL-CoA CARBOXYLASE		35% 1 mM
ACETYL-CoA HYDROLASE	NC	90% 200 μM
ADENYLATE CYCLASE		62% 5 mM
ADENYLATE CYCLASE		*I*50 = 120 μM
ADP-RIBOSE PYROPHOSPHATASE		75% 4 mM
ALANINE DEHYDROGENASE		
ALCOHOL DEHYDROGENASE	C	*K*i = 1.5 μM
ALCOHOL DEHYDROGENASE	C	*K*i = 18 μM
D-AMINO-ACID OXIDASE		
AMINOBUTYRALDEHYDE DEHYDROGENASE		
ARACHIDONATE 15-LIPOXYGENASE		46% 200 μM
ARGINIIE KINASE		
(+)-BORNEOL DEHYDROGENASE		
CITRATE (si)-SYNTHASE		
3-DEHYDROQUINATE SYNTHASE		
DIHYDROLIPOAMIDE DEHYDROGENASE		
2-ENOATE REDUCTASE		> 250 μM
2-ENOATE REDUCTASE		
L-ERYTHRO-3,5-DIAMINOHEXANOATE DEHYDROGENASE	C	*K*i = 4 μM
FORMATE DEHYDROGENASE		100% 250 μM
FRUCTOSE-6-PHOSPHATE PHOSPHOKETOLASE		
FRUCTURONATE REDUCTASE		
GLUCOSE-6-PHOSPHATE 1-DEHYDROGENASE		*K*i = 360 μM
GLUTAMATE SYNTHASE (NADPH)		
GLUTATHIONE REDUCTASE (NADPH)		
GLYCERALDEHYDE-3-PHOSPHATE DEHYDROGENASE (NADP)		23% 400 μM
GLYCEROL-3-PHOSPHATE DEHYDROGENASE (NAD)		
15-HYDROXYPROSTAGLANDIN DEHYDROGENASE (NAD)		
IMP DEHYDROGENASE		*K*i = 400 μM
IMP DEHYDROGENASE 1		*K*i = 102 μM
IMP DEHYDROGENASE 2		*K*i = 90 μM
ISOCITRATE DEHYDROGENASE		
L-LACTATE DEHYDROGENASE		
MALATE DEHYDROGENASE		
MALONATE-SEMIALDEHYDE DEHYDROGENASE (ACYLATING)		
2-METHYLCITRATE LYASE		
METHYLISOCITRATE LYASE		
METHYLMALONATE-SEMIALDEHYDE DEHYDROGENASE (ACYLATING)		
3-METHYL-2-OXOBUTANOATE DEHYDROGENASE (LIPOAMIDE)	C	*K*i = 8 μM
NAD-DINITROGEN-REDUCTASE ADP-D-RIBOSYLTRANSFERASE		
NADH PEROXIDASE		
NAD KINASE	C	*K*i = 910 μM
NAD NUCLEOSIDASE	C	*K*i = 3 μM
NADPH-CYTOCHROME-c2 REDUCTASE	C	*K*i = 55 μM
5′-NUCLEOTIDASE	C	49% 100 μM
D-OCTOPINE DEHYDROGENASE	NC	
OXOGLUTARATE DEHYDROGENASE (LIPOAMIDE)		
PHOSPHOENOLPYRUVATE CARBOXYKINASE		
PHOSPHOENOLPYRUVATE CARBOXYKINASE (ATP)		
6-PHOSPHO-β-GLUCOSIDASE A		*K*i = 2.2 mM
PHOSPHOKETOLASE		
PHOSPHORYLASE		
PREPHENATE DEHYDROGENASE		
PROPIONYL-CoA HYDROLASE		

PYRUVATE CARBOXYLASE		
PYRUVATE DEHYDROGENASE (LIPOAMIDE)		
[PYRUVATE DEHYDROGENASE (LIPOAMIDE)]-PHOSPHATASE		
RETINAL DEHYDROGENASE		
RIBOSE-PHOSPHATE PYROPHOSPHOKINASE		
SORBITOL-6-PHOSPHATE 2-DEHYDROGENASE		
SUCCINATE-SEMIALDEHYDE DEHYDROGENASE		
TAGATURONATE REDUCTASE		
UDP-N-ACETYLGLUCOSAMINE DEHYDROGENASE	C	*K*i = 50 μM
UDP-N-ACETYLGLUCOSAMINE 4-EPIMERASE		
UDP-GLUCOSE 6-DEHYDROGENASE		
UDP-GLUCOSE 4-EPIMERASE		
UDP-GLUCURONATE DECARBOXYLASE	C	*K*i = 2 μM
UDP-GLUCURONATE 5′-EPIMERASE		
URONATE DEHYDROGENASE		
β-NADH		
ACYL-CoA HYDROLASE		
NADH + CYANIDE		
NITRATE REDUCTASE (NADH)		
NADP		
ACETYL-CoA C-ACYLTRANSFERASE		
ACYL-CoA DEHYDROGENASE (NADP)		
ADENYLATE CYCLASE		75% 5 mM
ALCOHOL DEHYDROGENASE (NADP)		
ALDEHYDE REDUCTASE		
BILIVERDIN REDUCTASE		
CARBONYL REDUCTASE (NADPH)		40% 100 μM
CARBONYL REDUCTASE (NADPH)		99% 1 mM
CDP-DIACYLGLYCEROLPYROPHOSPHATASE		28% 3.3 mM
CHOLESTENONE 5α-REDUCTASE		
CHOLESTENONE 5β-REDUCTASE		
CINNAMOYL-CoA REDUCTASE	MIX	
3′,5′-CYCLIC-NUCLEOTIDE PHOSPHODIESTERASE		
CYCLOHEXANONE MONOOXYGENASE		
CYTOCHROME-b5 REDUCTASE		
DIHYDROFOLATE REDUCTASE		
DIHYDROPICOLINATE REDUCTASE		
DIHYDROPTERIDINE REDUCTASE	C	*K*i = 3.2 μM
2,5-DIKETO-D-GLUCONATE REDUCTASE	C	*K*i = 26 μM
ECDYSONE 20-MONOOXYGENASE		
ENDOPEPTIDASE TREPONEMA DENTICOLA ARCC35405		*K*i = 250 μM
ENOYL-[ACYL-CARRIER-PROTEIN] REDUCTASE (NADPH, B SPECIFIC)		
trans-2-ENOYL-CoA REDUCTASE (NADPH)		
ERYTHRULOSE REDUCTASE		
FAD PYROPHOSPHATASE		
FARNESYL-DIPHOSPHATE FARNESYLTRANSFERASE		
GLUCOSE-6-PHOSPHATE 1-EPIMERASE		
GLUTAMATE DEHYDROGENASE (NADP)		
GLUTAMATE SYNTHASE (NADH)		
GLUTATHIONE DEHYDROGENASE (ASCORBATE)		49% 100 μM
GLUTATHIONE PEROXIDASE		30% 700 μM
2-HYDROXY-4-CARBOXYMUCONATE-6-SEMIALDEHYDE DEHYDROGENASE	C	
15-HYDROXYPROSTAGLANDIN DEHYDROGENASE (NADP)		
20α-HYDROXYSTEROID DEHYDROGENASE		
21-HYDROXYSTEROID DEHYDROGENASE (NADP)		
myo-INOSITOL OXYGENASE		
[ISOCITRATE DEHYDROGENASE (NADP)] KINASE		
(–)-LIMONENE 3-MONOOXYGENASE		
(–)-LIMONENE 6-MONOOXYGENASE		
(–)-LIMONENE 7-MONOOXYGENASE		
METHYLENETETRAHYDROFOLATE CYCLOHYDROLASE		*K*i = 500 μM
[3-METHYL-2-OXOBUTANOATE DEHYDROGENASE (LIPOAMIDE)] KINASE		

NAD-DINITROGEN-REDUCTASE ADP-D-RIBOSYLTRANSFERASE		
NADH PEROXIDASE		77% 3 mM
NAD NUCLEOSIDASE		
NAD(P)H DEHYDROGENASE (QUINONE)		
NADPH-FERRIHEMOPROTEIN REDUCTASE	C	Ki = 2 μM
NADPH-FERRIHEMOPROTEIN REDUCTASE		
NAD(P) TRANSHYDROGENASE		
NICOTINAMIDASE	NC	Ki = 465 μM
NICOTINAMIDE PHOSPHORIBOSYLTRANSFERASE		
NITRATE REDUCTASE (NADH)	C	Ki = 11 mM
NITRITE REDUCTASE (NAD(P)H)		
4-NITROPHENYLPHOSPHATASE		60% 10 mM
5′-NUCLEOTIDASE	C	16% 100 μM
D-OCTOPINE DEHYDROGENASE	U	
3-OXOACYL-[ACYL-CARRIER-PROTEIN] REDUCTASE		
2-OXOALDEHYDE DEHYDROGENASE (NAD)		
15-OXOPROSTAGLANDIN Δ13 REDUCTASE		80% 2 mM
PHOSPHATE BUTYRYLTRANSFERASE		
PHOSPHODIESTERASE		30% 200 μM
PROSTAGLANDIN E2 9-KETOREDUCTASE		
PROSTAGLANDIN E2 9-REDUCTASE		
PYRROLINE-5-CARBOXYLATE REDUCTASE		> 160 μM
SERRATIA MARCESCENS NUCLEASE		
SORBOSE DEHYDROGENASE (NADP)	C	Ki = 130 μM
STEROID 5α-REDUCTASE		
Δ(24)-STEROL C-METHYLTRANSFERASE		
SULFITE REDUCTASE (NADPH)	C	Ki = 20 μM
THIOREDOXIN REDUCTASE (NADPH)		
THIOREDOXIN REDUCTASE (NADPH)		
UDP-N-ACETYLGLUCOSAMINE 4-EPIMERASE		
UDP-GLUCURNATE 4-EPIMERASE		

β-NADP

MALATE DEHYDROGENASE (OXALACETATE-DECARBOXYLATING) (NADP)	NC	Ki = 57 μM

NADP 2′,3′-DIALDEHYDE

MALATE DEHYDROGENASE (OXALACETATE-DECARBOXYLATING) (NADP)	IR	

NADPH

trans-ACENAPHTHENE-1,2-DIOL DEHYDROGENASE		
ACETOIN DEHYDROGENASE		> 150 μM
ACETYL-CoA C-ACETYLTRANSFERASE		
ACETYL-CoA CARBOXYLASE		32% 1mM
ALANINE DEHYDROGENASE		
D-AMINO-ACID OXIDASE		
CITRATE (si)-SYNTHASE		
DIHYDROPYRIMIDINE DEHYDROGENASE (NADP)		
FORMALDEHYDE DEHYDROGENASE (GLUTATHIONE)		
FRUCTOSE-6-PHOSPHATE PHOSPHOKETOLASE		
GLUCOSE 1-DEHYDROGENASE (NADP)	C	Ki = 38 μM
GLUCOSE-6-PHOSPHATE 1-DEHYDROGENASE	C	Ki = 10 μM
GLUCOSE-6-PHOSPHATE 1-DEHYDROGENASE	C	Ki = 4 μM
GLUCOSE-6-PHOSPHATE 1-DEHYDROGENASE		Ki = 220 μM
GLUTAMATE-5-SEMIALDEHYDE DEHYDROGENASE		
GLUTAMATE SYNTHASE (NADH)		
GLUTAMATE SYNTHASE (NADPH)		
GLUTATHIONE PEROXIDASE		30% 260 μM
GLUTATHIONE REDUCTASE (NADPH)		
GLYCERALDEHYDE-3-PHOSPHATE DEHYDROGENASE (NADP)		52% 400 μM
GLYCERALDEHYDE-3-PHOSPHATE DEHYDROGENASE (NADP)	C	Ki = 30 μM
ω-HYDROXYFATTY ACID: NADP OXIDOREDUCTASE		> 2 mM
[HYDROXYMETHYLGLUTARYL-CoA REDUCTASE (NADPH)] KINASE		
[HYDROXYMETHYLGLUTARYL-CoA REDUCTASE (NADPH)]-PHOSPHATASE	IR	
15-HYDROXYPROSTAGLANDIN DEHYDROGENASE (NADP)		
15-HYDROXYPROSTAGLANDIN DEHYDROGENASE (NADP)		Ki = 10.7 nM

15-HYDROXYPROSTAGLANDIN-I DEHYDROGENASE (NADP)		
ISOCITRATE DEHYDROGENASE (NADP)		
[ISOCITRATE DEHYDROGENASE (NADP)] KINASE		
2-METHYLCITRATE LYASE		
METHYLELETETRAHYDROFOLATE DEHYDROGENASE (NADP)		
METHYLISOCITRATE LYASE		
MYCOSERATE SYNHTASE		
NADH PEROXIDASE		
NITRATE REDUCTASE (NADH)		
D-NOPALINE DEHYDROGENASE		
5′-NUCLEOTIDASE	C	32% 100 μM
OXOGLUTARATE DEHYDROGENASE (LIPOAMIDE)		
15-OXOPROSTAGLANDIN Δ13 REDUCTASE		
PHOSPHATE BUTYRYLTRANSFERASE		
PHOSPHOGLUCONATE DEHYDROGENASE (DECARBOXYLATING)		
PHOSPHOKETOLASE		
PHOSPHOPROTEIN PHOSPHATASE		
Δ1PIPERIDINE-2-CARBOXYLATE RDUCTASE		> 150 μM
PYRROLINE-5-CARBOXYLATE REDUCTASE		> 130 μM
RIBOSE-PHOSPHATE PYROPHOSPHOKINASE		
RIBULOSE-BISPHOSPHATE CARBOXYLASE		
RUBREDOXIN-NAD REDUCTASE		
NaF		
ACYLCARNITINE HYDROLASE		
ACYLGLYCEROL LIPASE		40% 1 mM
CALF THYMUS RIBONUCLEASE H		10% 20 mM
CEPHALOSPORIN-C DEACETYLASE		
DIHYDROCOUMARIN HYDROLASE		20% 3.3 mM
EXORIBONUCLEASE H		
STEROL ESTERASE		
NAFAZATROM		
ARACHIDONATE 5-LIPOXYGENASE		*I*50 = 2.5 μM
ARACHIDONATE 5-LIPOXYGENASE		5 μM
ARACHIDONATE 12-LIPOXYGENASE		
LIPOXYGENASE		*I*50 = 1 μM
NAFCILLIN		
NEPRILYSIN		*K*i = 59 μM
NAFENOPIN		
LONG-CHAIN-FATTY-ACID CoA-LIGASE	C	*K*i = 390 μM
NAFOXIDINE		
PROTEIN KINASE C	NC	
NAFTIFINE		
SQUALENE MONOOXYGENASE		*I*50 = 144 μM
SQUALENE MONOOXYGENASE		*I*50 = 340 nM
SQUALENE MONOOXYGENASE		*I*50 = 340 nM
SQUALENE MONOOXYGENASE		*I*50 = 1.1 μM
SQUALENE MONOOXYGENASE	NC	*K*i = 1 μM
NAGSTATIN		
α-N-ACETYLGALACTOSAMINIDASE		*I*50 = 19 μg/ml
β-N-ACETYLHEXOSAMINIDASE	C	*K*i = 17 nM
β-N-ACETYLHEXOSAMINIDASE		*I*50 = 2 ng/ml
β-N-ACETYLHEXOSAMINIDASE		*I*50 = 4 ng/ml
NaH_2PO_4		
L-IDURONIDASE		
NaI		
GLUTAMINE-tRNA LIGASE		

NAJA KAOUTHIA PHOSPHOLIPASE A2 INHIBITOR
PHOSPHOLIPASE A2-1 — *K*i = 6.5 nM
PHOSPHOLIPASE A2-1 — *K*i = 10 nM
PHOSPHOLIPASE A2-1 — *K*i = 3.4 nM
PHOSPHOLIPASE A2-2

NAJA NAJA KAOUTHIA PHOSPHOLIPASE A2 INHIBITOR
PHOSPHOLIPASE A2

NAJA NAJA NAJA TRYPSIN INHIBITOR
TRYPSIN — *K*i = 3.5 pM

NAKIJIQUINONE C
PROTEIN KINASE C — *I*50 = 23 µM
PROTEIN TYROSINE KINASE cerb-2 — *I*50 = 26 µM
PROTEIN TYROSINE KINASE EGFR — *I*50 = 170 µM

NAKIJIQUINONE D
PROTEIN KINASE C — *I*50 = 220 µM
PROTEIN TYROSINE KINASE cerb-2 — *I*50 = 29 µM
PROTEIN TYROSINE KINASE EGFR — *I*50 > 400 µM

NALED
ACETYLCHOLINESTERASE
ACETYLESTERASE
CARBOXYLESTERASE — *K*ic = 8.5 nM

NALIDIXIC ACID
CYTOCHROME P450 LA2 — 67% 500 µM
DNA GYRASE
DNA TOPOISOMERASE (ATP-HYDROLYSING) — *I*50 = 650 µg/ml
GLYCINE-tRNA LIGASE
LEUCINE-tRNA LIGASE
RNA-DIRECTED DNA POLYMERASE

NALLANIN
PROTEIN KINASE A — *I*50 = 33 µM
PROTEIN KINASE C — *I*50 = 248 µM

NALOXONE
MORPHINE 6-DEHYDROGENASE

N-[4-[.N-[(2-AMINO-3,4-DIHYDRO-4-THIO-6-QUINAZOLINYL)-METHYL]PROP-2-YNYLAMINO]BENZOYL]-L-GLUTAMIC ACID
THYMIDYLATE SYNTHASE — *I*50 = 19 nM

1-NAPHTHALDEHYDE
TYROSINE-ESTER SULFOTRANSFERASE — C — *K*i = 29 µM

2-NAPHTHALDEHYDE
TYROSINE-ESTER SULFOTRANSFERASE — C — *K*i = 89 µM

NAPHTHALENE
AMINE OXIDASE MAO-A — *I*50 = 29 µM
CHOLINESTERASE — *K*i = 450 µM

1-NAPHTHALENEACETIC ACID
3-METHYLENEOXINDOLE REDUCTASE B

NAPHTHALENE 1,7-BIS[α,α-DIFLUOROMETHYLENEPHOSPHONATE]
PHOSPHOPROTEIN PHOSPHATASE 2A — *I*50 = 481 µM
PROTEIN TYROSINE-PHOSPHATASE 1B — *I*50 = 118 µM
PROTEIN TYROSINE-PHOSPHATASE CD45 — *I*50 = 27 µM

1,2-NAPHTHALENEDIOL
LACTOYLGLUTATHIONE LYASE — *I*50 = 34 µM

2,3-NAPHTHALENEDIOL
CATECHOL OXIDASE

NAPHTHALENEDIOL-1,5-DIPHOSPHATE		
FRUCTOSE-BISPHOSPHATE ALDOLASE	C	*K*i = 380 nM
NAPHTHALENE-1,2-DIONE		
trans-1,2-DIHYDROBENZENE-1,2-DIOL DEHYDROGENASE		
2-NAPHTHALENE SULPHONATE		
THIOSULFATE SULFURTRANSFERASE		
(3S)-(NAPHTHALENE-2-SULPHONYLAMINO)-1-[2R-(4-AMIDINOPHENYL)-1-PYRIDINOCARBONYLETHYL]-2-PYRROLIDINONE		
THROMBIN		*I*50 = 1.6 nM
Nα-(NAPHTHALENESULPHONYLGLYCYL)-4-AMIDINO-D,L-PHENYLALANINEPIPERIDINE		
α-THROMBIN		
2-NAPHTHALENETHIOL		
ARACHIDONATE 5-LIPOXYGENASE		*I*50 = 230 nM
NAPHTHALENE-1,3,6-TRISULPHONIC ACID		
PHOSPHOGLYCERATE KINASE	NC	*K*i = 5.5 mM
(E)-4-[2-(1-NAPHTHALENYL)ETHENYL]-1-PROPYL-PYRIDINIUM BROMIDE		
ACETYLCHOLINESTERASE		*I*50 = 18 μM
CHOLINE O-ACETYLTRANSFERASE		*I*50 = 190 nM
N-(2-NAPHTHALENYLSULPHONYL)GLYCINE		
ALDEHYDE REDUCTASE		*I*50 = 400 nM
N-(1-NAPHTHALENYL-THIOXOMETHYL)-N-METHYLGLYCINE		
ALDEHYDE REDUCTASE		*I*50 = 730 nM
1,8-NAPHTHALIMIDE		
NAD(P)-ARGININE ADP-RIBOSYLTRANSFERASE		*I*50 = 20 μM
α-NAPHTHOFLAVONE		
AROMATASE		*K*i = 200 nM
CYTOCHROME P-448		
CYTOCHROME P450 1A1		*K*i = 2.3 nM
CYTOCHROME P450 1A1	MIX	*K*i = 10 nM
CYTOCHROME P450 1A1	MIX	*K*i = 10 nM
CYTOCHROME P450 1A2		*K*i = 13 nM
CYTOCHROME P450 1A2		*K*i = 1.6 nM
CYTOCHROME P450 2A6	C	*K*i = 12 μM
EPOXIDE HYDROLASE		95% 100 μM
LEUKOTRIENE-B4 20-MONOOXYGENASE		
UNSPECIFIC MONOOXYGENASE		
β-NAPHTHOFLAVONE		
CYTOCHROME P450 6D1		*I*50 = 330 nM
EPOXIDE HYDROLASE		89% 100 μM
1,4-NAPHTHOHYDROQUINONE		
NAD(P)H DEHYDROGENASE (QUINONE)		
1-NAPHTHOIC ACID		
MONOPHENOL MONOOXYGENASE	C	*K*i = 140 μM
TYROSINE-ESTER SULFOTRANSFERASE IV	C	*K*i = 16 μM
2-NAPHTHOIC ACID		
MONOPHENOL MONOOXYGENASE	C	*K*i = 13 μM
PHENYLALANINE AMMONIA-LYASE		*K*i = 10 μM
PHENYLALANINE AMMONIA-LYASE		71% 1 mM
TYROSINE-ESTER SULFOTRANSFERASE IV	C	*K*i = 41 μM
NAPHTHO-[2,3-d]-IMIDAZOLE-4,-9-DIONE		
HYPOXANTHINE PHOSPHORIBOSYLTRANSFERASE		82% 100 μM
α-NAPHTHOL		
AMINE OXIDASE (FLAVIN-CONTAINING)		

D-AMINO-ACID OXIDASE
LIPOXYGENASE

β-NAPHTHOL
AMINE OXIDASE (FLAVIN-CONTAINING)

1-NAPHTHOL
ACETYLINDOXYL OXIDASE
UDP-GLUCURONOSYLTRANSFERASE — $Ki = 690\ \mu M$

2-NAPHTHOL
ACETYLINDOXYL OXIDASE

NAPHTHOMYCIN
PHOSPHODIESTERASE

2-NAPHTHOQUINOLINE
STIZOLOBATE SYNTHASE — 37% 100 μM
STIZOLOBINATE SYNTHASE — 43% 100 μM

NAPHTHOQUINONE
NAD(P)-ARGININE ADP-RIBOSYLTRANSFERASE — $I50 = 100\ \mu M$
NITRITE REDUCTASE (NAD(P)H)

α-NAPHTHOQUINONE
QUERCETIN 2,3-DIOXYGENASE

1,2-NAPHTHOQUINONE
15-HYDROXYPROSTAGLANDIN DEHYDROGENASE (NAD) — IR
NITRIC-OXIDE SYNTHASE I — $I50 = 12\ \mu M$

1,4-NAPHTHOQUINONE
NAD ADP-RIBOSYLTRANSFERASE — $I50 = 250\ \mu M$
NITRIC-OXIDE SYNTHASE I — $I50 = 26\ \mu M$

1,4-NAPHTHOQUINONE-2,3-BIS(PROPIONIC ACID)
TRYPANOTHIONE REDUCTASE — $I50 = 100\ \mu M$

1,4-NAPHTHOQUINONE-2-SULPHONIC ACID
NITRIC-OXIDE SYNTHASE I — $I50 = 44\ \mu M$

1-[(NAPHTHO[1,2-b]THIOPHEN-4-YL)METHYL]-1H-IMIDAZOLE
AMINE OXIDASE MAO — REV

1-NAPHTHOYL-ARG-CF_3
TRYPSIN — $Ki = 200\ nM$
TRYPTASE — $Ki = 900\ nM$

2-NAPHTHOYL-L-PHENYLALANYL-D,L-HOMOCYSTEINE
PEPTIDYLGLYCINE MONOOXYGENASE — $I50 = 10\ nM$

1-(4-NAPHTHYLACETYLOXYBENZOYLOXYACETYL)-4-ISOPROPYLPIPERAZINE
CHYMOTRYPSIN — $I50 = 8\ nM$

1-NAPHTHYLAMINE
CHOLINE SULFOTRANSFERASE

2-NAPHTHYLAMINE
AMINOPEPTIDASE — $Ki = 620\ \mu M$
AMINOPEPTIDASE B — $Ki = 10\ \mu M$
METHIONYL AMINOPEPTIDASE — $Ki = 10\ \mu M$

NAPHTHYL ARG CF3
TRYPSIN — $Ki = 200\ nM$

2-NAPHTHYLCYCLOPROPYLAMINE
AMINE OXIDASE MAO

(NAPHTH-2-YL)DIFLUOROMETHYLPHOSPHONIC ACID
PHOSPHOPROTEIN PHOSPHATASE 2A — 90% 200 μM

PHOSPHOPROTEIN PHOSPHATASE 1B		88% 200 μM
PROTEIN TYROSINE-PHOSPHATASE 1B		88% 200 μM
N[(2S,3R,4S)-1-(1-NAPHTHYL)-3,4-DIHYDROXY-6-METHYL-2-HEPTYL]BOC-CHE-ILENH$_2$		
ENDOTHELIN CONVERTING ENZYME		*I*50 = 23 nM
(1-NAPHTHYL)ETHYLACETAMIDE		
ARALKYLAMINE N-ACETYLTRANSFERASE		*I*50 = 8 μM
2-(1-NAPHTHYL)ETHYL URIDINEDIPHOSPHATE		
UDP-GLUCURONOSYLTRANSFERASE	C	
α-NAPHTHYLFLAVONE		
ESTRADIOL 2-HYDROXYLASE		
ESTRADIOL 16α-HYDROXYLASE		
S-(1-NAPHTHYLMETHYL)GLUTATHIONE		
GLUTATHIONE TRANSFERASE	C	*K*i = 2.5 μM
α-NAPHTHYL-N-METHYLPYRIDIN-4-YL-METHANE		
AMINE OXIDASE MAO-A		*K*i = 5.5 μM
AMINE OXIDASE MAO-B		*K*i = 850 μM
α-NAPHTHYLPHOSPHATE		
GLYCEROL-1,2-CYCLIC-PHOSPHATE PHOSPHODIESTERASE		51% 250 μM
NAPHTHYLPHTHALMIC ACID		
AUXIN CARRIER		
α-NAPHTHYLQUINOLINE		
CATECHOL 2,3-DIOXYGENASE	C	
Nα-2-NAPHTHYLSULPHONYL-L-3-AMIDINOPHENYLALANYL-4-METHYLPIPERIDIDE		
THROMBIN		*K*i = 2.5 nM
TRYPSIN		*K*i = 42 nM
N$^{\alpha}$(2-NAPHTHYLSULPHONYL)-3-AMIDINOPHENYLALANYL-4-(METHYLSULPHONYL)PIPERAZIDE HCl		
COAGULATION FACTOR Xa		*K*i = 22 μM
PLASMIN		*K*i = 12 μM
THROMBIN		*K*i = 3.3 nM
TRYPSIN		*K*i = 190 nM
Nα-2-NAPHTHYLSULPHONYL-GLYCYL-D-4-AMIDINOPHENYLALANYLPIPERIDIDE		
THROMBIN		*K*i = 2.1 nM
TRYPSIN		*K*i = 210 nM
Nα-(p-NAPHTHYLSULPHONYLGLYCYL)-4′-AMIDINOPHENYLAMINE PIPERIDINE		
ACROSIN		*K*i = 2.9 μM
COAGULATION FACTOR Xa		*K*i = 7.9 μM
COAGULATION FACTOR XIIa		*K*i = 500 μM
KALLIKREIN PLASMA		*K*i = 14 μM
KALLIKREIN TISSUE		*K*i = 93 μM
PLASMIN		*K*i = 30 μM
THROMBIN		*K*i = 6 nM
TRYPTASE		*K*i = 45 μM
VENOMBIN A		*K*i = 1.7 μM
Nα-(2-NAPHTHYLSULPHONYLGLYCYL)-4-AMIDOPHENYLALANYLPROLINE		
THROMBIN		*K*i = 510 nM
Nα-(2-NAPHTHYLSULPHONYLGLYCYL)D,L-p-AMINOPHENYLALANYLPIPERIDINE		
COAGULATION FACTOR Xa		*K*i = 7.9 μM
PLASMIN		*K*i = 30 μM
THROMBIN		*K*i = 6.6 nM
THROMBIN		*K*i = 6 nM
TRYPSIN		*K*i = 690 nM

NAPHTHYLVINYLPYRIDINE

Enzyme	Type	Value
CHOLINE O-ACETYLTRANSFERASE	NC	$I50 = 1.8\ \mu M$

4-(1-NAPHTHYLVINYL)PYRIDINE

Enzyme	Type	Value
CHOLINE O-ACETYLTRANSFERASE		$I50 = 10\ \mu M$
CHOLINE O-ACETYLTRANSFERASE		
LYSO PLATELET ACTIVATING FACTOR ACETYLTRANSFERASE		$I50 = 43\ \mu M$

β-NAPHTOQUINONE-4-SULPHONIC ACID

Enzyme	Type	Value
NICOTINAMIDE-NUCLEOTIDE ADENYLYLTRANSFERASE		

NAPROXEN

Enzyme	Type	Value
ARACHIDONATE 5-LIPOXYGENASE		$I50 > 100\ \mu M$
CYCLOOXYGENASE		$I50 = 2\ \mu M$
CYCLOOXYGENASE 1		$I50 = 1.1\ \mu M$
CYCLOOXYGENASE 1		$I50 = 4.8\ \mu M$
CYCLOOXYGENASE 2		$I50 = 36\ \mu M$
CYCLOOXYGENASE 2		$I50 = 28\ \mu M$
DIHYDRODIOL DEHYDROGENASE 2		96% 100 μM
DIHYDRODIOL DEHYDROGENASE 4		24% 100 μM
3α-HYDROXYSTEROID DEHYDROGENASE (B-SPECIFIC)		$I50 = 100\ \mu M$
3α-HYDROXYSTEROID/DIHYDRODIOL DEHYDROGENASE 2	U	$Ki = 1.1\ \mu M$
PHOSPHORIBOSYLAMINOIMIDAZOLECARBOXAMIDE FORMYLTRANSFERASE		
PROSTAGLANDIN-ENDOPEROXIDE SYNTHASE		$I50 = 85\ nM$
PROSTAGLANDIN-ENDOPEROXIDE SYNTHASE 1		$I50 = 600\ nM$
PROSTAGLANDIN-ENDOPEROXIDE SYNTHASE 2		$I50 = 2\ \mu M$
PROSTAGLANDIN H ENDOPEROXIDE SYNTHASE 1		$I50 = 24\ \mu M$
PROSTAGLANDIN H ENDOPEROXIDE SYNTHASE 2		$I50 = 209\ \mu M$
TYROSINE-ESTER SULFOTRANSFERASE IV	C	$Ki = 130\ \mu M$
UDP-GLUCURONOSYLTRANSFERASE	C	$Ki = 172\ \mu M$

(–)R NAPROXEN

Enzyme	Type	Value
LONG-CHAIN-FATTY-ACID CoA-LIGASE	MIX	$Ki = 650\ \mu M$

(+)S NAPROXEN

Enzyme	Type	Value
LONG-CHAIN-FATTY-ACID CoA-LIGASE	MIX	$Ki = 540\ \mu M$

NAPTHOL BLUE-BLACK

Enzyme	Type	Value
ELASTASE CERCARIAL		$Ki = 6\ \mu M$

1-NAPTHYLACETIC ACID

Enzyme	Type	Value
TYROSINE-ESTER SULFOTRANSFERASE IV	C	$Ki = 200\ \mu M$

2-NAPTHYLACETIC ACID

Enzyme	Type	Value
TYROSINE-ESTER SULFOTRANSFERASE IV	C	$Ki = 410\ \mu M$

NARINGENIN

Enzyme	Type	Value
ARACHIDONATE 5-LIPOXYGENASE		$I50 = 16\ \mu M$
ARACHIDONATE 12-LIPOXYGENASE		$I50 = 40\ \mu M$
ARYL SULFOTRANSFERASE P		$I50 = 11\ \mu M$
CHALCONE ISOMERASE		$Ki = 17\ \mu M$
CHLORIDE PEROXIDASE		
CYCLOOXYGENASE		$I50 = 100\ \mu M$
CYTOCHROME-C PEROXIDASE		
CYTOCHROME P450		$I50 = 100\ \mu M$
CYTOCHROME P450 1A2	C	$Ki = 29\ \mu M$
FLAVONE 4-REDUCTASE		
GLUTATHIONE TRANSFERASE		$I50 = 66\ \mu M$
11β-HYDROXYSTEROID DEHYDROGENASE		$I50 = 336\ \mu M$
11β-HYDROXYSTEROID DEHYDROGENASE		$I50 = 496\ \mu M$
11β-HYDROXYSTEROID DEHYDROGENASE		$I50 = 336\ \mu M$
11β-HYDROXYSTEROID DEHYDROGENASE 1		47% 1.2 μM
IODIDE PEROXIDASE		$I50 = 2.7\ \mu M$
LACTOPEROXIDASE		
MYELOPEROXIDASE		
NAD(P)H DEHYDROGENASE (QUINONE)	C	$Ki = 3.4\ \mu M$
NARINGENIN-CHALCONE SYNHTASE		

PEROXIDASE
PROTEINASE HIV-1 — $I50 = 165\ \mu M$
PROTEIN KINASE CASEIN KINASE G — 5% 2.5 μM
PROTEIN TYROSINE KINASE p40 — $I50 = 138\ \mu M$

NARINGENINCHALCONE
NARINGENIN-CHALCONE SYNHTASE

NARINGIN
ARACHIDONATE 5-LIPOXYGENASE — $I50 > 500\ \mu M$
ARYL SULFOTRANSFERASE P — $I50 = 265\ \mu M$
CYCLOOXYGENASE — $I50 = 320\ \mu M$
GLUTATHIONE TRANSFERASE — $I50 = 180\ \mu M$
11β-HYDROXYSTEROID DEHYDROGENASE — $I50 = 2.3$ mM
IODIDE PEROXIDASE — $I50 = 13\ \mu M$
PROTEINASE HIV-1 — $I50 = 220\ \mu M$
PROTEIN KINASE C — $I50 = 27\ \mu M$

NASCENT CAT-85 5-MER PEPTIDE
PEPTIDYLTRANSFERASE

Na_2SO_4
CYCLAMATE SULFOHYDROLASE — 57% 10 mM
L-IDURONIDASE

NAT 042152
STEROL O-ACYLTRANSFERASE

NAT-13-THIAPROSTAGLANDIN F2α
15-HYDROXYPROSTAGLANDIN DEHYDROGENASE (NAD)

NAZLININ
AMINE OXIDASE DIAMINE

NAZUMAMIDE A
THROMBIN — $I50 = 2.8\ \mu g/ml$

NB 598
FARNESYL-DIPHOSPHATE FARNESYLTRANSFERASE
SQUALENE MONOOXYGENASE — $I50 = 16$ nM
SQUALENE MONOOXYGENASE — C — $Ki = 410$ pM
SQUALENE MONOOXYGENASE — $I50 = 4$ nM
SQUALENE MONOOXYGENASE — $I50 = 2$ nM
SQUALENE MONOOXYGENASE — $I50 = 20$ nM

mNBA
KYNURENINE 3-MONOOXYGENASE — $I50 = 774$ nM

NBQ 59
DNA TOPOISOMERASE (ATP-HYDROLYSING)

.N-tert-BUTYL-2-(2(R)HYDROXY-4-PHENYL-3(S)-((N-((5-DIMETHYLAMINO)-6-CHLORO-2-PYRAZINYL)CARBONYL)-2(S)-(3(R)TETRAHYDRO-FURANOYL)GLYCINYL)AMINO)BUTY
PROTEINASE HIV-1 — $I50 = 60$ pM

NC 1300
Na^+/K^+-EXCHANGING ATPASE — $I50 = 20\ \mu M$
PROTON PUMP

NC 1300-B
H^+/K^+-EXCHANGING ATPASE — $Ki = 31\ \mu M$

NCO 700
CALPAIN — 65% 180 μM

Nd^{3+}
DEOXYRIBONUCLEASE — 33% 200 μM
GLUCOSE-6-PHOSPHATE 1-DEHYDROGENASE — 100% 830 μM

ISOCITRATE DEHYDROGENASE		63% 830 μM
L-LACTATE DEHYDROGENASE		74% 830 μM
N{4-[3-2,4-DIAMINO-7H-PYRROLO[2,3-d]PYRIMIDIN-YL)PROPYL]BENZOYL}(GLUTAMIC ACID)$_6$		
PHOSPHORIBOSYLAMINOIMIDAZOLECARBOXAMIDE FORMYLTRANSFERASE		*K*i = 70 nM
PHOSPHORIBOSYLGLYCINAMIDE FORMYLTRANSFERASE		*K*i = 14 μM
NEBULARINE		
ADENOSINE DEAMINASE		*K*i = 16 μM
ADENOSYLHOMOCYSTEINASE	C	*K*i = 1.4 mM
ADENOSYLHOMOCYSTEINASE		*K*i = 18 μM
XANTHINE OXIDASE	NC	85% 20 μM
NECTRISINE		
β-N-ACETYLGLUCOSAMINIDASE		
α-GLUCOSIDASE		
β-GLUCOSIDASE		
β-MANNOSIDASE		
NEEM		
METHIONINE γ-LYASE		
NEF		
MAP KINASE		
PROTEIN TYROSINE KINASE (Lck)		
NELFINAVIR MESYLATE		
CYTOCHROME P450 3A4		*K*i = 4.8 μM
NEMESULIDE		
PROSTAGLANDIN-ENDOPEROXIDE SYNTHASE 1		*I*50 = 70 μM
PROSTAGLANDIN-ENDOPEROXIDE SYNTHASE 2		*I*50 = 1.3 μM
NEOASTILBIN		
ALDEHYDE REDUCTASE		
NEOCARZINOSTATIN CHROMOPHORE		
PROTEIN KINASE CASEIN KINASE II		*I*50 = 60 nM
PROTEIN KINASE MICROTUBULIN ASSOCIATED		*I*50 = 900 nM
PROTEIN KINASE NUCLEAR cAMP INDEPENDENT		*I*50 = 600 nM
NEOCUPROINE		
AMINE DEHYDROGENASE		
AMINE OXIDASE (COPPER-CONTAINING)		
CYSTEAMINE DIOXYGENASE		
2,3-DIHYDROXYBENZOATE 2,3-DIOXYGENASE		79% 100 μM
INDOLE 2,3-DIOXYGENASE		
LACCASE		
STIZOLOBATE SYNTHASE		57% 100 μM
STIZOLOBINATE SYNTHASE		50% 100 μM
TAURINE DEHYDROGENASE		
NEOGRIFOLIN		
MONOPHENOL MONOOXYGENASE		*I*50 = 25 μM
NEOHESPERIDIN		
TRIACYLGLYCEROL LIPASE		*I*50 = 46 μg/ml
7-NEOHESPERIDOSYLLUTEOLIN		
XANTHINE OXIDASE		*I*50 = 12 μM
NEOLACTOTETRAOSYLCERAMIDE		
NEOLACTOSYLCERAMIDE α-2,3-SIALYLTRANSFERASE		
NEOMYCIN		
GENTAMICIN 3′-N-ACETYLTRANSFERASE		
HAMMERHEAD RIBOZYME		
ORNITHINE DECARBOXYLASE		*I*50 = 50 μM

PHOSPHATIDYLINOSITOL-BISPHOSPHATASE
1-PHOSPHATIDYLINOSITOL-4,5-BISPHOSPHATE PHOSPHODIESTERASE
1-PHOSPHATIDYLINOSITOL-4,5-BISPHOSPHATE PHOSPHODIESTERASE
1-PHOSPHATIDYLINOSITOL 3-KINASE
PHOSPHOLIPASE C — *I*50 = 100 μM

NEOMYCIN B
RIBOZYME HAMMERHEAD — 100% 500 μM

NEOMYCIN SULPHATE
1-PHOSPHATIDYLINOSITOL 3-KINASE — *I*50 = 9 μM

1-NEOPENTYL-1-[(4-CHLOROPHENYL)SULPHONYL]-3-n-PROPYLUREA
ALDEHYDE DEHYDROGENASE (NAD) — 100% 1 mM

NEOSTIGMINE
ACETYLCHOLINESTERASE — *I*50 = 35 nM
ACETYLCHOLINESTERASE — *I*50 = 7.3 nM
ACETYLSPERMIDINE DEACETYLASE
CHOLINESTERASE

NEPICASTAT
DOPAMINE β-MONOOXYGENASE — *I*50 = 8.5 nM
DOPAMINE β-MONOOXYGENASE — *I*50 = 9 nM

NEPLANOCIN A
ADENOSYLHOMOCYSTEINASE — 95% 200 nM
ADENOSYLHOMOCYSTEINASE — C — *K*i = 2 nM
ADENOSYLHOMOCYSTEINASE — SSI — *K*i = 3.8 nM

NERVE GROWTH FACTOR γ
PROTEINASE NEXIN-2 — *K*i = 9.1 nM

NERYLPHOSPHATE
ISOPENTENYL-DIPHOSPHATE Δ-ISOMERASE — 58% 5 mM

NERYLPYROPHOSPHATE
ISOPENTENYL-DIPHOSPHATE Δ-ISOMERASE

NETROPSIN
DNA POLYMERASE
DNA TOPOISOMERASE
DNA TOPOISOMERASE (ATP-HYDROLYSING) — *I*50 = 12 μM
RNA POLYMERASE

NEURINE
CHOLINE SULFOTRANSFERASE

NEUROENDOCRINE PEPTIDE 7B2
PROHORMONE CONVERTASE PC2

NEUROMEDIN N
NEUROLYSIN — *I*50 = 1.8 μM

NEUROPEPTIDE Y
ADENYLATE CYCLASE — *I*50 = 10 μM
3′,5′-CYCLIC-NUCLEOTIDE PHOSPHODIESTERASE (Ca^{2+}/CALMODULIN) — *I*50 = 500 nM
TYROSINE 3-MONOOXYGENASE — NC — *I*50 = 80 pM

NEUROPEPTIDE Y PEPTIDE (22-36)
3′,5′-CYCLIC-NUCLEOTIDE PHOSPHODIESTERASE (Ca^{2+}/CALMODULIN) — *I*50 = 50 nM

NEUROSERPIN
PLASMIN
t-PLASMINOGEN ACTIVATOR
u-PLASMINOGEN ACTIVATOR

NEUROSPORA CRASSA ENDO-EXONUCLEASE INHIBITOR
ENDO-EXONUCLEASE

NEUROTOXIN
PROTEIN KINASE C — $I50 \approx 100\ \mu M$

NEUTROPHIL CATHEPSING INHIBITOR
CATHEPSIN G

NEUTROPHIL ELASTASE INHIBITOR
ELASTASE NEUTROPHIL

NEVADENSIN
PROTEIN TYROSINE KINASE p40 — $I50 = 50\ \mu g/ml$

NEVIRAPINE
RNA-DIRECTED DNA POLYMERASE — $I50 = 175$ nM
RNA-DIRECTED DNA POLYMERASE — $I50 = 84$ nM

.N-(FLUOROMETHYL)SACCHARIN
ELASTASE LEUKOCYTE — IR

NH_2-ALA-LEU-TYR-LEU-COOH
ATROLYSIN A — C — $Ki = 250\ \mu M$

NH_2-GLY-PHECHN_2
DIPEPTIDYL-PEPTIDASE 1

NHOH-SUCC-PFR
ASTACIN — $Ki = 16\ \mu M$
MEPRIN A — $Ki = 146\ \mu M$

NH_2-PHE-ALA-$CH_2S^+(CH_3)_2$
DIPEPTIDYL-PEPTIDASE 1

N-(H-PHE-PRO)-O-(4-NITROBENZOYL)HYDROXYLAMINE
DIPEPTIDYL-PEPTIDASE 4 — IR

Ni^{2+}
N-ACETYLGALACTOSAMIDINE β-1,3-N-ACETYLGLUCOSAMINYLTRANSFERASE
N-ACETYLGLUCOSAMINE-6-PHOSPHATE DEACETYLASE
N-ACETYL-γ-GLUTAMYL-PHOSPHATE REDUCTASE
α-N-ACETYLNEURAMINATE α-2,8-SIALYLTRANSFERASE
ACETYLORNITHINE DEACETYLASE — 73% 200 μM
O-ACETYLSERINE (THIOL)-LYASE
ACYLAGMATINE AMIDASE — 85% 10 mM
ACYLGLYCERONE-PHOSPHATE REDUCTASE
N-ACYLHEXOSAMINE OXIDASE
ADENOSINETRIPHOSPHATASE ECTO — 68% 500 μM
AGMATINE DEIMINASE
β-ALA-ARG-HYDROLASE — 62% 400 μM
ALAKN-1-OL DEHYDROGENASE (ACCEPTOR)
ALANINE DEHYDROGENASE
ALDOSE 1-DEHYDROGENASE
2-ALKYNOL-1-OL DEHYDROGENASE
ALLANTOATE DEIMINASE
AMINE OXIDASE (COPPER-CONTAINING)
AMINO-ACID N-ACETYLTRANSFERASE
D-AMINO-ACID OXIDASE
AMINOCARBOXYMUCONATE-SEMIALDEHYDE DEHYDROGENASE
AMINOMUCONATE-SEMIALDEHYDE DEHYDROGENASE
AMINOPEPTIDASE — 87% 1 mM
AMINOPEPTIDASE B — 75% 17 μM
AMINOPEPTIDASE B
AMP DEAMINASE
AMP-THYMIDINE KINASE
α-AMYLASE
β-AMYLASE
ANHYDROTETRACYCLINE MONOOXYGENASE
ARABINOSE ISOMERASE

ARGINASE		
ARYLACETONITRILASE		80% 1 mM
ARYL-ACYLAMIDASE		
ARYLAMINE N-ACETYLTRANSFERASE		
ARYLSULFATE SULFOTRANSFERASE		
ARYL SULFOTRANSFERASE	NC	33% 600 μM
L-ASCORBATE OXIDASE		
L-ASCORBATE PEROXIDASE		
ASPARAGINASE		
ASPARTATE CARBAMOYLTRANSFERASE		
ASPERGILLUS ALKALINE PROTEINASE		
ATP PHOSPHORIBOSYLTRANSFERASE		
BENZYL-2-METHYL-HYDROXYBUTYRATE DEHYDROGENASE		
BIS(5′-ADENOSYL)-TRIPHOSPHATASE		*I*50 = 40 μM
γ-BUTYROBETAINE DIOXYGENASE		
Ca^{2+} CARRIER	C	
CARBOSINE SYNTHASE		
3-CARBOXY-cis,cis-MUCONATE CYCLOISOMERASE		100% 100 μM
CARNITINE 3-DEHYDROGENASE		
2-CHLOROBENZOATE 1,2-DIOXYGENASE		
CHOLINESTERASE		
CHORISMATE MUTASE		
dCTP DEAMINASE		80% 4 mM
2′,3′-CYCLIC-NUCLEOTIDE 2′-PHOSPHODIESTERASE		42% 1 mM
2′,3′-CYCLIC-NUCLEOTIDE 2′-PHOSPHODIESTERASE		17% 200 μM
3′,5′-CYCLIC-NUCLEOTIDE PHOSPHODIESTERASE		
CYSTINYL AMINOPEPTIDASE		
CYTOSINE DEAMINASE		
3-DEHYDROQUINATE SYNTHASE		
DEOXYHYPUSINE MONOOXYGENASE		38% 5 μM
DEOXYRIBONUCLEASE γ		*I*50 = 150 μM
DEOXYRIBONUCLEASE II		64% 5 mM
DESULFOGLUCINOLATE SULFOTRANSFERASE		
DETHIOBIN SYNTHASE		
DEXTRANASE		86% 11 mM
DEXTRANSUCRASE	C	*K*i = 6.5 mM
α-DEXTRIN ENDO-1,6-α-GLUCOSIDASE		
DIHYDROXYPHENYLALANINE TRANSAMINASE		
2,5-DIKETO-D-GLUCONATE REDUCTASE		95% 500 μM
DIMETHYLHISTIDINE N-METHYLTRANSFERASE		
DIPEPTIDASE		50% 1 mM
DIPEPTIDYL-PEPTIDASE 3		
DIPEPTIDYL-PEPTIDASE 4		
DNA (CYTOSINE-5-)-METHYLTRANSFERASE		
DOLICHYL-DIPHOSPHATE-POLYPHOSPHATE PHOSPHOTRANSFERASE		
trans-2-ENOYL-CoA REDUCTASE (NAD)		
ESTRONE SULFOTRANSFERASE		
FAD PYROPHOSPHATASE		
Fe^{3+} REDUCTASE		
FORMALDEHYDE DEHYDROGENASE		
FORMALDEHYDE TRANSKETOLASE		
FORMATE DEHYDROGENASE		
α-L-FUCOSIDASE		
GALACTINOL-RAFFINOSE GALACTOSYLTRANSFERASE		
α-GALACTOSIDASE		
β-GALACTOSIDASE		
GALACTOSIDE 2-L-FUCOSYLTRANSFERASE		100% 10 mM
GDP-MANNOSE 4,6-DEHYDRATASE		*I*50 = 8 mM
GIBBERELLIN 3β-DIOXYGENASE		
GLIOCLADIUM PROTEINASE		
GLOBOSIDE α-N-ACETYLGALACTOSAMINYLTRANSFERASE		
GLUCAN ENDO-1,3-β-GLUCOSIDASE		
GLUCAN 1,3-β-GLUCOSIDASE		
GLUCAN 1,4-α-MALTOTETRAOHYDROLASE		

GLUCONATE DEHYDRATASE		
GLUCONATE 5-DEHYDROGENASE		
GLUCOSAMINE-PHOSPHATE N-ACETYLTRANSFERASE		
GLUCOSE 1-DEHYDROGENASE		
GLUCOSE 1-DEHYDROGENASE (NADP)		
GLUCOSE-6-PHOSPHATE 1-DEHYDROGENASE		
α-GLUCOSIDASE		
β-GLUCOSIDASE		80% 1 mM
D-GLUCURONATE DEHYDRATASE		
β-GLUCURONIDASE		11% 2 mM
β-GLUCURONIDASE		34% 1 mM
GLUTAMATE DEHYDROGENASE (NAD(P))		
GLUTAMATE DEHYDROGENASE (NADP)		
GLUTAMATE-5-SEMIALDEHYDE DEHYDROGENASE		
GLUTAMATE SYNTHASE (FERREDOXIN)		
GLUTAMATE SYNTHASE (NADPH)		
GLUTAMIN-(ASPARAGIN-)ASE		
GLUTAMYL AMINOPEPTIDASE		89% 1 mM
GLUTATHIONE PEROXIDASE		26% 5 mM
GLUTATHIONE REDUCTASE (NADPH)		
GLYCINE C-ACETYLTRANSFERASE		
GLYCINE ACYLTRANSFERASE		
GLYCINE BENZOYLTRANSFERASE		
GLYCINE DEHYDROGENASE (DECARBOXYLATING)		
GLYCOCYAMINASE		
GLYCOGEN SYNTHASE (CASEIN) KINASE	C	*I*50 = 284 μM
GLYCOLIPID 3-α-MANNOSYLTRANSFERASE		
GMP REDUCTASE		
dGTAPASE MUT	NC	*K*i = 900 μM
2-HALOACID DEHALOGENASE		43% 1 mM
3-HEXULOSE PHOSPHATE SYNTHASE		
HEXULOSE 6-PHOSPHATE SYNTHASE		64% 1 mM
HIPPURATE HYDROLASE		
HISTIDINOL DEHYDROGENASE		54% 500 μM
HISTONE ACYLTANSFERASE		
HURAIN		
3-HYDROXYANTHRANILATE 3,4-DIOXYGENASE		20% 1 mM
5-HYDROXYFURANOCOUMARIN 5-O-METHYLTRANSFERASE		
8-HYDROXYFURANOCOUMARIN 8-O-METHYLTRANSFERASE		
2-HYDROXYPYRIDINE 5-MONOOXYGENASE		
HYOSCYAMINE (6S)-DIOXYGENASE		
HYOSCYAMINE (6S)-DIOXYGENASE		
INULIN FRUCTOTRANSFERASE (DEPOLYMERIZING, DIFRUCTOFURANOSE-1,2′: 2′,1-DIANHYDRIDE-		
3-ISOPROPYLMALATE DEHYDROGENASE		
KALLIKREIN TISSUE		
KYNURENINE-OXOGLUTARATE AMINOTRANSFERASE		
LACTALDEHYDE DEHYDROGENASE		
D-LACTALDEHYDE DEHYDROGENASE 1		
L-LACTATE DEHYDROGENASE		
LACTOSYLCERAMIDE α-2,3-SIALYLTRANSFERASE		
LEUCYL AMINOPEPTIDASE		
MALATE DEHYDROGENASE		57% 33 mM
MALONATE-SEMIALDEHYDE DEHYDROGENASE		
MANNOSE ISOMERASE		
MANNOSE-6-PHOSPHATE ISOMERASE		
α-MANNOSIDASE		
2-METHYLCITRATE SYNTHASE		
4-METHYLGLUTAMINE SYNTHETASE		
N^6-METHYL-LYSINE OXIDASE		
MONOSIALOGANGLIOSIDE SIALYLTRANSFERASE		
MONOTERPENOL β-GLUCOSYLTRANSFERASE		
MUCOROPEPSIN		
NEOLACTOSYLCERAMIDE α-2,3-SIALYLTRANSFERASE		
NICOTINATE-NUCLEOTIDE PYROPHOSPHORYLASE (CARBOXYLATING)		40% 100 μM

NITRILASE		83% 1 1mM
5′-NUCLEOTIDASE		
OLIGO-1,6-GLUCOSIDASE		
2-OXOALDEHYDE DEHYDROGENASE (NAD)		
8-OXO-dGTPASE MTH1		*I*50 = 801 μM
8-OXO-dGTPASE MUT		*I*50 = 1.5 mM
PANTOATE-β-ALANINE LIGASE		
PEPTIDOGLYCAN GLUCOSYLTRANSFERASE		
PERILLYL-ALCOHOL DEHYDROGENASE		
PEROXIDASE		*I*50 = 5.3 mM
PHENYLALANINE AMMONIA-LYASE		75% 800 μM
PHOSPHATASE ALKALINE	NC	*K*i = 1.4 mM
1-PHOSPHATIDYLINOSITOL 4-KINASE		
PHOSPHODIESTERASE		
1-PHOSPHOFRUCTOKINASE		
6-PHOSPHOFRUCTOKINASE		
β-PHOSPHOGLUCOMUTASE		29% 200 μM
PHOSPHOGLYCERATE MUTASE		20% 1 mM
PHOSPHOGLYCOLATE PHOSPHATASE		*I*50 = 1 mM
6-PHOSPHO-3-KETOHEXULOSE ISOMERASE		
PHOSPHOLIPASE C		
PHOSPHOPYRUVATE HYDRATASE		
3-PHYTASE		
PLASMIN		
PROCOLLAGEN N-ENDOPEPTIDASE		42% 100 μM
PROCOLLAGEN GLUCOSYLTRANSFERASE		
PROLYL OLIGOPEPTIDASE		97% 1 mM
PROSTAGLANDIN F SYNTHASE		
PROTEINASE SERINE CANDIDA LIPOLYTICA		
PROTEIN-DISULFIDE REDUCTASE (GLUTATHIONE)		
PROTEIN KINASE CASEIN KINASE		
PROTEIN TYROSINE-PHOSPHATASE		99% 1 mM
PROTEIN TYROSINE-PHOSPHATASE (Mg^{2+} DEPENDENT)		48% 250 μM
PSEUDOMONAS AEROGINOSA NEUTRAL PROTEINASE		100% 1 mM
PUTRESCINE OXIDASE		
PYRANOSE OXIDASE		
PYRIDOXINE DEHYDROGENASE		
PYRIMIDINE-DEOXYNUCLEOSIDE 2′-DIOXYGENASE		
PYROPHOSPHATE-FRUCTOSE-6-PHOSPHATE 1-PHOSPHOTRANSFERASE		
PYRUVATE KINASE		
QUEUINE tRNA-RIBOSYLTRANSFERASE		
L-RHAMNOSE ISOMERASE		
RIBONUCLEASE α		97% 10 mM
RIBONUCLEASE H		*I*50 = 200 μM
RIBONUCLEOSIDE-TRIPHOSPHATE REDUCTASE		
tRNA(CYTOSINE-5-) METHYLTRANSFERASE		
tRNA(GUANINE-N^7-) METHYLTRANSFERASE		
tRNA(URACIL-5-) METHYLTRANSFERASE		
SARCOSINE OXIDASE		
SECONDARY-ALCOHOL OXIDASE		
SERINE CARBOXYPEPTIDASE		
SERRALYSIN		
SORGHUM ASPARTIC PROTEINASE		38% 5 mM
STEROID 5α-REDUCTASE 1		*K*i = 176 μM
SULFATE ADENYLYLTRANSFERASE		
L-THREONINE 3-DEHYDROGENASE		
THYMIDINE KINASE		
TISSUE-ENDOPEPTIDASE DEGRADING COLLAGENASE-SYNTHETIC-SUBSTRATE		100% 100 μM
TRIACYLGLYCEROL LIPASE		
TRIMETHYLAMINE-N-OXIDE REDUCTASE		
TYROSINE 3-MONOOXYGENASE		
UDP-N-ACETYLGLUCOSAMINE PYROPHOSPHORYLASE		
UDP-N-ACETYLMURAMATE-ALANINE LIGASE		
UDP-GLUCOSE-1-PHOSPHATE URIDYLYLTRANSFERASE		

UDP-GLUCURONOSYLTRANSFERASE		
URATE OXIDASE		
UREASE		
UREIDOGLYCOLATE LYASE		
URIDINE NUCLEOSIDASE		
XAA-HIS DIPEPTIDASE		100% 400 μM
XAA-PRO DIPEPTIDASE		
XAA-TRP AMINOPEPTIDASE		
XYLOSE ISOMERASE		
NIALAMIDE		
AMINE OXIDASE MAO		
NICARDEPINE		
Δ5 DESATURASE	U	*K*i = 62 μM
LINOLEOYL-CoA DESATURASE	U	*K*i = 44 μM
NICARDIPINE		
BUFURALOL-1′-HYDROXYLASE (CYT P-450dbl)	C	*I*50 = 2 μM
3′,5′-CYCLIC-NUCLEOTIDE PHOSPHODIESTERASE	C	*K*i = 13 μM
$NiCl_2$		
CYSTEINYL-GLYCINE DIPEPTIDASE		
DIHYDROCOUMARIN HYDROLASE		54% 3.3 mM
MANNITOL-1-PHOSPHATASE		
NICLOSAMIDE		
OXIDATIVE PHOSPHORYLATION UNCOUPLER		
NICOTIANA ALATA PROTEINASE INHIBITOR C1		
CHYMOTRYPSIN		
NICOTIANA ALATA PROTEINASE INHIBITORS		
CHYMOTRYPSIN		
TRYPSIN		
NICOTIANA ALATA PROTEINASE INHIBITOR T1		
TRYPSIN		
NICOTIANA ALATA PROTEINASE INHIBITOR T3		
TRYPSIN		
NICOTIANA ALATA PROTEINASE INHIBITOR T4		
TRYPSIN		
NICOTIANA ALATA TRYPSIN INHIBITOR T1		
TRYPSIN		
NICOTIANA ALATA TRYPSIN INHIBITOR T2		
TRYPSIN		
NICOTIANA ALATA TRYPSIN INHIBITOR T3		
TRYPSIN		
NICOTIANA ALATA TRYPSIN INHIBITOR T4		
TRYPSIN		
NICOTIANA ATLANTA PROTEINASE INHIBITOR T2		
TRYPSIN		
NICOTIANAMINE		
GELATINASE A		*I*50 = 200 nM
GELATINASE B		*I*50 = 230 nM
PEPTIDYL-DIPEPTIDASE A		*I*50 = 260 μM
NICOTIANA TABACUM TRYPSIN/CHYMOTRYPSIN INHIBITOR		
CHYMOTRYPSIN		
TRYPSIN		

NICOTIFLORIN		
ARACHIDONATE 5-LIPOXYGENASE		*I*50 = 120 μM
NICOTIN		
LYCOPENE CYCLASE		*I*50 = 40 μM
NICOTINALDEHYDE		
NICOTINAMIDASE		*K*i = 69 μM
NICOTINAMIDE		
ADP-RIBOSYL CYCLASE		*I*50 = 40 μM
ADP-RIBOSYL CYCLASE		*I*50 = 1.5 mM
CARBONATE DEHYDRATASE I		61% 10 μM
CHYMOTRYPSIN		53% 40 mM
ECDYSONE 20-MONOOXYGENASE		
NAD ADP-RIBOSYLTRANSFERASE		72% 100 μM
NAD ADP-RIBOSYLTRANSFERASE		*I*50 = 74 μM
NAD ADP-RIBOSYLTRANSFERASE	C	*K*i = 5.7 μM
NAD ADP-RIBOSYLTRANSFERASE		63% 50 μM
NAD ADP-RIBOSYLTRANSFERASE		*I*50 = 30 μM
NAD ADP-RIBOSYLTRANSFERASE		*I*50 = 31 μM
NAD-DINITROGEN-REDUCTASE ADP-D-RIBOSYLTRANSFERASE		
NAD NUCLEOSIDASE	NC	*K*i = 2.1 mM
NAD(P)-ARGININE ADP-RIBOSYLTRANSFERASE		*I*50 = 3.4 mM
NAD(P) NUCLEOSIDASE	NC	
NMN NUCLEOSIDASE		
TRYPSIN		40% 60 mM
UNSPECIFIC MONOOXYGENASE		
C-NICOTINAMIDE ADENINEDINUCLEOTIDE		
ALCOHOL DEHYDROGENASE		*K*i = 1.1 nM
NICOTINAMIDE 2′-DEOXYADENINEDINUCLEOTIDE		
NAD KINASE	C	*K*i = 390 μM
NICOTINAMIDE 2′-DEOXY-2′-FLUOROARABINOSIDE ADENINEDINUCLEOTIDE		
NAD(P) NUCLEOSIDASE	REV	
NICOTINAMIDE NUCLEOSIDE		
NICOTINAMIDE PHOSPHORIBOSYLTRANSFERASE	C	
NICOTINAMIDE RIBONUCLEOTIDE		
NICOTINAMIDE-NUCLEOTIDE ADENYLYLTRANSFERASE		
NICOTINATE RIBONUCLEOTIDE		
NICOTINAMIDE-NUCLEOTIDE ADENYLYLTRANSFERASE		
NICOTINATE PHOSPHORIBOSYLTRANSFERASE	C	*K*i = 35 μM
NICOTINATE PHOSPHORIBOSYLTRANSFERASE	C	*K*i = 40 μM
NICOTINE		
CAPSANTHIN CAPSORUBIN SYNHTHASE β CYLASE		*I*50 = 30 μM
CAPSANTHIN CAPSORUBIN SYNHTHASE k CYLASE		*I*50 = 20 μM
CYTOCHROME P450 2A6	C	*K*i = 24 μM
HISTAMINE N-METHYLTRANSFERASE	C	
STEROID 11β-MONOOXYGENASE	C	*K*i = 9.9 μM
STEROID 21-MONOOXYGENASE	C	*K*i = 110 μM
SUPEROXIDE DISMUTASE		*I*50 = 810 μM
(S)NICOTINE		
ARYLAMINE N-METHYLTRANSFERASE		C
NICOTINIC ACID		
β-N-ACETYLGALACTOSAMINIDASE		*K*i = 11 mM
AMINE OXIDASE MAO-B		*K*i = 4 mM
D-AMINO-ACID OXIDASE		
IMIDAZOLEACETATE 4-MONOOXYGENASE		
LEGHEMOGLOBIN REDUCTASE		

NICOTINIC ACID HYDRAZIDE
NICOTINAMIDASE $I50 = 5\ \mu M$

NICOTINIC ACID MONONUCLEOTIDE
3-HYDROXYANTHRANILATE 3,4-DIOXYGENASE
NICOTINATE-NUCLEOTIDE PYROPHOSPHORYLASE (CARBOXYLATING)

NICOTINOHYDROXAMIC ACID
UREASE $I50 = 41\ \mu M$
UREASE $I50 = 2.9\ \mu M$
UREASE $I50 = 4.1\ \mu M$

NICOTINYLALANINE
KYNURENINASE $I50 = 800\ \mu M$
KYNURENINE 3-MONOOXYGENASE $I50 = 900\ \mu M$

NIDROXYZONE
GLUTATHIONE REDUCTASE (NADPH) 30% 20 μM

NIFLUNAMIC ACID
PROSTAGLANDIN-ENDOPEROXIDE SYNTHASE 1 $I50 = 16\ \mu M$
PROSTAGLANDIN-ENDOPEROXIDE SYNTHASE 2 $I50 = 100$ nM

NIFURTIMOX
GLUTATHIONE REDUCTASE (NADPH) $Ki = 30\ \mu M$
TRYPANOTHIONE REDUCTASE $I50 = 200\ \mu M$

NIGERAN TETRASACCHARIDES
MYCODEXTRANASE $Ki = 4.8$ mM

NIGERAN TRISACCHARIDES
MYCODEXTRANASE $Ki = 30$ mM

NIGEROSE
D-GLUCANSUCRASE $Ki = 15$ mM
GLUCOSIDASE II $I50 = 1$ mM
MYCODEXTRANASE $Ki = 60$ mM

NIGRANOIC ACID
DNA POLYMERASE α $I50 = 425\ \mu M$
DNA POLYMERASE β $I50 = 17\ \mu M$
RNA-DIRECTED DNA POLYMERASE $I50 = 159\ \mu M$
RNA-DIRECTED DNA POLYMERASE $I50 = 357\ \mu M$
RNA POLYMERASE $I50 = 188\ \mu M$

NIK 247
CHOLINESTERASE $I50 = 1.3\ \mu M$

NIKKOMYCIN
CHITIN SYNTHASE C

NIKKOMYCIN X
CHITIN SYNTHASE $I50 = 100$ ng/ml

NIKKOMYCIN Z
CHITIN SYNTHASE $I50 = 300$ ng/ml
CHITIN SYNTHASE 91% 10 μM
CHITIN SYNTHASE Chs 1 $Ki = 210$ nM
CHITIN SYNTHASE Chs 2 $Ki = 890\ \mu M$
CHITIN SYNTHASE Chs 3 $Ki = 1.2\ \mu M$

NILE BLUE A
MYOSIN-LIGHT-CHAIN KINASE
PROTEIN KINASE A
PROTEIN KINASE C
PROTEIN KINASE (Ca^{2+} DEPENDENT)

NIMBIN

ECDYSONE 20-MONOOXYGENASE		*I*50 = 1 mM

NIMESULIDE

3′,5′-CYCLIC-NUCLEOTIDE PHOSPHODIESTERASE IV		*I*50 = 39 μM
CYCLOOXYGENASE 1		*I*50 = 300 μM
CYCLOOXYGENASE 2		*I*50 = 14 μM

NINADATE

HYDROXYMETHYLBILANE SYNTHASE		

NINHYDRIN

ACONITATE HYDRATASE		*I*50 = 300 μM
XANTHINE OXIDASE		

$NISO_4$

CYCLAMATE SULFOHYDROLASE		80% 10 mM

NITECAPONE

CATECHOL O-METHYLTRANSFERASE		*K*i = 700 pM
CATECHOL O-METHYLTRANSFERASE		*K*i = 1 nM

NITIDINE

DNA TOPOISOMERASE		

NITIDINE CHLORIDE

DNA LIGASE I		*I*50 = 69 μM
RNA-DIRECTED DNA POLYMERASE		*I*50 = 18 μM
RNA-DIRECTED DNA POLYMERASE		*I*50 = 19 μM

NITRATE

ACETOACETATE DECARBOXYLASE		
ACETYLENECARBOXYLATE HYDRATASE		
N-ACETYLGALACTOSAMINE-6-SULFATASE		56% 17 mM
ACONITATE HYDRATASE		
ACYLAMINOACYL-PEPTIDASE		
ADENYLOSUCCINATE SYNTHASE		
4-AMINOBENZOATE 1-MONOOXYGENASE		
AMP DEAMINASE		
L-ASCORBATE OXIDASE		
ASPARAGINE-tRNA LIGASE		
CARBAMOYL-PHOSPHATE SYNTHASE (AMMONIA)	C	
CARBONATE DEHYDRATASE		
CARBOXY-cis,cis-MUCONATE CYCLASE		
CATALASE		
CREATINE KINASE	C	*K*i = 20 μM
CYANATE HYDROLASE		
2-DEHYDRO-3-DEOXYGLUCARATE ALDOLASE		
FORMATE DEHYDROGENASE		100% 250 μM
FUMARATE HYDRATASE		
GLUCOKINASE	NC	
GLUCOSE OXIDASE		
γ-GLUTAMYL HYDROLASE		
GLUTATHIONE TRANSFERASE β	NC	*K*i = 260 μM
ISOCITRATE LYASE		
LACTOYLGLUTATHIONE LYASE		
MALONATE-SEMIALDEHYDE DEHYDRATASE		
NADH DEHYDROGENASE		
NITROETANE REDUCTASE		
PHOSPHOGLUCOMUTASE		
PHOTINUS-LUCIFERIN 4-MONOOXYGENASE (ATP-HYDROLYSING)		
PYRUVATE DEHYDROGENASE (LIPOAMIDE)		
SALICYLATE 1-MONOOXYGENASE		
SULFATE ADENYLYLTRANSFERASE		*I*50 = 192 μM
SULFATE ADENYLYLTRANSFERASE	C	*K*i = 1.2 mM
SULFITE REDUCTASE (NADPH)		

NITRATE REDUCTASE INHIBITOR
NITRATE REDUCTASE

NITRIC OXIDE
ACONITATE HYDRATASE
ALCOHOL DEHYDROGENASE
ALDEHYDE DEHYDROGENASE $I50 = 150\ \mu M$
CASPASES
CATALASE $Ki = 180\ nM$
CREATINE KINASE
CYCLOOXYGENASE
CYTOCHROME-C3 HYDROGENASE
CYTOCHROME-C OXIDASE
CYTOCHROME-C PEROXIDASE
CYTOCHROME-C PEROXIDASE
CYTOCHROME OXIDASE
CYTOCHROME P450 2E1
DNA LIGASE
FORMAMIDOPYRIMIDINE-DNA GLYCOSIDASE $I50 \approx 120\ \mu M$
GLYCERALDEHYDE-3-PHOSPHATE DEHYDROGENASE (PHOSPHORYLATING)
HYDROGEN DEHYDROGENASE
LIPOXYGENASE
METHIONINE SYNTHASE VIT B12 DEPENDENT
NITRIC-OXIDE SYNTHASE
NITROGENASE
PHOSPHOLIPASE D
PROTEIN TYROSINE KINASE EGFR
PSEUDOMONAS CYTOCHROME OXIDASE
RIBONUCLEOSIDE-DIPHOSPHATE REDUCTASE IR
TRYPTOPHAN 5-MONOOXYGENASE IR
UPTAKE HYDROGENASE C $Ki = 12\ \mu M$

NITRIC OXIDE DONORS
Na^+/K^+-EXCHANGING ATPASE

NITRILACARB
ACETYLCHOLINESTERASE

NITRILOACETAE
CARBON-MONOXIDE DEHYDROGENASE

NITRILOACETIC ACID
ISOCITRATE DEHYDROGENASE (NAD)

NITRILOTRIACETIC ACID
ALTRONATE DEHYDRATASE 69% 3 mM
MANGANESE PEROXIDASE
MANNONATE DEHYDRATASE 70% 3mM

NITRITE
ACONITATE HYDRATASE
ADENYLOSUCCINATE SYNTHASE
ARGINASE NC $Ki = 4.9\ mM$
CATALASE
CYANATE HYDROLASE
CYTOCHROME-C3 HYDROGENASE
HYDROXYLAMINE REDUCTASE (NADH)
METHANOL-5-HYDROXYBENZIMIDAZOLYLCOBAMIDE Co-METHYLTRANSFERASE
NITRATE REDUCTASE (NADH)
NITRATE REDUCTASE (NAD(P)H)
NITRITE REDUCTASE (NAD(P)H)
PSEUDOMONAS CYTOCHROME OXIDASE
PYRUVATE SYNTHASE

p-NITROACETOPHENONE
ALDEHYDE DEHYDROGENASE (NAD) $Ki = 93\ \mu M$

NITROACRYLIC ACID
ADENYLOSUCCINATE LYASE

3-NITROACRYLIC ACID
SUCCINATE DEHYDROGENASE SSI

3-NITROALANINE
ALANINE TRANSAMINASE SSI
ASPARTATE TRANSAMINASE SSI

3-NITRO-5α-ANDROST-3-ENE-17β-DIISOPROPYLCARBOXAMIDE
STEROID 5α-REDUCTASE C Ki = 50 nM

p-NITROANILINE
AMINOPEPTIDASE Ki = 2.7 mM
AMINOPEPTIDASE B Ki = 440 μM
METHIONYL AMINOPEPTIDASE Ki = 1.3 mM

L-N^G-NITROARGININE
NITRIC-OXIDE SYNTHASE I I50 = 17 nM
NITRIC-OXIDE SYNTHASE II I50 = 3.5 μM
NITRIC-OXIDE SYNTHASE III I50 = 90 nM

L-N^G-NITROARGININE METHYLESTER
NITRIC-OXIDE SYNTHASE I I50 = 79 nM
NITRIC-OXIDE SYNTHASE I I50 = 660 nM
NITRIC-OXIDE SYNTHASE II I50 = 11 μM
NITRIC-OXIDE SYNTHASE III I50 = 6.5 μM
NITRIC-OXIDE SYNTHASE III I50 = 985 nM

L-N^g-NITRO ARGININE METHYLESTER
NITRIC-OXIDE SYNTHASE I50 = 870 nM

N^G-NITROARGININE METHYLESTER
ORNITHINE DECARBOXYLASE Ki = 20 μM
TYROSINE-ARGININE LIGASE I50 = 2.3 mM

NG-NITRO-L-ARGININE METHYLESTER
NITRIC-OXIDE SYNTHASE IR I50 = 300 μM

N^G-NITROARGININE
ARGINASE I50 = 27 mM
NADPH DEHYDROGENASE 53% 3 mM
NITRIC-OXIDE SYNTHASE Kd = 15 nM
NITRIC-OXIDE SYNTHASE I50 = 800 nM
NITRIC-OXIDE SYNTHASE I Ki = 200 nM
NITRIC-OXIDE SYNTHASE I Ki = 25 nM
NITRIC-OXIDE SYNTHASE I Kd = 15 nM
NITRIC-OXIDE SYNTHASE I Ki = 90 nM
NITRIC-OXIDE SYNTHASE II Ki = 8.7 μM
NITRIC-OXIDE SYNTHASE II Ki = 670 nM
NITRIC-OXIDE SYNTHASE II I50 = 20 μM
NITRIC-OXIDE SYNTHASE III I50 = 2 μM
NITRIC-OXIDE SYNTHASE III I50 = 300 μM
NITRIC-OXIDE SYNTHASE III Kd = 39 nM

N^G-NITRO-L-ARGININE
NITRIC-OXIDE SYNTHASE I I50 = 800 nM

NG-NITRO-L-ARGININE
NITRIC-OXIDE SYNTHASE I50 = 3 μM
NITRIC-OXIDE SYNTHASE IR I50 = 80 μM

N{[(4-(4-NITROAROYLAMINO))PHENYL]SULPHONYL-N-PHENYL-GLYCINE
ALDEHYDE REDUCTASE I50 = 280 nM

p-NITROBENZALDEHYDE
ALCOHOL DEHYDROGENASE (NADP)

3-NITROBENZAMIDE

NAD ADP-RIBOSYLTRANSFERASE		*I*50 = 160 μM
NAD ADP-RIBOSYLTRANSFERASE		71% 50 μM

p-NITROBENZAMIDINE

COMPLEMENT SUBCOMPONENT C1s	C	

3-NITROBENZENEBORONIC ACID

SUBTILISIN		*K*i = 10 μM

p-NITROBENZENEDIAZONIUM FLUOROBORATE

RHIZOPUSPEPSIN		

5-(4-NITROBENZENESULPHONYLAMIDO)-1,3,4-THIADIAZOLE-2-SULPHONAMIDE

CARBONATE DEHYDRATASE I		*I*50 = 2 nM
CARBONATE DEHYDRATASE II		*I*50 = 100 pM
CARBONATE DEHYDRATASE IV		*I*50 = 3 nM

trans-1-p-NITROBENZENESULPHONYL-3-ETHYL-4-ETHOXYCARBONYLAZETIDINONE

ELASTASE LEUKOCYTE		*I*50 = 10 μg/ml

4-NITROBENZENESULPHONYLFLUORIDE

EPOXIDE HYDROLASE		47% 1 mM
EPOXIDE HYDROLASE		94% 1 mM

3-[4-NITRO-1,2-BENZISOTHIAZOL-1,1-DIOXIDE-3(2H)-ON-2-YL]ACETIC ACID

ALDEHYDE REDUCTASE		*I*50 = 5.5 μM

S-7-NITROBENZOFURAN-4-YL-CoA

NAD(P) TRANSHYDROGENASE (AB-SPECIFIC)		

S-7-NITROBENZOFURAN-4-YL-DEPHOSPHO-CoA

NAD(P) TRANSHYDROGENASE (AB-SPECIFIC)		

3-NITROBENZOIC ACID

D-AMINO-ACID OXIDASE		
GLUTAMATE DEHYDROGENASE (NAD(P))	C	*K*i = 3.4 mM
GLUTAMATE DEHYDROGENASE (NAD(P))		

4-NITROBENZOIC ACID

BENZOATE 4-MONOOXYGENASE		
PROPIONATE-CoA LIGASE	C	*K*i = 4.7 mM

N-(7-NITROBENZ-2-OXA-1,3-DIAZOL-4-YL)COLCEMIDE

ALCOHOL DEHYDROGENASE	MIX	*K*i = 32 μM

α-N-(p-NITROBENZOXYCARBONYL)-L-ARGINYLCHLOROMETHYLKETONE

AGKISTRODON SERINE PROTEINASE		

S-p-NITROBENZOXYCARBONYLGLUTATHIONE

HYDROXYACYLGLUTATHIONE HYDROLASE	C	*K*i = 1.1 μM
HYDROXYACYLGLUTATHIONE HYDROLASE	C	*K*i = 6.5 μM
LACTOYLGLUTATHIONE LYASE		*K*i = 3.1 μM
LACTOYLGLUTATHIONE LYASE		*K*i = 3.1 μM

(m-NITROBENZOYL)ALANINE

KYNURENINASE		*I*50 = 100 μM
KYNURENINE 3-MONOOXYGENASE		*I*50 = 900 nM

(S)(+)-(m-NITROBENZOYL)ALANINE

KYNURENINASE		*I*50 = 60 μM
KYNURENINE 3-MONOOXYGENASE		*I*50 = 500 nM

(S)-(m-NITROBENZOYL)ALANINE

KYNURENINE 3-MONOOXYGENASE		

N-(3-NITROBENZOYL)GLUCOSAMINE

HEXOKINASE	C	*K*i = 45 μM

m-NITROBENZYLALANINE

KYNURENINASE		$I50 = 100\ \mu M$
KYNURENINE 3-MONOOXYGENASE		$I50 = 100$ nM

4-NITROBENZYL-4-GUANIDINOBENZOIC ACID

PLASMINOGEN ACTIVATOR

4-(4-NITROBENZYLIDENE)AMINOETHYLBENZENESULPHONAMIDE

CARBONATE DEHYDRATASE I		$Ki = 10\ \mu M$
CARBONATE DEHYDRATASE II		$Ki = 20$ nM
CARBONATE DEHYDRATASE IV		$Ki = 7$ nM

3-NITROBENZYLIDENE-N-HYDROXY-N′-AMINOGUANIDINE

RIBONUCLEOSIDE-DIPHOSPHATE REDUCTASE		$Ki = 77\ \mu M$
RIBONUCLEOSIDE-DIPHOSPHATE REDUCTASE		$Ki = 13\ \mu M$

6-(4-NITROBENZYLMERCAPTO)PURINE RIBONUCLEOSIDE

ADENOSINE TRANSPORT		$I50 = 20$ nM

N-(3-NITROBENZYL)-1,8-NAPHTHALIMIDE

THYMIDYLATE SYNTHASE		35% 5 μM

4-NITRO-2-BENZYL-1-NAPHTHOL

ARACHIDONATE 5-LIPOXYGENASE		$I50 = 48\ \mu M$

N-α-NITROBENZYLOXYCARBONYLARGINYLCHLOROMETHYLKETONE

CLOSTRIPAIN

p-NITROBENZYLSELENOSULPHATE

GLUTATHIONE REDUCTASE (NADPH)

p-NITROBENZYLSULPHONAMIDE

DIHYDROPYRIMIDINASE	C	$Ki = 600\ \mu M$

S(p-NITROBENZYL)-6-THIOGUANOSINE

NUCLEOSIDE TRANSPORT

NITROBENZYLTHIOINOSINE

ADENOSINE TRANSPORT (NBTI INSENSITIVE 1)		$I50 = 1\ \mu M$
ADENOSINE TRANSPORT (NBTI SENSITIVE)		$I50 = 300$ pM
NUCLEOSIDE TRANSPORT		$I50 = 35$ nM

6[(4-NITROBENZYL)THIO]-9-β-D-RIBOFURANOSYLPURINE

NUCLEOSIDE TRANSPORT

7-[(NITROBENZYL)-THIO]-3-(β-D-RIBOFURANOSYL)PYRAZOLO[4,3-d]PYRIMIDINE

NUCLEOSIDE TRANSPORT		$I50 = 18$ nM

N-(p-NITROBENZYOYL)-N-(PHENYL)-GLYCINE

ALDEHYDE REDUCTASE		$I50 = 2.6\ \mu M$

NITROBLUE TETRAZOLIUM

HYOSCYAMINE (6S)-DIOXYGENASE
HYOSCYAMINE (6S)-DIOXYGENASE
NITRIC-OXIDE SYNTHASE
NITROGENASE
PROCOLLAGEN-LYSINE 5-DIOXYGENASE
PROCOLLAGEN-PROLINE, 2-OXOGLUTARATE-4-DIOXYGENASE

3-NITRO-4-BROMOBENZENBORONIC ACID

TRIACYLGLYCEROL LIPASE		$Ki = 163$ nM

9-NITROCAMPTOTHECIN

DNA TOPOISOMERASE

m-NITROCARBOBENZOXYCARBONYLGLUTATHIONE

HYDROXYACYLGLUTATHIONE HYDROLASE		$Ki = 9\ \mu M$

o-NITROCARBOBENZOXYCARBONYLGLUTATHIONE

HYDROXYACYLGLUTATHIONE HYDROLASE		*K*i = 15 µM

p-NITROCARBOBENZOXYCARBONYLGLUTATHIONE

HYDROXYACYLGLUTATHIONE HYDROLASE		*K*i = 6.5 µM

S-p-NITROCARBOBENZOXY-GLUTATHIONE

HYDROXYACYLGLUTATHIONE HYDROLASE		*K*i = 6.5 µM

2-NITRO-4-CARBOXYPHENYL-N,N-DIPHENYLCARBAMATE

CHYMOTRYPSIN	IR	
ENVELYSIN		

4-NITROCATECHOL

CATECHOL O-METHYLTRANSFERASE		
CATECHOL OXIDASE		
LIPOXYGENASE		
PROTOCATECHUATE 3,4-DIOXYGENASE	C	*K*i = 1 µM
PROTOCATECHUATE 4,5-DIOXYGENASE		

p-NITROCATECHOL

3,4-DIHYDROXYPHENYLACETATE 2,3-DIOXYGENASE		*K*i = 14 µM

NITROCATECHOLSULPHATE

IDURONATE 2-SULFATASE		*I*50 = 100 µM

p-NITROCATECHOLSULPHATE

CEREBROSIDE-SULFATASE		
N-SULFOGLUCOSAMINE SULFOHYDROLASE		100% 20 mM

3′-NITRO-4′-CHLOROBENZYLIDENE-N-HYDROXY-N′-AMINOGUANIDINE

RIBONUCLEOSIDE-DIPHOSPHATE REDUCTASE		*I*50 = 20 µM

4-NITROCHOLESTEROL

β-GLUCURONIDASE	MIX	*K*i = 180 µM

3-NITROCOUMARIN

1-PHOSPHATIDYLINOSITOL-4,5-BISPHOSPHATE PHOSPHODIESTERASE		*I*50 = 208 µM
PHOSPHOLIPASE C		*I*50 = 16 µg/ml

2-NITRO-p-CRESOL

VANILLYL-ALCOHOL OXIDASE	C	

2-NITRO-1-(4,5-DIMETHOXY-2-NITROPHENYL)ETHYL

β-GALACTOSIDASE		

NITROFEN

PROTOPORPHYRINOGEN OXIDASE		

5′-O-NITRO-5′-FLUOROURIDINE

URIDINE KINASE		

5-NITRO 2-FURALDEHYDE

GLUTATHIONE REDUCTASE (NADPH)		

5-NITRO-2-FURALDEHYDE SEMICARBAZONE

GLUTATHIONE REDUCTASE (NADPH)		33% 20 µM

2-NITROFURAN

GLUTATHIONE REDUCTASE (NADPH)		26% 20 µM

NITROFURANTOIN

GLUTATHIONE REDUCTASE (NADPH)		*I*50 = 20 µM
GLUTATHIONE REDUCTASE (NADPH)	U	*K*i = 100 µM
TRYPANOTHIONE REDUCTASE	NC	*K*i = 40 µM
XI CRYSTALLIN	C	*K*i = 50 µM

2-(5-NITRO-2-FURNYLMETHYLIDENE)-n,n′-[1,4-PIPERAZINEDIYLBIS(1,3-PROPANEDIYL)]BISHYDRAZINECARBOXIMIDAMIDE HBR

TRYPANOTHIONE REDUCTASE	NC	*I*50 = 500 nM

5-NITROFUROATE		
GLUTAMATE DEHYDROGENASE (NAD(P))		
5-NITRO-2-FUROIC ACID		
GLUTAMATE DEHYDROGENASE (NAD(P))	C	*K*i = 170 μM
NITROFURYLACRYLAMIDE		
PROTEINASE ASPARTIC VARIOTI ASPARCTIC		
1-(5-NITRO-2-FURYL)-2-PHENYLSULPHINYL-2-FURYLCARBONYL ETHYLENE		
Na^+/K^+-EXCHANGING ATPASE		
1-(5-NITRO-2-FURYL)-2-PHENYLSULPHONYL-2-FURYLCARBONYL ETHYLENE		
Na^+/K^+-EXCHANGING ATPASE		*K*i = 13 μM
NITROGEN		
NITROGENASE	C	
NITROGEN MUSTARD		
GLUTATHIONE REDUCTASE (NADPH)		
NITROGLYCERIN		
ALDEHYDE DEHYDROGENASE (NAD) E2		*K*i = 44 μM
3-NITRO-4-HYDROXYBENZOIC ACID		
PROTOCATECHUATE 4,5-DIOXYGENASE		
5-NITRO-8-HYDROXYQUINOLINE		
FORMATE DEHYDROGENASE		
NG-NITRO-L-(N5-[IMINO(NITROAMINO)METHYL]-L-ORNITHINE		
NITRIC-OXIDE SYNTHASE		
7-NITROINDAZOLE		
AMINE OXIDASE MAO-B		
NITRIC-OXIDE SYNTHASE		*I*50 = 900 nM
NITRIC-OXIDE SYNTHASE		
NITRIC-OXIDE SYNTHASE		*I*50 = 470 nM
NITRIC-OXIDE SYNTHASE I		*I*50 = 4.8 μM
NITRIC-OXIDE SYNTHASE II		*I*50 = 6.9 μM
NITRIC-OXIDE SYNTHASE III		*I*50 = 2.1 μM
NITROMETHANE		
HISTIDINE AMMONIA-LYASE		
5-NITRO-2-METHYL FURAN		
GLUTATHIONE REDUCTASE (NADPH)		23% 20 μM
1-NITRO-2-NAPHTHOL		
N-HYDROXYARYLAMINE O-ACETYLTRANSFERASE		
1-NITRO-2-NAPTHOL-3,6-DISULPHONIC ACID		
CYSTATHIONINE γ-LYASE		55% 2 mM
5-NITRONORVALINE		
ALANINE TRANSAMINASE	SSI	
ASPARTATE TRANSAMINASE	SSI	
2-NITROORCINOL		
ORCINOL 2-MONOOXYGENASE		
3-NITRO-2-OXAZOLIDINONE		
ALDEHYDE DEHYDROGENASE (NAD)	C	*K*i = 3.5 μM
1-NITRO-3-OXO-BUTANE		
PORPHOBILINOGEN SYNTHASE		*K*i = 18 μM
2-NITRO-6(5H)-PHENANTHRIDINONE		
NAD ADP-RIBOSYLTRANSFERASE		*I*50 = 350 nM
NAD(P)-ARGININE ADP-RIBOSYLTRANSFERASE		*I*50 = 83 μM

N-(2-p-NITROPHENETHYL)CHLOROFLUOROACETAMIDE
CYTOCHROME P450 2B1

N-(2-p-NITROPHENETHYL)DICHLOROACETAMIDE
CYTOCHROME P450 3A1
CYTOCHROME P450 3A2

2-NITROPHENOL
AMIDASE
CATECHOL 2,3-DIOXYGENASE
MONOPHENOL MONOOXYGENASE C
2-NITROPHENOL 2-MONOOXYGENASE

3-NITROPHENOL
Ca^{2+}-TRANSPORTING ATPASE
PROTOCATECHUATE 3,4-DIOXYGENASE

4-NITROPHENOL
β-N-ACETYLGALACTOSAMINIDASE Ki = 37 μM
ALCOHOL DEHYDROGENASE
L-ASCORBATE OXIDASE Ki = 4.6 mM
CATECHOL OXIDASE C Ki = 62 μM
CATECHOL OXIDASE
Ca^{2+}-TRANSPORTING ATPASE
CHOLINE SULFOTRANSFERASE
CYCLOOXYGENASE I50 = 250 μM
CYTOCHROME P450 2A6 C Ki = 46 μM
β-GLUCOSIDASE C Ki = 1.5 mM
INDOLE-3-ACETALDEHYDE OXIDASE
LACCASE
MINOXIDIL SULFOTRANSFERASE 55% 500 nM
MONOPHENOL MONOOXYGENASE C
OLIGO-1,6-GLUCOSIDASE 100% 2 mM
PHOSPHATASE ACID NC Ki = 80 mM
3-PHYTASE
UDP-GLUCURONOSYLTRANSFERASE

p-NITROPHENYL-2-ACETAMIDO-2-DEOXY-3-O-β-D-GALACTOPYRANOSYL-β-GALACTOPYRANOSIDE
MUCINAMINYLSERINE MUCINAMINIDASE

4-NITROPHENYL O(2-ACETAMIDO-2-DEOXY-β-D-GLUCOPYRANOSYL)-(1- > 2)-O-(4,6-DI-O-METHYL-α-DMANNOPYRANOSYL)-(1- > 2)-β-D-GLUCOPYRANOSIDE
N-ACETYLGLUCOSAMINYLTRANSFERASE

4-NITROPHENYL O(2-ACETAMIDO-2-DEOXY-β-D-GLUCOPYRANOSYL)-(1- > 2)-O-(6-O-METHYL-α-D-
N-ACETYLGLUCOSAMINYLTRANSFERASE

p-NITROPHENYLACETIC ACID
CARBOXYPEPTIDASE A C Ki = 2.5 mM

4-NITROPHENYLACYLMETHYLPHOSPHONATES
TRYPSIN IR

p-NITROPHENYL-N-[succinyl-L-ALANYL-L-ALANYL-L-PROPYLMETHYL]-N-ISOPROPYL CARBAMATE
ELASTASE LEUKOCYTE I50 = 460 nM

4-NITROPHENYL-3-AMIDINOPHENYLMETHANE SULPHONATE
CHYMOTRYPSIN Ki = 13 μM
KALLIKREIN PLASMA Ki = 19 μM
PLASMIN Ki = 42 μM
THROMBIN Ki = 53 μM
TRYPSIN Ki = 9 μM

4-NITROPHENYL-4-AMIDINOPHENYLMETHANE SULPHONATE
CHYMOTRYPSIN Ki = 36 μM
KALLIKREIN PLASMA Ki = 47 μM
PLASMIN Ki = 50 μM

THROMBIN	IR	
TRYPSIN		Ki = 70 μM
p-NITROPHENYLAMIDRAZONE		
NUCLEOSIDE HYDROLASE		Ki = 10 μM
2-(3-NITROPHENYLAMINO)-3-NITRO-2H-CHROMENE		
1-PHOSPHATIDYLINOSITOL-4,5-BISPHOSPHATE PHOSPHODIESTERASE		I50 = 10 μM
4-NITROPHENYL-4-AMINOPHENYLMETHANESULPHONATE		
THROMBIN	IR	
p-NITROPHENYLARSONATE		
SUBTILISIN		
p-NITROPHENYL[[N(BENZYLOXYCARBONYL)AMINO]METHYL]PHOSPHONATE		
β-LACTAMASE		
p-NITROPHENYL-N-BUTYL CARBAMATE		
STEROL ESTERASE	IRR	
STEROL ESTERASE (NEUTRAL)		I50 = 250 μM
p-NITROPHENYL-N-BUTYLCARBAMATE		
STEROL ESTERASE		Ki = 2.6 μM
S-(2-NITROPHENYL)CYSTEINE		
KYNURENINASE	C	Ki = 100 μM
S-(2-NITROPHENYL)CYSTEINE S, S-DIOXIDE		
KYNURENINASE	C	Ki = 2.3 μM
p-NITROPHENYLDIMETHYLCARBAMATE		
CARBOXYLESTERASE		
p-NITROPHENYL-1,5-DITHIO-α-L-FUCOPYRANOSIDE		
α-L-FUCOSIDASE		Ki = 3.3 μM
p-NITROPHENYL-N-DODECYL CARBAMATE		
STEROL ESTERASE (NEUTRAL)		83% 250 μM
1-NITRO-2-PHENYLETHENE		
β-GALACTOSIDASE	IR	
N-(4-NITROPHENYLETHYLPHOSPHONYL)GLY-L-PRO-L-2-AMINOHEXANOIC ACID		
COLLAGENASE CLOSTRIDIUM HISTOLYTICUM		Ki = 5 nM
p-NITROPHENYL-β-.GALACTOSIDE		
β-MANNOSIDASE		
p-NITROPHENYL-α-D-GLUCOPYRANOSIDE		
β-GLUCOSIDASE		49% 15 mM
p-NITROPHENYL-α-D-GLUCOSE		
CYCLOMALTODEXTRIN GLUCANOTRANSFERASE		
p-NITROPHENYL-α-GLUCOSIDE		
MANNOSYL-OLIGOSACCHERIDE GLUCOSIDASE		65% 5 mM
SUCROSE α-GLUCOSIDASE	C	
p-NITROPHENYL-α-D-GLUCOSIDE		
GLUCOSIDASE II		76% 1 mM
GLUCOSIDASE II		I50 = 1 mM
p-NITROPHENYL-β-GLUCOSIDE		
THIOGLUCOSIDASE		
p-NITROPHENYL-β-D-GLUCOSIDE		
GLUCOSIDASE II		39% 6 mM
α,α-TREHALASE		Ki = 2.5 mM

(2S,3S)-3-(4-NITROPHENYL)GLYCIDOL
EPOXIDE HYDROLASE $I50 = 1.6\ \mu M$

4-NITROPHENYL 4-GUANIDINOBENZOIC ACID
CHYMASE $I50 = 6.3\ \mu M$
PROTEINASE In $I50 = 102\ \mu M$
TRYPTASE $I50 = 1$ pM

4-NITROPHENYL-4-GUANIDINOBENZOIC ACID
AGKISTRODON SERINE PROTEINASE
COMPLEMENT SUBCOMPONENT C1r
COMPLEMENT SUBCOMPONENT C1s
DIPEPTIDYL-PEPTIDASE 2
ENVELYSIN
PROTEINASE ALFA ALFA $Ki = 430$ nM
TRYPSIN $Ki = 420$ nM

p-NITROPHENYLHYDRAZINE
ALCOHOL DEHYDROGENASE (ACCEPTOR)

p-NITROPHENYL-α-MANNOPYRANOSIDE
MANNOSYL-GLYCOPROTEIN ENDO-β-N-ACETYLGLUCOSAMINIDASE 70% 30 mM

p-NITROPHENYL-α-D-MANNOPYRANOSIDE
β-MANNOSIDASE C

p-NITROPHENYL-β-D-MANNOPYRANOSIDE
β-GLUCOSIDASE

p-NITROPHENYLMANNOSE
UDP-N-ACETYLGLUCOSAMINE-LYSOSOMAL-ENZYME N-ACETYLGLUCOSAMINEPHOSPHOTRANSFERASE

p-NITROPHENYL α-D-MANNOSIDE
GLUCOSIDASE II 79% 6 mM

p-NITROPHENYL β-D-MANNOSIDE
GLUCOSIDASE II 51% 6 mM

4-NITROPHENYL 4-METHOXYPHENACYL METHYLPHOSPHONATE ESTER
THROMBIN IR

p-NITROPHENYL-N-[METHOXYSUCCINYL(N_δ-BENZOYL-L-ORNITHYL-L-ALANYL-L-PORLYLMETHYL]-N-IISOPROPYLCARBAMATE
ELASTASE LEUKOCYTE $Ki = 190$ nM

p-NITROPHENYL-N-[METHOXYSUCCINYL(N_β-CARBOBENZOXY-dl-PORLYLMETHYL]-N-ISOPROPYLCARBAMATE
ELASTASE LEUKOCYTE $Ki = 3.6\ \mu M$

p-NITROPHENYLOCTANOATE
ALCOHOL DEHYDROGENASE

p-NITROPHENYL-N-OCTYL CARBAMATE
STEROL ESTERASE IRR
STEROL ESTERASE (NEUTRAL) $I50 = 100\ \mu M$

N-p-NITROPHENYL-OXAMIC ACID
SIALIDASE

4-[[[4-(trans-2-NITROPHENYL)PHENYL]OXY]SULPHONYL]BENZOIC ACID N-METHYLAMIDE
PROTEIN KINASE C $I50 = 290\ \mu M$
PROTEIN TYROSINE KINASE EGFR $I50 = 400$ nM

2-NITROPHENYL PHENYLSULPHONE
RNA-DIRECTED DNA POLYMERASE $I50 = 51\ \mu M$

p-NITROPHENYLPHOSPHATE
PROTEIN TYROSINE-PHOSPHATASE

p-NITROPHENYLPHOSPHONATE
PHOSPHATASE ALKALINE C

1-(4-NITROPHENYL)-2-PROPEN-1-ONE		
3α-HYDROXYSTEROID DEHYDROGENASE (B-SPECIFIC)	IR	*K*i = 4 μM
7α-(4′-NITRO)PHENYLPROPYLANDROSTA-1,4-DIENE-3,17-DIONE		
AROMATASE		*I*50 = 64 nM
1-(4-NITROPHENYL)-2-PROPYN-1-ONE		
3α-HYDROXYSTEROID DEHYDROGENASE (B-SPECIFIC)	IR	*K*i = 6 μM
N-(3-NITROPHENYL)-4-(3-PYRIDYL)-2-PYRIMIDINAMINE		
PROTEIN KINASE Cα		*I*50 = 790 nM
PROTEIN KINASE Cβ1		*I*50 = 4.8 μM
PROTEIN KINASE Cβ2		*I*50 = 3.3 μM
PROTEIN KINASE Cγ		*I*50 = 3 μM
PROTEIN KINASE Cn		*I*50 = 9.2 μM
N-(p-NITROPHENYL)-D-RIBOAMIDRAZONE		
INOSINE-URIDINE NUCLEOSIDASE		Kd = 2 nM
p-NITROPHENYL β-D-RIBOPYRANOSIDE		
XYLAN 1,4-β-XYLOSIDASE		*K*i = 34 μM
p-NITROPHENYL N-(SUCCINYL-ALANYL-ALANAL-L-PROLYLMETHYL)-N-ISOPROPYLCARBAMATE		
ELASTASE LEUKOCYTE		
p-NITROPHENYLSULPHATE		
STERYL-SULFATASE		*I*50 = 7 mM
N-SULFOGLUCOSAMINE SULFOHYDROLASE		20% 5 mM
2-NITROPHENYLSULPHENYLCHLORIDE		
LACTOSE SYNTHASE		
p-NITROPHENYL-β-D-THIOGALACTOPYRANOSIDE		
β-GALACTOSIDASE		
p-NITROPHENYL α-D XYLOPYRANOSIDE		
XYLAN 1,4-β-XYLOSIDASE		*K*i = 300 μM
p-NITROPHENYL-β-D-XYLOPYRANOSIDE		
β-GLUCOSIDASE		
3-NITROPHTHALHYDRAZIDE		
NAD ADP-RIBOSYLTRANSFERASE		*I*50 = 72 μM
4-NITROPHTHALHYDRAZIDE		
NAD ADP-RIBOSYLTRANSFERASE		*I*50 = 510 μM
3-NITROPROPIONIC ACID		
FUMARATE HYDRATASE		
ISOCITRATE LYASE		*I*50 = 40 μM
SUCCINATE DEHYDROGENASE	SSI	
NITROPRUSSIDE		
LIPOXYGENASE 2	C	*K*i = 525 μM
3-NITROPYRIDINE-2-THIONE		
NICOTINATE-NUCLEOTIDE PYROPHOSPHORYLASE (CARBOXYLATING)	C	*K*i = 126 mM
NITROSALICYLALDEHYDE		
ACETOACETATE DECARBOXYLASE		*K*i = 300 nM
4-NITROSALICYLALDEHYDE		
PYRIDOXAMINE-PYRUVATE TRANSAMINASE		
N-NITROSALICYLALDEHYDE		
ALANINE RACEMASE	C	*K*i = 40 μM
3-NITROSALICYLAMIDE		
NAD ADP-RIBOSYLTRANSFERASE		*I*50 = 1.6 mM

5-NITROSALICYLIC ACID

GENTISATE 1,2-DIOXYGENASE		$Ki = 800\ \mu M$

3-NITROSALICYL-N-TETRADECYLAMIDE

SUCCINATE DEHYDROGENASE (UBIQUINONE)	NC	$Ki = 420$ nM

S-NITROSO-N-ACETYLPENICILLAMINE

LIPOXYGENASE 2	C	$Ki = 710\ \mu M$

NITROSOADAMANTANE

ALDEHYDE DEHYDROGENASE (NAD)		$I50 = 8.6\ \mu M$

NITROSOBENZENE

ALDEHYDE DEHYDROGENASE (NAD)		$I50 = 2.5\ \mu M$
NICOTINAMIDE-NUCLEOTIDE ADENYLYLTRANSFERASE	IR	31% 1.3 mM

6-NITROSO-1,2-BENZOPYRONE

NAD ADP-RIBOSYLTRANSFERASE	IR	$Ki = 40\ \mu M$

S-NITROSOGLUTATHIONE

GLUTATHIONE REDUCTASE (NADPH)		$Ki = 500\ \mu M$
GLUTATHIONE TRANSFERASE 1-1		$Ki = 130\ \mu M$
GLUTATHIONE TRANSFERASE 1-2		$Ki = 127\ \mu M$
GLUTATHIONE TRANSFERASE 3-3		$Ki = 13\ \mu M$

N-NITROSO-N-((7-METHOXYCOUMARIN-4-YL)METHYL-N′-ISOBUTYRYLALANINAMIDE

α-CHYMOTRYPSIN	IR	

N-NITROSO-N-(1-NAPHTHYLMETHYL)-N′-ISOBUTYRYLALANINE

α-CHYMOTRYPSIN	IR	

N-NITROSO-β-PHENYL-β-LACTAM

β-LACTAMASE		$I50 = 10$ mM

NITROSO-R-SALT

FORMATE DEHYDROGENASE		

4-NITROSTYRENE

3α-HYDROXYSTEROID DEHYDROGENASE (B-SPECIFIC)	IR	$I50 = 250\ \mu M$

2-NITRO-5-(THIOCYANATE)BENZOIC ACID

3-OXOACID CoA-TRANSFERASE		

2-NITRO-5-THIOCYANATOBENZOIC ACID

NADPH-FERRIHEMOPROTEIN REDUCTASE		

2-NITRO-5-THIOCYANOBENZOIC ACID

DEOXYRIBONUCLEASE I	IR	
LANOSTEROL SYNTHASE	IR	$Ki = 16$ mM
TRIACYLGLYCEROL LIPASE		

p-NITROTHIOPHENYL-β-D-GLUCOPYRANOSIDE

β-GLUCOSIDASE	C	$Ki = 1.5$ mM

2-(2-NITRO-4-TRIFLUOROMETHYLBENZOYL)-CYCLOHEXANE-1,3-DIONE

4-HYDROXYPHENYLPYRUVATE DIOXYGENASE		$I50 = 40$ nM

4-NITROURACIL

(R)-3-AMINO-2-METHYLPROPIONATE-PYRUVATE TRANSAMINASE	C	86% 1 mM

5-NITROURACIL

NAD ADP-RIBOSYLTRANSFERASE		$I50 = 430\ \mu M$
UDP-GLUCURONOSYLTRANSFERASE		85% 1 mM
URIDINE PHOSPHORYLASE	C	$Ki = 56\ \mu M$

NITROUS OXIDE

Ca^{2+}-TRANSPORTING ATPASE		
CYTOCHROME OXIDASE		
5-METHYLTETRAHYDROFOLATE-HOMOCYSTEINE S-METHYLTRANSFERASE	IRR	

NITROXYL
ALDEHYDE DEHYDROGENASE
ALDEHYDE DEHYDROGENASE (NAD)

NIZATIDINE

ACETYLCHOLINESTERASE	NC	*K*i = 5.1 μM
CHOLINESTERASE	MIX	*K*i = 220 μM

NK 104
HYDROXYMETHYLGLUTARYL-CoA REDUCTASE (NADPH)

NK 109
DNA TOPOISOMERASE (ATP-HYDROLYSING)

u-NKAI
Na^+/K^+-EXCHANGING ATPASE

NMN

ADP-RIBOSE PYROPHOSPHATASE	59% 4 mM
NAD NUCLEOSIDASE	

NNT24

TRYPTASE TL-2	85% 10 μM

NOBOTANIN

POLY(ADP-RIBOSE) GLYCOHYDROLASE	*I*50 = 15 μM

NOBOTANIN B

POLY(ADP-RIBOSE) GLYCOHYDROLASE	*I*50 = 4.4 μM

NOBOTANIN E

POLY(ADP-RIBOSE) GLYCOHYDROLASE	*I*50 = 1.8 μM

NOBOTANIN K

POLY(ADP-RIBOSE) GLYCOHYDROLASE	*I*50 = 380 pM

NOCARDIZIN A
D-ALANYL-D-ALANINE CARBOXYPEPTIDASE

iNOC-GLN-VAL-VAL-ALA-ALA-pNA
PAPAIN

NOCODAZOLE

PROTEIN TYROSINE KINASE (Fyn)	65% 100 μM
PROTEIN TYROSINE KINASE (Lck)	90% 100 μM

NODULARIN

PHOSPHOPROTEIN PHOSPHATASE	*I*50 = 700 pM
PHOSPHOPROTEIN PHOSPHATASE 1	*I*50 = 3 nM
PHOSPHOPROTEIN PHOSPHATASE 2A	*I*50 = 900 pM
PHOSPHOPROTEIN PHOSPHATASE 2A	*I*50 = 1 nM

NOFORMYCIN

NITRIC-OXIDE SYNTHASE II	C	*K*i = 1.3 μM

NOGALAMYCIN

HELICASE	*K*i = 1 μM
HELICASE	*I*50 = 360 nM
HELICASE II	*I*50 = 420 nM
RNA POLYMERASE	

NOJIRIMYCIN

α-AMYLASE	
CELLOBIOSE PHOSPHORYLASE	
DEXTRANSUCRASE	
β-FRUCTOFURANOSIDASE	
α-GLUCOSIDASE	
SUCROSE α-GLUCOSIDASE	*K*i = 127 nM
XYLAN 1,4-β-XYLOSIDASE	

NOJIRIMYCIN-6-PHOSPHATE
PHOSPHOGLUCOMUTASE

NOJIRITETRAZOLE
α-GLUCOSIDASE — Ki = 1.3 mM
β-GLUCOSIDASE — Ki = 1.4 μM
β-GLUCOSIDASE — Ki = 1.5 μM

2-NO$_2$-6-ME-SULPHOANILIDE
ACETOLACTATE SYNTHASE II — Ki = 70 nM

2-NO$_2$-6-ME-SULPHONANILIDE
ACETOLACTATE SYNTHASE I — Ki = 220 μM
ACETOLACTATE SYNTHASE III — Ki = 200 μM

NONACTIN
NH4+/K^+-TRANSPORT MITOCHONDRIA

10-NONADECENOIC ACID
H^+/K^+-EXCHANGING ATPASE — I50 = 4.9 μM

NONANAMIDE
EPOXIDE HYDROLASE — I50 = 19 μM

NONANOIC ACID
EPOXIDE HYDROLASE — I50 = 377 μM

(E)-3-NONENAL
GLUTATHIONE TRANSFERASE μ3-3 — Ki = 12 μM
GLUTATHIONE TRANSFERASE μ3-4 — Ki = 10 μM
GLUTATHIONE TRANSFERASE μ4-4 — Ki = 23 μM
GLUTATHIONE TRANSFERASE α1-1 — Ki = 51 μM
GLUTATHIONE TRANSFERASE α1-2 — Ki = 60 μM

trans-2-NONENAL
GLUCOSE-6-PHOSPHATE 1-DEHYDROGENASE

2-(NON-8-ENYL)-5-(HEX-5-ENYL)PYRROLIDINE
ACETYLCHOLINESTERASE — Ki = 2.5 μM

2-(NON-8-ENYL)-5-(HEX-5-ENYL)PYRROLINE
ACETYLCHOLINESTERASE — Ki = 13 μM

NONH-COCH(BENZYL)CO-ALA-GLY-NH$_2$
THERMOLYSIN — Ki = 660 nM

NONIDENT P40
CHLORAMPHENICOL O-ACETYLTRANSFERASE
CHOLESTENONE 5α-REDUCTASE
DOLICHYL-PHOSPHATE β-D-MANNOSYLTRANSFERASE
1,3-β-GLUCAN SYNTHASE
1-PHOSPHATIDYLINOSITOL 3-KINASE
STEROL O-ACYLTRANSFERASE
dTMP KINASE — 22% 0.4 %

NONIONIC DETERGENTS
ALDEHYDE OXIDASE

2-NONYL-5-(HEX-5-ENYL)PYRROLIDINE
ACETYLCHOLINESTERASE — Ki = 1.5 μM

2-NONYL-5-(HEX-5-ENYL)PYRROLINE
ACETYLCHOLINESTERASE — Ki = 5 μM

2-n-NONYL-4-HYDROXYQUINOLINE-N-OXIDE
CYTOCHROME bo3 UBIQUINOL OXIDASE
FUMARATE REDUCTASE (NADH)

1-NONYLIMIDAZOLE
THROMBOXANE A SYNTHASE

NONYLPHENOL
Ca^{2+}-TRANSPORTING ATPASE
Ca^{2+}-TRANSPORTING ATPASE 95% 100 μM
H^{+}-TRANSPORTING ATPASE 79% 25 μM
H^{+}-TRANSPORTING ATPASE 95% 100 μM

3-NONYLTHIO-1,1,1-TRIFLUORO-2-PROPANONE
JUVENILE HORMONE ESTERASE *I*50 = 38 nM

NORADRENALIN
DIHYDROPTERIDINE REDUCTASE
LIPOXYGENASE 1

NORADRENALINE
PROTEIN TYROSINE KINASE *I*50 = 100 μM

19-NORANDROSTENEDIONE
AROMATASE C *K*i = 21 nM

5′-NORARISTEROMYCIN
ADENOSYLHOMOCYSTEINASE *K*i = 11 nM
ADENOSYLHOMOCYSTEINASE *K*i = 11 nM

NORATROPINEN-ACETIC ACID
HYOSCYAMINE (6S)-DIOXYGENASE

5-NORBORNEN-2-YLMETHYLENE-N-HYDROXY-N′-AMINOGUANIDINE
RIBONUCLEOSIDE-DIPHOSPHATE REDUCTASE *K*i = 150 μM
RIBONUCLEOSIDE-DIPHOSPHATE REDUCTASE *K*i = 14 μM

NORCANTHARIDIN
PHOSPHOPROTEIN PHOSPHATASE 2A

NORCARNITINE
CARNITINE O-PALMITOYLTRANSFERASE

NORDAZEPAM
CYTOCHROME P450 2C19

2′-NORDEOXYGUANOSINE
PURINE-NUCLEOSIDE PHOSPHORYLASE *K*i = 30 μM

2′-NORDEOXYGUANOSINE 3′-DIPHOSPHATE
PURINE-NUCLEOSIDE PHOSPHORYLASE *K*i = 18 nM

2′-NORDEOXYGUANOSINE 5′-DIPHOSPHATE
PURINE-NUCLEOSIDE PHOSPHORYLASE *K*i = 9 nM

2′-NORDEOXYGUANOSINE 5′-MONOPHOSPHATE
PURINE-NUCLEOSIDE PHOSPHORYLASE *K*i = 2.2 μM

NOR-5,10-DIDEAZA-5,6,7,8-TETRAHYDROFOLIC ACID
PHOSPHORIBOSYLGLYCINAMIDE FORMYLTRANSFERASE *K*i = 630 nM

NORDIHYDROGUAIARETIC ACID
1-ALKYLGLYCEROPHOSPHOCHOLINE O-ACETYLTRANSFERASE *I*50 = 290 μM
D-AMINO-ACID OXIDASE
ARACHIDONATE 5-LIPOXYGENASE *I*50 = 500 nM
ARACHIDONATE 5-LIPOXYGENASE *I*50 = 800 nM
ARACHIDONATE 5-LIPOXYGENASE *I*50 = 2.4 μM
ARACHIDONATE 8-LIPOXYGENASE
ARACHIDONATE 12-LIPOXYGENASE *I*50 = 3.7 μM
ARACHIDONATE 12-LIPOXYGENASE *I*50 = 2.5 μM
AROMATASE *I*50 = 11 μM
L-ASCORBATE OXIDASE
CYCLOOXYGENASE *I*50 = 38 μM

CYTOCHROME P450 CATALYZED METABOLISM OF ARACHIDONIC ACID		*I*50 = 15 μM
EPOXIDE HYDROLASE		
7-ETHOXYRESORUFIN O-DEETHYLASE		*I*50 = 70 μM
FORMATE-TETRAHYDROFOLATE LIGASE		*I*50 = 240 μM
HYDROPEROXIDE ISOMERASE		68% 2.5 mM
LIPOXYGENASE		*I*50 = 1.1 μM
MYELOPEROXIDASE		*I*50 = 16 μM
NADH DEHYDROGENASE (UBIQUINONE)		88% 100 nM
NAD(P)-ARGININE ADP-RIBOSYLTRANSFERASE		*I*50 = 100 μM
PHOSPHOLIPASE A2 (HIGH MW)		*I*50 = 45 μM
PHOSPHOLIPASE A2 (LOW MW)		*I*50 = 220 nM
SUCCINATE COENZYM Q REDCUTASE		
TRANSACYLASE (COA INDEPENDENT)		*I*50 = 4 μM
UNSPECIFIC MONOOXYGENASE	C	*K*i = 95 μM

NOREPINEPHRINE

ADENYLATE CYCLASE		*I*50 = 180 μM
ARYLAMINE N-METHYLTRANSFERASE	C	
CARBOSINE SYNTHASE		
CATECHOL O-METHYLTRANSFERASE		
DIACYLGLYCEROL CHOLINEPHOSPHOTRANSFERASE		
DIHYDROPTERIDINE REDUCTASE		
DOPAMINE β-MONOOXYGENASE		
ETHANOLAMINEPHOSPHOTRANSFERASE		
LIPOXYGENASE 1	IR	
PHENYLALANINE 4-MONOOXYGENASE		
PHENYLETHANOLAMINE N-METHYLTRANSFERASE		
PYRIDOXAL KINASE		
SULFINOALANINE DECARBOXYLASE		
TYROSINE 3-MONOOXYGENASE		
TYROSINE TRANSAMINASE		

NORETHINDRONE

3α-HYDROXYSTEROID DEHYDROGENASE (B-SPECIFIC)		*I*50 = 9 μM

NORETHINDRONE ACETIC ACID

BUFURALOL-1′-HYDROXYLASE (CYT P-450dbl)	C	*I*50 = 100 μM

NORETHISTERONE

ESTRADIOL 2α-HYDROXYLASE		
3β-HYDROXY-Δ5-STEROID DEHYDROGENASE		

NORETHISTERONE ACETATE

3β-HYDROXY-Δ5-STEROID DEHYDROGENASE		

NORFLOXACIN

CYTOCHROME P450 1A	C	
CYTOCHROME P450 3A4	C	*K*i = 2.3 mM
CYTOCHROME P450 LA2		56% 500 μM
DNA TOPOISOMERASE (ATP-HYDROLYSING)		*K*i = 1 μg/ml

NORFLUOXETINE

CYTOCHROME P450 1A2		*K*i = 16 μM
CYTOCHROME P450 2C9		*K*i = 17 μM
CYTOCHROME P450 2D6		*K*i = 1.3 μM
CYTOCHROME P450 2D6		*K*i = 190 nM

NORHAMAN

CYTOCHROME P450 11		
CYTOCHROME P450 17		
HISTAMINE N-METHYLTRANSFERASE		*I*50 = 23 μM
INDOLEAMINE-PYRROLE 2,3-DIOXYGENASE	NC	*K*i = 100 μM
NAD ADP-RIBOSYLTRANSFERASE		*I*50 = 4.7 mM
TRYPTOPHAN 2,3-DIOXYGENASE	C	

NORISOGUAIACIN

FORMATE-TETRAHYDROFOLATE LIGASE		

NORJIRIMYCIN
β-GLUCOSIDASE

NORLAPACHOL
LACTOYLGLUTATHIONE LYASE C $Ki = 113\ \mu M$

NORLEUCINE
ALANINE TRANSAMINASE C $Ki = 30\ mM$
LEUCYL AMINOPEPTIDASE C
METHIONINE-tRNA LIGASE C $Ki = 100\ mM$
ORNITHINE CARBAMOYLTRANSFERASE
SACCHAROPINE DEHYDROGENASE (NAD, L-LYSINE FORMING)

D-NORLEUCINE
PHENYLALANINE DEHYDROGENASE

D,L-NORLEUCINE
AMINOACYLASE

L-NORLEUCINE
METHIONINE γ-LYASE

D,L-NORLEUCINE METHYLSTER
NEPENTHES ASPARTIC PROTEINASE

NORNUCIFERINE
PROTEIN TYROSINE-PHOSPHATASE CD45 $I50 = 5.3\ \mu M$

NOROFLOXACIN
DNA TOPOISOMERASE (ATP-HYDROLYSING) $I50 = 1.3\ \mu g/ml$

(R)NORRETICULINE
(S)-TETRAHYDROPROTOBERBERINE OXIDASE $I50 = 100\ \mu M$

(S)-NORRETICULINE
RETICULINE OXIDASE

NORSALSOLINOL
AMINE OXIDASE MAO-A MIX $Ki = 65\ \mu M$
AMINE OXIDASE MAO-B NC $Ki = 699\ \mu M$

19-NORTESTOSTERONE
AROMATASE C $Ki = 85\ nM$
STEROID Δ-ISOMERASE C $Ki = 5.2\ \mu M$

19-NORTESTOSTERONE-17-BROMOACETATE
3β(or 20α)HYDROXYSTEROID DEHYDROGENASE

19-NOR-2′,3′α-TETRAHYDROFURAN-2′-SPIRO-17-(4-METHYL-4-AZA-5α-ANDROST-5-ENE-3-ONE)
STEROID 5α-REDUCTASE $I50 = 10\ nM$
STEROID 5α-REDUCTASE $I50 = 10\ \mu M$

NORVALINE
ALANINE TRANSAMINASE C $Ki = 70\ mM$
2-AMINOHEXANOATE TRANSAMINASE
ARGININOSUCCINATE SYNTHASE
ORNITHINE CARBAMOYLTRANSFERASE $Ki = 430\ \mu M$
ORNITHINE-OXO-ACID TRANSAMINASE C
PUTRESCINE CARBAMOYLTRANSFERASE

L-NORVALINE
D-AMINO-ACID OXIDASE
GLYCINE AMIDINOTRANSFERASE C

O-(L-NORVALYL-5)ISOUREA
AMINO-ACID N-ACETYLTRANSFERASE

L-NORVALYL-N3-(4-METHOXYFUMAROYL)-L-2,3-DIAMINOPROPIONIC ACID
GLUTAMINE-FRUCTOSE-6-PHOSPHATE AMINOTRANSFERASE (ISOMERIZING) $I50 = 165\ \mu M$

Inhibitor / Enzyme	Type	Value
NORWOGONIN		
NADH DEHYDROGENASE (EXOGENOUS)		*I*50 = 70 μM
NADH DEHYDROGENASE (UBIQUINONE)		
NORZIMELDINE		
CYTOCHROME P450 2D6		*K*i = 5 μM
NOSTOPEPTIN A		
CHYMOTRYPSIN		*I*50 = 1.4 μg/ml
ELASTASE		*I*50 = 1.3 μg/ml
NOSTOPEPTIN B		
CHYMOTRYPSIN		*I*50 = 1.6 μg/ml
ELASTASE		*I*50 = 11 μg/ml
NOVOBIOCIN		
CDP-GLYCEROL GLYCEROPHOSPHOTRANSFERASE		
CDPRIBITOL RIBITOLTRANSFERASE		
DNA TOPOISOMERASE		*I*50 = 180 μM
DNA TOPOISOMERASE (ATP-HYDROLYSING)	C	*K*i = 10 nM
GLYCINE-tRNA LIGASE		
LEUCINE-tRNA LIGASE		
NAD ADP-RIBOSYLTRANSFERASE		*I*50 = 2.2 mM
NAD(P)-ARGININE ADP-RIBOSYLTRANSFERASE		*I*50 = 280 μM
NAD(P)-ARGININE ADP-RIBOSYLTRANSFERASE		*I*50 = 300 μM
REVERSE GYRASE		
NP 176 (R,R)		
CALPAIN		95% 190 μM
1151 NP		
PROTEIN TYROSINE KINASE (INSULIN RECEPTOR)		
NPC 15-437		
PROTEIN KINASE C (CYTOSOLIC)		*I*50 = 43 μM
PROTEIN KINASE C (MEMBRANE)		*I*50 = 18 μM
NPC 15437		
PROTEIN KINASE C	C	*K*i = 12 μM
PROTEIN KINASE C		*I*50 = 22 μM
PROTEIN KINASE C	C	*K*i = 5 μM
PROTEIN KINASE Cα	C	*K*i = 12 μM
Na_2PdCl_6		
β-AMYLASE		
NQ-Y15		
THROMBOXANE A SYNTHASE		
NRM		
OPHELINE KINASE		
NS 398		
CYCLOOXYGENASE 1		*I*50 = 200 μM
CYCLOOXYGENASE 2		*I*50 = 3.8 μM
CYCLOOXYGENASE 2		*I*50 = 10 μM
CYCLOOXYGENASE 2		*I*50 = 10 nM
PROSTAGLANDIN ENDOPEROXIDASE		*I*50 = 11 μM
PROSTAGLANDIN-ENDOPEROXIDE SYNTHASE		
PROSTAGLANDIN-ENDOPEROXIDE SYNTHASE 1		*I*50 = 75 μM
PROSTAGLANDIN-ENDOPEROXIDE SYNTHASE 2		*I*50 = 1.8 μM
NS 504		
α-AMYLASE		
NS 2028		
GUANYLATE CYCLASE	IR	*I*50 = 30 nM
GUANYLATE CYCLASE	IR	*I*50 = 200 nM

NbS_2
CARBONYL REDUCTASE (NADPH) 85% 100 μM

NSC 613
3′,5′-CYCLIC-NUCLEOTIDE PHOSPHODIESTERASE III *I*50 = 380 μM
3′,5′-CYCLIC-NUCLEOTIDE PHOSPHODIESTERASE IV *I*50 = 40 nM

NSC 101327
DNA TOPOISOMERASE *I*50 = 38 μM

NSC 158393
DNA TOPOISOMERASE *I*50 = 10 μM
INTEGRASE
PROTEINASE HIV-1 *I*50 = 1.7 μM

NSC 163501
ASPARTATE-AMMONIA LIGASE 95% 3 mM

NSC 287474
RNA-DIRECTED DNA POLYMERASE *I*50 = 360 nM

NSC 308795
DIHYDROPTEROATE SYNTHASE *I*50 = 60 μM

NSC 312180
PROTEINASE HIV-1 *I*50 = 320 nM

NSC 314622
DNA TOPOISOMERASE

NSC 324362
INTEGRASE *I*50 = 1.1 μM

NSC 38049
INTEGRASE *I*50 = 600 ng/ml

NSC 401005
GLUTATHIONE REDUCTASE (NADPH) *I*50 = 42 μM
THIOREDOXIN REDUCTASE (NADPH) *I*50 = 280 nM

NSC 617145
GLUTATHIONE REDUCTASE (NADPH) *I*50 = 125 μM
THIOREDOXIN REDUCTASE (NADPH) *I*50 = 250 nM

NSC 624231
RNA-DIRECTED DNA POLYMERASE *I*50 = 19 μM

NSC 625487
RNA-DIRECTED DNA POLYMERASE *I*50 = 520 nM

NSC 642710
INTEGRASE *I*50 = 1.7 μg/ml

NSC 662305
DNA TOPOISOMERASE (ATP-HYDROLYSING)

NSC 665517
DNA TOPOISOMERASE (ATP-HYDROLYSING)

NSC 665564
DIHYDROOROTATE DEHYDROGENASE *I*50 = 300 nM

NSC 666724
INTEGRASE *I*50 = 3.8 μM

NSD 1015
AMINE OXIDASE MAO

NSD 1034
THREONINE ALDOLASE 54% 5 mM

NSD 1055
AROMATIC-L-AMINO-ACID DECARBOXYLASE
HISTIDINE DECARBOXYLASE 83% 100 μM
THREONINE ALDOLASE 72% 5 mM

NaS_2O_3
CELLULASE

NSP 804
3′,5′-CYCLIC-NUCLEOTIDE PHOSPHODIESTERASE III *I*50 = 3 μM

NSP 805
3′,5′-CYCLIC-NUCLEOTIDE PHOSPHODIESTERASE III *I*50 = 300 nM

NU 1025
NAD ADP-RIBOSYLTRANSFERASE *I*50 = 440 nM

NUARIMOL
14α-METHYLSTEROL DEMETHYLASE

NUCLEAR LYSOZYME INHIBITOR
LYSOZYME

NUCLEAR PROTEIN PHOSPHATASE INHIBITOR
PHOSPHOPROTEIN PHOSPHATASE 1 *I*50 = 9.9 pM

NUCLEAR PROTEIN PHOSPHATASE 1 INHIBITOR
PHOSPHOPROTEIN PHOSPHATASE 1

NUCLEAR PROTEIN PHOSPHATASE INHIBITOR 1a
PHOSPHOPROTEIN PHOSPHATASE 1

NUCLEAR PROTEIN PHOSPHATASE INHIBITOR 1b
PHOSPHOPROTEIN PHOSPHATASE 1

NUCLEOCIDIN
ADENOSYLHOMOCYSTEINASE *K*i = 204 μM

NUCLEOSIDE TRIPHOSPHATES
GUANYLATE CYCLASE

NU/ICRF 500
DNA TOPOISOMERASE (ATP-HYDROLYSING)

NUPERCAINE
Ca^{2+}-TRANSPORTING ATPASE

NYSTATIN
CHITIN SYNTHASE

NZ 105
3′,5′-CYCLIC-NUCLEOTIDE PHOSPHODIESTERASE C *K*i = 30 μM

NZ 314
ALDEHYDE REDUCTASE

O

17-[(2-O-ADAMYNTYL)CARBAMOYL]-6-AZA-ANDROST-4ENE-ONE
STEROID 5α-REDUCTASE 1 *K*i = 6.9 nM
STEROID 5α-REDUCTASE 2 *K*i = 40 pM

OBOVATOL
STEROL O-ACYLTRANSFERASE *I*50 = 42 μM

OBSCUROLIDE A4
3′,5′-CYCLIC-NUCLEOTIDE PHOSPHODIESTERASE (Ca^{2+}/CALMODULIN) *I*50 = 2 mM

OBSCUROLIDES B2α
3′,5′-CYCLIC-NUCLEOTIDE PHOSPHODIESTERASE (Ca^{2+}/CALMODULIN) *I*50 = 2.5 mM

OBSCUROLIDES B3
3′,5′-CYCLIC-NUCLEOTIDE PHOSPHODIESTERASE (Ca^{2+}/CALMODULIN) *I*50 = 1.5 mM

OBSCUROLIDES B4
3′,5′-CYCLIC-NUCLEOTIDE PHOSPHODIESTERASE (Ca^{2+}/CALMODULIN) *I*50 = 400 μM

OBSCUROLIDES C2α
3′,5′-CYCLIC-NUCLEOTIDE PHOSPHODIESTERASE (Ca^{2+}/CALMODULIN) *I*50 = 10 mM

OBSCUROLIDES C2β
3′,5′-CYCLIC-NUCLEOTIDE PHOSPHODIESTERASE (Ca^{2+}/CALMODULIN) *I*50 = 12 mM

OBSCUROLIDES D2
3′,5′-CYCLIC-NUCLEOTIDE PHOSPHODIESTERASE (Ca^{2+}/CALMODULIN) *I*50 = 15 mM

1-OCATADECYL-2-O-METHYL-rac-3-GLYCEROPHOSPHO-myo-INOSITOL
1-PHOSPHATIDYLINOSITOL 3-KINASE *I*50 = 96 μM

OCHRACEOLIDE A
FARNESYL PROTEINTRANSFERASE *I*50 = 1 μg/ml

OCHRACEOLIDE B
FARNESYL PROTEINTRANSFERASE *I*50 = 700 ng/ml

OCN^-
FORMATE DEHYDROGENASE

14-OCTACOSANOL
AROMATASE 11% 15 μM

10,12-OCTADECADIENOATE
LIPOXYGENASE C

7,9-OCTADECADIYNOIC ACID
HYDROXYMETHYLGLUTARYL-CoA REDUCTASE (NADPH) *I*50 = 5 μM

OCTADECA-9,12-DIYNOIC ACID
LIPOXYGENASE IR

1-O-OCTADECANOYLPHOSPHONYL-MYO-INOSITOL
1-PHOSPHATIDYLINOSITOL PHOSPHODIESTERASE *I*50 = 100 μM

OCTADECATETRAENOIC ACID
STEROID 5α-REDUCTASE

cis-13-OCTADECENAL
ACETYLCHOLINESTERASE C *K*i = 500 μM

(E)-5-OCTADECEN-7,9-DIYNOIC ACID
HYDROXYMETHYLGLUTARYL-CoA REDUCTASE (NADPH) *I*50 = 500 nM

(Z)-5-OCTADECEN-7,9-DIYNOIC ACID
HYDROXYMETHYLGLUTARYL-CoA REDUCTASE (NADPH) *I*50 = 1.5 μM

Δ9-OCTADECENOIC ACID
Na^+/K^+-EXCHANGING ATPASE

OCTADEC-cis-9-EN-12-YNOIC ACID
PHOSPHOLIPASE A2 *I*50 = 100 μM
THROMBOXANE A SYNTHASE *I*50 = 10 μM

(E)-7-OCTADECEN-9-YNOIC ACID
HYDROXYMETHYLGLUTARYL-CoA REDUCTASE (NADPH) *I*50 = 2 μM

OCTADEC-trans-11-EN-9-YNOIC ACID
PHOSPHOLIPASE A2 *I*50 = 100 μM

(Z)-7-OCTADECEN-9-YNOIC ACID
HYDROXYMETHYLGLUTARYL-CoA REDUCTASE (NADPH) — *I*50 = 3 μM

OCTADECYLAMINE
MYOSIN-LIGHT-CHAIN KINASE — *I*50 = 11 μM
1-PHOSPHATIDYLINOSITOL PHOSPHODIESTERASE

3-(4-OCTADECYLBENZOYL)ACRYLIC ACID
PHOSPHOLIPASE A2 — IR — *I*50 = 70 nM

n-OCTADECYLDIMETHYLSULPHONIUM BROMIDES
PHOSPHOLIPASE C — *K*i = 250 μM

1-OCTADECYL-2-METHYL-rac-GLYCERO-3-PHOSPHOCHOLINE
1-PHOSPHATIDYLINOSITOL-4,5-BISPHOSPHATE PHOSPHODIESTERASE — *I*50 = 9.6 μM

1-O-OCTADECYL-2-O-METHYL-MC-GLYCERO-3-PHOSPHOCHOLINE
PHOSPHATIDYLINOSITOL PHOSPHOLIPASE Cγ — *I*50 = 400 nM
PHOSPHOLIPASE D (GLYCOSYLPHOSPHATIDYLINOSITOL SPECIFIC) — *I*50 = 38 μM
PHOSPHOLIPASE D (GLYCOSYLPHOSPHATIDYLINOSITOL SPECIFIC) — *I*50 = 175 μM

1-O-OCTADECYL-2-O-METHYL-rac-GLYCERO-3-PHOSPHOCHOLINE
CDP: PHOSPHOCHOLINE CYTIDYLTRANSFERASE — *I*50 = 8 μM
1-PHOSPHATIDYLINOSITOL 3-KINASE — *I*50 = 35 μM

rac-1-O-OCTADECYL-2-O-METHYLGLYCEROPHOSPHOCHOLINE
PROTEIN KINASE C — *I*50 = 12 μM

3-OCTADECYLOXY-2-ACETOXYPROPYLPHOSPHONOMyo-INOSITOL
1-PHOSPHATIDYLINOSITOL-4,5-BISPHOSPHATE PHOSPHODIESTERASE — *I*50 = 76 μM

OCTADECYLPHOSPHOCHOLINE
PHOSPHOLIPASE D — *I*50 = 6.4 μM

OCTADECYLPHOSPHODITHIONYL-1-myo-INOSITOL
1-PHOSPHATIDYLINOSITOL-4,5-BISPHOSPHATE PHOSPHODIESTERASE — *I*50 = 100 μM

1-OCTADECYLPHOSPHOTHIONYL-myo-INOSITOL
1-PHOSPHATIDYLINOSITOL PHOSPHODIESTERASE

OCTADECYLSULPHATE
MYOSIN-LIGHT-CHAIN KINASE

3-(1-OCTADECYLTHIO)PROPAN-1,2-DIOL
GLYCERYL-ETHER MONOOXYGENASE — 64% 500 μM

9-OCTADECYNOIC ACID
HYDROXYMETHYLGLUTARYL-CoA REDUCTASE (NADPH) — *I*50 = 5 μM

17-OCTADECYNOIC ACID
FATTY ACID ω-HYDROXYLASE — IR — *I*50 < 100 nM

2,3-OCTADIENYL-CoA
ACYL-CoA DEHYDROGENASE

trans-2-OCTAHEXAHYDROPYRIMIDINE-4,6-DICARBOXYLIC ACID
DIHYDROOROTASE — *K*i = 12 μM

2,3,4,5,6,7,8,9-OCTAHYDRO-2(S)MERCAPTO-3-OXO-1H-4-BENZAZACYCLOUNDECINE-5(S)-CARBOXYLIC ACID
NEPRILYSIN — *I*50 = 900 pM
THERMOLYSIN — *I*50 = 68 nM

OCTAMOYL-CoA
FATTY ACID SYNTHASE

n-OCTANETHIOL
LIPOXYGENASE 1 — NC

OCTANOIC ACID
ALCOHOL DEHYDROGENASE

ALDEHYDE REDUCTASE
[3-METHYL-2-OXOBUTANOATE DEHYDROGENASE (LIPOAMIDE)] KINASE
PROPIONATE-CoA LIGASE C *K*i = 58 μM
PYRUVATE KINASE

OCTANOL
BUTYRATE-CoA LIGASE
DEOXYRIBOSE-PHOSPHATE ALDOLASE

2-OCTANOL
CYTOCHROME P450 2E1 *K*i = 3 μM

n-OCTANOL
β-GLUCOSIDASE C *K*i = 40 mM
LIPOXYGENASE 1 C

OCTANOYL-CoA
ACETOACETATE-CoA LIGASE NC *K*i = 17 μM
METHYLMALONYL-CoA EPIMERASE 91% 2 mM
OXOGLUTARATE DEHYDROGENASE (LIPOAMIDE) *I*50 = 20 μM

N-β-OCTANOYLLYSYL-URACIL POLYOXIN C
CHITIN SYNTHASE *I*50 = 500 ng/ml

N-n-OCTANOYLNORNICOTINE
AROMATASE C *K*i = 650 μM

(N-OCTANOYLTHIO)GLYCOLLECITHINE
PHOSPHOLIPASE A2 *I*50 = 50 μM

4-OCTANOYLTHIOPROLYLPYRROLIDINE
PROLYL OLIGOPEPTIDASE *I*50 = 1 nM

OCTANOYL-D-THREO-p-NITRO-1-PHENYL-2-AMINO-1,3-PROPANEDIOL
2-HYDROXYACYLSPHINGOSINE 1β-GALACTOSYLTRANSFERASE

(E)-2-OCTENAL
GLUTATHIONE TRANSFERASE μ3-3 *K*i = 32 μM
GLUTATHIONE TRANSFERASE μ3-4 *K*i = 26 μM
GLUTATHIONE TRANSFERASE μ4-4 *K*i = 28 μM
GLUTATHIONE TRANSFERASE α1-1 *K*i = 1.9 μM
GLUTATHIONE TRANSFERASE α1-2 *K*i = 11 μM

4-OCTENOIC ACID
ACETYL-CoA C-ACYLTRANSFERASE IR

OCTENOYL-CoA
ACYL-CoA DEHYDROGENASE (MEDIUM CHAIN)

2,3-trans-OCTENOYL-CoA
FATTY ACID SYNTHASE

OCTIMBATE
STEROL O-ACYLTRANSFERASE *I*50 = 30 μM

OCTIVARIN
NADH DEHYDROGENASE (UBIQUINONE) *K*i = 800 pM

OCTOPAMINE
ARYLSULFATASE
DIHYDROPTERIDINE REDUCTASE
γ-GLUTAMYLHISTAMINE SYNTHASE

OCTOPINE
ALANOPINE DEHYDROGENASE
D-OCTOPINE DEHYDROGENASE

D-glycero-D-ido-OCTULOSE-1,8-DIPHOSPHATE
FRUCTOSE-BISPHOSPHATE ALDOLASE *K*i = 125 μM

n-OCTYL-6-O-[2-O-(2-ACETAMIDO-2-DEOXY-β-D-GLUCOPYRANOSYL)-6-DEOXY-α-D-MANNOPYRANOSYL]-β-D-GLUCOPYRANOSIDE
(N-ACETYLNEURAMINYL)-GALACTOSYLGLUCOSYLCERAMIDE N-ACETYLGALACTOSAMINYLTRANSFER
α-1,3(6)-MANNOSYLGLYCOPROTEIN β-1,6-N-ACETYLGLUCOSAMINYLTRANSFERASE C *K*i = 60 μM

OCTYLAMINE
GUANYLATE CYCLASE *K*i = 362 μM
TYROSYL PROTEIN SULFOTRANSFERASE 22% 200 μM

1-n-OCTYL-AMINOGLUTETHIMIDE
AROMATASE *I*50 = 1.6 μM
CHOLESTEROL MONOOXYGENASE (SIDE-CHAIN-CLEAVING) *I*50 = 19 μM

3,n-OCTYLAMINO-16,17,18,19,28,29-HEXAHYDRORIFAMYCIN
RNA-DIRECTED DNA POLYMERASE *I*50 = 16 nM
RNA-DIRECTED DNA POLYMERASE *I*50 = 640 nM
RNA-DIRECTED DNA POLYMERASE *I*50 = 6 nM

6-[4-n-OCTYLANILINO]URACIL
URACIL-DNA GLYCOHYDROLASE *I*50 = 8 μM

OCTYLBORONIC ACID
CUTINASE C *K*i = 5 μM

OCTYLCARBAMOYL ESEROLINE
ACETYLCHOLINESTERASE *I*50 = 15 nM
CHOLINESTERASE *I*50 = 4 nM

OCTYL-CoA
ACYL-CoA DEHYDROGENASE

OCTYL GALLATE
DOPAMINE SULFOTRANSFERASE 68% 6.7 μM
17α-ETHINYLOESTRADIOL SULFOTRANSFERASE 62% 6.7 μM
NADH OXIDASE (CN INSENITIVE) *I*50 = 70 nM
NAPHTHOL SULFOTRANSFERASE 44% 6.7 μM
SUCCINATE OXIDASE (CN INSENITIVE) *I*50 = 300 nM

OCTYL-β-GLUCOPYRANOSIDE
PHOSPHATIDYLGLYCEROPHOSPHATASE

1-O-n-OCTYL-β-D-GLUCOPYRANOSIDE
AMINE OXIDASE (FLAVIN-CONTAINING)

OCTYL D-GLUCOPYRANOSIDE
CHOLESTENONE 5α-REDUCTASE

OCTYL GLUCOSE
GLYCEROL-3-PHOSPHATE DEHYDROGENASE (NAD)

OCTYLGLUCOSIDE
CARNITINE O-PALMITOYLTRANSFERASE 1
DIACYLGLYCEROL KINASE
DOLICHYL-DIPHOSPHOOLIGOSACCHARIDE-PROTEIN GLYCOSYLTRANSFERASE
β-GALACTOSIDE α-2,3-SIALYLTRANSFERASE
1,3-β-GLUCAN SYNTHASE
IGA-SPECIFIC METALLOENDOPEPTIDASE 1 *I*50 = 0.1 %
IGA-SPECIFIC METALLOENDOPEPTIDASE 2 *I*50 = 0.1 %
PHOSPHATIDYLETHANOLAMINE N-METHYLTRANSFERASE
1-PHOSPHATIDYLINOSITOL 3-KINASE

OCTYL-β-GLUCOSIDE
2-ACYLGLYCEROL O-ACYLTRANSFERASE
MANNOSYL-OLIGOSACCHARIDE GLUCOSIDASE 90% 10 mM

OCTYL β-D-GLUCOSIDE
LIPID-A-DISACCHARIDE SYNTHASE

OCTYL-β-D-GLUCOSIDE
GLYCOPROTEIN O-FATTY-ACYLTRANSFERASE

S-n-OCTYLGLUTATHIONE
GLUTATHIONE TRANSFERASE
LACTOYLGLUTATHIONE LYASE $I50 = 59\ \mu M$

β-OCTYLGLYCOPYRANOSIDE
DOLICHOL O-ACYLTRANSFERASE

OCTYLGUANIDINE
H^+-TRANSPORTING ATPASE

2-OCTYL-4H-1,3,2-BENZODIOXAPHOSPHORIN
NEUROPATHY TARGET ESTERASE $I50 = 120\ pM$

2-OCTYL-4H-1,3,2-BENZODIOXAPHOSPHORIN-2-OXIDE
NEUROPATHY TARGET ESTERASE

L-1-O-OCTYL-2-(HEPTYLPHOSPHONYL)-sn-GLYCEROL-3-PHOSPHOETANOLAMINE
PHOSPHOLIPASE A2 (NON PANCREATIC SECRETED 14 KD) $I50 = 200\ nM$

L-1-OCTYL-2-HEPTYLPHOSPHONYL-SN-GLYCERO-3-PHOSPHOETHANOLAMINE
PHOSPHOLIPASE A2

OCTYLISOCYANATE
LEUKOCYTE-MEMBRANE NEUTRAL ENDOPEPTIDASE

O-OCTYL-O-(p-NITROPHENYL)METHYLPHOSPHONATE
CUTINASE

N^2-(p-n-OCTYLPHENYL)-2′-DEOXYGUANOSINE-5′-TRIPHOSPHATE
DNA POLYMERASE β $I50 = 100\ \mu M$
DNA POLYMERASE α $I50 = 5\ \mu M$
DNA POLYMERASE δ $I50 = 17\ \mu M$
DNA POLYMERASE T4 $I50 = 2.8\ \mu M$

1-OCTYL-2-(PHOSPHONOHEPTYL)-sn-GLYCERO-3-PHOSPHOETHANOLAMINE
PHOSPHOLIPASE A2

(OCTYLSULPHONYL)FLUOROMETHANESULPHONAMIDE
CARBONATE DEHYDRATASE II $I50 = 3\ nM$

3-n-OCTYLTHIO-1,1-DIFLUORO-2-PROPANONE
JUVENILE HORMONE ESTERASE $I50 = 971\ pM$

3-OCTYLTHIO-1,1,1-TRIFLUORO-2-PROPANONE
JUVENILE HORMONE ESTERASE $I50 = 23\ nM$

3-OCTYLTHIO-1,1,1-TRIFLUOROPROPAN-2-ONE
ANTENNAL ESTERASE $Ki = 25\ \mu M$
ESTERASE ANTENNAL $I50 = 80\ nM$

1-OCTYL-[1-(3,3,3-TRIFLUOROPROPAN-2,2-DIHYDROXY)]SULPHONE
JUVENILE HORMONE ESTERASE $I50 = 1.2\ nM$

N-OCTYL-β-VALIENAMINE
α-GLUCOSIDASE $I50 = 17\ \mu M$
GLUCOSYLCERAMIDASE $I50 = 30\ nM$

2-OCTYNOIC ACID
ACYL-CoA DEHYDROGENASE (NADP)

2-OCTYNOYL-CoA
ACYL-CoA DEHYDROGENASE
ACYL-CoA DEHYDROGENASE (MEDIUM CHAIN) SS

3-OCTYNOYL-CoA
ACYL-CoA DEHYDROGENASE

OCYTYL-β-GLUCOSIDE
DOLICHYL-PHOSPHATE β-D-MANNOSYLTRANSFERASE

3-OCYTYLTHIO-1,1,1-TRIFLUOROBUTAN-2-ONE
JUVENILE HORMONE ESTERASE — *I*50 = 880 pM
α-NAPHTHYLACETATE ESTERASE — *I*50 = 3.9 μM

3-OCYTYLTHIO-1,1,1-TRIFLUOROPROPAN-2-OL
JUVENILE HORMONE ESTERASE — *K*i = 950 pM
α-NAPHTHYLACETATE ESTERASE — *I*50 = 76 μM

OENOTHEIN A
AROMATASE — 70% 50 μM
STEROID 5α-REDUCTASE — *I*50 = 1.2 μM

OENOTHEIN B
AROMATASE — 33% 50 μM
POLY(ADP-RIBOSE) GLYCOHYDROLASE — *I*50 = 3.8 μM
POLY(ADP-RIBOSE) GLYCOHYDROLASE — *I*50 = 4.8 μM
STEROID 5α-REDUCTASE — *I*50 = 22 μM
STEROID 5α-REDUCTASE — *I*50 = 440 nM

β-OESTRADIOL
UDP-GLUCURONOSYLTRANSFERASE — C — *K*i = 25 μM

17β-OESTRADIOL
PROSTAGLANDIN E2 9-KETOREDUCTASE

OESTRADIOL SULPHATE
17β-HYDROXYSTEROID DEHYDROGENASE — C — *K*i = 240 μM

OESTRONE
UDP-GLUCURONOSYLTRANSFERASE — NC — *K*i = 5 μM

OESTRONE-3O-SULPHAMATE
OESTRONE SULFATASE — *I*50 = 50 nM

OF 4949-III
AMINOPEPTIDASE B — *I*50 = 40 μg/ml
PEPTIDYL-DIPEPTIDASE A — *I*50 = 70 μg/ml

OF 4949-IV
AMINOPEPTIDASE B — *I*50 = 20 μg/ml

OFLOXACIN
CYTOCHROME P450 LA2 — 12% 500 μM
DNA GYRASE

OFLOXAZIN
DNA TOPOISOMERASE (ATP-HYDROLYSING) — *K*i = 900 ng/ml

OK 1035
DNA PROTEIN KINASE — *I*50 = 8 μM
PROTEIN KINASE A — *I*50 = 390 μM

OKADAIC ACID
CALCINEURIN — *I*50 = 5 μM
MYOSIN-LIGHT-CHAIN PHOSPHATASE — *K*i = 250 nM
PHOSPHATASE ALKALINE
PHOSPHOPROTEIN PHOSPHATASE
PHOSPHOPROTEIN PHOSPHATASE 1 — *I*50 = 50 nM
PHOSPHOPROTEIN PHOSPHATASE 1 — *K*i = 145 nM
PHOSPHOPROTEIN PHOSPHATASE 1 — *I*50 = 160 nM
PHOSPHOPROTEIN PHOSPHATASE 1 — *I*50 = 10 nM
PHOSPHOPROTEIN PHOSPHATASE 1 — *K*i = 153 nM
PHOSPHOPROTEIN PHOSPHATASE 1 — *I*50 = 3.4 nM
PHOSPHOPROTEIN PHOSPHATASE 2 — *I*50 = 200 pM
PHOSPHOPROTEIN PHOSPHATASE 2A — *I*50 = 70 pM

PHOSPHOPROTEIN PHOSPHATASE 2A — *K*i = 34 pM
PHOSPHOPROTEIN PHOSPHATASE 2A — *I*50 = 1 nM
PHOSPHOPROTEIN PHOSPHATASE 2A — *I*50 = 2.6 nM
PHOSPHOPROTEIN PHOSPHATASE 2A — *I*50 = 4 nM
PHOSPHOPROTEIN PHOSPHATASE 2A — *I*50 = 40 nM
PHOSPHOPROTEIN PHOSPHATASE 2A — *K*i = 30 pM
PHOSPHOPROTEIN PHOSPHATASE (POLYCATION MODULATED) — *I*50 = 500 nM

OKADAIC ACID 9,10 EPISULPHIDE

PHOSPHOPROTEIN PHOSPHATASE 2A — *K*i = 47 pM

OKANIN

MYOSIN-LIGHT-CHAIN KINASE

OKY 046

THROMBOXANE A SYNTHASE — *I*50 = 4.5 μM
THROMBOXANE A SYNTHASE — *I*50 = 11 nM

OKY 46

THROMBOXANE A SYNTHASE

OKY 1581

THROMBOXANE A SYNTHASE — *I*50 = 150 nM
THROMBOXANE A SYNTHASE — *I*50 = 3 nM

OLEACEIN

PEPTIDYL-DIPEPTIDASE A — *I*50 = 26 μM

OLEANDRIGENIN

Na^+/K^+-EXCHANGING ATPASE — *I*50 = 1.2 μM

OLEANDRIN

Na^+/K^+-EXCHANGING ATPASE — *I*50 = 620 nM

OLEAN-12-ENE-11-OXO-3β,30-DIOL

11β-HYDROXYSTEROID DEHYDROGENASE

OLEANOLIC ACID

ADENOSINE DEAMINASE — *K*i = 950 nM
ARACHIDONATE 15-LIPOXYGENASE — *I*50 = 265 μM
AROMATASE — 12% 41 μM
C3 CONVERTASE
DNA LIGASE I — *I*50 = 216 μM
ELASTASE — *I*50 = 5.1 μM
ELASTASE LEUKOCYTE — *I*50 = 6.4 μM
HYALURONIDASE — *I*50 = 300 μM
11β-HYDROXYSTEROID DEHYDROGENASE — *I*50 = 150 nM
PHOSPHOLIPASE A2
PROTEINASE HIV-1 — *I*50 = 1 μM
PROTEIN KINASE A — *I*50 = 12 μM
PROTEIN KINASE C — *I*50 = 250 μM
PROTEIN KINASE (Ca^{2+} DEPENDENT) — *I*50 = 112 μM

OLEIC ACID

ADENYLATE KINASE
CHITIN SYNTHASE
CHOLATE-CoA LIGASE — %=% 750 μM
DNA METHYLTRANSFERASE ECO RI. — *I*50 = 16 μM
DNA POLYMERASE α — *I*50 = 38 μM
DNA POLYMERASE β — *I*50 = 32 μM
DOLICHOL ESTERASE
ELASTASE LEUKOCYTE — *K*i = 9 μM
ELASTASE SPUTUM — *K*i = 15 μM
β-GALACTOSIDE α-2,6-SIALYLTRANSFERASE
1,3-β-GLUCAN SYNTHASE — *I*50 = 88 μM
GLYCEROL-3-PHOSPHATE O-ACYLTRANSFERASE
LIPASE (HORMONE SENSITIVE) — NC — *I*50 = 500 nM

LIPOPROTEIN LIPASE		19% 80 μM
MULTICATALYTIC ENDOPEPTIDASE COMPLEX (CHYMOTRYPSIN LIKE ACTIVITY		
MULTICATALYTIC ENDOPEPTIDASE COMPLEX (PEPTIDYLGLUTAMYLPEPTIDE LI		
MYOSIN-LIGHT-CHAIN KINASE		
NAD ADP-RIBOSYLTRANSFERASE		*I*50 = 82 μM
NAD(P)-ARGININE ADP-RIBOSYLTRANSFERASE		*I*50 = 200 μM
Na^+/K^+-EXCHANGING ATPASE		
OXOGLUTARATE DEHYDROGENASE		
PHOSPHOLIPASE D (GLYCOSYLPHOSPHATIDYLINOSITOL SPECIFIC)		*I*50 = 220 μM
PHOSPHOPROTEIN PHOSPHATASE 1M		*I*50 = 250 μM
PLASMANYLETHANOYLAMINE DESATURASE		
PROSTAGLANDIN-ENDOPEROXIDE SYNTHASE		*I*50 = 2 mM
PROTEIN KINASE A		
PROTEIN KINASE A		*I*50 = 10 μM
PROTEIN KINASE C		37% 100 μM

cis-OLEIC ACID

PHOSPHOLIPASE A2 (Ca^{2+} DEPENDENT)		*I*50 = 1.5 μM

OLEIC ACID (8-HYDROXYPYRENE-1,3,6-TRISULPHONATE) ESTER

ELASTASE LEUKOCYTE		*K*i = 58 nM

1-OLEOYL-2-ACETYL-sn-GLYCEROL

1-ALKYL-2-ACETYLGLYCEROL O-ACYLTRANSFERASE

OLEOYL-CoA

PANTOTHENATE KINASE

N-OLEOYLDOPAMINE

ARACHIDONATE 5-LIPOXYGENASE		*I*50 = 7.5 nM

N-OLEOYLETHANOLAMINE

CERAMIDASE ALAKLINE

OLEUROPEIN

ARACHIDONATE 5-LIPOXYGENASE		*I*50 = 11 μM
ARACHIDONATE 12-LIPOXYGENASE		*I*50 = 9.4 μM

1-OLEYL-2-ACETYL-GLYCEROL

PHOSPHOLIPASE C		*I*50 = 5 μM

OLEYLAMINE

MYOSIN-LIGHT-CHAIN KINASE		*I*50 = 6 μM
PROTEIN KINASE (Ca^{2+} DEPENDENT)		*I*50 = 5 μM

OLEYL-CoA

ACETYL-CoA CARBOXYLASE		*K*i = 44 nM
ALKYLGLYCEROPHOSPHATE 2-O-ACETYLTRANSFERASE		
1-ALKYLGLYCEROPHOSPHOCHOLINE O-ACETYLTRANSFERASE		
GLUCOSE-6-PHOSPHATASE		46% 50 μM
3-HYDROXYDECANOYL-[ACYL-CARRIER-PROTEIN] DEHYDRATASE		60% 10 μM
LIPASE (HORMONE SENSITIVE)	NC	*I*50 = 100 nM
PHOSPHATIDYLCHOLINE-DOLICHOL O-ACYLTRANSFERASE		
PROTEIN KINASE (Ca^{2+} INDEPENDENT)		53% 1 μM
TRIACYLGLYCEROL LIPASE		*I*50 = 21 μM

1-OLEYL-2-DOCOSAHEXAENOYL-sn-GLYCERO-3-PHOSPHOCHOLINE

ARACHIDONATE 5-LIPOXYGENASE		*I*50 = 4 μM

OLEYLOXYETHYLPHOSPHOCHOLINE

PHOSPHOLIPASE A2		*I*50 = 14 μM

OLIGO(dG)

DNA TOPOISOMERASE (ATP-HYDROLYSING)

OLIGOMYCIN

ADENOSINETRIPHOSPHATASE		*I*50 = 400 ng/ml
ALDEHYDE OXIDASE		

APYRASE		
CALCIDOL 1-MONOOXYGENASE		
H^+-TRANSPORTING ATP SYNTHASE		400 ng/g
Na^+/K^+-EXCHANGING ATPASE		
NUCLEOSIDE-TRIPHOSPHATASE		45% 50 μg/ml

OLIGORIBONUCLEOTIDE LOW MW		
α,α-TREHALOSE-PHOSPHATE SYNTHASE		

OLIGOSACCHARIDES		
KERATAN-SULFATE ENDO-1,4-β-GALACTOSIDASE		

OLIVANIC ACID		
β-LACTAMASE		*I*50 = 400 ng/ml

OLIVOMYCIN		
RNA POLYMERASE	C	*K*i = 95 ng/ml

OLOMOUCINE		
CDC2 KINASE		*I*50 = 6 μM
CYCLIN A KINASE	C	
CYCLIN B KINASE	C	
CYCLIN DEPENDENT KINASE 1		*I*50 = 7 μM
CYCLIN DEPENDENT KINASE 2		*I*50 = 7 μM
CYCLIN DEPENDENT KINASE 2		
CYCLIN DEPENDENT KINASE 5		*K*i = 8 μM
CYCLIN DEPENDENT PROTEIN KINASE 5		*I*50 = 7 μM
CYCLIN DEPENDENT PROTEIN KINASES		
CYCLIN KINASE	C	
ERK1/MAP KINASE	C	
KINASE p35	C	
MAP KINASE		*I*50 = 30 μM
PROTEIN KINASE AMP STIMULATED		*I*50 = 230 μM
PROTEIN KINASE ARABIDOPSIS SHAGGY-RELATED		*I*50 = 130 μM
PROTEIN KINASE Cγ		*I*50 = 800 μM
PROTEIN KINASE Cn		*I*50 = 930 μM
PROTEIN TYROSINE KINASE EGFR		*I*50 = 440 μM

OLPRINONE		
3′,5′-CYCLIC-NUCLEOTIDE PHOSPHODIESTERASE III		

OLTIPRAZ		
CYTOCHROME P450 1A2		
RNA-DIRECTED DNA POLYMERASE		

OMBUIN		
NADH DEHYDROGENASE (EXOGENOUS)		*I*50 = 50 μM

OMEPRAZOLE		
CYTOCHROME P450 3A	NC	*K*i = 84 μM
CYTOCHROME P450 2C9	C	*K*i = 40 μM
CYTOCHROME P450 2C19	C	*K*i = 3.1 μM
CYTOCHROME P450 2D6		*K*i = 240 μM
DEXTROMETHROPHAN O-DEMETHYLASE		
7-ETHOXYCOUMARIN O-DEETHYLASE		*I*50 = 69 μM
H^+/K^+-EXCHANGING ATPASE		*I*50 = 470 nM
H^+/K^+-EXCHANGING ATPASE		*I*50 = 2.4 μM
H^+/K^+-EXCHANGING ATPASE		*I*50 = 50 μM
H^+-TRANSPORTING ATPASE		
Na^+/K^+-EXCHANGING ATPASE		*I*50 = 3.3 μM
PAPAIN		*I*50 = 17 μM
PYRUVATE DECARBOXYLASE	C	*K*i = 42 μM
TRANSKETOLASE		60% 200 μM
UREASE		*I*50 = 5.4 μM
UREASE		*I*50 = 10 μM

OMETHOATE
ACETYLCHOLINESTERASE

OMP
OROTIDINE-5′-PHOSPHATE DECARBOXYLASE — $Ki = 3.3\ \mu M$

ONCHIDAL
ACETYLCHOLINESTERASE — IR — Kd = 300 μM

ONCHOCYSTATIN
CATHEPSIN B — $Ki = 170$ nM
HISTOLYSAIN — $Ki = 70$ pM
PROTEINASE CACNORHABDITIS ELEGANS — $Ki = 25$ pM
PROTEINASES CYSTEINE

ONCOMODULIN
GLUTATHIONE REDUCTASE (NADPH) — $I50 = 5\ \mu M$

ONIDET P 40
CHLORAMPHENICOL O-ACETYLTRANSFERASE — $I50 = 20\ \mu M$

ONO-1
ALDEHYDE REDUCTASE

ONO-2
ALDEHYDE REDUCTASE

ONO-3
ALDEHYDE REDUCTASE

ONO 1301
THROMBOXANE A SYNTHASE

ONO 2235
ALDEHYDE REDUCTASE — $I50 = 10$ nM
ALDEHYDE REDUCTASE — $I50 = 26$ nM

ONO 3307
CHYMOTRYPSIN — $Ki = 47\ \mu M$
KALLIKREIN PLASMA — $Ki = 290$ nM
KALLIKREIN TISSUE — $Ki = 3.6\ \mu M$
PLASMIN — $Ki = 310$ nM
THROMBIN — $Ki = 180$ nM
TRYPSIN — $Ki = 48$ nM

ONO 3805
STEROID 5α-REDUCTASE 1 — $Ki = 27$ nM
STEROID 5α-REDUCTASE 1 — $Ki = 1$ nM
STEROID 5α-REDUCTASE 2 — $Ki = 500$ pM
STEROID 5α-REDUCTASE 2 — $Ki = 40$ nM

ONO 5046
CATHEPSIN G — $I50 = 7.2\ \mu M$
α-CHYMOTRYPSIN — $I50 = 9.5\ \mu M$
ELASTASE LEUKOCYTE — C — $Ki = 200$ nM
ELASTASE PANCREATIC — $I50 = 1.3\ \mu M$

ONO 9902
NEPRILYSIN

ONONIN
ALDEHYDE DEHYDROGENASE (NAD) 2 — C — $Ki = 10\ \mu M$

OOCYTE INHIBITOR
3′,5′-CYCLIC-NUCLEOTIDE PHOSPHODIESTERASE (CALMODULIN)

3(4-OO-3H, 5H-PYRROLO[3,2-d]PYRIMIDIN-7-YL)-3-(3-CHLOROPHENYL)PROPANENITRILE
PURINE-NUCLEOSIDE PHOSPHORYLASE — $I50 = 11$ nM

OOLONGHOMOBISFLAVAN
NADH DEHYDROGENASE $I50 = 800$ nM

OOLONGHOMOBISFLAVAN A
NADH DEHYDROGENASE $I50 = 500$ nM
NADH DEHYDROGENASE $I50 = 3.5$ μM
NADH DEHYDROGENASE $I50 = 3.1$ μM
NADH DEHYDROGENASE $I50 = 2.9$ μM

OOSPOREIN
DNA POLYMERASE $I50 > 700$ μM
DNA POLYMERASE I

OOSPORIN
DNA POLYMERASE $I50 = 610$ μM
DNA POLYMERASE $I50 = 75$ μM

OPC 13013
3′,5′-CYCLIC-NUCLEOTIDE PHOSPHODIESTERASE LOW Km $I50 = 730$ nM

OPC 13135
3′,5′-CYCLIC-NUCLEOTIDE PHOSPHODIESTERASE (cycAMP)
3′,5′-CYCLIC-NUCLEOTIDE PHOSPHODIESTERASE LOW Km $I50 = 185$ nM

OPC 3689
3′,5′-CYCLIC-NUCLEOTIDE PHOSPHODIESTERASE LOW Km $I50 = 70$ nM

OPC 3911
3′,5′-CYCLIC-NUCLEOTIDE PHOSPHODIESTERASE (cycAMP) cycGMP INHIBITED $I50 = 50$ nM
3′,5′-CYCLIC-NUCLEOTIDE PHOSPHODIESTERASE (cycGMP SENSITIVE) $I50 = 30$ nM
3′,5′-CYCLIC-NUCLEOTIDE PHOSPHODIESTERASE III $Ki = 5$ nM
3′,5′-CYCLIC-NUCLEOTIDE PHOSPHODIESTERASE III $I50 = 40$ nM
3′,5′-CYCLIC-NUCLEOTIDE PHOSPHODIESTERASE III $I50 = 60$ nM
3′,5′-CYCLIC-NUCLEOTIDE PHOSPHODIESTERASE III $I50 = 54$ nM
3′,5′-CYCLIC-NUCLEOTIDE PHOSPHODIESTERASE LOW Km $I50 = 42$ nM
3′,5′-CYCLIC-NUCLEOTIDE PHOSPHODIESTERASE (RI) $I50 = 130$ μM

OPHTHALMIC ACID
FORMALDEHYDE DEHYDROGENASE (GLUTATHIONE)
LACTOYLGLUTATHIONE LYASE C $Ki = 950$ μM

OPIPRAMOL
C8- > C7 STEROL ISOMERASE $Ki = 17$ nM

O_3PNHPO_3
INORGANIC PYROPHOSPHATASE

OPPOSSUM SERUM METALLOPROTEINASE INHIBITOR
ATROLYSIN

OPSOPYRROLE DICARBOXYLATE
HYDROXYMETHYLBILANE SYNTHASE

OR 462
CATECHOL O-METHYLTRANSFERASE $Ki = 23$ nM
CATECHOL O-METHYLTRANSFERASE $I50 = 18$ nM
TYROSINE 3-MONOOXYGENASE $I50 = 10$ μM

OR 486
CATECHOL O-METHYLTRANSFERASE $I50 = 12$ nM

OR 611
CATECHOL O-METHYLTRANSFERASE $Ki = 14$ nM
CATECHOL O-METHYLTRANSFERASE $I50 = 10$ nM
CATECHOL O-METHYLTRANSFERASE $I50 = 63$ nM

ORANGE A
GALACTOSYLCERAMIDE SULFOTRANSFERASE

ORANGE I		
AZOBENZENE REDUCTASE		
ORANGE II		
AZOBENZENE REDUCTASE		
ORANGE MX-2R		
FERRIC-CHELATE REDUCTASE		*I*50 = 250 µM
ORCINOL		
MONOPHENOL MONOOXYGENASE	C	
OREGANIC ACID		
FARNESYL PROTEINTRANSFERASE	C	*K*i = 4.5 nM
FARNESYL PROTEINTRANSFERASE	C	*K*i = 4.5 nM
GERANYLGERANYL PROTEINTRANSFERASE		*I*50 = 60 µM
ORG 20241		
3′,5′-CYCLIC-NUCLEOTIDE PHOSPHODIESTERASE III		*I*50 = 50 µM
3′,5′-CYCLIC-NUCLEOTIDE PHOSPHODIESTERASE IV		*I*50 = 6.3 µM
3′,5′-CYCLIC-NUCLEOTIDE PHOSPHODIETSTERASE III		*I*50 = 40 µM
ORG 30029		
3′,5′-CYCLIC-NUCLEOTIDE PHOSPHODIESTERASE III		*I*50 = 24 µM
3′,5′-CYCLIC-NUCLEOTIDE PHOSPHODIESTERASE IV		*I*50 = 16 µM
3′,5′-CYCLIC-NUCLEOTIDE PHOSPHODIESTERASE IV		*I*50 = 25 µM
3′,5′-CYCLIC-NUCLEOTIDE PHOSPHODIETSTERASE III		*I*50 = 25 µM
3′,5′-CYCLIC-NUCLEOTIDE PHOSPHODIETSTERASE IV		*I*50 = 16 µM
3′,5′-CYCLIC-NUCLEOTIDE PHOSPHODIETSTERASE V		*I*50 = 40 µM
ORG 30365		
AROMATASE		*I*50 = 90 nM
ORG 33201		
AROMATASE		*I*50 = 2.2 nM
ORG 9935		
3′,5′-CYCLIC-NUCLEOTIDE PHOSPHODIESTERASE IV		*I*50 = 25 µM
3′,5′-CYCLIC-NUCLEOTIDE PHOSPHODIESTERASE V		*I*50 = 12 µM
3′,5′-CYCLIC-NUCLEOTIDE PHOSPHODIESTERASE V		*I*50 = 9.8 µM
ORGANIC ACIDS		
KYNURENINE-OXOGLUTARATE AMINOTRANSFERASE		
ORIENTIN		
PROLYL OLIGOPEPTIDASE		*I*50 = 39 ppm
XANTHINE OXIDASE		
ORNITHINE		
ACETYLORNITHINE TRANSAMINASE		
AGMATINASE	NC	*K*i = 6 mM
5-AMINOVALERATE TRANSAMINASE		
ARGINASE	C	*K*i = 1.1 mM
ARGINASE		76% 25 mM
ARGINASE	C	*K*i = 1 mM
D-ARGINASE		
ARGININE DEIMINASE		
ARGININE-tRNA LIGASE		
GLUTAMATE SYNTHASE (FERREDOXIN)		
HISTIDINE-tRNA LIGASE		
INOSAMINE-PHOSPHATE AMIDINOTRANSFERASE		
KYNURENINASE		
LYSINE 2,3-AMINOMUTASE		
LYSINE (ARGININE) CARBOXYPEPTIDASE		
D-ORNITHINE 4,5-AMINOMUTASE		
ORNITHINE CARBAMOYLTRANSFERASE		> 10 mM
SACCHAROPINE DEHYDROGENASE (NADP, L-LYSINE FORMING)		

Inhibitor / Enzyme	Type	Value
D-ORNITHINE		
ORNITHINE DECARBOXYLASE	C	$Ki = 1.7$ mM
L-ORNITHINE		
AMIDINOTRANSFERASE		
ARGINASE		
ARGININE DECARBOXYLASE	C	$Ki = 3.8$ mM
ARGININE DEIMINASE	C	$Ki = 1$ mM
ARGININE RACEMASE	NC	
CARBOXYPEPTIDASE B	C	$Ki = 15$ mM
GLUTAMATE-5-SEMIALDEHYDE DEHYDROGENASE		
GLYCINE AMIDINOTRANSFERASE	C	$Ki = 500$ μM
D-LYSINE 5,6-AMINOMUTASE		
PROTEINASE I ACHROMOBACTER		$Ki = 22.5$ mM
PYRROLINE 5-CARBOXYLATE SYNTHASE		$I50 = 500$ μM
SACCHAROPINE DEHYDROGENASE (NAD, L-LYSINE FORMING)		
XAA-ARG DIPEPTIDASE		
ORN-P(O)(NH_2)$NHSO_3H$		
ORNITHINE CARBAMOYLTRANSFERASE		
OROBOL		
ARACHIDONATE 15-LIPOXYGENASE		$Ki = 44$ μM
1-PHOSPHATIDYLINOSITOL 4-KINASE		$I50 = 800$ nM
PROTEIN TYROSINE KINASE EGFR		$I50 = 3$ μg/ml
OROTIC ACID		
ACYLPHOSPHATASE		
CARBAMOYL-PHOSPHATE PHOSPHATASE	NC	
DIHYDROOROTASE	C	$Ki = 81$ μM
DIHYDROOROTASE		$Ki = 860$ μM
DIHYDROOROTATE DEHYDROGENASE		$Ki = 630$ μM
DIHYDROOROTATE DEHYDROGENASE	C	$Ki = 15$ μM
DIHYDROOROTATE OXIDASE	C	
OROTATE PHOSPHORIBOSYLTRANSFERASE		
OROTIDINE-5′-MONOPHOSPHATE		
ADENYLOSUCCINATE SYNTHASE		81% 10 mM
OROTIDYLATE		
OROTATE PHOSPHORIBOSYLTRANSFERASE		
ORPHENADRINE CITRATE		
ALDEHYDE OXIDASE		$I50 = 115$ μM
ORTHOVANADATE		
GLUTAMATE DEHYDROGENASE	C	$Ki = 40$ μM
PROTEIN TYROSINE-PHOSPHATASE		$I50 = 10$ μM
PROTEIN TYROSINE-PHOSPHATASE CD45		$I50 = 92$ μg/ml
ORYZACYSTATIN I		
CATHEPSIN H		$Ki = 790$ nM
CATHEPSIN H		$Ki = 78$ nM
CATHEPSIN L		$Ki = 5.1$ nM
PAPAIN		$Ki = 32$ nM
PAPAIN		$Ki = 30$ nM
ORYZACYSTATIN II		
CATHEPSIN B		$Ki = 8.2$ μM
CATHEPSIN H		$Ki = 10$ nM
CATHEPSIN H		$Ki = 250$ nM
CATHEPSIN L		$Ki = 39$ nM
PAPAIN		$Ki = 260$ nM
PAPAIN		$Ki = 830$ nM
ORYZALIM		
H^+-TRANSPORTING ATPASE		

OSCILLAMIDE Y

CHYMOTRYPSIN

OSCILLAPEPTIN

CHYMOTRYPSIN		*I*50 = 2.2 μg/ml
ELASTASE		*I*50 = 300 ng/ml

OSCILLAPEPTIN G

TYROSINASE

OSCILLATORIN

CHYMOTRYPSIN

OSIRISYNES

Na^+/K^+-EXCHANGING ATPASE

OSTRUTHIN

CYTOCHROME P450 1A1	MIX	*K*i = 121 nM
CYTOCHROME P450 2B1		*I*50 = 1.3 μM

OUABAGENIN

PROTEIN KINASE A		45% 400 μM

OUABAIN

Na^+/K^+-EXCHANGING ATPASE		*I*50 = 3 nM
Na^+/K^+-EXCHANGING ATPASE		*I*50 = 8.1 μM
Na^+/K^+-EXCHANGING ATPASE		*K*i = 30 nM
Na^+/K^+-EXCHANGING ATPASE	NC	*I*50 = 2 μM
Na^+/K^+-EXCHANGING ATPASE		*K*i = 27 nM
Na^+/K^+-EXCHANGING ATPASE		*I*50 = 300 nM
Na^+/K^+-EXCHANGING ATPASE		*I*50 = 32 nM
Na^+/K^+-EXCHANGING ATPASE		*I*50 = 220 nM
Na^+/K^+-EXCHANGING ATPASE		*I*50 = 120 nM
4-NITROPHENYLPHOSPHATASE		
PROTEIN KINASE A		43% 400 μM

OUDEMANSIN

CYTOCHROME bc1
NADH DEHYDROGENASE (UBIQUINONE)
UBIQUINOL-CYTOCHROME-C REDUCTASE

OVALICIN

METHIONYL AMINOPEPTIDASE 2		*I*50 = 400 pM

OVINE ARTICULAR CARTILAGE SERINE PROTEINASE INHIBITOR 6 Kd

CATHEPSIN G
CHYMOTRYPSIN
ELASTASE LEUKOCYTE
KALLIKREIN TISSUE
TRYPSIN

OVINE ARTICULAR CARTILAGE SERINE PROTEINASE INHIBITOR 58 Kd

CATHEPSIN G
CHYMOTRYPSIN
ELASTASE LEUKOCYTE
ELASTASE PANCREATIC
PLASMIN
TRYPSIN

OVINE UTERINE SERPIN 1

PEPSIN

OVOCYSTATIN

LEGUMAIN		*K*i = 5 nM

OVO-INHIBITOR CHICKEN

ELASTASE LEUKOCYTE

Inhibitor / Enzyme	Type	Value
OVO-INHIBITOR HEN		
CHYMOTRYPSIN		
OVOMACROGLOBULIN		
PEPSIN	C	$Ki = 13\ \mu M$
RENIN		
OVOMUCIN		
CHYMOSIN		
OVOMUCOID		
TRYPSIN		
OVOMUCOID (CHICKEN)		
ACROSIN		
OVOMUCOID DUCK		
CHYMOTRYPSIN		
ELASTASE LEUKOCYTE		
ELASTASE PANCREATIC		
HCG		
TRYPSIN		
7-OXABICYCLOHEPTYLPROSTANOIC ACIDS		
CYCLOOXYGENASE	IR	
OXACILLIN		
D-ALANYL-D-ALANINE CARBOXYPEPTIDASE	W	$I50 = 740\ \mu M$
N-7-OXADECYLDEOXYNOJIRIMYCIN		
α-GLUCOSIDASE		$I50 = 280\ nM$
1H-[1,2,4]OXADIAZOLO[4,3-a]QUINOXALIN-1-ONE		
GUANYLATE CYCLASE		$I50 = 10\ nM$
GUANYLATE CYCLASE		
OXADIAZON		
PROTOPORPHYRINOGEN OXIDASE		$I50 = 40\ nM$
10-OXA-5,8-DIDEAZAFOLIC ACID		
PHOSPHORIBOSYLAMINOIMIDAZOLECARBOXAMIDE FORMYLTRANSFERASE	C	$Ki = 2.9\ \mu M$
(2S,4S)-3-OXA-2-ETHYL-4-[[4-(1,3-DIOXABUTYL)PIPERIDIN-1-YL]CARBONYL]-5-PHENYLPENTANAMID		
RENIN		$I50 = 900\ pM$
γ-(L-OXAGLUTAMYL)-L-CYSTEINYL-GLYCINE		
γ-GLUTAMYLTRANSFERASE	NC	$Ki = 250\ \mu M$
OXALACETIC ACID		
2-(ACETAMIDOMETHYLENE)SUCCINATE HYDROLASE	C	$Ki = 24\ \mu M$
ACETOACETATE DECARBOXYLASE		
ACETYLPYRUVATE HYDROLASE	C	
ACONITATE HYDRATASE	C	$Ki = 700\ \mu M$
ADENYLOSUCCINATE SYNTHASE	C	$Ki = 500\ \mu M$
ALANINE DEHYDROGENASE		
ALDEHYDE REDUCTASE		
4-AMINOBUTYRATE TRANSAMINASE		
AMINOLEVULINATE TRANSAMINASE		
ASPARAGINE SYNTHASE (GLUTAMINE-HYDROLYSING)		
ASPARTATE 1-DECARBOXYLASE	C	$Ki = 810\ \mu M$
ASPARTATE TRANSAMINASE	C	$Ki = 50\ \mu M$
CITRATE (pro-3s)-LYASE	IR	
CYANATE HYDROLASE		$I50 = 6\ \mu M$
DECYLHOMOCITRATE SYNTHASE	C	$Ki = 600\ \mu M$
FRUCTOSE-BISPHOSPHATE ALDOLASE		
GLUTAMATE DECARBOXYLASE	C	$Ki = 810\ \mu M$
GLUTAMATE DEHYDROGENASE (NAD(P))		
GLUTAMATE SYNTHASE (NADH)	C	$Ki = 5\ mM$

Inhibitor / Enzyme	Type	Value
GLUTAMATE SYNTHASE (NADH)		
GLUTAMATE SYNTHASE (NADPH)		
GUANYLATE CYCLASE		
4-HYDROXY-2-OXOGLUTARATE ALDOLASE	C	Ki = 220 µM
ISOCITRATE DEHYDROGENASE (NAD)		
ISOCITRATE DEHYDROGENASE (NADP)		
ISOCITRATE LYASE		
KYNURENINE-OXOGLUTARATE AMINOTRANSFERASE		19% 1 mM
D-LACTATE DEHYDROGENASE (CYTOCHROME)	C	Ki = 10 µM
L-LACTATE DEHYDROGENASE		
LACTATE-MALATE TRANSHYDROGENASE		
LYSINE 2-MONOOXYGENASE		
D-MALATE DEHYDROGENASE (DECARBOXYLATING)		
MALATE DEHYDROGENASE (OXALACETATE-DECARBOXYLATING) (NADP)		
METHYLMALONYL-CoA CARBOXYLTRANSFERASE		Ki = 70 µM
N-METHYL-2-OXOGLUTARAMATE HYDROLASE	C	
NICOTINATE-NUCLEOTIDE PYROPHOSPHORYLASE (CARBOXYLATING)		
ORNITHINE-OXO-ACID TRANSAMINASE		
OXOGLUTARATE DEHYDROGENASE (LIPOAMIDE)		
PANTOTHENASE		Ki = 360 µM
PHOSPHOENOLPYRUVATE CARBOXYKINASE (ATP)		
PHOSPHOENOLPYRUVATE CARBOXYLASE		
PHOSPHOENOLPYRUVATE DECARBOXYLASE		
PHOSPHOGLUCONATE DEHYDROGENASE (DECARBOXYLATING)		
PROCOLLAGEN-PROLINE 3-DIOXYGENASE		
SUCCINATE DEHYDROGENASE		
SUCCINATE DEHYDROGENASE (UBIQUINONE)		
TARTRATE DEHYDROGENASE		
THIOSULFATE SULFURTRANSFERASE		
XAA-PRO DIPEPTIDASE		Ki = 160 µM
OXALGLYCINE		
ISOCITRATE DEHYDROGENASE (NAD)		
ISOCITRATE DEHYDROGENASE (NADP)		
N-OXALGLYCINE		
PROCOLLAGEN-PROLINE, 2-OXOGLUTARATE-4-DIOXYGENASE	C	I50 = 540 nM
OXALIC ACID		
ACETOLACTATE SYNTHASE		
ACETYLPYRUVATE HYDROLASE	C	
N-ACYLHEXOSAMINE OXIDASE		
ACYL-LYSINE DEACYLASE		
AGAVAIN		81% 10 mM
CARBONATE DEHYDRATASE III		
CARBOXYPEPTIDASE A		
CYANATE HYDROLASE		I50 = 4 µM
DEHYDROGLUCONATE DEHYDROGENASE		
FUMARYLACETOACETASE		
GLUCONATE 2-DEHYDROGENASE	C	Ki = 2.9 mM
GLUTAMATE SYNTHASE (NADPH)		
GLYOXYLATE REDUCTASE (NADP)		Ki = 36 µM
D-2-HYDROXY-ACID DEHYDROGENASE		
(S)-2-HYDROXY-ACID OXIDASE		
[HYDROXYMETHYLGLUTARYL-CoA REDUCTASE (NADPH)]-PHOSPHATASE		
4-HYDROXY-2-OXOGLUTARATE ALDOLASE		73% 40 mM
HYDROXYPYRUVATE REDUCTASE		Ki = 7 µM
7α-HYDROXYSTEROID DEHYDROGENASE		
myo-INOSITOL OXYGENASE		
ISOCITRATE LYASE	NC	Ki = 2 mM
ISOCITRATE LYASE	C	Ki = 5.1 µM
KYNURENINE-OXOGLUTARATE AMINOTRANSFERASE		
D-LACTATE DEHYDROGENASE (CYTOCHROME)	C	Ki = 1.6 µM
L-LACTATE DEHYDROGENASE (CYTOCHROME)		
L-LACTATE DEHYDROGENASE		80% 200 µM

LACTATE 2-MONOOXYGENASE		
LIGNIN PEROXIDASE	NC	
MALATE DEHYDROGENASE (OXALACETATE-DECARBOXYLATING) (NADP)	NC	
MALATE DEHYDROGENASE (OXALACETATE-DECARBOXYLATING) (NADP)		
MALATE SYNTHASE	C	Ki = 19 μM
MALYL-CoA LYASE	C	Ki = 160 μM
MEMBRANE ALANINE AMINOPEPTIDASE		
METHYLMALONYL-CoA CARBOXYLTRANSFERASE		Ki = 1.8 μM
NUCLEOTIDASE		
OXALATE OXIDASE		
OXALOACETATE DECARBOXYLASE		Ki = 3.8 μM
3-OXOACID CoA-TRANSFERASE		
2-OXOALDEHYDE DEHYDROGENASE (NAD)		
PANTOTHENASE		Ki = 310 μM
PHOSPHATASE ACID		
PHOSPHOENOLPYRUVATE CARBOXYKINASE (GTP)	C	Ki = 4.6 μM
PHOSPHOENOLPYRUVATE DECARBOXYLASE	C	Ki = 173 μM
PHOSPHOENOLPYRUVATE-PROTEIN KINASE		
PHOSPHOGLYCERATE MUTASE		
PHOTINUS-LUCIFERIN 4-MONOOXYGENASE (ATP-HYDROLYSING)		
3-PHYTASE		
PROCOLLAGEN-PROLINE, 2-OXOGLUTARATE-4-DIOXYGENASE		
PSEUDOMONAS AEROGINOSA NEUTRAL PROTEINASE		89% 200 μM
PYRUVATE CARBOXYLASE	NC	Ki = 70 μM
PYRUVATE, ORTHOPHOSPHATE DIKINASE		Ki = 10 μM
PYRUVATE, WATER DIKINASE		
THERMOLYSIN		
XAA-PRO DIPEPTIDASE		Ki = 210 μM
cis-OXALIC ACID		
4-HYDROXYPHENYLPYRUVATE DIOXYGENASE		
OXALIC BIS(2-HYDROXY-1-NAPHTHYLMETHYLENE)HYDRAZIDE)		
PROTEINASE CYSTEINE MALARIAL		I50 = 6 μM
OXALOACETIC ACID		
1-AMINOCYCLOPROPANE-1-CARBOXYLATE OXIDASE	C	Ki = 240 μM
ASPARTATE-AMMONIA LIGASE		
ATP CITRATE (pro-S)-LYASE		
D-2-HYDROXY-ACID DEHYDROGENASE		
MALATE DEHYDROGENASE		
OXALOMALIC ACID		
ACONITATE HYDRATASE	C	Ki = 1 μM
ACONITATE HYDRATASE	C	Ki = 23 μM
OXALURIC ACID		
PUTRESCINE CARBAMOYLTRANSFERASE		
S-OXALYL-N-ACETYLCYSTEAMINE		
MALATE DEHYDROGENASE (OXALACETATE-DECARBOXYLATING) (NADP)		
OXALYLALANINE		
PROCOLLAGEN-PROLINE, 2-OXOGLUTARATE-4-DIOXYGENASE	C	Ki = 40 μM
S-OXALYL-CoA		
MALATE DEHYDROGENASE (OXALACETATE-DECARBOXYLATING) (NADP)		0
β-N-OXALYL-L-α,β-DIAMINOPROPIONIC ACID		
TYROSINE TRANSAMINASE	U	Ki = 8.4 mM
5-(–)OXALYLETHYL-NADH		
ALANINE DEHYDROGENASE		
ALCOHOL DEHYDROGENASE		
GLYCERATE DEHYDROGENASE		
GLYCEROL DEHYDROGENASE (NADP)		
D-LACTATE DEHYDROGENASE		

Inhibitor / Enzyme	Type	Value
L-LACTATE DEHYDROGENASE		
MALATE DEHYDROGENASE		
S-OXALYLGLUTATHIONE		
PHOSPHORYLASE PHOSPHATASE		
OXALYLGLYCINE		
GLUTAMATE DEHYDROGENASE (NAD(P))	C	
PROCOLLAGEN-PROLINE, 2-OXOGLUTARATE-4-DIOXYGENASE	C	*K*i = 1.9 μM
PROCOLLAGEN-PROLINE, 2-OXOGLUTARATE-4-DIOXYGENASE	NC	*K*i = 400 nM
D,L-4-OXALYSINE		
LYSINE-tRNA LIGASE	C	
OXAMIC ACID		
ACETOLACTATE SYNTHASE		
ALANOPINE DEHYDROGENASE	MIX	*K*i = 6.4 mM
DEHYDROGLUCONATE DEHYDROGENASE		
GLUCONATE 2-DEHYDROGENASE	C	*K*i = 600 μM
L-LACTATE DEHYDROGENASE		10 μM
LACTATE RACEMASE		
MALATE SYNTHASE		
D-OCTOPINE DEHYDROGENASE		
PANTOTHENASE		
OXAMICETIN		
PEPTIDYLTRANSFERASE		
OXAMNIQUINE		
CYTOCHROME P450 2D6	C	*K*i = 22 μM
D-OXAMYCIN		
D-ALANINE TRANSAMINASE		
OXAMYL		
ACETYLCHOLINESTERASE		
OXANILIC ACIDS		
ALDEHYDE REDUCTASE		
OXANTEL		
FUMARATE REDUCTASE (NADH)	C	
4-OXAOCTANOYL-CoA		
ACYL-CoA DEHYDROGENASE		
α-OXAPHENAZINE		
(S)-2-HYDROXY-ACID OXIDASE		
7-OXA-PROSTAGLANDIN		
15-HYDROXYPROSTAGLANDIN DEHYDROGENASE (NAD)		*K*i = 500 nM
L-2-OXATHIAZOLIDINE-4-CARBOXYLATE		
5-OXOPROLINASE (ATP-HYDROLYSING)	C	
OXAZEPAM		
UDP-GLUCURONOSYLTRANSFERASE	C	*K*i = 188 μM
OXAZINE		
MYOSIN-LIGHT-CHAIN KINASE		
PROTEIN KINASE A		
PROTEIN KINASE C		
PROTEIN KINASE (Ca^{2+} DEPENDENT)		
OXAZOLIDINES		
ALCOHOL DEHYDROGENASE (NADP)		
OXETANOCIN-ATP		
DNA POLYMERASE		*K*i = 550 nM

DNA POLYMERASE α		$Ki = 4.3\ \mu M$
DNA POLYMERASE γ		$Ki = 60\ \mu M$
RNA-DIRECTED DNA POLYMERASE		$Ki = 850$ nM

OXETANOCIN-GTP

DNA POLYMERASE		$Ki = 550$ nM
DNA POLYMERASE α		$Ki = 4.5\ \mu M$
DNA POLYMERASE γ		$Ki = 33\ \mu M$
RNA-DIRECTED DNA POLYMERASE		$Ki = 900$ nM

OXETANOCINTRIPHOSPHATE

DNA POLYMERASE		$Ki = 150$ nM
DNA POLYMERASE α		$Ki = 2.2\ \mu M$

OXFENDAZOLE

FUMARATE REDUCTASE (NADH)		$I50 = 58\ \mu M$

2-OXGLUTARIC ACID

LYSINE 2-MONOOXYGENASE

5(S)Trans-5,6-OXIDO-7,9-trans-EICOSADIENOIC ACID

LEUKOTRIENE A4 HYDROLASE	IR	

3-(6S,7S)-OXIDOSQUALENE

SQUALENE MONOOXYGENASE		$I50 = 6.7\ \mu M$

OXILINIC ACID

DNA TOPOISOMERASE (ATP-HYDROLYSING)		$I50 = 475\ \mu g/ml$

OXINDOLE

AMINE OXIDASE MAO-B		$I50 = 73\ \mu M$
PHOSPHOLIPASE A2		$I50 = 380\ \mu M$

OXINDOLYL-L-ALANINE

TRYPTOPHANASE

2′R-32-OXIRANYLLANOST-7-EN-3β-OL

LANOSTEROL C14 DEMETHYLASE	C	$Ki = 620$ nM

OXMETIDINE

NADH DEHYDROGENASE (UBIQUINONE)		$I50 = 3.4\ \mu M$

4-(OXOACETYL)PHENOXYACETIC ACID

GLYCINE C-ACETYLTRANSFERASE

2-OXO ACIDS

D-AMINO-ACID OXIDASE	C	
SACCHAROPINE DEHYDROGENASE (NAD, L-LYSINE FORMING)	NC	

2-OXOADIPIC ACID

4-AMINOBUTYRATE TRANSAMINASE		
HOMOISOCITRATE DEHYDROGENASE		
KYNURENINE-OXOGLUTARATE AMINOTRANSFERASE		
[3-METHYL-2-OXOBUTANOATE DEHYDROGENASE (LIPOAMIDE)] KINASE		
PROCOLLAGEN-PROLINE, 2-OXOGLUTARATE-4-DIOXYGENASE		
PYRIMIDINE-DEOXYNUCLEOSIDE 2′-DIOXYGENASE		38% 2.5MM
PYRUVATE CARBOXYLASE		

3-OXOADIPIC ACID

PYRIMIDINE-DEOXYNUCLEOSIDE 2′-DIOXYGENASE		$I50 = 2.5$ mM

OXOALLOPURINOL

PURINE-NUCLEOSIDE PHOSPHORYLASE		$Ki = 810\ \mu M$

19-OXOANDROST-4-ENE-3,6,17-TRIONE

AROMATASE	C	$Ki = 7.5\ \mu M$

24-OXO-25AZACYCLOARTANOL

Δ(24)-STEROL C-METHYLTRANSFERASE	NC	$Ki = 170$ nM

3-OXO-4-AZA STEROIDES
3β-HYDROXY Δ5 STEROID DEHYDROGENASE/3-KETO-Δ5-STEROID ISOMERASE

4-OXO-4H-BENZOPYREN-2-CARBOXYLIC ACID
CARBONYL REDUCTASE (NADPH) | | 60% 100 μM

2-OXO-1,4-BISPHOSPHONOBUTANE
GLYCERALDEHYDE-3-PHOSPHATE DEHYDROGENASE (PHOSPHORYLATING) | | *K*i = 21 mM
PHOSPHOGLYCERATE KINASE | | *K*i = 84 μM

2-OXO-3-BUTYNOIC ACID
(S)-2-HYDROXY-ACID OXIDASE

2-OXOBUTYRIC ACID
ACETOLACTATE SYNTHASE
ACETOLACTATE SYNTHASE | C | *K*i = 5 mM
D-AMINO-ACID OXIDASE
2-DEHYDRO-3-DEOXY-L-ARABINONATE DEHYDRATASE
(S)-2-HYDROXY-ACID OXIDASE
4-HYDROXY-2-OXOGLUTARATE ALDOLASE
[3-METHYL-2-OXOBUTANOATE DEHYDROGENASE (LIPOAMIDE)] KINASE

2-OXOCAPROIC ACID
[3-METHYL-2-OXOBUTANOATE DEHYDROGENASE (LIPOAMIDE)] KINASE

5-[1-OXO-1-(4-CARBO-t-BUTOXYPHENYL)METHYL]-3-THIOPHENESULPHONAMIDE
CARBONATE DEHYDRATASE II | | *I*50 = 3 nM

3-[1′-OXO-7′-CARBOXYHEPTYL]-4-HYDROXY-6-OCTYL-2-PYRONE
ELASTASE SPUTUM | MIX | *K*i = 6 μM

α-OXO-1-CARBOXY-4-TETRAHYDROTHIOPYRANPROPIONIC ACID
PREPHENATE DEHYDRATASE | | *K*i = 20 mM

7-OXOCHOLESTEROL
CHOLESTEROL 7α-MONOOXYGENASE | C |

2-OXODENCANOIC ACID
ALDEHYDE REDUCTASE

2-OXO-3-DEOXYOCTONOIC ACID
2-DEHYDRO-3-DEOXYGLUCARATE ALDOLASE | C |

7-OXO-24xi(28)-DIHYDROCYCLOEUCALENOL
CYCLOEUCALENOL CYCLO-ISOMERASE | | *I*50 = 500 nM

7-OXO-24,25-DIHYDROLANOSTEROL
LANOSTEROL C14 DEMETHYLASE | C | *K*i = 200 nM

7-OXO-24(25)-DIHYDRO-29-NORLANOSTEROL
OBTUSIFOLIOL 14α-DEMETHYLASE | C | *K*i = 950 nM

7,11-OXO-24xi(24′)-DIHYDRO-OBTUSIFOLIOL
OBTUSIFOLIOL 14α-DEMETHYLASE | C | *K*i = 1.5 μM

7-OXO-24xi(24′)-DIHYDRO-OBTUSIFOLIOL
OBTUSIFOLIOL 14α-DEMETHYLASE | C | *K*i = 800 nM

N-(1-OXO-3,3-DIPHENYLPROPYL)-2-PROLINE-[trans(4-AMINOCYCLOHEXYL)METHYL]AMIDE
COAGULATION FACTOR Xa | | *K*i = 980 μM
PLASMIN | | *K*i = 744 μM
THROMBIN | | *K*i = 2 nM
TRYPSIN | | *K*i = 350 nM

N-(1-OXODODECYL)-4α,10-DIMETHYL-8-AZA-trans-DECAL-β-OL
LANOSTEROL SYNTHASE | | *I*50 = 110 nM

2-OXO-D-GLUCONATE
2-DEHYDRO-3-DEOXYGLUCARATE ALDOLASE | C |
GLUCONATE 2-DEHYDROGENASE | C | *K*i = 123 mM

2-OXOGLUTACONIC ACID		
ASPARTATE TRANSAMINASE	IR	
2-OXOGLUTARATE OXIME		
GLUTAMATE DEHYDROGENASE (NAD(P))		
α-OXOGLUTARIC ACID		
5-AMINOLEVULINATE SYNTHASE		
KYNURENINE 3-MONOOXYGENASE		
2-OXOGLUTARIC ACID		
ACETOACETATE DECARBOXYLASE		
ACETOIN DEHYDROGENASE		
ALANOPINE DEHYDROGENASE	UC	*K*i = 1.3 mM
ALDEHYDE REDUCTASE	NC	*K*i = 5 mM
2-AMINOADIPATE AMINOTRANSFERASE	NC	
5-AMINOLEVULINATE SYNTHASE	C	
ASPARAGINE SYNTHASE (GLUTAMINE-HYDROLYSING)		
ASPARTATE-AMMONIA LIGASE		
CARBAMOYL-PHOSPHATE SYNTHASE (AMMONIA)		*K*i = 3.6 mM
CITRATE (re)-SYNTHASE		31% 1 mM
CITRATE (si)-SYNTHASE		
CYANATE HYDROLASE		*I*50 = 15 mM
CYSTATHIONINE γ-LYASE		
FRUCTOSE-BISPHOSPHATE ALDOLASE		
GLUCONATE 2-DEHYDROGENASE		
[GLUTAMATE-AMMONIA-LIGASE] ADENYLYLTRANSFERASE		*I*50 = 800 μM
GLUTAMATE DECARBOXYLASE	C	*K*i = 900 μM
GLUTAMATE DECARBOXYLASE		50 = 9 mM
GLUTAMATE DEHYDROGENASE (NAD(P))	C	*K*i = 730 μM
GLUTAMATE SYNTHASE (NADH)		
GLUTAMINASE		*I*50 = 100 μM
HOMOISOCITRATE DEHYDROGENASE		
4-HYDROXY-4-METHYL-2-OXOGLUTARATE ALDOLASE	C	*K*i = 1.5 mM
4-HYDROXY-2-OXOGLUTARATE ALDOLASE		
myo-INOSITOL OXYGENASE		
ISOCITRATE DEHYDROGENASE		
ISOCITRATE DEHYDROGENASE (NADP)		
[ISOCITRATE DEHYDROGENASE (NADP)] KINASE		
ISOCITRATE LYASE	NC	*K*i = 1.4 mM
ISOCITRATE LYASE	NC	*K*i = 6.8 mM
KYNURENINE-OXOGLUTARATE AMINOTRANSFERASE		
L-LACTATE DEHYDROGENASE (CYTOCHROME)		
MALATE DEHYDROGENASE (OXALACETATE-DECARBOXYLATING) (NADP)		
METHYLGLUTAMATE DEHYDROGENASE		
NICOTINATE-NUCLEOTIDE PYROPHOSPHORYLASE (CARBOXYLATING)		
D-NOPALINE DEHYDROGENASE		
PHOSPHOENOLPYRUVATE CARBOXYKINASE (ATP)		
PHOSPHOENOLPYRUVATE DECARBOXYLASE		
PORPHOBILINOGEN SYNTHASE		
PYRIDOXAMINE-OXALOACETATE TRANSAMINASE		
PYRUVATE CARBOXYLASE		
PYRUVATE KINASE		
SACCHAROPINE DEHYDROGENASE (NAD, L-LYSINE FORMING)		
THREONINE DEHYDRATASE	C	*K*i = 148 mM
THYMINE DIOXYGENASE		
TYROSINE 2,3-AMINOMUTASE		32% 1 mM
XAA-PRO DIPEPTIDASE		*K*i = 12 μM
3-OXOGLUTARIC ACID		
ACETOACETATE DECARBOXYLASE		
HYOSCYAMINE (6S)-DIOXYGENASE		
PANTOTHENASE		
PROCOLLAGEN-PROLINE, 2-OXOGLUTARATE-4-DIOXYGENASE		
PYRIMIDINE-DEOXYNUCLEOSIDE 2′-DIOXYGENASE		27% 2.5 mM

Inhibitor / Enzyme	Type	Value
S-(3-OXOHEXADECYL)-CoA		
GLYCYLPEPTIDE N-TETRADECANOYLTRANSFERASE		
GLYCYLPEPTIDE N-TETRADECANOYLTRANSFERASE	C	$Ki = 250$ nM
3-OXOHEXANOYL-CoA		
HYDROXYMETHYLGLUTARYL-CoA SYNTHASE	MIX	$Ki = 7.1$ µM
2-OXO-5-(1-HYDROXY-2,4,6-HEPTATRIYNYL)-1,3-DIOXOLONE-4-HEPTANOIC ACID		
ACETYL-CoA C-ACETYLTRANSFERASE		
2-OXO-β-HYDROXYISOVALERATE		
KETOL-ACID REDUCTOISOMERASE		$I50 = 1$ mM
2-OXO-3-HYDROXYISOVALERIC ACID		
3-ISOPROPYLMALATE DEHYDROGENASE		
2-OXO-4-HYDROXY-3-METHYLBUTYRIC ACID γ-LACTONE		
2-DEHYDROPANTOYL-LACTONE REDUCTASE (A SPECIFIC)		
2-OXO-4-(4-HYDROXYPHENYL)BUTANOIC ACID		
4-HYDROXYPHENYLPYRUVATE DIOXYGENASE		
2-OXO-4-HYDROXYVALERIC ACID		
2-DEHYDRO-3-DEOXY-L-ARABINONATE DEHYDRATASE		
D,L-2-OXO-3-HYDROXYVALERIC ACID		
2-DEHYDRO-3-DEOXY-L-ARABINONATE DEHYDRATASE	C	$Ki = 160$ µM
2-OXOISOCAPROIC ACID		
KYNURENINE 3-MONOOXYGENASE		
[3-METHYL-2-OXOBUTANOATE DEHYDROGENASE (LIPOAMIDE)] KINASE		$Ki = 480$ µM
2-OXOISOPENTANOIC ACID		
GLUTAMINE-PYRUVATE TRANSAMINASE		
2-OXOISOVALERIC ACID		
2-DEHYDROPANTOYL-LACTONE REDUCTASE (A SPECIFIC)		
3-METHYL-2-OXOBUTANOATE DEHYDROGENASE (LIPOAMIDE)	C	$Ki = 15$ mM
OXOGLUTARATE DEHYDROGENASE (LIPOAMIDE)	C	$Ki = 500$ µM
OXOLINIC ACID		
DNA TOPOISOMERASE		$I50 = 460$ µM
DNA TOPOISOMERASE (ATP-HYDROLYSING)	C	$Ki = 40$ µM
GLYCINE-tRNA LIGASE		
OXOMALONATE		
PROCOLLAGEN-PROLINE, 2-OXOGLUTARATE-4-DIOXYGENASE		
OXOMALONIC ACID		
ISOCITRATE DEHYDROGENASE (NADP)		
2-OXOMALONIC ACID		
OXALOACETATE DECARBOXYLASE		
PANTOTHENASE		
PYRIDOXAMINE-OXALOACETATE TRANSAMINASE		
2-OXO-β-METHYLVALERIC ACID		
[3-METHYL-2-OXOBUTANOATE DEHYDROGENASE (LIPOAMIDE)] KINASE		
2-OXO-4-METHYLVALERIC ACID		
PYRUVATE DECARBOXYLASE		33% 5 mM
OXONIC ACID		
URATE OXIDASE		
3-OXOOCTANOYL-CoA		
FATTY ACID SYNTHASE		

3-(1′-OXOOCTYL)-4-HYDROXY-6-OCTYL-2-PYRONE
ELASTASE SPUTUM $Ki = 4.6\ \mu M$

11-OXO-18α-OLEANA-2,12-DIEN-30-OIC ACID
Na^+/K^+-EXCHANGING ATPASE $I50 = 8\ \mu M$

2-OXOPANTONIC ACID
2-DEHYDROPANTOYL-LACTONE REDUCTASE (A SPECIFIC)

2-OXOPENTADECYL-CoA
GLYCYLPEPTIDE N-TETRADECANOYLTRANSFERASE $I50 = 7\ nM$

S-(2-OXOPENTADECYL)-CoA
GLYCYLPEPTIDE N-TETRADECANOYLTRANSFERASE C $Ki = 28\ nM$

2-OXOPENTANOIC ACID
ALDEHYDE REDUCTASE

2-OXO-4-PENTENOIC ACID
GLUTAMATE DECARBOXYLASE C $Ki = 2.4\ \mu M$

3-OXO-4-PENTENOYL-CoA
ACETYL-CoA C-ACYLTRANSFERASE
CARNITINE O-ACETYLTRANSFERASE IR

2-OXO-4-PHENYL-3-BUTYNOIC ACID
PYRUVATE DECARBOXYLASE

3-OXO-3-PHENYLPROPENE
MANDELONITRILE LYASE IR

3-OXO-3-PHENYLPROPYNE
MANDELONITRILE LYASE IR

2-OXOPIMELIC ACID
PYRIMIDINE-DEOXYNUCLEOSIDE 2′-DIOXYGENASE 26% 2.5 mM

3-OXO-1,4-PIPERAZINE ACETIC ACID ESTERS N-SUBSTITUTED
ASPARTATE CARBAMOYLTRANSFERASE

L-5-OXOPROLINE
γ-GLUTAMYLCYCLOTRANSFERASE PRO
1-PYRROLINE-5-CARBOXYLATE DEHYDROGENASE

OXOPROPANE SULPHONATE
ACETOACETATE DECARBOXYLASE

2-OXOPROPANOL
RNA-DIRECTED DNA POLYMERASE

17β-(1-OXOPROP-2-YNYL)ANDROST-4-ENE-3-ONE
ESTRADIOL-17α-DEHYDROGENASE

7-OXOPROSTANOIC ACID
15-HYDROXYPROSTAGLANDIN DEHYDROGENASE (NAD)

OXOPURINAL-7-RIBOTIDE
OROTIDINE-5′-PHOSPHATE DECARBOXYLASE

OXOPURINOL
OROTATE PHOSPHORIBOSYLTRANSFERASE

11-OXO-11H-PYRIDO[2,1b]QUINAZOLINE-2-CARBOXYLIC ACID
ALDEHYDE REDUCTASE

6-OXOPYRIMIDINE N1-ACETIC ACIDS
ALDEHYDE REDUCTASE

4-OXO-4H-QUINOLINE-2-CARBOXYLIC ACID
ALDEHYDE REDUCTASE

2-OXOSUCCINAMIC ACID
SERINE-GLYOXYLATE TRANSAMINASE 45% 10 mM

2-OXOSUCCINIC ACID
PROCOLLAGEN-PROLINE, 2-OXOGLUTARATE-4-DIOXYGENASE

2-OXO-1,2,3,6-TETRAHYDROPYRIMIDINE-4,6-DICARBOXYLIC ACID
DIHYDROOROTASE C *K*i = 740 nM

(4R)-2-OXO-6-THIOXOHEXAHYDROPYRIMIDINE-4-CARBOXYLIC ACID
DIHYDROOROTASE C *K*i = 850 nM

2-OXOVALERIC ACID
ALDEHYDE REDUCTASE NC *K*i = 1.3 mM
PYRIMIDINE-DEOXYNUCLEOSIDE 2′-DIOXYGENASE

OXOVANADIUM(IV)
RIBONUCLEASE C *K*i = 60 μM

OXYCARBOXIN
SUCCINATE DEHYDROGENASE *I*50 = 8 μM

7-OXYCHOLESTEROL
CHOLESTEROL 7α-MONOOXYGENASE

OXYCILLIN
AMINOPEPTIDASE (HUMAN LIVER) NC *K*i = 1.6 mM

OXYDEMETON
ACETYLESTERASE

OXYDEMETON-METHYL
ACETYLCHOLINESTERASE

OXYDIGLYCOLIC ACID
GLUTAMATE DEHYDROGENASE (NAD(P))

OXYFLOURFEN
PROTOPORPHYRINOGEN OXIDASE

OXYGEN
DIMETHYLMALEATE HYDRATASE
GLUCONATE DEHYDRATASE
2-HYDROXYGLUTARATE DEHYDROGENASE
NITROGENASE
D-ORNITHINE 4,5-AMINOMUTASE
D-PROLINE REDUCTASE (DITHIOL)
TYROSINE 3-MONOOXYGENASE > 100 μM

OXYIMINO-Δ3-CEPHALOSPORINS
D-ALANYL-D-ALANINE CARBOXYPEPTIDASE

2-OXYJUGLONE
NADH DEHYDROGENASE (UBIQUINONE) *K*i = 30 μM

OXYLUCIFERIN
RENILLA-LUCIFERIN 2-MONOOXYGENASE

3-OXY-5-OXOPROLINE
5-OXOPROLINASE (ATP-HYDROLYSING)

4-OXY-5-OXOPROLINE
5-OXOPROLINASE (ATP-HYDROLYSING)

OXYPHENBUTAZONE
3α-HYDROXYSTEROID DEHYDROGENASE (A SPECIFIC)
3α-HYDROXYSTEROID DEHYDROGENASE (B-SPECIFIC) C *K*i = 70 μM
HYPOXANTHINE PHOSPHORIBOSYLTRANSFERASE

OXYPURINOL
OROTIDINE-5′-PHOSPHATE DECARBOXYLASE
XANTHINE DEHYDROGENASE
XANTHINE OXIDASE

1-OXYPURINOL-5′-PHOSPHATE
OROTIDINE-5′-PHOSPHATE DECARBOXYLASE — *K*i = 4 nM

7-OXYPURINOL-5′-PHOSPHATE
OROTIDINE-5′-PHOSPHATE DECARBOXYLASE — *K*i = 800 nM

3-OXYPURINOLRIBOSIDE MONOPHOSPHATE
OROTIDINE-5′-PHOSPHATE DECARBOXYLASE — *K*i = 3 nM

9-OXYPURINOLRIBOSIDE MONOPHOSPHATE
OROTIDINE-5′-PHOSPHATE DECARBOXYLASE — *K*i = 60 nM

OXYTETRACYCLINE
ANHYDROTETRACYCLINE MONOOXYGENASE

OXYTHIAMIN
THIAMIN PYROPHOSPHOKINASE — C — *K*i = 10 mM

OXYTHIAMINE
THIAMIN-PHOSPHATE KINASE

OXYTHIAMINE PYROPHOSPHATE
TRANSKETOLASE
TRANSKETOLASE

4′-OXYTHIAMINE PYROPHOSPHATE
PYRUVATE DEHYDROGENASE (LIPOAMIDE) — *I*50 = 6 nM

OXYTOCIN
ACETYLSEROTONIN O-METHYLTRANSFERASE
PROTEIN-DISULFIDE REDUCTASE (GLUTATHIONE)

OZAGREL
CYTOCHROME P450

P

Pγ
TRANSDUCIN — *K*i = 92 nM

P 21
CYCLIN DEPENDENT KINASE

P 335
THROMBIN — *K*i = 640 pM

P 498
THROMBIN

P 500
THROMBIN

P 596
THROMBIN — *K*i = 46 fM

P 10358
ACETYLCHOLINESTERASE

P 3622
FARNESYL-DIPHOSPHATE FARNESYLTRANSFERASE — *I*50 = 750 nM

P21

CYCLIN DEPENDENT KINASE 2	NC	Ki = 17 nM
CYCLIN DEPENDENT KINASE 3		
CYCLIN DEPENDENT KINASE 4		
CYCLIN DEPENDENT KINASE 6		
PROTEIN KINASE p38		
PROTEIN KINASE STRESS ACTIVATED β	NC	Ki = 75 nM

PACHYMIC ACID

PHOSPHOLIPASE A2		I50 = 2.9 mM

PACIFASTIN

CHYMOTRYPSIN
TRYPSIN

PACIFIC WHITING CATHEPSIN L INHIBITOR

CATHEPSIN L

PACILOQUINONE A

PROTEIN KINASE Cα		I50 = 69 μM
PROTEIN TYROSINE KINASE s-Src		I50 = 2 μM

PACILOQUINONE C

PROTEIN TYROSINE KINASE s-Src		I50 = 9 μM

PACILOQUINONE D

PROTEIN KINASE Cα		I50 = 5.5 μM
PROTEIN KINASE Cα		I50 = 69 μM
PROTEIN TYROSINE KINASE s-Src		I50 = 35 μM

(2S,3S)-PACLOBUTRAZOL

ent-KAURENE OXIDASE	C	Ki = 70 nM
ent-KAURENE OXIDASE		I50 = 23 nM

2′-P-ADP-RIBOSE

SERRATIA MARCESCENS NUCLEASE

PAECILOQUINONE A

PROTEIN TYROSINE KINASE (v-ABL)		I50 = 590 nM
PROTEIN TYROSINE KINASE EGFR		I50 = 11 μM
PROTEIN TYROSINE KINASE (MK CELLS)		I50 = 35 μM
PROTEIN TYROSINE KINASE c-Src		I50 = 2 μM
PROTEIN TYROSINE KINASE v-abl		I50 = 590 nM

PAECILOQUINONE B

PROTEIN TYROSINE KINASE (v-ABL)		I50 = 83 μM
PROTEIN TYROSINE KINASE EGFR		I50 = 21 μM
PROTEIN TYROSINE KINASE v-abl		I50 = 83 μM

PAECILOQUINONE C

PROTEIN TYROSINE KINASE (v-ABL)		I50 = 560 nM
PROTEIN TYROSINE KINASE EGFR		I50 = 6.7 μM
PROTEIN TYROSINE KINASE (MK CELLS)		I50 = 10 μM
PROTEIN TYROSINE KINASE v-abl		I50 = 560 nM

PAECILOQUINONE D

PROTEIN KINASE Cδ		I50 = 62 μM
PROTEIN TYROSINE KINASE (v-ABL)		I50 = 11 μM
PROTEIN TYROSINE KINASE EGFR		I50 = 8 μM
PROTEIN TYROSINE KINASE (MK CELLS)		I50 = 16 μM
PROTEIN TYROSINE KINASE c-Src		I50 = 35 μM
PROTEIN TYROSINE KINASE v-abl		I50 = 11 μM

PAECILOQUINONE E

PROTEIN TYROSINE KINASE EGFR		I50 = 38 μM
PROTEIN TYROSINE KINASE (MK CELLS)		I50 = 18 μM
PROTEIN TYROSINE KINASE c-Src		I50 = 51 μM
PROTEIN TYROSINE KINASE v-abl		I50 = 220 μM

PAECILOQUINONE F

PROTEIN TYROSINE KINASE (MK CELLS)		*I*50 = 40 μM
PROTEIN TYROSINE KINASE c-Src		*I*50 = 36 μM
PROTEIN TYROSINE KINASE v-abl		*I*50 = 3.6 μM

PAG PROTEIN

PROTEIN TYROSINE KINASE c-abl

PAIM I

α-AMYLASE

PALASONIN

PHOSPHOPROTEIN PHOSPHATASE 1		*I*50 = 656 nM
PHOSPHOPROTEIN PHOSPHATASE 2A		*I*50 = 120 nM

PALATINOSE

CYCLOMALTODEXTRIN GLUCANOTRANSFERASE		
OLIGO-1,6-GLUCOSIDASE		*K*i = 70 mM

PALBINONE

3α-HYDROXY DEHYDROGENASE		*I*50 = 46 nM
3α-HYDROXYSTEROID DEHYDROGENASE (B-SPECIFIC)		*I*50 = 46 nM

PALINAVIR

CATHEPSIN D		*I*50 = 100 μM
GASTRICSIN		*I*50 = 45 μM
PESIN		*I*50 = 33 μM
PROTEINASE HIV-1		*K*i = 31 pM
PROTEINASE HIV-2		*K*i = 130 pM

PALMATINE

TYROSINE 3-MONOOXYGENASE	C	*K*i = 670 μM

PALMITIC ACID

CYCLOOXYGENASE		*I*50 = 20 μM
1,3-β-GLUCAN SYNTHASE		*I*50 = 20 μM
GLUCOKINASE		*I*50 = 940 μM
GLUTATHIONE TRANSFERASE		*I*50 = 423 μM
GLUTATHIONE TRANSFERASE		*I*50 = 28 μM
GLUTATHIONE TRANSFERASE	C	*K*i = 14 μM
15-HYDROXYPROSTAGLANDIN DEHYDROGENASE (NAD)	C	
LYSOPHOSPHOLIPASE	C	*K*i = 37 μM
NAD(P)-ARGININE ADP-RIBOSYLTRANSFERASE		*I*50 = 16 μM
Na^+/K^+-EXCHANGING ATPASE		
PROSTAGLANDIN E2 9-KETOREDUCTASE		
TRIHYDROXYCOPROSTANOYL-CoA SYNTHETASE	NC	

PALMITOLEIC ACID

GLYCEROL-3-PHOSPHATE O-ACYLTRANSFERASE		
NAD ADP-RIBOSYLTRANSFERASE		*I*50 = 95 μM
NAD(P)-ARGININE ADP-RIBOSYLTRANSFERASE		*I*50 = 200 μM
PROTEIN KINASE A		*I*50 = 20 μM
STEROID 5α-REDUCTASE		

6-PALMITOYLASCORBIC ACID

LIPOXYGENASE 1	C	*K*i = 3 μM

(+)PALMITOYLCARNITINE

CARNITINE O-PALMITOYLTRANSFERASE

PALMITOYLCARNITINE

PALMITOYL-CoA HYDROLASE	C	
PROTEIN KINASE C		*I*50 = 34 μM

D,L-PALMITOYLCARNITINE

PROTEIN KINASE C		80% 50 μM
PROTEIN KINASE C		

Inhibitor / Enzyme	Type	Value
L-PALMITOYLCARNITINE		
LYSOPHOSPHOLIPASE	C	*K*i = 10 μM
PALMITOYLCHOLINE		
CARNITINE O-PALMITOYLTRANSFERASE		
PALMITOYL-CoA		
ACETATE-CoA LIGASE		
ACETOACETATE-CoA LIGASE	NC	*K*i = 9.8 μM
ACETYL-CoA CARBOXYLASE		*K*i = 130 nM
ACETYL-CoA CARBOXYLASE		*I*50 = 90 nM
ACYLGLYCEROL LIPASE		
ACYLGLYCERONE-PHOSPHATE REDUCTASE	C	
1-ACYLGLYCEROPHOSPHOCHOLINE O-ACYLTRANSFERASE		
2-ACYLGLYCEROPHOSPHOCHOLINE O-ACYLTRANSFERASE		
ADENOSINE DEAMINASE		
ADN-CARRIER MITOCHONDRIA		
1-ALKYLGLYCEROPHOSPHOCHOLINE O-ACETYLTRANSFERASE		85% 10 μM
AMP DEAMINASE		
ARYLAMINE N-ACETYLTRANSFERASE		
ATP CITRATE (pro-S)-LYASE		
BUTYRYL-CoA DEHYDROGENASE		
CARNITINE O-ACETYLTRANSFERASE		
CARNITINE O-PALMITOYLTRANSFERASE		
CITRATE (si)-SYNTHASE		
DECYLCITRATE SYNTHASE		
DECYLHOMOCITRATE SYNTHASE		
DIHYDROLIPOAMIDE S-ACETYLTRANSFERASE		
DYNEIN ATPASE		> 20 μM
ENOYL-[ACYL-CARRIER-PROTEIN] REDUCTASE (NADH)		
ENOYL-[ACYL-CARRIER-PROTEIN] REDUCTASE (NADPH, B SPECIFIC)		
ETHANOLAMINEPHOSPHOTRANSFERASE		
GLUCOKINASE	C	*K*i = 3.5 μM
GLUCOSE-6-PHOSPHATASE		47% 50 μM
GLUCOSE-6-PHOSPHATE 1-DEHYDROGENASE	C	*K*i = 600 nM
GLUCOSE-6-PHOSPHATE 1-DEHYDROGENASE		
GLUTAMINASE		
GLYCEROL-3-PHOSPHATE O-ACYLTRANSFERASE		> 80 μM
GLYCERONE-PHOSPHATE O-ACYLTRANSFERASE		> 600 μM
HEXOKINASE	C	
4-HYDROXYBENZOATE NONAPRENYLTRANSFERASE		
3-HYDROXYDECANOYL-[ACYL-CARRIER-PROTEIN] DEHYDRATASE		70% 10 μM
HYDROXYMETHYLGLUTARYL-CoA SYNTHASE		*K*i = 500 nM
MALONYL-CoA DECARBOXYLASE	NC	*K*i = 2.3 μM
NAD(P) TRANSHYDROGENASE (AB-SPECIFIC)	C	*K*i = 150 nM
OXOGLUTARATE DEHYDROGENASE (LIPOAMIDE)		*I*50 = 25 μM
PANTOTHENATE KINASE		
PHOSPHATE ACETYLTRANSFERASE		
PHOSPHATIDATE CYTIDILYLTRANSFERASE		
PROTEIN KINASE (Ca^{2+} INDEPENDENT)		44% 1 μM
SERINE C-PALMITOYLTRANSFERASE		
TRIACYLGLYCEROL LIPASE		
UDP-GLUCURONOSYLTRANSFERASE		*I*50 = 20 μM
PALMITOYL DEPHOSPHO-CoA		
ACETYL-CoA CARBOXYLASE		*K*i = 260 nM
PALMITOYLDIHYDROXYACETONEPHOSPHATE		
ACYLGLYCERONE-PHOSPHATE REDUCTASE	C	
ALKYLGLYCERONE-PHOSPHATE SYNTHASE		
PALMITOYL-1,N6 ETHENO-CoA		
ACETYL-CoA CARBOXYLASE		*K*i = 15 nM

1-PALMITOYL-sn-GLYCEROL-3-PHOSPHATE
1-ACYLGLYCEROL-3-PHOSPHATE O-ACYLTRANSFERASE
GLYCEROL-3-PHOSPHATE O-ACYLTRANSFERASE

1-PALMITOYL-sn-GLYCERO-3-PHOSPHOCHOLINE
β-GALACTOSIDE α-2,3-SIALYLTRANSFERASE

PALMITOYL-INOSINO-CoA
ACETYL-CoA CARBOXYLASE — $K\text{i} = 14$ nM

PALMITOYL-KETO-CoA
ACETYL-CoA CARBOXYLASE — $K\text{i} = 21$ nM

PALMITOYL-2-LYSO-sn-GLYCERO-3-PHOSPHOCHOLINE
1-ALKYLGLYCEROPHOSPHOCHOLINE O-ACETYLTRANSFERASE

2-(R)-N-PALMITOYLNORLEUCINOL-1-PHOSPHOGLYCOL
PHOSPHOLIPASE A2

PALMITOYL-4′-PHOSPHOPANTETHEINE
ACETYL-CoA CARBOXYLASE — $K\text{i} = 650$ nM

PALMITOYL-SACCHARIN
ELASTASE LEUKOCYTE — $I50 = 2$ μM

PALMITOYL-THR-VAL-SER-TYR-GLU-LEU
PROTEINASE HIV-1 — $I50 = 500$ nM
PROTEINASE HIV-2 — $I50 = 1.3$ μM

PALMITOYLTRIFLUOROMETHYLKETONE
PHOSPHOLIPASE A2 (Ca^{2+} DEPENDENT) — $I50 = 45$ μM
PHOSPHOLIPASE A2 (Ca^{2+} INDEPENDENT) — $I50 = 3.8$ μM
PHOSPHOLIPASE A2 (CYTOSOLIC) — $I50 = 300$ nM

PALMITOYL VANILLYLAMIDE
NADH DEHYDROGENASE (UBIQUINONE) — $I50 = 60$ μM

N-PALMITYL-p-AMINOBENZOIC ACID
STEROL O-ACYLTRANSFERASE — $I50 = 13$ μM

PALMITYL-CoA
CARNITINE O-OCTANOYLTRANSFERASE
DIACYLGLYCEROL CHOLINEPHOSPHOTRANSFERASE
GLUTAMINASE (PHOSPHATE ACTIVATED)
GLYEROL-3-PHOSPHATE DEHYDROGENASE (NAD(P))

2′-O-PALMTIOYL-cycAMP
ADENYLATE CYCLASE — C — $K\text{i} = 10$ μM

PAMODRONATE
FARNESYL-DIPHOSPHATE FARNESYLTRANSFERASE — $I50 = 12$ μM

PANAXYNDIOL
STEROL O-ACYLTRANSFERASE — $I50 = 45$ μM

PANAXYNDOL
STEROL O-ACYLTRANSFERASE — $I50 = 80$ μM

PANAXYNDTRIOL
STEROL O-ACYLTRANSFERASE — $I50 = 69$ μM

PANAXYNOL
ARACHIDONATE 5-LIPOXYGENASE — $I50 = 2$ μM
ARACHIDONATE 12-LIPOXYGENASE — $I50 = 4$ μM
ARACHIDONATE 12-LIPOXYGENASE — $I50 = 67$ μM
ARACHIDONATE 12-LIPOXYGENASE — $I50 = 1$ μM
STEROL O-ACYLTRANSFERASE — $I50 = 94$ μM

(–) PANCHICIN D
TRIACYLGLYCEROL LIPASE

PANCLICIN A
TRIACYLGLYCEROL LIPASE — *I*50 = 2.9 μM

PANCLICIN B
TRIACYLGLYCEROL LIPASE — *I*50 = 2.6 μM

PANCLICIN C
TRIACYLGLYCEROL LIPASE — *I*50 = 620 nM

PANCLICIN D
TRIACYLGLYCEROL LIPASE — *I*50 = 660 nM

PANCLICIN E
TRIACYLGLYCEROL LIPASE — *I*50 = 890 nM

PANCREATIC β-CELL PROLYL OLIGOPEPTDIASE INHIBITOR
PROLYL OLIGOPEPTIDASE

PANCREATIC SECRETED TRYPSIN INHIBITOR (BOVINE)
TRYPSIN — *K*i = 17 nM
TRYPSIN — *K*i = 145 nM

PANCREATIC SECRETED TRYPSIN INHIBITOR (OSTRICH)
TRYPSIN — *K*i = 8 nM
TRYPSIN — *K*i = 140 nM

PANCREATIC SECRETORY TRYPSIN INHIBITOR
β-TRYPSIN — Kd = 1.7 nM

PANCURONIUM
ACETYLCHOLINESTERASE
CHOLINESTERASE
HISTAMINE N-METHYLTRANSFERASE — C — *K*i = 2.7 μM

PANITOL
α-AMYLASE

PANNORIN
HYDROXYMETHYLGLUTARYL-CoA REDUCTASE (NADPH) — *I*50 = 160 μM

PANOSIALIN D
α-GLUCOSIDASE
β-GLUCOSIDASE
α-MANNOSIDASE

PANTETHEINE
CYSTEAMINE DIOXYGENASE
PANTETHEINASE — *K*i = 2.4 μM
PANTOTHENATE KINASE

D-PANTETHEINE
ACETYL-CoA CARBOXYLASE — C — *K*i = 13 nM

PANTETHEINE-4′-PHOSPHATE
PANTOTHENATE KINASE

D-PANTETHEINE-4′-PHOSPHATE
ACETYL-CoA CARBOXYLASE — C — *K*i = 4 nM

D-PANTETHINE
PHOSPHORYLASE PHOSPHATASE

PANTOPRAZOLE
H^+/K^+-EXCHANGING ATPASE — *I*50 = 6.8 μM
PAPAIN — *I*50 = 37 μM

Inhibitor / Enzyme	Type	Value
PANTOTHENATE		
ACETYL-CoA CARBOXYLASE	C	Ki = 5.4 nM
D-PANTOTHENATE		
PANTOTHENATE KINASE		> 500 μM
L-PANTOTHENATE		
PANTOTHENATE KINASE		
PANTOTHENATE-4′-PHOSPHATE		
ACETYL-CoA CARBOXYLASE	C	Ki = 3.8 nM
PANTOTHENOYLCYSTEINE-4′-PHOSPHATE		
PANTOTHENATE KINASE		
PANTOTHENYLALCOHOL		
PANTOTHENATE KINASE		
PAPAIN PRO REGION		
CATHEPSIN B		
CATHEPSIN L		
PAPAVERINE		
ACETYLCHOLINESTERASE		Ki = 19 μM
3′,5′-CYCLIC-GMP PHOSPHODIESTERASE		I50 = 21 μM
3′,5′-CYCLIC-NUCLEOTIDE PHOSPHODIESTERASE (cycAMP)		I50 = 2 μM
3′,5′-CYCLIC-NUCLEOTIDE PHOSPHODIESTERASE (cycAMP) (Ca^{2+}/CALMODULIN)		I50 = 23 μM
3′,5′-CYCLIC-NUCLEOTIDE PHOSPHODIESTERASE (cycAMP) LOW Km		I50 = 410 nM
3′,5′-CYCLIC-NUCLEOTIDE PHOSPHODIESTERASE F-I		Ki = 1.2 μM
3′,5′-CYCLIC-NUCLEOTIDE PHOSPHODIESTERASE F-II		Ki = 4.5 μM
3′,5′-CYCLIC-NUCLEOTIDE PHOSPHODIESTERASE F-III		Ki = 270 nM
3′,5′-CYCLIC-NUCLEOTIDE PHOSPHODIESTERASE (cycGMP)		I50 = 16 μM
3′,5′-CYCLIC-NUCLEOTIDE PHOSPHODIESTERASE (cycGMP)		I50 = 4 μM
3′,5′-CYCLIC-NUCLEOTIDE PHOSPHODIESTERASE (cycGMP) (Ca^{2+}/CALMODULIN)		I50 = 11 μM
3′,5′-CYCLIC-NUCLEOTIDE PHOSPHODIESTERASE I		I50 = 50 μM
3′,5′-CYCLIC-NUCLEOTIDE PHOSPHODIESTERASE II		I50 = 53 μM
3′,5′-CYCLIC-NUCLEOTIDE PHOSPHODIESTERASE III		I50 = 3.7 μM
3′,5′-CYCLIC-NUCLEOTIDE PHOSPHODIESTERASE IV		I50 = 3.1 μM
15-HYDROXYPROSTAGLANDIN DEHYDROGENASE (NAD)		
PROTEIN KINASE C		I50 = 1.8 mM
PAPAYA PRO REGION		
CATHEPSIN L		
GLYCYL ENDOPEPTIDASE		
PAPIN PRO REGION		
PAPAIN		
PAPP-A/proMBB COMPLEX		
CATHEPSIN G		KI = 1 μM
ELASTASE LEUKOCYTE	C	KI = 10 nM
PAPRIKA TRYPSIN INHIBITOR		
CHYMOTRYPSIN		Ki = 47 nM
PRONASE		Ki = 59 nM
TRYPSIN		Ki = 590 pM
PAPULACANDIN		
DIGUANYLATE CYCLASE		Ki = 70 μM
PAPULACANDIN B		
1,3-β-GLUCAN SYNTHASE		80% 10 μg/ml
1,3-β-GLUCAN SYNTHASE	NC	Ki = 72 μM
1,3-β-GLUCAN SYNTHASE		
PAPULACANDIN BID		
1,3-β-GLUCAN SYNTHASE		I50 = 40 μM

PARALDEHYDE
ALKANAL MONOOXYGENASE (FMN -LINKED)

PARAMYOSIN
MYOSIN ATPASE

PARAOXON

ACETYLCHOLINESTERASE		86% 100 nM
ACETYLESTERASE		
ACYLGLYCEROL LIPASE		
ALKYLAMIDASE		
ARYLFORMAMIDASE		92% 10 mM
CARBOXYLESTERASE		
CHOLINESTERASE		51% 5 pg/ml
CUTINASE		
ESTERASE		100% 10 mM
PHOSPHATIDYLCHOLINE-STEROL O-ACYLTRANSFERASE		
PHOSPHOLIPASE A2		71% 1 μg/ml
SIALATE -9-O-ACETYLESTERASE	IR	
SIALATE O-ACETYLESTERASE		*I*50 = 12 μM
TRIACYLGLYCEROL LIPASE ACID		64% 100 μM

PARA-PARI MI
ALDEHYDE REDUCTASE

PARAQUAT
GLUTATHIONE REDUCTASE (NADPH)

PARATHION

ACETYLCHOLINESTERASE		
Ca^{2+}/Mg^{2+}-TRANSPORTING ATPASE		*K*i = 36 μM
Na^{+}/K^{+}-EXCHANGING ATPASE		*K*i = 71 μM
PHOSPHATASE ACID		

PARAZOANTHUS AXINELLAE ACETYLCHOLINESTERASE INHIBITOR

ACETYLCHOLINESTERASE	C	*K*i = 4 μM

PARBENDAZOLE

FUMARATE REDUCTASE (NADH)		*I*50 = 120 μM

PARGYLINE

ACETYLSPERMIDINE DEACETYLASE		
ALDEHYDE DEHYDROGENASE		*I*50 = 300 μM
ALDEHYDE DEHYDROGENASE (NAD)		
AMINE OXIDASE (COPPER-CONTAINING)	UC	*K*i = 2.5 mM
AMINE OXIDASE MAO		
AMINE OXIDASE MAO-A		*K*i = 15 μM
AMINE OXIDASE MAO-B		*K*i = 1.8 μM
AMINE OXIDASE MAO-I	SSI	
DIMETHYLANILINE MONOOXYGENASE (N-OXIDE FORMING)	C	*K*i = 9 μM
HISTAMINE N-METHYLTRANSFERASE		*K*i = 126 μM

4-(2-PARIDYLAZO)RESORCINOL
4-METHOXYBENZOATE MONOOXYGENASE (O-DEMETHYLATING)

PAROMAMINE
AMINOGLYCOSIDE $N^{6'}$-ACETYLTRANSFERASE
KANAMYCIN 6′-N-ACETYLTRANSFERASE

PAROMOMYCIN
KANAMYCIN 6′-N-ACETYLTRANSFERASE

PAROMOMYCIN I
GENTAMICIN 3′-N-ACETYLTRANSFERASE

PAROXETINE

CYTOCHROME P450 1A2		*K*i = 5.5 μM
CYTOCHROME P450 2C9		*K*i = 35 μM

CYTOCHROME P450 2D6		Ki = 150 nM
CYTOCHROME P450 2D6		Ki = 3.2 μM
NITRIC-OXIDE SYNTHASE		

PAROXON

CARBOXYLESTERASE A
CARBOXYLESTERASE B

PARS INTERCEREBRALIS MAJOR PEPTIDE C

PROTEINASES

PARSOMYCIN

RNA-DIRECTED DNA POLYMERASE

PATROSELINIC ACID

STEROID 5α-REDUCTASE

PATULIN

CALF THYMUS RIBONUCLEASE H		
FARNESYL PROTEINTRANSFERASE		I50 = 290 μM

PAZELLIPTINE

DNA TOPOISOMERASE (ATP-HYDROLYSING)

Pb^{2+}

GUANINE DEAMINASE A

Pb^{2+}

N-ACETYL-β-GLUCOSAMINIDASE		25% 100 μM
β-N-ACETYLHEXOSAMINIDASE		
ADENOSINE DEAMINASE		
ADENOSYLMETHIONINE-8-AMINO-7-OXONONANOATE TRANSAMINASE		
ADENYLOSUCCINATE SYNTHASE		69% 200 μM
ALAKN-1-OL DEHYDROGENASE (ACCEPTOR)		
ALANINE DEHYDROGENASE		
ALANINE TRANSAMINASE		
D-ALANYL-D-ALANINE CARBOXYPEPTIDASE		
ALDOSE 1-EPIMERASE		20% 13 μM
ALLANTOIN RACEMASE		
AMINO-ACID N-ACETYLTRANSFERASE		
L-AMINOADIPATE-SEMIALDEHYDE DEHYDROGENASE		
5-AMINOLEVULINATE DEHYDROGENASE		
5-AMINOPENTANAMIDASE		100% 5 mM
AMINOPEPTIDASE		89% 1 mM
AMINOPEPTIDASE B		95% 100 μM
AMINOPEPTIDASE B		90% 17 μM
α-AMYLASE		
β-AMYLASE		
ANTHRANILATE SYNTHASE		
ARYL-ALCOHOL OXIDASE		
ARYLSULFATE SULFOTRANSFERASE		
ASPERGILLUS ALKALINE PROTEINASE		
AUREOLYSIN		
BACILLOLYSIN		86% 100 μM
(+)-BORNEOL DEHYDROGENASE		
BRANCHED-CHAIN-AMINO-ACID TRANSAMINASE		
CARBAMATE KINASE		
CARBOXYPEPTIDASE A	C	Ki = 48 μM
CARBOXYPEPTIDASE A		
CATHEPSIN D		
Ca^{2+}-TRANSPORTING ATPASE		I50 = 95 nM
CELLULASE		24% 2 mM
CHOLESTEROL 7α-MONOOXYGENASE		
COLLAGENASE CLOSTRIDIUM HISTOLYTICUM		
CREATINASE		15% 1 mM
3′,5′-CYCLIC-NUCLEOTIDE PHOSPHODIESTERASE V	C	

CYSTINYL AMINOPEPTIDASE		
CYTIDYLATE KINASE		
CYTOCHROME-C PEROXIDASE		
DESULFOGLUCINOLATE SULFOTRANSFERASE		
DEXTRANASE		
α-DEXTRIN ENDO-1,6-α-GLUCOSIDASE		
DIHYDROLIPOAMIDE DEHYDROGENASE		
DIPEPTIDASE		90% 1 mM
DIPEPTIDYL-PEPTIDASE 2		50% 1 mM
DIPEPTIDYL-PEPTIDASE 3		
ENDO-1,4-β-XYLANASE		
ETHANOLAMINE-PHOSPHATE PHOSPHO-LYASE		75% 100 μM
EXOPOLYPHOSPHATASE		
FERROCHELATASE		
FORMATE DEHYDROGENASE		
β-FRUCTOFURANOSIDASE		
α-L-FUCOSIDASE		
β-GALACTOSIDASE		
β-GALACTOSIDE α-2,6-SIALYLTRANSFERASE		
GLUCAN ENDO-1,3-α-GLUCOSIDASE		100% 20 mM
GLUCAN ENDO-1,3-β-GLUCOSIDASE		
GLUCAN 1,6-α-GLUCOSIDASE		
GLUCOKINASE		
α-GLUCOSIDASE		
β-GLUCOSIDASE		
GLUTAMINASE		
γ-GLUTAMYLTRANSFERASE		
GLYCEROL-2-PHOSPHATASE		
GLYCINE HYDROXYMETHYLTRANSFERASE		62% 100 μM
GTP CYCLOHYDROLASE I		
GUANYLATE CYCLASE		
2-HALOACID DEHALOGENASE		16% 1 mM
HISTIDINE AMMONIA-LYASE		> 1mM
INULIN FRUCTOTRANSFERASE (DEPOLYMERIZING, DIFRUCTOFURANOSE-1,2′: 2′,3′-DIANHYDRIDE		
LEUCYL AMINOPEPTIDASE		*I*50 = 50 μM
LEVANASE		100% 50 mM
LOTUS ASPARTIC PROTEINASE		
LUTEOLIN-7O-GLUCURONIDE 7-O-GLUCURONOSYLTRANSFERASE		40% 750 μM
MANNAN ENDO-1,4-β-MANNOSIDASE		
α-MANNOSIDASE		
2-METHYLCITRATE SYNTHASE		
MYTILIDASE		
NADH PEROXIDASE		
Na^+/K^+-EXCHANGING ATPASE	NC	*I*50 = 80 μM
NEOLACTOSYLCERAMIDE α-2,3-SIALYLTRANSFERASE		
NITRATE REDUCTASE (NADH)		
NITRIC-OXIDE SYNTHASE (Ca^{2+} STIMULATED)		*I*50 = 17 nM
NITRILE HYDRATASE		49% 1 mM
4-NITROPHENYLPHOSPHATASE		
NUCLEOSIDE PHOSPHOTRANSFERASE		
OLIGO-1,6-GLUCOSIDASE		90% 2 mM
PALMITOYL-CoA HYDROLASE		
PANTOTHENATE KINASE		
PAPAIN		
PEROXIDASE		*I*50 = 257 μM
PHOSPHATASE ACID		100% 1 mM
PHOSPHATIDYLCHOLINE-DOLICHOL O-ACYLTRANSFERASE		
PHOSPHOADENYLYLSULFATASE		46% 200 μM
PHOSPHOGLYCERATE MUTASE		12% 1 mM
3-PHOSPHOGLYCERATE PHOSPHATASE		
PHOSPHOLIPASE A2	C	
PHOSPHOLIPASE C		
3-PHOSPHOSHIKIMATE 1-CARBOXYVINYLTRANSFERASE		
POLYGALACTURONASE		

PORPHOBILINOGEN SYNTHASE	NC	
PROCOLLAGEN N-ENDOPEPTIDASE		
PROCOLLAGEN GALACTOSYLTRANSFERASE		
PROLYL AMINOPEPTIDASE		
PROTEIN-GLUTAMINE γ-GLUTAMYLTRANSFERASE		
PROTEIN KINASE C		
PROTOCATECHUATE 3,4-DIOXYGENASE		
PSEUDOMONAS AEROGINOSA ALKALINE PROTEINASE		
QUEUINE tRNA-RIBOSYLTRANSFERASE		
RIBONUCLEASE α		99% 10 mM
SARCOSINE OXIDASE		
SECONDARY-ALCOHOL OXIDASE		
SERINE-SULFATE AMMONIA-LYASE		
SUCCINATE-SEMIALDEHYDE DEHYDROGENASE		
TESTOSTERONE 17β-DEHYDROGENASE (NADP)		
THIAMIN-DIPHOSPHATE KINASE		
TRIACYLGLYCEROL LIPASE ALKALINE		
TRIPEPTIDE AMINOPEPTIDASE		100% 1 mM
UDP-N-ACETYLGLUCOSAMINE 2-EPIMERASE		90% 1 mM
URATE OXIDASE		
UREIDOGLYCOLATE LYASE		
dUTP PYROPHOSPHATASE		
XYLAN ENDO-1,3-XYLOSIDASE		
D-XYLULOSE REDUCTASE		

P-BENZAMIDINE

NITRIC-OXIDE SYNTHASE	C	*K*i = 120 μM

(E,E)[α,β(nP)]-8-O-(3-BENZOYLBENZOYL)-3,7-DIMETHYL-2,6-OCTADIEN-1-DIPHOSPHATE

PROTEIN FARNESYLPHOSPHATE TRANSFERASE	C	*K*i = 380 nM

PBS 1 URACIL DNA GLYCOSYLASE INHIBITOR

DNA GLYCOSYLASE		

PC 57

NEPRILYSIN		*K*i = 1.4 nM
PEPTIDYL-DIPEPTIDASE A		*K*i = 200 pM

PCA 4230

3′,5′-CYCLIC-NUCLEOTIDE PHOSPHODIESTERASE V		

$PhCH_2CH_2CO$-VAL-VAL-ASN-$(CH_3)_2$-ASP-LEU-OH

RIBONUCLEOTIDE REDUCTASE		*I*50 = 110 nM

8-(P-CHLOROBENZENESULPHONAMIDO)-2-[5-(3-PYRIDYL)PENTYL]OCTANOIC ACID

THROMBOXANE A SYNTHASE		*I*50 = 4 nM

PCPI 23

PAPAIN		*K*i = 320 pM

Pd^{2+}

CELLULOSE 1,4-β-CELLOBIOSIDASE I		90% 100 μM
FORMALDEHYDE DEHYDROGENASE		
PANTOATE-β-ALANINE LIGASE		

PD 089828

PROTEIN TYROSINE KINASE EGFR		*I*50 = 450 nM
PROTEIN TYROSINE KINASE FGFR		*I*50 = 130 nM
PROTEIN TYROSINE KINASE PDGFR		*I*50 = 1.1 μM
PROTEIN TYROSINE KINASE c-Src		*I*50 = 220 nM

PD 113271

DNA TOPOISOMERASE (ATP-HYDROLYSING)		

PD 114631

DNA TOPOISOMERASE (ATP-HYDROLYSING)		

PD 116124
PURINE-NUCLEOSIDE PHOSPHORYLASE C $Ki = 410$ nM

PD 116245
DNA TOPOISOMERASE (ATP-HYDROLYSING)

PD 116250
DNA TOPOISOMERASE (ATP-HYDROLYSING)

PD 123244-15
HYDROXYMETHYLGLUTARYL-CoA REDUCTASE (NADPH)

PD 125754
ENDOTHIAPEPSIN $Ki = 16$ μM
RENIN $I50 = 22$ nM

PD 125967
ENDOTHIAPEPSIN $Ki = 242$ nM
RENIN $I50 = 170$ nM

PD 128042
STEROL O-ACYLTRANSFERASE $I50 = 200$ ng/ml

PD 128763
NAD ADP-RIBOSYLTRANSFERASE $I50 = 100$ nM
NAD ADP-RIBOSYLTRANSFERASE $I50 = 360$ nM

PD 129337
STEROL O-ACYLTRANSFERASE $I50 = 17$ nM

PD 130327
ASPARTIC PROTEINASE A

PD 130328
ENDOTHIAPEPSIN $Ki = 110$ nM

PD 130693
ENDOTHIAPEPSIN $Ki = 69$ nM

PD 132002
RENIN $I50 = 36$ nM

PD 133450
ENDOTHIAPEPSIN $Ki = 6$ nM

PD 134390
PROTEINASE HIV-1 $I50 = 2$ nM

PD 134672
RENIN $I50 = 360$ pM

PD 134922
PROTEINASE HIV-1 $I50 = 15$ nM

PD 134967-15
HYDROXYMETHYLGLUTARYL-CoA REDUCTASE (NADPH) $I50 = 29$ nM
HYDROXYMETHYLGLUTARYL-CoA REDUCTASE (NADPH) $I50 = 18$ nM
HYDROXYMETHYLGLUTARYL-CoA REDUCTASE (NADPH) $I50 = 13$ nM
HYDROXYMETHYLGLUTARYL-CoA REDUCTASE (NADPH) $I50 = 102$ nM

PD 135023-15
HYDROXYMETHYLGLUTARYL-CoA REDUCTASE (NADPH) $I50 = 115$ nM
HYDROXYMETHYLGLUTARYL-CoA REDUCTASE (NADPH) $I50 = 266$ nM
HYDROXYMETHYLGLUTARYL-CoA REDUCTASE (NADPH) $I50 = 721$ nM
HYDROXYMETHYLGLUTARYL-CoA REDUCTASE (NADPH) $I50 = 19$ nM

PD 138387
PROSTAGLANDIN H2 SYNTHASE 2 $I50 = 220$ nM

PD 141955
PURINE-NUCLEOSIDE PHOSPHORYLASE $Ki = 67$ nM

PD 145709
PROTEIN TYROSINE KINASE FGFR

PD 146176
ARACHIDONATE 5-LIPOXYGENASE $I50 = 540$ nM
ARACHIDONATE 15-LIPOXYGENASE MIX $Ki = 197$ nM

PD 150606
CALCINEURIN CaM $Ki = 13$ µM
CALPAIN µ $Ki = 210$ nM
CALPAIN m $Ki = 370$ nM
CATHEPSIN B $Ki = 128$ µM
THERMOLYSIN $Ki = 204$ µM

PD 151746
CALPAIN I $Ki = 260$ nM
CALPAIN II $Ki = 5.3$ µM

PD 153035
PROTEIN TYROSINE KINASE EGFR $I50 = 25$ pM

PD 158780
PROTEIN TYROSINE KINASE EGFR $I50 = 80$ pM

PD 161570
PROTEIN TYROSINE KINASE EGFR 1 $I50 = 3.7$ µM
PROTEIN TYROSINE KINASE FGFR 1 $I50 = 40$ nM
PROTEIN TYROSINE KINASE PDGFRβ $I50 = 262$ nM

PD 162628
PROTEIN KINASE EGFR $I50 = 6.7$ µM

PD 166285
MAP KINASE $I50 = 5$ µM
PROTEIN KINASE C $I50 = 23$ µM
PROTEIN TYROSINE KINASE EGFR $I50 = 88$ nM
PROTEIN TYROSINE KINASE FGFR 1 $I50 = 39$ nM
PROTEIN TYROSINE KINASE PDGFR $I50 = 98$ nM
PROTEIN TYROSINE KINASE PDGFRβ $I50 = 98$ nM
PROTEIN TYROSINE KINASE c-Src $I50 = 8.4$ nM

PD 166866
PROTEIN KINASE EGFR $I50 = 50$ µM
PROTEIN TYROSINE KINASE FGFR 1 $I50 = 52$ nM

PD 69185
ENDOTHELIN CONVERTING ENZYME 1 $I50 = 900$ nM
GELATINASE A 13% 300 µM
NEPRILYSIN 25% 200 µM
STROMELYSIN 19% 300 µM

PD 83176
FARNESYL PROTEINTRANSFERASE $I50 = 20$ nM
GERANYLGERANYL PROTEINTRANSFERASE $Ki = 8.7$ µM
GERANYLGERANYL PROTEINTRANSFERASE $I50 = 1.3$ µM

PD 89828
MAP KINASE $I50 = 7.1$ µM
PROTEIN KINASE C $I50 = 50$ µM
PROTEIN KINASE EGFR $I50 = 450$ nM
PROTEIN TYROSINE KINASE EGFR C $I50 = 5.5$ µM
PROTEIN TYROSINE KINASE FGFR C $I50 = 150$ nM
PROTEIN TYROSINE KINASE PDGFRβ $I50 = 180$ nM
PROTEIN TYROSINE KINASE SRC C $I50 = 1.8$ µM

PD 98059
EXTRACELLULAR REGULATED PROTEIN KINASE
MAP KINASE KINASE
MAP KINASE p42
MAP KINASE p44

PD 99560
PROTEINASE HIV-1 C $Ki = 1\ \mu M$

PDE 132301-2
STEROL O-ACYLTRANSFERASE $I50 = 52$ nM

PEARL MILLET CYSTEINE PROTEINASE INHIBITOR
PAPAIN

PEA SEED TRYPSIN INHIBITOR I
TRYPSIN

PEA SEED TRYPSIN INHIBITOR II
TRYPSIN

PEA SEED TRYPSIN INHIBITOR III
TRYPSIN

PEA SEED TRYPSIN INHIBITOR IVa
TRYPSIN

PEA SEED TRYPSIN INHIBITOR IVb
TRYPSIN

PEA SEED TRYPSIN INHIBITORS I-V
TRYPSIN

PEA SEED TRYPSIN INHIBITOR V
TRYPSIN

PECTIC ACID
HYALURONIDASE

PECTIN
HYALURONIDASE
PECTINESTERASE
POLYGALACTURONASE C

PECTOLINARIGENIN
PROLYL OLIGOPEPTIDASE $I50 = 8.1$ ppm

PECTOLINARIN
PROLYL OLIGOPEPTIDASE $I50 = 71$ ppm

PEDUNCULAGIN
CARBONATE DEHYDRATASE $I50 = 550$ nM

PEFABLOC
ACETYLHYDROLASE PAF DEGRADING

PEFLOXACIN
CYTOCHROME P450 LA2 22% 500 μM
DNA GYRASE $I50 = 50$ ng/ml
DNA TOPOISOMERASE (ATP-HYDROLYSING) $I50 = 64$ ng/ml
DNA TOPOISOMERASE IV $I50 = 1\ \mu g/ml$

N-PELARGONOYLDOPAMINE
ARACHIDONATE 5-LIPOXYGENASE $I50 = 45$ nM

PELDESINE
PURINE-NUCLEOSIDE PHOSPHORYLASE $I50 = 36$ nM

PELRAGONIDIN
ANTHOCYANIDIN 3-O-GLUCOSYLTRANSFERASE

PELRINONE
3′,5′-CYCLIC-NUCLEOTIDE PHOSPHODIESTERASE III

PELTATOSIDE
DNA TOPOISOMERASE IV — $I50 = 20\ \mu g/ml$

PENAZETIDINE A
PROTEIN KINASE C

PENCICLOVIR TRIPHOSPHATE
DNA POLYMERASE — C — $Ki = 5.8\ \mu M$
DNA POLYMERASE α — C — $Ki = 175\ \mu M$

PENDUCULAGIN
NADH DEHYDROGENASE — $I50 = 4.4\ \mu M$

PENICILLAMINE
D-AMINO-ACID OXIDASE
AMINOLEVULINATE TRANSAMINASE
CYSTATHIONINE γ-LYASE
GLYCINE HYDROXYMETHYLTRANSFERASE
METHIONINE γ-LYASE
MYELOPEROXIDASE — $I50 = 100\ \mu M$
MYELOPEROXIDASE — 55% 50 μM
VALINE DECARBOXYLASE

D-PENICILLAMINE
4-AMINOBUTYRATE TRANSAMINASE
8-AMINO-7-OXONONAOATE SYNTHASE
AMINOPEPTIDASE B
3-CHLORO-D-ALANINE DEHYDROCHLORINASE — 30% 10 mM
DIAMINOPIMELATE DECARBOXYLASE
GELATINASE B — $I50 = 3\ \mu g/ml$
GELATINASE B — $I50 = 100\ \mu M$
GLUTAMATE-1-SEMIALDEHYDE 2,1-AMINOMUTASE
GLUTATHIONE PEROXIDASE
GLYCINE HYDROXYMETHYLTRANSFERASE — 16% 300 μM
LYSINE 2,3-AMINOMUTASE — 68% 10 mM
NITRILE HYDRATASE — 62% 1 mM
PHOSPHATIDYLSERINE DECARBOXYLASE — $I50 = 5\ mM$
t-PLASMINOGEN ACTIVATOR — $I50 = 18\ \mu g/ml$
u-PLASMINOGEN ACTIVATOR — $I50 = 14\ \mu g/ml$
u-PLASMINOGEN ACTIVATOR
PYRIDOXAL KINASE — 20% 100 μM

D,L-PENICILLAMINE
ADENOSYLMETHIONINE-8-AMINO-7-OXONONANOATE TRANSAMINASE
GLUTAMATE DECARBOXYLASE — C
GLUTAMATE DECARBOXYLASE — NC — $Ki = 200\ \mu M$
HOMOSERINE O-ACETYLTRANSFERASE

L-PENICILLAMINE
AMINODEOXYGLUCONATE DEHYDRATASE — 85% 1 mM
5-AMINOLEVULINATE SYNTHASE
AMINOLEVULINATE TRANSAMINASE — 72% 1 mM
ARGININE RACEMASE — 28% 5 mM
L-3-CYANOALANINE SYNTHASE — 56% 10 mM
GLYCINE HYDROXYMETHYLTRANSFERASE — 100% 300 μM
KYNURENINE-OXOGLUTARATE AMINOTRANSFERASE
NITRILE HYDRATASE — 39% 1 mM
ORNITHINE-OXO-ACID TRANSAMINASE
PHOSPHATIDYLSERINE DECARBOXYLASE — $I50 = 2.9\ mM$
SERINE-GLYOXYLATE TRANSAMINASE — 23% 1 mM
SULFINOALANINE DECARBOXYLASE

PENICILLAMINYL-VAL-ILE-MET
FARNESYL PROTEINTRANSFERASE — $I50 = 100$ nM

PENICILLANATE
D-ALANYL-D-ALANINE CARBOXYPEPTIDASE

PENICILLANIC ACID
β-LACTAMASE RTEM — IR

PENICILLANIC ACID SULPHONE
β-LACTAMASE — SSI
β-LACTAMASE RTEM — IR

PENICILLIC ACID
GLUTATHIONE TRANSFERASE

PENICILLIN
D-ALANYL-D-ALANINE CARBOXYPEPTIDASE
D-AMINO-ACID OXIDASE
CDP-GLYCEROL GLYCEROPHOSPHOTRANSFERASE
RIBONUCLEASE PANCREATIC

L-PENICILLINAMIN
VALINE-tRNA LIGASE — C — $Ki = 1.6$ mM

PENICILLIN DIMERS
PROTEINASE HIV-1

PENICILLIN G
D-ALANYL-D-ALANINE CARBOXYPEPTIDASE — $Ki = 75$ nM
D-ALANYL-D-ALANINE CARBOXYPEPTIDASE — $I50 = 7.6$ μM
β-LACTAMASE — $Ki = 20$ μM
MURAMOYL-PENTAPEPTIDE CARBOXYPEPTIDASE — C — $Ki = 16$ nM

PENICILLIN V
D-ALANYL-D-ALANINE CARBOXYPEPTIDASE — $I50 = 1$ μM

PENIDIAMIDE
PHOSPHOLIPASE A2 — $I50 = 30$ μM

PENTABROMOPROPEN-2-YL DIBROMOACETATE
ALDEHYDE REDUCTASE

PENTABROMOPROPEN-2-YL TRIBROMOACETATE
ALDEHYDE REDUCTASE

PENTACARBOXYLATE PORPHYRIN III
COPROPORPHYRINOGEN OXIDASE

PENTACHLOROPHENOL
ACETYLCHOLINESTERASE — NC
ARYL SULFOTRANSFERASE
ATP PHOSPHORIBOSYLTRANSFERASE
CHITIN SYNTHASE — 22% 250 μM
CYTOCHROME-b5 REDUCTASE
EPOXIDE HYDROLASE — $I50 = 35$ μM
GLUTATHIONE TRANSFERASE — $I50 = 24$ μM
N-HYDROXYARYLAMINE O-ACETYLTRANSFERASE
SULFOTRANSFERASE
TYROSINE-ESTER SULFOTRANSFERASE

15-PENTADECANOLIDE
PROTEIN KINASE A — $I50 = 20$ μM

PENTADECAN-2-ONE
ALDEHYDE DEHYDROGENASE (NADP)

N-PENTADECYLHYDROXYLAMINE
ARACHIDONATE 5-LIPOXYGENASE — $I50 = 500\ \mu M$
CYCLOOXYGENASE — $I50 = 80\ \mu M$

4-PENTADECYLPYRIDINE
POLYPHENOL OXIDASE — C — $Ki = 370\ \mu M$

2,4-PENTADIENOYL-CoA
ACETYL-CoA C-ACYLTRANSFERASE

3,4-PENTADIENOYL-CoA
ACYL-CoA DEHYDROGENASE — SSI

2,4-PENTADIONE
DIHYDROFOLATE REDUCTASE

PENTA-O-ETHYLQUERCETIN
3′,5′-CYCLIC-NUCLEOTIDE PHOSPHODIESTERASE (cycAMP) — $I50 = 8\ \mu M$
3′,5′-CYCLIC-NUCLEOTIDE PHOSPHODIESTERASE (cycGMP) — $I50 = 13\ \mu M$
3′,5′-CYCLIC-NUCLEOTIDE PHOSPHODIESTERASE (cycGMP) (Ca^{2+}/CALMODULIN) — $I50 = 189\ \mu M$

1,1,1,3,3-PENTAFLUORO-6,6-DIMETHYLHEPTAN-2-ONE
ACETYLCHOLINESTERASE — $Ki = 11$ nM
CHOLINESTERASE — $Ki = 61$ nM

PENTAFLUOROPHENYLPYRUVIC ACID
4-HYDROXYPHENYLPYRUVATE DIOXYGENASE — C — $Ki = 14\ \mu M$

PENTAFLUOROPROPIOPHENONE
STEROL ESTERASE — C — $Ki = 90\ \mu M$

PENTAGALLOYLGLUCOSE
α-GLUCOSIDASE — $I50 = 340$ nM
POLY(ADP-RIBOSE) GLYCOHYDROLASE — $I50 = 19\ \mu M$
XANTHINE OXIDASE — U — $Ki = 6.7\ \mu M$

1,2,3,4,6-PENTA-O-GALLOYLGLUCOSE
α-GLUCOSIDASE — $I50 = 1.7\ \mu M$

1,2,3,4,6-PENTA-O-GALLOYL-β-D-GLUCOSE
INTEGRASE — $I50 = 28\ \mu g/ml$
Na^+/K^+-EXCHANGING ATPASE — $I50 = 2.5\ \mu M$
XANTHINE OXIDASE — $I50 = 3.2\ \mu M$

1,2,3,4,6-PENTA-O-GALLOYL-D-GLUCOSE
NADH DEHYDROGENASE — $I50 = 6.8\ \mu M$
NADH DEHYDROGENASE — $I50 = 2.8\ \mu M$
NADH DEHYDROGENASE — $I50 = 600$ nM
NADH DEHYDROGENASE — $I50 = 180$ nM
NADH DEHYDROGENASE — $I50 = 3\ \mu M$

PENTA-O-GALLOYLGLUCOSE
ALDEHYDE REDUCTASE — $I50 = 70$ nM

PENTALENOLACTONE
GLYCERALDEHYDE-3-PHOSPHATE DEHYDROGENASE (PHOSPHORYLATING) — IR — $Ki = 2.2\ \mu M$
GLYCERALDEHYDE-3-PHOSPHATE DEHYDROGENASE (PHOSPHORYLATING) — IR — $I50 = 3.6\ \mu M$
GLYCERALDEHYDE-3-PHOSPHATE DEHYDROGENASE (PHOSPHORYLATING) — IR — $Ki = 12\ \mu M$
GLYCERALDEHYDE-3-PHOSPHATE DEHYDROGENASE (PHOSPHORYLATING) — $I50 = 500$ nM

21,21-PENTAMETHYLENE-4-AZA-5α-PREG-1-EN-3-ONE
STEROID 5α-REDUCTASE — $I50 = 300$ pM
STEROID 5α-REDUCTASE — $I50 = 1.5\ \mu M$

PENTAMETHYLENE-BIS-OXY-o-PHENYLENE-BIS[TRIMETHYLAMMONIUM IODIDE]
ACETYLCHOLINESTERASE — $I50 = 8.9\ \mu M$
CHOLINESTERASE — $I50 = 1.6$ mM

8,8′-(PENTAMETHYLENEDIOXY)BIS[1-ETHYLQUINOLINIUM IODIDE]		
ACETYLCHOLINESTERASE		*I*50 = 63 nM
CHOLINESTERASE		*I*50 = 35 μM
(3E,7E,11E,15E)-3,7,12,16,20-PENTAMETHYL-3,7,11,15,19-HENEICOSAPENTAEN-1-OL		
SQUALENE MONOOXYGENASE		*I*50 = 40 μM
2,2,5,7,8-PENTAMETHYL-6-HYDROXYCHROMANE		
PHOSPHOLIPASE C		
(5E,9E,13E,17E)-5,9,14,18,22-PENTAMETHYL-5,9,13,21-TRICOSAPENTAEN-1-OL		
SQUALENE MONOOXYGENASE		*I*50 = 4 μM
PENTAMIDINE		
ADENOSYLMETHIONINE DECARBOXYLASE		60% 10 mM
ADENOSYLMETHIONINE DECARBOXYLASE		*K*i = 50 mM
ADENOSYLMETHIONINE DECARBOXYLASE		*K*i = 19 μM
AMINE OXIDASE (COPPER-CONTAINING)	NC	*K*i = 3 μM
AMINE OXIDASE (POLYAMINE)	C	*K*i = 7.6 μM
Ca^{2+}/Mg^{2+}-TRANSPORTING ATPASE (CALMODULIN)		
CHOLINESTERASE	NC	*I*50 = 6.7 μM
DIAMINE N-ACETYLTRANSFERASE		
NITRIC-OXIDE SYNTHASE I	NC	
NITRIC-OXIDE SYNTHASE II		
OXIDATIVE PHOSPHORYLATION UNCOUPLER		
PEPTIDASE ALKALINE		*I*50 = 1.5 mM
SPERMIDINE/SPERMINE ACETYLTRANSFERASE	C	*K*i = 2.4 μM
TRYPSIN	C	*K*i = 2.3 μM
TRYPTASE	C	*K*i = 1.2 μM
PENTAMIDINE ISOTHIONATE		
GUANIDINOBENZOATASE	C	*K*i = 239 nM
PENTANEDIALDEHYDE BIS(GUANYLHYDRAZONE)		
ADENOSYLMETHIONINE DECARBOXYLASE		30% 1 mM
2,4-PENTANEDIONE		
20α-HYDROXYSTEROID DEHYDROGENASE		
(S)-TETRAHYDROPROTOBERBERINE OXIDASE		*I*50 = 100 μM
PENTANE-2,5-DIONE		
NAD(P) TRANSHYDROGENASE (AB-SPECIFIC)		
PENTANE-3-ONE		
ACETOIN DEHYDROGENASE		
PENTANOIC ACID		
20α-HYDROXYSTEROID DEHYDROGENASE		
MANDELONITRILE LYASE		*K*i = 91 μM
ORNITHINE-OXO-ACID TRANSAMINASE		
trans-2-PENTANOIC ACID		
D-AMINO-ACID OXIDASE		
1-PENTANOL		
ALDEHYDE DEHYDROGENASE (NAD)	NC	500 μM
CYTOCHROME-C OXIDASE	C	*K*i = 13 mM
n-PENTANOL		
CATECHOL 2,3-DIOXYGENASE		*K*i = 580 μM
2-PENTANONE		
CATECHOL 2,3-DIOXYGENASE		*K*i = 150 μM
PENTAQUINE		
PROTEIN KINASE A		*I*50 = 310 μM
PROTEIN KINASE A		32% 400 μM
PROTEIN KINASE (Ca^{2+} DEPENDENT)		*I*50 = 276 μM

PENTARIC ACID
2-DEHYDRO-3-DEOXYGLUCARATE ALDOLASE

(±)-PENTAZOCINE
C8- > C7 STEROL ISOMERASE $Ki = 1\ \mu M$

trans-2-PENTENAL
GLUCOSE-6-PHOSPHATE 1-DEHYDROGENASE

4-PENTENOIC ACID
ACETYL-CoA C-ACYLTRANSFERASE
FATTY ACID OXYDATION

1-PENTEN-3-ONE
GLUCOSE-6-PHOSPHATE 1-DEHYDROGENASE

3-PENTENOYLPANTETHEINE
BUTYRYL-CoA DEHYDROGENASE

3-PENTENOYL-S-PANTETHEINE 11-PIVALATE
ACETYL-CoA C-ACETYLTRANSFERASE

4-PENTHYLPYRAZOLE
ALCOHOL DEHYDROGENASE $Ki = 800$ pM

PENTOBARBITAL
Ca^{2+}-TRANSPORTING ATPASE $I50 = 11$ mM
Mg^{2+}-TRANSPORTING ATPASE $I50 = 17$ mM
Na^+/K^+-EXCHANGING ATPASE $I50 = 17$ mM

PENTOLAMINE
PROTEIN KINASE C 92% 200 μM

PENTOPRILAT
PEPTIDYL-DIPEPTIDASE A $I50 = 82$ nM

PENTOSAN POLYSULPHATE
ELASTASE LEUKOCYTE
GUANIDINOBENZOATASE NC $Ki = 680$ nM
PROTEIN KINASE C C $Ki = 32\mu g/ml$

PENTOSTATIN
ADENOSINE DEAMINASE

PENTOXYFYLLINE
3′,5′-CYCLIC-NUCLEOTIDE PHOSPHODIESTERASE (cGMP INHIBITABLE) $I50 = 157\ \mu M$
3′,5′-CYCLIC-NUCLEOTIDE PHOSPHODIESTERASE I $I50 = 164\ \mu M$
3′,5′-CYCLIC-NUCLEOTIDE PHOSPHODIESTERASE III $I50 = 50\ \mu M$
3′,5′-CYCLIC-NUCLEOTIDE PHOSPHODIESTERASE IV $I50 = 112\ \mu M$
3′,5′-CYCLIC-NUCLEOTIDE PHOSPHODIESTERASE IVa $I50 = 331\ \mu M$
3′,5′-CYCLIC-NUCLEOTIDE PHOSPHODIESTERASE IVb $I50 = 295\ \mu M$
3′,5′-CYCLIC-NUCLEOTIDE PHOSPHODIESTERASE IVc $I50 = 661\ \mu M$
3′,5′-CYCLIC-NUCLEOTIDE PHOSPHODIESTERASE IVd $I50 = 230\ \mu M$

3-PENTOYL-CoA
ACETYL-CoA C-ACETYLTRANSFERASE

2-(2-(PENTOYLOXY)PHENYL)-1H-IMIDAZO[4,5b]PYRIDINE
3′,5′-CYCLIC-NUCLEOTIDE PHOSPHODIESTERASE IV $I50 = 6\ \mu M$
3′,5′-CYCLIC-NUCLEOTIDE PHOSPHODIESTERASE V $I50 = 1\ \mu M$

n-PENTYLAMINE
PEPTIDASE ALKALINE $I50 = 9.3$ mM
SPERMIDINE SYNTHASE $I50 = 3.6\ \mu M$

1-n-PENTYL-AMINOGLUTETHIMIDE
AROMATASE $I50 = 6.6\ \mu M$
CHOLESTEROL MONOOXYGENASE (SIDE-CHAIN-CLEAVING) $I50 = 73\ \mu M$

N-nPENTYL-9-AMINO-1,2,3,4-TETRAHYDROACRIDINE		
ACETYLCHOLINESTERASE		Ki = 70 nM
6β-PENTYLANDROST-4-ENE-3,17-DIONE		
AROMATASE		Ki = 2.8 nM
N-PENTYL-1,3-DIAMINOPROPANE		
SPERMINE SYNTHASE	C	Ki = 109 nM
PENTYLENETETRAZOLE		
ASPARTATE TRANSAMINASE		
GLUTAMATE DEHYDROGENASE (NAD(P))		
S-n-PENTYLGLUTATHIONE		
GLUTATHIONE TRANSFERASE		
S-(n-PENTYL)HOMOCYSTEINE SULPHOXIMINE		
GLUTAMATE-CYSTEINE LIGASE		
4(5)-n-PENTYLOXYOXALYLAMINO-4(5)-CARBOXYAMIDO-1H-1,2,3-TRIAZOLE		
XANTHINE OXIDASE		I50 = 320 nM
2-PENTYLPUTRESCINE		
ORNITHINE DECARBOXYLASE	C	Ki = 500 μM
4-PENTYLPYRAZOLE		
ALCOHOL DEHYDROGENASE		Ki = 8 nM
n-PENTYL-1-THIO-β-D-XYLANOPYRANOSIDE		
XYLAN 1,4-β-XYLOSIDASE		
2-PENTYNOYL-CoA		
ACYL-CoA DEHYDROGENASE	SSI	
3-PENTYNOYL-S-PANTETHEIN		
BUTYRYL-CoA DEHYDROGENASE	SSI	
3-PENTYNOYLPANTETHEINE		
GLUTARYL-CoA DEHYDROGENASE		
N,N′-1,3-PEOPANEDIYLBIS(3,4-DIHYDROXYBENZAMIDE)		
CATECHOL O-METHYLTRANSFERASE		Ki = 300 nM
PEPS		
PROTEIN KINASE C		
PEPSIN INHIBITOR FROM ASCARIS		
CATHEPSIN D		
PEPSINOSTREPTIN		
PEPSIN		
PEPSTATIN		
CANDIDAPEPSIN		Ki = 2.9 nM
CATHEPSIN D		Ki = 105 pM
CATHEPSIN D		
CHYMOSIN		Ki = 220 nM
GASTRICSIN		
HUMICOLIN		
MUCOROPEPSIN		
NEPENTHES ASPARTIC PROTEINASE		
NEPRILYSIN		14% 200 ng/ml
PENICILLOPEPSIN		
PEPSIN		Ki = 45 pM
PROLINE CARBOXYPEPTIDASE		
PROTEINASE ASPARTIC ASPERGILLUS NIGER VAR.MARCOSPORUS		
PROTEINASE ASPARTIC ASPERGILLUS SAITOI		
PROTEINASE ASPARTIC PLASMODIUM		
PROTEINASE ASPARTIC VARIOTI ASPARCTIC		

PROTEINASE BOAR EPIDIDYMAL SPERM		60% 10 μM
PROTEINASE HIV		$I50 = 2$ μM
PROTEINASE HIV-1	C	$Ki = 360$ nM
PROTEINASE HTLV-1		$Ki = 80$ μM
PROTEINASES ASPARTIC		
RENIN		$Ki = 380$ nM
RENIN		$Ki = 13$ μM
RHIZOPUSPEPSIN		
SACCHAROPEPSIN		
YEAST PROTEINASE B		36% 500 μM

PEPSTATIN A

CANDIDAPEPSIN		$Ki = 6$ nM
CATHEPSIN D		
ENDOPEPTIDASE TREPONEMA DENTICOLA ARCC35405		$Ki = 71$ μM
ENDOTHELIN CONVERTING ENZYME		
ENDOTHIAPEPSIN		$Ki = 500$ pM
LACTOSYLCERAMIDE α-2,3-SIALYLTRANSFERASE		72% 1 μM
PRO-OPIOMELANOCORTIN CONVERTING ENZYME		100% 100 μM
PROTEINASE CANDIDA PARAPRILOSIS		$Ki = 600$ pM
PROTEINASE CANDIDA TROPICALIS		$Ki = 12$ nM
PROTEINASE HUMAN T-CELL LEUKAEMIA VIRUS 1		

PEPTICINNAMIN

FARNESYL PROTEINTRANSFERASE		$I50 = 1.1$ μM

PEPTICINNAMIN A

FARNESYL PROTEINTRANSFERASE		$I50 = 650$ nM

PEPTICINNAMIN B

FARNESYL PROTEINTRANSFERASE		$I50 = 200$ nM

PEPTICINNAMIN C

FARNESYL PROTEINTRANSFERASE		$I50 = 100$ nM

PEPTICINNAMIN D

FARNESYL PROTEINTRANSFERASE		$I50 = 1$ μM

PEPTICINNAMIN E

FARNESYL PROTEINTRANSFERASE		$I50 = 300$ nM

PEPTICINNAMIN F

FARNESYL PROTEINTRANSFERASE		$I50 = 500$ nM

PEPTIDE ALDEHYDES

PROTEINASES CYSTEINE		
PROTEINASES SERINE		

PEPTIDE CHLOROMETHYLKETONES

CATHEPSIN G	IR	
CHYMOSIN	IR	
CHYMOTRYPSIN	IR	
ELASTASE LEUKOCYTE	IR	
ELASTASE PANCREATIC	IR	
KALLIKREIN	IR	
PLASMINOGEN ACTIVATOR	IR	
PROLYL OLIGOPEPTIDASE	IR	
TRYPSIN	IR	
TRYPTASE	IR	

PEPTIDE FROM CRAYFISH MG 23000

PHOSPHORYLASE		$Ki = 10$ μG/ml

PEPTIDE FROM KIWI

PECTINESTERASE		$Ki = 220$ pM

PEPTIDE LIKE COMPOUNDS

INDOLEAMINE N-METHYLTRANSFERASE		

PEPTIDE MG 10000
CARBOXYPEPTIDASE A

PEPTIDE NUCLEIC ACIDS
TELOMERASE

PEPTIDE PROLYL BORONIC ACIDS
PROTEINASE IgA1

PEPTIDES
THROMBIN

PEPTIDES FROM ACTINOMADURA
PEPTIDYL-DIPEPTIDASE A

PEPTIDES FROM BOTHROPS JARARACA VENOM
PEPTIDYL-DIPEPTIDASE A

PEPTIDES FROM SOY BEAN
α-AMYLASE

PEPTIDES FROM WHEAT GERM
α-AMYLASE

PEPTIDES WITH C TERMINAL GLYCINE
PEPTIDYLGLYCINE MONOOXYGENASE

PEPTIDE YY
ADENYLATE CYCLASE — *I*50 = 14 μM

N-PEPTIDYL-O-ACYLHYDROXYLAMINES
PROTEINASES CYSTEINE

PEPTIDYL-4-AMINO-5-PHENYL-2-PENTYL BROMIDE
CHYMOTRYPSIN — IR

PEPTIDYL CHLOROMETHYLKETONES
PROTEINASES CYSTEINE — IR

PEPTIDYL DIAZOKETONES
CATHEPSIN L

PEPTIDYL DIAZOMETHANES
CALPAIN II

PEPTIDYL DIAZOMETHYLKETONES
CATHEPSIN B
PROTEINASES CYSTEINE

PEPTIDYL FLUOROMETHYLKETONES
PROTEINASES SERINE

PEPTIDYLNITRILES
PAPAIN

PEPTIDYL SULPHONIUM SALTS
CATHEPSIN B — IR
PAPAIN — IR

PERACETIC ACID
1-ACYLGLYCEROPHOSPHOCHOLINE O-ACYLTRANSFERASE — 70% 10 μM

PERAZIN
AMINE OXIDASE MAO — C — *K*i = 19 μM

PERCHLORATE
CARBONATE DEHYDRATASE
DIHYDROFOLATE REDUCTASE
GLUTATHIONE TRANSFERASE β — NC — *K*i = 190 μM
NITRATE REDUCTASE

NITRATE REDUCTASE (CYTOCHROME)		
PEPTIDE α-N-ACETYLTRANSFERASE		
PHOSPHOGLUCOMUTASE		
SUCCINATE DEHYDROGENASE		
SULFATE ADENYLYLTRANSFERASE	C	*K*i =
SULFATE ADENYLYLTRANSFERASE		*I*50 = 18 μM
PERFLUORODECALIN		
CYTOCHROME P450		
PERFLUORODECANOIC ACID		
ENOYL-CoA HYDRATASE		*K*i = 5 μM
3-HYDROXYACYL-CoA DEHYDROGENASE		*K*i = 5 μM
LONG-CHAIN-FATTY-ACID CoA-LIGASE		*I*50 = 80 μM
LONG-CHAIN-FATTY-ACID CoA-LIGASE		*I*50 = 30 μM
PERFLUOROOCTANOIC ACID		
OXIDATIVE PHOSPHORYLATION UNCOUPLER		
PERFLUOROSUCCINIC ACID		
ASPARTATE TRANSAMINASE		
3-OXOACID CoA-TRANSFERASE		
PHOSPHOENOLPYRUVATE CARBOXYKINASE (GTP)		
N-PERFLUORPELARGONOYLDOPAMINE		
ARACHIDONATE 5-LIPOXYGENASE		*I*50 = 66 nM
PERHEXILINE		
CARNITINE O-PALMITOYLTRANSFERASE 1		*I*50 = 77 μM
SPHINGOMYELIN PHOSPHODIESTERASE		82% 10 μM
PERHEXILINE MALEATE		
Ca^{2+}/Mg^{2+}-TRANSPORTING ATPASE		*I*50 = 100 μM
Na^{+}/K^{+}-EXCHANGING ATPASE		*I*50 = 50 μM
PERILC ACID METHYLESTER		
FARNESYL PROTEINTRANSFERASE		
GERANYLGERANYL PROTEINTRANSFERASE		
PERILLOSIDE A		
ALDEHYDE REDUCTASE	C	*K*i = 140 μM
PERILLOSIDE C		
ALDEHYDE REDUCTASE	C	*K*i = 230 μM
PERILLYL ALCOHOL		
FARNESYL PROTEINTRANSFERASE		14% 1 mM
GERANYLGERANYL PROTEINTRANSFERASE I		62% 1 mM
GERANYLGERANYL PROTEINTRANSFERASE II		72% 1 mM
PERINDOPRIL		
PEPTIDYL-DIPEPTIDASE A		
PERINDOPRILAT		
PEPTIDYL-DIPEPTIDASE A		*I*50 = 1.5 nM
PERIODATE		
PHOSPHOGLUCONATE DEHYDROGENASE (DECARBOXYLATING)		
PROTEIN KINASE C		
m-PERIODATE		
DEXTRANASE		
PERIPLANETA AMERICANA TREHALASE INHIBITOR		
α,α-TREHALASE		*I*50 = 5.4 μg
PERMANGANATE		
PHOSPHOGLUCONATE DEHYDROGENASE (DECARBOXYLATING)		
SIALIDASE		

PEROXIDASE
β-FRUCTOFURANOSIDASE
4-NITROPHENOL 2-MONOOXYGENASE
PHENOLL 2-MONOOXYGENASE

PEROXINITRITE
ACONITATE HYDRATASE

PEROXIVANADIUM COMPOUNDS
PROTEIN TYROSINE PHOSPHATASE

PEROXYNITRITE

ACONITATE HYDRATASE	IR	
CREATINE KINASE	IR	
NADH DEHYDROGENASE (UBIQUINONE)		
NITRIC-OXIDE SYNTHASE I		*I*50 = 15 μM
NITRIC-OXIDE SYNTHASE II		*I*50 = 10 μM
NITRIC-OXIDE SYNTHASE III		*I*50 = 28 μM
PROSTAGLANDIN-I SYNTHASE		*I*50 = 50 nM
SUCCINATE DEHYDROGENASE		
TYROSINE 3-MONOOXYGENASE	IR	

PERPHENAZINE

ACETYLCHOLINESTERASE		*I*50 = 100 μM
GLUTAMATE DEHYDROGENASE		*K*i = 610 μM
INOSITOL-1,4,5-TRISPHOSPHATE 5-PHOSPHATASE		53% 1 mM
PYRUVATE DEHYDROGENASE (LIPOAMIDE)		50% 267 μM

PERSICARIN

XANTHINE OXIDASE		*I*50 = 35 μM

PERUVIN

AROMATASE		*I*50 = 65 μM

PERVANADATE

PROTEIN TYROSINE-PHOSPHATASE		
PROTEIN TYROSINE-PHOSPHATASE 1B	IR	

PERYLENEQUINOID PIGMENTS
PROTEIN KINASE C

PETROSELENIC ACID

PROTEIN KINASE A		*I*50 = 20 μM

PETROSOLIC ACID

DNA POLYMERASE		*I*50 = 6.2 μM
RIBONUCLEASE H		*I*50 = 40 μM
RNA-DIRECTED DNA POLYMERASE		*I*50 = 1.2 μM

PETROSYNOL

DNA POLYMERASE		*I*50 = 36 μM
RNA-DIRECTED DNA POLYMERASE		*I*50 = 16 μM

PEYSSONOL A

DNA POLYMERASE α		*I*50 = 63 μM
RNA-DIRECTED DNA POLYMERASE		*I*50 = 21 μM
RNA-DIRECTED DNA POLYMERASE		*I*50 = 6.4 μM

PEYSSONOL B

DNA POLYMERASE α		*I*50 = 17 μM
DNA POLYMERASE β		*I*50 = 48 μM
RNA-DIRECTED DNA POLYMERASE		*I*50 = 15 μM
RNA-DIRECTED DNA POLYMERASE		*I*50 = 4.3 μM

PFIFEEEP

PROTEINASE HIV		70% ≈ 80 μM

P′GBDBIG

C1 ESTERASE		$I50 = 34\ \mu M$
α-CHYMOTRYPSIN		$I50 = 1$ mM
COMPLEMENT SUBCOMPONENT C1r (ESTEROLYTIC ACTIVITY)		$I50 = 4.4\ \mu M$
KALLIKREIN PLASMA		$I50 = 320$ nM
PLASMIN		$I50 = 190$ nM
THROMBIN		$I50 = 50\ \mu M$
TRYPSIN		$I50 = 34$ nM

PGBx

PHOSPHOLIPASE A2

PGH2

PROSTACYCLIN SYNTHASE

PHASEOLIN

CHALCONE ISOMERASE		$Ki = 50\ \mu M$

PHASEOLINONE

RNA POLYMERASE

PHASEOLOTOXIN

ORNITHINE CARBAMOYLTRANSFERASE

PHASEOLUS VULGARIS α-AMYLASE INHIBITOR

α-AMYLASE
α-AMYLASE

PHASEOLUS VULGARIS α-AMYLASE INHIBITOR 1

α-AMYLASE

PHASEOLUS VULGARIS α-AMYLASE INHIBITOR 2

α-AMYLASE
α-AMYLASE

PHASEOLUS VULGARIS POLYGALACTURONASE INHIBITOR

POLYGALACTURONASE
POLYGALACTURONASE

PHASEOLUS VULGARIS POLYGALACTURONASE INHINBITING PROTEIN

POLYGALACTURONASE

PHATALIC ACID

BENZOYLFORMATE DECARBOXYLASE		$Ki = 98\ \mu M$
SERINE CARBOXYPEPTIDASE		

PHCH$_2$CF$_2$COCO-ALA-LEU-ARG-OME

CHYMOTRYPSIN		$Ki = 190$ nM

D-PHE-ALA-CHNH$_2$

BOTHROPAIN	IR	$Ki = 10$ nM

(PHE-ALA-GLU)n

PROTEIN TYROSINE KINASE (INSULIN RECEPTOR)		$I50 = 1\ \mu M$

PHE-ALA-LYS-ARG-CH$_2$CL

PLASMIN	IR	$Ki = 3.7$ nM
PROTEINASE KEX-2	IR	$Ki = 3.7$ nM
THROMBIN	IR	$Ki = 3.7$ nM
TRYPSIN	IR	$Ki = 3.7$ nM

PHE-ARG

PEPTIDYL-DIPEPTIDASE A		$Ki = 810\ \mu M$

PHEBESTIN

AMINOPEPTIDASE B		$I50 = 9\ \mu g/ml$
GLUTAMYL AMINOPEPTIDASE		$I50 = 9\ \mu g/ml$
MEMBRANE ALANINE AMINOPEPTIDASE		$Ki = 30$ nM

[PHE-22]BIG ENDOTHELIN-1[19-37]
ENDOTHELIN CONVERTING ENZYME

PHEpsi[$CH_2CH(OH)$]PHE
PROTEINASE HIV-1 | | *I*50 = 800 fM

D-PHE-O-ETHYLESTER
SERINE CARBOXYPEPTIDASE | | *K*i = 5.3 mM

PHE-O-ETHYLESTER
SERINE CARBOXYPEPTIDASE | | *K*i = 8.3 mM

PHE-GLN-VAL-VAL-CYS(NPYS)-GLY-NH_2
CALPAIN

L-PHE-GLY-GLY
PHOSPHATASE ALKALINE

PHELLOPTERIN
O-DEMETHYLASE (CYT P450) | IR | *I*50 = 32 μM

PHENACAMIDE
ALCOHOL DEHYDROGENASE (NADP)

PHENACETIN

ACETYLSPERMIDINE DEACETYLASE		
ALCOHOL DEHYDROGENASE (NADP)	NC	
ALDEHYDE REDUCTASE	NC	*K*i = 3.4 mM
AMIDASE		
4-AMINOBUTYRATE TRANSAMINASE	NC	
CARBONYL REDUCTASE (NADPH)	NC	
SUCCINATE-SEMIALDEHYDE DEHYDROGENASE	NC	

PHENALAMID A1
RNA-DIRECTED DNA POLYMERASE | | *I*50 = 386 μM

PHENALAMID A2
NADH DEHYDROGENASE (UBIQUINONE)

PHENAMIDINE ISOTHIONATE
AMINE OXIDASE (COPPER-CONTAINING) | NC | *K*i = 4 μM

PHENANTHRAQUINONE
NITRIC-OXIDE SYNTHASE I | NC | *K*i = 9.6 μM

PHENANTHRENE

AMINE OXIDASE MAO-A	*I*50 = 21 μM
MYOSIN-LIGHT-CHAIN KINASE	28% 333 μM
PROTEIN KINASE A	33% 400 μM
PROTEIN KINASE A	*I*50 = 310 μM
PROTEIN KINASE C	26% 333 μM

PHENANTHRENEQUINONE

MILTPAIN	*K*i = 8 nM
MYOSIN-LIGHT-CHAIN KINASE	*I*50 = 6 μM
PROSTAGLANDIN F SYNTHASE	
PROTEIN KINASE A	*I*50 = 8 μM
PROTEIN KINASE A	*I*50 = 49 μM
PROTEIN KINASE C	*I*50 = 65 μM
PROTEIN KINASE (Ca^{2+} DEPENDENT)	*I*50 = 56 μM

9,10-PHENANTHRENE QUINONE

GLUTATHIONE REDUCTASE (NADPH)	U	*K*i = 2 μM
TRYPANOTHIONE REDUCTASE	U	*K*i = 2.5 μM

PHENANTHRIDINE

MYOSIN-LIGHT-CHAIN KINASE	*I*50 = 44 μM
PROTEIN KINASE A	*I*50 = 364 μM

PROTEIN KINASE A		$I50 = 110\ \mu M$
PROTEIN KINASE C		29% 333 μM
PROTEIN KINASE (Ca^{2+} DEPENDENT)		$I50 = 275\ \mu M$
6(5H)-PHENANTHRIDINONE		
MYOSIN-LIGHT-CHAIN KINASE		$I50 = 53\ \mu M$
NAD ADP-RIBOSYLTRANSFERASE		
PROTEIN KINASE A		$I50 = 62\ \mu M$
PROTEIN KINASE A		$I50 = 10\ \mu M$
PROTEIN KINASE (Ca^{2+} DEPENDENT)		$I50 = 312\ \mu M$
9-PHENANTHROL		
MYOSIN-LIGHT-CHAIN KINASE		$I50 = 10\ \mu M$
PROTEIN KINASE A		$I50 = 82\ \mu M$
PROTEIN KINASE A		$I50 = 10\ \mu M$
PROTEIN KINASE C		$I50 = 293\ \mu M$
PROTEIN KINASE (Ca^{2+} DEPENDENT)		$I50 = 84\ \mu M$
o-PHENANTHROLIINE		
(S)-6-HYDROXYNICOTINE OXIDASE		
2,10-PHENANTHROLINE		
myo-INOSITOL OXYGENASE		
4,7-PHENANTHROLINE		
MYOSIN-LIGHT-CHAIN KINASE		$I50 = 44\ \mu M$
PROTEIN KINASE A		$I50 = 16\ \mu M$
PROTEIN KINASE A		$I50 = 300\ \mu M$
PROTEIN KINASE C		20% 333 μM
PROTEIN KINASE (Ca^{2+} DEPENDENT)		$I50 = 57\ \mu M$
o-PHENANTHROLINE-2-CARBOXYLIC ACID		
SIGNAL PEPTIDASE I		
o-PHENANTHROLINE + IODOACETIC ACID		
HISTIDINOL DEHYDROGENASE		
m-PHENANTHROLINE		
CATECHOL 2,3-DIOXYGENASE	C	
PYRIDOXAL DEHYDROGENASE		$I50 = 2\ \mu M$
PYRIDOXINE 4-OXIDASE		
o-PHENANTHROLINE		
ACE SECRETASE		100% 5 mM
ACETOLACTATE DECARBOXYLASE		
N-ACETYL-β-GLUCOSAMINIDASE		25% 1 mM
N-ACETYLNEURAMINATE LYASE		40% 100 μM
N-ACETYLNEURAMINATE MONOOXYGENASE		100% 1 mM
ACONITATE HYDRATASE		$Ki = 500\ \mu M$
ADENOSINE DEAMINASE		
AEROMONOLYSIN		
AGAVAIN		100% 1 mM
ALANINE TRANSAMINASE		
ALCOHOL DEHYDROGENASE		$Ki = 1\ mM$
ALCOHOL DEHYDROGENASE		$Ki = 10\ \mu M$
ALCOHOL DEHYDROGENASE		$Ki = 8\ \mu M$
ALCOHOL DEHYDROGENASE (ACCEPTOR)		
ALCOHOL DEHYDROGENASE (NAD(P))		
ALCOHOL OXIDASE		
ALDEHYDE DEHYDROGENASE (NADP)		
ALDEHYDE DEHYDROGENASE (PYRROLOQUINOLINE-QUINONE)		15% 500 μM
ALDEHYDE REDUCTASE		
ALTRONATE DEHYDRATASE		81% 3 mM
AMINE OXIDASE (COPPER-CONTAINING)	NC	$Ki = 31\ \mu M$
AMINE OXIDASE (COPPER-CONTAINING)		
AMINE OXIDASE (COPPER-CONTAINING) I	NC	$Ki = 2.2\ mM$
AMINE OXIDASE (COPPER-CONTAINING) II	NC	$Ki = 3.1\ mM$

D-AMINO-ACID OXIDASE		
AMINOBUTYRALDEHYDE DEHYDROGENASE		
AMINOPEPTIDASE		Ki = 70 μM
AMINOPEPTIDASE		94% 1 mM
AMINOPEPTIDASE		96% 1 mM
AMINOPEPTIDASE		Ki = 200 μM
AMINOPEPTIDASE		
AMINOPEPTIDASE B		
AMINOPEPTIDASE B		Ki = 15 μM
AMINOPEPTIDASE (HUMAN LIVER)	C	Ki = 28 μM
AMINOPEPTIDASE Y		95% 1 mM
AMP DEAMINASE		
AMP-THYMIDINE KINASE		
ANTHRANILATE 2,3 DIOXYGENASE		
ANTHRANILATE 3-MONOOXYGENASE		
ANTHRANILATE 3-MONOOXYGENASE (DEAMINATING)		
D-ARABINITOL DEHYDROGENASE		
ARGININE 2-MONOOXYGENASE		
ASTACIN		
AUREOLYSIN		
BACILLOLYSIN		
BACTERIAL LEUCYL AMINOPEPTIDASE		
BENZOATE 4-MONOOXYGENASE		70% 50 μM
BETAINE-ALDEHYDE DEHYDROGENASE		
BIOTIN-CoA LIGASE		67% 10 mM
(R,R)-BUTANEDIOL DEHYDROGENASE		
CALF THYMUS RIBONUCLEASE H		
CARBONYL REDUCTASE (NADPH)		17% 1 mM
CARBONYL REDUCTASE (NADPH)		30% 1 mM
6-CARBOXYHEXANOATE-CoA LIGASE		
CARBOXYPEPTIDASE A		
CARBOXYPEPTIDASE B		
β-CAROTENE 15,15′-DIOXYGENASE		
CATECHOL 1,2-DIOXYGENASE		
CATECHOL 1,2-DIOXYGENASE 1		
CATECHOL 1,2-DIOXYGENASE 2		
CATECHOL 2,3-DIOXYGENASE	C	
CATECHOL OXIDASE		
CATHEPSIN H		
CELLULOSE 1,4-β-CELLOBIOSIDASE		84% 10 mM
2-CHLOROBENZOATE 1,2-DIOXYGENASE		
4-CHLOROPHENYLACETATE 3,4-DIOXYGENASE		
CHOLESTANETETRAOL 26-DEHYDROGENASE		
CINNAMOYL-CoA REDUCTASE		
CINNAMYL-ALCOHOL DEHYDROGENASE	NC	I50 = 380 μM
COCCOLYSIN		
COLLAGENASE		
CORTICOSTEROID SIDE-CHAIN-ISOMERASE		94% 200 μM
3′,5′-CYCLIC-NUCLEOTIDE PHOSPHODIESTERASE II		
CYSTEINE DIOXYGENASE		
CYSTINYL AMINOPEPTIDASE		
CYTIDINE DEAMINASE		
CYTOCHROME-b5 REDUCTASE		
CYTOSOL NON-SPECIFIC DIPEPTIDASE		
3-DEHYDROQUINATE SYNTHASE		
DEOXYHYPUSINE MONOOXYGENASE		100% 10 μM
DETHIOBIN SYNTHASE		
α-DEXTRIN ENDO-1,6-α-GLUCOSIDASE		
2,4-DICHLOROPHENOL 6-MONOOXYGENASE		
DIETHYL-2-METHYL-3-OXOSUCCINATE DEHYDROGENASE		
cis-1,2-DIHYDROBENZENE-1,2-DIOL DEHYDROGENASE		
trans-1,2-DIHYDROBENZENE-1,2-DIOL DEHYDROGENASE 1		$K$is = 20 μM
trans-1,2-DIHYDROBENZENE-1,2-DIOL DEHYDROGENASE 2		$K$is = 270 μM
DIHYDROOROTASE		

DIHYDROOROTASE		93% 18 mM
DIHYDROOROTATE DEHYDROGENASE		
4,5-DIHYDROOXYPHTHALATE DECARBOXYLASE		
DIHYDROPICOLINATE REDUCTASE		
DIHYDROPYRIMIDINASE		
2,3-DIHYDROXYINDOLE 2,3-DIOXYGENASE		
3,4-DIHYDROXYPHENYLACETATE 2,3-DIOXYGENASE	C	
2,5-DIHYDROXYPYRIDINE 5,6-DIOXYGENASE		
3,4-DIHYDROXY-9,10-SECOANDROSTA-1,3,5810)TRIENE-9,17-DIONE 4,5-OXIDOREDUCTASE		
DIPEPTIDASE		46% 1 mM
DIPEPTIDYL-PEPTIDASE 3		
DIPEPTIDYL-PEPTIDASE 3		100% 1 mM
DIPEPTIDYL-PEPTIDASE 4		
DNA NUCLEOTIDYLEXOTRANSFERASE		
DNA POLYMERASE		
DOPACHROME TAUTOMERASE		31% 500 µM
ENDOTHELIN CONVERTING ENZYME		*I*50 = 25 µM
ENDOTHELIN CONVERTING ENZYME		*I*50 = 34 µM
ENTEROBACTER RIBONUCLEASE		
ENVELYSIN		
FAD PYROPHOSPHATASE		
FERREDOXIN-NAD REDUCTASE		
FORMALDEHYDE DEHYDROGENASE		
FORMALDEHYDE DEHYDROGENASE (GLUTATHIONE)		
FORMALDEHYDE TRANSKETOLASE		
FORMATE DEHYDROGENASE		
FORMYLMETHIONINE DEFORMYLASE		
FRUCTOSE-BISPHOSPHATE ALDOLASE		100% 250 µM
GALACTONOLACTONE DEHYDROGENASE		
GELATINASE A		*I*50 = 11 µM
GELATINASE A		100% 1 mM
GELATINASE B		*I*50 = 17 µM
GERANIOL DEHYDROGENASE		
GLUCAN ENDO-1,3-β-GLUCOSIDASE		
GLUCOKINASE		
GLUCOSE-6-PHOSPHATASE		*K*i = 20 mM
GLUCURONATE ISOMERASE		90% 1 mM
GLUTAMATE DEHYDROGENASE (NAD(P))		
GLUTAMATE SYNTHASE (NADPH)		
GLUTAMYL AMINOPEPTIDASE		67% 100 nM
GLUTAMYL AMINOPEPTIDASE		
γ-GLUTAMYLCYCLOTRANSFERASE		93% 250 µM
γ-GLUTAMYL HYDROLASE		
GLUTATHIONE DEHYDROGENASE (ASCORBATE)		34% 1 mM
GLYCEROL-1,2-CYCLIC-PHOSPHATE PHOSPHODIESTERASE		
GLYCEROL-3-PHOSPHATE DEHYDROGENASE		
GLYCERYL-ETHER MONOOXYGENASE		68% 1 mM
GLYCINE HYDROXYMETHYLTRANSFERASE		20% 100 µM
GLYCOCYAMINASE		
GLYCYLPEPTIDE N-TETRADECANOYLTRANSFERASE		
γ-GUANIDINOBUTYRATE DEHYDROGENASE		
GUANOSINE-3′,5′-BIS(DIPHOSPHATE) 3′-PYROPHOSPHATASE		
HIPPURATE HYDROLASE		
HOMOACONITATE HYDRATASE		86% 300 mM
HOMOCARNOSINASE		98% 500 µM
HOMOCARNOSINASE		98% 500 µM
HYDROGENASE II		
D-2-HYDROXY-ACID DEHYDROGENASE		
(R)-3-HYDROXYACID ESTER DEHYDROGENASE		
(S)-3-HYDROXYACID ESTER DEHYDROGENASE		
3-HYDROXYBENZOATE 4-MONOOXYGENASE		
HYDROXYLAMINE REDUCTASE (NADH)		
HYDROXYMETHYLGLUTARYL-CoA REDUCTASE (NADPH)		
6-HYDROXYNICOTINATE REDUCTASE		

(R)-6-HYDROXYNICOTINE OXIDASE		
4-HYDROXYPHENYLACETATE 3-MONOOXYGENASE		
(R)-4-HYDROXYPHENYLLACTATE DEHYDROGENASE		
4-HYDROXYPHENYLPYRUVATE DIOXYGENASE		
4-HYDROXYPROLINE EPIMERASE		33% 320 µM
3α-HYDROXYSTEROID DEHYDROGENASE (B-SPECIFIC)		*I*50 = 9.1 µM
HYOSCYAMINE (6S)-DIOXYGENASE		
L-IDITOL 2-DEHYDROGENASE		
IGA-SPECIFIC METALLOENDOPEPTIDASE 1		*I*50 = 3.5 mM
IGA-SPECIFIC METALLOENDOPEPTIDASE 2		*I*50 = 35 mM
INDANOL DEHYDROGENASE	C	*K*i = 65 µM
INDOLEPYRUVATE C-METHYLTRANSFERASE		
INTERSTITIAL COLLAGENASE		100% 1 mM
ISOLEUCINE-tRNA LIGASE		
2-ISOPROPYLMALATE SYNTHASE		
β-LACTAMASE		
D-LACTATE DEHYDROGENASE (CYTOCHROME)		90% 3.5 mM
D-LACTATE DEHYDROGENASE (CYTOCHROME c-553)		
LACTOYLGLUTATHIONE LYASE		
LEUCOLYSIN		40% 50 µM
LEUCYL AMINOPEPTIDASE		*I*50 = 7 mM
LEUKOTRIENE A4 HYDROLASE		*I*50 = 700 µM
LINOLEATE ISOMERASE		
LIPOXYGENASE		
LYSINE 2,3-AMINOMUTASE		
LYSINE (ARGININE) CARBOXYPEPTIDASE		1 mM
MAGNESIUM PROTOPORPHYRIN O-METHYLTRANSFERASE		
MANDELATE 4-MONOOXYGENASE		80% 500 µM
MANGANESE PEROXIDASE		
MANNONATE DEHYDRATASE		75% 3 mM
MANNOSE-6-PHOSPHATE ISOMERASE		
MEMBRANE ALANINE AMINOPEPTIDASE		
METALLO-ENDOPEPTIDASE QG		
METALLOPROTEINASES	IR	
METHANE MONOOXYGENASE		58% 100 µM
METHANOL DEHYDROGENASE		
METHANOL OXIDASE		
METHIONYL AMINOPEPTIDASE		*K*i = 50 µM
4-METHOXYBENZOATE MONOOXYGENASE (O-DEMETHYLATING)		
(R)-2-METHYLMALATE DEHYDRATASE		99% 1 mM
MICROBIAL METALLOPROTEINASE		
MICROCOCCUS CASEOLYTICUS NEUTRAL PROTEINASE		
MILTPAIN		*K*i = 1 µM
MORPHINE 6-DEHYDROGENASE		
MULTICATALYTIC ENDOPEPTIDASE COMPLEX		
MYOSIN-LIGHT-CHAIN KINASE		*I*50 = 65 µM
NAD ADP-RIBOSYLTRANSFERASE		
NADH DEHYDROGENASE (UBIQUINONE)	U	
NADH PEROXIDASE		19% 3 mM
NAPHTHALEN 1,2-DIOXYGENASE		
NEPRILYSIN		95% 1 mM
NEUROLYSIN		95% 1 mM
NEUROLYSIN		*I*50 = 10 µM
NICOTINATE DEHYDROGENASE		
NICOTINE DEHYDROGENASE		
NITRATE REDUCTASE		
NITRATE REDUCTASE (CYTOCHROME)		
NITRIC-OXIDE REDUCTASE		
NITRILE HYDRATASE		78% 5 mM
NITRITE REDUCTASE		
NITRITE REDUCTASE (NAD(P)H)		
NITROGENASE		
NITROUS OXIDE REDUCTASE		
NUCLEOTIDE PYROPHOSPHATASE		

OCTANOL DEHYDROGENASE	
OLIGOPEPTIDASE M	
OXALATE OXIDASE	
(R)-OXOESTER OXIDO REDUCTASE	60% 5 mM
(S-3)-OXOESTER OXIDO REDUCTASE	60% 5 mM
PANTOATE-β-ALANINE LIGASE	
PARATHION HYDROLASE	79% 1 mM
PENICILLIUM ROQUEFORTI NEUTRAL PROTEINASE	
PEPTIDE DEFORMYLASE	*I*50 = 200 μM
PEPTIDYL-DIPEPTIDASE A	
PEPTIDYL-LYS METALLOENDOPEPTIDASE	*K*i = 3.9 mM
PERILLYL-ALCOHOL DEHYDROGENASE	
PEROXIDASE	
PHENYLALANINE 4-MONOOXYGENASE	
PHOSPHATASE ACID PURPLE	
PHOSPHATASE ALKALINE	
PHOSPHODIESTERASE	
PHOSPHOGLYCERATE MUTASE	
PHOSPHOLIPASE C	
PHTHALATE 4,5-DIOXYGENASE	
PLASMANYLETHANOYLAMINE DESATURASE	
POLYNUCLEOTIDE ADENYLYLTRANSFERASE	
PORPHOBILINOGEN SYNTHASE	89% 1 mM
PROCATECHUATE DECARBOXYLASE	*I*50 = 1 mM
PROCOLLAGEN N-ENDOPEPTIDASE	100% 1 mM
PROCOLLAGEN-PROLINE, 2-OXOGLUTARATE-4-DIOXYGENASE	
PROLYL OLIGOPEPTIDASE	
PROPANEDIOL DEHYDRATASE	85% 1 mM
1,3-PROPANEDIOL DEHYDROGENASE	
PROTEINASE 2A HRV2	*I*50 = 3.2 mM
PROTEINASE ASTACUS	
PROTEINASE IgA	
PROTEINASE SEPIA	
PROTEINASES METALLO	
PROTEIN KINASE A	18% 400 μM
PROTEIN KINASE A	*I*50 = 180 μM
PROTEIN KINASE C	20% 333 μM
PROTEIN KINASE (Ca^{2+} DEPENDENT)	11% 333 μM
PSEUDOMONAS AEROGINOSA ALKALINE PROTEINASE	97% 5 mM
PSEUDOMONAS AEROGINOSA NEUTRAL PROTEINASE	
PTEROYLGLUTAMATE HYDROLASE	
PYRANOSE OXIDASE	
PYRIDOXAL DEHYDROGENASE	*I*50 = 260 μM
PYRIDOXINE 4-OXIDASE	
4-PYRIDOXOLACTONASE	42% 200 μM
PYROGLUTAMYL-PEPTIDASE I	
PYRUVATE CARBOXYLASE	
QUERCETIN 2,3-DIOXYGENASE	
RETICULINE OXIDASE	
RHAMNULOSE-1-PHOSPHATE ALDOLASE	
tRNA ADENYLYLTRANSFERASE	
RNA-DIRECTED DNA POLYMERASE	
SACCHAROPINE DEHYDROGENASE (NADP, L-GLUTAMATE FORMING)	
SALICYLATE 1-MONOOXYGENASE	
SARCOSINE OXIDASE	47% 2 mM
SEPIA PROTEINASE	85% 1 mM
L-SERINE DEHYDRATASE	
SERRALYSIN	
STEAROYL-CoA DESATURASE	
STEROID 9α-MONOOXYGENASE	
STIZOLOBATE SYNTHASE	89% 100 μM
STIZOLOBINATE SYNTHASE	92% 100 μM
STREPTOMYCES GRISEUS METALLO-ENDOPEPTIDASE I	98% 1 mM
STREPTOMYCES GRISEUS METALLO-ENDOPEPTIDASE II	100% 1 mM

SUCCINATE DEHYDROGENASE (UBIQUINONE)		
SULFITE DEHYDROGENASE		
TETANUS TOXIN L CHAIN (PEPDIDOLYTIC ACITIVITY)		100% 1 mM
THERMOLYSIN		*K*i = 40 μM
THIMET OLIGOPEPTIDASE		93% 250 μM
L-THREONINE 3-DEHYDROGENASE		
TISSUE-ENDOPEPTIDASE DEGRADING COLLAGENASE-SYNTHETIC-SUBSTRATE		96% 10 μM
TOLUENE DIOXYGENASE		
TRIACETATE-LACTONASE		
TRIPEPTIDE AMINOPEPTIDASE		52% 500 μM
peptide-TRYPTOPHAN 2,3-DIOXYGENASE		
TRYPTOPHAN 5-MONOOXYGENASE		
TRYPTOPHAN-tRNA LIGASE		
TYROSINE 3-MONOOXYGENASE		*K*i = 2 μM
URIDINE NUCLEOSIDASE		
dUTP PYROPHOSPHATASE		
VALINE-tRNA LIGASE		
VENOM EXONUCLEASE		
XAA-ARG DIPEPTIDASE		
XAA-HIS DIPEPTIDASE		71% 100 μM
XAA-PRO DIPEPTIDASE		
XANTHINE DEHYDROGENASE		
p-PHENANTHROLINE		
COLUMBAMINE OXIDASE		
PYRIDOXAL DEHYDROGENASE		*I*50 = 2 μM
PYRIDOXINE 4-OXIDASE		
6(5H)PHENATHRIDINONE		
NAD(P)-ARGININE ADP-RIBOSYLTRANSFERASE		39% 100 μM
L-PHENAYLLACTATIC ACID		
CARBOXYPEPTIDASE A		
PHENAZINE		
MILTPAIN		
PHENAZINE ETHOSULPHATE		
L-PIPECOLATE OXIDASE		
PHENAZINE METHOSULPHATE		
PROGESTERONE MONOOXYGENASE		
STEROID 11β-MONOOXYGENASE		
(±)-PHENAZOCINE		
PROTEIN TYROSINE KINASE EGFR		*I*50 = 12 μM
PHENBARBITONE		
CARBONYL REDUCTASE (NADPH)		
PHENCYCLIDINE		
CYTOCHROME P450 2B1	IR	*K*i = 3.8 μM
CYTOCHROME P450 2B4		
PHENELZINE		
AMINE OXIDASE (COPPER-CONTAINING)		
AMINE OXIDASE MAO-A	IR	
AMINE OXIDASE MAO-B		
AMINE OXIDASE (SEMICARBAZIDE SENSITIVE)		
AROMATIC-L-AMINO-ACID DECARBOXYLASE		
BUFURALOL-1′-HYDROXYLASE (CYT P-450dbl)	C	*I*50 = 100 μM
TAURINE DEHYDROGENASE		
TYROSINE TRANSAMINASE		
PHENETHYL ISOTHIOCYANATE		
ALDEHYDE DEHYDROGENASE (NAD) (LOW Km)		*K*i = 1.4 μM
CYTOCHROME P450 2E1		

PHENETHYL ISOTHIOCYANATE-CYSTEINE		
N-NITROSODIMETHYLAMINE DEMETHYLASE		$I50 = 8.3\ \mu M$
PHENETHYL ISOTHIOCYANATE-GSH		
N-NITROSODIMETHYLAMINE DEMETHYLASE		$I50 = 4\ \mu M$
PHENFORMIN		
AMINE OXIDASE (COPPER-CONTAINING)	NC	$Ki = 900\ \mu M$
PHE-NH_2		
DIPEPTIDYL-PEPTIDASE 1		$Ki = 1$ mM
PHENIDIONE		
NAD(P)H DEHYDROGENASE (QUINONE)		
PHENIDONE		
ARACHIDONATE 5-LIPOXYGENASE		$I50 = 24\ \mu M$
ARACHIDONATE 5-LIPOXYGENASE		$I50 = 2.1\ \mu M$
ARACHIDONATE 12-LIPOXYGENASE		$I50 = 67\ \mu M$
ARACHIDONATE 12-LIPOXYGENASE		$I50 = 100$ nM
ARACHIDONATE 12-LIPOXYGENASE		$I50 = 21\ \mu M$
ARACHIDONATE 15-LIPOXYGENASE		$I50 = 400$ nM
CYCLOOXYGENASE		$I50 = 94\ \mu M$
CYCLOOXYGENASE		31% 75 μM
CYCLOOXYGENASE		$I50 = 89\ \mu M$
PHENIRAMINE		
ALDEHYDE OXIDASE		$I50 = 216\ \mu M$
PHENITIDINE		
AMIDASE		
PHENOBARBITAL		
ALCOHOL DEHYDROGENASE (NADP)		
ALDEHYDE REDUCTASE	NC	$Ki = 1.6$ mM
ALDEHYDE REDUCTASE		52% 1 mM
ALDEHYDE REDUCTASE 2		$I50 = 4.9$ mM
4-AMINOBUTYRATE TRANSAMINASE	NC	
CARBONYL REDUCTASE (NADPH)		24% 5 mM
GLUCURONATE REDUCTASE		
GLUCURONONOLATONE REDUCTASE		
γ-GLUTAMYLTRANSFERASE		
GLYCEROL DEHYDROGENASE (NADP)		
myo-INOSITOL OXYGENASE		
PHOSPHATASE ALKALINE	MIX	$Ki = 40$ mM
PHOSPHATASE ALKALINE	U	$Ki = 200$ mM
PHOSPHATASE ALKALINE		
PHOSPHATASE ALKALINE	MIX	$Ki = 10$ mM
PHOSPHATASE ALKALINE	MIX	$Ki = 55$ mM
SUCCINATE-SEMIALDEHYDE DEHYDROGENASE	NC	
PHENOBARBITONE		
ALDEHYDE REDUCTASE		
PHENOL		
ACYL-CoA OXIDASE		
ARACHIDONATE 5-LIPOXYGENASE		$I50 = 5.6$ mM
ARACHIDONATE 15-LIPOXYGENASE		$I50 = 14$ mM
L-ASCORBATE OXIDASE		$Ki = 6$ mM
CARBONATE DEHYDRATASE	C	
CARBONATE DEHYDRATASE		
CATECHOL O-METHYLTRANSFERASE		
CHOLINE SULFOTRANSFERASE		
CYTOCHROME P450 2E1	NC	$Ki = 970\ \mu M$
RENILLA-LUCIFERIN 2-MONOOXYGENASE		
SUBTILISIN		
TYROSINE PHENOL-LYASE	NC	$Ki = 36\ \mu M$

PHENOLPHTHALEIN
GLUTATHIONE TRANSFERASE
THYMIDYLATE SYNTHASE — $Ki = 4.4\ \mu M$

PHENOLPHTHALEIN DISULPHATE
IDURONATE 2-SULFATASE — $I50 = 2$ mM

PHENOL RED
GLUTATHIONE TRANSFERASE
1-D-myo-INOSITOL-TRISPHOSPHATE 3-KINASE — $I50 = 1.1$ mM
PHOSPHATIDYLINOSITOL 3-PHOSPHATASE — $I50 = 100\ \mu M$

PHENOLS
D-AMINO-ACID OXIDASE

PHENOLTETRABROMOPHENOLPHTHALEIN
GLUTATHIONE TRANSFERASE

PHENOLTETRABROMOPHENOLPHTHALEIN DISULPHONATE
GLUTATHIONE TRANSFERASE

PHENOLTHYMOLPHTHALEIN
THYMIDYLATE SYNTHASE — $I50 = 7\ \mu M$

PHENOTHIAZINE
3′,5′-CYCLIC-NUCLEOTIDE PHOSPHODIESTERASE

PHENOTHIAZINES
1-AMINOCYCLOPROPANE-1-CARBOXYLATE SYNTHASE

PHENOTHIAZINE SULPHONAMIDE
PYRUVATE DEHYDROGENASE (LIPOAMIDE)

PHENOXAN
NADH DEHYDROGENASE (UBIQUINONE) — $I50 = 5.8$ nM
RNA-DIRECTED DNA POLYMERASE — $I50 = 376\ \mu M$

PHENOXAZINE
ARACHIDONATE 5-LIPOXYGENASE — $I50 = 20$ nM

PHENOXAZINE-2-PROPENOIC ACID ETHYLESTER
ARACHIDONATE 5-LIPOXYGENASE — $I50 = 35$ nM

7-(PHENOXYACETAMIDO)-3-DESACETOXYCEPHALOSPORAMIC ACID
D-ALANYL-D-ALANINE CARBOXYPEPTIDASE

PHENOXYACETATE
BENZOYLFORMATE DECARBOXYLASE — $Ki = 5.5$ mM

2-[(PHENOXYACETYL)AMINO]-3,5-DICHLOROBENZOIC ACID
3α-HYDROXYSTEROID DEHYDROGENASE (B-SPECIFIC) — 63% 500 μM

PHENOXYAMINOCYCLOTRIPHOSPHATRIENES
UREASE

1-(2-PHENOXYETHYL)-1H-IMIDAZOLE
CYCLOOXYGENASE — $I50 = 67\ \mu M$
STEROID 11β-MONOOXYGENASE — 72% 10 μM
THROMBOXANE A SYNTHASE — $I50 = 660$ nM

PHENOXY-4H-1,3,2-BENZODIOXAPHOSPHORIN-2-OXIDE
GLUTATHIONE TRANSFERASE — IR

PHENOXYL RADICAL
NADPH-FERRIHEMOPROTEIN REDUCTASE — IR

1-PHENOXYMETHYL-3,4-DIHYDROISOQUINOLINE
SIALIDASE

PHENOXYMETHYLPENICILLIN
D-ALANYL-D-ALANINE CARBOXYPEPTIDASE

5′-[4-(PHENOXYSULPHONYL)BENZOYL]ADENOSINE
PHOSPHORYLASE KINASE — *I*50 = 80 μM
PROTEIN TYROSINE KINASE p60v-abl — *I*50 = 170 μM

PHENPROCOUMON
PROTEINASE HIV-1 — *K*i = 1 μM

17β-PHENPROPYLANDROST-4-ENE-3,17-DIONE
AROMATASE — *K*i = 13 nM

PHENSERINE
ACETYLCHOLINESTERASE

PHENTHOATE
ACETYLCHOLINESTERASE

PHENYETHYL-β-D-THIOGALACTOPYRANOSIDE
β-GALACTOSIDASE — *K*i = 1.5 μM

PHENYLACETALDEHYDE
INDOLE-3-ACETALDEHYDE OXIDASE

PHENYLACETALDOXIME
INDOLEACETALDOXIME DEHYDRATASE — C — *K*i = 22 nM

PHENYLACETAMIDINE
TRYPSIN — C — *K*i = 15.1 mM

PHENYL 5-ACETAMIDO-3,5-DIDEOXY-D-GLYCERO-α-D-GALACTO-2-SELENO-NONULOPYRANOSIDONIC ACID
SIALIDASE — *K*i = 1 mM

PHENYLACETAMIDOMETHANE BORONIC ACID
β-LACTAMASE — *K*i = 256 μM
β-LACTAMASE I

N[1]-PHENYLACETAMIDRAZON HI
LIPOXYGENASE — *I*50 = 35 nM

PHENYLACETIC ACID
CARBOXYPEPTIDASE A — C — *K*i = 390 μM
CATECHOL OXIDASE ppo — *I*50 = 1.3 mM
β-(9-CYTOKININ)-ALANINE SYNTHASE
DIPHOSPHOMEVALONATE DECARBOXYLASE — 28% 2.5 mM
2-ENOATE REDUCTASE
(S)-2-HYDROXY-ACID OXIDASE — C — *K*i = 2.5 mM
IMIDAZOLEACETATE 4-MONOOXYGENASE
INDOLE-3-ACETALDEHYDE OXIDASE
[3-METHYL-2-OXOBUTANOATE DEHYDROGENASE (LIPOAMIDE)] KINASE
PENICILLIN AMIDASE — C — *K*i = 5.3 mM
PHENYLALANINE 2-MONOOXYGENASE
PYRUVATE CARBOXYLASE
SERINE CARBOXYPEPTIDASE
TYROSINE TRANSAMINASE

PHENYLACETYLALANINE
NEPRILYSIN — *I*50 = 84 μM
THERMOLYSIN — *I*50 = 266 μM

PHENYLACETYL-ARG-pNA
KALLIKREIN TISSUE — *K*i = 400 nM

PHENYLACETYLATROPINE
HYOSCYAMINE (6S)-DIOXYGENASE

Inhibitor / Enzyme	Type	Value
PHENYLACETYL-CoA		
GLYCINE ACYLTRANSFERASE	C	$Ki = 2\ \mu M$
GLYCINE BENZOYLTRANSFERASE	C	
N6-PHENYLADENOSINE		
ADENOSINE KINASE	C	$Ki = 800$ nM
DEOXYCYTIDINE KINASE		
PHENYLALANINE		
ALANINE AMINOPEPTIDASE	NC	$I50 = 3.1$ mM
AMINOPEPTIDASE B	NC	$I50 = 25$ mM
ARGARATINE γ-GLUTAMYLTRANSFERASE		
ASPARAGINE-OXO-ACID TRANSAMINASE		
BENZOATE 4-MONOOXYGENASE		
CHORISMATE MUTASE 1		55% 170 μM
CHORISMATE MUTASE 3		50% 170 μM
2-DEHYDRO-3-DEOXYPHOSPHOHEPTONATE ALDOLASE		$Ki = 270\ \mu M$
2-DEHYDRO-3-DEOXYPHOSPHOHEPTONATE ALDOLASE		87% 1 mM
2-DEHYDRO-3-DEOXYPHOSPHOHEPTONATE ALDOLASE	C	$Ki = 7\ \mu M$
DIHYDROPTERIDINE REDUCTASE		
DIPHOSPHOMEVALONATE DECARBOXYLASE		20% 2.5 mM
GLUTAMATE SYNTHASE (NADH)		
4-HYDROXYPHENYLPYRUVATE DIOXYGENASE		
LEUCYL AMINOPEPTIDASE	NC	$I50 = 25$ mM
MONOPHENOL MONOOXYGENASE		
Na^+/K^+-EXCHANGING ATPASE		21% 1.2 mM
PEPTIDYL-DIPEPTIDASE A		100% 500 μM
PHENYLALANINE AMMONIA-LYASE		
PHOSPHATASE ALKALINE		72% 5 mM
PHOSPHATASE ALKALINE	U	$Ki = 610\ \mu M$
3-PHYTASE	C	
PREPHENATE DEHYDRATASE		
PREPHENATE DEHYDROGENASE		
PROTEIN TYROSINE-PHOSPHATASE		
PYRUVATE KINASE		
TRITRIRACHIUM ALKALINE PROTEINASE		$Ki = 2.7$ mM
TRYPTOPHAN 5-MONOOXYGENASE		
TYROSINE PHENOL-LYASE		
D-PHENYLALANINE		
AMINOACYLASE		76% 100 mM
CARBOXYPEPTIDASE A		$Ki = 120\ \mu M$
CARBOXYPEPTIDASE A		
NEPRILYSIN		
PHENYLALANINE AMMONIA-LYASE		24% 4 mM
PHENYLALANINE AMMONIA-LYASE	C	$Ki = 670\ \mu M$
PHENYLALANINE AMMONIA-LYASE		$Ki = 750\ \mu M$
PHENYLALANINE AMMONIA-LYASE	C	$Ki = 840\ \mu M$
PHENYLALANINE AMMONIA-LYASE		39% 500 μM
PHENYLALANINE DEHYDROGENASE		
PHOSPHATASE ALKALINE	U	$Ki = 48$ mM
SERINE CARBOXYPEPTIDASE		$Ki = 24$ mM
L-PHENYLALANINE		
CARBOXYPEPTIDASE A	C	$Ki = 5$ mM
CATECHOL OXIDASE	C	$Ki = 2$ mM
CHORISMATE MUTASE		$Ki = 37\ \mu M$
GLUTAMINE-PHENYLPYRUVATE TRANSAMINASE		90% 100 μM
(S)-2-HYDROXY-ACID OXIDASE		
LEUCYL AMINOPEPTIDASE	C	
PHENYLALANINE 4-MONOOXYGENASE		> 80 μM
PREPHENATE DEHYDRATASE		
PRETYROSINE DEHYDROGENASE	C	$I50 = 70\ \mu M$
PYRUVATE KINASE (L TYP)		
SERINE CARBOXYPEPTIDASE		$Ki = 3.9$ mM

TRYPTOPHANASE	C	Ki = 3.2 mM
TYROSINE 3-MONOOXYGENASE		
PHENYLALANINEPHOSPHONATE		
CARBOXYPEPTIDASE A		KI = 1 μM
PHENYLALANINOL		
TRITRIRACHIUM ALKALINE PROTEINASE		Ki = 400 μM
L-PHENYLALANINOL-AMP		
PHENYLALANINE-tRNA LIGASE	C	Ki = 1 μM
PHENYLALANINE-tRNA LIGASE	C	Ki = 2.5 μM
L-PHENYLALANINOL		
PHENYLALANINE-tRNA LIGASE	C	Ki = 6 μM
PHENYLALANYLCHLOROMETHYLKETONE		
TYROSYL AMINOPEPTIDASE		80% 1 mM
N-PHENYLALANYL-O-4-NITROBENZYL HYDROXAMATE		
CATHEPSIN H	IR	
LEUCYL AMINOPEPTIDASE		Ki = 40 μM
MEMBRANE ALANINE AMINOPEPTIDASE		Ki = 102 μM
PHENYLALKYLOXYRANECARBOXYLATE		
CARNITINE O-PALMITOYLTRANSFERASE		
2-PHENYLALLYLAMINE		
DOPAMINE β-MONOOXYGENASE	SSI	
3-PHENYL-4-AMINOBUTANOIC ACID		
4-AMINOBUTYRATE TRANSAMINASE		
γ-PHENYL-α-AMINOBUTYRIC ACID		
TRYPTOPHANASE		
N-PHENYL-(2-AMINOETHYL)CARBAMATE		
AMINE OXIDASE MAO-B	IR	Ki = 1 μM
1-PHENYL-1-AMINOMETHYLETHENE		
DOPAMINE β-MONOOXYGENASE	SSI	
1-PHENYL-1-(AMINOMETHYL)ETHENE HCl		
DOPAMINE β-MONOOXYGENASE	SSI	
7-PHENYL-1,4,6-ANDROSTATRIENE-3,17-DIONE		
AROMATASE	SSI	Ki = 172 nM
PHENYLARSENOXIDE		
TRYPANOTHIONE REDUCTASE		89% 50 μM
PHENYLARSINE OXIDE		
CASPASES		
DIHYDROLIPOAMIDE DEHYDROGENASE		
DIPHOSPHOMEVALONATE DECARBOXYLASE	IR	98% 25 μM
FLAVOCYTOCHROME b		
GLUCOSE TRANSPORT INSULIN STIMULATED		
3-HYDROXYBUTYRATE DEHYDROGENASE		
MORPHINE 6-DEHYDROGENASE		
NADPH DEHYDROGENASE (QUINONE)		
NAD(P) TRANSHYDROGENASE (AB-SPECIFIC)		
NUCLEOSIDE-TRIPHOSPHATASE		
1-PHOSPHATIDYLINOSITOL 3-KINASE		I50 = 20 μM
PROTEIN TYROSINE-PHOSPHATASE CD45		23% 100 μM
trans-N-(4-PHENYLAZO)BENZOYL)AMINOACETALDEHYDE		
PAPAIN	C	Ki = 2.1 μM

trans-N-(4-PHENYLAZO)PHENYL)AMINOBENZENEBORONIC ACID

CHYMOTRYPSIN	C	
SUBTILISIN	C	

N-(1-PHENYL-2-BENZIMIDAZOLYL)-N′-(2,6-DICHLOROPHENYL)UREA

STEROL O-ACYLTRANSFERASE		*I*50 = 11 nM

N-(4-PHENYLBENZOYL)-sn-(2(3-AMIDINOBENZYL)-3-(trans-STYRYL)-β-ALANINE METHYL ESTER

COAGULATION FACTOR Xa		*I*50 = 110 nM

3-PHENYL-1,5-BISTHIOGLUTARIMIDE

DOPAMINE β-MONOOXYGENASE	C	*K*i = 48 nM

N-[3-PHENYL-4,5-BIS[(TRIFLUOROMETHYL)IMINO]-2-THIAZOLIDINYLIDENE]BENZENEAMINE

CHOLINE O-ACETYLTRANSFERASE		

PHENYLBORIC ACID

UREASE	C	*K*i = 12 mM

PHENYLBORONIC ACID

CUTINASE	C	*K*i = 115 μM
LIPOPROTEIN LIPASE	C	*K*i = 3.6 mM
STEROL ESTERASE	C	
TRIACYLGLYCEROL LIPASE	C	*K*i = 75 μM

PHENYLBUTANOYL-PRO-ALA-THIAZOLE

PROLYL OLIGOPEPTIDASE		*I*50 = 5 nM

PHENYLBUTANOYL-PRO-ARG-THIAZOLE

PROLYL OLIGOPEPTIDASE		*I*50 = 28 μM
TRYPSIN		*I*50 = 230 nM

PHENYLBUTANOYL-PRO-LYS-THIAZOLE

PROLYL OLIGOPEPTIDASE		*I*50 = 7.8 μM
THROMBIN		*I*50 = 46 μM
TRYPSIN		*I*50 = 2.4 μM

PHENYLBUTANOYL-PRO-PRO-THIAZOLE

PROLYL OLIGOPEPTIDASE		*I*50 = 4.4 nM

PHENYLBUTANOYL-PRO-VAL-THIAZOLE

PROLYL OLIGOPEPTIDASE		*I*50 = 7.7 μM

2[[1[[1(4-PHENYLBUTANOYL)2(S)PYRROLIDINYL]CARBONYL2(S)PYRROLIDINYL]CARBOXYL]THIAZOLINE

PROLYL OLIGOPEPTIDASE		*I*50 = 3.8 nM

PHENYLBUTAZONE

ADENOSINE DEAMINASE		*K*i = 54 μM
D-AMINO-ACID OXIDASE		
CARBONYL REDUCTASE (NADPH)	NC	
CYCLOOXYGENASE		*I*50 = 79 μM
CYCLOOXYGENASE 1		*I*50 = 16 μM
CYTOCHROME P450 2C	C	*K*i = 572 μM
ELASTASE LEUKOCYTE		
11β-HYDROXYSTEROID DEHYDROGENASE		*I*50 = 1.4 mM
PROSTAGLANDIN-ENDOPEROXIDE SYNTHASE		*I*50 = 7 μM
UDP-GLUCURONOSYLTRANSFERASE		
URATE-RIBONUCLEOTIDE PHOSPHORYLASE		
URATE-RIBONUCLEOTIDE PHOSPHORYLASE	C	*K*i = 1.8 mM

4-PHENYL-3-BUTENOIC ACID

GLY-X CARBOXYPEPTIDASE		100% 10 mM
PEPTIDYLGLYCINE MONOOXYGENASE	SSI	*K*i = 1 μM

4-PHENYL-1-BUTYNE

CYTOCHROME P450 2b-10	SSI	*K*i = 3.9 μM

3-PHENYLBUTYRIC ACID
CARBOXYPEPTIDASE A C $Ki = 1.13$ mM

4-PHENYLBUTYRIC ACID
CARBOXYPEPTIDASE A

10-PHENYLBUTYRYLANTHRALIN
ARACHIDONATE 5-LIPOXYGENASE $I50 = 300$ nM

1-{N-(4-PHENYLBUTYRYL)-THIOPROLYL}-PYRROLIDINE
PROLYL OLIGOPEPTIDASE $I50 = 67$ μM

4-PHENYLBUTYRYL-THIOPROLYLPYRROLIDINE
PROLYL OLIGOPEPTIDASE $I50 = 67$ nM

7-[(N-PHENYLCARBAMOYL)AMINO]-4-CHLORO-3-METHOXYCOUMARIN
ELASTASE PANCREATIC

N-(N-[4-(4-PHENYLCARBOXAMIDOPHENYL)BUTYRYL]PROLYL)PYROLIDONE
PROLYL OLIGOPEPTIDASE C $I50 = 6.3$ nM
PROLYL OLIGOPEPTIDASE C $I50 = 4.4$ nM
PROLYL OLIGOPEPTIDASE C $I50 = 3.2$ nM

3-(1-PHENYL-4-CARBOXYPYRAZOL-5-YL)-6-CHLORO-1,2,3-BENZOTRIAZIN-4(3H)-ONE
3α-HYDROXYSTEROID DEHYDROGENASE (B-SPECIFIC) $I50 = 8$ μM

4′-PHENYLCHALCONE
GLUTATHIONE TRANSFERASE

4′-PHENYLCHALCONE OXIDE
GLUTATHIONE TRANSFERASE

4-PHENYLCHALCONE OXIDE
EPOXIDE HYDROLASE $I50 = 600$ nM

PHENYL-β-D-CHITOBIOSIDE
LYSOZYME

(E)-2-PHENYL-3-CHLOROALLYLAMINE
AMINE OXIDASE VASCULAR SSI

2-PHENYLCHROMONE (FLAVONE)
NAD ADP-RIBOSYLTRANSFERASE $I50 = 22$ μM

PHENYLCYCLIDINE
NITRIC-OXIDE SYNTHASE SSI $Ki = 4.9$ mM

1-PHENYLCYCLOBUTYLAMINE
AMINE OXIDASE IR

1-PHENYLCYCLOPENTANE CARBOXYLIC ACID 2-DIETHYL AMINOETHYL ESTER
ACETYLCHOLINESTERASE $I50 = 2.5$ mM
CHOLINESTERASE $I50 = 5$ μM

PHENYLCYCLOPROPYLAMINE
BUFURALOL-1′-HYDROXYLASE (CYT P-450dbl) NC $I50 = 17$ μM

1-PHENYLCYCLOPROPYLAMINE
AMINE OXIDASE MAO

S-PHENYLCYSTEINE
KYNURENINASE C $Ki = 700$ μM

S-PHENYLCYSTEINE S, S-DIOXIDE
KYNURENINASE C $Ki = 3.9$ μM

1-PHENYL-2-DECANOYLAMINO-3-MORPHOLINO-1-PROPANOL
CERAMIDE GLUCOSYLTRANSFERASE 90% 80 μM

D-threo-1-PHENYL-2-DECANOYLAMINO-3-MORPHOLINO-1-PROPANOL
CERAMIDE GLUCOSYLTRANSFERASE 20% 5 μM
UDP-GLUCOSE: CERAMIDE GLUCOSYLTRANSFERASE

threo-1-PHENYL-2-DECANOYLAMINO-3-MORPHOLINO-1-PROPANOL
CERAMIDE GLUCOSYLTRANSFERASE

N2-PHENYL-2′-DEOXYGUANOSINE
THYMIDINE KINASE C *K*i = 300 nM

p-PHENYLDEOXYMANNONOJIRIMICIN
β-N-ACETYLHEXOSAMINIDASE 50% 1 mM

1(S)-PHENYL-1,4-DIDEOXY-1,4-IMINORIBITOL
NUCLEOSIDE HYDROLASE *K*i = 170 nM

7-PHENYL-2,3-DIHYDROSPIRO[4H-1-BENZOPYRAN-4,4′-IMIDAZOLIDINE]-2′,5′-DIONE
ALDEHYDE REDUCTASE

1-PHENYL-3-DIMETHYLAMINOPROPYL CARBODIIMIDE
THYMIDYLATE SYNTHASE IR

3-PHENYL-6,7-DIMETHYLQUINOXALINE
PROTEIN TYROSINE KINASE PDGFR *I*50 = 400 nM

S,S′-(1,3-PHENYLENEBIS(1,2-ETHANEDIYL))BISISOTHIOUREA
NITRIC-OXIDE SYNTHASE II *K*i = 47 pM
NITRIC-OXIDE SYNTHASE III *K*i = 9 nM

m-PHENYLENEDIAMINE
AMINE OXIDASE (COPPER-CONTAINING)

o-PHENYLENEDIAMINE
AMINE OXIDASE (COPPER-CONTAINING)
THIOL OXIDASE

p-PHENYLENEDIAMINE
ORNITHINE DECARBOXYLASE 92% 1 mM

2-PHENYLETHANEBORONIC ACID
CHYMOTRYPSIN *K*i = 40 μM

PHENYLETHANOLAMINE
THIOL S-METHYLTRANSFERASE *I*50 = 6.9 mM

PHENYLETHYLAMINE
DIPHOSPHOMEVALONATE DECARBOXYLASE 20% 2.5 mM
PHENYLALANINE DEHYDROGENASE
RENIN
THIOL S-METHYLTRANSFERASE *I*50 = 780 μM
TYROSINE TRANSAMINASE

β-PHENYLETHYLAMINE
PHENYLALANINE AMMONIA-LYASE 41% 500 μM

1-PHENYLETHYLAMINE
PHENYLALANINE RACEMASE (ATP-HYDROLYSING) C *K*i = 260 μM

2-PHENYLETHYLAMINE
PHENYLALANINE RACEMASE (ATP-HYDROLYSING) *K*i = 150 μM

PHENYLETHYLBIGUANIDE
PYRUVATE KINASE
UBIQUINOL-CYTOCHROME-C REDUCTASE

N-PHENYLETHYLENEDIAMINE
DOPAMINE β-MONOOXYGENASE *K*i = 1.1 mM

PHENYLETHYLISOCYANATE

CYTOCHROME P450 2D6		

1-PHENYLETHYL-3-METHYLXANTHINE

PHOSPHATASE ALKALINE		46% 80 μM

N-(PHENYLETHYLPHOSPHONYL)-GLY-L-PRO-L-AMINOHEXANOIC ACID

NEUROLYSIN		*K*i = 900 pM
THIMET OLIGOPEPTIDASE		*K*i = 7.5 nM

PHENYLETHYL-β-D-THIOGALACTOPYRANOSIDE

β-GALACTOSIDASE		*K*i = 1.5 μM
LACTASE		*K*i = 1.2 mM

2-PHENYLETHYLUREIDOPURINE RIBOSIDE

ADENOSINE KINASE		

(E)-2-PHENYL-3-FLUOROALLYLAMINE

AMINE OXIDASE MAO-B		*I*50 = 2.5 μM

PHENYL β-D-GLUCOPYRANOSIDE

XYLAN 1,4-β-XYLOSIDASE		*K*i = 75 mM

PHENYL-α-D-GLUCOSIDE

GLUCAN 1,6-α-GLUCOSIDASE		
4-α-GLUCANOTRANSFERASE	C	*K*i = 10 mM
α-GLUCOSIDASE	C	*K*i = 400 μM
OLIGO-1,6-GLUCOSIDASE		58% 10 mM

β-PHENYLGLUCOSIDE

THIOGLUCOSIDASE		

PHENYL-β-GLUCOSIDE

α,α-TREHALASE		

PHENYL-β-D-GLUCOSIDE

CELLULASE		
4-α-GLUCANOTRANSFERASE	C	*K*i = 30 mM
β-GLUCOSIDASE	C	*K*i = 5.7 mM

(R)-(+)-PHENYLGLYCERATE

MANDELATE RACEMASE	C	*K*i = 1.3 mM

PHENYLGLYCINAMIDE

AMINE OXIDASE (COPPER-CONTAINING)		*I*50 = 18 mM
PHOSPHATASE ALKALINE	UC	*K*i = 50 mM

L-PHENYLGLYCINE

PHENYLALANINE DEHYDROGENASE		

PHENYLGLYOXAL

ACETYL-CoA CARBOXYLASE		
ACETYL-CoA HYDROLASE		
ALCOHOL DEHYDROGENASE (NADP)		
4-AMINOBUTYRATE TRANSAMINASE	IR	
ARYL SULFOTRANSFERASE		
BISPHOSPHOGLYCERATE PHOSPHATASE		
CARBONYL REDUCTASE (NADPH)		
DIAMINE N-ACETYLTRANSFERASE		
DI-trans,poly-cis-DECAPRENYLCISTRANSFERASE		
DIHYDROFOLATE REDUCTASE		
ENOYL-[ACYL-CARRIER-PROTEIN] REDUCTASE (NADH)	IR	
ENOYL-[ACYL-CARRIER-PROTEIN] REDUCTASE (NADPH, B SPECIFIC)		
trans-2-ENOYL-CoA REDUCTASE (NAD)		
GERANYLTRANSFERASE		
GLUCURONATE REDUCTASE		
GLUTAMATE DECARBOXYLASE	IR	
GLUTATHIONE REDUCTASE (NADPH)		

GLUTATHIONE TRANSFERASE
GLYCINE C-ACETYLTRANSFERASE
H^+-TRANSPORTING ATPASE
HYALURONOGLUCURONIDASE
4-HYDROXYBENZOATE 3-MONOOXYGENASE
3-HYDROXYBUTYRATE DEHYDROGENASE
15-HYDROXYPROSTAGLANDIN DEHYDROGENASE (NAD)
D-LACTALDEHYDE DEHYDROGENASE
LACTATE 2-MONOOXYGENASE
L-LYSINE 6-TRANSAMINASE
LYSOPHOSPHOLIPASE
MALATE DEHYDROGENASE (OXALACETATE-DECARBOXYLATING) (NADP)
MALONYL-CoA DECARBOXYLASE
METHYLCROTONYL-CoA CARBOXYLASE
METHYLENETETRAHYDROFOLATE REDUCTASE (NADPH)
NEPRILYSIN
NITRATE REDUCTASE (NADPH)
3-OXOACYL-[ACYL-CARRIER-PROTEIN] REDUCTASE
3-OXOACYL-[ACYL-CARRIER-PROTEIN] REDUCTASE
PANTOATE DEHYDROGENASE
PHENYLALANINE RACEMASE (ATP-HYDROLYSING)
PHOSPHATASE ACID — C — *K*i = 360 μM
PHOSPHATASE ALKALINE
PHOSPHOGLYCERATE MUTASE — *K*i = 50 μM
3-PHOSPHOSHIKIMATE 1-CARBOXYVINYLTRANSFERASE
PYRIDOXAMINE-PHOSPHATE OXIDASE
PYRUVATE DEHYDROGENASE (CYTOCHROME)
RIBONUCLEASE T1
RIBULOSE-BISPHOSPHATE CARBOXYLASE
RNA-DIRECTED DNA POLYMERASE
SIALATE -9-O-ACETYLESTERASE
SULFATE ADENYLYLTRANSFERASE
THIOSULFATE SULFURTRANSFERASE
TRANSKETOLASE
UROPORPHYRINOGEN-III SYNTHASE

PHENYLGLYOXAL BIS(GUANYLHYDRAZONE)
ADENOSYLMETHIONINE DECARBOXYLASE — *I*50 = 65 μM
AMIN OXODASE DIAMINE — *K*i = 120 nM

PHENYL 6-GUANIDINO-1-NAPHTOATE
TRYPSIN — *I*50 = 50 nM

PHENYLGUANINE
dTMP KINASE

N2-PHENYL-2′-GUANINE
THYMIDINE KINASE — C — *I*50 = 8 μM

PHENYLHYDRAZINE
ADENOSYLMETHIONINE-8-AMINO-7-OXONONANOATE TRANSAMINASE
ADENOSYLMETHIONINE DECARBOXYLASE
ALANINE TRANSAMINASE
D-ALANINE TRANSAMINASE
ALCOHOL DEHYDROGENASE — C — *K*i = 393 μM
ALCOHOL OXIDASE
AMINE OXIDASE (COPPER-CONTAINING)
AMINE OXIDASE (FLAVIN-CONTAINING)
AMINE OXIDASE MAO — SSI
AMINE OXIDASE (POLYAMINE)
4-AMINOBUTYRATE TRANSAMINASE
(2-AMINOETHYL)PHOSPHONATE-PYRUVATE TRANSAMINASE
8-AMINO-7-OXONONAOATE SYNTHASE
ARALKYLAMINE DEHYDROGENASE
ARGININE DECARBOXYLASE — 40% 50 mM
AROMATIC-AMINO-ACID TRANSAMINASE

ARYLACETONITRILASE		52% 1 mM
ASPARTATE 1-DECARBOXYLASE		
ASPARTATE 4-DECARBOXYLASE		
BRANCHED-CHAIN-AMINO-ACID TRANSAMINASE		
BROMELAIN		
CATECHOL OXIDASE		
3-CHLORO-D-ALANINE DEHYDROCHLORINASE		77% 10 mM
CYSTATHIONINE γ-LYASE		
2-DEHYDROPANTOATE 2-REDUCTASE		
DOPAMINE β-MONOOXYGENASE	SSI	
FORMATE DEHYDROGENASE		
GLUCOSE OXIDASE		
GLUTAMATE DECARBOXYLASE		
GLYCERATE DEHYDROGENASE		
GLYCINE C-ACETYLTRANSFERASE		
GLYCINE HYDROXYMETHYLTRANSFERASE		
GLYCINE TRANSAMINASE		
GLYOXYLATE REDUCTASE		
HISTIDINE AMMONIA-LYASE		
HISTIDINE DECARBOXYLASE		
4-HYDROXYGLUTAMATE TRANSAMINASE		
HYDROXYPYRUVATE REDUCTASE		77% 1 mM
INULINASE		
KYNURENINE-OXOGLUTARATE AMINOTRANSFERASE		81% 100 μM
L-LACTATE DEHYDROGENASE	C	Ki = 43 mM
LIGNIN PEROXIDASE	IR	
LIPOXYGENASE 1		
LIPOXYGENASE 2		
MANGANESE PEROXIDASE		
METHANOL OXIDASE		
MONOPHENOL MONOOXYGENASE		
NITRILASE		45% 1 1mM
NITRILE HYDRATASE		100% 1 mM
ORNITHINE-OXO-ACID TRANSAMINASE		
PAPAIN		
PHENYLALANINE AMMONIA-LYASE		
PHOSPHATIDYLSERINE DECARBOXYLASE		
PHOSPHOPANTOTHENOYLCYSTEINE DECARBOXYLASE		43% 10 mM
D-PROLINE REDUCTASE (DITHIOL)		
PROPANEDIOL DEHYDRATASE		
PROTEIN-LYSINE 6-OXIDASE		
SARCOSINE OXIDASE		
SERINE-GLYOXYLATE TRANSAMINASE		20% 1 mM
SPERMIDINE DEHYDROGENASE		
THETIN-HOMOCYSTEINE S-METHYLTRRANSFERASE		
TYROSINE 2,3-AMINOMUTASE		97% 1 mM
UROCANATE HYDRATASE		
XANTHINE OXIDASE		

6-(PHENYLHYDRAZINO)URACIL

DNA POLYMERASE III		

D,L-PHENYLHYDROXAMATE

BACTERIAL LEUCYL AMINOPEPTIDASE		Ki = 800 nM

3-PHENYL-7-HYDROXYFLAVONE

ALDEHYDE DEHYDROGENASE (NAD) 1	C	Ki = 1.5 μM
ALDEHYDE DEHYDROGENASE (NAD) 2	C	Ki = 1 μM

1-PHENYLIMIDAZOLE

NITRIC-OXIDE SYNTHASE (CALMODULIN DEPENDENT)		I50 = 25 μM

2-PHENYLIMIDAZOLE

NITRIC-OXIDE SYNTHASE (CALMODULIN DEPENDENT)		I50 = 160 μM

4-PHENYLIMIDAZOLE

AMINE OXIDASE (COPPER-CONTAINING)		$I50 = 1.2$ mM
α-AMYLASE		$Ki = 7.1$ mM
β-GLUCOSIDASE	C	$Ki = 6.2$ μM
INDOLEAMINE-PYRROLE 2,3-DIOXYGENASE	NC	$Ki = 4.4$ μM
NITRIC-OXIDE SYNTHASE (CALMODULIN DEPENDENT)		$I50 = 640$ μM

2-PHENYLINDANE-1,3-DIONE

ARACHIDONATE 5-LIPOXYGENASE		$I50 = 15$ μM

PHENYLIODOAETIC ACID

LOMBRICINE KINASE

4-PHENYL-5(E)-(IODOMETHYLIDENE)TETRAHYDRO-2-FURANONE

α-CHYMOTRYPSIN	C	$Ki = 49$ μM

PHENYLISOCYANIDE

STEROID 21-MONOOXYGENASE

PHENYLISOPROPYLHYDRAZINE

ALCOHOL DEHYDROGENASE

PHENYLISOTHIOCYANATE

EPOXIDE HYDROLASE		34% 1 mM
EPOXIDE HYDROLASE		54% 1 mM
PHOSPHATE CARRIER MITOCHONDRIA		

PHENYLLACTIC ACID

DIHYDROPTERIDINE REDUCTASE		
DIIODOTYROSINE AMINOTRANSFERASE		
DIPHOSPHOMEVALONATE DECARBOXYLASE		45% 2.5 mM
[3-METHYL-2-OXOBUTANOATE DEHYDROGENASE (LIPOAMIDE)] KINASE		
Na^{+}/K^{+}-EXCHANGING ATPASE		
PHENYLALANINE AMMONIA-LYASE	C	
STEAROYL-CoA DESATURASE		

β-PHENYLLACTIC ACID

SERINE CARBOXYPEPTIDASE		$Ki = 5.7$ mM

L-β-PHENYLLACTIC ACID

(S)-2-HYDROXY-ACID OXIDASE

N-PHENYLMALEIMIDE

EPOXIDE HYDROLASE		97% 1 mM
EPOXIDE HYDROLASE		60% 1 mM

PHENYL-α-MALTOSIDE

OLIGO-1,6-GLUCOSIDASE		$Ki = 300$ μM

PHENYLMATHANESULPHONYL FLUORIDE

CATHEPSIN G	IR	
CHYMOTRYPSIN	IR	

PHENYLMERCURIC ACETATE

α-AMYLASE
FORMATE DEHYDROGENASE
INORGANIC PYROPHOSPHATASE
ISOAMYLASE

PHENYLMERCURICACETATE

β-FRUCTOFURANOSIDASE

PHENYLMERCURIC ACID

GLUTATHIONE TRANSFERASE

PHENYLMERCURIC CHLORIDE

α-AMYLASE

PHENYLMERCURIC NITRATE
GLUCAN ENDO-1,3-β-GLUCOSIDASE 39% 200 μM
GLUCAN 1,3-β-GLUCOSIDASE

PHENYLMETHANE
STEROL ESTERASE

PHENYLMETHANESULPHONAMIDE
BIOTINIDASE 99% 100 nM

O-[[(1r)-n-[-n-(PHENYLMETHOXYCARBONYL)-L-ALANYL]-1-AMINOETHYL]HYDROXYPHOSPHINYL-L-3-PHENYL LACTIC ACID
CARBOXYPEPTIDASE A

(2S,4S,5S)2[[[(PHENYLMETHOXY)CARBONYL]L-VALINYL]AMINO5[[(1,1-DIMETHYLETHOXY)CARBONYL]AMINO]-1,6-DIPHENYL-4-HYDROXYHEXANE
PROTEINASE HIV-1 *I*50 = 3.1 nM

5-[4-(PHENYLMETHOXY)PHENYL]-2-(2-CYANOMETHYL)TETRAZOLE
AMINE OXIDASE MAO-A *I*50 = 86 μM
AMINE OXIDASE MAO-B *I*50 = 2 nM

(2R,4S,5S,1′S)-2-PHENYLMETHYL-4-HYDROXY-5-(t-BOC)AMINO-6-PHENYLHEXANOYL-N-(1′-IMIDAZO-2-YL)-2′-METHYLPROPANAMIDE
PROTEINASE HIV-1 *K*i = 18 nM

3-[1-(PHENYLMETHYL)-4-PIPERIDINYL]-1-(2,3,4,5-TETRAHYDRO-1H-1-BENZAZEPIN-8-YL)-1-PROPANONE FUMARATE
ACETYLCHOLINESTERASE *I*50 = 98 nM

PHENYLMETHYLSULPHONYLAZIDE
PENICILLIN AMIDASE *K*i = 7 μM

PHENYLMETHYLSULPHONYLCHLORIDE
PENICILLIN AMIDASE

PHENYLMETHYLSULPHONYLFLUORIDE
[ACYL-CARRIER-PROTEIN] S-MALONYLTRANSFERASE
AGKISTRODON SERINE PROTEINASE
ALDEHYDE DEHYDROGENASE (NAD) 50% 100 μM
AMINOPEPTIDASE 56% 1 mM
AMINOPEPTIDASE Y 31% 1 mM
ARYLESTERASE
ARYLFORMAMIDASE 100% 1 mM
ARYLFORMAMIDASE
BIOTINIDASE
CARBONATE DEHYDRATASE
CARBOXYLESTERASE 75% 1 mM
CARBOXYLESTERASE A *I*50 = 100 nM
CARBOXYLESTERASE B *I*50 = 100 μM
CATHEPSIN B
CATHEPSIN L 68% 100 μM
CHOLINE O-ACETYLTRANSFERASE
CHYMASE IR
CHYMOTRYPSIN
COAGULATION FACTOR XIIa
COMPLEMENT SUBCOMPONENT C1r
CUCUMISIN 45% 500 μM
DIPEPTIDASE
DIPEPTIDYL-PEPTIDASE 2 51% 1 mM
DIPEPTIDYL-PEPTIDASE 3
DIPEPTIDYL-PEPTIDASE 4
DIPEPTIDYL-PEPTIDASE 4 47% 1 mM
ELASTASE 98% 1 mM
ELASTASE LEUKOCYTE
ELASTASE PANCREATIC IR
ELASTASE PANCREATIC II 50% 1 mM
ENVELYSIN
EPOXIDE HYDROLASE 27% 1 mM
ESTERASE 45% 10 mM

EUPHORBAIN		
FATTY ACID SYNTHASE		
FICAIN		
FORMALDEHYDE DEHYDROGENASE		
β-GALACTOSIDASE		
γ-GLUTAMYLTRANSFERASE	IR	
γ-GLUTAMYLTRANSFERASE	IR	
HUMICOLIN		
HYDROGENASE II		
HYDROXYBUTYRATE-DIMER HYDROLASE		94% 100 μM
IGA-SPECIFIC METALLOENDOPEPTIDASE 1		*I*50 = 3 mM
INORGANIC PYROPHOSPHATASE		42% 1 mM
INTERSTITIAL COLLAGENASE		
LEUKOCYTE-MEMBRANE NEUTRAL ENDOPEPTIDASE		100% 1 mM
LIPOAMIDASE		36% 50 μM
LIPOPROTEIN LIPASE	QUI	*K*i = 1 μM
LYSOPHOSPHOLIPASE		
MALATE DEHYDROGENASE		
NITRIC-OXIDE REDUCTASE		
OLEYL-[ACYL-CARRIER-PROTEIN] HYDROLASE		
PALMITOYL-CoA HYDROLASE		
PANTOTHENASE		
PAPAIN		
PENICILLIN AMIDASE		*K*i = 2.2 μM
PHORBOL-DIESTER HYDROLASE		*I*50 = 3 μM
PHOSPHATIDYLCHOLINE-RETINOL O-ACYLTRANSFERASE		
1-PHOSPHOFRUCTOKINASE	IR	
PROLINE CARBOXYPEPTIDASE		
PROLINE-β-NAPHTHYLAMIDASE		67% 100 μM
PROLYL AMINOPEPTIDASE		
PROTEINASE ALKALINE		50% 5 mM
PROTEINASE ASPARTIC PLASMODIUM		
PROTEINASE I ACHROMOBACTER		
PROTEINASE PORPHYROMONAS GINGIVALIS TRYPSIN LIKE		
PROTEINASE SERINE ARTHROBACTER		
PROTEINASE SERINE DICHELOBACTER NODOSUS BASIC		100% 1 mM
PROTEINASE SERINE PROMONOCYTOC U-937		89% 2 mM
PROTEINASES SERINE		
PROTEINASE STARFISH SPERM ACROSOME		
α-PROTEINASE TENEBRIO		
PROTEINASE THERMUS Rt41A		92% 1 mM
RETINOL ISOMERASE		
RETINYL-PALMITATE ESTERASE		
SCOPULARIOPSIS PROTEINASE		
SEPIA PROTEINASE		
SEPIAPTERIN DEAMINASE		
SERINE CARBOXYPEPTIDASE		
SINAPOYLGLUCOSE-CHOLINE O-SINAPOYLTRANSFERASE		
STEROL ESTERASE		
SUBTILISIN		
TENEBRIO α-PROTEINASE		
TETANUS TOXIN L CHAIN (PEPDIDOLYTIC ACITIVITY)		100% 1 mM
THERMOMYCOLIN		
THERMOPHILIC STREPTOMYCES SERINE PROTEINASE		
THROMBIN	IR	
TRIACYLGLYCEROL LIPASE		
TRIPEPTIDE AMINOPEPTIDASE		29% 100 μM
TRIPEPTIDYL PEPTIDASE		90% 200 μM
TRYPSIN		95% 1 mM
TRYPTASE		63% 10 μg/ml
TRYPTASE TL-2		55% 10 μM
UCA PUGILATOR COLLAGENOLYTIC PROTEINASE		
XAA-HIS DIPEPTIDASE		38% 1 mM
YEAST PROTEINASE B		100% 1 mM

1-PHENYL-2-MYRISTOYLAMINO-3-MORPHOLINO-1-PROPANOL
GLYCOSYLCERAMIDE SYNTHASE

4-PHENYL-N-(2-NAPHTHALENEYLSULPHONYLOXY)-2-AZETIDINONE
β-LACTAMASE TEM-1β | IR

9-PHENYLNONANOIC ACID
CYTOCHROME P450 4A1

N-PHENYLOXAMIC ACID
SIALIDASE

2-PHENYLOXAZOLO-[5,4-d]PYRIMIDINE-7-ONE
HYPOXANTHINE PHOSPHORIBOSYLTRANSFERASE | C | Ki = 84 μM

2-PHENYL-PENEMS
D-ALANYL-D-ALANINE CARBOXYPEPTIDASE

4-[3(5-PHENYLPENTOXY)-4-METHOXYPHENYL]-2-METHYLBENZOIC ACID
3′,5′-CYCLIC-NUCLEOTIDE PHOSPHODIESTERASE IV | | I50 = 410 nM

5-PHENYL-1-PENTYNE
CYTOCHROME P450 2b-10 | SSI | Ki = 250 nM

1-PHENYLPERHYDRO-1,2,4-TRIAZIN-3-ONE
ARACHIDONATE 5-LIPOXYGENASE | | I50 = 21 μM

PHENYLPHOSPHATE
GLYCEROL-1,2-CYCLIC-PHOSPHATE PHOSPHODIESTERASE | | 58% 250 μM
RIBONUCLEASE PANCREATIC

PHENYLPHOSPHONATE
PHOSPHATASE ALKALINE | C

PHENYL-2-O-[2-PHOSPHONOETHYL]-β-D-GALACTOPYRANOSIDE GUANOSINE-5′-PHOSPHATE ANHYDRIDE
GALACTOSIDE 2-L-FUCOSYLTRANSFERASE | C | Ki = 16 μM

PHENYLPHOSPHORODIAMIDATE
UREASE | C | Ki = 95 nM

N-PHENYLPHOSPHORYL-L-PHENYLALANINE
CARBOXYPEPTIDASE A | | Ki = 2.1 μM

PHENYLPHOSPHORYL-PHE-PHE
THERMOLYSIN | | Ki = 3.8 μM

(R2) 3-PHENYLPROPANAL
α-CHYMOTRYPSIN | | Ki = 8.4 μM
SUBTILISIN | | Ki = 530 μM

N-3-PHENYLPROPANOYL-N-PHENYLETHYL)GLYCYLBOROLYSINE
THROMBIN

3-PHENYLPROPARGYLAMINE
DOPAMINE β-MONOOXYGENASE | SSI

3-PHENYLPROPENE
DOPAMINE β-MONOOXYGENASE | | Ki = 3.6 mM

1-PHENYL-2-PROPEN-1-ONE
3α-HYDROXYSTEROID DEHYDROGENASE (B-SPECIFIC) | IR | I50 = 91 μM

2-PHENYLPROP-2-ENYLAMINE
DOPAMINE β-MONOOXYGENASE | | Ki = 4.7 mM

PHENYLPROPIOLIC ACID
ACYL-CoA DEHYDROGENASE (NADP)
PHENYLALANINE AMMONIA-LYASE | | 95% 500 μM
PHENYLALANINE AMMONIA-LYASE | | 94% 1 mM
PHENYLALANINE AMMONIA-LYASE | NC | Ki = 24 μM

PHENYLALANINE AMMONIA-LYASE	C	$Ki = 3.8\ \mu M$
PHENYLALANINE AMMONIA-LYASE		61% 500 μM
PHENYLALANINE AMMONIA-LYASE		96% 1 mM
2-PHENYLPROPIONIC ACID		
CARBOXYPEPTIDASE A	C	$Ki = 62\ \mu M$
3-PHENYLPROPIONIC ACID		
CARBOXYPEPTIDASE A	C	$Ki = 100\ \mu M$
CARBOXYPEPTIDASE A		83% 1 mM
CARBOXYPEPTIDASE B		
CARBOXYPEPTIDASE C		
2-ENOATE REDUCTASE		
2-ENOATE REDUCTASE		68% 38 mM
PHENYLALANINE DEHYDROGENASE		
SERINE CARBOXYPEPTIDASE		$Ki = 4.5$ mM
3-(4-O-PHENYL)PROPIONOHYDROXAMIC ACID		
UREASE		$I50 = 87\ \mu M$
UREASE		$I50 = 1.8\ \mu M$
UREASE		$I50 = 13\ \mu M$
10-ω-PHENYLPROPIONYL-1,8-DIHYDROXY-9(10H)-ANTHRACENONE		
ARACHIDONATE 5-LIPOXYGENASE		$I50 = 300$ nM
ARACHIDONATE 12-LIPOXYGENASE		$I50 = 20\ \mu M$
β-PHENYLPROPIONYL-L-PHENYLALANINE		
THERMOLYSIN		
3-PHENYLPROPIONYL-PRO-BORO-L-LYS-OH		
THROMBIN		$Ki = 800$ pM
N-(3-PHENYLPROPIONYL)VAL-ALA-ASP-PHENYLBUTYLKETONE		
INTERLEUKIN CONVERTING ENZYME		$Ki = 42$ nM
7α-PHENYLPROPYLANDROSTA-1,4-DIENE-3,17-DIONE		
AROMATASE	SSI	$Ki = 7.7$ nM
S-n-PHENYLPROPYLGLUTATHIONE		
GLUTATHIONE TRANSFERASE		
4-(3′-PHENYLPROPYL)IMIDAZOLE		
β-GLUCOSIDASE		$Ki = 2.1\ \mu M$
N-(3-PHENYLPROPYLPHOSPHONYL)-GLY-PRO-L-2-AMINOHEXANOIC ACID		
COLLAGENASE CLOSTRIDIUM HISTOLYTICUM		$Ki = 60$ nM
PHENYLPROPYNAL		
β-LACTAMASE		
1-PHENYL-1-PROPYNE		
DOPAMINE β-MONOOXYGENASE	SSI	$Ki = 2.5$ mM
1-PHENYL-3-PYRAZOLIDONE		
ARACHIDONATE 5-LIPOXYGENASE		$I50 = 1\ \mu M$
2-PHENYLPYRIDINE		
β-GLUCOSIDASE	C	$Ki = 1.1$ mM
4-PHENYLPYRIDINE		
β-GLUCOSIDASE	C	$Ki = 4.4$ mM
2′-PHENYLPYRIMIDINE-5-SPIROCYCLOPROPANE-2,4,6-(1H, 3H, 5H)-TRIONE		
OROTATE REDUCTASE (NADH)		
PHENYLPYRUVIC ACID		
ACETOLACTATE SYNTHASE	C	$Ki = 110\ \mu M$
ALCOHOL DEHYDROGENASE (NADP)		
ALDEHYDE DEHYDROGENASE (NAD)	U	$Ki = 1.3$ mM

D-AMINO-ACID OXIDASE	C	Ki = 150 μM
CARBONYL REDUCTASE (NADPH)		
DIHYDROPTERIDINE REDUCTASE		
DIPHOSPHOMEVALONATE DECARBOXYLASE		50% 2.5 mM
4-HYDROXYPHENYLPYRUVATE DIOXYGENASE		
D-LACTATE DEHYDROGENASE (CYTOCHROME)		
L-LACTATE DEHYDROGENASE (CYTOCHROME)		
L-LACTATE DEHYDROGENASE		
3-METHYL-2-OXOBUTANOATE DEHYDROGENASE (LIPOAMIDE)	C	Ki = 1.7 mM
[3-METHYL-2-OXOBUTANOATE DEHYDROGENASE (LIPOAMIDE)] KINASE		
PHENYLALANINE AMMONIA-LYASE		Ki = 3 mM
PHENYLALANINE AMMONIA-LYASE		47% 500 μM
PYRUVATE CARRIER MITOCHONDRIA		
STEAROYL-CoA DESATURASE		
TYROSINE 2,3-AMINOMUTASE		40% 1 mM

3-PHENYLPYRUVIC ACID
CATHEPSIN D

2-PHENYLQUINAZOLINE-4,5,8(3HTRIONE
PURINE-NUCLEOSIDE PHOSPHORYLASE — IR

N-[(2-PHENYLQUINOLIN-4-YL)THIOCARBONYL]-N-METHYLGLYCINE
ALDEHYDE DEHYDROGENASE (NAD) — 91% 100 nM

N[1-PHENYL2(S)[[N(2QUINOLYLCARBONYL)ASPARAGINYL]AMINO]3(R)HYDROXYBUTAN4YL]PHE 2(R)HYDROXY-1(S)INDANYL)AMIDE
PROTEINASE HIV-1 — I50 = 5.4 nM

5-(PHENYLSELENYL)URACIL
URIDINE PHOSPHORYLASE — I50 = 4.8 μM

β-PHENYLD,L-SERINE
PHENYLALANINE AMMONIA-LYASE — 22% 6 mM

L-PHENYLSERINE
PHENYLALANINE AMMONIA-LYASE — 26% 1 mM

PHENYLSUCCINIC ACID
DICARBOXYLATE CARRIER MITOCHONDRIA

DL-PHENYLSUCCINIC ACID
CARBOXYPEPTIDASE A — C — Ki = 200 μM

2-(4-PHENYLSULPHONAMIDE)-3-(4-FLUOROPHENYL)INDENE
CYCLOOXYGENASE 1 — I50 = 7 nM
CYCLOOXYGENASE 2 — I50 = 5 nM

S-(PHENYLSULPHONYL)GLUTATHIONE
GLUTATHIONE TRANSFERASE 3-3 — 41% 100 μM

8-PHENYLTHEOPHYLLINE
3′,5′-CYCLIC-NUCLEOTIDE PHOSPHODIESTERASE (cycAMP)

1-PHENYL-3-(2-THIAZOLYL)-2-THIOUREA
DOPAMINE β-MONOOXYGENASE

N-PHENYL-N′-(2-THIAZOLYL)-THIOUREA
DOPAMINE β-MONOOXYGENASE

(3S,5R,6Z)6[1[[2(2-PHENYL-1,3,4-THIDIAZOL-5-YL)THIOETHYL-1,2,3-TRIATOL-4-YL]METHYLENE]PENICILLANATE 1,1-DIOXIDE
β-LACTAMASE — I50 = 4.6 μM
β-LACTAMASE — I50 = 400 nM
β-LACTAMASE TEM-2 — I50 = 50 nM

4-(PHENYLTHIO)-4-ANDROSTENEDIONE
AROMATASE — Ki = 35 μM

4-(PHENYLTHIO)-4-ANDROSTENE-3,17-DIONE
AROMATASE

(E)-3-[6-(4′-PHENYLTHIOBENZYL)-7-PHENYL-2,3-DIHYDRO-1H-PYRROLIZIN-5-YL]ACRYLIC ACID
ARACHIDONATE 5-LIPOXYGENASE — *I*50 = 2.8 μM
CYCLOOXYGENASE — *I*50 = 5.5 μM

PHENYLTHIOCYANATE
INDOLEACETALDOXIME DEHYDRATASE — 90% 100 nM

PHENYL-3-D-THIOGALACTOSIDE
β-GALACTOSIDASE — 90% 43 mM
β-GALACTOSIDASE — 1 mM

3-PHENYLTHIO-1-TRIFLUORO-2-PROPANONE
JUVENILE HORMONE ESTERASE — *I*50 = 32 μM

PHENYLTHIOUREA
AROMATIC-L-AMINO-ACID DECARBOXYLASE
MONOPHENOL MONOOXYGENASE

1-PHENYL-2-THIOUREA
CATECHOL OXIDASE

PHENYL 1-THIO-β-D-XYLOPYRANOSIDE
XYLAN 1,4-β-XYLOSIDASE — *K*i = 3.8 mM

S-PHENYL-p-TOLUENETHIOSULPHONATE
TRIOSE-PHOSPHATE ISOMERASE

2-PHENYL-6-(3-TRIFLUOROMETHYLPHENYL)-4-THIOMETHYLPYRIDINE
PHYTOENE DESATURASE

1-PHENYL-3-TRIMETHYLAMINOPROPYLCARBODIIMIDE
THYMIDYLATE SYNTHASE

L-PHENYL-2,3,7-TRITHIADECANE
LIPOXYGENASE — *I*50 = 9 μM

PHENYLUREA
UREASE

(7-PHENYLUREIDO)-4-CHLORO-3-(2-ISOTHIOUREIDOETHOXY)ISOCOUMARIN
GRANZYME A

PHENYTOIN
CYTOCHROME P450 2C — C — *K*i = 109 μM

PHEOPHORBIDE
MAGNESIUM PROTOPORPHYRIN O-METHYLTRANSFERASE
Mg^{2+} CHELATASE — *K*i = 920 μM

PHE-PHE
ELASTASE — *K*i = 1.5 mM

D-PHE-D-PHE(2E,2R)-2-AMINO-4-PHENYL-3-BUTENOIC ACID TRIFLUOROACETATE
PEPTIDYLGLYCINE MONOOXYGENASE — *I*50 = 300 μM

PHE-PHE-ARG-CHLOROMETHYLKETONE
LYS-GINGIVAIN — 85 % 4.4. μM

D-PHE-D-PHE METHYLESTER
CHYMOTRYPSIN — C — *K*i = 370 μM

D-PHE-L-PHE METHYLESTER
CHYMOTRYPSIN — C — *K*i = 90 μM

D-PHE-PHE-D-VINYLGLYCINE
PEPTIDYLGLYCINE MONOOXYGENASE — SSI — *K*i = 80 μM

D-PHE-PHE-L-VINYLGLYCINE
PEPTIDYLGLYCINE MONOOXYGENASE — C — *I*50 = 320 μM

D.PHE-L-PHE-D-VINYLGLYCINE		
PEPTIDYLGLYCINE MONOOXYGENASE	IR	Ki = 20 μM
157-PHE-psi(PO_2CH_2)LEU-PRO-NH_2		
THIMET OLIGOPEPTIDASE		Ki = 66 μM
(S)PHE(R)^P($OPH)_2$		
DIPEPTIDYL-PEPTIDASE 4	SSI	I50 = 70 μM
D-PHE-PRO-p-AMIDINIBENZYLAMINE		
PLASMIN		Kd = 38 μM
t-PLASMINOGEN ACTIVATOR		Kd = 250 μM
u-PLASMINOGEN ACTIVATOR		Kd = 588 μM
THROMBIN		Kd = 1.5 nM
TRYPSIN		Kd = 189 nM
H-D-PHE-PRO-ARG-CF_3		
THROMBIN		Ki = 1 nM
D-PHE-PRO-ARG CHLOROMETHYLKETONE		
CANCER PROCOAGULANT		100% 200 μM
COAGULATION FACTOR Xa		
KALLIKREIN		
PLASMIN		
PLASMINOGEN ACTIVATOR		
THROMBIN		I50 = 25 nM
TRYPSIN	IR	
D-PHE-PRO-ARG-CN		
THROMBIN	C	Ki = 700 nM
D-PHE-PRO-ARG-H		
PLASMIN		I50 = 2.4 μM
PROTEIN Ca		I50 = 4.2 μM
THROMBIN		I50 = 50 nM
THROMBIN		Ki = 75 nM
TRYPSIN		I50 = 220 nM
H-D-PHE-PRO-ARG-H		
THROMBIN		Ki = 7.2 nM
H-D-PHE-PRO-ARGINAL		
THROMBIN		
D-PHE-PRO-ARG-THIAZOLE		
u-PLASMINOGEN ACTIVATOR		I50 = 15 μM
THROMBIN		I50 = 1.5 nM
TRYPSIN		I50 = 4.2 nM
H-D-PHE-PROBORO-ARGOH		
COAGULATION FACTOR Xa		Ki = 8.2 nM
KALLIKREIN PLASMA	IR	Ki = 600 pM
KALLIKREIN PLASMA		Ki = 6 nM
PLASMIN		Ki = 2.3 nM
α-PLASMINOGEN ACTIVATOR (TISSUE TYPE)	IR	Ki = 2.3 nM
THROMBIN		Ki < 4 pM
[D-PHE-PRO-BOROBpgOPin)OC(CH_2)COGGHir		
THROMBIN		Ki = 600 pM
D-PHE-PRO-D,L-(4′-GUANIDO)PHE-TRIFLUOROMETHYLKETONE		
CANCER PROCOAGULANT		59% 100 μM
D-PHE-PRO-HOMOAGMATINE		
THROMBIN		
H-D-PHE-PRO-LYS-CF_3		
THROMBIN		Ki = 1 nM

D-PHE-PRO-D,L-LYS-TRIFLUOROMETHYLKETONE		
CANCER PROCOAGULANT		75% 100 µM
(H)-D-PHE-PRO-LYSYL-N-METHYLCARBOXAMIDE		
PLASMIN		*K*i = 1.2 µM
THROMBIN		*K*i = 1.7 nM
TRYPSIN		*K*i = 12 nM
(R,S)-PHE-PRO		
PROLYL OLIGOPEPTIDASE	C	*K*i = 8.4 mM
PHE-SER		
PHENYLALANINE AMMONIA-LYASE		26% 4 mM
PHETHIOL		
MEMBRANE ALANINE AMINOPEPTIDASE		*K*i = 5 nM
PHE-THRE-LEU-ASP-ALA-ASP-PHE		
RIBONUCLEOSIDE-DIPHOSPHATE REDUCTASE		
PHE-TYR		
PEPTIDYL-DIPEPTIDASE A		*I*50 = 3.7 µM
Z-D-PHE-VAL-BOROBPGOPIN		
α-THROMBIN		*I*50 = 3 nM
PHEVALIN		
CALPAIN		
PHLORETIN		
ALDOSE 1-EPIMERASE	C	*K*i = 22 µM
GLUCOSE TRANSPORT		
GLUTATHIONE TRANSFERASE		*I*50 = 68 µM
HYALURONOGLUCOSAMINIDASE		
LIPOXYGENASE		77% 30 mM
PHOSPHORYLASE		
PHLORETINPHOSPHATE		
HYALURONOGLUCURONIDASE		100 ng/ml
PHOSPHATASE ALKALINE		25 ng/ml
PHLORIDZIN		
ALDOSE 1-EPIMERASE	C	*K*i = 90 µM
GLUCOSE-6-PHOSPHATASE		
β-GLUCOSIDASE		*K*i = 2.9 µM
GLUTATHIONE TRANSFERASE		*I*50 = 204 µM
GLYCOGEN (STARCH) SYNTHASE		
LACTASE	C	*K*i = 180 µM
PHOSPHATASE		
PHOSPHORYLASE		
α,α-TREHALASE		*K*i = 2.4 mM
PHLORIZIN		
CELLOBIOSE PHOSPHORYLASE		
LACTASE		*K*i = 370 µM
α,α-TREHALASE		21% 2 mM
PHLOROGLUCINE		
ADENOSINE DEAMINASE		*I*50 = 130 µg/ml
PHLOROGLUCINOL		
ARYL SULFOTRANSFERASE P		*I*50 = 240 µM
IODIDE PEROXIDASE		
PHLOROGLUCINOL CARBOXYLIC ACID		
DIPHOSPHOMEVALONATE DECARBOXYLASE		

PHNYTOIN

THYMIDYLATE SYNTHASE		$I50 = 950\ \mu M$

PHO 81

CYCLIN DEPENDENT KINASE PHO85/PHO80		$Ki = 1$ nM

PHO81p

PROTEIN KINASE PHO85p		

PHOSALONE

ACETYLCHOLINESTERASE		

PHOSFOLAN

ACETYLCHOLINESTERASE		

PHOSMET

ACETYLCHOLINESTERASE		

PHOSPAHTIDYLSERINE

ACETYLCHOLINESTERASE		

PHOSPHAMIDON

ACETYLCHOLINESTERASE		
3-PHYTASE		

PHOSPHATE

ACETATE-CoA LIGASE		
ACETOLACTATE SYNTHASE		
ACETYLCHOLINESTERASE		
ACETYL-CoA HYDROLASE	NC	$Ki = 40\ \mu M$
N-ACETYLGALACTOSAMINE-4-SULFATASE		
N-ACETYLGALACTOSAMINE-6-SULFATASE		99% 17 mM
N-ACETYLGALACTOSAMINE-6-SULFATASE		95% 1 mM
N-ACETYLGLUCOSAMINE KINASE		
N-ACETYLGLUCOSAMINE-1-PHOSPHODIESTER N-ACETYLGLUCOSAMINIDASE		82% 33 mM
N-ACETYLGLUCOSAMINE-6-SULFATASE		95% 17 mM
N-ACETYLNEURAMINATE MONOOXYGENASE		
ACID-CoA LIGASE (GDP-FORMING)		
ACYLPHOSPHATASE		$I50 = 20$ mM
ACYL-PHOSPHATE PHOSPHOTRANSFERASE		
ADENOSINE-PHOSPHATE DEAMINASE		$Ki = 83$ mM
ADENYLATE CYCLASE		
ADENYLOSUCCINATE LYASE	C	$Ki = 4$ mM
ADENYLOSUCCINATE SYNTHASE	C	$Ki = 5$ mM
ADENYLYLSULFATASE		
β-ADRENERGIC-RECEPTOR KINASE		
AGAVAIN		63% 10 mM
β-ALA-ARG-HYDROLASE		80% 50 mM
ALDOSE-1-PHOSPHATE NUCLEOTIDYLTRANSFERASE		55% 4.5 mM
ALLANTOICASE		
AMIDOPHOSPHORIBOSYLTRANSFERASE		
AMINO-ACID N-ACETYLTRANSFERASE		
AMINOPEPTIDASE B		70% 17 mM
AMINOPEPTIDASE Y		44% 1 mM
(R)AMINOPROPANOL DEHYDROGENASE		
AMP DEAMINASE		$Ki = 8$ mM
AMP DEAMINASE		76% 8 mM
AMP DEAMINASE	C	
AMP NUCLEOSIDASE		
AMP NUCLEOSIDASE		
ANDROSTENOLONE SULFATASE	NC	$Ki = 800\ \mu M$
ARYLSULFATASE		
ARYLSULFATASE A		
ASPARTATE CARBAMOYLTRANSFERASE		
ASPERGILLUS NUCLEASE S1		
ATP CITRATE (pro-S)-LYASE		

ATP DEAMINASE		
BENZON NUCLEASE		
3′(2′),5′-BISPHOSPHATE NUCLEOTIDASE		25% 1 mM
γ-BUTYROBETAINE DIOXYGENASE		
CALCIDOL 1-MONOOXYGENASE		
CARBAMOYL-PHOSPHATE SYNTHASE (AMMONIA)	NC	
CARBOSINE SYNTHASE		
CARBOXY-cis,cis-MUCONATE CYCLASE		
3-CARBOXY-cis,cis-MUCONATE CYCLOISOMERASE		
CARBOXYPEPTIDASE A		
CARBOXYPEPTIDASE B		
CDP-DIACYLGLYCEROL-INOSITOL 3-PHOSPHATIDYLTRANSFERASE		
CEREBROSIDE-SULFATASE		$Ki = 9.2$ mM
CHITINASE		
CHOLINE-PHOSPHATE CYTIDYLYLTRANSFERASE		
CHONDROITIN ABC LYASE		
CHONDROITIN B LYASE		$I50 = 50$ mM
CHONDRO-4-SULFATASE		
CITRATE (si)-SYNTHASE		
CoA-GLUTATHIONE REDUCTASE (NADPH)		
CREATINE KINASE	C	$Ki = 10$ μM
CYCLAMATE SULFOHYDROLASE		
3′,5′-CYCLIC-NUCLEOTIDE PHOSPHATASE (IBMX INSENSITIVE)		
2′,3′-CYCLIC-NUCLEOTIDE 3′-PHOSPHODIESTERASE		
2′,3′-CYCLIC-NUCLEOTIDE 3′-PHOSPHODIESTERASE		
CYSTEINE-tRNA LIGASE		
CYTOCHROME-C OXIDASE		
CYTOCHROME-b5 REDUCTASE		
2-DEHYDRO-3-DEOXYPHOSPHOHEPTONATE ALDOLASE		
2-DEHYDRO-3-DEOXYPHOSPHOOCTONATE ALDOLASE		60% 10 mM
3-DEHYDROQUINATE SYNTHASE		
DEOXYRIBONUCLEASE II		
DEOXYRIBONUCLEASE IV (PHAGE T4-INDUCED)		80% 50 mM
DIACYLGLYCEROL KINASE		
DIHYDRODIPICOLINATE REDUCTASE		
DIHYDROOROTASE		51% 200 mM
DIPEPTIDASE	C	$Ki = 2.8$ mM
DIPHOSPHOMEVALONATE DECARBOXYLASE		
DNA α-GLUCOSYLTRANSFERASE		
DNA NUCLEOTIDYLEXOTRANSFERASE		
DNA POLYMERASE β		
DOLICHYL-PHOSPHATASE		75% 10 mM
ENDOPEPTIDASE TREPONEMA DENTICOLA ARCC35405		$Ki = 17$ mM
ENDOPOLYPHOSPHATASE		
ENDO-1,4-β-XYLANASE		$I50 = 100$ mM
ETHANOLAMINE-PHOSPHATE PHOSPHO-LYASE	C	$Ki = 1.3$ mM
EXOPOLYPHOSPHATASE		
FAD PYROPHOSPHATASE		
FARNESYL-DIPHOSPHATE FARNESYLTRANSFERASE		2% 100 mM
FOLYLPOLYGLUTAMATE SYNTHASE		97% 50 mM
FORMATE C-ACETYLTRANSFERASE		
FORMATE-TETRAHYDROFOLATE LIGASE	C	$Ki = 7$ mM
FORMATE-TETRAHYDROFOLATE LIGASE	MC	$Ki = 120$ mM
FRUCTOSE-BISPHOSPHATASE		
FRUCTOSE-BISPHOSPHATE ALDOLASE		$Ki = 920$ μM
FUMARATE REDUCTASE (NADH)		
GALACTONOLACTONE DEHYDROGENASE		
GALACTOSE-1-PHOSPHATASE		100% 10 mM
β-GALACTOSIDE α-2,3-SIALYLTRANSFERASE		
GERANYL-DIPHOSPHATE CYCLASE		89% 20 mM
GERANYLGERANYL-DIPHOSPHATE GERANYLGERANYLTRANSFERASE		
GLUCOKINASE	NC	
GLUCOMANNAN 4-β-MANNOSYLTRANSFERASE		
GLUCOSE-1,6-BISPHOSPHATE SYNTHASE		

GLUCOSE-1-PHOSPHATASE		54% 10 mM
GLUCOSE-6-PHOSPHATASE	C	$Ki = 20$ mM
GLUCOSE-1-PHOSPHATE ADENYLYLTRANSFERASE		$I50 = 22$ μM
GLUCOSE-1-PHOSPHATE ADENYLYLTRANSFERASE		$I50 = 260$ μM
GLUCOSE-1-PHOSPHATE ADENYLYLTRANSFERASE		$Ki = 580$ μM
GLUCOSE-6-PHOSPHATE 1-DEHYDROGENASE	C	$Ki = 100$ mM
GLUCOSE-6-PHOSPHATE 1-EPIMERASE		
GLUCOSE-6-PHOSPHATE ISOMERASE	C	$Ki = 1.7$ mM
GLUTAMATE-AMMONIA LIGASE		48% 35 mM
[GLUTAMATE-AMMONIA-LIGASE] ADENYLYLTRANSFERASE		$I50 = 20$ mM
GLUTAMATE-CYSTEINE LIGASE		
GLUTAMATE DECARBOXYLASE	C	
GLUTAMATE DEHYDROGENASE (NAD(P))		
GLUTAMATE DEHYDROGENASE (NADP)		
GLUTAMATE KINASE		
GLUTAMATE-5-SEMIALDEHYDE DEHYDROGENASE		
GLUTAMINASE	C	
γ-GLUTAMYL KINASE		
γ-GLUTAMYLTRANSFERASE		
GLUTATHIONE PEROXIDASE		
GLUTATHIONE REDUCTASE (NADPH)		
GLYCERALDEHYDE-3-PHOSPHATE DEHYDROGENASE (NADP)	C	$Ki = 2.4$ mM
GLYCERALDEHYDE-3-PHOSPHATE DEHYDROGENASE (NADP)		20% 10 mM
GLYCEROL-1,2-CYCLIC-PHOSPHATE PHOSPHODIESTERASE		
GLYCEROL-1-PHOSPHATASE		70% 10 mM
GLYCEROL-3-PHOSPHATE DEHYDROGENASE (NAD)		
GLYCEROPHOSPHOCHOLINE PHOSPHODIESTERASE		68% 10 mM
GLYCINE DEHYDROGENASE (CYTOCHROME)		> 100 mM
GLYCINE DEHYDROGENASE (DECARBOXYLATING)		
GLYCOGEN (STARCH) SYNTHASE D		
GLYCOGEN-SYNTHASE-D PHOSPHATASE		
GLYCOSULFATASE		
GLYEROL-3-PHOSPHATE DEHYDROGENASE (NAD(P))		> 100 mM
GTP CYCLOHYDROLASE I		
GTP CYCLOHYDROLASE I	U	
GUANIDINODEOXY-SCYLLO-INOSITOL-4-PHOSPHATASE		
HEXOKINASE		
HEXOKINASE 2		$Ki = 5.3$ mM
HISTIDINOL PHOSPHATASE	C	$Ki = 1.3$ mM
HOMOGENTISATE 1,2-DIOXYGENASE		
4-HYDROXYBENZOATE 3-MONOOXYGENASE		
HYDROXYLAMINE REDUCTASE (NADH)		
[HYDROXYMETHYLGLUTARYL-CoA REDUCTASE (NADPH)]-PHOSPHATASE		
2-HYDROXYPYRIDINE 5-MONOOXYGENASE		
HYDROXYPYRUVATE REDUCTASE		
HYPOTAURINE DEHYDROGENASE		
IDURONATE 2-SULFATASE		$I50$ 1 mM
IMIDAZOLEGLYCEROL-PHOSPHATE DEHYDRATASE	C	$Ki = 1.3$ mM
IMP DEHYDROGENASE		
INDOLE-3-ACETATE β-GLUCOSYLTRANSFERASE		50% 100 mM
INORGANIC PYROPHOSPHATASE	C	
myo-INOSITOL-1(or 4)-MONOPHOSPHATASE		38% 10 mM
myo-INOSITOL-1(or 4)-MONOPHOSPHATASE		$Ki = 100$ μM
ISOCITRATE LYASE		
ISOPENTENYL-DIPHOSPHATE Δ-ISOMERASE		60% 5 mM
3-ISOPROPYLMALATE DEHYDRATASE		
KYNURENINE-OXOGLUTARATE AMINOTRANSFERASE		
LACTATE 2-MONOOXYGENASE		
LACTOSE SYNTHASE		
LYSINE N-ACETYLTRANSFERASE		> 100 μM
MALEATE DEHYDRATASE		
MALTOSE SYNTHASE		
MANDELATE RACEMASE		
MANNOSE-6-PHOSPHATE ISOMERASE		$Ki = 86$ mM

MANNOTETRAOSE 2-α-N-ACETYLGLUCOSAMINYLTRANSFERASE		> 40 mM
METHIONINE ADENOSYLTRANSFERASE	NC	
METHYLGLYOXAL SYNTHASE		90% 40 mM
5-METHYLTHIORIBOSE KINASE		85% 5 mM
MONODEHYDROASCORBATE REDUCTASE (NADH)		
MONOTERPENYL-PYROPHOSPHATASE		*I*50 = 11 mM
MYOSIN-LIGHT-CHAIN KINASE		
NADH PEROXIDASE		60% 3 mM
NAD(P) TRANSHYDROGENASE		*K*i = 1.7 mM
NAD(P) TRANSHYDROGENASE (B-SPECIFIC)		
NEPRILYSIN		20% 4 mM
NICOTINATE-NUCLEOTIDE PYROPHOSPHORYLASE (CARBOXYLATING)		
NITROGENASE		> 30 mM
NUCLEOSIDE-DIPHOSPHATASE		
NUCLEOSIDE-TRIPHOSPHATASE		
3′-NUCLEOTIDASE		*I*50 = 5.9 mM
5′-NUCLEOTIDASE		80% 20 mM
5′-NUCLEOTIDASE	NC	*K*i = 19 μM
ORNITHINE CARBAMOYLTRANSFERASE	C	*K*i = 2 mM
OROTATE PHOSPHORIBOSYLTRANSFERASE		
OROTATE REDUCTASE (NADH)		
OROTIDINE-5′-PHOSPHATE DECARBOXYLASE		*K*i = 22 mM
OXOGLUTARATE DEHYDROGENASE (LIPOAMIDE)		
PEPTIDYL-DIPEPTIDASE A		
PEPTIDYLGLYCINE MONOOXYGENASE		
PHENYLALANINE 4-MONOOXYGENASE		
PHOSPHATASE ACID	C	*K*i = 2 mM
PHOSPHATASE ACID		59% 10 mM
PHOSPHATASE ACID PURPLE		
PHOSPHATASE ALKALINE		*K*i = 17 μM
PHOSPHATASE ALKALINE	C	*K*i = 0.6-10 μM
PHOSPHATASE ALKALINE (PHOSPHODIESTERASE ACTIVITY)	C	*K*i = 2.3 mM
PHOSPHATIDYLGLYCEROPHOSPHATASE		
PHOSPHODIESTERASE		59% 1 mM
PHOSPHOENOLPYRUVATE CARBOXYKINASE (PYROPHOSPHATE)		*K*i = 1.2 mM
PHOSPHOENOLPYRUVATE DECARBOXYLASE		
β-PHOSPHOGLUCOMUTASE		20% 10 mM
PHOSPHOGLUCONATE DEHYDRATASE		*K*i = 2.5 mM
PHOSPHOGLUCONATE DEHYDROGENASE (DECARBOXYLATING)		
PHOSPHOGLUCOSE ISOMERASE		*K*i = 8 mM
PHOSPHOGLYCERATE KINASE		
PHOSPHOGLYCOLATE PHOSPHATASE	C	*K*i = 2.6 mM
PHOSPHONOACETALDEHYDE HYDROLASE		
PHOSPHOPANTOTHENATE-CYSTEINE LIGASE		100% 80 mM
PHOSPHOPENTOMUTASE		50% 50 mM
PHOSPHOPROTEIN PHOSPHATASE		91% 50 mM
PHOSPHOPYRUVATE HYDRATASE	NC	*K*i = 6.4 mM
PHOSPHORIBOSYLFORMYLGLYCINAMIDINE SYNTHASE		
PHOSPHORIBULOKINASE		
PHOTINUS-LUCIFERIN 4-MONOOXYGENASE (ATP-HYDROLYSING)		
3-PHYTASE		*K*i = 28 μM
POLY(A)-SPECIFIC RIBONUCLEASE		22% 10 mM
POLYNUCLEOTIDE ADENYLYLTRANSFERASE		
POLYNUCLEOTIDE 3′-PHOSPHATASE		
PROCOLLAGEN N-ENDOPEPTIDASE		100% 10 mM
PROPANEDIOL-PHOSPHATE DEHYDROGENASE		
PROTEIN TYROSINE-PHOSPHATASE		70% 10 mM
PROTEIN TYROSINE-PHOSPHATASE ACID		*K*i = 2 mM
PROTEIN TYROSINE-PHOSPHATASE (Mg^{2+} DEPENDENT)		39% 5 mM
PURINE NUCLEOSIDASE		
PURINE-NUCLEOSIDE PHOSPHORYLASE		
PUTRESCINE CARBAMOYLTRANSFERASE	C	*K*i = 4 mM
PYROPHOSPHATE-SERINE PHOSPHOTRANSFERASE	C	
PYRROLINE-5-CARBOXYLATE REDUCTASE		

PYRROLINE 5-CARBOXYLATE SYNTHASE		*I*50 = 10 mM
PYRUVATE DECARBOXYLASE	C	
[PYRUVATE DEHYDROGENASE (LIPOAMIDE)]-PHOSPHATASE		
PYRUVATE KINASE		
PYRUVATE KINASE (L TYP)		
PYRUVATE, ORTHOPHOSPHATE DIKINASE		
QUERCETIN-3-SULFATE 4′-SULFOTRANSFERASE		
RIBONUCLEASE PANCREATIC		
RIBOSE-PHOSPHATE PYROPHOSPHOKINASE		
RIBULOSE-BISPHOSPHATE CARBOXYLASE	C	*K*i = 4.2 mM
RIBULOSE-BISPHOSPHATE CARBOXYLASE	NC	*K*i = 23 mM
tRNA(CYTOSINE-5-) METHYLTRANSFERASE		
RNA-DIRECTED DNA POLYMERASE		
mRNA GUANYLYLTRANSFERASE		
RNA NUCLEOTIDYLTRANSFERASE		
SEDOHEPTULOSE-BISPHOSPHATASE	C	*K*i = 2.2 mM
D-SERINE DEHYDRATASE		
SERINE-tRNA LIGASE		
SERINE-SULFATE AMMONIA-LYASE	NC	*K*i = 5 mM
STEROID 11β-MONOOXYGENASE		
STREPTOMYCIN-6-PHOSPHATASE		
SUCCINATE-CoA LIGASE (GDP-FORMING)	U	
SUCCINATE-SEMIALDEHYDE DEHYDROGENASE (NAD(P))		
O-SUCCINYLHOMOSERINE (THIOL)-LYASE		
SUCROSE PHOSPHORYLASE	C	*K*i = 250 mM
SULFATE ADENYLYLTRANSFERASE		> 5 mM
SULFITE DEHYDROGENASE		
N-SULFOGLUCOSAMINE-6-SULFATASE		22% 200 µM
SUPEROXIDE DISMUTASE (CU, ZN)		
TETRAHYDROXYPTERIDINE CYCLOISOMERASE		52% 18 mM
THERMOLYSIN		
THIAMIN PYROPHOSPHOKINASE		
THIOSULFATE SULFURTRANSFERASE		*K*i = 60 mM
THREONINE SYNTHASE		40% 50 mM
THYMIDINE-TRIPHOSPHATASE		*K*i = 31 mM
THYMINE DIOXYGENASE		
TRANSALDOLASE		
TRANSKETOLASE		
TREHALOSE-PHOSPHATASE		*K*i = 3.5 mM
α,α-TREHALOSE-PHOSPHATE SYNTHASE		
TRIOSE-PHOSPHATE ISOMERASE		*K*i = 6 mM
TRIOSE-PHOSPHATE ISOMERASE		*K*i = 6 mM
TRYPTOPHAN 2-MONOOXYGENASE		
TYROSINE TRANSAMINASE	C	*K*i = 28 µM
UDP-GLUCOSE-1-PHOSPHATE URIDYLYLTRANSFERASE	C	*K*i = 3.7 mM
UDP-GLUCOSE-1-PHOSPHATE URIDYLYLTRANSFERASE	C	
URACIL PHOSPHORIBOSYLTRANSFERASE		
URATE OXIDASE		
UREASE	C	
URIDINE PHOSPHORYLASE		
XAA-ARG DIPEPTIDASE		
XAA-HIS DIPEPTIDASE		70% 50 mM
XAA-TRP AMINOPEPTIDASE		
XANTHOTOXOL O-METHYLTRANSFERASE		
XYLOGLUCAN 6-XYLOSETRANSFERASE		
L-XYLULOSE REDUCTASE		

D,L-α-glycerol-1-PHOSPHATE

6-PHOSPHO-β-GLUCOSIDASE A		*K*i = 14 mM

PHOSPHATIDIC ACID

N-ACETYLLACTOSAMINE SYNTHASE		
ARACHIDONYL-DIACYLGLYCEROL KINASE		
CAM KINASE		
DNA POLYMERASE β		

DNA POLYMERASE γ
DOLICHYL-PHOSPHATASE — 96% 1 mM
ETHANOLAMINEPHOSPHOTRANSFERASE
GLYCEROL-3-PHOSPHATE O-ACYLTRANSFERASE
GLYCERONE-PHOSPHATE O-ACYLTRANSFERASE
1-PHOSPHATIDYLINOSITOL 3-KINASE
PHOSPHOLIPASE D (GLYCOSYLPHOSPHATIDYLINOSITOL SPECIFIC)
PROTEIN-DISULFIDE REDUCTASE (GLUTATHIONE)
PROTEIN TYROSINE KINASE (INSULIN RECEPTOR)

D-PHOSPHATIDYLCHOLIN

PHOSPHOLIPASE A2

PHOSPHATIDYLCHOLINE

Enzyme	Type	Value
ACYLGLYCERONE-PHOSPHATE REDUCTASE		
CARNITINE O-PALMITOYLTRANSFERASE		
DOLICHYL-PHOSPHATASE		68% 100 μM
β-GLUCOSIDASE		
GLYCEROL-3-PHOSPHATE DEHYDROGENASE (NAD)		
PHOSPHATIDYLETHANOLAMINE N-METHYLTRANSFERASE	C	56% 300 μM
1-PHOSPHATIDYLINOSITOL-4,5-BISPHOSPHATE PHOSPHODIESTERASE		
PHOSPHATIDYL N-METHYLETHANOLAMINE METHYLTRANSFERASE	C	38% 300 μM
SPHINGOMYELIN PHOSPHODIESTERASE		*K*i = 200 μM
UDP-N-ACETYLGLUCOSAMINE-DOLICHYL-PHOSPHATE N-ACETYLGLUCOSAMINEPHOSPHOTRANSFERA		

PHOSPHATIDYLCHOLINE DIFLUOROKETONES

PHOSPHATIDYLCHOLINE-STEROL O-ACYLTRANSFERASE

PHOSPHATIDYLCHOLINE (EGG)

PHOSPHOLIPASE D (GLYCOSYLPHOSPHATIDYLINOSITOL SPECIFIC) — *I*50 = 250 μM

PHOSPHATIDYLCHOLINES UNSATURATED

PROTEIN KINASE C

PHOSPHATIDYLETHANOL

1-PHOSPHATIDYLINOSITOL-4,5-BISPHOSPHATE PHOSPHODIESTERASE — *I*50 = 26 μM

PHOSPHATIDYLETHANOLAMINE

Enzyme	Type	Value
N-ACETYLLACTOSAMINE SYNTHASE		
DOLICHYL-PHOSPHATASE		71% 100 μM
ETHANOLAMINEPHOSPHOTRANSFERASE		
GLYCERONE-PHOSPHATE O-ACYLTRANSFERASE		> 1.5 mM
INDOLE-3-ACETATE β-GLUCOSYLTRANSFERASE		
PHOSPHATIDATE PHOSPHATASE		
1-PHOSPHATIDYLINOSITOL-4,5-BISPHOSPHATE PHOSPHODIESTERASE		
UDP-N-ACETYLGLUCOSAMINE-DOLICHYL-PHOSPHATE N-ACETYLGLUCOSAMINEPHOSPHOTRANSFERA		

PHOSPHATIDYLGLYCEROL

Enzyme	Type	Value
N-ACETYLLACTOSAMINE SYNTHASE		
N-ACETYLMURAMOYL-L-ALANINE AMIDASE		
ACYLGLYCERONE-PHOSPHATE REDUCTASE		
CDP-DIACYLGLYCEROL-SERINE O-PHOSPHATIDYLTRANSFERASE		
CHLOROPHYLLASE		
DNA TOPOISOMERASE		
GLYCEROL-3-PHOSPHATE O-ACYLTRANSFERASE		> 1.5 mM
α-MANNOSIDASE		
PHOSPHATIDATE PHOSPHATASE		75% 700 μg/ml

PHOSPHATIDYLINOSITOL

Enzyme	Type	Value
N-ACETYLNEURAMINATE MONOOXYGENASE		
CAM KINASE		
DNA POLYMERASE β	C	*K*i = 16 μM
DNA POLYMERASE γ		
DOLICHYL-PHOSPHATASE		
DOLICHYL-PHOSPHATE-MANNOSE-PROTEIN MANNOSYLTRANSFERASE		
β-GLUCOSIDASE		90% 500 μg/ml
GLYCEROL-3-PHOSPHATE O-ACYLTRANSFERASE		

GLYCERONE-PHOSPHATE O-ACYLTRANSFERASE		
HEXOKINASE		
α-MANNOSIDASE		
ORNITHINE DECARBOXYLASE		
PHOSPHATIDYLCHOLINE-DOLICHOL O-ACYLTRANSFERASE		
1-PHOSPHATIDYLINOSITOL PHOSPHODIESTERASE		93% 5 mM
PHOSPHOLIPASE D1a		*I*50 = 28 μM
PHOSPHOLIPASE D1b		*I*50 = 7.4 μM
PHOSPHOLIPASE D2		*I*50 = 7.2 μM
PROTEIN TYROSINE-PHOSPHATASE		*I*50 = 6 μM
STEROL ESTERASE		
PHOSPHATIDYLINOSITOL-3,4-DIPHOSPHATE		
GLUCOSE-6-PHOSPHATASE		*K*i = 5 μM
PHOSPHATIDYLINOSITOL-4,5-DIPHOSPHATE		
ARACHIDONYL-DIACYLGLYCEROL KINASE		
GLUCOSE-6-PHOSPHATASE		*K*i = 4.7 μM
PHOSPHATIDYLINOSITOL 3-KINASE (PROTEIN-SERINE KINASE ACTIVITY)		
SPHINGOMYELIN PHOSPHODIESTERASE		54% 5 μM
PHOSPHATIDYLINOSITOL (EGG)		
PHOSPHOLIPASE D (GLYCOSYLPHOSPHATIDYLINOSITOL SPECIFIC)		*I*50 = 60 μM
PHOSPHATIDYLINOSITOL-4-PHOSPHATE		
DNA POLYMERASE β		
DNA POLYMERASE α		
DNA POLYMERASE δ		
1-PHOSPHATIDYLINOSITOL 4-KINASE		
PHOSPHATIDYLINOSITOL 3-KINASE (PROTEIN-SERINE KINASE ACTIVITY)		
PHOSPHOLIPASE D (GLYCOSYLPHOSPHATIDYLINOSITOL SPECIFIC)		*I*50 = 20 μM
PHOSPHATIDYLINOSITOL-3,4,5-TRIPHOSPHATE		
GLUCOSE-6-PHOSPHATASE		*K*i = 1.7 μM
PHOSPHATIDYLSERINE		
N-ACETYLLACTOSAMINE SYNTHASE		
CHYMASE		
DNA POLYMERASE β		
β-GLUCOSIDASE		90% 500 μg/ml
GLUTAMATE DEHYDROGENASE (NAD(P))		
GLYCEROL-3-PHOSPHATE O-ACYLTRANSFERASE		
GLYCERONE-PHOSPHATE O-ACYLTRANSFERASE		> 1.5 mM
HEXOKINASE		
NITRIC-OXIDE SYNTHASE		
ORNITHINE DECARBOXYLASE		
PHOSPHATIDYLCHOLINE-DOLICHOL O-ACYLTRANSFERASE		
1-PHOSPHATIDYLINOSITOL-4,5-BISPHOSPHATE PHOSPHODIESTERASE		
1-PHOSPHATIDYLINOSITOL 4-KINASE		
1-PHOSPHATIDYLINOSITOL 4-KINASE III		64% 500 g/l
PROTEIN TYROSINE-PHOSPHATASE		*I*50 = 3.7 μM
SPHINGOMYELIN PHOSPHODIESTERASE		
STEROL ESTERASE		
PHOSPHATIDYLSERINE INOSITOL MICROSOMAL		
GLYCEROL-3-PHOSPHATE O-ACYLTRANSFERASE		
2-PHOSPHINOCYCLOHEXANECARBOXYLIC ACID		
XAA-PRO DIPEPTIDASE		*K*i = 24 μM
PHOSPHINOTHRICIN		
GLUTAMATE-AMMONIA LIGASE		
D,L-PHOSPHINOTHRICIN		
GLUTAMATE-AMMONIA LIGASE	C	*K*i = 25 μM
GLUTAMATE-AMMONIA LIGASE	C	*K*i = 1.1 μM
GLUTAMATE-AMMONIA LIGASE 1		*K*i = 4 μM
GLUTAMATE-AMMONIA LIGASE 2		*K*i = 1.5 μM

PHOSPHINOTHRICINS CYCLIC ANALOUGES
GLUTAMATE-AMMONIA LIGASE

PHOSPHITE
PHOSPHOENOLPYRUVATE DECARBOXYLASE | C | *K*i = 3.6 mM

3′PHOSPHOADENOSINE 5-PHOSPHATE
3′-PHOSPHOADENYLSULFATE 3-PHOSPHATASE | | 55% 13 mM

3′-PHOSPHOADENOSINE-5′-PHOSPHOSULPHATE
ARYL SULFOTRANSFERASE
SULFATE ADENYLYLTRANSFERASE | | *I*50 = 200 μM
SULFATE ADENYLYLTRANSFERASE

3′-PHOSPHOADENYLYLSULPHATE
TYROSINE-ESTER SULFOTRANSFERASE

2′-PHOSPHO-3-AMINOPYRIDINE DINUCLEOTIDE PHOSPHATE
20α-HYDROXYSTEROID DEHYDROGENASE

5-PHOSPHO-D-ARABINOHYDROXAMIC ACID
GLUCOSE-6-PHOSPHATE ISOMERASE

PHOSPHOCARPIN B1
CHYMOTRYPSIN

PHOSPHOCHOLINE
CHOLINE KINASE

PHOSPHOCREATINE
THIAMIN-PHOSPHATE PYROPHOSPHORYLASE

N-PHOSPHO-N′-(2,6-DICHLOROPHENYL)ETHYL GUANIDINE
CREATINE KINASE | | *K*i = 2 mM

PHOSPHODIEPRYL 03
NEUROLYSIN | | *K*i = 90 pM

PHOSPHODIEPRYL 08
NEUROLYSIN | | *K*i = 400 pM
THIMET OLIGOPEPTIDASE | | *K*i = 400 pM

PHOSPHOENOL ACETYLPHOSPHONATE
PHOSPHOENOLPYRUVATE CARBOXYLASE | C | *K*i = 2.2 mM
PHOSPHOPYRUVATE HYDRATASE | C | *K*i = 2.2 mM

PHOSPHOENOL ACETYLPHOSPHONATE METHYLESTER
PHOSPHOPYRUVATE HYDRATASE | C | *K*i = 27 mM

PHOSPHOENOL-2-OXOPENTANOATE
UDP-N-ACETYLGLUCOSAMINE 1-CARBOXYVINYLTRANSFERASE

PHOSPHOENOLPYRUVIC ACID
ADENYLATE KINASE
ADENYLOSUCCINATE SYNTHASE
ALANINE DEHYDROGENASE
FRUCTOSE-BISPHOSPHATASE
FRUCTOSE-BISPHOSPHATE ALDOLASE | | *I*50 = 2.6 mM
FRUCTOSE-6-PHOSPHATE PHOSPHOKETOLASE
1,3-β-GLUCAN SYNTHASE
GLUCOSE-1,6-BISPHOSPHATE SYNTHASE
GLUCOSE-1-PHOSPHATE ADENYLYLTRANSFERASE | | *I*50 = 47 μM
GLUCOSE-6-PHOSPHATE 1-DEHYDROGENASE
GLUCOSE-6-PHOSPHATE ISOMERASE | C | *K*i = 1.1 mM
GLUCOSE-6-PHOSPHATE ISOMERASE | | *K*i = 1.1 mM
GLUTAMATE DEHYDROGENASE
GLUTAMATE DEHYDROGENASE (NAD(P))
GLYCERALDEHYDE-3-PHOSPHATE DEHYDROGENASE (PHOSPHORYLATING)

GLYCEROL-3-PHOSPHATE DEHYDROGENASE		
GUANYLATE CYCLASE		
ISOCITRATE DEHYDROGENASE (NADP)		
[ISOCITRATE DEHYDROGENASE (NADP)] KINASE		
ISOCITRATE LYASE	NC	Ki = 910 µM
ISOCITRATE LYASE	UC	Ki = 630 µM
L-LACTATE DEHYDROGENASE		
MALATE SYNTHASE		
METHYLGLYOXAL SYNTHASE		65% 1.5 mM
5-METHYLTHIORIBOSE KINASE		45% 30 µM
NADH KINASE		57% 30 mM
NADH PEROXIDASE		73% 3 mM
OXALOACETATE DECARBOXYLASE		I50 = 28 nM
PHOSPHATIDYLSERINE DECARBOXYLASE		
PHOSPHOENOLPYRUVATE CARBOXYKINASE (PYROPHOSPHATE)		Ki = 170 µM
PHOSPHOENOLPYRUVATE DECARBOXYLASE		
6-PHOSPHOFRUCTOKINASE		
6-PHOSPHOFRUCTO-2-KINASE		
6-PHOSPHO-β-GALACTOSIDASE		
PHOSPHOGLUCONATE DEHYDROGENASE (DECARBOXYLATING)	C	Ki = 150 µM
PHOSPHOGLUCONATE DEHYDROGENASE (DECARBOXYLATING)	C	Ki = 540 µM
6-PHOSPHO-β-GLUCOSIDASE A		Ki = 8.1 mM
6-PHOSPHO-β-GLUCOSIDASE B		Ki = 720 µM
PHOSPHOGLYCERATE MUTASE		
PHOSPHOKETOLASE		
PHOSPHORIBULOKINASE		I50 = 850 µM
PYRUVATE, WATER DIKINASE		
THIAMIN-PHOSPHATE PYROPHOSPHORYLASE		
TRIOSE-PHOSPHATE ISOMERASE		Ki = 3 mM
TRIOSE-PHOSPHATE ISOMERASE		Ki = 1.3 mM
TRIOSE-PHOSPHATE ISOMERASE		Ki = 900 µM
TRIOSE-PHOSPHATE ISOMERASE		Ki = 500 µM
XAA-PRO DIPEPTIDASE		Ki = 8.5 nM
XAA-PRO DIPEPTIDASE		Ki = 300 nM

PHOSPHOENOLPYRUVIC ACID + CYANOBORHYDRIDE

3-PHOSPHOSHIKIMATE 1-CARBOXYVINYLTRANSFERASE		

4-PHOSPHOERYTHRONATE

RIBOSE-5-PHOSPHATE ISOMERASE	C	Ki = 4.4 µM
RIBULOSE-BISPHOSPHATE CARBOXYLASE	C	Ki = 40 µM
RIBULOSE-BISPHOSPHATE CARBOXYLASE	C	Ki = 70 µM
RIBULOSE-BISPHOSPHATE OXYGENASE		Ki = 40 µM

PHOSPHOETHANOLAMINE

CHOLINE-PHOSPHATE CYTIDYLYLTRANSFERASE	C	
ETHANOLAMINE KINASE		
PHOSPHATASE ALKALINE		

PHOSPHOGLUCONIC ACID

GLUTAMINE-FRUCTOSE-6-PHOSPHATE AMINOTRANSFERASE (ISOMERIZING)	C	
GLYCEROL-3-PHOSPHATE DEHYDROGENASE (NAD)		

2-PHOSPHOGLUCONIC ACID

4-NITROPHENYLPHOSPHATASE		40% 10 mM

6-PHOSPHOGLUCONIC ACID

FRUCTOSE-BISPHOSPHATE ALDOLASE		Ki = 92 µM
FRUCTOSE-BISPHOSPHATE ALDOLASE		I50 = 4.2 mM
GLUCOSE-6-PHOSPHATE ISOMERASE	C	Ki = 5 µM
GLUCOSE-6-PHOSPHATE ISOMERASE		Ki = 15 µM
GLYCERALDEHYDE-3-PHOSPHATE DEHYDROGENASE (NADP) (PHOSPHORYLATING)		
GLYCERALDEHYDE-3-PHOSPHATE DEHYDROGENASE (PHOSPHORYLATING)		
myo-INOSITOL-1-PHOSPHATE SYNTHASE		90% 5 mM
ISOCITRATE LYASE	NC	Ki = 1.3 mM
L-LACTATE DEHYDROGENASE		

LEVANSUCRASE		
MANNOSE-6-PHOSPHATE ISOMERASE		
4-NITROPHENYLPHOSPHATASE		55% 10 mM
PHOSPHOENOLPYRUVATE CARBOXYKINASE (GTP)		*I*50 = 40 μM
PHOSPHORIBULOKINASE		
RIBOSE-5-PHOSPHATE ISOMERASE	NC	*K*i = 7 mM
RIBOSE-5-PHOSPHATE ISOMERASE		
RIBULOSE-BISPHOSPHATE CARBOXYLASE		
3-PHOSPHOGLYCERALDEHYDE		
ADENYLATE KINASE		
PHOSPHOGLYCERATE		
PHOSPHORIBULOKINASE		
PYROPHOSPHATE-FRUCTOSE-6-PHOSPHATE 1-PHOSPHOTRANSFERASE		
2-PHOSPHOGLYCERATE		
ADENYLOSUCCINATE SYNTHASE		
BISPHOSPHOGLYCERATE MUTASE		60% 2 mM
BISPHOSPHOGLYCERATE PHOSPHATASE		60% 2 mM
2-DEHYDRO-3-DEOXYPHOSPHOHEPTONATE ALDOLASE		
INORGANIC PYROPHOSPHATASE		
PHOSPHOENOLPYRUVATE-PROTEIN KINASE		
6-PHOSPHOFRUCTOKINASE		
TRIOSE-PHOSPHATE ISOMERASE		*K*i = 15 μM
2-PHOSPHO-D-GLYCERATE		
HYDROXYPYRUVATE REDUCTASE		
PHOSPHOGLYCERATE MUTASE	C	
3-PHOSPHOGLYCERATE		
ADENYLOSUCCINATE SYNTHASE		
BISPHOSPHOGLYCERATE MUTASE		60% 2 mM
BISPHOSPHOGLYCERATE PHOSPHATASE		*K*i = 250 μM
BISPHOSPHOGLYCERATE PHOSPHATASE		60% 2 mM
[GLUTAMATE-AMMONIA-LIGASE] ADENYLYLTRANSFERASE		
GLUTAMATE SYNTHASE (FERREDOXIN)		
GLYCERALDEHYDE-3-PHOSPHATE DEHYDROGENASE (NADP)	MIX	*K*i = 10 mM
GLYCERATE KINASE		
ISOCITRATE DEHYDROGENASE (NADP)		
ISOCITRATE LYASE	C	*K*i = 800 μM
L-LACTATE DEHYDROGENASE		
METHYLGLYOXAL SYNTHASE		
PHOSPHOENOLPYRUVATE-PROTEIN KINASE		
6-PHOSPHOFRUCTOKINASE		
PHOSPHOGLUCOMUTASE		
PHOSPHOGLYCERATE KINASE		
PHOSPHOPYRUVATE HYDRATASE	C	*K*i = 450 μM
RIBOSE-5-PHOSPHATE ISOMERASE	C	*K*i = 1.2 mM
RIBULOSE-BISPHOSPHATE CARBOXYLASE	C	*K*i = 9.5 mM
STARCH (BACTERIAL GLYCOGEN) SYNTHASE	C	*K*i = 240 μM
D(+)-2-PHOSPHOGLYCERATE		
XAA-PRO DIPEPTIDASE		*K*i = 130 μM
D-3-PHOSPHOGLYCERATE		
RIBOSE-5-PHOSPHATE ISOMERASE	C	*K*i = 1.3 mM
D,L-2-PHOSPHOGLYCERATE		
TRIOSE-PHOSPHATE ISOMERASE		*K*i = 4.1 mM
TRIOSE-PHOSPHATE ISOMERASE		*K*i = 7.7 mM
TRIOSE-PHOSPHATE ISOMERASE		*K*i = 600 μM
D,L-3-PHOSPHOGLYCERATE		
TRIOSE-PHOSPHATE ISOMERASE		*K*i = 6.9 mM
TRIOSE-PHOSPHATE ISOMERASE		*K*i = 1.1 mM
TRIOSE-PHOSPHATE ISOMERASE		*K*i = 510 μM
XAA-PRO DIPEPTIDASE		*K*i = 71 μM

D-2-PHOSPHOGLYCERIC ACID
GLYCEROL-3-PHOSPHATE DEHYDROGENASE

D-3-PHOSPHOGLYCERIC ACID
GLYCEROL-3-PHOSPHATE DEHYDROGENASE

PHOSPHOGLYCERIDES
CHYMASE
PHOSPHATIDYLINOSITOL DEACETYLASE

PHOSPHOGLYCOLIC ACID
GLYCERALDEHYDE-3-PHOSPHATE DEHYDROGENASE (NADP)
GLYCEROL-3-PHOSPHATE DEHYDROGENASE
PHOSPHOENOLPYRUVATE DECARBOXYLASE C $Ki = 9\ \mu M$
6-PHOSPHOFRUCTOKINASE
PYROPHOSPHATE-FRUCTOSE-6-PHOSPHATE 1-PHOSPHOTRANSFERASE
TRIOSE-PHOSPHATE ISOMERASE $Ki = 6\ \mu M$

2-PHOSPHOGLYCOLIC ACID
PHOSPHOGLYCERATE MUTASE
PHOSPHOPYRUVATE HYDRATASE C
TRIOSE-PHOSPHATE ISOMERASE $Ki = 14\ \mu M$
TRIOSE-PHOSPHATE ISOMERASE $Ki = 27\ \mu M$
TRIOSE-PHOSPHATE ISOMERASE $Ki = 30\ \mu M$

PHOSPHOGLYCOLIC HYDROXAMATE
TRIOSE-PHOSPHATE ISOMERASE

O-PHOSPHO-D,L-HOMOSERINE
HOMOSERINE KINASE

2-α-PHOSPHOHYDANTOCIDIN
ADENYLOSUCCINATE SYNTHASE

5′-PHOSPHOHYDANTOCIDIN
ADENYLOSUCCINATE SYNTHASE C $Ki = 22\ nM$

PHOSPHOHYDROXYPYRUVIC ACID
GLYCERALDEHYDE-3-PHOSPHATE DEHYDROGENASE (NADP) 30% 100 μM
HYDROXYPYRUVATE REDUCTASE

3-PHOSPHOHYDROXYPYRUVIC ACID
PHOSPHOGLYCERATE MUTASE

PHOSPHOINOSITOL-4,5-DIPHOSPHATE
PROTEIN TYROSINE KINASE NC $I50 = 40\ \mu M$

PHOSPHOINOSITOL-4-PHOSPHATE
PROTEIN TYROSINE KINASE NC $I50 = 80\ \mu M$

6-PHOSPHO-2-KETO-3-DEOXY-D-GLUCONATE
2-DEHYDRO-3-DEOXYGLUCARATE ALDOLASE C

PHOSPHOLACTIC ACID
XAA-PRO DIPEPTIDASE $Ki = 5\ nM$

D-PHOSPHOLACTIC ACID
PHOSPHOENOLPYRUVATE DECARBOXYLASE C $Ki = 9\ \mu M$
PHOSPHOPYRUVATE HYDRATASE C $Ki = 350\ \mu M$

L-PHOSPHOLACTIC ACID
PHOSPHOENOLPYRUVATE DECARBOXYLASE C $Ki = 2\ \mu M$
PHOSPHOPYRUVATE HYDRATASE

PHOSPHOLAMBAN
Ca^{2+}-TRANSPORTING ATPASE

PHOSPHOLAMBAN(1-25)
Ca^{2+}-TRANSPORTING ATPASE

N-PHOSPHO-L-LEUCYL-L-TRYPTOPHAN
ASPERGILLUS ORYZAE NEUTRAL PROTEINASE

PHOSPHOLIPASE A
ECDYSONE 20-MONOOXYGENASE

PHOSPHOLIPASE C
ECDYSONE 20-MONOOXYGENASE

PHOSPHOLIPIDS
N-ACETYLMURAMOYL-L-ALANINE AMIDASE

PHOSPHOLIPIDS ACIDIC
Ca^{2+}-TRANSPORTING ATPASE

PHOSPHOMONOMETHYLETHANOLAMINE
CHOLINE-PHOSPHATE CYTIDYLYLTRANSFERASE — C

PHOSPHOMYCIN
PHOSPHATASE ACID

PHOSPHONATE ANALOGUE OF FRUCTOSE-2,6-BISPHOSPHATE
FRUCTOSE-BISPHOSPHATASE

PHOSPHONATE SUBSTRATE ANALOGA
1-PHOSPHATIDYLINOSITOL PHOSPHODIESTERASE

PHOSPHONOACETIC ACID
DNA POLYMERASE — Ki = 1 μM
FORMATE-TETRAHYDROFOLATE LIGASE
RNA-DIRECTED RNA POLYMERASE
XAA-PRO DIPEPTIDASE — Ki = 580 nM

3-(PHOSPHONOACETYLAMIDO)-ALANINE
GLUTAMATE-5-SEMIALDEHYDE DEHYDROGENASE

N-(PHOSPHONOACETYL)-L-ASPARTIC ACID
ASPARTATE CARBAMOYLTRANSFERASE — C — Ki = 26 nM
ASPARTATE CARBAMOYLTRANSFERASE — Ki = 27 nM

5-N-PHOSPHONOACETYL-L-ORNITHINE
ORNITHINE CARBAMOYLTRANSFERASE

PHOSPHONOALANINE
PHOSPHOSERINE PHOSPHATASE

4-PHOSPHONOBUTYRIC ACID
XAA-PRO DIPEPTIDASE — Ki = 2.4 μM

4-PHOSPHONOBUTYRONITRILE
PHOSPHOGLYCERATE KINASE

2-PHOSPHONOCYCLOHEXANECARBOXYLIC ACID
XAA-PRO DIPEPTIDASE — Ki = 93 nM

N-(PHOSPHONODIFLUOROACETYL)-L-ASPARTIC ACID
ASPARTATE CARBAMOYLTRANSFERASE — I50 = 5 μM

Z-3-(2-PHOSPHONOETHEN-1-METHYL)PYRIDINE-2-CARBOXYLIC ACID
O-SUCCINYLHOMOSERINE (THIOL)-LYASE — C — Ki = 45 μM

5-(PHOSPHONOETHYL)-5-DEOXYQUINATE
3-DEHYDROQUINATE SYNTHASE — Ki = 30 μM

PHOSPHONOFLUORIDIC ACID METHYL-5,8,11,14-EICOSATETRAENYL ESTER
PHOSPHOLIPASE A2

PHOSPHONOFORMIC ACID
RNA-DIRECTED DNA POLYMERASE — 98% 100 μM
XAA-PRO DIPEPTIDASE — Ki = 120 μM

1-PHOSPHONO-all-trans-GERANYLGERANIOL		
GERANYLGERANYL PROTEINTRANSFERASE		Ki = 720 nM
6-PHOSPHONOGLUCONATE		
TRIOSE-PHOSPHATE ISOMERASE		
PHOSPHONOGLYCOLIC ACID		
XAA-PRO DIPEPTIDASE		Ki = 88 nM
9-(7-PHOSPHONOHEPTYL)ADENINE		
5′-METHYLTHIOADENOSINE PHOSPHORYLASE		Ki = 16 μM
9-(6-PHOSPHONOHEXYL)GUANINE		
GUANYLATE KINASE	C	Ki = 150 μM
9-(2-PHOSPHONOMETHOXYETHYL)GUANINE DIPHOSPHATE		
DNA POLYMERASE β		Ki = 59 μM
DNA POLYMERASE α		Ki = 550 μM
2-(PHOSPHONOMETHYL)ACRYLIC ACID		
XAA-PRO DIPEPTIDASE		Ki = 13 nM
5-(PHOSPHONOMETHYL)-5-DEOXYQUINATE		
3-DEHYDROQUINATE SYNTHASE		Ki = 55 nM
N-(PHOSPHONOMETHYL)GLYCINE		
5-AMINOLEVULINATE SYNTHASE		
2-KETO-3-DEOXY-HEPTANAOTE SYNTHASE	C	86% 1 mM
XAA-PRO DIPEPTIDASE		Ki = 23 μM
S-PHOSPHONOMETHYLHOMOCYSTEINE		
GLUTAMATE-AMMONIA LIGASE	C	
S-PHOSPHONOMETHYLHOMOCYSTEINE SULPHONE		
GLUTAMATE-AMMONIA LIGASE	C	Ki = 1.4. mM
S-PHOSPHONOMETHYLHOMOCYSTEINE SULPHOXIDE		
GLUTAMATE-AMMONIA LIGASE	C	
2-(PHOSPHONOMETHYL)PENTANEDIOIC ACID		
DIPEPTIDASE N-ACYLATED α-LINKED ACIDIC		
3-(PHOSPHONOOXY)-4-HYDROXY-5-[N-(PHOSPHONOMETHYL-2-OXOETHYL)AMINO]-1-CYCLOHEXENE-1-CARBOXYLIC ACID (3α,4α,5β)		
3-PHOSPHOSHIKIMATE 1-CARBOXYVINYLTRANSFERASE		Ki = 7.4 μM
5-[(PHOSPHONOOXY)METHYL]-5-DEOXYQUINATE		
3-DEHYDROQUINATE SYNTHASE		Ki = 30 nM
3-(PHOSPHONOOXY)-5-QUINATE		
3-DEHYDROQUINATE SYNTHASE		Ki = 55 μM
9-(5-PHOSPHONOPENTYL)GUANINE		
PURINE-NUCLEOSIDE PHOSPHORYLASE	C	Ki = 400 nM
PURINE-NUCLEOSIDE PHOSPHORYLASE	C	Ki = 82 nM
PURINE-NUCLEOSIDE PHOSPHORYLASE	C	Ki = 320 nM
PURINE-NUCLEOSIDE PHOSPHORYLASE	C	Ki = 905 nM
D,L-4-PHOSPHONO-2-(PHOSPHONOMETHYL)BUTANOIC ACID		
PHOSPHOGLYCERATE MUTASE (COFACTOR DEPENDENT)	C	Ki = 800 μM
3-PHOSPHONOPROPIONIC ACID		
XAA-PRO DIPEPTIDASE		Ki = 72 nM
NN((R)-1-PHOSPHONOPROPYL)(S)LEUCYL)(S)3-INDOLYLALANINE-N-METHYLAMIDE		
COLLAGENASE		$I50$ = 50 nM
[S(E,E)]3PHOSPHONO-2[(4,8,12-TRIMETHYL-1-OXO-3,7,11-TRIDECATRIENYL)AMINO]PROPANOIC ACID		
FARNESYL PROTEINTRANSFERASE		$I50$ = 50 nM

9-[o-(2-PHOSPHONOVINYL)PHENYLDIETHYLESTER]GUANINE

PURINE-NUCLEOSIDE PHOSPHORYLASE		Ki = 3.2 nM
PURINE-NUCLEOSIDE PHOSPHORYLASE		Ki = 3.2 nM

9-(2-PHOSPHONYLMETHOXYETHOXYPROPYL)ADENINE DIPHOSPHATE

DNA POLYMERASE α		Ki = 1.2 μM
DNA POLYMERASE β		Ki = 70 μM
DNA POLYMERASE γ		Ki = 970 nM

9-(2-PHOSPHONYLMETHOXYETHOYL)ADENINE DIPHOSPHATE

DNA POLYMERASE α		Ki = 1.2 μM
DNA POLYMERASE β		Ki = 70 μM
DNA POLYMERASE γ		Ki = 970 nM
RNA-DIRECTED DNA POLYMERASE		Ki = 12 nM

9-(2-PHOSPHONYLMETHOXYETHYL)HYPOXANTHINE

RIBOSE-PHOSPHATE PYROPHOSPHOKINASE		Ki = 41 μM

PHOSPHOPEPTIDE MG 25000

ORNITHINE DECARBOXYLASE	NC	

N-PHOSPHO-L-PHENYLALANYL-L-ARGININE

ASPERGILLUS ORYZAE NEUTRAL PROTEINASE		
PSEUDOMONAS AEROGINOSA NEUTRAL PROTEINASE		

N-(5′-PHOSPHOPYRIDOXYL)GLUTAMIC ACID

GLUTAMATE DECARBOXYLASE	C	Ki = 10 μM

PHOSPHORAMIDON

ATROLYSIN A		Ki = 330 μM
COCCOLYSIN		
COLLAGENASE		I50 = 92 μM
ELASTASE		Ki = 60 nM
ENDOPEPTIDASE TREPONEMA DENTICOLA ARCC35405		Ki = 220 μM
ENDOTHELIN CONVERTING ENZYME		I50 = 2 μM
ENDOTHELIN CONVERTING ENZYME	C	Ki = 200 nM
ENDOTHELIN CONVERTING ENZYME	C	Ki = 3 μM
ENDOTHELIN CONVERTING ENZYME		I50 = 3.5 μM
ENDOTHELIN CONVERTING ENZYME		I50 = 490 nM
ENDOTHELIN CONVERTING ENZYME		I50 = 400 nM
ENDOTHELIN CONVERTING ENZYME M1		I50 = 1 μM
ENDOTHELIN CONVERTING ENZYME M2		I50 = 300 pM
NEPRILYSIN		I50 = 40 nM
NEPRILYSIN		I50 = 71 nM
NEPRILYSIN		I50 = 40 nM
NEPRILYSIN		Ki = 20 nM
NEPRILYSIN		I50 = 4 nM
NEPRILYSIN		Ki = 3.4 nM
NEPRILYSIN		I50 = 100 nM
PEPTIDYL-DIPEPTIDASE A		I50 = 3.7 μM
PROCOLLAGEN N-ENDOPEPTIDASE		40% 100 μM
PROTEINASE NEUTRAL BACILLUS SUBTILIS	C	Ki = 6.5 μM
PROTEINASES METALLO		
STREPTOMYCES GRISEUS METALLO-ENDOPEPTIDASE I		100% 10 μM
STREPTOMYCES GRISEUS METALLO-ENDOPEPTIDASE II		100% 10 μM
TETANUS TOXIN LIGHT CHAIN		
THERMOLYSIN		Ki = 28 nM
THERMOLYSIN		

1-(5′-PHOSPHO-β-D-RIBOFURANOSYL)BARBITURIC ACID

OROTIDINE-5′-PHOSPHATE DECARBOXYLASE		Ki = 8.8 pM

5-PHOSPHORIBONATE

RIBOSE-5-PHOSPHATE ISOMERASE		

D-PHOSPHORIBONIC ACID

RIBOSE-5-PHOSPHATE ISOMERASE	C	Ki = 119 μM

5-PHOSPHO-α-D-RIBOSE
ATP PHOSPHORIBOSYLTRANSFERASE

PHOSPHORIBOSE DIPHOSPHATE
NICOTINATE PHOSPHORIBOSYLTRANSFERASE
OROTATE PHOSPHORIBOSYLTRANSFERASE

5-PHOSPHO-α-D-RIBOSE-1-DIPHOSPHATE
RIBOSE-PHOSPHATE PYROPHOSPHOKINASE C

5-PHOSPHO-α-D-RIBOSE-1-PHOSPHATE
NICOTINATE-NUCLEOTIDE PYROPHOSPHORYLASE (CARBOXYLATING)

5′-PHOSPHORIBOSYLAMINE
RIBOSE-5-PHOSPHATE-AMMONIA LIGASE

PHOSPHORIBOSYLAMINOIMIDAZOLECARBOXAMIDE
ADENOSYLHOMOCYSTEINASE 97% 1 mM

PHOSPHORIBOSYLAMINOIMIDAZOLECARBOXAMIDE RIBOSIDE
ADENOSINE DEAMINASE C *K*i = 540 μM
ADENOSYLHOMOCYSTEINASE *K*i = 70 μM

PHOSPHORIBOSYL-AMP
PROCOLLAGEN-PROLINE, 2-OXOGLUTARATE-4-DIOXYGENASE *K*i = 16 nM

N-(5-PHOSPHO-D-RIBOSYL)ANTHRNILIC ACID
ANTHRANILATE PHOSPHORIBOSYLTRANSFERASE

5-PHOSPHORIBOSYL 1-α-DIPHOSPHATE PHOSPHONATE ANALOGUES
OROTATE PHOSPHORIBOSYLTRANSFERASE

PHOSPHORIBOSYLFORMYLGLYCINEAMIDINE
PHOSPHORIBOSYLFORMYLGLYCINAMIDINE SYNTHASE

5-PHOSPHORIBOSYL-1-PYROPHOSPHATE
PURINE-NUCLEOSIDE PHOSPHORYLASE *K*i = 5.2 μM

PHOSPHO-RIBOSYLPYRROPHOSPHATE
CARBAMOYL-PHOSPHATE SYNTHASE (GLUTAMINE-HYDROLYSING) SSI

1-(5-PHOSPHORIBOSYL-α-D-RIBOSYL)ATP
ATP PHOSPHORIBOSYLTRANSFERASE

PHOSPHORIC ACID 3,3-DIFLUORO-4-OXOEICOSYL-2-AMINOETHYLESTER
PHOSPHOLIPASE A2 *I*50 = 70 μM

PHOSPHORIC ACID, 3,3-DIFLUORO-4-OXOEICOSYL 2-AMINOETHYL ESTER
PHOSPHOLIPASE A2 *I*50 = 70 μM

PHOSPHORIC ACID, 3,3-DIFLUORO-4-OXOEICOSYL 2-(TRIMETHYLAMINO)ETHYL ESTER
PHOSPHOLIPASE A2 *I*50 = 700 μM

3′,5′-PHOSPHORODITHIOATE
PROTEIN KINASE A2 *K*i = 4 μM

PHOSPHORODITHIOATE DEOXYCYTIDINE 28 MER
RNA-DIRECTED DNA POLYMERASE *I*50 = 1.5 nM

PHOSPHOROTHIOATE OLIGONUCLEOTIDES
ENDONUCLEASE KSP 632-I
RNA-DIRECTED DNA POLYMERASE

PHOSPHORTHIATE OLIGOCYTIDINE(28)
DNA POLYMERASE α *K*i = 120 nM
DNA POLYMERASE β *K*i = 550 nM
DNA POLYMERASE δ *K*i = 31 nM
DNA POLYMERASE γ *K*i = 47 nM
RIBONUCLEASE H1 *K*i = 70 nM
RIBONUCLEASE H2 *K*i = 450 nM

Inhibitor / Enzyme	Type	Value
N-PHOSPHORYL-ALANYL-PROLINE		
PEPTIDYL-DIPEPTIDASE A		*I*50 = 10 nM
PHOSPHORYL-ALA-PRO		
PEPTIDYL-DIPEPTIDASE A		*K*i = 1.4 nM
PHOSPHORYLCHOLINE		
ETHANOLAMINE KINASE		
PHOSPHOSERINE PHOSPHATASE		*I*50 = 225 μM
trans-4-PHOSPHORYLCYCLOHEXYL-1-(1,2-DIO-n-HEXADECYL-sn-GLYCER-3-YL-PHOSPHATE		
1-PHOSPHATIDYLINOSITOL 4-KINASE		*K*i = 194 μM
Nα-PHOSPHORYL-GLY-PRO-ALA AMIDE		
COLLAGENASE		*I*50 = 1.5 mM
PEPTIDYL-DIPEPTIDASE A		*K*i = 45 μM
Nα-PHOSPHORYL-GLY-PRO-ALA		
COLLAGENASE		*I*50 = 780 μM
PEPTIDYL-DIPEPTIDASE A		*K*i =4.2 μM
Nα-PHOSPHORYL-GLY-PRO-AMIDE		
COLLAGENASE		*I*50 = 3 mM
PEPTIDYL-DIPEPTIDASE A		*K*i =3.4 μM
Nα-PHOSPHORYL-GLY-PRO		
PEPTIDYL-DIPEPTIDASE A		*K*i = 50 nM
PHOSPHORYLHOMOCHOLINE		
1-ACYLGLYCEROPHOSPHOCHOLINE O-ACYLTRANSFERASE		
N-PHOSPHORYL-LEU-PHE		
AMINOPEPTIDASE SOLUBLE		*I*50 = 100 μM
ELASTASE PSEUDOMONAS AERUGINOSA		*K*i = 260 nM
NEPRILYSIN		*I*50 = 300 pM
THERMOLYSIN		*K*i = 19 nM
Nα-PHOSPHORYL-LEU-PHE		
PEPTIDYL-DIPEPTIDASE A		*K*i = 70 nM
N-PHOSPHORYL-LEU-TRP		
ELASTASE PSEUDOMONAS AERUGINOSA		*K*i = 26 nM
THERMOLYSIN	C	*K*i = 2 nM
THERMOLYSIN		*K*i = 2 nM
L-2-PHOSPHORYLOXY-3-PHENYLPROPIONIC ACID		
CARBOXYPEPTIDASE A	C	*K*i = 140 nM
N-PHOSPHORYL-PHE		
CARBOXYPEPTIDASE A		*K*i = 5 μM
PHOSPHOSERINE		
CHOLINE O-ACETYLTRANSFERASE	C	*K*i = 750 μM
GLYCINE TRANSPORT MITOCHONDRIA		
4-NITROPHENYLPHOSPHATASE		35% 10 mM
PHOSPHOGLYCERATE PHOSPHATASE		66% 7 mM
3-PHOSPHOSERINE		
GLUTAMATE SYNTHASE (FERREDOXIN)		
O-PHOSPHOSERINE		
GLYCINE HYDROXYMETHYLTRANSFERASE		
O-PHOSPHO-L-SERINE		
SERINE-SULFATE AMMONIA-LYASE	NC	*K*i = 6 mM
SERINE-SULFATE AMMONIA-LYASE	NC	*K*i = 6 mM
PHOSPHOTHREONINE		
4-NITROPHENYLPHOSPHATASE		46% 10 mM

ORNITHINE DECARBOXYLASE	NC	
THREONINE SYNTHASE		35% 10 mM
PHOSPHOTYROSINE		
ORNITHINE DECARBOXYLASE	NC	
PROTEIN TYROSINE-PHOSPHATASE		57% 400 μM
PHOTERCIN B		
1-PHOSPHATIDYLINOSITOL-4,5-BISPHOSPHATE PHOSPHODIESTERASE		42% 2 μM
PHOXIM		
ACETYLCHOLINESTERASE		
o-PHTHALALDEHYDE		
ALCOHOL DEHYDROGENASE	IR	
Ca^{2+}-TRANSPORTING ATPASE		
DEXTRANSUCRASE		
GLUTATHIONE REDUCTASE (NADPH)	IR	
HEXOKINASE		
MALATE DEHYDROGENASE	IR	
6-PHOSPHOFRUCTO-2-KINASE	IR	
SUCCINATE-SEMIALDEHYDE DEHYDROGENASE		
PHTHALAMIDE		
NAD ADP-RIBOSYLTRANSFERASE		$I50 = 1$ mM
1-2H-PHTHALAZINE HYDRAZONE		
DOPAMINE β-MONOOXYGENASE		$K$is = 5.7 μM
1(2H)-PHTHALAZINONE		
NAD ADP-RIBOSYLTRANSFERASE		$I50 = 12$ μM
PHTHALEINE DYES		
GLUTAMINASE		
PHTHALIC ACID		
ASPARTATE TRANSAMINASE		
4,5-DIHYDROOXYPHTHALATE DECARBOXYLASE		
2,3-DIKETO-L-GULONATE DECARBOXYLASE		90% 40 mM
DIPHOSPHOMEVALONATE DECARBOXYLASE		
3-HYDROXYANTHRANILATE 3,4-DIOXYGENASE		
KYNURENINE-OXOGLUTARATE AMINOTRANSFERASE		10% 1 mM
NICOTINATE-NUCLEOTIDE PYROPHOSPHORYLASE (CARBOXYLATING)		
XAA-PRO DIPEPTIDASE		Ki = 1.2 μM
o-PHTHALIC ACID		
GLUCOSE OXIDASE		
3-[N-(PHTHALIMIDOMETHYL)AMINO]-5-ETHYL-6-METHYLPYRIDIN-2(1H)-ONE		
PROTEINASE HIV-1		$I50 = 30$ nM
PHTHALONIC ACID		
2-OXOGLUTARATE CARRIER MITOCHONDRIA	C	Ki = 30 μM
α-PHTHALOYLGLUTAMINE		
GMP SYNTHASE (GLUTAMINE-HYDROLYSING)		
1(2H)PHTHALZINONEISULPHITE		
NAD(P)-ARGININE ADP-RIBOSYLTRANSFERASE		$I50 = 510$ μM
PHTHICOL		
LACTOYLGLUTATHIONE LYASE	C	Ki = 28 μM
PHYLLAMYCIN B		
DNA POLYMERASE α		$I50 = 289$ μM
RNA-DIRECTED DNA POLYMERASE		$I50 = 3.5$ μM
PHYLPA		
DNA POLYMERASE β		

DNA POLYMERASE α
DNA POLYMERASE δ

PHYSARIUM LYSOPHOSPHATIDIC ACID
DNA POLYMERASE α

(–)-PHYSOSTIGMINE
ACETYLCHOLINESTERASE — *I*50 = 31 nM

PHYSOSTIGMINE
ACETYLCHOLINESTERASE — *I*50 = 33 nM
ACETYLCHOLINESTERASE — *I*50 = 20 nM
AMIDASE
CARBOXYLESTERASE I
CHOLINESTERASE — *I*50 = 28 nM
HISTAMINE N-METHYLTRANSFERASE — *I*50 = 20 μM
HISTAMINE N-METHYLTRANSFERASE — *I*50 = 8 μM

PHYSOSTIGMINES SUBSTITUTED
ACETYLCHOLINESTERASE

PHYSOSTIGMINE SULPHATE
ACETYLCHOLINESTERASE — *I*50 = 91 nM
ACETYLCHOLINESTERASE — *I*50 = 50 nM

PHYTIC ACID
BISPHOSPHOGLYCERATE MUTASE — 20% 2 mM

PHYTOCHELATIN
GLUTATHIONE γ-GLUTAMYLTRANSFERASE

PHYTOHEMAGGLUTININ
CYTOCHROME-b5 REDUCTASE

PHYTOSPHINGOSINE
2-ACYLGLYCEROL O-ACYLTRANSFERASE
PHOSPHATIDATE PHOSPHATASE

PHYTYL DIPHOSPHATE
MEVALONATE KINASE

PIBENZIMOL
DNA TOPOISOMERASE

PICEATANNOL
MYOSIN-LIGHT-CHAIN KINASE — *I*50 = 12 μM
PROTEIN KINASE A — *I*50 = 3 μM
PROTEIN KINASE C — *I*50 = 8 μM
PROTEIN KINASE (Ca^{2+} DEPENDENT) — *I*50 = 19 μM
PROTEIN TYROSINE KINASE p40 — C — *K*i = 15 μM
PROTEIN TYROSINE KINASE p56lck — *I*50 = 66 μM

PICEID
PROTEIN KINASE C — *I*50 = 200 μg/ml
PROTEIN TYROSINE KINASE p56lckC — *I*50 = 200 μg/ml

cis-PICEID
PROTEIN KINASE C — *I*50 = 200 μg/ml
PROTEIN TYROSINE KINASE p56lckC — *I*50 = 500 μg/ml

α-PICOLINAMIDE
NAD ADP-RIBOSYLTRANSFERASE — *I*50 = 250 μM

β-PICOLINE
THROMBOXANE A SYNTHASE — 66% 100 μM

γ-PICOLINE
THROMBOXANE A SYNTHASE — 68% 100 μM

PICOLINIC ACID
D-AMINO-ACID OXIDASE
DEOXYHYPUSINE MONOOXYGENASE
3-HYDROXYANTHRANILATE 3,4-DIOXYGENASE C
NICOTINATE N-METHYLTRRANSFERASE
NICOTINATE-NUCLEOTIDE PYROPHOSPHORYLASE (CARBOXYLATING)
PHOSPHOENOLPYRUVATE CARBOXYKINASE (GTP)
PYRIDOXAL KINASE

α-PICOLINIC ACID
ACONITATE HYDRATASE C Ki = 24 mM
ALTRONATE DEHYDRATASE 76% 3 mM
DIHYDROPICOLINATE REDUCTASE
MANNONATE DEHYDRATASE 63% 3 mM

PICOLOINIC ACID
GLUCOSE-6-PHOSPHATASE 13% 4 mM

PICOPRAZOLE
H^+/K^+-EXCHANGING ATPASE I50 = 3.3 μM

PICOTAMIDE
THROMBOXANE A SYNTHASE

PICRAMIDE
GLUTATHIONE REDUCTASE (NADPH) I50 = 1 μM

PICRYL CHLORIDE
GLUTATHIONE REDUCTASE (NADPH) I50 = 15 μM

PIERICIDIN
NADH DEHYDROGENASE (UBIQUINONE) I50 = 6 nM
UBIQUINOL-CYTOCHROME-C REDUCTASE I50 = 300 nM

PIERICIDIN A
GLYCEROL-3-PHOSPHATE DEHYDROGENASE
NADH DEHYDROGENASE (UBIQUINONE) Ki = 1 nM
UBIQUINOL OXIDASE CYTb558-d COMPLEX I50 = 15 μM
UBIQUINOL OXIDASE CYTb562-o COMPLEX I50 = 2 μM

PIEROTOXIN
4-AMINOBUTYRATE TRANSAMINASE

PIG AORTA PROTEIN PHOSPHATASE-1 INHIBITOR
MYOSIN-LIGHT-CHAIN PHOSPHATASE

PIG AORTA PRPTEIN PHOSPHATASE-1 INHIBITOR
PHOSPHOPROTEIN PHOSPHATASE 1

PIG BRAIN RIBONUCLEASE INHIBITOR
RIBONUCLEASE B Ki = 1 nM
RIBONUCLEASE PANCREATIC Ki = 1 nM

PIGEON PEA AMYLASE INHIBITORS
α-AMYLASE

PIGEON PEA PROTEINASE INHIBITOR
TRYPSIN C Ki = 153 nM

PIGEON PEA TRYPSIN INHIBITOR
CHYMOTRYPSIN
TRYPSIN

PIG LEUKOCYTE PROTEINASE INHIBITOR
CATHEPSIN B Ki = 335 nM
CATHEPSIN H Ki = 125 nM
CATHEPSIN L Ki = 67 pM
CATHEPSIN S Ki = 46 pM
PAPAIN Ki = 190 pM

PIG LIVER RIBONUCLEASE INHIBITOR
RIBONUCLEASE

PIG PLASMA CARBONATE DEHYDRATASE INHIBITOR
CARBONATE DEHYDRATASE

PIG PLASMA PLASMA KALLIKREIN INHIBITOR
KALLIKREIN PLASMA

PIG PLASMA α-PROTEINASE INHIBITOR PI1
CHYMOTRYPSIN
TRYPSIN

PIG PLASMA α-PROTEINASE INHIBITOR PI2
CHYMOTRYPSIN
TRYPSIN

PIG PLASMA α-PROTEINASE INHIBITOR PI3
CHYMOTRYPSIN

PIG PLASMA α-PROTEINASE INHIBITOR PI4
CHYMOTRYPSIN

PIG PLASMA α-PROTEINASE INHIBITOR PO1A
PAPAIN

PIG PLASMA α-PROTEINASE INHIBITOR PO1B
PAPAIN

PIG SERUM ELASTASE INHIBITOR
ELASTASE

PIG TESTICULAR TRYPSIN INHIBITOR A

TRYPSIN	C	*K*i = 15 nM

PIG TESTICULAR TRYPSIN INHIBITOR B

TRYPSIN	NC	*K*i = 15 nM

PIG TESTICULAR TRYPSIN INHIBITOR Cnb

TRYPSIN	C	*K*i = 220 nM

PIG TESTIS RIBONUCLEASE INHIBITOR
RIBONUCLEASE

PIGUAMERIN

KALLIKREIN PLASMA		*K*i = 1 nM
KALLIKREIN TISSUE		*K*i = 5 nM
TRYPSIN		*K*i = 5 nM

PIG URINE Na^+/K^+-ATPASE INHIBITOR

Na^+/K^+-EXCHANGING ATPASE		72% 195 nM
Na^+/K^+-EXCHANGING ATPASE		*I*50 = 90 nM

PILOCARPINE

CYTOCHROME P450 2A6	C	*K*i = 6 μM
CYTOCHROME P450 2A6	C	*K*i = 4 μM

PIMELATE

4-HYDROXY-2-OXOGLUTARATE ALDOLASE		57% 40 mM

PIMELIC ACID

4-AMINOBUTYRATE TRANSAMINASE		
GLUTAMATE DECARBOXYLASE		
KYNURENINE-OXOGLUTARATE AMINOTRANSFERASE		53% 1 mM

PIMOBENDAN

3′,5′-CYCLIC-NUCLEOTIDE PHOSPHODIESTERASE		*I*50 = 560 nM
3′,5′-CYCLIC-NUCLEOTIDE PHOSPHODIESTERASE (cycAMP)		21% 300 μM
3′,5′-CYCLIC-NUCLEOTIDE PHOSPHODIESTERASE (cycAMP) III		*I*50 = 700 nM

3′,5′-CYCLIC-NUCLEOTIDE PHOSPHODIESTERASE (cycAMP) IV		*I*50 = 34 µM
3′,5′-CYCLIC-NUCLEOTIDE PHOSPHODIESTERASE (Ca^{2+}/CALMODULIN)		*I*50 = 1.3 µM
3′,5′-CYCLIC-NUCLEOTIDE PHOSPHODIESTERASE (cycGMP)		26% 300 µM
3′,5′-CYCLIC-NUCLEOTIDE PHOSPHODIESTERASE (cGMP INHIBITABLE)		*I*50 = 3.7 µM
3′,5′-CYCLIC-NUCLEOTIDE PHOSPHODIESTERASE I		*I*50 = 131 µM
3′,5′-CYCLIC-NUCLEOTIDE PHOSPHODIESTERASE I		*I*50 = 233 µM
3′,5′-CYCLIC-NUCLEOTIDE PHOSPHODIESTERASE I		*I*50 = 295 µM
3′,5′-CYCLIC-NUCLEOTIDE PHOSPHODIESTERASE II		*I*50 = 242 µM
3′,5′-CYCLIC-NUCLEOTIDE PHOSPHODIESTERASE II		*I*50 = 112 µM
3′,5′-CYCLIC-NUCLEOTIDE PHOSPHODIESTERASE II		*I*50 = 221 µM
3′,5′-CYCLIC-NUCLEOTIDE PHOSPHODIESTERASE III		*I*50 = 880 nM
3′,5′-CYCLIC-NUCLEOTIDE PHOSPHODIESTERASE III		*I*50 = 2.4 µM
3′,5′-CYCLIC-NUCLEOTIDE PHOSPHODIESTERASE IV		*I*50 = 151 µM
3′,5′-CYCLIC-NUCLEOTIDE PHOSPHODIESTERASE IV (ROLIPRAM INSENSITIVE)	C	*K*i = 65 nM
PIMOZIDE		
3′,5′-CYCLIC-NUCLEOTIDE PHOSPHODIESTERASE		*I*50 = 7 µM
NITRIC-OXIDE SYNTHASE		*I*50 = 19 µM
PIN		
NITRIC-OXIDE SYNTHASE I		
PINANYL-N,N,N-TRIMETHYLAMINOMETHANE BORONATE		
BETAINE-HOMOCYSTEINE S-METHYLTRANSFERASE		
PINAPPLE STEM CYSTEINE PROTEINASE INHIBITOR		
PROTEINASES CYSTEINE		
PIN A PROTEIN		
PROTEINASE LON	NC	
PINDOLOL		
LIPOPROTEIN LIPASE		*I*25 = 16 mM
PINEAPPLE PROTEINASE INHIBITOR		
BROMELAIN		
D,L-PIPECOLIC ACID		
SACCHAROPINE DEHYDROGENASE (NADP, L-LYSINE FORMING)		
PIPECURONIUM		
HISTAMINE N-METHYLTRANSFERASE	C	*K*i = 45 µM
PIPEMIDIC ACID		
CYTOCHROME P450 LA2		60% 500 µM
DNA TOPOISOMERASE (ATP-HYDROLYSING)		*I*50 = 650 µg/ml
PIPERALIN		
CHOLESTENOL Δ-ISOMERASE		
PIPERASTATIN A		
PEPTIDYL-DIPEPTIDASE A		*I*50 = 42 µM
SERINE-TYPE CARBOXYPEPTIDASE	C	*K*i = 64 nM
PIPERASTATIN B		
PEPTIDYL-DIPEPTIDASE A		*I*50 = 40 µg/ml
SERINE-TYPE CARBOXYPEPTIDASE		*K*i = 55 nM
PIPERAZINE		
Na^+/K^+-EXCHANGING ATPASE		*I*50 = 600 µM
PIPERAZINE-N,N′-BIS(2-ETHANESULPHOINIC ACID)		
L-ASCORBATE OXIDASE		
PIPERIDINE		
CYTOCHROME P450 2E1	NC	*K*i = 3.1 mM
Δ^1-PIPERIDINE-2-CARBOXYLIC ACID		
D-AMINO-ACID OXIDASE		

3-(1-PIPERIDINO)-1-HYDROXYPROPYLIDENE-1,1-BISPHOSPHONIC ACID
FARNESYL-DIPHOSPHATE FARNESYLTRANSFERASE — *I*50 = 311 nM

20 PIPERIDIN-2-YL-5α-PREGNAN-3β,20-DIOL
Δ(24)-STEROL C-METHYLTRANSFERASE — *I*50 = 20 nM

2-PIPERIDONE-6-CARBOXYLIC ACID
5-OXOPROLINASE (ATP-HYDROLYSING)

PIPERONYL BUTOXIDE
CYTOCHROME P450 1A
CYTOCHROME P450 2B
CYTOCHROME P450 2C
CYTOCHROME P450 2C19
CYTOCHROME P450 2D
CYTOCHROME P450 6D1 — *I*50 = 430 nM
ECDYSONE 20-MONOOXYGENASE

PIPES BUFFER
CREATINE KINASE

P58IPK
PROTEIN KINASE R

PIRIMICARB
ACETYLCHOLINESTERASE

PIRIMIPHOS-ETHYL
ACETYLCHOLINESTERASE

PIRIPROST
GLUTATHIONE TRANSFERASE

PIRITREXIM
AMIDOPHOSPHORIBOSYLTRANSFERASE — *K*i = 66 μM

PIROGALLOL
PROTEIN TYROSINE KINASE — *I*50 = 12 μM

PIROXICAM
CYCLOOXYGENASE 1 — *I*50 = 13 μM
ELASTASE LEUKOCYTE
PROSTAGLANDIN-ENDOPEROXIDE SYNTHASE — *I*50 = 300 nM
PROSTAGLANDIN-ENDOPEROXIDE SYNTHASE 1 — *I*50 = 24 μM
PROSTAGLANDIN-ENDOPEROXIDE SYNTHASE 2 — *I*50 = 240 μM
PROSTAGLANDIN H ENDOPEROXIDE SYNTHASE 1 — *I*50 = 104 μM

PIROXIMONE
3′,5′-CYCLIC-NUCLEOTIDE PHOSPHODIESTERASE (cycAMP) LOW Km — *I*50 = 17 μM
3′,5′-CYCLIC-NUCLEOTIDE PHOSPHODIESTERASE (cycGMP) — *I*50 = 1.8 mM

PISUM SATIVUM TRYPSIN INHIBITOR
TRYPSIN

PITUITARY ADENYLATE CYLCASE ACTIVATING PROTEIN
CALMODULIN

N-(PIVALOYL)GLYCINOHYDROXAMIC ACID
UREASE

2-PIVALOYL-1,3-INDANEDIONE
NAD(P)H DEHYDROGENASE (QUINONE)

PIV-HIS-PRO-PHE-HIS-LEU-psi[CH(OH)CH_2]LEU-TYR-TYR-SER-NH_2
RENIN — *I*50 = 210 pM

PKC(19-31)
PROTEIN KINASE C — *I*50 = 750 nM

PKC INHIBITOR FROM PHYTOHAEMAGGLUTININ STIMULATED PERIPHERIAL BLOOD MONONUCLEAR CELLS
PROTEIN KINASE C

PKIα
PROTEIN KINASE A

PKIβ
PROTEIN KINASE A

PKSI
KALLIKREIN PLASMA

PKSI 527
KALLIKREIN PLASMA — Ki = 810 nM
KALLIKREIN PLASMA — Ki = 810 nM
PLASMIN — Ki = 390 μM
PLASMIN — Ki = 390 μM
u-PLASMINOGEN ACTIVATOR — I50 = 350 μM
u-PLASMINOGEN ACTIVATOR — Ki = 200 μM
TRYPSIN — I50 = 12 μM

PLACENTA APYRASE INHIBITOR
APYRASE

PLACENTA INHIBITOR
CARBONYL REDUCTASE (NADPH)

PLACENTA RIBONUCLEASE INHIBITOR
RIBONUCLEASE — Ki = 1 fM
RIBONUCLEASE PANCREATIC — Ki < 1 pM

PLASMA INHIBITOR
Na^+/K^+-EXCHANGING ATPASE — NC — I50 = 800 nM

PLASMA Na^+/K^+ TRANSPORTING ATPASE INHIBITOR
Na^+/K^+-EXCHANGING ATPASE

α2-PLASMIN INHIBITOR
PROTEINASES SERINE

PLASMINOGEN ACTIVATOR INHIBITOR-1
PLASMINOGEN RECEPTOR BOUND

PLASMINOGEN ACTIVATOR INHIBITOR-2
PLASMINOGEN RECEPTOR BOUND

PLASMINOSTATIN
SUBTILISIN BPN — Ki = 25 pM
TRYPSIN — Ki = 21 pM

PLASMINOSTREPTIN
STREPTOMYCES GRISEUS METALLO-ENDOPEPTIDASE I — 100% 100 nM
STREPTOMYCES GRISEUS METALLO-ENDOPEPTIDASE II — 100% 100 nM
SUBTILISIN BPN′ — Ki = 25 pM
TRYPSIN — Ki = 21 pM

PLASTOQUINONE 9
γ-GLUTAMYL CARBOXYLASE — I50 = 39 μM

PLATANETIN
NADH DEHYDROGENASE (EXOGENOUS) — I50 = 2 μM
NADH OXIDASE EXTERNAL — I50 = 4 μM
NADH OXIDASE (ROTENONE INSENSITIVE) — I50 = 20 μM
NADPH OXIDASE — I50 = 5 μM
NADPH OXIDASE ROTENONE INSENSITIVE — I50 = 14 μM

PLATANIN
NADH DEHYDROGENASE (EXOGENOUS) — I50 = 15 μM

PLATELET ACTIVATING FACTOR		
1-ALKYLGLYCEROPHOSPHOCHOLINE O-ACETYLTRANSFERASE	IR	
1,3-β-GLUCAN SYNTHASE		$I50 = 325\ \mu M$
Na^+/K^+-EXCHANGING ATPASE		
PLATELET RIBONUCLEASE INHIBITOR		
RIBONUCLEASE		$Ki = 130$ pM
PLENINDONE		
NAD(P)H DEHYDROGENASE (QUINONE)	C	$Ki = 110$ nM
P-LEU-psi($COCH_2$)(R,S)PHE-OMe		
NEPRILYSIN		$I50 = 15$ nM
PEPTIDYL-DIPEPTIDASE A		$I50 = 6.7\ \mu M$
THERMOLYSIN		$I50 = 1.3\ \mu M$
P-LEU-psi($COCH_2$)(R,S)TRP-OMe		
NEPRILYSIN		$I50 = 30$ nM
PEPTIDYL-DIPEPTIDASE A		$I50 = 1.5\ \mu M$
P-LEU-PHE-OH		
NEPRILYSIN		$I50 = 300$ pM
P-LEU-PHE-OMe		
NEPRILYSIN		$I50 = 9$ nM
PEPTIDYL-DIPEPTIDASE A		$I50 = 4.5\ \mu M$
THERMOLYSIN		$I50 = 1\ \mu M$
PLEUROTUS OSTREATUS SERINE PROTEINASE INHIBITORS		
PROTEINASES SERINE		
P-LEU-TRP-OH		
THERMOLYSIN		$I50 = 330$ nM
PLICACETIN		
PEPTIDYLTRANSFERASE		
PLIPASTATINS		
PHOSPHOLIPASE A2		
PHOSPHOLIPASE C		
PHOSPHOLIPASE D		
PLUMBAGIN		
Ca^{2+}-TRANSPORTING ATPASE		
β-LACTAMASE		
NAD ADP-RIBOSYLTRANSFERASE		
NITRIC-OXIDE SYNTHASE I		$I50 = 27\ \mu M$
PMDF		
PROTEINASE DICHELOBACTER NODOSUS EXTRACELLULAR		100% 1 mM
PMN CYTOSOLIC LEUKOCYTE ELASTASE INHIBITOR		
CATHEPSIN G		
ELASTASE LEUKOCYTE		
PMP		
FORMATE-TETRAHYDROFOLATE LIGASE		
PMP-C		
α-CHYMOTRYPSIN		$Ki = 200$ pM
ELASTASE LEUKOCYTE		$Ki = 120$ nM
PMP-D2		
α-CHYMOTRYPSIN		$Ki = 1.5\ \mu M$
PMPS		
CHOLESTANETRIOL 26-MONOOXYGENASE		
SULFITE REDUCTASE (NADPH)		

PMS 832
PHOSPHOLIPASE A2-2 — $Ki = 4.1\ \mu M$

PNEUMOCANDIN B0
1,3-β-GLUCAN SYNTHASE

P-NITROPHENYL-P′-GUANIDINOBENZOATE
PROTEINASE SERINE PROMONOCYTOC U-937 — 36% 1 μM

PNU 103017
PROTEINASE HIV-1

PNU 140690
PROTEINASE HIV-1 — $I50 < 10$ pM
PROTEINASE HIV-2 — $I50 < 1$ nM

PNU 142721
RNA-DIRECTED DNA POLYMERASE — $I50 = 20$ nM

PNU 157706
STEROID 5α-REDUCTASE 1 — $I50 = 3.9$ nM
STEROID 5α-REDUCTASE 2 — $I50 = 1.8$ nM

POCA-CoA
CARNITINE O-PALMITOYLTRANSFERASE 1

PODOCARPIC ACID PHOSPHATE
PROTEIN TYROSINE-PHOSPHATASE — $Ki = 4.9\ \mu M$

PODOSCYPHIC ACID
RNA-DIRECTED DNA POLYMERASE — $I50 = 20\ \mu g/ml$

POFLOXACIN
DNA TOPOISOMERASE (ATP-HYDROLYSING) — $I50 = 550\ \mu g/ml$
DNA TOPOISOMERASE (ATP-HYDROLYSING) — $I50 = 90\ \mu g/ml$

POLIMYXIN B
GLUCOSE TRANSPORT

POLIMYXIN E
GLUCOSE TRANSPORT

POLY(A)
ADENYLATE CYCLASE — $I50 = 450$ nM
ADP-RIBOSE GLYCOHYDROLASE — C — $Ki = 5.5\ \mu M$
PEPTIDYLTRANSFERASE

POLY(2-ACRYLAMIDO-2-METHYL-1-PROPANESULPHONIC ACID)
RNA-DIRECTED DNA POLYMERASE — $I50 = 2.4\ \mu g/ml$

POLY(ACRYLATE)
DNA POLYMERASE α — $I50 = 1.7\ \mu g/ml$

POLY(ADP-RIBOSE)
PHOSPHODIESTERASE — 33% 1 mM

POLY(ADP-RIBOSE)18
PROCOLLAGEN-PROLINE, 2-OXOGLUTARATE-4-DIOXYGENASE — NC — $Ki = 1.5$ nM

POLYALTHIDIN
NADH CYTOCHROME C REDUCTASE

POLY(ANETHOL SULPHONIC ACID)
COMPLEMENT SUBCOMPONENT C1r

POLY(ANETHOLSULPHONIC ACID)
RNA-DIRECTED DNA POLYMERASE — $I50 = 40$ ng/ml

POLYARGININE
HISTONE ACYLTANSFERASE

MYOSIN II HEAVY CHAIN KINASE A
PROTEIN KINASE G
RNA-DIRECTED RNA POLYMERASE

POLY-ASP
PROCOLLAGEN N-ENDOPEPTIDASE — 81% 10 μg/ml

POLY(L-ASPARTATE)42KDa
DNA POLYMERASE α — $I50$ = 1.5 μg/ml
DNA POLYMERASE β — $I50$ = 130 μg/ml

POLYASPARTIC ACID
β-ADRENERGIC-RECEPTOR KINASE — $I50$ = 1.3 μM

POLYCARBOPHIL
TRYPSIN
TRYPSIN

POLY[2′-O-(2,4-DINITROPHENYL)]POLY-A
3′,5′-CYCLIC-NUCLEOTIDE PHOSPHODIESTERASE I — $I50$ = 8 μM
3′,5′-CYCLIC-NUCLEOTIDE PHOSPHODIESTERASE II — $I50$ = 5.7 μM
RIBONUCLEASE B — $I50$ = 500 nM
RIBONUCLEASE H — $I50$ = 140 nM
RIBONUCLEASE S — $I50$ = 80 nM
RIBONUCLEASE T1 — $I50$ = 2.5 μM
RIBONUCLEASE T2 — $I50$ = 5.3 μM

POLYETHYLENE SULPHONATE
PHOSPHOGLUCONATE 2-DEHYDROGENASE — $I50$ < 10 nM
RNA POLYMERASE

POLY(2′-FLUORO-2′-DEOXYADENYLIC ACID)
RNA-DIRECTED DNA POLYMERASE — 94% 1 μg/ml

POLY(G)
ADP-RIBOSE GLYCOHYDROLASE — C — Ki = 2.8 μM
EXORIBONUCLEASE II

POLYGALACTURONASE INHIBITING PROTEIN
POLYGALACTURONASE

POLY(GALACTURONIC ACID)
HYALURONIDASE
PECTINESTERASE
PECTIN METHYLESTERASE — Ki = 30 μg/ml

POLY(GLU3,PHE1)
RNA POLYMERASE

POLYGLUTAMATE
POLYGALACTURONASE

POLY(L-GLUTAMATE)36KDa
DNA POLYMERASE α — $I50$ = 67 μg/ml
DNA POLYMERASE β — $I50$ = 660 μg/ml

POLY(GLUTAMIC ACID)
β-ADRENERGIC-RECEPTOR KINASE — $I50$ = 2 μM
HISTONE ACYLTANSFERASE

POLY(L-GLUTAMIC ACID)
N-ACETYLLACTOSAMINE SYNTHASE
ORNITHINE DECARBOXYLASE
PROTEINASE C1

POLY(GLU1,TYR1)
RNA POLYMERASE

POLY(I)
ADP-RIBOSE GLYCOHYDROLASE

POLYIXIN D
CHITIN SYNTHASE C $Ki = 6\ \mu M$

POLY(LYSINE)
β-ADRENERGIC-RECEPTOR KINASE $I50 = 69\ \mu M$
GUANYLATE CYCLASE $Ki = 685\ \mu M$
HISTONE ACYLTANSFERASE
LOW DENSITY LIPOPROTEIN KINASE
LYSOZYME
MYOSIN II HEAVY CHAIN KINASE A $Ki = 1.7\ \mu g/ml$
PHOSPHOPROTEIN PHOSPHATASE
PHOSPHORYLASE PHOSPHATASE
RNA-DIRECTED RNA POLYMERASE

POLY(L-LYSINE)
GLYCOGEN SYNTHASE KINASE 3 $I50 = 100\ nM$
INOSITOL-1,3,4-TRISPHOSPHATE KINASE C $I50 = 200\ \mu M$
PHOSPHORYLASE PHOSPHATASE
PROTEIN KINASE C U $Ki = 360\ \mu M$
RIBONUCLEASE U2

POLY(L-MALATE)12KDa
DNA POLYMERASE α $I50 = 300\ \mu g/ml$
PRIMASE $I50 = 300\ \mu g/ml$

POLY(L-MALATE)15KDa
DNA POLYMERASE α $I50 = 9\ ng/ml$
DNA POLYMERASE β $I50 = 44\ \mu g/ml$

POLY(L-MALATE)
DNA POLYMERASE α C $Ki = 10\ ng/ml$

POLYMORPHONUCLEAR LEUKOCYTE COLLAGENASE INHIBITOR
COLLAGENASE NEUTROPHIL

POLYMYXIN B
CALMODULIN-LYSINE N-METHYLTRANSFERASE $I50 = 298\ \mu M$
GLYCEROL-3-PHOSPHATE O-ACYLTRANSFERASE
MYOSIN-LIGHT-CHAIN KINASE C $Ki = 17\ \mu M$
PROTEIN KINASE C $I50 = 17\ \mu M$
PROTEIN KINASE C C $Ki = 1.8\ \mu M$

POLYNUCLEOTIDES
RNA POLYMERASE

POLY(ORNITHINE)
RNA-DIRECTED RNA POLYMERASE

POLY-D,L-ORNITHINE
α,α-TREHALOSE-PHOSPHATE SYNTHASE

POLYOXIN B
CHITIN SYNTHASE $I50 = 800\ ng/ml$

POLYOXIN D
CHITIN SYNTHASE C $Ki = 1.4\ \mu M$
CHITIN SYNTHASE 57% 13 μM
CHITIN SYNTHASE 55% 13 μM
CHITIN SYNTHASE 40% 50 μM
CHITIN SYNTHASE Chs 3 $Ki = 8.3\ \mu M$

POLYOXYETHYLATED CHOLESTEROL
STEROL O-ACYLTRANSFERASE

POLY(9)OXYETHYLENELAURYL ETHER
MONOPHENOL MONOOXYGENASE

POLYOZELLIN
PROLYL OLIGOPEPTIDASE — *K*i = 25 μM

POLYPHENOLS
α-ACID OXIDASE
HYALURONATE LYASE

POLYPHLORETINPHOSPHATE
CYSTINYL AMINOPEPTIDASE
15-HYDROXYPROSTAGLANDIN DEHYDROGENASE (NAD) — *K*i = 19 μg/ml
PHOSPHATASE ALKALINE

POLY-(L-PROLINE)
PROCOLLAGEN-PROLINE, 2-OXOGLUTARATE-4-DIOXYGENASE

POLYSACCHARIDE SULPHATES
ELASTASE LEUKOCYTE

POLY(4-STYRENESULPHONIC ACID)
RNA-DIRECTED DNA POLYMERASE — *I*50 = 60 ng/ml

POLYTRIPHOSPHATE
FRUCTOSE-BISPHOSPHATE ALDOLASE — Kd = 25 μM

POLY(U)
PEPTIDYLTRANSFERASE

POLY(URIDINE 5′-p-STYRENESULPHONATE)
GALACTOSYL TRANSFERASE — 75% 120 μM

POLYVINYLPYRROLIDONE
MONOPHENOL MONOOXYGENASE

POLYVINYLSULPHATE
ADENYLATE CYCLASE
DNA POLYMERASE α — *I*50 = 400 ng/ml
HEPARIN LYASE
INOSITOL-1,3,4-TRISPHOSPHATE KINASE — C — *I*50 = 250 μM
POLYNUCLEOTIDE ADENYLYLTRANSFERASE
tRNA(ADENINE-N^6)-METHYLTRANSFERASE

POLY(VINYLSULPHONIC ACID)
RNA-DIRECTED DNA POLYMERASE — *I*50 = 1.9 μg/ml

POMBUS
CARBONATE DEHYDRATASE II — *K*i = 22 nM
CARBONATE DEHYDRATASE IV — *K*i = 475 nM

PONALRESTAT
ALCOHOL DEHYDROGENASE (NADP) — *I*50 = 2.4 μM
ALDEHYDE REDUCTASE — *I*50 = 25 nM
ALDEHYDE REDUCTASE 1 — NC — *K*i = 60 μM
ALDEHYDE REDUCTASE 2 — MIX — *K*i = 7.7 μM
3α-HYDROXYSTEROID DEHYDROGENASE (B-SPECIFIC) — *I*50 = 59 μM

POPOLOHUANONE E
DNA TOPOISOMERASE (ATP-HYDROLYSING) — *I*50 = 400 nM

PORCINE KIDNEY H^+ TRANSPORTING ATPASE INHIBITOR
H^+-TRANSPORTING ATPASE (VACUOLAR)

PORCINE LEUKOCYTE PLASMIN/TRYPSIN INHIBITOR
PLASMIN — *K*i = 168 pM
TRYPSIN — *K*i = 16 pM

PORCINE LIVER RIBONUCLEASE INHIBITOR
RIBONUCLEASE

PORCINE PLASMA CARBONATE DEHYDRATASE II INHIBITOR
CARBONATE DEHYDRATASE I $K_i = 50\ \mu M$
CARBONATE DEHYDRATASE I $K_i = 20\ \mu M$
CARBONATE DEHYDRATASE II $K_i = 600\ pM$
CARBONATE DEHYDRATASE II $K_i = 2.9\ nM$
CARBONATE DEHYDRATASE II $K_i = 1\ nM$
CARBONATE DEHYDRATASE II $K_i = 10\ nM$
CARBONATE DEHYDRATASE III $K_i = 23\ \mu M$
CARBONATE DEHYDRATASE III $K_i = 70\ \mu M$
CARBONATE DEHYDRATASE IV $K_i = 150\ nM$

PORCION BLUE MX-R
PYRUVATE KINASE

PORCION RED
GALACTOSYLCERAMIDE SULFOTRANSFERASE

PORCION RED HE3B
ALCOHOL DEHYDROGENASE (NADP) C

PORCION RED MX-5B
ALCOHOL DEHYDROGENASE (NADP)

PORCION SCARLET MX-G
ALCOHOL DEHYDROGENASE (NADP)

PORPHOBILINOGEN
UROPORPHYRINOGEN DECARBOXYLASE 35% 8 µM

POSTSTATIN
CATHEPSIN B $I50 = 2.1\ \mu g/ml$
ELASTASE LEUKOCYTE $I50 = 110\ \mu g/ml$
PROLYL OLIGOPEPTIDASE $I50 = 30\ ng/ml$

POTASSIUMCHLORIDE
ELASTASE PANCREATIC
PHOSPHOLIPASE A1 82% 150 mM

POTASSIUMTETRACHLOROPLATINAT
DIHYDROPTERIDINE REDUCTASE IRR

POTATO ASPARTIC PROTEINASE INHIBITORS
PROTEINASES ASPARTIC

POTATO ATPASE INHIBITOR
H^+-TRANSPORTING ATPASE

POTATO CARBOXYPEPTIDASE INHIBITOR
ACETYLCHOLINESTERASE
CARBOXYPEPTIDASE A $K_i = 5\ nM$
CARBOXYPEPTIDASE B $K_i = 50\ nM$
CHOLINESTERASE

POTATO CARBOXYPEPTIDASE INHIBITOR I
CARBOXYPEPTIDASE A

POTATO CATHEPSIN D INHIBITOR
CATHEPSIN D
TRYPSIN

POTATO CHYMOTRYPSIN INHIBITOR
CHYMOTRYPSIN

POTATO CHYMOTRYPSIN INHIBITOR I
CHYMOTRYPSIN

POTATO INHIBITOR
ASPERGILLUS ALKALINE PROTEINASE
EUPHORBAIN

POTATO INHIBITOR I
PROTEINASE DF — *K*i = 660 pM

POTATO INHIBITOR IIa
PROTEINASE DF — *K*i = 9.2 nM

POTATO INHIBITOR IIb
PROTEINASE DF — *K*i = 5.7 nM

POTATO INVERTASE INHIBITOR
β-FRUCTOFURANOSIDASE

POTATO KUNITZ INHIBITOR-1
TRYPSIN

POTATO KUNITZ INHIBITOR-2
CHYMOTRYPSIN
SUBTILISIN
TRYPSIN

POTATO MULTICYSTATIN
PAPAIN — *K*i = 2 μM

POTATO PROTEINASE INHIBITOR
CHYMOPAPAIN
FICAIN
GLIOCLADIUM PROTEINASE
PAPAIN

POTATO TRYPSIN/CHYMOTRYPSIN INHIBITOR
CHYMOTRYPSIN
TRYPSIN

PP 1
PROTEIN TYROSINE KINASE Hck — *I*50 = 20 nM
PROTEIN TYROSINE KINASE p56lck — *I*50 = 5 nM
PROTEIN TYROSINE KINASE Src — *I*50 = 170 nM
PROTEIN TYROSINE KINASE p59fymT — *I*50 = 6 nM

PP 36
DOLICHYL-PHOSPHATE β-GLUCOSYLTRANSFERASE — 68% 50 μM

PP 37
DOLICHYL-PHOSPHATE β-GLUCOSYLTRANSFERASE — 80% 50 μM

PP 63
PROTEIN TYROSINE KINASE (INSULIN RECEPTOR)

PP 55A
DOLICHYL-PHOSPHATE β-GLUCOSYLTRANSFERASE — 82% 50 μM

PP 50B
UDP-GLUCURONOSYLTRANSFERASE — C — *K*i = 35 μM

P60v-src PEPTIDE(137-157)
PROTEIN TYROSINE KINASE EGFR — *I*50 = 7.5 μM
PROTEIN TYROSINE KINASE p60v-src — *I*50 = 7.5 μM

Pb(5-PIVALOYLAMIDO-1,3,4-THIDIAZOLE-2-SULPHONAMIDE)$_2$
CARBONATE DEHYDRATASE — *I*50 = 4 nM

P68 PROTEIN KINASE INHIBITOR
PROTEIN TYROSINE KINASE P68

Pr^{3+}

ATP PYROPHOSPHATASE		37% 400 μM
GLUCOSE-6-PHOSPHATE 1-DEHYDROGENASE		80% 400 μM

Pr^{3+}

GLUCOSE-6-PHOSPHATE 1-DEHYDROGENASE		91% 830 μM
ISOCITRATE DEHYDROGENASE		21% 830 μM
L-LACTATE DEHYDROGENASE		78% 830 μM

PR 109

STEROL O-ACYLTRANSFERASE		*I*50 = 6 nM

PRAETOLOL

LIPOPROTEIN LIPASE		

PRAVASTATIN

CYTOCHROME P450 2C9		*I*50 = 70 μM
HYDROXYMETHYLGLUTARYL-CoA REDUCTASE (NADPH)		*I*50 = 275 nM
HYDROXYMETHYLGLUTARYL-CoA REDUCTASE (NADPH)		*I*50 = 40 nM
HYDROXYMETHYLGLUTARYL-CoA REDUCTASE (NADPH)		*I*50 = 108 nM
HYDROXYMETHYLGLUTARYL-CoA REDUCTASE (NADPH)		*I*50 = 264 nM
HYDROXYMETHYLGLUTARYL-CoA REDUCTASE (NADPH)		*I*50 = 239 nM
HYDROXYMETHYLGLUTARYL-CoA REDUCTASE (NADPH)		*I*50 = 11 nM

PREDNISOLONE

CALPAIN		70% 4.1 mM
3α-HYDROXYSTEROID DEHYDROGENASE (A SPECIFIC)		
3α-HYDROXYSTEROID DEHYDROGENASE (B-SPECIFIC)		52% 10 μM
3α-HYDROXYSTEROID DEHYDROGENASE (B-SPECIFIC)	C	*K*i = 18 μM

PREDNISONE

3α-HYDROXYSTEROID DEHYDROGENASE (B-SPECIFIC)	C	*K*i = 18 μM

PREG-4-ENE-3,20,BISGUANYLHYDRAZONE

Na^+/K^+-EXCHANGING ATPASE		*I*50 = 8.7 μM

PREG-4-ENE-3,20-BISGUANYLHYDRAZONE

Na^+/K^+-EXCHANGING ATPASE		100% 100 μM

PREGNA-5,16-DIEN-3β-OL-20-ONE OXIME

STERIOD 17α-MONOOXYGENASE		*I*50 = 16 nM
STEROID 17α-HYDROXYLASE/17,20 LYASE		*I*50 = 16 nM

PREGNANCY ZONE PROTEIN

PROTEINASE		

5α-PREGNANDIONE

11β-HYDROXYSTEROID DEHYDROGENASE 2		*I*50 = 100 nM

5β-PREGNANDIONE

11β-HYDROXYSTEROID DEHYDROGENASE 2		*I*50 = 500 nM

5β-PREGNAN-3,20-DIONE

CARBONYL REDUCTASE (NADPH)		

5α-PREGNAN-3β-OL-20-ONE

trans-1,2-DIHYDROBENZENE-1,2-DIOL DEHYDROGENASE		66% 1 μM
3α(or 17β)-HYDROXYSTEROID DEHYDROGENASE		

5β-PREGNAN-3β-OL-20-ONE

trans-1,2-DIHYDROBENZENE-1,2-DIOL DEHYDROGENASE		80% 1 μM
3α(or 17β)-HYDROXYSTEROID DEHYDROGENASE		

5-PREGNENE-3β.20α-DIOL

4-ANDROSTENE-3,17-DIONE MONOOXYGENASE		

4-PREGNENE-3,20-DIONE

3α-HYDROXYSTEROID DEHYDROGENASE (B-SPECIFIC)		47% 1 μM

4-PREGNENE-3-ONE-20β-CARBOXALDEHYDE		
17α-HYDROXYPROGESTERONE ALDOLASE	C	*K*i = 410 nM
STEROID 17α-MONOOXYGENASE	C	*K*i = 8.5 μM
STEROID 5α-REDUCTASE		*K*i = 16 nM
PREGN-5-EN-β-OL		
STEROL O-ACYLTRANSFERASE		
PREGNENOLONE		
4-ANDROSTENE-3,17-DIONE MONOOXYGENASE		
CHOLESTEROL 7α-MONOOXYGENASE		
ESTRONE SULFOTRANSFERASE		89% 10 μM
STEROID 17α-MONOOXYGENASE		
STEROL O-ACYLTRANSFERASE		
PREGNENOLONE-3-O-METHYLTHIOPHOSPHONATE		
OESTRONE SULFATASE		*K*i = 1.4 μM
PREGNENOLONE MONOSULPHATE		
GLUTATHIONE TRANSFERASE		
PREMETHRIN		
NADH DEHYDROGENASE (UBIQUINONE)		*I*50 = 730 nM
8-PRENYLLUTEONE		
1-PHOSPHATIDYLINOSITOL PHOSPHODIESTERASE		*I*50 = 20 μM
PREPHENIC ACID		
ANTHRANILATE SYNTHASE		
AROMATIC-AMINO-ACID TRANSAMINASE		
PREPHENATE AMINOTRANSFERASE		> 1 mM
PRESICANIDINE A		
3′,5′-CYCLIC-NUCLEOTIDE PHOSPHODIESTERASE		*I*50 = 250 μM
PRESICANIDINE B 3-O-β-D-GLUCOPYRANOSIDE		
3′,5′-CYCLIC-NUCLEOTIDE PHOSPHODIESTERASE		*I*50 = 183 μM
PRE-α-TRYPSIN INHIBITOR		
TRYPSIN		
PREUSSOMERIN G		
FARNESYL PROTEINTRANSFERASE		*I*50 = 1.2 μM
PREUSSOMERIN H		
FARNESYL PROTEINTRANSFERASE		*I*50 = 12 μM
PREUSSOMERIN I		
FARNESYL PROTEINTRANSFERASE		*I*50 = 17 μM
PREVOTELLA INTERMEDIA PROTEINASE INHIBITOR		
CHYMOTRYPSIN		
PAPAIN		
PRONASE		
PROTEINASE K		
PROTEINASE TYP IX		
SUBTILISIN		
TRYPSIN		
TRYPSIN LIKE ACTIVITIY PSEUDOMONAS GINGIVALIS		
PRIMAQUINE		
ALCOHOL DEHYDROGENASE		
AMINOPYRINE DEMETHYLASE	NC	*K*i = 310 μM
AMINOPYRINE-N-DEMETHYLASE	C	*K*i = 27 μM
ANILINE HYDROXYLASE	NC	*K*i = 33 μM
p-CHLORO-N-METHYLANILINE-N-DEMETHYLASE (CYT P 450)	C	*K*i = 347 μM
CYTOCHROME P450 2D6	C	*K*i = 23 μM
GLUCOSE-6-PHOSPHATE 1-DEHYDROGENASE		*I*50 = 600 μM

GLYCEROL-3-PHOSPHATE DEHYDROGENASE (NAD(P))		$I50 = 490\ \mu M$
HEXOBARBITAL OXIDASE (CYT P 450)	NC	$Ki = 44\ \mu M$
INTEGRASE		$I50 = 15\ \mu M$
MALATE DEHYDROGENASE		$I50 = 470\ \mu M$
MYOSIN-LIGHT-CHAIN KINASE		$I50 = 49\ \mu M$
p-NITROANISOLE-O-DEMETHYLASE (CYT P-450)	NC	$Ki = 33\ \mu M$
PROTEIN KINASE A		21% 400 μM
PROTEIN KINASE A		$I50 = 284\ \mu M$
PROTEIN KINASE (Ca^{2+} DEPENDENT)		$I50 = 290\ \mu M$
PRIMAQUINE PHOSPHATE		
HISTAMINE N-METHYLTRANSFERASE		$I50 = 1.5$ mM
PRIME RNASE INHIBITOR		
PANCREATIC RIBONUCLEASE		
RIBONUCLEASE B		
RIBONUCLEASE C		
PRIMETIN		
NADH DEHYDROGENASE (EXOGENOUS)		$I50 = 30\ \mu M$
PRIMULIN		
CHITIN SYNTHASE		
PRO-L-ALA-D-GLU(CH_2Cl)-OH		
meso-DIAMINOPIMELATE ADDING ENZYME		
PROBARGIDE		
PYRIDOXAL KINASE		50% 100 μM
PROBENECID		
UDP-GLUCURONOSYLTRANSFERASE	C	$Ki = 900\ \mu M$
PROBESTIN		
AMINOPEPTIDASE B		$I50 = 74\ \mu M$
GLUTAMYL AMINOPEPTIDASE		$I50 = 20\ \mu M$
LEUCYL AMINOPEPTIDASE		$I50 = 170$ nM
MEMBRANE ALANINE AMINOPEPTIDASE		$I50 = 890\ \mu M$
MEMBRANE ALANINE AMINOPEPTIDASE		$I50 = 50$ nM
XAA-TRP AMINOPEPTIDASE		$I50 = 5\ \mu M$
PRO-BORO-OH		
PROLYL OLIGOPEPTIDASE 1 (IgA SPECIFIC)		$Ki = 16$ nM
PROLYL OLIGOPEPTIDASE 2 (IgA SPECIFIC)		$Ki = 4$ nM
PRO-BORO-PRO		
DIPEPTIDYL-PEPTIDASE 4		$I50 = 19$ nM
[L-Δ3PRO9]-BPP9a		
PEPTIDYL-DIPEPTIDASE A		$I50 = 200$ pM
des-PRO^3-BRADIKININ		
PEPTIDYL-DIPEPTIDASE A		$Ki = 4.5$ pM
PROCAINAMIDE		
DNA (CYTOSINE-5-)-METHYLTRANSFERASE		
Na^+/K^+-EXCHANGING ATPASE		$I50 = 14$ mM
PROCAINE		
ACETYLCHOLINESTERASE		$I50 = 194$ mM
ACETYLCHOLINESTERASE		$I50 = 400\ \mu M$
ACETYLCHOLINESTERASE		$I50 = 710\ \mu M$
1-ACYLGLYCEROPHOSPHOCHOLINE O-ACYLTRANSFERASE		
STEROL O-ACYLTRANSFERASE		25% 5 mM
PROCAINE HCL		
ACETYLCHOLINESTERASE		$I50 = 380\ \mu M$

PROCHLORAZ		
AROMATASE		*I*50 = 700 nM
14α-METHYLSTEROL DEMETHYLASE		
METHYLTETRAHYDROPROTOBERBERINE 14-MONOOXYGENASE		
PROCHLORPERAZINE		
PYRUVATE DEHYDROGENASE (LIPOAMIDE)		59% 267 μM
SPHINGOMYELIN PHOSPHODIESTERASE		46% 10 μM
PROCION BLUE MX-R		
PHOTINUS-LUCIFERIN 4-MONOOXYGENASE (ATP-HYDROLYSING)		
PROCION RED		
HYDROGENASE		
PROCION RED HE3B		
ASPARTATE CARBAMOYLTRANSFERASE	IR	
PROCYANIDIN B1		
RNA-DIRECTED DNA POLYMERASE		100% 1 mg/ml
TRIACYLGLYCEROL LIPASE		*I*50 = 42 μM
PROCYANIDIN B2		
RNA-DIRECTED DNA POLYMERASE	C	85% 1 mg/ml
PROCYANIDIN B3		
RNA-DIRECTED DNA POLYMERASE		72% 1 mg/ml
PROCYANIDIN C1		
RNA-DIRECTED DNA POLYMERASE		95% 1 mg/ml
PROCYANIDIN TETRAMER		
NADH DEHYDROGENASE		*I*50 = 1.5 μM
NADH DEHYDROGENASE		*I*50 = 4.5 μM
NADH DEHYDROGENASE		*I*50 = 6 μM
NADH DEHYDROGENASE		*I*50 = 9 μM
PROTEIN KINASE A		*I*50 = 1.4 μM
PROTEIN KINASE C		*I*50 = 600 nM
PROTEIN KINASE (Ca^{2+} DEPENDENT)		*I*50 = 600 nM
PRODELPHINIDIN B-2 3,3′-DI-O-GALLATE		
NADH DEHYDROGENASE		*I*50 = 9.3 μM
NADH DEHYDROGENASE		*I*50 = 7 μM
NADH DEHYDROGENASE		*I*50 = 1.2 μM
PRODIPINE		
BUFURALOL-1′-HYDROXYLASE (CYT P-450dbl)	C	*K*i 5 nM
DIPEPTIDYL-PEPTIDASE 4		*I*50 = 4.5 μM
PROFENOFOS		
ACETYLCHOLINESTERASE		
PROFILIN		
1-PHOSPHATIDYLINOSITOL-4,5-BISPHOSPHATE PHOSPHODIESTERASE		
PROFLAVIN		
CYTOCHROME-b5 REDUCTASE		72% 50μg/ml
NUCLEOSIDE-TRIPHOSPHATASE		
RNA-DIRECTED DNA POLYMERASE		
tRNA(GUANINE-N^2-) METHYLTRANSFERASE		
RNA POLYMERASE		
THROMBIN		
PROFLAVIN HEMISULPHATE		
STEROID 5α-REDUCTASE		*I*50 = 930 nM
PROFLAVIN SULPHATE		
tRNA ADENYLYLTRANSFERASE		100% 1 mM
tRNA CYTIDYLTRANSFERASE		100% 1 mM

Inhibitor / Enzyme	Type	Value
PROGABIDE		
EPOXIDE HYDROLASE		$Ki = 4.4\ \mu M$
EPOXIDE HYDROLASE		$I50 = 19\ \mu M$
PROGESTERONE		
ACYL-CoA DEHYDROGENASE (MEDIUM CHAIN)		20% 20 μM
ALDEHYDE DEHYDROGENASE (NAD)		30% 190 μM
ALDEHYDE DEHYDROGENASE (NAD(P))		
ALDEHYDE OXIDASE		65% 67 μM
4-ANDROSTENE-3,17-DIONE MONOOXYGENASE		
AROMATASE	C	$Ki = 32\ \mu M$
BUFURALOL-1′-HYDROXYLASE (CYT P-450dbl)		$I50 = 194\ \mu M$
CARBONYL REDUCTASE (NADPH)		
CHOLESTENONE 5α-REDUCTASE		
CORTISOL SULFOTRANSFERASE		
CORTISONE β-REDUCTASE		
ESTRADIOL 17β-DEHYDROGENASE		
ESTRONE SULFOTRANSFERASE		65% 10 μM
GLUCOSE-6-PHOSPHATE 1-DEHYDROGENASE		
3β-HYDROXYSTEROID DEHYDROGENASE	C	$Ki = 500$ nM
3β-HYDROXY-Δ5-STEROID DEHYDROGENASE		
11β-HYDROXYSTEROID DEHYDROGENASE		$I50 = 13$ nM
20α-HYDROXYSTEROID DEHYDROGENASE		
3α-HYDROXYSTEROID DEHYDROGENASE (B-SPECIFIC)		$I50 = 7\ \mu M$
3α-HYDROXYSTEROID DEHYDROGENASE (B-SPECIFIC)		
INDANOL DEHYDROGENASE	C	$Ki = 1.4\ \mu M$
PROSTAGLANDIN E2 9-KETOREDUCTASE		
5α-REDUCTASE	C	$Ki = 700$ nM
RETINOL FATTY-ACYLTRANSFERASE		90% 200 μM
STEROID 17α-MONOOXYGENASE		
STEROID 5α-REDUCTASE		$I50 = 100$ nM
STEROID 5α-REDUCTASE		$I50 = 1.5\ \mu M$
STEROID 5α-REDUCTASE		$I50 = 10\ \mu M$
STEROL O-ACYLTRANSFERASE		
STEROL O-ACYLTRANSFERASE		
STEROL O-ACYLTRANSFERASE	C	$Ki = 25\ \mu M$
PROGESTERONE DERIVATIVES		
Na^+/K^+-EXCHANGING ATPASE		
PRO-GLY-ARG CHLOROMETHYLKETONE		
PLASMINOGEN ACTIVATOR	IR	$Ki = 68\ \mu M$
PROGUANIL		
HISTAMINE N-METHYLTRANSFERASE		$I50 = 490\ \mu M$
PRO-HIS-PRO-PHE-HIS-PHE-PHE-VAL-TYR-LYS		
RENIN		
PRO-ILE		
NEUROLYSIN		$Ki = 90\ \mu M$
NEUROLYSIN		$Ki = 90\ \mu M$
OLIGOPEPTIDASE M		$Ki = 540\ \mu M$
SERINE-TYPE CARBOXYPEPTIDASE		$Ki = 5$ mM
PROLINAL		
TRIPEPTIDE AMINOPEPTIDASE		
PROLINE		
ARGINASE	C	$Ki = 21$ mM
ARGINASE		39% 40 mM
ARGINASE	C	$Ki = 4.5$ mM
BETAINE-ALDEHYDE DEHYDROGENASE		
GLUTAMATE-5-SEMIALDEHYDE DEHYDROGENASE		
GLUTAMATE SYNTHASE (NADH)		
MEMBRANE ALANINE AMINOPEPTIDASE		

Inhibitor / Enzyme	Type	Value
L-PIPECOLATE OXIDASE		
PROLINE DIPEPTIDASE	C	$Ki = 310\ \mu M$
PYRUVATE KINASE		
D,L-PROLINEAMIDE		
PROLINE-tRNA LIGASE		
L-PROLINE		
γ-GLUTAMYL KINASE		$I50 = 5$ mM
1-PYRROLINE-5-CARBOXYLATE DEHYDROGENASE		
PYRROLINE-5-CARBOXYLATE REDUCTASE	C	
L-PROLYL-D-ALANINE		
DIPEPTIDYL-PEPTIDASE 4	C	$Ki = 960\ \mu M$
L-PROLYL-L-ALANINE		
DIPEPTIDYL-PEPTIDASE 4	C	$Ki = 270\ \mu M$
3-PROLYLPHOSPHOENOLPYRUVIC ACID		
2-DEHYDRO-3-DEOXYPHOSPHOHEPTONATE ALDOLASE		
5′-O-[N-(L-PROLYL)-SULPHAMOYL]ADENOSINE		
PROLINE-tRNA LIGASE		$Ki = 600$ pM
PROLINE-tRNA LIGASE		$Ki = 4.3$ nM
PROMAZINE		
AMINE OXIDASE MAO	C	$Ki = 124\ \mu M$
3′,5′-CYCLIC-NUCLEOTIDE PHOSPHODIESTERASE		$I50 = 110\ \mu M$
1,3-β-GLUCAN SYNTHASE		$I50 = 400\ \mu M$
NITRIC-OXIDE SYNTHASE		$I50 = 36\ \mu M$
PYRUVATE DEHYDROGENASE (LIPOAMIDE)		37% 267 μM
PROMECARB		
ACETYLCHOLINESTERASE		$Ki = 11$ nM
CYCLOOXYGENASE		
PRO-MET		
GLUTAMINECYCLOTRANSFERASE		$Ki = 510\ \mu M$
NEUROLYSIN		$Ki = 287\ \mu M$
PROMETHACINE HCl		
ACETYLCHOLINESTERASE		$I50 = 2$ mM
CHOLINESTERASE		$I50 = 35\ \mu M$
PROMETHACINE METHOSULPHATE		
ACETYLCHOLINESTERASE		$I50 = 1.6$ mM
CHOLINESTERASE		$I50 = 16\ \mu M$
PROMETHAZINE		
ALCOHOL SULFOTRANSFERASE		31% 300 μM
3′,5′-CYCLIC-NUCLEOTIDE PHOSPHODIESTERASE		$I50 = 340\ \mu M$
CYTOCHROME P450 2D6	MIX	$Ki = 3.8\ \mu M$
1,3-β-GLUCAN SYNTHASE		$I50 = 500\ \mu M$
HISTAMINE N-METHYLTRANSFERASE		$I50 = 224\ \mu M$
NITRIC-OXIDE SYNTHASE		$I50 = 72\ \mu M$
NUCLEOSIDE-DIPHOSPHATASE		15% 1 mM
PROMETZINE		
AMINE OXIDASE MAO	C	$Ki = 31\ \mu M$
PRO-NVL-TYR-LYS-ARG-CH_2CL		
PROTEINASE KEX-2	IR	$Ki = 3.7$ nM
S-(PROPACHLOR)-GLUTATHIONE		
GLUTATHIONE TRANSFERASE		
PROPAFENONE		
CYTOCHROME P450 (O-DEMETHYLATION)		$Ki = 580$ nM

PROPAMIDINE
ADENOSYLMETHIONINE DECARBOXYLASE — *K*i = 334 μM

PROPAMIDINE ISOTHIONATE
AMINE OXIDASE (COPPER-CONTAINING) — NC — *K*i = 8 μM

D,L-PROPAN-1,2-DIOL
(R)AMINOPROPANOL DEHYDROGENASE

PROPANEDIALDEHYDE BIS(GUANYLHYDRAZONE)
ADENOSYLMETHIONINE DECARBOXYLASE — 15% 1 mM

1,3-PROPANEDIOL
PEPTIDE DEFORMYLASE — *K*i = 2.9 μM

PROPANETRICARBOXYLIC ACID
ISOCITRATE DEHYDROGENASE (NADP)

PROPANOL
ALDEHYDE DEHYDROGENASE (NAD)
β-GALACTOSIDASE

n-PROPANOL
FORMALDEHYDE DEHYDROGENASE

PROPANOLOL
ALCOHOL SULFOTRANSFERASE — 46% 300 μM
1,3-β-GLUCAN SYNTHASE — *I*50 = 2 mM
MULTI DRUG RESISTANCE PUMP

D,L-PROPANOLOL
PROSTAGLANDIN-ENDOPEROXIDE SYNTHASE

PROPAPHOS
ACETYLCHOLINESTERASE — *I*50 = 96 μM
ACETYLCHOLINESTERASE — *I*50 = 6.2 μM

PROPARGYLAMINE
AMINE OXIDASE PLASMA — SSI

6α-PROPARGYLANDROST-4-ENE-3,17-DIONE
AROMATASE — SSI — *K*i = 10 nM

10-PROPARGYL ANDROSTENEDIONE
AROMATASE

N^{ω}-PROPARGYL ARGININE
NITRIC-OXIDE SYNTHASE I — *K*i = 57 nM
NITRIC-OXIDE SYNTHASE II — *K*i = 180 μM
NITRIC-OXIDE SYNTHASE III — *K*i = 85 μM

N^{ω}-PROPARGYL-L-ARGININE
NITRIC-OXIDE SYNTHASE I — *K*i = 430 nM
NITRIC-OXIDE SYNTHASE II — *K*i = 620 nM

10-PROPARGYL-5-DEAZAAMINOPTERIN
DIHYDROFOLATE REDUCTASE — *K*i = 5.9 pM

N10 PROPARGYLDIDEAZAFOLIC ACID
THYMIDYLATE SYNTHASE — *K*i = 1 nM

N10-PROPARGYL-5,8-DIDEAZAFOLIC ACID
DIHYDROFOLATE REDUCTASE — *I*50 = 910 nM
NADPH OXIDASE
THYMIDYLATE SYNTHASE — *I*50 = 14 nM

10-PROPARGYLESTR-4-ENE-3,17-DIONE
AROMATASE — *I*50 = 36 nM

PROPARGYLGLYCINE

D-AMINO-ACID OXIDASE	IR	
CYSTATHIONINE β-LYASE	SSI	
CYSTATHIONINE γ-LYASE		
HOMOCYSTEINE DESULFHYDRASE		
SARCOSINE OXIDASE	SSI	
O-SUCCINYLHOMOSERINE (THIOL)-LYASE	SSI	
THREONINE SYNTHASE		

D,L-PROPARGYLGLYCINE

O-SUCCINYLHOMOSERINE (THIOL)-LYASE	IR	*K*i = 45 μM

N-PROPARGYLGUANIDINE

NITRIC-OXIDE SYNTHASE I	IR	

PROPENTOFYLLINE

3′,5′-CYCLIC-NUCLEOTIDE PHOSPHODIESTERASE (cGMP INHIBITABLE)		*I*50 = 97 μM
3′,5′-CYCLIC-NUCLEOTIDE PHOSPHODIESTERASE I		*I*50 = 32 μM
3′,5′-CYCLIC-NUCLEOTIDE PHOSPHODIESTERASE III		*I*50 = 19 μM
3′,5′-CYCLIC-NUCLEOTIDE PHOSPHODIESTERASE IV		*I*50 = 21 μM

PROPENYLPROPENYLSULPHIDE

LIPOXYGENASE		*K*i = 49 μM

1-PROPENYLPROPYLSULPHIDE

LIPOXYGENASE	C	

PROPEPTIDE α-LYTIC PROTEINASE

α-LYTIC ENDOPEPTIDASE		*K*i = 100 pM

PROPEPTIN

PROLYL OLIGOPEPTIDASE		*I*50 = 220 nM
PROLYL OLIGOPEPTIDASE	C	*K*i = 700 nM
PROLYL OLIGOPEPTIDASE		*I*50 = 350 nM

PROPEPTIN T

PROLYL OLIGOPEPTIDASE		*I*50 = 220 nM
PROLYL OLIGOPEPTIDASE		*I*50 = 430 nM

PROPERICIAZINE

AMINE OXIDASE MAO	C	*K*i = 23 μM

PROPETAMPHOS

ACETYLCHOLINESTERASE		

PRO-PHE

NEUROLYSIN		*K*i = 900 μM

cyclo[D-PRO-PHE-ALA-TRP-ARG-TYR]

α-AMYLASE		*K*i = 14 μM

PRO-PHE-ARG-CH_2Cl_2

KALLIKREIN PLASMA		
COAGULATION FACTOR XIIa		

PRO-PHE-ARG-CHN_2

HISTOLYSAIN	REV	kI 1.5 μM

PRO-PHE-HAR-CHLOROMETHYL KETONE

CATHEPSIN B	IR	

D-H-PRO-(N^{α}PHENETHYL)GLY-ARG-H

THROMBIN		

cyclo[PRO-PHE-PHE-LYS(GLX)-TRP-PHE]

TRIOSE-PHOSPHATE ISOMERASE		*I*50 = 3 μM

PRO-L-PHEpsi(PO_2CH_2)GLY-PRO

NEUROLYSIN		*K*i = 4 nM

PRO-PHE-psi(PO_2CH_2)LEU-PRO-NH_2

NEUROLYSIN		Ki = 12 nM
PEPTIDYL-DIPEPTIDASE A		Ki = 150 nM

PROPICONAZOLE

OBTUSIFOLIOL 14α-DEMETHYLASE		I50 = 2 μM

PROPIDIUM

ACETYLCHOLINESTERASE		Ki = 130 nM
ENDONUCLEASE I		

PROPIOCONAZOLE

14α-METHYLSTEROL DEMETHYLASE		

PROPIOLIC ACID

ACYL-CoA DEHYDROGENASE (NADP)		

PROPIONALDEHYDE

ARYL-ALDEHYDE DEHYDROGENASE		
DEOXYRIBOSE-PHOSPHATE ALDOLASE		
FORMALDEHYDE DEHYDROGENASE		50% 10 μM
PYRUVATE DECARBOXYLASE		

3-N-PROPIONAMIDOBENZAMIDE

NAD ADP-RIBOSYLTRANSFERASE		

PROPIONIC ACID

ACETATE-CoA LIGASE		
ACETOLACTATE SYNTHASE		
D-AMINO-ACID TRANSAMINASE	C	Ki = 3.5 mM
4-AMINOBUTYRATE TRANSAMINASE		
GUANIDINOBUTYRASE	C	
HEXOSE OXIDASE		
3-HYDROXYBUTYRATE DEHYDROGENASE		
4-HYDROXY-2-OXOGLUTARATE ALDOLASE		22% 40 mM
HYDROXYPYRUVATE REDUCTASE		
20α-HYDROXYSTEROID DEHYDROGENASE		
L-LYSINE 6-TRANSAMINASE		
MANDELONITRILE LYASE		Ki = 2.3 mM
PROSTAGLANDIN-ENDOPEROXIDE SYNTHASE		I50 = 141 mM
PROSTAGLANDIN-ENDOPEROXIDE SYNTHASE		
SARCOSINE OXIDASE		
SUCCINATE-CoA LIGASE (ADP-FORMING)		
β-UREIDOPROPIONASE		Ki = 160 μM
β-UREIDOPROPIONASE	C	Ki = 300 μM

Nα-PROPIONYL-L-ALA-BUTHIONINE SULPHOXIMINE

meso-DIAMINOPIMELATE-ADDING ENZYME		

PROPIONYL-ALA-PRO-CYCLOHEXYLAMIDE

ELASTASE PANCREATIC		Ki = 10 μM

PROPIONYL-CoA

AMINO-ACID N-ACETYLTRANSFERASE	C	Ki = 710 μM
ARYLAMINE N-ACETYLTRANSFERASE		
ASPARTATE N-ACETYLTRANSFERASE		
BUTYRYL-CoA DEHYDROGENASE		
CARBAMOYL-PHOSPHATE SYNTHASE (AMMONIA)		Ki = 5 mM
HYDROXYMETHYLGLUTARYL-CoA SYNTHASE	C	Ki = 5 μM
MALONYL-CoA DECARBOXYLASE		Ki = 300 μM
MALONYL-CoA DECARBOXYLASE	NC	Ki = 280 μM
MALYL-CoA LYASE	C	Ki = 400 μM
METHYLMALONYL-CoA CARBOXYLTRANSFERASE		Ki = 49 μM
METHYLMALONYL-CoA EPIMERASE		83% 2 mM
OXOGLUTARATE DEHYDROGENASE (LIPOAMIDE)	NC	Ki = 280 μM
PANTOTHENATE KINASE		
PYRUVATE DEHYDROGENASE (LIPOAMIDE)	C	Ki = 3.8 μM

Inhibitor / Enzyme	Type	Value
PROPIONYLMETHIONINE		
FORMYLMETHIONINE DEFORMYLASE	C	Ki = 170 μM
PROPIONYLPROMAZINE		
1,3-β-GLUCAN SYNTHASE		I50 = 400 μM
PROPIONYLPYRIDINE		
UDP-GLUCURONATE DECARBOXYLASE	C	Ki = 13 μM
PROPIOXANTHIN A		
DIPEPTIDYL-DIPEPTIDASE 3		Ki = 13 nM
DIPEPTIDYL-PEPTIDASE 3		
PROPIOXANTHIN B		
DIPEPTIDYL-DIPEPTIDASE 3		Ki = 5.9 nM
DIPEPTIDYL-PEPTIDASE 3		
PROPIOXATIN A		
DIPEPTIDYL-PEPTIDASE 3		Ki = 110 nM
DIPEPTIDYL-PEPTIDASE 3		Ki = 13 nM
PROTEINASE METALLO ASCIDIAN HEMOCYTE I		75% 10 μM
PROTEINASE METALLO ASCIDIAN HEMOCYTE II		72% 10 μM
PROTEINASE SERRATIA MARCESCENS		89% 1.1 nM
THERMOLYSIN		10% 1.1 nM
PROPIOXATIN B		
DIPEPTIDYL-PEPTIDASE 3		Ki = 110 nM
PROPOXUR		
ACETYLCHOLINESTERASE		Ki = 700 nM
D-PROPOXYPHENE		
ALDEHYDE OXIDASE		I50 = 16 μM
ALDEHYDE OXIDASE		I50 = 16 μM
p-PROPOXYPHENOL		
RIBONUCLEOTIDE REDUCTASE		I50 = 300 μM
PROPRANOLOL		
ALCOHOL DEHYDROGENASE		
LIPOPROTEIN LIPASE		I25 = 2 mM
LIPOPROTEIN LIPASE	NC	Ki = 550 μM
PHOSPHATIDATE PHOSPHATASE		
PHOSPHOLIPASE A		I50 = 200 μM
PROTEIN KINASE C	C	Ki = 120 nM
XANTHINE OXIDASE	NC	Ki = 440 μM
D,L-PROPRANOLOL		
HEXOKINASE		63% 50 μM
HEXOKINASE		60% 40 μM
D-H-PRO-PRO-ARG-H		
THROMBIN		
PRO-PRO-$CH_2N^{(}CH_3)_3$		
DIPEPTIDYL-PEPTIDASE 4	IR	
n-PROPYLALCOHOL		
UNDECAPRENYL-DIPHOSPHATASE		
PROPYLAMINE		
PROTEINASE I ACHROMOBACTER		Ki = 250 μM
n-PROPYLAMINE		
LYSYL ENDOPEPTIDASE		Ki = 250 μM
Na^+/K^+-EXCHANGING ATPASE		I50 = 60 μM
PEPTIDYL-LYS METALLOENDOPEPTIDASE	C	Ki = 8 mM
SPERMIDINE SYNTHASE		I50 = 35 μM

1-n-PROPYL-AMINOGLUTETHIMIDE
AROMATASE *I*50 = 5 µM
CHOLESTEROL MONOOXYGENASE (SIDE-CHAIN-CLEAVING) *I*50 = 220 µM

6α-n-PROPYLANDROST-4-ENE-3,17-DIONE
AROMATASE *K*i = 6.7 µM

N-PROPYL-3-BROMOCINNAMAMIDE
EPOXIDE HYDROLASE *I*50 = 48 µM

1-n-PROPYL-1-[(4-CHLOROPHENYL)SULPHONYL]-3-n-PROPYLUREA
ALDEHYDE DEHYDROGENASE (NAD) 91% 1 mM

3-n-PROPYLCIANIDANOL
CYTOCHROME P450 *I*50 = 50 µM

S-PROPYL-L-CYSTEINE
ALLIIN LYASE *K*i = 260 µM

N-PROPYL-1,3-DIAMINOPROPANE
SPERMINE SYNTHASE C *K*i = 121 nM

α-PROPYLDIHYDROXYPHENYLACETAMIDE
TYROSINE 3-MONOOXYGENASE

PROPYLENEDIPHOSPHONIC ACID
XAA-PRO DIPEPTIDASE *K*i = 13 µM

PROPYLENEGLYCOL
CYTOCHROME P450 2E1
3(or 17)β-HYDROXYSTEROID DEHYDROGENASE

5-PROPYLENE-6-METHOXYCOUMARIN
CYTOCHROME P450 1A1 *I*50 = 77 µM
CYTOCHROME P450 2B1 *I*50 = 14 µM

PROPYLGALLATE
ARACHIDONATE 5-LIPOXYGENASE *I*50 = 5 µM
ARACHIDONATE 12-LIPOXYGENASE *I*50 = 20 nM
L-ASCORBATE OXIDASE
CYCLOOXYGENASE *I*50 = 30 µM
Δ5 DESATURASE NC *K*i = 26 µM
Δ5 DESATURASE NC *K*i = 26 µM
DOPAMINE SULFOTRANSFERASE 59% 6.7 µM
17α-ETHINYLOESTRADIOL SULFOTRANSFERASE 46% 6.7 µM
GLUCOSE-6-PHOSPHATE 1-DEHYDROGENASE 100% 1.6 µM
LACCASE
LINOLEOYL-CoA DESATURASE NC *K*i = 170 µM
LINOLEOYL-CoA DESATURASE NC *K*i = 170 µM
LIPOXYGENASE 1 nM
TYROSINE 3-MONOOXYGENASE *I*50 = 1 µM

n-PROPYLGALLATE
1-AMINOCYCLOPROPANE-1-CARBOXYLATE SYNTHASE
ARACHIDONATE 5-LIPOXYGENASE *K*i = 124 µM
LIPOXYGENASE
RETINAL OXIDASE

S-n-PROPYLGLUTATHIONE
GLUTATHIONE TRANSFERASE

4-n-PROPYLHEPTANOIC ACID
SUCCINATE-SEMIALDEHYDE DEHYDROGENASE

6-PROPYL-4-HYDROXYPYRIMIDINE-2-SULPHINATE
GLUTATHIONE TRANSFERASE

6-PROPYL-4-HYDROXYPYRIMIDINE-2-SULPHONATE		
GLUTATHIONE TRANSFERASE		
PROPYLIODIDE		
CARBON-MONOXIDE DEHYDROGENASE		
5-METHYLTETRAHYDROFOLATE-HOMOCYSTEINE S-METHYLTRANSFERASE		
n-PROPYL ISOCYANATE		
ALDEHYDE DEHYDROGENASE (NAD)		100% 100 μM
N-n-PROPYLISOQUINOLIUM		
OXOGLUTARATE DEHYDROGENASE (LIPOAMIDE)		*I*50 = 3 mM
1-n-PROPYL-3-METHYLXANTHINE		
PHOSPHATASE ALKALINE		46% 80 μM
p-PROPYLOXYPHENOL		
RIBONUCLEOTIDE REDUCTASE R2		*I*50 = 700 nM
PROPYLPHENYLALANYLGLYCINE		
CROTALUS ADAMANTEUS SERINE PROTEINASE		
N-PROPYL, N-PROPYLTHIOCARBAMATE SULPHOXIDE		
ALDEHYDE DEHYDROGENASE (NAD) (LOW Km)		*I*50 = 38 μM
N-PROPYL PROTOPORPHYRIN		
FERROCHELATASE		100% 400 nM
2-PROPYLPUTRESCINE		
ORNITHINE DECARBOXYLASE	C	*K*i = 2 mM
6-PROPYL-2-SELENOURACIL		
THYROXINE DEIODINASE		*I*50 = 200 nM
THYROXINE DEIODINASE 1		
5′n-PROPYLTHIOADENOSINE		
ADENOSYLHOMOCYSTEINE NUCLEOSIDASE		99% 3.2 μM
6-PROPYLTHIOURACIL		
THYROXINE DEIODINASE 1		
6-PROPYL-2-THIOURACIL		
THYROXINE DEIODINASE		70% 1 μM
THYROXINE DEIODINASE		*I*50 = 400 nM
6-n-PROPYLTHIOURACIL		
IODOTHYRONINE DEIODINASE I		*K*i = 220 nM
6-n-PROPYL-2-THIOURACIL		
PEROXIDASE		
PEROXIDASE		
PROPYNAL		
ALCOHOL OXIDASE		
2-PROPYN-1-AL		
METHANOL OXIDASE		
2-PROPYN-1-OL		
METHANOL OXIDASE		
2-[N-(2-PROPYNYL)AMINOMETHYL]-1-METHYLINDOLE		
AMINE OXIDASE MAO-A	IR	*I*50 = 200 nM
AMINE OXIDASE MAO-B	IR	*I*50 = 300 nM
4-(1-PROPYNYL)BIPHENYL		
CYTOCHROME P450 1A2	SSI	*K*i = 13 nM
(5-PROPYNYL)URACIL		
DIHYDROPYRIMIDINE DEHYDROGENASE (NADP)		*K*i = 180 μM

Inhibitor / Enzyme	Type	Value
PRORENIN PROPART PEPTIDES		
CATHEPSIN E		
CHYMOSIN		
ENDOTHIAPEPSIN		
GASTRICSIN		
PEPSIN		
RENIN		
PROSTACYCLIN		
3α-HYDROXYSTEROID DEHYDROGENASE (B-SPECIFIC)		*I*50 = 300 μM
PROSTAGLANDIN A1		
3α-HYDROXYSTEROID DEHYDROGENASE (B-SPECIFIC)	C	*K*i = 3.1 μM
3α-HYDROXYSTEROID DEHYDROGENASE (B-SPECIFIC)		74% 20 μM
LEUKOTRIENE-B4 20-MONOOXYGENASE		
PROSTAGLANDIN A2		
9-HYDROXYPROSTAGLANDIN DEHYDROGENASE	NC	*K*i = 3 μM
3α-HYDROXYSTEROID DEHYDROGENASE (B-SPECIFIC)		*I*50 = 4.5 μM
PROSTAGLANDIN A2α		
3α-HYDROXYSTEROID DEHYDROGENASE (A SPECIFIC)		
PROSTAGLANDIN A GSH ADDUCT		
15-HYDROXYPROSTAGLANDIN DEHYDROGENASE (NADP)	C	*K*i = 580 nM
PROSTAGLANDIN B		
15-HYDROXYPROSTAGLANDIN DEHYDROGENASE (NAD)		*K*i = 60 μM
PROSTAGLANDIN B1		
3α-HYDROXYSTEROID DEHYDROGENASE (B-SPECIFIC)	C	*K*i = 800 nM
PROSTAGLANDIN-A1 Δ-ISOMERASE		
PROSTAGLANDIN B2		
9-HYDROXYPROSTAGLANDIN DEHYDROGENASE	NC	*K*i = 8 μM
15-HYDROXYPROSTAGLANDIN DEHYDROGENASE (NADP)		
3α-HYDROXYSTEROID DEHYDROGENASE (B-SPECIFIC)		*I*50 = 20 μM
PROSTAGLANDIN B1 OLYGOMERS		
PHOSPHOLIPASE A2		*I*50 = 5 μM
PROSTAGLANDIN D2		
9-HYDROXYPROSTAGLANDIN DEHYDROGENASE	NC	*K*i = 45 μM
3α-HYDROXYSTEROID DEHYDROGENASE (B-SPECIFIC)		41% 20 μM
PROSTAGLANDIN F SYNTHASE		
PROSTAGLANDIN E		
STEROL O-ACYLTRANSFERASE		*I*50 = 500 nM
PROSTAGLANDIN E1		
ADENYLATE CYCLASE		
3α-HYDROXYSTEROID DEHYDROGENASE (B-SPECIFIC)	C	*K*i = 7.5 μM
MORPHINE 6-DEHYDROGENASE		
PROSTAGLANDIN E2		
9-HYDROXYPROSTAGLANDIN DEHYDROGENASE	C	
3α-HYDROXYSTEROID DEHYDROGENASE (B-SPECIFIC)		50% 20 μM
3α-HYDROXYSTEROID DEHYDROGENASE (B-SPECIFIC)		*I*50 = 105 μM
Na^+/K^+-EXCHANGING ATPASE		
PROSTAGLANDIN F		
STEROL O-ACYLTRANSFERASE		*I*50 = 500 nM
PROSTAGLANDIN F1α		
3α-HYDROXYSTEROID DEHYDROGENASE (B-SPECIFIC)	C	*K*i = 12 μM
PROSTAGLANDIN F2a		
3α-HYDROXYSTEROID DEHYDROGENASE (B-SPECIFIC)		40% 20 μM

13-cisPROSTAGLANDIN F2α
15-HYDROXYPROSTAGLANDIN DEHYDROGENASE (NAD)

PROSTAGLANDIN G1
PROSTACYCLIN SYNTHASE

PROSTAGLANDIN H1
THROMBOXANE A SYNTHASE SSI $Ki = 28\ \mu M$

PROSTAGLANDINS
CYSTINYL AMINOPEPTIDASE
3α-HYDROXYSTEROID DEHYDROGENASE (A SPECIFIC)

PROSTANOIC ACID
15-HYDROXYPROSTAGLANDIN DEHYDROGENASE (NAD)

PROSTIGMINE
ACETYLCHOLINESTERASE
CHOLINESTERASE
CHOLINE SULFOTRANSFERASE

PROTAMINE
LYSINE (ARGININE) CARBOXYPEPTIDASE $I50 = 320$ nM
PHOSPHOPROTEIN PHOSPHATASE 1
PHOSPHORYLASE PHOSPHATASE
PROTEIN KINASE C $I50 = 800\ \mu M$
RNA-DIRECTED DNA POLYMERASE

PROTAMINE SULPHATE
ACYLGLYCEROL LIPASE
AMP DEAMINASE
LIPOPROTEIN LIPASE
PROTEIN TYROSINE KINASE I $I50 = 170\ \mu g/ml$
PROTEIN TYROSINE KINASE II $I50 = 220\ \mu g/ml$
TRYPTOPHAN SYNTHASE

PROTEIN 14-3-3
NITRATE REDUCTASE $I50 = 2\ \mu g/ml$

α1-PROTEINASE INHIBITOR
ACROSIN
CATHEPSIN G $I50 = 6.5$ nM
CATHEPSIN G
CHYMASE
CHYMOTRYPSIN
α-CHYMOTRYPSIN $I50 = 5.7$ nM
COAGULATION FACTOR Xa
COAGULATION FACTOR XIa
ELASTASE LEUKOCYTE $Ki = 100$ fM
ELASTASE PANCREATIC
ELASTASE PANCREATIC $I50 = 21$ nM
KALLIKREIN
KALLIKREIN PLASMA
KALLIKREIN TISSUE
KALLIKREIN URINE
MYELOBLASTIN
PLASMIN
PLASMINOGEN ACTIVATOR
THROMBIN
TRYPSIN $I50 = 180$ nM
TRYPTASE
TRYPTASE
TRYPTASE TL-2 13% $10\ \mu M$

α2-PROTEINASE INHIBITOR
ELASTASE LEUKOCYTE
MINIPLASMIN
PLASMIN

PROTEINASE INHIBITOR 6
- PLASMIN
- u-PLASMINOGEN ACTIVATOR
- PROTEIN Ca
- PROTEIN Ca
- THROMBIN
- TRYPSIN

PROTEINASE INHIBITOR 8
- CHYMOTRYPSIN — Ki = 1.7 nM
- COAGULATION FACTOR Xa — Ki = 272 pM
- SUBTILISIN — Ki = 8.4 pM
- THROMBIN — Ki = 350 pM
- TRYPSIN — Ki < 3.8 nM

PROTEINASE INHIBITOR 9
- GRANZYME B
- SUBTILISIN A — K = 3.6 pM

PROTEINASE K INHIBITOR 3
- PROTEINASE K

PROTEINASE NEXIN
- COAGULATION FACTOR Xa
- t-PLASMINOGEN ACTIVATOR
- u-PLASMINOGEN ACTIVATOR

PROTEINASE NEXIN-1
- ACROSIN — Ki = 1.4 nM
- PROTEINASE TLE
- PROTEIN Ca
- THROMBIN
- TRYPSIN — Ki = 220 nM

PROTEINASE NEXIN-2
- CHYMASE — Ki = 20 nM
- CHYMOTRYPSIN — Ki = 1.6 nM
- α-CHYMOTRYPSIN — Ki = 820 nM
- COAGULATION FACTOR IXa — Ki = 1.9 nM
- COAGULATION FACTOR XIa — Ki = 29 pM
- MULTICATALYTIC ENDOPEPTIDASE COMPLEX — Ki = 130 nM
- PLASMIN — Ki = 29 nM
- PLASMIN — Ki = 630 nM
- TRYPSIN — Ki = 420 pM

PROTEIN C INHIBITOR
- ACROSIN — Ki = 46 pM
- CHYMOTRYPSIN
- COAGULATION FACTOR Xa
- COAGULATION FACTOR XIa
- COAGULATION FACTOR XIIa
- KALLIKREIN PLASMA
- PLASMINOGEN ACTIVATOR
- PLASMINOGEN ACTIVATOR-2
- t-PLASMINOGEN ACTIVATOR
- u-PLASMINOGEN ACTIVATOR
- PROTEIN Ca
- THROMBIN
- TRYPSIN

PROTEIN 21 kD 7B2
- PRO-OPIOMELANOCORTIN CONVERTING ENZYME 2

PROTEIN INHIBITOR
- GANGLIOSIDE GALACTOSYLTRANSFERASE
- 2-HYDROXYACYLSPHINGOSINE 1β-GALACTOSYLTRANSFERASE
- PLASMINOGEN ACTIVATOR
- PROTEINASE ASPERGILLUS ACID

PROTEIN INHIBITOR ASCARIS LUMBRICOIDES
CHYMOTRYPSIN C $Ki = 47$ nM

PROTEIN INHIBITOR DEPENDENT ON FRUCTOSE 2,6-BISPHOSPHATE
HEXOKINASE

PROTEIN INHIBITOR OF NITRIC OXIDE SYNTHASE
NITRIC-OXIDE SYNTHASE I $I50 = 18$ μM

PROTEIN 14-3-3 ISOFORM BRAIN
PROTEIN KINASE C $I50 = 800$ nM

PROTEIN KINASE A II REGULATORY SUBUNIT
PHOSPHORYLASE PHOSPHATASE

PROTEIN KINASE INHIBITOR α
PROTEIN KINASE A $Ki = 220$ pM

PROTEIN KINASE INHIBITOR β1
PROTEIN KINASE A $Ki = 7.1$ nM

PROTEIN KINASE INHIBITOR β2
PROTEIN KINASE A

PROTEIN KINASE INHIBITOR γ
PROTEIN KINASE A $Ki = 73$ pM
PROTEIN KINASE A $Ki = 444$ pM

PROTEIN KINASE INHIBITOR PEPTIDE(5-24)
CAM KINASE $I50 = 76$ μM
MYOSIN-LIGHT-CHAIN KINASE $I50 = 450$ μM
PROTEIN KINASE A $Ki = 2.3$ nM
PROTEIN KINASE G $Ki = 100$ μM

PROTEIN KINASE INHIBITOR PEPTIDE (5-24 AMIDE)
PROTEIN KINASE A

PROTEIN KINASE R p58 KD INHIBITOR
PROTEIN KINASE R

γ-PROTEIN LAMDA PHAGE
EXODEOXYRIBONUCLEASE Y

PROTEIN PHOSPHATASE 2A INHIBITOR
PHOSPHOPROTEIN PHOSPHATASE 2A

PROTEIN PHOSPHATASE INHIBITOR 1 (MAMMALIAN)
PHOSPHOPROTEIN PHOSPHATASE 1 $I50 = 2$ nM

PROTEIN S
COAGULATION FACTOR Xa $Ki = 490$ nM

PROTEINS
PHOSPHOLIPASE A

PROTEINS FROM MACROPHAGES
STEROL ESTERASE

PROTHIOFOS
ACETYLCHOLINESTERASE

PROTHIONINE SULPHOXIME
GLUTAMATE-CYSTEINE LIGASE

PROTHOATE
ACETYLCHOLINESTERASE

PRO-THRE-HIS-ILE-LYS-TRP-GLY-ASP
PEPTIDYL-DIPEPTIDASE A

PRO-THRE-SERNH2
PROTEINASE IgA1 — $I50 = 5\ \mu M$

PROTOBERINES
(S)-TETRAHYDROPROTOBERBERINE OXIDASE

PROTOCATECHUALDEHYDE
PROTOCATECHUATE 3,4-DIOXYGENASE — C — $Ki = 14\ \mu M$
PROTOCATECHUATE 4,5-DIOXYGENASE

PROTOCATECHUATE
4-HYDROXYBENZOATE 3-MONOOXYGENASE

PROTOCATECHUIC ACID
ARACHIDONATE 5-LIPOXYGENASE — $I50 = 160\ \mu M$
ARACHIDONATE 12-LIPOXYGENASE — $I50 = 170\ \mu M$
BENZOATE 4-MONOOXYGENASE
CATECHOL OXIDASE — $Ki = 480\ \mu M$
CATECHOL OXIDASE ppo — $I50 = 1.7\ mM$
4,5-DIHYDROOXYPHTHALATE DECARBOXYLASE
DIPHOSPHOMEVALONATE DECARBOXYLASE
DOPAMINE SULFOTRANSFERASE — 56% 6.7 μM
SHIKIMATE 5-DEHYDROGENASE — C — $Ki = 200\ \mu M$
SHIKIMATE 5-DEHYDROGENASE — C — $Ki = 860\ \mu M$

PROTOCATECHUIC ACID METHYLESTER
PROTOCATECHUATE 3,4-DIOXYGENASE — C — $Ki = 400\ \mu M$

PROTOCHLOROPHYLLIDE A
MAGNESIUM PROTOPORPHYRIN O-METHYLTRANSFERASE

PROTODISCOIN
XANTHINE OXIDASE — 49% 100 μM

PROTOHEM
FERROCHELATASE — 20% 20 μM

PROTOHEMIN
GLUTATHIONE TRANSFERASE — C — $Ki = 32\ nM$

PROTOLICHESTERINIC ACID
DNA LIGASE I — $I50 = 20\ \mu M$
RNA-DIRECTED DNA POLYMERASE — $I50 = 24\ \mu M$

PROTOPORPHYRIN
GLUTATHIONE TRANSFERASE — C — $Ki = 110\ nM$
SUCCINATE-CoA LIGASE (ADP-FORMING)
TRYPTOPHAN 2,3-DIOXYGENASE — C — $Ki = 500\ nM$

PROTOPORPHYRIN COBALT
HEME OXYGENASE (DECYCLIZING) — C — $Ki = 82\ nM$

PROTOPORPHYRIN Fe^{2+}
MAGNESIUM PROTOPORPHYRIN O-METHYLTRANSFERASE

PROTOPORPHYRIN Fe^{3+}
MAGNESIUM PROTOPORPHYRIN O-METHYLTRANSFERASE

PROTOPORPHYRIN IRON
5-AMINOLEVULINATE SYNTHASE

PROTOPORPHYRIN IX
5-AMINOLEVULINATE DEHYDROGENASE — $Ki = 500\ \mu M$
5-AMINOLEVULINATE SYNTHASE — $I50 = 250\ nM$
GLUTATHIONE TRANSFERASE
LACTOYLGLUTATHIONE LYASE — C — $Ki = 200\ \mu M$
NITRIC-OXIDE SYNTHASE I — $I50 = 800\ nM$
NITRIC-OXIDE SYNTHASE II — $I50 = 4\ \mu M$
NITRIC-OXIDE SYNTHASE III — $I50 = 5\ \mu M$

PROTOPORPHYRIN IX IRON
5-AMINOLEVULINATE SYNTHASE — $I50 = 250$ nM

PROTOPORPHYRIN IX ZINC
NITRIC-OXIDE SYNTHASE I — IR
NITRIC-OXIDE SYNTHASE II — IR
NITRIC-OXIDE SYNTHASE III — IR

PROTOPORPHYRIN-MAGNESIUM
FERROCHELATASE — 75% 20 μM

PROTOPORPHYRIN TIN
HEME OXYGENASE (DECYCLIZING) — 97% 3.3 μM
HEME OXYGENASE (DECYCLIZING) — 76% 500 nM
HEME OXYGENASE (DECYCLIZING) — C — $K\text{i} = 33$ nM
HEME OXYGENASE (DECYCLIZING) — $K\text{i} = 17$ nM

PROTOPORPHYRIN ZINC
HEME OXYGENASE (DECYCLIZING) — C — $K\text{i} = 130$ nM

cyc(PRO-TYR-PRO-VAL)
STEROL O-ACYLTRANSFERASE — $I50 = 1.5$ mM

cyclo(PRO-TYR-PRO-VAL)
MONOPHENOL MONOOXYGENASE — $I50 = 1.5$ mM

PRP 101821
7-DEHYDROCHOLESTEROL REDUCTASE — $I50 = 1$ μM
FARNESYL-DIPHOSPHATE FARNESYLTRANSFERASE — $I50 = 1$ nM
Δ5,7-STEROL Δ7-REDUCTASE — $I50 = 1$ μM

PRUNETIN
ALCOHOL DEHYDROGENASE γ2-γ2
ALDEHYDE DEHYDROGENASE
ALDEHYDE DEHYDROGENASE (NAD) 1 — C — $K\text{i} = 150$ nM
ALDEHYDE DEHYDROGENASE (NAD) 2 — C — $K\text{i} = 6$ μM
PROTEIN TYROSINE KINASE
PROTEIN TYROSINE KINASE EGFR — $I50 = 4.2$ μg/ml

PS 28
CATHEPSIN B — $I50 = 20$ μM
CATHEPSIN B LIKE CYSTEINE PROTEINASE — $I50 = 10$ μM
PAPAIN — $I50 = 20$ μM

PSEUDOCARBA-NAD
NAD NUCLEOSIDASE — C — $K\text{i} = 7$ μM

PSEUDOHYPERICIN
PROTEIN KINASE A — $I50 > 50$ μM
PROTEIN KINASE C — $I50 = 30$ μM

PSEUDOMONAS AERUGINOSA ALLANTOINASE INHIBITOR
ALLANTOINASE

PSEUDOMONAS CARBOXYDOVORANS CARBON MONOXIDE DEHYDROGENASE INHIBITOR
CARBON-MONOXIDE DEHYDROGENASE

PSEUDOMONAS PHASEOLICA CHLOROSIS IDUCING TOXIN
ORNITHINE CARBAMOYLTRANSFERASE

PSEUDOMONIC ACID
ISOLEUCINE-tRNA LIGASE — 90% 100 μg/ml

PSEUDOMONIC ACID A
ISOLEUCINE-tRNA LIGASE

PSEUDONAJA TEXTILIS PLASMIN INHIBITOR
PLASMIN — $K\text{i} = 150$ nM
TRYPSIN — $K\text{i} = 300$ nM

PSEUDOSTELLARIN A

MONOPHENOL MONOOXYGENASE		$I50 = 131\ \mu M$

PSEUDOSTELLARIN B

MONOPHENOL MONOOXYGENASE		$I50 = 187\ \mu M$

PSEUDOSTELLARIN C

MONOPHENOL MONOOXYGENASE		$I50 = 63\ \mu M$

PSEUDOSTELLARIN D

MONOPHENOL MONOOXYGENASE		$I50 = 100\ \mu M$

PSEUDOSTELLARIN E

MONOPHENOL MONOOXYGENASE		$I50 = 175\ \mu M$

PSEUDOSTELLARIN F

MONOPHENOL MONOOXYGENASE		$I50 = 50\ \mu M$

PSEUDOSTELLARIN G

MONOPHENOL MONOOXYGENASE		$I50 = 75\ \mu M$

PSEUDOURIDINE

UDP-N-ACETYLGLUCOSAMINE PYROPHOSPHORYLASE

PSEUOMONAS PUTIDA HEAT LABILE INHIBITOR

NAD(P) NUCLEOSIDASE

PSEUROTIN A

CHITIN SYNTHASE		60% 232 μM
CHITIN SYNTHASE	C	$Ki = 93\ \mu M$

PSICOFURANINE

GMP SYNTHASE	IR	$Ki = 130$ nM
GMP SYNTHASE (GLUTAMINE-HYDROLYSING)		
NAD SYNTHASE	MIX	$Ki = 600\ \mu M$

D-PSICOSE

KETOHEXOKINASE

Ph-SO_2-GLY-PRO[4-AmPhGLY)^P(OPh)$_2$

GRANZYME A

PSORALEN

CYTOCHROME P450 6D1		$I50 = 4.7\ \mu M$
O-DEMETHYLASE (CYT P450)	IR	$I50 = 34\ \mu M$
MYOSIN-LIGHT-CHAIN KINASE		$I50 = 267\ \mu M$

PSPI 21

CHYMOTRYPSIN		$Ki = 1.8$ nM
ELASTASE LEUKOCYTE		$Ki = 800$ pM
TRYPSIN		$Ki = 1.5$ nM

PSPI 22

CHYMOTRYPSIN		$Ki = 25$ nM
TRYPSIN		$Ki = 580$ pM

PSYCHOSINE

CYTOCHROME-C OXIDASE		$I50 = 200\ \mu M$
GALACTOSYLGALACTOSYLGLUCOSYLCERAMIDASE	NC	$Ki = 740\ \mu M$
PROTEIN TYROSINE SULFOTRANSFERASE		
TYROSYL PROTEIN SULFOTRANSFERASE		92% 200 μM

PSYCHOTRINE DIHYDROGEN OXALATE

RNA-DIRECTED DNA POLYMERASE		$I50 = 9\ \mu M$
RNA-DIRECTED DNA POLYMERASE		$I50 = 28\ \mu M$

Pt^{2+}

MALATE DEHYDROGENASE

PT 523
DIHYDROFOLATE REDUCTASE — $I50 = 12$ nM

PTERIDINE
ADENOSINE DEAMINASE — C — $Ki = 23$ µM
ADENOSINE DEAMINASE — C — $Ki = 120$ µM

PTERIDINES
DIHYDROLIPOAMIDE DEHYDROGENASE

PTERIN
QUEUINE tRNA-RIBOSYLTRANSFERASE
SEPIAPTERIN DEAMINASE

PTERIN-6-ALDEHYDE
XANTHINE OXIDASE

PTERIN-6-CARBOXYLIC ACID
QUEUINE tRNA-RIBOSYLTRANSFERASE

PTEROATE
DIHYDROFOLATE REDUCTASE

PTEROYL-GLU-GLU-GLU
γ-GLUTAMYL HYDROLASE — $Ki = 4.5$ µM

PTEROYL-α-GLUTAMIC ACID
5-METHYLTETRAHYDROPTEROYLTRIGLUTAMATE-HOMOCYSTEINE METHYLTRANSFERASE — C

PTEROYL-γ-GLUTAMYL-GLUTAMIC ACID
5-METHYLTETRAHYDROPTEROYLTRIGLUTAMATE-HOMOCYSTEINE METHYLTRANSFERASE — C

PTEROYLORNITHINE
FOLYLPOLYGLUTAMATE SYNTHASE

L-PTEROYLPENTAGLUTAMIC ACID
PHOSPHORIBOSYLAMINOIMIDAZOLECARBOXAMIDE FORMYLTRANSFERASE — C — $Ki = 88$ nM

PTERULINIC ACID
NADH DEHYDROGENASE (UBIQUINONE)

PTERULONE
NADH DEHYDROGENASE (UBIQUINONE)

PTILODENE
ARACHIDONATE 5-LIPOXYGENASE

PTILOMYCIN A
Ca^{2+}-TRANSPORTING ATPASE — $I50 = 2$ µM
Na^{+}/K^{+}-EXCHANGING ATPASE — $I50 = 2$ µM

PTP
BUFURALOL-1′-HYDROXYLASE (CYT P-450dbl) — C — $I50 = 12$ µM

PUBERULONIC ACID
myo-INOSITOL-1(or 4)-MONOPHOSPHATASE — $I50 = 10$ µM

PUERANIN
ALDEHYDE DEHYDROGENASE 2 — $Ki = 15$ µM

PULCHELLALACTAM
PROTEIN TYROSINE-PHOSPHATASE CD45 — $I50 = 124$ µg/ml

PULEGONE
ACETYLCHOLINESTERASE — $Ki = 580$ µM

PUMPKIN CYSTEINE PROTEINASE INHIBITOR
FICAIN — $Ki = 106$ nM
PAPAIN — $Ki = 115$ nM

PUNICACORTEIN C
RNA-DIRECTED DNA POLYMERASE — $I50 = 5\ \mu M$

PUNICALAGIN
CARBONATE DEHYDRATASE — $I50 = 230$ nM

PUNICALIN
CARBONATE DEHYDRATASE — $I50 = 1\ \mu M$
RNA-DIRECTED DNA POLYMERASE — $I50 = 8\ \mu M$

PURINE
ADENINE DEAMINASE — C — $Ki = 220\ \mu M$
ADENOSINE DEAMINASE — C — $Ki = 2.6$ mM
RIBOSE-5-PHOSPHATE-AMMONIA LIGASE
XANTHINE OXIDASE — C — $I50 = 14\ \mu M$

PURINE-6-ALDEHYDE
XANTHINE OXIDASE

PURINE DERIVATIVES
URATE OXIDASE

PURINE RIBONUCLEOSIDE
ADENOSINE DEAMINASE — C — $Ki = 37\ \mu M$
ADENOSINE DEAMINASE — C — $Ki = 9.3\ \mu M$

9-PURINERIBONUCLEOSIDE
ADENOSINE DEAMINASE — C — $Ki = 8.8\ \mu M$

PURINE RIBONUCLEOTIDES
AMIDOPHOSPHORIBOSYLTRANSFERASE

PURINERIBOSE-5′-TRIPHOSPHATE
THREONINE-tRNA LIGASE

PURINE RIBOSIDE
ADENINE DEAMINASE — $Ki = 8.1\ \mu M$
ADENOSINE DEAMINASE — C — $Ki = 8\ \mu M$
ADENOSINE KINASE — C — $Ki = 1\ \mu M$
ADENOSYLHOMOCYSTEINASE — $Ki = 570\ \mu M$
PURINE-NUCLEOSIDE PHOSPHORYLASE

PURINERIBOSIDE-5′-TRIPHOSPHATE
ISOLEUCINE-tRNA LIGASE — $Ki = 2.1\ \mu M$

PURINES
URATE-RIBONUCLEOTIDE PHOSPHORYLASE

PUROMYCIN
AMINOPEPTIDASE — 87% 10 μM
AMINOPEPTIDASE — 45% 10 μM
AMINOPEPTIDASE B
AMINOPEPTIDASE (HUMAN LIVER)
ARYL-ACYLAMIDASE
3′,5′-CYCLIC-NUCLEOTIDE PHOSPHODIESTERASE — C
CYTOCHROME-C LYSINE N-METHYLTRANSFERASE
CYTOSOL ALANYL AMINOPEPTIDASE
O-DEMTETHYLPUROMYCIN O-METHYLTRANSFERASE
DIPEPTIDYL-PEPTIDASE 2 — $Ki = 19\ \mu M$
DIPEPTIDYL-PEPTIDASE 2 — 57% 1 mM
DIPEPTIDYL-PEPTIDASE 3 — 100% 2 mM
DIPEPTIDYL-PEPTIDASE 3
GLUTAMYL AMINOPEPTIDASE
LEUCYL AMINOPEPTIDASE
LEUCYLTRANSFERASE
MEMBRANE ALANINE AMINOPEPTIDASE
NEPRILYSIN — 36% 1 mM
PEPTIDYLTRANSFERASE

Inhibitor / Enzyme	Type	Value
PYROGLUTAMYL-PEPTIDASE I		
TETRAPEPTIDE DIPEPTIDASE		100% 25 μM
TRITRIRACHIUM ALKALINE PROTEINASE		*K*i = 2.8 mM
PUROMYCIN AMINONUCLEOSIDE		
DIPEPTIDYL-PEPTIDASE 2		*K*i = 30 μM
PURPACTIN A		
STEROL O-ACYLTRANSFERASE		*I*50 = 121 μM
PURPACTIN B		
STEROL O-ACYLTRANSFERASE		*I*50 = 121 μM
PURPACTIN C		
STEROL O-ACYLTRANSFERASE		*I*50 = 126 μM
PURPURIN		
7-ETHOXYCOUMARIN O-DEETHYLASE		*I*50 = 180 μM
MYOSIN-LIGHT-CHAIN KINASE		*I*50 = 25 μM
PROTEIN KINASE A		*I*50 = 4 μM
PROTEIN KINASE C		*I*50 = 19 μM
PROTEIN KINASE Cα		*I*50 = 20 μM
PROTEIN KINASE Cδ		*I*50 = 23 μM
PROTEIN KINASE Cγ		*I*50 = 23 μM
PROTEIN KINASE (Ca^{2+} DEPENDENT)		*I*50 = 14 μM
PROTEIN TYROSINE KINASE (v-Abl)		*I*50 = 5.2 μM
PROTEIN TYROSINE KINASE EGFR		*I*50 = 20 μM
PROTEIN TYROSINE KINASE s-Src		*I*50 = 64 μM
RNA-DIRECTED DNA POLYMERASE 1		*I*50 = 18 μM
PURPUROGALLIN		
CATECHOL O-METHYLTRANSFERASE		
GLUTATHIONE TRANSFERASE		
LACTOYLGLUTATHIONE LYASE		*K*i = 9 μM
PHOSPHOGLYCERATE KINASE		*I*50 = 1 μM
PURPURONE		
ATP CITRATE (pro-S)-LYASE		*I*50 = 7 μM
PUTRESCINE		
ACETYLSPERMIDINE DEACETYLASE	C	*K*i = 250 μM
ACETYLSPERMIDINE DEACETYLASE		*K*i = 250 μM
ADENYLATE CYCLASE		25% 10 μM
AMINO-ACID N-ACETYLTRANSFERASE		
ARGININE DECARBOXYLASE	C	*K*i = 7.25 mM
ARGININE DEIMINASE	UC	*K*i = 2.8 mM
CARBAMOYL-PHOSPHATE SYNTHASE (GLUTAMINE-HYDROLYSING)		18% 800 μM
DEOXYHYPUSINE SYNTHASE	C	*I*50 = 91 μM
DNA (CYTOSINE-5-)-METHYLTRANSFERASE		
GLUCOSE OXIDASE		
NITRIC-OXIDE SYNTHASE		*I*50 = 212 μM
4-NITROPHENYLPHOSPHATASE		
ORNITHINE DECARBOXYLASE	C	*K*i = 3.3 mM
ORNITHINE DECARBOXYLASE	C	*K*i = 1 mM
ORNITHINE-OXO-ACID TRANSAMINASE		
PHOSPHOLIPASE C		
POLYNUCLEOTIDE ADENYLYLTRANSFERASE		
PROTEIN-GLUTAMINE γ-GLUTAMYLTRANSFERASE		
PROTEIN TYROSINE KINASE I		*I*50 = 11 mM
PROTEIN TYROSINE KINASE II		*I*50 = 27 mM
PUTRESCINE OXIDASE		> 10 mM
PYRIDOXAL KINASE		
RIBONUCLEASE PANCREATIC		
tRNA(ADENINE-N^1-) METHYLTRANSFERASE		
tRNA ADENYLYLTRANSFERASE		
tRNA CYTIDYLTRANSFERASE		100% 1 mM

Inhibitor / Enzyme	Type	Value
RNA-DIRECTED RNA POLYMERASE		
tRNA(URACIL-5-) METHYLTRANSFERASE		
SPERMINE SYNTHASE		$Ki = 1.7$ mM
PYCNIDIONE		
STROMELYSIN		$I50 = 31$ μM
PYRACARBOLID		
SUCCINATE DEHYDROGENASE		$I50 = 5.5$ μM
PYRAN COPOLYMER		
RNA-DIRECTED DNA POLYMERASE		
1-PYRANEBUTYRYL-CoA		
CARNITINE O-PALMITOYLTRANSFERASE		
PYRAZINAMIDE		
NICOTINATE N-METHYLTRRANSFERASE		
PYRAZINE		
CYTOCHROME P450 2E1	U	$Ki = 200$ μM
PYRAZINE-2-CARBOXYLIC ACID		
NICOTINATE PHOSPHORIBOSYLTRANSFERASE		
PYRAZOFURIN 5′-PHOSPHATE		
OROTIDINE-5′-PHOSPHATE DECARBOXYLASE		$Ki = 5$ nM
PHOSPHORIBOSYLAMINOIMIDAZOLECARBOXAMIDE FORMYLTRANSFERASE		46% 200 μM
PYRAZOLE		
ALCOHOL DEHYDROGENASE		$Ki = 22$ nM
ALCOHOL DEHYDROGENASE		$Ki = 220$ μM
ALCOHOL DEHYDROGENASE-2		
ALDEHYDE REDUCTASE		33% 1 mM
CARBONYL REDUCTASE (NADPH)		
CARBONYL REDUCTASE (NADPH) 1		15% 10 mM
CARBONYL REDUCTASE (NADPH) 2		14% 10 mM
CARBONYL REDUCTASE (NADPH) 3		19% 10 mM
FORMALDEHYDE DISMUTASE	C	$Ki = 480$ μM
PYRAZOLE-1-CARBOXAMIDE		
NITRIC-OXIDE SYNTHASE I		$I50 = 200$ nM
NITRIC-OXIDE SYNTHASE II		$I50 = 200$ nM
NITRIC-OXIDE SYNTHASE III		$I50 = 200$ nM
3,5-PYRAZOLE DICARBOXYLIC ACID		
GLUTAMATE DECARBOXYLASE		$Ki = 13$ μM
PYRAZOMYCIN		
ADENOSYLHOMOCYSTEINASE	C	$Ki = 9.9$ mM
PYRENEGLYOXAL		
ELASTASE LEUKOCYTE		
PYRIDABEN		
NADH DEHYDROGENASE (UBIQUINONE)		$I50 = 4$ nM
PYRIDAZINE		
CYTOCHROME P450 2E1	NC	$Ki = 2.3$ mM
PYRIDINDOLOL		
β-GALACTOSIDASE		$I50 = 1.8$ μg/ml
PYRIDINE		
CATECHOL 2,3-DIOXYGENASE	C	
CATECHOL OXIDASE		$Ki = 13$ mM
CYTOCHROME P450 2E1	U	$Ki = 400$ nM
DIMETHYLANILINE-N-OXIDE ALDOLASE	C	$Ki = 2$ μM
β-FRUCTOFURANOSIDASE		

Inhibitor / Enzyme		
HISTIDINE DECARBOXYLASE		
THROMBOXANE A SYNTHASE		71% 100 μM
4-PYRIDINE ACETIC ACID (1S,2S,3S,5S)ISOPINOCAMPHENYL ESTER		
AROMATASE		$I50$ = 120 nM
17α-HYDROXYPROGESTERONE ALDOLASE		$I50$ = 260 nM
STEROID 17α-MONOOXYGENASE		$I50$ = 280 nM
PYRIDINE-4-ALDEHYDE		
PYRIDOXAMINE-PYRUVATE TRANSAMINASE		
3-PYRIDINEALDEHYDEADENINEDINUCLEOTIDE		
NITRATE REDUCTASE (NADH)	C	Ki = 740 μM
PYRIDINE-2-ALDEHYDE-2-PYRIDYLHYDRAZONE		
RIBONUCLEOTIDE REDUCTASE		$I50$ = 1.8 μM
PYRIDINE-2-ALDEHYDE-2′-QUINOLYLHYDRAZONE		
RIBONUCLEOTIDE REDUCTASE		$I50$ = 3 μM
PYRIDINE-2-CARBINOL		
CHOLINE SULFOTRANSFERASE		
2-PYRIDINECARBOXYALDEHYDE		
AMINE OXIDASE MAO-B		Ki = 1.9 mM
4-PYRIDINECARBOXYALDEHYDE		
AMINE OXIDASE MAO-B		Ki = 653 μM
2-PYRIDINE CARBOXYLIC ACID		
DIHYDROPICOLINATE REDUCTASE		
PYRIDINE-2-CARBOXYLIC ACID		
PROCOLLAGEN-LYSINE 5-DIOXYGENASE		
PROCOLLAGEN-PROLINE 3-DIOXYGENASE		
PROCOLLAGEN-PROLINE, 2-OXOGLUTARATE-4-DIOXYGENASE		
PYRIDINE-3-CARBOXYLIC ACID		
PROCOLLAGEN-PROLINE 3-DIOXYGENASE		
PROCOLLAGEN-PROLINE, 2-OXOGLUTARATE-4-DIOXYGENASE		
PYRIDINE 4-CARBOXYLIC ACID		
PROCOLLAGEN-PROLINE 3-DIOXYGENASE		
PYRIDINE-4-CARBOXYLIC ACID		
PROCOLLAGEN-PROLINE, 2-OXOGLUTARATE-4-DIOXYGENASE		
PYRIDINE DICARBOXYLIC ACID		
6β-HYDROXYHYOSCYAMINE EPOXIDASE		
PYRIDINE-2,3-DICARBOXYLIC ACID		
HYOSCYAMINE (6S)-DIOXYGENASE		
PROCOLLAGEN-PROLINE 3-DIOXYGENASE		
PROCOLLAGEN-PROLINE, 2-OXOGLUTARATE-4-DIOXYGENASE		
XAA-PRO DIPEPTIDASE		Ki = 440 μM
PYRIDINE-2,4-DICARBOXYLIC ACID		
DEOXYHYPUSINE MONOOXYGENASE		
FLAVONE SYNTHASE	C	Ki = 1.8 μM
HYOSCYAMINE (6S)-DIOXYGENASE		
HYOSCYAMINE (6S)-DIOXYGENASE	C	Ki = 9 μM
NARINGENIN, 2-OXOGLUTARATE 3-DIOXYGENASE	C	Ki = 1.2 μM
PEPTIDE-ASPARTATE β-DIOXYGENASE		$I50$ = 2 μM
PROCOLLAGEN-LYSINE 5-DIOXYGENASE		
PROCOLLAGEN-PROLINE 3-DIOXYGENASE		Ki = 3 μM
PROCOLLAGEN-PROLINE, 2-OXOGLUTARATE-4-DIOXYGENASE	C	Ki = 5 μM
PROCOLLAGEN-PROLINE, 2-OXOGLUTARATE-4-DIOXYGENASE		Ki = 8 μM
PROCOLLAGEN-PROLINE, 2-OXOGLUTARATE-4-DIOXYGENASE	C	Ki = 2 μM

Inhibitor / Enzyme	Type	Value
PYRIDINE-2,5-DICARBOXYLIC ACID		
AMINE OXIDASE MAO-B		*K*i = 1.8 mM
DEOXYHYPUSINE MONOOXYGENASE		
NARINGENIN, 2-OXOGLUTARATE 3-DIOXYGENASE	C	*K*i = 40 μM
NARINGENIN, 2-OXOGLUTARATE 3-DIOXYGENASE		40% 10 μM
OXOGLUTARATE DEHYDROGENASE (LIPOAMIDE)		
PROCOLLAGEN-LYSINE 5-DIOXYGENASE		
PROCOLLAGEN-PROLINE 3-DIOXYGENASE		
PROCOLLAGEN-PROLINE, 2-OXOGLUTARATE-4-DIOXYGENASE	C	*K*i = 800 nM
PROCOLLAGEN-PROLINE, 2-OXOGLUTARATE-4-DIOXYGENASE		*I*50 = 49 μM
PYRIDINE-2,6-DICARBOXYLIC ACID		
AMINE OXIDASE MAO-B		*K*i = 7.5 mM
DIHYDROPICOLINATE REDUCTASE		
GLUTAMATE DECARBOXYLASE		90% 1 mM
GLUTAMATE DEHYDROGENASE (NAD(P))	C	
NICOTINATE-NUCLEOTIDE PYROPHOSPHORYLASE (CARBOXYLATING)		
PROCOLLAGEN-PROLINE, 2-OXOGLUTARATE-4-DIOXYGENASE		
PYRIDINE-3,4-DICARBOXYLIC ACID		
PROCOLLAGEN-PROLINE, 2-OXOGLUTARATE-4-DIOXYGENASE		
PYRIDINE-3,5-DICARBOXYLIC ACID		
GLUTAMATE DEHYDROGENASE (NAD(P))	C	
PROCOLLAGEN-PROLINE, 2-OXOGLUTARATE-4-DIOXYGENASE		
PYRIDINE-2,4-DICARBOXYLIC ACID DIETHYLESTER		
PROCOLLAGEN-PROLINE, 2-OXOGLUTARATE-4-DIOXYGENASE	C	*K*i = 9 mM
PYRIDINE-2,4-DICARBOXYLIC AID		
OXOGLUTARATE DEHYDROGENASE (LIPOAMIDE)		
PYRIDINE-2,4-DIOL		
2,5-DIHYDROXYPYRIDINE 5,6-DIOXYGENASE		
2β-[(4-PYRIDINIUMMETHYL)-1,2,3-TRIAZOL-1-YL]METHYL-6,6-DIHYDROPENICILLANATE-1,1-DIOXIDE		
CEFOTAXIMASE		*I*50 = 1.6 μM
β-LACTAMASE		*I*50 = 1.6 μM
β-LACTAMASE (R-TEM)		*I*50 = 10 nM
PENICILLINASE		*I*50 = 100 nM
2,6-PYRIDINODICARBOXYLATE		
DIHYDROPICOLINATE REDUCTASE		
2(1H)-PYRIDINONE HYDRAZONE		
DOPAMINE β-MONOOXYGENASE		
(5Z)-7-[3-endo-(3-PYRIDINYL)BICYCLO[2.2.1]HEPT-2-endo-YL]HEPTANOIC ACID		
THROMBOXANE A SYNTHASE		*I*50 = 7 nM
(5Z)-7-[3-endo-(3-PYRIDINYL)BICYCLO[2.2.1]HEPT-2-exo-YL]HEPT-5-ENOIC ACID		
THROMBOXANE A SYNTHASE		*I*50 = 54 nM
N-(3-PYRIDINYLMETHYL)FARNESYLAMINE		
FARNESYL-DIPHOSPHATE FARNESYLTRANSFERASE		*I*50 = 50 nM
PYRIDOBENZO[b,f]1,5-OXAZOCYN-6-ONE		
RNA-DIRECTED DNA POLYMERASE		64% 350 nM
PYRIDOGLUTETHIMIDE		
AROMATASE		*I*50 = 2 μM
2-1H-PYRIDONE HYDRAZONE		
DOPAMINE β-MONOOXYGENASE		*K*is = 1.9 μM
PYRIDOSTIGMINE		
ACETYLCHOLINESTERASE	NC	*K*i = 3 nM
CHOLINESTERASE		*I*50 = 851 nM

PYRIDOXAL

O-ACETYLHOMOSERINE (THIOL)-LYASE	C	
O-ACETYLSERINE (THIOL)-LYASE		Ki = 122 μM
ALANINE RACEMASE	C	Ki = 320 μM
ARGININE DECARBOXYLASE		
β-FRUCTOFURANOSIDASE		
GLUTAMATE DEHYDROGENASE (NAD(P))		
GLYCOLALDEHYDE DEHYDROGENASE		
PHOSPHOPANTOTHENOYLCYSTEINE DECARBOXYLASE		
PYRIDOXAL KINASE		

PYRIDOXAL-L-ALANINE

PYRIDOXAMINE-PYRUVATE TRANSAMINASE	C	Ki = 180 nM

PYRIDOXAL-γ-AMINOBUTYRATE IMINE

PYRIDOXAL KINASE

PYRIDOXAL-L-ISOLEUCINE

PYRIDOXAMINE-PYRUVATE TRANSAMINASE	C	Ki = 120 μM

4-PYRIDOXALLACTONE

PYRIDOXAL DEHYDROGENASE	C	Ki = 700 μM

PYRIDOXALOLIGOPHOSPHATES

POLYRIBONUCLEOTIDE NUCLEOTIDYLTRANSFERASE

PYRIDOXAL-5′-PHOSPHATE

ACETYL-CoA CARBOXYLASE		
ACETYLSEROTONIN O-METHYLTRANSFERASE		
ACYLPHOSPHATASE	C	Ki = 320 mM
ADENYLATE CYCLASE		
β-ADRENERGIC-RECEPTOR KINASE		I50 = 900 μM
ALCOHOL DEHYDROGENASE		
ALCOHOL DEHYDROGENASE (NADP)		
ARYL SULFOTRANSFERASE	IR	Ki = 23 μM
ASPARTATE TRANSAMINASE		
CARBONATE DEHYDRATASE		
DEXTRANSUCRASE		
DNA LIGASE I		
DNA POLYMERASE		
DNA POLYMERASE α		
DNA POLYMERASE β		
DNA POLYMERASE γ		
β-FRUCTOFURANOSIDASE		
FRUCTOSE-BISPHOSPHATE ALDOLASE		
GLUCOSE-6-PHOSPHATASE		
GLUCOSE-6-PHOSPHATE 1-DEHYDROGENASE		
GLUCURONATE REDUCTASE		
GLUTAMATE DEHYDROGENASE		
GLUTAMATE DEHYDROGENASE (NAD(P))		
GLUTAMATE DEHYDROGENASE (NADP)		
GLUTAMATE-1-SEMIALDEHYDE 2,1-AMINOMUTASE		
GLUTAMATE SYNTHASE (NADPH)		
ω-HYDROXYFATTY ACID: NADP OXIDOREDUCTASE		
12α-HYDROXYSTEROID DEHYDROGENASE		
20α-HYDROXYSTEROID DEHYDROGENASE		
myo-INOSITOL-1-PHOSPHATE SYNTHASE		42% 1 mM
LACTOYLGLUTATHIONE LYASE		
LEUCINE DEHYDROGENASE		
MALATE SYNTHASE		
MALONYL-CoA DECARBOXYLASE		
MALONYL-CoA SYNTHETASE		
METHANOL-5-HYDROXYBENZIMIDAZOLYLCOBAMIDE Co-METHYLTRANSFERASE		
MEVALONATE KINASE		
NAD(P) TRANSHYDROGENASE (AB-SPECIFIC)		
NAD(P) TRANSHYDROGENASE (B-SPECIFIC)		

NEURAMINIDASE		
NITRATE REDUCTASE (NADH)		
NUCLEOSIDE-DIPHOSPHATASE	C	*K*i = 590 μM
NUCLEOSIDE-DIPHOSPHATASE	C	*K*i = 590 μM
3-OXOACYL-[ACYL-CARRIER-PROTEIN] REDUCTASE		
PHENOLL 2-MONOOXYGENASE		
PHOSPHOGLUCONATE DEHYDROGENASE (DECARBOXYLATING)	C	*K*i = 83 μM
PHOSPHOGLUCONATE DEHYDROGENASE (DECARBOXYLATING)	C	*K*i = 135 μM
PHOSPHOGLYCERATE DEHYDROGENASE		
PHOSPHOMEVALONATE KINASE		
PHOSPHOPANTOTHENOYLCYSTEINE DECARBOXYLASE		
PHOSPHORIBULOKINASE		
PORPHOBILINOGEN SYNTHASE	C	*K*i = 120 μM
PROTEIN KINASE CASEIN KINASE I	C	*I*50 = 1.5 mM
PROTEIN KINASE CASEIN KINASE II	C	*I*50 = 400 μM
PYRIDOXAL KINASE		
PYRIDOXAMINE-OXALOACETATE TRANSAMINASE		
PYRIDOXAMINE-OXALOACETATE TRANSAMINASE	C	
PYRIDOXAMINE-PHOSPHATE OXIDASE	C	*K*i = 3 μM
PYRIDOXINE 4-OXIDASE	C	*K*i = 2 μM
RIBONUCLEASE H		
RNA-DIRECTED DNA POLYMERASE		
RNA POLYMERASE I		
RNA POLYMERASE II		
SACCHAROPINE DEHYDROGENASE (NAD, L-LYSINE FORMING)		
SERINE-SULFATE AMMONIA-LYASE		
SUCCINATE-SEMIALDEHYDE DEHYDROGENASE		
THYMIDYLATE SYNTHASE	C	*K*i = 900 nM
TYPE II SITE SPECIFIC DEOXYRIBONUCLEASE		
TYROSINE TRANSAMINASE	C	*K*i = 1 mM
UROPORPHYRINOGEN-III SYNTHASE		

PYRIDOXALPHOSPHATE -6-AZOPHENYL-2′,4′-DISULPHONIC ACID

ADENOSINETRIPHOSPHATASE ECTO		*I*50 = 100 μM

PYRIDOXAL-5′-PHOSPHATE OXIME

PYRIDOXAMINE-PHOSPHATE OXIDASE	C	*K*i = 2 μM

PYRIDOXAL-L-TYROSINE

PYRIDOXAMINE-PYRUVATE TRANSAMINASE	C	*K*i = 36 μM

PYRIDOXAL-L-VALINE

PYRIDOXAMINE-PYRUVATE TRANSAMINASE	C	*K*i = 10 μM

PYRIDOXAMINE

ALANINE RACEMASE	C	*K*i = 590 μM
β-FRUCTOFURANOSIDASE		
PYRIDOXAL DEHYDROGENASE	C	*K*i = 1 mM
PYRIDOXINE 5′-O-β-D-GLUCOSYLTRANSFERASE		

PYRIDOXAMINEPHOSPHATE

PYRIDOXAMINE-OXALOACETATE TRANSAMINASE		

PYRIDOXAMINE-5′-PHOSPHATE

PHOSPHOGLYCERATE DEHYDROGENASE		
PYRIDOXAMINE-OXALOACETATE TRANSAMINASE	C	
PYRIDOXAMINE-PHOSPHATE OXIDASE		
TYROSINE TRANSAMINASE	C	*K*i = 1 mM

4-PYRIDOXIC ACID

PYRIDOXAL DEHYDROGENASE		
PYRIDOXAL DEHYDROGENASE	C	*K*i = 1 mM
PYRIDOXAMINE-PHOSPHATE OXIDASE		

5-PYRIDOXIC ACID

3-HYDROXY-2-METHYLPYRIDINE CARBOXYLATE DIOXYGENASE		

PYRIDOXINE

ALANINE RACEMASE	C	Ki = 23 mM
β-FRUCTOFURANOSIDASE		
β-GLUCOSIDASE	C	Ki = 3.2 mM
PHOSPHORYLASE b		
PYRIDOXAL DEHYDROGENASE		
PYRIDOXAMINE-PYRUVATE TRANSAMINASE	C	Ki = 52 μM
PYRIDOXINE 5′-O-β-D-GLUCOSYLTRANSFERASE		

PYRIDOXINE-5′-PHOSPHATE

GLYCOLALDEHYDE DEHYDROGENASE
PYRIDOXAMINE-PHOSPHATE OXIDASE

PYRIDOXOL

PYRIDOXAL DEHYDROGENASE	C	Ki = 800 μM
PYRIDOXAMINE-PYRUVATE TRANSAMINASE		Ki = 52 μM

PYRIDOXO 4,5 LACTONE

PYRIDOXAL DEHYDROGENASE	C	Ki = 700 μM

5-PYRIDOXOLACTONE

4-PYRIDOXOLACTONASE		Ki = 48 μM

4-PYRIDOXYC ACID PHOSPHATE

PYRIDOXAMINE-PHOSPHATE OXIDASE

PYRIDOXYL L-ALANINE

PYRIDOXAMINE-PYRUVATE TRANSAMINASE		Ki = 180 nM

PYRIDOXYL AMINO ACIDS

PYRIDOXAMINE-PYRUVATE TRANSAMINASE

2-[5-(2-PYRIDOXY)-PHENYLTHIO]-4,5-DIPHENYLIMIDAZOLE

STEROL O-ACYLTRANSFERASE		I50 = 10 nM

PYRIDYLAMINATED ACARBOSE

GLUCAN 1,4-α-GLUCOSIDASE		Ki = 500 nM

17-(3-PYRIDYL)ANDROSTA-5,16-DIEN-3β-OL

STEROID 17α-HYDROXYLASE/17,20 LYASE		I50 = 1 nM

4-(2-PYRIDYLAZO)RESORCINOL

ALCOHOL DEHYDROGENASE	IR	

4-[4′(2-PYRIDYL)BIPHENYL]BUTANE-α-PHOSPHONOSULPHONIC ACID

FARNESYL-DIPHOSPHATE FARNESYLTRANSFERASE		I50 = 13 nM

3-PYRIDYLCARBOXALDEHYDE

NICOTINATE PHOSPHORIBOSYLTRANSFERASE

PYRIDYLDISULPHIDE

CALPAIN

17-(3-PYRIDYL)ESTRA-1,3,5[10],16-TETRAEN-3-OL

AROMATASE		I50 = 1.8 μM
17α-HYDROXYPROGESTERONE ALDOLASE		I50 = 1.8 nM
STEROID 17α-MONOOXYGENASE		I50 = 2.6 nM
STEROID 5α-REDUCTASE		I50 = 10 μM

3-(4-PYRIDYLETHENYL)INDOLE

TRYPTOPHAN 2,3-DIOXYGENASE		Ki = 30 nM

(S)-1-(4-PYRIDYL)ETHYL 1-ADAMANTANECARBOXYLATE

STEROID 17α-MONOOXYGENASE/17,20 LYASE		I50 = 1.8 nM

11[2(3-PYRIDYL)ETHYLIDENE]6,11-DIHYDRODIBENZ[b,e]OXEPIN-2-CARBOXYLIC ACID

THROMBOXANE A SYNTHASE		I50 0 14 nM

7-[(Z)-(2′-PYRIDYL)METHYLENE]CEPHALOSPORANIC ACID SULPHONE
β-LACTAMASE — *I*50 = 5 μM

5-PYRIDYLMETHYLENE-N-HYDROXY-N′-AMINOGUANIDINE
RIBONUCLEOSIDE-DIPHOSPHATE REDUCTASE — *K*i = 53 μM
RIBONUCLEOSIDE-DIPHOSPHATE REDUCTASE — *K*i = 543 μM

N-(3-PYRIDYLMETHYL)FARNESYLAMINE
FARNESYL-DIPHOSPHATE FARNESYLTRANSFERASE — *I*50 = 4 nM

(E)-3-[4-(3-PYRIDYLMETHYL)PHENYL]-2-METHYLACRYLIC ACID
THROMBOXANE A SYNTHASE — *I*50 = 3 nM

PYRILAMINE MALEATE
ALDEHYDE OXIDASE — *I*50 = 510 μM

PYRILIUM SALTS
ACETYLCHOLINESTERASE
CHOLINESTERASE

PYRIMETHAMINE
DIHYDROFOLATE REDUCTASE — *I*50 1.8 μM
HISTAMINE N-METHYLTRANSFERASE — *I*50 = 26 μM
PROTEIN KINASE A — 13% 400 μM
PROTEIN KINASE A — 40% 400 μM
PROTEIN KINASE (Ca^{2+} DEPENDENT) — 22% 333 μM

PYRIMIDINE
CYTOCHROME P450 2E1 — U — *K*i = 170 μM
URATE-RIBONUCLEOTIDE PHOSPHORYLASE

4-PYRIMIDINE CARBOXALDEHYDE
AMINE OXIDASE MAO-B — C

PYRIMIDINE-5-SPIROCYCLOPROPANE-2,4,6-(1H, 3H, 5H)-TRIONE
OROTATE REDUCTASE (NADH)

(PYRIMIDINYL-4-OXY)ACETIC ACIDS
ALDEHYDE REDUCTASE

PYRIMIDO-[1,6-a]BENZIMIDAZOLES
DNA TOPOISOMERASE (ATP-HYDROLYSING)

(+)PYRIPYROPENE A
STEROL O-ACYLTRANSFERASE

PYRIPYROPENE A
STEROL O-ACYLTRANSFERASE — *I*50 = 58 nM

PYRIPYROPENE B
STEROL O-ACYLTRANSFERASE — *I*50 = 117 nM

PYRIPYROPENE C
STEROL O-ACYLTRANSFERASE — *I*50 = 53 nM

PYRIPYROPENE D
STEROL O-ACYLTRANSFERASE — *I*50 = 268 nM

PYRIPYROPENE E
STEROL O-ACYLTRANSFERASE

PYRIPYROPENE L
STEROL O-ACYLTRANSFERASE — *I*50 = 270 nM

PYRIPYROPENE M
STEROL O-ACYLTRANSFERASE — *I*50 = 3.8 μM

PYRIPYROPENE N
STEROL O-ACYLTRANSFERASE — *I*50 = 48 μM

PYRIPYROPENE O		
STEROL O-ACYLTRANSFERASE		*I*50 = 11 μM
PYRIPYROPENE P		
STEROL O-ACYLTRANSFERASE		*I*50 = 44 μM
PYRIPYROPENE Q		
STEROL O-ACYLTRANSFERASE		*I*50 = 40 μM
PYRIPYROPENE R		
STEROL O-ACYLTRANSFERASE		*I*50 = 78 μM
PYRITHIAMIN		
THIAMIN PYROPHOSPHOKINASE	C	*K*i = 85 μM
THIAMIN PYROPHOSPHOKINASE	C	*K*i = 1.2 μM
PYRITHIAMINE		
THIAMIN OXIDASE		
THIAMIN-PHOSPHATE KINASE		
THIAMIN PYROPHOSPHOKINASE		
THIAMIN-TRIPHOSPHATASE		
PYRITHIAMINE PHOSPHATE		
THIAMIN-PHOSPHATE KINASE		
PYRITHIOLAC		
ACETOLACTATE SYNTHASE		
PYRIZINOSTATIN		
PROLYLGLUTAMYL PEPTIDASE	UC	*I*50 = 1.8 μg/ml
PYR-LYS-TRP-ALA-PRO		
PEPTIDYL-DIPEPTIDASE A		*K*i = 500 μM
PYROCATECHOL		
ADENOSINE DEAMINASE		*I*50 = 20 μg/ml
TYROSINE PHENOL-LYASE	NC	*K*i = 460 μM
PYROGALLOL		
ALDEHYDE DEHYDROGENASE (NAD)	NC	*K*i = 1 mM
CATECHOL 1,2-DIOXYGENASE		
CATECHOL O-METHYLTRANSFERASE		10 μM
3-DEMETHYLUBIQUINONE-9 3-O-METHYLTRANSFERASE		29% 2 mM
GLUCOSE-6-PHOSPHATE 1-DEHYDROGENASE		100% 1.6 μM
LACTOYLGLUTATHIONE LYASE		24% 4 mM
LIPOXYGENASE		*I*50 = 30 μg/ml
PROTOCATECHUATE 4,5-DIOXYGENASE	C	90% 1 mM
PYROGALLOL OXIDATION PRODUCT		
LIPOXYGENASE		*I*50 = 12 μg/ml
PYRO-GLU-ASN-TRP		
ATROLYSIN C		*K*i = 3.5 μM
PYROGLUTAMINE		
PYROGLUTAMYL-PEPTIDASE I	C	*I*50 = 2 mM
PYROGLUTAMYL-AMINO ACID		
PYROGLUTAMYL-PEPTIDASE I		
PYROGLUTAMYL-LEU-ARG-CHO		
PLASMIN		*I*50 = 1.3 μg/ml
PYRONIN B		
PROTEINASE HIV-1		*I*50 = 17 μM
PYROPHOSPHATE		
2-(ACETAMIDOMETHYLENE)SUCCINATE HYDROLASE		
ACETATE-CoA LIGASE		

ACETATE KINASE		
ACETOLACTATE SYNTHASE		
ACETYL-CoA CARBOXYLASE		33% 1 mM
N-ACETYLGALACTOSAMINE-6-SULFATASE		
N-ACETYLGLUCOSAMINE KINASE		
N-ACETYLNEURAMINATE CYTIDYLYLTRANSFERASE		
ACONITATE HYDRATASE		
ACYL-LYSINE DEACYLASE		
ADENINE PHOSPHORIBOSYLTRANSFERASE		
ADENYLATE CYCLASE		$Ki = 180\ \mu M$
ADENYLATE CYCLASE		
ADENYLATE CYCLASE	C	$Ki = 135\ \mu M$
ADENYLYL-[GLUTAMATE-AMMONIA LIGASE] HYDROLASE		100% 20 mM
ADENYLYLSULFATASE	C	$Ki = 140\ \mu M$
ADENYLYLSULFATASE		
ADP DEAMINASE		
ALDEHYDE DEHYDROGENASE (NADP)		
ALDEHYDE DEHYDROGENASE (PYRROLOQUINOLINE-QUINONE)		15% 500 μM
L-AMINOADIPATE-SEMIALDEHYDE DEHYDROGENASE		
AMINOPEPTIDASE Y		93% 1 mM
AMP DEAMINASE		
ANTHRANILATE PHOSPHORIBOSYLTRANSFERASE		
ARGININE-tRNA LIGASE		$Ki = 120\ \mu M$
ARGININOSUCCINATE SYNTHASE	NC	$Ki = 220\ \mu M$
ARYL SULFOTRANSFERASE	NC	$Ki = 700\ \mu M$
ASPARAGINE SYNTHASE (GLUTAMINE-HYDROLYSING)		
ASPARTATE CARBAMOYLTRANSFERASE		
ASPARTATE 4-DECARBOXYLASE		
ASPERGILLUS NUCLEASE S1		
ASPULVINONE DIMETHYLALLYLTRANSFERASE		
ASPULVINONE DIMETHYLALLYLTRANSFERASE	MIX	$Ki = 18\ \mu M$
ATP DEAMINASE		
ATP PHOSPHORIBOSYLTRANSFERASE		
BISPHOSPHOGLYCERATE MUTASE		30% 2 mM
BISPHOSPHOGLYCERATE PHOSPHATASE		70% 20 mM
CALF THYMUS RIBONUCLEASE H		
CARBOXYPEPTIDASE B		
CDP-DIACYLGLYCEROL-INOSITOL 3-PHOSPHATIDYLTRANSFERASE		
CHOLATE-CoA LIGASE		
CHOLATE-CoA LIGASE	NC	
CITRAMALATE SYNTHASE		
CMP-N-ACYLNEURAMINATE PHOSPHODIESTERASE		82% 6.5 mM
COB(I)ALAMIN ADENOSYLTRANSFERASE		
CREATINE KINASE	C	$Ki = 10\ \mu M$
3′,5′-CYCLIC-NUCLEOTIDE PHOSPHODIESTERASE		
CYTOCHROME-b5 REDUCTASE	C	
2-DEHYDRO-3-DEOXYGLUCARATE ALDOLASE		
DEOXYCYTIDINE KINASE		
DICARBOXYLATE-CoA LIGASE		
trans-1,2-DIHYDROBENZENE-1,2-DIOL DEHYDROGENASE		
DIMETHYLALLYLtransTRANSFERASE		
DNA NUCLEOTIDYLEXOTRANSFERASE		
DNA POLYMERASE β		
DOLICHYL-PHOSPHATASE		
ENDOPEPTIDASE TREPONEMA DENTICOLA ARCC35405		$Ki = 240\ \mu M$
ETHANOLAMINE-PHOSPHATE CYTIDYLYLTRANSFERASE		
ETHANOLAMINE-PHOSPHATE PHOSPHO-LYASE		
FAD PYROPHOSPHATASE		
FARNESYL-DIPHOSPHATE FARNESYLTRANSFERASE		$I50 = 500\ \mu M$
FARNESYL PROTEINTRANSFERASE		$Ki = 1.4\ mM$
FARNESYLtransTRANSFERASE		
FMN ADENYLYLTRANSFERASE		
FRUCTOSE-BISPHOSPHATE ALDOLASE	C	$Ki = 5\ mM$
GERANYL-DIPHOSPHATE CYCLASE	C	$Ki = 100\ \mu M$

GLUCOSAMINE-6-PHOSPHATE ISOMERASE		
GLUCOSE-6-PHOSPHATASE		
GLUCOSE-1-PHOSPHATE CYTIDYLYLTRANSFERASE	NC	$Ki = 340\ \mu M$
GLUTAMATE-AMMONIA LIGASE		$I50 = 10$ mM
[GLUTAMATE-AMMONIA-LIGASE] ADENYLYLTRANSFERASE		$I50 = 5$ mM
GLYCERATE KINASE		
GLYCEROL-3-PHOSPHATE CYTIDILYLTRANSFERASE		
GLYCEROPHOSPHOCHOLINE PHOSPHODIESTERASE		71% 6 mM
GLYCINE DEHYDROGENASE (CYTOCHROME)		> 10 mM
GLYCOGEN-SYNTHASE-D PHOSPHATASE		
GLYCOSULFATASE		
GMP SYNTHASE	C	$Ki = 420\ \mu M$
GTP CYCLOHYDROLASE II		52% 100 μM
GUANYLATE CYCLASE		$I50 = 700\ \mu M$
GUANYLATE CYCLASE		64% 2 mM
HISTIDINE AMMONIA-LYASE		
HOMOGENTISATE 1,2-DIOXYGENASE		
H^+-TRANSPORTING PYROPHOSPHATASE		
3-HYDROXYMETHYLCEPHEM CARBAMOYLTRANSFERASE		85% 5 mM
[HYDROXYMETHYLGLUTARYL-CoA REDUCTASE (NADPH)]-PHOSPHATASE		
HYPOXANTHINE PHOSPHORIBOSYLTRANSFERASE		
INDOLE-3-ACETATE β-GLUCOSYLTRANSFERASE		
INORGANIC PYROPHOSPHATASE	C	$Ki = 100$ nM
myo-INOSITOL-1-PHOSPHATE SYNTHASE		
ISOPENTENYL-DIPHOSPHATE Δ-ISOMERASE		
LACTOSE SYNTHASE		
LEUCYL AMINOPEPTIDASE		
MANDELATE RACEMASE		
MANGANESE PEROXIDASE		
METHIONINE ADENOSYLTRANSFERASE	C	$Ki = 71$ mM
METHYLGLYOXAL SYNTHASE	C	$Ki = 95\ \mu M$
METHYLGLYOXAL SYNTHASE		80% 20 mM
(R)-2-METHYLMALATE DEHYDRATASE		
MONOTERPENYL-PYROPHOSPHATASE		$I50 = 600\ \mu M$
MYOSIN-LIGHT-CHAIN PHOSPHATASE		
NADH PEROXIDASE		90% 3 mM
NAD(P) TRANSHYDROGENASE		
NAD(P) TRANSHYDROGENASE (B-SPECIFIC)		
NICOTINAMIDE-NUCLEOTIDE ADENYLYLTRANSFERASE		
NICOTINATE-NUCLEOTIDE PYROPHOSPHORYLASE (CARBOXYLATING)		
NICOTINATE PHOSPHORIBOSYLTRANSFERASE		
NUCLEOSIDE-DIPHOSPHATASE	C	100% 1 mM
NUCLEOSIDE-TRIPHOSPHATASE		
3′-NUCLEOTIDASE		
OROTATE PHOSPHORIBOSYLTRANSFERASE		
PANTOTHENATE KINASE		
PEPTIDYL-DIPEPTIDASE A		100% 50 mM
PHENYLALANINE RACEMASE (ATP-HYDROLYSING)		> 2mM
PHOSPHATASE ALKALINE		
PHOSPHATE ACETYLTRANSFERASE		
PHOSPHATIDATE CYTIDILYLTRANSFERASE		
PHOSPHOAMIDASE		
PHOSPHOENOLPYRUVATE CARBOXYKINASE (PYROPHOSPHATE)		
6-PHOSPHOFRUCTO-2-KINASE		
PHOSPHOGLUCONATE DEHYDROGENASE (DECARBOXYLATING)	C	$Ki = 470\ \mu M$
PHOSPHOGLUCONATE DEHYDROGENASE (DECARBOXYLATING)	C	$Ki = 800\ \mu M$
PHOSPHOGLYCERATE MUTASE	C	
PHOSPHOPROTEIN PHOSPHATASE		76% 1 mM
PHOSPHOPROTEIN PHOSPHATASE SMP II		95% 1 mM
PHOSPHORYLASE PHOSPHATASE		
PHOTINUS-LUCIFERIN 4-MONOOXYGENASE (ATP-HYDROLYSING)		
α-PINENE CYCLASE	C	$Ki = 100\ \mu M$
POLYGALACTURONATE 4-α-GALACTURONOSYLTRANSFERASE		
POLYNUCLEOTIDE ADENYLYLTRANSFERASE		

Inhibitor / Enzyme	Type	Value
POLYNUCLEOTIDE ADENYLYLTRANSFERASE	NC	
POLYNUCLEOTIDE 3′-PHOSPHATASE		
POLYRIBONUCLEOTIDE SYNTHASE (ATP)		72% 12 mM
PROTEIN TYROSINE-PHOSPHATASE		70% 5 mM
PROTEIN TYROSINE-PHOSPHATASE (Mg^{2+} DEPENDENT)		47% 5 mM
PURINE-NUCLEOSIDE PHOSPHORYLASE		Ki = 330 μM
PUTRESCINE CARBAMOYLTRANSFERASE	C	Ki = 60 μM
PYRIDOXAL KINASE	C	Ki = 1.7 mM
PYROPHOSPHATE-FRUCTOSE-6-PHOSPHATE 1-PHOSPHOTRANSFERASE	C	
PYRUVATE DEHYDROGENASE (LIPOAMIDE)		
PYRUVATE, ORTHOPHOSPHATE DIKINASE		
QUINATE O-HYDROXYCINNAMOYLTRANSFERASE		
RHODOPSIN KINASE	C	Ki = 24 μM
RIBULOSE-PHOSPHATE EPIMERASE		82% 20 mM
tRNA CYTIDYLTRANSFERASE		100% 1 mM
RNA-DIRECTED RNA POLYMERASE		
mRNA GUANYLYLTRANSFERASE		I50 1 μM
tRNA ISOPENTENYLTRANSFERASE		
RNA URIDYLYLTRANSFERASE		
SERINE-tRNA LIGASE		
SHIKIMATE O-HYDROXYCINNAMOYLTRANSFERASE		
STREPTOMYCIN KINASE		
SUCCINYL-CoA HYDROLASE		
SULFATE ADENYLYLTRANSFERASE		I50 = 2.5 mM
THIAMIN-PHOSPHATE KINASE		
THIAMIN-PHOSPHATE PYROPHOSPHORYLASE		
THYMIDINE-TRIPHOSPHATASE		Ki = 270 μM
TRANSALDOLASE		
TRICHODIENE SYNTHASE		
TRIHYDROXYCOPROSTANOYL-CoA SYNTHETASE	NC	
UDP-GLUCOSE-1-PHOSPHATE URIDYLYLTRANSFERASE	NC	
URIDINE PHOSPHORYLASE		
PYROPHOSPHATE + FLUORIDE		
4-METHYLGLUTAMINE SYNTHETASE		
PYROSTATIN A		
β-N-ACETYLHEXOSAMINIDASE	C	Ki = 1.7 μM
PYROSTATIN B		
β-N-ACETYLHEXOSAMINIDASE	C	Ki = 2 μM
(S)-N-PYROVYL-1-(1-NAPHTHYL)ETHYLAMINE		
α-CHYMOTRYPSIN		Ki = 27 μM
SUBTILISIN		Ki = 37 μM
PYROXICAM		
CYCLOOXYGENASE 1		I50 = 18 μM
PYRROLE		
PROLINE-tRNA LIGASE		
PYRROLE-2-CARBOXYLATE		
D-AMINO-ACID OXIDASE	C	
4-HYDROXYPROLINE EPIMERASE	C	Ki = 16 mM
PROLINE RACEMASE		
δ-1-PYRROLE-2-CARBOXYLIC ACID		
PROLINE RACEMASE		50% 360 μM
2-PYRROLECARBOXYLIC ACID		
SARCOSINE OXIDASE		
PYRROLIDINE		
PROLINE-tRNA LIGASE		

PYRROLIDINE-2-PHOSPHONIC ACID		
XAA-PRO DIPEPTIDASE	C	*K*i = 950 μM

3-PYRROLIDINOL		
4-HYDROXYPROLINE EPIMERASE	C	*K*i = 90 mM

2-PYRROLIDINONE		
PROLINE-tRNA LIGASE		

3-PYRROLINE		
PROLINE-tRNA LIGASE		

Δ1-PYRROLINE-2-CARBOXYLATE		
1-PYRROLINE-5-CARBOXYLATE DEHYDROGENASE		

PYRROLOQUINOLINE QUINONE		
ALDEHYDE REDUCTASE		*I*50 = 61 μM

4-PYRROL-1-YL-PHENYLACETIC ACID HYDRAZIDE		
AMINE OXIDASE (COPPER-CONTAINING)	NC	*K*i = 500 nM

PYRUS COMMUNIS POLYGALACTURONASE INHIBITOR		
POLYGALACTURONASE		

PYRUVIC ACID		
ACETOACETATE DECARBOXYLASE		
ACETYLPYRUVATE HYDROLASE	C	
ALANINE DEHYDROGENASE		
4-AMINOBUTYRATE TRANSAMINASE		
5-AMINOLEVULINATE SYNTHASE	NC	
AMINOLEVULINATE TRANSAMINASE		
ANTHRANILATE SYNTHASE		
ASPARAGINE SYNTHASE (GLUTAMINE-HYDROLYSING)		
ASPARTATE-AMMONIA LIGASE		
CARBONATE DEHYDRATASE		
2-DEHYDRO-3-DEOXY-L-ARABINONATE DEHYDRATASE	C	*K*i = 4.6 mM
FRUCTOSE-BISPHOSPHATE ALDOLASE		
GLUCONATE 2-DEHYDROGENASE	C	*K*i = 18 mM
GLUTAMATE DEHYDROGENASE		
GLUTAMATE SYNTHASE (NADPH)		
GLYCERATE KINASE		
GLYCOSYLCERAMIDASE		
GUANYLATE CYCLASE		78% 5 mM
HEXOSE OXIDASE		
D-2-HYDROXY-ACID DEHYDROGENASE		
(S)-2-HYDROXY-ACID OXIDASE		
3-HYDROXYBUTYRATE DEHYDROGENASE		
4-HYDROXY-4-METHYL-2-OXOGLUTARATE ALDOLASE	C	
HYDROXYPHENYLPYRUVATE REDUCTASE	C	*K*i = 200 μM
HYDROXYPYRUVATE REDUCTASE		
myo-INOSITOL-1-PHOSPHATE SYNTHASE		
[ISOCITRATE DEHYDROGENASE (NADP)] KINASE		
ISOCITRATE LYASE		
KYNURENINE 3-MONOOXYGENASE		
D-LACTATE DEHYDROGENASE (CYTOCHROME)	NC	
L-LACTATE DEHYDROGENASE		*K*i = 15 mM
LACTATE-MALATE TRANSHYDROGENASE		
LACTATE 2-MONOOXYGENASE	C	*K*i = 1.1 mM
MALATE DEHYDROGENASE (OXALACETATE-DECARBOXYLATING) (NADP)		
MALATE SYNTHASE		
METHYLMALONYL-CoA CARBOXYLTRANSFERASE		
[3-METHYL-2-OXOBUTANOATE DEHYDROGENASE (LIPOAMIDE)] KINASE		
3-METHYL-2-OXOBUTANOATE HYDROXYMETHYLTRANSFERASE		38% 5 mM
D-OCTOPINE DEHYDROGENASE		
OXOGLUTARATE DEHYDROGENASE (LIPOAMIDE)		
PHOSPHOENOLPYRUVATE CARBOXYKINASE (GTP)	C	*K*i = 9 mM

Enzyme	Type	Value
PHOSPHOENOLPYRUVATE DECARBOXYLASE		
PORPHOBILINOGEN SYNTHASE	C	
PROCOLLAGEN-LYSINE 5-DIOXYGENASE		
PROCOLLAGEN-PROLINE, 2-OXOGLUTARATE-4-DIOXYGENASE		
PROLINE DEHYDROGENASE	C	*K*i = 3.3 mM
PYRUVATE DEHYDROGENASE (LIPOAMIDE)	IR	
[PYRUVATE DEHYDROGENASE (LIPOAMIDE)] KINASE	UC	*K*i = 325 μM
PYRUVATE KINASE	C	*K*i = 1 mM
PYRUVATE SYNTHASE	IR	
L-SERINE DEHYDRATASE	NC	
THIOSULFATE SULFURTRANSFERASE		
THREONINE DEHYDRATASE		
TYROSINE 2,3-AMINOMUTASE		20% 1 mM

Q

Inhibitor / Enzyme	Type	Value
Q 35		
DNA GYRASE		*I*50 = 2.5 μg/ml
Q 12713		
1-PHOSPHATIDYLINOSITOL-4,5-BISPHOSPHATE PHOSPHODIESTERASE		*I*50 = 8.5 μM
Q 8251		
PROTEINASE HIV-1		*I*50 = 720 pM
QA 208-199		
ARACHIDONATE 5-LIPOXYGENASE		*I*50 = 400 nM
QFS 12		
FURIN		*K*i = 4.6 μM
QS-10.MOR		
PHOSPHOLIPASE A2		*I*50 = 28 μM
QUERCETAGETIN		
INTEGRASE		*I*50 = 790 nM
NADH OXIDASE		*I*50 = 177 nmole
PROTEINASE HIV-1		*I*50 = 1 mM
PROTEIN KINASE C		*I*50 = 2 μM
RNA-DIRECTED DNA POLYMERASE		*K*i = 460 nM
SUCCIN OXIDASE		*I*50 = 104 nmole
QUERCETIN		
ADENOSINE DEAMINASE		*I*50 = 50 μg/ml
ADENOSINE DEAMINASE		*I*50 = 26 μM
ALCOHOL DEHYDROGENASE (NADP)		
ALDEHYDE REDUCTASE		*I*50 = 6.6 μM
ALDEHYDE REDUCTASE		*I*50 = 600 μM
ALDEHYDE REDUCTASE 2		*I*50 = 600 nM
AMINOGLYCOSIDE PHOSPHOTRANSFERASE (3′)-IIIa	C	*K*i = 126 μM
ANTHOCYANIDIN 3-O-GLUCOSYLTRANSFERASE		
ARACHIDONATE 5-LIPOXYGENASE		*I*50 = 3.5 μM
ARACHIDONATE 5-LIPOXYGENASE		*I*50 = 1 μM
ARACHIDONATE 5-LIPOXYGENASE		*I*50 = 400 nM
ARACHIDONATE 12-LIPOXYGENASE		*I*50 = 400 nM
ARACHIDONATE 15-LIPOXYGENASE 1		*I*50 = 98 μM
ARYL SULFOTRANSFERASE P	NC	*K*i = 100 nM
CAFFEATE O-METHYLTRANSFERASE		
CAM KINASE		
CARBONYL REDUCTASE (NADPH)		32% 10 μM
CARBONYL REDUCTASE (NADPH)		83% 10 μM
CATECHOL O-METHYLTRANSFERASE	NC	*I*50 ≈ 5 μM
Ca^{2+}-TRANSPORTING ATPASE		*I*50 = 10 μM
CHALCONE ISOMERASE		*K*i = 4.1 μM

CHLORDECONE REDUCTASE		
CHLORIDE PEROXIDASE		
3′,5′-CYCLIC-NUCLEOTIDE PHOSPHODIESTERASE (cycAMP)		*I*50 = 23 μM
3′,5′-CYCLIC-NUCLEOTIDE PHOSPHODIESTERASE (CALMODULIN INDEPENDENT)		*I*50 = 231 μM
3′,5′-CYCLIC-NUCLEOTIDE PHOSPHODIESTERASE (Ca^{2+}/CALMODULIN)		*I*50 = 78 μM
3′,5′-CYCLIC-NUCLEOTIDE PHOSPHODIESTERASE (cycGMP)		*I*50 = 20 μM
3′,5′-CYCLIC-NUCLEOTIDE PHOSPHODIESTERASE (cycGMP) (Ca^{2+}/CALMODULIN)		*I*50 = 57 μM
CYCLOOXYGENASE		*I*50 = 16 μM
CYTOCHROME-C PEROXIDASE		
CYTOCHROME P450 1A1		*I*50 = 350 nM
CYTOCHROME P450 2B1		
2-DEHYDROPANTOYL-LACTONE REDUCTASE (A SPECIFIC)		
DIACYLGLYCEROL KINASE		
DNA GYRASE		*I*50 = 2 μg/ml
DNA TOPOISOMERASE IV		*I*50 = 1 μg/ml
ETHOXY RESORUFIN DEALKYLASE	C	
β-FRUCTOFURANOSIDASE		*I*50 = 290 μM
β-GALACTOSIDASE		*K*i = 10 μM
α-GLUCOSIDASE		*K*i = 8.5 μM
GLUTATHIONE DEHYDROGENASE (ASCORBATE)		55% 20 μM
GLUTATHIONE REDUCTASE (NADPH)		39% 100 μM
GLUTATHIONE TRANSFERASE	NC	*K*i = 60 μM
GLUTATHIONE TRANSFERASE		*I*50 = 19 μM
GLUTATHIONE TRANSFERASE		73% 100 μM
GLYCEROL DEHYDROGENASE (NADP)		
HEXOKINASE		70% 8 μG/ml
H^+/K^+-EXCHANGING ATPASE	C	*I*50 = 2.3 μM
H^+-TRANSPORTING ATPASE		
H^+-TRANSPORTING ATP SYNTHASE		
HYALURONIDASE PH20		35% 50 μM
HYALURONOGLUCOSAMINIDASE		
6-HYDROXYMYELINE SYNTHASE		28% 1 mM
11β-HYDROXYSTEROID DEHYDROGENASE		*I*50 = 192 μM
L-IDITOL 2-DEHYDROGENASE	MIX	*K*i = 58 μM
L-IDITOL 2-DEHYDROGENASE		*I*50 = 45 μM
INTEGRASE		*I*50 = 23 μM
IODIDE PEROXIDASE		*I*50 = 2.4 μM
LACTOPEROXIDASE		
LACTOYLGLUTATHIONE LYASE		*K*i = 9 μM
LIPOXYGENASE		93% 30 mM
LYSO PLATELET ACTIVATING FACTOR ACETYLTRANSFERASE		*I*50 = 80 μM
MALATE DEHYDROGENASE (NADP)		*K*i = 830 nM
MONOPHENOL MONOOXYGENASE		
MYELOPEROXIDASE		*I*50 = 3.5 μM
MYELOPEROXIDASE		
MYOSIN-LIGHT-CHAIN KINASE		
NADH DEHYDROGENASE (EXOGENOUS)		*I*50 = 400 μM
NADH OXIDASE		*I*50 = 145 nmole
Na^+/K^+-EXCHANGING ATPASE		*I*50 = 5.5 μM
NUCLEOSIDE-TRIPHOSPHATASE		
PEROXIDASE		
PHENOL OXIDASE	NC	
PHENYLALANINE AMMONIA-LYASE		
PHOSPHATIDYLINOSITOL KINASE		*I*50 = 100 μM
1-PHOSPHATIDYLINOSITOL 3-KINASE	UC	*I*50 = 3.8 μM
1-PHOSPHATIDYLINOSITOL 3-KINASE		*I*50 = 10 μM
1-PHOSPHATIDYLINOSITOL 3-KINASE	C	*K*i = 880 nM
1-PHOSPHATIDYLINOSITOL 4-KINASE		*I*50 = 6 μM
1-PHOSPHATIDYLINOSITOL-4-PHOSPHATE 5-KINASE	C	*I*50 = 200 nM
PHOSPHOENOLPYRUVATE CARBOXYLASE		*K*i = 170 nM
PHOSPHOLIPASE A2		*I*50 = 2 μM
PHOSPHORYLASE KINASE		*I*50 = 50 μM
PROLYL OLIGOPEPTIDASE		*I*50 = 0.2 ppm
PROSTAGLANDIN E2 9-REDUCTASE		

PROSTAGLANDIN F SYNTHASE		
PROTEINASE HIV-1		*I*50 = 36 μM
PROTEIN KINASE C		*I*50 = 1 μM
PROTEIN KINASE C		*I*50 = 40 μM
PROTEIN KINASE CASEIN KINASE G	C	*K*i = 220 nM
PROTEIN KINASE CASEIN KINASE I		*I*50 = 45 μM
PROTEIN KINASE CASEIN KINASE II		*I*50 = 100 nM
PROTEIN TYROSINE KINASE		*I*50 = 13 μM
PROTEIN TYROSINE KINASE EGFR		*I*50 = 410 nM
PROTEIN TYROSINE KINASE EGFR		*I*50 = 50 μM
PROTEIN TYROSINE KINASE HPK40		*I*50 = 200 μM
PROTEIN TYROSINE KINASE I		
PROTEIN TYROSINE KINASE IIA		
PROTEIN TYROSINE KINASE IIB	C	*I*50 = 10 μM
PROTEIN TYROSINE KINASE III		*I*50 10 μM
PROTEIN TYROSINE KINASE (Mg^{2+} DEPENDENT)		90% 20 μM
PROTEIN TYROSINE KINASE p56lck		*I*50 = 8 μg/ml
PROTEIN TYROSINE KINASE p60v-src	C	*I*50 = 50 μM
PROTEIN TYROSINE KINASE ptabl50	C	*K*i = 3.7 μM
PROTEIN TYROSINE KINASE pp60c-Src		*I*50 = 600 μM
PROTEIN TYROSINE KINASE pp60c-Src		*I*50 = 4.4 μg/ml
PYRUVATE KINASE		
RNA-DIRECTED DNA POLYMERASE		*I*50 = 625 μM
RNA-DIRECTED DNA POLYMERASE		*I*50 = 646 μM
RNA-DIRECTED DNA POLYMERASE		*K*i = 80 nM
SUCCIN OXIDASE		*I*50 = 715 nmole
TRYPSIN		*K*i = 2 μM
TYROSINASE		*I*50 = 22 μg/ml
XANTHINE OXIDASE		*I*50 = 7.2 μM
XANTHINE OXIDASE		*I*50 = 383 nM
XANTHINE OXIDASE	MIX	Kei = 74 nM
XYLOSE REDUCTASE	NC	*K*i = 650 μM

QUERCETIN-3,7-DIGALACTOSIDE

β-GALACTOSIDASE		*K*i = 10 μM

QUERCETIN 3-O-(2″,6″-DIGALLOYL)-β-D-GALACTOPYRANOSIDE

INTEGRASE		*I*50 = 24 μg/ml

QUERCETIN-3-O-GAL

PROLYL OLIGOPEPTIDASE		*I*50 = 0.1 ppm

QUERCETIN 3-O-β-D-GALACTOPYRANOSIDE

INTEGRASE		*I*50 = 65 μg/ml

QUERCETIN 3-O-2″-GALLOYL)-α-L-ARABINOPYRANOSIDE

INTEGRASE		*I*50 = 18 μg/ml

QUERCETIN 3-O-2″-GALLOYL)-β-D-GALACTOPYRANOSIDE

INTEGRASE		*I*50 = 28 μg/ml

QUERCETIN 3-O-β-D-GLUCOPYRANOSIDE

RNA-DIRECTED DNA POLYMERASE		*I*50 = 50 μM

QUERCETIN 3-O-α-L-RHAMNOPYRANOSIDE

INTEGRASE		*I*50 = 75 μg/ml

QUERCETIN-3-SULPHATE

FLAVONOL 3-SULFOTRANSFERASE	NC	

QUERCITRIN

ACETOIN DEHYDROGENASE		
ALCOHOL DEHYDROGENASE (NADP)		*I*50 = 32 μM
ALCOHOL DEHYDROGENASE (NADP)		
ALDEHYDE REDUCTASE		84% 100 μM
ALDEHYDE REDUCTASE		*I*50 = 10 μM
ALDEHYDE REDUCTASE	NC	*K*i = 780 nM

ALDEHYDE REDUCTASE	NC	*K*i = 1.9 µM
CARBONYL REDUCTASE (NADPH)		
CARBONYL REDUCTASE (NADPH) 1		85% 100 µM
CARBONYL REDUCTASE (NADPH) 2		90% 100 µM
CARBONYL REDUCTASE (NADPH) 3		100% 100 µM
trans-1,2-DIHYDROBENZENE-1,2-DIOL DEHYDROGENASE		
trans-1,2-DIHYDROBENZENE-1,2-DIOL DEHYDROGENASE		70% 100 µM
trans-1,2-DIHYDROBENZENE-1,2-DIOL DEHYDROGENASE 2		87% 100 µM
trans-1,2-DIHYDROBENZENE-1,2-DIOL DEHYDROGENASE 3		64% 100 µM
trans-1,2-DIHYDROBENZENE-1,2-DIOL DEHYDROGENASE 4		59% 100 µM
GLUCURONATE REDUCTASE		*I*50 = 23 µM
GLUTATHIONE REDUCTASE (NADPH)		33% 100 µM
GLUTATHIONE TRANSFERASE		67% 100 µM
INTEGRASE		*I*50 = 60 µM
a,β-KETOALKENE DOUBLE BOUND REDUCTASE		68% 10 µM
KETONE REDUCTASES		
MORPHINE 6-DEHYDROGENASE		
PEPTIDYL-DIPEPTIDASE A		*I*50 = 670 µM
PROSTAGLANDIN F SYNTHASE		
PROTEIN KINASE C		*I*50 = 6 µM
RNA-DIRECTED DNA POLYMERASE		56% 1 mg/ml
QUERCITRIN 2″-O-GALLATE		
RNA-DIRECTED DNA POLYMERASE		93% 1 mg/ml
QUINACILLIN		
β-LACTAMASE RTEM	IR	
QUINACRINE		
ACETYLSPERMIDINE DEACETYLASE		
ALDEHYDE OXIDASE		83% 100 µM
AMINE DEHYDROGENASE		
AMINE OXIDASE (POLYAMINE)		80% 100 µM
BENZOATE 4-MONOOXYGENASE		
γ-BUTYROBETAINE DIOXYGENASE		
Ca^{2+}-TRANSPORTING ATPASE		
CHOLEST-5-ENE-3β,7α-DIOL 3β-DEHYDROGENASE		1 mM
CYCLOHEXYLAMINE OXIDASE		
ETHANOLAMINE OXIDASE		
GLYCINE DEHYDROGENASE (CYTOCHROME)		
HISTAMINE N-METHYLTRANSFERASE		84% 10 µM
LEGHEMOGLOBIN REDUCTASE		
LYSOPHOSPHOLIPASE		*I*50 = 700 µM
MYOSIN-LIGHT-CHAIN KINASE		*I*50 = 75 µM
NICOTINE DEHYDROGENASE		
PHOSPHATASE ALKALINE		
1-PHOSPHATIDYLINOSITOL-4,5-BISPHOSPHATE PHOSPHODIESTERASE		42% 9 µM
PHOSPHOLIPASE A2		*I*50 = 97 µM
PHOSPHOLIPASE A2		*I*50 = 17 µM
L-PIPECOLATE DEHYDROGENASE		
POLYAMINE OXIDASE	NC	
PROTEIN KINASE A		14% 400 µM
PROTEIN KINASE A		33% 400 µM
PROTEIN KINASE C		*I*50 = 300 µM
PROTEIN KINASE (Ca^{2+} DEPENDENT)		*I*50 = 270 µM
PROTOCHLOROPYLI'LIDE REDUCTASE		
PYRUVATE DEHYDROGENASE (NADP)		
SPERMIDINE DEHYDROGENASE		*I*50 = 200 µM
SPHINGOMYELIN PHOSPHODIESTERASE		61% 10 µM
THIAMIN OXIDASE		
QUINACRINEHYDROXYTOLUENE		
RETINAL OXIDASE		
QUINACRINE MUSTARD		
H^+-TRANSPORTING ATP SYNTHASE		

Inhibitor / Enzyme	Type	Value
QUINALDIC ACID		
ACONITATE HYDRATASE	NC	$Ki = 27$ mM
MONODEHYDROASCORBATE REDUCTASE (NADH)		
QUINALDINE		
AMINE OXIDASE MAO-A	C	$Ki = 14$ μM
AMINE OXIDASE MAO-B	NC	$Ki = 633$ μM
QUINALDINIC ACID		
PYRUVATE OXIDASE		
QUINALIZARIN		
7-ETHOXYCOUMARIN O-DEETHYLASE		$I50 = 63$ μM
QUINALIZERIN		
MYOSIN-LIGHT-CHAIN KINASE		$I50 = 53$ μM
PROTEIN KINASE A		$I50 = 2$ μM
PROTEIN KINASE C		$I50 = 4$ μM
PROTEIN KINASE (Ca^{2+} DEPENDENT)		$I50 = 65$ μM
QUINALPHOS		
ACETYLCHOLINESTERASE		
QUINAPRIL		
11β-HYDROXYSTEROID DEHYDROGENASE		
QUINAPRILAT		
PEPTIDYL-DIPEPTIDASE A		$I50 = 500$ pM
PEPTIDYL-DIPEPTIDASE A		
XAA-PRO AMINOPEPTIDASE		$I50 = 266$ μM
QUINATE		
QUINATE O-HYDROXYCINNAMOYLTRANSFERASE		
QUINAZARIN		
PROTEIN KINASE (Ca^{2+} DEPENDENT)		92% 160 μM
QUINAZOLINE		
AMINE OXIDASE MAO-A	C	$Ki = 169$ μM
AMINE OXIDASE MAO-B	NC	$Ki = 52$ μM
QUINAZOLINE DERIVATIVES		
THYMIDYLATE SYNTHASE		
2,4-(1H,3H)-QUINAZOLINEDIONE (BENZOYLENEUREA)		
NAD ADP-RIBOSYLTRANSFERASE		$I50 = 8.1$ μM
QUINIDINE		
ACETYLCHOLINESTERASE		
CHOLINESTERASE		57% 2 μg/ml
CYTOCHROME P450 2D1		
CYTOCHROME P450 2D6	C	$Ki = 70$ nM
CYTOCHROME P450 2D6		$Ki = 27$ nM
CYTOCHROME P450 2D6	C	$Ki = 400$ nM
CYTOCHROME P450 2D6	C	$Ki = 400$ nM
CYTOCHROME P450 2D6		$Ki = 80$ nM
CYTOCHROME P450 IID1 (DEBRISOQUINE)		$Ki = 5.9$ μM
DEBRISOQUINE 4-MONOOXYGENASE		$I50 = 3.6$ μM
DEBRISOQUINE 4-MONOXYGENASE		$I50 = 137$ μM
HEME POLYMERASE		$I50 = 90$ μM
HISTAMINE N-METHYLTRANSFERASE		$I50 = 8$ μM
MULTI DRUG RESISTANCE PUMP		
Na^+/K^+-EXCHANGING ATPASE		$I50 = 560$ μM
QUINIDINE N-OXIDE		
CYTOCHROME P450 2D6		$Ki = 430$ nM

QUINIDINESULPHATE		
CHOLINESTERASE		20 µM
QUININE		
ALDEHYDE DEHYDROGENASE (PYRROLOQUINOLINE-QUINONE)		15% 500 µM
AMINE DEHYDROGENASE		
BENZOATE 4-MONOOXYGENASE		
CYCLOHEXYLAMINE OXIDASE		
CYTOCHROME P450 2D6		*K*i = 1.8 µM
CYTOCHROME P450 2D6		*K*i = 4.6 µM
CYTOCHROME P450 (O-DEMETHYLATION)		*K*i = 70 nM
DEBRISOQUINE 4-MONOOXYGENASE		*I*50 = 223 µM
DEBRISOQUINE 4-MONOOXYGENASE		*I*50 = 2.4 µM
GLUCONATE 2-DEHYDROGENASE		
HEME POLYMERASE		*I*50 = 300 µM
H^+-TRANSPORTING ATP SYNTHASE F0F1		
LYSOPHOSPHOLIPASE		*I*50 = 2.6 mM
MYOSIN-LIGHT-CHAIN KINASE		20% 333 µM
PROTEIN KINASE A		29% 400 µM
PROTEIN KINASE A		34% 400 µM
PROTEIN KINASE (Ca^{2+} DEPENDENT)		20% 400 µM
QUININE SULPHATE		
D-AMINO-ACID OXIDASE		
QUININONE		
CYTOCHROME P450 2D6		*K*i = 720 nM
QUINIZARIN		
7-ETHOXYCOUMARIN O-DEETHYLASE		*I*50 = 22 µM
MYOSIN-LIGHT-CHAIN KINASE		*I*50 = 26 µM
PROTEIN KINASE A		*I*50 = 20 µM
PROTEIN KINASE C		*I*50 = 24 µM
QUINOIDE 2,4-DIAMINO-DIHYDROPTEROYLGLUTAMIC ACID		
DIHYDROPTERIDINE REDUCTASE	C	*K*i = 23 µM
QUINOLINE		
AMINE OXIDASE MAO-A	C	*K*i = 32 µM
AMINE OXIDASE MAO-A		23% 75 µM
AMINE OXIDASE MAO-B	NC	*K*i = 1.3 mM
AMINE OXIDASE MAO-B		14% 75 µM
CATECHOL 2,3-DIOXYGENASE	C	
PYRIDOXAL DEHYDROGENASE		
QUERCETIN 2,3-DIOXYGENASE		
1R-QUINOLINE-2-CARBONYL-ASN-PHE-psi-[CH(OH)CH_2N]PRO-O-t-Bu		
PROTEINASE HIV-1		*K*i = 33 nM
2-QUINOLINE CARBOXYLIC ACID		
DOPAMINE β-MONOOXYGENASE		*K*is = 140 µM
2-QUINOLINEMETHYLENE-N-HYDROXY-N′-AMINOGUANIDINE		
RIBONUCLEOSIDE-DIPHOSPHATE REDUCTASE		*K*i = 37 µM
RIBONUCLEOSIDE-DIPHOSPHATE REDUCTASE		*K*i = 20 µM
QUINOLINES		
ALDEHYDE OXIDASE		
QUINOLINIC ACID		
AMINE OXIDASE MAO-B	C	*K*i = 900 µM
3-HYDROXYANTHRANILATE 3,4-DIOXYGENASE	NC	
PHOSPHOENOLPYRUVATE CARBOXYKINASE (GTP)		
PYRIDOXAL KINASE		
8-QUINOLINOL		
GLYCEROL DEHYDROGENASE		90% 500 µM

o-QUINOLINOL
MICROCOCCUS CASEOLYTICUS NEUTRAL PROTEINASE

(2R,3S,11′S,8′S)3[(2-QUINOLINYLCARBAMYL)AMINO]2-PROPYLMETHYL]CARBONYL]AMINO-4-PHENYL-1[11′[7′,10′-DIOXO-8′[1-METHYLPROPYL)2′-OXA-6′,9′-DIAZABICYCLO[11.2.2]HEPTADECA-13′,15′,16′
PROTEINASE HIV-1 — *K*i = 300 pM

(2R,3S,11′S,8′S)3[(2-QUINOLINYLCARBONYL)AMINO]2-PROPYLMETHYL]CARBONYL]AMINO-4-PHENYL-1[11′[7′,10′-DIOXO-8′-(1-METHYLPROPYL)2′-OXA-6′,9′-DIAZABICYCLO[11.2.2]HEPTADECA-13′,15′,16′-TRIENE]AMINO]BUTAN-2-OL
PROTEINASE HIV-1 — *K*i = 300 pM

QUINOLI-2-YLCARBAMOYL ASN-[PHE-HEA-PRO[-ILE-PHE-OME
PROTEINASE HIV-1 — *K*i = 100 pM

QUINOLONES
DNA TOPOISOMERASE (ATP-HYDROLYSING)

QUINONES
ALDEHYDE OXIDASE

QUINOXALINYL-IMIDAZOLIDINEDIONE
ALDEHYDE REDUCTASE

QUINOXAPEPTIN A

Enzyme	Type	Value
DNA POLYMERASE α		*I*50 = 2.6 μM
DNA POLYMERASE β		*I*50 = 615 nM
DNA POLYMERASE δ		*I*50 = 494 nM
DNA POLYMERASE γ		*I*50 = 1.8 μM
RNA-DIRECTED DNA POLYMERASE	NC	*K*i = 18 nM
RNA-DIRECTED DNA POLYMERASE		*I*50 = 40 nM

QUINOXAPEPTIN B

Enzyme	Type	Value
RNA-DIRECTED DNA POLYMERASE	NC	*K*i = 22 nM
RNA-DIRECTED DNA POLYMERASE		*I*50 = 100 nM

QUISQALATE
N-ACETYL-L-ASPARTYL-L-GLUTAMATE PEPTIDASE — NC — *K*i = 3.2 μM

QUIZALOFOP
ACETYL-CoA CARBOXYLASE

R

R 830
CYCLOOXYGENASE — *I*50 = 500 μM
CYCLOOXYGENASE — *I*50 = 20 μM

R 40519
AMINE OXIDASE MAO-A — *I*50 = 50 nM
AMINE OXIDASE MAO-B — *I*50 = 55 μM

R 51975
NUCLEOSIDE TRANSPORT — *I*50 = 589 nM

R 53531
NUCLEOSIDE TRANSPORT — *I*50 = 893 nM

R 5421
SCRAMBLASE

R 57974
ADENOSINE TRANSPORT — *I*50 = 64 nM

R 59022
DIACYLGLYCEROL KINASE — *I*50 = 6.7 μM

R 59949
DIACYLGLYCEROL KINASE — $I50 = 125$ nM

R 68070
THROMBOXANE A SYNTHASE — $I50 = 3$ nM

R 68151
ARACHIDONATE 5-LIPOXYGENASE

R 70380
NUCLEOSIDE TRANSPORT — $I50 = 268$ nM

R 73335
NUCLEOSIDE TRANSPORT — $I50 = 242$ nM

R 73796
NUCLEOSIDE TRANSPORT — $I50 = 414$ nM

R 75231
NUCLEOSIDE TRANSPORT — $I50 = 268$ nM

R 76713

Enzyme	Type	Value
AROMATASE	C	$Ki = 430$ pM
AROMATASE	C	$Ki = 1.3$ nM
AROMATASE		$I50 = 3$ nM
AROMATASE		$Ki = 100$ pM
17α-HYDROXYPROGESTERONE ALDOLASE		
STEROID 17α-MONOOXYGENASE		

R 79595
3′,5′-CYCLIC-NUCLEOTIDE PHOSPHODIESTERASE III — $I50 = 38$ nM

R 80122

Enzyme	Type	Value
3′,5′-CYCLIC-NUCLEOTIDE PHOSPHODIESTERASE I		$I50 = 656$ μM
3′,5′-CYCLIC-NUCLEOTIDE PHOSPHODIESTERASE I		$I50 = 1$ mM
3′,5′-CYCLIC-NUCLEOTIDE PHOSPHODIESTERASE II		$I50 = 126$ μM
3′,5′-CYCLIC-NUCLEOTIDE PHOSPHODIESTERASE II		$I50 = 144$ μM
3′,5′-CYCLIC-NUCLEOTIDE PHOSPHODIESTERASE III		$I50 = 64$ nM
3′,5′-CYCLIC-NUCLEOTIDE PHOSPHODIESTERASE III		$I50 = 35$ nM
3′,5′-CYCLIC-NUCLEOTIDE PHOSPHODIESTERASE III		$I50 = 17$ nM
3′,5′-CYCLIC-NUCLEOTIDE PHOSPHODIESTERASE III		$I50 = 36$ nM
3′,5′-CYCLIC-NUCLEOTIDE PHOSPHODIESTERASE IV		$I50 = 104$ μM
3′,5′-CYCLIC-NUCLEOTIDE PHOSPHODIESTERASE IV		$I50 = 23$ μM

R 81267

Enzyme	Type	Value
3′,5′-CYCLIC-NUCLEOTIDE PHOSPHODIESTERASE I		$I50 = 570$ μM
3′,5′-CYCLIC-NUCLEOTIDE PHOSPHODIESTERASE II		$I50 = 664$ μM
3′,5′-CYCLIC-NUCLEOTIDE PHOSPHODIESTERASE III		$I50 = 290$ nM
3′,5′-CYCLIC-NUCLEOTIDE PHOSPHODIESTERASE IV		$I50 = 194$ μM

R 82150

Enzyme	Type	Value
RNA-DIRECTED DNA POLYMERASE	NC	$Ki = 700$ nM
RNA-DIRECTED DNA POLYMERASE		$I50 = 240$ nM

R 82913
RNA-DIRECTED DNA POLYMERASE — $I50 = 10$ nM

R 83355
ARACHIDONATE 5-LIPOXYGENASE

R 83839
AROMATASE — C — $Ki = 18$ nM

R 83842

Enzyme	Type	Value
AROMATASE		
AROMATASE	C	$Ki = 700$ pM

R 86183
RNA-DIRECTED DNA POLYMERASE — $I50$ = 50 nM

R 87366
PROTEINASE HIV — Ki = 1 nM

R 89439
RNA-DIRECTED DNA POLYMERASE — $I50$ = 200 nM

R 90385
RNA-DIRECTED DNA POLYMERASE

R 95845
RNA-DIRECTED DNA POLYMERASE

RA6
ACETYLCHOLINESTERASE

RA7
ACETYLCHOLINESTERASE

RA15
ACETYLCHOLINESTERASE

RABBIT LIVER ACAT INHIBITOR
STEROL O-ACYLTRANSFERASE

RABBIT SKELETAL MUSCLE cycAMP PROTEIN KINASE INHIBITOR
PROTEIN KINASE A — U — Ki = 490 pM
PROTEIN KINASE A — C — Ki = 2 nM

RABBIT SKELETAL MUSCLE MEMBRANE PROTEIN KINASE A INHIBITOR
PROTEIN KINASE A — $I50$ = 30 nM

RABDOSIIN
DNA TOPOISOMERASE

RABEPRAZOLE
H^+/K^+-EXCHANGING ATPASE
UREASE — IR — Ki = 140 nM

RAC-X 65
PEPTIDYL-DIPEPTIDASE A — Ki = 10 pM

RADICICOL
PROTEIN TYROSINE KINASE p60v-src — $I50$ = 100 ng/ml

RADICININ
PROTEINASE 3C — $I50$ = 500 μM

RADICIOL
PROTEIN TYROSINE KINASE SRC

RADIOSUMIN
PLASMIN — $I50$ = 6.2 μg/ml
THROMBIN — $I50$ = 88 μg/ml
TRYPSIN — $I50$ = 140 ng/ml

RAFFINOSE
α-GALACTOSIDASE — C — Ki = 170 mM
INULINASE

RAGI BIFUNTIONAL INHIBITOR
α-AMYLASE — Ki = 11 nM
TRYPSIN — Ki = 1.2 nM

RAMIPRIL
PEPTIDYL-DIPEPTIDASE A — $I50$ = 5 nM

RAMIPRILAT

PEPTIDYL-DIPEPTIDASE A		$I50 = 2$ nM
XAA-PRO AMINOPEPTIDASE		$I50 = 12$ μM
XAA-PRO AMINOPEPTIDASE		100% 1 mM
XAA-PRO AMINOPEPTIDASE		100% 1mM

RAMIPRIL DIACID

PEPTIDYL-DIPEPTIDASE A		$I50 = 590$ pM

RANA CATESBEIANA CYSTEINE PROTEINASE INHIBIOTR

PROTEINASES CYSTEINE

RANITIDINE

ACETYLCHOLINESTERASE	U	$Ki = 2.1$ μM
CHOLINESTERASE	NC	$Ki = 61$ μM
HISTAMINE N-METHYLTRANSFERASE		$I50 = 1.4$ mM
UREASE		73% 6.4 μM

RANITIDINECIMETIDINE

ALCOHOL DEHYDROGENASE	C	

RAPAMYCIN

KINASE S6 (70 kDa p70rsk)		
PEPTIDYLPROLYL ISOMERASE		$Ki = 330$ pM

RAPESEED TRYPSIN INHIBITOR

α-CHYMOTRYPSIN		$Ki = 410$ nM
TRYPSIN		
β-TRYPSIN		$Ki = 300$ pM

RASPBERRY POLYGALACTURONASE INHIBITOR

POLYGALACTURONASE

RAT BRAIN CARBOXYPEPTIDASE A INHIBITOR

CARBOXYPEPTIDASE A		$I50 = 2$ nM
CARBOXYPEPTIDASE A		$I50 = 16$ nM
CARBOXYPEPTIDASE A-1		$I50 = 3.2$ nM
CARBOXYPEPTIDASE A2		$I50 = 3.5$ nM
CARBOXYPEPTIDASE B		$I50 = 194$ nM

RAT BRAIN CERAMIDE GLUCOSYLTRANSFERASE INHIBITOR

CERAMIDE GLUCOSYLTRANSFERASE

RAT BRAIN CMP: N-ACETYLNEURAMINIC ACID LACTOYL CERAMIDE SIALYLTRANSFERASE INHIBITOR

LACTOSYLCERAMIDE α-2,3-SIALYLTRANSFERASE

RAT BRAIN INHIBITOR

LACTOSYLCERAMIDE α-2,6-SIALYLTRANSFERASE

RAT BRAIN Na^+/K^+-TRANSPORTING ATPASE INHIBITOR

Na^+/K^+-EXCHANGING ATPASE

RAT BRAIN Na^+/K^+-TRANSPORTING ATPASE INHIBITOR 13 KD

Ca^{2+}/Mg^{2+}-TRANSPORTING ATPASE
Ca^{2+}-TRANSPORTING ATPASE
Na^+/K^+-EXCHANGING ATPASE

RAT BRAIN PHOSPHOLIPASE D INHIBITORS

PHOSPHOLIPASE D

RAT BRAIN PROTEINASE INHIBITOR

α-CHYMOTRYPSIN
THROMBIN
TRYPSIN

RAT CALCIDOL MONOOXYGENASE INHIBITOR

CALCIDOL 1-MONOOXYGENASE

RAT CHONDROSARCOMA METALLOPROTEINASE INHIBITOR
COLLAGENASE

RAT HEART ACE INHIBITOR
PEPTIDYL-DIPEPTIDASE A C

RAT HEART INHIBITOR
PEPTIDYL-DIPEPTIDASE A

RAT KALLIKREIN BINDING PROTEIN
KALLIKREIN TISSUE

RAT KIDNEY PHOSPHOLIPASE INHIBITOR
PHOSPHOLIPASE A1

RAT LIVER ACETYL-CoA CARBOXYLASE INHIBITOR
ACETYL-CoA CARBOXYLASE

RAT LIVER AMP-PROTEIN KINASE INHIBITOR
PROTEIN KINASE AMP

RAT LIVER CYTOSOL CHOLESTEROL ESTERASE INHIBITOR
CARBOXYLESTERASE
HORMON SENSITIVE LIPASE
STEROL ESTERASE $I50 = 100\ \mu g$

RAT LIVER INOSITOL-1,3,4,5-TETRAKISPHOSPHATE 3-PHOSPHATASE INHIBITOR
INOSITOL-1,3,4,5-TETRAKISPHOSPHATE 3-PHOSPHATASE

RAT LIVER INSULIN DEGRADING ENZYME INHIBITOR
INSULYSIN C $Ki = 180\ nM$

RAT LIVER MITOCHONDRIA ATPASE INHIBITOR
H^+-TRANSPORTING ATP SYNTHASE

RAT LIVER PROTEINASE INHIBITOR
FICAIN
PAPAIN

RAT LIVER SERINE PROTEAS INHIBITORS
PROTEINASES SERINE

RAT MILK PEPTIDASE INHIBITORS
PEPTIDASES

RAT OVARIES PROTEIN KINASE C INHIBITOR
PROTEIN KINASE C C

RAT PROTEINASE INHIBITOR
CATHEPSIN B

RAT SMALL INTESTINE FUCOSYLTRANSFERASE INHIBITOR
α(1- > 3)FUCOSYLTRANSFERASE
GALACTOSIDE 2-L-FUCOSYLTRANSFERASE

RAT TESTES CHOLESTEROL ESTERASE INHIBITOR PROTEIN
STEROL ESTERASE

RAT VAGINAL RIBONUCLEASE INHIBITOR
RIBONUCLEASE

RAWSONOL
IMP DEHYDROGENASE $I50 = 10\ \mu M$

Rb^+
THETIN-HOMOCYSTEINE S-METHYLTRRANSFERASE

Rb^+
α-AMYLASE

RB 104

NEPRILYSIN		*K*i = 30 pM

RB 105

NEPRILYSIN		*K*i = 1.7 nM
PEPTIDYL-DIPEPTIDASE A		*K*i = 4.5 nM

RB 106

NEPRILYSIN		*K*i = 1.6 nM
PEPTIDYL-DIPEPTIDASE A		*K*i = 350 pM

RB 38A

NEPRILYSIN		*K*i = 900 pM

RCAM-LYSOZYM

PROTEIN TYROSINE KINASE (INSULIN RECEPTOR)		*K*i = 450 nM

RD 3-4082

PROTEINASE HEPATITIS C VIRUS		*I*50 = 5.8 µM

RD 4-2025

RNA-DIRECTED DNA POLYMERASE		*I*50 = 110 nM

RD 4-6250

CHYMOTRYPSIN		*I*50 = 790 ng/ml
ELASTASE		*I*50 = 5.9 µg/ml
PROTEINASE HEPATITIS C VIRUS		*I*50 = 2.3 µg/ml
TRYPSIN		*I*50 = 730 ng/ml

REACTIVE BLACK 5

PROTEINASE HIV-1		*I*50 = 100 µM

REACTIVE BLUE

ADENOSINETRIPHOSPHATASE ECTO		*I*50 = 32 µM
PROTEIN KINASE	NC	*K*i = 6 µM

REACTIVE BLUE 2

CHOLINE O-ACETYLTRANSFERASE		
2-HYDROXY-4-CARBOXYMUCONATE-6-SEMIALDEHYDE DEHYDROGENASE	C	*K*i = 72 µM

REACTIVE GREEN 19-DEXTRAN

L-LACTATE DEHYDROGENASE		

REACTIVE ORANGE 16

PROTEINASE HIV-1		*I*50 = 23 µM

REACTIVE RED

3′,5′-CYCLIC-NUCLEOTIDE PHOSPHODIESTERASE Iβ		*I*50 = 12 µM
3′,5′-CYCLIC-NUCLEOTIDE PHOSPHODIESTERASE II		*I*50 = 27 µM
3′,5′-CYCLIC-NUCLEOTIDE PHOSPHODIESTERASE III		*I*50 = 29 µM
3′,5′-CYCLIC-NUCLEOTIDE PHOSPHODIESTERASE IV		*I*50 = 34 µM

REACTIVE RED 120

Ca^{2+}-TRANSPORTING ATPASE		

REACTIVE RED 4-DEXTRAN

3′,5′-CYCLIC-NUCLEOTIDE PHOSPHODIESTERASE		
L-LACTATE DEHYDROGENASE		

REACTIVE YELLOW

3′,5′-CYCLIC-NUCLEOTIDE PHOSPHODIESTERASE Iβ		*I*50 = 1.2 µM
3′,5′-CYCLIC-NUCLEOTIDE PHOSPHODIESTERASE II		*I*50 = 1.4 µM
3′,5′-CYCLIC-NUCLEOTIDE PHOSPHODIESTERASE III		*I*50 = 3.3 µM
3′,5′-CYCLIC-NUCLEOTIDE PHOSPHODIESTERASE IV		*I*50 = 4.7 µM

REBECCAMYCIN

DNA TOPOISOMERASE		
DNA TOPOISOMERASE (ATP-HYDROLYSING)		
PROTEIN TYROSINE KINASE pp60c-Src		*I*50 = 55 µg/ml

RECOVERIN
RHODOPSIN KINASE — $I50 = 1\ \mu M$

RECOVERIN-Ca^{2+}
RHODOPSIN KINASE

RED BEET β-FRUCTOFURANOSIDASE INHIBITOR
β-FRUCTOFURANOSIDASE

RED H-8B
FERRIC-CHELATE REDUCTASE — $I50 = 8\ \mu M$

RED H-3BN
FERRIC-CHELATE REDUCTASE
FERRIC-CHELATE REDUCTASE — $I50 = 10\ \mu M$

REDOXAL
DIHYDROOROTATE DEHYDROGENASE — $Ki = 330$ nM

REDUVIIN
THROMBIN

REGUCALCIN
3′,5′-CYCLIC-NUCLEOTIDE PHOSPHODIETERASE (Ca^{2+}/CALMODULIN)
PROTEIN KINASE C

REGULATORY PROTEIN
HEXOKINASE

RENIN BINDING PROTEIN HOG KIDNEY
RENIN

RENIN INHIBITORS
PEPSIN A

RENTIAPRIL
PEPTIDYL-DIPEPTIDASE A — $I50 = 2$ nM
XAA-TRP AMINOPEPTIDASE — $I50 = 1.6\ \mu M$

RESERPINE
NAD ADP-RIBOSYLTRANSFERASE — $I50 = 790\ \mu M$

RESINIFERATOXIN
NADH OXIDASE

RESISTOMYCIN
RNA-DIRECTED DNA POLYMERASE — $I50 = 21\ \mu M$

RESOKAEMPFEROL
PROTEIN TYROSINE KINASE p56lck — $I50 = 5\ \mu g/ml$

RESORCINOL
CYTOCHROME-C PEROXIDASE
IODIDE PEROXIDASE
LACTOPEROXIDASE — 82% 250 μM
LACTOYLGLUTATHIONE LYASE — 24% 10 mM
MONOPHENOL MONOOXYGENASE — C
MONOPHENOL MONOOXYGENASE — 34% 800 μM
PYROGALLOL HYDROXYLTRANSFERASE
THYROID PEROXIDASE — 87% 250 μM

RESORCINOLDIPHOSPHATE
FRUCTOSE-BISPHOSPHATE ALDOLASE — C — $Ki = 880$ nM
FRUCTOSE-BISPHOSPHATE ALDOLASE — $Ki = 40\ \mu M$

RESORCYLIC ACID
DIPHOSPHOMEVALONATE DECARBOXYLASE

α-RESORCYLIC ACID
CATECHOL OXIDASE — Ki = 300 μM

β-RESORCYLIC ACID
CATECHOL OXIDASE — Ki = 420 μM

RESVERATROL
PROTEIN KINASE C — I50 = 40 μg6ml
PROTEIN TYROSINE KINASE p56lckC — I50 = 40 μg/ml
RIBONUCLEOTIDE REDUCTASE — I50 = 100 μM

cis-RESVERATROL
PROTEIN KINASE C — I50 = 30 μg/ml
PROTEIN TYROSINE KINASE p56lckC — I50 = 50 μg/ml

RESVERATROL-O^4-β-GLUCOSIDE
PROTEIN KINASE C — I50 = 3 μg/ml
PROTEIN TYROSINE KINASE p56lckC — I50 = 200 μg/ml

RETICULINE
DEBRISOQUINE HYDROXYLASE — I50 = 4 μM

RETICULOL
3′,5′-CYCLIC-NUCLEOTIDE PHOSPHODIESTERASE

RETIN
CARBONYL REDUCTASE (NADPH)

RETINAL
PROTEIN KINASE C — Ki = 10 μM

13-cis RETINAL
GLUTATHIONE TRANSFERASE — I50 = 215 μM

all-trans RETINAL
NAD ADP-RIBOSYLTRANSFERASE — I50 = 450 μM

all-trans-RETINAL
PHOSPHOLIPASE A2 — I50 = 6 μM

RETINOIC ACID
DNA TOPOISOMERASE (ATP-HYDROLYSING)
ESTRONE SULFOTRANSFERASE
GLUTATHIONE TRANSFERASE — C — Ki = 30 μM
NADH OXIDASE
STEROID 5α-REDUCTASE — I50 = 20 μM

9-cis RETINOIC ACID
3(or 17)β-HYDROXYSTEROID DEHYDROGENASE — Ki = 35 μM
3(or 17)β-HYDROXYSTEROID DEHYDROGENASE — Ki = 4.1 μM
all trans RETINOIC ACID HYDROXYLASE — C

13-cis RETINOIC ACID
GLUTATHIONE TRANSFERASE — I50 = 39 μM
3(or 17)β-HYDROXYSTEROID DEHYDROGENASE — Ki = 27 μM
3(or 17)β-HYDROXYSTEROID DEHYDROGENASE — Ki = 2.4 μM
all trans RETINOIC ACID HYDROXYLASE — C

13-cis-RETINOIC ACID
PHOSPHOLIPASE A2 — I50 = 15 μM
THIOREDOXIN REDUCTASE (NADPH) — SSI

all-trans RETINOIC ACID
GLUTATHIONE TRANSFERASE — NC — Ki = 41 μM

all-trans-RETINOIC ACID
DIFFERIC-TRANSFERRIN REDUCTASE — 41% 1 μM
DIFFERRIC-TRANSFERRIN REDUCTASE
3(or 17)β-HYDROXYSTEROID DEHYDROGENASE — Ki = 21 μM

PHOSPHOLIPASE A2		*I*50 = 10 μM
RETINOL O-FATTY-ACYLTRANSFERASE		
RETINOL		
GLUTATHIONE TRANSFERASE	C	*K*i = 10 μM
13-cis RETINOL		
GLUTATHIONE TRANSFERASE		*I*50 = 31 μM
all-trans-RETINOL		
GLUTATHIONE TRANSFERASE		*I*50 = 35 μM
PHOSPHOLIPASE A2		*I*50 = 165 μM
all-trans-RETINYL ACETATE		
GLUTATHIONE TRANSFERASE		*I*50 = 160 μM
all-trans-RETINYL-α-BROMOACETIC ACID		
LECITHIN RETINOL ACYLTRANSFERASE		
RETINYL MONOPHOSPHATE		
DOLICHYL-PHOSPHATASE		
all-trans-RETINYL PALMITATE		
GLUTATHIONE TRANSFERASE		*I*50 = 153 μM
RETRO-INVERSO PKC(19-31)		
PROTEIN KINASE C		*I*50 = 31 μM
RETRO-INVERSO SER 25 PKC(19-31)		
PROTEIN KINASE C	C	*I*50 = 5.5 μM
RETROJUSTICIDIN		
DNA POLYMERASE α		*I*50 = 989 μM
RNA-DIRECTED DNA POLYMERASE		*I*50 = 5.5 μM
RETROTHIORPHAN		
NEPRILYSIN		*K*i = 6 nM
REV 5367		
ARACHIDONATE 5-LIPOXYGENASE		*I*50 = 3 μM
ARACHIDONATE 15-LIPOXYGENASE		*I*50 = 3 μM
REV 5741		
ARACHIDONATE 5-LIPOXYGENASE		*I*50 = 8 μM
ARACHIDONATE 12-LIPOXYGENASE	C	*K*i = 68 μM
CYCLOOXYGENASE		30% 200 μM
REV 5827		
ARACHIDONATE 5-LIPOXYGENASE		*I*50 = 7 μM
ARACHIDONATE 12-LIPOXYGENASE	C	*K*i = 291 μM
REV 5875		
ARACHIDONATE 5-LIPOXYGENASE		*I*50 = 8 μM
REV 5901		
ARACHIDONATE 5-LIPOXYGENASE		
ARACHIDONATE 5-LIPOXYGENASE		*I*50 = 9 μM
ARACHIDONATE 5-LIPOXYGENASE	C	*I*50 = 6 μM
ARACHIDONATE 12-LIPOXYGENASE	C	*K*i = 8 μM
REV 5965		
ARACHIDONATE 5-LIPOXYGENASE		*I*50 = 17 μM
REV 6080		
ARACHIDONATE 5-LIPOXYGENASE		*I*50 = 14 μM
REY α-AMYLASE INHIBITOR		
α-AMYLASE		

dRF
PROTEIN KINASE (dsRNA DEPENDENT)

RG 12561
HYDROXYMETHYLGLUTARYL-CoA REDUCTASE (NADPH)

RG 13022
PROTEIN TYROSINE KINASE EGFR — *I*50 = 3 μM

RG 13291
PROTEIN TYROSINE KINASE PDGFR

RG 14467
PROTEIN TYROSINE KINASE EGFR

RG 14921
PROTEIN TYROSINE KINASE EGFR

RG 50864
ALDEHYDE DEHYDROGENASE (NAD) — *I*50 = 20 μM
OXIDATIVE PHOSPHORYLATION UNCOUPLER — E50 = 50 μM

RG 12561-Na
HYDROXYMETHYLGLUTARYL-CoA REDUCTASE (NADPH) — *I*50 = 3.4 nM

RH 5348
PROTOPORPHYRINOGEN OXIDASE — *K*i = 600 nM
PROTOPORPHYRINOGEN OXIDASE — *K*i = 50 nM

RHAMNETIN
INTEGRASE — *I*50 = 62 μM
NADH DEHYDROGENASE (UBIQUINONE)

L-RHAMNITOL-1-PHOSPHATE
RHAMNULOSE-1-PHOSPHATE ALDOLASE — C

L-RHAMNO-1,4-LACTONE
NARINGINASE — C — *K*i = 220 μM

3-O-α-L-RHAMNOPYRANOSYL(1- > 2)β-D-GALACTOPYRANOSYL(1- > 2)β-D-GLUCOPYRANOSYL SOYASAPOGENOL B 22-O-α-D-GLUCO-PYRANISIDE
DIGUANYLATE CYCLASE — NC — *I*50 = 5 μM

3βO-{β-D-RHAMNOPYRANOSYL(1- > 2)O-β-D-GLUCOPYRANOSYL(1- > 4)[O-βD-GLUCOPYRANOSYL]α-L-ARABINOSYL}16α-HYDROXY-13β,28
PHOSPHOLIPASE D — *I*50 = 2 μM

N-(α-RHAMNOPYRANOSYLOXYHYDROXYPHOSPHINYL)-L-LEUCINYL-L-HISTIDINE
THERMOLYSIN — C — *K*i = 330 nM

L-RHAMNOSE
α-L-RHAMNOSIDASE — C

RHAMNOSYL BUFALIN
Na^+/K^+-EXCHANGING ATPASE — *I*50 = 15 nM

3β-O-RHAMNOSYLCHLORMADINOL ACETATE
Na^+/K^+-EXCHANGING ATPASE — 16% 10 μM

3β-RHAMNOSYLOXY-5β,14β-ANDROSTAN-14-OL
Na^+/K^+-EXCHANGING ATPASE — 71% 30 μM

3β-RHAMNOSYLOXY-5β,14β-PREGNAN-14,20β-DIOL
Na^+/K^+-EXCHANGING ATPASE — 96% 100 μM

3β-RHAMNOSYL-5β-PREGNAN-20-ONE
Na^+/K^+-EXCHANGING ATPASE — 74% 20 μM

RHC 80267
DIACYLGLYCEROL LIPASE

RHEIN

7-ETHOXYCOUMARIN O-DEETHYLASE		*I*50 = 42 μM
GLUTATHIONE REDUCTASE (NADPH)	U	*K*i = 2.6 μM
NADH DEHYDROGENASE (UBIQUINONE)		0.3 mM
TRYPANOTHIONE REDUCTASE	U	*K*i = 80 μM

RHIZOBITOXINE

CYSTATHIONINE β-LYASE

RHODAMIN B

GLUTATHIONE TRANSFERASE

RHODAMINE 6G

H^+-TRANSPORTING ATP SYNTHASE F1		*I*50 = 75 μM

RHODAMINES

H^+-TRANSPORTING ATP SYNTHASE

RHODANIDE

DIHYDROFOLATE REDUCTASE		
GLUTATHIONE TRANSFERASE β	NC	*K*i = 110 μM

RHODIZONIC ACID

LACTOYLGLUTATHIONE LYASE

RHODNIIN

THROMBIN		*K*i = 200 fM

RHODOMYCIB A

PHOSPHOLIPASE C γ1 (PHOSPHATIDYLINOSITOL SPECIFIC)		*I*50 = 47 μM

RHOMBENONE

FARNESYL PROTEINTRANSFERASE		*I*50 = 2.3 μM

RHUS JAVANICA PROTEIN TYROSIN KINASE INHIBITOR

PROTEIN TYROSINE KINASE EGFR

RIBAVIRIN

DEOXYADENOSINE KINASE

RIBAVIRIN-5′-MONOPHOSPHATE

GDP-KINASE		
HYPOXANTHINE PHOSPHORIBOSYLTRANSFERASE		
IMP DEHYDROGENASE	C	*K*i = 580 μM
IMP DEHYDROGENASE	C	*K*i = 120 nM
IMP DEHYDROGENASE		*I*50 = 170 nM
IMP DEHYDROGENASE	C	*K*i = 800 nM

RIBAVIRIN PHOSPHATE

IMP DEHYDROGENASE 1	C	*K*i = 650 nM
IMP DEHYDROGENASE 2	C	*K*i = 390 nM

α-RIBAZOLE-3′-PHOSPHATE

5-METHYLTETRAHYDROFOLATE-HOMOCYSTEINE S-METHYLTRANSFERASE

RIBITOL

L-ARABINOSE ISOMERASE	C	*K*i = 350 mM
L-ARABINOSE ISOMERASE		
RIBOSE ISOMERASE		20% 50 μM

RIBITOL-1,5-DIPHOSPHATE

FRUCTOSE-BISPHOSPHATE ALDOLASE		*K*i = 20 μM

RIBOCITRIN

DEXTRANSUCRASE

RIBOFLAVIN

5-AMINO-6-(5PHOSPHORIBOSYLAMINO)URACIL REDUCTASE
GALACTONOLACTONE DEHYDROGENASE

GLUTAMATE RACEMASE		43% 100 μM
NADH DEHYDROGENASE		
PHOSPHORYLASE b		$I50 = 18$ μM
RIBOFLAVIN SYNTHASE		
STEAROYL-CoA DESATURASE		
RIBOFLAVIN-5′-DIPHOSPHATE		
FMN ADENYLYLTRANSFERASE		
RIBOFLAVINE		
CYTOCHROME-b5 REDUCTASE	C	$Ki = 20$ μM
PHOSPHORYLASE b		
RIBOFLAVINEHEMISULPHATE		
STEROID 5α-REDUCTASE		$I50 = 5$ μM
RIBOFLAVINE 5-SULPHATE		
D-AMINO-ACID OXIDASE		
L-AMINO-ACID OXIDASE		10 μM
RIBOFLAVIN-5′-MONOPHOSPHATE		
D-AMINO-ACID OXIDASE		
RIBOFLAVIN PHOSPHATE		
myo-INOSITOL OXYGENASE		
RIBOFLAVIN 5′-PHOSPHATE		
FMN ADENYLYLTRANSFERASE	C	$Ki = 9$ μM
RIBOFLAVIN-5′-PHOSPHATE		
RIBOFLAVINASE	C	
RIBOFLAVIN REDUCED		
ALKANAL MONOOXYGENASE (FMN -LINKED)		
1-(β-D-RIBOFURANOSYL)-5,6-DICHLOROBENZIMIDAZOLE		
PROTEIN KINASE CASEIN KINASE I		$Ki = 60$ μM
PROTEIN KINASE CASEIN KINASE IIA		$Ki = 175$ μM
PROTEIN KINASE CASEIN KINASE IIB		$Ki = 175$ μM
5-β-D-RIBOFURANOSYLNICOTINAMIDEADENINEDINUCLEOTIDE		
ALCOHOL DEHYDROGENASE	C	$Ki = 4$ nM
GLUTAMATE DEHYDROGENASE (NAD(P))	C	$Ki = 15$ μM
L-LACTATE DEHYDROGENASE	C	$Ki = 188$ μM
MALATE DEHYDROGENASE	C	$Ki = 410$ μM
5-β-D-RIBOFURANOSYLPICOLINAMIDEADENINEDINUCLEOTIDE		
ALCOHOL DEHYDROGENASE	C	$Ki = 21$ μM
GLUTAMATE DEHYDROGENASE (NAD(P))	C	$Ki = 50$ μM
9-(β-D-RIBOFURANOSYL)-PURINE		
ADENOSINE DEAMINASE	C	$Ki = 37$ μM
1-β-D-RIBOFURANOSYL-1,2,4-TRIAZOLE-3-CARBOXAMIDINE		
IMP DEHYDROGENASE		
PURINE-NUCLEOSIDE PHOSPHORYLASE		$Ki = 4$ μM
PURINE-NUCLEOSIDE PHOSPHORYLASE		$Ki = 30$ μM
RIBONOLACTONE		
NUCLEOSIDE HYDROLASE		$Ki = 90$ μM
D-RIBONO-γ-LACTONE		
KETOHEXOKINASE		
RIBONUCLEASE		
PROTEIN-DISULFIDE REDUCTASE (GLUTATHIONE)		
RIBONUCLEASE/ANGIOGENIN INHIBITOR		
RIBONUCLEASE		

RIBONUCLEASE L INHIBITOR		
RIBONUCLEASE L		
β-D-RIBOPYRANOSIDES		
XYLAN 1,4-β-XYLOSIDASE		
RIBOSE		
FORMALDEHYDE TRANSKETOLASE		
α-GLUCOSIDASE		
URIDINE NUCLEOSIDASE	C	*K*i = 7.2 mM
RIBOSE-5-DIPHOSPHATE		
GLUCOSE 1-DEHYDROGENASE		
L-RIBOSE		
β-GALACTOSIDASE		*K*i = 240 μM
D-RIBOSE-5-PHAOSPHATE		
GLYCERALDEHYDE-3-PHOSPHATE DEHYDROGENASE (NADP)		
RIBOSE-1-PHOSPHATE		
5′-METHYLTHIOADENOSINE PHOSPHORYLASE	C	46% 2.2 mM
PURINE-NUCLEOSIDE PHOSPHORYLASE	NC	*K*i = 361 μM
RIBOSE-PHOSPHATE PYROPHOSPHOKINASE		
THYMIDINE PHOSPHORYLASE		
URIDINE PHOSPHORYLASE		
RIBOSE-5-PHOSPHATE		
FRUCTOSE-BISPHOSPHATE ALDOLASE		*I*50 = 5.3 mM
GLUCOSE-6-PHOSPHATE ISOMERASE		*K*i = 500 μM
GLYCERATE KINASE		
IMP DEHYDROGENASE		
OROTATE PHOSPHORIBOSYLTRANSFERASE		
PHOSPHORIBOKINASE		> 2 mM
PHOTINUS-LUCIFERIN 4-MONOOXYGENASE (ATP-HYDROLYSING)		
RIBOSE-5-PHOSPHATE ADENYLYLTRANSFERASE	NC	*K*i = 4 mM
URIDINE NUCLEOSIDASE		
D-RIBOSE-1-PHOSPHATE		
TRIOSE-PHOSPHATE ISOMERASE		*K*i = 4.8 mM
D-RIBOSE-5-PHOSPHATE		
NMN NUCLEOSIDASE		
RIBOSOMAL PROTEIN PHOSPHATASE 1 INHIBITOR		
PHOSPHOPROTEIN PHOSPHATASE 1		*K*i = 20 nM
RIBOSTAMYCIN		
GENTAMICIN 3′-N-ACETYLTRANSFERASE		
1-RIBOSYL-ALLOPURINOL-5′-PHOSPHATE		
IMP DEHYDROGENASE 1	C	*K*i = 670 μM
IMP DEHYDROGENASE 2	C	*K*i = 1.9 mM
1-RIBOSYLOXIPURINOL-5-PHOSPHATE		
OROTIDINE-5′-PHOSPHATE DECARBOXYLASE		*K*i = 20 nM
RIBOSYL-RIBOSYL-ADENINE		
PROCOLLAGEN-PROLINE, 2-OXOGLUTARATE-4-DIOXYGENASE		
RIBOSYL-RIBOSYL-HYPOXANTHINE		
PROCOLLAGEN-PROLINE, 2-OXOGLUTARATE-4-DIOXYGENASE		
RIBOSYLTHYMIDINE		
URIDINE NUCLEOSIDASE	C	*K*i = 70 μM
1-RIBOSYLURACIL		
URATE-RIBONUCLEOTIDE PHOSPHORYLASE		

RIBOVIRIN
- DEOXYGUANOSINE KINASE

RIBULOSE-1,5-DIPHOSPHATE
- GLUCOSE-6-PHOSPHATE 1-DEHYDROGENASE
- GLYCERATE KINASE
- PHOSPHORIBULOKINASE
- RIBOSE-5-PHOSPHATE ISOMERASE 1 mM

L-RIBULOSE
- KETOHEXOKINASE

RIBULOSEPHOSPHATE
- PHOSPHORIBULOKINASE

RIBULOSE-5-PHOSPHATE
- GLUCOSE-6-PHOSPHATE ISOMERASE $K_i = 50\ \mu M$
- PHOSPHOGLUCONATE DEHYDROGENASE (DECARBOXYLATING)
- URIDINE NUCLEOSIDASE

RICE SUBTILISIN/α-AMYLASE INHIBITOR
- α-AMYLASE

RICE TRYPSIN INHIBITOR
- TRYPSIN

RIDOGREL
- THROMBOXANE A SYNTHASE $I_{50} = 5$ nM

RIFAMPICIN
- COLLAGENASE XI $I_{50} = 13\ \mu M$
- POLYNUCLEOTIDE ADENYLYLTRANSFERASE
- RNA-DIRECTED DNA POLYMERASE $I_{50} < 1$ nM
- RNA-DIRECTED DNA POLYMERASE $I_{50} = 2$ nM
- RNA-DIRECTED DNA POLYMERASE $I_{50} = 200$ nM
- RNA POLYMERASE

RIFAMYCIN
- DNA POLYMERASE
- POLYNUCLEOTIDE ADENYLYLTRANSFERASE
- RNA-DIRECTED DNA POLYMERASE
- RNA POLYMERASE

RIFAMYCIN B
- RIFAMYCIN-B OXIDASE > 2 mM

RIFAMYCIN DERIVATIVE AF/05
- RNA-DIRECTED DNA POLYMERASE
- RNA POLYMERASE I
- RNA POLYMERASE II

RIFAMYCIN DERIVATIVE AF/13
- RNA-DIRECTED DNA POLYMERASE
- RNA POLYMERASE I C
- RNA POLYMERASE II C

RIFAMYCIN SV
- POLYRIBONUCLEOTIDE NUCLEOTIDYLTRANSFERASE

RIPOSTATIN A
- RNA POLYMERASE

RIPOSTATIN B
- RNA POLYMERASE

RISEDRONATE
- PYROPHOSPHATE-FRUCTOSE-6-PHOSPHATE 1-PHOSPHOTRANSFERASE $K_i = 300\ \mu M$

RISHIRILIDE B

GLUTATHIONE TRANSFERASE	C	$Ki = 9.4\ \mu M$

RISPERIDONE

CYTOCHROME P450 2D6

RISTOCETIN

CDPRIBITOL RIBITOLTRANSFERASE

RISTOMYCIN

PHOSPHO-N-ACETYLMURAMOYL-PENTAPEPTIDE-TRANSFERASE

RITANSERIN

CYTOCHROME P450 2D6		$Ki = 1.8\ \mu M$

RITODRINE

ALCOHOL SULFOTRANSFERASE		13% 300 μM

RITONAVIR

CYTOCHROME P450 3A		$I50 = 70$ nM
CYTOCHROME P450 3A4		$Ki = 19$ nM
CYTOCHROME P450 2C9		$I50 = 4.2\ \mu M$
CYTOCHROME P450 2C9		$I50 = 8\ \mu M$
CYTOCHROME P450 2D6		$I50 = 2.5\ \mu M$
17α-ETHINYLESTRADIOL 2-HYDROXYLASE		$I50 = 2\ \mu M$
PROTEINASE HIV		
PROTEINASE HIV-1		

RIWrYWAV

CATHEPSIN B		$Ki = 300$ nM
CATHEPSIN L		$Ki = 180$ nM
CRUZIPAIN		$Ki = 110$ nM
KALLIKREIN		$Ki = 10$ mM

RK 682

PROTEIN TYROSINE-PHOSPHATASE		$I50 = 1\ \mu M$
PROTEIN TYROSINE-PHOSPHATASE CD45		$I50 = 54\ \mu M$
PROTEIN TYROSINE-PHOSPHATASE VHR		$I50 = 2\ \mu M$

RK 1409

PROTEIN KINASE C

RK 1409B

PROTEIN KINASE C		$I50 = 400$ nM

RK 286C

PROTEIN KINASE C		$I50 = 40$ nM

RK 286D

PROTEIN KINASE C

R13K0 (LAI)

PROTEINASE SERINE PROMONOCYTOC U-937		80% 58 nM

RL 44004

NEPRILYSIN		$Ki = 200$ nM

RM 3

PHOSPHOLIPASE A2

RMI 18341

5α-REDUCTASE	IR	$Ki = 35$ nM

RMI 12330A

ADENYLATE CYCLASE		$Ki = 620\ \mu M$
3′,5′-CYCLIC-NUCLEOTIDE PHOSPHODIESTERASE (cycAMP) HIGH Km	NC	$Ki = 1.5$ mM
3′,5′-CYCLIC-NUCLEOTIDE PHOSPHODIESTERASE (cycAMP) LOW Km	NC	$Ki = 3.5$ mM

RN 521-61-9
ELASTASE LEUKOCYTE 100% 65 μM

RN 100902-07-6
RENIN *I*50 = 1 nM

RN 102994-14-9
AROMATASE *I*50 = 6.4 μM

RN 106585-70-7
DIHYDROFOLATE REDUCTASE *K*i = 11 nM

RN 112711-08-7
ALDEHYDE REDUCTASE *I*50 = 60 nM

RN 122280-72-2
3′,5′-CYCLIC-NUCLEOTIDE PHOSPHODIESTERASE *I*50 = 36 nM

RN 123411-33-6
RENIN *I*50 = 1 nM

RN 125313-77-1
PROTEIN KINASE A *I*50 = 8.4 μM
PROTEIN KINASE C *I*50 = 110 nM

RN 125830-36-6
DNA TOPOISOMERASE (ATP-HYDROLYSING) *I*50 = 5 μM

RN 127142-43-2
RIBONUCLEOSIDE-DIPHOSPHATE REDUCTASE

RN 1313068-27-4
CYCLOOXYGENASE *I*50 = 1.1 μM

RN 136034-25-8
β-GLUCOSIDASE C *K*i = 70 μM

RN 139347-01-6
DIHYDROFOLATE REDUCTASE *I*50 = 400 nM
FOLYLPOLYGLUTAMATE SYNTHASE *K*i = 570 μM
THYMIDYLATE SYNTHASE *I*50 = 55 μM

RN 139565-32-5
ELASTASE LEUKOCYTE 100% 62 μM

RN 140112-65-8
STEROL O-ACYLTRANSFERASE *I*50 = 7 nM

RN 140413-27-0
RNA-DIRECTED DNA POLYMERASE *I*50 = 12 nM

RN 140697-02-5
CARBONATE DEHYDRATASE *I*50 = 230 nM

RN 140926-41-6
ALDEHYDE REDUCTASE *I*50 = 2.1 nM

RN 142867-54-7
ALDEHYDE DEHYDROGENASE (NAD) *I*50 = 25 μM

RN 143620-93-3
ARACHIDONATE 5-LIPOXYGENASE *I*50 = 1.1 μM
LIPOXYGENASE *I*50 = 40 μM

RN 143621-14-1
ARACHIDONATE 5-LIPOXYGENASE *I*50 = 60 nM

RN 19228-39-8
ELASTASE LEUKOCYTE 100% 55 μM

RNA
N-ACETYL-β-GLUCOSAMINIDASE
DNA (CYTOSINE-5-)-METHYLTRANSFERASE 100% 10 μg
EXORIBONUCLEASE H
β-GALACTOSIDASE
β-GLUCOSIDASE
β-GLUCURONIDASE
γ-GLUTAMYL HYDROLASE $I50 = 1$ μg/ml
HISTONE ACYLTANSFERASE
HISTONE-LYSINE N-METHYLTRANSFERASE
NAD PYROPHOSPHATASE
PHOSPHATASE
PHOSPHODIESTERASE

5 S RNA
RIBONUCLEASE P

RNA APTAMERS
PROTEIN KINASE Cβ2

tRNA
DEOXYRIBONUCLEASE I 100% 67 μM
DEOXYRIBONUCLEASE V 13% 67 μM
RESTRICTION ENDONUCLEASE EcoRI C
RIBONUCLEASE P

RO 02-0683
CHOLINESTERASE IR $Ki = 32$ μM

RO 8-8717
PHOSPHOLIPASE A2 $I50 = 90$ μM

RO 09-1470
LANOSTEROL C14 DEMETHYLASE $I50 = 300$ ng/ml
LANOSTEROL C14 DEMETHYLASE $I50 = 5$ μM

RO 09-1679
COAGULATION FACTOR Xa $I50 = 3.3$ μM
THROMBIN $I50 = 34$ μM
TRYPSIN $I50 = 40$ nM

RO 11-2933
MULTI DRUG RESISTANCE PUMP

RO 12-7310
PHOSPHOLIPASE A2 $I50 = 86$ μM

RO 13-6298
PHOSPHOLIPASE A2 $I50 = 400$ μM

RO 13-7410
PHOSPHOLIPASE A2 $I50 = 12$ μM

RO 14-3899
PHOSPHOLIPASE A2 $I50 = 9$ μM

RO 14-6258
PHOSPHOLIPASE A2 $I50 = 140$ μM

RO 14-9572
PHOSPHOLIPASE A2 $I50 = 6.6$ μM

RO 15-4513
ALCOHOL DEHYDROGENASE Kd = 345 μM

RO 16-3177
AMINE OXIDASE MAO-A $I50 = 20$ μM

RO 16-6327
AMINE OXIDASE MAO-B — Ki = 180 nM

RO 16-6354
AMINE OXIDASE MAO-A — I50 = 20 μM
AMINE OXIDASE MAO-B — I50 = 7 μM

RO 16-6491
AMINE OXIDASE MAO-A — I50 = 200 μM
AMINE OXIDASE MAO-B — I50 = 30 nM

RO 17-5380
H^+/K^+-EXCHANGING ATPASE — Ki = 20 μM

RO 17-8604
AMINE OXIDASE MAO-A — I50 = 60 μM
AMINE OXIDASE MAO-B — I50 = 2 μM

RO 18-0503
AMINE OXIDASE MAO-B — I50 = 2 μM

RO 18-1236
AMINE OXIDASE MAO-B — I50 = 3 μM

RO 18-3185
AMINE OXIDASE MAO-B — I50 = 4 μM

RO 18-4426
AMINE OXIDASE MAO-B — I50 = 4 μM

RO 18-5364
H^+/K^+-EXCHANGING ATPASE — I50 = 56 nM

RO 18-5920
AMINE OXIDASE MAO-A — I50 = 70 μM

RO 19-0002
AMINE OXIDASE MAO-A — I50 = 60 μM

RO 19-0337
AMINE OXIDASE MAO-B — I50 = 30 μM

RO 19-3704
PHOSPHOLIPASE A2 — I50 = 7 μM

RO 19-6327
AMINE OXIDASE MAO-A — I50 = 300 μM
AMINE OXIDASE MAO-B — I50 = 30 nM
AMINE OXIDASE MAO-B — I50 = 37 nM

RO 19-7241
AMINE OXIDASE MAO-B — I50 = 85 μM

RO 20-1724
3′,5′-CYCLIC-NUCLEOTIDE PHOSPHODIESTERASE
3′,5′-CYCLIC-NUCLEOTIDE PHOSPHODIESTERASE 4A — I50 = 1.5 μM
3′,5′-CYCLIC-NUCLEOTIDE PHOSPHODIESTERASE (cycAMP) — I50 = 16 μM
3′,5′-CYCLIC-NUCLEOTIDE PHOSPHODIESTERASE (cycAMP) — I50 = 2.5 μM
3′,5′-CYCLIC-NUCLEOTIDE PHOSPHODIESTERASE (cycAMP) — I50 = 3 μM
3′,5′-CYCLIC-NUCLEOTIDE PHOSPHODIESTERASE (cycAMP) — I50 = 600 μM
3′,5′-CYCLIC-NUCLEOTIDE PHOSPHODIESTERASE (cycAMP) — I50 = 270 μM
3′,5′-CYCLIC-NUCLEOTIDE PHOSPHODIESTERASE (cycAMP) — I50 = 26 μM
3′,5′-CYCLIC-NUCLEOTIDE PHOSPHODIESTERASE (cycAMP) — I50 = 300 μM
3′,5′-CYCLIC-NUCLEOTIDE PHOSPHODIESTERASE (cycGMP) — I50 = 422 μM
3′,5′-CYCLIC-NUCLEOTIDE PHOSPHODIESTERASE (cycGMP) — 25% 2 mM
3′,5′-CYCLIC-NUCLEOTIDE PHOSPHODIESTERASE (cycGMP) — I50 = 380 μM
3′,5′-CYCLIC-NUCLEOTIDE PHOSPHODIESTERASE (cycGMP) (Ca^{2+}/CALMODULIN) — I50 = 394 μM
3′,5′-CYCLIC-NUCLEOTIDE PHOSPHODIESTERASE (cycGMP) (Ca^{2+}/CALMODULIN) — I50 = 940 μM

3′,5′-CYCLIC-NUCLEOTIDE PHOSPHODIESTERASE Ia — *I*50 = 50 μM
3′,5′-CYCLIC-NUCLEOTIDE PHOSPHODIESTERASE Ib — *I*50 = 447 μM
3′,5′-CYCLIC-NUCLEOTIDE PHOSPHODIESTERASE II (cycGMP STIMULATED) — *I*50 = 631 μM
3′,5′-CYCLIC-NUCLEOTIDE PHOSPHODIESTERASE III — *I*50 = 50 μM
3′,5′-CYCLIC-NUCLEOTIDE PHOSPHODIESTERASE III — *K*i = 62 μM
3′,5′-CYCLIC-NUCLEOTIDE PHOSPHODIESTERASE III — *I*50 = 190 μM
3′,5′-CYCLIC-NUCLEOTIDE PHOSPHODIESTERASE III — *I*50 = 316 μM
3′,5′-CYCLIC-NUCLEOTIDE PHOSPHODIESTERASE III — *I*50 = 120 μM
3′,5′-CYCLIC-NUCLEOTIDE PHOSPHODIESTERASE III — *I*50 = 1.1 mM
3′,5′-CYCLIC-NUCLEOTIDE PHOSPHODIESTERASE III (cycGMP INHIBITED) — *I*50 = 132 μM
3′,5′-CYCLIC-NUCLEOTIDE PHOSPHODIESTERASE III (cycGMP INHIBITED) — *I*50 = 1 μM
3′,5′-CYCLIC-NUCLEOTIDE PHOSPHODIESTERASE IV — *K*i = 3.8 μM
3′,5′-CYCLIC-NUCLEOTIDE PHOSPHODIESTERASE IV — *I*50 = 3 μM
3′,5′-CYCLIC-NUCLEOTIDE PHOSPHODIESTERASE IV — *I*50 = 4.7 μM
3′,5′-CYCLIC-NUCLEOTIDE PHOSPHODIESTERASE IV — *I*50 = 18 μM
3′,5′-CYCLIC-NUCLEOTIDE PHOSPHODIESTERASE IV — *I*50 = 1.7 μM
3′,5′-CYCLIC-NUCLEOTIDE PHOSPHODIESTERASE IVa — *I*50 = 5.9 μM
3′,5′-CYCLIC-NUCLEOTIDE PHOSPHODIESTERASE IVb — *I*50 = 5.6 μM
3′,5′-CYCLIC-NUCLEOTIDE PHOSPHODIESTERASE IVc — *I*50 = 22 μM
3′,5′-CYCLIC-NUCLEOTIDE PHOSPHODIESTERASE IVd — *I*50 = 5.6 μM
3′,5′-CYCLIC-NUCLEOTIDE PHOSPHODIESTERASE LOW Km — *I*50 = 190 μM
3′,5′-CYCLIC-NUCLEOTIDE PHOSPHODIESTERASE V (cycGMP) — *I*50 = 467 μM

RO 22-9194
THROMBOXANE A SYNTHASE — *I*50 = 12 μM

RO 23-1717
PHOSPHOLIPASE A2 — *I*50 = 56 μM

RO 23-8358
PHOSPHOLIPASE A2

RO 24-0553
ARACHIDONATE 5-LIPOXYGENASE — *I*50 = 6 nM

RO 31-2201
PEPTIDYL-DIPEPTIDASE A — *I*50 = 600 pM

RO 31-2848
PEPTIDYL-DIPEPTIDASE A — *I*50 = 1.4 nM

RO 31-4493
PHOSPHOLIPASE A2 — *I*50 = 25 μM

RO 31-4639
PHOSPHOLIPASE A2 — *I*50 = 10 μM

RO 31-4724
CASEINASE — *I*50 = 26 nM
COLLAGENASE — *I*50 = 9 nM
COLLAGENASE — *I*50 = 10 nM
GELATINASE — *I*50 = 7.7 nM
STROMELYSIN — *K*i = 26 nM

RO 31-6045
PROTEIN KINASE A — *I*50 = 100 μM

RO 31-7467
CASEINASE — *I*50 = 209 nM
COLLAGENASE — *I*50 = 18 nM
COLLAGENASE — *I*50 = 17 nM
GELATINASE — *I*50 = 239 nM

RO 31-7549
CAM KINASE — *I*50 = 15 μM
CAM KINASE — *I*50 = 15 μM
PHOSPHORYLASE KINASE — *I*50 = 840 nM

PROTEIN KINASE A	$I50 = 5.1\ \mu M$
PROTEIN KINASE A	$I50 = 4.2\ \mu M$
PROTEIN KINASE C	$I50 = 80$ nM
PROTEIN KINASE C	$I50 = 48$ nM

RO 31-8161

CAM KINASE	$I50 = 14\ \mu M$
PROTEIN KINASE A	$I50 = 3.3\ \mu M$
PROTEIN KINASE C	$I50 = 30$ nM

RO 31-8220

CAM KINASE	$I50 = 17\ \mu M$
MAPKAP KINASE 1β	$I50 = 3$ nM
PROTEIN KINASE A	$I50 = 1.5\ \mu M$
PROTEIN KINASE C	$I50 = 10$ nM
PROTEIN KINASE C	$I50 = 5$ nM
PROTEIN KINASE C (CYTOSOLIC)	$I50 = 48$ nM
PROTEIN KINASE C (MEMBRANE)	$I50 = 2.2$ nM
RIBOSOMAL PROTEIN S6 KINASE p70	$I50 = 15$ nM

RO 31-8425

CAM KINASE	$I50 = 19\ \mu M$
PHOSPHORYLASE KINASE	$I50 = 1.3\ \mu M$
PROTEIN KINASE A	$I50 = 2.8\ \mu M$
PROTEIN KINASE A	$I50 = 1.2\ \mu M$
PROTEIN KINASE C	$I50 = 4.8$ nM
PROTEIN KINASE C	$I50 = 7.6$ nM

RO 31-8830

PHOSPHORYLASE KINASE	$I50 = 11\ \mu M$
PROTEIN KINASE A	$I50 = 8.5\ \mu M$
PROTEIN KINASE C	$I50 = 42$ nM

RO 31-8959

PROTEINASE FELINE IMMUNODEFICIENCY VIRUS	$Ki = 67\ \mu M$
PROTEINASE HIV	$Ki = 100$ pM
PROTEINASE HIV-1	$Ki = 100$ pM
PROTEINASE HIV-2	$Ki = 120$ pM
PROTEINASE HTLV-1	$Ki = 3.7\ \mu M$

RO 31-9790

COLLAGENASE	$I50 = 5$ nM
GELATINASE A	$I50 = 8$ nM
GELATINASE B	$I50 = 12$ nM
INTERSTITIAL COLLAGENASE	$I50 = 10$ nM
STROMELYSIN	$I50 = 470$ nM
STROMELYSIN 1	$I50 = 700$ nM

RO 32-0432

MYOSIN-LIGHT-CHAIN KINASE	$I50 = 11\ \mu M$
PHOSPHORYLASE KINASE	$I50 = 16\ \mu M$
PROTEIN KINASE A	$I50 = 22\ \mu M$
PROTEIN KINASE C	$I50 = 17$ nM
PROTEIN KINASE Cβ	$I50 = 37$ nM
PROTEIN KINASE Cα	$I50 = 8.8$ nM
PROTEIN KINASE Cβ1	$I50 = 28$ nM
PROTEIN KINASE Cβ2	$I50 = 31$ nM
PROTEIN KINASE Cγ	$I50 = 37$ nM

RO 32-0557

PROTEIN KINASE C	$I50 = 6.8$ nM
PROTEIN KINASE Cβ	$I50 = 48$ nM
PROTEIN KINASE Cα	$I50 = 2.9$ nM
PROTEIN KINASE Cβ1	$I50 = 5.3$ nM
PROTEIN KINASE Cβ2	$I50 = 6.8$ nM
PROTEIN KINASE Cγ	$I50 = 5.4$ nM

RO 32-2313

THYMIDINE KINASE		*I*50 = 1.8 nM
THYMIDINE KINASE		*I*50 = 190 pM

RO 32-3555

COLLAGENASE 2		*K*i = 4.4 nM
COLLAGENASE 3		*K*i = 3.4 nM
COLLAGENASE FIBROBLAST		*I*50 = 3 nM
COLLAGENASE NEUTROPHIL		*I*50 = 4 nM
GELATINASE A		*K*i = 154 nM
GELATINASE B		*K*i = 59 nM
MICROBIAL COLLAGENASE 1		*K*i = 3 nM
STROMELYSIN 1		*K*i = 527 nM

RO 40-7592

CATECHOL O-METHYLTRANSFERASE		*K*i = 6.4 nM
CATECHOL O-METHYLTRANSFERASE	C	*K*i = 30 nM
CATECHOL O-METHYLTRANSFERASE		*K*i = 4.3 nM

RO 41-0770

AMINE OXIDASE MAO-A		*I*50 = 200 nM
AMINE OXIDASE MAO-B		*I*50 = 5 μM

RO 41-0960

CATECHOL O-METHYLTRANSFERASE		*I*50 = 42 nM
CATECHOL O-METHYLTRANSFERASE		*I*50 = 16 nM

RO 41-1010

AMINE OXIDASE MAO-A		*I*50 = 60 nM
AMINE OXIDASE MAO-B		*I*50 = 200 nM

RO 41-1049

AMINE OXIDASE MAO-A	SSI	*I*50 = 20 nM
AMINE OXIDASE MAO-A		*K*i = 170 nM
AMINE OXIDASE MAO-B	SSI	*I*50 = 12 μM

RO 42-5892

CATHEPSIN D		*I*50 = 35 μM
PEPSIN		*I*50 = 240 μM
RENIN		*I*50 = 3.6 μM
RENIN		*I*50 = 800 pM

RO 46-2825

DNA GYRASE

RO 46-5934

ACETYLCHOLINESTERASE		*I*50 = 38 nM

RO 46-6240

THROMBIN		*K*i = 300 pM
THROMBIN		*K*i = 270 pM
TRYPSIN		*K*i = 1.9 μM

RO 46-6962

DNA GYRASE

RO 46-7864

DNA GYRASE

RO 47-3359

DNA GYRASE

RO 48-1220

β-LACTAMASE RTEM-1		*I*50 = 1.1 μM
β-LACTAMASE TEM-1		*I*50 = 80 nM

RO 48-8071

LANOSTEROL SYNTHASE		*I*50 = 6.5 nM
SQUALENE-HOPENE CYCLASE		

RO 48-8077

SQUALENE-HOPENE CYCLASE	NC	*K*i = 6.6 nM
SQUALENE-LANOSTEROL CYCLASE		

RO 61-8048

KYNURENINE 3-MONOOXYGENASE		*I*50 = 37 nM

RO 55615

DIHYDROPTEROATE SYNTHASE		*I*50 = 34 μM

RO 80122

3′,5′-CYCLIC-NUCLEOTIDE PHOSPHODIESTERASE		

ROBINETIN

ADENOSINE DEAMINASE		*I*50 = 40 μg/ml
INTEGRASE		*I*50 = 5.9 μM
NADH DEHYDROGENASE (UBIQUINONE)		
PROTEIN TYROSINE KINASE p40		*I*50 = 7 μg/ml
XANTHINE OXIDASE		*I*50 = 4.3 μM

ROBUSTIC ACID

PROTEIN KINASE A		*I*50 = 10 μM

ROCANIC ACID

HISTIDINE DECARBOXYLASE	C	*K*i = 2.1 mM

ROEMERINE

PROTEIN TYROSINE-PHOSPHATASE CD45		*I*50 = 107 μM

ROGLETIMIDE

AROMATASE		*I*50 = 63 μM

ROLINETIN

INTEGRASE		*I*50 = 5.9 μM

ROLINIASTATIN 2

NADH DEHYDROGENASE (UBIQUINONE)		*I*50 = 510 pM

(+)ROLIPRAM

3′,5′-CYCLIC-NUCLEOTIDE PHOSPHODIESTERASE IV		*I*50 = 570 nM

(–)ROLIPRAM

3′,5′-CYCLIC-NUCLEOTIDE PHOSPHODIESTERASE IV		*I*50 = 200 nM

ROLIPRAM

3′,5′-CYCLIC-NUCLEOTIDE PHOSPHODIESTERASE		*I*50 = 800 nM
3′,5′-CYCLIC-NUCLEOTIDE PHOSPHODIESTERASE		
3′,5′-CYCLIC-NUCLEOTIDE PHOSPHODIESTERASE 46		*I*50 = 1.6 μM
3′,5′-CYCLIC-NUCLEOTIDE PHOSPHODIESTERASE (cycAMP)		*I*50 = 5 μM
3′,5′-CYCLIC-NUCLEOTIDE PHOSPHODIESTERASE (cycAMP)		*I*50 = 700 nM
3′,5′-CYCLIC-NUCLEOTIDE PHOSPHODIESTERASE (cycAMP)		*I*50 = 3 μM
3′,5′-CYCLIC-NUCLEOTIDE PHOSPHODIESTERASE (cycAMP)		*I*50 = 3 μM
3′,5′-CYCLIC-NUCLEOTIDE PHOSPHODIESTERASE (cycAMP) (Ca^{2+} INDEPENDENT)		*I*50 = 490 μM
3′,5′-CYCLIC-NUCLEOTIDE PHOSPHODIESTERASE (cycAMP) III		*I*50 = 257 μM
3′,5′-CYCLIC-NUCLEOTIDE PHOSPHODIESTERASE (cycAMP) IV		*I*50 = 1.5 μM
3′,5′-CYCLIC-NUCLEOTIDE PHOSPHODIESTERASE 4B2B		
3′,5′-CYCLIC-NUCLEOTIDE PHOSPHODIESTERASE (cycGMP)		*I*50 = 281 μM
3′,5′-CYCLIC-NUCLEOTIDE PHOSPHODIESTERASE (cycGMP)		*I*50 = 246 μM
3′,5′-CYCLIC-NUCLEOTIDE PHOSPHODIESTERASE (cycGMP) (Ca^{2+}/CALMODULIN)		*I*50 = 2.7 mM
3′,5′-CYCLIC-NUCLEOTIDE PHOSPHODIESTERASE (cycGMP) (Ca^{2+}/CALMODULIN)		*I*50 = 493 μM
3′,5′-CYCLIC-NUCLEOTIDE PHOSPHODIESTERASE (cycGMP STIMULATED)		*I*50 = 417 μM
3′,5′-CYCLIC-NUCLEOTIDE PHOSPHODIESTERASE (cycGMP STIMULATED)		*I*50 = 435 μM
3′,5′-CYCLIC-NUCLEOTIDE PHOSPHODIESTERASE Ia		*I*50 = 50 μM
3′,5′-CYCLIC-NUCLEOTIDE PHOSPHODIESTERASE Ib		*I*50 = 813 μM
3′,5′-CYCLIC-NUCLEOTIDE PHOSPHODIESTERASE IB		*I*50 = 812 μM
3′,5′-CYCLIC-NUCLEOTIDE PHOSPHODIESTERASE II		*I*50 = 126 μM
3′,5′-CYCLIC-NUCLEOTIDE PHOSPHODIESTERASE II (cycGMP STIMULATED)		*I*50 = 126 μM
3′,5′-CYCLIC-NUCLEOTIDE PHOSPHODIESTERASE III		*I*50 = 158 μM

3′,5′-CYCLIC-NUCLEOTIDE PHOSPHODIESTERASE III (cycGMP INHIBITED) *I*50 = 159 μM
3′,5′-CYCLIC-NUCLEOTIDE PHOSPHODIESTERASE III (cycGMP INHIBITED) *I*50 = 107 μM
3′,5′-CYCLIC-NUCLEOTIDE PHOSPHODIESTERASE IV *I*50 = 794 nM
3′,5′-CYCLIC-NUCLEOTIDE PHOSPHODIESTERASE IV *K*i = 230 nM
3′,5′-CYCLIC-NUCLEOTIDE PHOSPHODIESTERASE IV *I*50 = 65 nM
3′,5′-CYCLIC-NUCLEOTIDE PHOSPHODIESTERASE IV *I*50 = 1.3 μM
3′,5′-CYCLIC-NUCLEOTIDE PHOSPHODIESTERASE IV *I*50 = 1.3 μM
3′,5′-CYCLIC-NUCLEOTIDE PHOSPHODIESTERASE IV *I*50 = 1.5 μM
3′,5′-CYCLIC-NUCLEOTIDE PHOSPHODIESTERASE IV *I*50 = 600 nM
3′,5′-CYCLIC-NUCLEOTIDE PHOSPHODIESTERASE IVa *I*50 = 2.2 μM
3′,5′-CYCLIC-NUCLEOTIDE PHOSPHODIESTERASE IVa *I*50 = 300 nM
3′,5′-CYCLIC-NUCLEOTIDE PHOSPHODIESTERASE IVa *I*50 = 900 nM
3′,5′-CYCLIC-NUCLEOTIDE PHOSPHODIESTERASE IVb *K*i = 85 nM
3′,5′-CYCLIC-NUCLEOTIDE PHOSPHODIESTERASE IVb *I*50 = 404 μM
3′,5′-CYCLIC-NUCLEOTIDE PHOSPHODIESTERASE IVb *I*50 = 201 nM
3′,5′-CYCLIC NUCLEOTIDE PHOSPHODIESTERASE IVc *I*50 = 15 μM
3′,5′-CYCLIC-NUCLEOTIDE PHOSPHODIESTERASE IVc *I*50 = 1.5 μM
3′,5′-CYCLIC-NUCLEOTIDE PHOSPHODIESTERASE IVc *I*50 = 2.7 μM
3′,5′-CYCLIC-NUCLEOTIDE PHOSPHODIESTERASE IVd *I*50 = 1.7 μM
3′,5′-CYCLIC-NUCLEOTIDE PHOSPHODIESTERASE IVd *I*50 = 1.6 μM
3′,5′-CYCLIC-NUCLEOTIDE PHOSPHODIESTERASE IVd *I*50 = 135 nM
3′,5′-CYCLIC-NUCLEOTIDE PHOSPHODIESTERASE V *I*50 = 631 μM
3′,5′-CYCLIC-NUCLEOTIDE PHOSPHODIESTERASE V (cycGMP) *I*50 = 83 μM
3′,5′-CYCLIC-NUCLEOTIDE PHOSPHODOESTERASE HUMAN EOSINOPHYL *I*50 = 3.3 μM

(R)-ROLIPRAM
3′,5′-CYCLIC-NUCLEOTIDE PHOSPHODOESTERASE 4A *I*50 = 103 nM
3′,5′-CYCLIC-NUCLEOTIDE PHOSPHODOESTERASE 4B *I*50 = 182 nM
3′,5′-CYCLIC-NUCLEOTIDE PHOSPHODOESTERASE 4C *I*50 = 23 nM
3′,5′-CYCLIC-NUCLEOTIDE PHOSPHODOESTERASE 4D *I*50 = 219 nM

ROLLIMEMBRIN
NADH DEHYDROGENASE (UBIQUINONE) *I*50 = 330 pM

ROLLINIASTATIN 1
NADH DEHYDROGENASE (UBIQUINONE) *K*i = 300 pM
NADH DEHYDROGENASE (UBIQUINONE) *I*50 = 580 pM

ROLLINIASTATIN 2
NADH DEHYDROGENASE (UBIQUINONE) *K*i = 600 pM

R-ROLPRAM
3′,5′-CYCLIC-NUCLEOTIDE PHOSPHODIESTERASE 4A *I*50 = 204 nM

ROSCOVITINE
CDC2 KINASE *I*50 = 200 nM
CYCLIN DEPENDENT KINASE 1 *I*50 = 650 nM
CYCLIN DEPENDENT KINASE 2 *I*50 = 700 nM
CYCLIN DEPENDENT KINASE CDC/CYCLIN B *I*50 = 650 nM
CYCLIN DEPENDENT KINASE CDK4/p35 *I*50 = 200 nM
CYCLIN DEPENDENT KINASE CDK/CYCLIN A *I*50 = 700 nM
CYCLIN DEPENDENT KINASE CDK2/CYCLIN E *I*50 = 700 nM
MAP KINASE *I*50 = 30 nM
PROTEIN TYROSINE KINASE EGFR *I*50 = 70 μM

ROSE BENGAL
ALCOHOL DEHYDROGENASE *K*i = 3 μM
ASPARTATE KINASE
CARBONATE DEHYDRATASE
DEOXYGUANOSINE KINASE
α-GLUCOSIDASE
GLUTATHIONE TRANSFERASE
GLUTATHIONE TRANSFERASE 4-4 *I*50 = 150 nM
GLUTATHIONE TRANSFERASE A1-1 *I*50 = 500 nM
GLUTATHIONE TRANSFERASE M1a-1a *I*50 = 1 μM
GLUTATHIONE TRANSFERASE P1-1 *I*50 = 8 μM

1-D-myo-INOSITOL-TRISPHOSPHATE 5-KINASE		
1-D-myo-INOSITOL-TRISPHOSPHATE 6-KINASE		
LEUKOTRIENE-C4 SYNTHASE		*I*50 = 50 μM
NITRITE REDUCTASE (NAD(P)H)		
RNA POLYMERASE		100% 100 μM
ROSEMANOL		
PROTEINASE HIV-1		*I*50 = 600 μg/ml
ROSMARINIC ACID		
ADENYLATE CYCLASE		
ARACHIDONATE 5-LIPOXYGENASE		*I*50 = 6.2 mM
ARACHIDONATE 12-LIPOXYGENASE		*I*50 = 6.4 mM
COMPLEMENT C3 CONVERTASE		
HYDROXYPHENYLPYRUVATE REDUCTASE	C	*K*i = 210 μM
INTEGRASE		*I*50 = 9 μM
ROSMARINATE SYNTHASE		
TYROSINE TRANSAMINASE		
ROSMARINIC ACID METHYL ESTER		
ARACHIDONATE 5-LIPOXYGENASE		*I*50 = 600 μM
ARACHIDONATE 12-LIPOXYGENASE		*I*50 = 400 μM
ROTENONE		
CALCIDOL 1-MONOOXYGENASE		
GLYCEROL-3-PHOSPHATE DEHYDROGENASE		
HYDROXYLAMINE REDUCTASE (NADH)		
4-METHOXYBENZOATE MONOOXYGENASE (O-DEMETHYLATING)		
NADH DEHYDROGENASE (UBIQUINONE)		*I*50 = 25nmol/mg
NADH DEHYDROGENASE (UBIQUINONE)		*I*50 = 19 nM
NADH DEHYDROGENASE (UBIQUINONE)		*K*i = 4 nM
NADH DEHYDROGENASE (UBIQUINONE)		*I*50 =1.3mmol/mg
NADH DEHYDROGENASE (UBIQUINONE) 1		*I*50 = 21 μM
NADH DEHYDROGENASE (UBIQUINONE) 1		*I*50 = 11 μM
NITRITE REDUCTASE (NAD(P)H)		
PROTEIN KINASE CASEIN KINASE G		8% 2.5 μM
SQUALENE MONOOXYGENASE		
ROTTLERIN		
CAM KINASE III		*I*50 = 5.3 μM
ELONGATION FACTOR-2 KINASE		*I*50 = 5.3 μM
PROTEIN KINASE A		*I*50 = 78 μM
PROTEIN KINASE Cβ		*I*50 = 100 μM
PROTEIN KINASE Cα		*I*50 = 30 μM
PROTEIN KINASE Cβ		*I*50 = 42 μM
PROTEIN KINASE Cδ		*I*50 = 6 μM
PROTEIN KINASE Cδsp		*I*50 = 3 μM
PROTEIN KINASE Cγ		*I*50 = 40 μM
PROTEIN KINASE CASEIN KINASE II		*I*50 = 30 μM
PROTEIN KINASE Cn		*I*50 = 82 μM
PROTEIN KINASE Cx		*I*50 = 100 μM
RP 53801		
PROTEIN KINASE C		*K*i = 3.6 mM
PROTEIN TYROSINE KINASE p60v-src	NC	*I*50 = 22 μM
RP 64477		
STEROL O-ACYLTRANSFERASE		*I*50 = 194 nM
STEROL O-ACYLTRANSFERASE		*I*50 = 89 nM
STEROL O-ACYLTRANSFERASE		*I*50 = 32 nM
RP 64966		
LEUKOTRIENE A4 HYDROLASE		*I*50 = 1.5 μM
RP 70676		
STEROL O-ACYLTRANSFERASE		*I*50 = 25 nM
STEROL O-ACYLTRANSFERASE		*I*50 = 21 nM
STEROL O-ACYLTRANSFERASE		*I*50 = 44 nM

RP 73163

STEROL O-ACYLTRANSFERASE		$I50$ = 245 nM
STEROL O-ACYLTRANSFERASE		$I50$ = 86 nM
STEROL O-ACYLTRANSFERASE		$I50$ = 370 nM

RP 73401

3′,5′-CYCLIC-NUCLEOTIDE PHOSPHODIESTERASE I		$I50$ = 44 µM
3′,5′-CYCLIC-NUCLEOTIDE PHOSPHODIESTERASE II		$I50$ = 72 µM
3′,5′-CYCLIC-NUCLEOTIDE PHOSPHODIESTERASE III		$I50$ = 148 µM
3′,5′-CYCLIC-NUCLEOTIDE PHOSPHODIESTERASE IV		$I50$ = 1.2 nM
3′,5′-CYCLIC-NUCLEOTIDE PHOSPHODIESTERASE IVa		$I50$ = 8 nM
3′,5′-CYCLIC-NUCLEOTIDE PHOSPHODIESTERASE IVa		$I50$ = 200 pM
3′,5′-CYCLIC-NUCLEOTIDE PHOSPHODIESTERASE IVb		$I50$ = 4 nM
3′,5′-CYCLIC-NUCLEOTIDE PHOSPHODIESTERASE IVb		$I50$ = 800 pM
3′,5′-CYCLIC-NUCLEOTIDE PHOSPHODIESTERASE IVb		$I50$ = 2.3 nM
3′,5′-CYCLIC-NUCLEOTIDE PHOSPHODIESTERASE IVc		$I50$ = 2 nM
3′,5′-CYCLIC-NUCLEOTIDE PHOSPHODIESTERASE IVd		$I50$ = 1.5 nM
3′,5′-CYCLIC-NUCLEOTIDE PHOSPHODIESTERASE IVd		$I50$ = 200 pM
3′,5′-CYCLIC-NUCLEOTIDE PHOSPHODIESTERASE IVd		$I50$ = 3 nM
3′,5′-CYCLIC-NUCLEOTIDE PHOSPHODIESTERASE V		$I50$ = 22 µM
3′,5′-CYCLIC-NUCLEOTIDE PHOSPHODOESTERASE 4A		$I50$ = 500 pM
3′,5′-CYCLIC-NUCLEOTIDE PHOSPHODOESTERASE 4B		$I50$ = 500 pM
3′,5′-CYCLIC-NUCLEOTIDE PHOSPHODOESTERASE 4C		$I50$ = 1 nM
3′,5′-CYCLIC-NUCLEOTIDE PHOSPHODOESTERASE 4D		$I50$ = 100 pM
3′,5′-CYCLIC-NUCLEOTIDE PHOSPHODOESTERASE HUMAN EOSINOPHYL		$I50$ = 9 nM

pRb2/p130

CYCLIN DEPENDENT KINASE 2

pRb/p107

CYCLIN DEPENDENT KINASE 2

RPI 856 A

CATHEPSIN D		$I50$ = 880 nM
PEPSIN		$I50$ = 1.9 µM
PROTEINASE HIV-1		$I50$ = 37 nM
PROTEINASE HTLV-1		$I50$ = 27 nM

RPI 856A

PROTEINASE HIV-1		$I50$ = 37 nM

RPI 856 B

CATHEPSIN D		$I50$ = 8.6 µM
PROTEINASE HIV-1		$I50$ = 260 nM
PROTEINASE HTLV-1		$I50$ = 81 nM

RPI 856 C

CATHEPSIN D		$I50$ = 2 µM
PEPSIN		$I50$ = 4.8 µM
PROTEINASE HIV-1		$I50$ = 55 nM
PROTEINASE HTLV-1		$I50$ = 92 nM

RPI 856 D

PROTEINASE HIV-1		$I50$ = 280 nM
PROTEINASE HTLV-1		$I50$ = 200 nM

RPR 107393

FARNESYL-DIPHOSPHATE FARNESYLTRANSFERASE		$I50$ = 800 pM

RPR 110993

PROTEIN TYROSINE KINASE CSF-1R		$I50$ = 180 nM

RPR 113228

FARNESYL PROTEINTRANSFERASE	C	Ki = 400 nM
FARNESYL PROTEINTRANSFERASE		
GERANYLGERANYL PROTEINTRANSFERASE		$I50$ = 59 µM

RPR 113829
- FARNESYL PROTEINTRANSFERASE — *I*50 = 1.8 nM

RPR 108514A
- PROTEIN TYROSINE KINASE CSF-1R — *I*50 = 500 nM
- PROTEIN TYROSINE KINASE EGFR — *I*50 = 4 μM
- PROTEIN TYROSINE KINASE PDGFR — *I*50 = 15 μM

RPR 108518A
- PROTEIN TYROSINE KINASE EGFR — *I*50 = 500 nM
- PROTEIN TYROSINE KINASE p56lck — *I*50 = 500 nM
- PROTEIN TYROSINE KINASE p60v-src — *I*50 = 500 nM
- PROTEIN TYROSINE KINASE PDGFR — *I*50 = 20 μM

RPR II
- ACETYLCHOLINESTERASE — *I*50 = 410 μM

RPR V
- ACETYLCHOLINESTERASE — *I*50 = 750 μM

RS 130830
- COLLAGENASE 3 — *I*50 = 590 nM
- COLLAGENASE FIBROBLAST — *I*50 = 520 pM
- GELATINASE A — *I*50 = 220 pM
- GELATINASE B — *I*50 = 580 pM
- MATRILYSIN — *I*50 = 1.2 nM
- STROMELYSIN 1 — *I*50 = 9.3 nM

RS 14203
- 3′,5′-CYCLIC-NUCLEOTIDE PHOSPHODOESTERASE 4A — *I*50 = 1.1 nM
- 3′,5′-CYCLIC-NUCLEOTIDE PHOSPHODOESTERASE 4B — *I*50 = 400 pM
- 3′,5′-CYCLIC-NUCLEOTIDE PHOSPHODOESTERASE 4C — *I*50 = 3.3 nM
- 3′,5′-CYCLIC-NUCLEOTIDE PHOSPHODOESTERASE 4D — *I*50 = 250 pM

RS 21607
- AROMATASE — *K*i = 7.6 nM
- CHOLESTEROL 7α-MONOOXYGENASE — *K*i = 1.6 μM
- CHOLESTEROL MONOOXYGENASE (SIDE-CHAIN-CLEAVING) — *K*i = 18 μM
- LANOSTEROL C14 DEMETHYLASE — *K*i = 840 pM
- LANOSTEROL C14 DEMETHYLASE — *K*i = 2.5 nM
- LANOSTEROL C14 DEMETHYLASE — *K*i = 1.4 nM
- STEROID 17α-MONOOXYGENASE/17,20 LYASE — *K*i = 446 nM

RS 2232
- AMINE OXIDASE MAO-A — *K*i = 54 nM

RS 25344
- 3′,5′-CYCLIC-NUCLEOTIDE PHOSPHODIESTERASE 4D3
- 3′,5′-CYCLIC-NUCLEOTIDE PHOSPHODIESTERASE IV — *I*50 = 300 pM
- 3′,5′-CYCLIC-NUCLEOTIDE PHOSPHODOESTERASE 4A — *I*50 = 11 nM
- 3′,5′-CYCLIC-NUCLEOTIDE PHOSPHODOESTERASE 4B — *I*50 = 1.9 nM
- 3′,5′-CYCLIC-NUCLEOTIDE PHOSPHODOESTERASE 4C — *I*50 = 250 pM
- 3′,5′-CYCLIC-NUCLEOTIDE PHOSPHODOESTERASE 4D — *I*50 = 100 pM

RS 25560-198
- DOPAMINE β-MONOOXYGENASE — *I*50 = 18 nM
- DOPAMINE β-MONOOXYGENASE — *I*50 = 25 nM

RS 39066
- COLLAGENASE — *I*50 = 6.9 nM
- MATRILYSIN — *I*50 = 230 pM
- STROMELYSIN — *I*50 = 800 pM

RS 5186
- THROMBOXANE A SYNTHASE

RS 61980
- DIHYDROOROTATE DEHYDROGENASE — *K*i = 2.7 μM

Inhibitor / Enzyme	Type	Value
RS 82856		
3′,5′-CYCLIC-NUCLEOTIDE PHOSPHODIESTERASE II		$I50 = 60$ pM
3′,5′-CYCLIC-NUCLEOTIDE PHOSPHODIESTERASE IV	MIX	$Ki = 500$ pM
RS 8359		
AMINE OXIDASE MAO		
AMINE OXIDASE MAO-A		$I50 = 520$ nM
AMINE OXIDASE MAO-A		
AMINE OXIDASE MAO-B		$I50 \approx 900$ µM
RS 93522		
3′,5′-CYCLIC-NUCLEOTIDE PHOSPHODIESTERASE		
RU 44004		
NEPRILYSIN		
RU 69296		
ENDOTHELIN CONVERTING ENZYME		$I50 = 25$ nM
RU 69738		
ENDOTHELIN CONVERTING ENZYME		$I50 = 20$ nM
RUBRATOXIN B		
4-NITROPHENYLPHOSPHATASE		
$RuCl_3$		
MALEATE DEHYDRATASE		
RUFIGFALLOL		
PHOSPHOGLYCERATE KINASE		$I50 = 800$ nM
RUGOSIN D		
POLY(ADP-RIBOSE) GLYCOHYDROLASE		$I50 = 6.1$ µM
XANTHINE OXIDASE		$I50 = 3.1$ µM
RUM 1		
CYCLIN DEPENDENT KINASE CDC2/CDC		
RUSCOGENIN		
ELASTASE		$I50 = 120$ µM
RUSTMICIN		
INOSITOLCERAMIDE SYNTHASE		$I50 = 70$ pM
RUTHENIUM RED		
Ca^{2+} CARRIER MITOCHONDRIA		
Ca^{2+}-TRANSPORTING ATPASE		$I50 = 25$ µM
MYOSIN ATPASE (ACTIN ACTIVATED Mg^{2+})	C	$Ki = 4.4$ µ
RUTIN		
ADENOSINE DEAMINASE		$I50 = 305$ µM
ALDEHYDE REDUCTASE		
ARACHIDONATE 5-LIPOXYGENASE		$I50 = 12$ µM
ARACHIDONATE 5-LIPOXYGENASE		$I50 = 45$ µM
ARYL SULFOTRANSFERASE P		$I50 = 200$ µM
CARBONYL REDUCTASE (NADPH)		60% 1 µM
CYCLOOXYGENASE		$I50 = 450$ µM
CYTOCHROME P450 1A1		$I50 = 40$ µM
DNA TOPOISOMERASE IV		$I50 = 1$ µg/ml
GLUTATHIONE DEHYDROGENASE (ASCORBATE)		54% 10 µM
GLUTATHIONE REDUCTASE (NADPH)		22% 100 µM
GLUTATHIONE TRANSFERASE		45% 100 µM
GLUTATHIONE TRANSFERASE		$I50 = 35$ µM
IODIDE PEROXIDASE		$I50 = 41$ µM
MYELOPEROXIDASE		$I50 = 10$ µM
PHOSPHOLIPASE A2		
PHOSPHOLIPASE A2-2		
PROSTAGLANDIN F SYNTHASE		

PROTEINASE HIV-1 — $I50 = 500\ \mu M$
PROTEIN KINASE C — $I50 = 32\ \mu M$
PROTEIN KINASE CASEIN KINASE G — 11% 2.5 μM
RUTININ OXIDASE — $I50 = 52\ \mu M$
SEPIAPTERIN REDUCTASE

RUTINETIN
ARACHIDONATE 12-LIPOXYGENASE — $I50 = 7\ \mu M$

3-RUTINOSYLKAEMPFEROL
XANTHINE OXIDASE — $I50 = 15\ \mu M$

RUTIN SULPHATE
MYELOPEROXIDASE — $I50 = 85\ \mu M$

RUTOSIDE
ADENOSINE DEAMINASE — $I50 = 80\ \mu g/ml$

RWJ 50215
THROMBIN — $Ki = 1.2\ \mu M$

RWJ 50353
PLASMIN — $Ki = 12\ \mu M$
t-PLASMINOGEN ACTIVATOR — $Ki = 3.3\ \mu M$
PROTEIN Ca — $Ki = 19\ \mu M$
STREPTOKINASE — $Ki = 6.3\ \mu M$
THROMBIN — $Ki = 160\ pM$
TRYPSIN — $Ki = 16\ nM$

RWJ 63556
CYCLOOXYGENASE 2 — $I50 = 1.9\ \mu M$

S

S 17162
ENDOTHELIN CONVERTING ENZYME

S 18326
COAGULATION FACTOR Xa — $I50 = 420\ nM$
FACTOR I — $I50 = 84\ nM$
KALLIKREIN — $I50 = 25\ nM$
PLASMIN — $I50 = 124\ nM$
t-PLASMINOGEN ACTIVATOR — $I50 = 15\ nM$
u-PLASMINOGEN ACTIVATOR — $I50 = 936\ nM$
PROTEIN Ca — $I50 = 30\ nM$
THROMBIN — $I50 = 4\ nM$

S 1924
H^+/K^+-EXCHANGING ATPASE

S 21402
NEPRILYSIN — $Ki = 1.7\ nM$
PEPTIDYL-DIPEPTIDASE A — $Ki = 4.2\ nM$

S 21402-1
NEPRILYSIN — $Ki = 1.7\ nM$
PEPTIDYL-DIPEPTIDASE A — $Ki = 1.7\ nM$

S 23142
PROTOPORPHYRINOGEN OXIDASE — $I50 = 100\ pM$

S 2467
HYDROXYMETHYLGLUTARYL-CoA REDUCTASE (NADPH)

S 266056
FARNESYL-DIPHOSPHATE FARNESYLTRANSFERASE — $I50 = 55\ \mu M$

FARNESYL-DIPHOSPHATE FARNESYLTRANSFERASE — $I50 = 14\ \mu M$
FARNESYL-DIPHOSPHATE FARNESYLTRANSFERASE — $I50 = 48\ \mu M$

S 267338
FARNESYL-DIPHOSPHATE FARNESYLTRANSFERASE — $I50 = 1\ \mu M$
FARNESYL-DIPHOSPHATE FARNESYLTRANSFERASE — $I50 = 68$ nM
FARNESYL-DIPHOSPHATE FARNESYLTRANSFERASE — $I50 = 24$ nM

S 2720
RNA-DIRECTED DNA POLYMERASE — $I50 = 140$ nM

S 2864
RENIN — $I50 = 380$ pM

S 3337
H^+/K^+-EXCHANGING ATPASE

S 58035
STEROL O-ACYLTRANSFERASE

S 862033
RENIN

S 863390
RENIN

S B2864
RENIN — $I50 = 380$ pM

S 9780L
PEPTIDYL-DIPEPTIDASE A

SA 152
α-THROMBIN — IR — $Ki = 700\ \mu M$

SA 898
NEPRILYSIN — $Ki = 18\ \mu M$
PEPTIDYL-DIPEPTIDASE A — $Ki = 170$ nM
THIMET OLIGOPEPTIDASE — C — $Ki = 9.1$ nM

SA 6541
LEUKOTRIENE A4 HYDROLASE

D-SACCHARIC ACID
β-GLUCURONIDASE

SACCHARIC ACID 1,4-LACTONE
β-GLUCURONIDASE — $I50 = 80\ \mu g/ml$

1,4-D-SACCHARIC ACID LACTONE
β-GLUCURONIDASE — 100% 10 mM

SACCHARIN
DOPAMINE SULFOTRANSFERASE — 44% 6.7 μM
GLUCOSE-6-PHOSPHATASE

SACCHARO 1,4-LACTONE
β-GLUCURONIDASE — C — $Ki = 170$ nM

D-SACCHAROLACTONE
L-IDURONIDASE — $Ki = 690\ \mu M$

SACCHAROMYCES CEREVISIAE P40 PROTEIN
PROTEIN KINASE p34CDC28

SACCHAROMYCES CEREVISIAE PROTEINASE INHIBITOR
SERINE CARBOXYPEPTIDASE

SACCHAROPINE
SACCHAROPINE DEHYDROGENASE (NADP, L-LYSINE FORMING)

SAE 9

AROMATASE		$I50 = 550$ nM
CHOLESTEROL MONOOXYGENASE (SIDE-CHAIN-CLEAVING)		$I50 = 100$ μM

SAFINGOL

PROTEIN KINASE C		$I50 = 38$ μM

SAFRANIN

GLUTATHIONE REDUCTASE (NADPH)	NC	$Ki = 453$ μM

SAFRAZINE

AMINE OXIDASE MAO

SAINTOPIN

DNA TOPOISOMERASE
DNA TOPOISOMERASE (ATP-HYDROLYSING)

SALANNIN

ECDYSONE 20-MONOOXYGENASE		25% 10 μM

SALBOSTATIN

ALDEHYDE REDUCTASE		$I50 = 680$ μM
α,α-TREHALASE	C	$Ki = 180$ nM

SALFREDIN A3

ALDEHYDE REDUCTASE

SALFREDIN A4

ALDEHYDE REDUCTASE

SALFREDIN A7

ALDEHYDE REDUCTASE

SALFREDIN B11

ALDEHYDE REDUCTASE

SALFREDIN C1

ALDEHYDE REDUCTASE

SALFREDIN C2

ALDEHYDE REDUCTASE

SALFREDIN C3

ALDEHYDE REDUCTASE

SALICIN

CYCLOMALTODEXTRIN GLUCANOTRANSFERASE		
β-GLUCOSIDASE	C	$Ki = 6.2$ mM
SUCROSE SYNTHASE		
THIOGLUCOSIDASE		

SALICORTIN

β-GLUCOSIDASE	SSI	

SALICOYL CYCLIC PHOSPHATE

β-LACTAMASE

SALICYLALDOXIME

ANTHRANILATE 2,3 DIOXYGENASE		
ANTHRANILATE 3-MONOOXYGENASE (DEAMINATING)		
L-ASCORBATE OXIDASE		
CATECHOL OXIDASE		
CYCLOHEXYLAMINE OXIDASE		
CYSTEAMINE DIOXYGENASE		
CYTOCHROME-C OXIDASE		
2,3-DIHYDROXYBENZOATE 2,3-DIOXYGENASE		68% 100 μM
3-HYDROXYBENZOATE 4-MONOOXYGENASE		
HYDROXYLAMINE REDUCTASE (NADH)		
LACCASE		

SECONDARY-ALCOHOL OXIDASE		
STIZOLOBATE SYNTHASE		
STIZOLOBINATE SYNTHASE		
SALICYLHYDROXAMIC ACID		
CATECHOL OXIDASE		*I*50 = 200 nM
DIHYDROOROTATE DEHYDROGENASE		80% 1 mM
LACTOPEROXIDASE		Kd = 70 µM
LIPOXYGENASE		90% 1 mM
MONOPHENOL MONOOXYGENASE		95% 100 µM
MYELOPEROXIDASE		*I*50 = 5 µM
MYELOPEROXIDASE		
PEROXIDASE		Kd = 2 µM
PROCOLLAGEN-PROLINE, 2-OXOGLUTARATE-4-DIOXYGENASE		
URATE OXIDASE		
XANTHINE DEHYDROGENASE		
SALICYLIC ACID		
ACETOIN DEHYDROGENASE		
ACYL-CoA SYNTHETASE (MEDIUM CHAIN)		*K*i = 37 µM
ALCOHOL DEHYDROGENASE		*K*i = 1.3 mM
D-AMINO-ACID OXIDASE		
ARYLAMINE N-ACETYLTRANSFERASE		
ARYL SULFOTRANSFERASE P		*I*50 = 40 µM
L-ASCORBATE PEROXIDASE		*I*50 = 78 µM
BENZOATE 4-MONOOXYGENASE		
CATALASE		32% 1 mM
CATALASE		57% 1 mM
CATALASE		47% 1 mM
CATALASE		51% 1 mM
CATALASE		66% 1 mM
CATECHOL OXIDASE		*K*i = 870 µM
CREATINE KINASE		
4,5-DIHYDROOXYPHTHALATE DECARBOXYLASE		
GENTISATE 1,2-DIOXYGENASE		*K*i = 3.2 mM
β-GLUCOSIDASE	MIX	*K*i = 61 mM
GLYCEROL DEHYDRATASE		
3-HYDROXYANTHRANILATE 3,4-DIOXYGENASE	NC	*K*i = 98 µM
3α-HYDROXYSTEROID DEHYDROGENASE (B-SPECIFIC)	NC	*K*i = 420 µM
MANDELATE RACEMASE		
6-METHYLSALYCILATE DECARBOXYLASE	NC	
MONOPHENOL MONOOXYGENASE		23% 800 µM
NAD(P)H DEHYDROGENASE (QUINONE)		
NICOTINATE N-METHYLTRRANSFERASE		
NITRITE REDUCTASE (NAD(P)H)		
PECTATE LYASE		*K*i = 200 µM
PHOSPHOGLYCERATE KINASE		
PHOSPHOLIPASE A2		*I*50 = 18 mM
PHYLLOQUINONE EPOXIDE REDUCTASE		*I*50 = 330 µM
PROPIONATE-CoA LIGASE	C	*K*i = 1.5 mM
o-PYROCATECHUATE DECARBOXYLASE	C	
THYROXINE DEIODINASE		
TYROSINE-ESTER SULFOTRANSFERASE IV	C	*K*i = 67 µM
UDP-GLUCURONOSYLTRANSFERASE	C	*K*i = 367 µM
XANTHINE OXIDASE		
SALMINE		
RNA-DIRECTED RNA POLYMERASE		
(R) SALSINOL		
CATECHOL O-METHYLTRANSFERASE	C	*K*i = 92 µM
(S) SALSINOL		
CATECHOL O-METHYLTRANSFERASE	C	*K*i = 81 µM

SALSOLIDINE

CATECHOL O-METHYLTRANSFERASE	C	$Ki = 190\ \mu M$

(R)-(+)-SALSOLIDINE

AMINE OXIDASE MAO-A	C	$Ki = 6\ \mu M$

(+)-(R)-SALSOLINE

AMINE OXIDASE MAO-A	C	$Ki = 77\ \mu M$

SALSOLINOL

DEBRISOQUINE 4-MONOOXYGENASE	C	$Ki = 430\ \mu M$
TYROSINE 3-MONOOXYGENASE		

(+)-(R)-SALSOLINOL

AMINE OXIDASE MAO-A	C	$Ki = 31\ \mu M$

(R)-SALSOLINOL

AMINE OXIDASE MAO-A	C	$Ki = 38\ \mu M$
TRYPTOPHAN 5-MONOOXYGENASE		$I50 = 20\ \mu M$

R-SALSOLINOL

AMINE OXIDASE MAO-A	C	$Ki = 76\ \mu M$
AMINE OXIDASE MAO-B	NC	$Ki = 68\ \mu M$

(S)-SALSOLINOL

TRYPTOPHAN 5-MONOOXYGENASE		$I50 = 20\ \mu M$

SALVIANOLIC ACID

H^+/K^+-EXCHANGING ATPASE		$I50 = 520\ \mu M$

SALYRGAN-SODIUM

METHYLGLUTACONYL-CoA HYDRATASE

SAMPATRILAT

NEPRILYSIN		$Ki = 8\ nM$
PEPTIDYL-DIPEPTIDASE A		$Ki = 1.2\ nM$

SAN 54817

NADH DEHYDROGENASE (UBIQUINONE)

SANDARACOPIMARIC ACID

ARACHIDONATE 15-LIPOXYGENASE

SANDOSTATIN

PROTEIN TYROSINE KINASE

SANDOZ COMPOUND 57-118

STEROL O-ACYLTRANSFERASE	C	

SANDOZ COMPOUND 58-035

STEROL O-ACYLTRANSFERASE

SANGIVAMYCIN

β-ADRENERGIC-RECEPTOR KINASE	C	$I50 = 67\ \mu M$
PROTEIN KINASE A		$I50 = 50\ \mu M$
PROTEIN KINASE C	C	$Ki = 15\ \mu M$
PROTEIN KINASE (NUCLEAR)		
RHODOPSIN KINASE	C	$Ki = 180\ nM$

SANGIVAMYCINAMIDINE

PROTEIN KINASE (NUCLEAR)

SANGIVAMYCINAMIDOXIME

PROTEIN KINASE (NUCLEAR)

SANGUIIN H-2

NADH DEHYDROGENASE		$I50 = 6.5\ \mu M$
NADH DEHYDROGENASE		$I50 = 4.2\ \mu M$

SANGUIIN H-6

NADH DEHYDROGENASE		$I50 = 1.5\ \mu M$
NADH DEHYDROGENASE		$I50 = 7.1\ \mu M$
NADH DEHYDROGENASE		$I50 = 1.9\ \mu M$
NADH DEHYDROGENASE		$I50 = 4.1\ \mu M$

SANGUIIN H-11

NADH DEHYDROGENASE		$I50 = 1.7\ \mu M$
NADH DEHYDROGENASE		$I50 = 550$ nM
NADH DEHYDROGENASE		$I50 = 1.2\ \mu M$
NADH DEHYDROGENASE		$I50 = 1\ \mu M$
NADH DEHYDROGENASE		$I50 = 700$ pM

SANGUINARINE

AMINE OXIDASE I		$Ki = 600\ \mu M$
AMINE OXIDASE II		$Ki = 500\ \mu M$
ARACHIDONATE 5-LIPOXYGENASE		$I50 = 400$ nM
ARACHIDONATE 12-LIPOXYGENASE		$I50 = 13\ \mu M$
AROMATIC-L-AMINO-ACID DECARBOXYLASE	IR	$Ki = 120\ \mu M$
Ca^{2+}/Mg^{2+}-TRANSPORTING ATPASE		
PROTEIN KINASE A		$I50 = 6\ \mu M$

SANGUINARINE NITRATE

DNA LIGASE I		$I50 = 322\ \mu M$

SANOCRISIN

INTERSTITIAL COLLAGENASE	NC	$Ki = 60$ nM

SAPONIN C GLEDITSIA

Na^+/K^+-EXCHANGING ATPASE		$I50 = 400\ \mu M$

SAPONIN C GLEDITSIA 3-O-GLYCOSIDE METHYLESTER

Na^+/K^+-EXCHANGING ATPASE		$I50 = 25\ \mu M$

SAQUAYAMYCIN A

FARNESYL PROTEINTRANSFERASE		$I50 = 1.8\ \mu M$

SAQUAYAMYCIN B

FARNESYL PROTEINTRANSFERASE		$I50 = 1.8\ \mu M$

SAQUAYAMYCIN C

PROTEIN FARNESYLTRANSFERASE		$I50 = 1.4\ \mu M$

SAQUAYAMYCIN D

FARNESYL PROTEINTRANSFERASE		$I50 = 1.5\ \mu M$

SAQUAYAMYCIN E

FARNESYL PROTEINTRANSFERASE		$I50 = 1.8\ \mu M$

SAQUAYAMYCIN F

FARNESYL PROTEINTRANSFERASE		$I50 = 2\ \mu M$

SAQUINAVIR

CYTOCHROME P450 3A4		$Ki = 3\ \mu M$

SARCOSINE

ALANINE DEHYDROGENASE	C	$Ki = 4.3$ mM
4-HYDROXYPROLINE EPIMERASE	C	$Ki = 19$ mM
PROLINE RACEMASE		

SARCOSYL

IGA-SPECIFIC METALLOENDOPEPTIDASE 1		$I50 = 0.1$ %
IGA-SPECIFIC METALLOENDOPEPTIDASE 2		$I50 = 0.1$ %
PHORBOL-DIESTER HYDROLASE		

SARIN

ACETYLCHOLINESTERASE	IR	
CARBOXYLESTERASE		
CHOLINESTERASE		

SARMESIN

3′,5′-CYCLIC-NUCLEOTIDE PHOSPHODIESTERASE (cycGMP) (Ca^{2+}/CALMODULIN)		

R(+)-SATERINONE

3′,5′-CYCLIC-NUCLEOTIDE PHOSPHODIESTERASE I		*I*50 = 37 μM
3′,5′-CYCLIC-NUCLEOTIDE PHOSPHODIESTERASE II		*I*50 = 119 μM
3′,5′-CYCLIC-NUCLEOTIDE PHOSPHODIESTERASE III		*I*50 = 10 nM
3′,5′-CYCLIC-NUCLEOTIDE PHOSPHODIESTERASE IV		*I*50 = 40 nM

S(–)-SATERINONE

3′,5′-CYCLIC-NUCLEOTIDE PHOSPHODIESTERASE I		*I*50 = 27 μM
3′,5′-CYCLIC-NUCLEOTIDE PHOSPHODIESTERASE II		*I*50 = 159 μM
3′,5′-CYCLIC-NUCLEOTIDE PHOSPHODIESTERASE III		*I*50 = 50 nM
3′,5′-CYCLIC-NUCLEOTIDE PHOSPHODIESTERASE IV		*I*50 = 70 nM

SAXITOXIN

AMINE OXIDASE (COPPER-CONTAINING)	NC	*K*i = 140 μM

S100Bββ

PROTEIN KINASE C		*I*50 = 10 μM

SB 200517

ATP CITRATE (pro-S)-LYASE		

SB 202190

MAP KINASE p38		*I*50 = 40 nM
MAP KINASE p38	UC	*K*i = 63 μM
MAP KINASE p38		*I*50 = 50 nM
MAP KINASE p38β		*I*50 = 10 μM
MAP KINASE p38β2		*I*50 = 80 nM

SB 202235

ARACHIDONATE 5-LIPOXYGENASE		*I*50 = 2.3 μM

SB 202742

β-LACTAMASE		*I*50 = 17 mg/ml
β-LACTAMASE		*I*50 = 41 mg/ml
β-LACTAMASE		*I*50 = 110 mg/ml
β-LACTAMASE		*I*50 = 10 μG/ML
β-LACTAMASE ECOXA1		*I*50 = 80 mg/ml
β-LACTAMASE EC(PSE4)		*I*50 = 10 mg/ml
β-LACTAMASE ECTEM1		*I*50 = 5 mg/ml

SB 203238

PROTEINASE HIV-1		*K*i = 430 nM

SB 203347

PHOSPHOLIPASE A2 (14KD)		*I*50 = 500 nM
PHOSPHOLIPASE A2 (85KD)		*I*50 = 20 μM

SB 203386

PROTEINASE HIV-1		

SB 203580

MAP KINASE p38	UC	*K*i = 84 μM
MAP KINASE p38β		*I*50 = 7 μM
MAP KINASE p38β2		*I*50 = 100 nM

SB 204144

PROTEINASE HIV-1	C	*K*i = 2.8 Nm

SB 206343

PROTEINASE HIV		*K*i = 600 nM
PROTEINASE HIV-1		*K*i = 600 pM

SB 206718

MAP KINASE p38		*I*50 = 640 nM

SB 207499

3′,5′-CYCLIC-NUCLEOTIDE PHOSPHODIESTERASE IV	*I*50 = 94 nM
3′,5′-CYCLIC-NUCLEOTIDE PHOSPHODOESTERASE 4A	*I*50 = 50 nM
3′,5′-CYCLIC-NUCLEOTIDE PHOSPHODOESTERASE 4B	*I*50 = 41 nM
3′,5′-CYCLIC-NUCLEOTIDE PHOSPHODOESTERASE 4C	*I*50 = 134 nM
3′,5′-CYCLIC-NUCLEOTIDE PHOSPHODOESTERASE 4D	*I*50 = 9.6 nM
3′,5′-CYCLIC-NUCLEOTIDE PHOSPHODOESTERASE HUMAN EOSINOPHYL	*I*50 = 170 nM

SB 210661

ARACHIDONATE 5-LIPOXYGENASE

SB 212021

β-LACTAMASE	*I*50 = 19 μM
PEPTIDYL-DIPEPTIDASE A	*I*50 = 55 μM

SB 212305

β-LACTAMASE	*I*50 = 1 μM
PEPTIDYL-DIPEPTIDASE A	*I*50 = 68 μM

SB 216357

PROCESSING PEPTIDASE

SB 216477

PHOSPHOLIPASE A2

SB 220025

MAP KINASE p38	*I*50 = 60 nM
PROTEIN KINASE A	*I*50 = 30 μM
PROTEIN KINASE C	*I*50 = 2.9 μM
PROTEIN TYROSINE KINASE p56lck	*I*50 = 3.5 μM

SB 227931

MAP KINASE p38	*I*50 = 46 nM

Sc^{3+}

ACETYLCHOLINESTERASE	30% 100 μM
ADENYLATE CYCLASE	34% 5 μM
CHOLINESTERASE	80% 1 mM
DIHYDROPTERIDINE REDUCTASE	54% 7.3 μM
FERROXIDASE	90% 100 μM

SC 0051

4-HYDROXYPHENYLPYRUVATE DIOXYGENASE	*I*50 = 45 nM

SC 558

CYCLOOXYGENASE 1	*I*50 = 18 μM
CYCLOOXYGENASE 2	*I*50 = 93 nM

SC 903

GELATINASE B	*K*i = 2.6 nM
INTERSTITIAL COLLAGENASE	*K*i = 2.8 nM
STROMELYSIN 1	*K*i = 24 nM

SC 31769

STEROL O-ACYLTRANSFERASE

SC 37698

ELASTASE NEUTROPHIL

SC 39026

ELASTASE LEUKOCYTE	*I*50 = 500 nM
ELASTASE PANCREATIC	57 % 10 μM

SC 40827

COLLAGENASE	*I*50 = 1.5 μM
METALLOPROTEINASE UTERINE SMALL	*I*50 = 750 nM

SC 44463
GELATINASE A — $I50 = 18$ nM
INTERSTITIAL COLLAGENASE — $I50 = 1.2$ nM
METALLOPROTEINASE UTERINE SMALL — $I50 = 10$ nM

SC 45662
ARACHIDONATE 5-LIPOXYGENASE — $I50 = 68$ μM
ARACHIDONATE 5-LIPOXYGENASE — $I50 = 3.5$ μM

SC 46944
RENIN — $I50 = 5$ nM

SC 47557
RENIN — $I50 = 78$ nM

SC 47563
RENIN — $I50 = 1$ nM

SC 48272
RENIN — $I50 = 700$ nM

SC 52151
PROTEINASE HIV-1 — $I50 = 6$ nM

SC 5251
PROTEINASE HIV-1 — $Ki = 3$ nM

SC 56525
RENIN — $I50 = 1.2$ nM

SC 57461
LEUKOTRIENE A4 HYDROLASE — $Ki = 23$ nM

SC 57666
CYCLOOXYGENASE 2 — $I50 = 30$ nM

SC 58125
CYCLOOXYGENASE 2 — $I50 = 90$ nM

SC 58231
CYCLOOXYGENASE 1 — $I50 = 18$ μM
CYCLOOXYGENASE 1 — $I50 = 18$ μM
CYCLOOXYGENASE 2 — $I50 = 15$ nM
CYCLOOXYGENASE 2 — $I50 = 15$ nM

SC 58272
PROTEIN N-MYRISTOYLTRANSFERASE — $I50 = 56$ nM
PROTEIN N-MYRISTOYLTRANSFERASE — $I50 = 14$ μM

SC 58451
CYCLOOXYGENASE 1 — $I50 = 5.4$ μM
CYCLOOXYGENASE 1 — $I50 = 5.4$ μM
CYCLOOXYGENASE 1 — $I50 = 140$ nM
CYCLOOXYGENASE 2 — $I50 = 8$ nM
CYCLOOXYGENASE 2 — $I50 = 8$ nM
CYCLOOXYGENASE 2 — $I50 = 1.1$ nM

SC 59383
PROTEIN N-MYRISTOYLTRANSFERASE — $I50 = 1.5$ μM
PROTEIN N-MYRISTOYLTRANSFERASE — $I50 = 810$ μM

SC 41661A
ARACHIDONATE 5-LIPOXYGENASE — $I50 = 17$ μM
ARACHIDONATE 5-LIPOXYGENASE — $I50 = 300$ nM

SC 55389A
PROTEINASE HIV

SCALARADIAL

PHOSPHOLIPASE A2		*I*50 = 70 pM
PHOSPHOLIPASE A2 (14KD)		*I*50 = 70 pM
PHOSPHOLIPASE A2 (85KD)		*I*50 = 20 nM
PHOSPHOLIPASE A2 (SECRETORY)		99% 3 μM

SCH 207278

FARNESYL PROTEINTRANSFERASE		*I*50 = 3.5 μM
GERANYLGERANYL PROTEINTRANSFERASE I		*I*50 = 70 μM

SCH 28080

FLIPASE		*I*50 = nM
H^+/K^+-EXCHANGING ATPASE		66% 18 μM
H^+/K^+-EXCHANGING ATPASE		*K*i = 1 μM
H^+/K^+-EXCHANGING ATPASE	C	*K*i = 24 nM
H^+/K^+-EXCHANGING ATPASE		*I*50 = 12 μg/ml
4-NITROPHENYLPHOSPHATASE	C	*K*i = 275 nM

SCH 32615

NEPRILYSIN		*K*i = 20 nM

SCH 34826

NEPRILYSIN

SCH 37137

GLUCOSAMINE-FRUCTOSE-6-PHOSPHATE AMINOTRANSFERASE (ISOMERIZING)

SCH 39304

STEROL 14α-DEMETHYLASE

SCH 39370

NEPRILYSIN		*I*50 = 15 nM
NEPRILYSIN		*I*50 = 11 nM

SCH 40120

ARACHIDONATE 5-LIPOXYGENASE		*I*50 = 7 μM
ARACHIDONATE 5-LIPOXYGENASE		*I*50 = 8 μM
ARACHIDONATE 5-LIPOXYGENASE		*I*50 = 4 μM

SCH 42354

NEPRILYSIN

SCH 42495

NEPRILYSIN

SCH 44342

FARNESYL PROTEINTRANSFERASE		*I*50 = 280 nM

SCH 47112

PROTEIN KINASE C		*I*50 = 1.7 nM

SCH 47890

ATROLYSIN C		*K*i = 520 nM

SCH 49209

GERANYLGERANYL PROTEINTRANSFERASE		*I*50 = 25 nM

SCH 49211

PHOSPHOLIPASE D		*I*50 = 11 μM

SCH 49212

PHOSPHOLIPASE D		*I*50 = 12 μM

SCH 51688

3′,5′-CYCLIC-NUCLEOTIDE PHOSPHODIESTERASE (cycGMP)

SCH 51866

3′,5′-CYCLIC-NUCLEOTIDE PHOSPHODIESTERASE I		*I*50 = 70 nM
3′,5′-CYCLIC-NUCLEOTIDE PHOSPHODIESTERASE V		*I*50 = 60 nM

SCH 53823
PHOSPHOLIPASE A2 — $I50 = 24\ \mu M$
PHOSPHOLIPASE D

SCH 53825
PHOSPHOLIPASE A2 — $I50 = 19\ \mu M$
PHOSPHOLIPASE D

SCH 53827
PHOSPHOLIPASE A2 — $I50 = 17\ \mu M$
PHOSPHOLIPASE D — $I50 = 17\ \mu M$

SCH 56580
FARNESYL PROTEINTRANSFERASE — $I50 = 40$ nM

SCH 58450
FARNESYL PROTEINTRANSFERASE — $I50 = 29\ \mu M$
GERANYLGERANYL PROTEINTRANSFERASE — $I50 = 740\ \mu M$

SCH 58540
FARNESYL PROTEINTRANSFERASE — $I50 = 29\ \mu M$

SCH 65576
PROTEINASE CYTOMEGALO VIRUS — $I50 = 9.8\ \mu g/ml$

SCH 68631
PROTEINASE HEPATITIS C VIRUS NS3

SCHINOL
PHOSPHOLIPASE A2

SCHINUS TEREBINTHIFOLIUS PLA2 INHIBITORS
PHOSPHOLIPASE A2

SCHISTOCERCA GREGARIA OVARIES SERINE PROTEINASE INHIBITOR
CHYMOTRYPSIN
ELASTASE PANCREATIC
TRYPSIN

SCHISTOCERCA GREGARIA PROTEINASE INHIBITOR 1
α-CHYMOTRYPSIN — 100% 10 μM
TRYPSIN — 100% 10 μM

SCHISTOCERCA GREGARIA PROTEINASE INHIBITOR 2
α-CHYMOTRYPSIN — 100% 10 μM
ELASTASE PANCREATIC — 100% 10 μM

SCHISTOCERCA GREGARIA PROTEINASE INHIBITOR 3
α-CHYMOTRYPSIN — 100% 10 μM

SCHISTOCERCA GREGARIA PROTEINASE INHIBITOR 4
α-CHYMOTRYPSIN — 100% 10 μM

SCHISTOCERCA GREGARIA PROTEINASE INHIBITOR 5
α-CHYMOTRYPSIN — 100% 10 μM

SCHISTOCERCA GREGARIA SERINE PROTEINASE INHIBITOR 1
PROTEINASES SERINE

SCHISTOCERCA GREGARIA SERINE PROTEINASE INHIBITOR 2
PROTEINASES SERINE

SCHISTOCERCA GREGARIA SERINE PROTEINASE INHIBITOR 3
PROTEINASES SERINE

SCHISTOSOMA MANSONI PROTEINASE INHIBITOR
COAGULATION FACTOR XIIa

(±)γ-SCHIZANDRIN
3′,5′-CYCLIC-NUCLEOTIDE PHOSPHODIESTERASE (cycAMP) — $I50 = 19\ \mu M$

SCHIZOLOBIUM PARAHYBA CHYMOTRYPSIN INHIBITOR
α-CHYMOTRYPSIN — $Ki = 59$ nM

SCHIZOLOBIUM PARAHYBA TRYPSIN INHIBITOR
CHYMOTRYPSIN — $Ki = 58$ nM

SCHIZOSTATIN
FARNESYL-DIPHOSPHATE FARNESYLTRANSFERASE — C — $Ki = 450$ nM
FARNESYL-DIPHOSPHATE FARNESYLTRANSFERASE — $I50 = 840$ nM

.S-(p-CHLOROPHENACYL)GLUTATHIONE
HYDROXYACYLGLUTATHIONE HYDROLASE — $Ki = 400\ \mu M$

SCHRADAN
ACETYLCHOLINESTERASE

SCN^-
4-AMINOBENZOATE 1-MONOOXYGENASE
FORMATE DEHYDROGENASE
NADH DEHYDROGENASE
PEPTIDE α-N-ACETYLTRANSFERASE
PHOTINUS-LUCIFERIN 4-MONOOXYGENASE (ATP-HYDROLYSING)
SALICYLATE 1-MONOOXYGENASE
SULFITE REDUCTASE (NADPH)
THYMIDINE KINASE

SCOPADULCIC ACID METHYLESTER
H^+/K^+-EXCHANGING ATPASE — 81% 100 μM

SCOPADULCIOL
H^+/K^+-EXCHANGING ATPASE — 45% 100 μM

SCOPADULIC ACID B
H^+/K^+-EXCHANGING ATPASE — $I50 = 28\ \mu M$

SCOPAFUNGIN
H^+/K^+-EXCHANGING ATPASE — $I50 = 36\ \mu g/ml$

SCOPARIC ACID A
β-GLUCURONIDASE — $I50 = 6.8\ \mu M$

SCOPARIC ACID C
β-GLUCURONIDASE — $I50 = 10\ \mu M$

SCOPOLETIN
PHOSPHOGLUCONATE DEHYDROGENASE (DECARBOXYLATING)

SCOPOLIN
PHOSPHOGLUCONATE DEHYDROGENASE (DECARBOXYLATING)

SCREROTIORIN
PHOSPHOLIPASE A2

SCRIP
RENIN — $I50 = 16$ nM

SCUTELLARIN
NADH DEHYDROGENASE (EXOGENOUS) — $I50 = 500\ \mu M$
XANTHINE OXIDASE — $I50 = 13\ \mu M$

SCYPHOSTATIN
SPHINGOMYELIN PHOSPHODIESTERASE (ACID) — $I50 = 49\ \mu M$
SPHINGOMYELIN PHOSPHODIESTERASE (NEUTRAL) — $I50 = 1\ \mu M$

SCYTONEMIN
PHOSPHOLIPASE A2 — $I50 = 4.4\ \mu M$

SD 146
PROTEINASE HIV-1 — $Ki = 24$ pM

SD 894
PROTEINASE HIV

SDZ 87-469
SQUALENE MONOOXYGENASE — $I50 = 20$ nM
SQUALENE MONOOXYGENASE — $I50 = 16$ μM
SQUALENE MONOOXYGENASE — $I50 = 10$ nM

SDZ 115-358
THROMBIN — $Ki = 70$ pM

SDZ 283471
PROTEINASE HIV-1 — $Ki = 4.3$ nM

SDZ 63370
HYDROXYMETHYLGLUTARYL-CoA REDUCTASE (NADPH) — $I50 = 34$ nM

SDZ-CPI 975
CARNITINE O-PALMITOYLTRANSFERASE 1

SDZ ENA 713
ACETYLCHOLINESTERASE — $Ki = 1.4$ μM
ACETYLCHOLINESTERASE G1 — $I50 = 3.2$ μM
ACETYLCHOLINESTERASE G4 — $I50 = 13$ μM

SDZ PRI 053
PROTEINASE HIV-1 — $Ki = 9.5$ nM

SDZ SBA 586
SQUALENE MONOOXYGENASE — $I50 = 8$ nM
SQUALENE MONOOXYGENASE — $I50 = 39$ nM
SQUALENE MONOOXYGENASE — $I50 = 11$ μM
SQUALENE MONOOXYGENASE — $I50 = 1.9$ μM

Se^{2+}
PROTEIN-DISULFIDE REDUCTASE (GLUTATHIONE)

Se^{4+}
GLUTAMINE-FRUCTOSE-6-PHOSPHATE AMINOTRANSFERASE (ISOMERIZING) — 90% 10 μM

SE 205
COLLAGENASE FIBROBLAST — $I50 = 1.2$ nM
GELATINASE B — $Ki = 1.8$ nM
INTERSTITIAL COLLAGENASE — $Ki = 1.2$ nM
STROMELYSIN 1 — $Ki = 33$ nM
TNFα CONVERTING ENZYME — $I50 = 1.2$ μM

SEBACATE
4-HYDROXY-2-OXOGLUTARATE ALDOLASE — 62% 40 mM

SEBAIC ACID
GLUTATHIONE TRANSFERASE

SECALONIC ACID D
MYOSIN-LIGHT-CHAIN KINASE — $I50 = 60$ μM
PROTEIN KINASE A — $I50 = 12$ μM
PROTEIN KINASE C — $I50 = 13$ μM
PROTEIN KINASE (Ca^{2+} DEPENDENT) — $I50 = 67$ μM

(4R)-5,10-SECOESTRA-4,5-DIENE-3,10,17-TRIONE
5α-REDUCTASE — NC — $Ki = 5.5$ μM

(3R,S)-1,10-SECO-5α-ESTR-1-YNE-3,17β-DIOL
3α-HYDROXYSTEROID DEHYDROGENASE (B-SPECIFIC) — IR

Inhibitor / Enzyme	Type	Value
5,10-SECOESTR-4-YNE-3,10,17-TRIONE		
3β-HYDROXY-Δ5-STEROID DEHYDROGENASE	IR	
SECOISOLARICIRESINOL		
AROMATASE		11% 409 μM
SECOLOGANIN		
17-O-DEACETYLVINDOLINE O-ACETYLTRANSFERASE		
SECO-MEVINIC ACID		
HYDROXYMETHYLGLUTARYL-CoA REDUCTASE (NADPH)		*I*50 = 8 nM
SECONAL		
NADH DEHYDROGENASE (UBIQUINONE)		
5,10-SECO-19-NOR-5-CHOLESTYN-3,10-DIONE		
CHOLESTEROL OXIDASE		
(4R)-5,10-SECO-19-NORPREGNA-4,5-DIENE-3,10,20-TRIONE		
5α-REDUCTASE	NC	*K*i = 980 nM
SECOSTEROID		
STEROID 5α-REDUCTASE		
STEROID 5β-REDUCTASE	NC	
SECRETIN		
PEPTIDYL-DIPEPTIDASE A		
SECRETORY LEUKOCYTE PROTEINASE INHIBITOR		
CHYMASE		*K*i = 30 nM
CHYMASE 1		*K*i = 300 pM
CHYMASE 1		*K*i = 9 μM
CHYMOTRYPSIN		*K*i = 71 pM
PROTEINASES SERINE		
SECRETORY LEUKOPROTEINASE INHIBITOR		
CATHEPSIN G		*I*50 = 5.5 nM
CHYMOTRYPSIN		*I*50 = 10 nM
ELASTASE LEUKOCYTE		*I*50 = 10 nM
TRYPSIN		*I*50 = 330 nM
SEDOHEPTULOSE-1,7-BISPHOSPHATE		
GLYCEROL-3-PHOSPHATE DEHYDROGENASE (NAD)		
D-SEDOHEPTULOSE		
KETOHEXOKINASE		
SEDOHEPTULOSE-1,7-DIPHOSPHATE		
2-DEHYDRO-3-DEOXYPHOSPHOHEPTONATE ALDOLASE		
FRUCTOSE-BISPHOSPHATASE	C	*K*i = 29 μM
FRUCTOSE-BISPHOSPHATASE	C	*K*i = 400 μM
RIBOSE-5-PHOSPHATE ISOMERASE	C	*K*i = 600 μM
RIBULOSE-BISPHOSPHATE CARBOXYLASE		
SEDOHEPTULOSE-7-PHOSPHATE		
GLUCOSE-6-PHOSPHATE ISOMERASE		*K*i = 8.6 μM
GLYCERALDEHYDE-3-PHOSPHATE DEHYDROGENASE (NADP)		
TRANSALDOLASE		
SEF 19		
AROMATASE		*I*50 = 5.3 nM
SEL 2711		
COAGULATION FACTOR Xa		*K*i = 3 nM
THROMBIN		*K*i = 40 μM
SELEGILINE		
AMINE OXIDASE MAO-A		*I*50 = 1.4 μM
AMINE OXIDASE MAO-B		*I*50 = 6 nM

SELENALYSINE
HOMOCITRATE SYNTHASE

β-SELENAPROLINE
PROLINE-tRNA LIGASE

SELENATE
N-ACETYLMANNOSAMINE KINASE 19% 1 mM
CARBONATE DEHYDRATASE III
6-PHOSPHOFRUCTO-2-KINASE
RIBOSE-5-PHOSPHATE ADENYLYLTRANSFERASE
UDP-N-ACETYLGLUCOSAMINE 2-EPIMERASE 82% 1 mM

SELENAZOFURIN
IMP DEHYDROGENASE

SELENAZOLE-4-CARBOXAMIDE ADENINE DINUCLEOTIDE
IMP DEHYDROGENASE NC *K*i = 55 nM

SELENAZOLE-4-CARBOXAMIDE ADENINEDINULCEOTIDE
IMP DEHYDROGENASE 1 *K*i = 33 nM
IMP DEHYDROGENASE 2 *K*i = 21 nM

SELENITE
PROSTAGLANDIN-H2 D-ISOMERASE NC *K*i = 10 μM
UDP-N-ACETYLGLUCOSAMINE 2-EPIMERASE 91% 1 mM

SELENIUM(IV)OXIDE
UDP-N-ACETYLGLUCOSAMINE 2-EPIMERASE 82% 1 mM

SELENOCYSTEAMINE
ACETYLSEROTONIN O-METHYLTRANSFERASE *I*50 = 13 μM
GLUTAMATE-CYSTEINE LIGASE

D,L-SELENOCYSTEINE
L-SERINE DEHYDRATASE *I*50 = 7 μM

6-SELENOGUANOSINE-5′-MONOPHOSPHATE
GUANYLATE KINASE

SELENOHOMOCYSTEINE LACTONE
PROTEIN-LYSINE 6-OXIDASE *K*i = 8.3 μM

SELENO-L-METHIONINE
METHIONINE ADENOSYLTRANSFERASE *K*i = 510 μM

SELENOPHENFURIN
IMP DEHYDROGENASE

SELENOUREA
UREASE

SEMICARBAZIDE
ACETYL-CoA C-ACYLTRANSFERASE
ACETYLSPERMIDINE DEACETYLASE
ADENOSYLMETHIONINE-8-AMINO-7-OXONONANOATE TRANSAMINASE
ADENOSYLMETHIONINE DECARBOXYLASE
ALANINE-GLYOXYLATE TRANSAMINASE 25% 1 mM
ALANINE TRANSAMINASE
AMINE DEHYDROGENASE
AMINE OXIDASE (COPPER-CONTAINING)
AMINE OXIDASE (POLYAMINE)
AMINE OXIDASE (SEMICARBAZIDE SENSITIVE) IR *K*i = 85 μM
1-AMINOCYCLOPROPANE-1-CARBOXYLATE SYNTHASE
AMINODEOXYGLUCONATE DEHYDRATASE 57% 1 mM
8-AMINO-7-OXONONAOATE SYNTHASE
ARALKYLAMINE DEHYDROGENASE
ARGININE DECARBOXYLASE 42% 50 mM

AROMATIC-L-AMINO-ACID DECARBOXYLASE	95% 1 mM
AROMATIC-AMINO-ACID-GLYOXYLATE AMINOTRANSFERASE	92% 1 mM
ASPARTATE 4-DECARBOXYLASE	
BRANCHED-CHAIN-AMINO-ACID TRANSAMINASE	
CEREBROSIDE-SULFATASE	44% 13 mM
CHOLINE DEHYDROGENASE	
CYSTATHIONINE γ-LYASE	
CYTOCHROME-C OXIDASE	
2-DEHYDROPANTOATE 2-REDUCTASE	
DIAMINE TRANSAMINASE	100% 100 μM
DIAMINOBUTYRATE-PYRUVATE TRANSAMINASE	58% 33 μM
DIAMINOPIMELATE EPIMERASE	
ETHANOLAMINE OXIDASE	
FORMALDEHYDE DISMUTASE	
FORMATE DEHYDROGENASE	
GLUCAN ENDO-1,3-β-GLUCOSIDASE	
GLUCOSE OXIDASE	
GLUTAMATE DECARBOXYLASE	
GLUTAMINE-PHENYLPYRUVATE TRANSAMINASE	
GLYCINE C-ACETYLTRANSFERASE	
GLYCINE HYDROXYMETHYLTRANSFERASE	
GLYCOSULFATASE	
GLYOXYLATE REDUCTASE	
HISTIDINE DECARBOXYLASE	
HISTIDINOL-PHOSPHATE TRANSAMINASE	
erythro-3-HYDROXYASPARTATE DEHYDRATASE	55% 2 mM
HYDROXYLAMINE REDUCTASE	
KYNURENINE-GLYOXYLATE TRANSAMINASE	
LYSINE DECARBOXYLASE	100% 1 mM
METHIONINE γ-LYASE	
NITRILE HYDRATASE	77% 1 mM
OXALATE OXIDASE	
PHENYLALANINE AMMONIA-LYASE	14% 4 mM
PHENYLALANINE (HISTIDINE) TRANSAMINASE	12% 1 mM
PROTEIN-LYSINE 6-OXIDASE	
SERINE-GLYOXYLATE TRANSAMINASE	31% 1 mM
THERMOPHILIC STREPTOMYCES SERINE PROTEINASE	
THIAMIN OXIDASE	
THREONINE ALDOLASE	100% 1 mM
THREONINE DEHYDRATASE	
TRIMETHYLAMINE-N-OXIDE REDUCTASE	
TRYPTOPHAN 2-MONOOXYGENASE	
TYROSINE 2,3-AMINOMUTASE	25% 1 mM
UROCANATE HYDRATASE	53% 20 mM
UROCANATE HYDRATASE	52% 100 μM
VALINE DECARBOXYLASE	
XANTHINE OXIDASE	

SEMICOCHLIODINOL A

CATHEPSIN D	*I*50 = 2.5 μM
PROTEINASE HIV-1	*I*50 = 370 nM
PROTEIN TYROSINE KINASE EGFR	*I*50 = 20 μM

SEMICOCHLIODINOL B

CATHEPSIN D	*I*50 = 4.9 μM
PROTEIN TYROSINE KINASE EGFR	*I*50 = 60 μM

SEMINAL PLASMA PROTEINASE INHIBITOR

CATHEPSIN B
CATHEPSIN H
FICAIN
PAPAIN

SEMPERVINE

CYTOCHROME P450 2D6	*K*i = 9.7 μM

SEMPERVIRINE
BUFURALOL-1′-HYDROXYLASE (CYT P-450dbl) — *I*50 = 4 μM

SENECIOYLDITHRANOL
GLUCOSE-6-PHOSPHATE 1-DEHYDROGENASE

SEOCL$_2$
PEROXIDASE — *I*50 = 11 mM

SEPHAGENIN A
RNA-DIRECTED DNA POLYMERASE — *I*50 = 49nM

SEPHAGENIN B
RNA-DIRECTED DNA POLYMERASE — *I*50 = 74 nM

SEPIAPTERIN
GTP CYCLOHYDROLASE A2

L-SEPIAPTERIN
GTP CYCLOHYDROLASE I — *I*50 = 35 μM
NITRIC-OXIDE SYNTHASE — *I*50 = 10 μM

SEQUOYITOL
MYOINOSITOL D-3-HYDROXY KINASE — C — *K*i = 271 μM

SEQUTERPE LACTONES
IMP DEHYDROGENASE

SER-ASN-VAL-L-3-(4′-(1′,1″-BIPHENYLYL)ALA-ALA-OBzl
MYOSIN-LIGHT-CHAIN KINASE — *I*50 = 1 μM

SER-ASN-VAL-3-IODO-TYR-ALA-OBzl
MYOSIN-LIGHT-CHAIN KINASE — *I*50 = 1 μM

SER-ASN-VAL-(1-NAPHTHYLALANINE)-ALA-OBZL
MYOSIN ATPASE
MYOSIN-LIGHT-CHAIN KINASE

SERATRODAST
CYTOCHROME P450 2C9 — *I*50 = 60 μM
CYTOCHROME P450 2C19 — *I*50 = 50 μM

SERICIN
TYROSINASE

SERICYSTATIN
CATHEPSIN B
PAPAIN

SERINE
ALANINE DEHYDROGENASE
1-AMINOCYCLOPROPANE-1-CARBOXYLATE DEAMINASE
8-AMINO-7-OXONONAOATE SYNTHASE
ASPARAGINE-OXO-ACID TRANSAMINASE
ASPARAGINE SYNTHASE (GLUTAMINE-HYDROLYSING)
ASPARTATE 1-DECARBOXYLASE — *K*i = 730 μM
CYSTEINE LYASE
GLUTAMATE SYNTHASE (FERREDOXIN)
GLUTAMATE SYNTHASE (NADH)
GLUTAMATE SYNTHASE (NADPH)
γ-GLUTAMYLTRANSFERASE
GLYCINE DEHYDROGENASE
GLYCINE HYDROXYMETHYLTRANSFERASE
GLYCINE-OXALOACETATE TRANSAMINASE
ORNITHINE-OXO-ACID TRANSAMINASE
PHOSPHOGLYCERATE DEHYDROGENASE
PHOSPHOGLYCERATE PHOSPHATASE
PHOSPHOGLYCOLATE PHOSPHATASE

1-PYRROLINE-5-CARBOXYLATE DEHYDROGENASE		
UDP-N-ACETYLMURAMATE-ALANINE LIGASE		

SERINE-BORATE

D-ALANINE γ-GLUTAMYLTRANSFERASE		

D-SERINE + BORATE

γ-GLUTAMYLTRANSFERASE		$Ki = 170\ \mu M$

L-SERINE + BORATE

γ-GLUTAMYLTRANSFERASE		$Ki = 20\ \mu M$

D-SERINE

D-ALANINE-ALANYL-POLY(GLYCEROPHOSPHATE)LIGASE	C	$Ki = 34$ mM
D-ALANINE-POLY(PHOSPHORIBITOL)LIGASE		$Ki = 34$ mM
D-ALANINE TRANSAMINASE		
AMINODEOXYGLUCONATE DEHYDRATASE		$Ki = 63\ \mu M$
ASPARTATE 1-DECARBOXYLASE	C	$Ki = 160\ \mu M$
GLUTAMATE DECARBOXYLASE	C	$Ki = 160\ \mu M$
PHOSPHOSERINE PHOSPHATASE		$Ki = 27$ mM
PYRUVATE KINASE		
L-SERINE DEHYDRATASE		
SERINE-GLYOXYLATE TRANSAMINASE	C	$Ki = 1.6$ mM
THREONINE DEHYDRATASE	C	$Ki = 45$ mM
UDP-N-ACETYLMURAMATE-ALANINE LIGASE		

D,L-SERINE

(R)AMINOPROPANOL DEHYDROGENASE		

DL-SERINE HYDROXAMIC ACID

PROTEINASE ASTACUS		$I50 = 2.6$ mM

L-SERINE

D-ALANINE 2-HYDROXYMETHYLTRANSFERASE	C	$Ki = 5.6$ mM
AMINODEOXYGLUCONATE DEHYDRATASE		$Ki = 4.7$ mM
ASPARTATE 1-DECARBOXYLASE	C	$Ki = 730\ \mu M$
CYSTATHIONINE γ-LYASE	C	$Ki = 13$ mM
CYSTATHIONINE γ-LYASE	C	$Ki = 13$ mM
GLUTAMATE-AMMONIA LIGASE		
GLUTAMATE DECARBOXYLASE	C	$Ki = 730\ \mu M$
GLYCINE DEHYDROGENASE (CYTOCHROME)	NC	$Ki = 4.4$ mM
GLYCINE-tRNA LIGASE	NC	$Ki = 52$ mM
HOMOSERINE DEHYDROGENASE		
HOMOSERINE KINASE		17% 20 mM
PHOSPHOGLYCERATE DEHYDROGENASE	NC	
PHOSPHOSERINE PHOSPHATASE		$Ki = 650\ \mu M$
D-SERINE DEHYDRATASE		
SERINE-ETHANOLAMINEPHOSPHATE PHOSPHODIESTERASE	C	

D,L-SERINE PHOSPHATE

XAA-PRO DIPEPTIDASE		$Ki = 510\ \mu M$

L-SERINE-O-PHOSPHATE

THREONINE SYNTHASE		

D-SERINE SULPHATE

SERINE-SULFATE AMMONIA-LYASE	C	$Ki = 18$ mM

L-SERINE-O-SULPHATE

ASPARTATE TRANSAMINASE	SSI	
GLUTAMATE DECARBOXYLASE	SSI	

SERINE-O-SULPHATE

ASPARTATE TRANSAMINASE		

SERINOL

D-SERINE DEHYDRATASE	IR	

SEROTONIN
ACETYLCHOLINESTERASE
AROMATIC-L-AMINO-ACID DECARBOXYLASE
ARYL-ACYLAMIDASE
CHOLINESTERASE
DIHYDROPTERIDINE REDUCTASE
HISTAMINE N-METHYLTRANSFERASE
PHENYLETHANOLAMINE N-METHYLTRANSFERASE 75% 50 μM
PYRIDOXAL KINASE
TYROSINE TRANSAMINASE

SERP 1
PROTEINASES SERINE

SERPENTINE
CYTOCHROME P450 2D6 *K*i = 2.2 μM

SERPIN CrmA
INTERLEUKIN 1β CONVERTING ENZYME *K*i = 400 fM

SERPIN II (RAT)
TRYPTASE *K*i = 1 nM

SERRATIA MARCESCENS METALLOPROTEINASE INHIBITOR
PROTEINASE SERRATIA MARCESCENS

SERTRALINE
CYTOCHROME P450 1A2 *K*i = 8.8 μM
CYTOCHROME P450 2C9 *K*i = 33 μM
CYTOCHROME P450 2D6 *K*i = 700 nM

SERUM ALBUMINE
PRENYL-PYROPHOSPHATEASE

SERUM ALBUMINE BOVINE
N-ACETYLGLUCOSAMINE-6-SULFATASE 78% 0.5%
PROCOLLAGEN N-ENDOPEPTIDASE 56% 2%

SERUM ALBUMIN HUMAN
GALACTOSYLGALACTOSYLGLUCOSYLCERAMIDASE

SERUM PROTEINS
PHOSPHOLIPASE A1

SER-VAL-ALA-LYS-LEU-GLU-LYS
PEPTIDYL-DIPEPTIDASE A *I*50 = 82 μM

N-(D,L-SERYL)-N′-(2,3,4-TRIHYDROXYBENZYL)HYDRAZINE
AROMATIC-L-AMINO-ACID DECARBOXYLASE IR

(+)SESAMIN
Δ5 DESATURASE *K*i = 155 μM
Δ5 DESATURASE 87% 28 μM

SESAMIN
Δ5 DESATURASE *K*i = 155 μM

SESAMINOL
Δ5 DESATURASE 78% 28 μM

SESAMOL
Δ5 DESATURASE 21% 28 μM

SESAMOLIN
Δ5 DESATURASE 81% 28 μM

SESELIN
CYTOCHROME P450 1A1 *I*50 = 67 μM
CYTOCHROME P450 2B1 *I*50 = 7.7 μM

SET		
PHOSPHOPROTEIN PHOSPHATASE 2A		$I50 = 2$ nM
SETHOXYDIM		
ACETYL-CoA CARBOXYLASE	NC	$Ki = 1.9$ μM
ACETYL-CoA CARBOXYLASE		$I50 = 6$ μM
ACETYL-CoA CARBOXYLASE		$I50 = 5.6$ μM
PROPIONYL-CoA CARBOXYLASE		$I50 = 3.5$ μM
SETRALINE		
CYTOCHROME P450 1A2		$I50 = 70$ μM
CYTOCHROME P450 2D6		$Ki = 1.5$ μM
SEZOLAMIDE		
CARBONATE DEHYDRATASE II		$I50 = 540$ pM
SF 2370		
PROTEIN KINASE C		$I50 = 92$ nM
PROTEIN KINASE C		$I50 = 240$ nM
SF 86327		
SQUALENE MONOOXYGENASE	NC	$Ki = 30$ nM
SFK 525A		
ALDEHYDE OXIDASE		
SG 210		
ALDEHYDE REDUCTASE		$I50 = 9.5$ nM
SH 489		
AROMATASE		$I50 = 700$ nM
SaH 57-118		
STEROL O-ACYLTRANSFERASE		$I50 = 5.8$ μM
SaH 58-035		
STEROL O-ACYLTRANSFERASE		$I50 = 57$ nM
SHAARGROCKOL B		
RNA-DIRECTED DNA POLYMERASE		$I50 = 8.5$ μM
SHAARGROCKOL C		
RNA-DIRECTED DNA POLYMERASE		$I50 = 3.3$ μM
SH-COMPOUNDS		
ALANINE-tRNA LIGASE		
SHEEP BRAIN PROTEIN KINASE C INHIBITORY PROTEIN		
PROTEIN KINASE C		$I50 = 1.7$ μM
SHEEP BRAIN PROTEIN KINASE INHIBITOR 1 (KCIP-1)		
PROTEIN KINASE C		
SHEEP BRAIN PROTEIN KINASE INHIBITOR 2 (KCIP-2)		
PROTEIN KINASE C		
SHEEP LUNG PROTEINASE INHIBITOR		
ELASTASE LEUKOCYTE		
ELASTASE PANCREATIC		
SHEEP PLACENTA AROMATASE INHIBITOR		
AROMATASE		
SHERMILAMINE B		
DNA TOPOISOMERASE (ATP-HYDROLYSING)		$I90 = 118$ μM
SHERMILAMINE C		
DNA TOPOISOMERASE (ATP-HYDROLYSING)		$I90 = 138$ μM

SHIKIMATE-3-PHOSPHATE
SHIKIMATE KINASE

SHIKIMATE-5-PHOSPHATE
SHIKIMATE KINASE

SHIKIMIC ACID
PHOSPHOENOLPYRUVATE CARBOXYLASE C $Ki = 22\ \mu M$

SHIKONIN
DNA TOPOISOMERASE $I50 = 208\ \mu M$

SHISTOSOMA MANSONI PROTEINASE INHIBITOR
ELASTASE LEUKOCYTE
ELASTASE PANCREATIC
PROTEINASE SHISTOSOMA MANSONI 28KD

SHORT PEPTIDES FROM THE TERMINAL SEGMENT OF HIV PROTEINASE
PROTEINASE HIV-1

SHOWDOMYCIN
1,3-β-GLUCAN SYNTHASE
UDP-N-ACETYLGLUCOSAMINE-DOLICHYL-PHOSPHATE N-ACETYLGLUCOSAMINEPHOSPHOTRANSFERA

SH REAGENTS
PROTEINASE CATHEPSIN S LIKE

SIALIC ACID
NEURAMINIDASE $I50 = 2$ mM

SIALOSYLPARAGLOBOSIDE
DNA POLYMERASE α $I50 = 50\ \mu M$

SIASTAIN B
EXO-α-SIALIDASE $I50 = 150\ \mu M$
β-GLUCURONIDASE $I50 = 39\ \mu M$

SIASTATIN
SIALIDASE $I50 = 3\ \mu g/ml$

SIASTATIN B
EXO-α-SIALIDASE $I50 = 3\ \mu g/ml$
EXO-α-SIALIDASE $I50 = 6.3\ \mu g/ml$
EXO-α-SIALIDASE $I50 = 50\ \mu g/ml$
β-GLUCURONIDASE $I50 = 16\ \mu g/ml$

SIC 1
CYCLIN DEPENDENT KINASE CDC28/CLB $Ki < 1$ nM

SICOS AUSTRALIS TRYPSIN INHIBITOR I
TRYPSIN

SICOS AUSTRALIS TRYPSIN INHIBITOR II
TRYPSIN

SICOS AUSTRALIS TRYPSIN INHIBITOR III
TRYPSIN

SIGUAZODAN
3′,5′-CYCLIC-NUCLEOTIDE PHOSPHODIESTERASE III $I50 = 800$ nM

SILDENAFIL
3′,5′-CYCLIC-NUCLEOTIDE PHOSPHODIESTERASE I $I50 = 260$ nM
3′,5′-CYCLIC-NUCLEOTIDE PHOSPHODIESTERASE III $I50 = 65\ \mu M$
3′,5′-CYCLIC-NUCLEOTIDE PHOSPHODIESTERASE V $I50 = 3.6$ nM

SILIBINI-GDIHEMISUCCINATE
ADENOSINE DEAMINASE $I50 = 50\ \mu g/ml$

Inhibitor / Enzyme	Type	Value
SILIBININ		
XANTHINE DEHYDROGENASE		
XANTHINE OXIDASE		
SILICOTUNGSTIC ACID		
RNA-DIRECTED DNA POLYMERASE		
SILK WORM CHYMOTRYPSIN INHIBITOR		
α-CHYMOTRYPSIN		Kd < 1 nM
SILVEX		
LONG-CHAIN-FATTY-ACID CoA-LIGASE	MIX	*K*i = 2.9 mM
SILYBIN		
β-GLUCURONIDASE		*I*50 = 120 μg/ml
HYDROXYMETHYLGLUTARYL-CoA REDUCTASE (NADPH)		
SILYCHRISTIN		
TRYPSIN		*K*i = 40 μM
SIMVASTATIN		
CYTOCHROME P450 2C9		*I*50 = 35 μM
HYDROXYMETHYLGLUTARYL-CoA REDUCTASE (NADPH)	C	*I*50 = 940 pM
HYDROXYMETHYLGLUTARYL-CoA REDUCTASE (NADPH)		*I*50 = 3 nM
17-OXOSTEROID OXIDOREDUCTASE		
STEROL O-ACYLTRANSFERASE	C	*K*i = 12 μM
SINAPIC ACID		
ADENOSINE DEAMINASE		*I*50 = 60 μg/ml
CELLULASE		92% 830 μM
SINAPSIS ARVENIS TRYPSIN INHIBITOR		
TRYPSIN		*K*i = 7 μM
SINEFUNGIN		
1-AMINOCYCLOPROPANE-1-CARBOXYLATE SYNTHASE		
CALMODULIN-LYSINE N-METHYLTRANSFERASE		*I*50 = 41 μM
CATECHOL O-METHYLTRANSFERASE		*I*50 = 800 μM
CYCLOPROPANE-FATTY-ACYL-PHOSPHOLIPID SYNTHASE		
DEMETHYLMACROCIN O-METHYLTRANSFERASE		
ECO RI ADENINE DNA METHYLASE		*K*i = 10 nM
GUANIDINOACETATE N-METHYLTRRANSFERASE		
HISTAMINE N-METHYLTRANSFERASE		*I*50 = 30 μM
HISTONE-LYSINE N-METHYLTRANSFERASE		*K*i = 3.5 μM
MACROCIN O-METHYLTRANSFERASE		
MAGNESIUM PROTOPORPHYRIN O-METHYLTRANSFERASE		*I*50 = 500 nM
METHYLTHIOADENOSINE NUCLEOSIDASE	C	*K*i = 122 μM
PHENYLETHANOLAMINE N-METHYLTRANSFERASE		*I*50 = 50 μM
PROTEIN-ARGININE N-METHYLTRANSFERASE		*K*i = 3.5 μM
PROTEIN-GLUTAMATE METHYLTRANSFERASE		*K*i = 220 nM
PROTEIN-GLUTAMATE O-METHYLTRANSFERASE		
mRNA(GUANINE-N7-)-METHYLTRANSFERASE		*K*i = 150 nM
mRNA(NUCLEOSIDE-2′O-)-METHYLTRANSFERASE		*K*i = 75 nM
SPERMIDINE SYNTHASE		*I*50 = 100 μM
SPERMIDINE SYNTHASE		20% 1 mM
SPERMINE SYNTHASE		82% 1 mM
THIOETHER S-METHYLTRANSFERASE		
SINENSETIN		
ARACHIDONATE 15-LIPOXYGENASE 1		*I*50 = 114 μM
β-SITOSTEORL		
CYCLOOXYGENASE		*I*50 = 24 mM
SITOSTEROL		
11β-HYDROXYSTEROID DEHYDROGENASE		*I*50 = 1.4 mM
STEROL O-ACYLTRANSFERASE		

STEROL O-ACYLTRANSFERASE		
Δ(24)-STEROL C-METHYLTRANSFERASE	C	*K*i = 26 μM
β-SITOSTEROL-3-O-β-D-GLUCOPYRANOSIDE		
PROLYL OLIGOPEPTIDASE		*I*50 = 28 ppm
β-SITOSTEROL-β-D-GLYCOSIDE		
XANTHINE OXIDASE		46% 100 μM
SITOSTYROL GLUCOSIDE		
STERYL-β-GLUCOSIDASE		
SJA 6017		
CALPAIN II		*I*50 = 80 nM
SKATOLE		
TRYPTOPHAN 2′-DIOXYGENASE		
SKB 207499		
3′,5′-CYCLIC-NUCLEOTIDE PHOSPHODIESTERASE IVa		*I*50 = 149 nM
3′,5′-CYCLIC-NUCLEOTIDE PHOSPHODIESTERASE IVb		*I*50 = 117 nM
3′,5′-CYCLIC-NUCLEOTIDE PHOSPHODIESTERASE IVc		*I*50 = 612 nM
„3′,5′-CYCLIC-NUCLEOTIDE PHOSPHODIESTERASE IVd		*I*50 = 39 nM
SKELETAL MUSCLE AMP-PROTEIN KINASE INHIBITOR		
PROTEIN KINASE AMP STIMULATED		
SKF 525		
PHENOL O-METHYLTRANSFERASE		
SKF 102698		
DOPAMINE β-MONOOXYGENASE		*K*i = 40 nM
SKF 104864		
DNA TOPOISOMERASE		*I*50 = 56 nM
SKF 10497		
LANOSTEROL C14 DEMETHYLASE		*I*50 = 2 nM
SKF 105561		
ARACHIDONATE 5-LIPOXYGENASE		*I*50 = 3 μM
PROSTAGLANDIN-H SYNTHASE		*I*50 = 100 μM
SKF 105657		
5α-REDUCTASE 1		*I*50 = 7.5 μM
5α-REDUCTASE 2		*I*50 = 8 nM
STEROID 5α-REDUCTASE		*I*50 = 350 nM
STEROID 5α-REDUCTASE		*I*50 = 700 pM
STEROID 5α-REDUCTASE 1		*K*i = 425 nM
STEROID 5α-REDUCTASE 2		*K*i = 1 nM
SKF 107457		
PROTEINASE HIV-1		*K*i = 18 nM
PROTEINASE SIMIAN IMMUNODEFICIENCY VIRUS		
SKF 107461		
PROTEINASE HIV-1		
SKF 108361		
PROTEINASE HIV-1		*K*i = 800 nM
SKF 108738		
PROTEINASE HIV-1		*K*i = 1 nM
SKF 29661		
PHENYLETHANOLAMINE N-METHYLTRANSFERASE		*K*i = 550 nM
SKF 36914		
INTERSTITIAL COLLAGENASE	NC	*K*i = 48 μM

SKF 38393

PHOSPHOPROTEIN PHOSPHATASE 2A		*I*50 = 50 µM
TRYPTOPHAN 5-MONOOXYGENASE		*I*50 = 4.4 µM

SKF 45905

ARACHIDONATE 5-LIPOXYGENASE		*I*50 ≈ 3 µM
PHOSPHOLIPASE A2 (85KD)		*I*50 = 3 µM
TRANSACYLASE (COA INDEPENDENT)		*I*50 = 6 µM

SKF 64139

PHENYLETHANOLAMINE N-METHYLTRANSFERASE		*K*i = 200 nM
PHENYLETHANOLAMINE N-METHYLTRANSFERASE		*I*50 = 10 nM

SKF 7698

PHENYLETHANOLAMINE N-METHYLTRANSFERASE		

SKF 80544

INTERSTITIAL COLLAGENASE	NC	*K*i = 1.5 mM

SKF 86607

PHENYLETHANOLAMINE N-METHYLTRANSFERASE		*I*50 = 33 nM

SKF 91488

HISTAMINE N-METHYLTRANSFERASE		
HISTAMINE N-METHYLTRANSFERASE		

SKF 94120

3′,5′-CYCLIC-NUCLEOTIDE PHOSPHODIESTERASE (cycGMP STIMULATED)		*I*50 = 558 µM
3′,5′-CYCLIC-NUCLEOTIDE PHOSPHODIESTERASE (cycGMP STIMULATED)		*I*50 = 643 µM
3′,5′-CYCLIC-NUCLEOTIDE PHOSPHODIESTERASE III		*I*50 = 12 µM
3′,5′-CYCLIC-NUCLEOTIDE PHOSPHODIESTERASE III		*I*50 = 11 µM
3′,5′-CYCLIC-NUCLEOTIDE PHOSPHODIESTERASE III		*I*50 = 14 µM
3′,5′-CYCLIC-NUCLEOTIDE PHOSPHODIESTERASE III (cycAMP)		*I*50 = 3.7 µM
3′,5′-CYCLIC-NUCLEOTIDE PHOSPHODIESTERASE III (cycAMP)		*I*50 = 11 µM

SKF 94836

3′,5′-CYCLIC-NUCLEOTIDE PHOSPHODIESTERASE (cycAMP) LOW Km		*K*i = 3 µM
3′,5′-CYCLIC-NUCLEOTIDE PHOSPHODIESTERASE Ia		*I*50 = 31 µM
3′,5′-CYCLIC-NUCLEOTIDE PHOSPHODIESTERASE III		*I*50 = 1.4 µM

SKF 95601

H^+/K^+-EXCHANGING ATPASE		

SKF 95654

3′,5′-CYCLIC-NUCLEOTIDE PHOSPHODIESTERASE III		*I*50 = 700 nM

SKF 96067

H^+/K^+-EXCHANGING ATPASE		*K*i = 1 µM
H^+/K^+-EXCHANGING ATPASE	C	*K*i = 390 nM
Na^+/K^+-EXCHANGING ATPASE		*I*50 = 60 µM

SKF 96356

H^+/K^+-EXCHANGING ATPASE	C	*K*i = 71 nM

SKF 97574

H^+/K^+-EXCHANGING ATPASE		*K*i = 460 nM

SKF 98625

ARACHIDONATE 5-LIPOXYGENASE		*I*50 ≈ 3 µM
TRANSACYLASE (COA INDEPENDENT)		*I*50 = 9 µM

SKF 525A

ALDEHYDE OXIDASE		*I*50 = 2.6 µM
ALKANAL MONOOXYGENASE (FMN -LINKED)		
BENZENE 1,2-DIOXYGENASE		85% 1 mM
BENZOATE 4-MONOOXYGENASE		
CYTOCHROME P450 1A		
CYTOCHROME P450 2A		

CYTOCHROME P450 3A	
CYTOCHROME P450 3A1/2	$I50 = 130\ \mu M$
CYTOCHROME P450 2B	
CYTOCHROME P450 2B1	$I50 = 110\ \mu M$
CYTOCHROME P450 2C	
CYTOCHROME P450 2C11	$I50 = 180\ \mu M$
CYTOCHROME P450 2C19	
CYTOCHROME P450 2D	
CYTOCHROME P450 2E	
ECDYSONE 20-MONOOXYGENASE	
ESTRADIOL 2-HYDROXYLASE	
ESTRADIOL 16α-HYDROXYLASE	
ISOPENTENYL-DIPHOSPHATE Δ-ISOMERASE	98% 1 mM
(–)-LIMONENE 3-MONOOXYGENASE	
(–)-LIMONENE 6-MONOOXYGENASE	
(–)-LIMONENE 7-MONOOXYGENASE	
MENTHOL MONOOXYGENASE	
PROGESTERONE 11α-MONOOXYGENASE	
THIOL S-METHYLTRANSFERASE	
THIOPURINE S-METHYLTRANSFERASE	

SKF 3301A

ISOPENTENYL-DIPHOSPHATE Δ-ISOMERASE	99% 1 mM

(R)-SKF 95654

3′,5′-CYCLIC-NUCLEOTIDE PHOSPHODIESTERASE III	$I50 = 400$ nM

SKIN ELASTASE INHIBITOR

ELASTASE LEUKOCYTE	$Ki = 20$ pM

SLPI

SUBTILISIN BPN	$Ki = 30$ pM
TRYPSIN	$Ki = 690$ pM

Sm^{3+}

PROTEIN KINASE (Ca^{2+} DEPENDENT)

SM 10888

ACETYLCHOLINESTERASE
CHOLINESTERASE

Sn^{2+}

ACETOLACTATE DECARBOXYLASE	70% 400 μM
ALDEHYDE DEHYDROGENASE (NADP)	
L-AMINOADIPATE-SEMIALDEHYDE DEHYDROGENASE	
5-AMINOPENTANAMIDASE	100% 5 mM
α-L-ARABINOFURANOSIDASE	46% 1 mM
ARACHIDONATE 12-LIPOXYGENASE	
ARGININE DEIMINASE	82% 1 mM
ARYLACETONITRILASE	87% 1 mM
ASPARAGINE SYNTHASE (GLUTAMINE-HYDROLYSING)	
CATECHOL OXIDASE	
CELLULASE	
CERAMIDASE	
CEREBROSIDE-SULFATASE	100% 20 mM
CHITINASE	
DIHYDROFOLATE REDUCTASE	
DIHYDROPYRIMIDINASE	100% 10 mM
DIMETHYLHISTIDINE N-METHYLTRANSFERASE	
GLUCONATE 5-DEHYDROGENASE	
α-GLUCOSIDASE	
GLYCEROL-3-PHOSPHATE CYTIDILYLTRANSFERASE	
GTP CYCLOHYDROLASE I	
2-HYDROXYGLUTARATE DEHYDROGENASE	
3-HYDROXYMETHYLCEPHEM CARBAMOYLTRANSFERASE	
KYNURENINE-OXOGLUTARATE AMINOTRANSFERASE	

MANNITOL 2-DEHYDROGENASE (NADP)
NITRILASE 64% 1 1mM
4-NITROPHENOL 2-MONOOXYGENASE
OLIGO-1,6-GLUCOSIDASE
PECTINESTERASE
3-PHOSPHOGLYCERATE PHOSPHATASE
PHOSPHOLIPASE D 32% 1 mM
POLYGALACTURONASE
PORPHOBILINOGEN SYNTHASE
SACCHAROPINE DEHYDROGENASE (NAD, L-LYSINE FORMING)
TRIACYLGLYCEROL LIPASE
TRIACYLGLYCEROL LIPASE ALKALINE
XYLOSE ISOMERASE

SN 38
DNA TOPOISOMERASE *I*50 = 400 nM

SN 18071
DNA TOPOISOMERASE *I*50 = 150 μM

SN 5949
UBIQUINOL-CYTOCHROME-C REDUCTASE

SN 6131
DNA TOPOISOMERASE *I*50 = 50 μM

SN 6132
DNA POLYMERASE I
DNA TOPOISOMERASE *I*50 = 46 μM
RNA POLYMERASE

SN 6999
DNA POLYMERASE I
DNA TOPOISOMERASE *I*50 = 23 μM
RNA POLYMERASE

SNA 60-367-2
AROMATASE *I*50 = 63 μM

SNA 8073 B
PROLYL OLIGOPEPTIDASE *I*50 = 8.9 μM
PROLYL OLIGOPEPTIDASE *I*50 = 14 μM

SNA 8073B
PROLYL OLIGOPEPTIDASE *I*50 = 8.9 μM

SNK 860
ALDEHYDE REDUCTASE *I*50 = 35 nM

S_2O_3-
SULFATE ADENYLYLTRANSFERASE *I*50 = 1 mM

SODIUM(6R)-6-ACETONYLPENICILLANATE S,S-DIOXIDE
β-LACTAMASE *I*50 = 1.4 μM
β-LACTAMASE *I*50 = 13 nM
β-LACTAMASE *I*50 = 24 μM

SODIUM BISULPHITE
XYLOSE REDUCTASE *K*i = 39 mM

SODIUM BOROHYDRIDE
ADENOSYLMETHIONINE DECARBOXYLASE
DIAMINOBUTYRATE-PYRUVATE TRANSAMINASE
FORMALDEHYDE DISMUTASE
PHOSPHOPANTOTHENOYLCYSTEINE DECARBOXYLASE

SODIUM CHLORIDE
N-ACETYLGALACTOSAMINE-6-SULFATASE 44% 20 mM

ELASTASE PANCREATIC		$Ki = 70$ mM
β-GALACTOSIDE α-2,3-SIALYLTRANSFERASE		80% 200 mM
GLUCURONATE 2-SULFATASE		$I50 = 75$ mM
ISOCITRATE DEHYDROGENASE (NAD)		> 50 mM
α-MANNOSIDE β1-2-N-ACETYLGLUCOSAMINYLTRANSFERASE		$I50 = 100$ mM
MANNOTETRAOSE 2-α-N-ACETYLGLUCOSAMINYLTRANSFERASE		$I50 = 300$ mM
RIBONUCLEASE V		53% 150 mM
SODIUM CITRATE		
LYSOZYME		> 100 mM
SODIUM CYANOBOROHYDRIDE		
ADENOSYLMETHIONINE DECARBOXYLASE		
SODIUMDIHYDROGENPHOSPHATE		
GLUCURONATE 2-SULFATASE		$I50 = 250$ μM
SODIUM DOCOSANEDIOATE		
PROTEINASE HIV-1		$I50 = 12$ μM
SODIUM PENTOSAN POLYSULPHATE		
STROMELYSIN		
SOLANACEOUS GLYCOALKALOIDE		
ACETYLCHOLINESTERASE		
CHOLINESTERASE		
SOLANESYL DIPHOSPHATE		
trans-OCTAPRENYLTRANSFERASE		
α-SOLANINE		
CHOLINESTERASE		
SOLANUM TUBEROSUM PECTIN METHYLESTERASE INHIBITOR		
PECTINESTERASE		
SOLANUM TUBEROSUM TRYPSIN/CHYMOTRYPSIN INHIBITOR		
CHYMOTRYPSIN		
TRYPSIN		
SOLUFLAZINE		
ADENOSINE TRANSPORT		$I50 = 42$ nM
NUCLEOSIDE TRANSPORT		$I50 = 315$ nM
NUCLEOSIDE TRANSPORT		$I50 = 11$ μM
SOMAN		
ACETYLCHOLINESTERASE		
CARBOXYLESTERASE		
CHOLINESTERASE		
SOMATOSTATIN		
PROTEIN TYROSINE KINASE		
SORANGICIN A		
RNA POLYMERASE		
SORBINIL		
ALCOHOL DEHYDROGENASE (NADP)		$I50 = 1.8$ μM
ALDEHYDE REDUCTASE		$I50 = 36$ nM
ALDEHYDE REDUCTASE		$I50 = 120$ nM
ALDEHYDE REDUCTASE	NC	$Ki = 140$ μM
ALDEHYDE REDUCTASE		$I50 = 1.3$ μM
ALDEHYDE REDUCTASE 2		$I50 = 110$ μM
4-AMINOBUTYRATE TRANSAMINASE	NC	
trans-1,2-DIHYDROBENZENE-1,2-DIOL DEHYDROGENASE		
trans-1,2-DIHYDROBENZENE-1,2-DIOL DEHYDROGENASE 2		44% 10 μM
trans-1,2-DIHYDROBENZENE-1,2-DIOL DEHYDROGENASE 3		63% 10 μM
trans-1,2-DIHYDROBENZENE-1,2-DIOL DEHYDROGENASE 4		63% 10 μM

Inhibitor / Enzyme	Type	Value
GLUCURONATE REDUCTASE		
SUCCINATE-SEMIALDEHYDE DEHYDROGENASE	NC	
SORBITOL		
GLUCOSE-6-PHOSPHATE ISOMERASE		
L-IDITOL 2-DEHYDROGENASE		
MANNONATE DEHYDRATASE		> 49 mM
THIOGLUCOSIDASE	C	*K*i = 9 mM
α,α-TREHALASE		
XYLOSE ISOMERASE	C	
D-SORBITOL		
MANNONATE DEHYDRATASE	C	*K*i = 9 mM
XYLOSE ISOMERASE		
SORBITOL-6-PHOSPHATE		
GLUCOSE-6-PHOSPHATE ISOMERASE		*K*i = 25 μM
MANNOSE-6-PHOSPHATE ISOMERASE		*K*i = 2.6 mM
SORBITOL-6-PHOSPHATE 2-DEHYDROGENASE		
D-SORBITOL-6-PHOSPHATE		
myo-INOSITOL-1-PHOSPHATE SYNTHASE		
SORBOSE		
1,3-β-GLUCAN SYNTHASE	U	*K*i = 70 mM
L-SORBOSE 1,6-DIPHOSPHATE		
FRUCTOSE-BISPHOSPHATE ALDOLASE	C	*K*i = 130 μM
L-SORBOSE-1,6-DIPHOSPHATE		
FRUCTOSE-BISPHOSPHATE ALDOLASE	C	*K*i = 44 μM
FRUCTOSE-BISPHOSPHATE ALDOLASE		*K*i = 130 μM
L-SORBOSE		
FRUCTOKINASE		
KETOHEXOKINASE		
MANNOKINASE		
SUCROSE PHOSPHORYLASE	C	*K*i = 170 mM
SUCROSE PHOSPHORYLASE	C	*K*i = 67 mM
SORBOSE-6-PHOSPHATE		
FRUCTOSE-2,6-BISPHOSPHATASE		
L-SORBOSE 1-PHOSPHATE		
FRUCTOSE-BISPHOSPHATE ALDOLASE	C	*K*i = 200 μM
SORBYL DITHRANOL		
GLUCOSE-6-PHOSPHATE 1-DEHYDROGENASE		
SORGHUM BICOLOR CYSTEINE PROTEINASE INHIBITOR		
PAPAIN		
D.SORITOL		
MANNONATE DEHYDRATASE	C	*K*i = 9 mM
D,L-SOTALOL		
ACETYLCHOLINESTERASE		*K*i = 310 μM
CHOLINESTERASE		*K*i = 510 μM
SOY BEAN BOWMAN BIRK INHIBITOR		
ELASTASE		71% 2 mg/ml
TRYPSIN		85% 2 mg/ml
SOY BEAN CYSTEIN EPROTEINASE INHIBITOR		
PROTEINASES CYSTEINE		
SOY BEAN CYSTEINE PROTEINASE INHIBITORS		
BROMELAIN		
FICAIN		
PAPAIN		

SOY BEAN INHIBITOR
ELASTASE LEUKOCYTE

SOY BEAN LOW MOLECULAR COMPONENT
5-AMINOLEVULINATE DEHYDROGENASE

SOY BEAN PHOSPHATASE INHIBITOR
PHOSPHATASE ACID

SOY BEAN POLYGALACTURONASE INHIBITOR
POLYGALACTURONASE

SOY BEAN PROTEINASE INHIBITOR BOWMAN BIRK

CATHEPSIN G		$Ki = 1.2$ nM
CHYMASE ATYPICAL		$Ki = 410$ nM
CHYMOTRYPSIN		
ELASTASE LEUKOCYTE		$Ki = 2$ nM
TRYPSIN		$Ki = 1.6$ nM

SOY BEAN PROTEINASE INHIBITOR KUNITZ

TRYPSIN		$Ki = 600$ pM

SOY BEAN TRYPSIN INHIBITOR

ACROSIN		
CATHEPSIN G		
CATHEPSIN R		
CHYMASE		$I50 = 34$ nM
CHYMASE		
CHYMOTRYPSIN		
α-CHYMOTRYPSIN		$Ki = 500$ nM
CLOSTRIPAIN		
ELASTASE PANCREATIC II		90% 100 μg/ml
ENDOPEPTIDASE BRACHIONUS PLICATILIS		100% 2.5 μM
HURAIN		
KALLIKREIN PLASMA		$Ki = 4$ nM
KALLIKREIN TISSUE	NC	$Ki = 52$ μM
LEUKOCYTE-MEMBRANE NEUTRAL ENDOPEPTIDASE		97% 1 mg/ml
MYELOBLASTIN		
PLASMIN		$Ki = 370$ nM
PLASMIN		
PLASMIN		$Ki = 2.1$ nM
PROTEINASE ALKALINE		
PROTEINASE BOAR EPIDIDYMAL SPERM		100% 1 mM
PROTEINASE DF		$Ki = 3$ nM
PROTEINASE SERINE PROMONOCYTOC U-937		100% 1 μM
PROTEINASE STARFISH SPERM ACROSOME		
SEPIA PROTEINASE		
TONIN		$Ki = 13$ μM
TRYPSIN		$Ki = 10$ nM
TRYPSIN		$Ki = 350$ nM
TRYPSIN LIKE ACTIVITY		42% 125 ng/ml
TRYPTASE		54% 250 μg/ml
YEAST PROTEINASE B		36% 100 μg/ml

SOY BEAN TRYPSIN INHIUBITOR
UCA PUGILATOR COLLAGENOLYTIC PROTEINASE

SOY CYSTATIN
PAPAIN

SOY PROTEINASE INHIBITOR 2-II
CATHEPSIN G
CHYMOTRYPSIN
ELASTASE LEUKOCYTE
TRYPSIN

SP 346
PROTEINASE EIAV — $I50 = 1.7\ \mu M$
PROTEINASE HIV-1 — $I50 = 1.4\ \mu M$
PROTEINASE HIV-2 — $I50 = 600$ nM

SP 88226
PROTEINASE SERINE PROMONOCYTOC U-937 — 60% 110 nM

SPAI
Na^+/K^+-EXCHANGING ATPASE — C — $I50 = 12\ \mu M$

SPAI-1
Na^+/K^+-EXCHANGING ATPASE — $I50 = 18\ \mu M$
Na^+/K^+-EXCHANGING ATPASE — $I50 = 10\ \mu M$
Na^+/K^+-EXCHANGING ATPASE — $I50 = 18\ \mu M$
Na^+/K^+-EXCHANGING ATPASE — $I50 = 12\ \mu M$
Na^+/K^+-EXCHANGING ATPASE — $I50 = 22\ \mu M$

SPAI-2
Na^+/K^+-EXCHANGING ATPASE

SPAI-3
Na^+/K^+-EXCHANGING ATPASE

SPANDS
PROTEINASE HIV-1 — $I50 = 110\ \mu M$

SPARFLOXACIN
DNA GYRASE — $I50 = 30\ \mu g/ml$
DNA TOPOISOMERASE (ATP-HYDROLYSING)
DNA TOPOISOMERASE IV — $I50 = 7.4\ \mu g/ml$

SPARSOMYCIN
PEPTIDYLTRANSFERASE

SPERM ACROSIN INHIBITOR
ACROSIN

SPERMIDINE
ACETYLSPERMIDINE DEACETYLASE — C — $Ki = 55\ \mu M$
ACETYLSPERMIDINE DEACETYLASE — $Ki = 55\ \mu M$
ADENYLATE CYCLASE — 72% 10 μM
β-ADRENERGIC-RECEPTOR KINASE — $I50 = 2.6$ mM
AMINO-ACID N-ACETYLTRANSFERASE
ARGININE DECARBOXYLASE — C — $Ki = 7.2$ mM
ARGININE DEIMINASE — UC — $Ki = 4.3\ \mu M$
ARGINYLTRANSFERASE — C
CALF THYMUS RIBONUCLEASE H
CALPAIN II — $I50 = 50$ mM
CARBAMOYL-PHOSPHATE SYNTHASE (GLUTAMINE-HYDROLYSING)
CYTOSINE-DNA METHYLASE HaeI — 42% 20 mM
CYTOSINE-DNA METHYLASE HaeIII — 74% 20 mM
CYTOSINE-DNA METHYLASE HpaII — 78% 20 mM
CYTOSINE-DNA METHYLASE SssI — 88% 20 mM
DIHYDROFOLATE REDUCTASE — > 12 mM
DNA (CYTOSINE-5-)-METHYLTRANSFERASE
HISTONE ACYLTANSFERASE
LYSINE 2-MONOOXYGENASE
METHIONINE ADENOSYLTRANSFERASE — 30% 10 mM
NITRIC-OXIDE SYNTHASE — $I50 = 140\ \mu M$
4-NITROPHENYLPHOSPHATASE
ORNITHINE DECARBOXYLASE
PEPTIDASE ALKALINE — $I50 = 9.1$ mM
PHOSPHOLIPASE C
POLYNUCLEOTIDE ADENYLYLTRANSFERASE
PROTEIN-GLUTAMINE γ-GLUTAMYLTRANSFERASE
PROTEIN KINASE CASEIN KINASE N42

PROTEIN TYROSINE KINASE I		$I50 = 2.5$ mM
PROTEIN TYROSINE KINASE II		$I50 = 4.5$ mM
PROTEIN TYROSINE KINASE (Zn^{2+} DEPENDENT)		40% 5 mM
PROTEIN TYROSINE-PHOSPHATASE		
PUTRESCINE OXIDASE		
RIBONUCLEASE U2		
tRNA(ADENINE-N^1-) METHYLTRANSFERASE		
tRNA ADENYLYLTRANSFERASE		
tRNA CYTIDYLTRANSFERASE		100% 1 mM
tRNA(CYTOSINE-5-) METHYLTRANSFERASE		
[RNA-POLYMERASE]SUBUNIT KINASE		
tRNA(URACIL-5-) METHYLTRANSFERASE		
SPERMIDINE SYNTHASE		

SPERMIDINEPANTOTHEINE 4′-PHOSPHATE

PHOSPHOPANTOTHENOYLCYSTEINE DECARBOXYLASE	C	$Ki = 430$ μM

SPERMINE

ACETYLSPERMIDINE DEACETYLASE	C	$Ki = 36$ μM
ACETYLSPERMIDINE DEACETYLASE		$Ki = 36$ μM
ADENOSYLMETHIONINE DECARBOXYLASE	NC	$Ki = 500$ μM
ADENYLATE CYCLASE		50% 10 μM
β-ADRENERGIC-RECEPTOR KINASE		$I50 = 990$ μM
ARGININE DECARBOXYLASE	C	$Ki = 4.4$ mM
ARGINYLTRANSFERASE	C	
BIS(5′-ADENOSYL)-TRIPHOSPHATASE		
CALF THYMUS RIBONUCLEASE H		
CALPAIN II		$I50 = 25$ mM
CARBAMOYL-PHOSPHATE SYNTHASE (GLUTAMINE-HYDROLYSING)		
Ca^{2+}-TRANSPORTING ATPASE		
3′,5′-CYCLIC-NUCLEOTIDE PHOSPHODIESTERASE		
3′,5′-CYCLIC-NUCLEOTIDE PHOSPHODIESTERASE (cycAMP) (Ca^{2+}/CALMODULIN)	NC	$Ki = 1.1$ mM
DIHYDROFOLATE REDUCTASE		> 12 mM
DNA (CYTOSINE-5-)-METHYLTRANSFERASE		
DNA POLYMERASE		$I50 = 250$ μM
GLYCOGEN SYNTHASE KINASE 3		$I50 = 700$ μM
INOSITOL-1,3,4-TRISPHOSPHATE KINASE	C	$Ki = 12$ mM
INOSITOL-1,4,5-TRISPHOSPHATE 5-PHOSPHATASE		
ISOLEUCINE-tRNA LIGASE		
NITRIC-OXIDE SYNTHASE		$I50 = 56$ μM
4-NITROPHENYLPHOSPHATASE		50% 30 μM
ORNITHINE DECARBOXYLASE		
PEPTIDASE ALKALINE		$I50 = 8.5$ mM
PEPTIDYLTRANSFERASE		$Ki = 84$ μM
PHOSPHODIESTERASE		
PHOSPHOLIPASE C		
PHOSPHOPROTEIN PHOSPHATASE		
POLYNUCLEOTIDE ADENYLYLTRANSFERASE		
POLYNUCLEOTIDE ADENYLYLTRANSFERASE		
PROTEIN KINASE A		40% 2 mM
PROTEIN KINASE C		
PROTEIN KINASE C		22% 500 μM
PROTEIN KINASE CASEIN KINASE N42		
PROTEIN TYROSINE KINASE I		$I50 = 350$ nM
PROTEIN TYROSINE KINASE II		$I50 = 430$ nM
PROTEIN TYROSINE KINASE (Zn^{2+} DEPENDENT)		70% 5 mM
PROTEIN TYROSINE-PHOSPHATASE CD45		93% 2 mM
PUTRESCINE OXIDASE		
RIBONUCLEASE U2		
tRNA ADENYLYLTRANSFERASE		
tRNA CYTIDYLTRANSFERASE		100% 1 mM
tRNA(CYTOSINE-5-) METHYLTRANSFERASE		
RNA-DIRECTED RNA POLYMERASE		
tRNA(GUANINE-N^1-) METHYLTRANSFERASE		

SPHINGANINE
2-ACYLGLYCEROL O-ACYLTRANSFERASE
PHOSPHATIDATE PHOSPHATASE
PHOSPHOLIPASE A2

D-(+)-threo-SPHINGANINE
SPHINGANINE KINASE

L-(−)-erythro-SPHINGANINE
SPHINGANINE KINASE

L-(−)-threo-SPHINGANINE
SPHINGANINE KINASE

SPHINGOFUNGIN B
SERINE C-PALMITOYLTRANSFERASE *I*50 = 1 ng/ml
SERINE C-PALMITOYLTRANSFERASE *I*50 = 20 nM
SERINE C-PALMITOYLTRANSFERASE *I*50 = 2.5 nM

SPHINGOFUNGIN C
SERINE C-PALMITOYLTRANSFERASE *I*50 = 100 nM

SPHINGOFUNGIN E
SERINE C-PALMITOYLTRANSFERASE *I*50 = 7.2 nM

SPHINGOFUNGIN F
SERINE C-PALMITOYLTRANSFERASE *I*50 = 57 nM

SPHINGOFUNGINS
SERINE C-PALMITOYLTRANSFERASE

SPHINGOMYELIN
LECITHIN-CHOLESTEROL ACYLTRANSFERASE (PHOSPHOLIPASE A2 ACTIVITY)
PHOSPHATIDYLCHOLINE-STEROL O-ACYLTRANSFERASE
PHOSPHATIDYLCHOLINE-STEROL O-ACYLTRANSFERASE
TYROSYL PROTEIN SULFOTRANSFERASE 20% 200 μM

SPHINGOMYELINES
1-PHOSPHATIDYLINOSITOL PHOSPHODIESTERASE

SPHINGOSINE
2-ACYLGLYCEROL O-ACYLTRANSFERASE
CAM KINASE *I*50 = 27 μM
CERAMIDASE
CHOLINE-PHOSPHATE CYTIDYLYLTRANSFERASE *I*50 25 MOL%
3′,5′-CYCLIC-NUCLEOTIDE PHOSPHODIESTERASE (Ca^{2+}/CALMODULIN) *I*50 = 200 μM
ELONGATION FACTOR-2 KINASE *I*50 = 20 μM
β-GALACTOSIDASE
2-HYDROXYACYLSPHINGOSINE 1β-GALACTOSYLTRANSFERASE
MYOSIN-LIGHT-CHAIN KINASE *I*50 = 200 μM
Na^+/K^+-EXCHANGING ATPASE
PHOSPHATIDATE PHOSPHATASE *I*50 = 240 μM
PHOSPHOLIPASE A2
PROTEIN KINASE A *I*50 = 98 μM
PROTEIN KINASE C *K*i = 53 μM
PROTEIN KINASE C *I*50 = 87 μM
PROTEIN KINASE Cμ
PROTEIN KINASE (Ca^{2+} DEPENDENT) *I*50 = 1 mM
PROTEIN KINASE (FS STIMULATED)
PROTEIN TYROSINE KINASE (INSULIN RECEPTOR) *I*50 = 145 μM
PROTEIN TYROSINE SULFOTRANSFERASE
TYROSYL PROTEIN SULFOTRANSFERASE *I*50 = 150 μM

(2R,2S)-D-erythro-4-trans-SPHINGOSINE
PROTEIN KINASE C C *I*50 = 2.8 MOL%

(2R,2S)-L-erythro-4-trans-SPHINGOSINE
PROTEIN KINASE C C *I*50 = 3.3 MOL%

(2R,3R)-D-threo-SPHINGOSINE
PROTEIN KINASE C — $I50 = 2.8$ mol%

(2R,3S)-L-erythro-SPHINGOSINE
PROTEIN KINASE C — $I50 = 3.3$ mol%

(2S,3R)-D-erythro-SPHINGOSINE
PROTEIN KINASE C — $I50 = 2.8$ mol%

D-SPHINGOSINE
β-ADRENERGIC-RECEPTOR KINASE
GLUCOSYLCERAMIDASE — 59% 300 μM
MYOSIN-LIGHT-CHAIN KINASE — $I50 = 6$ μM

D-erythroSPHINGOSINE
GLUCOSYLCERAMIDASE — NC — $Ki = 50$ μM

D-(+)-erythro-cis-SPHINGOSINE
PROTEIN KINASE C — $I50 = 7$ mol%

D-threo-SPHINGOSINE
SPHINGOSINE KINASE — $Ki = 0.46$ MOL%

L-threo-SPHINGOSINE
SPHINGOSINE KINASE — $Ki = 0.4$ MOL%

SPI 2.1
PROTEINASE DES(1-3) INSULIN LIKE GROWTH FACTOR GENERATING

SPI 447
H^+/K^+-EXCHANGING ATPASE — $I50 = 4.2$ μM

SPINACH LEAF NITRATE REDUCTASE INHIBITOR
NITRATE DEHYDROGENASE NADH KINASE

SPINACH NITRATE REDUCTASE INHIBITOR
NITRATE REDUCTASE

SPINAPIC ACID
CHORISMATE MUTASE 1 — 30% 170 μM
CHORISMATE MUTASE 2 — 35% 170 μM

SPIRAPRILAT
PEPTIDYL-DIPEPTIDASE A — $I50 = 1$ nM
XAA-PRO AMINOPEPTIDASE — $I50 = 560$ μM

SPIRO[9H-FLUORENE-9,4′-ISOTHIAZOLIDIN]-3′-ONE 1′,1′-DIOXIDE
GLUCURONATE REDUCTASE — $I50 = 4.5$ μM

SPIROHYDANTOINYLISATIN-1-ACETC ACID
ALDEHYDE REDUCTASE — $I50 = 150$ nM

17-(SPIRO-γ-LACTONE)ESTRONE
17β-HYDROXYSTEROID DEHYDROGENASE 2

SPIRONE
BUFURALOL-1′-HYDROXYLASE (CYT P-450dbl) — $I50 = 11$ μM

SPIRONOLACTONE
CARBONATE DEHYDRATASE I — 25% 100 μM
CARBONATE DEHYDRATASE II — 30% 100 μM
CARBONATE DEHYDRATASE IV — 35% 100 μM

SPIROPENTANE ACETIC ACID
ACYL-CoA DEHYDROGENASE

SPIROSTA-5,25(27)DIENE-1β,3β-DIOL(NEORUSCOGENIN)-1-O-{O-α-L-RHAMNOPYRANOSYL(- > 2)-α-L-ARABINOPYRANOSIDE
3′,5′-CYCLIC-NUCLEOTIDE PHOSPHODIESTERASE — $I50 = 84$ μM

Inhibitor / Enzyme	Type	Value
(25R,s)-5α-SPIROSTAN-3β-OL TETRASACCHARIDE		
3′,5′-CYCLIC-NUCLEOTIDE PHOSPHODIESTERASE		
Na^+/K^+-EXCHANGING ATPASE		
SPORAMIN		
TRYPSIN		
SPOROSTATIN		
3′,5′-CYCLIC-NUCLEOTIDE PHOSPHODIESTERASE	NC	*I*50 = 41 µg/ml
SPR 210		
ALDEHYDE REDUCTASE		*I*50 = 9.5 nM
SQ 20006		
3′,5′-CYCLIC-NUCLEOTIDE PHOSPHODIESTERASE (cycAMP)		*I*50 = 13 µM
3′,5′-CYCLIC-NUCLEOTIDE PHOSPHODIESTERASE (cycGMP)		*I*50 = 47 µM
PROTEIN KINASE Cα		*I*50 = 270 µM
PROTEIN KINASE Cβ2		*I*50 = 900 µM
PROTEIN KINASE Cγ		*I*50 = 160 µM
SQ 20009		
3′,5′-CYCLIC-NUCLEOTIDE PHOSPHODIESTERASE		*I*50 = 4.6 µM
PROTEIN KINASE A		> 200 µM
RHODOPSIN KINASE		
SQ 20881		
PEPTIDYL-DIPEPTIDASE A		*I*50 = 18 nM
SQ 21611		
3′,5′-CYCLIC-NUCLEOTIDE PHOSPHODIESTERASE		
SQ 22536		
ADENYLATE CYCLASE		*I*50 = 310 µM
ADENYLATE CYCLASE		*I*50 = 390 µM
ADENYLATE CYCLASE		*I*50 = 13 µM
ADENYLATE CYCLASE		*I*50 = 10 µM
3′,5′-CYCLIC-NUCLEOTIDE PHOSPHODIESTERASE		
SQ 26333		
PEPTIDYL-DIPEPTIDASE A		*I*50 = 8 nM
SQ 26771		
4-AMINOBUTYRATE TRANSAMINASE		*K*i = 22 mM
SQ 27519		
PEPTIDYL-DIPEPTIDASE A		*I*50 = 11 nM
SQ 28603		
NEPRILYSIN		*I*50 = 20 nM
PEPTIDYL-DIPEPTIDASE A		*I*50 = 30 µM
SQ 29072		
NEPRILYSIN		*K*i = 26 nM
NEPRILYSIN	C	*K*i = 19 nM
SQ 29852		
PEPTIDYL-DIPEPTIDASE A		*I*50 = 36 nM
SQ 31429		
MYOSIN-LIGHT-CHAIN KINASE		
SQ 31511		
MYOSIN-LIGHT-CHAIN KINASE		*K*i = 24 µM
SQ 32056		
CATHEPSIN E		*I*50 = 5 nM
SQ 32285		
CATHEPSIN E		*I*50 = 95 nM

SQ 32328		
MYOSIN-LIGHT-CHAIN KINASE		
SQ 32429		
MYOSIN ATPASE		
MYOSIN-LIGHT-CHAIN KINASE		
SQ 32732		
MYOSIN-LIGHT-CHAIN KINASE		
SQ 32970		
CATHEPSIN D		*K*i = 73 μM
CHYMOSIN		*K*i = 1.1 mM
ENDOTHIAPEPSIN		*K*i = 20 μM
PEPSIN		*K*i = 113 μM
RENIN		*K*i = 32 μM
SQ 33600		
HYDROXYMETHYLGLUTARYL-CoA REDUCTASE (NADPH)		
SQ 33800		
RENIN		*I*50 = 350 pM
SQ 65442		
3′,5′-CYCLIC-NUCLEOTIDE PHOSPHODIESTERASE		*I*540 = 10 μM
2,3,22,23-SQALENE DIEPOXIDE		
SQUALENE MONOOXYGENASE	NC	*I*50 = 16 μM
SQUALESTATIN		
FARNESYL-DIPHOSPHATE FARNESYLTRANSFERASE		*I*50 = 12 nM
SQUALESTATIN 1		
FARNESYL-DIPHOSPHATE FARNESYLTRANSFERASE	C	*K*i = 1.6 nM
FARNESYL-DIPHOSPHATE FARNESYLTRANSFERASE		*I*50 = 15 nM
SQUALESTATIN 2		
FARNESYL-DIPHOSPHATE FARNESYLTRANSFERASE		*I*50 = 15 nM
FARNESYL-DIPHOSPHATE FARNESYLTRANSFERASE		*I*50 = 32 nM
FARNESYL-DIPHOSPHATE FARNESYLTRANSFERASE		*I*50 = 5 nM
SQUALESTATIN 3		
FARNESYL-DIPHOSPHATE FARNESYLTRANSFERASE		*I*50 = 6 nM
FARNESYL-DIPHOSPHATE FARNESYLTRANSFERASE		*I*50 = 60 nM
FARNESYL-DIPHOSPHATE FARNESYLTRANSFERASE		*I*50 = 38 nM
SQUALESTATIN 4		
FARNESYL-DIPHOSPHATE FARNESYLTRANSFERASE		*I*50 = 43 nM
SQUALESTATIN DERIVATIVES		
FARNESYL-DIPHOSPHATE FARNESYLTRANSFERASE		
SQUALESTATIN H1		
FARNESYL-DIPHOSPHATE FARNESYLTRANSFERASE		*I*50 = 26 nM
SQUALESTATIN S1		
FARNESYL-DIPHOSPHATE FARNESYLTRANSFERASE		*I*50 = 12 nM
SQUAMOCIN		
NADH DEHYDROGENASE (UBIQUINONE)		*K*i = 400 pM
SQUAMOUS CELL CARCINOMA ANTIGEN		
CATHEPSIN L	NC	*K*i = 64 pM
CHYMOTRYPSIN		
PAPAIN		
SQUAMOUS CELL CARCINOMA ANTIGEN 1		
CATHEPSIN K		75% 700 nM
CATHEPSIN L		93% 86 nM

CATHEPSIN S		99% 1 μM
PAPAIN		91% 360 nM
SQUAMOUS CELL CARCINOMA ANTIGEN 2		
CATHEPSIN G		
CHYMASE		
SQUARIC ACID DERIVATIVES		
LACTOYLGLUTATHIONE LYASE		
SQUASH ASPARTIC PROTEINASE INHIBITOR 1		
PEPSIN		*K*i = 3 nM
PROTEINASE ASPARTIC GLOMERELLA EINGULAT SECRETED		*K*i = 34 nM
SQUASH ASPARTIC PROTEINASE INHIBITOR 2		
PEPSIN		*K*i = 3.8 nM
PROTEINASE ASPARTIC GLOMERELLA EINGULAT SECRETED		*K*i = 15 nM
SQUASH ASPARTIC PROTEINASE INHIBITOR 3		
PEPSIN		*K*i = 2.3 nM
SQUIBB 32377		
FARNESYL-DIPHOSPHATE FARNESYLTRANSFERASE		*I*50 = 30 μM
Sr^{2+}		
D-AMINO-ACID OXIDASE		
ASPARAGINE SYNTHASE (GLUTAMINE-HYDROLYSING)		
ASPERGILLUS ALKALINE PROTEINASE		
CALCIDOL 1-MONOOXYGENASE		
α-DEXTRIN ENDO-1,6-α-GLUCOSIDASE		
GLYCINE C-ACETYLTRANSFERASE		
GTP CYCLOHYDROLASE I		
METHYLASPARTATE AMMONIA-LYASE		
5-METHYLTETRAHYDROFOLATE-HOMOCYSTEINE S-METHYLTRANSFERASE		
NAD(P) TRANSHYDROGENASE (AB-SPECIFIC)		
Na^+/K^+-EXCHANGING ATPASE		
NICOTINATE-NUCLEOTIDE PYROPHOSPHORYLASE (CARBOXYLATING)		
OXALATE OXIDASE		
PANTOATE-β-ALANINE LIGASE		
PHOSPHOLIPASE A2	C	
PHOSPHOLIPASE A2		
PHOSPHOPYRUVATE HYDRATASE		
3-PHYTASE		90% 500 μM
POLYGALACTURONASE		
RENILLA-LUCIFERIN 2-MONOOXYGENASE		
RIBOFLAVIN KINASE		
SIALIDASE		
THYMIDINE KINASE		
TRITRIRACHIUM ALKALINE PROTEINASE		
SR 25477		
THROMBIN		
SR 26831		
ELASTASE LEUKOCYTE		*I*50 = 80 nM
ELASTASE PANCREATIC II		*I*50 = 4.8 μM
SR 31747		
CHOLESTENOL Δ-ISOMERASE		*I*50 = 600 nM
SR 33557		
SPHINGOMYELIN PHOSPHODIESTERASE		31% 10 μM
SR 41476		
PROTEINASE HIV		*I*50 = 7 μM
SR 42128		
RENIN	C	*K*i = 2 nM

SR 43845
RENIN

SR 48692
CARBOXYPEPTIDASE A

SR 95191
AMINE OXIDASE MAO-A | IR | *I*50 = 140 μM

SR 95531
AMINE OXIDASE MAO-A

SRC 416
PROTEIN TYROSINE KINASE pp60c-Src | | *K*i = 700 μM

SRI 20
PEPTIDYL-DIPEPTIDASE A | | *I*50 = 13 nM

SRI 62834
PHOSPHATIDYLINOSITOL PHOSPHOLIPASE Cγ | | *I*50 = 2.3 μM
PHOSPHOLIPASE D (GLYCOSYLPHOSPHATIDYLINOSITOL SPECIFIC) | | *I*50 = 50 μM

1S,2S,4aR,SS,8S,8a.S, 4′R,6′R)-6′-[1,2,4a,5,6,7,8,8a-OCTAHYDRO-2-METHYL-8-[(2″,2″-DIMETHYLBUTYRYL)OXY]-6-(E)PROP-1-ENYL)-NAPHTHALENYL]ETHYL]TETRAHYDRO-4-′-HYDROXY-2′H-PYRAN-2′-ONE
HYDROXYMETHYLGLUTARYL-CoA REDUCTASE (NADPH) | | *I*50 = 3 nM

(2.S, 3S)-L-threo-SPHINGOSINE
PROTEIN KINASE C | | *I*50 = 2.2 mol%

ST 43
NEPRILYSIN | | *K*i = 1.5 nM

ST 271
OXIDATIVE PHOSPHORYLATION UNCOUPLER | | E50 = 28 μM
PROTEIN TYROSINE KINASE EGFR | | *I*50 = 1.1 μM

ST 280
Na^+/K^+-EXCHANGING ATPASE | | *I*50 = 74 μM
PROTEIN TYROSINE KINASE EGFR | | *I*50 = 440 nM
PROTEIN TYROSINE KINASE EGFR | | *I*50 = 440 nM

ST 458
PROTEIN TYROSINE KINASE EGFR | | *I*50 = 440 nM
PROTEIN TYROSINE KINASE EGFR | | *I*50 = 440 nM

ST 633
PROTEIN TYROSINE KINASE EGFR | | *I*50 = 250 nM

ST 638
ALDEHYDE DEHYDROGENASE (NAD) | | *I*50 = 6 μM
PROTEIN TYROSINE KINASE EGFR | | *I*50 = 370 nM
PROTEIN TYROSINE KINASE EGFR | | *I*50 = 850 nM

ST 642
PROTEIN TYROSINE KINASE EGFR | | *I*50 = 850 nM
PROTEIN TYROSINE KINASE EGFR | | *I*50 = 850 nM

STACHYBOTRYDIAL
STEROL ESTERASE | | *I*50 = 60 μM

STACHYOSE
α-GALACTOSIDASE

STANOZOLOL
STEROID 17α-MONOOXYGENASE | C | *K*i = 6.3 μM

STAPHYLOCOCCUS TANABEENSIS PROTEINASE INHIBITOR
CALPAIN
PAPAIN

STATIL

GLUCURONATE REDUCTASE

STAUROSPORINE

Ca^{2+} DEPENDENT PROTEIN KINASE α		*I*50 = 110 nM
Ca^{2+} DEPENDENT PROTEIN KINASE β		*I*50 = 120 nM
Ca^{2+} DEPENDENT PROTEIN KINASE γ		*I*50 = 70 nM
CAM KINASE		*I*50 = 40 nM
CAM KINASE II	NC	*I*50 = 20 nM
CYCLIN DEPENDENT KINASE 1		*I*50 = 3 nM
CYCLIN DEPENDENT KINASE 2		*I*50 = 4 nM
CYCLIN DEPENDENT KINASE 5		*K*i = 39 nM
DNA TOPOISOMERASE (ATP-HYDROLYSING)		
F2A KINASE		*I*50 = 5 nM
F2B KINASE		*I*50 = 4 nM
F2C KINASE		*I*50 = 5 nM
MAP KINASE		*I*50 = 20 nM
MYOSIN-LIGHT-CHAIN KINASE		*I*50 = 21 nM
NADPH OXIDASE		
PHOSPHATIDYLINOSITOL KINASE		*I*50 = 26 μg/ml
PHOSPHORYLASE KINASE		*I*50 = 520 pM
PHOSPHORYLASE KINASE		*I*50 = 5 pM
PHOSPHORYLASE KINASE		*I*50 = 3 nM
PROLYL OLIGOPEPTIDASE		*I*50 = 750 nM
PROTEIN KINASE A		*I*50 = 8.2 nM
PROTEIN KINASE A		*I*50 = 120 nM
PROTEIN KINASE C		*I*50 = 5 nM
PROTEIN KINASE C		*I*50 = 3 nM
PROTEIN KINASE C		*I*50 = 9 nM
PROTEIN KINASE C		*K*i = 5.5 nM
PROTEIN KINASE Cβ		*I*50 = 160 nM
PROTEIN KINASE Cα		*I*50 = 58 nM
PROTEIN KINASE Cβ		*I*50 = 65 nM
PROTEIN KINASE Cβ1		*I*50 = 9 μM
PROTEIN KINASE Cβ2		*I*50 = 3 μM
PROTEIN KINASE Cδ		*I*50 = 330 nM
PROTEIN KINASE Cγ		*I*50 = 49 nM
PROTEIN KINASE (Ca^{2+} DEPENDENT)		*I*50 = 8 nM
PROTEIN KINASE C (CYTOSOLIC)		*I*50 = 16 nM
PROTEIN KINASE Ceta		*I*50 = 10 μM
PROTEIN KINASE C (MEMBRANE)		*I*50 = 1.3 nM
PROTEIN KINASE Cxi		*I*50 = 220 nM
PROTEIN KINASE Czeta		*I*50 = 1.3 mM
PROTEIN KINASE G		*I*50 = 8.5 nM
PROTEIN TYROSINE KINASE		*I*50 = 70 μM
PROTEIN TYROSINE KINASE		*I*50 = 6.4 nM
PROTEIN TYROSINE KINASE EGFR		*I*50 = 10 nM
PROTEIN TYROSINE KINASE FynT		*I*50 = 5 nM
PROTEIN TYROSINE KINASE (Fyn)		*I*50 = 5 nM
PROTEIN TYROSINE KINASE HPK40		*I*50 = 20 nM
PROTEIN TYROSINE KINASE (IGF RECEPTOR)		
PROTEIN TYROSINE KINASE (INSULIN RECEPTOR)		*I*50 = 60 nM
PROTEIN TYROSINE KINASE JAK2		*I*50 = 40 μM
PROTEIN TYROSINE KINASE (Lck)		*I*50 = 100 nM
PROTEIN TYROSINE KINASE p56lck		*I*50 = 200 nM
PROTEIN TYROSINE KINASE NGFR		*K*i = 600 pM
PROTEIN TYROSINE KINASE p56lck		*I*50 = 3 nM
PROTEIN TYROSINE KINASE c-Src		*I*50 = 10 nM
PROTEIN TYROSINE KINASE pp60c-Src		*I*50 = 8 ng/ml
PROTEIN TYROSINE KINASE pp60c-Src		*I*50 = 90 nM
PROTEIN TYROSINE KINASE pp60v-Src		*I*50 = 6.4 nM
PROTEIN TYROSINE KINASE v-abl		*I*50 = 80 nM
PROTEIN TYROSINE KINASE v-Src		*I*50 = 6 nM
PROTEIN TYROSINE KINASE ZAP70		*I*50 = 20 μM
RIBOSOMAL PROTEIN S6 KINASE		*I*50 = 5 nM

STAUROSPORINE AGLYCONE		
PROTEIN KINASE C		$I50 = 310$ nM
STAUROSPORINE CGP 41251		
PROTEIN KINASE C		
STEARIC ACID		
DNA METHYLTRANSFERASE ECO RI.		$I50 = 133$ μM
β-GALACTOSIDE α-2,6-SIALYLTRANSFERASE		
1,3-β-GLUCAN SYNTHASE		$I50 = 100$ μM
GLUTATHIONE TRANSFERASE		$I50 = 22$ μM
GLUTATHIONE TRANSFERASE	C	$Ki = 19$ μM
GLUTATHIONE TRANSFERASE	C	$Ki = 1.1$ μM
GLUTATHIONE TRANSFERASE		$I50 = 349$ μM
15-HYDROXYPROSTAGLANDIN DEHYDROGENASE (NAD)	C	
NAD(P)-ARGININE ADP-RIBOSYLTRANSFERASE		$I50 = 6.1$ μM
PEPTIDYL-DIPEPTIDASE A		
PROTEIN KINASE A		$I50 = 500$ μM
PROTEIN KINASE (Ca^{2+} DEPENDENT)		$I50 = 100$ μM
STEARIDONIC ACID		
ARACHIDONATE 5-LIPOXYGENASE		
STEAROYL-CoA		
2-ACYLGLYCEROPHOSPHOCHOLINE O-ACYLTRANSFERASE		
ETHANOLAMINE KINASE		
GLUCOSE-6-PHOSPHATASE		40% 50 μM
GLUCOSE-6-PHOSPHATE 1-DEHYDROGENASE	C	$Ki = 400$ nM
GLUTAMINASE		
GLUTAMINASE (PHOSPHATE ACTIVATED)		
[PYRUVATE DEHYDROGENASE (LIPOAMIDE)] KINASE		
N-STEAROYLDOPAMINE		
ARACHIDONATE 5-LIPOXYGENASE		$I50 = 16$ nM
STEAROYL VANILLYLAMIDE		
NADH DEHYDROGENASE (UBIQUINONE)		$I50 = 95$ μM
STEARYLAMINE		
2-ACYLGLYCEROL O-ACYLTRANSFERASE		
PHOSPHOLIPASE A2		
PHOSPHOLIPASE D (GLYCOSYLPHOSPHATIDYLINOSITOL SPECIFIC)		$I50 = 130$ μM
PROTEIN KINASE C		$I50 = 2.4$ mol%
PROTEIN TYROSINE SULFOTRANSFERASE		
TYROSYL PROTEIN SULFOTRANSFERASE		56% 200 μM
STEARYL-CoA		
ACETYL-CoA CARBOXYLASE		$Ki = 1.3$ nM
STEFIN A		
CATHEPSIN B		$Ki = 2.4$ μM
CATHEPSIN B		$Ki = 2$ nM
CATHEPSIN B		$Ki = 1.9$ nM
CATHEPSIN H		$Ki = 69$ pM
CATHEPSIN H		$Ki = 400$ pM
CATHEPSIN L		$Ki = 20$ pM
CATHEPSIN L		$Ki = 29$ pM
CATHEPSIN S		$Ki = 53$ pM
CRUZIPAIN		Kd = 21 pM
PAPAIN		$Ki = 27$ pM
PAPAIN		$Ki = 174$ pM
STEFIN A (HUMAN)		
ACTINIDAIN		$Ki = 130$ pM
CATHEPSIN B		$Ki = 2.7$ nM
CATHEPSIN H		$Ki = 72$ pM
CATHEPSIN L		$Ki = 2.1$ pM

CATHEPSIN S $Ki = 27$ pM
PAPAIN $Ki = 10$ pM

STEFIN B
CATHEPSIN B $Ki = 84$ nM
CATHEPSIN B $Ki = 19$ mM
CATHEPSIN H $Ki = 27$ pM
CATHEPSIN L $Ki = 9$ pM
CATHEPSIN S $Ki = 90$ pM
CRUZIPAIN Kd = 60 pM
PAPAIN $Ki = 128$ pM
PAPAIN $Ki = 124$ pM

STEFIN B (BOVINE)
CATHEPSIN L $Ki = 9$ pM
CATHEPSIN S $Ki = 8$ pM
CATHEPSIN S $Ki = 8.4$ pM
PAPAIN $Ki = 120$ pM
PAPAIN $Ki = 120$ pM

STEFIN B (HUMAN)
ACTINIDAIN $Ki = 52$ pM
CATHEPSIN B $Ki = 74$ nM
CATHEPSIN H $Ki = 15$ pM
CATHEPSIN L $Ki = 230$ pM
CATHEPSIN S $Ki = 27$ pM
PAPAIN $Ki = 120$ pM

STEFIN B (SHEEP)
CATHEPSIN L $Ki = 50$ pM
PAPAIN $Ki = 40$ pM

STEFIN C (BOVINE THYMUS)
CATHEPSIN B $Ki > 100$ nM
CATHEPSIN L
PAPAIN $Ki < 180$ pM

STEFIN D1
CATHEPSIN B $Ki = 335$ nM
CATHEPSIN H $Ki = 125$ nM
CATHEPSIN L $Ki = 67$ pM
CATHEPSIN S $Ki = 46$ pM
PAPAIN $Ki = 190$ pM

STEFIN D2
CATHEPSIN B $Ki = 195$ nM
CATHEPSIN H $Ki = 102$ nM
CATHEPSIN L $Ki = 161$ pM
PAPAIN $Ki = 42$ pM

STEMPHONE
DIACYLGLYCEROL KINASE $I50 = 3.3$ µM

STENOPHYLLANIN
NADH DEHYDROGENASE $I50 = 1.4$ µM
NADH DEHYDROGENASE $I50 = 1.1$ µM

STEORID 17β-CARBOXYLIC ACID
STEROID 5α-REDUCTASE $Ki = 1.1$ µM

STERCOBILIN
PROTEINASE HIV-1 $Ki = 4$ µM

STERCULIC ACID
STEAROYL-CoA DESATURASE

STEREUM HIRSUTUM GLYOXALASE INHIBITOR
LACTOYLGLUTATHIONE LYASE

STEROIDAL A RING FUSED PYRAZOLES
3(or 17)β-HYDROXYSTEROID DEHYDROGENASE

STEROIDAL SAPONINS
3′,5′-CYCLIC-NUCLEOTIDE PHOSPHODIESTERASE
3′,5′-CYCLIC-NUCLEOTIDE PHOSPHODIESTERASE

STEROIDES
CORTISONE α-REDUCTASE
NADH DEHYDROGENASE (UBIQUINONE) 40% 50 μM

3α-STEROIDES
ALCOHOL DEHYDROGENASE

STEVASTELLIN A
PROTEIN TYROSINE-PHOSPHATASE VHR *I*50 = 2.7 μM

STEVASTELLIN A3
PROTEIN TYROSINE-PHOSPHATASE VHR *I*50 = 3.6 μM

STEVASTELLIN B
PROTEIN TYROSINE-PHOSPHATASE VHR *I*50 = 20 μM

STEVASTELLIN B3
PROTEIN TYROSINE-PHOSPHATASE VHR *I*50 = 14 μM

STEVASTELLIN C3
PROTEIN TYROSINE-PHOSPHATASE VHR *I*50 = 16 μM

STEVASTELLIN D3
PROTEIN TYROSINE-PHOSPHATASE VHR *I*50 = 1.7 μM

STEVASTELLIN E3
PROTEIN TYROSINE-PHOSPHATASE VHR *I*50 = 14 μM

STEVASTELLIN H
PROTEIN TYROSINE-PHOSPHATASE VHR *I*50 = 24 μM

STEVASTELLIN P
PROTEIN TYROSINE-PHOSPHATASE CD45 *I*50 = 160 μM

STEVIOL
ADN-CARRIER MITOCHONDRIA NC

STIBOGLUCONATE
DNA TOPOISOMERASE

STICHODACTYLA HELIANTHUS PROTEINASE INHIBITORS
PROTEINASES

STICTIC ACID
INTEGRASE *I*50 = 4.4 μM

STIGMASTEROL
11β-HYDROXYSTEROID DEHYDROGENASE *I*50 = 2 mM
XANTHINE OXIDASE 51% 100 μM

STIGMASTEROL-β-D-GLYCOSIDE
XANTHINE OXIDASE 58% 100 μM

STIGMATELLIN
CYTOCHROME bc1
MENADIOL: NITRATE OXIDOREDUCTASE *I*50 = 250 nM
MENAQUINOL OXIDASE 61% 10 μM
NADH DEHYDROGENASE (UBIQUINONE) 39% 3.2 μM
PLASTOQUINOL-PLASTOCYANIN REDUCTASE
UBIQUINOL-CYTOCHROME-C2 REDUCTASE

Inhibitor / Enzyme	Type	Value
STILBAMIDINE ISOTHIONATE		
AMINE OXIDASE (COPPER-CONTAINING)	NC	*K*i = 12 μM
STILBENE HYDROXAMIC ACID		
ARYLAMINE N-ACETYLTRANSFERASE	SSI	*K*i = 0.24/min
STILBESTROL		
trans-1,2-DIHYDROBENZENE-1,2-DIOL DEHYDROGENASE		75% 10 μM
trans-1,2-DIHYDROBENZENE-1,2-DIOL DEHYDROGENASE 2		52% 10 μM
trans-1,2-DIHYDROBENZENE-1,2-DIOL DEHYDROGENASE 3		54% 10 μM
trans-1,2-DIHYDROBENZENE-1,2-DIOL DEHYDROGENASE 4		43% 10 μM
3α(or 17β)-HYDROXYSTEROID DEHYDROGENASE		
3α-HYDROXYSTEROID DEHYDROGENASE (B-SPECIFIC)		39% 1 μM
STREPTAVIDIN		
UREASE (ATP-HYDROLYSING)		
STREPTIDINE		
AMINE OXIDASE (COPPER-CONTAINING)	NC	*K*i = 900 μM
STREPTOCOCCUS SOBRINUS DEXTRANASE INHIBITOR		
DEXTRANASE		
STREPTOCOCCUS SOBRINUS ENDODEXTRANASE INHIBITOR		
ENDODEXTRANASE		
STREPTOLYDIGIN		
DNA NUCLEOTIDYLEXOTRANSFERASE		
RNA POLYMERASE	NC	
STREPTOMYCES ALKALINE PROTEINASE INHIBITOR		
PROTEINASE ALKALINE		
STREPTOMYCES β-AMYLASE INHIBITOR		
β-AMYLASE		
STREPTOMYCES BIKINIENSIS SUBTILISIN INHIBITOR (SIL 15)		
SUBTILISIN BNP′		*K*i = 27 pM
STREPTOMYCES CACOI PROTEINASE INHIBITOR		
SUBTILISIN BNP		*K*i = 28 pM
TRYPSIN		*K*i = 55 nM
STREPTOMYCES CLAVULINGERS β-LACTAMASE INHIBITOR		
β-LACTAMASE		*K*i = 18 pM
STREPTOMYCES CORUCHORUSHII α-AMYLASE INHIBITOR		
α-AMYLASE		
STREPTOMYCES LIVIDANS PROTEINASE INHIBITOR		
SUBTILISIN BPN		
SUBTILISIN BPN′		*K*i = 30 pM
TRYPSIN		*K*i = 1 nM
STREPTOMYCES METALLOPROTEINASE INHIBITOR		
STREPTOMYCES GRISEUS METALLO-ENDOPEPTIDASE I		99% 100 nM
STREPTOMYCES GRISEUS METALLO-ENDOPEPTIDASE II		99% 100 nM
STREPTOMYCES NANIWAENSIS PEPSIN INHIBITOR		
RHODOTORULAPEPSIN		
STREPTOMYCES NIGRESCEUS METALLOPROTEINASE INHIBITOR		
THERMOLYSIN		*K*i = 10 pM
STREPTOMYCES NITROSPOREUS α-AMYLASE INHIBITOR		
α-AMYLASE		
STREPTOMYCES PHOSPHOLIPASE C INHIBITOR		
3′,5′-CYCLIC-NUCLEOTIDE PHOSPHODIESTERASE I		*I*50 = 8.4 μg

1-PHOSPHATIDYLINOSITOL PHOSPHODIESTERASE — *I*50 = 2.3 μg
PHOSPHOLIPASE C — *K*i = 9.5 nM
SPHINGOMYELIN PHOSPHODIESTERASE — *I*50 = 23 μg

STREPTOMYCES RIBONUCLEASE INHIBITOR SaI14
RIBONUCLEASE

STREPTOMYCES RIBONUCLEASE INHIBITOR SaI20
RIBONUCLEASE

STREPTOMYCES SP 22 PAPAIN INHIBITOR
PAPAIN

STREPTOMYCES SP 22 TRYPSIN INHIBITOR
TRYPSIN

STREPTOMYCES SUBTILISIN INHIBITOR
CHYMOTRYPSIN
LYSYL ENDOPEPTIDASE
PROTEINASE SERINE ASPERGILLUS FUMIGATUS — *K*i = 1 nM
STREPTOMYCES GRISEUS METALLO-ENDOPEPTIDASE I — 100% 100 nM
STREPTOMYCES GRISEUS METALLO-ENDOPEPTIDASE II — 100% 100 nM
SUBTILISIN
TRYPSIN

STREPTOMYCES SUBTILISIN INHIBITOR LIKE 2
SUBTILISIN BPN′ — *K*i = 29 pM
TRYPSIN — *K*i = 590 pM

STREPTOMYCES SUBTILISIN INHIBITOR LIKE 3
SUBTILISIN BPN′ — *K*i = 46 pM
TRYPSIN — *K*i = 410 pM

STREPTOMYCES SUBTILISIN INHIBITOR LIKE 4
SUBTILISIN BPN′ — *K*i = 12 pM
TRYPSIN — *K*i = 1.1 nM

STREPTOMYCES TRYPSIN INHIBITOR
PRONASE E — 45% 10 μg
SUBTILISIN — 65% 10 μg
TRYPSIN — 93% 10 μg

STREPTOMYCIN
D-AMINO-ACID OXIDASE
D-AMINO-ACID OXIDASE
ASPARTYLTRANSFERASE
STREPTOMYCIN KINASE

STREPTONIGRIN
RNA-DIRECTED DNA POLYMERASE

STREPTOTHRICIN B
PHOSPHOLIPASE C γ1 (PHOSPHATIDYLINOSITOL SPECIFIC) — *I*50 = 6.7 μM

STREPTOVARICIN
RNA-DIRECTED DNA POLYMERASE
RNA POLYMERASE

STREPTOVERTICILLIUM ANTICOAGULANT I
α-CHYMOTRYPSIN
COAGULATION FACTOR XIIa — *K*i = 53 nM
KALLIKREIN PLASMA — *K*i = 7.2 nM
THERMOLYSIN
TRYPSIN

STREPTOVERTICILLIUM ANTICOAGULANT II
PROTEINASE STREPTOMYCES GRISEUS ALKALINE
PROTEINASE TITRACHIUM ALBUM
SUBTILISIN

STREPTOVERTICILLIUM ANTICOAGULANT III
PROTEINASE STREPTOMYCES GRISEUS ALKALINE
PROTEINASE TITRACHIUM ALBUM
SUBTILISIN

STREPTTOMYCES CLAVULIGERIUS β-LACTAMASE INHIBITORY PROTEIN
β-LACTAMASE

STRICTININ
RNA-DIRECTED DNA POLYMERASE — $I50 = 87$ nM

STROBILURIN
CHITIN SYNTHASE

STROBILURIN A
NADH DEHYDROGENASE (UBIQUINONE)
UBIQUINOL-CYTOCHROME-C REDUCTASE

STROBILURIN B
NADH DEHYDROGENASE (UBIQUINONE)
UBIQUINOL-CYTOCHROME-C REDUCTASE

D-STROMBINE
D-OCTOPINE DEHYDROGENASE

STROPHANTHIDIN

Ca^{2+}-TRANSPORTING ATPASE	C	$Ki = 250$ μM
H^+/K^+-EXCHANGING ATPASE	C	$Ki = 160$ μM
H^+-TRANSPORTING ATPASE	C	$Ki = 90$ μM

STYLOGERANIDINE 1
CHITINASE

STYLOGERANIDINE 2
CHITINASE

STYLOGERANIDINE 3
CHITINASE

STYRENE OXIDE
MANDELATE RACEMASE

STYRENE-7,8-OXIDE
EPOXIDE HYDROLASE

trans-STYRENE OXIDE
EPOXIDE HYDROLASE

trans-STYRYLACETIC ACID

PEPTIDYLGLYCINE MONOOXYGENASE	IR	$Ki = 600$ nM

trans-STYRYLTHIOACETIC ACID

PEPTIDYLGLYCINE MONOOXYGENASE	SSI	$Ki = 100$ μM

SU 4885
CORTICOSTERONE 18-MONOOXYGENASE

SU 9055
CORTICOSTERONE 18-MONOOXYGENASE

SUAM 1221

PROLYL OLIGOPEPTIDASE	C	$I50 = 4.7$ nM
PROLYL OLIGOPEPTIDASE	C	$I50 = 7.2$ nM
PROLYL OLIGOPEPTIDASE	C	$I50 = 9$ nM

SUBACATE
KYNURENINE-OXOGLUTARATE AMINOTRANSFERASE

Inhibitor / Enzyme		
SUBARATE		
4-HYDROXY-2-OXOGLUTARATE ALDOLASE		62% 40 mM
SUBEROGORGIN		
ACETYLCHOLINESTERASE		*I*50 = 93 μM
SUBEROYLANILIDE		
HISTONE DEACETYLASE		
SUBSTANCE P		
MEMBRANE ALANINE AMINOPEPTIDASE		*K*i = 450 nM
PEPTIDYL-DIPEPTIDASE A		
6-SUBSTITUTED (9-HYDROXYALKYL)PURINE		
ADENOSINE DEAMINASE	C	
SUBSTITUTED ISOCOUMARINS		
COAGULATION FACTOR Xa		
COAGULATION FACTOR XIa		
COAGULATION FACTOR XIIa		
KALLIKREIN PLASMA		
KALLIKREIN TISSUE		
THROMBIN		
TRYPSIN		
5′-SUBSTITUTED THYMIDINE ANALOGUES		
THYMIDINE KINASE		
SUBSTRATE		
D-AMINO-ACID TRANSAMINASE	C	*K*i = 43 mM
ARALKYLAMINE DEHYDROGENASE		
LACTALDEHYDE DEHYDROGENASE		
METHYLMALONYL-CoA DECARBOXYLASE		
SUBSTRATE ANALOGUES		
D-ALANYL-D-ALANINE CARBOXYPEPTIDASE		
SUBTILISIN INHIBITOR LIKE 1		
SUBTILISIN BPN		*K*i = 28 pM
TRYPSIN		*K*i = 55 nM
SUBTILISIN INHIBITOR LIKE 2		
SUBTILISIN		
SUBTILISIN BPN		*K*i = 29 pM
TRYPSIN		*K*i = 590 pM
SUBTILISIN INHIBITOR LIKE 3		
SUBTILISIN BPN		*K*i = 46 pM
TRYPSIN		*K*i = 410 pM
SUBTILISIN INHIBITOR LIKE 4		
SUBTILISIN BPN		*K*i = 12 pM
TRYPSIN		*K*i = 1.1 nM
SUBTILISIN INHIBITOR LIKE 5		
SUBTILISIN		
SUBTILISIN INHIBITOR LIKE 7		
SUBTILISIN		
SUBTILISIN INHIBITOR LIKE 10		
SUBTILISIN BPN		*K*i = 160 pM
TRYPSIN		*K*i = 160 pM
SUBTILISIN INHIBITOR LIKE 13		
SUBTILISIN BPN		*K*i = 25 pM
TRYPSIN		*K*i = 990 pM

SUBTILISIN INHIBITOR LIKE 14

SUBTILISIN BPN		*K*i = 53 pM
TRYPSIN		*K*i = 350 pM

SUBTILISIN LIKE INHIBITOR 8

CHYMOTRYPSIN		*K*i = 11 nM
SUBTILISIN BPN′		*K*i = 92 pM

SUCC-AP-$NHNH_2$

SUCCINYLDIAMINOPIMELATE TRANSAMINASE		*I*50 = 500 nM

SUCCINAMIC ACID

D-AMINO-ACID TRANSAMINASE	C	*K*i = 28 mM

SUCCINAMIDE

ASPARAGINASE		

SUCCINATE SEMIALDEHYDE

ALCOHOL DEHYDROGENASE (NADP)		
D-AMINO-ACID OXIDASE		
γ-BUTYROBETAINE DIOXYGENASE		
2-HYDROXY-3-OXOPROPIONATE REDUCTASE		
MALONATE-SEMIALDEHYDE DEHYDROGENASE	NC	*K*i = 560 μM

SUCCINIC ACID

ADENINE PHOSPHORIBOSYLTRANSFERASE		
ADENOSINE-PHOSPHATE DEAMINASE		*K*i = 18 mM
ADENYLOSUCCINATE SYNTHASE	C	*K*i = 1.8 mM
ALANINE TRANSAMINASE		
D-ALANINE TRANSAMINASE		
ALANOPINE DEHYDROGENASE	NC	*K*i = 22 mM
AMINO-ACID N-ACETYLTRANSFERASE		
5-AMINOLEVULINATE SYNTHASE		
ARGININOSUCCINATE LYASE		
ASPARTATE 1-DECARBOXYLASE	C	*K*i = 730 μM
ASPARTATE 4-DECARBOXYLASE		68% 5.6 mM
ASPARTATE TRANSAMINASE	C	*K*i = 18 mM
ATP DEAMINASE		
CARBOXYPEPTIDASE A	C	*K*i = 400 μM
CYANATE HYDROLASE		*I*50 = 4 mM
CYTOCHROME-b5 REDUCTASE		
DEHYDROGLUCONATE DEHYDROGENASE		*K*i = 175 mM
3-DEHYDROQUINATE DEHYDRATASE		*K*i = 74 mM
DEOXYRIBONUCLEASE II		
2,3-DIKETO-L-GULONATE DECARBOXYLASE		90% 40 mM
FUMARATE HYDRATASE	C	*K*i = 52 mM
FUMARATE REDUCTASE (NADH)		
GIBBERELLIN 2β-DIOXIGENASE I		
GLUTAMATE-AMMONIA LIGASE		
GLUTAMATE DECARBOXYLASE	C	*K*i = 730 μM
GLUTAMATE DEHYDROGENASE (NAD(P))		
GLUTAMATE DEHYDROGENASE (NADP)		
GLUTAMATE FORMIMINOTRANSFERASE		
GLUTAMATE SYNTHASE (NADPH)		
ISOCITRATE LYASE	NC	*K*i = 7 mM
ISOCITRATE LYASE	NC	*K*i = 290 μM
LYSINE 2-MONOOXYGENASE		
MALATE DEHYDROGENASE (OXALACETATE-DECARBOXYLATING) (NADP)	C	*K*i = 20 mM
MALONYL-CoA SYNTHETASE	NC	*K*i = 47 mM
2-METHYLENEGLUTARATE MUTASE	C	*K*i = 1.8 mM
METHYLITACONATE Δ-ISOMERASE	C	*K*i = 1.8 mM
NICOTINATE-NUCLEOTIDE PYROPHOSPHORYLASE (CARBOXYLATING)		
3-OXOACID CoA-TRANSFERASE	C	*K*i = 1 mM
OXOGLUTARATE DEHYDROGENASE (LIPOAMIDE)		*K*i = 1.2 μM
OXOGLUTARATE DEHYDROGENASE (LIPOAMIDE)		
PHOSPHATIDYLSERINE DECARBOXYLASE		

PHOSPHOENOLPYRUVATE CARBOXYKINASE (PYROPHOSPHATE)		
PHOSPHOENOLPYRUVATE CARBOXYLASE		
PHOSPHOENOLPYRUVATE DECARBOXYLASE		
PROCOLLAGEN-PROLINE 3-DIOXYGENASE		
PROCOLLAGEN-PROLINE, 2-OXOGLUTARATE-4-DIOXYGENASE		
SERINE-SULFATE AMMONIA-LYASE	C	*K*i = 35 mM
SUCCINATE-HYDROXYMETHYLGLUTARATE CoA-TRANSFERASE	C	*K*i = 1.5 mM
TAUROPINE DEHYDROGENASE		
TAUROPINE DEHYDROGENASE	NC	*K*i = 10 mM
THIOSULFATE SULFURTRANSFERASE		
TRIMETHYLYSINE DIOXYGENASE		
TRIOSE-PHOSPHATE ISOMERASE		*K*i = 1.4 mM
UROCANATE HYDRATASE	C	
XAA-PRO DIPEPTIDASE		*K*i = 176 μM

SUCCINIC ANHYDRIDE

INORGANIC PYROPHOSPHATASE		47% 25 mM

SUCCINIMIDES

ALCOHOL DEHYDROGENASE (NADP)

N-SUCCINIMIDYL-3-(2-PYRIDYLTHIO)PROPIONATE

ELECTRON-TRANSFERRING-FLAVOPROTEIN DEHYDROGENASE

SUCCINYL-ALA-ALA-PRO-PHE-CH_2Cl

PROTEINASE NUCLEAR SCAFFOLD		*K*i = 56 nM

3-SUCCINYLAMINOBENZAMIDE

NAD ADP-RIBOSYLTRANSFERASE		91% 50 μM

(2S,6RS)-2-(N-(SUCCINYL)AMINO-6-HYDRAZINOHEPTANE-1,7-DIOIC ACID

SUCCINYLDIAMINOPIMELATE TRANSAMINASE		*K*i = 22 nM

SUCCINYLCHOLINE

CHOLINESTERASE	C	*K*i = 20 μM
CHOLINESTERASE		

SUCCINYL-CoA

HYDROXYMETHYLGLUTARYL-CoA SYNTHASE		*I*50 = 200 μM
MALONYL-CoA DECARBOXYLASE		*K*i = 580 μM
METHYLMALONYL-CoA EPIMERASE		61% 2 mM
3-OXOACID CoA-TRANSFERASE	C	*K*i = 1.9 mM
OXOGLUTARATE DEHYDROGENASE (LIPOAMIDE)		
PYRUVATE KINASE		
SUCCINATE-CoA LIGASE (GDP-FORMING)	C	*K*i = 20 μM

SUCCINYL-GLN-VAL-VAL-ALA-ALA-pNITROANILIDE

PAPAIN		*I*50 = 59 μM

N-SUCCINYL-GLUTAMIC ACID

N-ACETYL-L-ASPARTYL-L-GLUTAMATE PEPTIDASE	C	*K*i = 1.2 μM

O-SUCCINYL-D,L-HOMOSERINE

O-ACETYLHOMOSERINE (THIOL)-LYASE		31% 30 mM

SUCCINYLOACETONE

PORPHOBILINOGEN SYNTHASE		
PORPHOBILINOGEN SYNTHASE	IR	

SUCCINYLPHOSPHONATE

OXOGLUTARATE DEHYDROGENASE (LIPOAMIDE)

O-SUCCINYLSERINE

O-SUCCINYLHOMOSERINE (THIOL)-LYASE

SUCROSE

AMYLOSUCRASE		> 100 mM
DEXTRANSUCRASE		

FLAVONE 4-REDUCTASE		
α-GALACTOSIDASE		
GLUCAN ENDO-1,3-β-GLUCOSIDASE		
α-GLUCOSIDASE	C	Ki = 37 mM
β-GLUCOSIDASE		51% 15 mM
GLYCEROL DEHYDROGENASE		
GUANYLATE CYCLASE		
INULINASE	NC	Ki = 2 mM
PECTINESTERASE		
PROCOLLAGEN GLUCOSYLTRANSFERASE		
PROPIONATE-CoA LIGASE		
SUCROSE-PHOSPHATASE		
THIOGLUCOSIDASE		
α,α-TREHALASE		
URATE OXIDASE		
SUC-TYR-D-LEU-D-VAL-BENZOYLANILINE		
ELASTASE LEUKOCYTE		Ki = 60 μM
SUFINDAC SULPHIDE		
CATHEPSIN G		Ki = 150 μM
ELASTASE LEUKOCYTE		Ki = 10 μM
SUFONAZO III		
PROTEINASE HIV-1		I50 = 290 μM
SUGAR ACIDS		
2-DEHYDRO-3-DEOXYGLUCARATE ALDOLASE		
SUGAR BEET β-FRUCTOFURANOSIDASE INHIBITOR		
β-FRUCTOFURANOSIDASE		
SUGAR PHOSPHATES		
myo-INOSITOL-1(or 4)-MONOPHOSPHATASE		
SULBACTAM		
β-LACTAMASE		I50 = 110 nM
β-LACTAMASE		I50 = 26 μM
β-LACTAMASE		I50 = 8 μM
β-LACTAMASE		Ki = 800 nM
β-LACTAMASE		I50 = 255 μg/ml
β-LACTAMASE		I50 = 2.8 μM
β-LACTAMASE BRO-1		I50 = 20 nM
β-LACTAMASE CAZ-3		I50 = 2.5 μM
β-LACTAMASE DJP-1		I50 = 210 nM
β-LACTAMASE MJ-1		I50 = 40 μM
β-LACTAMASE OXA-1		I50 = 4.7 μM
β-LACTAMASE OXA-2		I50 = 140 nM
β-LACTAMASE OXA-4		I50 = 16 μM
β-LACTAMASE OXA-5		I50 = 18 μM
β-LACTAMASE OXA-6		I50 = 51 μM
β-LACTAMASE OXA-7		I50 = 40 μM
β-LACTAMASE PSE-2		I50 = 37 μM
β-LACTAMASE PSE-4		I50 = 3.7 μM
β-LACTAMASE SHV-1		I50 = 17 μM
β-LACTAMASE SHV-2		I50 = 2.8 μM
β-LACTAMASE SHV-3		I50 = 2.7 μM
β-LACTAMASE SHV-5		I50 = 630 nM
β-LACTAMASE TEM-1		I50 = 6.1 μM
β-LACTAMASE TEM-2		I50 = 8.7 μM
β-LACTAMASE TEM-3		I50 = 30 nM
β-LACTAMASE TEM-5		I50 = 1.2 μM
β-LACTAMASE TEM-6		I50 = 450 nM
β-LACTAMASE TEM-7		I50 = 620 nM
β-LACTAMASE TEM-9		I50 = 900 nM
β-LACTAMASE TEM-10		I50 = 340 nM

β-LACTAMASE TEM-E1		$I50$ = 640 nM
β-LACTAMASE TEM-E2		$I50$ = 1.6 μM
β-LACTAMASE TEM-E3		$I50$ = 200 nM
β-LACTAMASE TEM-E4		$I50$ = 790 nM
β-LACTAMASE TLE-1		$I50$ = 5.5 μM
SULCOTRIONE		
4-HYDROXYPHENYLPYRUVATE DIOXYGENASE	C	Ki = 9.8 nM
SULFADIAZINE		
CYTOCHROME P450 2C	C	Ki = 1.7 mM
DIHYDROPTEROATE SYNTHASE		Ki = 89 μM
SULFADIMETHOXINE		
CYTOCHROME P450 2C	C	Ki = 377 μM
SULFADIMIDINE		
CYTOCHROME P450 2C	C	Ki = 476 μM
SULFADOXINE		
CYTOCHROME P450 2C	C	Ki = 205 μM
DIHYDROPTEROATE SYNTHASE		Ki = 39 μM
SULFA DRUGS		
DIHYDROPTEROATE SYNTHASE		
SULFAMERAZINE		
CYTOCHROME P450 2C	C	Ki = 972 μM
SULFAMETHAZINE		
LACTOPEROXIDASE		Ki = 110 μM
THYROID PEROXIDASE		Ki = 420 μM
SULFAMETHOXAZOLE		
CYTOCHROME P450 2C	C	Ki = 902 μM
DIHYDROPTEROATE SYNTHASE		Ki = 36 μM
SULFAMOXOLE		
CYTOCHROME P450 2C	C	Ki = 401 μM
DIHYDROPTEROATE SYNTHASE		Ki = 7 μM
(p-O-SULFAMOYL)-N-TRIDECANOYL TYRAMINE		
ESTRONE SULFATASE		$I50$ = 61 nM
N-(4-SULFAMYLBENZOYL)BENZYLAMINE		
CARBONATE DEHYDRATASE		Kd 1.1NM
SULFANILAMIDE		
CARBONATE DEHYDRATASE I		Ki = 50 μM
CARBONATE DEHYDRATASE II		Ki = 1 μM
CARBONATE DEHYDRATASE III		Ki > 5 mM
CARBONATE DEHYDRATASE IV		Ki = 26 μM
CYTOCHROME P450 2C	C	Ki = 4.5 mM
SULFAPHENAZOLE		
CYTOCHROME P450 2C	C	Ki = 558 μM
CYTOCHROME P450 2C9	C	Ki = 300 nM
CYTOCHROME P450 TB	C	Ki = 110 nM
SULFAPYRIDINE		
CYTOCHROME P450 2C	C	Ki = 799 μM
GLUTATHIONE TRANSFERASE μ		$I50$ = 85 μM
GLUTATHIONE TRANSFERASE α		$I50$ = 145 μM
GLUTATHIONE TRANSFERASE pi		$I50$ = 55 μM
SULFAQUINOXALINE		
DIHYDROPTEROATE SYNTHASE		Ki = 15 μM

SULFASALAZINE		
GLUTATHIONE TRANSFERASE		
GLUTATHIONE TRANSFERASE μ		*I*50 = 200 nM
GLUTATHIONE TRANSFERASE α		*I*50 = 28 μM
GLUTATHIONE TRANSFERASE pi		*I*50 = 13 μM
SULFATROXAZOLE		
CYTOCHROME P450 2C	C	*K*i = 892 μM
SULFAZECINE		
D-ALANYL-D-ALANINE CARBOXYPEPTIDASE		
SULFIRCIN		
PROTEIN TYROSINE-PHOSPHATASE 1B		*I*50 = 30 μM
PROTEIN TYROSINE-PHOSPHATASE CDC25A		*I*50 = 7.8 μM
PROTEIN TYROSINE-PHOSPHATASE VHR		*I*50 = 4.7 μM
SULFISOMIDINE		
CYTOCHROME P450 2C	C	*K*i = 1.3 mM
SULFISOXAZOLE		
CYTOCHROME P450 2C	C	*K*i = 1.9 mM
SULFMETURON		
ACETOLACTATE SYNTHASE		
SULFOMETURON		
ACETOLACTATE SYNTHASE		
SULFORAPHANE		
CYTOCHROME P450 2E1	C	*K*i = 37 μM
SULFOTEP		
ACETYLCHOLINESTERASE		
SULINDAC		
CATHEPSIN G		
ELASTASE LEUKOCYTE		
(E/Z) SULINDAC		
3α-HYDROXYSTEROID DEHYDROGENASE (B-SPECIFIC)		*I*50 = 6.1 μM
SULINDAC N-METHYL HYDROXAMATE		
ARACHIDONATE 5-LIPOXYGENASE		*I*50 = 1 μM
SULINDAC SULPHIDE		
CYCLOOXYGENASE 1		*I*50 = 1.3 μM
CYCLOOXYGENASE 2		*I*50 = 51 μM
CYCLOOXYGENASE 2		*I*50 = 1.2 nM
PROSTAGLANDIN-ENDOPEROXIDE SYNTHASE 1		*I*50 = 500 nM
PROSTAGLANDIN-ENDOPEROXIDE SYNTHASE 2		*I*50 = 14 μM
SULMAZOLE		
3′,5′-CYCLIC-NUCLEOTIDE PHOSPHODIESTERASE		*I*50 = 140 μM
3′,5′-CYCLIC-NUCLEOTIDE PHOSPHODIESTERASE (cycAMP)		*I*50 = 286 μM
3′,5′-CYCLIC-NUCLEOTIDE PHOSPHODIESTERASE (cycGMP)		*I*50 = 320 μM
3′,5′-CYCLIC-NUCLEOTIDE PHOSPHODIESTERASE III		35% 100 μM
3′,5′-CYCLIC-NUCLEOTIDE PHOSPHODIESTERASE IV		31% 100 μM
3′,5′-CYCLIC-NUCLEOTIDE PHOSPHODIESTERASE V		*I*50 = 21 μM
SULPHADIAZINE		
PROTEIN KINASE A		49% 400 μM
PROTEIN KINASE A		23% 400 μM
SULPHAGUANIDINE		
GUANIDINOBENZOATASE	C	*K*i = 1.1 μM

5′-(4-SULPHAMOYLBENZOYL)ADENOSINE

PHOSPHORYLASE KINASE		$I50 = 120\ \mu M$
PROTEIN TYROSINE KINASE p60v-abl		$I50 = 120\ \mu M$

3-SULPHAMOYLOXY-17β-(N-HEPTYLCARBAMOYL)ESTRA-1,3,5(10)-TRIENE

ESTRONE SULFATASE		$I50 = 500$ pM

(p-O-SULPHAMOYL)-N-TETRADECANOYLTYRAMINE

ESTRONE SULFATASE		$I50 = 56$ nM
STEROL SULFATASE		

SULPHANILAMIDE

CARBONATE DEHYDRATASE		

SULPHAPHENAZOLE

CYTOCHROME P450 2C		
CYTOCHROME P450 2C9		$I50 = 800$ nM
CYTOCHROME P450 2C9	C	$Ki = 300$ nM

SULPHAQUINOXALINE

NAD(P)H DEHYDROGENASE (QUINONE)		$I50 = 10\ \mu M$
VITAMIN K EPOXIDE REDUCTASE	NC	$Kis = 1\ \mu M$
VITAMIN K EPOXIDE REDUCTASE		$I50 = 8\ \mu M$
VITAMIN K QUINONE REDUCTASE		

SULPHASALAZINE

PHOSPHOGLYCERATE KINASE		
15-PROSTAGLANDINE DEHYDROGENASE		
THIOPURINE S-METHYLTRANSFERASE		$I50 = 78\ \mu M$

SULPHATE

ACETOLACTATE SYNTHASE		
ACETYLENECARBOXYLATE HYDRATASE		
N-ACETYLGALACTOSAMINE-4-SULFATASE		
N-ACETYLGALACTOSAMINE-6-SULFATASE		
N-ACETYLGLUCOSAMINE-6-SULFATASE		
ACYLCARNITINE HYDROLASE		
ADENINE PHOSPHORIBOSYLTRANSFERASE		
ADENYLATE KINASE		
ADENYLOSUCCINATE SYNTHASE		
AMIDOPHOSPHORIBOSYLTRANSFERASE		
AMINO-ACID N-ACETYLTRANSFERASE		
AMP DEAMINASE		
α-AMYLASE		44% 10 mM
ANDROSTENOLONE SULFATASE	NC	$Ki = 400\ \mu M$
ARYLSULFATASE	C	
ARYLSULFATASE A		
ARYLSULFATASE B	NC	
BISPHOSPHOGLYCERATE MUTASE		81% 50 mM
CARBAMOYL-PHOSPHATE SYNTHASE (AMMONIA)	C	
CARBONATE DEHYDRATASE III		
CEREBROSIDE-SULFATASE		
CHOLINESUFATASE		25% 10 mM
CREATINE KINASE	C	$Ki = 6\ \mu M$
CYANATE HYDROLASE		$I50 = 3.3$ mM
CYTOCHROME-C OXIDASE		
2-DEHYDRO-3-DEOXYPHOSPHOHEPTONATE ALDOLASE		40% 60 mM
DEOXYRIBONUCLEASE II		
ETHANOLAMINE-PHOSPHATE PHOSPHO-LYASE		49% 20 mM
GLUCOKINASE	NC	
GLUCOSE-1-PHOSPHATE ADENYLYLTRANSFERASE		$Ki = 450\ \mu M$
GLUCOSE-6-PHOSPHATE 1-EPIMERASE		
GLUCURONATE 2-SULFATASE		$I50 = 80\ \mu M$
[GLUTAMATE-AMMONIA-LIGASE] ADENYLYLTRANSFERASE		$I50 = 5$ mM
γ-GLUTAMYL HYDROLASE		
GLUTATHIONE PEROXIDASE		

GLYCOGEN (STARCH) SYNTHASE		
GLYCOPROTEIN N-PALMITOYLTRANSFERASE		
GLYCOSULFATASE		34% 5 mM
GTP CYCLOHYDROLASE I		
HYDROXYPYRUVATE REDUCTASE		
IDURONATE 2-SULFATASE		Ki = 66 μM
IDURONATE 2-SULFATASE		I50 = 1 mM
myo-INOSITOL-1(or 4)-MONOPHOSPHATASE	C	47% 17 mM
ISOCITRATE LYASE		
KYNURENINE-OXOGLUTARATE AMINOTRANSFERASE		
MALONATE-SEMIALDEHYDE DEHYDRATASE		
MALONYL-CoA DECARBOXYLASE		
5′-NUCLEOTIDASE		37% 20 mM
PALMITOYL-CoA HYDROLASE		
PEPTIDE α-N-ACETYLTRANSFERASE		
PHOSPHATASE ACID	C	Ki = 17 mM
PHOSPHATE ACETYLTRANSFERASE		
PHOSPHATIDYLSERINE DECARBOXYLASE		
PHOSPHOENOLPYRUVATE CARBOXYKINASE (PYROPHOSPHATE)	C	Ki = 18 mM
6-PHOSPHOFRUCTO-2-KINASE		
PHOSPHOGLUCONATE DEHYDROGENASE (DECARBOXYLATING)		
PHOSPHOGLYCERATE KINASE		
PHOSPHOPENTOMUTASE		85% 50 mM
PHOTINUS-LUCIFERIN 4-MONOOXYGENASE (ATP-HYDROLYSING)		
POLYNUCLEOTIDE ADENYLYLTRANSFERASE		
POLYNUCLEOTIDE 3′-PHOSPHATASE		
POLYPHOSPHATE KINASE		
PROPANEDIOL-PHOSPHATE DEHYDROGENASE		
PYRUVATE KINASE (L TYP)		
RIBONUCLEASE P		
RIBULOSE-PHOSPHATE EPIMERASE		57% 100 mM
SIALIDASE ACID ACTIVATED		
SUCROSE-PHOSPHATE SYNTHASE		
SULFATE ADENYLYLTRANSFERASE		
SULFITE REDUCTASE (NADPH)		
N-SULFOGLUCOSAMINE-6-SULFATASE		43% 200 μM
THIOSULFATE SULFURTRANSFERASE		Ki = 36 mM
THIOSULFATE-THIOL SULFURTRANSFERASE		
THREONINE SYNTHASE	C	
TRANSALDOLASE		
TRANSKETOLASE		
TRIOSE-PHOSPHATE ISOMERASE		Ki = 4.5 mM
TRIOSE-PHOSPHATE ISOMERASE		Ki = 8 mM
TRIOSE-PHOSPHATE ISOMERASE		Ki = 5.7 mM
XAA-HIS DIPEPTIDASE		35% 3 mM

SULPHATED GALACTOSAMINOGLYCANS
CHODROTIN 6-SULFOTRANSFERASE

SULPHATED GLYCOSAMINOGLYCANS
GLYCOSAMINOGLYCAN SULFOTRANSFERASES

SULPHATED POLYSACCHARIDES
ACROSIN

SULPHATHIAZOLE
IODIDE PEROXIDASE

SULPHATHIOZOLE
DIHYDROPTEROATE SYNTHASE

SULPHHYDRYL REAGENTS
CATHEPSIN B
CATHEPSIN H
CATHEPSIN L
N5-FORMYL TETRAHYDROFOLATE CYCLODEHYDRASE

FRUCTOSE-BISPHOSPHATE ALDOLASE
GLUCOSE-6-PHOSPHATASE
GLUTAMINE-PHENYLPYRUVATE TRANSAMINASE
GLYCEROL-3-PHOSPHATE O-ACYLTRANSFERASE
HYDROXYMETHYLGLUTARYL-CoA REDUCTASE (NADPH)
LYSOSOMAL CARBOXYPEPTIDASE B
TYROSINE TRANSAMINASE

SULPHIDE

ACONITATE HYDRATASE		
AMINOPEPTIDASE (HUMAN LIVER)	MIX	*K*i = 76 μM
ATROLYSIN A		
L-ASCORBATE OXIDASE		
CARBONATE DEHYDRATASE		*K*i = 3 μM
CARBOXYPEPTIDASE A		
CYSTATHIONINE γ-LYASE		*K*i = 7.5 μM
CYSTEINE-tRNA LIGASE		
CYTOCHROME-C OXIDASE		100 μM
FORMATE DEHYDROGENASE		
L-GALACTONOLACTONE OXIDASE		
3-HYDROXYBUTYRATE DEHYDROGENASE		
IRON-CYTOCHROME-C REDUCTASE		
LACTOPEROXIDASE	SSI	
LEUCYL AMINOPEPTIDASE		
MALEYLACETOACETATE ISOMERASE		
MEMBRANE ALANINE AMINOPEPTIDASE		
NITRATE REDUCTASE		
NITRATE REDUCTASE (CYTOCHROME)		
PEROXIDASE		*I*50 = 53 μM
SULFATE ADENYLYLTRANSFERASE		
SULFITE REDUCTASE (NADPH)		
THIOSULFATE-THIOL SULFURTRANSFERASE		
XAA-HIS DIPEPTIDASE		87% 300 μM

SULPHIDEAMINE

NITRITE REDUCTASE (NAD(P)H)

SULPHINPYRAZONE

UDP-GLUCURONOSYLTRANSFERASE

SULPHITE

2-(ACETAMIDOMETHYLENE)SUCCINATE HYDROLASE		
N-ACETYLGALACTOSAMINE-6-SULFATASE		99% 17 mM
N-ACETYLGLUCOSAMINE-6-SULFATASE		
L-AMINOADIPATE-SEMIALDEHYDE DEHYDROGENASE		
AROMATIC α-KETOACID DEHYDROGENASE		30% 1 mM
ARYLSULFATASE	C	
ARYLSULFATASE A	C	*K*i = 2 μM
L-ASCORBATE OXIDASE		
CARBONATE DEHYDRATASE		
CATECHOL OXIDASE		
CELLULASE		
CEREBROSIDE-SULFATASE		*K*i = 12 μM
CREATINE KINASE		
CYANATE HYDROLASE		*I*50 = 120 μM
DOPAMINE β-MONOOXYGENASE	IR	
FERREDOXIN-NITRITE REDUCTASE		24% 10 mM
FORMATE DEHYDROGENASE		
FORMATE DEHYDROGENASE (NADP)		
L-GALACTONOLACTONE OXIDASE		
3-HYDROXYBUTYRATE DEHYDROGENASE		
HYDROXYLAMINE REDUCTASE (NADH)		
LACCASE		
MALATE DEHYDROGENASE (OXALACETATE-DECARBOXYLATING) (NADP)		
METHANOL-5-HYDROXYBENZIMIDAZOLYLCOBAMIDE Co-METHYLTRANSFERASE		

Inhibitor / Enzyme	Type	Value
NITRATE REDUCTASE		
NITRATE REDUCTASE (CYTOCHROME)		
NITRITE REDUCTASE (NAD(P)H)		
PEPTIDYLGLYCINE MONOOXYGENASE	IR	
PEROXIDASE		$I50 = 1.1$ mM
PSEUDOMONAS CYTOCHROME OXIDASE		
PYROPHOSPHATE-FRUCTOSE-6-PHOSPHATE 1-PHOSPHOTRANSFERASE		
STAPHYLOCOCCAL CYSTEINE PROTEINASE		
SULFOACETALDEHYDE LYASE		
N-SULFOGLUCOSAMINE-6-SULFATASE		92% 200 μM
THIOSULFATE DEHYDROGENASE		
THIOSULFATE SULFURTRANSFERASE		
THIOSULFATE-THIOL SULFURTRANSFERASE		
UROCANATE HYDRATASE		
SULPHITE IRON		
PEPTIDYLGLYCINE MONOOXYGENASE	IR	
2-SULPHOBENZOIC ACID		
TRIACYLGLYCEROL LIPASE		
SULPHOBETAINE		
CARNITINE O-PALMITOYLTRANSFERASE		
SULPHOBROMOPHTHALEIN		
GLUTATHIONE TRANSFERASE A1-1		$I50 = 25$ μM
GLUTATHIONE TRANSFERASE M1a-1a		$I50 = 5$ μM
GLUTATHIONE TRANSFERASE P1-1		$I50 = 20$ μM
ISOCITRATE DEHYDROGENASE (NAD)		
LEUKOTRIENE-C4 SYNTHASE		
SULPHOCARNITINE		
CARNITINE O-PALMITOYLTRANSFERASE		
S-SULPHOCYSTEINE		
GLUTAMATE-CYSTEINE LIGASE	IRR	
S-SULPHOGLUTATHIONE		
GLUTATHIONE TRANSFERASE		
S-SULPHOHOMOCYSTEINE		
GLUTAMATE-CYSTEINE LIGASE	IRR	
SULPHOMETURON METHYL		
ACETOLACTATE SYNTHASE		$Ki = 65$ nM
ACETOLACTATE SYNTHASE		92% 1 μM
ACETOLACTATE SYNTHASE I		$Ki = 950$ μM
ACETOLACTATE SYNTHASE II		$Ki = 1.8$ μM
ACETOLACTATE SYNTHASE III		$Ki = 190$ μM
SULPHONAMIDE		
DIHYDROPTEROATE SYNTHASE		
6-SULPHONAMIDE URACIL		
DIHYDROOROTATE DEHYDROGENASE		71% 1 mM
6-SULPHONAMIDO-3-CHLORO-3H-1,3,4-THIDIAZOLO[2,3-C]1,2,4-THIADIAZOLE		
CARBONATE DEHYDRATASE		$I50 = 5.6$ nM
6β-SULPHONAMIDOPENICILLANIC ACIDS		
β-LACTAMASE		
SULPHONYLFLUORIDE		
ELASTASE PANCREATIC		
STEROL ESTERASE		
2-(4-SULPHONYLMETHYLPHENYL)-3-(4-FLUOROPHENYL)INDENE		
CYCLOOXYGENASE 2		$I50 = 1$ nM

Inhibitor / Enzyme	Type	Value
SULPHONYL UREA		
ARGINASE		
SULPHONYLUREA		
ACETOLACTATE SYNTHASE		
2(p-SULPHOPHENYLAZO)-1,8-DIHYDROXY-3,6-NAPHTHALENE-DISULPHONIC ACID		
PHOSPHOGLYCERATE KINASE	C	$Ki = 148\ \mu M$
p-SULPHOPHENYLISOTHIOCYANATE		
PHOSPHATE CARRIER MITOCHONDRIA		
3-SULPHOPROPANOIC ACID		
3-OXOACID CoA-TRANSFERASE		
3-(1-SULPHOPROPYL)-7-MERCAPTOHEXANOYL-L-THREONINEPHOSPHATE		
COENZYME-M-7-MERCAPTOHEPTANOYLTHREONINE-PHOSPHATE-HETEROSULFIDE HYDROGENASE		
β-SULPHOPYRUVIC ACID		
PHOSPHOENOLPYRUVATE CARBOXYKINASE (GTP)	C	$Ki = 1\ \mu M$
6-SULPHO-O-α-D-QUINOVOPYRANOSYLDIACYLGLYCEROL		
Na^+/K^+-EXCHANGING ATPASE		
SULPHOQUINOVOSYLDIACYLGLYCEROL		
α-GLUCOSIDASE	C	$Ki = 2.9\ \mu M$
α-GLUCOSIDASE	C	$Ki = 2.9\ \mu M$
O-SULPHO-D-SERINE		
D-AMINO-ACID TRANSAMINASE		
SULPHO-N-SUCCINIMIDYL MYRISTIC ACID		
FATTY ACID TRANSPORT	IR	
SULPHO-N-SUCCINIMIDYL OLEIC ACID		
FATTY ACID TRANSPORT	IR	
FATTY ACID TRANSPORT		
SULPHUR		
ADENYLATE KINASE		$I50 = 400\ \mu M$
ADENYLATE KINASE		$I50 = 140\ \mu M$
ADENYLATE KINASE		$I50 = 3.6\ \mu M$
ADENYLATE KINASE		$I50 = 2\ \mu M$
ADENYLATE KINASE		$I50 = 17\ \mu M$
ADENYLATE KINASE		$I50 = 3\ \mu M$
SULPHUR ELEMENTAL		
L-SERINE DEHYDRATASE		$I50 = 6\ \mu M$
SULPROFOS		
ACETYLCHOLINESTERASE		
SULVANINE		
THROMBIN		$I50 = 27\ \mu M$
TRYPSIN		$I50 = 12\ \mu M$
SUPEROXIDE DISMUTASE		
GALACTOSE OXIDASE		
INDOLEAMINE-PYRROLE 2,3-DIOXYGENASE		
INDOLE 2,3-DIOXYGENASE		
2-NITROPROPNAE DIOXYGENASE		
TRYPTOPHAN 2,3-DIOXYGENASE		
SUPPRESCINS		
NUCLEOTIDE TRIPHOSPHATASES (ECTO)		
SUPROFEN		
CARBONYL REDUCTASE (NADPH)		62% 1 mM
PROSTAGLANDIN-ENDOPEROXIDE SYNTHASE 1		$I50 = 500\ nM$
PROSTAGLANDIN-ENDOPEROXIDE SYNTHASE 2		$I50 = 2\ \mu M$

SURAMIN

Enzyme	Type	Value
N-ACETYLGALACTOSAMINE-4-SULFATASE		36% 5 mM
β-N-ACETYLHEXOSAMINIDASE		60% 100 μM
ADENOSINETRIPHOSPHATASE ECTO		*I*50 = 100 μM
ADENOSINETRIPHOSPHATASE ECTO		*I*50 = 50 μM
BIS(5′-ADENOSYL)-TRIPHOSPHATASE	C	*K*i = 300 nM
BIS(5′-NUCLEOSYL)-TETRAPHOSPHATASE (ASYMMETRICAL)	C	*K*i = 5 μM
CATHEPSIN B		
Ca^{2+}-TRANSPORTING ATPASE		
CEREBROSIDE-SULFATASE		58% 5 mM
CYCLIN DEPENDENT KINASE 1		*I*50 = 4 μM
DIADENOSINEPOLYPHOSPHATE HYDROLASE ECTO NEUTRAL	C	*K*i = 1.8 μM
DIHYDROFOLATE REDUCTASE		*I*50 = 38 μM
DNA POLYMERASE α		
DNA POLYMERASE α	C	*K*i = 350 nM
DNA PRIMASE	C	*K*i = 2,6 μM
DNA TOPOISOMERASE (ATP-HYDROLYSING)		*I*50 = 100 μg/ml
DNA TOPOISOMERASE (ATP-HYDROLYSING)		*I*50 = 5 μM
ELASTASE LEUKOCYTE		
FOLYLPOLYGLUTAMATE SYNTHASE		*I*50 = 900 nM
β-FRUCTOFURANOSIDASE		
FRUCTOSE-BISPHOSPHATE ALDOLASE		*I*50 = 17 μM
FRUCTOSE-BISPHOSPHATE ALDOLASE		*I*50 = 44 μM
α-GALACTOSIDASE		45% 100 μM
β-GALACTOSIDASE		71% 100 μM
GLUCOSE-6-PHOSPHATE ISOMERASE		*I*50 = 60 μM
GLUCOSE-6-PHOSPHATE ISOMERASE	C	*K*i = 360 μM
β-GLUCURONIDASE		75% 100 μM
GLYCERALDEHYDE-3-PHOSPHATE DEHYDROGENASE (PHOSPHORYLATING)		*I*50 = 23 μM
GLYCERALDEHYDE-3-PHOSPHATE DEHYDROGENASE (PHOSPHORYLATING)		*I*50 = 3 μM
GLYCERALDEHYDE-3-PHOSPHATE DEHYDROGENASE (PHOSPHORYLATING)		*I*50 = 74 μM
GLYCERALDEHYDE-3-PHOSPHATE DEHYDROGENASE (PHOSPHORYLATING)		*I*50 = 74 mM
GLYCERALDEHYDE-3-PHOSPHATE DEHYDROGENASE (PHOSPHORYLATING)		*I*50 = 418 μM
GLYCEROL KINASE		*I*50 = 10 μM
GLYCEROL KINASE		*I*50 = 130 μM
GLYCEROL KINASE		*I*50 = 140 μM
GLYCEROL-3-PHOSPHATE DEHYDROGENASE (NAD)		*I*50 = 10 μM
GLYCEROL-3-PHOSPHATE DEHYDROGENASE (NAD)		*I*50 = 40 μM
HEPARANASE	MIX	*K*i = 48 μM
HEPARAN N-SULFATASE		55% 1 mM
HEXOKINASE		*I*50 = 210 μM
HEXOKINASE		*I*50 = 220 μM
HEXOKINASE		*I*50 = 24 μM
H^{+}-TRANSPORTING ATPASE		*I*50 = 50 μM
HYALURONOGLUCOSAMINIDASE		90% 1 mM
IDURONATE 2-SULFATASE		100% 50 μM
L-IDURONIDASE		46% 1 mM
INTEGRASE		*I*50 = 200 nM
METHYLENETETRAHYDROFOLATE CYCLOHYDROLASE		
Mg^{2+} TRANSPORTING ATPASE		*I*50 = 200 nM
Na^{+}/K^{+}-EXCHANGING ATPASE		
PHOSPHATASE ACID		75% 1 mM
1-PHOSPHATIDYLINOSITOL 3-KINASE		
PHOSPHATIDYLINOSITOL PHOSPHOLIPASE Cγ		*I*50 = 218 μM
6-PHOSPHOFRUCTOKINASE		*I*50 = 3 μM
6-PHOSPHOFRUCTOKINASE		*I*50 = 600 nM
PHOSPHOGLUCONATE DEHYDROGENASE (DECARBOXYLATING)	C	*K*i = 20 μM
PHOSPHOGLUCONATE DEHYDROGENASE (DECARBOXYLATING)	C	*K*i = 15 μM
PHOSPHOGLYCERATE KINASE	C	*K*i = 130 μM
PHOSPHOGLYCERATE KINASE	C	*K*i = 8 μM
PHOSPHOGLYCERATE KINASE	C	*K*i = 55 μM
PHOSPHOGLYCERATE KINASE		*I*50 = 80 μM
PHOSPHOGLYCERATE KINASE	C	*K*i = 20 μM
PHOSPHOGLYCERATE KINASE	C	*K*i = 250 μM

PHOSPHOLIPASE Cγ (PHOSPHATIDYLINOSITOL SPECIFIC)		$I50 = 63\ \mu M$
PHOSPHOLIPASE D		$I50 = 15\ \mu M$
PROTEINASE 3		
PROTEIN KINASE		
PROTEIN KINASE A		$I50 = 656\ \mu M$
PROTEIN KINASE A		$I50 = 10\ \mu M$
PROTEIN KINASE C	C	$Ki = 10\ \mu M$
PROTEIN KINASE C		$I50 = 50\ \mu M$
PROTEIN TYROSINE KINASE EGFR		$I50 = 70\ \mu M$
PROTEIN TYROSINE KINASE EGFR		$I50 = 70\ \mu M$
PROTEIN TYROSINE KINASE HPK40		$I50 = 400\ \mu M$
PROTEIN TYROSINE KINASE p56lck		$I50 = 200\ \mu M$
PROTEIN TYROSINE KINASE pp60c-Src		$I50 = 200\ \mu M$
PROTEIN TYROSINE-PHOSPHATASE 1B		$Kis = 4\ \mu M$
PROTEIN TYROSINE-PHOSPHATASE CD45	IR	$I50 = 80\ \mu M$
PROTEIN TYROSINE PHOSPHATASE VHR		
PROTEIN TYROSINE-PHOSPHATASE YERSINIA		
RIBONUCLEASE H		100% 5 mM
RNA-DIRECTED DNA POLYMERASE	C	$Ki = 540$ nM
RNA POLYMERASE	C	$Ki = 700$ nM
SIALIDASE		100% 50 μg/ml
SIALIDASE		62% 50 μg/ml
STEROID 5α-REDUCTASE 1	NC	$Ki = 2.4\ \mu M$
THYMIDYLATE SYNTHASE		$I50 = 87\ \mu M$
TRIOSE-PHOSPHATE ISOMERASE		$Ki = 100\ \mu M$
TRIOSE-PHOSPHATE ISOMERASE		$Ki = 100\ \mu M$
TRIOSE-PHOSPHATE ISOMERASE		$Ki = 600\ \mu M$
UREASE	C	

SURFACTANTS

CYCLOPROPANE-FATTY-ACYL-PHOSPHOLIPID SYNTHASE

SURFACTIN

PHOSPHATASE ALKALINE		$I50 = 70\ \mu M$
PHOSPHOLIPASE A2		$I50 = 8.5\ \mu M$

SURMANIN

PROTEIN TYROSINE-PHOSPHATASE		$I50 = 90\ \mu M$

SW 3a

H^+-TRANSPORTING PYROPHOSPHATASE		83% 10 μM

SW 26

H^+-TRANSPORTING PYROPHOSPHATASE		91% 10 μM

SW 26b

H^+-TRANSPORTING PYROPHOSPHATASE		100% 10 μM

SWAINSONINE

α-MANNOSIDASE		$I50 = 100$ nM
α-MANNOSIDASE	C	$Ki = 70$ nM
MANNOSYL-OLIGOSACCHARIDE 1,3-1,6-α-MANNOSIDASE		$I50 = 100$ nM
MANNOSYL-OLIGOSACCHARIDE 1,3-1,6-α-MANNOSIDASE		$I50 = 1\ \mu M$

L-SWAINSONINE

NARINGINASE		$Ki = 450$ nM

SWARTZIA PICKELLI TRYPSIN INHIBITOR

CHYMOTRYPSIN		$Ki = 650$ nM
KALLIKREIN PLASMA		$Ki = 4$ nM
TRYPSIN		$Ki = 10$ nM

SWEAT GLAND CYSTEINE PROTEINASE INHIBITOR

PROTEINASES CYSTEINE	NC	

SWEET POTATO LEAVES TRYPSIN INHIBITOR

TRYPSIN

SWERTIFRANCHESIDE
DNA LIGASE I — $I50 = 11\ \mu M$
RNA-DIRECTED DNA POLYMERASE
RNA-DIRECTED DNA POLYMERASE — $I50 = 43\ \mu M$

SWERTISIN
XANTHINE OXIDASE

SYLIBIN
GLUTATHIONETRANSFERASE 1-1 — C — $Ki = 8\ \mu M$
GLUTATHIONE TRANSFERASE 2-2 — NC — $Ki = 41\ \mu M$
GLUTATHIONE TRANSFERASE 3-3 — NC — $Ki = 800$ nM
GLUTATHIONETRANSFERASE 3-4 — $I50 = 1\ \mu M$
GLUTATHIONE TRANSFERASE 4-4 — NC — $Ki = 500$ nM
HYALURONOGLUCOSAMINIDASE

SYLL
AROMATASE — $I50 = 95\ \mu M$

SYN 1012
β-LACTAMASE

SYNEFUNGIN
ADENOSYLHOMOCYSTEINASE — $Ki = 6\ \mu M$

SYNERGIMYCIN
PEPTIDYLTRANSFERASE

α-SYNUCLEIN
PHOSPHOLIPASE D1 — $Ki > 50\ \mu M$
PHOSPHOLIPASE D2 — $Ki = 10$ nM

β-SYNUCLEIN
PHOSPHOLIPASE D2

SYRINGETIN
PROTEIN TYROSINE KINASE p56lck — $I50 = 20\ \mu g/ml$

SYRINGIC ACID
ARACHIDONATE 5-LIPOXYGENASE — $I50 = 240\ \mu M$
THIOPURINE S-METHYLTRANSFERASE

SYRINGOMYCIN E
H^+-TRANSPORTING ATPASE

SYRINGOPEPTIN 22-A
H^+-TRANSPORTING ATPASE — 53% 10 μM

SYRINGOPEPTIN 25-A
H^+-TRANSPORTING ATPASE — 57% 10 μM

SYSTOSE
TRIACYLGLYCEROL LIPASE

T

T 330
H^+/K^+-EXCHANGING ATPASE — $I50 = 75\ \mu M$

T 440
3′,5′-CYCLIC-NUCLEOTIDE PHOSPHODIESTERASE
3′,5′-CYCLIC-NUCLEOTIDE PHOSPHODIESTERASE II — $I50 = 23\ \mu M$
3′,5′-CYCLIC-NUCLEOTIDE PHOSPHODIESTERASE III — $I50 = 49\ \mu M$
3′,5′-CYCLIC-NUCLEOTIDE PHOSPHODIESTERASE IV — $I50 = 71$ nM
3′,5′-CYCLIC-NUCLEOTIDE PHOSPHODIESTERASE V — $I50 = 67\ \mu M$

Inhibitor / Enzyme	Type	Value
T 614		
CYCLOOXYGENASE 2		$I50 \approx 7.7$ µg/ml
T 0757		
ARACHIDONATE 5-LIPOXYGENASE		
T 757		
ARACHIDONATE 5-LIPOXYGENASE		
T 776		
H^+/K^+-EXCHANGING ATPASE		
T 794		
AMINE OXIDASE MAO-A		
T 799		
ARACHIDONATE 5-LIPOXYGENASE		
T 2591		
STEROL O-ACYLTRANSFERASE		$I50 = 260$ nM
STEROL O-ACYLTRANSFERASE		$I50 = 4.1$ µM
STEROL O-ACYLTRANSFERASE		$I50 = 4.6$ µM
T 30177		
INTEGRASE		$I50 = 79$ nM
T 30678		
INTEGRASE		$I50 = 98$ nM
9TA		
PROTEIN TYROSINE KINASE p56lck		$I50 = 2.1$ µM
PROTEIN TYROSINE KINASE c-Src		$I50 = 120$ nM
TA 6366		
PEPTIDYL-DIPEPTIDASE A		
TA 3037A		
GLUTATHIONE TRANSFERASE	C	$Ki = 4.9$ µM
TABACCO INVERTASE INHIBITOR		
β-FRUCTOFURANOSIDASE		
TABACCO TRYPSIN INHIBITOR		
CHYMOTRYPSIN		
TRYPSIN		
TABANAMIN		
THROMBIN		
TABERSONINE		
17-O-DEACETYLVINDOLINE O-ACETYLTRANSFERASE		
TABTOXIN		
GLUTAMATE-AMMONIA LIGASE		
TABTOXININE β-LACTAME		
GLUTAMATE-AMMONIA LIGASE	IRR	$Ki = 1$ µM
TABUN		
ACETYLCHOLINESTERASE		
CARBOXYLESTERASE		
CHOLINESTERASE		
TACRINE		
ACETYLCHOLINESTERASE	REV	$I50 = 130$ nM
ACETYLCHOLINESTERASE		$I50 = 358$ nM
ACETYLCHOLINESTERASE	NC	$I50 = 10$ nM
ACETYLCHOLINESTERASE		$I50 = 1.3$ µM
ACETYLCHOLINESTERASE		$Ki = 50$ nM

ACETYLCHOLINESTERASE		$I50 = 240$ nM
AMINE OXIDASE MAO-A	C	$Ki = 13$ μM
AMINE OXIDASE MAO-B	C	30% 100 μM
CHOLINESTERASE		$I50 = 400$ nM
CHOLINESTERASE		$I50 = 100$ nM
CHOLINESTERASE		$I50 = 47$ nM
CHOLINESTERASE		$I50 = 5.2$ μM
HISTAMINE N-METHYLTRANSFERASE	C	$Ki = 35$ nM
HISTAMINE N-METHYLTRANSFERASE	MIX	$I50 = 200$ nM

TAENIAESTATIN
PROTEINASE TAENIA TAENIAEFORMIS

D-TAGATOSE
KETOHEXOKINASE

TAGATOSE-6-PHOSPHATE
PYROPHOSPHATE-FRUCTOSE-6-PHOSPHATE 1-PHOSPHOTRANSFERASE

TAGATURONATE
FRUCTURONATE REDUCTASE
TAGATURONATE REDUCTASE

TAGETITOXIN
RNA POLYMERASE

TAK 147

ACETYLCHOLINESTERASE		$I50 = 51$ nM

TALOPEPTIN
PEPTIDYL-DIPEPTIDASE A
PSEUDOMONAS AEROGINOSA NEUTRAL PROTEINASE

TAMOXIFEN

β-ADRENERGIC-RECEPTOR KINASE		
Ca^{2+}-TRANSPORTING ATPASE		46% 25 μM
CYTOCHROME P450 2A1		$I50 = 192$ μM
CYTOCHROME P450 3A1/2		$I50 = 104$ μM
CYTOCHROME P450 2B1		$I50 = 69$ μM
CYTOCHROME P450 2C11		$I50 = 173$ μM
MYOSIN-LIGHT-CHAIN KINASE		77% 50 μM
PROTEIN KINASE C	NC	80% 100 μM
PROTEIN KINASE C		$I50 = 194$ μM
SPHINGOMYELIN PHOSPHODIESTERASE		79% 10 μM

cis-TAMOXIFEN

3′,5′-CYCLIC-NUCLEOTIDE PHOSPHODIESTERASE (Ca^{2+}/CALMODULIN)		$I50 = 6.5$ μM
Na^+/K^+-EXCHANGING ATPASE		$Ki = 6$ μM
Na^+/K^+-EXCHANGING ATPASE		$Ki = 4.2$ μM

trans-TAMOXIFEN

3′,5′-CYCLIC-NUCLEOTIDE PHOSPHODIESTERASE (Ca^{2+}/CALMODULIN)		$I50 = 6.8$ μM
Na^+/K^+-EXCHANGING ATPASE		$Ki = 7$ μM
Na^+/K^+-EXCHANGING ATPASE		$Ki = 4.5$ μM

TAMOXIFEN TRIMETHYLAMMONIUMIODIDE

PROTEIN KINASE C		$I50 = 19$ μM

TAN 931

AROMATASE		$I50 = 162$ μM
AROMATASE	U	$Ki = 40$ μM

TAN 999
PROTEIN KINASE C

TAN 1831

FARNESYL PROTEINTRANSFERASE		$I50 = 30$ μM

TAN 1496 A
DNA TOPOISOMERASE

TAN 1518 A
DNA TOPOISOMERASE

TAN 1518 B
DNA TOPOISOMERASE

TAN 1496 C
DNA TOPOISOMERASE

TAN 1496 E
DNA TOPOISOMERASE

TANNIC ACID

ALDEHYDE REDUCTASE		*I*50 = 1.8 μg/ml
DOPAMINE SULFOTRANSFERASE		54% 6.7 μM
17α-ETHINYLOESTRADIOL SULFOTRANSFERASE		43% 6.7 μM
α-GLUCOSIDASE		32% 250 μM
GLUTATHIONE TRANSFERASE		*I*50 = 1 μM
H^+/K^+-EXCHANGING ATPASE	C	*I*50 = 29 nM
HYALURONIDASE PH20		10% 50 μM
PECTINESTERASE		
PROSTAGLANDIN-ENDOPEROXIDE SYNTHASE		
SUCRASE		49% 250 μM

TANNIN

CELLULASE	
ELASTASE LEUKOCYTE	*I*50 = 300 ng/ml
GLYCOSYLTRANSFERASE	
LACCASE	
PECTIN ESTERASE	
PROTEIN KINASE C	

TANNINS
MYOSIN-LIGHT-CHAIN KINASE
PROTEIN KINASE A
PROTEIN KINASE C
PROTEIN KINASE (Ca^{2+} DEPENDENT)

TANNINS CONDENSED
MYOSIN-LIGHT-CHAIN KINASE
PROTEIN KINASE A
PROTEIN KINASE C

TANSINONE I
ALDEHYDE REDUCTASE — *I*50 = 4.8 μM

TANSINONE IIA
ALDEHYDE REDUCTASE — *I*50 = 1.1 μM

TAPI-2
ACE SECRETASE — *I*50 = 18 μM

D-TARTARIC ACID
GLUCARATE DEHYDRATASE

(+)TARTRATE
GLUCARATE DEHYDRATASE — 58% 1 mM

TARTRATE

3-DEHYDROQUINATE DEHYDRATASE	*K*i = 21 mM
4-HYDROXY-2-OXOGLUTARATE ALDOLASE	63% 40 mM
NUCLEOTIDASE	82% 1 mM
NUCLEOTIDASE	80% 10 μM
PHOSPHATASE ACID	*I*50 = 1.6 mM
PHOSPHOENOLPYRUVATE CARBOXYKINASE (ATP)	

PHOSPHOENOLPYRUVATE DECARBOXYLASE		Ki = 3.5 mM
3-PHOSPHOGLYCERATE PHOSPHATASE	IRR	
PHOSPHOPROTEIN PHOSPHATASE		38% 10 mM
PHOTINUS-LUCIFERIN 4-MONOOXYGENASE (ATP-HYDROLYSING)		
3-PHYTASE		
PROTEIN TYROSINE-PHOSPHATASE ACID		I50 = 335 µM
D(+)TARTRATE		
FUMARATE HYDRATASE	C	Ki = 25 mM
D-TARTRATE		
HYDROXYPYRUVATE REDUCTASE		
L+ TARTRATE		
PHOSPHATASE ACID	NC	Ki = 102 mM
L-TARTRATE		
HYDROXYPYRUVATE REDUCTASE		
meso-TARTRATE		
D-ASPARTATE OXIDASE		64% 10 mM
D-MALATE DEHYDROGENASE (DECARBOXYLATING)		
TARTRATE DEHYDROGENASE		
TARTRAZINE		
DOPAMINE SULFOTRANSFERASE		94% 6.7 µM
17α-ETHINYLOESTRADIOL SULFOTRANSFERASE		17% 6.7 µM
TARTRONATE		
PHOSPHOGLYCERATE KINASE		Ki = 38 mM
TARTRONATE SEMIALDEHYDE PHOSPHATE		
PHOSPHOPYRUVATE HYDRATASE		
TARTRONIC ACID		
3-HYDROXYBUTYRATE DEHYDROGENASE		
TAS 103		
DNA TOPOISOMERASE		I50 = 2 µM
DNA TOPOISOMERASE (ATP-HYDROLYSING)		I50 = 6.5 µM
TAURINE		
GLUTAMATE DEHYDROGENASE		
PANTOATE-β-ALANINE LIGASE		52% 1 mM
PHOSPHATIDYLCHOLINE METHYLTRANSFERASE		
TAURINE CHLORAMINE		
NITRIC-OXIDE SYNTHASE		56% 1 mM
TAUROCHENODEOXYCHOLIC ACID		
PROTEIN KINASE A		46% 400 µM
TAUROCHOLIC ACID		
CARBONATE DEHYDRATASE	NC	Ki = 821 µM
CARBONATE DEHYDRATASE I	NC	Ki = 2.4 mM
CARBONATE DEHYDRATASE II	NC	Ki = 714 µM
CHOLATE-CoA LIGASE		
DIACYLGLYCEROL O-ACYLTRANSFERASE		
DOLICHOL ESTERASE		> 0.06%
α-GALACTOSIDASE		
GALACTOSYLCERAMIDASE		> 0.3%
GLUCOSE-6-PHOSPHATASE		
β-GLUCOSIDASE		
GLUTATHIONE TRANSFERASE		
GLYCINE N-CHOLOYLTRANSFERASE		
GLYCOLIPID 2-α-MANNOSYLTRANSFERASE		
2-HYDROXYACYLSPHINGOSINE 1β-GALACTOSYLTRANSFERASE		
β-MANNOSIDASE		

PLASMANYLETHANOYLAMINE DESATURASE		
PROTEIN KINASE A		47% 400 μM
RETINOL FATTY-ACYLTRANSFERASE		100% 0.5%
RETINOL O-FATTY-ACYLTRANSFERASE		
STEROL 3β-GLUCOSYLTRANSFERASE		
STERYL-SULFATASE		90% 1 mg/ml
TRIACYLGLYCEROL LIPASE		
TAURODEOXYCHOLIC ACID		
BILE-SALT SULFOTRANSFERASE		
CYTOCHROME-b5 REDUCTASE		
GLUTATHIONE TRANSFERASE		
GLYCINE N-CHOLOYLTRANSFERASE		
PROTEIN KINASE A		46% 400 μM
TAUROSPONGEN A		
DNA POLYMERASE β		*K*i = 1.7 μM
RNA-DIRECTED DNA POLYMERASE		*K*i = 1.7 μM
TAUROURSODEOXYCHOLIC ACID		
UDP-GLUCURONOSYLTRANSFERASE	C	*K*i = 201 μM
TAUTOMYCIN		
PHOSPHOPROTEIN PHOSPHATASE 1		*K*i = 160 pM
PHOSPHOPROTEIN PHOSPHATASE 1		*I*50 = 1 nM
PHOSPHOPROTEIN PHOSPHATASE 1		*I*50 = 700 pM
PHOSPHOPROTEIN PHOSPHATASE 1		*K*i = 480 pM
PHOSPHOPROTEIN PHOSPHATASE 1		*I*50 = 32 nM
PHOSPHOPROTEIN PHOSPHATASE 2A		*I*50 = 650 pM
PHOSPHOPROTEIN PHOSPHATASE 2A		*K*i = 400 pM
PHOSPHOPROTEIN PHOSPHATASE 2A		*K*i = 286 nM
PHOSPHOPROTEIN PHOSPHATASE 2A		*I*50 = 32 nM
PHOSPHOPROTEIN PHOSPHATASE (CHICKEN GIZZARD ACTOMYOSIN)		*I*50 = 6 nM
TAXIFOLIN		
CYTOCHROME P450		*I*50 = 460 μM
GLUTATHIONE TRANSFERASE		*I*50 = 600 μM
NADH OXIDASE		*I*50 = 173 nmole
PROTEIN KINASE CASEIN KINASE G		9% 2.5 μM
PROTEIN TYROSINE KINASE p40		*I*50 = 300 μg/ml
RNA-DIRECTED DNA POLYMERASE		
SUCCIN OXIDASE		*I*50 = 220 nmole
TAZOBACTAM		
β-LACTAMASE		*I*50 = 2.3 μM
β-LACTAMASE BRO-1		*I*50 = 20 nM
β-LACTAMASE CAZ-3		*I*50 = 60 nM
β-LACTAMASE DJP-1		*I*50 = 20 nM
β-LACTAMASE MJ-1		*I*50 = 430 nM
β-LACTAMASE OXA-1		*I*50 = 1.4 μM
β-LACTAMASE OXA-2		*I*50 = 10 nM
β-LACTAMASE OXA-4		*I*50 = 5.6 μM
β-LACTAMASE OXA-5		*I*50 = 250 nM
β-LACTAMASE OXA-6		*I*50 = 1.7 μM
β-LACTAMASE OXA-7		*I*50 = 610 μM
β-LACTAMASE PSE-2		*I*50 = 940 μM
β-LACTAMASE PSE-4		*I*50 = 100 nM
β-LACTAMASE SHV-1		*I*50 = 140 nM
β-LACTAMASE SHV-2		*I*50 = 130 nM
β-LACTAMASE SHV-3		*I*50 = 100 nM
β-LACTAMASE SHV-5		*I*50 = 80 nM
β-LACTAMASE TEM-1		*I*50 = 40 nM
β-LACTAMASE TEM-2		*I*50 = 50 nM
β-LACTAMASE TEM-3		*I*50 = 10 nM
β-LACTAMASE TEM-5		*I*50 = 280 nM
β-LACTAMASE TEM-6		*I*50 = 170 nM

β-LACTAMASE TEM-7		$I50$ = 180 nM
β-LACTAMASE TEM-9		$I50$ = 340 nM
β-LACTAMASE TEM-10		$I50$ = 80 nM
β-LACTAMASE TEM-E1		$I50$ = 20 nM
β-LACTAMASE TEM-E2		$I50$ = 50 nM
β-LACTAMASE TEM-E3		$I50$ = 60 nM
β-LACTAMASE TEM-E4		$I50$ = 40 nM
β-LACTAMASE TLE-1		$I50$ = 50 nM

Tb^{3+}

PROTEIN-GLUTAMINE γ-GLUTAMYLTRANSFERASE

N-((T-BUTOXYCARBONYL)ALA-PRO-PHE-)-O-NITROBENZOYLHYDROXYLAMINE

SUBTILISIN

TCNR-MONOPHOSPHATE

IMP DEHYDROGENASE

TDP

α-1,3-MANNOSYL-GLYCOPROTEIN β-1,2-N-ACETYLGLUCOSAMINYLTRANSFERASE		
5′-NUCLEOTIDASE	C	55% 100 μM
RIBOSE-PHOSPHATE PYROPHOSPHOKINASE		
SUCROSE SYNTHASE	SUB	Ki = 4 mM
dTDPGLUCOSE 4,6-DEHYDRATASE		
THIAMIN-TRIPHOSPHATASE		55% 7.5 mM
UDP-GLUCURNATE 4-EPIMERASE		95% 3 mM

dTDP

dCTP DEAMINASE	NC	
dTDP-DIHYDROSTREPTOSE-STREPTIDINE-6-PHOSPHATE DIHYDROSTREPTOSYLTRANSFERASE		80% 1 mM

dTDP-6-DEOXYGLUCOSE

dTDPGLUCOSE 4,6-DEHYDRATASE	C	

TDP-GLUCOSE

CMP-N-ACYLNEURAMINATE PHOSPHODIESTERASE		74% 2.9 mM
1,3-β-GLUCAN SYNTHASE		
PHOSPHORYLASE		
UDP-GLUCOSE-1-PHOSPHATE URIDYLYLTRANSFERASE		Ki = 2 mM
UDP-GLUCURNATE 4-EPIMERASE		83% 3 mM

TDP-RHAMNOSE

UDP-GLUCOSE-1-PHOSPHATE URIDYLYLTRANSFERASE		Ki = 520 μM

Te^{4+}

DOLICHYLPHOSPHATE-GLUCOSE PHOSPHODIESTERASE	C	Ki = 700 nM

TEA

GLYCEROL-3-PHOSPHATE DEHYDROGENASE (NAD)

TEBUFELONE

ARACHIDONATE 5-LIPOXYGENASE		$I50$ = 3 μM
CYCLOOXYGENASE 1		$I50$ = 250 nM
CYCLOOXYGENASE 2		$I50$ = 100 nM

TEBUFENPYRAD

NADH DEHYDROGENASE (UBIQUINONE)

psi-TECTORIGENIN

PROTEIN TYROSINE KINASE EGFR		$I50$ = 100 nM

TEI 5624

CATHEPSIN G		$I50$ = 1 μM
α-CHYMOTRYPSIN		$I50$ = 6.3 μM
ELASTASE LEUKOCYTE		Ki = 6.9 nM
PLASMIN		$I50$ = 110 μM
THROMBIN		$I50$ = 130 μM
TRYPSIN		$I50$ = 1.4 μM

TEI 6344

CATHEPSIN G		$I50 = 900$ nM
α-CHYMOTRYPSIN		$I50 = 7.2$ μM
ELASTASE LEUKOCYTE		$Ki = 16$ nM
PLASMIN		$I50 = 30$ μM
THROMBIN		$I50 = 49$ μM
TRYPSIN		$I50 = 420$ nM

TEI 6522

STEROL O-ACYLTRANSFERASE		$I50 = 13$ nM

TEI 6620

STEROL O-ACYLTRANSFERASE		$I50 = 9$ nM
STEROL O-ACYLTRANSFERASE		$I50 = 20$ nM

TEI 6720

XANTHINE OXIDASE		$I50 = 1.8$ nM
XANTHINE OXIDASE		$I50 = 1.4$ nM
XANTHINE OXIDASE		$I50 = 2.2$ nM

TEICHOIC ACID GLYCAN COMPLEX

N-ACETYLMURAMOYL-L-ALANINE AMIDASE

TELLIMAGRANDIN

CARBONATE DEHYDRATASE		$I50 = 320$ nM

TELLIMAGRANDIN II

XANTHINE OXIDASE		$I50 = 3.1$ μM

TELLINAGRANDIN

POLY(ADP-RIBOSE) GLYCOHYDROLASE		$I50 = 12$ μM

TEMAFLOXACIN

CYTOCHROME P450 LA2		10% 500 μM
DNA TOPOISOMERASE (ATP-HYDROLYSING)		$I50 = 30$ μg/ml
DNA TOPOISOMERASE (ATP-HYDROLYSING)		$I50 = 300$ μg/ml

TEMAZEPAM

UDP-GLUCURONOSYLTRANSFERASE	C	$Ki = 217$ μM

TEMEPHOS

ACETYLCHOLINESTERASE

TEMOCAPRILATE

PEPTIDYL-DIPEPTIDASE A

TEMPOROMANDIBULAR JOINT DISC CELLS PROTEINASE INBITORS

PROTEINASE TEMPOROMANDIBULAR JOINT DISC CELLS

TENDAMISTAT

α-AMYLASE

TENEVEL

PEPTIDYLTRANSFERASE

TENIDAP

CYCLOOXYGENASE		$I50 = 20$ nM
PROSTAGLANDIN H ENDOPEROXIDE SYNTHASE 1		$I50 = 2.1$ μM
PROSTAGLANDIN H ENDOPEROXIDE SYNTHASE 2		$I50 = 5.2$ μM
XANTHINE OXIDASE	NC	$Ki = 34$ μM

TENIPOSIDE

DNA TOPOISOMERASE (ATP-HYDROLYSING)

TENTOXIN

AF1 ATPASE		
H^+-TRANSPORTING ATP SYNTHASE		
H^+-TRANSPORTING ATP SYNTHASE F1		10 nM

TEPARY BEAN α-AMYLASE INHIBITOR
α-AMYLASE

TEPOXALIN
ARACHIDONATE 5-LIPOXYGENASE $I50 = 150$ nM
CYCLOOXYGENASE $I50 = 4.6$ μM
CYCLOOXYGENASE $I50 = 2.9$ μM
CYCLOOXYGENASE $I50 = 100$ nM
PROSTAGLANDIN H SYNTHASE (PEROXYDASE ACTIVITY) $I50 = 4$ μM

TEPP
ACETYLCHOLINESTERASE

TEPROTIDE
PEPTIDYL-DIPEPTIDASE A

TERBINAFINE
SQUALENE MONOOXYGENASE $I50 = 4.1$ μM
SQUALENE MONOOXYGENASE NC $Ki = 30$ nM
SQUALENE MONOOXYGENASE NC $Ki = 30$ nM
SQUALENE MONOOXYGENASE $I50 = 18$ nM
SQUALENE MONOOXYGENASE C $Ki = 77$ nM
SQUALENE MONOOXYGENASE $I50 = 40$ nM

TERBUFOS
ACETYLCHOLINESTERASE

TERBUTALINE
CHOLINESTERASE

TEREPHTHALIC ACID
4,5-DIHYDROOXYPHTHALATE DECARBOXYLASE
SERINE-SULFATE AMMONIA-LYASE C $Ki = 18$ mM

TERFEROL
3′,5′-CYCLIC-NUCLEOTIDE PHOSPHODIESTERASE (cycAMP) $I50 = 820$ nM
3′,5′-CYCLIC-NUCLEOTIDE PHOSPHODIESTERASE (cycGMP)

TERGITOL NP 40
DIACYLGLYCEROL O-ACYLTRANSFERASE

TERGITOL T-7
NITRIC-OXIDE REDUCTASE

TERGITOL T 15-S-9
NITRIC-OXIDE REDUCTASE

(S)-(+)-TERINDOLE
AMINE OXIDASE MAO-A $I50 = 20$ nM
AMINE OXIDASE MAO-B $I50 = 6.2$ μM

TERLAKIREN
RENIN $I50 = 300$ pM

TERPENDOLE A
STEROL O-ACYLTRANSFERASE $I50 = 15$ μM

TERPENDOLE B
STEROL O-ACYLTRANSFERASE $I50 = 27$ μM

TERPENDOLE C
STEROL O-ACYLTRANSFERASE $I50 = 2.1$ μM

TERPENDOLE D
STEROL O-ACYLTRANSFERASE $I50 = 3.2$ μM

TERPENDOLE J
STEROL O-ACYLTRANSFERASE $I50 = 39$ μM

TERPENDOLE K
STEROL O-ACYLTRANSFERASE | | *I*50 = 38 µM

TERPENDOLE L
STEROL O-ACYLTRANSFERASE | | *I*50 = 32 µM

α-TERPENYL CATION
GERANYL-DIPHOSPHATE CYCLASE

(1,1′,4′,1″)-TERPHENYL-3,3″-DICARBAMIDINIUM DIACETATE
COAGULATION FACTOR Xa | | *K*i = 3.8 µM
PLASMIN | | *K*i = 2.1 µM
THROMBIN | | *K*i = 4.8 µM
TRYPSIN | | *K*i = 1.3 µM

α-TERPINEOL
GERANIOL DEHYDROGENASE

TERRITREM A
ACETYLCHOLINESTERASE

TERRITREM B
ACETYLCHOLINESTERASE | | *I*50 = 7.6 nM
ACETYLCHOLINESTERASE | | *I*50 = 260 nM

TERRITREM C
ACETYLCHOLINESTERASE | | *I*50 = 6.8 nM

7-TERTAHYDROBIOPTERIN
PHENYLALANINE 4-MONOOXYGENASE | | > 50% 1µM

TERTAHYDROECHINOCANDIN B
1,3-β-GLUCAN SYNTHASE | | *I*50 = 850 nM

α-TERTHIENYL
ACETYLCHOLINESTERASE
SUPEROXIDE DISMUTASE

TERTRAGALLOYLGLUCOSE
SUCCINATE DEHYDROGENASE (UBIQUINONE) | C | *K*i = 1.5 µM

Δ9-TERTRAHYDROCANNABINOL
ADENYLATE CYCLASE | | *I*50% 50 µM

TERTRAHYDROFOLIC ACID
FORMIMINOTETRAHYDROFOLATE CYCLODEAMINASE | C | *K*i = 4 µM

Δ1-TESTOLACTONE
AROMATASE | | *K*i = 20 µM

TESTOSTERONE
ALDEHYDE DEHYDROGENASE (NAD(P))
4-ANDROSTENE-3,17-DIONE MONOOXYGENASE
CARBONYL REDUCTASE (NADPH)
CHOLESTENONE 5α-REDUCTASE
trans-1,2-DIHYDROBENZENE-1,2-DIOL DEHYDROGENASE | | 48% 1 µM
ESTRADIOL 17β-DEHYDROGENASE | C |
3β-HYDROXYSTEROID DEHYDROGENASE | C | *K*i = 300 nM
3α-HYDROXYSTEROID DEHYDROGENASE (B-SPECIFIC) | | *I*50 = 25 µM
21-HYDROXYSTEROID DEHYDROGENASE (NAD) | C | *K*i = 14 µM
STEROID 21-MONOOXYGENASE
STEROL O-ACYLTRANSFERASE
UDP-GLUCURONOSYLTRANSFERASE
UDP-GLUCURONOSYLTRANSFERASE | C | *K*i = 250 µM

TETCYCLACIN
FLAVONOID 3′-MONOOXYGENASE

TETCYCLASIS
ent-KAURENE OXIDASE
OBTUSIFOLIOL 14α-DEMETHYLASE — *I*50 = 9 μM

TETHYA INGALLI PROTEINASE INHIBITOR
TRYPSIN — *I*50 = 65 nM

N,N2,N3,N4-TETRAACETYLCHITOTETRAPENTOASE
CHITOBIASE — *K*i = 14 μM

5′[[[(2″,3″,4″,6‴-TETRA-O-ACETYL 6″-ETHYNYL)α-D-MANNOPYRANOSYL OXY]CARBONYL]AMINO]]SULPHONYL-2′,3′-ISOPROPYLIDENE N2 ACETYL GUANOSINE
GDP-MANNOSE 6-DEHYDROGENASE — 75% 100 μM

TETRA-p-AMIDINOPHENOXYPROPANE
PROTEINASES SERINE

TETRABORATE
STERYL-SULFATASE — 83% 200 μM

TETRABROMO-2-AZABENZIMIDAZOLE
PROTEIN KINASE CASEIN KINASE I — *K*i = 39 μM
PROTEIN KINASE CASEIN KINASE II — *K*i = 640 nM
PROTEIN KINASE CASEIN KINASE II — *K*i = 580 nM
PROTEIN KINASE CASEIN KINASE I (27 KD) — *K*i = 76 μM
PROTEIN KINASE CASEIN KINASE I (45 KD) — *K*i = 120 μM

4,5,6,7-TETRABROMO-2-AZABENZIMIDAZOLE
PROTEIN KINASE CASEIN KINASE II — *K*i = 700 nM
PROTEIN KINASE 60S — *K*i = 100 nM

3,3′,5,5′-TETRABROMO-4,4′-DIHYDROXYBENZOPHENONE
GLUTATHIONE TRANSFERASE

TETRABROMOPHENOL BLUE
GLUTATHIONE TRANSFERASE
1-D-myo-INOSITOL-TRISPHOSPHATE 3-KINASE — *I*50 = 12 μM
PHOSPHATIDYLINOSITOL 3-PHOSPHATASE — *I*50 = 25 μM

TETRABROMOPHENOLPHTHALEIN
GLUTATHIONE TRANSFERASE
THYMIDYLATE SYNTHASE — *I*50 = 5 μM

TETRABROMOPHENOLPHTHALEIN ETHYL ESTER
GLUTATHIONE TRANSFERASE

TETRABROMOPHENOLPHTHALEIN ISOPROPYL ESTER
GLUTATHIONE TRANSFERASE

3,4,5,6-TETRABROMOPHENOLSULPHONEPHTHALEIN
1-D-myo-INOSITOL-TRISPHOSPHATE 3-KINASE — *I*50 = 105 μM

TETRABROMOSULPHOPHTHALEIN
GLUTATHIONE TRANSFERASE 11-11 — *I*50 = 12 μM

TETRACAINE
ACETYLCHOLINESTERASE — *I*50 = 180 μM
ACETYLCHOLINESTERASE — *I*50 = 44 μM
ACETYLCHOLINESTERASE — *I*50 = 120 μM
ACETYLCHOLINESTERASE — *I*50 = 5 μM
1-ACYLGLYCEROPHOSPHOCHOLINE O-ACYLTRANSFERASE
PROTEIN KINASE C — 61% 200 μM
STEROL O-ACYLTRANSFERASE — *I*50 = 1.3 mM

1,2,3,4-TETRACARBOXYCYCLOPENTANE
ACONITATE HYDRATASE — C — *K*i = 15 mM

2,3,7,8-TETRACHLOROBENZO-p-DIOXIN
ESTRADIOL 2-HYDROXYLASE — *K*i = 8 nM

TETRACHLORO-1,4-BENZOQUINONE

GLUTATHIONE TRANSFERASE μ	IR	
GLUTATHIONE TRANSFERASE π	IR	
GLUTATHIONE TRANSFERASE 1-1	IR	93%
GLUTATHIONE TRANSFERASE 1-2	IR	89%
GLUTATHIONE TRANSFERASE 2-2	IR	72%
GLUTATHIONE TRANSFERASE 3-3	IR	85%
GLUTATHIONE TRANSFERASE 3-4	IR	84%
GLUTATHIONE TRANSFERASE 4-4	IR	92%
GLUTATHIONE TRANSFERASE psi	IR	

2,2′,5,5′-TETRACHLOROBIPHENYL

TYROSINE 3-MONOOXYGENASE		33% 100 μM

TETRACHLOROETHYLENE

AMMONIA MONOOXYGENASE		*I*50 = 5 μM
METHANE MONOOXYGENASE	C	*K*i = 90 μM

2,3,5,6-TETRACHLORO4-METHOXYPHENOL

CHITIN SYNTHASE

2,3,5,6-TETRACHLORO-4-PYRIDINOL

VITAMIN K-EPOXIDE REDUCTASE (WARFARIN INSENSITIVE)
VITAMIN K-EPOXIDE REDUCTASE (WARFARIN SENSITIVE)

TETRACHLORO-4-PYRIDINOL

PHYLLOQUINONE MONOOXYGENASE (2,3-EPOXIDIZING)

TETRACHLOROVINPHOS

ACETYLCHOLINESTERASE

TETRACINE

Ca^{2+}-TRANSPORTING ATPASE

TETRACOSANOIC ACID

Na^{+}/K^{+}-EXCHANGING ATPASE

TETRACOSANOYL-CoA

ACETYL-CoA CARBOXYLASE		*K*i = 150 nM

TETRACYCLASIS

5-O-(4-COUMARYOYL)-D-QUINATE 3′-MONOOXYGENASE

TETRACYCLIN

RIBOZYME HAMMERHEAD		56% 500 μM

TETRACYCLINE

ANHYDROTETRACYCLINE MONOOXYGENASE		
GELATINASE A		96% 50 μg/ml
GTP PYROPHOSPHOKINASE 1		
GUANOSINE-3′,5′-BIS(DIPHOSPHATE) 3′-PYROPHOSPHATASE		
NUCLEOTIDE PYROPHOSPHOKINASE		

(6E,12E)TETRADECADIENE-8,10-DIYNE-1,3-DIOL

Na^{+}/K^{+}-EXCHANGING ATPASE		40% 1 mM
XANTHINE OXIDASE		*I*50 = 50 μM

(6E,12E)TETRADECADIENE-8,10-DIYNE-1,3-DIOL DIACETATE

Na^{+}/K^{+}-EXCHANGING ATPASE		40% 1 mM
XANTHINE OXIDASE		*I*50 = 100 μM

(6E,12E)-TETRADECADIENE-8,10-DSIYNE-1,3-DIOL MONOACETATE

Na^{+}/K^{+}-EXCHANGING ATPASE		*I*50 = 170 μM

7,9-TETRADECADIYNOIC ACID

HYDROXYMETHYLGLUTARYL-CoA REDUCTASE (NADPH)		*I*50 = 3 μM

TETRADECANE

LIPOXYGENASE		*I*50 = 4 μM

TETRADECANEDIDIC ACID
GLUTATHIONE TRANSFERASE

2-(R)-TETRADECANOYLAMINOHEXANOL-1-PHOSPHOCHOLINE
PHOSPHOLIPASE A2 — Ki = 15 μM
PHOSPHOLIPASE A2 — Ki = 3.5 μM
PHOSPHOLIPASE A2 — Ki = 25 μM

2-(R)-TETRADECANOYLAMINOHEXANOL-1-PHOSPHOGLYCOL
PHOSPHOLIPASE A2 — Ki = 200 nM
PHOSPHOLIPASE A2 — Ki = 700 nM
PHOSPHOLIPASE A2 — Ki = 500 nM

(R)-2-TETRADECANOYLAMINO-HEXANOYL-1-PHOSPHOCHOLINE
PHOSPHOLIPASE A2

12-O-TETRADECANOYLPHORBOL-13-ACETIC ACID
HYDROXYACYLGLUTATHIONE HYDROLASE

(4E,6E,12E)TETRADECATRIENE-8,10-DIYNE-1,3-DIOL
XANTHINE OXIDASE — I50 = 50 μM

(4E,6E,12E)TETRADECATRIENE-8,10-DIYNE-1,3-DIOL DIACETATE
XANTHINE OXIDASE — I50 = 32 μM

2R,3R-1-TETRADECENYL-2-PALMITOYLAMINO-3-MORPHOLINO-1-PROPANOL
CERAMIDE GLUCOSYLTRANSFERASE — 73% 5 μM

cis-7-TETRADECENYL-1-YL ACETATE
ACETYLCHOLINESTERASE — C — Ki = 2.6 mM

TETRADECYLAMINE
MYOSIN-LIGHT-CHAIN KINASE — I50 = 12 μM

TETRADECYLBORONIC ACID
CUTINASE — C — Ki = 15 μM

N-TETRADECYL-N,N-DIMETHYL-3-AMINOPROPYLSULPHONATE
PHOSPHOLIPASE A2 — I50 = 7.5 μM

TETRADECYLGLYCIDYL-CoA
CARNITINE O-PALMITOYLTRANSFERASE A — IR
CARNITINE O-PALMITOYLTRANSFERASE 1 — I50 = 60 nM

2-TETRADECYLGLYCIDYL-CoA
CARNITINE O-PALMITOYLTRANSFERASE — IR — Ki = 270 nM

4-TETRADECYLOXYBENZAMIDINE
PHOSPHOLIPASE A2 — I50 = 3 μM

5-(TETRADECYLOXY)-2-FUROIC ACID
ACETYL-CoA CARBOXYLASE
OXIDATIVE PHOSPHORYLATION UNCOUPLER
TRICARBOXYLATE CARRIER MITOCHONDRIA

5-(TETRADECYLOXY)-2-FUROYL-CoA
ACETYL-CoA CARBOXYLASE
ADN-CARRIER MITOCHONDRIA

TETRADECYLSPHINGOSINE
GLUCOSYLCERAMIDASE — I50 = 78 μM

2-TETRADECYLTHIOACETYL-CoA
ACYL-CoA OXIDASE — C — Ki = 2.6 μM

TETRADECYLTHIOACRYLIC ACID
FATTY ACID β OXIDATION

3-TETRADECYLTHIOPROPIONYL-CoA		
ACYL-CoA OXIDASE	C	*K*i = 240 nM
TETRADECYLTRIMETHYLAMMONIUM BROMIDE		
MYOSIN-LIGHT-CHAIN KINASE		
1-PHOSPHATIDYLINOSITOL PHOSPHODIESTERASE		
TETRADIFON		
H^+-TRANSPORTING ATP SYNTHASE		*I*50 = 50 nM
TETRADYMOL		
OXIDATIVE PHOSPHORYLATION UNCOUPLER		
TETRAETHYLTHIURAM DISULPHIDE		
L-ASCORBATE OXIDASE		
CHOLESTEROL 7α-MONOOXYGENASE		
XANTHINE OXIDASE		
1,3,4,6-TETRA-O-GALLOYL-β-D-GLUCOPYRANOSIDE		
RNA-DIRECTED DNA POLYMERASE		*I*50 = 86 μM
TETRAGALLOYLGLUCOSE		
POLY(ADP-RIBOSE) GLYCOHYDROLASE		*I*50 = 24 μM
1,2,3,6-TETRA-O-GALLOYL-GLUCOSE		
DNA POLYMERASE α		45% 418 nM
DNA POLYMERASE β		92% 13 μM
DNA POLYMERASE γ		99% 42 μM
α-GLUCOSIDASE		*I*50 = 3.4 μM
1,2,3,6-TETRA-O-GALLOYL-D-GLUCOSE		
NADH DEHYDROGENASE		*I*50 = 800 nM
1,2,4,6-TETRA-O-GALLOYL-β-D-GLUCOSE		
XANTHINE OXIDASE		*I*50 = 8.1 μM
TETRAHYDROALSTOMINE		
CYTOCHROME P450 2D6		*K*i = 5 μM
TETRAHYDROBIOPTERIN		
DIHYDROPTERIDINE REDUCTASE		> 50 μM
GTP CYCLOHYDROLASE A2		
GTP CYCLOHYDROLASE I	NC	*K*i = 16 μM
PHENYLALANINE 4-MONOOXYGENASE		
PHENYLALANINE 4-MONOOXYGENASE		
QUEUINE tRNA-RIBOSYLTRANSFERASE		
(6R)-5,6,7,8-TETRAHYDRO-L-BIOPTERIN		
GTP CYCLOHYDROLASE I		*I*50 = 40 μM
GTP CYCLOHYDROLASE I		*K*i = 16 μM
NITRIC-OXIDE SYNTHASE		*I*50 = 10 μM
(4R,5R,6R)-TETRAHYDRO-1,3-BIS[[3-(N-METHYLAMINO)PHENYL]METHYL]-5-HYDROXY-4-[2-PHENYLETHYL-2-(1H)-PYRIMIDINONE		
PROTEINASE HIV-1		*K*i = 18 pM
Δ9-TETRAHYDROCANNABINOL		
1-ACYLGLYCEROPHOSPHOCHOLINE O-ACYLTRANSFERASE		55% 50 μM
N-(LONG-CHAIN-ACYL)ETHANOLAMINE DEACYLASE		31% 160 μM
LYSOPHOSPHATIDYLETHANOLAMINE ACYL-CoA ACYLTRANSFERASE		53% 50 μM
TETRAHYDRO-β-CARBOLINE		
DEBRISOQUINE HYDROXYLASE		*I*50 = 50 μM
TETRAHYDROCERULENIN		
EROTHRONOLIDE SYNTHASE		
3,3′,4,4′-TETRAHYDROCHALCONE		
XANTHINE OXIDASE		

TETRAHYDROCORTISOL
CORTISOL SULFOTRANSFERASE

D,L-2-[(5,6,7,8-TETRAHYDRO-5-DEAZAPTEROYL)AMINO]-4-PHOSPHONOBUTABOIC ACID
FOLYLPOLYGLUTAMATE SYNTHASE — $Ki = 3.2\ \mu M$
PHOSPHORIBOSYLGLYCINAMIDE FORMYLTRANSFERASE — C — $Ki = 47$ nM

TETRAHYDRODEOXYURIDINE
dCMP DEAMINASE — 24% 1.4 mM

3,4,5,6-TETRAHYDRODEOXYURIDINE
DEOXYCYTIDINE DEAMINASE

TETRAHYDRODEOXYURIDINE 5′-MONOPHOSPHATE
dCMP DEAMINASE — C — $Ki = 20$ nM

(4α,5α,6β,7β)3,3′-[[TETRAHYDRO-5,6-DIHYDROXY-2-OXO-4,7-BIS(PHENYLMETHYL)1H-1,3-DIAZEPINE-1,3(2H)DIYL]BIS(METHYLENE)]BIS[N(5-METHYL-2-PYRDINYL)BENZAMIDE
PROTEINASE HIV-1 — $Ki = 10$ nM

(4α,5α,6β,7β)3,3′[[TETRAHYDRO-5,6-DIHYDROXY-2-OXO-4,7-BIS(PHENYLMETHYL)1H-1,3-DIZEPINE-1,3(2H)DIYL]BIS(METHYLENE)]BIS[N(5-METHYL-2-PYRIDINYL)BENZAMIDE
PROTEINASE HIV-1 — $Ki = 20$ pM

(+)TETRAHYDROFOLIC ACID
FORMYLTETRAHYDROFOLATE DEHYDROGENASE — C — $Ki = 10\ \mu M$

(–)TETRAHYDROFOLIC ACID
FORMYLTETRAHYDROFOLATE DEHYDROGENASE — C — $Ki = 1\ \mu M$

TETRAHYDROFOLIC ACID
FORMIMINOTETRAHYDROFOLATE CYCLODEAMINASE — C — $Ki = 4\ \mu M$
GUANINE DEAMINASE — C — $Ki = 260\ \mu M$
INDOLEACETALDOXIME DEHYDRATASE — 100% 100 μM
myo-INOSITOL OXYGENASE

(±)TETRAHYDROFOLIC ACID
GTP CYCLOHYDROLASE I — $I50 = 130\ \mu M$

TETRAHYDROFURAN
ACETYL-CoA CARBOXYLASE
EPOXIDE HYDROLASE
PROLINE-tRNA LIGASE

2′,3′α-TETRAHYDROFURAN-2′-SPIRO-17-(4,7α-DIMETHYL-4-AZA-5α-ANDROSTAN-3-ONE)
STEROID 5α-REDUCTASE — $I50 = 1\ \mu M$
STEROID 5α-REDUCTASE — $I50 = 5$ nM

2′,3′α-TETRAHYDROFURAN-2′-SPIRO-17-(4,7β-DIMETHYL-4-AZA-5α-ANDROSTAN-3-ONE)
STEROID 5α-REDUCTASE — $I50 = 250$ nM

2′,3′α-TETRAHYDROFURAN-2′-SPIRO-17-(4-METHYL-4-AZA-5α-ANDROSTAN-3-ONE)
STEROID 5α-REDUCTASE — $I50 = 1.3\ \mu M$
STEROID 5α-REDUCTASE — $I50 = 3$ nM
STEROID 5α-REDUCTASE — $I50 = 100$ nM

2′,3′α-TETRAHYDROFURAN-2′-SPIRO-17-(4-METHYL-4-AZA-5α-ANDROST-5-ENE-3-ONE)
STEROID 5α-REDUCTASE — $I50 = 2$ nM

2,3,4,5-TETRAHYDRO-1H-BENZAZEPINE
PHENYLETHANOLAMINE N-METHYLTRANSFERASE — $Ki = 3.3\ \mu M$

(6R,S)TETRAHYDROHOMOFOLATEPENTAGLUTAMATE
PHOSPHORIBOSYLGLYCINAMIDE FORMYLTRANSFERASE — $I50 = 290$ nM

(6R,S)TETRAHYDROHOMOFOLIC ACID
PHOSPHORIBOSYLGLYCINAMIDE FORMYLTRANSFERASE — $I50 = 1\ \mu M$

(R*)N[[1,2,3,4-TETRAHYDRO-2[N[2(1H-IMIDAZOL-4-YL)ETHYL-3-ISOQUINOLINYL]METHIONINE([IMIDAZOL-4-YLETHYL]VAL-TIC-MET
GERANYLGERANYL PROTEINTRANSFERASE — *I*50 = 234 nM

(R*)N[[1,2,3,4-TETRAHYDRO-2[N[2(1H-IMIDAZOL-4-YL)ETHYL]VAL-3-ISOQUINOLINYL]CARBONYL]METHIONINE([IMIDAZOL-4-YLETHYL]VAL-TIC-MET
FARNESYL PROTEINTRANSFERASE — *I*50 = 790 pM

4-(5,6,7,8-TETRAHYDROIMIDAZO[1,5-a]PYRIDIN-5-YL)BENZONITRILE HCL
AROMATASE — *I*50 = 1.8 nM

2,3,5,6-TETRAHYDROIMIDAZO[2,1-b]THIAZOLE
AMINE OXIDASE (COPPER-CONTAINING) — *I*50 = 720 μM
PHOSPHATASE ALKALINE — UC — *K*i = 3.7 mM

TETRAHYDROISOQUINOLINE
AMINE OXIDASE MAO-A — C — *K*i = 55 μM
AMINE OXIDASE MAO-A — *I*50 = 90 μM
AMINE OXIDASE MAO-B — *I*50 = 48 μM
AMINE OXIDASE MAO-B — NC — *K*i = 57 μM

1,2,3,4-TETRAHYDROISOQUINOLINE
AMINE OXIDASE MAO-A — C — *K*i = 210 μM
AMINE OXIDASE MAO-B — C — *K*i = 15 μM
PHENYLETHANOLAMINE N-METHYLTRANSFERASE — *K*i = 10 μM

(1,2,3,4-TETRAHYDROISOQUINOLIN-2-YL)CARBONYL-LEU-D,L-PHECH$_2$F
CALPAIN I — *I*50 = 100 nM

TETRAHYDROLIPSTATIN
LIPOPROTEIN LIPASE — *I*50 = 6 nM
STEROL ESTERASE
TRIACYLGLYCEROL LIPASE — IR

9-(TETRAHYDRO-5-METHYL-2-FURYL)ADENINE
ADENYLATE CYCLASE — *I*50 = 8 μM

1,2,3,4-TETRAHYDRO-2-METHYL-4,6,7-ISOQUINOLINETRIOL
TYROSINE 3-MONOOXYGENASE — *I*50 = 4 μM

(+)-(S)-1,2,3,4-TETRAHYDRO-1-NAPHTHOL
TYROSINE-ESTER SULFOTRANSFERASE IV — C

1-[1[(S)-2-(1,2,3,4-TETRAHYDRONAPHTHYL)ACETYL]PROLYL]PROLINAL
PROLYL OLIGOPEPTIDASE — *I*50 = 450 pM

1,4,5,6-TETRAHYDRONICOTINAMIDE ADENINE DINUCLEOTIDE PHOSPHATE
ALCOHOL DEHYDROGENASE (NADP)

TETRAHYDRONORHARMANE
AMINE OXIDASE MAO-A — *I*50 = 8 μM

1,2,3,4-TETRAHYDRO-4-OXO-7-CHLORO-2-NAPHTHYL) PYRIDINE
STEROID 17α-MONOOXYGENASE — C — *K*i = 860 nM

TETRAHYDROPAPAVERINE
BUFURALOL-1′-HYDROXYLASE (CYT P-450dbl) — C — *I*50 = 39 μM

TETRAHYDROPENTALENOLACTONE
GLYCERALDEHYDE-3-PHOSPHATE DEHYDROGENASE (PHOSPHORYLATING) — IR

2,3,5,6-TETRAHYDRO-6-PHENYL(+-)IMIDAZO[2,1-b]THIAZOLE
PHOSPHATASE ALKALINE

3α,5β-TETRAHYDROPROGESTERONE
11β-HYDROXYSTEROID DEHYDROGENASE

TETRAHYDROPTEROATE
FOLYLPOLYGLUTAMATE SYNTHASE

TETRAHYDROPTEROYLGLUTAMATE
FORMYLTETRAHYDROFOLATE DEHYDROGENASE $Ki = 1\ \mu M$

(R)-TETRAHYDROPTEROYLGLUTAMATE
FORMIMINOTETRAHYDROFOLATE CYCLODEAMINASE

TETRAHYDROPTEROYLHEXA-γ-GLUTAMATE
FORMYLTETRAHYDROFOLATE DEHYDROGENASE

N^{α}-(TETRAHYDROPTEROYL)-L-ORNITHINE
FOLYLPOLYGLUTAMATE SYNTHASE $Ki = 200$ nM

1-[2-(3,4,5,6-TETRAHYDROPYRIDYL)]-1,3-PENTADIENE
INDOLEAMINE N-METHYLTRANSFERASE

(4R,5R)-1,4,5,6-TETRAHYDROPYRIMIDINE
β-GALACTOSIDASE $Ki = 80\ \mu M$

(4R,5S)-1,4,5,6-TETRAHYDROPYRIMIDINE
β-GLUCOSIDASE $Ki = 30\ \mu M$

5,6,7,8-TETRAHYDRO-9-(1′-D-RIBITYL)ISOALLOXAZINE
RIBOFLAVIN SYNTHASE

6,7,8,9-TETRAHYDRO[1,2,5]THIADIAZOLO[3,4-h]-ISOQUINOLINE
PHENYLETHANOLAMINE N-METHYLTRANSFERASE

TETRAHYDROTHIAMINE DIPHOSPHATE
PYRUVATE DEHYDROGENASE (LIPOAMIDE)

TETRAHYDROTHIAMINE PYROPHOSPHATE
PYRUVATE DEHYDROGENASE (LIPOAMIDE) $I50 = 46$ nM

TETRAHYDROTHOIPHEN
PROLINE-tRNA LIGASE

TETRAHYDRO-d-UMP
dCMP DEAMINASE

TETRAHYDROURIDINE
CYTIDINE DEAMINASE C $Ki = 51$ nM
CYTIDINE DEAMINASE $Ki = 220$ nM

2,4,5,6-TETRAHYDROURIDINE
CYTIDINE DEAMINASE $Ki = 240$ nM

3,4,5,6-TETRAHYDROURIDINE
CYTIDINE DEAMINASE C $Ki = 1O9$ nM

TETRAHYDROURIDINE 5′-MONOPHOSPHATE
dCMP DEAMINASE 44% 40 μM

2,2′,4,4′-TETRAHYDROXYBENZOPHENONE
XANTHINE OXIDASE $I50 = 39\ \mu M$

7,8,3′,4′-TETRAHYDROXYFLAVON
PROTEIN KINASE C $I50 = 1\ \mu M$

E.2,3,5,4′-TETRAHYDROXYSTILBENE-2O-β-D-GLUCOPYRANOSIDE
Ca^{2+}-TRANSPORTING ATPASE $I50 = 240\ \mu M$

1,3,5,6-TETRAHYDROXYXANTHONE
PEPTIDYL-DIPEPTIDASE A-I $I50 = 69\ \mu M$

1,3,6,7-TETRAHYDROXYXANTHONE
PEPTIDYL-DIPEPTIDASE A-I $I50 = 530\ \mu M$

3,4,5,6-TETRAHYDROXYXANTHONE
PEPTIDYL-DIPEPTIDASE A-I $I50 = 239\ \mu M$

3,4,6,7-TETRAHYDROXYXANTHONE		
PEPTIDYL-DIPEPTIDASE A-I		*I*50 = 35 μM
3′,3″,5′,5″-TETRAIODOPHENOLSULPHTHALEIN		
1-D-myo-INOSITOL-TRISPHOSPHATE 3-KINASE		*I*50 = 35 μM
PHOSPHATIDYLINOSITOL 3-PHOSPHATASE		*I*50 = 220 μM
TETRAIODOPHENOLTHYMOLPHTHALEIN		
THYMIDYLATE SYNTHASE		*I*50 = 3 μM
3′,3″,5′,5″-TETRAIODOPHENYLSULPHONEPHTHALEIN		
Ca^{2+}-TRANSPORTING ATPASE		*I*50 = 1.1 μM
TETRAIODOTHYROACETIC ACID		
15-HYDROXYPROSTAGLANDIN DEHYDROGENASE (NAD)		
THYROXINE DEIODINASE		*I*50 = 200 nM
TETRAISOPROPYLPYROPHOSPHORAMIDE		
CHOLINESTERASE		10 μM
meso-TETRAKIS(N-METHYL-4-PYRIDINIUM)DIPHENYL PORPHYRIN-Co		
ADENOSYLMETHIONINE DECARBOXYLASE		
DIAMINE N-ACETYLTRANSFERASE		*K*i = 120 nM
ORNITHINE DECARBOXYLASE		
meso-TETRAKIS(N-METHYL-4-PYRIDINIUM)DIPHENYL PORPHYRIN		
DIAMINE N-ACETYLTRANSFERASE		*K*i = 64 nM
POLYAMINE OXIDASE		*K*i = 395 nM
meso-TETRAKIS(N-METHYL-4-PYRIDINIUM)DIPHENYL PORPHYRIN-Ni		
ADENOSYLMETHIONINE DECARBOXYLASE		
DIAMINE N-ACETYLTRANSFERASE		*K*i = 6 nM
2,4,2′,4′-TETRAKIS(PIVALYLOXY)BENZOPHENONE		
ELASTASE LEUKOCYTE		*I*50 = 27 nM
meso-TETRAKIS(N-TRIMETHYLANILINIUM) PORPHYRIN		
DIAMINE N-ACETYLTRANSFERASE		*K*i = 24 nM
POLYAMINE DECARBOXYLASE		*K*i = 140 nM
TETRAMETHYLAMMONIUM		
ACETYLCHOLINESTERASE		*K*i = 4 mM
TRIMETHYLAMINE DEHYDROGENASE		
N,N,N′,N′-TETRAMETHYLENEDIAMINE		
ACETYLCHOLINESTERASE		*I*50 = 2.8 mM
TETRAMETHYLENE GLUTARIC ACID		
ACETOIN DEHYDROGENASE		
ALCOHOL DEHYDROGENASE (NADP)		
ALDEHYDE REDUCTASE	UC	*I*50 = 2.4 μM
ALDEHYDE REDUCTASE		83% 10 μM
ALDEHYDE REDUCTASE		*I*50 = 10 μM
CARBONYL REDUCTASE (NADPH)		40% 1 mM
GLUCURONATE REDUCTASE		*I*50 = 31 μM
meso-TETRA[(p-METHYLETHYLSULPHONIO)PHENYL] PORPHYRIN		
DIAMINE N-ACETYLTRANSFERASE		*K*i = 90 nM
POLYAMINE OXIDASE		*K*i = 1.3 μM
2,5,9,11-TETRAMETHYL-5H-INDOLO-[2,3-b]QUINOLINE		
DNA TOPOISOMERASE (ATP-HYDROLYSING)		
(4E,8E,12E)-5,9,13,17-TETRAMETHYL-4,8,12,16-OCTADECATETRAEN-1-OL		
SQUALENE MONOOXYGENASE		*I*50 = 100 μM
([[(2,2,5,5-TETRAMETHYL-1-OXY-3-PYRROLINDINYL)CARBAMOYL]METHYL]THIOAMP		
AMP NUCLEOSIDASE		

3,6,7,3′-TETRAMETHYLQUERCETAGETIN
METHYLQERCETAGETIN 6-O-METHYLTRANSFERASE

TETRAMETHYLSCUTELLAREIN
ARACHIDONATE 15-LIPOXYGENASE 1 | | *I*50 = 110 μM

TETRAMETHYLTHIURAM DISULPHIDE
GLUTATHIONE REDUCTASE (NADPH) | IR |

L-TETRAMISOLE
PHOSPHATASE ALKALINE

TETRANITROMETHANE
ACHROMOBACTER IOPHAGUS COLLAGENASE
ARYLSULFATASE A | IR | *K*i = 2 μM
L-ASPARTATE OXIDASE
CATHEPSIN D
CROTALUS ADAMANTEUS SERINE PROTEINASE
2-DEHYDRO-3-DEOXYPHOSPHOHEPTONATE ALDOLASE | IR |
DODECENOYL-CoA Δ-ISOMERASE
FORMIMINOGLUTAMATE DEIMINASE
β-D-FUCOSIDASE | IR |
GLUCAN 1,3-β-GLUCOSIDASE
GLUCONOLACTONASE
H^+-TRANSPORTING ATP SYNTHASE
3-HYDROXYDECANOYL-[ACYL-CARRIER-PROTEIN] DEHYDRATASE
LACTATE 2-MONOOXYGENASE
LACTOYLGLUTATHIONE LYASE
D-LYSINE 5,6-AMINOMUTASE
MALATE SYNTHASE
MUCOROPEPSIN
PEPSIN A
POLYGALACTURONASE
PREPHENATE DEHYDROGENASE
PROLYL AMINOPEPTIDASE
STEAROYL-CoA DESATURASE
THROMBIN

TETRAPHENYLBORON
ADN-CARRIER MITOCHONDRIA
PHOSPHATE CARRIER MITOCHONDRIA
SUCCINATE CARRIER MITOCHONDRIA

TETRAPHENYLRESORCINOL DIPHOSPHATE
CARBOXYLESTERASE

TETRAPLATIN
ADENYLATE CYCLASE | | *I*50 = 10 μM

TETRA-n-PROPYLAMMONIUM CHLORIDE
ALCOHOL SULFOTRANSFERASE | | *I*50 = 17 μM

O,O,O′,O′-TETRAPROPYL DITHIOPYROPHOSPHATE
ACETYLCHOLINESTERASE

TETRARIC ACID
2-DEHYDRO-3-DEOXYGLUCARATE ALDOLASE

4,5,8,9-TETRATHIADODECA-1,11-DIENE
LIPOXYGENASE | | *I*50 = 25 μM

4,8,9,13-TETRATHIAHEXADECA-1,15-DIENE
LIPOXYGENASE | | *I*50 = 2 μM

4,5,9,10-TETRATHIATRIDECA-1,12-DIENE
LIPOXYGENASE | | *I*50 = 8 μM

1,5,9,10-TETRATHIATRIDECANE
LIPOXYGENASE — $I50 = 8\ \mu M$

TETRATHIONATE
1-AMINOCYCLOPROPANE-1-CARBOXYLATE DEAMINASE
ASCLEPAIN
BROMELAIN
CREATINASE
FICAIN
FORMALDEHYDE DEHYDROGENASE
GLYCERALDEHYDE-3-PHOSPHATE DEHYDROGENASE (PHOSPHORYLATING)
PREPHENATE DEHYDRATASE
PYROGLUTAMYL-PEPTIDASE I
STREPTOPAIN

TETRAZOLE
CARBONATE DEHYDRATASE

α-TETRAZOLE
GLUTAMATE DEHYDROGENASE (NAD(P))

γ-TETRAZOLE
GLUTAMATE DEHYDROGENASE (NAD(P))

TETRINDOLE
AMINE OXIDASE MAO-A — C — $Ki = 400\ nM$
AMINE OXIDASE MAO-B — C — $Ki = 110\ \mu M$

TETRNITROMETHANE
CYCLOMALTODEXTRIN GLUCANOTRANSFERASE

TETRODOXIN
RUBREDOXIN-NAD REDUCTASE

TETROXOPRIM
DIHYDROFOLATE REDUCTASE

TFP
3′,5′-CYCLIC-NUCLEOTIDE PHOSPHODIESTERASE (Ca^{2+}/CALMODULIN) — $I50 = 8\ \mu M$

TGG-II-23-A
RNA-DIRECTED DNA POLYMERASE — $I50 = 136\ \mu M$

THALIPORPHINE
INTERLEUKIN CONVERTING ENZYME

THAPSIC ACID
GLUTATHIONE TRANSFERASE

THAPSIGARGICIN
Ca^{2+}-TRANSPORTING ATPASE

THAPSIGARGIN
Ca^{2+}-TRANSPORTING ATPASE
Ca^{2+}-TRANSPORTING ATPASE — 48% 30 μM
Ca^{2+}-TRANSPORTING ATPASE — 81% 30 μM
Ca^{2+}-TRANSPORTING ATPASE — $I50 = 170\ pM$
H^{+}-TRANSPORTING ATPASE — 88% 30 μM
H^{+}-TRANSPORTING ATPASE — 75% 30 μM

THAPSIVILLOSIN A
Ca^{2+}-TRANSPORTING ATPASE

THEAFLAVIN
COLLAGENASE IV

THEAFLAVIN DIGALLATE
α-GLUCOSIDASE — 62% 100 μM
SUCRASE — 42% 100 μM

THEAFLAVIN 3,3′-DI-O-GALLATE
NADH DEHYDROGENASE $I50 = 800$ nM
NADH DEHYDROGENASE $I50 = 3.8$ μM
NADH DEHYDROGENASE $I50 = 7.9$ μM
NADH DEHYDROGENASE $I50 = 8.8$ μM

THEAFLAVIN 3′-O-GALLATE
NADH DEHYDROGENASE $I50 = 600$ nM
NADH DEHYDROGENASE $I50 = 5.4$ μM
NADH DEHYDROGENASE $I50 = 3.8$ μM

THEAFLAVIN MONOGALLATE A
α-GLUCOSIDASE 48% 100 μM

THEAFLAVIN MONOGALLATE B
α-GLUCOSIDASE 61% 100 μM

THEANINE
GLUTAMINE-tRNA LIGASE

D,L-THEANINE
GLUTAMATE SYNTHASE (NADPH)

THEASINENSIN A
NADH DEHYDROGENASE $I50 = 2.1$ μM
NADH DEHYDROGENASE $I50 = 4$ μM
NADH DEHYDROGENASE $I50 = 7$ μM

THEOBROMINE
3′,5′-CYCLIC-NUCLEOTIDE PHOSPHODIESTERASE
NAD ADP-RIBOSYLTRANSFERASE

2-THEONYLFLUORIDE
SUCCINATE DEHYDROGENASE (UBIQUINONE)

THEONYLTRIFLUOROACETONE
DIHYDROOROTATE DEHYDROGENASE $Ki = 20$ μM
4-HYDROXYPHENYLPYRUVATE DIOXYGENASE
NAD(P)H DEHYDROGENASE (QUINONE)
PHOSPHATIDYLCHOLINE DESATURASE
STEAROYL-CoA DESATURASE
SUCCINATE DEHYDROGENASE (UBIQUINONE)
UBIQUINOL-CYTOCHROME-C REDUCTASE

1-(2-THEONYL)-3,3,3,-TRIFLUOROACETONE
MONODEHYDROASCORBATE REDUCTASE (NADH)

1-(2-THEONYL)-3,3,3-TRIFLUOROACETONE
GLYCEROL-3-PHOSPHATE DEHYDROGENASE

THEOPHYLLINE
3′,5′-CYCLIC-GMP PHOSPHODIESTERASE $I50 = 47$ μM
3′,5′-CYCLIC-NUCLEOTIDE PHOSPHODIESTERASE (cycAMP) $I50 = 210$ μM
3′,5′-CYCLIC-NUCLEOTIDE PHOSPHODIESTERASE (cycAMP) $I50 = 2.8$ mM
3′,5′-CYCLIC-NUCLEOTIDE PHOSPHODIESTERASE (cycAMP) C $I50 = 120$ μM
3′,5′-CYCLIC-NUCLEOTIDE PHOSPHODIESTERASE (cycAMP) LOW Km $I50 = 96$ μM
3′,5′-CYCLIC-NUCLEOTIDE PHOSPHODIESTERASE F-I $Ki = 323$ μM
3′,5′-CYCLIC-NUCLEOTIDE PHOSPHODIESTERASE F-II $Ki = 286$ μM
3′,5′-CYCLIC-NUCLEOTIDE PHOSPHODIESTERASE F-III $Ki = 45$ μM
3′,5′-CYCLIC-NUCLEOTIDE PHOSPHODIESTERASE (cycGMP) $I50 = 271$ μM
3′,5′-CYCLIC-NUCLEOTIDE PHOSPHODIESTERASE (cycGMP) $I50 = 630$ μM
3′,5′-CYCLIC-NUCLEOTIDE PHOSPHODIESTERASE I $I50 = 372$ μM
3′,5′-CYCLIC-NUCLEOTIDE PHOSPHODIESTERASE Ia $I50 = 42$ μM
3′,5′-CYCLIC-NUCLEOTIDE PHOSPHODIESTERASE IB $I50 = 280$ μM
3′,5′-CYCLIC-NUCLEOTIDE PHOSPHODIESTERASE II $I50 = 270$ μM
3′,5′-CYCLIC-NUCLEOTIDE PHOSPHODIESTERASE II $I50 = 620$ μM
3′,5′-CYCLIC-NUCLEOTIDE PHOSPHODIESTERASE II (cycGMP STIMULATED) $I50 = 270$ μM

3′,5′-CYCLIC-NUCLEOTIDE PHOSPHODIESTERASE III		*I*50 = 240 μM
3′,5′-CYCLIC-NUCLEOTIDE PHOSPHODIESTERASE III		*I*50 = 380 μM
3′,5′-CYCLIC-NUCLEOTIDE PHOSPHODIESTERASE III		*I*50 = 146 μM
3′,5′-CYCLIC-NUCLEOTIDE PHOSPHODIESTERASE III (cycAMP)		*K*i = 40 μM
3′,5′-CYCLIC-NUCLEOTIDE PHOSPHODIESTERASE III (cycGMP INHIBITED)		*I*50 = 162 μM
3′,5′-CYCLIC-NUCLEOTIDE PHOSPHODIESTERASE III (cycGMP INHIBITED)		*I*50 = 390 μM
3′,5′-CYCLIC-NUCLEOTIDE PHOSPHODIESTERASE IV		*I*50 = 158 μM
3′,5′-CYCLIC-NUCLEOTIDE PHOSPHODIESTERASE IV		*I*50 = 155 μM
3′,5′-CYCLIC-NUCLEOTIDE PHOSPHODIESTERASE IV		*I*50 = 295 μM
3′,5′-CYCLIC-NUCLEOTIDE PHOSPHODIESTERASE IV		*I*50 = 98 μM
3′,5′-CYCLIC-NUCLEOTIDE PHOSPHODIESTERASE IVa		*I*50 = 661 μM
3′,5′-CYCLIC-NUCLEOTIDE PHOSPHODIESTERASE IVb		*I*50 = 453 μM
3′,5′-CYCLIC-NUCLEOTIDE PHOSPHODIESTERASE IVb		*I*50 = 708 μM
3′,5′-CYCLIC-NUCLEOTIDE PHOSPHODIESTERASE IVc		*I*50 = 644 μM
3′,5′-CYCLIC-NUCLEOTIDE PHOSPHODIESTERASE IVd		*I*50 = 444 μM
3′,5′-CYCLIC-NUCLEOTIDE PHOSPHODIESTERASE V (cycGMP)		*I*50 = 630 μM
GLYCEROL-1,2-CYCLIC-PHOSPHATE PHOSPHODIESTERASE		
15-HYDROXYPROSTAGLANDIN DEHYDROGENASE (NAD)		*K*i = 5.7 mM
NAD ADP-RIBOSYLTRANSFERASE		*I*50 = 46 μM
NAD ADP-RIBOSYLTRANSFERASE		68% 1 mM
NAD(P)-ARGININE ADP-RIBOSYLTRANSFERASE		*I*50 = 2.8 mM
4-NITROPHENYLPHOSPHATASE		
NUCLEOSIDE-DIPHOSPHATE KINASE		
5′-NUCLEOTIDASE		56% 100 μM
PHOSPHATASE ALKALINE		21% 80 μM
PHOSPHATASE ALKALINE	UC	*K*i = 820 mM
PHOSPHATASE ALKALINE		97% 80 μM
PHOSPHATASE ALKALINE		*I*50 = 100 μM
PHOSPHATASE ALKALINE		43% 1 mM
PHOSPHATASE ALKALINE	U	*K*i = 200 μM
PHOSPHATASE ALKALINE	U	*K*i = 126 μM
PHOSPHATASE ALKALINE (PHOSPHODIESTERASE ACTIVITY)	U	*K*i = 370 μM
1-PHOSPHATIDYLINOSITOL 4-KINASE		
PURINE NUCLEOSIDASE		83% 10 mM
PYRIDOXAL KINASE		*K*i 13 μM
PYRIDOXAL KINASE		86% 100 μM
RHODOPSIN KINASE		
THYMIDINE KINASE		

THERIN

TRYPSIN		*K*i = 600 fM

THERMOLYSIN

PITRILYSIN		kI = 480 nM

THERMORUBIN

ALDEHYDE REDUCTASE		*I*50 = 8 nM

3-THIAAMINOLAEVULINIC ACID

PORPHOBILINOGEN SYNTHASE	IR	

THIABENAZOLE

FUMARATE REDUCTASE (NADH)		*I*50 = 420 μM

18-THIA-19-DEHYDRO-(R,S)-2,3-OXIDOSQUALENE

SQUALENE-HOPENE CYCLASE	SSI	*K*i = 31 nM

THIAMINE

L-ASCORBATE OXIDASE
PHOSPHORYLASE b
THIAMIN-PHOSPHATE KINASE
THIAMIN PYRIDINYLASE
THIAMIN PYRIDINYLASE

THIAMINE DISULPHIDE DERIVATIVES

BROMELAIN

THIAMINE PYROPHOSPHATE
[3-METHYL-2-OXOBUTANOATE DEHYDROGENASE (LIPOAMIDE)] KINASE
[PYRUVATE DEHYDROGENASE (LIPOAMIDE)] KINASE

THIAMINE THIAZOLONE-PYROPHOSPHATE
ACETOLACTATE SYNTHASE — *K*i = 300 nM
PYRUVATE DEHYDROGENASE (LIPOAMIDE)

THIAMPHENICOL
PYRIDOXAL KINASE — 31% 100 μM

THIANGAZOLE
NADH DEHYDROGENASE (UBIQUINONE)
RNA-DIRECTED DNA POLYMERASE — *I*50 = 252 μM

13-THIA-9(Z)-,11(E)-OCTADECADIENOIC ACID
LIPOXYGENASE 1 — C — *K*i = 30 μM

3-THIAOCTANOYL-CoA
ACYL-CoA DEHYDROGENASE

4-THIAOCTANOYL-CoA
ACYL-CoA DEHYDROGENASE

3-THIAOCTYL-CoA
ACYL-CoA DEHYDROGENASE

(6E)-8-THIA-2,3-OXIDOSQUALENE
LANOSTEROL SYNTHASE — *I*50 = 680 nM

18-THIA-2,3-OXIDOSQUALENE
OXIDOSQUALENE SYNTHASE — *I*50 = 50 nM

7-THIA-13-PROSTYNOIC ACID
15-HYDROXYPROSTAGLANDIN DEHYDROGENASE (NAD)

THIAZOCIN A
ALDEHYDE REDUCTASE — *I*50 = 455 nM

THIAZOCIN B
ALDEHYDE REDUCTASE — *I*50 = 220 nM

THIAZOFURIN
IMP DEHYDROGENASE

THIAZOLE ADENINE
IMP DEHYDROGENASE — UC — *K*i = 1.5 μM

THIAZOLE-4-CARBOXAMIDE ADENINEDINUCLEOTIDE
IMP DEHYDROGENASE — U — *K*i = 75 nM
IMP DEHYDROGENASE — U — *K*i = 130 nM
IMP DEHYDROGENASE 1 — *I*50 = 94 nM
IMP DEHYDROGENASE 2 — *I*50 = 101 nM

THIAZOLIDINE-4-CARBOXYLIC ACID
PYRROLINE-5-CARBOXYLATE REDUCTASE

R-4-THIAZOLIDINECARBOXYLIC ACID
XAA-PRO DIPEPTIDASE — C — *K*i = 160 μM

D-2-THIAZOLINE-2-CARBOXYLIC ACID
DOPAMINE β-MONOOXYGENASE

4-THIAZOLYLMETHOXYAMINE
HISTIDINE DECARBOXYLASE — 74% 100 μM

6-(5-THIAZOLYLMETHYL)-5,6,7,8-TETRAHYDRONAPHTHALENE-2-CARBOXYLATE SODIUM
THROMBOXANE A SYNTHASE — *I*50 = 140 nM

THIELAVIN A

PHOSPHOLIPASE A2-2	*I*50 = 29 μM
PHOSPHOLIPASE A2-2	*I*50 = 43 μM

THIELAVIN B

PHOSPHOLIPASE A2-2	*I*50 = 2.4 μM
PHOSPHOLIPASE A2-2	*I*50 = 1.3 μM

THIELAVIN C

PHOSPHOLIPASE A2-2	*I*50 = 2.1 μM
PHOSPHOLIPASE A2-2	*I*50 = 460 nM

THIELAVIN D

PHOSPHOLIPASE A2-2	*I*50 = 6.2 μM
PHOSPHOLIPASE A2-2	*I*50 = 1.1 μM

THIELAVIN E

PHOSPHOLIPASE A2-2	*I*50 = 9.3 μM
PHOSPHOLIPASE A2-2	*I*50 = 4.5 μM

(±)THIELOCIN A1β

PHOSPHOLIPASE A2	*I*50 = 21 nM
PHOSPHOLIPASE A2-2	*I*50 = 3.3 nM

THIELOCIN A1

PHOSPHOLIPASE A2 (NON PANCREATIC SECRETED)	*I*50 = 12 mM

THIELOCIN A1β

PHOSPHOLIPASE A2	*I*50 = 3.3 nM
PHOSPHOLIPASE A2-1	*I*50 = 140 μM
PHOSPHOLIPASE A2-2	*I*50 = 6.2 μM
PHOSPHOLIPASE A2-2	*I*50 = 12 μM
PHOSPHOLIPASE A2-2	*I*50 = 320 nM

THIELOCIN A2α

PHOSPHOLIPASE A2	*I*50 = 51 nM
PHOSPHOLIPASE A2-2	*I*50 = 310 nM

THIELOCIN A2β

PHOSPHOLIPASE A2	*I*50 = 38 nM
PHOSPHOLIPASE A2-2	*I*50 = 240 nM

THIELOCIN A3

PHOSPHOLIPASE A2	*I*50 = 32 nM
PHOSPHOLIPASE A2-2	*I*50 = 390 nM

THIELOCIN B1

PHOSPHOLIPASE A2	*I*50 = 7.8 nM
PHOSPHOLIPASE A2-2	*I*50 = 170 nM

THIELOCIN B2

PHOSPHOLIPASE A2	*I*50 = 70 nM
PHOSPHOLIPASE A2-2	*I*50 = 2.7 μM

THIELOCIN B3

PHOSPHOLIPASE A2	*I*50 = 12 nM
PHOSPHOLIPASE A2-1	*I*50 = 18 μM
PHOSPHOLIPASE A2-1	*I*50 = 2.8 μM
PHOSPHOLIPASE A2-2	*I*50 = 76 nM
PHOSPHOLIPASE A2-2	*I*50 = 76 nM
PHOSPHOLIPASE A2-2	*I*50 = 12 nM
PHOSPHOLIPASE A2 (NON PANCREATIC SECRETED)	*I*50 = 76 μM

THIENAMYCINE

β-LACTAMASE	SSI	

THIENOLOCIN A1β

PHOSPHOLIPASE A2	*I*50 = 63 μM

PHOSPHOLIPASE A2		$I50 = 21\ \mu M$
PHOSPHOLIPASE A2		$I50 = 33$ nM
PHOSPHOLIPASE A2		$I50 = 2\ \mu M$
PHOSPHOLIPASE A2		$I50 = 7.1\ \mu M$
THIENOTHIOPYRAN-2-SULPHONAMIDE		
CARBONATE DEHYDRATASE		
THIENYLALANINE		
2-DEHYDRO-3-DEOXYPHOSPHOHEPTONATE ALDOLASE		84% 1 mM
HISTIDINE AMMONIA-LYASE	C	Ki = 2.4 mM
PRETYROSINE DEHYDROGENASE		
β-2-THIENYLALANINE		
PREPHENATE DEHYDRATASE		
2-(2-THIENYL)ALLYLAMINE		
DOPAMINE β-MONOOXYGENASE	SSI	Ki = 35 μM
2-THIENYL-4,5-DIHYDRO-1H-THIENO-[3,2-e]BENZIMIDAZOLE		
H^+/K^+-EXCHANGING ATPASE		$I50 = 3\ \mu M$
3-(2-THIENYL)-4-(3,5-DIMETHYL-4-HYDROXYPHENYL)METHYLIDIENE-2-ISOXAZOLIN-5-ONE		
PROTEIN KINASE A		$I50 > 50\ \mu M$
PROTEIN KINASE C		$I50 = 2.5\ \mu M$
1-[L-(2-THIENYL)ETHYL]IMIDAZOLE-2(3H)-THIONE		
DOPAMINE β-MONOOXYGENASE		Ki = 26 nM
9-(3-THIENYLMETHYL)-9-DEAZAGUANINE		
PURINE-NUCLEOSIDE PHOSPHORYLASE		$I50 = 80$ nM
2-[4-(THIEN-2-YLMETHYL)PHENYL]PROPANOIC ACID		
CYCLOOXYGENASE		$I50 = 23\ \mu M$
4-(2-THIENYL)-2-OXO-3-BUTENOIC ACID		
[3-METHYL-2-OXOBUTANOATE DEHYDROGENASE (LIPOAMIDE)] KINASE		
4-(3-THIENYL)-2-OXO-3-BUTENOIC ACID		
3-METHYL-2-OXOBUTANOATE DEHYDROGENASE (LIPOAMIDE)	NC	Ki = 150 μM
2-THIENYLPYRUVIC ACID		
4-HYDROXYPHENYLPYRUVATE DIOXYGENASE	SSI	
3-THIENYLPYRUVIC ACID		
4-HYDROXYPHENYLPYRUVATE DIOXYGENASE	SSI	
[5-(3-THIENYL)TETRAZOL-1-YL]ACETIC ACID		
ALCOHOL DEHYDROGENASE (NADP)		$I50 = 2.4\ \mu M$
ALDEHYDE REDUCTASE		$I50 = 28$ nM
ALDEHYDE REDUCTASE		$I50 = 23$ nM
ALDEHYDE REDUCTASE		$I50 = 21$ nM
THIGLYCOLIC ACID		
GLUTAMATE DECARBOXYLASE		
UREIDOSUCCINASE		
THIGLYCOLIC ACIDCOLLIDINE		
MALEYLACETOACETATE ISOMERASE		60% 1 mM
THIMERASAL		
β-N-ACETYLHEXOSAMINIDASE		
THIMEROSAL		
LYSOPHOSPHOLIPASE		$I50 = 10\ \mu M$
4-NITROPHENYLPHOSPHATASE		
THIOACETAMIDE		
ALDEHYDE DEHYDROGENASE (PYRROLOQUINOLINE-QUINONE)		23% 1 mM
METHANE MONOOXYGENASE		92% 10 μM

2-THIOACETIC ACID
DIHYDROOROTASE

THIOAMOPHENICOL
PEPTIDYLTRANSFERASE

7-THIOARACHIDONIC ACID
ARACHIDONATE 5-LIPOXYGENASE

2-THIO-6-AZAURIDINE
CYTIDINE DEAMINASE $Ki = 290\ \mu M$

THIOBARBITURIC ACID
DIHYDROOROTATE DEHYDROGENASE 17% 1 mM
MEVALDATE REDUCTASE (NADPH)

THIOBENZAMIDE
NAD ADP-RIBOSYLTRANSFERASE $I50 = 620\ \mu M$

22-THIO-23,24-BISNOR-5-CHOLEN-3β-OL
CHOLESTEROL MONOOXYGENASE (SIDE-CHAIN-CLEAVING) $Ki = 700$ nM

N-(THIOCARBOXY)-L-APSARTIC ANHYDRIDE
ADENYLOSUCCINATE SYNTHASE

THIOCHOLINE
CHOLINE SULFOTRANSFERASE
GLYCEROPHOSPHOCHOLINE CHOLINEPHOSPHODIESTERASE (Zn^{2+}) $Ki = 2.6\ \mu M$

THIOCHROME PYROPHOSPHATE
PYRUVATE DEHYDROGENASE (LIPOAMIDE) C

L-THIOCITRULLINE
NITRIC-OXIDE SYNTHASE I $Ki = 1.2$ nM
NITRIC-OXIDE SYNTHASE I $Ki = 60\ \mu M$
NITRIC-OXIDE SYNTHASE II $Ki = 36\ \mu M$
NITRIC-OXIDE SYNTHASE II $Ki = 3.6$ nM

2-THIO-CTP
CTP SYNTHASE $Ki = 80\ \mu M$

THIOCYANATE
ADENYLOSUCCINATE SYNTHASE
ALCOHOL DEHYDROGENASE C
AMINE OXIDASE (COPPER-CONTAINING)
CHLORIDE PEROXIDASE
CITRATE (si)-SYNTHASE
CYANATE HYDROLASE
CYTIDYLATE KINASE
GUANYLATE CYCLASE
HYDROXYLAMINE REDUCTASE (NADH)
ISOCITRATE DEHYDROGENASE (NAD)
MANDELONITRILE LYASE
NITRATE REDUCTASE (NADH)
SUCCINATE DEHYDROGENASE
XANTHINE OXIDASE

2-THIOCYCLOHEXANONE
CYCLOHEXANONE MONOOXYGENASE

2-THIOCYTIDINE
α-2,6-SIALYLTRANSFERASE C $Ki = 150\ \mu M$

2,2′-THIODIACETIC ACID
GLUTAMATE DEHYDROGENASE (NAD(P))

1-THIO-2-(2′,4′-DICHLOROPHENYL)ACETIC ACID
DIHYDROXY ACID DEHYDRATASE $Ki = 100\ \mu M$

THIODIGLYCOLIC ACID
GLUTAMATE DEHYDROGENASE (NAD(P))
PYRIMIDINE-DEOXYNUCLEOSIDE 2′-DIOXYGENASE — 24% 2.5 mM
XAA-PRO DIPEPTIDASE — Ki = 15 μM

THIOFANOX
ACETYLCHOLINESTERASE

2-THIO-5-FLUORO-dUMP
THYMIDYLATE SYNTHASE — Ki = 79 nM

4-THIO-5-FLUORO-dUMP
THYMIDYLATE SYNTHASE — Ki = 60 nM

2-THIO-FMN
FMN ADENYLYLTRANSFERASE

5-THIO-D-FRUCTOSE
KETOHEXOKINASE — C

5-THIO-L-FUCOSE
α-L-FUCOSIDASE — Ki = 42 μM

2-THIOFdUMP
THYMIDYLATE SYNTHASE — C — Ki = 220 nM

4-THIOFdUMP
THYMIDYLATE SYNTHASE — C — Ki = 580 nM

THIOGALACTOSIDE
β-GALACTOSIDASE — C

THIO GAR DIDEAZAFOLIC ACID
PHOSPHORIBOSYLGLYCINAMIDE FORMYLTRANSFERASE — Kd = 250 pM

THIOGLUCOSE
HEXOKINASE

5-THIO-D-GLUCOSE
GLUCOSE TRANSPORT — C

THIOGLYCOLIC ACID
AGAVAIN — 76% 10 mM
ALLANTOINASE
AMINOPEPTIDASE B — 100% 17 mM
COLLAGENASE CLOSTRIDIUM HISTOLYTICUM
DIAMINOPIMELATE DEHYDROGENASE
ENVELYSIN
LEUCYL AMINOPEPTIDASE
MALEYLACETOACETATE ISOMERASE
PROLINE RACEMASE
PSEUDOMONAS AEROGINOSA NEUTRAL PROTEINASE
SERRALYSIN
SIALIDASE
THERMOLYSIN
UROCANATE HYDRATASE
VENOM EXONUCLEASE

p-THIOGLYCOLIC ACID
AMINOLEVULINATE TRANSAMINASE

6-THIOGUANINE
HYPOXANTHINE-GUANINE PHOSPHORIBOSYLTRANSFERASE — I50 = 13 μM
HYPOXANTHINE PHOSPHORIBOSYLTRANSFERASE
PROTEIN KINASE ECTO
PROTEIN KINASE N — Ki = 10 μM
QUEUINE tRNA-RIBOSYLTRANSFERASE

6-THIOGUANOSINE 5′-PHOSPHATE

GUANYLATE KINASE	C	

THIO-IMP

ADENYLOSUCCINATE LYASE		
IMP DEHYDROGENASE		

6-THIO-IMP

ADENYLOSUCCINATE SYNTHASE		35% 2 mM
GMP REDUCTASE		
IMP DEHYDROGENASE 1	C	Ki = 15 μM
IMP DEHYDROGENASE 2	C	Ki = 29 μM

6-THIOINOSINE

HYPOXANTHINE PHOSPHORIBOSYLTRANSFERASE		

THIOINOSINE 5′-PHOSPHATE

IMP DEHYDROGENASE		

THIOINOSINIC ACID

ADENYLOSUCCINATE LYASE		
PHOSPHORIBOSYLAMINOIMIDAZOLECARBOXAMIDE FORMYLTRANSFERASE	C	Ki = 120 μM

3-THIOLACTIC ACID

DIHYDROXY ACID DEHYDRATASE		Ki = 5 μM

THIOLACTOMYCIN

FATTY ACID SYNTHASE II		
3-OXOACYL-[ACYL-CARRIER-PROTEIN] SYNTHASE I		

5-THIOLATED POLYURIDYLIC ACID

RNA-DIRECTED DNA POLYMERASE		Ki = 31 nM

THIOL REAGENTS

PROTEIN KINASE C		
SUCCINATE DEHYDROGENASE		
SUCCINATE DEHYDROGENASE (UBIQUINONE)		

THIOLYSINE

HOMOCITRATE SYNTHASE		

THIOMALATE

GLUTAMATE DECARBOXYLASE		Ki = 12 μM

19-THIOMETHYL ANDROSTENEDIONE

AROMATASE		Ki = 2.4 nM

THIOMETON

ACETYLCHOLINESTERASE		

THIO NAD

NICOTINAMIDE PHOSPHORIBOSYLTRANSFERASE		

THIO-NAD

ALDEHYDE DEHYDROGENASE (NAD(P))		
NAD KINASE	C	Ki = 140 μM
NAD NUCLEOSIDASE		
PYRROLINE-5-CARBOXYLATE REDUCTASE		

THIO-NADP

PYRROLINE-5-CARBOXYLATE REDUCTASE		

THIONAZIN

ACETYLCHOLINESTERASE		

THIONICOTINAMIDE

NAD ADP-RIBOSYLTRANSFERASE		I50 = 1.8 mM
NICOTINAMIDE PHOSPHORIBOSYLTRANSFERASE	C	
UDP-GLUCURONATE DECARBOXYLASE	C	Ki = 16 μM

THIONICOTINAMIDE-ADENINEDINUCLEOTIDE PHOSPHATE		
MALATE DEHYDROGENASE (OXALACETATE-DECARBOXYLATING) (NADP)	C	Ki = 260 nM
THIONICOTINAMIDE DINUCLEOTIDE		
L-LACTATE DEHYDROGENASE		Ki = 21 µM
L-LACTATE DEHYDROGENASE		Ki = 47 µM
L-LACTATE DEHYDROGENASE		Ki = 59 µM
THIONICOTINAMIDE-NAD		
ADP-RIBOSE PYROPHOSPHATASE		76% 4 mM
RUBREDOXIN-NAD REDUCTASE		
8-THIO-2′-NORDEOXYGUANOSINE		
PURINE-NUCLEOSIDE PHOSPHORYLASE		Ki = 13 µM
(R)-1-THIOOCTYL-2-HEPTYLPHOSPHONYL-1-DEOXYGLYCERO-3-PHOSPHOGLYCOL		
PHOSPHOLIPASE A2		
6-THIOOROTIC ACID		
DIHYDROOROTASE	C	Ki = 82 µM
(18E)-20-THIO-2,3-OXIDOSQUALENE		
LANOSTEROL SYNTHASE		I50 = 320 nM
LANOSTEROL SYNTHASE		I50 = 200 nM
2-THIO-6-OXOPURINE		
URATE-RIBONUCLEOTIDE PHOSPHORYLASE		
THIOPENTAL		
Ca^{2+}-TRANSPORTING ATPASE		
THIOPHENE-3-CARBOXAMIDE		
NAD ADP-RIBOSYLTRANSFERASE		
2-THIOPHENECARBOXYLATE		
PROLINE RACEMASE		
2-THIOPHENECARBOXYLIC ACID		
SARCOSINE OXIDASE		
THIOPHENFURIN		
IMP DEHYDROGENASE		
THIOPHENOL		
ARACHIDONATE 5-LIPOXYGENASE		I50 = 2.5 µM
4-THIOPHENYL-4-ANDROSTADIENE-3,17-DIONE		
AROMATASE	C	Ki = 9.8 nM
4-THIOPHENYL-4-ANDROSTADINE-3,17-DIONE		
AROMATASE	C	Ki = 9.8 nM
4-THIOPHENYL-4-ANDROSTENE-3,17-DIONE		
AROMATASE	C	Ki = 33 nM
1-THIOPHENYL-β-D-GLUCOSIDE		
XYLAN 1,4-β-XYLOSIDASE		
2-THIO-4-PHENYLIMIDAZOLIDINE		
AMINE OXIDASE (COPPER-CONTAINING)		I50 = 18 mM
1-THIO-1-PHENYLMETHANE		
L-IDITOL 2-DEHYDROGENASE		Ki = 300 nM
1-THIOPHENYL-β-D-XYLOSIDE		
XYLAN 1,4-β-XYLOSIDASE		
THIOPHOSPHATE		
ALKALINE PHOSPHATASE	C	
PHOSPHATASE ALKALINE	C	Ki = 17 µM
PROTEIN TYROSINE-PHOSPHATASE		I50 = 470 µM

THIOPHOSPHATIDIC ACID
PHOSPHATIDATE CYTIDILYLTRANSFERASE
PHOSPHATIDATE PHOSPHATASE C $Ki = 60\ \mu M$

THIOPHOSPHORYLATED RCM-LYSOZYME
PROTEIN TYROSINE-PHOSPHATASE 1 $I50 = 230$ nM

L-THIOPROLINE
GLUTAMATE KINASE

L-THIOPROLYLTHIAZOLIDINE
DIPEPTIDYL-PEPTIDASE 4 $Ki = 20\ \mu M$
DIPEPTIDYL-PEPTIDASE 4 $Ki = 36\ \mu M$

6-THIOPURINE
THIOPURINE S-METHYLTRANSFERASE
THIOPURINE S-METHYLTRANSFERASE

(2′S)-10β-THIORANYLANDROST-4-ENE-3,17-DIONE
AROMATASE C $I50 = 22$ nM

THIOREDOXIN REDUCED HUMAN
MONOPHENOL MONOOXYGENASE

4-THIO-1-βD-RIBOFURANOSYLPYRAZOLO-(3,4-d)PYRIMIDINE
RIBOSYLPYRIMIDINE NUCLEOSIDASE $Ki = 35\ \mu M$

THIORIDAZINE
Ca^{2+}/Mg^{2+}-TRANSPORTING ATPASE
CYTOCHROME P450 2D6
INOSITOL-1,4,5-TRISPHOSPHATE 5-PHOSPHATASE 61% 1 mM
Na^{+}/K^{+}-EXCHANGING ATPASE
PYRUVATE DEHYDROGENASE (LIPOAMIDE) 48% 267 μM

THIORPHAN
ENDOTHELIN CONVERTING ENZYME $I50 = 53\ \mu M$
ENDOTHELIN CONVERTING ENZYME M2
NEPRILYSIN $Ki = 4.7$nM
NEPRILYSIN $I50 = 10$ nM
NEPRILYSIN
NEPRILYSIN $I50 = 34$ nM

2-THIORPHAN
NEPRILYSIN $I50 = 4$ nM

(R)-THIORPHAN
NEPRILYSIN $I50 = 1.7$ nM
PEPTIDYL-DIPEPTIDASE A $I50 = 4.8\ \mu M$
THERMOLYSIN $I50 = 13\ \mu M$

(S)-THIORPHAN
PEPTIDYL-DIPEPTIDASE A $I50 = 110$ nM
THERMOLYSIN $I50 = 5.9\ \mu M$

THIOSALICYLIC ACID
GENTISATE 1,2-DIOXYGENASE $Ki = 200\ \mu M$

THIOSANGIVAMYCIN
PROTEIN KINASE (NUCLEAR)

THIOSEMICARBAZIDE
ASPARTATE TRANSAMINASE
BRANCHED-CHAIN-AMINO-ACID TRANSAMINASE
GLUTAMATE DECARBOXYLASE
METHANE MONOOXYGENASE 85% 1 μM
PHOSPHATIDYLSERINE DECARBOXYLASE $I50 = 400\ \mu M$

THIOSEMICARBAZONE		
RNA-DIRECTED DNA POLYMERASE		
THIOSORBITOL		
L-IDITOL 2-DEHYDROGENASE		
THIOSTREPTON		
GUANOSINE-3′,5′-BIS(DIPHOSPHATE) 3′-PYROPHOSPHATASE		
6-THIOSUCROSE		
DEXTRANSUCRASE	C	Ki = 7.3 mM
D-GLUCANSUCRASE	C	Ki = 7.3 mM
SUCROSE-1,6-GLUCAN 3(6)-α-GLUCOSYLTRANSFERASE		
THIOSULPHATE		
N-ACETYLGALACTOSAMINE-6-SULFATASE		99% 17 mM
N-ACETYLGLUCOSAMINE-6-SULFATASE		78% 0.5%
CYANATE HYDROLASE		I50 = 1 mM
FORMALDEHYDE DISMUTASE		
PYROPHOSPHATE-FRUCTOSE-6-PHOSPHATE 1-PHOSPHOTRANSFERASE		
SULFATE ADENYLYLTRANSFERASE		
THIOSULFATE SULFURTRANSFERASE		
THIOSULPHITE		
DISPASE		
THIOTHIAZOLONE PYROPHOSPHATE		
PYRUVATE DEHYDROGENASE (LIPOAMIDE)	IR	
THIOTHIXENE		
INOSITOL-1,4,5-TRISPHOSPHATE 5-PHOSPHATASE		20% 1 mM
PYRUVATE DEHYDROGENASE (LIPOAMIDE)		40% 267 μM
4-THIO-dTMP		
THYMIDYLATE 5′-PHOSPHATASE		
2-THIO-UMP		
ASPARTATE CARBAMOYLTRANSFERASE		
2-THIO-dUMP		
THYMIDYLATE SYNTHASE	C	Ki = 300 nM
4-THIO-dUMP		
THYMIDYLATE SYNTHASE	C	Ki = 220 nM
THIOURACIL		
IODIDE PEROXIDASE		
2-THIOURACIL		
4-AMINOBUTYRATE TRANSAMINASE		Ki = 1.9 mM
PEROXIDASE		
THIOPURINE S-METHYLTRANSFERASE		
THYROXINE DEIODINASE		
URATE-RIBONUCLEOTIDE PHOSPHORYLASE		
THIOUREA		
ACETYLINDOXYL OXIDASE		
ALCOHOL DEHYDROGENASE		
ALCOHOL DEHYDROGENASE-2		
ALDEHYDE DEHYDROGENASE (PYRROLOQUINOLINE-QUINONE)		33% 1 mM
CATECHOL OXIDASE		
CHLORIDE PEROXIDASE		
γ-GLUTAMYL HYDROLASE		27% 50 mM
IODIDE PEROXIDASE		
LACCASE		
MANDELATE 4-MONOOXYGENASE		
METHANE MONOOXYGENASE		100% 10 μM
MONOPHENOL MONOOXYGENASE		

NITRITE REDUCTASE (NAD(P)H)		
PROPANEDIOL DEHYDRATASE		
UREA CARBOXYLASE		
UREASE	C	
XANTHINE DEHYDROGENASE		
2-THIO-UTP		
CTP SYNTHASE		$Ki = 250\ \mu M$
4-THIO-UTP		
CTP SYNTHASE		$Ki = 510\ \mu M$
2-THIOXANTHINE		
HYPOXANTHINE PHOSPHORIBOSYLTRANSFERASE	C	$Ki = 604\ \mu M$
XANTHINE PHOSPHORIBOSYLTRANSFERASE		
6-THIOXANTHINE		
HYPOXANTHINE PHOSPHORIBOSYLTRANSFERASE	C	$Ki = 190\ \mu M$
PURINE-NUCLEOSIDE PHOSPHORYLASE		$Ki = 1.1$ mM
XANTHINE PHOSPHORIBOSYLTRANSFERASE		
THIOXANTHONE		
AMINE OXIDASE MAO-A		$I50 = 3.4\ \mu M$
AMINE OXIDASE MAO-B		15% 10 μM
6-THIOXANTHOSINE-5′-MONOPHOSPHATE		
GMP SYNTHASE	C	$Ki = 5\ \mu M$
6-THIO-XMP		
GMP REDUCTASE		
THIPHENOL OXALATE		
4-HYDROXYPHENYLPYRUVATE DIOXYGENASE	C	$Ki = 150\ \mu M$
2[[[N[5(S)[[2THIPHENOXYMETHOXY)BENZOYL]AMINO]6CYCLOHEXYL3(R),4(R)DIHYDROXY2(R)ISOPROPYLHEXANOYL]ILE]-AMINO]METHYL]BENZIMIDAZOLE		
PROTEINASE HIV-1		$I50 = 5$ nM
THISTREPTON		
GTP PYROPHOSPHOKINASE 1		
THIZOL-4-CARBOXAMIDE ADENINEDINUCLEOTIDE		
IMP DEHYDROGENASE 2		$Kii = 110$ nM
THRENE RED VIOLET RH		
DIHYDROFOLATE REDUCTASE		$I50 = 200\ \mu M$
DIHYDROFOLATE REDUCTASE		$I50 = 6.9\ \mu M$
L-THREONIC ACID		
L(+)-TARTRATE DEHYDRATASE		22% 300 μM
THREONINE		
ALANINE AMINOPEPTIDASE	NC	$I50 = 25$ mM
ALANINE DEHYDROGENASE		
5-AMINOLEVULINATE SYNTHASE		
ASPARTATE-SEMIALDEHYDE DEHYDROGENASE		
GLUTAMATE-AMMONIA LIGASE		
GLUTAMATE SYNTHASE (FERREDOXIN)		
GLYCINE HYDROXYMETHYLTRANSFERASE		
PYRUVATE KINASE		
SERINE-PHOSPHOETHANOLAMINE SYNTHASE		
allo-D,L-THREONINE		
L-SERINE DEHYDRATASE	C	$Ki = 3$ mM
THREONINE DEHYDRATASE	C	$Ki = 70\ \mu M$
allo-L-THREONINE		
L-THREONINE 3-DEHYDROGENASE	C	1.2 mM

Inhibitor / Enzyme	Type	Value
D-THREONINE		
D-SERINE DEHYDRATASE	C	Ki = 3.4 mM
THREONINE DEHYDRATASE	C	Ki = 56 mM
D,L-allo-THREONINE		
HOMOSERINE DEHYDROGENASE		
DL-THREONINE HYDROXAMIC ACID		
PROTEINASE ASTACUS		I50 = 1.9 mM
THREONINE INHIBITORY PEPTIDE		
CAM KINASE		I50 = 40 μM
L-THREONINE		
ASPARTATE KINASE		94% 20 mM
HOMOSERINE DEHYDROGENASE		
HOMOSERINE KINASE	C	Ki = 4 mM
HOMOSERINE KINASE		34% 2 mM
L-THREONINE + L-LYSINE		
ASPARTATE KINASE		I50 = 400 μM
D,L-THREONINE-O-PHOSPHATE		
ETHANOLAMINE-PHOSPHATE PHOSPHO-LYASE		90% 10 mM
THREONINE SYNTHASE		
D-THREONOHYDROXAMIC ACID		
XYLOSE ISOMERASE		
XYLOSE ISOMERASE		Ki = 100 nM
D-THREOSE-2,4-DIPHOSPHATE		
GLYCERALDEHYDE-3-PHOSPHATE DEHYDROGENASE (PHOSPHORYLATING)	NC	Ki = 200 nM
THRE-PHE-GLN-ALA-ILE-PRO-LEU-ARG-GLU-ALA		
RETROVIRAL PROTEINASE		
THROMBIN APTAMER		
THROMBIN		
THROMBOMODULIN		
THROMBIN		
THROMBOMODULIN PEPTIDE(426-444)		
THROMBIN		I50 = 140 μM
THROMBOSPONDIN		
ELASTASE LEUKOCYTE		Kd = 17 nM
THROMBOSPONDIN 1		
CATHEPSIN G		Ki = 10 nM
THROMSTOP		
PLASMIN		I50 = 35 μg/ml
PROLYL OLIGOPEPTIDASE		I50 = 84 μg/ml
THROMBIN		I50 = 60 ng/ml
TRYPSIN		I50 = 650 ng/ml
β-THUJAPLICIN		
CATECHOL O-METHYLTRANSFERASE		
THYLAKOID INHIBITOR		
GLYCEROL-3-PHOSPHATE DEHYDROGENASE (NAD)		
THYMIDILIC ACID POLYMERS		
REC A PEPTIDASE		
THYMIDINE		
ADP-RIBOSE PYROPHOSPHATASE		
CYTIDINE DEAMINASE		Ki = 470 μM

CYTOSINE DEAMINASE		$I50 = 690\ \mu M$
DIHYDROOROTASE	C	$Ki = 70\ \mu M$
GLUCOSE TRANSPORT		$I50 = 90$ mM
GLUTAMATE-AMMONIA LIGASE		
NAD ADP-RIBOSYLTRANSFERASE		$I50 = 180\ \mu M$
NAD ADP-RIBOSYLTRANSFERASE		94% 1 mM
NAD NUCLEOSIDASE		
NAD(P)-ARGININE ADP-RIBOSYLTRANSFERASE		$I50 = 1.9$ mM
PYRIMIDINE-DEOXYNUCLEOSIDE 2′-DIOXYGENASE		
RIBONUCLEASE T2		
dTMP KINASE		$I50 = 23\ \mu M$
URATE-RIBONUCLEOTIDE PHOSPHORYLASE		
THYMIDINE DIMER		
VENOM EXONUCLEASE		
THYMIDINE 3′,5′-DIPHOSPHATE		
MICROCOCCAL NUCLEASE	C	$Ki = 540$ nM
THYMIDINE 5′-[α,β-IMIDO]TRIPHOSPHATE		
RNA-DIRECTED DNA POLYMERASE	C	$Ki = 2.4\ \mu M$
L-THYMIDINE		
THYMIDINE KINASE	C	$Ki = 2.2\ \mu M$
P1-(THYMIDINE-5′)-P5-(ADENOSINE-5′)-P1,P2-METHYLENEHEXAPHOSPHATE		
RIBONUCLEOSIDE-DIPHOSPHATE REDUCTASE		$I50 = 231\ \mu M$
dTMP KINASE		$I50 = 4\ \mu M$
P1-(THYMIDINE-5′)-P5-(ADENOSINE-5′)-P1,P2-METHYLENEPENTAPHOSPHATE		
RIBONUCLEOSIDE-DIPHOSPHATE REDUCTASE		$I50 = 128\ \mu M$
dTMP KINASE		$I50 = 6\ \mu M$
P1-(THYMIDINE-5′)-P4-(ADENOSINE-5′)-P1,P2-METHYLENETETRAPHOSPHATE		
RIBONUCLEOSIDE-DIPHOSPHATE REDUCTASE		$I50 = 296\ \mu M$
dTMP KINASE		$I50 = 340\ \mu M$
P1-(THYMIDINE-5′)-P3-(ADENOSINE-5′)-P1,P2-METHYLENETRIPHOSPHATE		
RIBONUCLEOSIDE-DIPHOSPHATE REDUCTASE		$I50 = 455\ \mu M$
dTMP KINASE		$I50 = 730\ \mu M$
THYMIDINERIBOSIDE		
CYTIDINE DEAMINASE		$Ki = 530\ \mu M$
THYMINE		
THYMIDINE PHOSPHORYLASE		
URACIL DEHYDROGENASE	C	$Ki = 38\ \mu M$
URATE-RIBONUCLEOTIDE PHOSPHORYLASE		
THYMOL BLUE		
1-D-myo-INOSITOL-TRISPHOSPHATE 3-KINASE		$I50 = 590\ \mu M$
PHOSPHATIDYLINOSITOL 3-PHOSPHATASE		$I50 = 950\ \mu M$
THYMUS DNASE INHIBITOR		
DEOXYRIBONUCLEASE		
THYROXINE		
ACYLPHOSPHATASE		
BISPHOSPHOGLYCERATE MUTASE		
CREATINE KINASE		
DIHYDROPTERIDINE REDUCTASE	NC	
GLUTAMATE DEHYDROGENASE (NAD(P))		51% 13 μM
GLUTATHIONE TRANSFERASE	C	$Ki = 6.6\ \mu M$
NADPH-CYTOCHROME-c2 REDUCTASE		
TYROSINE TRANSAMINASE		
L-THYROXINE		
15-HYDROXYPROSTAGLANDIN-D DEHYDROGENASE		
MYOSIN-LIGHT-CHAIN KINASE		$I50 = 31\ \mu M$

THYRSIFERYL 23 ACETATE
PHOSPHOPROTEIN PHOSPHATASE 2A — $I50 = 16\ \mu M$

THYSANONE
PROTEINASE 3C — $I50 = 13\ \mu g/ml$

TIAMULIN
CYTOCHROME P450 3A — $Ki = 6\ \mu M$

TIAZOFURIN
IMP DEHYDROGENASE — UC — $Ki = 72$ mM

TIBENELAST
3′,5′-CYCLIC-NUCLEOTIDE PHOSPHODIESTERASE (cycAMP) — $I50 = 388\ \mu M$
3′,5′-CYCLIC-NUCLEOTIDE PHOSPHODIESTERASE (cycAMP) — $I50 = 324\ \mu M$
3′,5′-CYCLIC-NUCLEOTIDE PHOSPHODIESTERASE (cycAMP) — $I50 = 597\ \mu M$
3′,5′-CYCLIC-NUCLEOTIDE PHOSPHODIESTERASE (cycAMP) — $I50 = 219\ \mu M$
3′,5′-CYCLIC-NUCLEOTIDE PHOSPHODIESTERASE (cycAMP) — $I50 = 202\ \mu M$
3′,5′-CYCLIC-NUCLEOTIDE PHOSPHODIESTERASE (cycAMP) — C — $Ki = 9.4\ \mu M$
3′,5′-CYCLIC-NUCLEOTIDE PHOSPHODIESTERASE (cycAMP) — $I50 = 267\ \mu M$

TIBO 82913
RNA-DIRECTED DNA POLYMERASE — $I50 = 50$ nM

TICK ANTICOAGULANT PEPTIDE
COAGULATION FACTOR Xa — $Ki = 500$ pM

TICK ANTICOAGULANT PROTEIN REC.
COAGULATION FACTOR Xa — $Ki = 180$ pM

TICK SALIVARY GLAND PROTEIN PHOSPHATASE INHIBITOR
PHOSPHOPROTEIN PHOSPHATASE

TICLOPIDINE
CYTOCHROME P450 2C9 — $Ki = 39\ \mu M$
CYTOCHROME P450 2C19 — $Ki = 3.7\ \mu M$

TICRYNAFEN
NAD(P)H DEHYDROGENASE (QUINONE) — $I50 = 3\ \mu M$
VITAMIN K EPOXIDE REDUCTASE — $I50 = 5$ mM

TIENILIC ACID
CYTOCHROME P450 2C9 — SSI
CYTOCHROME P450 2C10 — IR
GLUTATHIONE TRANSFERASE

TIGLYL-CoA
AMINO-ACID N-ACETYLTRANSFERASE — 60% 3 mM
2-METHYLACYL-CoA DEHYDROGENASE

TILUDRONATE
H^+-TRANSPORTING ATPASE (V-TYPE) — $I50 = 466$ nM
H^+-TRANSPORTING ATPASE (V-TYPE) — $I50 = 3.5\ \mu M$
PROTEIN TYROSINE-PHOSPHATASE

TILURONE
RNA-DIRECTED DNA POLYMERASE
RNA POLYMERASE — C

TIMERSAL
H^+-TRANSPORTING ATP SYNTHASE — $I50 = 2\ \mu M$

TIMOLOL
LIPOPROTEIN LIPASE

TIMOLOL MALEATE
Na^+/K^+-EXCHANGING ATPASE

TIMOPRAZOLE
- H^+/K^+-EXCHANGING ATPASE — $I50 = 15\ \mu M$

TIMP
- GELATINASE
- GELATINASE A — 100% 0.32 U

TIMP-1
- GELATINASE
- GELATINASE A — $Ki = 4.3$ nM
- GELATINASE B — $Ki = 8.5$ nM
- STROMELYSIN — $I50 = 25$ nM

TIMP-2
- COLLAGENASE
- GELATINASE
- GELATINASE A — $Ki = 7.9$ nM
- GELATINASE B — $Ki = 8.7$ nM
- MATRIN
- TRANSIN

TIMP-3
- PROTEINASES METALLO

TIMP-4
- GELATINASE A — $I50 = 3$ nM
- GELATINASE B — $I50 = 83$ nM
- INTERSTITIAL COLLAGENASE — $I50 = 19$ nM
- MATRILYSIN — $I50 = 8$ nM
- MATRIX METALLOPROTEINASES
- METALLOPROTEINASES
- STROMELYSIN 1 — $I50 = 45$ nM

TIOCONAZOLE
- AROMATASE — $I50 = 220$ nM
- AROMATASE — $I50 = 1.7\ \mu M$
- STEROID 17α-HYDROXYLASE/17,20 LYASE
- STEROID 17α-MONOOXYGENASE
- STEROID 21-MONOOXYGENASE

TIRON
- CATECHOL 1,2-DIOXYGENASE
- CATECHOL 2,3-DIOXYGENASE
- 2,5-DIHYDROXYPYRIDINE 5,6-DIOXYGENASE
- GENTISATE 1,2-DIOXYGENASE
- GLYCEROL-3-PHOSPHATE DEHYDROGENASE
- HYDROGENASE
- HYOSCYAMINE (6S)-DIOXYGENASE
- INDOLE 2,3-DIOXYGENASE
- LACTATE RACEMASE
- LINOLEOYL-CoA DESATURASE
- LIPOXYGENASE
- NADH PEROXIDASE — 65% 3 mM
- 2-NITROPROPNAE DIOXYGENASE
- XANTHINE DEHYDROGENASE

TISSUE FACTOR PATHWAY INHIBITOR
- COAGULATION FACTOR Xa — $Ki = 2$ nM

TISSUE FACTOR PATHWAY INHIBITOR DOMAIN 1
- CATHEPSIN G — $Ki = 200$ nM
- COAGULATION FACTOR VIIa — $Ki = 250$ nM
- PLASMIN — $Ki = 26$ nM

TISSUE FACTOR PATHWAY INHIBITOR DOMAIN 2
- CHYMOTRYPSIN — $Ki = 750$ pM
- COAGULATION FACTOR Xa — $Ki = 90$ nM
- TRYPSIN — $Ki = 100$ pM

TISSUE INHIBITOR OF METTALLO-PROTEINASES
METALLOPROTEINASES

TISSUE INHIBITOR PLASMA
β-GLUCURONIDASE

TL-1

AMINE OXIDASE MAO-A		*K*i = 40 μM
AMINE OXIDASE MAO-B		*K*i = 7.9 μM

TMC 2A

DIPEPTIDYL-PEPTIDASE 4		*I*50 = 8.1 μM

TMC 52A

CATHEPSIN L		*I*50 = 13 nM

TMC 2B

DIPEPTIDYL-PEPTIDASE 4		*I*50 = 17 μM

TMC 52B

CATHEPSIN L		*I*50 = 10 nM

TMC 2C

DIPEPTIDYL-PEPTIDASE 4		*I*50 = 20 μM

TMC 52C

CATHEPSIN L		*I*50 = 10 nM

TMC 52D

CATHEPSIN L		*I*50 = 6 nM

TMK 688

ARACHIDONATE 5-LIPOXYGENASE		*I*50 = 140 nM
CYCLOOXYGENASE		
UNSPECIFIC MONOOXYGENASE		*I*50 = 180 nM

TMK 777

ARACHIDONATE 5-LIPOXYGENASE		*I*50 = 17 nM
UNSPECIFIC MONOOXYGENASE		*I*50 = 10 nM

TMP

dCMP DEAMINASE	C	*K*i = 120 μM
DIHYDROOROTASE	C	*K*i = 74 μM
IMP DEHYDROGENASE		
MICROCOCCAL NUCLEASE	C	*K*i = 19 μM
THIAMIN-TRIPHOSPHATASE		12% 7.5 mM
THYMIDYLATE SYNTHASE	C	*K*i = 24 μM
URIDINE PHOSPHORYLASE		

TMP 153

STEROL O-ACYLTRANSFERASE		*I*50 = 6.4 nM
STEROL O-ACYLTRANSFERASE		*I*50 = 9 nM
STEROL O-ACYLTRANSFERASE		*I*50 = 2.3 nM

dTMP

dCTP DEAMINASE	NC	
NAD ADP-RIBOSYLTRANSFERASE		
NAD NUCLEOSIDASE		
PHENYLETHANOLAMINE N-METHYLTRANSFERASE	NC	*K*i = 7.6 μM
PHENYLETHANOLAMINE N-METHYLTRANSFERASE	C	*K*i = 24 μM
RIBONUCLEASE T2		*K*i = 78 μM
THYMIDINE-TRIPHOSPHATASE		*K*i = 7.4 mM
THYMIDYLATE SYNTHASE		*K*i = 16 μM
THYMIDYLATE SYNTHASE	NC	*K*i = 1 μM
THYMIDYLATE SYNTHASE	C	*K*i = 24 μM

TNP 351
DIHYDROFOLATE REDUCTASE

Inhibitor / Enzyme	Type	Value
PHOSPHORIBOSYLAMINOIMIDAZOLECARBOXAMIDE FORMYLTRANSFERASE		*K*i = 52 μM
PHOSPHORIBOSYLGLYCINAMIDE FORMYLTRANSFERASE		*K*i = 47 μM
TOBACCO SERINE PROTEINASE INHIBITOR		
PROTEINASES SERINE		
TOBRAMYCIN		
KANAMYCIN KINASE		
PHOSPHOLIPASE C γ1 (PHOSPHATIDYLINOSITOL SPECIFIC)		*I*50 = 9.7 μM
α-TOCOPHEROL		
L-ASCORBATE OXIDASE		
GLUCOSE-6-PHOSPHATE 1-DEHYDROGENASE		29% 32 μM
LIPOXYGENASE		
LIPOXYGENASE 1	IR	
PROTEIN KINASE C		
D,L-α-TOCOPHEROL		
ARACHIDONATE 5-LIPOXYGENASE		*K*i = 259 μM
α-TOCOPHEROL HEMISUCCINATE		
ACETYLCHOLINESTERASE		*I*50 = 1.7 μM
CHOLINESTERASE		*I*50 = 100 μM
α-TOCOPHEROLQUINONE		
PHYLLOQUINONE EPOXIDE REDUCTASE		*I*50 = 330 μM
TOCOTRIENOLS		
HYDROXYMETHYLGLUTARYL-CoA REDUCTASE		
3-TODOACETOLPHOSPHATES		
METHYLGLYOXAL SYNTHASE	IR	
TOLBUTAMIDE		
ACETOHEXAMIDE REDUCTASE		36% 200 μM
CARNITINE ACYLTRANSFERASE		*I*50 = 2.1 mM
FRUCTOSE-2,6-BISPHOSPHATASE		
PHOSPHOPROTEIN PHOSPHATASE		*K*i = 210 μM
PROTEIN KINASE A		
TOLCINATE		
SQUALENE MONOOXYGENASE		*I*50 = 120 nM
SQUALENE MONOOXYGENASE		*I*50 = 145 μM
TOLFENAMIC ACID		
ARACHIDONATE 5-LIPOXYGENASE		*I*50 = 10 μM
CYCLOOXYGENASE		*I*50 = 50 nM
TOLICICLATE		
SQUALENE MONOOXYGENASE		
TOLMETIN		
CYCLOOXYGENASE		*I*50 = 100 nM
3α-HYDROXYSTEROID DEHYDROGENASE (A SPECIFIC)		
3α-HYDROXYSTEROID DEHYDROGENASE (B-SPECIFIC)	C	*K*i = 29 μM
TOLNAFTATE		
SQUALENE MONOOXYGENASE		
SQUALENE MONOOXYGENASE		*I*50 = 215 μM
SQUALENE MONOOXYGENASE		*I*50 = 1 μM
TOLOXATONE		
AMINE OXIDASE MAO-A		*K*i = 1.8 μM
AMINE OXIDASE MAO-B		*K*i = 44 μM
R-TOLOXATONE		
AMINE OXIDASE MAO-A		*K*i = 1.5 μM

p-TOLOYL ARGININETRIFLUOROMETHYLKETONE
KALLIKREIN TISSUE | | *K*i = 35 μM

TOLRESTAT
ALCOHOL DEHYDROGENASE (NADP) | | *I*50 = 980 nM
ALDEHYDE REDUCTASE | | *I*50 = 31 nM
ALDEHYDE REDUCTASE | | 47% 5 nM

TOLUENE
BENZENE 1,2-DIOXYGENASE
CHOLINESTERASE | | *K*i = 6 mM
CYTOCHROME P450 2E1 | C | *K*i = 1.6 mM
RENILLA-LUCIFERIN 2-MONOOXYGENASE

TOLUENE-3,4-DITHIOL
ARACHIDONATE 5-LIPOXYGENASE | | *I*50 = 24 μM
ARACHIDONATE 12-LIPOXYGENASE | | *I*50 = 20 μM
CYCLOOXYGENASE | | *I*50 = 6.6 μM
ENDOTHELIN CONVERTING ENZYME | | *I*50 = 30 μM
LIPOXYGENASE | | *I*50 = 219 μM
QUERCETIN 2,3-DIOXYGENASE

p-TOLUENESULPHONAMIDE
CARBONATE DEHYDRATASE

N-TOLUENE-p-SULPHONYL-L-p-AMINOPHENYLALANINE
PREPHENATE DEHYDROGENASE

Nα-(4-TOLUENESULPHONYL)D,L-m-AMINOPHENYLALANYLPIPERIDINE
THROMBIN | | *K*i = 540 nM
THROMBIN | | *K*i = 340 nM
TRYPSIN | | *K*i = 1.2 μM

Nα-(4-TOLUENESULPHONYL)D,L-p-AMINOPHENYLALANYLPIPERIDINE
THROMBIN | | *K*i = 640 nM
THROMBIN | | *K*i = 1.3 μM
TRYPSIN | | *K*i = 64 μM

N-TOLUENESULPHONYL-L-ARGININE METHYL ESTER
THROMBIN

N-TOLUENESULPHONYL-L-ARGINYL-GLYCINE
THROMBIN

trans-1-p-TOLUENESULPHONYL-3-ETHYL-4-ETHOXYCARBONYLAZETIDINONE
ELASTASE LEUKOCYTE | | *I*50 = 10 μg/ml

p-TOLUENESULPHONYLFLUORIDE
AGKISTRODON SERINE PROTEINASE

N-TOLUENE-p-SULPHONYL-PHENYLALANINE
PREPHENATE DEHYDROGENASE

m-TOLUIC ACID
SERINE-SULFATE AMMONIA-LYASE | C | *K*i = 16 mM

o-TOLUIC ACID
MONOPHENOL MONOOXYGENASE | C | *K*i = 6.8 μM

TOLUIDINE
SPERMIDINE SYNTHASE | | *I*50 = 108 μM

m-TOLUIDINE
β-FRUCTOFURANOSIDASE

TOLUIDINONAPHTHALENE SULPHONIC ACID
PHOTINUS-LUCIFERIN 4-MONOOXYGENASE (ATP-HYDROLYSING)

2,6-TOLUIDINONAPHTHALENE SULPHONIC ACID
ALCOHOL DEHYDROGENASE — *K*i = 17 μM
PHOTINUS-LUCIFERIN 4-MONOOXYGENASE (ATP-HYDROLYSING)

4-TOLUOLSULPHONYLFLUORIDE
PROTEINASES SERINE

2-(m-TOLYL)-8-AZAADENOSINE
ADENINE DEAMINASE — *K*i = 8.5 μM

5-p-TOLYLAZOTROPOLONE
LACTOYLGLUTATHIONE LYASE — *I*50 = 24 μM

m-TOLYL METHYLCARBAMATE
ACETYLCHOLINESTERASE — *K*i = 24 μM

TOLYL SALIGENIN PHOSPHATE
CARBOXYLESTERASE

TOLYPOCLADIN
3′,5′-CYCLIC-NUCLEOTIDE PHOSPHODIESTERASE (CALMODULIN INDEPENDENT) — *I*50 = 83 μM
3′,5′-CYCLIC-NUCLEOTIDE PHOSPHODIESTERASE (Ca^{2+}/CALMODULIN) — C — *I*50 = 30 μM

TOMATO EXTENSIN PEROXIDASE INHIBITOR
EXTENSIN PEROXIDASE

TOMATO INVERTASE INHIBITOR
β-FRUCTOFURANOSIDASE

TOMATO POLYGALACTURONASE INHIBITOR
POLYGALACTURONASE

TOMATO ROOT GLUTAMINE SYNTHASE INHIBITOR
GLUTAMATE-AMMONIA LIGASE

TOPOSTIN
DNA TOPOISOMERASE

TOPOSTIN A
DNA TOPOISOMERASE

TOPOSTIN B
DNA TOPOISOMERASE

TOPOTECAN
DNA TOPOISOMERASE

TOPSENTIN
PHOSPHOLIPASE A2 — IR — *I*50 = 500 nM

TORBAFYLLINE
3′,5′-CYCLIC-NUCLEOTIDE PHOSPHODIESTERASE (cGMP INHIBITABLE) — *I*50 = 294 μM
3′,5′-CYCLIC-NUCLEOTIDE PHOSPHODIESTERASE I — *I*50 = 56 μM
3′,5′-CYCLIC-NUCLEOTIDE PHOSPHODIESTERASE III — *I*50 = 153 μM
3′,5′-CYCLIC-NUCLEOTIDE PHOSPHODIESTERASE IV — *I*50 = 177 μM

TORRESEA ACREANA TRYPSIN INHIBITOR 1
CHYMOTRYPSIN
COAGULATION FACTOR XIIa
TRYPSIN

TORRESEA ACREANA TRYPSIN INHIBITOR 2
CHYMOTRYPSIN
COAGULATION FACTOR XIIa
PLASMIN
TRYPSIN

TORRESEA CEARENSIS TRYPSIN INHIBITOR
CHYMOTRYPSIN — *K*i = 50 nM

COAGULATION FACTOR XIIf — $Ki = 1.5\ \mu M$
TRYPSIN — $Ki = 1.1$ nM

TORULASPORA DELBRUECKII PHOSPHOLIPASE B INHIBITOR
LYSOPHOSPHOLIPASE

TOSUFLOXACIN
DNA GYRASE — $I50 = 25\ \mu g/ml$
DNA TOPOISOMERASE IV — $I50 = 1.8\ \mu g/ml$

N-TOSYL.-l-ALANINE
AMINOACYLASE

TOSYLAMIDO-2-PHENYLETHYL CHLOROMETHYLKETONE
PHOTINUS-LUCIFERIN 4-MONOOXYGENASE (ATP-HYDROLYSING)

TOSYLAMIDO-2-PHENYLETHYLCHLOROMETHYLKETONE
ENTOMOPHTHORA COLLAGENOLYTIC PROTEINASE

L-1-TOSYLAMIDO-2-PHENYLETHYL CHLOROMETHYLKETONE
AMINOPEPTIDASE — 50% 1 mM
CATHEPSIN L
CHYMOTRYPSIN
γ-GLUTAMYLTRANSFERASE
METRIDIUM PROTEINASE A
PHORBOL-DIESTER HYDROLASE
PROLYL OLIGOPEPTIDASE
PROTEINASES SERINE

TOSYLARGININE
TRYPSIN

N-α-TOSYLARGINYLCHLOROMETHYLKETONE
CLOSTRIPAIN

p-TOSYLARSONATE
SUBTILISIN

TOSYLGLYCOLIC ACID
RIBONUCLEASE T1

Nα-TOSYLGLYCYL-3-AMIDINOPHENYLALANINE METHYLESTER
COAGULATION FACTOR Xa
THROMBIN

α-N-TOSYL-p-GUANIDINO-L-PHENYLALANINE
THROMBIN

N-α-p-TOSYL-L-HOMOARGININE METHYL ESTER
CLOSTRIPAIN — $Ki = 26\ \mu M$

N-TOSYLLYSINE CHLOROMETHYLKETONE
CATHEPSIN L — 50% 200 nM

N-TOSYL-L-LYSINE CHLOROMETHYLKETONE
ACROSIN
ADENYLATE CYCLASE — IR
AMINOACYLASE
AMINOPEPTIDASE — 21% 10 μM
AMINOPEPTIDASE B
ARYLSULFATE SULFOTRANSFERASE
BROMELAIN
CATHEPSIN B — 94% 100 μM
CATHEPSIN L
CLOSTRIPAIN
COMPLEMENT SUBCOMPONENT C1r
DIPEPTIDASE
DIPEPTIDYL-PEPTIDASE 4

ELASTASE PANCREATIC II		91% 500 μM
ENDOPEPTIDASE BRACHIONUS PLICATILIS		100% 1 mM
ENTEROPEPTIDASE		
ENTOMOPHTHORA COLLAGENOLYTIC PROTEINASE		
GINGIPAIN		100% 1 mM
GLUCOSE-6-PHOSPHATASE	IR	
LACTOSYLCERAMIDE α-2,3-SIALYLTRANSFERASE		73% 135 μM
PAPAIN		
PROTEINASE 2A HRV2		*I*50 = 80 μM
PROTEINASE ALFA ALFA		*K*i = 100 nM
PROTEINASE BOAR EPIDIDYMAL SPERM		97% 1 mM
PROTEINASE CATHEPSIN S LIKE		89% 1 μM
PROTEINASE I ACHROMOBACTER		
PROTEINASE II MYXOBACTER AL-1		
PROTEINASE In		34% 1 mM
PROTEINASE NS3		
PROTEINASE PORPHYROMONAS GINGIVALIS		
PROTEINASE PORPHYROMONAS GINGIVALIS TRYPSIN LIKE		
PROTEINASE SERINE PROMONOCYTOC U-937		31% 2 mM
PROTEINASES SERINE		
PROTEINASE STARFISH SPERM ACROSOME		
PROTEIN KINASE C		*I*50 = 1 mM
PROTEIN TYROSINE KINASE		
PYROGLUTAMYL-PEPTIDASE I	IR	*I*50 = 33 μM
SUBMANDIBULAR PROTEINASE A		
TENEBRIO α-PROTEINASE		
THROMBIN		
TRYPSIN		99% 500 μM
TRYPSIN		
TRYPTASE		*I*50 = 270 μM
UCA PUGILATOR COLLAGENOLYTIC PROTEINASE		

α-N-p-TOSYLLYSYLCHLOROMETHYLKETONE

PROTEINASE ASPARTIC PLASMODIUM		

TOSYLNORLEUCINYL-PHENYLALANINE

STREPTOPAIN		

TOSYLOXYPHTHALIMIDE

CHYMOTRYPSIN	IR	
TRYPSIN	IR	

N-TOSYL-PHE-ALA-CHLOROMETHYLKETONE

BROMELAIN		

N-TOSYL-L-PHENYLALANINE CHLOROMETHYLKETONE

ACROSIN		
CATHEPSIN G		
CATHEPSIN H		79% 10 μM
CATHEPSIN L		50% 1 mM
CHYMASE		*I*50 = 240 μM
DIPEPTIDASE		
ENVELYSIN		
LACTOSYLCERAMIDE α-2,3-SIALYLTRANSFERASE		67% 284 μM
PAPAIN		
PROLINE-β-NAPHTHYLAMIDASE		91% 200 μM
PROLYL OLIGOPEPTIDASE (THIOL DEPENDENT)		
PROTEIN KINASE C		*I*50 = 8 mM
TRIPEPTIDE AMINOPEPTIDASE		*I*50 = 90 μM
TRYPSIN		

TOSYLPHENYLCHLOROMETHYLKETONE

SERINE CARBOXYPEPTIDASE		

TOWEL GOURD TRYPSIN INHIBITOR I

TRYPSIN		

TOWEL GOURD TRYPSIN INHIBITOR II
TRYPSIN

TOXADOCIAL A
THROMBIN — $I50 = 6.5\ \mu g/ml$

TOXICOL A
RNA-DIRECTED DNA POLYMERASE — $I50 = 3.1\ \mu M$

TOXIUSOL
DNA POLYMERASE α — $I50 = 1.2\ \mu M$
DNA POLYMERASE β — $I50 = 900$ nM
RNA-DIRECTED DNA POLYMERASE — $I50 = 1.5\ \mu M$

TOYOCAMYCIN
CYCLIN DEPENDENT KINASE 2 — C — $Ki = 1.6\ \mu M$
CYCLIN DEPENT KINASE CDK2 — $Ki = 5.1\ \mu M$
1-PHOSPHATIDYLINOSITOL 3-KINASE — $I50 = 11\ \mu M$
1-PHOSPHATIDYLINOSITOL 4-KINASE — $I50 = 12\ \mu M$
PROTEIN KINASE A — $I50 = 2.5$ mM
PROTEIN KINASE CASEIN KINASE II — $I50 = 177\ \mu M$
PROTEIN KINASE (NUCLEAR)
RHODOPSIN KINASE — $I50 = 14\ \mu M$

TPCK
GINGIPAIN — 100% 1 mM
GLUCOSE-6-PHOSPHATASE — IR
ISOPRENYLATED PROTEIN ENDOPROTEASE — IR
PROTEINASE 2A HRV2 — $I50 = 1.3$ mM
PROTEINASE Ca^{2+} THIOL DROSOPHILA MELANOGASTER
SERINE-TYPE CARBOXYPEPTIDASE — 29% 100 μM

TPI
3′,5′-CYCLIC-NUCLEOTIDE PHOSPHODIESTERASE

TRACHYSPIC ACID
HEPARANASE — $I50 = 36\ \mu M$

TRALKOXYDIM
ACETYL-CoA CARBOXYLASE — $I50 = 1\ \mu M$

TRANDOLAPRIL
11β-HYDROXYSTEROID DEHYDROGENASE

TRANDOLPRILATE
PEPTIDYL-DIPEPTIDASE A — $Ki = 24$ pM

TRANEXAMIC ACID
PEPTIDYL-DIPEPTIDASE A
PLASMIN

TRANSPOSASE INHIBITORY PROTEINS
TRANSPOSASE

TRANYLCYPROMINE
AMINE OXIDASE (COPPER-CONTAINING)
AMINE OXIDASE (FLAVIN-CONTAINING) — IRR — 100 nM
AMINE OXIDASE MAO
CYTOCHROME P450 2A6 — C — $Ki = 40$ nM
PHENYLETHANOLAMINE N-METHYLTRANSFERASE — C
PROSTACYCLIN SYNTHASE
PROTEIN-LYSINE 6-OXIDASE — $Ki = 375\ \mu M$

TRAPIDIL
THROMBOXANE A SYNTHASE

TRAPOXIN
HISTONE DEACETYLASE — IR

TRASYLOL

CHYMOTRYPSIN		*K*i = 4.7 nM
LEUKOCYTE-MEMBRANE NEUTRAL ENDOPEPTIDASE		68% 1 mM
SUBMANDIBULAR PROTEINASE A		
TRYPSIN		

TRAUMATIC ACID

GLUTATHIONE TRANSFERASE

TRAZODONE

CYTOCHROME P450 2D6		*K*i = 9 μM

TREHALAMINE

α-GLUCOSIDASE		*I*50 = 210 μM
α-GLUCOSIDASE		*I*50 = 340 μM
OLIGO-1,6-GLUCOSIDASE		*I*50 = 500 μM
SUCRASE		*I*50 = 68 μM
SUCROSE α-GLUCOSIDASE		*I*50 = 160 μM
α,α-TREHALASE		*I*50 = 460 nM
α,α-TREHALASE		*I*50 = 290 μM
α,α-TREHALASE		*I*50 = 180 μM

TREHALOSE

α,α-TREHALOSE-PHOSPHATE SYNTHASE

TREHALOSEPHOSPHATE

GLUCOSE-1-PHOSPHATE ADENYLYLTRANSFERASE

TREHALOSE-6-PHOSPHATE

HEXOKINASE	C	*K*i = 40 μM
HEXOKINASE	C	*K*i = 5 μM
HEXOKINASE 1	C	*K*i = 200 μM
HEXOKINASE 2	C	*K*i = 40 μM

TREHALOSTATIN

α,α-TREHALASE

TREHAZOLIN

α-GLUCOSIDASE		*I*50 = 95 μM
α-GLUCOSIDASE		*I*50 = 190 μM
OLIGO-1,6-GLUCOSIDASE		*I*50 = 7.6 μM
SUCRASE		*I*50 = 170 μM
SUCROSE α-GLUCOSIDASE		*I*50 = 370 μM
α,α-TREHALASE		*I*50 = 27 nM
α,α-TREHALASE		*I*50 = 19 μM
α,α-TREHALASE		*I*50 = 5.5 nM
α,α-TREHALASE		*I*50 = 16 nM
α,α-TREHALASE		*I*50 = 3.7 nM
α,α-TREHALASE		

(1- > 4) TREHAZOLOID PSEUDOSACCHERIDE

α-GLUCOSIDASE	C	*K*i = 9.3 μM
β-GLUCOSIDASE	C	*K*i = 48 μM

TREQUINSIN

3′,5′-CYCLIC-NUCLEOTIDE PHOSPHODIESTERASE (cycAMP)		*I*50 = 400 nM
3′,5′-CYCLIC-NUCLEOTIDE PHOSPHODIESTERASE (cycAMP)		*I*50 = 200 nM
3′,5′-CYCLIC-NUCLEOTIDE PHOSPHODIESTERASE (cycAMP)		*I*50 = 200 nM
3′,5′-CYCLIC-NUCLEOTIDE PHOSPHODIESTERASE 4D3		
3′,5′-CYCLIC-NUCLEOTIDE PHOSPHODIESTERASE (cycGMP)		*I*50 = 2 μM
3′,5′-CYCLIC-NUCLEOTIDE PHOSPHODIESTERASE (cycGMP)		*I*50 = 5 μM
3′,5′-CYCLIC-NUCLEOTIDE PHOSPHODIESTERASE (cycGMP) (Ca^{2+}/CALMODULIN)		*I*50 = 13 μM
3′,5′-CYCLIC-NUCLEOTIDE PHOSPHODIESTERASE (cycGMP) (Ca^{2+}/CALMODULIN)		*I*50 = 4 μM
3′,5′-CYCLIC-NUCLEOTIDE PHOSPHODIESTERASE (cycGMP STIMULATED)		*I*50 = 600 nM
3′,5′-CYCLIC-NUCLEOTIDE PHOSPHODIESTERASE (cycGMP STIMULATED)		*I*50 = 4 μM

Inhibitor / Enzyme	Type	Value
TRESTATIN A		
α-AMYLASE	C	Ki = 700 nM
GLUCAN 1,4-α-GLUCOSIDASE	C	Ki = 66 nM
TRESTATIN B		
α-AMYLASE		
TRESTATIN C		
α-AMYLASE	C	Ki = 150 nM
GLUCAN 1,4-α-GLUCOSIDASE	C	Ki = 37 nM
TRESTATINS		
α-GLUCOSIDASE		
TRIABIN		
THROMBIN		
TRI-N-ACETYLCHITOTRIOSE		
LYSOZYME		
TRIACETYLOLEANDOMYCIN		
CYTOCHROME P450 3A		
TRIACSIN A		
LONG-CHAIN-FATTY-ACID CoA-LIGASE	C	Ki = 9 μM
TRIACSIN C		
LONG-CHAIN-FATTY-ACID CoA-LIGASE		I50 = 8.7 μM
LONG-CHAIN-FATTY-ACID CoA-LIGASE	IR	
TRIADIMENOL		
14α-METHYLSTEROL DEMETHYLASE		
TRIALLYLAMINE		
H^+/K^+-EXCHANGING ATPASE		Ki = 1.8 mM
TRIAMICINOLONE		
3α-HYDROXYSTEROID DEHYDROGENASE (B-SPECIFIC)		65% 10 μM
3,4,5-TRIAMINO-4H-TRIAZOLE		
RIBONUCLEOSIDE-DIPHOSPHATE REDUCTASE		I50 = 2.9 mM
TRIAMPTERENE		
DIHYDROFOLATE REDUCTASE		
TRIAZINATE		
DIHYDROFOLATE REDUCTASE		
1,3,5-TRIAZINE		
CYTOCHROME P450 2E1	NC	Ki = 600 μM
2-TRIAZINE-5-NITROFURAN		
GLUTATHIONE REDUCTASE (NADPH)		I50 = 4.3 μM
P^1-TRIAZOFURIN-P^2-ADENOSINE-5′-METHYLENE BISPHOSPHONATE		
IMP DEHYDROGENASE 1		I50 = 107 nM
IMP DEHYDROGENASE 2		I50 = 109 nM
1,2,3-TRIAZOLE		
CARBONATE DEHYDRATASE		
1,2,4-TRIAZOLE		
CARBONATE DEHYDRATASE		
CARBONATE DEHYDRATASE II		
FATTY-ACID PEROXIDASE		
1,2,4-TRIAZOLO[1,5A]PYRIMIDINES		
ACETOLACTATE SYNTHASE		

2α-(1,2,3-TRIAZOL-1-YL)METHYL-2β-METHYL-6,6-DIHYDROPENAM-3α-CARBOXYLATE-1,1-DIOXIDE SODIUM
β-LACTAMASE — $I50 = 5.1\ \mu M$

2β-[(1,2,3-TRIAZOL-1-YL)METHYL]-2α-METHYLPENAM-3α-CARBOXYLATE-1,1-DIOXIDE
β-LACTAMASE TEM2 — $I50 = 7\ \mu M$

TRIAZOPHOS
ACETYLCHOLINESTERASE

TRIBENZENESULPHONIC ACID
DIHYDROFOLATE REDUCTASE

TRIBOLIDE
Ca^{2+}-TRANSPORTING ATPASE

3,4,5-TRIBROMO-2-(2′,4′-DIBROMOPHENOXY)PHENOL
ARACHIDONATE 15-LIPOXYGENASE — $I50 = 7.4\ \mu M$
IMP DEHYDROGENASE — $I50 = 4\ \mu M$

TRI-n-BUTYLAMINE
3′-PHOSPHOADENOSINE 5′-PHOSPHOTRANSFERASE — 61% 300 μM

S,S,S-TRIBUTYL PHOSPHOROTRITHIOATE
CARBOXYLESTERASE

TRIBUTYLTIN
FERROCHELATASE — 70% 50 μM
H^+-TRANSPORTING ATP SYNTHASE

TRIBUTYLTIN ACETATE
GLUTATHIONE TRANSFERASE 4-4 — $I50 = 1.5\ \mu M$

TRIBUTYLTIN CHLORIDE
ADENOSINETRIPHOSPHATASE — 70% 500 μM
GLUTATHIONE TRANSFERASE

TRICARBALLYLIC ACID
ACONITATE HYDRATASE — C — $Ki = 4$ mM
GLUCOSE-1,6-BISPHOSPHATE SYNTHASE
2-METHYLCITRATE SYNTHASE — C
XAA-PRO DIPEPTIDASE — $Ki = 130\ \mu M$

1,2,3-TRICARBOXYCYCLOPENTENE-1
ACONITATE HYDRATASE — C — $Ki = 25$ mM

TRICETIN
PROTEIN KINASE C — $I50 = 1\ \mu M$

TRICHLOMETHYLOXIRANE
EPOXIDE HYDROLASE — $I50 = 4\ \mu M$

TRICHLORFON
ACETYLESTERASE
CARBOXYLESTERASE

TRICHLORMETHIAZIDE
THIOPURINE S-METHYLTRANSFERASE — $I50 = 1$ mM

TRICHLOROACETATE
CREATINE KINASE

TRICHLOROACETIC ACID
METHYLGUANINASE

TRICHLOROBENZOIC ACID
PECTATE LYASE — 35% 200 μM

TRICHLOROETHANOL
ALCOHOL DEHYDROGENASE
ALCOHOL DEHYDROGENASE-1

2,2,2-TRICHLOROETHYL-3,4-DICHLOROCARBANILIC ACID		
H^+-TRANSPORTING ATPASE		
TRICHLOROETHYLENE		
CYTOCHROME P450 2E1	IR	
2-TRICHLOROMETHYL-4(3H)QUINAZOLINONE		
NAD ADP-RIBOSYLTRANSFERASE		*I*50 = 2.2 mM
1-TRICHLOROMETHYL-1,2,3,4-TETRAHYDRO-β-CARBOLINE		
NADH DEHYDROGENASE		100% 800 μM
TRICHLORONATE		
ACETYLCHOLINESTERASE		
TRICHLOROPHENE OXIDE		
EPOXIDE HYDROLASE		
2,4,5-TRICHLOROPHENOL		
2,4-DICHLOROPHENOL 6-MONOOXYGENASE		
2,4,6-TRICHLOROPHENOL		
2,4-DICHLOROPHENOL 6-MONOOXYGENASE		
2,4,5-TRICHLOROPHENOXYACETIC ACID		
GLUTATHIONE TRANSFERASE		
LONG-CHAIN-FATTY-ACID CoA-LIGASE	MIX	*K*i = 3 mM
1,1,1,-TRICHLOROPROPANE-2,3-OXIDE		
EPOXIDE HYDROLASE		
1.1.1. TRICHLOROPROPENE OXIDE		
EPOXIDE HYDROLASE		
TRICHLOROPROPYLENE OXIDE		
EPOXIDE HYDROLASE		
2,6,8-TRICHLOROPURINE		
URATE OXIDASE		*K*i = 800 nM
5,7,8-TRICHLORO-1,2,3,4-TETRAHYDROISOQUINOLINE		
PHENYLETHANOLAMINE N-METHYLTRANSFERASE		
TRICHLORPHON		
ACETYLCHOLINESTERASE		
TRICHOSTATIN		
HISTONE DEACETYLASE		*I*50 = 3.8 nM
TRICHOSTATIN A		
HISTONE DEACETYLASE		*I*50 = 100 nM
TRICHOVIRIDIN		
MONOPHENOL MONOOXYGENASE		*I*50 = 6.6 μg/ml
TRICINE		
SHIKIMATE O-HYDROXYCINNAMOYLTRANSFERASE		
TRICIRIBINE PHOSPHATE		
AMIDOPHOSPHORIBOSYLTRANSFERASE		70% 1 mM
IMP DEHYDROGENASE	C	66% 1.2 mM
TRICOSANOIC ACID		
Na^+/K^+-EXCHANGING ATPASE		
TRI-o-CRESOL PHOSPHATE		
ACETYLCHOLINESTERASE		
ESTERASE NEUROTOXIC		

TRICRESYLPHOSPHATE
ESTERASE

1,1,3-TRICYANO-1-MAINO-1-PROPENE
CHLORIDE PEROXIDASE

TRICYCLAZOLE
TRIHYDROXYNAPHTHALENE REDUCTASE C *K*i = 15 nM
1,3,8-TRIHYDROXYNAPHTHALENE REDUCTASE

TRICYCLODECAN-9-YL XANTHOGENATE
PHOSPHOLIPASE C
PHOSPHOLIPASE D

TRICYCLOHEXYLTIN CHLORIDE
GLUTATHIONE TRANSFERASE

TRICYCLO[3.3.1.0 ^{2,7}]NONAN-4-OL
CHORISMATE MUTASE *I*50 = 500 μM

TRICYCLO[3.3.1.0 ^{2,7}]NON-3-ENE-DICARBOXYLIC ACID
CHORISMATE MUTASE *I*50 = 20 mM

TRICYCLO[3.3.1.0 ^{2,7}]NON-3-ENE-1,3-DIPHOSPHONIC ACID
CHORISMATE MUTASE *I*50 = 3 mM

TRIDECANOIC ACID
ALKANAL MONOOXYGENASE (FMN -LINKED)

4-[(2-TRIDECYNYLOXY)METHYL]PHENYLACETIC ACID
ARACHIDONATE 12-LIPOXYGENASE *I*50 = 50 nM

TRIDEGIN
PROTEIN-GLUTAMINE γ-GLUTAMYLTRANSFERASE *I*50 = 9.2 nM

TRIDEMETHYLTHIOCOLCHICINE
DNA TOPOISOMERASE (ATP-HYDROLYSING) *I*50 = 9 μM

TRIDEMOPRH
C8- > C7 STEROL ISOMERASE

TRIDEMORPH
CHOLESTENOL Δ-ISOMERASE *I*50 = 400 nM
CHOLESTENOL Δ-ISOMERASE *I*50 = 33 nM
CYCLOEUCALENOL CYCLO-ISOMERASE *I*50 = 400 nM
STEROL C8-C7 ISOMERASE
Δ5,7-STEROL Δ7-REDUCTASE *I*50 = 500 nM
Δ8,14-STEROL Δ14 REDUCTASE *I*50 = 25 μM
Δ8,14-STEROL Δ14 REDUCTASE *I*50 = 98 μM

1,2,5-TRIDEOXY-1,5-IMINO-D-ALLOHEXITOL
β-GALACTOSIDASE *I*50 = 3.1 μM
α-GLUCOSIDASE *I*50 = 500 μM
LACTASE *I*50 = 4 μM
OLIGO-1,6-GLUCOSIDASE *I*50 = 6.4 μM

1,4,6-TRIDEOXY-1,4-IMINO-D-MANNITOL
α-MANNOSIDASE *K*i = 1.3 μM
α-MANNOSIDASE *I*50 = 60 μM
α-MANNOSIDASE *I*50 = 300 μM

1,4,6-TRIDEOXY-1,4-IMINO-L-MANNITOL
NARINGINASE

(+)-3,7,8-TRIDEOXY-3,7-IMINO-D-THRO-L-galacto-OCTITOL
β-GALACTOSIDASE *K*i = 15 μM

2′,3′,5′-TRIDEOXYURIDINE-5′-TRIPHOSPHONATE
RNA-DIRECTED DNA POLYMERASE *I*50 = 370 μM

TRIDIMEFON		
OBTUSIFOLIOL 14α-DEMETHYLASE		*I*50 = 8 μM
(1R,2S)-TRIDIMENOL		
ent-KAURENE OXIDASE		*I*50 = 170 nM
TRIDIPHANEDIOL		
GLUTATHIONE TRANSFERASE		
S-(TRIDIPHANE)-GLUTATHIONE		
GLUTATHIONE TRANSFERASE		
TRIETHANOLAMINE		
CHOLINE SULFOTRANSFERASE		
CYSTATHIONINE γ-LYASE		
α-GALACTOSIDASE		
HYDROGEN DEHYDROGENASE		
TRIETHANOLAMMONIO-1-CHOLINE		
CHOLINE KINASE		
TRIETHYLAMINE		
3′-PHOSPHOADENOSINE 5′-PHOSPHOTRANSFERASE		52% 300 μM
TRIETHYLENETETRAAMINE		
AMINE OXIDASE (COPPER-CONTAINING)		*K*i = 570 nM
DNA NUCLEOTIDYLEXOTRANSFERASE		
SUPEROXIDE DISMUTASE		
TRIETHYLGERMANIUM CHLORIDE		
GLUTATHIONE TRANSFERASE		
TRIETHYLLEAD CHLORIDE		
GLUTATHIONE TRANSFERASE		
TRIETHYLPHOSPHINE GOLD		
6-PHOSPHOFRUCTOKINASE	IR	
PHOSPHOLIPASE C		*I*50 = 800 nM
TRIETHYLTIN BROMIDE		
GLUTATHIONE TRANSFERASE		
TRIFLIRHIZIN		
PROLYL OLIGOPEPTIDASE		*I*50 = 88 ppm
TRIFLUOPERAZINE		
ACETYLCHOLINESTERASE		*I*50 = 60 μM
ADENOSINETRIPHOSPHATASE		
β-ADRENERGIC-RECEPTOR KINASE		
1-ALKYLGLYCEROPHOSPHOCHOLINE O-ACETYLTRANSFERASE	IR	
AMINE OXIDASE DIAMINE	NC	*K*i = 2 nM
CALPAIN I		*I*50 = 130 μM
CAM KINASE II		40% 10 μM
Ca^{2+}-TRANSPORTING ATPASE		90% 14 μM
C8- > C7 STEROL ISOMERASE		*K*i = 500 nM
3′,5′-CYCLIC-NUCLEOTIDE PHOSPHODIESTERASE		*I*50 = 17 μM
3′,5′-CYCLIC-NUCLEOTIDE PHOSPHODIESTERASE (Ca^{2+}/CALMODULIN)		*I*50 = 14 μM
H^+-TRANSPORTING ATP SYNTHASE	C	
H^+-TRANSPORTING ATP SYNTHASE F1		
[HYDROXYMETHYLGLUTARYL-CoA REDUCTASE (NADPH)] KINASE		
INOSITOL-1,4,5-TRISPHOSPHATE 5-PHOSPHATASE		62% 1 mM
INOSITOL-1,4,5-TRISPHOSPHATE 5-PHOSPHATASE		
MYOSIN-LIGHT-CHAIN KINASE		*I*50 = 14 μM
NITRIC-OXIDE SYNTHASE		*I*50 = 7 μM
NITRIC-OXIDE SYNTHASE		*I*50 = 140 μM
NITRIC-OXIDE SYNTHASE II		
PHOSPHOLIPASE		
PHOSPHOLIPASE A2		*I*50 = 73 μM

PHOSPHOLIPASE A2		*I*50 = 42 μM
PHOSPHOPROTEIN PHOSPHATASE 1		
PHOSPHOPROTEIN PHOSPHATASE 2B		
PROTEIN KINASE C		51% 40 μM
PYRUVATE DEHYDROGENASE (LIPOAMIDE)		32% 267 μM
SPHINGOMYELIN PHOSPHODIESTERASE		51% 10 μM
TRYPANOTHIONE REDUCTASE	C	*K*i = 22 μM
TRYPANOTHIONE REDUCTASE		*K*i = 22 μM
XAA-METHYL-HIS DIPEPTIDASE		
TRIFLUOROACETAMIDOBENZAMIDOMETHANE BORONIC ACID		
β-LACTAMASE		*K*i = 34 μM
TRIFLUOROACETAMIDOMETHANEBORONIC ACID		
β-LACTAMASE I		
TRIFLUOROACETATE		
HALOACETATE DEHALOGENASE	C	*K*i = 2.1 mM
TRIFLUOROACETONE		
CYTOCHROME-b5 REDUCTASE		
TRIFLUOROACETOPHENONE		
STEROL ESTERASE	C	*K*i = 18 μM
2,2,2-TRIFLUOROACETOPHENONE		
ALDEHYDE DEHYDROGENASE (NAD)		*K*i = 7 μM
N-TRIFLUOROACETYLADRIAMYCIN-14-O-HEMIADIPATE		
CARNITINE O-PALMITOYLTRANSFERASE		30% 100 μM
N-TRIFLUOROACETYLADRIAMYCIN-14-VALERIC ACID		
CARNITINE O-PALMITOYLTRANSFERASE		70% 1 mM
TRIFLUOROACETYL-CYS-PHE-O-MET		
INTERSTITIAL COLLAGENASE		*I*50 = 63 nM
TRIFLUOROACETYL-LYS-ALA-NHC$_6$H$_4$-4-CF$_3$		
ELASTASE PANCREATIC		*K*i = 25 nM
TRIFLUOROACETYL-LYS-ALA-p(TRIFLUOROMETHYL)ANILIDE		
ELASTASE PANCREATIC		*K*i = 25 nM
TRIFLUOROACETYLPHENYLALANINE		
THERMOLYSIN		*I*50 = 490 μM
TRIFLUOROACETYLPHENYLALANYLGLYCINE		
NEPRILYSIN		*I*50 = 14 mM
TRIFLUOROACETYL-TRIPEPTIDES		
ELASTASE PANCREATIC		
TRIFLUOROALANINE		
ALANINE RACEMASE		
β-TRIFLUOROALANINE		
CYSTATHIONINE γ-LYASE	IR	
β,β,β-TRIFLUOROALANINE		
ALANINE RACEMASE	SSI	
TRIFLUORO-D,L-ALANINE		
CYSTATHIONINE β-LYASE		
3,3,3-TRIFLUORO-2-AMINOISOBUTYRIC ACID		
2,2-DIALKYLGLYCINE DECARBOXYLASE (PYRUVATE)	IR	
5-TRIFLUORO-2′-DEOXYURIDINE		
NAD ADP-RIBOSYLTRANSFERASE		*K*i = 1.6 μM

1,1,1-TRIFLUORO-6,6-DIMETHYLHEPTAN-2-ONE
ACETYLCHOLINESTERASE $Ki = 12$ nM
CHOLINESTERASE $Ki = 29$ nM

1,1,1-TRIFLUORODODEC-3-YN-2-ONE
JUVENILE HORMONE ESTERASE

2,2,2-TRIFLUOROETHANOL
ALCOHOL DEHYDROGENASE

1,1,1-TRIFLUORO-4-(p-HYDROXYPHENYL)BUTAN-2-ONE
CARBOXYLESTERASE IR $Ki = 6$ nM

5,5,5-TRIFLUORO-D,L-LEUCINE
2-ISOPROPYLMALATE SYNTHASE C

D,L-TRIFLUOROLEUCINE
2-ISOPROPYLMALATE SYNTHASE

20,20,20-TRIFLUOROLEUKOTRIENE E4
LEUKOTRIENE-B4 20-MONOOXYGENASE
LEUKOTRIENE-E4 20 MONOOXYGENASE

TRIFLUOROMETHANE SULPHONAMIDE
CARBONATE DEHYDRATASE II Kd = 2 nM

1-(M-4-TRIFLUOROMETHOXYPHENYL)AMINO-1-PHENYLMETHANEPHOSPHONIC ACID
NADH DEHYDROGENASE (UBIQUINONE) $Ki = 26$ μM

p-TRIFLUOROMETHOXYPHENYLHYDRAZONE
OXIDATIVE PHOSPHORYLATION UNCOUPLER 100 nM

2-TRIFLUOROMETHYL ALANINE
2,2-DIALKYLGLYCINE DECARBOXYLASE (PYRUVATE) SSI

5-TRIFLUOROMETHYLAMINOBUTYRIC ACID
O-SUCCINYLHOMOSERINE (THIOL)-LYASE SSI

4-TRIFLUOROMETHYLBENZOIC ACID
4-METHOXYBENZOATE MONOOXYGENASE (O-DEMETHYLATING)

7-p-TRIFLUOROMETHYLBENZYL-8-DEOXYHEMIGOSSYLIC ACID
L-LACTATE DEHYDROGENASE $Ki = 200$ nM
L-LACTATE DEHYDROGENASE $Ki = 13$ μM
L-LACTATE DEHYDROGENASE $Ki = 81$ μM

6-TRIFLUOROMETHYL-2-(1,1′-BIPHENYL-4-YL)-3-METHYL-4-QUINOLINE CARBOXYLIC ACID SODIUM SALT
DIHYDROOROTATE DEHYDROGENASE

3-(TRIFLUOROMETHYLCARBONYLAMINO)ESTRA-1,3,4-(10)TRIEN-17-ONE
ESTRONE SULFATASE $I50 = 8.7$ μM

5-(TRIFLUOROMETHYL)-2′-DEOXYURIDINE MONOPHOSPHATE
THYMIDYLATE SYNTHASE SSI

5-TRIFLUOROMETHYL-2′-DEOXYURIDINE TRIPHOSPHATE
DNA POLYMERASE C

Nα-[5-(TRIFLUOROMETHYL)-5,8-DIDEOXYAZAPTEROYL]-L-ORNITHINE
FOLYLPOLYGLUTAMATE SYNTHASE $Ki = 8.8$ nM

Nα-[5-(TRIFLUOROMETHYL)-5,8-DIDEOXYISOAZAPTEROYL]-L-ORNITHINE
FOLYLPOLYGLUTAMATE SYNTHASE $Ki = 11$ nM

3-TRIFLUOROMETHYL FARNESYLPYROPHOSPHATE
FARNESYL PROTEINTRANSFERASE C $Ki = 11$ nM

2-TRIFLUOROMETHYL GLUTAMIC ACID
GLUTAMATE DECARBOXYLASE SSI

S-TRIFLUOROMETHYL-L-HOMOCYSTEINE
METHIONINE ADENOSYLTRANSFERASE

TRIFLUOROMETHYLKETONES
JUVENILE HORMONE ESTERASE

N-([5-(TRIFLUORO-METHYL)-6-METHOXY-1-NAPHTHALENYL]-THIOXOMETHYL)-N-METHYLGLYCINE
ALDEHYDE REDUCTASE — *I*50 = 11 nM
GLUCURONATE REDUCTASE — *I*50 = 240 nM

2-TRIFLUOROMETHYL-5-METHYL-BENZOXAZINONE
CHYMOTRYPSIN — *K*i = 150 nM
ELASTASE LEUKOCYTE — *K*i = 29 nM
THROMBIN — *K*i = 5.1 µM

1-(TRIFLUOROMETHYL)PHENOXATHIN 10,10-DIOXIDE
AMINE OXIDASE MAO-A — *I*50 = 80 nM

(2S,3a,s,7a,s)1[[trans(R,S))2[3)TRIFLUOROMETHYL)PHENYL]CYCLOPROPAN-1-YL]CARBONYL]2(PYRROLIDIN-1-YLCARBONYL)PERHYDRO-INDOLE
PROLYL OLIGOPEPTIDASE — *I*50 = 900 pM

N-(4-TRIFLUOROMETHYLPHENYLETHYLPHOSPHONYL)-GLY-PRO-L-2-AMINOHEXANOIC ACID
COLLAGENASE CLOSTRIDIUM HISTOLYTICUM — *K*i = 24 nM

N2-[m-(TRIFLUOROMETHYL)PHENYL]GUANINE
THYMIDYLATE SYNTHASE — *I*50 = 150 nM
dTMP KINASE — *I*50 = 150 nM

3-[(4′-TRIFLUOROMETHYL)PHENYL]-5H-INDENO[1,2-c]PYRIDAZIN-5-ONE
AMINE OXIDASE MAO-B — *I*50 = 90 nM

1-(2-TRIFLUOROMETHYLPHENYL)IMIDAZOLE
NITRIC-OXIDE SYNTHASE I — *I*50 = 28 µM
NITRIC-OXIDE SYNTHASE II — *I*50 = 27 µM

10-(3′-TRIFLUOROMETHYLPHENYL)-3-METHYLFLAVIN
GLUTATHIONE REDUCTASE (NADPH) — 90% 31 µM

(3-TRIFLUOROMETHYLPHENYL)-(2-(1,3,4-THIADIAZOL-2-ONE-5-YL)PHENYL)AMINE
ARACHIDONATE 5-LIPOXYGENASE — 100% 16 µM

8-TRIFLUOROMETHYLPURINE
ADENOSINE DEAMINASE — C — *K*i = 170 µM

5-TRIFLUOROMETHYLTHIORIBOSE
5-METHYLTHIORIBOSE KINASE — C — *K*i = 7 µM

5-TRIFLUOROMETHYLdUMP
THYMIDYLATE SYNTHASE — *K*i = 39 nM

1,1,1-TRIFLUORO-10(Z)NONADECEN-2-ONE
OLEAMIDE HYDROLASE — *K*i = 1.2 nM

(5,5,5-TRIFLUORO-4-OXOPENTYL)TRIMETHYLAMMONIUM
ACETYLCHOLINESTERASE — *K*i = 60 pM
CHOLINESTERASE — *K*i = 70 nM

1,1,1-TRIFLUORO-4-PHENYLBUTAN-2-ONE
CARBOXYLESTERASE — IR — *K*i = 6.8 nM

9-[o-(1,2,2-TRIFLUORO-2-PHOSPHONOETHYL)PHENYLDIISOPROPYLESTER]GUANINE
PURINE-NUCLEOSIDE PHOSPHORYLASE — *K*i = 1.3 nM
PURINE-NUCLEOSIDE PHOSPHORYLASE — *K*i = 600 pM

TRIFLUOROTETRADECANONE
ACETYLESTERASE

TRIFLUPERIDOL
BUFURALOL-1′-HYDROXYLASE (CYT P-450dbl) C *I*50 = 500 nM
C8- > C7 STEROL ISOMERASE

TRIFLUPROMAZINE
1,3-β-GLUCAN SYNTHASE *I*50 = 100 μM
PYRUVATE DEHYDROGENASE (LIPOAMIDE) 48% 267 μM

TRIFOP
ACETYL-CoA CARBOXYLASE

TRIGALLOYLGLUCOSE
POLY(ADP-RIBOSE) GLYCOHYDROLASE *I*50 = 32 μM

1,2,3-TRI-O-GALLOYLGLUCOSE
α-GLUCOSIDASE *I*50 = 6.8 μM

1,2,6-TRI-O-GALLOYL-β-D-GLUCOSE
INTEGRASE *I*50 = 28 μg/ml

1,2,6-TRI-O-GALLOYL-D-GLUCOSE
NADH DEHYDROGENASE *I*50 = 1.8 μM

3,4,5-TRI-O-GALLOYLQUINIC ACID
DNA POLYMERASE α NC *K*i = 280 nM
DNA POLYMERASE β C *K*i = 44 μM
DNA POLYMERASE γ C *K*i = 7.5 μM

2,7,8-TRIHYDROXYANTHRAQUINONE
PROCOLLAGEN-PROLINE, 2-OXOGLUTARATE-4-DIOXYGENASE *K*i = 40 μM

3′,4′,6-TRIHYDROXYAURONE
XANTHINE OXIDASE *I*50 = 6.2 μM

1,2,4-TRIHYDROXYBENZENE
ALDEHYDE DEHYDROGENASE (NAD)

2,4,6-TRIHYDROXYBENZOIC ACID
GENTISATE 1,2-DIOXYGENASE *K*i = 2.3 mM

3,4,5-TRIHYDROXYBENZOIC ACID
GLUTAMATE DECARBOXYLASE *K*i = 13 μM
GLUTAMATE DECARBOXYLASE 12% 1 mM
PROCOLLAGEN-PROLINE, 2-OXOGLUTARATE-4-DIOXYGENASE

(2,3,4-TRIHYDROXYBENZYL)HYDRAZINE
AROMATIC-L-AMINO-ACID DECARBOXYLASE

2,3′,4′-TRIHYDROXYBENZYLIDENE-N-HYDROXY-N′-AMINOGUANIDINE
RIBONUCLEOSIDE-DIPHOSPHATE REDUCTASE *I*50 = 122 μM

2′,4′,5′-TRIHYDROXYBENZYLIDENE-N-HYDROXY-N′-AMINOGUANIDINE
RIBONUCLEOSIDE-DIPHOSPHATE REDUCTASE *I*50 = 188 μM

2′,4′,6′-TRIHYDROXYBENZYLIDENE-N-HYDROXY-N′-AMINOGUANIDINE
RIBONUCLEOSIDE-DIPHOSPHATE REDUCTASE *I*50 = 80 μM

3β,16β,17α-TRIHYDROXYCHOLEST-5-ENE-22-ONE 16-O-(2-O-3,4-DIMETHOXYBENZOYL-β-D-XYLOPYRANOSYL-(1- >)(2-O-ACETYL-α-L-ARABINO-PYRANOSIDE
3′,5′-CYCLIC-NUCLEOTIDE PHOSPHODIESTERASE *I*50 = 500 nM

3β,16β,17α-TRIHYDROXYCHOLEST-5-ENE-22-ONE 16-O-(2-O-4-METHOXYBENZOYL-β-D-XYLOPYRANOSYL)(1- >)(2-O-ACETYL-α-L-ARABINO-PYRANOSIDE
3′,5′-CYCLIC-NUCLEOTIDE PHOSPHODIESTERASE *I*50 = 55 μM

3′,5,7-TRIHYDROXY-4′,6-DIMETHOXYISOFLAVONE
CATECHOL O-METHYLTRANSFERASE

3′,5,7-TRIHYDROXY-4′,8-DIMETHOXYISOFLAVONE
AROMATIC-L-AMINO-ACID DECARBOXYLASE
CATECHOL O-METHYLTRANSFERASE

(2S)-3′,4′,7-TRIHYDROXYFLAVAN-(4α- > 8)CATECHIN
TRIACYLGLYCEROL LIPASE | | $I50$ = 5.5 μM

3,3′,4′TRIHYDROXYFLAVONE
PROTEIN KINASE C | | $I50$ = 2 μM

5,6,7-TRIHYDROXYFLAVONE
XANTHINE OXIDASE | | $I50$ = 2.4 μM

5,7,4′-TRIHYDROXY-8-METHOXYFLAVONE
SIALIDASE | | $I50$ = 55 μM

2-[(2,3,4-TRIHYDROXYPHENYL)METHYL]HYDRAZIDE
PROCOLLAGEN-LYSINE 5-DIOXYGENASE

2-(3′,4′,5′-TRIHYDROXYPHENYL)-3,5,7-TRIHYDROXY-BENZOPYRYLIUM
XANTHINE OXIDASE | | Kei = 70 nM

[1R-(1α,3α,4β,5α)]-1,3,4-TRIHYDROXY-5-(PHOSPHONOMETHYL)CYCLOHEXANE-1-CARBOXYLIC ACID
3-DEHYDROQUINATE SYNTHASE | | Ki = 7.3 nM

[1S-(1α,3β,4α,5β)]-1,3,4-TRIHYDROXY-5-(PHOSPHONOMETHYL)CYCLOHEXANE-1-CARBOXYLIC ACID
3-DEHYDROQUINATE SYNTHASE | | Ki = 5.4 nM

3,4′,5-TRIHYDROXYSTILBENE-4′-O-β-D-(2″-O-GALLOYL)GLUCOPYRANOSIDE
MONOPHENOL MONOOXYGENASE | | $I50$ = 6.7 μM

3,4′,5-TRIHYDROXYSTILBENE-4′-O-β-D-(6″-O-GALLOYL)GLUCOPYRANOSIDE
MONOPHENOL MONOOXYGENASE | | $I50$ = 15 μM

6,3′,4′-TRIHYDROXY-5,7,8-TRIMETHOXYFLAVONE
ALCOHOL DEHYDROGENASE (NADP) | | $I50$ = 230 nM
ALDEHYDE REDUCTASE | NC | Ki = 200 nM
ALDEHYDE REDUCTASE | NC | Ki = 26 nM

TRIIODOBENZOIC ACID
AUXIN CARRIER

TRIIODOTHYROACETIC ACID
ALCOHOL DEHYDROGENASE | | Ki = 2 μM
ALDEHYDE DEHYDROGENASE | | Ki = 9.8 μM
ALDEHYDE DEHYDROGENASE | | Ki = 5.7 μM
9-HYDROXYPROSTAGLANDIN DEHYDROGENASE | NC |

3,3′,5-TRIIODOTHYROACETIC ACID
15-HYDROXYPROSTAGLANDIN-D DEHYDROGENASE

TRIIODOTHYRONINE
ALCOHOL DEHYDROGENASE | | Ki = 6.7 μM
ALDEHYDE DEHYDROGENASE | | Ki = 12 μM
ALDEHYDE DEHYDROGENASE | | Ki = 7.4 μM
GLUTAMATE DEHYDROGENASE (NAD(P)) | | 60% 20 μM
NAD(P) TRANSHYDROGENASE (AB-SPECIFIC)

3,3′,5-TRIIODOTHYRONINE
15-HYDROXYPROSTAGLANDIN DEHYDROGENASE (NAD)

3,3,′5′-TRIIODO-L-THYRONINE
3′,5′-CYCLIC-NUCLEOTIDE PHOSPHODIESTERASE | C | Ki = 400 μM

3,5,3′-TRIIODOTHYRONINE
TYROSINE TRANSAMINASE | C | Ki = 38 μM

L-TRIIODOTHYRONINE
CREATINE KINASE

L-TRIIODOTHYROXINE
CREATINE KINASE

TRILFUORPERAZINE
PROTOCHLOROPYLI′LIDE REDUCTASE

TRILOBOLIDE
Ca^{2+}-TRANSPORTING ATPASE 97% 100 μM
Ca^{2+}-TRANSPORTING ATPASE 100% 100 μM
H^{+}-TRANSPORTING ATPASE 91% 100 μM
H^{+}-TRANSPORTING ATPASE 67% 100 μM

TRILOSTANE
3β-HYDROXY-Δ5-STEROID DEHYDROGENASE C *K*i = 4 nM

TRIMEPROPAZINE
1,3-β-GLUCAN SYNTHASE *I*50 = 400 μM

TRIMERESURUS FLAVOVIRIDIS ANTIHEMORRHAGIC FACTOR
PROTEINASE H2

TRIMERESURUS FLAVOVIRIDIS PHOSPHOLIPASE A2 INHIBITOR
PHOSPHOLIPASE A2
PHOSPHOLIPASE A2

TRIMESIC ACID
GLUTAMATE DEHYDROGENASE (NAD(P)) *K*i = 4 mM
SERINE-SULFATE AMMONIA-LYASE C *K*i = 82 mM

TRIMETAPHOSPHATE
COB(I)ALAMIN ADENOSYLTRANSFERASE

TRIMETAZIDINE
Na^{+}/K^{+}-EXCHANGING ATPASE

TRIMETHOPRIM
DIHYDROFOLATE REDUCTASE *K*i = 17 μM
DIHYDROFOLATE REDUCTASE *K*i = 1.4 nM
GLUTATHIONE SYNTHASE
HISTAMINE N-METHYLTRANSFERASE *I*50 = 8.6 mM
THYMIDYLATE SYNTHASE NC *K*i = 5 mM

6[(3′,4′,5′-TRIMETHOXY-N-METHYLANILINO)METHYL]PYRIDO[3,2-d]PYRIMIDINE
DIHYDROFOLATE REDUCTASE *I*50 = 130 nM
DIHYDROFOLATE REDUCTASE *I*50 = 26 nM
DIHYDROFOLATE REDUCTASE *I*50 = 4.7 nM

(Z)-N-(2,4,6-TRIMETHOXYPHENYL)9-OCTADECENAMIDE
STEROL O-ACYLTRANSFERASE *I*50 = 44 nM

3′,4′,5′-TRIMETHOXYTRICETIN
PROTEIN KINASE C *I*50 = 31 μM

TRIMETHYLAMINE
ANTHRANILATE SYNTHASE

TRIMETHYLAMINE-N-OXIDE
TRIMETHYLAMINE-OXIDE ALDOLASE

TRIMETHYLAMINE + PYROPHOSPHATE
TRICHODIENE SYNTHASE *K*i = 680 nM

2-(TRIMETHYLAMINO)ETHYL DIPHOSPHATE
ISOPENTENYL-DIPHOSPHATE Δ-ISOMERASE *K*i = 100 nM

N,N,N-TRIMETHYLAMINOMETHANEBORONATE
BETAINE-HOMOCYSTEINE S-METHYLTRANSFERASE C *K*i = 45 μM

3-TRIMETHYLAMMONIO-1-PROPANOL
CHOLINE KINASE

m-(N,N,N-TRIMETHYLAMMONIO)TRIFLUOROACETOPHENONE
ACETYLCHOLINESTERASE | | *K*i = 1.3 fM
ACETYLCHOLINESTERASE | | *K*i = 15 fM

TRIMETHYLAMMONIUM
CHOLINE SULFOTRANSFERASE

TRIMETHYLAMMONIUM-N-(BENZYLOXYCARBONYL)AMINOMETHYLPHOSPHONOFLUORIDATE
β-LACTAMASE | IR |

4-TRIMETHYLAMMONIUMBUTYRIC ACID
CARNITINE O-PALMITOYLTRANSFERASE

3-TRIMETHYLAMMONIUMETHYLCATECHOL
CHOLINE O-ACETYLTRANSFERASE

5-TRIMETHYLAMMONIUM-2-PENTANONE
ACETYLCHOLINESTERASE | C | *K*i = 35 μM

TRIMETHYLAPIGENIN
ARACHIDONATE 15-LIPOXYGENASE 1 | | *I*50 = 295 μM

3-[4-[(3,5,6,-TRIMETHYL-1,4-BENZOQUINON-2-YL)-3-PYRIDYLMETHYL]PHENYL]-2-METACRYLIC ACID
ARACHIDONATE 5-LIPOXYGENASE | | *I*50 = 100 μM
THROMBOXANE A SYNTHASE | | *I*50 = 520 nM

3-[4-[(3,5,6,-TRIMETHYL-1,4-BENZOQUINON-2-YL)-3-PYRIDYLMETHYL]PHENYL]-2-PROPIONIC ACID
ARACHIDONATE 5-LIPOXYGENASE | | *I*50 = 100 μM

(6β,8aβ)-2-(1,5,9-TRIMETHYL(E)-4,8-DECADIENYL)-1,2,3,5,6,8,8a-OCTAHYDRO-5,5,8a-TRIMETHYL-6-ISOQUINOLINOL
LANOSTEROL SYNTHASE | | *I*50 = 600 nM

4,4,10-β-TRIMETHYL-trans-DECAL-3β-OL
LANOSTEROL SYNTHASE | NC | *K*i = 13 μM

N-(1,5,9-TRIMETHYLDECYL)-4α,10-DIMETHYL-8-AZA-trans-DECAL-3β-OL
CHOLESTENOL Δ-ISOMERASE | | *I*50 = 200 nM
CYCLOARTENOL SYNTHASE | | *I*50 = 1 μM
CYCLOEUCALENOL CYCLO-ISOMERASE | | *I*50 = 25 nM
LANOSTEROL SYNTHASE | | *I*50 = 100 μM
LANOSTEROL SYNTHASE | | *I*50 = 2 μM
Δ8,14-STEROL Δ14 REDUCTASE | | *I*50 = 300 nM
STEROL Δ14-REDUCTASE | | *I*50 = 700 nM

3,7,11-TRIMETHYL DODECANOATE
MEVALONATE KINASE

N,N,N-TRIMETHYL-L-HISTIDINE
HISTIDINE METHYLTRANSFERASE | C |

3-(2,2,2-TRIMETHYLHYDRAZINIUM)PROPIONIC ACID
γ-BUTYROBETAINE DIOXYGENASE | NC |

2,3,5-TRIMETHYL-6-(12-HYDROXY-5,10-DODECADIYNYL)-1,4-BENZOQUINONE
ARACHIDONATE 5-LIPOXYGENASE

5-(1,3,3-TRIMETHYLINDOLINYL)-N-METHYLCARBAMATE
ACETYLCHOLINESTERASE | | Kd = 71 nM

N,N,,N-TRIMETHYL-1-METHYL-(2′,2′,4′,6′-TETRAMETHYLCYCLOHEX-3′-EN-1′-YL)PROP-2-ENYLAMMONIUM IODIDE
ent-KAUREN SYNTHASE | | *I*50 = 100 nM

(2R,4S/2S,4R)-2,6,6-TRIMETHYL-2-OXO-4-HEXADECYL-1,3-DIOXA-6-AZA-2-PHOSPHACYCLOOCTANE
PROTEIN KINASE C | | *I*50 = 90 μM

3-[2-(2,4,5-TRIMETHYLPHENYL)THIOETHYL]-4-METHYLSYDNONE

CYTOCHROME P450 2B1	IR	
CYTOCHROME P450 2C6	IR	
CYTOCHROME P450 2C11	IR	

3,5,6-TRIMETHYL-2-[1-(3-PYRIDYL)ETHYL]-1,4-BENZOQUINONE

ARACHIDONATE 5-LIPOXYGENASE		*I*50 = 9.5 μM
THROMBOXANE A SYNTHASE		*I*50 = 720 nM

2,3,5-TRIMETHYL-6-(3-PYRIDYLMETHYL)-1,4-BENZOQUINONE

ARACHIDONATE 5-LIPOXYGENASE		*I*50 = 360 nM
THROMBOXANE A SYNTHASE		*I*50 = 330 nM

TRIMETHYLSULPHONIUM CHLORIDE

TRIMETHYLAMINE DEHYDROGENASE

N-(4E,8E)-5,9,13-TRIMETHYL-4,8,12-TETRADECATRIEN-1-YLPYRIDINIUM

LANOSTEROL SYNTHASE

TRIMETHYLTIN CHLORIDE

GLUTATHIONE TRANSFERASE

7,3′.5′-TRI-O-METHYLTIRCETIN

DNA TOPOISOMERASE

4,8,12-TRIMETHYL-3,7,11-TRIDECATRIEN-1-PHOSPHONOFORMIC ACID

FARNESYL-DIPHOSPHATE FARNESYLTRANSFERASE	C	*K*i = 2.2 μM

TRIMIDOX

RIBONUCLEOSIDE-DIPHOSPHATE REDUCTASE		
RIBONUCLEOTIDE REDUCTASE		*I*50 = 50 μM

TRIMIPRAMINE

SPHINGOMYELIN PHOSPHODIESTERASE		39% 10 μM
TRYPANOTHIONE REDUCTASE		*I*50 = 297 μM

TRINITROBENZENE

GLUTATHIONE REDUCTASE (NADPH)		*I*50 = 1O μM

1,3,5-TRINITROBENZENE

GLUTATHIONE TRANSFERASE	C	*K*i = 10 μM

2,4,6-TRINITROBENZENE

S-FORMYLGLUTATHIONE HYDROLASE		
S-SUCCINYLGLUTATHIONE HYDROLASE		45% 1 mM

2,4,6,-TRINITROBENZENESULPHONATE + NBUTYLAMINE

THIOSULFATE SULFURTRANSFERASE	IR	

TRINITROBENZENESULPHONIC ACID

ADENYLYLSULFATE KINASE		
BISPHOSPHOGLYCERATE MUTASE	IR	
BISPHOSPHOGLYCERATE PHOSPHATASE	IR	
GLUTATHIONE REDUCTASE (NADPH)		*I*50 = 7 μM
PHOSPHOENOLPYRUVATE CARBOXYLASE		
PHOSPHOGLYCERATE MUTASE		
PYRUVATE CARBOXYLASE		
SULFATE ADENYLYLTRANSFERASE		
THIOGLUCOSIDASE		

2,4,6-TRINITROBENZENESULPHONIC ACID

ACETATE KINASE		
2-ACYLGLYCEROL O-ACYLTRANSFERASE		
4,5-DIHYDROOXYPHTHALATE DECARBOXYLASE		
GLUTATHIONE REDUCTASE (NADPH)		
HYDROXYACYLGLUTATHIONE HYDROLASE	IR	
15-HYDROXYPROSTAGLANDIN DEHYDROGENASE (NAD)		
INORGANIC PYROPHOSPHATASE		

myo-INOSITOL-1-PHOSPHATE SYNTHASE		80% 1 mM
LACTOSE SYNTHASE		
LACTOYLGLUTATHIONE LYASE		
ORNITHINE CARBAMOYLTRANSFERASE		
PHOSPHATIDYLCHOLINE-DOLICHOL O-ACYLTRANSFERASE		
TRIACYLGLYCEROL LIPASE ACID		51% 4 mM

TRINITROMETHANE
SULFATE ADENYLYLTRANSFERASE

TRINITROPHENOL
DIHYDROOROTATE DEHYDROGENASE

2′(3′)-O-(2,4,6-TRINITROPHENYL-ATP

1-PHOSPHATIDYLINOSITOL-4-PHOSPHATE 5-KINASE	C	*K*i = 55 μM

2,4,6-TRINITROPHENYL-2-DEOXY-2,2-DIFLUORO-α-GLUCOSIDE

α-GLUCOSIDASE	IR	

2,4,6-TRINITROPHENYL-2-DEOXY-2,2-DIFLUORO-α-MALTOSIDE

α-AMYLASE	IR	

TRINITROTOLUENE

GLUTATHIONE REDUCTASE (NADPH)		*I*50 = 30 μM

TRIOSTAM
GLUTATHIONE REDUCTASE (NADPH)
TRYPANOTHIONE REDUCTASE

TRIPARANOL

C8- > C7 STEROL ISOMERASE		
7-DEHYDROCHOLESTEROL REDUCTASE		
LANOSTEROL SYNTHASE		
Δ24 REDUCTASE		*K*i = 523 nM
Δ24(25) REDUCTASE		
STEROID Δ8- > 7 ISOMERASE		
Δ(24)-STEROL C-METHYLTRANSFERASE		

TRIPELENAMINE

CYTOCHROME P450 2D6	MIX	*K*i = 5.6 μM
HISTAMINE N-METHYLTRANSFERASE		*I*50 = 515 μM

TRIPHENYLACETIC ACID

UDP-GLUCURONOSYLTRANSFERASE	C	*K*i = 110 μM

4,4,4-TRIPHENYLBUTANOIC ACID

UDP-GLUCURONOSYLTRANSFERASE	C	*K*i = 61 μM

TRIPHENYLCHLORIDE

GLUTATHIONE TRANSFERASE III		*I*50 = 600 nM

2,2,2-(TRIPHENYL)ETHYLURIDINEDIPHOSPHATE

UDP-GLUCURONOSYLTRANSFERASE	C	

7,7,7-TRIPHENYLHEPTANOIC ACID

UDP-GLUCURONOSYLTRANSFERASE	C	*K*i = 5 μM

7,7,7-TRIPHENYLHEPTYL-UDP
UDP-GLUCURONOSYLTRANSFERASE

6,6,6-TRIPHENYLHEXANOIC ACID

UDP-GLUCURONOSYLTRANSFERASE	C	*K*i = 25 μM

TRIPHENYLMETHANE DYES
ARGINASE

9,9,9-TRIPHENYLNONANOIC ACID

UDP-GLUCURONOSYLTRANSFERASE	C	*K*i = 77 μM

8,8,8-TRIPHENYLOCTANOIC ACID
UDP-GLUCURONOSYLTRANSFERASE C $Ki = 55\ \mu M$

5,5,5-TRIPHENYLPENTANOIC ACID
UDP-GLUCURONOSYLTRANSFERASE C $Ki = 47\ \mu M$

TRIPHENYLPHOSPHATE
CARBOXYLESTERASE

TRIPHENYLPHOSPHINE
Ca^{2+}-TRANSPORTING ATPASE

TRIPHENYLPHOSPHITE
CARBOXYLESTERASE

3,3,3-TRIPHENYLPROPIONIC ACID
UDP-GLUCURONOSYLTRANSFERASE C $Ki = 75\ \mu M$

TRIPHENYLSULPHONIUMCHLORIDE
H^{+}-TRANSPORTING ATPASE

TRIPHENYLTIN
H^{+}-TRANSPORTING ATP SYNTHASE REV
LEUKOTRIENE-C4 SYNTHASE $I50 > 100\ \mu M$

TRIPHENYLTIN ACETATE
GLUTATHIONE TRANSFERASE

TRIPHENYLTIN CHLORIDE
GLUTATHIONE TRANSFERASE
GLUTATHIONE TRANSFERASE MGT $I50 = 500$ nM

TRIPHENYLTIN HYDROXIDE
GLUTATHIONE TRANSFERASE

TRIPHENYLTIN ISOTHIOCYANATE
GLUTATHIONE TRANSFERASE

TRIPHENYLTIN OXIDE
GLUTATHIONE TRANSFERASE

TRIPHENYLTIN SULPHIDE
GLUTATHIONE TRANSFERASE

TRIPHENYLTRICHLORIDE
GLUTATHIONE TRANSFERASE II $I50 = 900$ nM
GLUTATHIONE TRANSFERASE IV $I50 > 10\ \mu M$
GLUTATHIONE TRANSFERASE V $I50 > 10\ \mu M$

TRIPHOSPHATE
AMINOPEPTIDASE Y 100% 1 mM

TRIPOLYPHOSPHATE
ADENOSINE TETRAPHOSPHATASE $Ki = 20\ \mu M$
AMP DEAMINASE
COB(I)ALAMIN ADENOSYLTRANSFERASE
CREATINE KINASE C $Ki = 8\ \mu M$
CYSTEINE-tRNA LIGASE
GUANYLATE CYCLASE $I50 = 800\ \mu M$
METHIONINE ADENOSYLTRANSFERASE C $Ki = 14\ \mu M$
NICOTINATE PHOSPHORIBOSYLTRANSFERASE
POLYPHOSPHATE-GLUCOSE PHOSPHOTRANSFERASE C
PYROPHOSPHATE-FRUCTOSE-6-PHOSPHATE 1-PHOSPHOTRANSFERASE

TRI-n-PROPYLAMINE
3′-PHOSPHOADENOSINE 5′-PHOSPHOTRANSFERASE 45% 300 μM

2,4,6-TRIPYRIDYL-(2)-1,3,5-TRIAZINE
myo-INOSITOL OXYGENASE

TRIS

Enzyme	Type	Value
ACETOIN DEHYDROGENASE		
ACETYL-CoA C-ACYLTRANSFERASE		
ACETYLENECARBOXYLATE HYDRATASE		90% 50 mM
ADENOSYLHOMOCYSTEINE NUCLEOSIDASE		
ADENYLOSUCCINATE LYASE		> 100 mM
ALTERNASUCRASE		
4-AMINOBUTYRATE TRANSAMINASE	IR	
AMYLO-1,6-GLUCOSIDASE		
AMYLOSUCRASE		
ARABINOSE ISOMERASE		
L-ARABINOSE ISOMERASE		
ARYLAMINE GLUCOSYLTRANSFERASE		
ASPARTATE-SEMIALDEHYDE DEHYDROGENASE		
BIS-γ-GLUTAMYLCYSTEINE REDUCTASE (NADPH)		
BRANCHED-CHAIN-AMINO-ACID TRANSAMINASE		
CARBAMOYL-PHOSPHATE SYNTHASE (AMMONIA)		
CLOSTRIPAIN		
COLLAGENASE		
CREATINE KINASE		
CYSTATHIONINE γ-LYASE		
CYSTEINE-tRNA LIGASE		
CYTOCHROME-b5 REDUCTASE		
2-DEHYDRO-3-DEOXY-3-GLUCONATE 6-DEHYDROGENASE		
DEOXYCYTIDINE KINASE		
3-DEOXY-MANNO-OCTULOSONATE-8-PHOSPHATASE		
DEXTRANSUCRASE	C	*K*i = 9 mM
DIPEPTIDYL-PEPTIDASE 2		96% 50 mM
DIPEPTIDYL-PEPTIDASE 2		65% 50 mM
ECDYSONE 20-MONOOXYGENASE		
FARNESYL-DIPHOSPHATE FARNESYLTRANSFERASE		19% 100 mM
β-FRUCTOFURANOSIDASE		
β-D-FUCOSIDASE		
GALACTINOL-RAFFINOSE GALACTOSYLTRANSFERASE		
α-GALACTOSIDASE		
GLUCAN 1,3-α-GLUCOSIDASE		*I*50 = 50 mM
GLUCAN 1,4-α-GLUCOSIDASE	C	*K*i = 8.5 mM
GLUCOSAMINATE AMMONIA-LYASE		
α-GLUCOSIDASE	C	*K*i = 320 μM
α-GLUCOSIDASE		79% 5 mM
β-GLUCOSIDASE		*K*i = 500 μM
GLUTAMATE FORMIMINOTRANSFERASE		
GLUTAMATE METHYLTRANSFERASE		
GLYCEROL DEHYDROGENASE		
GLYCEROL-3-PHOSPHATE DEHYDROGENASE (NAD)		
GLYCERONE-PHOSPHATE O-ACYLTRANSFERASE		
GLYCINE DEHYDROGENASE (CYTOCHROME)		
GLYCINE-OXALOACETATE TRANSAMINASE		
GLYOXYLATE-L-ASPARTATE AMINOTRANSFERASE		
GLYOXYLATE DEHYDROGENASE (ACYLATING)		
GMP SYNTHASE (GLUTAMINE-HYDROLYSING)		
GUANOSINE PHOSPHORYLASE		
HOMOSERINE DEHYDROGENASE		
3-HYDROXYBUTYRYL-CoA EPIMERASE		
HYDROXYPYRUVATE REDUCTASE		
LACTASE		*K*i = 5.1 mM
LEVANASE	NC	*K*i = 6 mM
LEVANSUCRASE		
β-LYSINE 5,6-AMINOMUTASE		
L-LYSINE 6-TRANSAMINASE		
MALONATE-SEMIALDEHYDE DEHYDRATASE		
MALYL-CoA LYASE		50% 100 mM
MANNOSIDASE IA		
α-MANNOSIDASE IA		

MANNOSYL-OLIGOSACCHARIDE GLUCOSIDASE		60% 10 mM
MANNOTETRAOSE 2-α-N-ACETYLGLUCOSAMINYLTRANSFERASE		*I*50 = 60 mM
METHYLAMINE-GLUTAMATE METHYLTRANSFERASE		> 50 mM
MONODEHYDROASCORBATE REDUCTASE (NADH)		
Na^+/K^+-EXCHANGING ATPASE		*I*50 = 50 mM
NITROETANE REDUCTASE		
NUCLEOSIDE DEOXYRIBOSYLTRANSFERASE		
OLIGO-1,6-GLUCOSIDASE		93% 5 mM
OLIGO-1,6-GLUCOSIDASE		*K*i = 600 μM
OXOGLUTARATE DEHYDROGENASE (LIPOAMIDE)		
PHENYLALANINE 4-MONOOXYGENASE		
PHOSPHATE ACETYLTRANSFERASE		
PHOSPHOENOLPYRUVATE CARBOXYKINASE (PYROPHOSPHATE)		
6-PHOSPHOFRUCTOKINASE		
POLYGALACTURONASE		
PORPHOBILINOGEN SYNTHASE		
QUINATE O-HYDROXYCINNAMOYLTRANSFERASE		
D-SERINE DEHYDRATASE		
L-SERINE DEHYDRATASE		
SUCROSE α-GLUCOSIDASE	C	*K*i = 700 μM
SUCROSE SYNTHASE		
SULFATE ADENYLYLTRANSFERASE		
SULFITE REDUCTASE (FERREDOXIN)		
α,α-TREHALASE	C	*K*i = 3.8 mM
XYLAN 1,4-β-XYLOSIDASE		*K*i = 6.8 mM
XYLOGLUCAN 6-XYLOSETRANSFERASE		
XYLOSE ISOMERASE		
TRISAMINOMETHANE		
β-GALACTOSIDASE		
TRIS-MALEATE		
HOMOCARNOSINASE		48% 150 mM
α-MANNOSIDASE		94% 50 mM
SUCROSE-PHOSPHATE SYNTHASE		
TRISNORSQUALENE ALCOHOL		
SQUALENE MONOOXYGENASE	NC	*K*i = 4 μM
SQUALENE MONOOXYGENASE		*I*50 = 4 μM
TRISNORSQUALENE CYCLOPROPYLAMINE		
SQUALENE MONOOXYGENASE		*I*50 = 5 μM
SQUALENE MONOOXYGENASE	NC	*K*i = 30 nM
SQUALENE MONOOXYGENASE	NC	*K*i = 180 μM
TRISNORSQUALENE N-METHYLCYCLOPROPYLAMINE		
LANOSTEROL SYNTHASE		
TRISNORSQUALENEMETHYLHYDROXYLAMINE		
LANOSTEROL SYNTHASE		
SQUALENE MONOOXYGENASE		*I*50 = 13 μM
25,26,27-TRISNOR-24-TRIMETHYLAMMONIUM CYCLOARTANOL IODIDE		
Δ(24)-STEROL C-METHYLTRANSFERASE	NC	*K*i = 35 nM
TRISULFURON		
ACETOLACTATE SYNTHASE		
TRISULPHATED DISACCHERIDES		
MYOSIN ATPASE		
4,5,9-TRITHIADODECA-1,11-DIENE		
LIPOXYGENASE		*I*50 = 29 μM
4,5,9-TRITHIADODECA-1,11-DIENE 9-OXIDE		
LIPOXYGENASE		*I*50 = 21 μM

4,5,9-TRITHIADODECA-1,6,11-TRIENE
LIPOXYGENASE $I50 = 37\ \mu M$

2-(2′,3′,4′-TRITHIA-6′-HEPTENYL)-3,4-DIHYDRO-2H-THIOPYRAN
LIPOXYGENASE $I50 = 51\ \mu M$

TRITICUM AESTIVUM α-AMYLASE INHIBITOR
α-AMYLASE

TRITICUM AESTIVUM α-AMYLASE INHIBITOR II
α-AMYLASE C $Ki = 290$ pM

TRITON W-1339
LIPOPROTEIN LIPASE

TRITON WR 1339
DIACYLGLYCEROL O-ACYLTRANSFERASE
ETHANOLAMINEPHOSPHOTRANSFERASE

TRITON X 100
ACETYLCHOLINESTERASE 0.1%
ACETYLCHOLINESTERASE $I50 = 8.3$ mM
ACETYLGALACTOSAMINYL-O-GLYCOSYL-GLYCOPROTEIN β-1,3-N-ACETYLGLUCOSAMINYLTRANSFE
N-ACETYLGLUCOSAMINYLDIPHOSPHOUNDECAPRENOL N-ACETYL-β-D-MANNOSAMINYLTRANSFERAS
N-ACETYLNEURAMINATE 7-O(or 9-O) ACETYLTRANSFERASE > 0.02 %
ACYL[ACYL-CARRIER-PROTEIN]-UDP-N-ACETYLGLUCOSAMINE O-ACYLTRANSFERASE
1-O-ACYLCERAMIDE SYNTHASE
1-ACYLGLYCEROL-3-PHOSPHATE O-ACYLTRANSFERASE
1-ACYLGLYCEROPHOSPHOCHOLINE O-ACYLTRANSFERASE
β-ADRENERGIC-RECEPTOR KINASE
ALDEHYDE OXIDASE
1-ALKYLGLYCEROPHOSPHOCHOLINE O-ACETYLTRANSFERASE
ARYLESTERASE
Ca^{2+}/Mg^{2+}-TRANSPORTING ATPASE 0.1%
CARNITINE O-PALMITOYLTRANSFERASE 1
CATECHOL O-METHYLTRANSFERASE
CDP-DIACYLGLYCEROL-GLYCEROL-3-PHOSPHATE 3-PHOSPHATIDYLTRANSFERASE
CDP-DIACYLGLYCEROL-INOSITOL 3-PHOSPHATIDYLTRANSFERASE
CDP-DIACYLGLYCEROL-SERINE O-PHOSPHATIDYLTRANSFERASE
CHLORAMPHENICOL O-ACETYLTRANSFERASE $I50 = 24\ \mu M$
CHLORAMPHENICOL O-ACETYLTRANSFERASE
CHOLINESTERASE $I50 = 612\ \mu M$
CYTOCHROME-C OXIDASE
DIACYLGLYCEROL O-ACYLTRANSFERASE
DIACYLGLYCEROL KINASE
DIMETHYLALLYLtransTRANSFERASE
DOLICHOL ESTERASE > 0.06%
DOLICHOL KINASE > 0.5%
DOLICHYL-PHOSPHATE β-D-MANNOSYLTRANSFERASE
ECDYSONE 20-MONOOXYGENASE
trans-2-ENOYL-CoA REDUCTASE (NADPH)
FATTY ACID CoA REDUCTASE 100% 400 μM
GANGLIOSIDE GALACTOSYLTRANSFERASE
1,3-β-GLUCAN SYNTHASE
GLYCEROL-3-PHOSPHATE O-ACYLTRANSFERASE
GLYCEROL-3-PHOSPHATE DEHYDROGENASE (NAD)
GLYCOSAMINOGLYCAN GALACTOSYLTRANSFERASE
4-HYDROXYBENZOATE NONAPRENYLTRANSFERASE 85% 0.5%
3(or 17)β-HYDROXYSTEROID DEHYDROGENASE
7α-HYDROXYSTEROID DEHYDROGENASE
LEUCINE DEHYDROGENASE
LYSOPHOSPHOLIPASE 76% 1%
METHYLGLUTAMATE DEHYDROGENASE
NADH DEHYDROGENASE (UBIQUINONE) 95% 50 μg/ml
NITRIC-OXIDE REDUCTASE
NUATIGENIN 3-β-GLUCOSYLTRANSFERASE

PHOSPHATIDYLCHOLINE-DOLICHOL O-ACYLTRANSFERASE		
PHOSPHATIDYLCHOLINE-STEROL O-ACYLTRANSFERASE		
PHOSPHATIDYLETHANOLAMINE N-METHYLTRANSFERASE		
PHOSPHATIDYLGLYCEROPHOSPHATASE		
PHOSPHATIDYLINOSITOL-BISPHOSPHATASE		
1-PHOSPHATIDYLINOSITOL 4-KINASE I		
1-PHOSPHATIDYLINOSITOL PHOSPHODIESTERASE		
PHOSPHATIDYL N-METHYLETHANOLAMINE METHYLTRANSFERASE		
PHOSPHOLIPASE A1		
RETINYL-PALMITATE ESTERASE		67% 0.02%
SARSAPOGENIN 3-β-GLUCOSYLTRANSFERASE		
SIALIDASE		
SQUALENE MONOOXYGENASE		
STEROL O-ACYLTRANSFERASE		
TESTOSTERONE 17β-DEHYDROGENASE		
TRIACYLGLYCEROL LIPASE		
TYROSINE-ESTER SULFOTRANSFERASE		
UBIQUINOL-CYTOCHROME-C REDUCTASE		
UDP-GLUCURONOSYLTRANSFERASE		> 0.01%
UNDECAPRENL-PHOSPHATE MANNOSYLTRANSFERASE		
XANTHURENIC ACID: UDP GLUCOSYLTRANSFERASE		*I*50 = 0.06%
TROGLITAZONE		
MAP KINASE		*I*50 = 10 μM
TROLEANDOMYCIN		
CYTOCHROME P450 3A		
CYTOCHROME P450 2B		
ESTRADIOL 2/4-HYDROXYLASE		
NITRIC-OXIDE SYNTHASE		26% 100 μM
TROLOX		
LIPOXYGENASE 1	C	*K*i = 18 μM
TROLOX c		
ARACHIDONATE 5-LIPOXYGENASE		*I*50 = 170 μM
CYCLOOXYGENASE		*I*50 = 1 mM
TROMEXAN		
NAD(P)H DEHYDROGENASE (QUINONE)		
TROPOLONE		
D-AMINO-ACID OXIDASE		
1-AMINOCYCLOPROPANE-1-CARBOXYLATE OXIDASE		
CATECHOL O-METHYLTRANSFERASE		*I*50 = 64 μM
CATECHOL OXIDASE		*I*50 = 700 nM
DOPAMINE β-MONOOXYGENASE		
LACTOYLGLUTATHIONE LYASE		*I*50 = 152 μM
MONOPHENOL MONOOXYGENASE	C	*K*i = 220 nM
THIOPURINE S-METHYLTRANSFERASE		
TROPOLONE-3,4-DICARBOXYLIC ACID		
STIPITATONATE DECARBOXYLASE		
TROPONIN C		
PHOSPHORYLASE KINASE		
PROTEIN KINASE C		
TROPONIN I		
MYOSIN ATPASE		
TRP-ALA		
UBIQUITIN-PROTEIN LIGASE		
D-TRP-D-MET-D-FCL-L-GLA-NH_2		
FARNESYL PROTEINTRANSFERASE		*I*50 = 42 nM

TRP-NHOH

AMINOPEPTIDASE HUMAN BLOOD		$Ki = 3\ \mu M$

cyclo[TRP-PHE-D-PRO-PHE-PHE-LYS(Z)]

TRIOSE-PHOSPHATE ISOMERASE		32% 125 µM
TRIOSE-PHOSPHATE ISOMERASE		32% 125 µM
TRIOSE-PHOSPHATE ISOMERASE		32% 125 µM
TRIOSE-PHOSPHATE ISOMERASE		32% 125 µM
TRIOSE-PHOSPHATE ISOMERASE		$I50 = 3\ \mu M$

TRUE CHOLINESTERASE INHIBITOR

ACETYLCHOLINESTERASE

TRYPAN BLUE

HEPARANASE		$I50 = 320\ \mu M$
PHOSPHOGLUCONATE DEHYDROGENASE (DECARBOXYLATING)	C	$Ki = 8.7\ \mu M$
PHOSPHOGLUCONATE DEHYDROGENASE (DECARBOXYLATING)	C	$Ki = 4.7\ \mu M$
PROTEINASE HIV-1		$I50 = 2.6\ \mu M$

TRYPANOSOMA CRUZI PEPTIDE INHIBITOR

H^+-TRANSPORTING ATPASE

TRYPAN RED

SIALIDASE

TRYPSIN

THROMBIN		$Ki = 1.9\ \mu M$

TRYPSIN INHIBITOR II HUMAN SPERMA

ACROSIN

TRYPSIN INHIBITOR P25

TRYPSIN

TRYPSIN-SUBTILISIN INHIBITOR BASIC

SUBTILISIN
TRYPSIN

TRYPSTATIN

CHYMASE		$Ki = 24\ nM$
COAGULATION FACTOR Xa		$Ki = 120\ pM$
TRYPSIN		$Ki = 14\ nM$
TRYPTASE		$Ki = 360\ pM$
TRYPTASE		$Ki = 360\ pM$
TRYPTASE TL-2		92% 10 µM

TRYPTAMINE

AGMATINE DEIMINASE		
AMINE OXIDASE (COPPER-CONTAINING)		
ARYL-ACYLAMIDASE		
BUFURALOL-1′-HYDROXYLASE (CYT P-450dbl)		$I50 = 138\ \mu M$
17-O-DEACETYLVINDOLINE O-ACETYLTRANSFERASE		
β-GLUCOSIDASE	MIX	$Ki = 34\ mM$
HISTAMINE N-METHYLTRANSFERASE		
PYRIDOXAL KINASE		
TRYPTOPHAN 2,3-DIOXYGENASE		
TRYPTOPHAN-tRNA LIGASE	C	

TRYPTOPHAN

ACETYLCHOLINESTERASE		
ALANINE AMINOPEPTIDASE	NC	$I50 = 6.3\ mM$
ALANOPINE DEHYDROGENASE		
ARGINASE		26% 10 mM
ARGININOSUCCINATE SYNTHASE		
CHOLINESTERASE		
β-CYCLOPIAZONATE DEHYDROGENASE		
2-DEHYDRO-3-DEOXYPHOSPHOHEPTONATE ALDOLASE		

DIHYDROPTERIDINE REDUCTASE		
DOPACHROME TAUTOMERASE	C	*I*50 = 2 mM
GLUTAMATE-AMMONIA LIGASE		
[GLUTAMATE-AMMONIA-LIGASE] ADENYLYLTRANSFERASE		
GLUTAMATE DEHYDROGENASE (NADP)		
INDOLE 3-ACETALDEHYDE REDUCTASE (NADPH)		
PHENYLALANINE AMMONIA-LYASE		
1-PYRROLINE-5-CARBOXYLATE DEHYDROGENASE		
SACCHAROPINE DEHYDROGENASE (NADP, L-GLUTAMATE FORMING)		
SACCHAROPINE DEHYDROGENASE (NADP, L-LYSINE FORMING)		
TRITRIRACHIUM ALKALINE PROTEINASE		*K*i = 4 mM
D-TRYPTOPHAN		
INDOLEAMINE-PYRROLE 2,3-DIOXYGENASE	C	
PHENYLALANINE DEHYDROGENASE		
PREPHENATE DEHYDRATASE		
TRYPTOPHANDEHYDROBUTYRIDINE		
GLUTATHIONE TRANSFERASE	C	*K*i = 6 μM
TRYPTOPHAN HYDROXAMIC ACID		
TRYPTOPHAN-tRNA LIGASE	C	
L-TRYPTOPHAN HYDROXAMIC ACID		
PROTEINASE ASTACUS		*I*50 = 450 μM
L-TRYPTOPHAN		
ANTHRANILATE SYNTHASE		
CYSTATHIONINE γ-LYASE		
PHOSPHATASE ALKALINE		
PREPHENATE DEHYDRATASE		
PREPHENATE DEHYDROGENASE		
TRYPTOPHAN 5-MONOOXYGENASE		> 200 μM
TYROSINE 3-MONOOXYGENASE		
TRYPTOPHAN TETRAMATE		
β-CYCLOPIAZONATE DEHYDROGENASE		
TSAA 291		
5α-REDUCTASE	C	*K*i = 1.4 μM
TSAO-T		
RNA-DIRECTED DNA POLYMERASE		
TSETSE SALIVARY GLAND THROMBIN INHIBITOR		
THROMBIN		
TSETSE THROMBIN INHIBITOR		
THROMBIN		K = 584 fM
TSUSHIMYCIN		
PHOSPHO-N-ACETYLMURAMOYL-PENTAPEPTIDE-TRANSFERASE		
TTADFIASGRTGRRNAIHD (DERIVED FROM NATIVE INHIBITOR)		
PROTEIN KINASE A		
TTP		
N-ACETYLNEURAMINATE CYTIDYLYLTRANSFERASE		
CDP-DIACYLGLYCEROL-INOSITOL 3-PHOSPHATIDYLTRANSFERASE		
dCMP DEAMINASE	NC	*K*i = 100 μM
FLAVANONE 7-O-β-GLUCOSYLTRANSFERASE		
GLUCOKINASE		
GLUCOSE-1-PHOSPHATE CYTIDYLYLTRANSFERASE	pC	*K*i = 40 μM
GLUCOSE-1-PHOSPHATE THYMIDYLYLTRANSFERASE	C	*K*i = 100 μM
GLUCOSE-1-PHOSPHATE THYMIDYLYLTRANSFERASE	C	*K*i = 26 μM
GLUCURONOKINASE		
LACTOSE SYNTHASE		
NADH PEROXIDASE		95% 3 mM

Inhibitor / Enzyme	Type	Value
NAD(P) TRANSHYDROGENASE (B-SPECIFIC)		
5′-NUCLEOTIDASE	C	47% 100 µM
PYRUVATE CARBOXYLASE	NC	*K*i = 1.3 mM
RIBOSE-PHOSPHATE PYROPHOSPHOKINASE		
dTDPGLUCOSE 4,6-DEHYDRATASE		
THYMIDINE KINASE	NC	*K*i = 21 µM
URACIL PHOSPHORIBOSYLTRANSFERASE		
dTTP		
ARGININOSUCCINATE SYNTHASE		
dCMP DEAMINASE		
dCTP DEAMINASE	NC	
dCTP DEAMINASE		94% 200 µM
NICOTINATE-NUCLEOTIDE PYROPHOSPHORYLASE (CARBOXYLATING)		
REC A PEPTIDASE		
RIBONUCLEOSIDE-DIPHOSPHATE REDUCTASE		*I*50 = 19 µM
dTDP-DIHYDROSTREPTOSE-STREPTIDINE-6-PHOSPHATE DIHYDROSTREPTOSYLTRANSFERASE		46% 1 mM
dTMP KINASE		*I*50 = 690 µM
ddTTP		
RNA-DIRECTED DNA POLYMERASE		*I*50 = 150 µM
RNA-DIRECTED DNA POLYMERASE		98% 50 µM
TUBERCIDIN		
ADENOSINE KINASE		
ADENOSYLHOMOCYSTEINASE		*K*i = 6.8 mM
ADENOSYLHOMOCYSTEINASE		
METHYLENETETRAHYDROFOLATE CYCLOHYDROLASE		
TUBERCIDIN-5′-PHOSPHATE		
AMP NUCLEOSIDASE	C	*K*i = 51 µM
RHODOPSIN KINASE		*K*i = 3.6 µM
TUBERCIDIN-5′-TRIPHOSPHATE		
THREONINE-tRNA LIGASE		
S-TUBERCIDINYLHOMOCYSTEINE		
ADENOSYLHOMOCYSTEINE NUCLEOSIDASE		*K*i = 1.9 µM
DNA (CYTOSINE-5-)-METHYLTRANSFERASE		
mRNA(NUCLEOSIDE-2′O-)-METHYLTRANSFERASE		
S-TUBERCIDINYL-L-HOMOCYSTEINE		
THIOL S-METHYLTRANSFERASE	C	*K*i = 60 µM
S-TUBERCIDINYLMETHIONINE		
SPERMIDINE SYNTHASE		
SPERMINE SYNTHASE		
S-TUBERCIDINYL-L-METHIONINE		
THIOL S-METHYLTRANSFERASE	C	*K*i = 95 µM
TUBOCURARE		
HISTAMINE N-METHYLTRANSFERASE		
TUBOCURARINE		
ACETYLCHOLINESTERASE		*I*50 = 500 µM
CHOLINESTERASE		*I*50 = 5.6 mM
D-TUBOCURARINE		
HISTAMINE N-METHYLTRANSFERASE	C	*K*i = 8.2 µM
TUBULIN PEPTIDE T37 + 5		
FRUCTOSE-BISPHOSPHATE ALDOLASE		
TUMOR ASSOCIATED TRYPSIN INHIBITOR		
TRYPSIN		

TUNA MUSCLE ACE INHIBITOR
PEPTIDYL-DIPEPTIDASE A — $I50$ = 900 nM

TUNGSTATE
ARYLSULFATASE — C — Ki = 640 μM
2-FUROYL-CoA DEHYDROGENASE
GLUCOSE-6-PHOSPHATASE — Kis = 11 μM
PHOSPHATASE ACID — C — Ki = 230 μM
PHOSPHATASE ACID — C — Ki = 9.2 nM
PHOSPHATASE ALKALINE — C — Ki = 6.8 μM
PHOSPHOGLUCONATE DEHYDROGENASE (DECARBOXYLATING)
PROTEIN TYROSINE-PHOSPHATASE
PYROPHOSPHATE-FRUCTOSE-6-PHOSPHATE 1-PHOSPHOTRANSFERASE

5-TUNGSTO-2-ANIMONIATE
RNA-DIRECTED DNA POLYMERASE

TUNICAMYCIN
CELLULOSE SYNTHASE (UDP FORMING)
DOLICHYL-PHOSPHATE β-D-MANNOSYLTRANSFERASE
FERROXIDASE
GLUCOSE TRANSPORT
LACTOSE SYNTHASE
PHOSPHATIDYLCHOLINE-STEROL O-ACYLTRANSFERASE
PHOSPHO-N-ACETYLMURAMOYL-PENTAPEPTIDE-TRANSFERASE — $I50$ = 12 μg/ml
PROTEIN PALMITOYLTRANSFERASE
UDP-N-ACETYLGLUCOSAMINE-DOLICHYL-PHOSPHATE N-ACETYLGLUCOSAMINEPHOSPHOTRANSFERA — $I50$ = 2-8ng/ml
UDP-GLUCOSE TRANSGLUCOSYLASE — $I50$ = 1 μg/ml

TURANOSE
β-FRUCTOFURANOSIDASE
α-GLUCOSIDASE — C — Ki = 11 mM
α-GLUCOSIDASE

TURKEY OVOMUCOID
CHYMOTRYPSIN
TRYPSIN

TURKEY OVOMUCOID THIRD DOMAINE P1 GLU18
PROTEINASE STREPTOMYCES GRISEUS (GLU SPECIFIC)

TURMERONOL A
LIPOXYGENASE

TURMERONOL B
LIPOXYGENASE

TUROSTERIDE
STEROID 5α-REDUCTASE — $I50$ = 55 nM
STEROID 5α-REDUCTASE 1 — Ki = 108 nM
STEROID 5α-REDUCTASE 2 — Ki = 22 nM

TURQUOISE H-A
CHOLINE-PHOSPHATE CYTIDYLYLTRANSFERASE

TURQUOISE MX-G
CHOLINE-PHOSPHATE CYTIDYLYLTRANSFERASE

TWEEN
7α-HYDROXYSTEROID DEHYDROGENASE

TWEEN 20
1-ACYLGLYCEROPHOSPHOCHOLINE O-ACYLTRANSFERASE
β-ADRENERGIC-RECEPTOR KINASE
DIACYLGLYCEROL O-ACYLTRANSFERASE
DIACYLGLYCEROL CHOLINEPHOSPHOTRANSFERASE
ETHANOLAMINEPHOSPHOTRANSFERASE
FATTY ACID CoA REDUCTASE — 100% 400 μM

GLOBOSIDE α-N-ACETYLGALACTOSAMINYLTRANSFERASE
MYELIN-PROTEOLIPID O-PALMIOYLTRANSFERASE
PHOSPHATIDATE CYTIDILYLTRANSFERASE
STEROL O-ACYLTRANSFERASE

TWEEN 80
1-ACYLGLYCEROPHOSPHOCHOLINE O-ACYLTRANSFERASE
ALCOHOL DEHYDROGENASE (NAD(P))
CHOLESTEROL 7α-MONOOXYGENASE
ECDYSONE 20-MONOOXYGENASE
EPOXIDE HYDROLASE
GANGLIOSIDE GALACTOSYLTRANSFERASE
GLOBOSIDE α-N-ACETYLGALACTOSAMINYLTRANSFERASE
3(or 17)β-HYDROXYSTEROID DEHYDROGENASE
MANNOTETRAOSE 2-α-N-ACETYLGLUCOSAMINYLTRANSFERASE *I*50 = 1.5%
PHENYLALANINE 4-MONOOXYGENASE
PHOSPHATIDYLCHOLINE-DOLICHOL O-ACYLTRANSFERASE
RETINYL-PALMITATE ESTERASE 95% 0.02%
STEROL O-ACYLTRANSFERASE
TREHALOSE O-MYCOLYLTRANSFERASE

TWEEN 20X
PHOSPHATIDATE PHOSPHATASE 15% 3 mg/ml

TY 11345
H^+/K^+-EXCHANGING ATPASE *I*50 = 9.9 μM
Na^+/K^+-EXCHANGING ATPASE

(TYE-ALA-GLU)$_9$
PROTEIN TYROSINE KINASE pp60c-Src

(TYR-ALA-GLU)n
PROTEIN TYROSINE KINASE (INSULIN RECEPTOR)

TYR-ALA-GLY-ALA-VAL-VAL-ASN-ASP-LEU
RIBONUCLEOSIDE-DIPHOSPHATE REDUCTASE
RIBONUCLEOTIDE REDUCTASE

TYR-ALA-GLY-ALA-VAL-VAL-ASP-ASN-LEU
RIBONUCLEOSIDE-DIPHOSPHATE REDUCTASE

TYR-ALA-LEU-PRO-HIS-ALA
PEPTIDYL-DIPEPTIDASE A *I*50 = 9.8 μM

TYR-ALA-LYS-ARG-CH_2CL
PROTEINASE PORPHYROMONAS GINGIVALIS

TYRAMINE
ARYLSULFATASE
DIHYDROPTERIDINE REDUCTASE
HISTAMINE N-METHYLTRANSFERASE
PYRIDOXAL KINASE
RENILLA-LUCIFERIN 2-MONOOXYGENASE
TYROSINE-tRNA LIGASE C *K*i = 6 μM

TYR-ARG
DIPEPTIDYL-PEPTIDASE 3 *K*i = 25 μM

D-TYR-ARG-LYS-ARG-β-CYCLOHEXYLALANYL-VAL-GLN-LYS-ASP
PRO-OPIOMELANOCORTIN CONVERTING ENZYME *K*i = 4.4 nM

D-TYR-ARG-SER-LYS-ARG-β-ALA-VAL-GLN-LYS-ASP
FURIN *K*i = 800 nM
PRO-OPIOMELANOCORTIN CONVERTING ENZYME *K*i = 8.7 μM

D-TYR-ARG-SER-LYS-ARG-γ-AMINOBUTYRYL-VAL-GLN-LYS-ASP
FURIN *K*i = 2.2 μM
PRO-OPIOMELANOCORTIN CONVERTING ENZYME *K*i = 1.1 μM

D-TYR-ARG-SER-LYS-ARG-β-CYCLOHEXYLALANYL-VAL-GLN-LYS-ASP
FURIN | | *K*i = 1.3 μM

(TYR-GLU-ALA-GLY)n
PROTEIN TYROSINE KINASE

TYR-GLY-ALA-VAL-VAL-ASN-ASP-LEU
RIBONUCLEOTIDE REDUCTASE

TYR-GLY-ARG-CH_2Cl
MULTICATALYTIC ENDOPEPTIDASE COMPLEX (TRYPSIN LIKE ACTIVITY)

TYR-LEU
PEPTIDYL-DIPEPTIDASE A | C | *I*50 = 82 μM

D-TYR-LYS-GLU-ARG-SER-LYS-ARG-AIB-VAL-GLN-LYS-APS
FURIN | | *K*i = 8.7 μM
PRO-OPIOMELANOCORTIN CONVERTING ENZYME 1 | | *K*i = 1.2 μM

TYR-NHOH
AMINOPEPTIDASE HUMAN BLOOD | | *K*i = 40 μM
BACTERIAL LEUCYL AMINOPEPTIDASE | | *K*i = 83 μM
LEUCYL AMINOPEPTIDASE | | *K*i = 16 μM
MEMBRANE ALANINE AMINOPEPTIDASE | | *K*i = 19 μM

TYROMYCIN
CARBOXYPEPTIDASE A | | *I*50 = 60 μg/ml
LEUCYL AMINOPEPTIDASE | | *I*50 = 31 μg/ml
MEMBRANE ALANINE AMINOPEPTIDASE | | *I*50 = 41 μg/ml

TYROSINASE
β-FRUCTOFURANOSIDASE

TYROSINE
CHORISMATE MUTASE 1 | | 55% 170 μM
CHORISMATE MUTASE 3 | | 50% 170 μM
2-DEHYDRO-3-DEOXYPHOSPHOHEPTONATE ALDOLASE | C | *K*i = 900 nM
3-DEOXY-D-ARABINO-HEPTULOSONATE-7-PHOSPHATE SYNTHASE | | *K*i = 900 nM
DIHYDROPTERIDINE REDUCTASE
LEUCYL AMINOPEPTIDASE | C | *K*i = 13 mM
PHENYLALANINE AMMONIA-LYASE | | 54% 2 mM
PHENYLALANINE 4-MONOOXYGENASE
PREPHENATE DEHYDROGENASE
PREPHENATE DEHYDROGENASE (NADP)
PYRUVATE KINASE
TRITRIRACHIUM ALKALINE PROTEINASE | | *K*i = 860 μM
TRYPTOPHAN 5-MONOOXYGENASE | NC | *K*i = 19 μM

TYROSINE AMIDE
TYROSINE-tRNA LIGASE | C | *K*i = 8.1 μM

D-TYROSINE
PHENYLALANINE DEHYDROGENASE
PREPHENATE DEHYDROGENASE
PRETYROSINE DEHYDROGENASE

L-TYROSINE HYDROXAMIC ACID
PROTEINASE ASTACUS | | *I*50 = 174 μM

L-TYROSINE
CHORISMATE MUTASE | | *K*i = 31 μM
HISTIDINE AMMONIA-LYASE | | *K*i = 1.7 mM
PHENYLALANINE AMMONIA-LYASE | C | *K*i = 1.3 mM
PREPHENATE DEHYDROGENASE
PRETYROSINE DEHYDROGENASE
TYROSINE 3-MONOOXYGENASE | | > 100 μM

Inhibitor / Enzyme	Type	Value
m-TYROSINE		
AROMATIC-L-AMINO-ACID DECARBOXYLASE		
TYROSINE METHYL ESTER		
TYROSINE-tRNA LIGASE	C	*K*i = 17 μM
TYROSINE-4O-SULPHATE		
TYROSINE-ESTER SULFOTRANSFERASE	U	
L-TYROSINOL-AMP		
TYROSINE-tRNA LIGASE	C	*K*i = 29 nM
L-TYROSINOL		
TYROSINE-tRNA LIGASE	C	*K*i = 4.1 μM
TYROSINYL-ALANINE		
PEPTIDYL-DIPEPTIDASE A		*K*i = 60 μM
PROTEIN TYROSINE KINASE		
TYROSINYL-AMP		
TYROSINE-tRNA LIGASE		
TYROSTATIN		
PROTEINASE CARBOXYL (PEPSTATIN INSENSITIV)	C	*K*i 2.6 nM
PROTEINASES CARBOXYL		
TYROXINE		
ACYLPHOSPHATASE		
15-HYDROXYPROSTAGLANDIN DEHYDROGENASE (NAD)		
TYRPHOSTIN		
PROTEIN KINASE C		*I*50 = 114 μM
PROTEIN TYROSINE KINASE EGFR		*I*50 = 21 μM
TYRPHOSTIN 8		
GTP-ASE TRANSDUCIN		*I*50 = 45 μM
TYRPHOSTIN 23		
GTP-ASE TRANSDUCIN		*I*50 = 10 μM
GUANYLATE CYCLASE		*I*50 = 26 μM
6-PHOSPHOFRUCTOKINASE		*I*50 = 6.9 μM
TYRPHOSTIN 24		
PROTEIN TYROSINE KINASE EGFR		*I*50 = 6 μM
TYRPHOSTIN 25		
GTP-ASE TRANSDUCIN		*I*50 = 7 μM
PROTEIN TYROSINE KINASE		*I*50 = 150 μM
TYRPHOSTIN 46		
PROTEIN TYROSINE KINASE EGFR		*I*50 = 10 μM
PROTEIN TYROSINE KINASE EGFR		*I*50 = 20 μM
PROTEIN TYROSINE KINASE pp60c-Src		*I*50 = 1 mM
TYRPHOSTIN 47		
GTP-ASE TRANSDUCIN		*I*50 = 5 μM
TYRPHOSTIN 50		
PROTEIN TYROSINE KINASE		*I*50 = 100 μM
TYRPHOSTIN 537		
INTEGRASE		*I*50 = 1.9 μM
TYRPHOSTINE		
1-PHOSPHATIDYLINOSITOL 3-KINASE		*I*50 = 33 μM
TYRPHOSTIN RG 50864		
PROTEIN TYROSINE KINASE EGFR		*K*i = 850 μM

TYRPHOSTINS

PROTEIN KINASE A		$I50 > 1$ mM
PROTEIN KINASE C		$I50 > 1$ mM
PROTEIN TYROSINE KINASE EGFR		$I50 = 35$ μM
PROTEIN TYROSINE KINASE (INSULIN RECEPTOR)		$I50 = 1.2$ mM

TYR-TYR

DIPEPTIDYL-PEPTIDASE 3		$Ki = 5.8$ μM

TYR-TYR-(GLU)$_9$

PROTEIN KINASE CASEIN KINASE II	C	$Ki = 150$ μM

TZC 5665

3′,5′-CYCLIC-NUCLEOTIDE PHOSPHODIESTERASE III		$I50 = 210$ nM
3′,5′-CYCLIC-NUCLEOTIDE PHOSPHODIESTERASE IV		$I50 = 138$ μM

TZI

ARACHIDONATE 5-LIPOXYGENASE		

TZI 41127

ARACHIDONATE 5-LIPOXYGENASE		$I50 = 1.8$ μM
LIPOXYGENASE		$I50 = 19$ μM

TZP 4238

3α-HYDROXYSTEROID OXIDOREDUCTASE	C	$Ki = 54$ μM
STEROID 5α-REDUCTASE		$I50 = 200$ μM

U

U^{2+}

ADENYLYLSULFATE KINASE		

U 0126

MEK 1		$I50 = 72$ nM
MEK 2		$I50 = 58$ nM

U 0521

CATECHOL O-METHYLTRANSFERASE		$I50 = 4.9$ μM

U 103017

PROTEINASE HIV-1		$Ki = 800$ pM
PROTEINASE HIV-2		$Ki = 3.2$ nM

U 104489

RNA-DIRECTED DNA POLYMERASE		

U 140690

CATHEPSIN D		$Ki = 15$ μM
CATHEPSIN E		$Ki = 9$ μM
PEPSIN		$Ki = 2$ μM
PROTEINASE HIV-1		$Ki = 8$ pM
PROTEINASE HIV-2		$Ki = 1$ nM

U 23469

CYTOCHROME P450 3A1/2		$I50 = 133$ μM
CYTOCHROME P450 2B1		$I50 = 181$ μM
CYTOCHROME P450 2C11		$I50 = 168$ μM

U 31355

RNA-DIRECTED DNA POLYMERASE		$I50 = 56$ μM

U 3585

ARACHIDONATE 5-LIPOXYGENASE		$I50 = 9.5$ μM
PHOSPHOLIPASE A2		$I50 = 5.5$ μM
PHOSPHOLIPASE A2		$I50 = 4.1$ μM

U 46619
THROMBOXANE A SYNTHASE

U 56497
CASPASE-1 $I50 = 630\ \mu M$

U 57908
DIACYLGLYCEROL LIPASE $I50 = 4\ \mu M$
DIACYLGLYCEROL LIPASE NC $I50 = 3.5\ \mu M$
DIACYLGLYCEROL LIPASE $I50 = 4\ \mu M$
LIPOPROTEIN LIPASE $I50 = 1.1\ \mu M$
LIPOPROTEIN LIPASE $I50 = 1.5\ \mu M$

U 63557a
THROMBOXANE A SYNTHASE C

U 66858
ARACHIDONATE 5-LIPOXYGENASE $I50 = 1.1\ \mu M$

U 67154
3′,5′-CYCLIC-NUCLEOTIDE PHOSPHODIESTERASE

U 68244
ARACHIDONATE 5-LIPOXYGENASE $I50 = 820$ nM

U 73122
1-PHOSPHATIDYLINOSITOL-4,5-BISPHOSPHATE PHOSPHODIESTERASE $I50 = 101\ \mu M$
PHOSPHOLIPASE C
PHOSPHOLIPASE D $I50 = 78\ \mu M$

U 75875
PROTEINASE HIV-1 $Ki < 1$ nM
PROTEINASE HIV-2

U 75878
PROTEINASE HIV-1 $Ki < 1$ nM

U 78036
RNA-DIRECTED DNA POLYMERASE

U 817492
PROTEINASE HIV-1 $I50 = 73$ nM

U 84569
3′,5′-CYCLIC-NUCLEOTIDE PHOSPHODIESTERASE $I50 = 400$ nM
1-PHOSPHATIDYLINOSITOL-4,5-BISPHOSPHATE PHOSPHODIESTERASE $I50 = 800$ nM

U 86983
3′,5′-CYCLIC-NUCLEOTIDE PHOSPHODIESTERASE $I50 = 2\ \mu M$

U 88352
RNA-DIRECTED DNA POLYMERASE $I50 = 210$ nM

U 90152
RNA-DIRECTED DNA POLYMERASE

U 92163
PROTEINASE HIV-2

U 96988
CATHEPSIN D 52% 10 μM
CATHEPSIN E 72% 10 μM
PEPSIN 53% 10 μM
PROTEINASE HIV-1 $I50 = 38$ nM
PROTEINASE HIV-2 $I50 = 32$ nM

U 98017
MAP KINASE

U 9843
RNA-DIRECTED DNA POLYMERASE NC $Ki = 630$ nM

U 10029A
PHOSPHOLIPASE A2 REV $I50 = 250$ μM

U 18666A
CYCLOARTENOL SYNTHASE $I50 = 250$ nM
LANOSTEROL SYNTHASE $I50 = 800$ nM
Δ24 REDUCTASE $Ki = 157$ nM

U 19451 A
NITRIC-OXIDE SYNTHASE I $I50 = 54$ μM
NITRIC-OXIDE SYNTHASE II $I50 = 14$ μM

U 36557A
THROMBOXANE A SYNTHASE

U 74500A
NADPH OXIDASE SUPEROXIDE GENERATING

U 84645E
PROTEINASE HIV-1 C $Ki = 3.5$ μM

U 85549E
PROTEINASE HIV-1 C $Ki = 3.7$ μM

U 87201E
RNA-DIRECTED DNA POLYMERASE

U 88204E
RNA-DIRECTED DNA POLYMERASE

U 90152E
RNA-DIRECTED DNA POLYMERASE

U 78518F
NADPH OXIDASE SUPEROXIDE GENERATING

U 75875R
PROTEINASE HIV-1 $I50 < 1$ nM

U 90152S
RNA-DIRECTED DNA POLYMERASE $I50 = 1.1$ μM

348U87
RIBONUCLEOSIDE-DIPHOSPHATE REDUCTASE IR

1592U89
RNA-DIRECTED DNA POLYMERASE

1842U89
THYMIDYLATE SYNTHASE NC $Ki = 90$ pM

1843U89
THYMIDYLATE SYNTHASE $Ki = 90$ pM

1954U89
DIHYDROFOLATE REDUCTASE $Ki = 1.4$ pM

UBENIMEX
LEUCYL AMINOPEPTIDASE $I50 = 16$ μg/ml

UBIQUINONE 0
ACETOLACTATE SYNTHASE I $Ki = 42$ μM
ACETOLACTATE SYNTHASE II $Ki = 110$ μM
ACETOLACTATE SYNTHASE III $Ki = 78$ μM
GLUCOSE TRANSPORT C $Ki = 860$ μM

UBIQUINONE 5

ACETOLACTATE SYNTHASE I		$Ki = 330\ \mu M$
ACETOLACTATE SYNTHASE II		$Ki = 660\ \mu M$
ACETOLACTATE SYNTHASE III		$Ki = 440\ \mu M$

UBIQUINONE 9

γ-GLUTAMYL CARBOXYLASE		$I50 = 25\ \mu M$

UBIQUINONE 10

γ-GLUTAMYL CARBOXYLASE		$I50 = 36\ \mu M$
PHOSPHOLIPASE A2		33% 10 μM

UC 38

RNA-DIRECTED DNA POLYMERASE		$I50 = 800$ nM

UC 84

RNA-DIRECTED DNA POLYMERASE		$I50 = 4.3\ \mu M$

UC 781

RNA-DIRECTED DNA POLYMERASE

UCF1 A

FARNESYL PROTEINTRANSFERASE

UCF1 B

FARNESYL PROTEINTRANSFERASE

UCN 01

CYCLIN DEPENDENT KINASE 1		$I50 = 31$ nM
CYCLIN DEPENDENT KINASE 2		$I50 = 42$ nM
CYCLIN DEPENDENT KINASE 2		$I50 = 30$ nM
CYCLIN DEPENDENT KINASE 4		$I50 = 32$ nM
CYCLIN DEPENDENT KINASE 4		$I50 = 32$ nM
CYCLIN DEPENDENT KINASE 6		$I50 = 58$ nM
MAP KINASE		$I50 = 910$ nM
PROTEIN KINASE A		$I50 = 42$ nM
PROTEIN KINASE C		$I50 = 4.1$ nM
PROTEIN KINASE Cβ		$Ki = 25$ nM
PROTEIN KINASE Cβ		$I50 = 530$ nM
PROTEIN KINASE Cα		$I50 = 29$ nM
PROTEIN KINASE Cα		$Ki = 440$ pM
PROTEIN KINASE Cβ		$I50 = 34$ nM
PROTEIN KINASE Cβ		$Ki = 1.7$ nM
PROTEIN KINASE Cδ		$I50 = 590$ nM
PROTEIN KINASE Cδ		$Ki = 20$ nM
PROTEIN KINASE Cγ		$I50 = 30$ nM
PROTEIN KINASE Cγ		$Ki = 940$ pM
PROTEIN KINASE C (CYTOSOLIC)		$I50 = 18$ nM
PROTEIN KINASE C (MEMBRANE)		$I50 = 3.2$ nM
PROTEIN KINASE Cn		$Ki = 21$ nM
PROTEIN KINASE Cxi		$Ki = 3.8\ \mu M$

UCN 02

PROTEIN KINASE A		$I50 = 250$ nM
PROTEIN KINASE C		$I50 = 62$ nM
PROTEIN KINASE Cβ		$I50 = 1.2\ \mu M$
PROTEIN KINASE Cα		$I50 = 530$ nM
PROTEIN KINASE Cβ		$I50 = 700$ nM
PROTEIN KINASE Cδ		$I50 = 2.8\ \mu M$

UCN 1028A

PROTEIN KINASE C

UCSF 8

PEPSIN		$I50 = 700\ \mu M$
PROTEINASE HIV-1	C	$Ki = 15\ \mu M$
PROTEINASE HIV-2		$Ki = 100\ \mu M$
RENIN		$I50 = 1.4$ mM

UCSF 191

PROTEINASE HIV-1	IR	Ki = 5.7 μM
PROTEINASE HIV-2	IR	Ki = 156 μM

UD-CG 115

3′,5′-CYCLIC-NUCLEOTIDE PHOSPHODIESTERASE		
Na^+/K^+-EXCHANGING ATPASE		

UD-CG 212 Cl

3′,5′-CYCLIC-NUCLEOTIDE PHOSPHODIESTERASE (cycAMP)		*I*50 = 107 μM
3′,5′-CYCLIC-NUCLEOTIDE PHOSPHODIESTERASE (cycGMP)		*I*50 = 99 μM
3′,5′-CYCLIC-NUCLEOTIDE PHOSPHODIESTERASE I		*I*50 = 128 μM
3′,5′-CYCLIC-NUCLEOTIDE PHOSPHODIESTERASE II		*I*50 = 203 μM
3′,5′-CYCLIC-NUCLEOTIDE PHOSPHODIESTERASE III		*I*50 = 190 nM
3′,5′-CYCLIC-NUCLEOTIDE PHOSPHODIESTERASE I		*I*50 = 175 μM
3′,5′-CYCLIC-NUCLEOTIDE PHOSPHODIESTERASE II		*I*50 = 181 μM
3′,5′-CYCLIC-NUCLEOTIDE PHOSPHODIESTERASE III		*I*50 = 50 nM
3′,5′-CYCLIC-NUCLEOTIDE PHOSPHODIESTERASE IV		*I*50 = 41 μM

UDP

N-ACETYLGLUCOSAMINYLDIPHOSPHODOLICHOL N-ACETYLGLUCOSAMINYLTRANSFERASE		45% 1 mM
N-ACETYLGLUCOSAMINYLDIPHOSPHOUNDECAPRENOL N-ACETYL-β-D-MANNOSAMINYLTRANSFERAS		
N-ACETYLLACTOSAMINE SYNTHASE	C	
ADENYLOSUCCINATE SYNTHASE		
ALDOSE-1-PHOSPHATE NUCLEOTIDYLTRANSFERASE	C	Ki = 46 μM
ARYLAMINE GLUCOSYLTRANSFERASE		
CARBAMOYL-PHOSPHATE SYNTHASE (GLUTAMINE-HYDROLYSING)		
CDP-DIACYLGLYCEROL-INOSITOL 3-PHOSPHATIDYLTRANSFERASE		
CELLULOSE SYNTHASE (UDP FORMING)		
CHITIN SYNTHASE		77% 5 mM
CHITOBIOSYLDIPHOSPHODOLICHOL α-MANNOSYLTRANSFERASE		62% 1 mM
CMP-N-ACYLNEURAMINATE PHOSPHODIESTERASE		76% 860 μM
DEOXYCYTIDINE KINASE		
DEOXYGUANOSINE KINASE	C	Ki = 3 μM
1,2-DIACYLGLYCEROL 3-β-GALACTOSYLTRANSFERASE		
DOLICHYL-PHOSPHATE β-GLUCOSYLTRANSFERASE		
DOLICHYL-PHOSPHATE D-XYLOSYLTRANSFERASE		
FLAVONE APIOSYLTRANSFERASE		
FRUCTOKINASE		
FRUCTOSE-BISPHOSPHATASE		
FUCOSYLGALACTOSE α-N-ACETYLGALACTOSAMINYLTRANSFERASE		84% 50 μM
FUCOSYLGLYCOPROTEIN 3α-GALACTOSYLTRANSFERASE		
GALACTOLIPID GALACTOSYLTRANSFERASE		*I*50 = 8 mM
GALACTOSYLGALACTOSYLGLUCOSYLCERAMIDE β-D-ACETYLGALACTOSAMINYLTRANSFERASE		
GALLATE 1-β-GLUCOSYLTRANSFERASE		
GLOBOSIDE α-N-ACETYLGALACTOSAMINYLTRANSFERASE		
1,3-β-GLUCAN SYNTHASE	C	Ki = 100 μM
1,3-β-GLUCAN SYNTHASE	C	Ki = 1 mM
GLUCURONOKINASE		
sn-GLYCEROL-3-PHOSPHATE 2-α-GALACTOSYLTRANSFERASE		
GLYCOGENIN GLUCOSYLTRANSFERASE		78% 100 μM
GLYCOGEN (STARCH) SYNTHASE	C	
HEXOKINASE		
4-HYDROXYBENZOATE 4O-β-D-GLUCOSYLTRANSFERASE		
HYDROXYCINNAMATE 4-β-GLUCOSYLTRANSFERASE		
myo-INOSITOL OXYGENASE		
LACTOSE SYNTHASE		
LUTEOLIN-7O-DIGLUCURONIDE 4′-O-GLUCURONOSYLTRANSFERASE		
LUTEOLIN-7O-GLUCURONIDE 7-O-GLUCURONOSYLTRANSFERASE		
α-MANNOSIDE β1-2-N-ACETYLGLUCOSAMINYLTRANSFERASE		Ki = 45 μM
α-1,3-MANNOSYL-GLYCOPROTEIN β-1,2-N-ACETYLGLUCOSAMINYLTRANSFERASE		
MANNOSYL-GLYCOPROTEIN β-1,4-N-ACETYLGLUCOSAMINYLTRANSFERASE		
MANNOTETRAOSE 2-α-N-ACETYLGLUCOSAMINYLTRANSFERASE		*I*50 = 400 μM
[3-METHYL-2-OXOBUTANOATE DEHYDROGENASE (LIPOAMIDE)] PHOSPHATASE		*I*50 = 250 μM
NUATIGENIN 3-β-GLUCOSYLTRANSFERASE		

Enzyme	Type	Value
5′-NUCLEOTIDASE	C	58% 100 µM
OLIGONUCLEOTIDASE		39% 200 µM
OROTATE PHOSPHORIBOSYLTRANSFERASE		
PHOSPHOGLUCOMUTASE	C	
PHOSPHOPOLYPRENOL GLUCOSYLTRANSFERASE		
POLYGALACTURONATE 4-α-GALACTURONOSYLTRANSFERASE		
POLYPEPTIDE N-ACETYLGALACTOSAMINYLTRANSFERASE		
POLYRIBONUCLEOTIDE NUCLEOTIDYLTRANSFERASE	C	*K*i = 500 µM
POLYRIBONUCLEOTIDE NUCLEOTIDYLTRANSFERASE	C	*K*i = 2.5 mM
PROCOLLAGEN GALACTOSYLTRANSFERASE		
PROCOLLAGEN GLUCOSYLTRANSFERASE		
REC A PEPTIDASE		
RIBONUCLEOSIDE-DIPHOSPHATE REDUCTASE		*K*i = 29 µM
SARSAPOGENIN 3-β-GLUCOSYLTRANSFERASE		
SINAPATE 1-GLUCOSYLTRANSFERASE		
STEROL 3β-GLUCOSYLTRANSFERASE		
SUCROSE-PHOSPHATE SYNTHASE	C	*K*i = 3.6 mM
SUCROSE SYNTHASE	SUB	*K*i = 7.6 mM
SUCROSE SYNTHASE		
α,α-TREHALOSE-PHOSPHATE SYNTHASE		
UDP-N-ACETYLGLUCOSAMINE-DOLICHYL-PHOSPHATE N-ACETYLGLUCOSAMINEPHOSPHOTRANSFERA		
UDP-N-ACETYLGLUCOSAMINE-LYSOSOMAL-ENZYME N-ACETYLGLUCOSAMINEPHOSPHOTRANSFERASE		
UDP-GLUCOSE 4-EPIMERASE		*K*i = 35 µM
UDP-GLUCOSE SIALIC ACID 3-O-GLUCOSYLTRANSFERASE		*K*i = 250 µM
UDP-GLUCURONATE DECARBOXYLASE	C	*K*i = 200 µM
UDP-GLUCURONATE DECARBOXYLASE	C	*K*i = 200 µM
UDP-GLUCURONATE 5′-EPIMERASE		
UDP-GLUCURONOSYLTRANSFERASE	C	*K*i = 4.4 µM
UDP-GLUCURONOSYLTRANSFERASE	NC	*K*i = 13 µM
URACIL PHOSPHORIBOSYLTRANSFERASE		
XANTHURENIC ACID: UDP GLUCOSYLTRANSFERASE		98% 5 mM
1,4-β-D-XYLAN SYNTHASE		

UDP-N-ACETYLGLUCOSAMINE

Enzyme	Type	Value
N-ACETYLGLUCOSAMINE KINASE	NC	
N-ACETYLGLUCOSAMINE-1-PHOSPHODIESTER N-ACETYLGLUCOSAMINIDASE		72% 5 mM
CMP-N-ACYLNEURAMINATE PHOSPHODIESTERASE		44% 1 mM
GLUCOSAMINE-FRUCTOSE-6-PHOSPHATE AMINOTRANSFERASE (ISOMERIZING)		
GLUTAMINE-FRUCTOSE-6-PHOSPHATE AMINOTRANSFERASE (ISOMERIZING)	C	*K*i = 5 µM
GLYCOGENIN GLUCOSYLTRANSFERASE		89% 100 µM
UDP-GLUCURONOSYLTRANSFERASE		

UDP-N-ACETYLGLUCOSAMINE-6-PHOSPHATE

Enzyme	Type	Value
N-ACETYLGLUCOSAMINE KINASE	NC	

UDP-N-ACETYLMURAMIC ACID-L-ALA

Enzyme	Type	Value
UDP-N-ACETYLGLUCOSAMINE 1-CARBOXYVINYLTRANSFERASE		

UDP-N-ACETYLMURAMIC ACID

Enzyme	Type	Value
UDP-N-ACETYLGLUCOSAMINE 1-CARBOXYVINYLTRANSFERASE		

UDP-N-ACETYLMURAMYL-ALA-D-GLU-MESO-2,6-DIAMONOPIMELIC ACID-D-ALA-D-ALA

Enzyme	Type	Value
UDP-N-ACETYLGLUCOSAMINE 1-CARBOXYVINYLTRANSFERASE	C	*K*i = 2.5 mM

UDP-N-ACETYLMURAMYL-L-ALA-D-GLU-MESO-2,6-DIAMONOPIMELIC ACID

Enzyme	Type	Value
UDP-N-ACETYLGLUCOSAMINE 1-CARBOXYVINYLTRANSFERASE	C	*K*i = 1.5 mM

UDP-ARABINBOSE

Enzyme	Type	Value
UDP-GLUCOSE 6-DEHYDROGENASE		

dUDP

Enzyme	Type	Value
CARBAMOYL-PHOSPHATE SYNTHASE (GLUTAMINE-HYDROLYSING)		
DEOXYCYTIDINE KINASE		
5′-NUCLEOTIDASE	C	*I*50 = 100 µM
THYMIDINE-TRIPHOSPHATASE		*K*i = 46 µM
URACIL PHOSPHORIBOSYLTRANSFERASE		

Inhibitor / Enzyme		
UDP-6-DEOXYGALACTOSE		
UDP-GLUCOSE 4-EPIMERASE		
UDP-4-DEOXY-D-GLUCOSE		
UDP-GLUCOSE 4-EPIMERASE		Ki = 440 μM
UDP-DIALDEHYDE		
LACTOSYLCERAMIDE α-2,3-SIALYLTRANSFERASE		Ki = 220 μM
LACTOSYLCERAMIDE α-2,3-SIALYLTRANSFERASE		Ki = 29 μM
UDP-GALACTOSE		
(N-ACETYLNEURAMINYL)-GALACTOSYLGLUCOSYLCERAMIDE N-ACETYLGALACTOSAMINYLTRANSFER		
CMP-N-ACYLNEURAMINATE PHOSPHODIESTERASE		35% 1 mM
FUCOSYLGALACTOSE α-N-ACETYLGALACTOSAMINYLTRANSFERASE		
β-GALACTOSIDE α-2,3-SIALYLTRANSFERASE	C	Ki = 560 μM
GANGLIOSIDE GALACTOSYLTRANSFERASE		
GLYCOGENIN GLUCOSYLTRANSFERASE		85% 100 μM
UDP-GLUCOSE-HEXOSE-1-PHOSPHATE URIDYLYLTRANSFERASE		
UDP-GLUCOSE-1-PHOSPHATE URIDYLYLTRANSFERASE		
UDP-GALACTURONATE		
PROCOLLAGEN GALACTOSYLTRANSFERASE		
UDP-GALACTURONIC ACID		
CHITIN SYNTHASE		
CMP-N-ACYLNEURAMINATE PHOSPHODIESTERASE		79% 2.9 mM
UDP-GLUCOSE 6-DEHYDROGENASE		
UDP-GLC NAC		
β-GALACTOSIDE α-2,3-SIALYLTRANSFERASE		
o-UDP-GLCNAC ALDEHYDE		
METHYLCOENZYM M METHYLREDUCTASE		80% 220 μM
UDP-GLUCOSE		
N-ACETYLGLUCOSAMINE-1-PHOSPHODIESTER N-ACETYLGLUCOSAMINIDASE		25% 5 mM
N-ACETYLGLUCOSAMINYLDIPHOSPHODOLICHOL N-ACETYLGLUCOSAMINYLTRANSFERASE		45% 1 mM
ATP PYROPHOSPHATASE		
CARBAMOYL-PHOSPHATE SYNTHASE (GLUTAMINE-HYDROLYSING)		
CMP-N-ACYLNEURAMINATE PHOSPHODIESTERASE		70% 2.9 mM
β-GALACTOSIDE α-2,3-SIALYLTRANSFERASE		
GLUCOSE-1-PHOSPHATE ADENYLYLTRANSFERASE		
MALTOSE SYNTHASE		
PHOSPHORYLASE		
SUCROSE SYNTHASE		
α,α-TREHALOSE-PHOSPHATE SYNTHASE		
UDP-N-ACETYLGLUCOSAMINE 2-EPIMERASE		98% 360 μM
UDP-N-ACETYLGLUCOSAMINE-LYSOSOMAL-ENZYME N-ACETYLGLUCOSAMINEPHOSPHOTRANSFERASE		
UDP-GLUCOSE-HEXOSE-1-PHOSPHATE URIDYLYLTRANSFERASE		
UDP-GLUCOSE-1-PHOSPHATE URIDYLYLTRANSFERASE	C	
UDP-GLUCURNATE 4-EPIMERASE		100% 3 mM
UDP-GLUCURONATE DECARBOXYLASE	C	Ki = 2 mM
UDP-GLUCURONATE DECARBOXYLASE	C	Ki = 2 mM
UDP-GLUCURONATE		
PROCOLLAGEN GALACTOSYLTRANSFERASE		
UDP-GLUCOSE 6-DEHYDROGENASE	C	Ki = 500 μM
UDP-GLUCURONIC ACID		
CHITIN SYNTHASE		
CMP-N-ACYLNEURAMINATE PHOSPHODIESTERASE		76% 860 μM
1,4-β-D-XYLAN SYNTHASE		
UDP-D-GLUCURONIC ACID		
GLUCURONOKINASE		

UDP-HEXANOLAMINE
MANNOSYL-GLYCOPROTEIN β-1,4-N-ACETYLGLUCOSAMINYLTRANSFERASE
UDP-N-ACETYLGLUCOSAMINE-DOLICHYL-PHOSPHATE N-ACETYLGLUCOSAMINEPHOSPHOTRANSFERA

UDP-HEXYLAMINE
MANNOTETRAOSE 2-α-N-ACETYLGLUCOSAMINYLTRANSFERASE — *I*50 = 600 μM

UDP-MANNOSE
CHITIN SYNTHASE
CMP-N-ACYLNEURAMINATE PHOSPHODIESTERASE — 61% 2.9 mM

o-UDP
βN-ACETYLGLUCOSAMINYL-GLYCOPEPTIDE β-1,4-GALACTOSYLTRANSFERASE
METHYLCOENZYM M METHYLREDUCTASE — 62% 280 μM
UDP-GALACTOSE N-ACETYLGLUCOSAMINE β-4-GALACTOSYLTRANSFERASE

UDP-PYRIDOXAL
GLYCOGEN (STARCH) SYNTHASE

UDP-XYLOSE
CHITIN SYNTHASE
GLYCOGENIN GLUCOSYLTRANSFERASE — 97% 100 μM
UDP-N-ACETYLGLUCOSAMINE-DOLICHYL-PHOSPHATE N-ACETYLGLUCOSAMINEPHOSPHOTRANSFERA
UDP-GLUCOSE 6-DEHYDROGENASE — C — *K*i = 2.7 μM
UDP-GLUCURONATE DECARBOXYLASE

UDP-D-XYLULOSE
UDP-GLUCOSE 6-DEHYDROGENASE — FB — *K*i = 4-5 μM

(S) U 85548 E
PROTEINASE HIV-1

(S) simp U 85548 E
PROTEINASE HIV-1

UK 129485
RNA-DIRECTED DNA POLYMERASE — *I*50 = 190 nM

UK 34787
STEROID 11β-MONOOXYGENASE — *I*50 = 2 μM
THROMBOXANE A SYNTHASE — *I*50 = 15 nM

UK 35493
3′,5′-CYCLIC-NUCLEOTIDE PHOSPHODIESTERASE II (cycAMP) — *I*50 = 600 nM
3′,5′-CYCLIC-NUCLEOTIDE PHOSPHODIESTERASE III (cycAMP) — *I*50 = 900 nM

UK 37248
THROMBOXANE A SYNTHASE

UK 38485
THROMBOXANE A SYNTHASE — *I*50 = 12 μM

UK 61260
3′,5′-CYCLIC-NUCLEOTIDE PHOSPHODIESTERASE

UK 63831
NEPRILYSIN — *I*50 = 220 nM
PEPTIDYL-DIPEPTIDASE A — *I*50 = 20 nM

UK 69578
NEPRILYSIN — *K*i = 28 nM

UK 73967
NEPRILYSIN

UK 73976
NEPRILYSIN

ULINASTATIN
- CHYMOTRYPSIN
- ELASTASE
- TRYPSIN

UMBELLIFERONE

XANTHINE OXIDASE		$I50 = 44\ \mu M$

UMP

N-ACETYLGLUCOSAMINYLDIPHOSPHODOLICHOL N-ACETYLGLUCOSAMINYLTRANSFERASE		55% 1 mM
ADENYLOSUCCINATE SYNTHASE		47% 10 mM
D-AMINO-ACID OXIDASE	C	
CARBAMOYL-PHOSPHATE SYNTHASE (GLUTAMINE-HYDROLYSING)		
CDP-DIACYLGLYCEROLPYROPHOSPHATASE		26% 3.3 mM
CELLULOSE SYNTHASE (UDP FORMING)		
CMP-N-ACETYLNEURAMINATE MONOOXYGENASE		
CMP-N-ACYLNEURAMINATE PHOSPHODIESTERASE		26% 860 μM
DIHYDROOROTASE	C	Ki = 230 μM
DOLICHYL-PHOSPHATE β-GLUCOSYLTRANSFERASE	C	Ki = 62 μM
FRUCTOKINASE		
FUCOSYLGALACTOSE α-N-ACETYLGALACTOSAMINYLTRANSFERASE		83% 50 μM
GALACTOLIPID GALACTOSYLTRANSFERASE		$I50 = 10$ mM
1,3-β-GLUCAN SYNTHASE		
HEXOKINASE		
IMP DEHYDROGENASE		
LACTOSE SYNTHASE		
α-MANNOSIDE β1-2-N-ACETYLGLUCOSAMINYLTRANSFERASE		Ki = 220 μM
NUCLEOSIDE PHOSPHOACYLHYDROLASE		Ki = 8.5mM
1,3-β-OLIGIGOGLUCAN PHOSPHORYLASE		
OLIGONUCLEOTIDASE		45% 400 μM
OROTATE PHOSPHORIBOSYLTRANSFERASE		
OROTIDINE-5′-PHOSPHATE DECARBOXYLASE		Ki = 88 μM
OROTIDINE-5′-PHOSPHATE DECARBOXYLASE		Ki = 150 μM
POLYGALACTURONATE 4-α-GALACTURONOSYLTRANSFERASE		
POLYPEPTIDE N-ACETYLGALACTOSAMINYLTRANSFERASE		
PROCOLLAGEN GALACTOSYLTRANSFERASE		
PYRIMIDINE-NUCLEOSIDE PHOSPHORYLASE	C	Ki = 15 μM
RIBONUCLEASE T2		Ki = 450 μM
RIBOSE-5-PHOSPHATE ISOMERASE		
SARSAPOGENIN 3-β-GLUCOSYLTRANSFERASE		
SUCROSE SYNTHASE		
α,α-TREHALOSE-PHOSPHATE SYNTHASE		
UDP-N-ACETYLGLUCOSAMINE-DOLICHYL-PHOSPHATE N-ACETYLGLUCOSAMINEPHOSPHOTRANSFERA		
UDP-N-ACETYLGLUCOSAMINE 2-EPIMERASE		98% 360 μM
UDP-GLUCOSE 4-EPIMERASE		Ki = 50 μM
UDP-GLUCURNATE 4-EPIMERASE		100% 3 mM
UDP-GLUCURONATE DECARBOXYLASE	C	Ki = 2 mM
UDP-GLUCURONATE DECARBOXYLASE	C	Ki = 2 mM
UDP-GLUCURONOSYLTRANSFERASE		
UNDECAPRENYLPHOSPATE GALACTOSEPHOSPHOTRANSFERASE		56% 100 μM
URACIL PHOSPHORIBOSYLTRANSFERASE		
URIDINE NUCLEOSIDASE		
1,4-β-D-XYLAN SYNTHASE		

2′-UMP

2′,3′-CYCLIC-NUCLEOTIDE 2′-PHOSPHODIESTERASE	C	Ki = 2 mM
RIBONUCLEASE T1		
RIBONUCLEASE T2		Ki = 310 μM

3′-UMP

RIBONUCLEASE T1		
RIBONUCLEASE T2		Ki = 210 μM
RIBOSE-5-PHOSPHATE ISOMERASE		

5′-UMP

3′,5′-EXONUCLEASE		32% 5 mM

UMP-BUTANESULPHONIC ACID
POLYPEPTIDE N-ACETYLGALACTOSAMINYLTRANSFERASE

cyc-2′,3′-UMP
2′,3′-CYCLIC-NUCLEOTIDE 3′-PHOSPHODIESTERASE | C | *K*i = 185 μM

dUMP
dCMP DEAMINASE | C | *K*i = 700 μM
DEOXYCYTIDINE KINASE
DIHYDROOROTASE | C | *K*i = 250 μM
OROTATE PHOSPHORIBOSYLTRANSFERASE
PYRIMIDINE-DEOXYNUCLEOSIDE 2′-DIOXYGENASE
THYMIDINE-TRIPHOSPHATASE | | *K*i = 1.2 mM
THYMIDYLATE 5′-PHOSPHATASE
THYMIDYLATE SYNTHASE | | *K*i = 5.2 μM
dTMP KINASE | C | *K*i = 3 mM
dTMP KINASE | | *I*50 = 500 μM

UNDECANOIC ACID
20α-HYDROXYSTEROID DEHYDROGENASE

UNDECANOIC ACID BIS{[2(3H-IMIDAZOL-4-YL)ETHYL]AMIDE
GELATINASE A | | *I*50 = 29 μM
GELATINASE B | | *I*50 = 50 μM

UNDECANOYL VANILLYLAMIDE
NADH DEHYDROGENASE (UBIQUINONE) | | *I*50 = 3 μM

UNDECAPRENYLPYROPHOSPHATE
DOLICHYL-PHOSPHATASE | C | *K*i = 160 μM

5-n-UNDECENYL-6-HYDROXY-4,7-DIOXOBENZOTHIAZOLE
CYTOCHROME bo3 UBIQUINOL OXIDASE

5-UNDECYL-6-HYDROXY-4,7-DIOXOBENZOTHIAZOLE
PLASTOQUINOL-PLASTOCYANIN REDUCTASE
UBIQUINOL-CYTOCHROME-C REDUCTASE

5-(n-UNDECYL)-6-HYDROXY-4,7-OXOBENZOTHIAZOLE
SUCCINATE DEHYDROGENASE (UBIQUINONE)

N-UNDECYLOYLDOPAMINE
ARACHIDONATE 5-LIPOXYGENASE | | *I*50 = 17 nM

10-UNDECYNOIC ACID
CYTOCHROME P450 4A1
LAURIC ACID HYDROXYLASE | SSI |

10-UNDECYNYL SULPHATE SODIUM
LAURIC ACID HYDROXYLASE | SSI |

UNGINOL
PHOSPHOLIPASE A2 | | *I*50 = 104 μM

UNIVERSITY OF WISCONSIN SOLUTION
COLLAGENASE

UNSATURATED FATTY ACIDS
H^+/K^+-EXCHANGING ATPASE

4′,5′-UNSATURATED 5′-FLUOROADENOSINE NUCLEOSIDES
ADENOSYLHOMOCYSTEINASE

α,β-UNSATURATED STEROIDES
CORTISONE β-REDUCTASE

UO^{2+}
α-GLUCOSIDASE

UP 5180-47		
ALDEHYDE DEHYDROGENASE (NAD)		$I50 = 45$ nM
URACIL		
ALKANAL MONOOXYGENASE (FMN -LINKED)		
DIHYDROPYRIMIDINE DEHYDROGENASE (NADP)		
OROTATE PHOSPHORIBOSYLTRANSFERASE		
THYMINE DIOXYGENASE		
UDP-GLUCURONATE 5′-EPIMERASE		
URACIL-DNA GLYCOHYDROLASE	NC	$Ki = 200$ μM
URACIL-DNA GLYCOSYLASE		$Ki = 600$ μM
URATE-RIBONUCLEOTIDE PHOSPHORYLASE		
URIDINE PHOSPHORYLASE		
URACIL-5-CARBOXYLIC ACID		
ORATE REDUCTASE (NADPH)		
URACILY DNA GLYCOSYLASE INHIBITOR		
DNA GLYCOSYLASE		
URATE ANALOGUES		
URATE OXIDASE		
UREA		
N-ACETYL-β-GLUCOSAMINIDASE		
β-N-ACETYLHEXOSAMINIDASE	MIX	$Ki = 115$ mM
ACYLPHOSPHATASE		100% 5 mM
AGMATINASE		
AMIDASE		
AMIDASE	NC	$Ki = 500$ μM
AMIDINOASPARTASE	C	$Ki = 250$ μM
AMYLO-1,6-GLUCOSIDASE		
ARGININOSUCCINATE LYASE		
ARYLFORMAMIDASE		100% 2 mM
ASPARAGINE-tRNA LIGASE		
BENZOYLFORMATE DECARBOXYLASE		
BIOTIN-[METHYLCROTONYL-CoA-CARBOXYLASE] LIGASE		
CARBOXYPEPTIDASE B		
CATHEPSIN D		
CELLULOSE 1,4-β-CELLOBIOSIDASE		
CYTIDINE DEAMINASE		
β-(9-CYTOKININ)-ALANINE SYNTHASE		
DDT-DEHYDROCHLORINASE		100% 8 M
ETHANOLAMIN AMMONIA-LYASE		95% 1 mM
FORMAMIDASE		
β-FRUCTOFURANOSIDASE		
GLUTAMATE-AMMONIA LIGASE		
3-HYDROXYDECANOYL-[ACYL-CARRIER-PROTEIN] DEHYDRATASE		
HYDROXYMETHYLBILANE SYNTHASE		
INULINASE		
MANDELATE 4-MONOOXYGENASE		
MYCODEXTRANASE		
OLIGO-1,6-GLUCOSIDASE		
PHENOLL 2-MONOOXYGENASE		
PHOSPHOLIPASE A1		
POLYGALACTURONASE		
RIBONUCLEASE (POLY-(U)-SPECIFIC)		71% 1 M
L(+)-TARTRATE DEHYDRATASE		
UDP-GLUCURNATE 4-EPIMERASE		
UREASTIBAMINE		
DNA TOPOISOMERASE		
β-UREIDOISOBUTYRIC ACID		
β-UREIDOPROPIONASE		$Ki = 18$ μM

β-UREIDOPROPIONATE
β-UREIDOPROPIONASE

URETHAN
CARBONATE DEHYDRATASE I — *K*i = 3.6 mM
CARBONATE DEHYDRATASE II — *K*i = 77 mM
D-GLUTAMATE OXIDASE

URETHANE
ALKANAL MONOOXYGENASE (FMN -LINKED)
ARGININE DECARBOXYLASE

URIC ACID
ARGINASE — C — *K*i = 9 μM
PLASMINOGEN ACTIVATOR — *I*50 = 4 mM
XANTHINE OXIDASE — U — *K*i = 70 μM

URIC ACID RIBOSIDE
PURINE NUCLEOSIDASE

URIDINE
ADENOSINE KINASE — 40% 100 μM
CYTIDINE DEAMINASE — *K*i = 280 μM
DEOXYCYTIDINE DEAMINASE
DIHYDROOROTASE — C — *K*i = 300 μM
FUCOSYLGALACTOSE α-N-ACETYLGALACTOSAMINYLTRANSFERASE — 38% 2 mM
PURINE NUCLEOSIDASE
RIBONUCLEASE — C — *K*i = 8 mM
RIBONUCLEASE T2 — *K*i = 370 μM
THYMIDINE PHOSPHORYLASE
UDP-N-ACETYLGLUCOSAMINE 2-EPIMERASE — 98% 360 μM
UDP-GLUCURONATE 5′-EPIMERASE
URIDINE PHOSPHORYLASE
UTP-HEXOSE-1-PHOSPHATE URIDYLYLTRANSFERASE — 50% 400 μM

URIDINE 5′-DIPHOSPHATE BROMOACETOL
UDP-GLUCOSE 4-EPIMERASE — SI — *K*i = 110 μM

URIDINE-5-′DIPHOSPHATE CHLOROACETATE
UDP-GLUCOSE 6-DEHYDROGENASE — IR

URIDINE 5′-DIPHOSPHATE CHLOROACETOL
UDP-GLUCOSE 4-EPIMERASE — SI — *K*i = 110 μM

URIDINE DIPHOSPHOGLUCOSE
myo-INOSITOL OXYGENASE

[1-[(6-URIDINEDIPHOSPHO)HEXANAMIDO]ETHYL](2,4-DICARBOXYBUTYL)PHOSPHINATE
D-GLUTAMIC ACID ADDING ENZYME — *I*50 = 680 nM

URIDINE NUCLEOTIDES
CARBAMOYL-PHOSPHATE SYNTHASE (AMMONIA)

URIDINE 3′-PHOSPHATE
RIBONUCLEASE PANCREATIC — *K*i = 2.16 mM

URIDINE 5′-PHOSPHORIC(1-HEXADECANESULPHONIC)ANHYDRIDE
POLYPEPTIDE N-ACETYLGALACTOSAMINYLTRANSFERASE — C — *I*50 = 160 μM

URIFINE
OROTATE PHOSPHORIBOSYLTRANSFERASE

URINASTATIN
TRYPSIN

URINE TRYPSIN INHIBITOR
TRYPSIN

URINE TRYPSIN INHIBITOR (HUMAN)

ACROSIN		Ki = 12 nM
CATHEPSIN G		
α-CHYMOTRYPSIN		
ELASTASE LEUKOCYTE		
PLASMIN		Ki = 100 nM
PLASMINOGEN ACTIVATOR		

UROBILIN

PROTEINASE HIV-1		Ki = 6 μM

UROCANIC ACID

HISTIDINE AMMONIA-LYASE		Ki = 130 μM
HISTIDINE DECARBOXYLASE	C	Ki = 2.1 mM

UROPORPHYRIN

NITRIC-OXIDE SYNTHASE II		I50 = 25 μM
NITRIC-OXIDE SYNTHASE III		I50 = 20 μM

UROPORPHYRIN I

GLUTATHIONE TRANSFERASE		
UROPORPHYRINOGEN DECARBOXYLASE		46% 8 μM

UROPORPHYRIN III

UROPORPHYRINOGEN DECARBOXYLASE		51% 8 μM

UROPORPHYRINOGEN III

UROPORPHYRIN-III C-METHYLTRANSFERASE		> 2μM
UROPORPHYRINOGEN-III SYNTHASE		

UROXIN

Na^+/K^+-EXCHANGING ATPASE		

URSODEOXYCHOLIC ACID

trans-1,2-DIHYDROBENZENE-1,2-DIOL DEHYDROGENASE 1		Ki = 8.8 μM
trans-1,2-DIHYDROBENZENE-1,2-DIOL DEHYDROGENASE 2		Ki = 30 nM
DIHYDRODIOL DEHYDROGENASE 2		98% 100 μM
DIHYDRODIOL DEHYDROGENASE 4		10% 100 μM
GLYCINE N-CHOLOYLTRANSFERASE		
11β-HYDROXYSTEROID DEHYDROGENASE		I50 = 30 μM
3α-HYDROXYSTEROID/DIHYDRODIOL DEHYDROGENASE 2	U	Ki = 60 nM
PROTEINASE HIV-1		I50 = 70 μM
UDP-GLUCURONOSYLTRANSFERASE	C	Ki = 346 μM

URSOLIC ACID

ADENOSINE DEAMINASE		Ki = 9.7 μM
ARACHIDONATE 5-LIPOXYGENASE		80% 1 μM
ARACHIDONATE 5-LIPOXYGENASE		I50 = 300 μM
ARACHIDONATE 15-LIPOXYGENASE		60% 1 μM
ARACHIDONATE 15-LIPOXYGENASE		I50 = 300 μM
AROMATASE		30% 82 μM
CYCLOOXYGENASE		38% 1 μM
DNA LIGASE I		I50 = 216 μM
ELASTASE LEUKOCYTE	NC	Ki = 6.6 μM
ELASTASE LEUKOCYTE		87% 20 μM
PHOSPHOLIPASE A2		
PROTEINASE HIV-1		Ki = 3.4 μM
PROTEIN KINASE A		I50 = 9 μM
PROTEIN KINASE C		I50 = 106 μM
PROTEIN KINASE (Ca^{2+} DEPENDENT)		I50 = 71 μM

$(s^4dU)_{35}$

RNA-DIRECTED DNA POLYMERASE	C	Ki = 3.3 nM

UTERINE PLASMIN/TRYPSIN INHIBITOR

PLASMIN		
TRYPSIN		

UTEROFERRIN

UDP-GLUCOSE-GLYCOPROTEIN GLUCOSEPHOSPHOTRANSFERASE		

UTEROGLOBIN

PHOSPHOLIPASE A2		30% 2 nM

UTP

N-ACETYLGLUCOSAMINE KINASE		
N-ACETYLNEURAMINATE CYTIDYLYLTRANSFERASE		
ADENYLYLSULFATASE	C	K_i = 150 μM
ARGININOSUCCINATE SYNTHASE		
ARYL SULFOTRANSFERASE	NC	60% 600 μM
ASPARTATE CARBAMOYLTRANSFERASE		
BACILLUS SUBTILIS RIBONUCLEASE		54% 500 μM
CARBAMOYL-PHOSPHATE SYNTHASE (GLUTAMINE-HYDROLYSING)	C	
CDP-DIACYLGLYCEROL-INOSITOL 3-PHOSPHATIDYLTRANSFERASE		
CELLULOSE SYNTHASE (UDP FORMING)		
CHITIN SYNTHASE		
3′,5′-CYCLIC-NUCLEOTIDE PHOSPHODIESTERASE		
DIACYLGLYCEROL KINASE		
DIPEPTIDASE	C	K_i = 700 μM
DNA POLYMERASE β		K_i = 880 nM
FLAVANONE 7-O-β-GLUCOSYLTRANSFERASE		
1,3-β-GLUCAN SYNTHASE		
GLUTAMATE SYNTHASE (NADPH)		
GTP CYCLOHYDROLASE I		
GUANYLATE CYCLASE		
HYDROXYPYRUVATE REDUCTASE		
INOSINE NUCLEOSIDASE		
myo-INOSITOL OXYGENASE		
[ISOCITRATE DEHYDROGENASE (NADP)] KINASE		
LACTOSE SYNTHASE		
α-MANNOSIDE β1-2-N-ACETYLGLUCOSAMINYLTRANSFERASE		K_i = 135 μM
[3-METHYL-2-OXOBUTANOATE DEHYDROGENASE (LIPOAMIDE)] PHOSPHATASE		I_{50} = 100 μM
NADH PEROXIDASE		95% 3 mM
NAD PYROPHOSPHATASE		36% 66 μM
5′-NUCLEOTIDASE	C	30% 100 μM
OROTATE PHOSPHORIBOSYLTRANSFERASE		
PHOSPHATIDYLINOSITOL KINASE		70% 500 μM
PHOSPHOACETYLGLUCOSAMINE MUTASE		
PHOSPHOGLUCOMUTASE	C	
PHOSPHOPANTOTHENOYLCYSTEINE DECARBOXYLASE		
POLYGALACTURONATE 4-α-GALACTURONOSYLTRANSFERASE		
POLYNUCLEOTIDE ADENYLYLTRANSFERASE		
POLYPEPTIDE N-ACETYLGALACTOSAMINYLTRANSFERASE		
PROCOLLAGEN GLUCOSYLTRANSFERASE		
PROTEIN-GLUTAMINE γ-GLUTAMYLTRANSFERASE		
PURINE-NUCLEOSIDE PHOSPHORYLASE		K_i = 164 μM
PYRUVATE CARBOXYLASE	NC	K_i = 450 μM
PYRUVATE KINASE		
STEROL 3β-GLUCOSYLTRANSFERASE		
SUCROSE-PHOSPHATE SYNTHASE		
THIAMIN-PHOSPHATE PYROPHOSPHORYLASE		
THYMIDINE-TRIPHOSPHATASE		K_i = 8.6 mM
α,α-TREHALOSE-PHOSPHATE SYNTHASE		
UDP-GLUCOSE 4-EPIMERASE		K_i = 220 μM
UDP-GLUCURNATE 4-EPIMERASE		100% 3 mM
UDP-GLUCURONOSYLTRANSFERASE		
UMP KINASE		
URACIL PHOSPHORIBOSYLTRANSFERASE		
URIDINE KINASE		
UTP-HEXOSE-1-PHOSPHATE URIDYLYLTRANSFERASE		42% 400 μM
1,4-β-D-XYLAN SYNTHASE		

dUTP

CARBAMOYL-PHOSPHATE SYNTHASE (GLUTAMINE-HYDROLYSING)		
dCMP DEAMINASE	C	
dCTP DEAMINASE		83% 200 μM
DEOXYCYTIDINE KINASE		
GLUCOSE-1-PHOSPHATE THYMIDYLYLTRANSFERASE	C	*K*i = 260 μM
GLUCOSE-1-PHOSPHATE THYMIDYLYLTRANSFERASE	C	*K*i = 100 μM
5′-NUCLEOTIDASE	C	27% 100 μM
THYMIDINE-TRIPHOSPHATASE		*K*i = 14 μM
URACIL PHOSPHORIBOSYLTRANSFERASE		

UTTRONIN A

1-PHOSPHATIDYLINOSITOL 3-KINASE		*I*50 = 1.1 μM

UVAOL

ELASTASE LEUKOCYTE	C	*I*50 = 16 μM
ELASTASE LEUKOCYTE	C	*K*i = 16 μM
PROTEINASE HIV-1		*I*50 = 3.5 μM
PROTEIN KINASE A		*I*50 = 165 μM

V

cis-VACCENIC ACID

GLYCEROL-3-PHOSPHATE O-ACYLTRANSFERASE		
4-HYDROXYBENZOATE NONAPRENYLTRANSFERASE		
PROTEIN KINASE A		*I*50 = 6 μM

cis-VACCENOYL-CoA

GLYCEROL-3-PHOSPHATE O-ACYLTRANSFERASE

2-VAD-DCB

CASPASE-1

DVAL^{22}-BIG ENDOTHELIN-1[16-38]

ENDOTHELIN CONVERTING ENZYME		*K*i = 25 μM

VAL-BORO-PRO

DIPEPTIDYL-PEPTIDASE 4	IR	

VALCLAVAM

HOMOSERINE O-SUCCINYLTRANSFERASE

VAL-$COCH_2Cl$

PROTEINASE DICHELOBACTER NODOSUS EXTRACELLULAR		98% 1 mM

n-VALERALDEHYDE

FORMALDEHYDE DEHYDROGENASE

VALERIC ACID

4-HYDROXY-2-OXOGLUTARATE ALDOLASE		28% 40 mM

n-VALERIC ACID

GLUTAMATE DECARBOXYLASE

VALEROLACTAM

1-PYRROLINE-5-CARBOXYLATE DEHYDROGENASE

VALEROLACTAMATE

1-PYRROLINE-5-CARBOXYLATE DEHYDROGENASE

4-VALEROLACTIC ACID

1-PYRROLINE-5-CARBOXYLATE DEHYDROGENASE

δ-VALEROLACTONE

1-PYRROLINE-5-CARBOXYLATE DEHYDROGENASE

VALERYL-CoA

ACETYL-CoA HYDROLASE		
PYRUVATE CARBOXYLASE		*K*i = 80 μM

VALERYLSALICYLIC ACID

PROSTAGLANDIN-ENDOPEROXIDE SYNTHASE 1		

VALIDAMINE

α-GLUCOSIDASE		
α,α-TREHALASE	C	*K*i = 140 μM

VALIDAMYCIN

α,α-TREHALASE		*I*50 = 250 μM
α,α-TREHALASE		*I*50 = 1 nM

VALIDAMYCIN A

α,α-TREHALASE	C	*K*i = 47 nM

VALIDAMYCIN B

α,α-TREHALASE	C	*K*i = 190 nM

VALIDAMYCIN C

α,α-TREHALASE	C	*K*i = 2.8 μM

VALIDAMYCIN D

α,α-TREHALASE	C	*K*i = 3.2 nM

VALIDAMYCIN E

α,α-TREHALASE	C	*K*i = 140 nM

VALIDAMYCIN F

α,α-TREHALASE	C	*K*i = 330 nM

VALIDAMYCIN G

α,α-TREHALASE	C	*K*i = 190 nM

VALIDOXYLAMINE A

α,α-TREHALASE	C	*K*i = 430 pM
α,α-TREHALASE		*I*50 = 2.4 nM

VALIDOXYLAMINE B

α,α-TREHALASE	C	*K*i = 55 nM

VALIDOXYLAMINE G

α,α-TREHALASE	C	*K*i = 120 nM

VALIENAMINE

α-GLUCOSIDASE		

VALINE

ACETOLACTATE SYNTHASE III		85% 4 mM
ALANINE AMINOPEPTIDASE	NC	*I*50 = 6.3 mM
2-AMINOHEXANOATE TRANSAMINASE		
ARGINASE		*I*50 = 6.2 mM
ARGINASE	C	*K*i = 3.6 mM
ARGININOSUCCINATE SYNTHASE		
GLUTAMATE-AMMONIA LIGASE		
GLUTAMATE SYNTHASE (NADH)		
GLUTAMATE SYNTHASE (NADPH)		
LEUCYL AMINOPEPTIDASE	C	*K*i = 17 mM
ORNITHINE CARBAMOYLTRANSFERASE		
ORNITHINE-OXO-ACID TRANSAMINASE		
PHENYLALANINE DEHYDROGENASE		
1-PYRROLINE-5-CARBOXYLATE DEHYDROGENASE		
PYRUVATE KINASE		

D-VALINE

PHENYLALANINE DEHYDROGENASE		

Inhibitor / Enzyme	Type	Value
D,L-VALINE		
GLUTAMATE DEHYDROGENASE		
DL-VALINE HYDROXAMIC ACID		
PROTEINASE ASTACUS		*I*50 = 1.2 mM
L-VALINE		
ACETOLACTATE SYNTHASE	NC	*I*50 = 100 µM
HOMOSERINE KINASE		24% 20 mM
ISOLEUCINE-tRNA LIGASE	C	*K*i = 380 µM
3-ISOPROPYLMALATE DEHYDROGENASE	C	*K*i = 1.8 mM
3-METHYL-2-OXOBUTANOATE HYDROXYMETHYLTRANSFERASE		27% 5 mM
VALINEMETHYLKETONE		
ALANINE AMINOPEPTIDASE		*K*i = 550 nM
VALINOCTIN A		
FARNESYL PROTEINTRANSFERASE		*I*50 = 3.3 µM
VALINOCTIN B		
FARNESYL PROTEINTRANSFERASE		*I*50 = 1 µg/ml
L-VALINOL-AMP		
VALINE-tRNA LIGASE	C	*K*i = 29 nM
VALINOMYCIN		
UBIQUINOL OXIDASE		*I*50 = 100 nM
VALIOLAMINE		
α-GLUCOSIDASE		*I*50 = 2.2 µM
β-GLUCOSIDASE		
SUCROSE α-GLUCOSIDASE		*I*50 = 42 nM
VAL-LEU-ARG-ARG-ALA-SER-VAL-ALA		
TRIPEPTIDYL PEPTIDASE	C	*K*i = 1.5 µM
D-VAL-LEU-LYS-CHLORMETHYLKETONE		
CANCER PROCOAGULANT		92% 200 µM
VAL-D-LEU-PRO-PHE-PHE-VAL-D-LEU		
PEPSIN A		
VAL-PHE		
PEPTIDYL-DIPEPTIDASE A	C	*I*50 = 44 µM
VALPROIC ACID		
ALCOHOL DEHYDROGENASE (NADP)	NC	
ALDEHYDE REDUCTASE		62% 1 mM
ALDEHYDE REDUCTASE	NC	*K*i = 5.4 mM
4-AMINOBUTYRATE TRANSAMINASE	NC	
4-AMINOBUTYRATE TRANSAMINASE		52% 2 mM
CARBONYL REDUCTASE (NADPH)	NC	
EPOXIDE HYDROLASE		*K*i = 500 µM
FATTY ACID OXYDATION		
OXIDATIVE PHOSPHORYLATION UNCOUPLER		
PROSTAGLANDIN F SYNTHASE		
SUCCINATE-SEMIALDEHYDE DEHYDROGENASE	NC	
SUCCINATE TRANSPORT MITOCHONDRIA		
VAL-PRO-LEU		
DIPEPTIDYL-PEPTIDASE 4		
VALPROMIDE		
EPOXIDE HYDROLASE	C	*K*i = 160 µM
EPOXIDE HYDROLASE	C	*K*i = 110 µM
EPOXIDE HYDROLASE		*I*50 = 4 µM

VAL-THIAZOLID

DIPEPTIDYL-PEPTIDASE 4	TS	$Ki = 270$ nM

VAL-THR-VAL-ALA-PRO-VAL-HIS-ILE

CATHEPSIN G		$Ki = 1.4$ mM
ELASTASE LEUKOCYTE		$Ki = 120$ μM

VALYLARGININE

AMINOPEPTIDASE B		$Ki = 10$ μM

VAMIDOTHION

ACETYLCHOLINESTERASE		

VANADATE

N-ACETYLGLUCOSAMINE KINASE		99% 10 mM
N-ACETYLMANNOSAMINE KINASE		41% 1 mM
ADENOSINETRIPHOSPHATASE		$I50 = 12$ μM
ARYLSULFATASE	C	$Ki = 6$ μM
BISPHOSPHOGLYCERATE PHOSPHATASE		
Ca^{2+}-TRANSPORTING ATPASE		69% 100 μM
Ca^{2+}-TRANSPORTING ATPASE		61% 100 μM
Ca^{2+}-TRANSPORTING ATPASE		
DYNEIN ATPASE		$I50 = 100$ nM
FRUCTOSE-2,6-BISPHOSPHATASE	NC	$I50 = 150$ μM
GLUCOSE-6-PHOSPHATASE		$Ki = 1.4$ μM
GLYCEROL-1-PHOSPHATASE		81% 50 μM
GLYCEROPHOSPHOCHOLINE PHOSPHODIESTERASE		65% 50 μM
H^{+}/K^{+}-EXCHANGING ATPASE		
H^{+}-TRANSPORTING ATPASE		$I50 = 1.6$ mM
H^{+}-TRANSPORTING ATPASE		95% 100 μM
H^{+}-TRANSPORTING ATP SYNTHASE (CF1 ATPASE)		$I50 = 500$ μM
HYDROXYMETHYLBILANE SYNTHASE		
MONOTERPENYL-PYROPHOSPHATASE		$I50 = 500$ μM
MYOSIN ATPASE		
Na^{+}/K^{+}-EXCHANGING ATPASE		$Ki = 50$ nM
NITRATE REDUCTASE (NADH)		
4-NITROPHENYLPHOSPHATASE		
NUCLEOSIDE-TRIPHOSPHATASE		
PEROXIDASE		
PHOSPHATASE ACID	C	$Ki = 29$ μM
PHOSPHATASE ACID	C	$Ki = 210$ nM
PHOSPHATASE ACID		
PHOSPHATASE ALKALINE	C	$Ki = 130$ nM
PHOSPHATASE ALKALINE	C	$Ki = 22$ μM
PHOSPHATASE ALKALINE		36% 100 μM
1-PHOSPHATIDYLINOSITOL PHOSPHODIESTERASE		
PHOSPHOENOLPYRUVATE CARBOXYLASE		
6-PHOSPHOFRUCTOKINASE		
6-PHOSPHOFRUCTO-2-KINASE		
PHOSPHOGLUCOMUTASE	C	$Ki = 740$ nM
PHOSPHOGLUCOMUTASE	C	$Ki = 1.1$ μM
PHOSPHOGLUCONATE DEHYDROGENASE (DECARBOXYLATING)	C	$Ki = 70$ μM
PHOSPHOGLUCONATE DEHYDROGENASE (DECARBOXYLATING)	C	$Ki = 12$ μM
PHOSPHOGLYCERATE MUTASE		
PHOSPHOLIPASE C		
PHOSPHOPROTEIN PHOSPHATASE		
PHOSPHOPROTEIN PHOSPHATASE		26% 100 μM
3-PHYTASE		$Ki = 28$ μM
POLYNUCLEOTIDE ADENYLYLTRANSFERASE		
PROTEIN TYROSINE-PHOSPHATASE	NC	90% 100 μM
PROTEIN TYROSINE-PHOSPHATASE		81% 1 μM
PROTEIN TYROSINE-PHOSPHATASE β		$I50 = 30$ nM
PROTEIN TYROSINE-PHOSPHATASE ACID		$I50 = 53$ μM
PROTEIN TYROSINE-PHOSPHATASE 1B	C	$Ki = 380$ nM
PROTEIN TYROSINE-PHOSPHATASE (Mg^{2+} DEPENDENT)		69% 10 μM

PROTEIN TYROSINE-PHOSPHATASE MEG-1		*I*50 = 17 nM
PROTEIN TYROSINE-PHOSPHATASE sigma		*I*50 = 1 nM
SULFATE ADENYLYLTRANSFERASE (ADP)		
UDP-N-ACETYLGLUCOSAMINE 2-EPIMERASE		19% 1 mM
UTEROFERRIN (PURPLE PHOSPHATASE ACID)	NC	
deca-VANADATE		
ADENYLATE KINASE		
6-PHOSPHOFRUCTO-2-KINASE		*I*50 = 27 μM
TRANSDUCIN GUANOSINE TRIPHOSPHATASE		
VANADATE DIMER		
GLUCOSE-6-PHOSPHATE 1-DEHYDROGENASE	NC	*K*i = 160 μM
VANADATE-myoINOSITOL COMPLEX		
1-PHOSPHATIDYLINOSITOL PHOSPHODIESTERASE		
meta-VANADATE		
GLUCOSE-6-PHOSPHATE 1-DEHYDROGENASE	C	*K*i = 2.7 mM
GLUTATHIONE REDUCTASE (NADPH)		
VANADATE OLIGOMERS		
HEXOKINASE		
PHOSPHOGLUCONATE DEHYDROGENASE (DECARBOXYLATING)		
PHOSPHOGLYCERATE MUTASE		
PHOSPHORYLASE		
ortho-VANADATE		
APYRASE		
PROTEIN TYROSINE-PHOSPHATASE		100% 1 mM
PROTEIN TYROSINE-PHOSPHATASE CDC25B		89% 500 μM
VANADATE TETRAMER		
GLUCOSE-6-PHOSPHATE 1-DEHYDROGENASE	NC	
GLYCEROL-3-PHOSPHATE DEHYDROGENASE (NAD)	C	*K*i = 120 μM
VANADIUM (IV)		
PROTEIN KINASE A		*I*50 = 360 μM
VANADIUM OXIDE SULPHATE		
DOLICHYL-PHOSPHATASE		
VANADIUM OXOANIONS		
SUPEROXIDE DISMUTASE (CU, ZN)		
VANADIUM PENTOXIDE		
ISOPENTENYL-DIPHOSPHATE Δ-ISOMERASE		
VANADYLRIBONUCLEOSIDE		
POLYRIBONUCLEOTIDE SYNTHASE (ATP)		50% 400 μM
VANADYL SULPHATE		
GLUTAMATE DEHYDROGENASE	NC	*K*i = 147 μM
VANCOMYCIN		
CDP-GLYCEROL GLYCEROPHOSPHOTRANSFERASE		
PEPTIDOGLYCAN GLUCOSYLTRANSFERASE		
PHOSPHO-N-ACETYLMURAMOYL-PENTAPEPTIDE-TRANSFERASE		
VANICOSIDE B		
PROTEIN KINASE C		*I*50 = 31 μg/ml
VANILLIC ACID		
ARACHIDONATE 5-LIPOXYGENASE		*I*50 = 770 μM
PHENYLALANINE AMMONIA-LYASE 1		
PROTOCATECHUATE 3,4-DIOXYGENASE	C	*K*i = 3 mM
SHIKIMATE 5-DEHYDROGENASE	C	*K*i = 570 μM

VANILLIN

ARYL SULFOTRANSFERASE		99% 1 μM
DOPAMINE SULFOTRANSFERASE		100% 6.7 μM
17α-ETHINYLOESTRADIOL SULFOTRANSFERASE		72% 6.7 μM
MALONYL-CoA DECARBOXYLASE		
MANDELONITRILE LYASE		*K*i = 2.2 mM
MONOPHENOL MONOOXYGENASE		*I*50 = 4.1 mM
NAPHTHOL SULFOTRANSFERASE		91% 6.7 μM
SHIKIMATE 5-DEHYDROGENASE	C	*K*i = 93 μM

VANILLYLMANDELIC ACID

TYROSINE TRANSAMINASE

VANOXONIN

THYMIDYLATE SYNTHASE

VARDAX

3′,5′-CYCLIC-NUCLEOTIDE PHOSPHODIESTERASE (cycAMP)		*I*50 = 103 μM
3′,5′-CYCLIC-NUCLEOTIDE PHOSPHODIESTERASE (cycGMP)		*I*50 = 151 μM
3′,5′-CYCLIC-NUCLEOTIDE PHOSPHODIESTERASE (cycGMP) (Ca^{2+}/CALMODULIN)		*I*50 = 321 μM

VARIABILIN

PHOSPHOLIPASE A2		*I*50 = 76.9 μM
PHOSPHOLIPASE A2 SECRETORY		*I*50 = 6.9 μM

VASCULAR ANTICOAGULANT-α

COAGULATION FACTOR Xa		*I*50 = 5.5 nM
PHOSPHOLIPASE A2		*I*50 = 26 nM

VASCULAR ANTICOAGULANT-β

COAGULATION FACTOR Xa		*I*50 = 1.4 nM
PHOSPHOLIPASE A2		*I*50 = 29 nM

VASOPRESSIN

ACETYLSEROTONIN O-METHYLTRANSFERASE
PROTEIN-DISULFIDE REDUCTASE (GLUTATHIONE)

VECURONIUM

HISTAMINE N-METHYLTRANSFERASE	C	*K*i = 1.2 μM

VELENOIC ACID DILACTONE

XANTHINE OXIDASE		*I*50 = 760 nM

VELNACRINE

ACETYLCHOLINESTERASE

VELUTIN

DNA TOPOISOMERASE

VENLAFAXINE

CYTOCHROME P450 2D6		*K*i = 41 μM
CYTOCHROME P450 2D6		

VENTURICIDIN

ADENOSINETRIPHOSPHATASE		*I*50 = 470 ng/ml
H^+-TRANSPORTING ATP SYNTHASE	REV	
H^+-TRANSPORTING ATP SYNTHASE F0		

VERAPAMIL

H^+/K^+-EXCHANGING ATPASE		*I*50 = 15 μM
H^+-TRANSPORTING ATPASE (V-TYPE)		*I*50 = 398 μM
H^+-TRANSPORTING PYROPHOSPHATASE VACUOLAR		*I*50 = 700 μM
MYOSIN-LIGHT-CHAIN KINASE		44% 333 μM
PROTEIN KINASE C		85% 200 μM
PROTEIN KINASE (Ca^{2+} DEPENDENT)		*I*50 = 400 μM

VERBASCOSIDE

PROTEIN KINASE C	C	*K*i = 22 μM

VERBENONE
- ACETYLCHOLINESTERASE — C — K_i = 290 µM

VERCURONIUM
- ACETYLCHOLINESTERASE
- CHOLINESTERASE

VERMEERIN
- GLYCERALDEHYDE-3-PHOSPHATE DEHYDROGENASE (PHOSPHORYLATING) — 8% 3 mM
- HEXOKINASE — 69% 1 mM
- 6-PHOSPHOFRUCTOKINASE — 50% 800 µM

VERNOLATE SULPHOXIDE
- ALDEHYDE DEHYDROGENASE (NAD) (LOW Km) — I_{50} = 113 µM

VERONAL
- CLOSTRIPAIN
- D-GLUTAMATE OXIDASE

VERSENE
- N-ACETYLGLUCOSAMINE DEACYLASE
- PHOSPHATASE ALKALINE
- STIPITATONATE DECARBOXYLASE
- XAA-PRO AMINOPEPTIDASE

VERSICONOL
- PROTEIN TYROSINE KINASE EGFR — I_{50} = 21 µM
- PROTEIN TYROSINE KINASE (MK CELLS) — I_{50} = 22 µM
- PROTEIN TYROSINE KINASE v-abl — I_{50} = 23 µM

VERTINE
- LYSINE DECARBOXYLASE

VESNARINONE
- 3′,5′-CYCLIC-NUCLEOTIDE PHOSPHODIESTERASE (cycAMP) — I_{50} = 220 µM
- 3′,5′-CYCLIC-NUCLEOTIDE PHOSPHODIESTERASE (Ca^{2+}/CALMODULIN) — I_{50} = 40 µM
- 3′,5′-CYCLIC-NUCLEOTIDE PHOSPHODIESTERASE (cGMP INHIBITABLE) — I_{50} = 6.4 µM
- 3′,5′-CYCLIC-NUCLEOTIDE PHOSPHODIESTERASE (cGMP INHIBITABLE) — I_{50} = 65 µM
- 3′,5′-CYCLIC-NUCLEOTIDE PHOSPHODIESTERASE (cGMP INHIBITABLE) — K_i = 8.5 µM
- 3′,5′-CYCLIC-NUCLEOTIDE PHOSPHODIESTERASE (cGMP INHIBITABLE) — I_{50} = 11 µM
- 3′,5′-CYCLIC-NUCLEOTIDE PHOSPHODIESTERASE (cGMP INHIBITABLE) — I_{50} = 54 µM
- NAD ADP-RIBOSYLTRANSFERASE — 32% 500 µM

VICTORIN
- GLYCINE DECARBOXYLASE MULTIENZYM COMPLEX

VIGNAFURAN
- ARACHIDONATE 5-LIPOXYGENASE

VIGNA UNGUICULATA PROTEINASE INHIBITOR
- PAPAIN

VINAXANTHONE
- 1-PHOSPHATIDYLINOSITOL-4,5-BISPHOSPHATE PHOSPHODIESTERASE — I_{50} = 5 µM
- PHOSPHOLIPASE C — I_{50} = 5.4 µM

VINBLASTINE
- AMINE OXIDASE MAO-B — K_i = 77 µM
- BUFURALOL-1′-HYDROXYLASE (CYT P-450dbl) — I_{50} = 130 µM
- CYTOCHROME P450 2D6 — C — K_i = 42 µM

VINCAMINE
- BUFURALOL-1′-HYDROXYLASE (CYT P-450dbl) — I_{50} = 60 µM

VINCRISTINE
- AMINE OXIDASE MAO-B

VINDOLINE
17-O-DEACETYLVINDOLINE O-ACETYLTRANSFERASE

VINILACETIC ACID
D-AMINO-ACID OXIDASE

VINORELBINE
CYTOCHROME P450 2D6 — C — $Ki = 22\ \mu M$

VINPOCETINE
3′,5′-CYCLIC-NUCLEOTIDE PHOSPHODIESTERASE (cycAMP) (Ca^{2+}/CALMODULIN)
3′,5′-CYCLIC-NUCLEOTIDE PHOSPHODIESTERASE (Ca^{2+}/CALMODULIN) — NC — $Ki = 14\ \mu M$

4-VINYL-4-AMINOBUTYRRATE
4-AMINOBUTYRATE TRANSAMINASE

(±)α-VINYLARGININE
ARGININE DECARBOXYLASE — IR — $Ki = 1.8$ mM

2-VINYL DIHYDROSPHINGOSINE-1-PHOSPHATE
SPHINGOSINE-1-PHOSPHATE LYASE — $I50 = 2.4\ \mu M$

2-VINYL DOPA
AROMATIC-L-AMINO-ACID DECARBOXYLASE — SSI

VINYLGLYCINE
ALANINE RACEMASE — IR
L-AMINO-ACID OXIDASE
D-AMINO-ACID TRANSAMINASE
THREONINE SYNTHASE

2-VINYLINOSINIC ACID
IMP DEHYDROGENASE — IR

(±)α-VINYLLYSINE
LYSINE DECARBOXYLASE — IR — $Ki = 500\ \mu M$

18-VINYLPROGESTERONE
CYTOCHROME P450 11β — SSI
18-HYDROXYLASE — $I50 = 40$ nM
STEROID 18β-MONOOXYGENASE — $Ki = 5\ \mu M$

N-VINYLPROTOPORPHYRIN IX
FERROCHELATASE

2-VINYLPYRIDINE
CANCER PROCOAGULANT — 55% 100 μM

4-VINYLPYRIDINE
HYDROXYMETHYLGLUTARYL-CoA REDUCTASE

VIPERA AMMODYTES PHOSPHOLIPASE A2 INHIBITOR
PHOSPHOLIPASE A2

VIRAL PROTEIN p12
EXORIBONUCLEASE H

VIRENSIC ACID
INTEGRASE — $I50 = 4.6\ \mu M$

VIRGINIAMYCIN
PEPTIDYLTRANSFERASE

VIRIDIN
PROTEIN TYROSINE KINASE pp60 v-arc — $I50 = 30\ \mu M$

VIRIDIOFUNGIN A
SERINE C-PALMITOYLTRANSFERASE — $I50 = 3$ ng/ml

VIRIDIOFUNGIN B
SERINE C-PALMITOYLTRANSFERASE — $I50 = 3$ ng/ml

VIRIDIOFUNGIN C
SERINE C-PALMITOYLTRANSFERASE — $I50 = 3$ ng/ml

VIRUS ENCODED METHYLASE INHIBITOR
DNA (CYTOSINE-5-)-METHYLTRANSFERASE

VITAMIN A ACETATE
CATHEPSIN D — 22% 1 mM

VITAMIN B12
GLYCEROL DEHYDRATASE
RIBONUCLEASE PANCREATIC

VITAMIN D BINDING PROTEIN
CALCIDOL 1-MONOOXYGENASE

VITAMIN E
ARACHIDONATE 5-LIPOXYGENASE — $I50 = 63$ µM

VITAMIN K1
NAD ADP-RIBOSYLTRANSFERASE — $I50 = 520$ µM
NAD(P)-ARGININE ADP-RIBOSYLTRANSFERASE — $I50 = 1.9$ µM

VITAMIN K2
NAD(P)-ARGININE ADP-RIBOSYLTRANSFERASE — $I50 = 13$ µM

VITAMIN K5
NAD ADP-RIBOSYLTRANSFERASE — $I50 = 1.3$ mM

VITAMIN K EPOXIDE ANALOGUES
VITAMIN K-EPOXIDE REDUCTASE (WARFARIN SENSITIVE)

VITAMIN K QUINONE
VITAMIN K EPOXIDE REDUCTASE

VITELLO INHIBITOR
CHYMOTRYPSIN
PROTEINASE BACILLUS LICHENIFORMIS ALKALINE
TRYPSIN

VITRONECTIN
CHYMASE — $Ki = 2$ nM

VK 19911
PROTEIN KINASE MAP p38

V3 LOOP OF GP120
PROTEINASE MOLT CELLS

VM 26
DNA TOPOISOMERASE (ATP-HYDROLYSING)

VO^{2+}
PHOSPHATASE ALKALINE

V_2O_5
TYROSINE-ESTER SULFOTRANSFERASE

VO3-
PHOSPHATASE ALKALINE

VO^{2+}/NMP COMPLEXES
RIBONUCLEASE PANCREATIC

VOROZOLE
17α-HYDROXYPROGESTERONE ALDOLASE — $I50 = 1.8$ µM
STEROID 17β-MONOOXYGENASE — $I50 = 64$ µM

VP 16
DNA TOPOISOMERASE (ATP-HYDROLYSING) — $I50 = 60\ \mu M$

VP 16-213
DNA TOPOISOMERASE (ATP-HYDROLYSING)

VP26
DNA TOPOISOMERASE (ATP-HYDROLYSING)

VP 612L
1-PHOSPHATIDYLINOSITOL PHOSPHODIESTERASE — 92% 5 mM

VP 616L
1-PHOSPHATIDYLINOSITOL PHOSPHODIESTERASE — C

VX 478
PROTEINASE HIV-1 — $Ki = 600$ pM
PROTEINASE HIV-2 — $Ki = 19$ nM

VZ 564
ARACHIDONATE 5-LIPOXYGENASE

W

W 5
H^+-TRANSPORTING ATPASE (V-TYPE) — $I50 = 470\ \mu M$
PROTEIN KINASE (Ca^{2+} DEPENDENT) — $I50 = 500$ nM

W 7
CALMODULIN-LYSINE N-METHYLTRANSFERASE — $I50 = 95\ \mu M$
CAM KINASE II — $I50 = 15\ \mu M$
3′,5′-CYCLIC-NUCLEOTIDE PHOSPHODIESTERASE (Ca^{2+}/CALMODULIN) — $I50 = 48\ \mu M$
H^+-TRANSPORTING ATPASE (V-TYPE) — $I50 = 71\ \mu M$
H^+-TRANSPORTING PYROPHOSPHATASE VACUOLAR — $I50 = 191\ \mu M$
MYOSIN-LIGHT-CHAIN KINASE — $I50 = 38\ \mu M$
NITRIC-OXIDE SYNTHASE II
PHOSPHOLIPASE A2 — $I50 = 92\ \mu M$
PHOSPHOLIPASE A2 — $I50 = 144\ \mu M$
PROTEIN KINASE C — 50% 100 μM
PROTEIN KINASE (Ca^{2+} DEPENDENT) — $I50 = 150$ nM
SPHINGOMYELIN PHOSPHODIESTERASE

W 13
3′,5′-CYCLIC-NUCLEOTIDE PHOSPHODIESTERASE (Ca^{2+}/CALMODULIN) — $I50 = 84\ \mu M$
PHOSPHOLIPASE A2 — $I50 = 330\ \mu M$
PHOSPHOLIPASE A2 — $I50 = 140\ \mu M$

W 77
GLUTATHIONE TRANSFERASE — C — $Ki = 520\ \mu M$

3744W
PROTEIN KINASE bcr-abl — $I50\%$ 30 μM
PROTEIN KINASE Cβ2 — $I50\%$ 60 μM
PROTEIN TYROSINE KINASE bcr-abl — $I50\%$ 30 μM
PROTEIN TYROSINE KINASE erbB 2 RECEPTOR — 55% 50 μM
PROTEIN TYROSINE KINASE c-abl — 79% 50 μM
PROTEIN TYROSINE KINASE PDGFR — $I50 = 15$ nM

WA 8242A
PHOSPHOLIPASE A2-1 — $I50 = 1.5$ nM
PHOSPHOLIPASE A2-2 — $I50 = 250$ pM
PHOSPHOLIPASE A2 (SECRETORY)

WA 8242B
PHOSPHOLIPASE A2-1 — $I50 = 1.1$ nM

PHOSPHOLIPASE A2-2 $I50 = 140$ pM
PHOSPHOLIPASE A2 (SECRETORY)

WA 8242C
PHOSPHOLIPASE A2 (SECRETORY)

WARANGALONE
PROTEIN KINASE A $I50 = 3.5$ μM

WARFARIN
γ-GLUTAMYL CARBOXYLASE
NAD(P)H DEHYDROGENASE (QUINONE) $I50 = 20$ μM
PROTEINASE HIV $I50 = 18$ μM
PROTEINASE HIV-1 C $Ki = 3.3$ μM
THYROXINE DEIODINASE
VITAMIN K-EPOXIDE REDUCTASE (WARFARIN SENSITIVE) $I50 = 2$ μM

WAY 121509
ALDEHYDE REDUCTASE $I50 = 14$ nM

WAY 121520
ARACHIDONATE 5-LIPOXYGENASE $I50 = 10$ nM
ARACHIDONATE 5-LIPOXYGENASE $I50 = 4$ nM
PHOSPHOLIPASE A2 $I50 = 4$ μM

WAY 121898
STEROL ESTERASE $I50 = 650$ nM

WAY 122220
PHOSPHOLIPASE A2 $I50 = 15$ μM

WAY 124466
PEPTIDYLPROLYL ISOMERASE $Ki = 13$ nM

WAY-PDA 641
3′,5′-CYCLIC-NUCLEOTIDE PHOSPHODIESTERASE III $I50 = 15$ μM
3′,5′-CYCLIC-NUCLEOTIDE PHOSPHODIESTERASE IV $I50 = 420$ nM

WCPI-3
PAPAIN $Ki = 6$ nM

WF 3681
ALDEHYDE REDUCTASE $I50 = 250$ nM

WHEAT α-AMYLASE INHIBITOR 0.19
α-AMYLASE $Ki = 2.9$ nM

WHEAT α-AMYLASE INHIBITOR 0.28
α-AMYLASE $Ki = 1.3$ nM

WHEAT AMYLASE INHIBITOR (WRP 25)
α-AMYLASE $Ki = 90$ nM
α-AMYLASE 1 $Ki = 50$ nM
α-AMYLASE 1 $Ki = 980$ nM
α-AMYLASE 2 $Ki = 50$ nM
α-AMYLASE 2 $Ki = 20$ nM

WHEAT α-AMYLASE SUBTILISIN INHIBITOR
α-AMYLASE

WHEAT DIMERIC α-AMYLASE INHIBITOR
α-AMYLASE 78% 1 μg

WHEAT ENDOSPERM TRYPSIN INHIBITOR
TRYPSIN

WHEAT FLOUR α-AMYLASE INHIBITOR 1
α-AMYLASE 92% 12 μg

WHEAT FLOUR α-AMYLASE INHIBITOR 2
α-AMYLASE 95% 2 μg

WHEAT FLOUR α-AMYLASE INHIBITOR 3
α-AMYLASE 95% 1 μg

WHEAT FLOUR α-AMYLASE INHIBITOR 4
α-AMYLASE 95% 250 ng

WHEAT GERM AGLUTININ
CYTOCHROME-b5 REDUCTASE

WHEAT LIPASE INHIBITOR
TRIACYLGLYCEROL LIPASE

WHEAT MONOMERIC α-AMYLASE INHIBITOR
α-AMYLASE 64% 5 μg

WIN 41662
PROTEIN TYROSINE KINASE p56lck *I*50 = 30 μM
PROTEIN TYROSINE KINASE PDGFR *K*i = 15 nM

WIN 58237
3′,5′-CYCLIC-NUCLEOTIDE PHOSPHODIESTERASE I *I*50 850 nM
3′,5′-CYCLIC-NUCLEOTIDE PHOSPHODIESTERASE III *I*50 14 μM
3′,5′-CYCLIC-NUCLEOTIDE PHOSPHODIESTERASE IV *I*50 = 300 nM
3′,5′-CYCLIC-NUCLEOTIDE PHOSPHODIESTERASE V C *K*i = 170 nM

WIN 58993
3′,5′-CYCLIC-NUCLEOTIDE PHOSPHODIESTERASE III *K*i = 25 nM

WIN 61626
3′,5′-CYCLIC-NUCLEOTIDE PHOSPHODIESTERASE I *I*50 = 500 nM

WIN 61651
PROTEIN TYROSINE KINASE p56lck *I*50 = 18 μM
PROTEIN TYROSINE KINASE PDGFR *I*50 = 10 μM

WIN 61691
3′,5′-CYCLIC-NUCLEOTIDE PHOSPHODIESTERASE I *I*50 = 85 nM
3′,5′-CYCLIC-NUCLEOTIDE PHOSPHODIESTERASE III *I*50 = 290 nM
3′,5′-CYCLIC-NUCLEOTIDE PHOSPHODIESTERASE V *I*50 = 3.1 μM

WIN 61972
3′,5′-CYCLIC-NUCLEOTIDE PHOSPHODIESTERASE I *I*50 = 80 nM
3′,5′-CYCLIC-NUCLEOTIDE PHOSPHODIESTERASE V *I*50 = 820 nM

WIN 62005
3′,5′-CYCLIC-NUCLEOTIDE PHOSPHODIESTERASE III *K*i = 26 nM

WIN 62582
3′,5′-CYCLIC-NUCLEOTIDE PHOSPHODIESTERASE III

WIN 62785
ELASTASE LEUKOCYTE *K*i = 300 pM

WIN 63110
ELASTASE LEUKOCYTE *K*i = 7 pM

WIN 63291
3′,5′-CYCLIC-NUCLEOTIDE PHOSPHODIESTERASE III

WIN 63394
ELASTASE LEUKOCYTE *K*i = 23 pM

WIN 63759
ELASTASE LEUKOCYTE SSI *K*i = 13 pM

WIN 63995
ELASTASE LEUKOCYTE — *K*i = 270 pM

WIN 64593
DNA TOPOISOMERASE (ATP-HYDROLYSING) — *I*50 = 96 nM

WIN 64733
ELASTASE LEUKOCYTE — SSI — *K*i = 14 pM

WIN 65579
3′,5′-CYCLIC-NUCLEOTIDE PHOSPHODIESTERASE III — *I*50 = 3 μM
3′,5′-CYCLIC-NUCLEOTIDE PHOSPHODIESTERASE IV — *I*50 = 135 μM
3′,5′-CYCLIC-NUCLEOTIDE PHOSPHODIESTERASE V — *I*50 = 2 μM

WIN 65936
ELASTASE LEUKOCYTE — *K*i = 66 pM

WIN 67694
CASPASE-1

WIN 68123
ELASTASE LEUKOCYTE — *K*i = 380 pM

WIN 68769
ELASTASE LEUKOCYTE — *K*i = 22 pM

WINGED BEAN CHYMOTRYPSIN INHIBITOR
CHYMOTRYPSIN

WINGED BEAN KUNITZ CHYMOTRYPSIN INHIBITOR
CHYMOTRYPSIN

WOGONIN
XANTHINE OXIDASE — *I*50 = 176 μM

WOLFRAMATE
6-PHOSPHOFRUCTO-2-KINASE
TRIMETHYLAMINE-N-OXIDE REDUCTASE

WOODFRUTICOSIN
DNA TOPOISOMERASE (ATP-HYDROLYSING)

WOODWARD'S REAGENT K
ACYLAMINOACYL-PEPTIDASE
β-GLUCOSIDASE — IR
SULFITE OXIDASE
XYLOSE ISOMERASE

WORTMANNIN
DNA PROTEIN KINASE
MYOSIN-LIGHT-CHAIN KINASE — *I*50 = 170 nM
1-PHOSPHATIDYLINOSITOL 3-KINASE — *I*50 = 5 nM
1-PHOSPHATIDYLINOSITOL 3-KINASE — IR — *I*50 = 2 nM
1-PHOSPHATIDYLINOSITOL 3-KINASE — IR — *I*50 = 4.2 nM
1-PHOSPHATIDYLINOSITOL 3-KINASE — *I*50 = 1 nM
PHOSPHOLIPASE A2 — *I*50 = 2 nM
SERINE KINASE ASSOCIATED WITH 3-PHOSPHATIDYLINOSITOL KINASE

WR 1065
DNA TOPOISOMERASE (ATP-HYDROLYSING)

WR 99210
DIHYDROFOLATE REDUCTASE

WS 1358
DEHYDROPEPTIDASE

WS 1358A1
DEHYDROPEPTIDASE

WS 79089A

COLLAGENASE		$I50 = 102\ \mu M$
ENDOTHELIN CONVERTING ENZYME		$I50 = 730$ nM

WS 9659A

STEROID 5α-REDUCTASE

WS 75624B

ENDOTHELIN CONVERTING ENZYME		$I50$ 0 30 ng/ml

WS 79089B

COLLAGENASE		$I50 = 49\ \mu M$
ENDOTHELIN CONVERTING ENZYME	C	$Ki = 89$ nM
ENDOTHELIN CONVERTING ENZYME		$I50 = 140$ nM

WS 9659B

STEROID 5α-REDUCTASE

WS 79089C

COLLAGENASE		$I50 = 61\ \mu M$
ENDOTHELIN CONVERTING ENZYME		$I50 = 3.4\ \mu M$

WSP 1

AROMATASE		$Ki = 1\ \mu M$

WSP 3

AROMATASE		$I50 = 3\ \mu M$

WY 14643

1-ACYLGLYCEROPHOSPHOCHOLINE O-ACYLTRANSFERASE		
LONG-CHAIN-FATTY-ACID CoA-LIGASE	C	$Ki = 110$ nM
OXIDATIVE PHOSPHORYLATION UNCOUPLER		

WY 26769

Na^+/K^+-EXCHANGING ATPASE		$I50 = 42\ \mu M$

WY 26876

Na^+/K^+-EXCHANGING ATPASE		$I50 = 18\ \mu M$

WY 27569

THROMBOXANE A SYNTHASE		$I50 = 2.8\ \mu M$

WY 45911

ARACHIDONATE 5-LIPOXYGENASE		$I50 = 1.4\ \mu M$
ARACHIDONATE 12-LIPOXYGENASE		$I50 = 2.7\ \mu M$
CYCLOOXYGENASE		$I50 = 40\ \mu M$
LIPOXYGENASE		$I50 = 10\ \mu M$

WY 47288

ARACHIDONATE 5-LIPOXYGENASE		$I50 = 400$ nM
CYCLOOXYGENASE		$I50 = 6.3\ \mu M$

WY 22284A

ALCOHOL DEHYDROGENASE (NADP)

WY 50295 TROMETHAMINE

ARACHIDONATE 5-LIPOXYGENASE		$I50 = 21\ \mu M$
3′,5′-CYCLIC-NUCLEOTIDE PHOSPHODIESTERASE III		$I50 = 16\ \mu M$
3′,5′-CYCLIC-NUCLEOTIDE PHOSPHODIESTERASE IV		$I50 = 8.9\ \mu M$

X

Inhibitor / Enzyme	Type	Value
XANTHENE		
AMINE OXIDASE MAO-A		*I*50 = 8.9 μM
AMINE OXIDASE MAO-B		*I*50 = 23 μM
XANTHENE-9-CARBOXYLIC ACID		
AMINE OXIDASE MAO-A		*I*50 = 137 μM
XANTHINE		
HYPOXANTHINE PHOSPHORIBOSYLTRANSFERASE		
PTERIDINE OXIDASE		
URATE OXIDASE		
XANTHINE PHOSPHORIBOSYLTRANSFERASE		
XANTHINE-5′-MONOPHOSPHATE		
IMP DEHYDROGENASE	C	*K*i = 11 μM
IMP DEHYDROGENASE	C	*K*i = 67 μM
IMP DEHYDROGENASE	C	*K*i = 85 μM
XANTHINE RIBONUCLEOTIDE		
PYRIMIDINE-NUCLEOSIDE PHOSPHORYLASE	C	*K*i = 130 μM
URIDINE PHOSPHORYLASE	C	*K*i = 130 μM
XANTHOANGELOL		
H^+/K^+-EXCHANGING ATPASE	C	*I*50 = 1.8 μM
4-NITROPHENYLPHOSPHATASE (K^+ STIMULATED)		*I*50 = 1.3 μM
XANTHOGENATE		
CATECHOL OXIDASE		
CYCLOHEXYLAMINE OXIDASE		
LACCASE		
XANTHOHUMOL		
DIACYLGLYCEROL O-ACYLTRANSFERASE		*I*50 = 50 μM
XANTHOHUMOL B		
DIACYLGLYCEROL O-ACYLTRANSFERASE		*I*50 = 194 μM
XANTHOMATIN		
KYNURENINE 3-MONOOXYGENASE		
XANTHONE		
AMINE OXIDASE MAO-A		*I*50 = 840 nM
AMINE OXIDASE MAO-B		*I*50 = 122 μM
XANTHONE-2-CARBOXYLIC ACIDS		
ALDEHYDE REDUCTASE		
XANTHOPTERIN		
SEPIAPTERIN DEAMINASE		
XANTHOSINE		
DIHYDROOROTASE	C	*K*i = 800 μM
PURINE NUCLEOSIDASE		
XANTHOSINEDIPHOSPHATE		
ADENYLOSUCCINATE SYNTHASE		42% 10 mM
3-XANTHOSINE MONOPHOSPHATE		
OROTIDINE-5′-PHOSPHATE DECARBOXYLASE		*K*i = 450 nM
XANTHOSINE-5′-MONOPHOSPHATE		
ADENYLOSUCCINATE SYNTHASE	C	*K*i = 140 μM
IMP CYCLOHYDROLASE	C	*K*i = 120 nM
9-XANTHOSINE MONOPHOSPHATE		
OROTIDINE-5′-PHOSPHATE DECARBOXYLASE		*K*i = 6.3 μM

Inhibitor / Enzyme	Type	Value
XANTHOSINE-5-PHOSPHATE		
IMP DEHYDROGENASE 1		Ki = 80 μM
IMP DEHYDROGENASE 2		Ki = 94 μM
XANTHOTOXIN		
CYTOCHROME P450 1A1	C	Ki = 758 nM
CYTOCHROME P450 2B1		I50 = 1.3 μM
CYTOCHROME P450 6D1		I50 = 16 nM
O-DEMETHYLASE (CYT P450)	IR	I50 = 400 nM
XANTHURENIC ACID		
ACETYL-CoA CARBOXYLASE	NC	Ki = 160 μM
AMINE OXIDASE MAO-B		
myo-INOSITOL OXYGENASE		I50 = 190 μM
NAD ADP-RIBOSYLTRANSFERASE		
PYRIDOXAL KINASE		
XANTHYLETIN		
CYTOCHROME P450 2B1		I50 = 21 μM
XDP		
XANTHINE PHOSPHORIBOSYLTRANSFERASE		
XENON		
Ca^{2+}-TRANSPORTING ATPASE		
XENOPSIN		
NEUROLYSIN		Ki = 2.1 μM
XENOPUS LAEVIS OOCYTE INHIBITOR		
NAD ADP-RIBOSYLTRANSFERASE		
XENOPUS LAEVIS OVAR DNA TOPOISOMERASE INHIBITOR		
DNA TOPOISOMERASE		
XERULINIC ACID		
HYDROXYMETHYLGLUTARYL-CoA SYNTHASE		
XESTOCLYCAMINE		
PROTEIN KINASE C		I50 = 4 μg/ml
XESTOQUINOLIDE A		
PROTEIN TYROSINE KINASE pp60 v-arc		I50 = 80 μM
XESTOQUINONEN		
PROTEIN TYROSINE KINASE pp60c-Src	IR	I50 = 60 μM
5-X-3HYDROXY-4-METHOXYBENZOIC ACIDS		
CATECHOL O-METHYLTRANSFERASE		
XIAP		
CASPASE 3		
CASPASE 7		
CASPASE 9		
XIP		
Na^{+}/Ca^{2+} EXCHANGER		
XL 075		
PROTEINASE HIV-1		Ki = 3.4 nM
XM 323		
PROTEINASE HIV-1		Ki = 270 pM
XMEO 8		
CUSATIVIN		
XMP		
ADENYLOSUCCINATE SYNTHASE		42% 200 μM

GMP REDUCTASE		
HYPOXANTHINE PHOSPHORIBOSYLTRANSFERASE		
IMP CYCLOHYDROLASE	C	*K*i = 120 nM
IMP DEHYDROGENASE 2		*I*50 = 770 μM
URIDINE NUCLEOSIDASE		
XANTHINE PHOSPHORIBOSYLTRANSFERASE		
2′-dXMP		
GMP REDUCTASE		
XN 975		
PROTEINASE HIV-1		*K*i = 27 pM
X-PROTEIN		
TRYPTASE TL-2		61% 1 μM
XTP		
XANTHINE PHOSPHORIBOSYLTRANSFERASE		
XU 348		
PROTEINASE HIV-1		
XU 430		
PROTEINASE HIV-1		
XYLAMIDINE		
CYTOCHROME P450 2D6		*K*i = 220 nM
XYLAN		
ENDO-1,4-β-XYLANASE		
XYLARIC ACID		
INTERLEUKIN CONVERTING ENZYME	C	*K*i = 8 μM
XYLERYTHRIN		
XANTHINE OXIDASE		*I*50 = 1.8 μM
XYLIDYL BLUE		
PROTEINASE HIV-1		*I*50 = 40 μM
XYLITOL		
ARABINOSE ISOMERASE		
L-ARABINOSE ISOMERASE		
XYLOSE ISOMERASE	C	*K*i = 1.5 μM
D-XYLITOL		
XYLOSE ISOMERASE	C	*K*i = 2.7 mM
XYLITOL-1,5-DIPHOSPHATE		
FRUCTOSE-BISPHOSPHATE ALDOLASE		*K*i = 2.8 μM
FRUCTOSE-BISPHOSPHATE ALDOLASE		*K*i = 2.8 μM
XYLOBIOSE		
ENDO-1,4-β-XYLANASE		
XYLOCAINE		
15-HYDROXYPROSTAGLANDIN DEHYDROGENASE (NAD)		*I*50 = 50 μM
XYLOCOSONE		
GLUTATHIONE REDUCTASE (NADPH)	IR	
XYLONO-1,4-LACTONE		
XYLAN 1,4-β-XYLOSIDASE		15% 8.2 mM
α-D-XYLOPYRANOSYL β-D-FRUCTOFURANOSIDE		
D-GLUCANSUCRASE	C	*K*i = 2.8 mM

3O-{β-D-XYLOPYRANOSYL(1- > 2)O-β-D-GLUCOPYRANOSYL(1- > 4)[O-βD-GLUCOPYRANOSYL(1- > 2)]α-L-ARABINOSYL}16α-HYDROXY-13β,28EPOXYOLEANANE
PHOSPHOLIPASE D — $I50 = 3\ \mu M$

XYLOSE
ARABINOGALACTAN ENDO-1,4-β-GALACTOSIDASE
D-ARABINOSE 1-DEHYDROGENASE
ENDO-1,4-β-XYLANASE
α-GLUCOSIDASE
THIOGLUCOSIDASE

D-XYLOSE
CELLULOSE 1,4-β-CELLOBIOSIDASE
GLYCEROL DEHYDROGENASE
HEXOKINASE
XYLAN 1,4-β-XYLOSIDASE — $Ki = 26\ mM$

L-XYLOSE
ALDOSE 1-EPIMERASE

XYLULOSE-1,5-BISPHOSPHATE
RIBULOSE-BISPHOSPHATE CARBOXYLASE

D-XYLULOSE-1,5-BISPHOSPHATE
RIBULOSE-BISPHOSPHATE CARBOXYLASE

D-XYLULOSE
GLUCOKINASE — C — $Ki = 120\ mM$
KETOHEXOKINASE

XYLULOSE-1,5-DIPHOSPHATE
FRUCTOSE-BISPHOSPHATE ALDOLASE — $Ki = 50\ \mu M$

XYLULOSE-5-PHOSPHATE
GLUCOSE-6-PHOSPHATE ISOMERASE — $Ki = 700\ \mu M$

L-XYLULOSE-5-PHOSPHATE
GLUCOSE-6-PHOSPHATE ISOMERASE — $Ki = 700\ \mu M$

3,4-XYLYL METHYLCARBAMATE
ACETYLCHOLINESTERASE — $Ki = 8.1\ \mu M$

3,5-XYLYL METHYLCARBAMATE
ACETYLCHOLINESTERASE — $Ki = 3.7\ \mu M$

Y

Y^{3+}
ACETYLCHOLINESTERASE — 34% 100 μM
ALCOHOL DEHYDROGENASE — 82% 3 mM
ATP PYROPHOSPHATASE — 67% 400 μM
CATALASE — 58% 2 mM
DEOXYRIBONUCLEASE — 50% 200 μM
DIHYDROPTERIDINE REDUCTASE — 40% 3.7 μM
FERROXIDASE — 55% 100 μM
GLUCOSE-6-PHOSPHATE 1-DEHYDROGENASE — 57% 400 μM
ISOCITRATE DEHYDROGENASE — 80% 830 μM
L-LACTATE DEHYDROGENASE — 100% 830 μM
MALATE DEHYDROGENASE — 80% 830 μM
PLASMIN — 50% 37 μM
PLASMINOGEN ACTIVATOR — 50% 31 μM

Y 20811
THROMBOXANE A SYNTHASE

Y 208118
THROMBOXANE A SYNTHASE

Y 29794
PROLYL OLIGOPEPTIDASE C $Ki = 1$ nM

YAGAVVNDL
RIBONUCLEOSIDE-DIPHOSPHATE REDUCTASE $I50 = 28$ μM

YAKUCHINONE B
MONOPHENOL MONOOXYGENASE $I50 = 57$ μM

YARAKR-CMK
FURIN

YCQRTLREIQILLRF
MEK C $Ki = 840$ nM

YEAST HYDROXYMETHYLGLUTARYL CoA REDUCTASE INHIBITOR
HYDROXYMETHYLGLUTARYL-CoA REDUCTASE (NADPH)

YEAST PROTEINASE INHIBITOR
SACCHAROPEPSIN

YF 044P-D
CANDIDAPEPSIN $I50 = 640$ nM
CATHEPSIN D

YF 0200R-A
CANDIDAPEPSIN $I50 = 650$ μM

YF 0200R-B
CANDIDAPEPSIN $I50 = 620$ μM

YH 1885
H^+-TRANSPORTING ATPASE

YJA 20379-4
H^+/K^+-EXCHANGING ATPASE $I50 = 81$ μM

YJA 20379-5
H^+/K^+-EXCHANGING ATPASE $I50 = 31$ μM

YL 01869P
GELATINASE B $I50 = 1.6$ μM

YM 175
FARNESYL-DIPHOSPHATE FARNESYLTRANSFERASE C $Ki = 57$ nM
PROTEIN TYROSINE-PHOSPHATASE MEG-1 $I50 = 412$ μM

YM 511
AROMATASE $I50 = 120$ pM
AROMATASE $I50 = 400$ pM

YM 553
AROMATASE $I50 = 38$ pM

YM 750
STEROL O-ACYLTRANSFERASE $I50 = 180$ nM

YM 19020
Na^+/K^+-EXCHANGING ATPASE

YM 21095
RENIN

YM 26365
CATHEPSIN D $I50 = 17$ μM
RENIN $I50 = 2.9$ μM
RENIN $I50 = 4.5$ μM

YM 26567-1
PHOSPHOLIPASE A2-1 C $Ki = 12$ μM
PHOSPHOLIPASE A2-2 C $Ki = 1.6$ μM

YM 26734

PHOSPHOLIPASE A2-1		Ki = 6.8 µM
PHOSPHOLIPASE A2-2	C	Ki = 48 nM

YM 47141

CATHEPSIN G		Ki = 9.2 µM
α-CHYMOTRYPSIN		Ki = 1.3 µM
ELASTASE		
ELASTASE LEUKOCYTE		Ki = 210 nM

YM 47142

ELASTASE

YM 51084

CATHEPSIN L		I50 = 9.6 nM

YM 51085

CATHEPSIN L		I50 = 9.6 nM

YM 60828

COAGULATION FACTOR Xa		Ki = 1.3 nM
KALLIKREIN PLASMA		Ki = 400 nM
PLASMIN		Ki = 2.2 µM
t-PLASMINOGEN ACTIVATOR		Ki = 1.9 µM
TRYPSIN		Ki = 46 nM

YM 9429

3β-HYDROXYSTEROID Δ7-REDUCTASE		Ki = 40 µnM

YM 17E

STEROL O-ACYLTRANSFERASE		I50 = 44 nM

YOHIMBINE

CYTOCHROME P450 2D6		Ki = 180 nM

γ YOHIMBINE

CYTOCHROME P450 2D6		Ki = 31 nM

Y-c-[D-PHE(3-I)TYR-GSFG]KR-NH_2

PROTEIN TYROSINE KINASE p60c-Src		I50 = 130 nM

YS 980

PEPTIDYL-DIPEPTIDASE A		I50 = 3.5 nM
XAA-PRO AMINOPEPTIDASE		100% 1 mM
XAA-PRO AMINOPEPTIDASE		I50 = 20 µM
XAA-TRP AMINOPEPTIDASE		I50 = 18 µM

YS 3025

PROTEINASES SERINE

YT 18

ARACHIDONATE 5-LIPOXYGENASE		I50 = 7.5 µM
ARACHIDONATE 5-LIPOXYGENASE		I50 = 1.5 µM

YTR 830H

β-LACTAMASE

YVADcmk

CASPASE-1

3,4-YXILIDENE

β-FRUCTOFURANOSIDASE

3,5-YXILIDENE

β-FRUCTOFURANOSIDASE

YYGAKIYRPDKM

TRYPSIN		Ki = 9.4 µM

Z

Z 321
PROLYL OLIGOPEPTIDASE

Z-AA-PHE{PO_2^-O}PHE-O-(3-(4-PYRIDYL)PROPYL ESTER
PEPSIN | IR | *K*i = 340 pM

ZABICIPRILATE
PEPTIDYL-DIPEPTIDASE A | | *I*50 = 1.8 nM

Z-ALA-ALA-ALA
CARBOXYPEPTIDASE A | C | *K*i = 76 nM

Z-ALA-ALA-ALA-$^P(OPH)_2$
ELASTASE LEUKOCYTE

Z-ALA-ALA$(CO)_3OBnH_2O$
α-CHYMOTRYPSIN | | *K*i = 210 μM
ELASTASE LEUKOCYTE | | *K*i = 800 nM

Z-ALA-ALA-PHE-CH_3
THERMITASE | | *K*i = 460 nM

Z-ALA-ALA-PHE-CH_2CL
PROTEINASE SERINE DICHELOBACTER NODOSUS BASIC | | 100% 1 mM

Z-ALA-ALA-PHE-$COCH_2Cl$
PROTEINASE DICHELOBACTER NODOSUS EXTRACELLULAR | | 77% 1 mM

Z-ALA-ALA-PHE-H
α-CHYMOTRYPSIN | | *K*i = 2.2 μM
SUBTILISIN | | *K*i = 180 nM

Z-L-ALA-L-ALA-L-PRO-D,L-ALA-TRIFLUOROMETHYLKETONE
ELASTASE LEUKOCYTE | C | *K*i = 680 nM
α-LYTIC ENDOPEPTIDASE | C | *K*i = 5 μM

Z-ALA-ALA-PRO-CHN_2
PROLYL OLIGOPEPTIDASE | IR |
PROLYL OLIGOPEPTIDASE | C | *K*i = 1.7 μM

Z-ALA-ALA-PRO-VAL-TRIFLUOROMETHYLKETONE
ELASTASE LEUKOCYTE | | *K*i = 1 nM

Z-ALA-CH_2Cl
PROLIPOPROTEIN-SIGNAL PEPTIDASE

Z-ALA-GLY-PHE-CHLOROMETHYLKETONE
CHYMOTRYPSIN | IR |

Z-ALA-GLY^P-(O)PHE
CARBOXYPEPTIDASE A | | *K*i = 156 pM

Z-ALA-ILE BORONIC ACID
ELASTASE PANCREATIC | | *K*i = 300 nM

Z-D-ALA-LEU-PHE-(OCO-2,6-Fl(2)-Ph)
CALPAIN I

Z-ALA-PHE-CH_2Cl
PROTEINASE K

Z-ALA-PRO
PROLYL OLIGOPEPTIDASE

Z-β-AMINOCAPROYL-D-PHE
SERINE CARBOXYPEPTIDASE | | *K*i = 950 μM

Z-(AMINOMETHYL)PHOSPHONYL-LEU-ALA
INTERSTITIAL COLLAGENASE *K*i = 78 μM

Z-(AMINOMETHYL)PHOSPHONYL-LEU-ALA-GLY
INTERSTITIAL COLLAGENASE *K*i = 14 μM

N-Z-AMINOPHOSPHONAMIDIC ACID
D-ALANINE-D-ALANINE LIGASE

D,L-Z-2-AMINO-5-PHOSPHONO-3-PENTENOIC ACID
THREONINE SYNTHASE IR

Z(4-AMPHGLY)^{P}(OPH)$_2$
KALLIKREIN PLASMA

ZANAMIVIR
SIALIDASE *I*50 = 2 nM

ZAPRINAST
3′,5′-CYCLIC-NUCLEOTIDE PHOSPHODIESTERASE (cycAMP) *I*50 = 36 μM
3′,5′-CYCLIC-NUCLEOTIDE PHOSPHODIESTERASE (cycGMP)
3′,5′-CYCLIC-NUCLEOTIDE PHOSPHODIESTERASE I *I*50 = 16 μM
3′,5′-CYCLIC-NUCLEOTIDE PHOSPHODIESTERASE I *I*50 = 3.6 μM
3′,5′-CYCLIC-NUCLEOTIDE PHOSPHODIESTERASE I *I*50 = 5.6 μM
3′,5′-CYCLIC-NUCLEOTIDE PHOSPHODIESTERASE Ia *I*50 = 100 nM
3′,5′-CYCLIC-NUCLEOTIDE PHOSPHODIESTERASE Ib *I*50 = 112 μM
3′,5′-CYCLIC-NUCLEOTIDE PHOSPHODIESTERASE Ic *I*50 = 27 μM
3′,5′-CYCLIC-NUCLEOTIDE PHOSPHODIESTERASE IB *I*50 = 122 μM
3′,5′-CYCLIC-NUCLEOTIDE PHOSPHODIESTERASE II *I*50 = 32 μM
3′,5′-CYCLIC-NUCLEOTIDE PHOSPHODIESTERASE II *I*50 = 66 μM
3′,5′-CYCLIC-NUCLEOTIDE PHOSPHODIESTERASE II *I*50 = 18 μM
3′,5′-CYCLIC-NUCLEOTIDE PHOSPHODIESTERASE II *I*50 = 107 μM
3′,5′-CYCLIC-NUCLEOTIDE PHOSPHODIESTERASE II *I*50 = 155 μM
3′,5′-CYCLIC-NUCLEOTIDE PHOSPHODIESTERASE II (cycGMP STIMULATED) *I*50 = 155 μM
3′,5′-CYCLIC-NUCLEOTIDE PHOSPHODIESTERASE III 14% 10 μM
3′,5′-CYCLIC-NUCLEOTIDE PHOSPHODIESTERASE III (cycGMP INHIBITED) *I*50 = 74 μM
3′,5′-CYCLIC-NUCLEOTIDE PHOSPHODIESTERASE IV *I*50 = 25 μM
3′,5′-CYCLIC-NUCLEOTIDE PHOSPHODIESTERASE IV *I*50 = 100 μM
3′,5′-CYCLIC-NUCLEOTIDE PHOSPHODIESTERASE IV 13% 10 μM
3′,5′-CYCLIC-NUCLEOTIDE PHOSPHODIESTERASE V *I*50 = 427 nM
3′,5′-CYCLIC-NUCLEOTIDE PHOSPHODIESTERASE V *I*50 = 1.5 μM
3′,5′-CYCLIC-NUCLEOTIDE PHOSPHODIESTERASE V *I*50 = 1 μM
3′,5′-CYCLIC-NUCLEOTIDE PHOSPHODIESTERASE V *I*50 = 251 nM
3′,5′-CYCLIC-NUCLEOTIDE PHOSPHODIESTERASE V (cycGMP) *I*50 = 1.5 μM
3′,5′-CYCLIC-NUCLEOTIDE PHOSPHODIESTERASE VIII *I*50 = 24 μM

ZARAGOZIC ACID A
FARNESYL-DIPHOSPHATE FARNESYLTRANSFERASE *K*i = 78 pM
FARNESYL-DIPHOSPHATE FARNESYLTRANSFERASE *I*50 = 500 pM
FARNESYL PROTEINTRANSFERASE *I*50 = 2 μM
FARNESYL PROTEINTRANSFERASE *I*50 = 216 nM
GERANYLGERANYL PROTEINTRANSFERASE I *I*50 = 620 nM
GERANYLGERANYL PROTEINTRANSFERASE II *I*50 = 66 μM

ZARAGOZIC ACID A ANALOGUE
FARNESYL PROTEINTRANSFERASE 1 C *K*i = 1 nM
GERANYLGERANYL PROTEINTRANSFERASE I *I*50 = 1.7 μM
GERANYLGERANYL PROTEINTRANSFERASE II *I*50 = 17 μM

ZARAGOZIC ACID B
FARNESYL-DIPHOSPHATE FARNESYLTRANSFERASE *K*i = 29 pM
FARNESYL-DIPHOSPHATE FARNESYLTRANSFERASE *K*i = 15 pM
FARNESYL-DIPHOSPHATE FARNESYLTRANSFERASE *I*50 = 200 pM
FARNESYL-DIPHOSPHATE FARNESYLTRANSFERASE *I*50 = 200 pM
FARNESYL PROTEINTRANSFERASE *I*50 = 1 μM
FARNESYLPYROPHOSPHATASE *I*50 = 5 μM
GERANYLGERANYLPYROPHOSPHATASE *I*50 = 20 μM

ZARAGOZIC ACID C		
FARNESYL-DIPHOSPHATE FARNESYLTRANSFERASE		Ki = 45 pM
FARNESYL-DIPHOSPHATE FARNESYLTRANSFERASE		I50 = 500 pM
FARNESYL-DIPHOSPHATE FARNESYLTRANSFERASE		Ki = 45 pM
FARNESYL-DIPHOSPHATE FARNESYLTRANSFERASE		I50 = 2 μM
FARNESYL PROTEINTRANSFERASE		I50 = 150 nM
ZARAGOZIC ACID D		
FARNESYL-DIPHOSPHATE FARNESYLTRANSFERASE		I50 = 6 nM
FARNESYL PROTEINTRANSFERASE		I50 = 100 nM
ZARAGOZIC ACID D2		
FARNESYL-DIPHOSPHATE FARNESYLTRANSFERASE		I50 = 2 nM
FARNESYL PROTEINTRANSFERASE		I50 = 100 nM
ZARAGOZID ACID 3D		
FARNESYL PROTEINTRANSFERASE		I50 = 600 nM
ZARAGOZIG ACID A		
FARNESYL-DIPHOSPHATE FARNESYLTRANSFERASE		I50 = 1.3 nM
ZARDAVERINE		
3′,5′-CYCLIC-NUCLEOTIDE PHOSPHODIESTERASE Ib		I50 = 1 mM
3′,5′-CYCLIC-NUCLEOTIDE PHOSPHODIESTERASE II (cycGMP STIMULATED)		I50 = 447 μM
3′,5′-CYCLIC-NUCLEOTIDE PHOSPHODIESTERASE III (cycGMP INHIBITED)		I50 = 600 nM
3′,5′-CYCLIC-NUCLEOTIDE PHOSPHODIESTERASE III (cycGMP INHIBITED)	C	Ki = 450 nM
3′,5′-CYCLIC-NUCLEOTIDE PHOSPHODIESTERASE IV		Ki = 160 nM
3′,5′-CYCLIC-NUCLEOTIDE PHOSPHODIESTERASE IV		I50 = 1.7 μM
3′,5′-CYCLIC-NUCLEOTIDE PHOSPHODIESTERASE V (cycGMP)		I50 = 138 μM
Nα-Z-ARG-ARG-PRO-PHE-HIS-STA-ILE-HIS-Nβ-BOC-LYS METHYL ESTER		
RENIN		
Nα-Z-ARG-ARG-PRO-PHE-HIS-STA-ILE-HIS-N-βBOCLYS METHYL ESTER		
RENIN		
Z-ARG-ILE-PHE-H		
CATHEPSIN B	IR	Ki = 11 nM
PAPAIN	IR	Ki = 900 pM
Z-ARG-LEU-VAL-GLY-CHN_2		
CATHEPSIN B	IR	
PAPAIN	IR	
PROHORMONE THIOL PROTEINASE	IR	
Z-ARG TRIFLUOROMETHYLKETONE		
TRYPSIN		Ki = 600 nM
Z-ASN-PHE-psi[CH(OH)CH_2N]PRO-OBU^t		
PROTEINASE HIV		I50 = 140 nM
Z-ASP-CH_2OC(O)-2,6-DICHLOROBENZENE		
INTERLEUKIN 1β CONVERTING ENZYME	IR	
Z-ASP-FLUOROMETHYLKETONE		
CASPASES		
Z-t.-BUTYL-GLY-LEU-H		
CALPAIN I		I50 = 4 nM
ZD 1033		
AROMATASE		I50 = 15 nM
ZD 1542		
PROSTAGLANDIN-I SYNTHASE		I50 = 18 μM
THROMBOXANE A SYNTHASE		I50 = 16 nM
ZD 1694		
THYMIDYLATE SYNTHASE		

ZD 2138		
ARACHIDONATE 5-LIPOXYGENASE		*I*50 = 3.8 µM
ZD 5522		
ALDEHYDE REDUCTASE	NC	*K*i = 7.2 nM
ZD 7717		
ARACHIDONATE 5-LIPOXYGENASE		*I*50 = 40 nM
ZD 9331		
THYMIDYLATE SYNTHASE		*K*i = 400 pM
α-ZEARALENOL		
OXIDATIVE PHOSPHORYLATION UNCOUPLER		
β-ZEARALENOL		
OXIDATIVE PHOSPHORYLATION UNCOUPLER		
ZEARALENONE		
OXIDATIVE PHOSPHORYLATION UNCOUPLER		
ZEATIN		
INDOLE-3-ACETATE β-GLUCOSYLTRANSFERASE		
ZEATIN RIBOSIDE		
ADENOSINE NUCLEOSIDASE		
ZEAXANTHIN		
β-CAROTENE 15,15′-DIOXYGENASE	NC	*K*i = 7.8 µM
ZEBULARINE		
CYTIDINE DEAMINASE	C	*K*i = 816 nM
CYTIDINE DEAMINASE	C	*K*i = 360 nM
CYTIDINE DEAMINASE		*K*i = 2.3 µM
CYTIDINE DEAMINASE		*K*i = 38 µM
Z, E, E GERANYL-GERANYL DIPHOSPHATE		
DIMETHYLALLYLtransTRANSFERASE		43% 20 µM
ZENARESTAT		
ALDEHYDE REDUCTASE		*I*50 = 12 nM
14-3-3-ZETA		
MYOSIN II HEAVY CHAIN PROTEIN KINASE C		
ZG 1494α		
PLATELET ACTIVATING FACTOR ACETYLTRANSFERASE		*I*50 = 304 µM
Z-GGL-CHO		
MULTICATALYTIC ENDOPEPTIDASE COMPLEX (CHYMOTRYPSIN LIKE ACTIVITY		
Z-GLN-TYR		
PENICILLOPEPSIN		76% 1 mM
Z-GLU		
ENVELYSIN	C	*K*i = 2.4 mM
PAPAIN		
PENICILLOPEPSIN		50% 1 mM
Z-GLU-TYR		
PENICILLOPEPSIN		77% 1 mM
Z-GLY		
CARBOXYPEPTIDASE A	C	*K*i = 27 mM
Z-GLY-GLY-LEUCINAL		
MULTICATALYTIC ENDOPEPTIDASE COMPLEX		
MULTICATALYTIC ENDOPEPTIDASE COMPLEX (CHYMOTRYPSIN LIKE ACTIVITY		

Z-GLY-GLY-PHE-CH_2CL		
PROTEINASE SERINE DICHELOBACTER NODOSUS BASIC		100% 1 mM
Z-GLY-GLY-PHE-$COCH_2Cl$		
PROTEINASE DICHELOBACTER NODOSUS EXTRACELLULAR		86% 1 mM
Z-GLY-GLY^P-(O)PHE		
CARBOXYPEPTIDASE A		*K*i = 4.2 nM
Z-GLY-LEUOH		
COLLAGENASE CLOSTRIDIUM HISTOLYTICUM	C	*K*i = 1.5 Mm
Z-GLY-LEU-PHE CHLOROMETHYLKETONE		
CATHEPSIN G	IR	
PROTEINASES SERINE		
TRYPTASE	IR	
Z-GLY-D-PHE		
SERINE CARBOXYPEPTIDASE		*K*i = 4.5 mM
Z-GLY-PHE-GLY-SEMICARBAZONE		
CATHEPSIN B	IR	*K*i = 1 µM
PAPAIN	IR	*K*i = 9.2 nM
Z-GLY-PHE-NHO-NBZ		
THERMITASE		
Z-GLY^P-(O)PHE		
CARBOXYPEPTIDASE A		*K*i = 32 nM
Z-GLY-PRO-CHLOROMETHYLKETONE		
PROLYL OLIGOPEPTIDASE		*K*i = 70 µM
Z-GLY-PRO-DIAZOMETHYLKETONE		
PROLYL OLIGOPEPTIDASE		*I*50 < 60 nM
Z-GLY-PRO-LEU-NHOH		
COLLAGENASE CLOSTRIDIUM HISTOLYTICUM	C	*K*i = 200 µM
Z-GLY-PRO-LEUOH		
COLLAGENASE CLOSTRIDIUM HISTOLYTICUM	C	*K*i = 4.3 mM
Z-GLY-PRO-NHOH		
COLLAGENASE CLOSTRIDIUM HISTOLYTICUM		*K*i = 240 nM
Z-GLY-PRO-PHE-LEUCINAL		
MULTICATALYTIC ENDOPEPTIDASE COMPLEX (BrAAP)	C	*K*i = 1.5 µM
MULTICATALYTIC ENDOPEPTIDASE COMPLEX (CHYMOTRYPSIN LIKE ACTIVITY)	C	*K*i = 41 µM
MULTICATALYTIC ENDOPEPTIDASE COMPLEX (PGHP)	C	*K*i = 3.1 µM
MULTICATALYTIC ENDOPEPTIDASE COMPLEX (SNAAP)	C	*K*i = 2.3 µM
Z-HIS-TYR(OPO_4^{2-})SER(OBN)$TRPNH_2$		
FARNESYL PROTEINTRANSFERASE		*K*i = 2.4 nM
Z-HOMOCYS-LEU-OCH_3		
ENDOTHELIN CONVERTING ENZYME		*I*50 = 4 µM
Z-L-HOMOPHE-(R,S)-3-AMINO-2-OXOVALERYL-D-LEU-L-VAL		
CATHEPSIN B		*I*50 = 4.1 µg/ml
ELASTASE LEUKOCYTE		*I*50 = 34 µg/ml
PROLYL OLIGOPEPTIDASE		*I*50 = 4.7 ng/ml
N-Z-ILE-GLU(O-t-BUTYL)-ALA-LEUCINAL		
MULTICATALYTIC ENDOPEPTIDASE COMPLEX (CHYMOTRYPSIN LIKE ACTIVITY)		*I*50 = 250 nM
Z-ILE-GLU(OtBU)ALA-LEUCINALE		
MULTICATALYTIC ENDOPEPTIDASE COMPLEX (ACIDIC CHYMOTRYPSIN LIKE A		*I*50 = 250 nM
MULTICATALYTIC ENDOPEPTIDASE COMPLEX (NEUTRAL CHYMOTRYPSIN LIKE		*I*50 = 6.5 µM

Z-ILE-GLU-PRO-PHE-CO-GLU-ASP-ARG-OME
CHYMASE — *K*i = 100 nM
CHYMOTRYPSIN — *K*i = 40 μM

Z-ILE-GLU-PRO-PHE-CO_2-ME
CHYMASE — *K*i = 1 nM
CHYMOTRYPSIN — *K*i = 10 nM

Z-ILE-TYR
STREPTOPAIN — *K*i = 2.3 mM

ZILEUTON
ARACHIDONATE 5-LIPOXYGENASE — *I*50 = 10 μM
ARACHIDONATE 5-LIPOXYGENASE
ARACHIDONATE 5-LIPOXYGENASE — *I*50 = 500 nM
ARACHIDONATE 15-LIPOXYGENASE — 51% 500 μM
LIPOXYGENASE — *I*50 > 600 μM

ZIMELDINE
CYTOCHROME P450 2D6 — *K*i = 1.6 μM

ZINCON
AMINOBUTYRALDEHYDE DEHYDROGENASE
FORMATE DEHYDROGENASE

ZINC PROTOPORPHYRIN IX
NITRIC-OXIDE SYNTHASE

N-Z-ISATIN
CHYMOTRYPSIN — IR

ZK 114043
AROMATASE — *K*i = 55 μM

ZK 114863
AROMATASE — *K*i = 13 μM

ZK 138723
AROMATASE

ZK 62711
3′,5′-CYCLIC-NUCLEOTIDE PHOSPHODIESTERASE (cycAMP) — *I*50 = 490 μM
3′,5′-CYCLIC-NUCLEOTIDE PHOSPHODIESTERASE (cycGMP) — 25% 1.4 mM

ZK 98299
AROMATASE — *K*i = 19 μM

Z-LAG-$NHCH_2CN$
PROTEINASE ADENOVIRUS 2 — NC — *K*i = 16 μM

Z-LEU-ABU-$CONHCH_2CHOHC_6H_5$
CALPAIN II — *K*i = 15 nM

Z-LEU-ALA-CHN_2
BOTHROPAIN — IR — *K*i = 35 nM
CALPAIN II — IR
CATHEPSIN I — IR

Z-LEU-LEU-CHN_2
CALPAIN — IR

Z-LEU-LEU-CH_2N_2
CATHEPSIN L — IR

Z-LEU-LEU-H
CALPAIN — *I*50 = 1.2 μM
CALPAIN I — *I*50 = 8 nM
MULTICATALYTIC ENDOPEPTIDASE COMPLEX — *I*50 = 110 μM

Z-LEU-LEU-LEU-AL

DNA POLYMERASE α		95% 1 mM
DNA POLYMERASE β	NC	$Ki = 63\ \mu M$
DNA POLYMERASE γ		99% 1 mM
DNA POLYMERASE I		99% 1 mM
DNA POLYMERASE TdT		21% 1 mM

Z-LEU-LEU-LEU-CH_2CL

DNA POLYMERASE β		$I50 = 32\ \mu M$

Z-LEU-LEU-LEUCINAL

PROTEINASE NEUTRAL (Ca^{2+} ACTIVATED)

Z-LEU-LEU-LEU-H

CALPAIN		$I50 = 1.5\ \mu M$
MULTICATALYTIC ENDOPEPDTIDASE COMPLEX (BRANCHED CHAIN AMINO ACID		$Ki = 12$ nM
MULTICATALYTIC ENDOPEPDTIDASE COMPLEX (CHYMOTRYPSIN LIKE ACTIVIT		$Ki = 4$ nM
MULTICATALYTIC ENDOPEPTIDASE COMPLEX		$I50 = 850$ nM
MULTICATALYTIC ENDOPEPTIDASE COMPLEX 6S		

Z-LEU-LEU-LEU-VINYLSULPHONE

MULTICATALYTIC ENDOPEPTIDASE COMPLEX (CHYMOTRYPSIN LIKE ACTIVITY	IR	
MULTICATALYTIC ENDOPEPTIDASE COMPLEX (PEPTIDYL GLUTAMYL PEPTIDAS	IR	
MULTICATALYTIC ENDOPEPTIDASE COMPLEX (TRYPSIN LIKE ACTIVITY)	IR	

Z-LEU-LEU-NVAL-H

MULTICATALYTIC ENDOPEPTIDASE COMPLEX 6S

Z-LEU-LEU-PHE-CHN_2

HISTOLYSAIN	REV	$Ki = 1.5\ \mu M$

Z-LEU-LEU-PHE-COOEt

MULTICATALYTIC ENDOPEPTIDASE COMPLEX (CHYMOTRYPSIN LIKE ACTIVITY)		$Ki = 53\ \mu M$

Z-LEU-LEU-PHE-H

PROTEINASE IKB

Z-LEU-LEU-PHENYLALANINAL

MULTICATALYTIC ENDOPEPTIDASE COMPLEX (CHYMOTRYPSIN LIKE ACTIVITY)		$Ki = 460$ nM
MULTICATALYTIC ENDOPEPTIDASE COMPLEX (TRYPSIN LIKE ACTIVITY)		$Ki = 43\ \mu M$

Z-LEU-LEU-PRO-CHN_2

PROLYL OLIGOPEPTIDASE	C	$Ki = 73$ nM

Z-LEU-LEU-TYR-CHN_2

CALPAIN I	IR	

Z-LEU-LEU-TYR-NH_2

CATHEPSIN B	IR	
CATHEPSIN H	IR	
CATHEPSIN L	IR	
CATHEPSIN S	IR	

Z-LEU-LYS-CHN_2

HISTOLYSAIN	IR	

Z-LEU-MET-CHN_2

HISTOLYSAIN	IR	

Z-LEU-MET-H

CALPAIN

Z-LEU-NHOH

AMINOPEPTIDASE		$I50 = 7\ \mu M$
NEPRILYSIN		$I50 = 40$ nM
THERMOLYSIN		$Ki = 10\ \mu M$

Z-LEU-NVA-CONHCH$_2$-2-PYRIDYL

CALPAIN I		$Ki = 19$ nM

Z-LEU-D,L-PHECONHET

CALPAIN I		Ki = 74 nM
CALPAIN II		Ki = 35 nM

Z-LEU-PHE-COOH

CALPAIN I		Ki = 8.5 nM
CALPAIN II		Ki = 5.7 nM
CATHEPSIN B		Ki = 4.5 µM
PAPAIN		Ki = 7 µM

Z-LEU-PHE-L-(N,N-DIMETHYLGLUTAMINAL)

PROTEINASE 3C

Z-LEU-VAL-GLY-CHN_2

CATHEPSIN B	IR	
PAPAIN	IR	
PROHORMONE THIOL PROTEINASE	IR	

Z-LEU-VAL-GLY- DIAZOMETHYLKETONE

PROTEINASES CYSTEINE

Z-LEU-VAL-H

CALPAIN I		I50 = 4 nM

ZLIII133A

CATHEPSIN B		I50 = 20 µM
CATHEPSIN B LIKE CYSTEINE PROTEINASE		I50 = 500 nM
CRUZPAIN		I50 = 600 nM
PAPAIN		I50 = 50 µM

Z-LLF-CHO

MULTICATALYTIC ENDOPEPTIDASE COMPLEX (CHYMOTRYPSIN LIKE ACTIVITY

Z-LYS-CHN_2

CANCER PROCOAGULANT		98% 200 µM
CLOSTRIPAIN	IR	

Z-LYS-$CH_2S^+(CH_3)(CH_2C_6H_5)$

CLOSTRIPAIN	IR	Ki = 7.8 nM

Z-LYS-PRO

DIPEPTIDYL-PEPTIDASE 2

Z-LYS(Z)-VAL-PRO-VAL-TRIFLUOROMETHYLKETONE

ELASTASE LEUKOCYTE		Ki < 100 pM

ZM 211965

ARACHIDONATE 5-LIPOXYGENASE		I50 = 100 nM

ZM 218287

ARACHIDONATE 5-LIPOXYGENASE		I50 = 10 nM

(+)-(.Z)-2,3-METHANO-m-TYROSINE

AROMATIC-L-AMINO-ACID DECARBOXYLASE	C	Ki = 22 µM

Zn^+

HYDROXYPYRUVATE REDUCTASE		21% 1 mM

Zn^{2+}

4-ACETAMIDOBUTYRYL-CoA DEACETYLASE		30% 500 µM
ACETOIN DEHYDROGENASE		
ACETOLACTATE DECARBOXYLASE		58% 400 µM
N-ACETYLGALACTOSAMIDINE β-1,3-N-ACETYLGLUCOSAMINYLTRANSFERASE		
N-ACETYLGALACTOSAMIDINE β-1,6-N-ACETYLGLUCOSAMINYLTRANSFERASE		
N-ACETYLGALACTOSAMINE-6-SULFATASE		88% 17 mM
ACETYLGALACTOSAMINYL-O-GLYCOSYL-GLYCOPROTEIN β-1,6-N-ACETYLGLUCOSAMINYLTRANSFE		
N-ACETYLGLUCOSAMINE KINASE		99% 1 mM
N-ACETYL-β-GLUCOSAMINIDASE		

N-ACETYL-γ-GLUTAMYL-PHOSPHATE REDUCTASE		
β-N-ACETYLHEXOSAMINIDASE		
N-ACETYLLACTOSAMINE SYNTHASE		
N-ACETYLMANNOSAMINE KINASE		97% 1 mM
N-ACETYLNEURAMINATE CYTIDYLYLTRANSFERASE		
ACETYLORNITHINE DEACETYLASE		84% 200 µM
O-ACETYLSERINE (THIOL)-LYASE		
ACONITATE HYDRATASE m	C	*K*i = 2 µM
ACROSIN		39% 250 µM
ACTINOMYCIN LACTONASE		
ACYLAGMATINE AMIDASE		95% 10 mM
ACYLAMINOACYL-PEPTIDASE		
ACYLCARNITINE HYDROLASE		
ACYLGLYCERONE-PHOSPHATE REDUCTASE		
N-ACYLHEXOSAMINE OXIDASE		
ADENOSINE DEAMINASE		
ADENOSINE NUCLEOSIDASE		
ADENOSINE-PHOSPHATE DEAMINASE		*I*50 = 40 µM
ADENOSYLMETHIONINE DECARBOXYLASE		100% 500µM
ADENYLATE CYCLASE		
ADENYLOSUCCINATE SYNTHASE		
ADP DEAMINASE		
β-ADRENERGIC-RECEPTOR KINASE		
AGMATINE DEIMINASE		
β-ALA-ARG-HYDROLASE		67% 100 µM
ALAKN-1-OL DEHYDROGENASE (ACCEPTOR)		
ALANINE DEHYDROGENASE	NC	
ALANINE TRANSAMINASE		
D-ALANYL-D-ALANINE CARBOXYPEPTIDASE		*I*50 = 400 µM
ALCOHOL DEHYDROGENASE (ACCEPTOR)		
ALDEHYDE DEHYDROGENASE (NADP)		
ALDEHYDE DEHYDROGENASE (PYRROLOQUINOLINE-QUINONE)		20% 200 µM
ALDEHYDE REDUCTASE		
ALDOSE 1-DEHYDROGENASE		
ALDOSE 1-EPIMERASE		20% 13 µM
ALDOSE-1-PHOSPHATE NUCLEOTIDYLTRANSFERASE		100% 5 mM
ALDOSE-6-PHOSPHATE REDUCTASE (NADPH)		
ALGINATE LYASE		84% 1 mM
1-ALKYL-2-ACETYLGLYCEROL O-ACYLTRANSFERASE		
ALKYLGLYCEROPHOSPHOETHANOLAMINE PHOSPHODIESTERASE		> 10 mM
2-ALKYNOL-1-OL DEHYDROGENASE		
ALLANTOATE DEIMINASE		
ALLANTOIN RACEMASE		79% 800 µM
AMIDINOASPARTASE	C	
AMINE DEHYDROGENASE		
D-AMINO-ACID N-ACETYLTRANSFERASE		
AMINO-ACID N-ACETYLTRANSFERASE		
L-AMINO-ACID DEHYDROGENASE		
D-AMINO-ACID OXIDASE		
1-AMINOCYCLOPROPANE-1-CARBOXYLATE SYNTHASE		
5-AMINOPENTANAMIDASE		100% 500 µM
AMINOPEPTIDASE		100% 1 mM
AMINOPEPTIDASE		
AMINOPEPTIDASE B		30% 17 µM
AMINOPEPTIDASE B		90% 1 mM
AMINOPEPTIDASE Y		72% 500 µM
AMP DEAMINASE		23% 10 µM
α-AMYLASE		
β-AMYLASE		
4-ANDROSTENE-3,17-DIONE MONOOXYGENASE		
ANTHOCYANIDIN 3-O-GLUCOSYLTRANSFERASE		
ANTHRANILATE SYNTHASE		
AQUACOBALAMIN REDUCTASE (NADPH)		
α-L-ARABINOFURANOSIDASE		45% 10 mM

L-ARABINOSE ISOMERASE		
ARABINOSE-5-PHOSPHATE ISOMERASE		
ARACHIDONATE 5-LIPOXYGENASE		
ARALKYLAMINE N-ACETYLTRANSFERASE		
ARGARATINE γ-GLUTAMYLTRANSFERASE		
ARGINASE		
ARGINIIE KINASE		
ARGININE DECARBOXYLASE		
ARGININE DEIMINASE		32% 1 mM
AROMATIC-L-AMINO-ACID DECARBOXYLASE		
ARYLACETONITRILASE		83% 1 mM
ARYL-ACYLAMIDASE		
ARYLAMINE N-ACETYLTRANSFERASE		
ARYLESTERASE	NC	*K*i = 2.6 mM
ARYLFORMAMIDASE		100% 1 mM
ARYLSULFATASE		
ARYLSULFATE SULFOTRANSFERASE		
ARYL SULFOTRANSFERASE	NC	99% 600 μM
L-ASCORBATE OXIDASE		
L-ASCORBATE PEROXIDASE		
ASPARAGINE-AMMONIA LIGASE (ADP-FORMING)		40% 2 mM
ASPARAGINE SYNTHASE (GLUTAMINE-HYDROLYSING)		
ASPARTATE-AMMONIA LIGASE		100% 3 mM
ASPARTATE CARBAMOYLTRANSFERASE		
β-ASPARTYLDIPEPTIDASE		
ASPERGILLUS ALKALINE PROTEINASE		
ATP DEAMINASE		
ATP PHOSPHORIBOSYLTRANSFERASE		
BACILLUS SUBTILIS RIBONUCLEASE		25% 1 mM
BENZALDEHYDE DEHYDROGENASE (NADP)		
BENZOATE 4-MONOOXYGENASE		
BETAINE-ALDEHYDE DEHYDROGENASE		
BILE-SALT SULFOTRANSFERASE		
BIOTINIDASE		30% 1 mM
BIS(5′-ADENOSYL)-TRIPHOSPHATASE		*I*50 = 30 μM
BIS-γ-GLUTAMYLCYSTEINE REDUCTASE (NADPH)		
BIS(5′-NUCLEOSYL)-TETRAPHOSPHATASE (ASYMMETRICAL)		
BLASTICIDIN-S-DEAMINASE		
BRANCHED-CHAIN-AMINO-ACID TRANSAMINASE		
γ-BUTYROBETAINE DIOXYGENASE		
Ca^{2+}/Mg^{2+}-TRANSPORTING ATPASE	U	*K*i = 8 μM
N-CARBAMOYLSARCOSINE AMIDASE		38% 2 mM
CARBONATE DEHYDRATASE		
CARBOSINE SYNTHASE		
CARBOXYLESTERASE		
CARBOXYMETHYLHYDANTOINASE		
CARBOXYPEPTIDASE		49% 1 mM
CARBOXYPEPTIDASE A	C	*K*i = 24 μM
CARNITINE O-ACETYLTRANSFERASE		100% 500 μM
CARNITINE O-ACETYLTRANSFERASE		
CARNITINE 3-DEHYDROGENASE		
CARNITINE O-OCTANOYLTRANSFERASE		
κ-CARRAGEENASE		
CASPASE 3		*I*50 = 100 nM
CATECHOL 1,2-DIOXYGENASE		
CATHEPSIN B		90% 1 mM
CATHEPSIN D		
Ca^{2+}-TRANSPORTING ATPASE		50% 100 μM
CDP-DIACYLGLYCEROL-GLYCEROL-3-PHOSPHATE 3-PHOSPHATIDYLTRANSFERASE		
CDP-DIACYLGLYCEROL-INOSITOL 3-PHOSPHATIDYLTRANSFERASE		
CERAMIDASE		
CERAMIDE GLUCOSYLTRANSFERASE		
CHITINASE		
CHITIN SYNTHASE		

CHITOBIOSYLDIPHOSPHODOLICHOL α-MANNOSYLTRANSFERASE		
2-CHLOROBENZOATE 1,2-DIOXYGENASE		
CHOLESTENONE 5α-REDUCTASE		
CHOLESTEROL 7α-MONOOXYGENASE		
CHOLESTEROL OXIDASE		
CHOLINE O-ACETYLTRANSFERASE		
CHOLINE OXIDASE		
CHOLINE-PHOSPHATE CYTIDYLYLTRANSFERASE		
CHOLINESTERASE		*K*i = 56 μM
CHOLINESTERASE		
CHOLYLGLYCINE HYDROLASE		69% 5 mM
CHONDROITIN ABC LYASE		100% 1 mM
CHONDROITIN AC LYASE		*I*50 = 5 mM
CHYMOPAPAIN		
CLOSTRIDIUM HISTOLYTICUM AMINOPEPTIDASE		100% 1mM
COB(I)ALAMIN ADENOSYLTRANSFERASE		
COLLAGENASE CLOSTRIDIUM HISTOLYTICUM		
CORTICOSTERONE 18-MONOOXYGENASE		
CORTISOL SULFOTRANSFERASE		
4-COUMARATE-CoA LIGASE		
2-COUMARATE O-β-GLUCOSYLTRANSFERASE		77% 1 mM
CREATINASE		100% 1 mM
CREATINE KINASE		
CYANIDIN-3-RHAMNOSYLGLUCOSIDE 5-O-GLUCOSYLTRANSFERASE		
3′,5′-CYCLIC-GMP PHOSPHODIESTERASE		68% 100 μM
1,2-CYCLIC-INOSITOL-PHOSPHATE PHOSPHODIESTERASE		*I*50 = 10 μM
2′,3′-CYCLIC-NUCLEOTIDE 2′-PHOSPHODIESTERASE		94% 1 mM
2′,3′-CYCLIC-NUCLEOTIDE 2′-PHOSPHODIESTERASE		90% 200 μM
2′,3′-CYCLIC-NUCLEOTIDE 3′-PHOSPHODIESTERASE		82% 500 μM
3′,5′-CYCLIC-NUCLEOTIDE PHOSPHODIESTERASE		58% 2.5 mM
3′,5′-CYCLIC-NUCLEOTIDE PHOSPHODIESTERASE II		
CYCLOHEXYLAMINE OXIDASE		
CYCLOMALTODEXTRIN GLUCANOTRANSFERASE		
β-CYCLOPIAZONATE DEHYDROGENASE		
CYSTATHIONINE γ-LYASE		
CYSTEINE-tRNA LIGASE		
CYTIDINE DEAMINASE		36% 10 mM
CYTIDYLATE KINASE		
CYTOSINE DEAMINASE		
CYTOSOL NON-SPECIFIC DIPEPTIDASE		
2-DEHYDRO-3-DEOXYPHOSPHOHEPTONATE ALDOLASE		
2-DEHYDRO-3-DEOXYPHOSPHOOCTONATE ALDOLASE		34% 1 mM
2-DEHYDROPANTOYL-LACTONE REDUCTASE (A SPECIFIC)		
3-DEHYDROQUINATE DEHYDRATASE		68% 5 mM
DEOXYRIBONUCLEASE γ		*I*50 = 40 μM
DEOXYRIBONUCLEASE I		*I*50 = 5 μM
DEOXYRIBONUCLEASE II		50% 5 mM
DEOXYRIBONUCLEASE (PYRIMIDINE DIMER)		
DESULFOGLUCINOLATE SULFOTRANSFERASE		
DESULFOHEPARIN SULFOTRANSFERASE		
DEXTRANASE		92% 1.2 mM
DEXTRANSUCRASE	C	*K*i = 1 mM
α-DEXTRIN ENDO-1,6-α-GLUCOSIDASE		66% 10 mM
2,4-DICHLOROPHENOL 6-MONOOXYGENASE		
DIETHYL-2-METHYL-3-OXOSUCCINATE DEHYDROGENASE		
DIHYDRODIPICOLINATE SYNTHASE		
DIHYDROOROTASE		18% 200 μM
4,5-DIHYDROOXYPHTHALATE DECARBOXYLASE		
DIHYDROPYRIMIDINASE		*K*i = 23 μM
DIHYDROXYPHENYLALANINE TRANSAMINASE		
2,6-DIHYDROXYPYRIDINE 3-MONOOXYGENASE		
2,5-DIKETO-D-GLUCONATE REDUCTASE		95% 500 μM
DIMETHYLARGINASE		*K*i = 2 μM
DIMETHYLGLYCINE OXIDASE		

DIMETHYLHISTIDINE N-METHYLTRANSFERASE	
DINUCLEOSIDE TETRAPHOSPHATASE	
DIPEPTIDASE	40% 1 mM
DIPEPTIDASE	
DIPEPTIDYL-PEPTIDASE 2	93% 10 mM
DIPEPTIDYL-PEPTIDASE 3	
DIPEPTIDYL-PEPTIDASE 4	
DIPEPTIDYL-PEPTIDASE 4	98% 1 mM
DIPEPTIDYL-PEPTIDASE 4	
DISCADENINE SYNTHASE	
DISPASE	> 10 mM
DNA (CYTOSINE-5-)-METHYLTRANSFERASE	
DNA POLYMERASE	
DNA TOPOISOMERASE	
DOLICHOL KINASE	
DOLICHYL-DIPHOSPHOOLIGOSACCHARIDE-PROTEIN GLYCOSYLTRANSFERASE	
ENDO-β-GALACTOSIDASE C	56% 1 mM
ENDOGLYCERAMIDASE	
ENDO-1,4-β-XYLANASE	
trans-2-ENOYL-CoA REDUCTASE (NAD)	
ESTRADIOL 17β-DEHYDROGENASE	
ESTRONE SULFOTRANSFERASE	
ETHANOLAMINE-PHOSPHATE PHOSPHO-LYASE	94% 100 µM
EUPHORBAIN	
EXODEOXYRIBONUCLEASE III	
FAD PYROPHOSPHATASE	> 2mM
FARNESOL DEHYDROGENASE	
FERRIC REDUCTASE	*K*i = 25 µM
FERROCHELATASE	80% 100 µM
FLAVONE SYNTHASE	
FLAVONOL 3-O-GLUCOSYLTRANSERASE	> 500 µM
FORMALDEHYDE DEHYDROGENASE	
FORMALDEHYDE DEHYDROGENASE (GLUTATHIONE)	
FORMALDEHYDE DISMUTASE	
FORMALDEHYDE TRANSKETOLASE	
β-FRUCTOFURANOSIDASE	
FRUCTOSE-BISPHOSPHATASE	*I*50 = 400 nM
FRUCTOSE-BISPHOSPHATE ALDOLASE	
FRUCTOSE 5-DEHYDROGENASE	
α-L-FUCOSIDASE	
1,2-α-L-FUCOSIDASE	
FUMARATE REDUCTASE (NADH)	
GALACTINOL GALACTOSYLTRANSFERASE	
GALACTINOL-RAFFINOSE GALACTOSYLTRANSFERASE	
GALACTOKINASE	
GALACTOLIPID GALACTOSYLTRANSFERASE	
GALACTONATE DEHYDRATASE	100% 5 µM
α-GALACTOSIDASE	
β-GALACTOSIDASE	
GALACTOSIDE 2-L-FUCOSYLTRANSFERASE	100% 10 mM
GDP-MANNOSE 3,5-EPIMERASE	
GENTAMICIN 3′-N-ACETYLTRANSFERASE	
GERANIOL DEHYDROGENASE	
GERANYLGERANYL PROTEINTRANSFERASE	
GERANYLGERANYLPYROPHOSPHATASE	
GIBBERELLIN 2β-DIOXIGENASE I	
GIBBERELLIN 3β-DIOXYGENASE	
GLIOCLADIUM PROTEINASE	
GLOBOSIDE α-N-ACETYLGALACTOSAMINYLTRANSFERASE	
GLUCAN ENDO-1,3-α-GLUCOSIDASE	43% 20 mM
GLUCAN ENDO-1,3-β-GLUCOSIDASE	
GLUCAN 1,4-β-GLUCOSIDASE	
GLUCAN 1,6-α-GLUCOSIDASE	
GLUCAN 1,4-α-MALTOHEXAOSIDASE	80 % 1 mM

GLUCAN 1,4-α-MALTOTETRAOHYDROLASE		
GLUCONATE DEHYDRATASE		
GLUCONATE 2-DEHYDROGENASE		
GLUCOSAMINE-PHOSPHATE N-ACETYLTRANSFERASE		
GLUCOSE 1-DEHYDROGENASE (NADP)		
GLUCOSE-6-PHOSPHATE 1-DEHYDROGENASE		
α-GLUCOSIDASE		
β-GLUCOSIDASE		
D-GLUCURONATE DEHYDRATASE		
GLUCURONATE ISOMERASE		90% 10 μM
β-GLUCURONIDASE		25% 1 mM
GLUCURONONOLATONE REDUCTASE		
GLUTAMATE DECARBOXYLASE		14% 100 μM
GLUTAMATE DECARBOXYLASE	C	Ki = 55 μM
GLUTAMATE DECARBOXYLASE		I50 = 25 μM
GLUTAMATE DEHYDROGENASE		
GLUTAMATE DEHYDROGENASE (NAD(P))		
GLUTAMATE DEHYDROGENASE (NADP)		> 10 mM
GLUTAMATE FORMIMINOTRANSFERASE		
GLUTAMATE SYNTHASE (NADH)		
GLUTAMINE-FRUCTOSE-6-PHOSPHATE AMINOTRANSFERASE (ISOMERIZING)		95% 10 μM
GLUTAMYL AMINOPEPTIDASE		86% 1 mM
γ-GLUTAMYLTRANSFERASE		100% 5 mM
GLUTARYL-CoA DEHYDROGENASE		
GLUTATHIONE DEHYDROGENASE (ASCORBATE)		
GLUTATHIONE OXIDASE		
GLUTATHIONE PEROXIDASE		100% 5 mM
GLUTATHIONE REDUCTASE (NADPH)		Ki = 6.5 μM
GLUTATHIONE SYNTHASE		
GLUTATHIONE THIOLESTERASE		
GLUTATHIONE TRANSFERASE		
GLYCEROL DEHYDROGENASE		
GLYCEROL-1-PHOSPHATASE		42% 5 mM
GLYCEROL-2-PHOSPHATASE		
GLYCEROL-3-PHOSPHATE O-ACYLTRANSFERASE		
GLYCEROL-3-PHOSPHATE CYTIDILYLTRANSFERASE		
GLYCEROL-3-PHOSPHATE DEHYDROGENASE		
GLYCEROPHOSPHOCHOLINE PHOSPHODIESTERASE		
GLYCERYL-ETHER MONOOXYGENASE		
GLYCINE C-ACETYLTRANSFERASE		
GLYCINE ACYLTRANSFERASE		
GLYCINE BENZOYLTRANSFERASE		
GLYCINE DEHYDROGENASE (DECARBOXYLATING)		
GLYCINE-tRNA LIGASE		100% 6.7 mM
GLYCOCYAMINASE		
GLYCOGEN (STARCH) SYNTHASE		
GLYCOGEN SYNTHASE (CASEIN) KINASE	C	I50 = 65 μM
GLYCOGEN-SYNTHASE-D PHOSPHATASE		
GLYCOLATE DEHYDROGENASE		
GLYCOLIPID 3-α-MANNOSYLTRANSFERASE		
GLYCOPEPTIDE α-N-ACETYLGALACTOSAMINIDASE		66% 2.5 mM
GMP REDUCTASE		
GTP CYCLOHYDROLASE I		
GTP CYCLOHYDROLASE II		
GUANYLATE CYCLASE		95% 1 mM
2-HALOACID DEHALOGENASE		92% 1 mM
HEPARAN-α-GLUCOSAMINIDINE N-ACETYLTRANSFERASE		
HEPARITIN-SULFATE LYASE		1 mM
HEPARITIN SULFOTRANSFERASE		
HIPPURATE HYDROLASE		
HISTIDINE METHYLTRANSFERASE	C	Ki = 39 μM
HISTIDINOL DEHYDROGENASE		66% 500 μM
HISTIDINOL PHOSPHATASE		Ki = 1.1 μM
HISTONE ACYLTANSFERASE		

H^+/K^+-EXCHANGING ATPASE	
HOMOSERINE O-ACETYLTRANSFERASE	
HURAIN	
HYALURONATE LYASE	
HYDROXYANTHRAQUINONE GLUCOSYLTRANSFERASE	
4-HYDROXYBENZOATE 4O-β-D-GLUCOSYLTRANSFERASE	
4-HYDROXYBENZOATE 3-MONOOXYGENASE	
4-HYDROXYBENZOATE NONAPRENYLTRANSFERASE	46% 10 mM
5-HYDROXYFURANOCOUMARIN 5-O-METHYLTRANSFERASE	
8-HYDROXYFURANOCOUMARIN 8-O-METHYLTRANSFERASE	
2-HYDROXYGLUTARATE DEHYDROGENASE	
6β-HYDROXYHYOSCYAMINE EPOXIDASE	
4-HYDROXYMANDELATE OXIDASE	
HYDROXYMETHYLBILANE SYNTHASE	63% 500 μM
3-HYDROXYMETHYLCEPHEM CARBAMOYLTRANSFERASE	
(R)-4-HYDROXYPHENYLLACTATE DEHYDROGENASE	
4-HYDROXYPROLINE EPIMERASE	100% 1 mM
3β-HYDROXY-Δ5-STEROID DEHYDROGENASE	> 1 mM
3(or 17)β-HYDROXYSTEROID DEHYDROGENASE	
7α-HYDROXYSTEROID DEHYDROGENASE	
HYOSCYAMINE (6S)-DIOXYGENASE	
HYOSCYAMINE (6S)-DIOXYGENASE	
HYPOXANTHINE PHOSPHORIBOSYLTRANSFERASE	
IMIDAZOLEGLYCEROL-PHOSPHATE DEHYDRATASE	*I*50 = 1.1 μM
INDOLE 3-ACETALDEHYDE REDUCTASE (NADPH)	
INDOLEACETALDOXIME DEHYDRATASE	100% 1 mM
INDOLYLACETYL-MYO-INOSITOL GALACTOSYLTRANSFERASE	
INORGANIC PYROPHOSPHATASE	*K*i = 15 μM
myo-INOSITOL-1-PHOSPHATE SYNTHASE	
INULINASE	
ISOCITRATE DEHYDROGENASE (NADP)	
ISOCITRATE LYASE	
ISOFLAVONE 4′-O-METHYLTRANSFERASE	
ISOORIENTIN 3-METHYLTRANSFERASE	
3-ISOPROPYLMALATE DEHYDROGENASE	
2-ISOPROPYLMALATE SYNTHASE	
KERATAN-SULFATE ENDO-1,4-β-GALACTOSIDASE	100% 10 mM
KERATAN-SULFATE ENDO-1,4-β-GALACTOSIDASE	
KYNURENINE-OXOGLUTARATE AMINOTRANSFERASE	
LACTALDEHYDE DEHYDROGENASE	
β-LACTAMASE	
L-LACTATE DEHYDROGENASE	
LACTOYLGLUTATHIONE LYASE	100% 500 μM
LEUCINE-tRNA LIGASE	
LEUCYL AMINOPEPTIDASE	*I*50 = 170 μM
LEVANASE	
LICODIONE 2-O-METHYLTRANSFERASE	
LIPOAMIDASE	57% 500 μM
LYSOPHOSPHOLIPASE	98% 10 mM
LYSOZYME	
LYSYL AMINOPEPTIDASE	
LYSYL ENDOPEPTIDASE	*K*i = 580 μM
MACROLIDE KINASE	
MAGNESIUM PROTOPORPHYRIN O-METHYLTRANSFERASE	
MALATE DEHYDROGENASE	33% 33 mM
MALONATE-SEMIALDEHYDE DEHYDROGENASE	
MANDELONITRILE LYASE	
MANNAN ENDO-1,4-β-MANNOSIDASE	
MANNAN 1,2-(1,3)-α-MANNOSIDASE	
MANNITOL 2-DEHYDROGENASE	
MANNITOL 2-DEHYDROGENASE (NADP)	
MANNITOL-1-PHOSPHATE 5-DEHYDROGENASE	
MANNOSE ISOMERASE	
MANNOSE-1-PHOSPHATE GUANYLYLTRANSFERASE	

Enzyme	Type	Value
MANNOSE-6-PHOSPHATE ISOMERASE	C	$K$is = 6.4 μM
α-MANNOSIDASE		I50 = 1 mM
MEMBRANE ALANINE AMINOPEPTIDASE		> 100 μM
METHANE MONOOXYGENASE		
METHIONYL DIPEPTIDASE		100% 50 μM
4-METHOXYBENZOATE MONOOXYGENASE (O-DEMETHYLATING)		28% 100 μM
2-METHYLCITRATE LYASE		
2-METHYLCITRATE SYNTHASE		
METHYLENETETRAHYDROFOLATE CYCLOHYDROLASE		
METHYLENETETRAHYDROFOLATE DEHYDROGENASE (NAD)		
METHYLISOCITRATE LYASE		
N^6-METHYL-LYSINE OXIDASE		
METHYLMALONYL-CoA EPIMERASE		
MEVALONATE KINASE		
MICROCOCCAL NUCLEASE		
MONODEHYDROASCORBATE REDUCTASE (NADH)		
MONOTERPENOL β-GLUCOSYLTRANSFERASE		
MUCINAMINYLSERINE MUCINAMINIDASE		66% 2.5 mM
MUCOROPEPSIN		
NAD NUCLEOSIDASE		
NAD(P)-ARGININE ADP-RIBOSYLTRANSFERASE		I50 = 11 mM
NAD PYROPHOSPHATASE		91% 333 μM
NAD SYNTHASE		
Na^+/K^+-EXCHANGING ATPASE		
NARINGENIN-CHALCONE SYNHTASE		
NEOLACTOSYLCERAMIDE α-2,3-SIALYLTRANSFERASE		
NEUROLYSIN		
NICOTINATE GLUCOSYLTRANSFERASE		
NITRATE REDUCTASE (NADH)		
NITRIC-OXIDE SYNTHASE (Ca^{2+} STIMULATED)		46% 100 nM
NITRIC-OXIDE SYNTHASE I		I50 = 30 μM
NITRILASE		85% 1 1mM
NITROGENASE		
NMN NUCLEOSIDASE		
NUATIGENIN 3-β-GLUCOSYLTRANSFERASE		
NUCLEOSIDE PHOSPHOTRANSFERASE		
3′-NUCLEOTIDASE		
5′-NUCLEOTIDASE		
trans-OCTAPRENYLTRANSFERASE		
OLIGO-1,6-GLUCOSIDASE		96% 2 mM
ORNITHINE CARBAMOYLTRANSFERASE		
ORNITHINE DECARBOXYLASE		
OROTATE PHOSPHORIBOSYLTRANSFERASE		
OXALACETATE TAUTOMERASE		52% 100 μM
OXALOACETATE DECARBOXYLASE		54% 1.5 mM
2-OXOALDEHYDE DEHYDROGENASE (NADP) I		28% 5 mM
3-OXOSTEROID Δ^1-DEHYDROGENASE		
3-OXO-5α-STEROID Δ^4-DEHYDROGENASE		
PALMITOYL-CoA HYDROLASE		
PANTOATE-β-ALANINE LIGASE		
PAPAIN		
PARAOXON HYDROLASE	NC	Ki = 130 μM
PEPTIDE-N4-(N-ACETYL-β-GLUCOSAMINYL)ASPARAGINE AMIDASE		54% 10 mM
PEPTIDE-N4-(N-ACETYL-β-GLUCOSAMINYL)ASPARAGINE AMIDASE		20% 10 mM
PEPTIDE α-N-ACETYLTRANSFERASE		
PEPTIDE-ASPARTATE β-DIOXYGENASE		
PEPTIDOGLYCAN GLUCOSYLTRANSFERASE		
PERILLYL-ALCOHOL DEHYDROGENASE		
PHENOL β-GLUCOSYLTRANSFERASE		
PHENYLALANINE 2-MONOOXYGENASE		
PHLORETIN HYDROLASE		63% 1 mM
PHORBOL-DIESTER HYDROLASE		
PHOSPHATASE ACID		
PHOSPHATASE ACID		

PHOSPHATASE ACID		92% 10 mM
PHOSPHATASE ACID	C	$Ki = 2.5$ mM
PHOSPHATASE ALKALINE		
PHOSPHATASE ALKALINE (PHOSPHODIESTERASE ACTIVITY)	NC	$Ki = 210$ μM
PHOSPHATIDATE CYTIDILYLTRANSFERASE		
PHOSPHATIDYLCHOLINE-DOLICHOL O-ACYLTRANSFERASE		
PHOSPHATIDYLGLYCEROPHOSPHATASE		74% 1 mM
PHOSPHATIDYLINOSITOL DEACETYLASE		
1-PHOSPHATIDYLINOSITOL PHOSPHODIESTERASE		
PHOSPHOACETYLGLUCOSAMINE MUTASE		100% 500 μM
PHOSPHODIESTERASE		
PHOSPHOENOLPYRUVATE-PROTEIN KINASE		
6-PHOSPHOFRUCTOKINASE		
6-PHOSPHO-β-GALACTOSIDASE		
β-PHOSPHOGLUCOMUTASE		50% 200 μM
PHOSPHOGLUCONATE 2-DEHYDROGENASE	C	
PHOSPHOGLYCERATE KINASE		
PHOSPHOGLYCERATE MUTASE		76% 1 mM
3-PHOSPHOGLYCERATE PHOSPHATASE		
PHOSPHOGLYCOLATE PHOSPHATASE		$I50 = 500$ μM
6-PHOSPHO-3-KETOHEXULOSE ISOMERASE		
PHOSPHOLIPASE A1		
PHOSPHOLIPASE A2	NC	
PHOSPHOLIPASE C		
PHOSPHOLIPASE D		
PHOSPHOMANNOMUTASE		
PHOSPHOPENTOMUTASE		> 90% 1 mM
PHOSPHOPROTEIN PHOSPHATASE		
PHOSPHOPROTEIN PHOSPHATASE lambda		100% 1 μM
PHOSPHORYLASE		
PHOSPHORYLASE KINASE		
3-PHOSPHOSHIKIMATE 1-CARBOXYVINYLTRANSFERASE		
PHOTINUS-LUCIFERIN 4-MONOOXYGENASE (ATP-HYDROLYSING)		
PHTHALATE 4,5-DIOXYGENASE		
Δ1PIPERIDINE-2-CARBOXYLATE RDUCTASE		
PLASMIN		
POLYGALACTURONASE		
POLYNUCLEOTIDE ADENYLYLTRANSFERASE		
POLYPHOSPHATASE		
POLYPHOSPHATE KINASE		
PORPHOBILINOGEN SYNTHASE		> 1 mM
PRENYL-PYROPHOSPHATEASE		58% 500 μM
PROCOLLAGEN C-ENDOPEPTIDASE		$I50 = 61$ μM
PROCOLLAGEN N-ENDOPEPTIDASE		$I50 = 8.3$ μM
PROCOLLAGEN N-ENDOPEPTIDASE		92% 100 μM
PROCOLLAGEN GLUCOSYLTRANSFERASE		
PROCOLLAGEN-LYSINE 5-DIOXYGENASE		
PROCOLLAGEN-PROLINE, 2-OXOGLUTARATE-4-DIOXYGENASE		
PROLINE DEHYDROGENASE		
PROLYL AMINOPEPTIDASE		
PROLYL OLIGOPEPTIDASE		100% 100 μM
PROLYL OLIGOPEPTIDASE		74% 10 mM
PROSTAGLANDIN F SYNTHASE		
PROSTAGLANDIN-H2 D-ISOMERASE		
PROTEINASE ALKALINE		48% 1 mM
PROTEINASE ASPARTIC PLASMODIUM		
PROTEINASE HIV		$Ki = 700$ μM
PROTEINASE I ACHROMOBACTER		
PROTEINASE NEUTRAL BACILLUS SUBTILIS		
PROTEINASE SERINE CANDIDA LIPOLYTICA		
PROTEIN-DISULFIDE REDUCTASE (GLUTATHIONE)		
PROTEIN-DISULFIDE REDUCTASE (NAD(P)H)		
PROTEIN-GLUCOSYLGALACTOSYLHYDROXYLYSINE GLUCOSIDASE		44% 1 mM
PROTEIN-GLUCOSYLGALACTOSYLHYDROXYLYSINE GLUCOSIDASE		

PROTEIN-GLUTAMINE γ-GLUTAMYLTRANSFERASE		
PROTEIN KINASE C		
PROTEIN KINASE CASEIN KINASE		
PROTEIN TYROSINE KINASE (Mg^{2+} DEPENDENT)		100% 5 mM
PROTEIN TYROSINE-PHOSPHATASE	NC	40% 100 μM
PROTEIN TYROSINE-PHOSPHATASE CD45		87% 1 mM
PROTEIN TYROSINE-PHOSPHATASE (Mg^{2+} DEPENDENT)		100% 250 μM
PROTEIN TYROSINE SULFOTRANSFERASE		
PSEUDOMONAS AEROGINOSA NEUTRAL PROTEINASE		82% 1 mM
PTERIDIN DEAMINASE		
PYRIDOXINE DEHYDROGENASE		
4-PYRIDOXOLACTONASE		
PYRIMIDINE-DEOXYNUCLEOSIDE 2′-DIOXYGENASE		
PYRIMIDINE-5′-NUCLEOTIDE NUCLEOSIDASE		
PYROPHOSPHATE-FRUCTOSE-6-PHOSPHATE 1-PHOSPHOTRANSFERASE		
PYRROLINE-5-CARBOXYLATE REDUCTASE		
PYRUVATE DECARBOXYLASE		
PYRUVATE KINASE		
[PYRUVATE KINASE] PHOSPHATASE		
QUEUINE tRNA-RIBOSYLTRANSFERASE		
QUINATE O-HYDROXYCINNAMOYLTRANSFERASE		
RENILLA-LUCIFERIN 2-MONOOXYGENASE		
RENIN	NC	Ki = 24 μM
L-RHAMNOSE DEHYDROGENASE		75% 6.7 mM
L-RHAMNOSE ISOMERASE		
RHODOPSIN KINASE		I50 = 1 mM
RHODOTORULAPEPSIN		23% 5 mM
RIBONUCLEASE α		1005 10 mM
RIBONUCLEASE H		I50 = 50 μM
RIBONUCLEASE M5		
RIBONUCLEASE T1		
RIBONUCLEASE T2		
RIBOSE-5-PHOSPHATE ISOMERASE		
RIBOSYLPYRIMIDINE NUCLEOSIDASE		
tRNA(CYTOSINE-5-) METHYLTRANSFERASE		
RNA-DIRECTED RNA POLYMERASE		
mRNA(GUANINE-N7-)-METHYLTRANSFERASE		
tRNA(GUANINE-N^1-) METHYLTRANSFERASE		
mRNA GUANYLYLTRANSFERASE		> 10 mM
tRNA(URACIL-5-) METHYLTRANSFERASE		
SACCHAROPINE DEHYDROGENASE (NAD, L-GLUTAMATE FORMING)		
SACCHAROPINE DEHYDROGENASE (NAD, L-LYSINE FORMING)		
SALICYL ALCOHOL β-D-GLUCOSYLTRANSFERASE		60% 2 mM
SARCOSINE OXIDASE		93% 2 mM
SECONDARY-ALCOHOL OXIDASE		
SEDOHEPTULOSE-BISPHOSPHATASE		
SELENEPHOSPHATE SYNTHASE		
L-SERINE DEHYDRATASE		
SERINE-SULFATE AMMONIA-LYASE		
SHIKIMATE 5-DEHYDROGENASE		
SIALATE O-ACETYLESTERASE		
SIALIDASE		
O-SIALOGLYCOPROTEIN ENDOPEPTIDASE		
SORGHUM ASPARTIC PROTEINASE		30% 5 mM
SPHINGOMYELIN PHOSPHODIESTERASE		
SPLEEN EXONUCLEASE		
STAPHYLOCOCCAL CYSTEINE PROTEINASE		
STARCH SYNTHASE		
STEROID 5α-REDUCTASE		
STEROID 5α-REDUCTASE 1		Ki = 1.5 μM
STEROL 3β-GLUCOSYLTRANSFERASE		
SUBMANDIBULAR PROTEINASE A		
SUCCINATE-HYDROXYMETHYLGLUTARATE CoA-TRANSFERASE		
SUCCINATE-SEMIALDEHYDE DEHYDROGENASE		

SUCCINATE-SEMIALDEHYDE DEHYDROGENASE (NAD(P))
SUCROSE 1^{F}-FRUCTOSYLTRANSFERASE 91% 10 mM
SUCROSE SYNTHASE
SULFATE ADENYLYLTRANSFERASE (ADP)
SULFUR DIOXYGENASE
L(+)-TARTRATE DEHYDRATASE 61% 200 μM
TESTOSTERONE 17β-DEHYDROGENASE
TESTOSTERONE 17β-DEHYDROGENASE (NADP)
TETANUS TOXIN L CHAIN (PEPDIDOLYTIC ACITIVITY) 100% 1 mM
2,3,4,5-TETRAHYDROPYRIDINE-2-CARBOXYLATE N-SUCCINYLTRANSFERASE
THETIN-HOMOCYSTEINE S-METHYLTRRANSFERASE
THIAMINASE
THIAMIN-DIPHOSPHATE KINASE
THIAMIN PYRIDINYLASE II 100% 100 μM
THIAMIN-TRIPHOSPHATASE
THIOGLUCOSIDASE
THIOREDOXINE SYSTEMS
L-THREONINE 3-DEHYDROGENASE
THROMBOXANE A SYNTHASE
THYMIDINE KINASE
THYMIDINE-TRIPHOSPHATASE 99% 1 mM
TISSUE-ENDOPEPTIDASE DEGRADING COLLAGENASE-SYNTHETIC-SUBSTRATE 90% 100 μM
TOLUENE DIOXYGENASE
α,α-TREHALASE C 100% 100 μM
TREHALOSE-PHOSPHATASE 80% 600 μM
α,α-TREHALOSE-PHOSPHATE SYNTHASE
TRIACETATE-LACTONASE
TRIMETHYLYSINE DIOXYGENASE
TRIPEPTIDE AMINOPEPTIDASE 100% 1 mM
TRIPEPTIDE AMINOPEPTIDASE 100% 100 μM
TRYPSIN *K*i = 33 mM
TYPE II SITE SPECIFIC DEOXYRIBONUCLEASE 100% 2 mM
TYROSINE-ESTER SULFOTRANSFERASE
TYROSINE 3-MONOOXYGENASE
TYROSINE TRANSAMINASE
TYROSYL PROTEIN SULFOTRANSFERASE
UBIQUINOL OXIDASE CYTb558-d COMPLEX *I*50 = 60 μM
UBIQUINOL OXIDASE CYTb562-o COMPLEX *I*50 = 1 μM
UDP-N-ACETYLGLUCOSAMINE 2-EPIMERASE 46% 100 μM
UDP-N-ACETYLGLUCOSAMINE 4-EPIMERASE
UDP-N-ACETYLMURAMATE-ALANINE LIGASE
UDP-GLUCURONOSYLTRANSFERASE
URATE OXIDASE
UREASE
UREIDOGLYCOLATE LYASE
UREIDOGLYCOLLATE DEHYDROGENASE
β-UREIDOPROPIONASE 80% 1 mM
URIDINE NUCLEOSIDASE
URIDINE NUCLEOSIDASE 24% 100 μM
UROPORPHYRINOGEN-III SYNTHASE 26% 10 μM
VALINE-PYRUVATE TRANSAMINASE 20% 2 mM
XAA-ARG DIPEPTIDASE
XAA-HIS DIPEPTIDASE 43% 100 μM
XAA-TRP AMINOPEPTIDASE
XYLAN ENDO-1,3-XYLOSIDASE
XYLAN 1,4-β-XYLOSIDASE
D-XYLULOSE REDUCTASE
YEAST RIBONUCLEASE

Z-(2-NAPHTHYL)ALANINE-(1-NAPHTHYL)ALANINE-LEUCINAL
PROTEINASE IKB

ZNCL
DIHYDROCOUMARIN HYDROLASE 85% 3.3 mM

Zn^{2+} FORMATE
CYTOCHROME P450

2-(Z-NH(CH_2)$_2$CONH)C_6SO_2F
PROTEINASE LYMPHOCYTE GRANULE (LY-CHYMASE)
PROTEINASE LYMPHOCYTE GRANULE (MET-ASE)

Z-NHCH_2PO_2-LEU-LEU
THERMOLYSIN — *K*i = 9 nM

$ZNSO_4$
CYCLAMATE SULFOHYDROLASE — 59% 1 mM
MANNITOL-1-PHOSPHATASE

ZOFENOPRIL
PEPTIDYL-DIPEPTIDASE A — *I*50 = 80 nM

ZOFENOPRILAT
PEPTIDYL-DIPEPTIDASE A — *I*50 = 3 nM
XAA-PRO AMINOPEPTIDASE — *I*50 = 440 μM
XAA-TRP AMINOPEPTIDASE — *I*50 = 7 μM

ZOLEDRONATE
PYROPHOSPHATE-FRUCTOSE-6-PHOSPHATE 1-PHOSPHOTRANSFERASE — *K*i = 50 μM

ZOMEPIRAC
3α-HYDROXYSTEROID DEHYDROGENASE (A SPECIFIC)
3α-HYDROXYSTEROID DEHYDROGENASE (B-SPECIFIC) — C — *K*i = 37 μM
3α-HYDROXYSTEROID DEHYDROGENASE (B-SPECIFIC) — 41% 10 μM

ZOPOLRESTAT
ALDEHYDE REDUCTASE — NC — *K*i = 19 nM
3α-HYDROXYSTEROID DEHYDROGENASE (B-SPECIFIC) — *I*50 = 46 μM

(Z-ORN-VAL-CHApsi[H.E.]-ALA-NHBU
PROTEINASE HIV-1 — *I*50 = 8 nM

Z-PEPTIDES
ACYLAMINOACYL-PEPTIDASE — C

Z-PHE
PREPHENATE DEHYDROGENASE
SERINE CARBOXYPEPTIDASE — *K*i = 710 μM

Z-PHE-ALA-CH_2
PROTEINASE CYSTEINE LEISHMANIA MEXICANA

Z-PHE-ALA CHLOROMETHYLKETONE
CHYMOTRYPSIN

Z-PHE-ALA-CHN_2
CANCER PROCOAGULANT — 95% 200 μM
PROHORMONE THIOL PROTEINASE — IR
PROTEINASE CYSTEINE LEISHMANIA MEXICANE

Z-PHE-ALA-CHO
CATHEPSIN B — *K*i = 78 μM

Z-PHE-ALA-$CH_2S^+(CH_3)_2$
CATHEPSIN B — IR — *K*i = 780 nM

Z-PHE-ALA-DIAZOMETHYLKETONE
CATHEPSIN B

erythro-Z-PHE-ALA-EPOXIDE
CATHEPSIN B — *K*i = 600 μM
PAPAIN — *K*i = 320 μM

Z-PHE-ALA-H
CATHEPSIN B | C | $Ki = 21\ \mu M$

Z-PHE-ALA METHYLKETONE
CATHEPSIN B | IR | $Ki = 31\ \mu M$

Z-PHE-ALA MONOFLUOROMETHYLKETONE
CATHEPSIN B | IR | $Ki = 1.5\ \mu M$

Z-PHE-ALA TRIFLUOROMETHYLKETONE
CATHEPSIN B | IR | $Ki = 470\ \mu M$

Z-PHE-(AMINOMETHYL)PHOSPHONYL-LEU-ALA
INTERSTITIAL COLLAGENASE | | $Ki = 71\ \mu M$

Z-PHE-ARG-CHN_2
KALLIKREIN PLASMA
PROTEINASES CYSTEINE

Z-PHE CHLOROMETHYLKETONE
CHYMASE | | $Ki = 797\ \mu M$
CHYMOTRYPSIN | | $Ki = 109\ \mu M$
PROTEINASE THERMUS Rt41A | | 50% 300 μM
SERINE CARBOXYPEPTIDASE
THERMOMYCOLIN

Z-PHE-psi-(CH_2NH)PRO-O-TbU
PROTEINASE HIV-1 | | $Ki = 17\ \mu M$

Z-PHE-$CH_2S^+(CH_3)_2$
PAPAIN | IR |

Z-PHE-CYS(OBZL)CH_2OCO 2,6$(CF_3)_2$PH
CATHEPSIN L | IR | $Ki = 1.4$ nM
CATHEPSIN S | IR |

Z-PHE-DIAZOMETHYLKETONE
PAPAIN

Z-PHE-EPOXIDE
CATHEPSIN B | IR | $Ki = 367\ \mu M$
PAPAIN | IR | $Ki = 574\ \mu M$

Z-PHE-GLYCINENITRILE
CATHEPSIN B | SI | $Ki = 500$ nM

Z-PHE-GLY$(CO)_3$OBnH_2O
α-CHYMOTRYPSIN | | $Ki = 4.9\ \mu M$
ELASTASE LEUKOCYTE | | $Ki = 3\ \mu M$
ELASTASE PANCREATIC | | $Ki = 47\ \mu M$

Z-PHE-GLY-NHO-NBZ
CATHEPSIN L

Z-PHE-GLY^P-(O)PHE
CARBOXYPEPTIDASE A | | $Ki = 81$ pM

1-(N-Z-PHE)-2-IODOACETYL HYDRAZINE
CATHEPSIN B | IR | $Ki = 370$ nM
PAPAIN | IR | $Ki = 180$ nM

Z-D-PHE-D-LEU
SERINE CARBOXYPEPTIDASE | | $Ki = 130\ \mu M$

Z-PHE-LEU-LYSINOL
KALLIKREIN URINE | | $I50 = 26\ \mu M$
PLASMIN | | $I50 = 95\ \mu M$

Inhibitor / Enzyme	Type	Value
Z-PHE-LYS-$CH_2OCO(2,4,6\text{-}ME_3)PH$		
CATHEPSIN B	IR	
Z-PHE-LYS-$CH_2S^+(CH_3)_2$		
KALLIKREIN PLASMA	IR	$Ki = 60\ \mu M$
PLASMIN	IR	$Ki = 27\ \mu M$
TRYPSIN	IR	$Ki = 27\ \mu M$
Z-PHE-LYS-$CH_2S^+(CH_3)\text{-}CH_2PH$		
KALLIKREIN PLASMA	REV	$Ki = 1.9\ \mu M$
PLASMIN	REV	$Ki = 10\ \mu M$
TRYPSIN		
Z-PHE-$NH_2CH_2COCH_2S^+(CH_3)_2$		
PROTEIN-GLUTAMINE γ-GLUTAMYLTRANSFERASE	IR	
Z-PHE$NH_2CH_2COCH_2S^+(CH_3)_2$		
CATHEPSIN B	IR	
Z-PHE-$NH_2(CH_2)_3COCH_2S^+(CH_3)_2$		
PROTEIN-GLUTAMINE γ-GLUTAMYLTRANSFERASE	IR	
Z-PHE-$NHCH_2C(O)CO_2(n\text{-}C_6H_{13})$		
CATHEPSIN B		$Ki = 150$ nM
PAPAIN		$Ki = 800$ nM
Z-PHE-NHO-CO-4-NO_2PHE		
CATHEPSIN B	IR	$Ki = 12\ \mu M$
CATHEPSIN L	IR	$Ki = 44$ nM
CATHEPSIN S	IR	$Ki = 900$ nM
PAPAIN	IR	$Ki = 15\ \mu M$
Z-D-PHE-NHOH		
AMINOPEPTIDASE		$I50 = 1.5\ \mu M$
NEPRILYSIN		$I50 = 200$ nM
THERMOLYSIN		$Ki = 3\ \mu M$
Z-L-PHENYLALANYLBROMOMETHANE		
SUBTILISIN		
Z-L-PHENYLALANYLDIAZOMETHANE		
RHIZOPUSPEPSIN		
Z-PHENYLALANYLGLYOXAL		
CHYMOTRYPSIN	C	$Ki = 65\ \mu M$
erythro-Z-PHE-O-PN-THR-EPOXIDE		
CATHEPSIN B		$Ki = 10\ \mu M$
PAPAIN		$Ki = 30\ \mu M$
Z-PHE-OXAPROLINE-GLYCINE BENZYLESTER		
PROCOLLAGEN-PROLINE, 2-OXOGLUTARATE-4-DIOXYGENASE		
Z-L-PHE-(S)-2-OXO-2-(2-PYRROLIDINYL)ACETYL-D-PHE		
CATHEPSIN B		$I50 = 100\ \mu g/ml$
PROLYL OLIGOPEPTIDASE		$I50 = 650$ pg/ml
Z-PHE-PHE-CHN_2		
BOTHROPAIN	IR	$Ki = 35$ nM
CATHEPSIN L	IR	
PROHORMONE THIOL PROTEINASE	IR	
PROTEINASE CATHEPSIN S LIKE		100% 500 nM
PROTEINASE I PACIFIC WHITING		92% 100 μM
Z-PHE-PHE-DIAZOMETHYLKETONE		
PAPAIN		
Z-PHE-PHE-H		
CALPAIN II		$I50 = 104$ nM

CATHEPSIN B		$I50 = 70$ nM
CATHEPSIN L		$I50 = 740$ pM
Z-PHEpsi(PO_2CH_2)ALA-ARG-MET		
NEUROLYSIN		$Ki = 15$ pM
Z-PHE-psi-(PO_2CH_2)GLY-PRO-AHX		
COLLAGENASE		$Ki = 8$ nM
Z-PHE-psi(PO2CH_2)GLY-PRO-NLE		
COLLAGENASE		$Ki = 8$ nM
Z-(R,S)-PHE^{P}-^{O}-LEU-ALANH_2		
THERMOLYSIN		
Z-PHE-psi[PO_2NH]LEU-ALA		
THERMOLYSIN		$Ki = 68$ pM
Z-D-PHE-PROBORO-METHOXYPROPYLGLYCINEPINANDIOL		
PLASMIN		$Ki = 16$ μM
THROMBIN		$Ki = 8.9$ nM
TRYPSIN		$Ki = 1.1$ μM
Z-D-PHE-PRO-BORO-3-METHOXYPROPYLGLYCINE PINANEDIOL ESTER		
COAGULATION FACTOR Xa		$I50 = 1.7$ μM
KALLIKREIN		$I50 = 7.3$ μM
PLASMIN		$I50 = 3.2$ μM
u-PLASMINOGEN ACTIVATOR		$I50 = 3.6$ μM
Z-D-PHE-PRO-METHOXYPROPYLBOROGLYCINE		
COAGULATION FACTOR Xa	C	$Ki = 12$ μM
KALLIKREIN TISSUE	C	$Ki = 77$ μM
PLASMIN		
THROMBIN	C	$Ki = 22$ nM
Z-D-PHE-PRO-1-(3-METHOXYPROPYL)BOROGLYCINE ESTER		
THROMBIN		$Ki = 7$ nM
Z.PHE-PRO-PHE-^{P}(OPH)$_2$		
CHYMOTRYPSIN		
Z-PHE-THR(OBZL)CHN_2		
CATHEPSIN B	IR	
Z-PHE-TYR(t-BU)CHN_2		
CATHEPSIN L	IR	
CATHEPSIN S	IR	
Z-PHE-VAL^{P}-(O)-PHE		
CARBOXYPEPTIDASE A		$Ki = 30$ fM
N-Z-PRO-α-AZALYS p-NITROPHENYLESTER		
THROMBIN		
Z-PRO-LEU-GLY-HNOH		
GELATINASE A		$I50 = 2$ μM
GELATINASE B		$I50 = 870$ nM
N-Z-PRO-LEU-GLY-HYDROXAMATE		
COLLAGENASE		
Z-PRO-LEU-NHOH		
ENDOPEPTIDASE TREPONEMA DENTICOLA ARCC35405		$Ki = 12$ μM
Z-PROLINAL		
PROLYL OLIGOPEPTIDASE		$I50 = 1.3$ nM
PROLYL OLIGOPEPTIDASE		$I50 = 640$ pM

Inhibitor / Enzyme	Type	Value
Z-PROLINE		
XAA-PRO DIPEPTIDASE	C	*K*i = 90 µM
Z-PROLINOL		
PROLYL OLIGOPEPTIDASE		*I*50 = 175 µM
PROLYL OLIGOPEPTIDASE		*I*50 = 205 µM
N-Z-PROLYL-3-FLUOROPYRROLIDINE		
PROLYL OLIGOPEPTIDASE		*K*i = 800 pM
Z-PROLYL-MORPHOLINE		
PROLYL OLIGOPEPTIDASE	TS	*I*50 = 1 mM
Z-PROLYL-PIPERIDINE		
PROLYL OLIGOPEPTIDASE	TS	*I*50 = 1 mM
Z-PROLYL-PYROLIDINE		
PROLYL OLIGOPEPTIDASE	TS	*I*50 = 540 nM
Z-PROLYL-THIAZOLIDINE		
PROLYL OLIGOPEPTIDASE	TS	*I*50 = 4.6 nM
Z-PROLYL-THIOPROLINAL		
PROLYL OLIGOPEPTIDASE		*I*50 = 1.6 nM
Z-PROLYL-THIOPROLINOL		
PROLYL OLIGOPEPTIDASE	TS	*I*50 = 120 µM
Z-PRO-LYSINAL		
LYSYL ENDOPEPTIDASE	NC	*K*i = 40 µM
Z-PRO-LYS-PHE-psi(PO_2CH_2)-D,L-AL-PRO-ome		
ASTACIN		
Z-PRO-PHE-CHO		
CHYMOTRYPSIN		*K*i = 850 nM
Z-PRO-PRO-$CH_2N^+C_5H_5$		
PROLYL OLIGOPEPTIDASE	IR	*K*i = 9.3 nM
Z-PRO-PROLINAL		
DIPEPTIDYL-PEPTIDASE 4		64% 1.5 mM
PROLYL OLIGOPEPTIDASE	TS	*K*i = 3.4 nM
PROLYL OLIGOPEPTIDASE		*K*i = 350 pM
PROLYL OLIGOPEPTIDASE		*K*i = 3.7 nM
Z-PRO-PROLINOL		
PROLYL OLIGOPEPTIDASE		*K*i = 1 mM
PROLYL OLIGOPEPTIDASE	TS	*I*50 = 100 µM
Z-PRO-PYRROLIDINE		
PROLYL OLIGOPEPTIDASE		*K*i = 240 µM
Z-PROTEINS		
PHOSPHOLIPASE C		
Z-PRO-THIAZOLIDINE		
PROLYL OLIGOPEPTIDASE		*K*i = 39 nM
Z-PRO-THIOPROLINAL		
PROLYL OLIGOPEPTIDASE		*K*i = 3.5 nM
Z-PRO-THIOPROLINOL		
PROLYL OLIGOPEPTIDASE		*K*i = 1 mM
ZSY 27		
3′,5′-CYCLIC-NUCLEOTIDE PHOSPHODIESTERASE		
Z-THIOPROLINE-THIOPROLINAL		
PROLYL OLIGOPEPTIDASE		*K*i = 10 pM

Z-THIOPROLINE-THIOPROLINE
PROLYL OLIGOPEPTIDASE — *K*i = 37 μM

Z-THIOPROLINE-THIOPROLINOL
PROLYL OLIGOPEPTIDASE — *K*i = 1 mM

Z-THIOPROOXAZOLIDINE
PROLYL OLIGOPEPTIDASE — *K*i = 590 nM

Z-THIOPROPYRROLIDINE
PROLYL OLIGOPEPTIDASE — *K*i = 210 nM

Z-THIOPROTHIAZOLIDINE
PROLYL OLIGOPEPTIDASE — *K*i = 1.6 nM

Z-THIOPROTHIOPROLINAL
PROLYL OLIGOPEPTIDASE — *K*i = 350 pM

Z-THIOPROTHIOPROLINALDIMETHYLACETAL
PROLYL OLIGOPEPTIDASE — *K*i = 12 μM

Z-THIOPROTHIOPROLINE METHYL ESTER
PROLYL OLIGOPEPTIDASE — *K*i = 12 μM

Z-TRP-OH
STROMELYSIN — *K*i = 2.1 μM

Z-TYR-ALA-ASP-CHLOROMETHYLKETONE
INTERLEUKIN-β-CONVERTING ENZYME LIKE

Z-TYR-ALA-CH_2
PROTEINASE CYSTEINE LEISHMANIA MEXICANA

Z-TYR-ALA-CH_3
CATHEPSIN B — IR
CATHEPSIN L — IR
CATHEPSIN S — IR

Z-TYR-ALA-CH_2F
CATHEPSIN L — IR

Z-TYR(I)-AGLY-CH_2Cl
CATHEPSIN B — IR — *K*i = 54 nM

Z-TYR(I)-AGLY-CHN_2Cl
CALPAIN I — IR
CALPAIN II — IR

ZUCLOMIPHENE
C8- > C7 STEROL ISOMERASE

Z-VAD-FMK
CASPASE-1
PROTEINASE CPP32

Z-VAL-ALA-ASP-CH_2DCB
CASPASE-1 — IR

Z-VAL-ALA-ASP-CH_2DCPP
INTERLEUKIN 1β CONVERTING ENZYME

Z-VAL-ALA-ASP-CH_2-α((2,6-DICHLOROBENZOYL)OXY)METHYL KETONE
CASPASE-1 — IR

Z-VAL-ALA-ASP-CH_2-PTP
INTERLEUKIN CONVERTING ENZYME — IR

Z-VAL-ALA-ASP-FLUOROMETHYLKETONE
CASPASES
INTERLEUKIN-β-CONVERTING ENZYME LIKE

Z-VAL-ALA-ASP-H

CATHEPSIN B		*K*i = 1.4 μM
INTERLEUKIN CONVERTING ENZYME		*K*i = 11 nM

Z-VAL-LYSINAL

LYSYL ENDOPEPTIDASE	NC	*K*i = 6.5 μM

Z-VAL-(S)-2-OXO-2-(2-PYRROLIDINYL)ACETYLCYCLOHEXYLAMINE

PROLYL OLIGOPEPTIDASE

Z-VAL-PHE-H

CALPAIN		*K*i = 7 nM

Z-VAL-PORLINAL

PROLYL OLIGOPEPTIDASE		*I*50 = 4 ng/ml

Z-VAL-PRO-H

PROLYL OLIGOPEPTIDASE		*K*i = 2.4 nM

Z-VAL-PROLINAL

PROLYL OLIGOPEPTIDASE		*I*50 = 16 ng/ml
PROLYL OLIGOPEPTIDASE		*K*i = 9 nM

Z-VAL-PRO-VAL^{P}(OPhH-pCl)$_2$

CHYMOTRYPSIN
ELASTASE LEUKOCYTE
ELASTASE PANCREATIC

Z-VAL-TYR

PENICILLOPEPSIN		76% 1 mM

Z-VAL-VAL-NLECHN$_2$

CATHEPSIN L	IR	
CATHEPSIN S	IR	

Z-VVK-NHO-NBz

CATHEPSIN B	IR	*K*i = 12 μM
CATHEPSIN H	IR	*K*i = 830 μM
CATHEPSIN L	IR	*K*i = 610 nM
CATHEPSIN S	IR	*K*i = 94 nM
PLASMIN	IR	
t-PLASMINOGEN ACTIVATOR	IR	
THROMBIN	IR	
TRYPSIN	IR	

ZWITTERGENT

α-GLUCOSAMIDINE-N-ACETYLTRANSFERASE		100% 1%
HEPARAN-α-GLUCOSAMINIDINE N-ACETYLTRANSFERASE		

Glossary

Trivial name	Synonym or systematic name
A 3	N-(6-AMINOETHYL)-5-CHLORO-1-NAPHTHALENESULPHONAMIDE
A 2131	3-(5′-HYDROXYMETHYL-2′-FURYL)-1-BENZYLTHIENO(3,2-c)PYRAZOLE
A 60586	6-HYDROXY-N-METHYL-N-(2-[4-PHENYLPHENYL]ETHYL)-1,2,3,4-TETRAHYDRO-1-NAPHTHALENE METHAN-AMINE
A 61442	N-HYDROXY-N-METHYL-3-[4-2,4,6-TRIMETHYLPHENYL)PHENYL]PROPENAMIDE
A 62198	DIMETHYLACETYL-PHE-HIS-NHCH(CYCLOHEXYLMETHYLCH(OH)CH(OH)CH2N3
A 63162	N-HYDROXY-N-[1-PHENYLMETHOXYPHENYLETHYL]ACETAMIDE
A 64077	N-(1-(BENZO[b]THIEN-2-yl)ETHYL-N-HYDROXYUREA
A 64077	ZILEUTON
A 64662	ENALKIREN
A 69412	N-1-(FUR-3-YLETHYL)-N-HYDROXYUREA
A 771726	N-(4-TRIFLUOROMETHYLPHENYL-2-CYANO-3-HYDROXYCROTOAMIDETMOXIFEN CITRATE
A 78773	N[3-[5-(4-FLUOROPHENOXY)-2-FURANYL]-1-METHYL-2-PROPYNYL]-N-HYDROXYUREA
A 79175	R-(+)-N-[3-[5-(FLUOROPHENOXY)-2-FURANYL]-1-METHYL-2-PROPYNYL]-N-HYDROXYUREA
A 80987	YY 752
A 80914A	3,4a-DICHLORO-3,4,4a-10a-TETRAHYDRO-6,8-DIHYDROXY-2,2,7-TRIMETHYL-10a-[(2,2-DIMETHYL-3-CHLORO-6-METHYLENECYCLOHEXYL)METHYL]2H-NAPHTHOL[2,3,b]PYRAN5,10DIONE
A 25822B	15-AZA-24-METHYLENE-d-HOMOCHOLESTADIENE 3β-OL
A 9154C	5-[5′-DEOXY-5′-(C)-4′-,5′-DIDEHYDROADENOSYL-L-ORNITHINE
A 723 U	2-ACETYLPYRIDINE 4-(MORPHOLINOETHYL)THIOSEMICARBAZONE
A 1110U	2-ACETYLPYRIDINE 5-[(DIMETHYLAMINO)THIOCARBONYL]-THIOCARBONOHYDRAZONE
1233A	(E, E)-11-[3′-(HYDROXYMETHYL)-4′-OXO-2′-OXETANYL]-3,5,7-TRIMETHYL-2,4-UNDECADIENOIC ACID
AA	α-ALLENYLAGMATINE
AA 861	2-(12-HYDROXY-5,10-DODECADINYL)-3-5-6-TRIMETHYL-2,5-CYCLOHEXADIEN-1,4-DIONE
AAL 05	N-CYCLOHEXYL-N-METHYL-4-(1,2-DIHYDRO-2-OXO-6-QUINOYLOXY)VALERAMIDE
AAPCK	ALA-ALA-PHE-CHLOROMETHYLKETONE
AB 47	N-[(8-AMINO-1(s)-CARBOXYOCTYL]-L-ALANYL-L-PROLINE
ABEZYL 10-FORMYL-TRIDEAZAFOLATE	N-[7-2-AMINO-3,4-DIHYDRO-4-OXOQUINAZOLIN-6-YL)-6-FORMYL-1-OXOHEPTYL]-L-GLUTAMIC ACID
ABT 719	8-(3-(S)-AMINOPYRROLIDIN-1-YL)-CYCLOPROPYL-7-FLUORO-9-METHYL-4H-4-OXO-QUINOLIZINE-3-CARBOXYLIC ACID
ABT 761	(R)(7)-N-[3[5[(4-FLUOROPHENYL)METHYL]-2-THIENYL]-1-METHYL-2-PROPYL]N-HYDROXYUREA
ABYSSINONE V	4′-HYDROXY-3′,5′-DIPRENYLISOFLAVONE
β1AC	β1-ANTICOLLAGENASE
AC 3-1	1-[8-HYDROXY-2-METHYL, 2-(4-METHYLPENT-3-ENYL)CHROMENE], 3-[2,4-DIHYDROXYPHENYL]PROPAN-3-ONE
AC 3-2	4,2′,4′-TRIHYDROXY-8-GERANYLDIHYDROCHALCONE
AC 3-3	5,7,4′-TRIHYDROXY-8-GERANYLDIHYDROCHALCONE
AC 5-1	3,4,2′,4′-TETRAHYDROXY-2-GERANYLDIHYDROCHALCONE
AC 5-2	7,3′,4′-TRIHYDROXY-2′-GERANYLFLAVONE
ACARBOSE	BAY-g 5421
ACARBOSE	O-4,6-DIDEOXY-4-[[[1S-(1α,4α,5β,6α)]-4,5,6-TRIHYDORXY-3-(HYDROXYMETHYL)-2-CYCLOHEXEN-1-YL]AMINO]-α-D-GLUCOPYRANOSYL-(1->4)-O-α-D-GLUCOPYRANOSYL-(1->4)-D-GL
ACETAZOLAMIDE	DIAMOX
ACETAZOLAMIDE	N-[5-(AMINOSULPHONYL)-1,3,4-THIADIAZOL-2-YL]ACETAMIDE
ACETOIN	3-HYDROXY-2-BUTANONE
ACETORPHAN	N-(R,S)-3-ACETYLMERCAOTO-2-BENZYLPROPANOYL-GLYCINE BENZYLESTER
γ-ACETYLENIC GABA	4-AMINOHEX-5-YNOIC ACID
ACETYLPEPSTATIN	STREPOMYCES PEPSIN INHIBITOR
ACHPA	4(S)-AMINO-3(S)-HYDROXY-5-CYCLOHEXYLPENTANOIC ACID
α1-ACHY	α1-ANTICHYMOTRYPSIN
ACIFLUORFEN	5-[2-CHLORO-4-(TRIFLUOROMETHYL)PHENOXY]-2-NITROBENZOIC ACID
ACIVICIN	α-AMINO-3-CHLORO-4,5-DIHYDRO-5-ISOXAZOLACETIC ACID
ACP	ACYLCARRIER PROTEIN
ACYCLOTHYMIDINE	5-METHYL-1-(2′-HYDROXYETHOXYMETHYL)URACIL
ACYCLOURIDINE	1-(2′-HYDROXYETHOXYMETHYL)URACIL
ACYCLOVIR	9-[(2-HYDROXYETHOXY)METHYL]GUNANINE

Trivial name	Synonym or systematic name
AD6	8-MONOCHLORO-3-β-DIETHYLAMINOETHYL-4-METHYL-7-ETOXYCARBONYLMETHOXY COUMARIN
AD 5467	3,4-DIHYDRO-2,8-DISIOPROPYL-3-THIOXO-2H-1,4-BENZOXAZINE-4-ACETIC ACID
ADIPOSTATIN A	5-n-PENTADECYLRESORCINOL
ADIPOSTATIN B	5-ISOPENTADECYLRESORCINOL
ADN 138	8′-CHLORO-2′,3′-DIHYDROSPIRO[PYRROLIDINE-3,6′(5′h)-PYRROLO[1,2,3-de][1,4]BENZOXAZINE]-2,5,5′-TRIONE
ADRENALIN	(R)-4-[1-HYDROXY-2-(METHYLAMINO)ETHYL]-1,2-BENZENEDIOL
ADRIAMYCIN	DOXORUBICIN
AFK 108	1-[2-(2,4-DICHLOROPHENYL)-2-((2E)-3,7-DIMETHYLOCTA-2,6-DIENYLOXY)ETHYL]-1H-IMIDAZOLE
AFZELIN	KEMPFEROL-3-O-RHAMNOSIDE
AG 17	(3,5-DI-tert.-BUTYL-4-HYDROXYBENZILIDENE)MALONITRILE
AG 34	3-METHOXYL-5-(2,2-DICYANOEHTENYL)CATECHOL
AG 331	N^6-[4-N-MORPHOLINOSULPHONYL)BENZYL]N^6-METHYL-2,6-DIAMINO-BENZ[cd]INDOLE GLUCURONATE
AG 337	3,4-DIHYDRO-2-AMINO-6-METHYL-4-OXO-5-(4-PYRIDYLTHIO)QUINAZOLINE
AG 372	1,3-BIS-(DICYANOMEHTENYL)-2-(3,4-DIHYDROXYPHENYL)-METHENYLINDANE
AG 799	5-BENZYLTHIOMETHYLVANILLINE
AG 800	3-METHOXY-4-HYDROXY-5-BENZYLTHIOMETHYL-α-AMIDO(cis)-CINNAMONITRILE
AG 805	α-(3,4-DIHYDROXYBENZOYL)-5-(cis)-INDOLECINNAMONITRILE
AG 814	(2′,5′-DIHYDROXY)BENZYL)-3-AMINOBENZOIC ACID
AG 824	3,4-DIHYDROXY-5-BENZYLTHIOMETHYL-α-AMIDO-(cis)-CINNAMONITRILE
AG 826	(3′-4′-DIHYDROXY)BENZYL-3-AMINOBENZOIC ACID
AG 1112	3-[2-[(3-AMINO-4-CYANO)PYRAZOL-5-YL]-2-CYANO]ETHENYLINDOLE
AG 1343	NELFINAVIR MESYLATE
AG 1343	[3S, (3R*,4aR*,8aR*,2′S*,3′S*)]2[2′HYDROXY3′[(PHENYLTHIO)METHYL]4′AZA5′OXO5′-(2″-METHYL-3″-HYDROXYPHENYL)PENTYL]DECAHYDROISOQUINOLINE-3N-tBUTYLCARBOXAMIDE
AG 1343	[3S(3R*,4aR*,8aR+,2′S*,3′S*)]2[2′HYDROXY3′PHENYLTHIO-METHYL4 ′AZA5′OXO5′(2″METHYL3″HYDROXY-PHENYL) PENTHYL]DECAHYDROISOQUINOLOINE3Nt-BUTYLCARBOXAMIDE MSA
AG 1749	LANSOPRAZOLE
AG 1749	2-[[[3-METHYL-4-(2,2,2-TRIFLUOROETHOXY)-2-PYRIDYL]METHYLSULPHINYL]-1H-BENZIMIDAZOLE
AG 2034	4-[2-(AMINO-4-OXO-4,6,7,8-TETRAHYDRO-3H-PYRIMIDINO[5,4-6][1,4]THIAZIN-6-YL-(S)-ETHYL]-2,5-THIENOYL-L-GLUTAMIC ACID
AGARICIC ACID	2-HYDROXY-1,2,3-NONADECANETRICARBOXYLIC ACID
AGM 1470	O-(CHLOROACETYLCARBAMOYL)FUMAGILLOL
AGM 1470	TNP 470
AGMATINE	(4-AMINOBUTYL)-GUANIDINE
AGN 1135	N-PROPARGYL-1-AMINOINDANE
AH 21132	(±)cis-6-(p-ACETAMIDOPHENYL)-1,2,3,4,4a,10b-HEXAHYDRO-8,9-DIMETHOXY-2-METHYL-BENZO-[c][1,6]NAPHTYRI-DINE)
AH 21132	BENAFENTRINE
AH 2429	4-CHLORO-1-METHYL-5-(4-NITROPHENOXY)-3-TRIFLOUROMETHYL)-1H-PYRAZOLE
AH 2430	4-CHLORO-1-METHYL-3-(4-NITROPHENOXY)-5-TRIFLOUROMETHYL)-1H-PYRAZOLE
AH 2431	5((4-CHLORO-1-METHYL-5-(TRIFLUOROMETHYL)-1H-PYRAZOL-3-YL)OXY)-2-NITROBENZOIC ACID
AH 2432	5((4-CHLORO-1-METHYL-3-(TRIFLUOROMETHYL)-1H-PYRAZOL-5-YL)OXY)-2-NITROBENZOIC ACID
AHR 5333	1-[4-[3-[4-[BIS(4-FLUOROPHENYL)HYDROXYMETHYL]-1-PIPERIDINYL]PROPOXY]-3-METHOXYPHENYL]ETHA-NONE
AHR 9294	8-METHOXY-4-[(2-ISOPROPYLPHENYL)AMINO]-3-QUINOLINECARBOXYLATE ETHYLESTER
A4 INHIBITOR	ALZHEIMER AMYLOID PRECUSOR PROTEIN FRAGMENT
AJOENE	(E/Z)-4,5,9-TRITHIADODECA-1,6,11-TRIENE 9.OXIDE
AK 275	Z-LEU-ABU-CONH-CH_2CH_3
AL 03152	2,7-DIFLUORO-4-METHOXYSPIRO(9H-FLUORENE-9,4′-IMIDAZOLIDINE)-2′,5′-DIONE
AL 1567	ALCONIL
AL 1576	2,7-DIFLUOROSPIROFLUORENE-9,5′-IMIDAZOLIDINE-2′,4′-DIONE
AL 1576	2,7-DIFLUOROSPIRO(9H-FLUORENE-9,4′-IMIDAZOLINE)-2,5′-DIONE
ALADOTRILAT	(S)N(3(3,4-METHYLENEDIOXYPHENYL)-2-MERCAPTOMETHYL)-1-OXOPROPYL-(S)-ALANINE
ALANINE PHOSPHONATE	(1-AMINOMETHYL)PHOSPHONIC ACID
ALATRIOPRILAT	(S)-N-[3-(3,4-METHYLENEDIOXYPHENYL)-2-(MERCAPTOMETHYL)-1-OXOPROLYL](S)ALANINE
ALBIZZIIN	3-[(AMINOCARBONYL)AMINO]-L-ALANINE
ALDRIN	1,2,3,4,10,10-HEXACHLORO-1,4,4α,5,8,8a-HEXAHYDRO-1,4: 5,8-DIMETHANONAPHTHALENE

Trivial name	Synonym or systematic name
ALENDRONATE	4-AMINO-1-HYDROXYBUTYLIDENE-1,1-BISPHOSPHONIC ACID
ALLANTOIC ACID	BIS[(AMINOCARBONYL)AMINO]-ACETIC ACID
ALLANTOIN	(2,5-DIOXO-4-IMIDAZOLIDINYL)-UREA
ALLANTOXANIC ACID	1,4,5,6-TETRAHYDRO-4,6-DIOXO-1,3,5-TRIAZINE-2-CARBOXYLIC ACID
ALLOPURINOL	1,4-DIHYDRO-4-HYDROXYPYRAZOLO(3,4-D)-PYRIMIDIN-4-ONE
ALLOXANTHINE	1H-PYRAZOLO[3,4-d]PYRIMIDINE-4,6-DIOL
ALMOXATONE	(R)-3-{4-[(3-CHLOROPHENYL)METHOXY]PHENYL}-5-[(METHYLAMINO)METHYL]-2-OXAZOLIDINONE
ALO 1567	2-FLUORO-SPIRO(9h-FLUORENE-9,4′-IMIDOLIDINE)-2′,5′DIONE
ALO 1576	2,7-DIFLUORO-SPIRO(9h-FLUORENE-9,4′-IMIDOLIDINE)-2′,5′DIONE
ALO 1576	HOE 843
ALO 1576	IMRESTAT
ALO-1750	7-FLUORO-SPIRO(5h-INDENO[1,2b]PYRIDINE-5,3′-PYRROLIDINE)-2,5′-DIONE
ALO 4114	2,7-DIFLUORO-4,5-DIMETHOXYSPIRO-(9H-FLUORENE-9,4′-IMIDAZOLIDINE)-2′,5′-DIONE
ALRESTATIN	1,3-DIOXO-1H-BENZENE-(de)-ISOQUINOLIN-(3H) ACETIC ACID
AMA	S-(5′-DEOXY-5′-ADENOSYL)METHIOETHYLHYDROXYLAMINE
AMASTATIN	[S-(R*,S*)]-N-[N-[N-(3-AMINO-2-HYDROXY-5-METHYL-1-OXOHEXYL)-L-VALYL]-L-VALYL]-L-ASPARTIC ACID
AMDR	ADENOSINE 2′-MONOPHOSPHO-5′-DIPHOSPHORIBOSE
AMERCIN CYANAMID	2-p-AMINOBENZENESULFONAMIDO-1,3,4-THIAZOLE-5-SULPHONAMIDE
AMETHOPTERIN	N-[4-[[(2,4-DIAMINO-6-PTERIDINYL)METHYL]METHYLAMINO]BENZOYL]-L-GLUTAMIC ACID
AMFLUTIZOLE	4-AMINO-3[3-(TRIFLUOROMETHYL)PHENYL]-5-ISOTHIAZOLE CARBOXYLIC ACID
AMIDOX	3,4-DIHYDROXYBENZAMIDOXIME
AMIFLAMINE	(S)-4-(DIMETHYLAMINO)-α,2-DIMETHYL-BENZENEETHANAMINE
AMILORIDE	3,5-DIAMINO-N-(AMINIMINOMETHYL)-6-CHLOROPYRAZINECARBOXAMIDE
AMINOACYCLOTHYMIDINE	5-METHYL-1-(2′-AMINOETHOXYMETHYL)URACIL
AMINOCARNITINE	3-AMINO-4-TRIMETHYL-AMINOBUTYRATE
AMINOGLUTETHIMIDE	3-(4-AMINOPHENYL)-3-ETHYLPIPERIDINE-2,5-DIONE
o-AMINOGLYCOLIC ACID	CARBOXYMETHOXYLAMINE
AMINOPTERIN	N-[4-[[(2,4-DIAMINO-6-PTERIDINYL)METHYL]AMINO]BENZOYL]-L-GLUTAMIC ACID
AMITROLE	3-AMINO-1,2,4-TRIAZOLE
AMITRYPTILINE	3-(10,11-DIHYDRO-5H-DIBENZO[A, D]CYCLOHEPTEN-5-YLIDENE)-N, N-DIMETHYL-1-PROPANAMINE
AMO 1618	N, N, N, 2-TETRAMETHYL-5-(1-METHYLETHYL)-4-[(1-PIPERIDINYLCARBONYL)OXY]-BENZENAMINIUM CHLORIDE
AMOBARBITAL	5-ETHYL-5-(3-METHYLBUTYL)-2,4,6(1H, 3H, 5H)-PYRIMIDINETRIONE
AMP	ADENOSINE 5′-MONOPHOSPHATE
AMPD	4-AZA-4-METHYL-5α-PREGNANE-3,20-DIONE
AMPHETAMINE	α-METHYL-BENZENEETHANAMINE
AMPICILLIN	[2S-[2α,5α,6β(S*)]]-6-[(AMINOPHENYLACETYL)AMINO]-3,3-DIMETHYL-7-OXO-4-THIA-1-AZABICYCLO[3.2.0]HEPTANE-2-CARBOXYLIC ACID
AMP-PNP	ADENYL-5′-YLIMINODIPHOSPHATE
cAMPS	Rp-ADENOSINE-3′,5′-cyclicMONOPHOSPHOTHIOATE
AMRINONE	5-AMINO-(3,4′-BIPYRIDIN)-6(1H)-ONE
m-AMSA	4′-(9-ACRIDINYLAMINO) METHANE-SULPHONYL-m-ANISIDE
m-AMSA	N-[4-(ACRIDINYLAMINO)-3-METHOXYPHENYL]METHANESULPHONAMIDE
AMYLAMINE	1-PENTANAMINE
ANACARDIC ACID	6-[8(Z),11(Z),14-PENTADECATRIENYL]SALICYLIC ACID
ANACARDIC ACID A	6-TRIDECYLSALICYLIC ACID
ANACARDIC ACID B	6-[(8Z)-PENTADECENYL]SALICYLIC ACID
ANACARDIC ACID C	6-[(9Z, 12Z)-HEPTADECADIENYL]SALICYLIC ACID
ANACARDIC ACID D	6-[(8Z)-HEPTADECENYL]SALICYLIC ACID
ANASTTROZOLE	2,2′[5-(1H-1,2,4-TRIAZOL-1-YLMETHYL)-1.3-PHENYLENEBIS(2-METHYLPROPIONITRILE)
ANCYMIDOL	α-CYCLOPROPYL-α-[p-METHOXYPHENYL]-5-PYRIMIDINE METHYLALCOHOL
ANDROSTERONE	(3α,5α)-3-HYDROXY-ANDROSTAN-17-ONE
ANF	α-NAPHTHOFLAVONE
ANNONIN IV	ROLINIASTATIN 2
ANTHRALIN	1,8-DIHYDROXY-9[10H]-ANTHRACENONE
ANTHRANILIC ACID	2-AMINO-BENZOIC ACID

Trivial name	Synonym or systematic name
ANTHRANYLAMIDE	2-AMINOBENZAMIDE
β1-ANTICOLLAGENASE	TIMP
ANTILEUKOPROTEASE	SECRETORY LEKOPROTEASE INHIBITOR
ANTIMYCIN A1	3-METHYL-BUTANOIC ACID 3-[[3-(FORMYLAMINO)-2-HYDROXYBENZOYL]AMINO]-8-HEXYL-2,6-DIMETHYL-4,9-DIOXO-1,5-DIOXONAN-7-YL ESTER
ANTIMYCIN D	DACTINOMYCIN
ANTIPAIN	N2-[[(1-CARBOXY-2-PHENYLETHYL)AMINO]CARBONYL]-L-ARGINYL-N-[4-[(AMINOIMINOMETHYL)AMINO]-1-FORMYLBUTYL]-L-VALINAMIDE
α1-ANTITRYPSIN	α1-PROTEINASE INHIBITOR
ANTRYPOL	SURAMIN
AO 128	N-[2-HYDROXY-1-(HYDROXYMETHYL)ETHYL]VALIOLAMINE
AO 128	(+)1L-[1(OH),2,4,5/3]-5-[2-HYDROXY-1-(HYDROXYMETHYL)ETHYL]AMINO-1-C-(HYDROXYMETHYL)-1,2,3,4-CYCLO-HEXANETETROL
AO 128	VOGLIBOSE
AP5A	DI(ADENOSIN-5′-)PENTAPHOSPHATE (P1,P4)
APC 3328	N-[1S-(2-PHENYLETHYL)-3-PHENYLSULPHONYLALLYL]-4-METHYL-2R-PIPERAZINYL CARBONYLAMINO-VALERAMIDE
APHIDICOLIN	[3R-(3α,4α,4Aα,6Aβ,8β,9β,11Aβ,11Bβ)]-TETRADECAHYDRO-3,9-DIHYDROXY-4,11B-DIMETHYL-8,11A-METHANO-11-AH-CYCLOHEPTA[A]NAPHTHALENE-4,9-DIMETHANOL
APMSF	4-AMIDINOPHENYLMETHANESULPHONYLFLUORIDE
APOCYNIN	4-HYDROXY-3-METHOXYACETOPHENON
APROTININ	BOVINE PANCREATIC TRYPSIN INHIBITOR
APROTININ	PANCREATIC BASIC TRYPSIN INHIBITOR
APROTININ	PANCREATIC TRYPSIN INHIBITOR (KUNITZ)
APROTININ	TRASYLOL
APROTININ	TRYPSIN-KALLEKREIN INHIBITOR
APR-TC	ALLOPURINOL RIBOSIDE-3-THIOCARBOXAMIDE
APR-TC	4(5H)-OXO-1-β-D-RIBOFURANOSYLPYRAZOLO[3,4-d]PYRIMIDINE-3-THIOCARBOXAMIDE
APSTATIN	N-[(2S, 3R)-3-AMINO-2-HYDROXY-4-PHENYLBUTANOYL]-L-PROLYL-L-PROLYL-L-ALANINAMIDE
ARACHIDONIC ACID	(ALL-Z)-5,8,11,14-EICOSTETRAENOIC ACID
ARBOXIMIDAMIDE HBR	2-(5-NITRO-2-FURNYLMETHYLIDENE)-n,n′-[1,4-PIPERAZINEDIYLBIS(1,3-PROPANEDIYL)]BISHYDRAZINEC
ARBUTIN	HYDROQUINONE -O-β-D-GLUCOPYRANOSIDE
ARCAIN	N, N‴-1,4-BUTANEDIYLBIS-GUANIDINE
ARECAIDINE	1,2,5,6-TETRAHYDRO-1-METHYL-3-PYRIDINECARBOXYLIC ACID
ARGATROBAN	MD 805
ARGATROBAN	(2R,4R)-4-METHYL-1-[N2-[(3-METHYL-1,2,3,4-TETRAHYDRO-8-QUINOLINYL)SULPHONYL]ARGINYL]-2-PIPERIDINE CARBOXYLIC ACID
L-ARGININOSUCCINIC ACID	(S)-N-[[(4-AMINO-4-CARBOXYBUTYL)AMINO]IMINOMETHYL]-L-ASPARTIC ACID
ARGIPIDINE	ARGATROBAN
ARICEPT	E 2020
ARIMIDEX	ANASTROZOLE
ARIMIDEX	ZD 1033
ARL 67156	6-N, N-DIETHYL-D-β,γ-DIBROMOMETHYLENE-ATP
ARPHAMENINES A	2R,5S-5-AMINO-8-GUANIDINO-4-OXO-2-PHENYLMETHYLOCTANOIC ACID
ARPHAMENINES B	2R,5S-5-AMINO-8-GUANIDINO-4-OXO-2-(p-HYDROXYPHENYLMETHYL)OCTANOIC ACID
ARTEPARON	GLYCOSAMINOGLYCAN POLYSULPHATE
AS 183	2,4-DIMETHYL-2-HYDROXY-5-(1,2,5,7-TETRAMETHYLNONYL)-3(2H)FURANONE
AS 186a	PENICILLIDE
AS 186b	PURPACTIN A
ASI 222	3-β-O-(4-AMINO-4,6-DIDEOXY-β-D-GALACTOPYRANOSYL)DIGITOXIGENIN
ASPARENOMYCIN A	[5R-[3[R*(E)],5R*,6E]]-3-[[2-(ACETYLAMINO)ETHENYL]SULPHINYL]-6-(2-HYDROXY-1-METHYLETHYLIDENE)]-7-OXO-1-AZABICYCLO[3.2.0]HEPT-2-ENE-2-CARBOXYLIC ACID
ASPIRIN	2-(ACETYLOXY)-BENZOIC ACID
ASPISOL	LYSINEMONOACETYLSALICYLATE
ASTERRIQUINONE	2,5-BIS-[(1″,1″-DIMETHYL-2″-PROPENYL)-INDOL-3′-YL]-3,6-DIHYDROXY-1,4-BENZOQUINONE
AT-125	[S-(R*,R*)]-α-AMINO-3-CHLORO-4,5-DIHYDRO-5-ISOXAZOLEACETIC ACID
ATABRINE	QUINACRINE

Trivial name	Synonym or systematic name
ATAMESTANE	1-METHYL-ANDROSTA-1,4-DIENE-3,17-DIONE
ATAMESTANE	1-METHYL-1,4-ANDROSTADIENE-3,17-DIONE
ATEBRIN	QUINACRINE
AT III	ANTITHROMBIN III
ATORVASTATIN	CI 981
ATORVASTIN	LIPITER
ATRACTYLIGENIN	(2β,4α,15α)-2,15-DIHYDROXY-19-NORKAUR-16-EN-18-OIC ACID
ATRACTYLOSIDE	15-HYDROXY-2-[[2-O-(3-METHYL-1-OXOBUTYL)-3,4-DI-O-SULPHO-β-D-GLUCOPYRANOSYL]OXY]-19-NORKAUR-16-EN-18-OIC ACID DIPOTASSIUM SALT
ATRAZINE	2-CHLORO-4-(ETHYLAMINO)-6-(ISOPROPYLAMINO)-s-TRIAZINE
A723U	2-ACETYLPYRIDINE 4-(MORPHOLINOETHYL)THIOSEMICARBAZONE
AU 1421	(Z)-5-METHYL-2-[(1-NAPHTHYL)ETHENYL]-4-PIPERIDINOPYRIDINE
AURANOFIN	(1-THIO-β-D-GLUCOPYRANOSE-2,3,4,6-TETRAACETATO-S)(TRIETHYLPHOSPHINE)GOLD
AURIN	4-[BIS(4-HYDROXYPHENYL)METHYLENE]-2,5-CYCLOHEXADIEN-1-ONE
AURINTRICARBOXYLIC ACID	5-[(3-CARBOXY-4-HYDROXYPHENYL)(3-CARBOXY-4-OXO-2,5-CYCLOHEXADIEN-1-YLIDENE)METHYL]-2-HYDROXY-BENZOIC ACID
AUSTRALINE	(1R,2R,3R,7S,7aR)-3-(HYDROXYMETHYL)-1,2,7-TRIHYDROXYPYRROLIZIDINE
AY 22284	1,3-DIOXO-1H-BENZ[de]ISOQUINOLINE-2-(3H)ACETIC ACID
AY 27773	TOLRESTAT
AY 28768	PELRINONE
AY 31390	3,4-DIHYDRO-5-BROMO-2-METHYL-4-OXO-6[[(5-BROMO-3-PYRIDINYL)METHYL]AMINO]PYRIMIDINE HBr
AY 9944	1,4-BIS(2-CHLOROBENZYLAMINOMETHYL)CYCLOHEXANE
AZACHOLESTEROL	6-AZACHOLES-4-EN-3β-OL-7-ONE
8-AZA-1-DEAZAADENOSINE	7-AMINO-3-(β-D-RIBOFURANOSYL)-3H-1,2,3-TRIAZOLO[4,5-b]PYRIDINE
8-AZADECALIN	N-(1,5,9-TRIMETHYLDECYL)-4α,10-DIMETHYL-8-AZA-trans-DECAL-3β-OL
15-AZASTEROL	15-AZA-24-METHYLENE-D-HOMOCHOLESTA-8,14-DIEN-3β-OL
AZATOXIN	NSC 640737-M
AZATOXIN	5R, 11aS-1H, 6H, 3ONE-5,4,11,11a-TETRAHYDRO-5-(3,5-DIMETHOXY-4-HYDROXYPHENYL)OXAZOLO(3′,4′:1,6)PYRIDO(3,4-b)INDOLE
AZIRIDINE GLUTAMATE	2-(2-CARBOXYETHYL)AZIRIDINE-2-CARBOXYLIC ACID
AZIRIDIONO-DAP	2-(4-AMINO-4-CARBOXYBUTYL)AZIRIDINE-2-CARBOXYLIC ACID
B 428	4-IODO-BENZO[b]THIOPHENE-2-CARBOXYMIDINE
B 581	N-{2(S)-[2(R)-AMINO-3-MERCAPTOPROPYLAMINO]-3(S)-METHYLBUTYL}PHENYLALANYL-METHIONINE
B 807-27	2-[5-(4-CHLOROPHENYL)PENTYL]OXIRANE-2-CARBOXYLATE
B 807-27	POCA
B 827-33	ETOMOXIR
B 832-145	6-(3-ISOPROPOXY-4-METHOXYPHENYL)-3-(2H)PYRIDAZINONE
B 859-35	DEXNIGULDIPINE
B 956	N-{[8(R)-AMINO-2(S)-BENZYL-5(S)-ISOPROPYL-9-MERCAPTONONA-2,5-DIENOYL]}METHIONINE
B 956	PD 331
B 8301-078	2,2-DIFLUORO-6-[(4-METHOXY-3-METHYL-2-PYRIDYL)METHYLSULPHINYL]-5H-[1,3]-DIOXOLO[4,5-f]BENZIMIDAZOLE
B 9004-070	(cis-8,9-DIMETHOXY-2-METHYL-6-[4-p-TOLUENESULPHONAMIDOPHENYL]-1,2,3,4,4a,10b-HEXAHYDROBENZO-[c][1,6]-NAPHTHYRIDINE
B1-4	3,6-DIHYDROXY-2-[2′-(1″,1″-DIMETHYL-2″-PROPENYL)-INDOL-3′-YL]-5-[1′,7′-(1″,1″-DIMETHYLPROPANO)-INDOL-3′-YL]-1,4-BENZOQUINONE
B24	3,5-ETHOXY-4-AMINOMETHYLPYRIDINE
BAFILOMYCIN C1	L 681110A1
BAICALEIN	5,6,7-TRIHYDROXY-2-PHENYL-4H-1-BENZOPYRAN-4-ONE
BAICALIN	7-D-GLUCURONIC ACID -5,6-DIHYDROXYFLAVONE
BANA 113	3-GUANIDINO-4-(N-ACETYLAMINO)BENZOIC ACID
BANA 115	3,5-DIGUANIDINO-4-(N-ACETYLAMINO)BENZOIC ACID
BAPN	β-AMINOPROPIONITRILE
BARBITAL	5,5-DIETHYL-2,4,6(1H, 3H, 5H)-PYRIMIDINE TRIONE
BARBITURIC ACID	2,4,6-(1H, 3H, 5H)-PYRIMIDINE TRIONE
BATHOCUPROINE	2,9-DIMETHYL-4,7-DIPHENYL-1,10-PHENANTHROLINE

Trivial name	Synonym or systematic name
BATHOPHENANTHROLINE	4,7-DIPHENYL-1,10-PHENANTHROLINE
BATIMISTAT	BB 94
BAYER 205	SURAMIN
BAY G 5421	ACARBOSE
BAY G 6576	NAFAZATROM
BAY H 5595	1-DEOXYNOJIRIMYCIN
BAY M 1099	N-HYDROXYETHYL-1-DEOXYNOJIRIMYCIN
BAY M 1099	MIGLITOL
BAY O 1248	EMIGLITATE
BAY O 1248	N-[β-(4-ETHOXYCARBONYLPHENOXY)ETHYL]-1-DEOXYNOJIRIMYCIN
BAY W 1807	(–)(S)-3-ISOPROPYL-4-(2-CHLOROPHENYL)-1,4-DIHYDRO-1-ETHYL-2-METHYLPYRIDINE-3,5,6-TRICARBOXYLATE
BAY W 6228	CERIVASTATIN
BAY W 6228	(E)-(+)(3R, 5S)-7-(4-(4-FLUOROPHENYL)-2,6-DISIOPROPYL-5-(METHOXYMETHYL)PYRID-3-YL)-3,5-DIHYDROXY-HEPT-6-ENOATE
BAY X 1005	(R)-2-[4-QUINOLIN-2-YL-METHOXY)PHENYL]-2-CYCLOPENTYL ACETIC ACID
BB 87	N-4-HYDROXY-2R-ISOBUTYL-N1-{2-PHENYL-1S-[(PYRID-2-YLMETHYL)CARBAMOYL]ETHYL}SUCCINAMIDE
BB 94	BATIMASTAT
BB 94	4-(N-HYDROXYAMINO)-2R-ISOBUTYL-3S-(THIOPHEN-2-YLTHIOMETHYL)SUCCINYL-PHENYLALANYL-N-METHYLAMIDE
BB 250	N-4-HYDROXY-2R-ISOBUTYL-N1-(1S METHYLCARBAMOYL]-2-PHENYLETHYL)-3S-THIOPHENE-2-SULPHONYL-METHYL)SUCCINAMIDE
BB 2001	3R{1S(3-MORPHOLIN-4-YL-PROPYLCARBAMOYL)-2-PHENYLETHYLCARBAMOYL}6-PHENYLHEXENOIC ACID
BB 2014	3R-(1S-METHYLCARBAMOYL-2-PHENYLETHYLCARBAMOYL)-6-PHENYLHEXANOIC ACID
BB 2516	MARIMASTAT
BB 3003	3R-(1S-METHYLCARBAMOYL-2-PHENYLETHYLCARBAMOYL)NONADECANOIC ACID
BCNU	1,3-BIS(2-CHLOROETHYL)-1-NITROSOUREA
BCNU	CARMUSTINE
BD 40	10-[γ-DIETHYLAMINOPROPYLAMINO]-6-METHYL-5H-PYRIDO[3′,4′: 4,5]PYRROLO[2,3-G]ISOQUINOLINE
BD 40	NSC 303565
BD 40	PAZELLIPTINE
BD 40	SR 95225
BDF 8634	SATERINONE
BE 16627B	L-N-(HYDROXY-2-ISOBUTYLSUCCINAMOYL)SERYL-L-VALINE
BEFLAXATONE	3-[4,4,4-TRIFLUORO-3(R)-HYDROXYBUTOXYPHENYL](R)METHOXYOXAZOLIDIN-2-ONE
BEHENIC ACID	DOCOSANOIC ACID
BELACTIN A	4-[3-[(2-AMINO-5-CHLOROBENZOYL)AMINO]-1,1-DIMETHYL-2-OXOBUTYL]-3-METHYL-2-OXETANONE
BELACTIN B	4-[3-[[2-(β-GLUCOPYRANOSYLAMINO)-5-CHLOROBENZOYL)AMINO]-1,1-DIMETHYL-2-OXOBUTYL]-3-METHYL-2-OXETANONE
BEMORADAN	7(1,4,5,6-TETRAHYDRO-4-METHYL-6-OXO-3-PYRIDAZINYL)-2H-1,4-BENZOXAZIN-3(4H)-ONE
BENARTHIN	L-(2,3-DIHYDROBENZOYL)ARGINYL-L-THREONINE
BENASTATIN A	8,13-DIHYDRO-1,7,9,11-TETRAHYDROXY-13-DIMETHYL-8-OXO-3-PENTYLBENZO[a]NAPHTHACENE-2-CARBOXY-LIC ACID
BENASTATIN B	5,6,8,13-TETRAHYDRO-1,7,9,11-TETRAHYDROXY-13-DIMETHYL-8-OXO-3-PENTYLBENZO[a]NAPHTHACENE-2-CAR-BOXYLIC ACID
BENAZEPRILAT	L 155360
BENEXATE	BENZYL 2-[trans-4-(GUANIDINOMETHYL)CYCLOHEXYLCARBONYLOXY]BENZOATE HYDROCLORIDE CYCLO-DEXTRIN CLATHRATE
BENOXAPROFEN	2-(4-CHLOROPHENYL)-α-METHYL-5-BENZOXAZOLEACETIC ACID
BENSERAZIDE	2-[(2,3,4-TRIHYDROXYPHENYL)METHYL]HYDRAZIDE D, L-SERINE
BENZBROMARONE	2-ETHYL-3-(4-HYDROXY-3,5-DIBROMOBENZOYL)-BENZOFURAN
BENZIL	DIPHENYLETHANEDIONE
BENZOCAINE	4-AMINOBENZOIC ACID ETHYL ESTER
BENZOTHIADIAZOLE	BENZO(1,2,3)THIADIAZOLE-7-CARBOTHIOIC ACID S-METHYLESTER
5-BENZYLACYCLOURIDINE	5-BENZYL-1-(2′-HYDROXYETHOXYMETHYL)URACIL
5-BENZYLOXYBENZYL-URACIL	5-{[-(PHENYLMETHOXY)PHENYL]METHYL}URACIL

Trivial name	Synonym or systematic name
5-m-BENZYLOXYBENZYL-URACIL	5-(m-BENZYLOXYBENZYL)-1-(2′-HYDROXYETHOXYMETHYL)URACIL
BERENIL	4,4′-DIAZOAMINO-BIS-BENZAMIDINE
BESTATIN	[S-(R*,S*)]-N-(3-AMINO-2-HYDROXY-1-OXO-4-PHENYLBUTYL)-L-LEUCINE
BETAINE	N, N, N-TRIMETHYLAMMONIOACETIC ACID
BETAINE ALDEHYDE	N, N, N-TRIMETHYL-2-OXO-ETHANAMINIUM
BHT	3,5-DIBUTYL-4-HYDROXYTOLUENE
BIALAPHOS	L-PHOSPHINOTRICINYL-L-ALANYL-L-ALANINE
BIBB 515	1-(4-CHLOROBENZOYL)-4-((4-(2-OXAZOLIN-2-YL)BENZYLIDENE))PIPERIDINE
BIBX 79	trans-N-(4-CHLOROBENZOYL)-N-METHYL-(4-DIMETHYLAMINOMETHYLPHENYL)CYCLOHEXYLAMINE
BIFENOX	METHYL-5-(2,4-DICHLOROPHENOXY)-2-NITROBENZOIC ACID
BIKUNIN	LIGHT CHAIN OF INTER α-TRYPSIN INHIBITOR
BI-L 239	2,6-DIMETHYL-4-[2-(4-FLUOROPHENYL)ETHENYL]PHENOL
BIOCHANIN A	GENISTEIN-4′-METHYL ETHER
BIOPTERIN	[S-(R*,S*)]-2-AMINO-6-(1,2-DIHYDROXYPROPYL)-4(1H)-PTERIDINONE
BI-RG 587	NEVIRAPINE
BIRLANE	1-CHLORO-2-(2,4-DICHLOROPHENYL)-VINYLDIETHYL PHOSPHATE
BL 2401	(±)-3-[2-BENZYL-3-(PROPIONYLTHIO)PROPIONYL]AMINO-5-METHYLBENZOIC ACID
BL P2013	2β-(CHLOROMETHYL)-2α-METHYLPENAM-3α-CARBOXYLIC ACID
BM 13677	2-(3-PHENYLPROPOXYIMINO)BUTYRIC ACID
BM 15766	4-[2-[1-(4-CHLOROCINAMYL)PIPERAZIN.4.YL]BENZOIC ACID
BM 210955	1-HYDROXY-3-(METHYLPENTYLAMINO)PROPYLIDEN-1,1-BISPHONIC ACID
BM 41440	3-HEXADECYLMERCAPTO-2-METHOXYMETHYLPROPYL-1-PHOSPHOCHOLINE
BM 41440	ILOFOSINE
BM 42304	METHYL(3)CINNAMOYLHYDRAZONO-2-PROPIONATE
BMS 180742	SUCCINYL-PHE-GLU-PRO-ILE-PRO-GLU-GLU-TYR-CYCLOHEXYLALANINE-GLN
BMS 181162	4-(3′-CARBOXYPHENYL)-3,7-DIMETHYL-9-(2″,6″,6″-TRIMETHYL-1″-CYCLOHEXYL)-2Z, 4E, 6E, 8E-NONATETRA-ENOIC ACID
BMS 182657	S(R*,R*)2,3,4,5-TETRAHYDRO-3[(2-MERCAPTO-1-OXO-3-PHENYLPROPYL)AMINO]-2-OXO-1H-BENZOZEPINE-1-ACETIC ACID
BMS 185878	(E, E)N[N[N[3-HYDROXY(3,7,11-TRIMETHYL-1-OXO-2,6,10-DODECATRIENYL)PHOSPHINYL]-1-OXOPROPYL]VAL]-METH
BMY 20844	1,3-DIHYDRO-7,8-DIMETHYL-2H-IMIDAZO[4,5-b]QUINOLIN-2-ONE
BMY 21190	7-[4-(N-CYCLOHEXYL-N-METHYLAMINO)-4-OXOBUTOXY]-1,3-DIHYDRO-2H-IMIDAZO[4,5-b]QUINOLIN-2-ONE
BMY 30094	9-PHENYLNONANOHYDROXAMIC ACID
(–)BO 2367	(–)-7-[(1R*,2R*,6R*)-2-AMINO-8-AZABICYCLO[4.3.0.]-NON-3-EN-8-YL]-1-CYCLOPROPYL-6,8-DIFLUORO-1,4-DI-HYDRO-4-OXO-3-QUINOLINECARBOXYLIC ACID
BOF 4272	(±)-8-(3-METHOXY-4-PHENYLSULPHINYLPHENYL)PYRAZOLO[1,5-α]-1,3,5-TRIANZINE-4-OLATE
BOF 4272	SODIUM-(±)8-(3-METHOXY-4-PHENYLSULPHINYLPHENYL)PYRAZOLO[1,5-α]-1,3,5-TRIAZINE-4-OLEATE MONO-HYDRATE
BONGKREKIC ACID	[R-[R*,S*-(E, Z, Z, E, E, Z, E)]]-20-(CARBOXYMETHYL)-6-METHOXY-2,5,17-TRIMETHYL-2,4,8,10,14,18,20-DOCOSAHEP-TANEDIOIC ACID
BPP9a	BRADIKININ-POTENTIATING PEPTIDE 9a
BREDININ	4-CARBAMOYL-11-β-D-RIBOFURANOSYLIMIDOZOLIUM-5-OLATE
BREFELDIN A	[1R-(1R*,2E, 6S*,10E, 11AS*,13S*,14AR*)]-1,6,7,8,9,11A, 12,13,14,14A-DECAHYDRO-1,13-DIHYDROXY-6-METHYL-4H-CYCLOPENT[F]OXACYCLOTRIDECIN-4-ONE
BREQUINAR	DUP 785
BREQUINAR	6-FLUORO-2-(2′-FLUORO-1,1′-BIPHENYL-4-YL)-3-METHYL-4-QUINOLINECARBOXYLIC ACID
BREQUINAR	NSC 368390
BRL 30892	DENBUFYLLINE
BRL 39123	9-(4-HYDROXY-3-HYDROXYMETHYLBUT-1-YL)GUANINE
BRL 42715	C_6-(N1-METHYL-1,2,3-TRIAZOYLMETHYLENE)PENEM
BRL 42715	(5R)(Z)-6-1-METHYL-1,2,3-TRIAZOL-4-YLMETHYLENE)PENEM-3-CARBOXYLIC ACID
BRL 61063	1,3-DI(CYCLOPROPYLMETHYL)-8-AMINO]XANTHINE
BRL 8242	2-[BENZIMIDAZOLYL]-IMIDAZOLINE DIHYDROCHLORIDE
BROFAROMINE	4-(7-BROMO-5-METHOXY-2-BENZOFURANYL)PIPERIDINE
BROMCRESOL PURPLE	4,4′-(3H-2,1-BENZOXATHIOL-3-YLIDENE)BIS[2-BROMO-6-METHYLPHENOL]S, S-DIOXIDE

Trivial name	Synonym or systematic name
5-BROMOACYCLO-THYMIDINE	5-BROMO-1-(2′-HYDROXYETHOXYMETHYL)URACIL
BROMOCONDURITOL	(1,2,4/3)-1-BROMO-2,3,4-TRIHYDROXYCYCLOHEX-5-ENE
BROMOLEVAMISOLE	(–)6-(4-BROMOPHENYL)2,3,4,6-TETRAHYDROIMIDAZO[2,1-b]THIAZOLE
BROMOSULPHOPHTHALEIN	3,3′-(4,5,6,7-TETRABROMO-3-OXO-1(3H)-ISOBENZOFURANYLIDENE)BIS[6-HYDROXYBENZENESULPHONIC ACID] DISODIUM SALT
BROMTHYMOL BLUE	4,4′-(3H-2,1-BENZOXATHIOL-3-YLIDENE)BIS[2-BROMO-3-METHYL)PHENOL]S, S-DIOXIDE
B103U	4-HYDROXY-6-MERCAPTOPYRAZOLZ[3,4-d]PYRIMIDINE
BU AdATP	N2-(p-n-BUTYLANILINO)dATP
BUDDLENOID A	KAEMPFEROL 7-(6″-p-COUMAROYLGLUCOSIDE)
BUDDLENOID B	ISORHAMNETIN 7-(6″-p-COUMAROYLGLUCOSIDE)
BUDIPINE	1-t-BUTYL-4,4-DIPHENYLPIPERIDINE
BU 2743E	(2,3-DIHYDROXYBENZYL)-ALANYL-THREONINE
BU PdGTP	N2-(p-n-BUTYLPHENYL)dGTP
BUQUINERAN	N-BUTYL-N′-[1-(6,7-DIMETHOXY-4-QUINAZOLINYL)-4-PIPERIDINYL]-UREA
BUTHIONINE	S-BUTYLHOMOCYSTEINE
γ-BUTYROBETAINE	3-CARBOXY-N, N, N-TRIMETHYL-1-PROPANAMINIUM HYDROXIDE
BUTYROLACTONE I	α-OXO-β-(p-HYDROXY-m-3,3-DIMETHYLALLYLBENZYL)-γ-METHOXYCARBONYL-γ-BUTYROLACTONE
BW 175	2S,3R,4S)-4-{L-N-[(2S)3-ETHYLSULPHONYL-2-(1-NAPHTHYLMETHYL)PROPIONYL]NORLEUCYL}AMINO-5-CYCLO-HEXYL-1-MORPHOLINO-2,3-PENTANEDIOL
BW A4C	N-[(E)-3-(3-PHENYLOXYPHENYL)PROP-2-ENYL]ACETOHYDROXAMIC ACID
BW A137C	N-[4-(BENZYLOXY)BENZYL]ACETOHYDROXAMIC ACID
BW 284C51	1: 5-BIS(4-ALLYLDIMETHYLAMMONIUMPHENYL)PENTAN-3-ONE DIBROMIDE
BW 755C	DIHYDRO-1-[3-(TRIFLUOROMETHYL)PHENYL]-1H-PYRAZOL-3-AMINE
BW 301U	2,4-DIAMINO-6-(2,5-DIMETHOXYBENZYL)-5-METHYLPYRIDO[2,3-d]PYRIMIDINE
BW543U76	5-DACTHF
BW 1370U87	1-ETHYLPHENOXATHIIN-10,10-DIOXIDE
BW 1843U89	(S)-2-(5-(((1,2-DIHYDRO-3-METHYL-1-OXOBENZO[f]QUINOZOLIN-9-YL)METHYLAMINO)-1-OXO-2-ISOINDOLINYL)-GLUTARIC ACID
BZA-2B	CYSTEINYL-[(R)-3-METHYLAMINO-1-CARBOXYMETHYL-2-3-DIHYDRO-5-PHENYL-1H-1,4-BENZODIAZEPIN-2-ONE)METHIONINE
BZAR	BIS(CARBOBENZOXYCARBONYL-L-ARGININAMIDO)RHODAMINE
BZSA	L-BENZYLSUCCINATE
C 105	SER-VAL-ALA-LYS-LEU-GLU-LYS
C 107	ALA-LEU-PRO-HIS-ALA
C 111	GLY-VAL-TYR-PRO-HIS LYS
C 112	ILE-ARG-PRO-VAL-GLN
C1-1	3,6-DIHYDROXY-2-[2′-(1″,1″-DIMETHYL-2″-PROPENYL)-INDOL-3′-YL]-5-INDOL-3′-YL-1,4-BENZOQUINONE
284 C 51	1,5-BIS-(4-ALLYLDIMETHYL AMMONIUM PHENYL)PENTANE-3-ONE DIBROMIDE
447C88	N-HEPTYL-N′-(2,4-DIFLUORO-6-{2-[4-(2,2-DIMETHYLPROPYL)PHENYL]ETHYL}PHENYL)UREA
680C91	((E)-6-FLUORO-3-[2-(3-PYRIDYL)VINYL]-1H-INDOLE
776C	5-ETHYNYLURACIL
CA 030	N-(L-3-trans-ETHOXYCARBONYLOXIRANE-2-CARBONYL)-L-ISOLEUCYL-L-PROLINE
CA 074	N-(L-3-trans-PROPYLCARBAMOYLOXIRANE-2-CARBONYL)-L-ISOLEUCYL-L-PROLINE
CADAVERINE	1,5-PENTANEDIAMINE
CAFFEINE	3,7-DIHYDRO-1,3,7-TRIMETHYL-1H-PURINE-2,6-DIONE
CALCININE	A 23187
CALIMIDAZOLIUM	N-(6-AMINOHEXYL)-5-CHLORO-1-NAPHTHALENESULPHONAMIDE
CALPACTIN I	LIPOCORTIN II
CALPAIN INHIBITOR 1	N-ACETYL-LEUCYL-LEUCYL-NORLEUCINAL
CALPAIN INHIBITOR 1	AC-LEU-LEU-NORLEUCINAL
CALPAIN INHIBITOR 2	AC-LEU-LEU-MET-H
CALPAIN INHIBITOR 2	N-ACETYL-LEU-LEU-METHIONINAL
CALPEPTIN	BENZYLOXYCARBONYL-LEU-NORLEUCINAL
CAMONAGREL	FI 2845
CAMONAGREL	(+)-5-[2-IMIDAZOLE-1-ETHYLOXY]-1-INDANCARBOXYLIC ACID

Trivial name	Synonym or systematic name
CAMOSTAT	(N, N, -DIMETHYLCARBAMOYLMETHYL-p-(p-GUANIDINOBENZOYLOXY)PHENYLACETATE
CANAVANINE	O-[(AMINOIMINOMETHYL)AMINO]-L-HOMOSERINE
CANDOXATRILAT	cis-4-{[2-CARBOXY-3-(2-METHOXYETHOXY)PROLYL-1-CYCLOPENTANECARBONYLAMINE}-1-CYCLOHEXANE CARBOXYLIC ACID
CANTHARIDIN	exo,exo-2,3-DIMETHYL-7-OXABICYCLO[2.2.1]HEPTANE-2,3-DICARBOXYLIC ACID ANHYDRIDE
CANTHARIDIN	HEXAHYDRO-3a,7a-DIMETHYL-4,7-EPOXYISOBENZOFURAN-1,3-DIONE
CAPTAN	N-TRICHLOROMETHYLTHIO-4-CYCLOHEXIMIDE
CAPTOPRIL	(S)-1-(3-MERCAPTO-2-METHYL-1-OXOPROPYL-L-PROLINE
CAPTOPRIL	SQ 14225
CARACEMIDE	N-ACETYL-N, O-DI(METHYLCARBAMOYL)HYDROXYLAMINE
CARBAMAZEPINE	5H-DIBENZ[B, F]AZEPINE-5-CARBOXAMIDE
CARBAZERAN	1-(6,7-DIMETHOXY-1-PHTHALAZINYL)-4-PIPERIDINYL CARBAMIC ACID ETHYL ESTER
CARBENOXOLONE	3-(3-CARBOXY-1-OXOPROPOXY)-11-OXOOLEAN-12-EN-29-OIC ACID
CARBIDOPA	(S)-α-HYDRAZINO-3,4-DIHYDROXY-α-METHYLBENZENEPROPANOIC ACID
CARBONYLSALICYLAMIDE	CARSALAM
CARBOVIR	CARBOCYCLIC 2′,3′-DIDEHYDRO-2′,3′-DIDEOXYGUANOSINE TRIPHOSPHATE
CARBOVIR	NSC 614846
CARBOXYATRACTYLOSIDE	(2β,15α)-15-HYDROXY-2-[[2-O-(3-METHYL-1-OXOBUTYL)-3,4-DI-O-SULPHO-β-D-GLUCOPYRANOSYL]OXY]-KAUR-16-ENE-18,19-DIOIC ACID DIPOTASSIUM SALT
1-CARBOXYSALSOLINE	1-CARBOXY-1-METHYL-6-HYDROXY-7-METHOXY-1,2,3,4-TETRAHYDROISOQUINOLINE
CARDOL	5-[8(Z),11(Z),14-PENTADECATRIENYL]RESORCINOL
CARIPORIDE	HOE 642
CARIPORID MESILAT	(4-ISOPROPYL-3-METHANESULPHONYLBENZOYL)GUANIDINE METHANSULPHONAT
CARMUSTINE	1,3-BIS(2-CHLOROETHYL)-1-NITROSOUREA
CARNITINE	3-CARBOXY-2-HYDROXY-N, N, N-TRIMETHYL-1-PROPANAMINIUM HYDROXIDE
CARNOSINE	N-β-ALANYL-L-HISTIDINE
CARPROPAMIDE	[(1RS, 3SR)-2,2-DICHLORO-N-[(R)-1-(4-CHLOROPHENYL)ETHYL)-1-ETHYL-3-METHYLCYCLOPROPANECARBOX-AMIDE
CASEIN KINASE INHIBITOR 6	N-(2-AMINOETHYL)-ISOQUINOLINE-8-SULPHONAMIDE
CASEIN KINASE INHIBITOR 7	N-(2-AMINOETHYL)-5-CHLOROISOQUINOLINE-8-SULPHONAMIDE
CASEIN KINASE INHIBITOR 8	1-(5-CHLORO-8-ISOQUINOLINESULPHONYL)PIPERAZINE
CASSAINE	CASSAIC ACID DIMETHYLAMINOMETHYL ESTER
CASTANOSPERMINE	[1S-(1α,6β,7α,8β,8Aβ)]-OCTAHYDRO-1,6,7,8-INDOZOLIZINETETROL
CATECHIN(FLAVAN)	(2R-trans)-2-(3,4-DIHYDROXYPHENYL)-3,4-DIHYDRO-2H-1-BENZOPYRAN-3,5,7-TRIOL
CATECHOL	CATECHIN
CATECHOL(FLAVAN)	CATECHIN(FLAVAN)
CATECHOL(PHENOL)	1,2-BENZENEDIOL
CB 30-256	2-DESAMINO-2-METHYL-10-METHYL-2′,5′-DIFLUORO-5,8-DIDEAZAFOLIC ACID
CB 30900	N-[N-4[N-[3,4-DIHYDRO-2,7-DIMETHYL-4-OXO-6-QUINAZOLINYL]-N-PROP-2-YNYLAMINO]-2-FLUOROBENZYL]-L-γ-GLUTAMYL]-D-GLUTAMIC ACID
CB 3717	N-(4-{N-[(2-AMINO-4-HYDROXY-6-QUINAZOLINYL)METHYL]PROP-2-YNYLAMINO}BENZOYL-L-GLUTAMIC ACID
CB 3717	N10-PROPARGYL-5,8-DIDEAZAFOLIC ACID
CB 3717	PDDF
CB 3717	N10-PROPARGYL-5,8-DIDEAZAFOLIC ACID
CB 3804	2-DESAMINO-10-PROPARGYL-5,8-DIDEAZAFOLIC ACID
CB 3819	2-DESAMINO-2-METHYL-N10-PROPARGYL-5,8-DIDEAZAFOLIC ACID
CBDP	2-/O-CRESYL/4H-1-3-2-BENZODIOXAPHOSPHORIN-2-OXIDE
CCP	CARBONYLCYANIDE m-CHLORPHENYLHYDRAZONE
CD 840	LY 765527
CDRI COMPOUND 80/53	(N1-3-ACETYL-4-5-DIHYDRO-2-FURANYL)-N4-(METHOXY-8-QUNILINYL)1,4-PEPTANE-DIAMINE
CEFDINIR	CI 983
CELECOXIB	4-[5-(4-METHYLPHENYL)-3-(TRIFLUOROMETHYL)-1H-PYRAZOL-1-YL]BENZENESULPHONAMIDE
CEPAENE-1	trans-5-ETHYL-4,6,7-TRITHIA-2-DECENE-4-OXIDE
CEPAENE-2A	trans,trans-5-ETHYL-4,6,7-TRITHIA-2,8-DECADIENE-4-OXIDE
CEPHALOTHIN	(SR-trans)-3-[(ACETYLOXY)METHYL]-8-OXO-7-[(2-THIENYLACETYL)AMINO]-5-THIA-1-AZABICYCLO[4.2.0]OCT-2-ENE-2-CARBOXYLIC ACID
CEPHAMYCIN	7α-CYCLOSPORIN

Trivial name	Synonym or systematic name
CERANOPRIL	SQ 29852
CERIVASTATIN	BAY W 6228
CERULENIN	[2R-[2α,3α(4E, 7E)]]-3-(1-OXO-4,7-NONADIENYL)-OXIRANECORBOXAMIDE
CGA 201029	METHYL-1-(2,2-DIMETHYLINDAN-1YL)IMIDAZOLE-5-CARBOXYLIC ACID
CGA 201029	R 69020
CGA 214372	R(–)METHYL-1-(2,2-DIMETHYLINDAN-1-YL)IMIDAZOLE-5-CARBOXYLATE
CGP 28014	N-(2-PYRIDONE-6-YL)-N,N′-DI-n-PROPYPLFORMAMIDINE
CGP 28238	6-(2,4-DIFLUOROPHENOXY)-5-METHYLSULPHONYLAMINO-1-INDANONE
CGP 32349	4-HYDROXYANDROSTENDIONE
CGP 38560	N-(2(R)-BENZYL-3-tert.-BUTYLSULPHONYLPROPIONYL)HIS-CHAcVAL-n-BUTYLAMIDE METHANESULPHONATE SALT
CGP 41251	4′-N-BENZOYLSTAUROSPORINE
CGP 5211	4,5-DIAMINOPHTHALAMIDE
CGP 53153	N-2-(CYANO-2-PROPYL)-3-OXO-4-AZA-5α-ANDROST-1-ENE-17β-CARBOXAMIDE
CGP 53716	N-[4-METHYL-3(4-PYRIDIN-3-YLPYRIMIDIN-2-YLAMINO)PHENYL]BENZAMIDE
CGP 61755	5(S)-(BOC-AMINO)-4(S)-HYDROXY-6-PHENYL-2(R)-(2,3,4-TRIMETHOXYPHENYLMETHYL)HEXANOYL(L)-VAL-N-(2-METHOXYETHYL)AMIDE
CGP 11305A	BROFAROMINE
CGP 48664A	4-AMIDINOINDAN-1-ONE 2′-AMINDINOHYDRAZONE
CGS 12970	3-METHYL-2-(3-PYRIDYL)-1-INDOLEOCTANOIC ACID
CGS 13080	IMIDAZO-(1,5-2)PYRIDINE-5-HEXANOIC ACID
CGS 14831	BENAZEPRIL
CGS 15135	5-CHLORO-3-METHYL-2-(3-PYRIDYL)-1-INDOLEOCTANOIC ACID
CGS 16617	3-[(5-AMINO-1-CARBOXY-1S-PENTYL)AMINO]-2,3,4,5-TATRAHYDRO-2-OXO-3S-1H-1-BENZAZEPINE-1- ACETIC ACID
CGS 20267	4-[1-(CYANOPHENYL)-1-(1,2,4-TRIAZOLYL)METHYL]BENZONITRILE
CGS 20267	LETROZOLE
CGS 20267	4,4′-(1H-1,2,4-TRIAZOL-1-YL-METHYLENE)-BISBENZONITRILE
CGS 24128	N-[2-(PHOSPHONOMETHYLAMINO)-3-(4-BIPHENYLYL)PROPIONYL]-3-AMINOPROPIONIC ACID
CGS 24592	(S)-3-[N-[2-[(PHOSPHONOMETHYL)AMINO)-3,4-BIPHENYL)PROPIONYL]AMINO]PROPIONIC ACID
CGS 25015	α-[N-[1-OXO-3-THIO-2-(TRIFLUOROMETHYL)PROPYL]AMINO]-1-NAPHTHALENEPROPANOIC ACID
CGS 26129	N-[3-MERCAPTO-2(R)-[(2-METHYLPHENYL)METHYL]-1-OXOPROPYL]-L-METHIONINE
CGS 8515	METHYL 2-[3,4-DIHYDRO-3,4-DIOXO-1-NAPHTHALENYL]AMINO)BENZOIC ACID
CGS 16949A	FADROZOLE
CGS 16949A	4-(5,6,7,8-TETRAHYDROIMIDAZO[1,5a]-PYRIDIN-5-YL)BENZONITRILE MONOHYDROCHLORIDE
CGS 18320B	BIS-(p-CYANOPHENYL)-IMIDAZO-1-YL-METHANE HEMISUCCINATE
CGS 14796C	cis-1-[(4-[(1-IMIDAZOLYL)METHYL]CARBOXYMETHYL]IMIDAZOLE SUCCINATE
CHAPS	3-[(3-CHOLAMIDOPROPYL)DIMETHYLAMMONIO]-1-PROPANESULPHONATE
CHLOMETHOXYNIL	2,4-DICHLORO-1-(3-METHOXY-4-NITROPHENOXY)BENZENE
CHLORANILIC ACID	2,5-DICHLORO-3,6-DIHYDROXY-1,4-BENZOQUINONE
CHLORGYLIN	N-[3-(2,4-DICHLOROPHENOXY)PROPYL]-N-METHYL-2-PROPYN-1-AMINE
CHLORMADINONE ACETATE	17α-ACETOXY-6-CHLORO-3β-HYDROXY-PREGNA-4,6-DIEN-20-ONE
CHLORNITROFEN	2,4,6-TRICHLORO-1-(4-NITROPHENOXY)BENZENE
CHLOROGENIC ACID	[1S-(1α,3β,4α,5α)]-3-[[3-(3,4-DIHYDROXYPHENYL)-1-OXO-2-PROPENYL]OXY]-1,4,5-TRIHYDROXY-CYCLOHEXANE-CARBOXYLIC ACID
CHLORPROMAZINE	2-CHLORO-N, N-DIMETHYL-10H-PHENOTHIAZINE-10-PROPANAMINE
CHLORTETRACYCLINE	[4S-(4α,4Aα,5Aα,6β,12Aα)]-7-CHLORO-4-(DIMETHYLAMINO)-1,4,4A, 5,5A, 6,11,12A-OCTAHYDRO-3,6,10,12,12A-PENTAHYDROXY-6-METHYL-1,11-DIOXO-2-NAPHTHACENECARBOXAMIDE
CHOLECYSTOKININ RELEASING PEPT	PANCREATIC SECRTORY TRYPSIN INHIBITOR I
CHOLESTEROL	CHOLEST-5-EN-3β-OL
CHOLIC ACID	(3α,5β,7α,12α)-3,7,12-TRIHYDROXY-CHOLAN-24-OIC ACID
CHOLINE	2-HYDROXY-N, N, N-TRIMETHYL-ETHANAMINIUM
CHORISMIC ACID	TRANS-3-[(1-CARBOXYETHENYL)-OXY]-4-HYDROXY-1,5-CYCLOHEXADIENE-1-CARBOXYLIC ACID
CHROMOMYCIN A3	3B-O-(4-O-ACETYL-2,6-DIDEOXY-3-C-METHYL-α-L-ARABINO-HEXOPYRANOSYL)-7-METHYLOLIVOMYCIN D

Trivial name	Synonym or systematic name
CHYMOTRYPSIN INHIBITOR 2	BARLEY SERINE PROTEINASE INHIBITOR 2
CI 914	IMAZODAN
CI 930	4,5-DIHYDRO-6-[4-(1H-IMIDAZOL-1-YL)PHENYL]-5-METHYL-3(2H)PYRIDAZINONE
CI 930	IMAZODAN
CI 949	5-METHYL-3-(1-METHYLETHOXY)-1-PHENYL-N-1H-TETRAZOL-5-YL-1H-INDOLE-2-CARBOXAMIDE
CI 972	2,6-DIAMINO-3,5-DIHYDRO-7-(3-THIENYL)-4H-PYRROLO[3,2-d]PYRIMIDIN-4-ONE
CI 972	2,6-DIAMINO-3,5-DIHYDRO-7(3-THIENYLMETHYL)-4H-PYRROLO[3,2-d]PYRIMIDIN-2-ONE
CI 976	2,2-DIMETHYL-N-(2,4,6-TRIMETHOXYPHENYL)DODECANAMIDE
CI 976	PD 128042
CI 981	ATROVASTATIN
CI 981	[R(R*,R*)]-2-(4-FLUOROPHENYL)-β,δ-DIHYDROXY-5-(1-METHYLETHYL)-3-PHENYL-4-[(PHENYLAMINO)CARBONYL]-1H-PYRROLO-1-HEPTANOIC ACID
CI 986	5-[3,5-BIS(1,1-DIMETHYLETHYL)-4-HYDROXYPHENYL]-1,3,4-THIADIAZOLE-2(3H)-THIONE CHOLINE
CI 1000	PD 141955
CI 1011	[[2,4,6-TRIS(1-METHYLETHYL)PHENYL]ACETYL]SULPHAMIC ACID, 2,6-BIS(1-METHYLETHYL]PHENYL ESTER
CICLETANINE	3-(4-CHLOROPHENYL)-1,3-DIHYDRO-7-HYDROXY-6-METHYLFURO-[3,4-c]PYRIDINE
CILASTATIN	Z-S-[6-CARBOXY-6-(2,2-DIMETHYL-(S)-CYCLOPROPYL)CARBOXY)-AMINO-5-HEXENYL]-L-CYSTEINE
CILAZAPRIL	RO 31-2848/006
CILAZAPRILAT	RO 31-3113/000
CILOSTAMIDE	N-CYCLOHEXYL-N-METHYL-4-(1,2-DIHYDRO-2-OXO-6-QUINOYLOXY)BUTYRAMIDE
CIMOXATONE	3-{4-[(3-CYANOPHENYL)METHOXY]PHENYL}-5-(METHOXYMETHYL)-2-OXAZOLIDINONE
CINNAMIC ALCOHOL	3-PHENYL-2-PROPEN-1-OL
CIP 28	RAT TESTES CHOLESTERYLESTERASE INHIBITOR PROTEIN
CIRSILIOL	3′,4′,5′-TRIHYDROXY-6,7-DIMETHOXYFLAVONE
CISPLATIN	cis-DIAMMINEDICHLOROPLATINIUM
CITRIC ACID	2-HYDROXY-1,2,3-PROPANETRICARBOXYLIC ACID
CITRININ	(3R-trans-)4,6-DIHYDRO-8-HYDROXY-3,4,5-TRIMETHYL-6-OXO-3H-2-BENZOPYRAN-7-CARBOXYLIC ACID
L-CITRULLINE	N5-(AMINOCARBONYL)-L-ORNITHINE
CL 13475	2-p-AMINOBENZENESULFONAMIDO-1,3,4-THIAZOLE-5-SULPHONAMIDE
CL 277082	N′-(2,4-DIFLUOROPHENYL)-N-((4-2,2-DIMETHYLPROPYL)PHENYL)METHYL)N-HEPTYLUREA
CL 283546	N′-HEPTYL-N-{[4-(3-METHYLBUTYL)PHENYL]METHYL}-N′-(2,4,6-TRIFLUOROPHENYL)UREA
CL 283796	N′-(4-CHLORO-2,6-DIMETHYLPHENYL)-N-HEPTYL-N-{[4-3-METHYLBUTYL)PHENYL]METHYL}UREA
CL-ARA-ATP	2-CHLORO-2′-DEOXYADENOSINETRIPHOSPHATE
CLAVULANIC ACID	[2R-(2α,3Z, 5α)]-3-(2-HYDROXYETHYLIDENE)-7-OXO-4-OXA-1-AZABICYCLO[3.2.O.]HEPTANE-2-CARBOXYLIC ACID
CL-F-ARA-ATP	2-CHLORO-9-(2-DEOXY-2-FLUORO-β-D-ARABINOFURANOSYL)ADENOSINETRIPHOSPHATE
CLODRONATE	DICHLOROMETHYLENE BISPHOSPHONATE
CLOFIBRATE	2-(4-CHLOROPHENOXY)-2-METHYLPROPANOIC ACID ETHYL ESTER
CMI 206	trans-2[3-METHOXY-4-PROPOXY-5(N′-BUTYL-N′-HYDROXYUREIDYL)PHENYL]-5-(3,4,5-TRIMETHOXYPHENYL)-TETRAHYDROFURAN
62 C 47 MPA	1,5-BIS-(4-ALLYLTRIMETHYL AMMONIUM PHENYL)PENTANE-3-ONE DIBROMIDE
COFORMYCIN	3-β-D-ERYTHROPENTOFURANOSYL)-6,7,8-TRIHYDROIMIDAZO[4,5-4][1,3]DIAZEPIN-8-OL
COFORMYCIN	(R)-3,4,7,8-TETRAHYDRO-3-β-D-RIBOFURANOSYL-IMIDAZO[4,5-D][1,3]DIAZEPIN-8-OL
COGNEX	TACRINE
COLCHICINE	(S)-N-(5,6,7,9-TETRAHYDRO-1,2,3,1O-TETRAMETHOXY-9-OXOBENZA[A]HEPTALEN-7-YL)ACETAMIDE
COMPACTIN	MEVASTIN
COMPOUND 48/80	p-METHOXYPHENETHYLAMIN + FORMALDEHYDE
CONCANAMYCIN A	FOLIMYCIN
CONDURITOL	(1α,2β,3β,4α)-5-CYCLOHEXENE-1,2,3,4-TETROL
CONDURITOL B EPOXIDE	1,2-ANHYDRO-myo-INOSITOL
CONGOCIDINE	4-[[[(AMINOIMINOMETHYL)AMINO]-ACETYL]AMINO]-N-[5-[[(3-AMINO-3-IMINOPROPYL)AMINO]CARBONYL]-1-METHYL-1H-PYRROL-3-YL]-1-METHYL-1H-PYRROLE-2-CARBOXAMIDE
CONGO RED	3,3′-[[1,1′BIPHENYL]-4,4′-DIYLBIS-(AZO)BIS[4-AMINO-1-NAPHTHALENESULPHONIC ACID] DISODIUM SALT
COPRINE	N-(1-HYDROXYCYCLOPROPYL)-L-GLUTAMINE
CORDYCEPIN	3′-DEOXYADENOSINE

Trivial name	Synonym or systematic name
CORTISOL	(11β)-11,17,21-TRIHYDROXY-PREGN-4-ENE-3,20-DIONE
CORTISONE	17,21-DIHYDROXY-PREGN-4-ENE-3,11,20-TRIONE
COUMALIC ACID	2-OXO-2H-PYRAN-5-CARBOXYLIC ACID
p-COUMARIC ACID	3-(4-HYDROXYPHENYL)-2-PROPENOIC ACID
CP 105191	(S)-N-(2,4-BIS(METHYLTHIO)QUINOLIN-5-YL)-2-(HEXYLTHIO)DECANOIC AMIDE
CP 113818	(S)-N-(2,4-BIS(METHYLTHIO)6-METHYLPYRIDIN-3-YL)-2-(HEXYLTHIO)DECANOIC AMIDE
CP 45634	SORBINIL
CP 73850	ZOPOLRESTAT
CP 80633	(2′S)-5-[3-(2′-EXOBICYCLO[2.2.1]HEPTYLOXY-4-METHOXYPHENYL]TETRAHYDRO-2(1H)PYRIMIDONE
CP 80794	ISOPROPYL N-[N-(4-MORPHOLINOCARBONYL)-L-PHENYLALANINYL-S-METHYL-L-CYSTEINYL]-2-(R)HYDROXY-3(S)-AMINO-4-CYCLOHEXYBUTANOATE
CP 80794	TERLAKIREN
CP-83101	(3R*,5S*)-(E)-3,5-DIHYDROXY-9,9-DIPHENYL-6,8-NONADIENOIC ACID
CPC 405	9′-CHLOROPHENYL-erythro-9-(2-HYDROXY-3-NONYL)ADENINE
CPC 406	9′-PHTHALIMIDO-erythro-9-(2-HYDROXY-3-NONYL)ADENINE
CPG 48664	4-AMIDINOINDAN-1-ONE 2′-AMIDINOHYDRAZONE $(HCL)_2$
CPG 11305 A	BROFAROMINE
αCPI	α-CYSTEINE PROTEINASE INHIBITOR
CPI 17	PIG AORTA PROTEIN PHOSPHATASE-1 INHIBITOR
CPT 11	7-EHTYL-10-[4-(1-PIPERIDINO)-1-PIPERIDINO]CARBONYLOXY CAMPTOTHECIN
CPT 11	β-(5-IMINO-2-PYRROLIDINECARBOXAMIDO)PROPAMIDE
CPT 11	IRINOTECAN
CRC 220	4-METHOXY-2,3,6-TRIMETHYLPHENYLSULPHONYL-L-ASPARTYL-D-AMIDINOPHENYLALANINEPIPERIDIDE
CREATINE	N-(AMINOIMINOMETHYL)-N-METHYL-GLYCINE
CREPENYNIC ACID	OCTADEC-cis-9-EN-12-YNOIC ACID
CRIXIVAN	INDINAVIR
CROCONAZOLE	1-[1-[2-[(8-CHLOROBENZYL)OXYL]PHENYL-1-H-IMIDAZOLE
CROTOXYPHOS	DIMETHYL-cis-1-METHYL-2-(1-PHENYLETHOXYCARBAMYL)-VYNYL PHOSPHATE
CS 045	TROGLIZATONE
CS 514	PRAVASTATIN
CS 514	(+)SODIUM (3R,5R)-3,5-DIHYDROXY-7-[(1S,2S,6S,8S,8aR)-6-HYDROXY-2-METHYL-8-[(S)-2-METHYLBUTYRYL]-1,2,6,8-HEXAHYDRO-1-NAPHTHYL]HEPTANOATE
CS 518	2-(1-IMIDAZOLYLMETHYL)-4,5,6,7-TETRHYDROBENZO[b]-THIOPHENE-6-CARBOXYLATE SODIUM
CS 518	6-{2-[1-(1H)-IMIDAZOYL]METHYL-4,5-DIHYDROBENZO[b]THIOPHENE}CARBOXYLATE
CS 518	RS 5186
3116 CT	BIS(3-DIMETHYLAMINO-5-HYDROXYPHENOXY) 1,3-PROPANE DIMETHYLIODIDE
CUPFERRON	N-HYDROXY-N-NITROSO-BEZENAMINE AMMONIUM SALT
CUPRIZONE	BIS(CYCLOHEXYLIDENEHYDRAZIDE) ETHANEDIOIC ACID HYDRAZIDES
CURCUMIN	DIFERULOYL METHANE
CV 159	1,4-DIHYDRO-2,6-DIMETHYL-4-(3-NITROPHENYL)-3,5-PYRIDINE DICARBOXYLIC ACID METHYL 6-(5-PHENYL-3-PYRAZOLYLOXY)HEXYLESTER
CV 3317	DELAPRIL
CV 4151	(E)-7-PHENYL-7-(3-PYRIDYL)-6-HEPTENOIC ACID
CV 4151	ISBOGREL
CV 4151	(E)-7-PHENYL-7-(3-PYRIDYL)-6-HEPTENOIC ACID
CV 5875	(R)-3-[(S)-1-CARBOXY-5-(4-PIPERIDYL)PENTYL]AMINO-4-OXO-2,3,4,5-TETRAHYDRO-1,5-BENZOTHIAZEPINE-5-ACETIC ACID
CV 5975	(R)-3-[(S)-1-CARBOXY-5-(4-PIPERIDYL)PENTYL]AMINO-4-OXO-2,3,4,5-TETRAHYDRO-1,5-BENZOTHIAZEPINE-5-ACETIC ACID
CV 6504	2,3,5-TRIMETHYL-6-(3-PYRIDYLMETHYL)-1,4-BENZOQUINONE
CVT 313	2-[BIS(HYDROXYETHYL)AMINO]-6-(4-METHOXYBENZYLAMINO)-9-ISOPROPYLPURINE
γ-CYANOMETHOTHREXATE	N-(4-AMINO-4-DEOXYN10METHYLPTEROYL)-4-CYANOGLUTAMIC ACID
CYCLANDELATE	3,3,5-TRIMETHYLCYCLOHEXANYLMADELATE
CYCLOHEXYL-GLUTETHIMIDE	(+)-3-[4-AMINOPHENYL]-3-CYCLOHEXYLPIPERIDINE-2,6-DIONE
CYCLOOCTATIN	1,2,3,3a,4,5,7,8,9,9a,10,10a-DODECAHYDRO-3,4-DIHYDROXY-1-HYDROXYMETHYL-4,9a-DIMETHYL-7-(1-METHYL-ETHYL)-DICYCLOPENTA[a,d]CYCLOOCTENE

Trivial name	Synonym or systematic name
CYCLOPHELLITOL	[(1S, 2R, 3S, 4R, 5R, 6R)-5-HYDROXYMETHYL-7-OXABICYCLO[4,1,0)HEPTANE-2,3,4-TRIOL
CYCLOPHOSPHAMIDE	2-[BIS(2-CHLOROETHYL)AMINO]TETRAHYDRO-2H-1,3,2-OXAZAPHOSPHORINE-2-OXIDE
CYCLOPHOSTIN	NK 901093
CYCLOSERINE	(R)-4-AMINO-3-ISOXAZOLIDINONE
CYLANDELATE	TRIMETHYLCYCLOHEXANYLMANDELATE
CYPROMIN	ROLICYPRIN
CYSFLUORETIN	7,8,9,11-TETRAHYDRO-5,10-DIHYDROXY-4,9-DIMETHOXY-2-(4-ACETAMIDO-4-CARBOXY-2-THIABUTYL)-6H-BENZO[b]FLUOREN-6-ONE
L-CYSTATHIONINE	(R)-S-(2-AMINO-2-CARBOXYETHYL)-L-HOMOCYSTEINE
CYTOPLASMIC ANTI-PROTEINASE	PROTEINASE INHIBITOR 6
D 600	α-[3-[[2-(3,4-DIMETHOXYPHENYL)ETHYL]METHYLAMINO]PROPYL]-3,4,5-TRIMETHOXY-α-(1-METHYLETHYL)-BENZENEACETONITRILE
D 609	TRICYLCODECAN-9-YLXANTHATE
D 1694	ICI D 1694
D 1694	N-{5-[N-(3,4-DIHYDRO-2-METHYL-4-OXOQUINAZOLIN-6-YLMETHYL)-N-METHYLAMINO]-2-THEONYL}-L-GLUTAMIC ACID
D 19391	OCTADECYLPHOSPHOCHOLINE
D-1	3,6-DIHYDROXY-2,5-DIINDOL-3′-YL-1,4-BENZOQUINONE
5-DACTHF	5-DEAZAACYCLOTETRAHYDROFOLATE
5-DACTHF	N-[4-[[3-(2,4-DIAMINO-1,6-DIHYDRO-6-OXO-5-PYRIMIDINYL)PROPYL]AMINO]BENZOYL]-L-GLUTAMIC ACID
DALVASTATIN	(4R, 6S)-6-[(E)-2-[2-(4-FLUORO-3-METHYLPHENYL)-4,4,6,6-TETRAMETHYL-1-CYCLOHEXENYL]VINYL]-4-HYDROXY TETRAHYDROPYRAN-2-ONE
DAN	DIAZOACETYL NORLEUCIN METHYLETSER
DAPSONE	4,4′-DIAMINODIPHENYLSULPHONE
DAUNORUBICIN	8-ACETYL-10-[(3-AMINO-2,3,6-TRIDEOXY-α-L-LYXO-HEXOPYRANOSYL)OXY]-7,8,9,10-TETRAHYDRO-6,8,11-TRI-HYDROXY-1-METHOXY-5,12-NAPHTHACENEDIONE
DAUNORUBICIN	DAUNOMYCIN
DAZIP	2,3-DIMETHYL-8-[(4-AZIDOPHENYL)METHOXY]IMIDAZOLE-[1,2-a] PYRIDINE
DAZOXIBEN	4-[2-(1H-IMIDAZOL-1-YL)ETHOXY]-BENZOIC ACID
DCCD	N, N′-METHANETETRAYLBIS-CYCLOHEXANAMINE
DD 089	TETRA(2-HYDROXYPHENYL)PORPHYRIN
DDE	DICHLORODIPHENYLDICHLOROETHANE
3-DEAZA-SIBA	5′-DEOXYISOBUTYLTHIO-3-DEAZA ADENOSINE
DEBRISOQUIN	3,4-DIHYDRO-2(1H)-ISOQUINOLINECARBOXIMIDAMIDE
trans 1-DECALONE	trans DECAHYDRO-1-NAPHTHALENONE
DECOYININE	9-(6-DEOXY-β-D-ERYTHRO-HEX-5-EN-2-ULOFURANOSYL)-9H-PURIN-6-AMINE
DELAVIRDINE	U 90152
DEMEROL HCl	MEPERIDINE
DENBUFYLLINE	1,3-DI-n-BUTYL-7-(2-OXOPROPYL)XANTHINE
2′-DEOXY-8-AZA-1-DEAZA-ADENOSIN	7-AMINO-3-(2-DEOXY-β-D-erytro-PENTAFURANOSYL)-3H-1,2,3-TRIAZOLO[4,5-b]PYRIDINE
2-DEOXYCOFORMYCIN	(R)-3-(2-DEOXY-β-D-ERYTHRO-PENTOFURANOSYL)-3,6,7,8-TETRAHYDRO-IMIDAZO[4,5-D][1,3]DIAZEPIN-8-OL
DEOXYFUCONOJIRIMYCIN	1,5-DIDEOXY-1,5-IMINO-L-FUCITOL
2′-DEOXYGLUCOSYL-THYMINE	1-(2′-DEOXY-β-D-GLUCOPYRANOSYL)THYMINE
1-DEOXYMANNOJIRIMYCIN	1,5-DIDEOXY-1,5-IMINO-D-MANNITOL
DEOXYMANNONOJIRI-MYCIN	1,5-DIDEOXY-1,5-IMINO-D-MANNITOL
1-DEOXYNOJIRIMYCIN	1,5-DIDEOXY-1,5-IMINO-D-GLUCITOL
1-DEOXYNOJIRIMYCIN	MORANOLINE
DEPHOSTATIN	2-(N-METHYL-N-NITROSO)HYDROQUINONE
DEPRENYL	N, α-DIMETHYL-N-2-PROPYNYL-BENZENEETHANAMINE
(–)DEPRENYL	SELEGILINE
DEQUALINIUM	1,1′-(1,10-DECANEDIYL)BIS-[4-AMINO-2-METHYLQUINOLINIUM]
2-DESAMINOMETHO-THREXATE	N[4-[[(4-AMONOPTERIDIN-6-YL)METHYLAMINO]-BENZOYL]-L-GLUTAMIC ACID

Trivial name	Synonym or systematic name
DESULPHATE HIRUDIN	CGP 39393
DEXMEDETOMIDINE	A 85499
DEXMEDETOMIDINE	(±)-4-[1-(2,3-DIMETHYLPHENYL)ETHYL]-1H-IMIDAZOLE
DEXRAZOXANE	ICRF 187
DFMO	2,2-DIFLUOROMETHYLORNITHINE
DFP	DIISOPROPYL FLUOROPHOSPHATE
DG 35 VIII	t-BUTYL-3-ISOPROPYL-3-[(2S, 3S)-2-HYDROXY-3-(N-QUINALDOYL-L-ASPARAGINYL)AMINO-4-PHENYLBUTYLCARBAZATE
D-HAMAMELONIC ACID-2-PHOSPHATE	2-CARBOXYARABINITOL-1-PHOSPHATE
DIALIFOS	DIALIFOR
DIALLYL SULPHIDE	ALLYL SULFIDE
DIAMIDE	AZODICARBOXYLIC ACID BIS(DIMETHYLAMIDE)
DIAMOX	2-ACETYLAMINO-1,3,4-THIADIAZOLE-5-SULPHONAMIDE
DIBUCAINE	2-BUTHOXY-N-[2-(DIETHYLAMINO)ETHYL]-4-QUINOLINECARBOXAMIDE
DICHLORVOS	2,2-DICHLOROVINYLDIMETHYL PHOSPHATE
DICLOFENAC	O(2,6-DICHLOROANILINO(PHENYL ACETIC ACID
DICLOFOP	2-[4(2′,4′-DICHLOROPHENOXY)PHENOXY]PROPIONIC ACID
DICLRB	5,6-DIBROMO-1-(β-D-RIBOFURANOSYL)BENZIMIDAZOLE
DICUMAROL	3,3′-METHEYLENE-BIS[4-HYDROXY-2H-1-BEZOPYRAN-2-ONE]
DIDOX	3,4-DIHYDROXYBENZOHYDROXAMIC ACID
DIDOX	N, 3,4-TRIHYDROXYBENZAMIDE
DIECA	DIETHYLDITHIOCARBAMATE
DIFLUFENICAN	N-(2,4-DIFLUOROPHENYL)-2-[3-(TRIFLUROMETHYL)PHENOXY]-3-PYRIDINECARBOXAMIDE
DIGITOXIGENIN	17β-(BUT-2′-EN-4′-OLID-3′-YL)-5β,14β-ANDROSTANE-3β,14β-DIOL
DIGLYCOLAMIDIC ACID	IMINODIACETIC ACID
7,8-DIHYDROBIOPTERIN	2-AMINO-7,8-DIHYDRO-6-[(1S, 2R)-1,2-DIHYDROXYPROPYL]-1H-PTERIDIN-4-ONE
DIISOPROPYL FLUORO-PHOSPHATE	ISOFLUROPHATE
DIISOPROPYLFLUORO-PHOSPHATE	PHOSPHOROFLUORIDIC ACID BIS(1-METHYLETHYL) ESTER
DIM	1,4-DIDEOXY-1,4-IMINO-D-MANNITOL
DIMAZ	O, O-DIETHYL-S-2-ETHYLTHIOETHYL PHOSPHORODITHIOATE
DIPICOLINIC ACID	2,6-PYRIDINEDICARBOXYLIC ACID
DIPROTIN A	LEU-PRO-ILE
DIPROTIN B	VAL-PRO-LEU
DIPYRIDAMOLE	2,2′,2″,2‴-[(4,8-DI-1-PIPERIDINYLPYRIMIDO[5,4-D]PYRIMIDINE-2,6-DIYL)DINITRILO]TETRAETHANOL
DISTAMYCIN	STALLIMYCIN
DISTAMYCIN A	N-[5-[[(3-AMINO-3-IMINOPROPYL)AMINO]CARBONYL]-1-METHYL-1H-PYRROL-3-YL-4-[[[4-(FORMYLAMINO)-1-METHYL-1H-PYRROL-2-YL]CARBONYL]AMINO]-1-METHYL-1H-PYRROLE-2-CA
DISULFIRAM	TETRAETHYLTHIOPEROXYDICARBONIC DIAMIDE
DISULFOTON	O, O-DIETHYL-S-2-ETHYLTHIOETHYL PHOSPHORODITHIOATE
DISYTON	O, O-DIETHYL-S-2-ETHYLTHIOETHYL PHOSPHORODITHIOATE
DITHIZONE	PHENYLDIAZENECARBOTHIOIC ACID 2-PHENYLHYDRAZIDE
DIURON	DCMU
DIURON	3-(3,4-DICHLOROPHENYL)-1,1-DIMETHYLUREA
DMDP	2,5-DIHYDROXYMETHYL-3,4-DIHYDROXYPYRROLIDINE
DMI 1	8-METHYLPENTADECANOIC ACID
DMI 2	4″R, 6aR, 10S, 10aS-8-ACETYL-6a,10a-DIHYDROXY-2-METHOXY-12-METHYL-10[4′[3″,5″-DIMETHYL4(Z2,4DIMETHYL--Z2″,4″DIMETHYL-2″-HEPTENOYL)TETRAHYDROPYRAN-1″-YLOXY]5′METHYL
DMP 266	L 743 726
DMP 323	([4R-(4α,5α,6β,7β)]HEXAHYDRO-5,6-BIS(HYDROXY)-1,3-BIS([(4-HYDROXYMETHYL)PHENYL]METHYL)-4,7-BIS-(PHENYLMETHYL)-2H-1,3-DIAZEPIN-2-ONE
DMP 777	2-[4-[[(4-METHYL)PIPERAZIN-1-YL]CARBONYL]PHENOXY]-3,3-N-[1-(3,4-METHYLENEDIOXYPHENYL)BUTYL]-4-OXO-1-AZETIDINECARBOXAMIDE
DN 9693	1,5-DIHYDRO-7-(1-PIPERIDINYL)IMIDAZO[2,1-b]QUINAZOLINE-2(3H)-ONE DIHYDROCHLORIDE HYDRATE

Trivial name	Synonym or systematic name
5838 DNI	1,4,4a,5,12,12a-HEXAHYDRO-4,4a,11,12a-TETRAHYDROXY-3,8-DIMETHOXY-9-METHOXYCARBONYL-10-METHYL-1,5,12-TRIOXO NAPHTHACENE
DOLICHOL	2,3-DIHYDROPOLYPRENOL(C55-C100)
DON	6-DIAZO-5-OXO-L-NORLEUCINE
DONV	5-DIAZO-4-OXO L-NORVALINE
DOPACHROME	2,3,5,6,-TETRAHYDRO-5,6-DIOXO-1H-INDOLE-2-CARBOXYLIC ACID
DOPAMINE	4-(2-AMINOETHYL)-1,2-BENZENEDIOL
DORZOLAMIDE	(−)-(4S,6S)-4-ETHYLAMINO-5,6-DIHYDRO-6-METHYL-4H-THIENO[2,3-b]THIOPYRAN-2-SULPHONAMIDE 7,7-DIOXIDE
DORZOLAMIDE HCL	MK 507
DOXORUBICIN	10-[(3-AMINO-2,3,6-TRIDEOXY-α-L-LYXO-HEXOPYRANOSYL)OXY]-7,8,9,10-TETRAHYDRO-6,8,11-TRIHYDROXY-8-(HYDROXYACETYL)-1-METHOXY-5,12-NAPHTHACENEDIONE
DP 1904	6-(1-IMIDAZOLYLMETHYL)-5,6,7,8-TETRAHYDRONAPHTHALENE-2-CARBOXYLIC ACID
DPDS	DIPHENYLDISULFIDE
D-PENICILLAMINE	3-MERCAPTO-D-VALINE
DP-X 8405	4-FLUORO-α-(3-1,4-DIMETHYL-7-(1-METHYLETHYL)-1,2,3,4,4A-8,10,10A-OCTAHYDRO-1-PHENANTHRENYL)-METHYLAMINO)PROPYL)BENZENEMETHANOL
DT-DX 30 SE	E-6-(4-(2-(4-CHLOROBENZENESULPHONYLAMINO)ETHYL)PHENYL)-6-(3-PYRIDYL)-HEX-5-ENOIC ACID
DTNB	3,3′-DITHIOBIS[6-NITRO-BENZOIC ACID]
DU 14	p-O-SULPHAMOYL)-N-TETRADECANOYL TYRMAINE
DU 1777	(2S, 3aS, 7aS)-1-(N2-NICOTINYL-L-LYSYL-γ-D-GLUTAMYL)OCTAHYDRO-1H-INDOLE-2-CARBOXYLIC ACID
DU 6622	(HYDROXY-2-(HYDROXYMETHYL)-5-[7-(METHOXYCARBONYL)NAPHTHALEN-1-YL]PENTANOIC ACID 1,3-LACTONE
DULCITOL	GALACTITOL
DUP 128	N′-(2,4,-DIFLUOROPHENYL)-N-[5-(4,5-DIPHENYL-1H-IMIDAZOL-2-YLTHIO)PENTYL]-N-HEPTYLUREA
DUP 654	1-METHOXY-2-BENZYLNAPTHTHALENE
DUP 714	ACETYL-D-PHE-PRO-BORO-ARG-OH
DUP 785	BREQUINAR
DV 1006	4-(2-CARBOXYETHYL)PHENYL-trans-4-AMINOMETHYLCYCLOHEXANECARBOXYLATE HCl
DX 9065a	(+)-(2S)-2-[4[[(3S)-1-ACETIMIDOYL-3-PYRROLIDINYL]OXY]PHENYL]-3-[7-AMIDINO-2-NAPHTHYL]PROPANOIC ACID
DYL}AMINOBUTYRATE	ISOPROPYL4CYCLOHEXYL2HYDROXY3{N[2MORPHOLINOCARBONYLMETHYL3(1NAPHTHYL)PROPIONYL]-L-HIS
E 64	L-3-CARBOXY-trans-2,3-EPOXIPROPIONYL-LEUCINYLAMIDO-(4-GUANIDINO)BUTANE
E 64	EPOXYSUCCINYL-LEUCYL-AGMATINE
E 64	[2S-[2α,3β(R*9]]-3-[[[1-[[[4-[(AMINOIMINOMETHYL)AMINO]BUTYL]AMINO]CARBONYL]-3-METHYLBUTYL]AMINO]-CARBONYL]-OXYRANE CARBOXYLIC ACID
E 64c	LOXISTATIN
E 64c	(+)-(2S,3S)-3-{1-[N-(3-METHYLBUTYL)AMINO]LEUCYLCARBONYL}OXYRANE-2-CARBOXYLATE
E 64d	(+)-(2S,3S)-3-{1-[N-(3-METHYLBUTYL)AMINO]LEUCYLCARBONYL}OXYRANE-2-CARBOXYLIC ACID ETHYLESTER
E 600	PHOSPHORIC ACID DIETHYL 4-NITROPHENYL ESTER
E 1020	1,2-DIHYDRO-6-METHYL-2-OXO-5-(IMIDZO[1,2-a]PYRIDIN-6-YL)-3-PYRIDINE CARBONITRILE
E 1020	LOPRINONE
E 2011	(5R)-3-[2((1S)-3-CYANO-1-HYDROXYPROPYL)BENZTHIAZOL-6-YL]-5-METHOXYMETHYL-2-OXAZOLIDINE
E 2020	1-BENZYL-4-[(5,6-DIMETHOXY-1-INDANON)-2-YL]METHYLPIPERIDINE HCl
E 3040	6-HYDROXY-5,7-DIMETHYL-2-(METHYLAMINO)-4-(3-PYRIDYLMETHYL)BENZOTHIAZOLE
E 3123	4-[2-SUCCINIMIDEOTHYLTHIO]PHENYL-4-GUANIDINOBENZOATE-METHANESULPHONATE
E 3123	4-(2-SUCCINIMIDOETHYLTHIO)PHENYL 4-GUANIDINOBENZOATE
E 3810	LY 307640
E 3810	2-[{4-(3-METHOXYPROPOXY)-3-METHYLPYRIDIN-2-YL}-METHYLSULPHINYL]-1H-BENZIMIDAZOLE
E 3810	RABEPRAZOLE
E 3810	(±)-SODIUM 2-[[4-(3-METHOXYPROPOXY)-3-METHYLPYRIDIN-2-YL]-METHYLSULPHINYL]-1H-BENZIMIDAZOLE
E 4021	1-[6-CHLORO-4-(3,4-METHYLENEDIOXYBENZYL)AMINOQUINAZOLIN-2-YL]PYPERIDINE-4-CARBOXYLATE
E 5324	N-BUTYL-N′-[2-[3-(5-ETHYL-4-PHENYL-1H-IMIDAZOL-1-YL)PROPOXY]-6-METHYLPHENYL]UREA
E 6080	6-HYDROXY-2-(4-SULPHAMOYLBENZYLAMINO)-4,5,7-TRIMETHYLBENZOTHIAZOLE HCl
EA	α-ETHYNYLAGMATINE
EB 1053	3-(1-PYROLIDINO)-1-HYDROXYPROPYLIDENE-1,1-BISPHOSPHONIC ACID
EBELACTON A	3,11-DIHYDROXY-2,4,6,8,10,12-HEXAMETHYL-9-OXO-6-TETRADECEN-1,3-LACTONE

Trivial name	Synonym or systematic name
EBELACTON B	2-ETHYL-3,11-DIHYDROXY-4,6,8,10,12-PENTAMETHYL-)-9-OXO-6-TETRADECEN-1,3-LACTONE
EBELACTONE A	3,11-DIHYDROXY-2,4,6,8,10,12-HEXAMETHYL-9-OXO-6-TETRADECEN-1,3-LACTONE
EBSELEN	2-PHENYL-1,2-BENZOISOSELENAZOL-3-(2H)-ONE
EC 64c	EP 475
ECIGUANINE A	N-(2-AMIDINOETHYL)-2-AMINO-4-HYDROXY-7H-PYRROLO[2,3-d]PYRIMIDINE-5-CARBOXAMIDE
ECIGUANINE B	N-(3-AMINOPROPYL)-2-AMINO-4-HYDROXY-7H-PYRROLO[2,3-d]PYRIMIDINE-5-CARBOXAMIDE
ECOTIN	E.COLI TRYPSIN INHIBITOR
EDDnp	N-(2,4-DINITROPHENYL)ETHYLENEDIAMINE
EDIFENPHOS	HINOSAN
EDIFENPHOS	O-ETHYL S, S-DIPHENYLPHOSPHORODITHIOATE
EDTA	ETHYLENEDIAMINETETRAACETIC ACID
E-EPU	5-ETHYL-1-ETHOXYMETHYL-6-(PHENYLTHIO)URACIL
EFEGATRAN	D-METHYL-PHE-PRO-ARGINAL
EFEGATRON	ARGATROBAN
EGTA	ETHYLENEGLYCOL-BIS-(β-AMINOETHYLETHER)-N, N, N′,N′,-TETRAACETIC ACID
ELAIDIC ACID	(E)-9-OCTADECENOIC ACID
ELASNIN	3,5-DIBUTYL-6-(1-BUTYL-2-OXOHEPTYL)-4-HYDROXY-2H-PYRAN-2-ONE
ELASTATINAL	L-2-(2-AMINO-1,4,5,6-TETRAHYDRO-4-PYRIMIDINYL)-N-[[(1-CARBOXY-3-METHYLBUTYL)AMINO]CARBONYL]-GLYCYL-N1-(1-METHYL-2-OXOETHYL)-L-GLUTAMAMIDE
ELENOLIC ACID	[2S-(2α,3α,4β)]-3-FORMYL-3,4-DIHYDRO-5-(METHOXYCARBONYL)-2-METHYL-2H-PYRAN-4-ACETIC ACID
EMD 54622	5-[1-(3,4-DIMETHOXYBENZOYL)-4,4-DIMETHYL-1,2,3,4-TETRAHYDROCHINOLIN-6-YL]-6-METHYL-3,6-DIHYDRO-1,-3,4-THIAZIN-2-ONE
EMERIAMINE	(R)-3-AMINO-4-TRIMETHYL-AMINOBUTYRIC ACID
EMIGLITATE	N-[β-(ETHOXYCARBONYLPHENOXY)-ETHYL]-1-DEOXYNOJIRIMYCIN
EMULPHOGENE	POLYOXYETHYLENE ALCOHOLS
ENALAPRIL	MK 421
ENALAPRIL	(S)-1-[N-[1-(ETHOXYCARBONYL)-3-PHENYLPROPYL]-L-ALANYL]-L-PROLINE
ENALAPRILAT	MK 422
ENDOTHAL	7-OXYBICYCLO[2.2.1]HEPTANE-2,3-DICARBOXYLIC ACID
ENDOTHAL THIO-ANHYDRIDE	7-OXABICYCLO[2.2.1]HEPTANE-2,3-DICARBOXYLIC ACID THIOANHYDRIDE
ENNATIN D	CYCLO[D-α-HYDROXYISOVALERYL(D-HIV)-L-N-METHYLELUCINYL(L-ME-LEU)D-HIV-L-N-METHYLVALINYL-(L-ME-VAL)-D-HIV-L-ME-VAL
ENOXIMONE	1,3-DIHYDRO-4-METHYL-5-[4-(METHYLTHIO)BENZOYL]-2H-IMIDAZOL-2-ONE
ENOXIMONE	LY 186655
ENTACAPONE	(E)-2-CYANO-3-(3,4-DIHYDROXY-5-NITROPHENYL)-N, N-DIETHYL-2-PROPENAMIDE
EOSINE Y	2′,4′,5′,7′-TETRABROMO-3′,6′-DIHYDROXYSPIRO[ISOBENZOFURAN-1(3H),9′-[9H]XANTHEN]-3-ONE DISODIUM SALT
EP 459	L-trans-EPOXYSUCCINYLLEUCYLAMIDO(7-AMINO)HEPTANE
EP 475	[2S-[2α,3β(R*)]]-3-[[[3-METHYL-1-[[(3-METHYLBUTYL)AMINO]CARBONYL]BUTYL]AMINO]CARBONYL]-OXYRANE-CARBOXYLIC ACID
EPC-K1	L-ASCORBIC ACID 2-[3,4-DIHYDRO-2,5,7,8-TETRAMETHYL-2-(4,8,12-TRIMETHYLTRIDECYL)-2H-1-BENZOPYRAN-6-YL-HYDROGENPHOSPHATE] POTASSIUM SALT
EPIRUBICIN	4′-EPIDOXORUBICIN
EPISESAMIN	ASARININ
EPITESTOSTERONE	17α-HYDROXY-4-ANDROSTEN-3-DIONE
EPNP	1,2-EPOXY-3-(p-NITROPHENOXY)PROPANE
EPOSTANE	(4α,5α,17β)-4,5-EPOXY-3,17-DIHYDROXY-4,17-DIMETHYLANDROST-2-ENE-2-CARBONITRILE
EPRISTERIDE	17β-[[(1,1-DIMETHYLETHGYL)AMINO]CARBOXYL]ANDROSTA-3,5-DIENE-CARBOXYLIC ACID
EPTASTIGMINE	HEPTYLPHYSOSTIGMINE
EQUILENIN	3-HYDROXYESTRA-1,3,5,7,9-PENTAEN-17-ONE
ER 3826	N-(1-BENZYL-4-PIPERIDINYL)-4-[N-METHYL-N-[(E)-3-[4-(METHYLSULPHONYL)PHENYL-2-PROPENOYL]AMINO]-BENZENESULPHONAMIDE
ERGOSTEROL	(3β,22E)-ERGOSTA-5,7,22-TRIEN-3-OL
ERIOCHROME T	3-HYDROXY-4-[(1-HYDROXY-2-NAPHTHALENYL)AZO]-7-NITRO-1-NAPHTHALENESULPHONIC ACID MONO-SODIUM SALT
ERYCRYSTAGALLIN	3,9-DIHYDROXY-2,10-DIPRENYLPTEROCARP-6a-ENE

Trivial name	Synonym or systematic name
ERYTHRITOL	(R*,S*)-1,2,3,4-BUTANETETROL
ES 34	N-[2-(MERCAPTOMETHYL)-1-OXO-3-PHENYLPROPYL]-L-LEUCINE
ES 1005	BIS-[(1-NAPHTHYL)METHYL]ACETY-HISTIDYL-STATYL-LEUCYL-ε-LYSINOL DIHYDROCHLORIDE
ES 8891	N-MORPHOLINOACETYL-81-NAPHTHYL)-L-ALANYL-(4-THIAZOLYL)-L-ALANYL(3S, 4S)-4-AMINO-3-HYDROXY-5-CYCLOHEYLPENTANOYL-n-HEXYLAMIDE
ESCULETIN	6,7-DIHYDROXY-2H-1-BENZOPYRAN-2-ONE
ESERINE	PHYSOSTIGMIN
ESTRADIOL-17α	(17α)-ESTRA-1,3,5(10)-TRIENE-3,17-DIOL
ESTRADIOL-17β	(17β)-ESTRA-1,3,5(10)-TRIENE-3,17-DIOL
ET 18	1-OCTADECYL-2-METHYL-rac-GLYCERO-3-PHOSPHOCHOLINE
ETHACRYNIC ACID	[2,3-DICHLORO-4-(2-METHYLENE-1-OXOBUTYL)PHENOXY]ACETIC ACID
ETHANOLAMINE	2-AMINO-ETHANOL
ETHIDIUM	HOMIDIUM
ETHIDIUMBROMIDE	2,7-DIAMINO-10-ETHYL-9-PHENYLPHENANTHRIDINIUM BROMIDE
ETHOPROPAZINE	N, N-DIETHYL-α-METHYL-10H-PHENOTHIAZINE-10-ETHANAMINE
ETHOXYFORMIC ANHYDE	DIETHYL PYROCARBONATE
ETHOXYQUIN	1,2-DIHYDRO-6-ETHOXY-2,2,4-TRIMETHYLQUINOLINE
β-ETHYNYLTYRAMINE	β-ETHYNYL-4-HYDROXYBENZENEETHANOLAMINE HYDROCHLORIDE
ET-18-OCH_3	1-O-OCTADECYL-2-O-METHYL-rac-GLYCERO-3-PHOSPHOCHOLINE
ETOMOXIR	2-[6-(4-CHLOROPHENOXY)HEXYL]OXIRANE-2-CARBOXYLATE
ETYA	5,8,11,14-EICOSATETRAYNOIC ACID
EUCILAT	2-(1-SUCCINOYLOXYETHYL)-3-METHYL-5-(2-OXO-2,5-DIHYDRO-4-FURYL)BENZO(b)FURANNE MORPHOLINIUM SALT
EUGENOL	4-ALLYL-2-METHOXYPHENOL
EUPATORIN	3′,5′-DIHYDROXY-4′,6,7-TRIMETHOXYFLAVONE
EXEMESTANE	FCE 24304
EXEMESTANE	6-METHYLENEANDROSTA-1,4-DIENE-3,17-DIONE
EXTRINSIC PATHWAY INHIBITOR	LIPOPROTEIN ASSOCIATED COAGULATION INHBIBITOR
F 1394	(1S, 2S)-2-[3-(2,2-DIMETHYLPROPYL)-3-NONYLUREIDO]AMINOCYCLOHEXANE-1-YL 3-[N-(2,2,5,5-TETRAMETHYL-1,3-DIOXANE-4-CARBONYL)AMINO]PROPIONATE
F 1394	(1S, 2S)-2-[3-(2,2-DIMETHYLPROPYL)-3-NONYLUREIDO]CYCLOHEXANE-1-YL 3-[(4R)-N-(2,2,5,5,-TETRAMETHYL-1,3-DIOXANE-4-CARBONYL)AMINO]PROPIONATE
F 10863A	ZARAGOCID ACID 3D
F-244	12-HYDROXY-13-HYDROXYMETHYL-3,5,7-TRIMETHYLTETRADECA-2,4-DIENOIC ACID 12,14-LACTONE
FAD	FLAVIN ADENINE DINUCLEOTIDE
FAG	MACROPHAGE 23-KD STRESS PROTEIN
FAGARONINE	2-HYDROXY-3,8,9-TRIMETHOXY-5-METHYLBENZO[c]PHENANTHRIDINE
FAGOMINE	1,2,5-TRIDEOXY-1,5-IMINO-D-ARABINO-HEXITOL
3-epi-FAGOMINE	1,2,5-TRIDEOXY-1,5-IMINO-D-ALLO-HEXITOL
F-ARA-ATP	9-β-D-ARABINOFURANOSYL-2-FLUOROADENOSINETRIPHOSPHATE
FASULIDE	1-(5-ISOQUINOLINESULFONYL)HOMOPIPERAZINE HCL
FCCP	[[4-(TRIFLOUROMETHOXY)PHENYL]HYDRAZONO]-PROPANEDINITRILE
FCDP	n3-FUMARAMOYL-l-2,3-DIAMINOPROPIONIOC ACID
FCE 22178	5,6-DIHYDRO-7-(1H-IMIDAZOL-1-YL)NAPHTHALENE-2-CARBOXYLIC ACID
FCE 22178	ROLAFARGREL
FCE 24304	EXEMESTANE
FCE 24304	6-METHYLENEANDROSTA-1,4-DIENE-3,17-DIONE
FCE 24928	4-AMINOANDROSTA-1,4,6-TRIENE-3,17-DIONE
FCE 26073	1-(4-METHYL-3-OXO-4-AZA-5α-ANDROSTANE-17β-CARBONYL)-1,3-DIISOPROPYLUREA
FCE 26743	(S)-2-(4-(3-FLUOROBENZYLOXY)BENZYLAMINO)PROPIONAMIDE
FCE 27677	[(–)N-[2,6-BIS(1-METHYLETHYL)PHENYL]-N′-[(4R, 5R)-2-(4-DIMETHYLAMINOPHENYL)-4,5-DIMETHYLDIOXOLAN-2-YL]METHYLUREA
FCE 28260	(22R, S)-N-(1,1,1-TRIFLUORO-2-PHENYLPROP-2-YL)-3-OXO-4-AZA-5α-ANDROST-1-ENE-17β-CORBOXAMIDE
FDNB	2,4-FLUORODINITROBENZENE
FDP	n3-FUMAROYL-L-2,3-DIAMINOPROPANOIC ACID
FENPROPIDIN	N-[3-(p-tert-BUTYLPHENYL)-2-METHYLPROPYL]PIPERIDINE

Trivial name	Synonym or systematic name
FENPROPIMORPH	N-[3-(p-tert-BUTYLPHENYL)-2-METHYLPROPYL]cis-2,6-DIMETHYLMORPHOLINE
FIALURIDINE	1-(2′-DEOXY-2′-FLUORO-βD-ARABINOFURANOSYL)5-IODOURIDINE
FINASTERIDE	17β-(N-ter.-BUTYL)CARBAMOYL-4-AZA-5α-ANDROST-1-ENE-3-ONE
FISETIN	3,3′,4′,7-TETRAHYDROXYFLAVONE
FIT 276	N-(4-[2(R)-AMINO-3-MERCAPTOPROPYLAMINO-3-PHENYL]BENZOYL]METHIONINE METHYLESTER
FITC	FLUORESCEIN-5′-ISOTHIOCYANATE
FK 143	4-[3-[3-[BIS(4-ISOBUTYLPHENYL)METHYLAMINO]BENZOYL]-1H-INDOL-1-YL]BUTYRIC ACID
FK 706	2-[4-[[(S)-1-[[(RS)-3,3,3-TRIFLUORO-1-ISOPROPYL-2-OXOPROPYL]AMINOCARBONYL]PYRROLIDIN-1-YL]CARBONYL]-2-METHYLPROPYL]AMINOCARBONYL]BENZOYLAMINO]ACETIC ACID
FK 3311	4′-ACETYL-2′-(2,4-DIFLUOROPHENOXY)METHANESULPHONANILIDE
FLAVIANIC ACID	8-HYDROXY-5,7-DINITRO-2-NAPHTHALENESULPHONIC ACID
FLAVOPIRIDOL	(−)cis-5,7-DIHYDROXY-2-(2-CHLOROPHENYL)-8-[4-(3-HYDROXY-1-METHYL)PIPERIDINYL]-1H-1-BENZOPYREN-4-ONE
FLAVOPIRIDOL	L 86-8275
FLM 5011	2-HYDROXY-5-METHYLLAUROPHENONOXIM
FLOSULIDE	CGP 28238
FLOSULIDE	6-(2,4-DIFLUOROPHENOXY)-5-METHANESULPHONAMIDOINDAN-1-ONE
FLP 62064	N-(4-METHOXYPHENYL)-1-PHENYL-1H-PYRAZOLE-3-AMINE
FLUAZIFOP	2-[4-[[5-(TRIFLUOROMETHYL)-2-PYRIDINYL]OXY-PHENOXY]PROPANOIC ACID
FLUAZIFOP-BUTYL	BUTYL-2-[4-(5-TRIFLOUROMETHYL-2-PYRIDYLOXY)-PHENOXY]PROPIONIC ACID
FLUDARABINE	2-FLUOROADENOINEARABINOSIDE-5′-PHOSPHATE
FLUDARABINE PHSOPHATE	9β-ARABINOSYL-2-FLUOROADENINE PHOSPHATE
FLUDARABINE PHSOPHATE	FLUDARA
2FLUOROACETAMIDIE 1,2DIDEOXYNOJ IRIMYCIN	1,5-DIDEOXY-1,5-IMINO-N-FLUOROACETYLGLUCOSAMINOTOL
5-FLUOROACYCLO-THYMIDINE	5-FLUORO-1-(2′-HYDROXYETHOXYMETHYL)URACIL
6′βFLUOROARISTERO-MYCIN	(±)-(1α,2α,3β,4β,5β)-3-ADENIN-9-YL-4-FLUORO-5-(HYDROXYMETHYL)-1,2-CYCLOPENTANEDIOL
FLUOROFAMIDE	N-(DIAMINOPHOSPHINYL)-4-FLUOROBENZENEAMIDE
FLUOZINAM	3-CHLORO-N-(3-CHLORO-2,6-DINITRO-4-TRIFLUOROMETHYLPHENYL-5-TRIFLUOROMETHYL-2-PYRIDINAMINE
FMA	2-MONOFLUOROAGMATINE
FMDP	N3-(4-METHOXYFUMAROYL)-L-2,3-DIAMINOPROPIONIOC ACID
FMN	FLAVIN MONONUCLEOTIDE
FOLIMYCIN	CONCANAMYCIN A
FOMESAN	5-[2-CHLORO-4-(TRIFLUOROMETHYL)PHENOXY]-N-(METHYLSULPHONYL)-2-NITROBENZAMIDE
FORMASTANE	4-HYDROXYANDROST-4-ENE-3,17-DIONE
FOROXYMITHINE	(3S,6S)-3-[3-[N-[(N2-ACETYL-N5-FORMYL-N5-HYDROXY-L-ORNITHINYL)-L-SERYL]-N-HYDROXYAMINO]PROPYL]-2,5-PIPERAZINEDIONE
FOS 37	FUMAGILLOL
FOSINOPRIL	4-CYCLOHEXYL-[[[2-METHYL-1-(OXOPROPOXY)PROPOXY](4-PHENYLBUTYL)PHOSPHINYL]ACETYL]PROLINE
FOSIOPRILAT	SQ 27519
FOSTRIECIN	PD 110161
FOSTRIECIN	PHOSPHOTRIENIN
FOTEMUSTINE	DIETHYL(1-[3-(2-CHLOROETHYL)-3-NITROSOUREIDO)ETHYLPHOSPHONATE
FPL 67156	6-N, N-DIETHYL-D-β,γ-DIBROMOMETHYLENE-ATP
FR 110302	(+)-,2,-DIBUTYL-5-(2-QUINOLYLMETHOXY)-1,2,3,4-TETRAHYDRO-1-NAPHTHOL
FR 122047	1-[(4,5-BIS(4-METHOXYPHENYL)-2-THIAZOYL)CARBONYL]-4-METHYLPIPERAZINE HCl
FR 129169	N-(1,2-DIPHENYLETHYL)-2-OCTYLPHENYLACETAMIDE
FR 145237	N-BENZYL-N[3(4-CHLOROPHENYL)-5-METHYL-2-BENZOFURANYL]METHYL-N′-(2,4,6-TRIFLUOROPHENYL)UREA
FR 186054	N-BENZYL-N-[3-(PYRAZOL-3-YL)BENZYL]-N′-[2,4-BIS(METHYLTHIO)-6-METHYLPYRIDIN-3-YL]UREA
FR 74366	[3-(4-BROMO-2-FLUOROBENZYL)-7-CHLORO-2,4-DIOXO-1,2,3,4-TETRAHYDROQUINAZOLIN-1-YL] ACETIC ACID
FR 74366	FK 366
FR 74366	ZENARESTAT
FRONTALIN	1,5-DIMETHYL-6,8-DIOXOBICYCLO 3.2.1 OCTANE
FRUMIN	O, O-DIETHYL-S-2-ETHYLTHIOETHYL PHOSPHORODITHIOATE

Trivial name	Synonym or systematic name
FSBA	5′-p-FLUOROSULPHONYLBENZOYLADENOSINE
5′-FSBA	5′-p-FLUOROSULPHONYLBENZENE
FUMIGATOL	3-METHOXY-6-METHYL-1,2,4-BENZENETRIOL
FdUMP	5-FLUORO-2′-DEOXYURIDINEMONOPHOSPHATE
FURALTADONE	5-(4-MORPHOLINYLMETHYL)-3-[[(5-NITRO-2-FURANYL)METHYLENE]AMINO]-2-OXAZOLIDINONE
FURAZOLIDONE	3-[[(5-NITRO-2-FURANYL)METHYLENE]AMINO]-2-OXAZOLIDINONE
FUREGRELATE	5-(3′-PYRIDINYLMETHYL)BENZOFURAN-2-CARBOXYLIC ACID
FUROIC ACID	2-FURANCARBOXYLIC ACID
FUROSEMIDE	5-[AMINOSULPHONYL]-4-CHLORO-2-(2-FURANYLMETHYL)-AMINO]BENZOIC ACID
FUSARIC ACID	5-BUTYL-2-PYRIDINECARBOXYLIC ACID
FUSCIN	9,10-DIHYDRO-5-HYDROXY-4,8,8-TRIMETHYL-2H, 4H-BENZO[1,2-B: 4,3-C′]DIPYRAN-2,6(8H)-DIONE
FUT 175	6′-AMIDINO-2-NAPHTHYL p-GUANIDINOBENZOIC ACID
FUT 187	6-AMIDINO-2-NAPHTHYL[4-(4,5-DIHYDRO-1H-IMIDAZOL-2-YL)AMINO]BENZOATE DIMETHANE SULPHONATE
FUT 187	6-AMINO-2-NAPHTHYL-4-[(4,5-DIHYDRO-1H-IMIDAZOL-2-YL)AMINO]BENZOIC ACID
FUT 187	SEPINOSTAT MESILATE
FUT 5895	4-(β-AMIDINOMETHENYL)PHENYL-4-GUANIDINOBENZOATE DIMETHANESULPHONATE
FUT 5897	4-AMIDINO-2-BENZOYLPHENYL-4-GUANIDINOBENZOATE DIMETHANESULPHONATE
G 137	PICOTAMIDE
GABACULIN	5-AMINO-1,3-CYCLOHEXADIENE-1-CARBOXYLIC ACID
GABACULINE	5-AMINO-1,3-CYCLOHEXADIENYLCARBOXYLIC ACID
GABEXATE MESILATE	ε-GUANIDINOCAPROIC ACID p-CARBOXYETHYLPHENYLESTER
GALACTAL	1,5-ANHYDRO-2-DEOXY-LYXO-HEX-1-ENITOL
GALACTO-NOJIRIMYCIN	GALACTOSTATIN
GALBONOLIDE A	RUSTMICIN
GALLIC ACID	3,4,5-TRIHYDROXY-BENZOIC ACID
GALLIN	2-(3,4,5,6-TETRAHYDROXY-9H-XANTHEN-9-YL)BENZOIC ACID
GAMSA	AMSACRINE
GANA	4-GUANIDINO-NEUAC2EN
GEMCITABINE	2′-DEOXY-2′,2′-DIFLUOROCYTIDINE
GEMSA	DL-S-(GUANIDINOETHYL)MERCAPTOSUCCINIC ACID
GENTISIC ACID	2,5-DIHYDROXY-BENZOIC ACID
GERMANIN	SURAMIN
GG 167	5-ACETYLAMINO-2,6-ANHYDRO-4-GUANIDINO-3,4,5-TRIDEOXY-D-GLYCEROL-D-GALACTONON-2-ENOIC ACID
GG 167	2,3-DIDEHYDRO-2,4-DIDEOXY-4-GUANIDINYL-N-ACETYLNEURAMINIC ACID
GG 167	4-GUANIDINO-2,4-DIDEOXY-2,3-DEHYDRO-N-ACETYLNEURAMINIC ACID
GG 167	ZANAMVIR
GG 211	GI 147211
GGTI 287	N-4-[2(R)-AMINO-3-MERCAPTOPROPYL]AMINO-2-PHENYLBENZOYL-LEUCINE
GI 157669X	17β-(ISOBUTYL)CARBONYL-6-AZA-ANDROST-4-EN-3-ONE
GLISOXEPIDE	N-[2-[4-[[[[(HEXAHYDRO-1H-AZEPIN-1-YL)AMINO]CARBONYL]AMINO]SULPHONYL]PHENYL]ETHYL]-5-METHYL-3-ISOXAZOLECARBOXAMIDE
GLUCANA	L-4-GLUTAMYL-3-CARBOXY-4-NITROANILIDE
D-GLUCITOL	D-SORBITOL
GLUTACONIC ACID	2-PENTENEDIOIC ACID
L-GLUTAMATE 5-SEMI-ALDEHYDE	(S)-2-AMINO-5-OXO-PENTANOIC ACID
GLYCERONE	1,3-DIHYDROXYPROPAN-2-ONE
GLYCERRHETINIC ACID	3β-HYDROXY-11-OXOOLEAN-12-EN-30-OIC ACID
GLYCIDIC ACID	OXIRANECARBOXYLIC ACID
GLYCIDOL PHOSPHATE	OXYRANEMETHANOL PHOSPHATE
GLYCOLALDEHYDE	HYDROXY-ACETALDEHYDE
GLYCOLIC ACID	HYDROXY-ACETIC ACID
GLYCOPRILAT	(S)-N-[3-(3,4-METHYLENEDIOXYPHENYL)-2-(MERCAPTOMETHYL)-1-OXOPROLYL]GLYCINE
GLYCYRRHIZIC ACID	20β-CARBOXY-11-OXO-30-NOROLEAN-12-EN-3β-YL-2-O-β-D-GLUCOPYRANOSYL-α-D-GLUCOPYRANOSIDURONIC ACID
GLYOXYLIC ACID	OXO-ACETIC ACID

Trivial name	Synonym or systematic name
GLYPHOSATE	N-[PHOSPHONOMETHYL]-GLYCINE
GM 6001	GALARDIN
GM 6001	N-[2(R)-2-(HYDROXAMIDOCARBONYLMETHYL)-4-METHYLPENTANOYL]-L-TRYPTOPHANE METHYLAMIDE
cGMPS	Rp-GUANOSINE-3′,5′-cyclicMONOPHOSPHOTHIOATE
GOSSYPOL	1,1′,6,6′,7,7′-HEXAHYDRO-3,3′-DIMETHYL-5,5′-DIISOPROPYL-2,2′-BINAPHTHYL-8,8′-DIALDEHYDE
GOSSYPOL	1,1′,6,6′,7,7′-HEXAHYDROXY-3,3′-DIMETHYL-5,5′-BIS(1-METHYLETHYL)-[2,2′-BINAPHTHALENE]-8,8′-DICARBOXAL-DEHYDE
GP 1-515	4-AMINO-1-(5-AMINO-5-DEOXY-1-β-D-RIBOFURANOSYL)-3-BROMOPYRAZOLO[3,4-D]PYRIMIDINE
GPC	L-3-GLYCERYLPHOSPHORYLCHOLINE
GPE	L-3-GLYCERYLPHOSPHORYLETHANOLAMINE
GR 92754	6,7-DIBUTYL-2,4-DIAMINOPTERIDINE
GRAHAM'S SALT	METAPHOSPHORIC ACID, SODIUM SALT
GS 4071	(3.R, 4R, 5S)-4-ACETAMIDO-5-AMINO-3-(1-ETHYLPROPOXY)-1-CYCLOHEXENE-1-CARBOXYLIC ACID
G-STROPHANTHIN	OUBAIN
GUAIACOL	o-METHOXYPHENOL
GUAZATINE	N-N‴-(IMINODI-8,1-OCTANEDIYL)BIS-GUANIDINE
GUVACINE	1,2,5,6-TETRAHYDRO-3-PYRIDINECARBOXYLIC ACID
GW 3600	trans(R*,R*)(±)METHYLACERYL-4-[3-(CYCLOPENTOXY)-4-METHOXYPHENYL]-3-METHYL-1-PYRROLIDINE-CARBOXYLATE
GYKI 14166	D-PHE-PRO-ARG-H
GYKI 14766	D-METHYL-PHE-PRO-ARGINAL
H 7	1-(5-ISOQUINOLINESULPHONYL)-2-METHYLPIPERAZINE DIHYDROCHLORIDE
H 8	N-[2-(METHYLAMINO)ETHYL]-5-ISOQUINOLINESULPHONAMIDE DIHYDROCHLORIDE
H 9	N-(6-AMINOETHYL)-5-ISOQUINOLINESULPHONAMIDE
H 9	N-[2-AMINOETHYL]-5-ISOQUINOLINESULPHONAMIDE DIHYDROCHLORIDE
H 85	N-[2-(N-FORMYL-p-CHLOROCINNAMYLAMINO)ETHYL]-5-ISOQUINOLINESULPHONAMIDE
H 88	N[2-(CYNNAMYLAMINO)ETHYL]-5-ISOQUINOLINESULPHONAMIDE
H 89	N-[2-(p-BROMOCINNAMYLAMINO)ETHYL]-5-ISOQUINOLINESULFONAMIDE
H 142	PRO-HIS-PRO-PHE-HIS-LEUpsi[CH2NH]-VAL-ILE-HIS-LYS
H 149/94	2-{[2-(3-METHYL)PYRIDYLMETHYL]SULPHINYL}-5-METHOXYCARBONYL-6-BENZIMIDAZOLE
H 256	PRO-THR-GLU-PHEpsi[CH_2NH]-PHE-ARG-GLU
H 297	PRO-THR-GLU-PHEpsi[CH2NH]-NLE-ARG-LEU psi = {CH2NH IN PLACE OF CONH}
HA 1004	N-(2-GUANIDINOETHYL)-5-ISOQUINOLINESULPHONAMIDE HYDROCHLORIDE
HA 1077	1-(5-ISOQUINOLINESULPHONYL)-HOMOPIPERAZINE HCl
HAMAMELONIC ACID 2-PHOSPHATE	2-CARBOXY-D-ARABINITOL-1-PHOSPHATE
HAMAMELONIC ACID 2-PHOSPHATE	2-C-(HYDROXYMETHYL)-D-RIBONIC ACID 2-PHOSPHATE
HAMAMELOSE	2-C-(HYDROXYMETHYL)-D-RIBOSE
HARMALINE	4,9-DIHYDRO-7-METHOXY-1-METHYL-3H-PYRIDO[3,4-B]INDOLE
HARMAN	1-METHYL-β-CARBOLINE
HARMINE	7-METHOXY-1-METHYL-9H-PYRIDO[3,4-B]INDOLE
HBY 97	(S)-ISOPROPOXYCARBONYL-6-METHOXY-3-(METHYLTHIOMETHYL)-3,4-DIHYDRO-QUINOXALINE-2-(1H)-THIONE
HEMATIN	[7,12-DIETHENYL-3,8,13,17-TETRAMETHYL-21H, 23H-PORPHINE-2,18-DIPROPANOATO(4-)-N21,N22,N23,N24]-HYDROXY-FERRATE(2-)DIHYDROGEN
HEMIACYLCARNITINIUM	(2s,6r)-6-(CARBOXYMETHYL)-2-HYDROXY-2,4,4-TRIMETHYLMORPHOLINIUM Cl
HEMIN	CHLORO[7,12-DIETHENYL-3,8,13,17-TETRAMETHYL-21H, 23H-PORPHINE-2,18-DIPROPANOATO(4-)-N21,N22,N23,N24]FERRATE(2-)DIHYDROGEN
HEMIPALMITOYL-CARNITINIUM	6-(CARBOXYMETHYL)-2-HYDROXY-2-PENTADECYL-4,4-DIMETHYLMORPHOLINIUM Br
HEMIPALMITOYL-CARNITINIUM	(2s,6r: 2r,6s)-6-CARBOXYMETHYL-2-HYDROXY-2-PENTADECYL-4,4-DIMETHYLMORPHOLINIUM BROMIDE
HEMIPROPANOYL-CARNITINIUM	(2s,6r)-6-(CARBOXYMETHYL)-2-ETHYL-2-HYDROXY-4,4-DIMETHYLMORPHOLINIUM Br
HERCYNINE	α-CARBOXY-N, N, N-TRIMETHYL-1H-IMIDAZOLE-4-ETHANAMINIUM HYDROXIDE INNER SALT
15-HETE	15-HYDROXY-6,8,11,14-EICOSATETRAENOIC ACID

Trivial name	Synonym or systematic name
HEXAHYDROXYFLAVONE	MYRICETIN
HEXARIC ACID	2,3,4,5-TETRAHYDROXYHEXANEDIOIC ACID
HIPPURIC ACID	N-BENZOYL-GLYCINE
HISPIDIN	6-(3,4-DIHYDROXYSTYRYL)-4-HYDROXY-2-PYRONE
HISTARGIN	N-[(S)-1-CARBOXY-4-GUANIDINOBUTYL]-N′-(S)-1-CARBOXY-2-(4-IMIDAZOLYL)ETHYL]ETHYLENEDIAMINE
HL 725	TREQUINSIN
HOE 065	N-OCTYL-2-[N-[(S)-1-ETHOXYCARBONYL-3-PHENYLPROPYL]-L-ALANYL]-1S,3S,5S)-2-AZABICYCLO[3.3.0]OCTANE-3-CARBOXYLATE MALEATE SALT
HOE 077	2,4-PYRIDINE DICARBOXYLIC ACID BIS[(2-METHOXYETHYLAMIDE
HOE 288	3′S,5′S-1′[(S)-N-[(S)-1-CARBOXY-3-PHENYLPROPYL]ALANYL-SPIRO[BICYCLO[2,2,2,]OCTANE-2,3′-PYRROLIDINE]-5′-CABOXYLIC ACID ETHYLESTER
HOE 467	TENDAMISTAT
HOE 642	4-ISOPROPYL-3-METHYLSULPHONYLBENZOYLGUANIDINE METHANESULPHONATE
HOE 694	(3-METHANESULPHONYL-4-PIPERIDINOBENZOYL)GUANIDINE
HOE 704	2-DIMETHYLPHOSPHINOYL-2-HYDROXYACETIC ACID
HOE 731	SAVIPRAZOLE
HOE 843	2,7-DIFLUOROSPIRO(9H-FLUORENE-9,4′-IMIDAZOLINE)-2,5′-DIONE
α-HOMONOJIRIMYCIN	2,6-DIDEOXY-2,6-IMINO-D-GLYCERO-L-GULO-HEPTITOL
HP 029	9-AMINO-1,2,3,4-TETRAHYDROACRIDINE-1-OL
HP 029	VELNACRINE
HQNO	2-HEPTYL-4-QUINOLINOL-1-OXIDE
HSR 803	N-[4-[2-(DIMETHYLAMINO)ETHOXY]BENZYL]-3,4-DIEMTHOXYBENZAMIDE HCl
HTS	5-(3-HYDROXYBENZOYL)-2-THIOPHENYLSULPHONAMIDE
HUMAN PROTEINASE INHIBITOR 6	HUMAN PLACENTA SERINE PROTEINASE INHIBITOR
HYDANTOIN	2,4-IMIDAZOLIDINEDIONE
α-HYDRAZINO m-KRESOL	m-HYDROXYBENZYLHYDRAZINE
HYDROCINNAMIC ACID	BENZENPROPANOIC ACID
HYDROXYAKALONE	4-AMINO-1H-PYRAZOLO[3,4-d]PYRIMIDINE-3-ONE-6-OL
HYPOGLYCINE A	α-AMINO-2-METHYLENE-CYCLOPROPANEPROPANOIC ACID
IBMX	3-ISOBUTYL-1-METHYL-XANTHINE
IBUPROFEN	2-(4-ISOBUTYLPHENYL)PROPIONIC ACID
ICI 128436	PONALRESTAT
ICI 128436	STATIL
ICI 186756	(R,S)-N2-(3-CARBOXY-1-OXOPROPYL)-N6-[(PHENYLMETHOXY)CARBONYL]-L-LYSYL-L-VALYL-N-(1-FORMYL-2-METHYLPROPYL)-L-PROLINAMIDE
ICI 198583	2-DESAMINO-2-METHYL-N[1]0-PROPARGYL-5,8-DIDEAZAFOLIC ACID
ICI 200355	4-(4-BROMOPHENYLSULPHONYLCARBAMOYL)BENZOYL-L-VALYL-L-PROLINE 1 (RS)-(1-TRIFLUOROACETYL-2-METHYLPROLYL)AMIDE
ICI 207968	1,2-DIHYDRO-2-(3-PYRIDYLMETHYL)-3H-INDAZOL-3-ONE
ICI 207968	2-(3-PYRIDYLMETHYL)-INDAZOLINONE
ICI 207968	ZILEUTON
ICI 211965	1-[3-(NAPHTH-2-YLMETHOXY)PHENYL]-1-(THIAZOL-2-YL)PROPYL METHYL ETHER
ICI 216800	(+)1-METHOXY-6-(NAPHTH-2-YL-METHOXY)-1-(THIAZOL-2-YL)INDANE
ICI 63197	2-AMINO-6-METHYL-4-PROPYL-[1,2,4]-TRIAZOLO[1,5a]PYRIMIDIN-5(4H)-ONE
ICI D 1694	N-(5-[N-(3,4-DIHYDRO-2-METHYL-4-OXOQUINAZOLIN-6-YLMETHYL)-N-METHYLAMINO]2-THENOYL)-S-GLUTAMIC ACID
ICI D2138	6[[FLUORO-5-(4-METHOXY-3,4,5,6-TETRAHYDRO-2H-PYRAN-4-YL)PHENOXY]METHYL]-1-METHYLQUINOL-2-ONE
ICRF 154	4,4′-(1,2-ETHANEDIYL)-BIS-(2,6-PIPERAZINEDIONE)
ICRF 159	4,4′-(METHYLETHANE-1,2-DIYL)-BIS-(2,6-PIPERAZINEDIONE)
ICRF 187	[(+)-1,2-BIS(3,5-DIOXOPIPERAZINYL-1-YL)PROPANE
ICRF 193	BIS(2,6-DIOXOPIPERAZINE)BUTANE
ICRF 193	4,4′-(1,2-DIMETHYLETHANE-1,2-DIYL)-BIS-(2,6-PIPERAZINEDIONE)
IDARUBICIN	4-DEMETHOXYDAUNORUBICIN
IDRAPRIL	(+)-(1S, 2R)-2-[[N-(2-HYDROXYAMINO-2-OXOETHYL)-N-METHYLAMINO]CARBONYL]CYCLOHEXANE-1-CARBOXYLIC ACID
IFO	3-[4[3(1H-IMIDAZOL-1-YL)PROPOXY]PHENYL]-5-TRIFLUOROMETHYL-1,2,4-OXADIAZOLE

Trivial name	Synonym or systematic name
7-IHA	7-(1-IMIDAZOYL)HEPTANOIC ACID
IaI	INTER-α-TRYPSIN INHIBITOR
ILMOFOSINE	BM 41440
IMIDAPRIL	(4S)-1-METHYL-3-{(2S)-2-[N-((1S)-1-ETHOXYCARBONYL-3-PHENYLPROPYL)AMINO]PROPIONYL}-2-OXO-IMIDAZOLIDINE-4-CARBOXYLIC ACID HCl
IMIDAPRILAT	(4S)-3-{(2S)-2-[N-((1S)-1-CARBOXY-3-PHENYLPROPYL)AALANYL}-1-METHYL-2-OXO-IMIDAZOLIDINE-4-CARBOXYLIC ACID HCl
IMIPRAMINE	10,11-DIHYDRO-N, N-DIMETHYL-5H-DIBENZ[B, F]AZEPINE-5-PROPANAMINE
tIMP	THIOINOSINIC ACID
IMRESTAT	2,7-DIFLUOROSPIRO(9H-FLUORENE-9,4′-IMIDAZOLINE)-2,5′-DIONE
INDINAVIR	L 735524
INDINAVIR	[1α[αS*,γR*,δR*],2αN(2,3DIHYDRO2HYDROXY-1H-INDEN1YL)2[[(1,1DIMETHYLETHYL)AMINO]CARBONYL]γ-HYDROXYα(PHENYLMETHYL)4(3-PYRIDINYLMETHYL)1-PIPERAZINEPENTANAMIDE
INDOBUFEN	(±)(1,3-DIHYDRO-1-OXO-2H-ISOINDOL-2-YL)-α-ETHYLBENZAENEACETIC ACID
INDOLAPRIL	PD 109763-2
INDOLAPRILAT	PD 110021-0
INDOLIDAN	1,3-DIHYDRO-3,3-DIMETHYL-5-(1,4,5,6-TETRAHYDRO-6-OXO-3-PYRIDAZINYL)-2H-INDOL-2-ONE
INDOMETHACIN	1-(4-CHLOROBENZOYL)-5-METHOXY-2-METHYL-1H-INDOLE-3-ACETIC ACID
INDOSPICINE	(S)-2,7-DIAMINO-7-IMINOHEPTANOIC ACID
INOGATRAN	GLYCINE, N[2-[2[[[3[(AMINOIMINOMETHYL)AMINO]PROPYL]AMINO]CARBONYL]-1-PIPERIDINYL]-1-(CYCLOHEXYLMETHYL)-2-OXOETHYL[2S]]]
12IODE	12-IODO-cis-9-OCTADECENOIC ACID
5-IODOACYCLOTHYMIDINE	5-IODO-1-(2′-HYDROXYETHOXYMETHYL)URACIL
IPRONIAZID	4-PYRIDINECARBOXYLIC ACID 2-(1-METHYLETHYL)HYDRAZIDE
IPSDIENOL	2-METHYL-6-METHYLENE-2,7-OCTADIEN-4-OL
IPSENOL	2-METHYL-6-METHYLENE-7-OCTEN-4-OL
IPTG	ISOPROPYL-β-D-GALACTOPYRANOSIDE
ISATIN	INDOLE-2,3-DIONE
ISATIN	TRIBULIN
ISATOIC ACID	2-(CARBOXYAMINO)-BENZOIC ACID
ISOBONGKREKIC ACID	[R-[R*,S*-(E, E, Z, E, E, Z, E)]]-20-(CARBOXYMETHYL)-6-METHOXY-2,5,17-2,4,8,10,14,18,20-DOCOSAHEPTAENEDIOIC ACID
ISOCITRIC ACID	3-CARBOXY-3,4-DIDEOXY-PENTARIC ACID
ISOFLUROPHATE	DIISOPROPYL FLUOROPHOSPHATE
ISOLUMINOL	4-AMINOPHTHALYLHYDRAZIDE
ISONIAZID	4-PYRIDINECARBOXYLIC ACID HYDRAZIDE
ISONICOTINIC ACID	4-PYRIDINECARBOXYLIC ACID
ISO-OMPA	N, N′,N″,N‴-TETRAKIS(1-METHYLETHYL)-DIPHOSPHORAMIDE
ISOPHTHALIC ACID	1,3-BENZENEDICARBOXYLIC ACID
ITACONIC ACID	METHYLENE-BUTANEDIOIC ACID
ITAZIGREL	2-(TRIFLUOROMETHYL)-4,5-BIOS(p-METHOXYPHENYL)THIAZOLE
IY 80843	N-[2-(2-METHOXYPHENYL)ETHYL]-N′-[4-(IMIDAZOLE-4-YL)PHENYL]FORMAMIDE
IY 80845	N-[2-(4-METHOXYPHENYL)ETHYL]-N′-[4-(IMIDAZOLE-4-YL)PHENYL]FORMAMIDE
JMV 390-1	N-[3-(R, S)[(HYDROXYAMINO)CARBONYL]-2-BENZYL-1-OXOPROLYL]-L-ISOLEUCINE-L-LEUCINE
JTE 522	4-(4-CYLOHEXYL-2-METHYLOXAZOL-5-YL)-2-FLUOROBENZENESULPHONAMIDE
JTP 4819	(S)-2-[[(S)-2-(HYDROXYACETYL)-1-PYRROLIDINYL]CARBONYL]-N-(PHENYLMETHYL)-1-PYRROLIDINE-CARBOXAMIDE
JUGLONE	5-HYDROXY-1,4-NAPHTHOQUINONE
K 252a	(8R*,9S*,11S*)-(–)-9-HYDROXY-9-METHOXYCARBONYL-8-METHYL-2,3,9,10-TETRAHYDRO-8,11-EPOXY-1H, 8H, 121H-2,7b,11a-TRIAZADIBENZO[a,g]CYCLOOCTA[cde]TRIDEN-1-ONE
K 3920	INDOBUFEN
KB-R 7785	[4-(N-HYDROXYAMINO)-2(R)-ISOBUTYL-3(S)-METHYLSUCCINYL]-L-PHENYLGLYCINE-N-METHYLAMIDE
KBT 3022	ETHYL 2-[4,5-BIS(4-METHOXYPHENYL)THIAZOL-2-YL]PYRROL-1-YLACETATE
KDI 792	FK 070
KELATORPHAN	N-[3(R)-[(HYDROXYAMINO)CARBONYL]-2-BENZYL-1-OXOPROPYL]-L-ALANINE
KELLETIN I	ERYTHRITYL TETRAKIS(p-HYDROXYBENZOIC ACID)
KERATOLININ	CYSTATIN A HUMAN

Trivial name	Synonym or systematic name
KETOPROFEN	m-BENZOYLHYDROTROPIC ACID
KETOPROFEN	(R, S)-2-(3′-BENZOYLPHENYL)PROPANOIC ACID
KF 13218	(Z)-11-(5-CARBOXYPENTYLIDENE)-6-METHYL-5,11-DIHYDROPYRIDO[4,3-C][1]BENZAZEPIN-5(6H)ONE
KF 17828	2-BROMO-N-(2,6-DIISOPROPYLPHENYL)-6,11-DIHYDROBENZ[b,e]OXEPIN-11-CARBOXAMIDE
KF 18678	(E)4-{2-[[3-[1-[BIS(4-FLUOROPHENYL)METHYLINDOL-5-YL]-1-OXO-2-BUTENYL]AMINO]PHENOXY}BUTYRIC ACID
KF 19514	5-PHENYL-3-(3-PYRIDYL)METHYL-3H-IMIDAZO(4,5-c)(1,8)NAPTHTHYRIDIN-4(5H)ONE
KI 6783	(3,4-DIMETHOXY)-4-PHENOXY-6,7-DIMETHOXYQUINOLINE
KIH 200	2-(4′-HYDROXYBENZYLIDENE)-4-CYCLOPENTENE-13,-DIONE
KIH 201	2-(4′-HYDROXY-3′-METHOXYBENZYLIDENE)-4-CYCLOPENTENE-1,3-DIONE
KIH 202	2-(4′-HYDROXY-3′,5′-DIMETHOXYBENZYLIDENE)-4-CYCLOPENTENE-1,3-DIONE
KININOGEN	α-CYSTEINPROTEINASE INHIBITOR
KK 505	2-ETHYL-4-(3-PYRIDYL)-1(2H)-PHTHALAZINONE
K76 MONOCARBOXYLIC ACID	6,7-DIHYDROXY-2,5,5,8a-TERTAMETHYL-1,2,3,4,4a,5,6,7,8,8a-DECAHYDRONAPHTHALENE-1-SPIRO-2′-(7′-CARBOXYLATE-6′-FORMYL-4′-HYDROXY-2′,3′-DIHYDROBENZOFURAN
KN 62	1-[N,O-BIS(1,5-ISOQUINOLINESULPHONYL)-N-METHYL-L-TYROSYL]-4-PHENYLPIPERAZINE
KN 93	2-[N-(2-HYDROXYETHYL)-N-4-METHOXYBENZENESULPHONYL)]AMINO-N-(4-CHLOROCINNAMYL)-N-METHYLBENZYLAMINE
KNI 93	SER-APNS-PRO-ILE-VAL-NH_2
KNI 102	Z-ASN-(2S, 3S)-3-AMINO-2-HYDROXY-4-PHENYLBUTYRIC ACID-PRO-NHBUt
KNI 174	1-NAPHTHOXYACETYL-ASN-(2S, 3S)-3-AMINO-2-HYDROXY-4-PHENYLBUTYRIC ACID-L-5,5-DIMETHYLTHIAZOLIDINE-4-CARBOXYLIC ACID-NHBUt
KOJIC ACID	5-HYDROXY-2-(HYDROXYMETHYL)-γ-PYRONE
KP 363	BUTENAFINE
KS 501	2-(p-D-GALACTOFURANOSYLOXY)-6-HEPTYL-4-HYDROXYBENZOIC ACID 3-HEPTYL-5-HYDROXYPHENYL ESTER
KS 502	2-(p-D-GALACTOFURANOSYLOXY)-6-HEPTYL-4-HYDROXYBENZOIC ACID 4-CARBOXY-3-HEPTYL-5-HYDROXYPHENYL ESTER
KT 5926	(8R*,9S*,11S*)-(–)-9-HYDROXY-9-METHOXYCARBONYL-8-METHYL-14-n-PROPOXY-2,3,9,10-TETRAHYDRO-8,11-EPOXY, 1H, 8H, 11H, -2,7b,11a-TRIAZADIBENZYO[a,g]CYCLOOCTA[cde]TRINDEN-1-ONE
KTU 3616	CAPROPAMID
KUKOAMINE A	N1,N12-BIS(DIHYDROCAFFEOYL)SPERMINE
KW 3635	(E)-11-[2-(5,6-DIMETHYL-1-BENZIMIDAZOYL)ETHYLIDENE]6,11-DIHYDRODIBENZ[b,e]OXEPIN-2-CARBOXYLATE
KW 5092	{1-[2-[[[5-(PIPERIDINOMETHYL)-2-FURANYL]METHYL]AMINO]ETHYL-2-IMIDAZOLIDINYLIDENE}PROPANEDINITRILE FUMARATE
KYNURENIC ACID	4-HYDROXY-2-QUINOLINECARBOXYLIC ACID
KYNURENINE	α,2-DIAMINO-γ-OXO-BENZENEBUTANOIC ACID
KYOTORPHIN	N2-L-TYROSYL-L-ARGININE
L 86-8275	(–)cis-5,7-DIHYDROXY-2-(2-CHLOROPHENYL)-8-[4-(3-HYDROXY-1-METHYL)PIPERIDINYL]-4H-BENZOPYRAN-4-ONE
L 86-8275	FLAVOPIRIDOL
L 372460	N-(9-HYDROXY-9-FLUORENECARBOXY)PROLYL t-4-AMINOCYCLOHEXYLMETHYL AMIDE
L 612710	2′,3′α-TETRAHYDROFURAN-2′-SPIRO-17-(6-METHYLENE-4-ANDROSTEN-3-ONE
L 645066	(5α)-23-METHYL-4-AZA-21-NORCHOL-1-ENE-3,20-DIONE
L 645151	2-SULPHAMOYL-6-BENZOTHIAZOLYL-2,2-DIMETHYL PROPIONIC ACID
L 645390	17β-N,N-DIISOPROPYLCARBAMOYL-4-AZA-5α-ANDROSTAN-1-ONE
L 651580	METHYL 3-OXO-4-METHYL-4-AZA-5α-ANDROST-1-ENE-17β-CARBOXYLATE
L 651896	2,3-DIHYDRO-6-[3-(2-HYDROXYMETHYL)PHENYL-2-PROPENYL]-5-BENZOFURANOL
L 652117	trans-4-ETHOXYCARBONYL-3-ETHYL-1-(4-NITROPHENYLSULPHONYL)AZETIDIN-3-ONE
L 652343	3-HYDROXY-5-TRIFLUOROMETHYL-N-(2-(2-THIENYL)-2-PHENYL-ETHENYL)-BENZO(B)THIOPHENE-2-CARBOXAMIDE
L 652731	trans-2,5-BIS-(3,4,5-TRIMETHOXYPHENYL)TETRAHYDROFURAN
L 653180	9-{[(Z)-2-(HYDROXYMETHYL)CYCLOHEXYL]METHYL}GUANINE
L 654066	(5α)-23-METHYL-4-AZA-21-NORCHOL-1-ENE-3,20-DIONE
L 654066	MK 0963
L 656224	7-CHLORO-2-[(4-METHOXYPHENYL)METHYL]-4-HYDROXY-3-METHYL-5-PROPYL-BENZOFURAN

Trivial name	Synonym or systematic name
L 658758	3-(ACETOXYMETHYL)-2-([2(S)-CARBOXYPYROLIDINO)CARBONYL]-7α-METHOXY-5-OXO-5-THIA-1-AZA-BICYCLO[4.2.0]-OCT-2-ENE 5,5-DIOXIDE
L 659699	1233 A
L 659699	F 244
L 659699	(E,E)-11-[3-HYDROXYMETHYL)-4-OXO-2-OXETANYL]-3,5,7-TRIMETHYL-2,4-UNDECADIENOIC ACID
L 659699	2R-HYDROXYMETHYL-3R-HYDROXY-8R, 10,12-TRIMETHYL-(E, E)-10.12-TETRADECADIENDIOIC ACID 1,3-LACTONE
L 660039	MK 927
L 663536	3-[1-4-CHLOROBENZYL)-3-tert-BUTYLTHIO-5-ISOPROPYLINDOL-2-YL]-2,2-DIMETHYLPROPANOIC ACID
L 663536	MK 886
L 665829	MK 417
L 669262	iso-SIMVASTIN-6-ONE
L 671152	s,s-5,6-DIHYDRO-4h-4-ETHYLAMINO-6-METHYLTHIENO-[2,3b]THIOPYRAN-2-SULFONAMIDE-7,7-DIOXIDE
L 671152	MK 507
L 680833	[S(R*,S*)]-4-[(1-(((1-(4-METHYLPHENYL)BUTYL)AMINO)CARBONYL)-3,3-DIETHYL-4-OXO-2-AZETIDINYL)OXY]-BENZNE ACETIC ACID
L 683590	ASCOMYCIN
L 683845	[S(R*,S*)]-4-[(1-(((1-(5-BENZOFURANYL)BUTYL)AMINO)CARBONYL)-3,3-DIETHYL-4-OXO-2-AZETIDINYL)OXY]-BENZNE ACETIC ACID
L 685502	N-2(R)-HYDROXY-1(S)-INDANYL-5(S)-[(tetr.BUTYLOXYCARBONYL)AMINO]-4(S)-HYDROXY-6-PHENYL-2(R)-[[4-[2-(4-MORPHOLINYL)ETHOXY]PHENYL]METHYL]HEXANAMIDE
L 689065	3-[1-(4-CHLOROBENZYL)-4-METHYL-6-(5-PHENYLPYRIDIN-2-YLMETHOXY)-4,5-DIHYDRO-1H-THIOPYRANO(2,3,4-c,d]INDOL-2-YL]2,2-DIMETHYLPROPANOIC ACID
L 689502	N-[2-(R)-HYDROXY-1(S)-INDANYL]-5(S)-(1,1-DIMETHYLETHOXYCARBONYLAMINO)-4(S)-HYDROXY-6-PHENYL-2 (R)-{4-[2-(4-MORPHOLINYL)ETHOXY]PHENYL}METHYLHEXAMIDE
L 693612	trans-(–)-5,6-DIHYDRO-6-(3-METHOXYPROPYL)-4-PROPYLAMINO-4H-THIENO[2,3-b]THIOPYRAN-2-SULPHONAMIDE-7,7-DIOXIDE
L 694458	[S(R*,S*)]-2-[(4-METHYLPIPERAZIN-1-YL)CARBONYL]PHENOXY]-3,3-DIETHYL-N-[1-[3,4-METHYLENEDIOXY)-PHENYL]BUTYL-4-OXO-1-ACETIDINECARBOXAMIDE
L 694599	ZARAGOZIC ACID A
L 696229	3-[2-(1,3-BENZOXAZOL-2-YL)ETHYL]-5-ETHYL-6-METHYLPYRIDIN-2(1H)-ONE
L 696229	2-[2-(5-ETHYL-6-METHYL-2(1H)-PYRIDON-3-YL)ETHYL]BENZOXAZOLE
L 696418	N-[(R)CARBOXYETHYL]α(S)(2-PHENYLETHYL)GLYCINE-L-LEUCINE, N-PHENYLAMIDE
L 696474	18-DEHYDROXY CYTOCHALASIN H
L 697639	3-{[(4,7-DIMETHYL-1,3-BENZOXAZOL-2-YL)METHYL]AMINO}-5-ETHYL-6-METHYLPIRIN-2(1H)-ONE
L 697661	3-[2-(2-(4,7-DICHLOROBENZOXAZOLYL)METHYLAMINO]-5-ETHYL-6-METHYL-1H-PYRIDIN-2-ONE
L 699333	2[2[1(4-CHLOROBENZYL)-4-METHYL-6-[(5-PHENYLPYRIDIN-2-YL)METHOXY]4,5-DIHYDRO-1H-THIOPYRANO-[2,3,4-cd]INDOL-2-YL]ETHOXY]BUTANOIC ACID
L 735524	CRIXIVAN
L 735524	INDINAVIR
L 735524	MK 639
L 735524	N-[2(R)-HYDROXY-1(S)-INDANYL]-5-[[2(S)-tert.-BUTYLAMINOCARBONYL]-4-(3-PYRIDYLMETHYL)PIPERAZINO]-4(S)-HYDROXY-(2R)-PHENYLMETHYLPENTANAMIDE
L 735882	4-[N-BENZYL-3-(4CHLOROBENZYL)PIPERIDIN-3-YL]-2,4-DIOXOBUTANOIC ACID
L 738372	6-CHLORO(4S)CYCLOPROPYL-3,4-DIHYDRO-4-((2-PYRIDYL)ETHYNYL)QUINAZOLIN-2(1H)ONE
L 739010	[1S, 5R]-3-CYANO-1-(3FURYL)-6-{6-[3-(3α-HYDROXY-6,8-DIOXABICYXCLO[3.2.1]OCTANYL)]PYRIDIN-2-YLMETHOXYL}-NAPHTHALENE
L 739749	2(S)[2(S)[2(R)-AMINO-3-MERCAPTO]PROPYLAMINO-3(S)-METHYL]PENTYLOXY-3-PHENYLPROPIONYLMETHIONINESULPHONE METHYLESTER
L 739750	[2(S){2(S)[2(R)AMINO-3-MERCAPTO]PROPYLAMINO-3(S)METHYL}PENTOXY-3-PHENYLMETHIONINESULPHONE
L 739750	2(S)[2(S)[2(R)-AMINO-3-MERCAPTO]PROPYLAMINO-3(S)-METHYL]]PENTYLOXY-3-PHENYLPROPIONYLMETHIONINESULPHONE
L 741494	XYLARIC ACID
L 743726	6-CHLORO-4-CYCLOPROPYLETHNEYL-4-TRIFLUOROMETHYL-1,4-DIHYDRO-2H-BENZOXAZIN-2-ONE
L 745337	5-METHANESULPHOAMIDO-6-(2,4-DIFLUOROTHIOPHENYL)-1-INDANONE
L 746530	[1S, 5R]-2-CYANO-4-(3-FURYL)-7-{3-FLUORO-5-[3-(3α-HYDROXY-6,8-DIOXABICYCLO[3.2.1]-OCTANYL)]PHENOXYMETHYL}QUINOLINE

Trivial name	Synonym or systematic name
L 754394	N[2(R)HYDROXY-1(S)INDANYL]5[2(S)[(((1,1-DIMETHYLETHYL)AMINO)CARBONYL]4[(FURO[2,3-b]PYRIDIN-5-YL)-METHYL]PIPERAZIN1YL]4(S)HYDROXY-2(R)(PHENYLMETHYL)PENTAMIDE
L8027	2-ISOPROPYLINDOL-3-YL-3-PYRIDYL KETONE
β-LACTAMASE INHIBITOR PROTEIN	STREPTOMYCES CLAVULINGERS β-LACTAMASE INHIBITOR
LACTIC ACID	2-HYDROXY-PROPANOIC ACID
LACTITOL	4-O-β-D-GALACTOPYRANOSYL-D-GLUCITOL
LAMISIL	TERBINAFINE
LAMIVUDINE	DIDEOXYTHIACYTIDINE
LANNATE	METHYL N-{[(METHYLAMINO)CARBONYL]OXY}ETHANIMIDOTHIOATE
LAPACHOL	2-HYDROXY-3-(3-METHYL-2-BUTENYL)-1,4-NAPHTHALENEDIONE
β-LAPACHONE	3,4-DIHYDRO-2,2-DIMETHYL-2H-NAPHTHOL[1,2,b]PYRAN-5,6-DIONE
LATERITIN	4-METHYL-6-(1-METHYLETHYL)-3-PHENYLMETHYL-1,4-PERHYDROOXAZINE-2,5-DIONE
LAWSONE	2-HYDROXY-1,4-NAPHTHOQUINONEONE
LAZABEMIDE	RO 19-6327
LB 1350	(5S)-[(N-(ISOPROPYLOXYCARBONYL)-β-METHANESULPHONYL-L-VALINYL]AMINO]-(4R, 3S)-EPOXY-6-PHENYL-1-HEXANOYL]-[(2S)-(1-PHENYL-3-; ETHYL-1-OXO)BUTYLAMINO)AMIDE]
L-CANALINE	O-AMINO-L-HOMOSERINE
L-CARNITINE	[-]-β-HYDROXY-γ-TRIMETHYLAMINOBUTYRIC ACID
L-DOPA	3-HYDROXY-L-TYROSINE
LEMIPRAZOLE	((±)-2-[[2-(ISOBUTYLMETHYLAMINO)BENZYL]SULPHINYL]-1H-BENZIMIDAZOLE
LEUCINTHIOL	2-AMINO-4-METHYL-1-PENTANETHIOL
LEUCOVORIN	5-FORMYLTETRAHYDROFOLIC ACID
LEUKOTRIENE C4	[R-[R*,S*-(E, E, Z, Z)]]-N-[S-[1-(4-CARBOXY-1-HYDROXYBUTYL)-2,4,6,9-PENTADECATETRAENYL]-N-L-γ-GLUTAMYL-L-CYSTEINYL]-GLYCINE
LEUKOTRIENE D4	[R-[R*,S*-(E, E, Z, Z)]]-N-[S-[1-(4-CARBOXY-1-HYDROXYBUTYL)-2,4,6,9-PENTADECATETRAENYL]-L-CYSTEINYL]-GLYCINE
LEUKOTRIENE E4	[5S-[5R*,6S*(S*),7E, 9E, 11Z, 14Z]]-6-[(2-AMINO-2-CARBOXYETHYL)THIO]-5-HYDROXY-7,9,11,14-EICOSOTETRAENOIC ACID
LEUPEPTIN	N-ACETYL-LEUCYL-LEUCYL-ARGININALDEHYDE
5nor-L-FUCO-1-DEOXY-NOJIRIMYCIN	1,4-DIDEOXY-1,4-IMINO-D-ARABINITOL
LIAROZOLE	R 75251
LIBENZAPRIL	3[(5-AMINO-1-CARBOXY-1S-PENTYL)AMINO]2,3,4,5-TETRAHYDRO-2-OXO-3S-1H-1-BENZAZEPENE-1-ACETIC ACID
LIBENZAPRIL	CGS 16617
LIDOCAINE	2-(DIETHYLAMINO)-N-(2,6-DIMETHYLPHENYL)-ACETAMIDE
LIGNOCERIC ACID	TETRACOSANOIC ACID
LILLY 18947	2-[(3,5-DICHLORO[1,1′-DIPHENYL]-2-YL)OXY]-N, N-DIETHYLETANAMINE, HYDROBROMIDE
LILLY 51641	N-[2-(2-CHLOROPHENOXY)ETHYL]-CYCLOPROPANAMINE
LILLY 53325	2-[(3,5-DICHLORO-[1,1′-BIPHENYL]-2-YL)OXY]-ETHANAMINE, HYDROBROMIDE
LIMP	LARGE INHIBITOR OF MEMTALLOPROTEINASES
LINETASTINE	TMK 688
LINOLEIC ACID	(Z, Z)-9,12-OCTADECADIENOIC ACID
LINURON	N′-(3,4-DICHLOROPHENYL)-N-METHOXY-N-METHYLUREA
LIPOAMIDE	1,2-DITHIOLANE-3-PENTANAMIDE
LIPOCORTIN I	ANNEXIN I
LIPOIC ACID	1,2-DITHIOLANE-3-PENTANOIC ACID
LIPSTATIN	N-FORMYL-L-LEUCINE [2S, -[2α(1R*,3Z, 6Z),3β]]-1[(3-HEXYL-4-OXO-2-OXETANYL)METHYL]-3,6-DODECADIENYL ESTER
LISINOPRIL	MK 521
LISINOPRIL	1-[N2-(1-CARBOXY-3-PHENYLPROPYL)-L-LYSYL]-L-PROLINE
LITHOCHOLIC ACID	(3α,5β)-3-HYDROXY-CHOLAN-24-OIC ACID
LND 623	14β-AMINO 3β-RHAMNOSYL-20(R)-β-OL, 5β-PREGNAN
LOVASTATIN	(+)-(1S,3R,7S,8S,8aR)-1,2,3,7,8,8a-HEXAHYDRO-8-[2-[(2R, 4R)-TETRAHYDRO-4-HYDROXY-6-OXO-2H-PYRAN-2-YL]-ETHYL]-1-NAPHTHYL 2-METHYLBUTANOATE
LP 149	AC-NAPHTHYLALANINYL-VAL-STATINYL-GLU-NAPHTHYLALANINE-NH_2
LS 82556	(S)-3-N-(METHYLBENZYL)CARBAMOYL-5-PROPIONYL-2,6-LUTIDINE

Trivial name	Synonym or systematic name
LUMICHROME	7,8-DIMETHYL-BENZO[G]PTERIDINE-2,4(1H, 3H)-DIONE
LUMIFLAVINE	7,8,10-TRIMETHYL-BENZO[G]PTERIDINE-2,4(3H, 10H)-DIONE
LUMINOL	3-AMINOPHTHALHYDRAZIDE
LUTEOSKYRIN	(1S, 1′S, 2R, 2′R, 3S, 3′S, 9AR, 9′AR,)-8,8′-DIHYDROXY-RUGULOSIN
LY 181984	N-(4-METHYLPHENYLSULPHONYL)-N′-(4-CHLOROPHENYL)UREA
LY 186126	1,3-DIHYDRO-1,3,3-TRIMETHYL-5-(1,4,5,6-TETRAHYDRO-6-OXO-3-PYRIDAZINYL)-2H-INDOL-2-ONE
LY 186655	5,6-DIETHOXYBENZO(b)THIOPHENE-2-CARBOXYLYC ACID
LY 186655	TIBENELAST
LY 191704	trans-8-CHLORO-1,4,4a,5,6,10b-HEXAHYDRO-4-METHYLBENZO[f]QUINOLIN-3(2H)-ONE
LY 191704	8-CHLORO-4-METHYL-1,2,3,4,4a,5,6,10b-OCTAHYDROBENZO[f]QUINOLIN-3(2H)ONE
LY 195115	INDOLIDAN
LY 207320	6-METHYLENE-4-PREGENENE-3,20-DIONE
LY 231514	N-[4-[2-[2-AMINO-4,7-DIHYDRO-4-OXY-3H-PYRROLO[2,3-d]PYRIMIDINE-5-YL)ETHYL]BENZOYL]-L-GLUTAMIC ACID
LY 233569	N-HYDROXY-N-METHYL-3-[2-(METHYLTHIO)PHENYL]-2-PROPENAMIDE
LY 245769	(1S,2R)-5-{3-[1-HYDROXY-15,15,15-TRIFLUORO-2-(2-1H-TETRAZOL-5-YLETHYLTHIO)PENTADECA-3-(E),5(Z)-DIPHENYL}-1H-TETRAZOLE
LY 254155	LOMETREXOL
LY 294002	2-(4-MORPHOLINYL)-8-PHENYL-4H-1-BENZOPYRAN-4-ONE
LY 294468	D-METHYL-PHE-PRO-ARGINAL
LY 300046	N-(2-PYRIDYL)-N′-(5-BROMO-2-PYRIDYL)THIOUREA
LY 303366	ECHINOCANDIN B
LY 309887	6R-2′,5′-THIENYL-5,10-DIDEAZATETRAHYDROFOLIC ACID
LY 333531	(S)-3-[[(DIMETHYLAMINO)METHYL]-10,11,14,15-TETRAHYDRO-4,9: 16,21-DIMETHENO-1H, 13H-DIBENZO[e,k]-PYRROLO[3,4-h][1,4,13]OXO DIAZACYCLOHEXADECENE
LY 745337	6-[(2,4-DIFLUOROPHENYL)THIO]-5-METHANESULPHONAMIDE-1-INDANONE
LY 78335	2,3-DICHLORO-α-METHYLBENZYLAMINE
LY 806303	METHYL-3-(2-METHYL-1-OXOPROPOXY)[1]BENZOTHIENO[3,2-b]FURAN-2-CARBOXYLATE
LY 83583	6-ANILINO-5,8-QUINOLINEDIONE
LYSERGIC ACID	9,10-DIDEHYDRO-6-METHYL-ERGOLINE-8-CARBOXYLIC ACID
α2M	α2-MACROGLOBULIN
M 16209	1-(3-BROMOBENZO[b]FURAN-2-YLSULPHONYL)HYDANTOIN
M 16287	1-(3-CHLOROBENZO[b]FURAN-2-YLSULPHONYL)HYDANTOIN
M 79175	2-METHYL-6-FLUOROSPIROCHROMAN-4,5′-IMIDAZOLIDINE-2′,4′-DIONE
4-MA	N,N-DIETHYL-4-METHOXY-3-OXO-4-AZA-5-ANDROSTANE-17-CARBOXAMIDE
4-MA	17β-N, N-DIETHYLCARBAMOYL-4-METHYL-4-AZA-5α-ANDROSTAN-3-ONE
MACROCORTIN	LIPOMODULIN
MALEIC ACID	(Z)-2-BUTENEDIOIC ACID
MALIC ACID	HYDROXY-BUTANEDIOIC ACID
MALLOTOCHROMENE	8-ACETYL-5,7-DIHYDROXY-6-(3-ACETYL-2,4-DIHYDROXY-5-METHYL-6-METHOXYBENZYL)-2,2-DIMETHYL-CHROMENE
MALLOTOJAPONIN	3-(3,3-(DIMETHYLALLYL)-5-(3-(ACETYL-2,4-DIHYDROXY-5-METHYL-6-METHOXYBENZYL)-PHLORACETO-PHENONE
MALONIC ACID SEMI-ALDEHYDE	3-OXOPROPANOIC ACID
MALOTILATE	DIISOPROPYL-1,3-DITHIOL-2-YLIDENE MALONATE
MALTITOL	4-O-α-D-GLUCOPYRANOSYL-D-GLUCITOL
MANGOSTIN	1,3,6-TRIHYDROXY-7-METHOXY-2,8-BIS(3-METHYL-2-BUTENYL)-9H-XANTHEN-9-ONE
γ-MANGOSTIN	1,3,6,7-TETRHYDROXY-2,8-BIS(3-METHYL-2-BUTENYL)-9H-XANTHEN-9-ONE
MANOALIDE	3,7-DI(HYDROXYMEHYL-4-HYDROXY-11-METHYL-13-(2,6,6-TRIMETHYLCYCLOHEXENYL)-2,6,10-TRIDECATRI-ANOIC ACID-LACTONE
MAOEA	5′-DEOXY-5′-[n-METHYL-n-[2-AMINOOXY)ETHYL]AMINO]ADENOSINE
MAOPA	5′-DEOXY-5′-[n-METHYL-n-[3-(AMINOOXY)PROPYL]AMINO]ADENOSINE
α-MAPI	MICROBIAL ALKALINE PROTEASE INHIBITOR α
β-MAPI	MICROBIAL ALKALINE PROTEASE INHIBITOR β
MARGINOIC ACID	HEPTADECANOIC ACID
MAZ 525	2-[(1H-BENZIMIDAZOL-2-YL)SULPHINYLMETHYL]-4-DIMETHYLAMINO-5-PYRIMIDINE CARBOXYLATE

Trivial name	Synonym or systematic name
MB 22948	2-o-PROPOXYPHENYL-8-AZAPURINE-6-ONE
MB 22948	ZAPRINAST
MB 39279	5-AMINO-4-CYANO-1-(2,6-DICHLORO-4-TRIFLUOROMETHYLPHENYL)PYRAZOL
MB 35902A	2,4-DIAMINO-6-METHYL-5,3-(3-NITROPHENOXY)PROP-1′-YLOXYPYRIMIDINE
MCI 154	6-[4-(4′-PYRIDYL)AMINOPHENYL]-4,5-DIHYDRO-3(2H)-PYRIDAZINONE
MCI 9038	ARGATROBAN
MCI 9038	(2R,4R)-4-METHYL-1-[N2-[(3-METHYL-1,2,3,4-TERTAHYDRO-8-QUINOLINYL)SULPHONYL]-L-ARGINYL]-2-PIPERIDI-NECARBOXYLIC ACID
MCPA-CoA	METHYLENECYCLOPROPYLACETYL-CoA
MD 805	ARGATROBAN
MD 805	(2R,4R)-4-METHYL-1-[N2-[(3-METHYL-1,2,3,4-TERTAHYDRO-8-QUINOLINYL)SULPHONYL]-L-ARGINYL]-2-PIPERIDI-NECARBOXYLIC ACID
MD 230254	5-[4-(BENZYLOXY)PHENYL]-3-(2-CYANOETHYL)-1,3,4-OXADIAZOL-2(3H)-ONE
MD 240929	ALMOXATONE
MD 370503	BEFLOXATONE
MD 780515	CIMOXATONE
MDL 100173	[4S-[4α,7α(R),12bβ]]-7-[[2-(THIO-1-OXO-3-PHENYLPROPYL]AMINO]-1,2,3,4,6,7,8,12b-OCTAHYDRO-6-OXO-PYRIDO-[2,1-a][2]BENZAZEPINE-4-CARBOXYLIC ACID
MDL 101731	2′-DEOXY-2′-(FLUOROMETHYLENE)CYTIDINE
MDL 17043	1,3-DIHYDRO-4-METHYL-5-[4-(METHYLTHIO)-BENZOYL]-2H-IMIDAZOL-2-ONE
MDL 17043	ENOXIMONE
MDL 18962	10-PROPARGYL-4-ESTRENE-3,17-DIONE
MDL 19205	PIROXIMONE
MDL 201053	Z-L-PHE-L-ALA-CH_2F
MDL 25637	2,6-DIDEOXY-2,6-IMINO-7-O-(β-D-GLUCOPYRANOSYL)-D-GLYCERO-L-GLUCOHEPTITOL
MDL 27032	4-PROPYL-5-(4-PYRIDINYL)-2(3H)-OXAZOLONE
MDL 28170	Z-VAL-PHE-H
MDL 28574	6-O-BUTYLCASTANOSPERIMINE
MDL 28815	N-(1,5,9-TRIMETHYLDECYL)-4α,10-DIMETHYL-8-AZA-trans-DECAL-3β-OL
MDL 28842	MDL 28842
MDL 28842	(Z)-5′-FLUORO-4′,5′-DIDEHYDRO-5′-DEOXYADENOSINE
MDL 72145	(E)-2-(3,4-DIMETHOXYPHENYL)-3-FLUOROALLYLAMINE
MDL 72161	(E)-2-PHENYL-3-FLUOROALLYLAMINE
MDL 72274	(E)-2-PHENYL-3-CHLOROALLYLAMINE
MDL 72392	(E)-β-FLUOROMETHYLENE-m-TYRAMINE
MDL 72483	(4S)-4-AMINO-5,6-HEPTADIENOIC ACID
MDL 72527	N,N′-BIS(BUTA-2,3-DIENYL)BUTANE-1,4-DIAMINE
MDL 72912	5-FLUROMETHYLORNITHINE
MDL 73745	2,2,2-TRIFLUORO-1-(3-TRIMETHYLSILYL-PHENYL)ETHANONE
MDL 73756	D-PHE-PRO-ARG-CHLORMETHYLKETONE
MDL 73811	5′-{[(Z)-4-AMINO-2-BUTENYL]METHYLAMINO}-5′-DEOXYADENOSINE
MDL 73945	1,5-DIDEOXY-1,5-[(6-DEOXY-1-O-METHYL-6-α,D-GLUCOPYRANOSYL)IMINO]-D-GLUCITOL
MDL 74428	{2-[2-[(2-AMINO-1,6-DIHYDRO-6-OXO-9H-PURIN-9-YL)METHYL]PHENYL]ETHENYL}PHOSPHONIC ACID
MDL 12330A	N-(cis-2-PHENYLCYCLOPENTYL)AZACYCLO-TRIDECAN-2-IMINE HCl
MDL 72274A	(E)-2-PHENYL-3-CHLOROALLYLAMINE
MDL 72974A	(E)-4-FLUORO-β-FLUOROMETHYLENE BENZENE BUTANAMINE HCl
MDL 72974A	(E)-2-(4-FLUOROPHENETHYL)-3-FLUOROALLYLAMINE
MDL 72974A	MOFEGILINE
MECLOFENAMIC ACID	2-[(2,6-DICHLORO-3-METHYLPHENYL)AMINO]BENZOIC ACID
MEDIPINE	1-METHYL-4,4-DIPHENYLPIPERIDINE
MEDORINONE	5-METHYL-1,6-NAPHTHYRIDIN-2(1H)-ONE
MEGESTEROL ACETATE	17α-ACETOXY-6-METHYL-PREGNA-4,6-DIENE-3,20-DIONE
MELEZITOSE	O-α-D-GLUCOPYRANOSYL-[1->3]-O-β-D-FRUCTOFURANOSYL-[2->1]-α-D-GLUCOPYRANOSIDE
MELINAMIDE	N-(α-METHYLBENZYL)LINOLEAMIDE
MEN 10979	6,7-DIHYDRO-7-METHYL-12-ETHYLPYRIDO[2,3-b]PYRIDO[2,3-4,5]THIENO[2,3-f][1,4]DIAZEPIN-6[12H]THIONE
MENADIONE	2-METHYL-1,4-NAPHTHALENEDIONE

Trivial name	Synonym or systematic name
MENTAX	BUTENAFINE
MEPACRINE	QUINACRINE
MER 29	4-CHLORO-α-[4-[2-(DIETHYLAMINO)ETHOXY]PHENYL]-α-(4-METHYLPHENYL)-BENZENETHANOL
MERSALYL	[3-[[2-(CARBOXMETHOXY)BEZOYL]AMINO]-2-METHOXYPROPYL]HYDROXYMERCURY MONOSODIUM SALT
MESACONIC ACID	(E)-2-METHYL-2-BUTENEDIOIC ACID
MET 88	3-(2,2,2-TRIMETHYLHYDRAZINIUM)PROPIONIC ACID
METHICILLIN	2,6-DEMETHOXYPHENYLPENICILLIN
METHICILLIN	[2S-(2α,5α,6β)]-6-[(2,6-DIMETHOXYBENZOYL)AMINO]-3,3-DIMETHYL-7-OXO-4-THIA-1-AZABICYCLO[3.2.O]HEPTANE-2-CARBOXYLIC ACID
METHIMAZOLE	1,3-DIHYDRO-1-METHYL-2H-IMIDAZOLE-2-THIONE
METHOTREXATE	N-[4-[[(2,4-DIAMINO-6-PTERIDINYL)METHYL]METHYLAMINO]BENZOYL]-L-GLUTAMIC ACID
METHOXY E 3810	5-METHOXY-2-[{4-(3-METHOXYPROPOXY)-3-METHYLPYRIDIN-2-YL}-METHYLSULPHINYL]-1H-BENZIMIDAZOLE
2-METHYLCHROMONE	2-METHYL-1,4-PHTHALAZINEDIONE
METHYLDOPA	3-HYDROXY-α-METHYL-L-TYROSINE
γ-METHYLENEAMINOPTERIN	N-(4-AMINO-4-DEOXYPTEROYL)-4-METHYLENE-DL-GLUTAMIC ACID
γ-METHYLENEMETHOTHREXATE	N-(4-AMINO-4-DEOXY-N10-METHYLPTEROYL)-4-METHYLENE-DL-GLUTAMIC ACID
METHYLGLYOXAL	2-OXOPROPANAL
METHYLMALONICACID SEMIALDEHYDE	2-METHYL-3-OXOPROPANOIC ACID
35-METHYLOKADAIC ACID	DINOPHYSISTOXIN-1
35-METHYLOKADAIC ACID	DINOPHYSISTOXIN-1
METHYLPALMOXIRATE	2-TETRAGLYCYLGLYCIDATE
METHYLTHIOTUBERCIDIN	METHYLTHIO-7-DEAZAADENOSINE
METOPIRON	2-METHYL-1,2-DI-3-PYRIDINYL-1-PROPANONE
METRONIDAZOLE	2-METHYL-5-NITRO-1H-IMIDAZOLE-1-ETHANOL
METYRAPONE	2-METHYL-1,2-BIS-(3-PYRIDYL)-1-PROPANONE
MEVASTIN	2-METHYLBUTANOIC ACID 1,2,3,7,8,8A-HEXAHYDRO-7-METHYL-8-[2-(TETRAHYDRO-4-HYDROXY-6-OXO-2H-PYRAN-2-YL)ETHYL]-1-NAPHTHALENYL ESTER
MEVINPHOS	DIMETHYL-1-METHOXYCARBONYL-1-PROPEN-2-YL PHOSPHATE
MF 201	HEPTYLPHYSOSTIGMINE
MF 268	(3aS, 8aR)-1,2,3,3a,8,8a-HEXAHYDRO-1,3a,8-TRIMETHYLPYRROLO[2,3-b]INDOL-5-OL[8-(cis2,6-DIMETHGYLMORPHOLIN-4-YL)OCTYL CARBAMATE BITARTRATE
MFMA	MONOFLUORO(2)METHYLARGININE
Δ-MFMA	MONOFLUORO((E)-2)METHYLDEHYDROARGININE
MFT 279	3-[N-(2-CHLOROBENZYL)AMINO]-6-(1H-IMIDAZOL-1-YL)PYRIDAZINE
MGBG	METHYLGLYOXAL BIS(GUANYLHYDRAZONE)
MGP	METHYL(6-O)GLUCOSEPOLYSACCHARIDE
MHZPA	5′-DEOXY-5′-[n-METHYL-n-(3-HYDRAZINOPROPYL)AMINO]ADENOSINE
MIDIN-5-ONE	6-{2-[(4-FLUOROPHENYL)PHENYLMETHYLENE]-1-PIPERIDINYL}-ETHYL-7-METHYL 5H THIAZOLO[3,2a]PYRI
MILACEMIDE	2n-PENTYLAMINOACETAMIDE
MILDRONATE	3-(2,2,2-TRIMETHYLHYDRAZINIUM)PROPIONIC ACID
MILRINONE	1,6-DIHYDRO-2-METHYL-6-OXO-[3,4′-BIPYRIDINE]-5-CARBONITRILE
MIMOSINE	β-N(3-HYDROXY-4-PYRIDONE)-α-AMINOPROPIONIC ACID
MIMOSINE	(S)-α-AMINO-3-HYDROXY-4-OXO-1(4H)-PYRIDINEPROPANOIC ACID
MINOXIDIL	6-AMINO-1,2-DIHYDRO-1-HYDROXY-2-IMINO-4-PIPERIDINOPYRIMIDINE
MIPAFOX	BIS-MONOISOPROPYLAMINOFLUOROPHOSPHATE
MJ 33	1-HEXADECYL-3-TRIFLUOROETHYLGLYCERO-sn-2-PHOSPHOMETHANOL
MJ 133	1-HEXADECYL-3-(TRIFLUOROMETHYL)-sn-GLYCERO-2-PHOSPHOMETHANOL
MK 386	4,7β-DIMETHYL-4-AZA-5α-CHOLESTAN-3-ONE
MK 417	S(+),[5,6-DIHYDRO-4H-(ISOBUTYLAMINOTHIENO-(2,3-B)-THIOPYRAN-2-SULPHONAMIDE-7.7-DIOXIDE
MK 417	SEZOLAMIDE
MK 421	ENALAPRIL MALEATE
MK 507	(–)-(4S,6S)-4-ETHYLAMINO-5,6-DIHYDRO-6-METHYL-4H-THIENO[2,3-b]THIOPYRAN-2-SULPHONAMIDE 7,7-DIOXIDE
MK 0521	N-α-[(S)-1-CARBOXY-3-PHENYLPROPYL]-L-LYSYL-L-PROLINE

Trivial name	Synonym or systematic name
MK 538	PONALRESTAT
MK 591	3-[1-(4-CHLOROBENZYL)-3-(t-BUTHYLTHIO)-5-(QUINOLIN-2-YL-METHOXY)-INDOL-2-YL]-2,2-DIMETHYL PROPANOIC ACID
MK 639	INDINAVIR
MK 639	L 735524
MK 733	SIMVASTATIN
MK 886	3-[1-(4-CHLOROBENZYL)-3-tert-BUTYLTHIO-5-ISOPROPYLINDOL-2-YL]-2,2-DIMETHYLPROPANOIC ACID
MK 906	N-[1,1-DIMETHYL]-3-OXO-4-AZA-5α-ANDROST-1-ENE-17β-CARBOXAMIDE
MK 906	FINASTERIDE
MK 927	R(–),S(+)-5,6-DIHYDRO-4H-4(ISOBUTYLAMINO)THIENO(2,3-B)THIOPYRAN-2-SULPHONAMIDE-7,7-DIOXIDE
S-(+) MK 927	MK 417
ML 7	1-(5-IODONAPHTHALENE-1-SULPHONYL)-1H-HEXAHYDRO-1,4-DIAZEPINE HCl
ML 9	1-(5-CHLORONAPHTHALENE-1-SULPHONYL)-1H-HEXAHYDRO-1,4-DIAZEPINE HCl
ML 3000	2,2-DIMETHYL-6-(4-CHLOROPHENYL)-7-PHENYL, 2,3-DIHYDRO-1H-PYRROLIZINE-5-YL)ACETIC ACID
ML 236A	[1S-[1α(4S*,6S*),2α,8β,8aα]]-6-[2-(1,2,6,7,8,8a-HEXAHYDRO-8-HYDROXY-2-METHYL-1-NAPHTHALENYL)ETHYL]-TETRAHYDRO-4-HYDROXY-2H-PYRAN-2-ONE
ML 236B	COMPACTIN
ML 236B	MEVASTIN
MLR 52	4′-DEMETHYLAMINO-4′,5′-DIHYDROXYSTAUROSPORINE
MNBS	METHYL p-NITROBENZENE SULPHONATE
McN 3716	2-TETRAGLYCYLGLYCIDATE
MOCLOBEMIDE	p-CHLORO-N-[2-MORPHOLINOETHYL]BENZAMIDE
MONACOLIN	CS 514
MONACOLIN MB-530B	MEVINOLIN
MONOCROTOPHOS	PHOSPHORIC ACID DIMETHYL[1-METHYL-3-(METHYLAMINO]-3-OXO-n-PROPENYL)ESTER
MORANTEL	1,4,5,6-TETRAHYDRO-1-METHYXL-2-[2-(3-METHYL-2-THIENYL)ETHENYL]PYRIMIDINE
MORANYL	SURAMIN
MORIN	2-(2,4-DIHYDROXYPHENYL)-3,5,7-TRIHYDROXY-4H-1-BENZOPYRAN-4-ONE
MOTAZIPONE	4,5-DIHYDRO-6-(4-(1H-IMIDAZOL-1-YL)-2-THIENYL]-5-METHYL-3-(2H)PYRIDAZINONE
MOTUPORIN	NODULRAIN V
MOXIPRIL	2[1-ETHOXY(CARBONYL)-3-PHENYLPROPYL]AMINO-1-OXOPROPYL]-6,7-DIMETHOXY-1,2,3,4-TETRAHYDROISO-QUINOLOINE-3-CARBOXYLIC ACID
MPP+	1-METHYL-4-PHENYLPYRIDINIUM
MPTP	1-METHYL-4-PHENYL-1,2,3,6-TETRAHYDROPYRIDINE
MR 889	N-(2-OXO-2,3,4,5-)-TETRAHYDRO-3-THIENYL)-2-(2-THEONYLTHIO)PROPIONAMIDE
MR 889	n-[S(2-THIOPHENECARBONYL)-2-MERCAPTOPROPIONYL]HOMOCYSTEINE LACTONE
MR 889	2-[3-THIOPHENCARBOXYTHIO]-N-[DIHYDRO-2(3H)-THIOPHENONE-3-IL]-PROPIONAMIDE
MS 857	4-ACETYL-1-METHYL-7[4-PYRIDYL]-5,6,7,8-TETRAHYDRO-3(2H)-ISOQUINOLINONE
MST 16	4,4′-(1,2-ETHANEDIYL)-BIS-(1-ISOBUTOXYCARBONYLOXYMETHYL-2,6-PIPERAZINEDIONE)
MUSCIMOL	5-(AMINOMETHYL)-3(2H)-ISOXAZOLONE
MUSCUS PROTEINASE INHIBITOR	SECRETORY LEUKOCYTE PROTEINASE INHIBITOR
MY 5445	1-(3-CHLOROANILINO)-4-PHENYLPHTHALAZINE
MYCOPHENOLIC ACID	(E)-6-(1,3-DIHYDRO-4-HYDROXY-6-METHOXY-7-METHYL-3-OXO-5-ISOBENZOFURANYL)-4-METHYL-4-HEXENOIC ACID
MYOCRISIN	SODIUM GOLD THIOMALATE
MYRICETIN	3,5,7-TRIHYDROXY-2-(3,4,5-TRIHYDROXYPHENYL)-4H-1-BENZOPYRAN-4-ONE
MYRICITRIN	3-[(6-DEOXY-α-L-MANNOPYRANOSYL)OXY]-5,7-DIHYDROXY-2-(3,4,5-TRIHYDROXYPHENYL)-4H-1-BENZOPYRAN-4-ONE
MYRIOCIN	(2S, 3R, 4R)(E)-2-AMINO-3,4-DIHYDROXY-2-HYDROXYMETHYL-14-OXOECOS-6-ENOIC ACID
MYRYSTIC ACID	TETRADECANOIC ACID
MYXOTHIAZOL	7-[2′-(1,6-DIMETHYL-2,4-HEPTADIENYL)[2,4′-BITHIAZOL]-4-YL]-3,5-DIMETHYOXY-4-METHYL-2,6-HEPTADIEN-AMIDE
N 751	2-CHLORO-9-(3-[PIPERAZINYL]PROPYLIDENE)THIOXANTHENE
4N-62	1[n,o-BIS(5-ISOQUINOLINESULPHONYL)-n-METHYL-l-TYROSYL]-4-PHENYLPIPERAZINE
NA 382	N-ETHOXYCARBONYL-7-OXOSTAUROSPAORINE
N-ACETYLHYDROXYLAMIN	ACETOHYDROXAMIC ACID

Trivial name	Synonym or systematic name
NAD	NICOTINAMIDE DINUCLEOTIDE
NAFAMOSTAT MESILATE	FUT 175
NAFAZATROM	2,4-DIHYDRO-5-METHYL-2-[2-(2-NAPHTHALENYLOXY)ETHYL]-3H-PYRAZOL-3-ONE
NALIDIXIC ACID	1-ETHYL-1,4-DIHYDRO-7-METHYL-4-OXO-1,8-NAPHTHYRIDINE-3-CARBOXYLIC ACID
NAPROXEN	6-METHOXY-α-METHYLNAPHTHALENE-2-ACETIC ACID
NAZLININ	1-(4-BUTYLAMINO)-1,2,3,4-TETRAHYDRO-β-CARBOLINE
NB 598	(E)-N-ETHYL-N-(6,6-DIMETHYL-2-HEPTEN-4-YNYL)-3-[(3,3′-BITHIOPHEN-5-YL)METHOXY]BENZENEMETHAN-AMINE
NBD-Cl	7-CHLORO-4-NITROBENZOFURAZAN
NBD-CL	7-CHLORO-4-NITROBENZO-2-OXA-1,3-DIAZOLE
NBMPR	6[(4-NITROBENZYL)THIO]-9-β-D-RIBOFURANOSYLPURINE
NBQ 59	(1-PROPENYL)-3-NITROBENZIMIDAZOLO[3,2-a]QUINOLINIUM CHLORIDE
NBTF	7-[(NITROBENZYL)-THIO]-3-(β-D-RIBOFURANOSYL)PYRAZOLO[4,3-d]PYRIMIDINE
NBTGF	5-AMINO-7-[(NITROBENZYL)-THIO]-3-(β-D-RIBOFURANOSYL)PYRAZOLO[4,3-d]PYRIMIDINE
NBTI	NITROBENZYLTHIOINOSINE
NC 1300	2-[(2-DIMETHYLAMINOBENZYL)SULPHINYL]BENZIMIDAZOLE
NCDC	2-NITRO-4-CARBOXYPHENYL-N, N-DIPHENYL-CARBAMATE
NCS 665564	2-ACETYL-7-BROMO-1-PHENYL-1,2,3,4-TETRAHYDRO-β-CARBOLINE
NDGA	NORDIHYDROGUAIARETIC ACID
NEBULARINE	9-β-D-RIBOFURANOSYL-9H-PURINE
NEM	N-ETHYLMALEIMIDE
NEOCUPROINE	2,9-DIMETHYL-1,10-PHENANTHROLINE
NEOMYCIN A	2-DEOXY-4-O-(2,6-DIAMINO-2,6-DIDEOXY-α-D-GLUCOPYRANOSYL)-D-STREPTAMINE
NEOPTERIN	2-AMINO-6-(1,2,3-TRIHYDROXYPROPYL)-4(3H)-PTERIDINONE
NEOSTIGMINE	3-[[(DIMETHYLAMINO)CARBONYL]OXY]N, N, N, -TRIMETHYL-BENZENAMINIUM
NEPICASTAT	S-5-AMINOMETHYL-1-(5,7-DIFLUZORO-1,2,3,4-TETRAHYDRONAPHTHYL)-1,3-DIHYDROIMIDAZOLE-2-THIONE HCl
NEPLANOCIN A	[1S-(1α,2α,5β)]-5-(6-AMINO-9H-PURIN-9-YL)-3-(HYDROXYMETHYL)-3-CYCLOPENTENE-1,2-DIOL
N-ETHYLMALEIMIDE	1-ETHYL-1H-PYRROLE-2,5-DIONE
NETROPSIN	CONGOCIDINE
NEVIRAPINE	11-CYCLOPROPYL-5,11-DIHYDRO-4-METHYL-6H-DIPYRIDO[3,2-b: 2′,3′-e][1,4]DIAZEPINONE
NICOTIANAMINE	N-[N-(3-AMINO-3-CARBOXYPROPYL)-3-AMINO-3-CARBOXYPROPYL]AZETIDINE-2-CARBOXYLIC ACID
NICOTINE	(S)-3-(1-METHYL-2-PYRROLIDINYL)-PYRIDINE
NIDROXYZONE	1-(2-HYDROXYETHYL)-2-[(5-NITRO-2-FURANYL)METHYLENE]HYDRAZINECARBOXAMIDE
NIFEDIPINE	1,4-DIHYDRO-2,6-DIMETHYL-4-(2-NITROPHENYL)-3,5-PYRIDINEDICARBOXYLIC ACID DIMETHYL ESTER
NIK 247	9-AMINO-2,3,5,6,7,8-HEXAHYDRO-1H-CYCLOPPENTA-(b)-QUINOLINE MONOHYDRTAE HCl
NIKKOMYCIN B	5-[[2-AMINO-4-HYDROXY-4-(4-HYDROXYPHENYL)-3-METHYL-1-OXOBUTYL]AMINO]-1,5-DIDEOXY-1-(4-FORMYL-2-,3-DIHYDRO-2-OXO-1H-IMIDAZOL-1-YL)-HEXOFURANURIC ACID
NIMESULIDE	4-NITRO-2-PHENOXYMETHANESULFONANILIDE
NIPECOTIC ACID	3-PIPERIDINECARBOXYLIC ACID
NITECAPONE	3-(3,4-DIHYDROXY-5-NITROBENZYLIDENE)-2,4-PENTANEDIONE
NITROFEN	2,4-DICHLORO-1-(4-NITROPHENOXY)BENZENE
NITROFURANTOIN	1-[[(5-NITRO-2-FURANYL)METHYLENE]AMINO-2,4-IMIDAZOLIDINEDIONE
NITROFURAZONE	2-[(5-NITRO-2-FURANYL)METHYLENE]-HYDRAZINECARBOXAMIDE
NK 109	2,3-(METHYLENEDIOXY)-5-METHYL-7-HYDROXY-8-METHOXYBENZO[c]PHENANTHRIDINIUM HYDROGEN-SULPHATE HYDRATE
N-METHYLNORSALSOLINOL	2-METHYL-6,7-DIHYDROXY-1,2,3,4-TETRAHYDROISOQUINOLINE
NMN	NICOTINAMIDE ADENINE MONONUCLEOTIDE
NOJIRIMYCIN	5-AMINO-5-DEOXY-D-GLUCOPYRANOSE
2-NO_2-6-ME-SULPHON-ANILIDE	1,2,4-TRIAZOLO-(1,5-a)-2,4-DIMETHYL-3-(N-SUKPHONYL-(2-NITRO-6-METHYLANILINE))-1,5-PYRIMIDINE
NONACTIN	2,5,11,14,20,23,29,32-OCTAMETHYL-4,13,22,31,37,38,39,40-OCTAOXAPENTACYCLO[32.2.1.1 7,10.1 16,19.1 25,28]TETRA-CONTANE-3,12,21,30-TETRONE
NORADRENALIN	4-(2-AMINO-1-HYDROXYETHYL)-1,2-BENZENEDIOL
2′-NORDEOXYGUANOSINE	9-(1,3-DIHYDROXY-2-PROPOXYMETHYL)GUANINE
NORDIHYDROGUAIARETIC ACID	4,4′-[2,3-DIMETHYL-1,4-BUTANEDIYL]BIS-1,2-BENZENEDIOL

Trivial name	Synonym or systematic name
NORVALINE	2-AMINOPENTANOIC ACID
NORVIR	RITONAVIR
NORWOGONIN	5,7,8-TRIHYDROXYFLAVONE
NOVALDEX	TMOXIFEN CITRATE
NOVOBIOCIN	N-[7-[[3-O-(AMINOCARBONYL)-6-DEOXY-5-C-METHYL-4-O-METHYL-β-L-LYXO-HEXOPYRANOSYL]OXY]-4-HYDROXY-8-METHYL-2-OXO-2H-1-BENZOPYRAN-3-YL]-4-HYDROXY-3-(3-METHYL-2-BUTENYL)-BENZAMIDE
NPC 15437	2,6-DIAMINO-N-([1-(OXOTRIDECYL)-2-PIPERIDINYL]METHYL)-HEXANAMIDE
NQ-Y15	2-[(4-ACETYLPHENYL)AMINO]-3-CHLORO-1,4-NAPHTHALENEDIONE
NS 398	N-[2-CYCLOHEXYLOXY-4-NITROPHENYL]METHANESULPHONAMIDE
NS 2028	4H-8-BROMO-1,2,4-OXADIAZOLO(3,4-d)BENZ(b)(1,4)OXAZIN-1-ONE
NSC 11905	LAPACHOL
NSC 143095	PYRAZOFURIN
NSC 224131	PALA
NSC 242557	(3,5-DI-tert.-BUTYL-4-HYDROXYBENZILIDENE)MALONITRILE
NSC 249992	AMSACRINE
NSC 264137	2-METHYL-9-HYDROXY ELLIPTICINIUM
NSC 26980	MITOMYCIN
NSC 284356	MITINDOMIDE
NSC 336628	MERBARONE
NSC 336628	5-[N-PHENYLCARBOXAMIDO]-2-THIOBARBITURIC ACID
NSC 340847	2β-D-RIBOFURANOSYLSELENAZOLE-4-CARBOXAMIDE
NSC 368390	BREQUINAR
NSC 375575	CYCLOPENTYLCYTOSINE
NSC 51143	2,3-DIHYDRO-1H-IMIDAZOL(1,2-b)PYRAZOLE
NSC 51143	IMPY
NSC 609699	TOPOTECAN
NSC 624231	2-NITROPHENYLPHENYLSULPHONE
NSC 625487	1-(2′-,6′-DIFLUOROPHENYL)-1H, 3H-THIZOLO[3,4-a]BENZIMIDAZOLE
NSC 625487	THIAZOLOBENZIMIDAZOLE
NSC 649890	FLAVOPIRIDOL
NSD 1015	3-HYDROXYBENZYL HYDRAZINE
NSD 1055	5-[(AMINOOXY)METHYL]-2-BROMO-PHENOL
NU 1025	8-HYDROXY-2-METHYLQUINAZOLIN-4(3H)-ONE
OESTRONE	ESTRONE
OFLOXAZIN	(–)-9-FLUORO-3(S)-METHYL-10-(4-METHYL-1-PIPERAZINYL)-7-OXO-2,3-DIHYDRO-7H-PYRIDO[1,2,3-de]-1,4-BENZOXAZINE-6-CARBOXYLIC ACID
OK 1035	3-CYANO-5-(4-PYRIDYL)-6-HYDRAZONOMETHYL-2-PYRIDONE
OKY 046	[E]-3-[p-(1H-IMIDAZOL-1-YL-METHYL)-PHENYL]-2-PROPANOIC ACID
OKY 046	OZAGREL
OKY 1581	(E)-2-METHYL-3-[4-(3-PYRIDINYLMETHYL)PHENYL]-2-PROPENOIC ACID
OLEACEIN	2-(3,4-DIHYDROXYPHENYL)ETHYL-4-FORMYL-3-(2-OXOETHYL)-4E-HEXENOATE
OLEIC ACID	(Z)-9-OCTADECENOIC ACID
OLIVANIC ACID	3-[(2-AMINOETHENYL)THIO]-6-(1-HYDROXYETHYL)-7-OXO-1-AZABICYCLO[3.2.O]HEPT-2-ENE-2-CARBOXYLIC ACID
OLOMOUCINE	2-(2-HYDROXYETHYLAMINO)-6-BENZYLAMINO-9-METHYLPURINE
OLPRINONE	1,2-DIHYDRO-6-METHYL-2-OXO-5-(IMIDAZO[1,2-a]PYRIDIN-6-YL)-3-PYRIDINE CARBONITRILE
OLTIPRAZ	5-PYRAZINYL-4-METHYL-1,2-DITHIOLE-3-THIONE
OM 805	ARGATROBAN
ONO 1301	7,8-DIHYDRO-5-[(E)-[[a-(3-PYRIDYL)BENZYLIDENE]AMINOOXY]ETHYL-1-NAPHTHYLOXY]ACETIC ACID
ONO 2235	(E)-3-CARBOXYMETHYL-5-[(2E)-METHYL-3-PHENYLPROPENYLIDENE]RHODANINE
ONO 2235	EPALRESTAT
ONO 3307	4-SULPHAMOYL PHENYL-4-GUANIDINOBENZOATE METHANESULPHONATE
ONO 3805	4-[2-[4-(4-ISOBUTYLBENZYLOXY)-2,3-DIMETHYLBENZOYLAMINO]PHENOXY]BUTANOIC ACID
ONO 5046	N-[2-[4-(2,2-DIMETHYLPROPIONYLOXY)PEHNYLSULPHONYLAMINO]BENZOYL]AMINO ACETIC ACID
ONO 9302	EPRISTERIDE
OPC 13013	6-4(1-CYCLOHEXYL-5-TETRAZOLYL)BUTOXYL)-1,2,3,4-TETRAHYDRO-2-OXOQUINOLONE

Trivial name	Synonym or systematic name
OPC 13135	N-CYCLOHEXYL-N-(HYDROXYBUTYL)5-(6-1,2,3,4-TETRAHYDRO-2-OXOQUINOLYLOXY)BUTYRAMIDE
OPC 3689	CILOSTAMIDE
OPC 3689	N-CYCLOHEXYL-N-METHYL-4-(1,2-DIHYDRO-2-OXO-6-QUINOLYLOXY)BUTYRAMIDE
OPC 3911	N-CYCLOHEXYL-N-2-HYDROXYETHYL-4(6-(1,2-DIHYDRO-2-OXOQUINOLYLOXY))BUTYRAMIDE
OPHTALMIC ACID	L-γ-GLUTAMYL-L-α-AMINOBUTYRYL-GLYCINE
OPRIN	OPPOSSUM SERUM METALLOPROTEINASE INHIBITOR
OPSOPYRROLE	3-ETHYL-4-METHYL-1H-PYRROLE
OR 462	3-(3,4-DIHYDROXY-5-NITROBENZYLIDENE)-2,4-PENTANEDIONE
OR 462	NITECAPONE
OR 468	3,4-DINITROPYROCATECHOL
OR 611	(E)-2-CYANO-N, N-DIETHYL-3-(3,4-DIHYDROXY-5-NITROPHENYL)PROPENAMIDE
OR 611	ENTACAPONE
ORG 20241	N-HYDROXY-4-[3,4-DIMETHOXYPHENYL]-THIAZOLE-2-CARBOXIMIDAMIDE
ORG 30029	N-HYDROXY-5,6-DIMETHOXY-BENZO[b]-THIOPHENE-2-CARBOXIMIDAMIDE HCl
ORG 30365	19-MERCAPTOANDROST-4-ENE-3,17-DIONE
ORG 33201	3α-R)-trans-1-[(3α-ETHYL-9-(ETHYLTHIO)-2,3,3α,4,5,6-HEXAHYDRO-1H-PHENALEN-2-yl)METHYL]1H-IMIDAZOLE HCl
ORLISTAT	TETRAHYDROLIPSTATIN
OROBOL	5,7-DIHYDROXY-3-(3,4-HYDROXYPHENYL)-4H-1-BENZOPYRAN-4-ONE
OROTIC ACID	1,2,3,6-TETRAHYDRO-2,6-DIOXO-4-PYRIMIDINECARBOXYLIC ACID
ORYZALINA	4-(DIPROPYLAMINO)-3,5-DINITROBENZENE SULPHONAMIDE
OUABAIN	3-[(6-DEOXY-α-L-MANNOPYRANOSYL)OXY]-1,5,11α,14,19-PENTAHYDROXY-CARD-20(22)-ENOLIDE
OXADIAZON	5-tert-BUTYL-3-(2,4-DICHLORO-5-ISOPROPOXYPHENYL)-1,3,4-OXADIAZOLIN-2-ONE
OXALOACETIC ACID	2-OXOBUTANEDIOIC ACID
OXALURIC ACID	[(AMINOCARBONYL)AMINO]OXO-ACETIC ACID
OXAMIC ACID	AMINOOXO-ACETIC ACID
OXAMYCIN	CYCLOSERINE
OXANILIC ACID	OXO(PHENYLAMINO)-ACETIC ACID
OXANTEL	(E)-3-[2-[1,4,5,6-TETRAHYDRO-1-METHYL-2-PYRMIDINYL)ETHENYL]PHENOL
OXOLINIC ACID	5-ETHYL-5,8-DIHYDRO-8-OXO-1,3-DIOXOLO[4,5-G]QUINOLINE-7-CARBOXYLIC ACID
OXONIC ACID	1,4,5,6-TETRAHYDRO-4,6-DIOXO-1,3,5-TRIAZINE-2-CARBOXYLIC ACID
OXPHALIN	1-(1,3-DIHYDROXYBENZYLIDENE)2,4,6-TRIMETHYLANILIN
OXPHAMAN	1-(2,5-DIHYDROXYBENZYLIDENE)AMINOADAMANTAN
OXYFLUORFEN	2-CHLORO-1-(3-ETHOXY-4-NITROPHENOXY)-4-TRIFLUOROMETHYL)BENZENE
P 596	D-CYCLOHEXYLAMINE-PRO-ARG($CH_2NC_5H_4CH_2CO$)(GLY)$_4$ASP-TYR-GLU-PRO-ILE-PRO-GLU-GLU-ALA-CYCLOHEXYLALANINE-D-GLU
P 10358	1-[3-FLUORO-4-PYRIDINYL)AMINO]-3-METHYL-1H-INDOL-5-YL METHYLCARBAMATE
P 3622	3-(4-CHLOROPHENYL)-2-(4-DIETHYLAMINOETHOXYPHENYL)-A-PENTENONITRILE MONOHYDROGEN CITRATE
PACIFASTIN	CRAYFISH TRYPSIN INHIBITOR
PADAC	7-(THIENYL-2-ACETAMIDE)-3-[2-(4-N,N-DIMETHYLAMINOPHENYLAZO)PYRIDINIUM-METHYL]-3-CEPHAM-4-CARBOXYLIC ACID
PALA	N-(PHOSPHONOACETYL)-L-ASPARTIC ACID
PALMOXIRATE	2-TETRADECYLGLYCIDATE
PALMOXIRATE	TETRADECYL-2-OXIRANECARBOXYLIC ACID
PALO	5N-(PHOSPHONOACETYL)-L-ORNITHINE
PAMIDRONATE	3-AMINO-1-HYDROXYPROPYLIDENE-1,1-BISPHOSPHONIC ACID
PANTOIC ACID	(R)-2,4-DIHYDROXY-3,3-DIMETHYL-BUTANOIC ACID
PANTOTHENIC ACID	(R)-N-(2,4-DIHYDROXY-3,3-DIMETHYL-1-OXOBUTYL)-β-ALANINE
PAPULACANDIN B	1,1 6-ANHYDRO-1-C-[2,4-DIHYDROXY-6-(HYDROXYMETHYL)PHENYL]-4-O-[6-O-(8-HYDROXY-1-OXO-2,4,6-DECA-TRIENYL)-β-D-GALACTOPYRANOSYL]-α-D-GLUCOPYRANOSE
PARAOXON	DIETHYL 4-NITROPHENYLPHOSPHATE
PARAOXON	E 600
PARAOXON	PHOSPHORIC ACID DIETHYL 4-NITROPHENYLESTER
PARGYLINE	N-METHYL-N-2-PROPYNYL-BENZENEMETHANAMINE
PC 18	(2S)-1-[2(AMNIO)-4-(METHYLTHIO)]BUTANE THIOL
PC 57	N-[N-(2S)-2-MERCAPTO-3-METHYL-1-OXOBUTYL)-(S)-ISOLEUCYL](S)-TYROSINE

Trivial name	Synonym or systematic name
PCHB	p-CHLOROHYDROXYBENZOIC ACID
PCMB	p-MERCURIBENZOIC ACID
PCMPS	p-CHLOROMERCURIPHENYL SULPHONATE
PCO	5-PHENYL-3(N-CYCLOPROPYL)ETHYLAMINE-1,2,3-OXADIAZOLE
PD 089828	[[2-AMINO-6-(2,6-DICHLOROPHENYL)PYRIDO[2,3-d]PYRIMIDIN-7-YL]-3-t-BUTYLUREA
PD 123244-15	(±)-(R*,R*)-3,4-DIBROMO-β,δ-DIHYDROXY-2-(4-FLUROPHENYL)-5-(1-METHYLETHYL)-1H-PYRROL-1-HEPTANOIC ACID
PD 128042	CI 976
PD 128763	3,4-DIHYDRO-5-METHOXYISOQUINOLIN-1(2H)-ONE
PD 128763	3,4-DIHYDRO-5-METHYL-1(2H)ISOQUINOLINE
PD 132002	[1S-(1R*,2S*,3R*)]-3-[[1-(CYCLOHEXYLMETHYL)-2,3-DIHYDROXY-5-METHYLHEXYL]AMINO]-N-[N-(4-MORPHO-LINOSULPHONYL)PHENYLALANYL]-3-OXO-DL-ALANINE METHYLESTER
PD 132301-2	N-[2,6-BIS(1-METHYLETHYL)PHENYL]-N′-[[1-[4-(DIMETHYLAMINO)PHENYL]CYCLOPENTYL]METHYL]UREA
PD 134672	CI 992
PD 134672	[1S-(1R*,2S*,3R*)] N-(4-MORPHOLINYLSULPHONYL)-PHENYLALANYL-3-(2-AMINO-4-THIAZOLYL)-N-[1-(CYCLO-HEXYLMETHYL)-2,3-DIHYDROXY-5-METHYLHEXYL]-L-ALANINAMIDE
PD 141955	2-AMINO-3,5-DIHYDRO-7-(3-THIENYLMETHYL)-4H-PYRROLO[3,2-d]PYRIMIDIN-4-ONE
PD 141955	CI 1000
PD 141955	9-DEAZA-9-(3-THIENYLMETHYL)GUANINE
PD 153035	4-(3-BROMOANILO)-6,7-DIMETHXYQUINAZOLINE
PD 161570	1-tert-BUTYL-3-[6-(2,6-DICHLOROPHENYL)-2-(4-DIETHYLAMINOBUTYLAMINO)PYRIDO[2,3-d]PYRMIDIN-7-YL]-UREA
PD 166285	6-(2,6-DICHLOROPHENYL)-2-[4(2-DIETHYLAMINOETHOXY)PHENYLAMINO]-8-METHYL-8H-PYRIDO[2,3-d]PYRI-MIDIN-7-ONE
PD 83176	N-CARBOBENZOXY-HIS-L-(O-BENZYL)TYR-L-(O-BENZYL)SER-TRP-D-ALANINAMIDE
PD 98059	2′-AMINO-3-′-METHOXYFLAVONE
PD 98059	2-(2′-AMINO-3′-METHOXYPHENYL)OXANAPHTHALEN-4-ONE
PD 134298-38A	ATROVASTATIN
PD 143298-38A	CI 981
PEFABLOC	4-(2-AMINOETHYL)-BENZYLSULPHONYLFLUORIDE
PEFABLOC	5-METHANESULPHOAMIDO-6-(2,4-DIFLUOROTHIOPHENYL)-1-INDANONE
PELDESINE	BCX 34
PELRINONE	3,4-DIHYDRO-2-METHYL-4-OXO-6[(3-PYRIDINYLMETHYL)AMIN0]-5-PYRIMIDINECARBONITRILE HCl
PENCICLOVIR	9-(4-HYDROXY-3-HYDROXYMETHYLBUT-1-YL)GUANINE
PENTAGALLOYLGLUCOSE	3-O-DIGALLOYL-1,2,6-TRIGALLOYLGLUCOSE
PENTAMIDINE	4,4′-DIAMIDINOPHENOXYPENTANE
PENTARIC ACID	2,3,4-TRIHYDROXY-GLUTARIC ACID
PENTOPRIL	CGS 13945
PENTOPRILAT	CGS 13934
PENTOSTATIN	2′-DEOXYCOFORMYCIN
PENTOSTATIN	(8R)-HYDROXYL-2′-DEOXYCOFORMYCIN
PEPSTATIN	ISOVAL-VAL-VAL-STA-ALA-STA ; STA = (3S, 4S)-4-AMINO-3-HYDROXY-6-METHYLHEPTANOIC ACID
PEPSTATIN A	ISOVALERYL-VAL-VAL-STA-ALA-STA
PERINDOPRILAT	(2S,3aS,7aS)-1-[N[1(S)-CARBOXYBUTYL]-(S)-ALANYL]-2-CARBOXYPERHYDROINDOLE
P′GBDBIG	N,N-DIMETHYLAMINO-4-(4′-GUANIDINOBENZYLOXY)BENZYLCARBONYLOXYGLYCOLATE
PHASEOLOTOXIN	N6-(AMINOIMINOMETHYL)-N2-[N-[N5-[AMINO(SULPHAMINO)PHOSPHINYL]-L-ORNITHYL]-L-ALANYL]-L-LYSINE
PHENACEMIDE	N-(AMINOCARBONYL)BENZENEACETAMIDE
PHENACETIN	N-(4-ETHOXYPHENYL)ACETAMIDE
o-PHENANTHROLINE	1,10 PHENANTHROLINE
PHENELZINE	(2-PHENYLETHYL)HYDRAZINE
PHENFORMIN	N-(2-PHENYLETHYL)-IMIDODICARBONIMIDIC DIAMIDE
PHENIDONE	1-PHENYL-3-PYRAZOLIDINONE
PHENOBARBITAL	5-ETHYL-5-PHENYL-2,4,6-(1H, 3H, 5H)PYRIMIDINETRIONE
PHENOTHIAZINE	THIODIPHENYLAMINE
PHENYLETHYLALCOHOL	2-PHENYLETHANOL
PHENYLLACTATE	L-2-HYDROXY-3-PHENYLPROPIONIC ACID
o-PHENYLPHENOL	2-HYDROXYBIPHENYL

Trivial name	Synonym or systematic name
p-PHENYLPHENOL	4-HYDROXYBIPHENYL
PHENYLPROPIOLIC ACID	PHENYLPROPYNOIC ACID
PHENYLPROPYNAL	3-PHENYL-2-PROPYNAL
PHENYLSTATINE	4-AMINO-3-HYDROXY-5-PHENYLPENTANOIC ACID
PHETHIOL	1-AMINO-1-BENZYL-MERCAPTOETHANE
PHLORETIN	3-(4-HYDROXYPHENYL)-1-(2,4,6-TRIHYDROXYPHENYL)-1-PROPANONE
PHLORIDZIN	1-[2-(β-D-GLUCOPYRANOSYLOXY)-4,6-DIHYDROXYPHENYL]-3-(4-HYDROXYPHENYL)-1-PROPANONE
PHMB	p-HYDROXYMERCURIBENZOATE
PHOSPHAMIDON	1-CHLORO-1-N,N-DIETHYLCARBAMOYL-1-PROPEN-2-YLDIMETHYL PHOSPHATE
PHOSPHINOTRICIN	L-2-AMINO-4-HYDROXYMETHYLPHOSPHINYL)BUTANOIC ACID
PHOSPHINOTRICIN	GLUFOSINATE
PHOSPHODIERPRYL	N-(PHENYLPHOSPHONYL)-GLY-PRO-L-AMINOHEXANOIC ACID
PHOSPHOENOLPYRUVIC ACID	2-(PHOSPHONOOXY)-2-PROPENOIC ACID
PHOSPHORAMIDON	N-[N-[[(6-DEOXY-α-L-MANNOPYRANOSYL)OXY]HYDROXYPHOSPHINYL]-L-LEUCYL]-L-TRYPTOPHAN
PHTHALAZINE	2,3-BENZODIAZINE
PHTHALHYDRAZIDE	2,3-DIHYDRO-1,4-PHTHALAZINEDIONE
PHTHALIC ACID	1,2-BENZENEDICARBOXYLIC ACID
PHTHALONIC ACID	2-CARBOXY-α-OXO-BENZENEACETIC ACID
PHYLPA	SODIUM 1-O-[(9′S, 10′R)9′,10′-METHANOHEXADECANOYL]sn-GLYCEROL 2,3-CYCLIC PHOSPHATE
PHYSOSTIGMINE	(3AS-cis)-1,2,3,3A, 8,8A-HEXAHYDRO-1,3A, 8-TRIMETHYLPYRROLO[2,3-B]INDOL-5-OL METHYLCARBAMATE (ESTER)
PHYSOSTIGMINE	ESERINE
PICOTAMIDE	N,N′-BIS-(3-PICOYL)-4-METHOXY-ISOPHTHALAMIDE
PIERICIDIN A	[R-[R*,R*-(ALL-E)]]-2-(10-HYDROXY-3,7,9,11-TETRAMETHYL-2,5,7,11-TRIDECATETRAENYL)-5,6-DIMETHOXY-3-METHYL-4-PYRIDINOL
PIMELIC ACID	HEPTANEDIOIC ACID
PIMOBENDAN	4,5-DIHYDRO-6-2[4-METHOXYPHENYL]-1H-BENZIMIDAZOLE-5-YL-5-METHYL-3(2H)-PYRIDIZANONE
PINOCEMBRIN	5,7-DIHYDROXYFLAVONE
PIPECOLIC ACID	2-PIPERIDINECARBOXYLIC ACID
PIPERALIN	3-(2-METHYLPYRIDINE)PROPYL-3,4-DICHLOROBENZOIC ACID
PIPERASTATIN A	N-FORMYL-allo-ILE-HTRE-LEU-VAL-PIP-LEU-PIP
PIPERASTATIN B	N-FORMYL-VAL-THR-LEU-VAL-PIP-LEU-PIP
PIRIPROST	6,9-DEEPOXY-6,9-PHENYLIMINO)-6,8-PROSTAGLANDIN I1
PIRIPROST	U 60257
PIRITREXIM	BW 301U
PIROXICAM	4-HYDROXY-2-METHYLN-2-PYRIDINYL-2H-1,2-BENZOTHIAZINE-3-CARBOXAMIDE-1,1-DIOXIDE
PKSI 527	trans-4-AMINOMETHYLCYCLOHEXANECARBONYL-L-PHENYLALANINE 4-CARBOXYMETHYLANILIDE
PLACENTAL PROTEINASE INHIBITOR	PROTEINASE INHIBITOR 6
PLACENTAL THROMBIN INHIBITOR	PROTEINASE INHIBITOR 6
α2-PLASMIN INHIBITOR	α2-ANTIPLASMIN
PLASMINOGEN ACTIVATOR INHIBITOR	PROTEIN C INHIBITOR
PLATANETIN	3,5,7,8-TETRAHYDROXY-6-ISOPRENYL FLAVONE
PLOMESTANE	19 ACETYLEBICANDROSTENEDIONE
PLUMBAGIN	5-HYDROXY-2-METHYL-1,4-NAPHTHOQUINONE
PMS 832	1-(4′-METHOXYBENZOYL)-2n-TRIDECYLPIPERAZINE
PMSF	PHENYLMETHYLSULPHONYL FLUORIDE
PNEUMOCANDIN B0	L 743872
PNEUMOCANDIN B0	MK 0991
PNU 103017	4-CYANO-N-(3-CYCLOPROPYL-(5,6,7,8,9,10-HEXAHYDRO-4-HYDROXY-2-OXO-2H-CYCLOOCTA(b)PYRAN-3-YL)-METHYL)PHENYL)BENZENESULPHONAMIDE
PNU 140690	U 140690
POCA	2-[5-(4-CHLOROPHENYL)PENTYL]OXIRANE-2-CARBOXYLATE

Trivial name	Synonym or systematic name
PODOSCYPHIC ACID	(E)-4,5-DIOXO-2-HEXADECANOIC ACID
POL 647	FUROYL-LEU-TRP
POL 656	FUROYL-LEU-cycloTRP
POLYOXIN C	5-AMINO-1,5-DIDEOXY-1-[3,4-DIHYDRO-5-(HYDROXYMETHYL)-2,4-DIOXO-1(2H)-PYRIMIDINYL]-β-D-alloFURAN-URONIC ACID
PONALRESTAT STATIL	3-(-4-BROMO-2-FLUOROBENZYL-4-OXO-3-PHTHALAZINE-1-YLACETIC ACID
POSTSTATIN	VAL-VAL-3-AMINO-2-OXO-VALERYL-D-LEU-VAL
PP1	4-AMINO-5-(4-METHYLPHENYL)-7-(t-BUTYL)-PYRAZOLO[3,4-d]PYRIMIDINE
PP2	4-AMINO-5-(4-CHLOROPHENYL)-7-(t-BUTYL)-PYRAZOLO[3,4-d]PYRIMIDINE
PP 36	5′-[[N-(2-DECANOYLAMINO-3-HYDROXY-3-PHENYLPROPYLOXYCARBONYL)GLYCYL]AMINO]-5′-DEOXY-URIDINE
PP 37	5′-[[(2-DECANOYLAMINO-3-HYDROXY-3-PHENYLPROPYLOXYCARBONYL)]AMINO]SULPHONYL]-URIDINE
PP 55	5′-[[(2-DECANOYLAMINO-3-PHENYLPROPYLOXYCARBONYL)]AMINO]SULPHONYL]-URIDINE
PP50B	D-5′-O-[[(2-DECANOYLAMINO-3-PHENYLPROPYLOXYCARBONYL)AMINO]SULPHONYL]-2′,3′-O-ISOPROPYLIDE-NEURIDINE
PPESTATIN A	ISOVALERYLPEPSTATIN
PPPA	L-2-PHOSPHORYLOXY-3-PHENYLPROPIONIC ACID
PRAVASTIN	CS 514
PREPHENIC ACID	1-CARBOXY-4-HYDROXY-α-OXO-2,5-CYCLOHEXADIENE-1-PROPANOIC ACID
PRIMACRINE	QUINACRINE
PRODIAX	PONALRESTAT
PRODIPINE	DIPHENYL-1-(S)-PROLYLPYRROLIDINE-2(R, S)PHOSPHONATE
PRODIPINE	1-ISOPROPYL-4,4-DIPHENYLPIPERIDINE
PROFLAVINE	3,6-ACRIDINEDIAMINE
PROGESTERONE	PREGN-4-ENE-3,20-DIONE
PROLIXAN	AZAPROPAZONE
PROPARGYLGLYCINE	2-AMINO-4-PENTYNOIC ACID
PROPERICIAZINE	PERICYAZINE
PROPIOLIC ACID	2-PROPYNOIC ACID
PROSCOR	FINASTERIDE
PROTEINASE H IV 1	PNU 140690
PROTEINASE H IV 2	PNU 140690
PROTEINASE INHIBITOR 6	PLACENTAL PROTEINASE INHIBITOR
PROTEIN PHOSPHATASE 2A INHIBITOR	PUTATIVE CLASSII HUMAN HISTOCOMPATIBILITY LEUKOCYTE ASSOCIATED PROTEIN
PROTOCATECHUIC ACID	3,4-DIHYDROXYBENZOIC ACID
PSICOFURANINE	9β-D-PSICOFURANOSYL-9H-PURIN-6-AMINE
6-PTERIDYLALDEHYDE	6-PTERIDINECARBOXALDEHYDE
PTEROIC ACID	4-[[(2-AMINO-1,4-DIHYDRO-4-OXO-6-PTERIDINYL)METHYL]AMINO]BENZOIC ACID
PTILODENE	11-HYDROXY-16-OXO-5(Z),8(Z),12(E),14(E),17(E)-ICOSAPENTAENOIC ACID
PUG	D-GLUCONOHYDROXIMO-1,5-LACTONE-N-PHENYLURETHANE
PUTRESCINE	1,4-BUTANEDIAMINE
PYRAZOMYCIN	4-HYDROXY-3-β-D-RIBOFURANOSYL-1H-PYRAZOLE-5-CARBOXAMIDE
PYRIDOGLUTETHIMIDE	3-ETHYL-3-(4-PYRIDYL)PIPERIDINE-2,6-DIONE
PYRIDOGLUTETHIMIDE	ROGLETIMIDE
PYRIDOXAL	3-HYDROXY-5-(HYDROXYMETHYL)-2-METHYL-4-PYRIDINECARBOXALDEHYDE
PYRIDOXAMINE	4-(AMINOMETHYL)-5-HYDROXY-6-METHYL-3-PYRIDINEMETHANOL
4-PYRIDOXIC ACID	3-HYDROXY-5-(HYDROXYMETHYL)-2-METHYL-4-PYRIDINECARBOXYLIC ACID
PYRIDOXINE	5-HYDROXY-6-METHYL-3,4-PYRIDINEDIMETHANOL
PYRIMETHAMINE	5-(4-CHLOROPHENYL)-6-ETHYL-2,4-PYRIMIDINEDIAMINE
PYRITHIAMINE	1-[(4-AMINO-2-METHYL-5-PYRIMIDINYL)METHYL]-3-(2-HYDROXYETHYL)-2-METHYL-PYRIDINIUM BROMIDE
PYRIZINISTATIN	2,4,4a,8-TETRAHYDRO-2,6,8-TRIMETHYL-4a-(2-OXOPROPYL)PYRIMIDO[5,4-E]-1,2,4-TRIAZINE-3,5,7(6H)-TRIONE
PYROGALLOL	1,2,3-BENZENETRIOL
PYROSTATIN A	4-HYDROXY-2-IMINO-1-METHYLPYROLLIDINE-5-CARBOXYLIC ACID
PYROSTATIN B	2-IMINO-1-METHYLPYROLLIDINE-5-CARBOXYLIC ACID
Q 8111	RO 31-8959

Trivial name	Synonym or systematic name
QUERCETIN	2-(3,4-DIHYDROXYPHENYL)-3,5,7-TRIHYDROXY-4H-1-BENZOPYRAN-4-ONE
QUERCETIN	3,3′,4′,5,7-PENTAHYDROXYFLAVONE
QUERCITRIN	3-[(6-DEOXY-α-L-MANNOPYRANOSYL)-OXY]-2-(3,4-DIHYDROXYPHENYL)-5,7-DIHYDROXY-4H-1-BENZOPYRAN-4-ONE
QUERCITRIN	QUERCETIN-3-O-RHAMNOSIDE
QUINACRINE	N4-(6-CHLORO-2-METHOXY-9-ACRIDINYL)-N1,N1-DIETHYL-1,4-PENTANEDIAMINE DIHYDROCHLORIDE
QUINALDINE	2-METHYLQUINOLINE
QUINALDINIC ACID	2-QUINOLINECARBOXYLIC ACID
QUINAPRIL	PD 109452-2Q
QUINAPRILAT	PD 109548-1
QUINAZOLINE	1,3-BENZODIAZINE
R 830	[3,5-BIS(1,1-DIMETHYLETHYL)-4-HYDROXYPHENYL]2-THIENYLMETHANONE
R 24571	CALMIDAZOLIUM
R 40519	N-ETHYL-2,3-DIHYDROIMIDAZO[2,1-b]BENZOTHIAZOL-6-AMINE
R 51469	MIOFLAZINE
R 5421	ETHANIMIDOTHIOIC ACID N-[[N-BUTHYLTHIO-N-METHYLAMINO]CARBONYLOXY]METHYL ESTER
R 59022	6-[2-[4-[(4-FLUOROPHENYL)PHENYLMETHYLENE]-1-PIPERIDINYL]ETHYL]-7-METHYL-5H-THIAZOLO[3,2-a]PYRIMIDIN-5-ONE
R 59949	3-[2-[4-[BIS(4-FLUOROPHENYL)METHYLENE]-1-PIPERIDINYL]ETHYL]-2,3-DIHYDRO-2-THIOXO-4(1H)-QUINAZOLINONE
R 76713	6-[(4-CHLOROPHENYL)(1H-1,2,4-TRIAZOL-1-YL)METHYL]-1-METHYL-1H-BENZOTRIAZOLE
R 7904	LIDOFLAZINE
R 79595	N-CYCLOHEXYL-N-METHYL-2-[[[PHENYL(1,2,3,5-TETRAHYDRO-2-OXOIMIDAZO[2,1-b]QUINAZOLIN-7-YL)-METHYLENE]AMIN]OXY]ACETAMIDE
R 80122	(E)-N-CYCLOHEXYL-N-METHYL-2-[[[PHENYL(1,2,3,5-TETRAHYDRO-2-OXOIMIDAZO[2,1-b]QUINAZOLIN-7-YL)-METHYLENE]AMINO]OXY]ACETAMIDE
R 82150	(+)-(S)-4,5,6,7-TETRAHYDRO-5-METHYL-6-(3-METHYL-2-BUTENYL)IMIDAZO[4,5,1-j,k][1,4]BENZODIAZEPIN-2(1H)-THIONE
R 82913	(+)-(S)-4,5,6,7-TETRAHYDROIMIDAZO-9-CHLORO-5-METHYL-6-(3-METHYL-2-BUTENYL)-IMIDAZOLE[4,5,1-j,k][1,4]-BENZODIAZEPIN-2(1H)THIONE
R 82913	4,5,6,7-TETRAHYDRO-5(S)-METHYL-6-(3-METHYL-2-BUTENYL)-9-CHLOROIMIDAZO[4,5,1-j,k][1,4]BENZODIAZEPIN-2(1H)THIONE
R 82913	TIBO
R 83842	dextro-(6-[(4-CHLOROPHENYL)(1H-1,2,4-TRIAZOL-1-YL)METHYL]-1-METHYL-1H-BENZOTRIAZOLE
R 83842	VOROZOLE
R 85355	2-[2-[4-CHLOROPHENYL)-2-OXOETHYL]-2,4-DIHYDRO-4-[4-(4-HYDROXYPHENYL)-1-PIPERAZINYL]-5-METHYL-3H-1,2,4-TRIAZOL-3-ONE
R 87366	(2S, 3S)-3-[N-(QUINOXALINE-2-CARBONYL)-L-ASPARAGINYL]AMINO-2-HYDROXY-4-PHENYLBUTANOYL-L-PROLINE tert.-BUTYLAMIDE
R 90385	LOVIRIDE
RA6	N-ETHYL-3[1-(DIMETHYLAMINO)ETHYL]PHENYL CARBAMATE
RA7	N-ETHYL, N-METHYL-3[1-(DIMETHYLAMINO)ETHYL]PHENYL CARBAMATE
RA15	N-PROPYL-3[1-(DIMETHYLAMINO)ETHYL]PHENYL CARBAMATE
RABEPRAZOLE	E 3810
RAC-X 65	N-[1(S)-CARBOXY-3-CARBOXANILIDOPROPYL]-ALA-PRO
RAMIPRIL	HOE 498
RAPEPRAZOLE	2[[4-(3-METHOXYPROPOXY)-3-METHYLSULPHINYL]-1H-BENZIMIDAZOLE
RAZOXANE	ICFR 159
RB 105	N[2S, 3R)-2-(MERCAPTOMETHYL)-3-PHENYLBUTANOYL-L-ALANINE
RB 105	S 21402
RB 38A	N-[3(R)(HYDROXYAMINOCARBONYL)-2-BENZYL-1-OXOPROPYL]-L-PHENYLALANINE
RB 38B	(S,S)-N-[3-(HYDROXYAMINOCARBONYL)-2-BENZYL-1-OXOPROPYL]-L-PHENYLALANINE
RD 4-2025	4-(2,6-DICHLOROPHENYL)-1,2,5-THIDIAZOL-3-YL-N-METHYL, N-ETHYLCARBAMATE
REACTIVE BLUE 2	CIBACRONE BLUE 3GA
REDOXAL	2,2′-[3,3′-DIMETHOXY[1,1′-BIPHENYL]-4,4′-DIYL)DIIMINO]BIS-BENZOIC ACID
REMIKIREN	RO 42-5892
REV 5901	α-PENTYL-3-(2-QUINOLYLMETHOXY)BENZENE

Trivial name	Synonym or systematic name
REVLON 5901	2-[3-(1-HYDROXYHEXYL)PHENYOXYMETHYL]QUINOLINE
RG 12561	4α,6β(E)-(±)-6-[2-[2-(4-FLUORO-3-METHYLPHENYL)-4,4,6,6-TETRAMETHYL-1-CYCLOHEXEN-1-YL]ETHENYL]-TETRAHYDRO-4-HYDROXY-2H-PYRAN-2-ONE
RG 14620	(Z)-α-[(3,5-DICHLOROPHENYL)METHYLENE]-3-PYRIDYLACETONITRILE
RG 50864	α-CYANO-3,4-DIHYDROXYTHIOCIANNAMIDE
RG 6866	N-METHYL-4-BENZYLOXYPHENYLACETOHYDROXAMIC ACID
RG 80267	1,6-BIS(CYCLOHEXYLOXIMINOCARBONYLAMINO)HEXANE
RH 5348	METHYL-5-[2-CHLORO-5-(TRIFLUOROMETHYL)PHENOXY]-2-NITROBENZOIC ACID
RHAMNETIN	3,3′,4′,5-TETRAHYDROXYFLAVONE
RHC 80267	1,6-DI(O-(CARBAMOYL)CYCLOHEXANE OXIME)HEXANE
RHC 80267	U 57908
RHEIN	9,10-DIHYDRO-4,5-DIHYDROXY-9,10-DIOXO-2-ANTHRACENECARBOXYLIC ACID
RIBAVIRIN	1-β-D-RIBOFURANOSYL-1,2,4,TRIAZOLE-3-CARBOXAMIDE
RIBAVIRIN	VIRAZOLE
RIBITOL	ADONITOL
RICININE	1,2-DIHYDRO-4-METHOXY-1-METHYL-2-OXO-3-PYRIDINECARBONITRILE
RIDOGREL	(E)-5-[[[(3-PYRIDINYL)[3-(TRIFLUOROMETHYL)PHENYL]METHYLENE]AMINO]OXYPENTANOIC ACID
RIFAMPICIN	3-[[(4-METHYL-1-PIPERAZINYL)IMINO]METHYL]RIFAMYCIN
RIFAMPICIN	RIFAMPIN
RITONAVIR	ABT 538
RK 1409	7-OXOSTAUROSPORINE
RK 1409B	4′-DEMETHYLAMINO-4′-HYDROXY-3′-EPISTAUROSPORINE
RK 286C	4′-DEMETHYLAMINO-4′-HYDROXYSTAUROSPORINE
RMI 14514	5-TETRADECYLOXY-2-FURANCARBOXYLIC ACID
RMI 18341	(5α,20R)-4-DIAZO-21-HYDROXY-20-METHYLPREGNAN-3-ONE
RMI 12330A	MDL 12330A
RMI 12330A	N-(cis-2-PHENYLCYCLOPENTYL)AZACYCLOTRIDECAN-2-IMINE HCl
RO 1-2812	3,5-DINITROCATECHOL
RO 2-0683	4-[[(DIMETHYLAMINO)CARBONYL]OXY]-N, N, N-TRIMETHYL-[1,1′-BIPHENYL]-3-METHANAMINIUM BROMIDE
RO 4-4602	BESERAZIDE
RO 09-0154	TRESTATIN
RO 09-1437	CYCLOTHILIDINE
RO 09-1450	VINAXANTHONE
RO 11-1163	MOCLOBEMIDE
RO 11-8958	EPIROPRIM
RO 15-4513	8-AZIDO-5,6-DIHYDRO-5-METHYL-6-OXO-4H-IMIDAZO[1,5-a][1,4]BENZODIAZEPINE-3-CARBOXYLATE
RO 17-5380	(±)-5,7-DIHYDRO-2-[[(4-METHOXY-3,5-DIMETHYL-2-PYRIDINYL)METHYL]SULPHINYL]-5,5,7,7-TETRAMETHYL-INDENO[5,6-D]IMIDAZOL-6(1H)-ONE
RO 18-0647	TETRAHYDROLIPSTATIN
RO 18-5364	5,7-DIHYDRO-2-[[(4-METHOXY-3-METHYL-2-PYRIDINYL)METHYL]SULPHINYL]-5,5,7,7-TETRAMETHYL-INDENO[5,6-D]IMIDAZOL-6(1H)-ONE
RO 19-3704	3-4(R)-2-METHOXYCARBONYL)OXY-3-(OCTADECYLCARBAMOYL)OXYPROPOXY BUTHIAZOLIUM IODIDE
RO 19-6327	N-(2-AMINOETHYL-5-CHORO-2-PYRIDINE CARBOXAMIDE)
RO 19-6327	LAZABEMIDE
RO 19-6327	N-(2-AMINOETHYL)-5-CHLORO-2-PYRIDINECARBOXAMIDE
RO 20-1724	(±)-4-(3-BUTOXY-4-METHOXYBENZYL)-2-IMIDAZOLIDINONE
RO 31-4724	(N-[(N-[2-[(N-HYDROXYCARBAMOYL)METHYL]-4-METHYLVALERYL]L-LEUCYL]-L-ALANINE ETHYLESTER
RO 31-8220	3[1-[3-(AMININOTHIO)PROPYL]-1H-INDOLYL-3-YL]-3-(1-METHYL-1H-INDOLYL-3-YL)MALEIIMIDE METHANE SUL-PHONATE
RO 31-8959	INVIRASE
RO 31-8959	SAQUINAVIR
RO 32-0432	(S)-3[8(DIMETHYLAMINOMETHYL)-6,7,8,9-TETRAHYDROPYRIDO[1,2-a]INDOL-10-YL]-4-METHYL-3-INDOLYL)-1H-PYRROLE-2,5-DIONE HCl
RO 32-2313	1-[5-[2(S)-(2,4-DICHLORO-5-METHOXY)PROPIONAMIDO]-2,5-DIDEOXY-2-FLUORO-β-D-ARABINOFURANOSYL]5-ETHYLURACIL
RO 32-3555	3(R)-(CYCLOPENTYLMETHYL)-2R)-[(3,4,4-TRIMETHYL-2,5-DIOXO-1-IMIDAZOLINIDYL)METHYL]-4-OXO-4-PIPERI-DINOBUTYROHYDROXAMIC ACID

Trivial name	Synonym or systematic name
RO 40-7592	3,4-DIHYDROXY-4′-METHYL-5-NITROBENZOPHENONE
RO 40-7592	TOLCAPONE
RO 41-0960	3,4-DIHYDROXY-5-NITRO-2′-FLUOROBENZOPHENE
RO 41-1049	N-(2-AMINOETHYL)-5-(m-FLUOROPHENYL)-4-THIAZOLE CARBOXAMIDE
RO 42-5892	(S)-α[(S)-α-[(tertBUTYLSULPHONYL)METHYL]HYDROCYNNAMIDO]-N-[(1S,2R,3S)-1-(CYCLOHEXYLMETHYL)-3-CYCLOPROPYL-2,3-DIHYDROXYPROPYL]IMIDAZOLE-4-PROPIONAMIDE
RO 42-5892	REMIKIREN
RO 44-9375	CIPROKIREN
RO 46-6240	NAPSAGATRON
RO 48-8071	[4′-(6-ALLYLMETHYLAMINOHEXYLOXY)-2′-FLUOROPHENYL](4-BROMOPHENYL)METHANONE FUMARATE
RO 1724	D, L-4-(3-BUTYLOXY-4-METHYOXYBENZYL-2-IMIDAZOLIDONE
RO 80122	((E)-N-CYCLOHEXYL-N-METHYL-2-[[[PHENYL(1,2,3,5-TETRAHYDRO-2-OXOIMIDAZO[2,1-b]QUINAZOLIN-7-YL)-METHYLENE]AMINO]OXY]ACETAMIDE
RO 80122	SODIUM-(±)8-(3-METHOXY-4-PHENYLSULPHINYLPHENYL)PYRAZOLO[1,5-α]-1,3,5-TRIAZINE-4-OLEATE MINO-HYDRATE
ROBINETIN	3,3′,4′,5′,7-PENTAHYDROXYFLAVONE
ROGLETIMIDE	PYRIDOGLUTETHIMIDE
ROLIPRAM	4-(3′-CYCLOPENTYLOXY-4′-METHOXYPHENYL)-2-PYRROLIDONE
ROSCOVITINE	2-(1-ETHYL-2-HYDROXYETHYLAMINO)-6-BENZYLAMINO-9-ISOPROPYLPURINE
ROSE BENGAL	4,5,6,7-TETRACHLORO-3′,6′-DIHYDROXY-2′,4′,5′,7′-TETRAIODOSPIRO[ISOBENZOFURAN-1(3H),9′-[9H]XANTHEN-3-ONE DIPOTASSIUM SALT
ROTTLERIN	5,7-DIHYDROXY-2,2-DIMETHYL-6-(2,4,6-TRIHYDROXY-3-METHYL-5-ACETYLBENZYL)-8-CINNAMOYL-1,2-CHRO-MENE
ROTTLERIN	MALLOTOXIN
RP 866	QUINACRINE
RP 60475	INTOPLICINE
RP 64477	N-BUTYL-3-(p-DECYLBENZAMIDO)-4-(METHYLTHIO)BENZAMIDE
RP 73163	(S)-2-[5-(3,5-DIMETHYL-1-PYRAZOLYL)PENT-1-YL)SULPHINYL]-5,6-DIPHENYLIMIDAZOLE
RP 73401	3-(CYCLOPENTYLOXY)-N-(3,5-DICHLORO-4-PYRIDYL)-4-METHOXYBENZAMIDE
RPR 101821	trans-2-[4-(BENZOXAZOL-2-YL)-PHENYLMETHOXY]AMINO CYCLOHEXANE HCl
RPR 107393	3-HYDORXY-3-[4-(QUINOLIN-6-YL)PHENYL]-1-AZABICYCLO[2-2-2]OCTANE-DIHYDROCHLORIDE
RPR II	2-BUTENOIC ACID -3-(DIETHOXYPHOSPHINOTHIOYL)-METHYL ESTER
RPR V	2-BUTENOIC ACID -3-(DIETHOXYPHOSPHINOTHIOYL)-ETHYL ESTER
RS 21607	AZALANSTAT
RS 21607	(2S, 4S)cis-2[(1H-IMIDAZOL-1-YL)METHYL]-2-[2-(4-CHLOROPHENYL)ETHYL]-4-[[(4-AMINOPHENYL)THIO]METHYL]-1,3-DOXOLANE
RS 2232	(±)-4-(4-CYANOPHENYL)AMINO-6,7-DIHYDRO-5H-CYCLOPENTA[d]PYRIMIDINE
RS 25344	8-AZA-1-(3-NITROPHENYL)-3-(4-PYRIDYLMETHYL9-2,4-QUINAZOLINEDIONE
RS 25560-197	NEPICASTAT
RS 5186	6-[2-[1-(1H)-IMIDAZOYL]METHYL-4,5-DIHYDROBENZO[b]THIOPHENE]CARBOXYLATE
RS 59022	6-[2-[4-[(4-FLUOROPHENYL)PHENYLMETHYLENE]-1-PIPERIDINYL]ETHYL]-7-METHYL-5H-THIAZOLO[3,2-a]-PYRIMIDIN-5-ONE
RS 82856	N-CYCLOHEXYL-N-METHYL-4-(7-OXY-1,2,3,5-TETRAHYDROIMIDAZO[2,1-b]QUINAZOLIN-2-ONE)BUTYRAMIDE
RS 8359	(±)-4-(4-CYANOPHENYL)AMINO-6,7-DIHYDRO-7-HYDROXY-5H-CYCLOPENTA[d]PYRIMIDINE
RS 95191	3-2(MORPHOLINOETHYLAMINO)-4-CYANO-6-PHENYLPYRAZINE
6(R, S)LY 309887	LOMETREXOL
RU 44403	TRANDOLPRILAT
RWJ 63556	N-[5-(4-FLUOROPHENOXY)THIEN-2-YL]METHANE SULPHONAMIDE
S 10211	(3S)-2-[(2S)-N[(1S)-1-CARBOXY-3-PHENYLPROPYL]ALANYL]2-AZABICYCLO[2.2.2]-OCTANE-3-CARBOXYLIC ACID
S 10211	ZABICIPRILAT
S 17162	N-(2,3-DIHYDROXYPROPYLPHOSPHONYL)-(S)-LEU(S)-TRP-OH
S 1924	2-(5-METHYL-2-PICOLYLSULPHINYL)-1H-THIENO[3,4-d]IMIDAZOLE
S 21402	N[2S, 3R(2-MERCAPTOMETHYL-1-OXO-3-PHENYLBUTYL)-L-ALA
S 21402-1	(2S)-2-[(2S, 3R)-2-THIOMETHYL-3-PHENYLBUTANAMIDO]PROPIONIC ACID
S 23142	N-[4-CHLORO-2-FLUORO-5-PROPAGYLOXY]PHENYL-3,4,5,6-TETRAHYDROPHTHALIMIDE
S 2720	6-CHLORO-3,3-DIMETHYL-4-(ISOPROPENYLOXYCARBONYL)-3,4-DIHYDROQUINOXALIN-2(1H)-THIONE

Trivial name	Synonym or systematic name
S 2864	N-[N-(3-(4-AMINO-1-PIPERIDINYLCARBONYL)2(R)BENZYLPROPIONYL)HISTIDINYL](2s,3r,4s)-1-CYCLOHEXYL-3,4-DIHYDROXY-6(2-PYRIDYL)HEXANE-2-AMIDE
S 3337	2-(ETHYLAMINOETHYLSULPHINYL)-5,6-DIMETHOXYBENZIMIDAZOLE
S 3483	1-{[2-(4-CHLOROPHENYL)CYCLOPROPYL]METHOXY}-3,4-DIHYDROXY-5-{[3-(4-HYDROXYPHENYL)-1-OXO-2-PROPENYL]OXY}CYCLOHEXANECARBOXYLIC ACID
SA 898	N-[(2R, 4R)-2-(2-HYDROXYPHENYL)-3-(3-MERCAPTOPROPIONYL)-4-THIAZOLIDINECARBONYL]-L-PHENYLALANINE
SA 6541	S-(4-DIMETHYLAMINOBENZYL)-N-[(2S)-3-MERCAPTO-2-METHYLPROPIONYL]CYSTEINE
SACCHAROPINE	(S)-N-(5-AMINO-5-CARBOXYPENTYL)-L-GLUTAMIC ACID
SAE 9	2-N,N-DIMETHYLAMINO-4,6-BIS(1-H-IMIDAZOL-1-YL)-1,3,5-TRIAZINE
SAFINGOL	L-threo-DIHYDROSPHINGOSINE
SAFINGOL	(2S, 3S) 2-AMINO-1,3-OCTADECANEDIOL
SALICYLAMIDE	2-HYDROXYBENZAMIDE
SALSOLIDINE	1-METHYL-6,7-DIMETHOXY-1,2,3,4-TETRAHYDROISOQUINOLINE
SALSOLINOL	1-METHYL-6,7-DIHYDROXY-1,2,3,4-TETRAHYDROISOQUINOLINE
SAM	S-ADENOSYL-L-METHIONINE
SAMH	S-AQRISTEROMYCENYL-L-HOMOCYSTEINE
SANDOZ COMPOUND 58-035	3-[DECYLDIMETHYLSILYL]-N-[2-(4-METHYLPHENYL)-1-PHENYLETHYL)PROPANAMIDE
SANGIVAMYCIN	4-AMINO-5-CARBOXAMIDE-7-(β-D-RIBOFURANOSYL)PYRROLO[2,3-d]PYRIMIDINE
SANOCRISIN	SODIUM GOLD THIOSULPHATE
SAQUANVIR	RO 31-8959
SAQUINAVIR	RO 31-8959
SARCOSINE	N-METHYL-GLYCINE
SARIN	ISOPROPYL METHYLPHOSPHONOFLUORIDATE
SATERINONE	(±)-1,2-DIHYDRO-5-[4-[2-HYDROXY-3-[4-(2-METHOXYPHENYL)-1-PIPERAZINYL]PROPOXYL]PHENYL-6-METHYL-2-OXO-3-PYRIDINECARBONITRILE
SB 202742	6-(HEPTADECATRIEN-8′,11′,14′-YL)-2-HYDROXY BENZOIC ACID
SB 203347	2-[2-[3,5-BIS(TRIFLUOROMETHYL)SULPHONAMIDO]-4-TRIFLUOROMETHYLPHENOXY]BENZOIC ACID
SB 210661	(S)-N-HYDROXY-N-[2,3-DIHYDRO-6-(2,6-DIFLUOROPHENYLMETHOXY)-3-BENZOFURANYL]UREA
SC 0051	2-(2-CHLORO-4-METHANESULPHONYLBENZOYL)-1,3-CYCLOHEXANEDIONE
SC 28080	2-METHYL-8-(PHENYLMETHOXY)IMIDAZOL[1,2-a]PYRIDINE-3-ACETONITRILE
SC 39026	(±)2-CHLORO-4-(1-HYDROXYOCTADECYL)BENZOIC ACID
SC 40827	N-[3-N-(BENZYLOXYCARBONYL)AMINO-1-(R)CARBOXYPROPYL]-L-LEUCYL-O-METHYL-L-TYROSINE-N-METHYLAMIDE
SC 44463	N4-HYDROXY-N1-{1S-[(4-METHOXYPHENYL)METHYL]-2-(METHYLAMINO-2-OXOETHYL}-2R-(2-METHYLPROPYL)-BUTANE DIAMIDE
SC 45662	±[2S*-[[3,5-BIS(1,1-DIMETHYLETHYL)-4-HYDROXYPHENYL]THIO]1R*-METHYLPROPOXY]ACETIC ACID
SC 57461	N-METHYL-N-[3-[4-(PHENYLMETHYL)PHENOXY]PROPYL]-β-ALANINE
SC 58451	5-(FLUOROPHENYL)-6[4-(METHYLSULFONYL)PHENYL]SPIRO[2,4]-HEPT-5-ENE
SC 58635	4-[5-(4-METHYLPHENYL)-3-(TRIFLUOROMETHYL)-1H-PYRAZOL-1-YL]BENZENESULPHONAMIDE
SC 415661A	3-[3,5-BIS(1,1-DIMETHYLETHYL)-4-HYDROXYPHENYL]THIO]-N-METHYL-N-[2-(2-PYRIDINYL)-ETHYL]PROPANAMIDE
SC 41661A	3-{3,5-BIS(1,1-DIMETHYL)-4-HYDROXYPHENYL)THIOL]-N-METHYL-N-[2-(2-PHRIDINYLPROPANAMIDE
SCH 28080	3-CYANOMETHYL-2-METHYL-8-(PHENYLMETHOXY)IMIDAZO[1,2-a]PYRIDINE
SCH 32615	N-[L-(1-CARBOXY-2-PHENYL)ETHYL]-L-PHENYLALANYL-β-ALANINE
SCH 34826	(S)-N-[N-[1-[[(2,2-DIMETHYL-1,3-DIOXALAN-4-YL)METHOXY]CARBONYL]-2-PHENYLETHYL]-L-PHENYLALANYL]-β-ALANINE
SCH 37224	1-(1,2-DIHYDRO-4-HYDROXY-2-OXO-1-PHENYL-1,8-NAPHTHYRIDIN-3-YL)PYRROLIDINIUM, HYDROXIDE INNER SALT
SCH 39370	N-[N-[1-(S)-CARBOXY-3-PHENYLPROPYL]-(S)-PHENYLALANYL]-(S)-ISOSERINE
SCH 40120	{10-(3-CHLOROPHENYL)-6,8,9,10-TETRAHYDROBENZO[b][1,8]NAPHTHYRIDIN-5(7H)ONE
SCH 42354	N-[2(S)-(MERCAPTOMETHYL)-3-(2-METHYLPHENYL)-1-OXOPROLYL]-L-METHIONINE
SCH 42354	[(2S)3(4-METHYLPIPERAZIN-1-YL)SULPHONYL2(PHENYLMETHYL)PROPIONYL]N(1S, 2R, 3S)-1-CYCLOHEXYLMETHYL)-2,3-DIHYDROXY-5-METHYLHEXYL]-L-[3-(THIAZOL-4-YL)ALANINAMIDE
SCH 44342	1-(4-PYRIDYLACETYL)-4-(8-CHLORO-5,6-DIHYDRO-11H-BENZO-[5,6]CYCLOHEPTA(1,2-b)PYRIDIN-11-YLIDENE)-PIPERIDINE

Trivial name	Synonym or systematic name
SCH 51866	cis-5,6a,7,8,9,9a-HEXAHYDRO-2-[4-(TRIFLUOROMETHYL)PHENYLMETHYL]-5-METHYL-CYCLOPENT[4,5]IMIDAZO[2,1-b]PURIN-4(3H)ONE
SCOPADULCIOL	6β-BENZOYL-12-METHYL-13-OXO-9(12)α,9(12)b-DIHOMO-18-PRODOCARPANOL
SCOPARIC ACID A	6-BENZOYL-12-HYDROXYLABDA-8(17),13-DIEN-18-OIC ACID
SCOPARIC ACID C	6-BENZOYL-15-NOR-14-OXO-8(17)LABDEN-18-OIC ACID
SDI 158	2-HYDROXYMETHYL-4-(4-N,N-DIMETHYLAMINOSULPHONYL-1-PIPERAZINO)PYRIMIDINE
SDZENA713	-(S)-N-ETHYL-3[(1-DIMETHYL-AMINO)ETHYL]-N-METHYLPHENYLCARBAMATE
SECCOSTEROID	(4R)-5,10-SECO-19-NORPREGNA-4,5-DIENE-3,10,20-TRIONE
SEF 19	2-IMIDAZOL-1-YL)-4,6-DIMORPHORINO-1,3,5-TRIAZINE
SELENAZOFURIN	2β-D-RIBOFURANOSYLSELENAZOLE-4-CARBOXAMIDE
SEZOLAMIDE	MK 417
SF 6847	(3,5-DI-tert.-BUTYL-4-HYDROXYBENZILIDENE)MALONITRILE
SF 86327	TERBINAFINE
SH 489	1-METHYL-1,4-ANDROSTADIENE-3,17-DIONE
SHOWDOMYCIN	3-β-D-RIBOFURANOSYL-1H-PYRROLE-2,5-DIONE
SIBA	5′-DEOXY-5′-S-ISOBUTYL-THIOADENOSINE
SIMVASTATIN	(1S,3R,7S,8S,8aR)-1,2,3,7,8,8a-HEXAHYDRO-3,7-DIMETHYL-8-[2-[(2R, 4R)-TETRAHYDRO-4-HYDROXY-6-OXO-2H-PYRAN-2-YL]ETHYL]-1-NAPHTHYL 2,2-DIMETHYLBUTANOATE
SIMVASTATIN	MK 733
SINEFUNGIN	6,9-DIAMINO-1-(6-AMINO-9H-PURIN-9-YL)-1,5,6,7,8,9-HEXADEOXY-DECOFURANURONIC ACID
SJA 6017	N-(4-FLUOROPHENYLSULPHONYL)-L-VALYL-L-LEUCINAL
SKF 104864	TOPOTECAN
SKF 104976	3β-HYDROXYLANOST-8,15-DIEN-32CARBOXYLIC ACID
SKF 10561	2-(4-METHYLTHIOPHENYL)-3-(4-PYRIDYL)-6,7-DIHYDRO-[5H]-PYRROLO[1,2-α]IMIDAZOLE
SKF 105657	EPRISTERIDE
SKF 105657	17β-(N-tert.-BUTYLCARBOXAMIDO)ANDROST-3,5-DIENE-3-CARBOXYLIC ACID
SKF 107461	Z-ALA-ALA-PHE-PSI[CHOHCH$_2$]GLY-VAL-ONE
SKF 29661	1,2,3,4-TETRAHYDRO-7-ISOQUINOLINESULPHONAMIDE
SKF 36914	CHLORO(TRIETHYLPHOSPHINE)GOLD
SKF 38393	2,3,4,5-TETRAHYDRO-7,8-DIHYDROXYL-1H-3-BENZAZEPINE
SKF 45905	2-[2-[3-(4-CHLORO-3-TRIFLUOROMETHYLPHENYL)UREIDO]-4-TRIFLUOROMETHYLPHENOXY]-4,5-DICHLOROBENZENESULPHONIC ACID
SKF 45905	2-[2-(3,4-DICHLORO-3-(TRIFLUOROMETHYL)PHENYL)UREIDO]-4-(TRIFLUOROMETHYLPHENOXY)-4,5-DICHLOROBENZENE SULPHONIC ACID
SKF 64139	7,8-DICHLORO-1,2,3,4-TETRAHYDROISOQUINOLINE
SKF 80544	BIS(TRIETHYLPHOSPHINE)GOLD
SKF 92508	1-METHYL-3-[2-(5-METHYL-4-IMIDAZOYL)METHYLTHIOETHYL]THIOUREA
SKF 92508	METIAMIDE
SKF 94120	5-(4-ACETIMIDOPHENYL)PYRAZIN-(1H)-ONE
SKF 94836	2-CYANO-1-METHYL-3-[4-(4-METHYL-6-OXO-1,4,5,6-TETRAHYDROPYRIDAZINE-3-YL)PHENYL]GUANIDINE
SKF 94836	SIGUAZODAN
SKF 95601	2-[[(3-CHLORO-4-MORPHOLINO-2-PYRIDYL)METHYL]SULPHINYL]-5-METHOXY-(1H)-BENZIMIDAZOLE
SKF 95654	4,5-DIHYDRO-6-[4-(1,4-DIHYDRO-4-OXOPYRIDIN-1-YL)PHENYL]-5-METHYL-3(2H)PYRIDAZINONE
SKF 96067	3-BUTYRYL-4-(2-METHYLPHENYLAMINO)-8-METHOXYQUINOLINE
SKF 96356	1-(2-METHYLPHENYL)-4-(METHYLAMINO)-6-METHYL-2,3-DIHYDROPYRROLO[3,2-cQUINOLINE
SKF 97574	3-BUTYRYL-4-(2-METHYLAMINO)-8-(2-HYDROXYETHOXY)QUINOLINE
SKF 98625	DIETHYL 7-(3,4,5-TRIPHENYL-2-OXO-2,3-DIHYDROIMIDAZOL-1-YL)HEPTANEPHOSPHONATE
SKF 98625	DIETHYL 7-(3m4m5-TRIPHENYL-2-OXO-2,3-DIHYDROIMIDAZOL-1-YL)HEPTATINE PHOSPHONATE
SKF 525A	DIMETHYLAMINOMETHYL-2,2-DIPHENYLVALERATE
SKF 525A	α-PHENYL-α-PROPYL-BENZENEACETIC ACID 2-(DIETHYLAMINO)ETHYL ESTER, HYDROCHLORIDE
SKF 525A	PROADIFEN
SKF 3301A	2-[(2,2-DIPHENYLPENTYL)OXY]-N, N-DIMETHYL-ETHANAMINE HYDROCHLORIDE
SKF 6817 A2	2-(2-AMINOETHYL)-5,6-DICHLOROBENZIMIDAZOLE
SKF 92994 A2	2-[2-(5-METHYL-1-IMIDAZOLYLMETHYLTHIO)-ETHYLAMINO]-5-(3,4-METHYLENE-DIOXYBENZYL)-4-PYRIMIDONE
SKF 92994 A2	OXMETIDINE
SKF TRANS 385	trans-(+-)-2-PHENYLCYCLOPROPYLAMINE HYDROCHLORIDE

Trivial name	Synonym or systematic name
SN 390	QUINACRINE
SN 5949	2-HYDROXY-3-(2-METHYLOCTYL)-1,4-NAPHTHALENEDIONE
SNK 860	(2S, 4S)-6-FLUORO-2,3-DIHYDRO-2′,5′-DIOXOSPIRO[4H-1-BENZOPYRAN-4,4′-IMIDAZOLIDINE]-2-CARBOXAMIDE
SNK 860	(2S, 4S)-6-FLUORO-2′,5′-DIOXOSPIRO[CHROMAN-4,4′-IMIDAZOLIDINE]-2-CARBOXAMIDE
SOLVIREX	O, O-DIETHYL-S-2-ETHYLTHIOETHYL PHOSPHORODITHIOATE
SOMAN	METHYLPHOSPHONOFLUORIDIC ACID 1,2,2-TRIMETHYLPROPYLESTER
SORBINIL	(S)-6-FLUORO-2,3-DIHYDRO-SPIRO[4H-1-BEZOPYRAN-4,4′-IMIDAZOLIDINE]-2′,5′-DIONE
SP 346	VAL-SER-GLN-ASN-TYR-PIPECOLIC ACID-ILE-VAL-GLN-NH2
SPERMIDINE	N-(3-AMINOPROPYL)-1,4-BUTANEDIAMINE
SPERMINE	N, N′-BIS(3-AMINOPROPYL)-1,4-BUTANEDIAMINE
SPHINGOFUNGIN B	2S-AMINO-3R, 4R, 5S, 14-TETRAHYDROXYEICOS-6-ENOIC ACID
SPHINGOFUNGIN C	2S-AMINO-2S-ACETOXY-3R, 4R, 14-TRIHYDROXYEICOS-6-ENOIC ACID
SPI 447	3-AMINO-5-METHYL-2-(2-METHYL-3-THIENYL)IMIDAZO[1,2-α]THIENO[3,2-c]PYRIDINE
SPI 501	1-BUTYL-3(6,7-DIMETHOXY-2-NAPHTHYLSULPHONYL)AMINO-3(3-GUANIDINOPROPYL)2-PYRROLIDONE
SPIRAPRILAT	SCH 33861
SPR 210	2-[4-(4,5,7-TRIFLUOROBENZOTHIAZOL-2-YL)m,4-DIHYDRO-2H-1,4-BENZOTHIAZIN-2-YL]ACETIC ACID
SQ 14225	CAPTORPIL
SQ 14225	D-3-MERCAPTO-2-METHYL PROPANOYL-L-PROLINE
SQ 20009	1-ETHYL-4-[(1-METHYLETHYLIDENE)HYDRAZINO]-1H-PYRAZOLO[3,4-b]PYRIDINE-5-CARBOXYLIC ACID ETHYL ESTER MONO HYDROCHLORIDE
SQ 20881	TEPROTIDE
SQ 21611	9-BENZYLADENINE
SQ 22536	9-(TETRAHYDRO-2-FURYL)ADENINE
SQ 28133	ES 34
SQ 28555	4-CYCLOHEXYL-[[[2-METHYL-1-(OXOPROPOXY)PROPOXY](4-PHENYLBUTYL)PHOSPHINYL]ACETYL]PROLINE
SQ 28603	N-[2-(MERCAPTOMETHYL)-1-OXO-3-PHENYLPROPYL]-β-ALANINE
SQ 29072	7-[(2-MERCAPTOMETHYL)-1-OXO-3-PHENYLPROPYL]AMINOHEPTANOIC ACID
SQ 29852	(S)-1-[6-AMINO-2-[[HYDORXY(4-PHENYLBUTYL)-PHOSPHINYL]OXY]-1-OXOHEXYL]-L-PROLINE
SQ 31000	CS 514
SQ 31429	LYS-ARG-ARG-TRP-LYS-LYS-ASN-PHE-ILE-ALA-VAL
SQ 32238	LYS-ARG-ARG-TRP-LYS-LYS
SQ 32732	LYS-ARG-ARG-(1-NAPHTYLALANIE)-LYS-LYS
SQ 33600	(S)-4-[[[1-(4-FLUOROPHENYL)-3-1-METHYLETHYL)-1H-INDOL-2-YL]ETHYNYL]HYDROXYPHOSPHINYL-3-HYDRO-XYBUTANOIC ACID
SQUALESTATIN I	ZARAGOZIC ACID A
SR 26831	[[5-(2-CHLOROBENZYL-2-(TERBUTYLOXYCARBONYL)]-4,5,6,7-TETRAHYDROTHIENO(3,2-c)PYRIDINE]N-OXIDE
SR 31747	(Z)N-CYCLOHEXYL-N-ETHYL-3-(3-CHLORO-4-CYCLOHEXYLPHENYL)PROPEN-2-YLAMINE
(+)-S R 76713	VOROZOLE
SRI 62834	(±)-2-(HYDROXY[TETRAHYDRO-2-(OCTADECYLOXY)METHYLFURAN-2-YL]METHOXY)PHOPSHINYLOXY-N,N,N-TRIMETHYLETHANIMINIUM HYDROXIDE
ST 43	N-[2-(MERCAPTOMETHYL)-1-OXO-3-PHENYLPROPYL]-L-PHENYLALANINE
ST 638	α-CYANO-3-ETHOXY-4-HYDROXY-5-PHENYLTHIOMETHYLCINNAMAMIDE
STATIL	PONALRESTAT
STATINE	4(S)-AMINO-3(S)-HYDORXY-6-METHYLHEPTANOIC ACID
STEFFIN D1	PIG LEUKOCYTE CYSTEINE PROTEINASE INHIBITOR
STH	S-TUBERCIDINYL-L-HOMOCYSTEINE
STIPITATIC ACID	3,6-DIHYDROXY-5-OXO-1,3,6.CYCLOHEPTATRIENE-1-CARBOXYLIC ACID
STIPITATONIC ACID	4,7-DIHYDROXY-1H-CYCLOHEPTA[C]FURAN-1,3,6-TRIONE
STM	S-TUBERCIDINYL-L-METHIONINE
STREPTIDINE	N,N‴-(2,4,5,6-TETRAHYDROXY-1,3-CYCLOHEXANEDIYL)BISGUANIDE
SU 10603	7-CHLORO-3,4-DIHYDRO-2-(3-PYRIDINYL)-1(2H)-NAPHTHALENONE
SU 8000	3-(6-CHLORO-3-METHYL-1H-INDEN-2-YL)-PYRIDINE
SUAM 1221	4-PHENYLBUTYRYLPROLYLPYRROLIDINE
SUCCINIC ACID SEMI-ALDEHYDE	4-OXOBUTANOIC ACID
SUCCINYLACETONE	4,6-DIOXOHEPTANOIC ACID

Trivial name	Synonym or systematic name
SULBACTAM	(2S-cis)-3,3-DIMETHYL-7-OXO-4-THIA-1-AZABICYCLO[3.2.0]HEPTANE-2-CARBOXYLIC ACID 4,4-DIOXIDE
SULCOTRIONE	2-(2-CHLORO-4-METHYNESULPHONYLBENZOYL)-1,3-CYCLOHEXANEDIONE
SULFORAPHANE	1-ISOTHIOCYANATE-4-METHYLSULPHINYLBUTANE
SULINDAC	5-FLUORO-2-METHYL-1-[[4-(METHYLSULPHINYL)PHENYL]METHYLENE]-1H-INDENE-3-ACETIC ACID
SULINDAC	MK 231
SULPHANILAMIDE	4-AMINO-BENZENESULPHONAMIDE
SULPHAQUINOXALINE	N1-(2-QUINAXALINYL)SULPHANILAMIDE
SURAMIN	8,8′-[CARBONYLBIS[IMINO-3,1-PHENYLENECARBONYLIMINO(4-METHYL-3,1-PHENYLENE)CARBONYLIMINO]]-BIS-1,3,5-NAPHTHALENETRISULPHONIC ACID
SWAINSONINE	(1S, 2R, 8R, 8αR)-1,2,8-TRIHYDROXYOCTAHYDROXINDOLIZIDINE
SYNTHALIN	DECAMETHYLENEDIGUANIDINE
T 330	2-[(2-DIMETHYLAMINOBENZYL)SULFINYL]-1-METHYLPYRIDINE-2-YL)IMIDAZOLE
T 440	1-[1-(2-METHOXYETHYL)PYRID-2-ONE-4-YL]-2,3-BIS(HYDROXYMETHYL)-6,7-DIETHOXYNAPHTHALENE
T 614	3-FORMYLAMINO-7-METHYLSULPHONYLAMINO-6-PHENOXY-4H-1-BENZOPYRAN-4-ONE
T 794	(5R)-3-(6-(CYCLOPROPYLMETHOXY)-2-NAPHTHALENYL)-5-(METHOXYMETHYL)-2-OXAZOLIDONE
T 2591	1-(3-t-BUTYL-2-HYDROXY-5-METHOXYPHENYL)-3-(2-CYCLOHEXYLETHYL)-3-(4-DIMETHYLAMINOPHENYL)-UREA
TA 6366	(4S)-1-METHYL-3-{(2S)-2-[N-((1S)-1-ETHOXYCARBONYL-3-PHENYLPROPYL)AMINO]PROPIONYL}-2-OXO-IMIDAZO-LIDINE-4-CARBOXYLIC ACID
TA 3037A	(Z)-3-BENZILYDENE-3,4-DIHYDRO-2-OXO-2H-1,4-BENZOXAZINE-5-CARBOXYLIC ACID
TABTOXIN	4-(3-HYDROXY-2-OXO-3-AZETIDINYL)-L-2-AMINOBUTANOYL-L-THREONINE
TABTOXININE-β-LACTAM	2-AMINO-4-(-HYDROXY-2-OXOAZACYCLOBUTAN-3-YL)BUTANOIC ACID
TABUN	DIMETHYLPHOSPHOROAMIDOCYANIDIC ACID ETHYL ESTER
TACRINE	9-AMINO-1,2,3,4-TETRAHYDROACRIDINE
TACRINE	1,2,3,4-TETRAHYDRO-9-ACRIDINAMINE
TACRINE	TETRAHYDROAMINOACRIDINE
TAK 144	3-[1-(PHENYLMETHYL)-4-PIPERIDINYL]-1-(2,3,4,5-TETRAHYDRO-1H-1-BENZAZEPIN-8-YL)-1-PROPANONE FUMARATE
TAK 147	3-[1-(PHENYLMETHYL)-4-PIPERIDINYL]-1-(2,3,4,5-TETRAHYDRO-1H-1-BENZAZEPIN-8-YL)-1-PROPANONE FUMA-RATE
TAME	N-TOLUENESULPHONYL-L-ARGINE METHYL ESTER
TAS 103	6-[[2-(DIMETHYLAMINO)ETHYL]AMINO]-3-HYDROXY-7H-INDENO[2,1-c]QUINOLIN-7-ONE
TAURINE	2-AMINOETHANESULPHONIC ACID
TAUROCHOLIC ACID	2-[[(3α,5β,7α,12α)-3,7,12-TRIHYDROXY-24-OXOCHOLAN-24-YL]AMINO]-ETHANESULPHONIC ACID
TAUROLIPID A	2-(3-ACYLOXY-7,13-DIHYDROXYOCTADECANOYL)AMINOETHANESULPHONIC ACID
TAUROLIPID B	2-(3-ACYLOXY-2,7,13-TRIHYDROXYOCTADECANOYL)AMINOETHANESULPHONIC ACID
TAZOBACTAM	3-METHYL-7-OXO-3-(1H-1,2,3-TRIAZOL-1-YLMETHYL)-4-THIA-1-AZABICYCLO[3.2.0]-HEPTANE-2-CARBOXYLIC ACID, 4,4-DIOXIDE
TCNP	1β-D-RIBOFURANOSYL-1,2,4-TRIAZOLE-3-CAROXAMIDE
TEI 5624	7-(4-CHLOROPHENYLSULPHONYL-L-GLUTAMYL)AMINO-5-METHYL-2-ISOPROPYLAMINO-4H-3,1-BENZOXAZIN-4-ONE
TEI 6344	7-(4-CHLOROPHENYLSULPHONYL-L-LYSYL)AMINO-5-METHYL-2-ISOPROPYLAMINO-4H-3,1-BENZOXAZIN-4-ONE
TEI 6522	N-(7-METHOXY-4-OXOCHROMAN-8-YL)-2,2-DIMETHYLDODECANAMIDE
TEI 6620	N-[5-(DIMETHYLAMINO)2,2,4,6-TETRAMETHYL-2,3-DIHYDROBENZOFURAN-7-YL]-2,2-DIMETHYLDODECYL-AMIDE
TEI 6720	2-(3-CYANO-4-ISOBUTOXYPHENYL)-4-METHYL-5-THIAZOLECARBOXYLIC ACID
TENIPOSIDE	4′-DEMETHYLEPIPODOPHYLLOTOXINTHENYLIDENE-β-D-GLUCOSIDE
TENOX PG	3,4,5-TRIHYDROXY-BENZOIC ACID, PROPYLESTER
TENTOXIN	cyclo(LEUCYL-N-METHYL-(Z)-DEHYDROPHENYLALANYL-GLYCYL-N-METHYL-ALANYL)
TEPOXALIN	5-(4-CHLOROPHENYL)-N-HYDROXY-1-(4-METHOXYPHENYL)-N-METHYL-1H-PYRAZOLE-3-PROPANAMIDE
TEPOXALIN	ORF 20485
TEPOXALIN	RWJ 20485
TEPROTIDE	2-L-TRYPTOPHAN-3-DE-L-LEUCINE-4-DE-L-PROLINE-8-L-GLUTAMINE-BRADIKININ POTENTIATOR B
TERTHIENYL	2,2′: 5,2″-TERTHIPHENE
TETCYCLASIS	5-[4-CHLOROPHENYL]-3,4,5,9,10-PENTAZOTETRACYCLO-5,4,1,O2,6,O8,11-DODECA-3,9-DIENE
TETRACAINE	4-[BUTYLAMINO]BENZOIC ACID 2-[DIMETHYLAMINO]ETHYL ESTER
TETRADYMOL	10β-HYDROXYFURANOEREMOPHILANE

Trivial name	Synonym or systematic name
7-TETRAHYDROBIOPTERIN	2-AMINO-4-HYDROXY-7-[DIHYDROXYLPROPYL-(erythro)-5,6,7,8-TETRAHYDROPTERIN
TETRAHYDROLIPSTAIN	ORLISTAT
TETRAMISOLE	2,3,5,6-TETRAHYDRO-6-PHENYL(+-)IMIDAZO[2,1-b]THIAZOLE
TETRAPLATIN	NSC 363812
TETRAPLATIN	TETRACHLORO(D,L-trans)-1,2-DIAMINOCYCLOHEXANE PLATIUNIUM
TETRODOTOXIN	OCTAHYDRO-12-(HYDROXYMETHYL)-2-IMINO-5,9: 7,10A-DIMETHANO-10AH-[1,3]DIOXOCINO[6,5-D]PYRIDINE-4,7,10,11,12-PENTOL
TETROLIC ACID	2-BUTYNOIC ACID
TETROXOPRIM	5-[[3,5-DIMETHOXY-4-(2-METHOXYETHOXY)PHENYL]METHYL]-2,4-PYRIMIDINEDIAMINE
THEONYLTRIFLUORO-ACETONE	4,4,4-TRIFLUORO-1-(2-THIENYL)-1,3-BUTANEDIONE
THERMOZYMOCIDIN	MYRIOCIN
THIABENDAZOLE	2-[4-THIAZOLYL)-1H-BENZIMIDAZOLE
THIENAMYCIN	[5R-[5α,6α(R*)]]-3-[(2-AMINOETHYL)THIO]-6-(1-HYDROXYETHYL)-7-OXO-1-AZABICYCLO[3.2.0]HEPT-2-ENE-2-CARBOXCYLIC ACID
THIENOTHIOPYRAN-2-SULPHONAMIDE	(–)-(S,S)-4-ETHYLAMINO-5,6-DIHYDRO-6-METHYL-7,7-DIOXIDE-4H-THIENO(2,3-b)-THIOPYRAN-2-SULPHONAMIDE
THIENOTHIOPYRAN-2-SULPHONAMIDE	TRUSOPT
6-THIODIHYDROOROTATE	(4r)-2-OXO-6-THIOXOHEXAHYDROPYRIMIDINE-4-CARBOXYLIC ACID
THIORPHAN	DL-3-MERCAPTO-2-BENZYLPROPANOYL-GLYCINE
THYROXINE	O-(4-HYDROXY-3,5-DIIODOPHENYL)-3,5-DIIODO-L-TYROSINE
TIBA	2,3,5-TRIIODOBENZOIC ACID
TIC	(L)-1,2,3,4-TETRAHYDRO-3-ISOQUINOLINECARBOXYLIC ACID
TIGLYL-CoA	TRANS-2-METHYL-2-BUTENOIC ACID CoA ESTER
TILORONE	2,7-BIS[2-(DIETHYLAMINO)ETHOXY]-9H-FLUOREN-9-ONE
TIMP	TISSUE INHIBITOR OF METTALLOPROTEINASES
TIRON	4,5-DIHYDROXY-1,3-BENZENE DISULPHONIC ACID
TISSUE FACTOR PATHWAY INHIBITOR	LIPOPROTEIN ASSOCIATED COAGULATION INHIBITOR
TLCK	N-TOSYL-L-LYSINE CHLOROMETHYLKETONE
TMGA	TETRAMETHYLENEGLUTARIC ACID
TMK 688	1-{[5′-(3″-METHOXY-4″-ETHOXYCARBONYLOXYPHENYL)-2′,4′-PENTADIENOYL]AMINOETHYL}-4-DIPHENYLMETHOXYPIPERIDINE
TMK 688	1-{[5′-(3″-METHOXY-4″-HYDROXYPHENYL)-2′,4′-PENTADIENOYL]AMINOETHYL}-4-DIPHENYLMETHOXYPIPERIDINE
TMK 777	1-[{5′-(3″-METHOXY-4″-ETHOXYCARBONYLOXYPHENYL)-2′,4′-PENTADIENOYL}AMINOETHYL]-4-DIPHENYLMETHOXYPIPERIDINE
TMP 153	N-[4-(2-CHLOROPHENYL)-6,7-DIMETHYL-3-QUINOLYL]-N′-(2,4-DIFLUOROPHENYL)UREA
TNP 351	N{4-[3-2,4-DIAMINO-7H-PYRROLO[2,3-d]PYRIMIDIN-YL)PROPYL]BENZOYL}-GLUTAMIC ACID
TOFA	5-(TETRADECYLOXY)-2-FUROIC ACID
TOFYL-CoA	(5-TETRADECYLOXY)-2-FUROYL-CoA
TOGAMYCIN	4-AMINO-5-CYANO-7-(β-D-RIBOFURANOSYL)-PYRROLO[2,3-d]PYRIMIDINE
TOLAFENTRINE	(cis-8,9-DIMETHOXY-2-METHYL-6-[4-p-TOLUENESULPHONAMIDOPHENYL]-1,2,3,4,4a,10b-HEXAHYDROBENZO-[c][1,6]-NAPHTHYRIDINE
TOLCAPONE	3,4-DIHYDROXY-4′-METHYL-2-NITROBENZOPHENONE
TOLOXATONE	5-HYDROXYMETHYL-3-m-TOLYLOXAZOLIDIN-2-ONE
TOLRESTAT	ALREDASE
TOLRESTAT	N-([5-(TRIFLUORO-METHYL)-6-METHOXY-1-NAPHTHALENYL]-THIOXOMETHYL)-N-METHYLGLYCINE
TOLYPOCLADIN	3-METHYL-5,6,8-TRIHYDROXY-2-AZAANTHRAQUIONONE-(9,10)
TOMUDEX	ICI D1694
TOMUDEX	ZD 1694
TPCK	L-1-TOSYLAMIDO-2-PHENYL(ETHYL) CHLOROMETHYLKETONE
Trp-P-2	3-AMINO-1METHYL-5H-PYRIDO(4,3-b)INDOLE
TRANYLCYPROMINE	trans(+-)-2-PHENYL-CYCLOPROPANAMINE
TRASYLOL	APROTININ
TREHAZOLINN 2	TREHALOSTATIN

Trivial name	Synonym or systematic name
TREQUINSIN	9,10-DIMETHOXY-2-MESITYLIMINO-3-METHYL-3,4,6,7-TETRAHYDRO-2H-PYRIMIDO(6,1-a)ISOQUINOLIN-4-ONE
TRIACSIN A	1-HYDROXY-3(E, E, -2$'$,4$'$-UNDECADIENYLIDENE)TRIAZINE
TRIACSIN C	1-HYDROXY-3-(E, E, E, -2$'$,4$'$,7$'$-UNDECADIENYLIDINE)TRIAZENE
TRIARIMOL	α-(2,4-DICHLOROPHERNYL)-α-PHENYL-5-PYRIMIDINE METHANOL
TRIBULIN A	4-HYDROXYPHENYLETHANOL
TRICHOSTATIN A	7-[4-(DIMETHYLAMINO)PHENYL]-N-HYDROXY-4,6-DIMETHYL-7-OXO-2,4-HEPTADIENAMIDE
TRICIRIBINE PHOSPHATE	6-AMINO-4-METHYL-8-(β-D-RIBOFURANOSYL)PYRROLO[2,3,4-de]PYRIMIDO[4,5-c]PYRIDAZINE 5$'$-PHOSPHATE
TRIDEMORPH	2,6-DIMETHYL-N-TRIDECYLMORPHOLINE
TRIFLUOPERAZINE	10-[3-(4-METHYL-PIPERAZINYL)PROPYL]-2-(TRIFLUOROMETHYL)-10H-PHENOTHIAZINE
TRILOSTANE	(4α,5α,17β)-4,5-EPOXY-3,17-DIHYDROXY-ANDROST-2-ENE-2-CARBONITRILE
TRIMESIC ACID	1,3,5-BENZENTRICARBOXYLIC ACID
TRIMETAZIDINE	1-(2,3,4-TRIMETHOXYLBENZYL)PIPERAZINE DIHYDROCHLORIDE
TRIMETHOPRIM	5-[(3,4,5-TRIMETHOXYPHENYL)METHYL-2,4-PYRIMIDINEDIAMINE
TRIMIDOX	3,4,5-TRIHYDROXYBENZAMIDOXIME
TRIMIDOX	3,4,5-TRIHYDROXYBENZOHYDROXAMIDOXIME
TRIPARANOL	4-CHLORO-α-[4-[2-(DIETHYLAMINO)ETHOXY]PHENYL]-α-(4-METHYLPHENYL)BENZENEETHANOL
TRIS	2-AMINO-2-(HYDROMETHYL)-1,3-PROPANEDIOL
TROPOLONE	2-HYDROXY-2,4,6-CYCLOHEPTATRIEN-1-ONE
TRYPSIN/CHYMOTRYPSIN INHIBITOR	BOWMAN BIRK PROTEASE INHIBITOR
TRYPSIN/CHYMOTRYPSIN INHIBITOR	BOWMAN BIRK SOY BEAN PROTEASE INHIBITOR
TSAA 291	16β-ETHYL-17β-HYDROXY-4-ESTREN-3-ONE
TSAO-T	[2$'$,5$'$-BIS-O-(tetr-BUTYLDIMETHYLSILYL)-β-D-RIBOFURNAOSYL]-$'$-SPIRO-5$''$(4$''$-AMINO-1$''$,2$''$-OXATHIOLE-2$''$,24 DIOXIDE)THYMINE
TUBERCIDIN	7-DEAZAADENOSINE
S-TUBERCIDINYL-METHIONINE	(S)-7-DEAZA-ADENOSYL-L-METHIONINE
TURANOSE	3-O-α-D-GLUCOPYRANOSYL-D-FRUCTOSE
TUROSTERIDE	FCE 26073
TUROSTERIDE	1-[4-METHYL-3-OXO-4-AZA-5α-ANDROSTANE-17β-CARBONYL]-1,3-DIISOPROPYLUREA
TUT 7	MENOGARIL
TY 11345	($\pm$)-2-[(4-METHOXY-6,7,8,9-TETRAHYDRO-5H-CYCLOHEPTA[b]PYRIDIN-9-YL)SULPHINYL]-1H-BENZIMIDAZOLE SODIUM SALT
TY 11345	($\pm$)-2-[(4METHOXY-6,7,8,9-TETRAHYDRO-5H-CYCLOHEPTA[b]PYRIDIN-9-YL]SUPHINYL]-1H-BENZIMIDAZOLE
TYROMYCIN	1,16-BIS[4-METHYL-2,5-DIOXO-3-FURYLHEXADECANE
TYRPHOSTIN 1	[4-METHOXYBENZYLIDENE]MALONITRILE
TYRPHOSTIN 23	[3,4-DIHYDROXYBENZYLIDENE]MALONITRILE
TYRPHOSTIN 25	[3,4,5-TRIHYDROXYBENZYLIDENE]MALONITRILE
TYRPHOSTIN 46	3,4-DIHYDROXY-α-CYANOCINNAMAMIDE
TYRPHOSTIN 47	3,4-DIHYDROXY-α-CYANOTHIOCINNAMAMIDE
TYRPHOSTIN 51	2-AMINO-1,1,3-TRICYANO-4-[3$'$,4$'$,5$'$-TRIHYDROYXYPHENYL]BUTADIENE
TYRPHOSTIN 63	[4-HYDROXYBENZYL]MALONITRILE
TZC 5665	6-[4-[2-[3-(5-CHLORO-2-CYANOPHENOXY)-2-HYDROXYPROPYLAMINO]-2-METHYLPROPYLAMINO]PHENYL]-4,5-DIHYDRO-5-METHYL-3(2H) PYRIDAZINONE MONOETHYL MALEATE
TZI 41127	2-(4-HYDROXY-3,4-DIMETHYLPHENYL)-5-METHOXY-4-METHYLINDOLE
U 0126	1,4-DIAMINO-2,3-DICYANO-1,4-BIS[2-AMINOPHENYLTHIO]BUTADIENE
U 0521	3$'$,4$'$-DIHYDROXY-2-METHYLPROPIOPHENONE
U 103017	4-CYANO-N-(3-CYCLOPROPYL-(5,6,7,8,9,10-HEXAHYDRO-4-HYDROXY-2-OXO-2H-CYCLOOCTA(b)PYRAN-3-YL)-METHYL)PHENYL)BENZENESULPHONAMIDE
U 31355	4-AMINO-2-(BENZYLTHIO)-6-CHLOROPYRIMIDINE
U 46619	9,11-(METHANOEPOXY)-15(S)-HYDROXYPROSTA-5Z,13E-DIENOIC ACID
U 54701	9,11-IMINOEPOXYPROSTA-5,13-DIENOIC ACID
U 57908	CYCLOHEXANONE-O, O$'$-[(1,6-HEXADIENYL)-BIS(IMINOCARBONYL)]DIOXIME
U 57908	RHC 80267
U 60257	6,9-DEEPOXY-6,9(PHENYLIMINO)-Δ6,8-PROSTAGLANDIN I1

Trivial name	Synonym or systematic name
U 63557a	5-(3′-PYRIDINYLMETHYL)BENZOFURAN-2-CARBOXYLATE
U 71038	DITEKIREN
U 73122	1-[6-[[17β-3-METHOXYESTRA-1,3,5(10)-TRIEN-17-YL]AMINO]HEXYL]-1H-PYRROLE-2,5-DIONE
U 78036	1-(4-CHLOROBENZOYL)-1,2-DIHYDRO-2-QUINOLINECARBONITRILE
U 9843	ETHYLENESULPHONIC ACID TETRAMER
U 18666A	3β-(2-DIETHYLAMINOETHOXY)ANDROSTENONE
U 19451A	2-BENZYL-2-THIO-PSEUDOUREA
U 87201E	ATEVIRDINE
U 90152E	1-(5-METHANESULFONAMIDO-1H-INDOL-2-YL-CARBONYL)-4-[3-(1-METHYLAMINO)PYRIDINYL]PIPERAZINE
U 90152S	DELAVIRIDINE
U 90152S	1-(5-METHANESULFONAMIDO-1H-INDOL-2-YL-CARBONYL)-4-[3-(1-METHYLAMINO)PYRIDINYL]PIPERAZINE
348U87	2-ACETYLPYRIDINE 5-[(2-CHLOROANILO)-THIOCARBONYL]-THIOCARBONOHYDRAZONE
1843U89	GW 1843
UBENIMEX	BESTATIN
UBRETID	DISTIGMINE BROMIDE
UC 38	NSC 629243
UC 781	N-[4-CHLORO-3-(3-METHYL-2-BUTENYLOXY)PHENYL]-2-METHYL-3-FURACARBOTHIOAMIDE
UCF1C	MANUMYCIN
UCN 01	7-HYDROXY STAUROSPORIN
UCN 1028C	CALPHOSTIN C
UDP NAG	UDP-N-ACETYL-GLUCOSAMINE
UK 14275	BUQUINERAN
UK 31557	CARBAZERAN
UK 35493	6,7-DIMETHOXY-4-[4-(2-(ISOTHIAZOLIDNYL)ETHYL-1-PIPERIDINYL]QUINAZOLINE S, S DIOXIDE
UK 37248-01	DAZOXIBEN
UK 61260	6-(2,4-DIMETHYL-1H-IMIDAZOL-1YL)-8-METHYL-2(1H)QUINOLINE METHANESULPHONATE HYDRATE
UK 61260	NANTERINONE
UK 69578	(±)CANDOXATRILAT
UK 69578	cis-4-{[2-CARBOXY-3-(2-METHOXYETHOXY)PROLYL-1-CYCLOPENTANECARBONYLAMINE}-1-CYCLOHEXANE CARBOXYLIC ACID
UK 79300	CANDOXATRIL
UM 26	4′-DEMETHYLEPIPODOPHYLLOTOXIN 9-(4,6-O-THIENYLIDENE-β-D-GLUCOPYRANOSIDE
URIC ACID	7,9-DIHYDRO-1H-PURINE-2,6,8(3H)-TRIONE
URINASTATIN	URINARY TRYPSIN INHIBITOR
UROCANIC ACID	3-(1H-IMIDAZOL-4-YL)-2-PROPENOIC ACID
2-VAD-DCB	2-VALYL-ALANYL-3(S)-3-AMINO-4-OXO-5-(2,6-DICHLOROBENZOYLOXOPENTANOIC ACID
VALPROIC ACID	2-PROPYLPENTANOIC ACID
VANILLIC ACID	4-HYDROXY-3-METHOXY-BENZOIC ACID
VANILLIN	4-HYDROXY-3-METHOXYBENZALDEHYDE
VELNACRINE	9-AMINO-1,2,3,4-TETRAHYDRO-1-ACRIDINOL
VERAPAMIL	α-[3-[[2-(3,4-DIMETHOXYPHENYL)ETHYL]METHYLAMINO]PROPYL]-3,4-DIMETHOXY-α-(1-METHYLETHYL)-BENZENEACETONITRILE
VERBENONE	4,6,6-TRIMETHYLBICYCLO 3.1.1 HEPT-3-EN-2-ONE
VESNARINONE	3,4-DIHYDRO-6-[4-(3,4-DIMETHOXYBENZOYL)-1-PIERAZINYL]-2(1H)-QUINOLINONE
VIAGRA	SILDENAFIL
γ-VINYL GABA	4-AMINOHEX-5-ENOIC ACID
VK 14275	BUQUINERAN
VK 19911	4-(4-FLUOROPHENYL)-1-(4-PIPERIDINYL)-5-(4-PYRIDYL)IMIDAZOLE
VK 31557	CARBAZERAN
VM-26	TENIPOSIDE
VOLTAREN	DICLOFENAC
VP 16	ETOPOSIDE
VP 16-123	ETOPOSIDE
VP 616L	2-DEOXY-2-FLUORO-scyllo-INOSITOL-1-O-DODECYLPHOSPHONATE
VZ 564	N-HYDROXY-N-(6-METHOXY-3,4-DIHYDRO-NAPHTHYLMETHYL)UREA
W 7	CALMIDAZOLIUM

Trivial name	Synonym or systematic name
W 7	N-(6-AMINOHEXYL)-5-CHLORO-1-NAPHTHALENESULPHONAMIDE HYDROCHLORIDE
W 77	(S)-p-(2-AMINOETHYLOXY)-N-[2-(4-BENZYLOXYCARBONYLPIPERAZINYL)-1-(p-METHOXYBENZYL)ETHYL]-N-METHYLBENZENESULPHONAMIDE 2HCl
W-5	N-(6-AMINOHEXYL)-1-NAPHTHALENESULPHONAMIDE HYDROCHLORIDE
W-12	N-(6-AMINOBUTYL)-2-NAPHTHALENESULPHONAMIDE HYDROCHLORIDE
W-13	N-(6-AMINOBUTYL)-5-CHLORO-2-NAPHTHALENESULPHONAMIDE HYDROCHLORIDE
1400 W	N-[3-(AMINOMETHYL)BENZYL]ACETAMIDINE
WARFARIN	3α-PHENYL-β-ACETYLETHYL-4-HYDROXYCOUMARIN
WAY 121509	2[(4-BROMO-2-FLUOROPHENYL)METHYL]SPIRO[6-FLUOROISOQUINOLINE-4-(1H)3′-PYRROLIDI-NE]1,2′,3,5′(2H)TETRONE
WF 3681	3-(4-HYDROXY-5-OXO-3-PHENYL-2,5-DIHYDRO-2-FURYL)PROPIONIC ACID
WIN 41662	3-PHENYL-N1-[1-(4-PYRIDYL)PYRIMIDINE]HYDRAZONE
WIN 47203-2	MILRINONE
WIN 61651	1,4-DIHYDRO-7-(4-METHYL-1-PIPERIZINYL)-1-(4-(4-METHYL-1-PIPERIZINYL))PHENYL-4-OXO-3-QUINOLINE-CARBOXAMIDE
WIN 61691	8-(3-BUTYLSULPHONAMIDOPHENYL)-2,3-DIBROMOIMIDAZOL[1,2-d][1,2,4]TRIAZIN-5(6H)-ONE
WIN 62785	4(1-METHYLETHYL)-2-[(1-PHENYL-TETRAZOL-5-YLTHIO)METHYL]-1,2-BENZISOTHIAZOL-3(2H)-ONE 1,1-DIOXIDE
WIN 67694	Z-VAL-ALA-ASP-CH_2O(CO)[2,6-(Cl-2)]PH
WR 1065	2-[(AMINOPROPYL)AMINO]ETHANETHIOL
WS 1228A	TRIASCIN C
WS 7622A	FR 901277
WSP 3	3-(4′-AMINOPHENYL)PYRROLIDINE-2,5-DIONE
WY 14643	[4-CHLORO-6-(2,3-XYLININO)-2-PYRIMIDINYLTHIO]ACETIC ACID
WY 26769	2,3-DIHYDRO-2-(2-PYRIDINYL)THIAZOLO-[3,2-a]BEZIMIDAZOLE
WY 27569	1,4-DIHYDRO-2-[IMIDAZOL-1-YL-METHYL]-6-METHYL-4-[3-NITROPHENYL]PYRIDINE-3,5-DICARBOXYLIC ACID
WY 47288	2-[(1-NAPHTHALENYLOXY)METHYL]QUINOLINE
WY 50295	S-(+)-α-METHYL-6-(2-QUINOLINYLMETHOXY)-2-NAPHTHALENEACETIC ACID
WY 50295 SALT	TROMETHAMINE
WY 50295 T	S-(+)-αMETHYL-6-(2-QUINOLINYLMETHOXY)-2-NAPHTHALENEACETIC ACID 2-HYDROXY-1,1-BIS(HYDROXY-METHYL)ETHYLAMINE SALT
XANTHINE	3,7-DIHYDRO-1H-PURINE-2,6-DIONE
XANTHURENIC ACID	4,8-DIHYDROXY-2-QUINOLINECARBOXYLIC ACID
XIMENIC ACID	OCTADEC-trans-11-EN-9-YNOIC ACID
XM 323	[4R-(4α,5α,6β)]-HEXAHYDRO-5,6-BIS(HYDROXY)-1,3-BIS([4-(HYDROYMETHYL)PHENYL]METHYL)-4,7-BIS(PHE-NYLMETHYL)-2H-1,3-DIAZEPIN-2-ONE
XU 62320	FLUVASTATIN
Y 20811	4-[α-HYDROXY-5-(1-IMIDAZOYL)-2-METHYLBENZYL]-3,5-DIMETHYLBENZOIC ACID
Y 29794	2-(8-DIMETHYLAMINOOCTYLTHIO)-6-ISOPROPYL-3-PYRIDYL 2-THIENYL KETONE CITRATE
YH 1885	5,6-DIMETHYL-2-(4-FLUOROPHENYLAMINO)-4-(1-METHYL-1,2,3,4-TETRAHYDROISOQUINOLIN-2-YL PYRIMIDINE
YJA 20379-5	2-AMINO-4,5-DIHYDROPYRIDO[1,2-a]THIAZOLO[5,4-g]BENZIMIDAZOLE
YLS 820340	METHYL-5-[2-CHLORO-4-(TRIFLUOROMETHYL)PHENOXY]-2-CHLOROBENZOIC ACID
YM 175	CYCLOHEPTYLAMINOMETHYLENE-1,1-BISPHOSPHONIC ACID
YM 511	4-[(4-BROMOBENZYL)(4-CYANOPHENYLAMINO]-4H-1,2,4-TRIAZOLE
YM 553	5-[(4-CYANOPHENYL)(3,5-DIFLUOROBENZYL)AMINO]PYRIMIDINE
YM 190202	3-CYANOMETHYL-2-METHYL-8-[(3-METHYL-2-BUTENYL9OXY]IMIDAZO[1,2-α]PYRIDINE
YM 190202	YM 020
YM 21095	(2R, S),(3S)-3-[N|α-[1,4-DIOXO-4-MORPHOLINO-2-(1-NAPHTHYLMETHYL)BUTYL]-L-HISTIDYLAMINO]CYCLOHE-XYL-1-[(1-METHYL-5-TETRAZOYL)THIO]-2-BUTANOL
YM 26365	(3R)-3-[3[(1S)-1-CYCLOHEXYLMETHYL-2-HYDROXY-3-[(1-METHYL-5-TETRAZOLYL)THIO]PROPYL]UREIDO]-1-METHYL-5-PHENYL-2,3-DIHYDRO-1H-1,4-BENZODIAZEPIN-2-ONE
YM 26567-1	(+)trans-4-(3-DODECANOYL-2,4,6-TRIHYDROXYPHENYL)-7-HYDROXY-2-(4-HYDROXYPHENYL)CHROMAN
YM 26734	4-(3,5-DIDODECANOYL-2,4,6-TRIHYDROXYPHENYL)-7-HYDROXY-2-(4-HYDROXYPHENYL)CHROMAN
YM 51084	IVAL-TYR-VAL-PHE-H
YM 51085	IVAL-TYR-VAL-PHENYLALANINOL
YM 60828	[N-[4-[(1-ACETIMIDOYL-4-PIPERIDYL)OXY]PHENYL]-N-[(7-AMIDINO-2-NAPHTHYL)METHYL]SULPHAMOYL]-ACETIC ACID
YM 9429	cis-1-[4-p-MENTHAN-8-YLOXY)PHENYL]PIPERIDINE1-BISPHONIC ACID

Trivial name	Synonym or systematic name
YM 17E	1-3-BIS[[1-CYCLOHEPTYL-3-(p-DIMETHYLAMINOPHENYL)UREIDO]METHYL]BENZENE HCl
YRIMIDIN-5-ONE	6-[2-[4-[(4-FLUOROPHENYL)PHENYLMETHYLENE]-1-PYPERIDINYL]ETHYL]-7-METHYL-5h-THIAZOL[2,3-a]P
YS 980	[R-(R*,S*)]-3-(3-MERCAPTO-2-METHYL-1-OXOPROPYL)-4-THIAZOLIDINECARBOXYLIC ACID
YS 3025	3-[2-(2-THIOHENCARBOXYTHIO)]-PROPANOYL-4-THIAZOLIDIN CARBOXYLIC ACID
YT 18	2,3-DIHYDRO-2,4,6,7-TETRAMETHYL-2-[(4-PHENYL-1-PIPERAZINYL)METHYL]-5-BENZOFURANAMINE
YTR 830	TAZOBACTAM
YTR 830	2β-[(1,2,3-TRIAZOL-1-YL)METHYL]-2α-METHYLPENAM-3α-CARBOXYLATE-1,1-DIOXIDE
YTR 830H	(2S,3R,5S)-3-METHYL-7-OXO-3-(1H, 1,2,3-TRIAZOL-1-YL-METHYL)-THIA-1-AZABICYCLO[3.2.0]HEPTANE-2-CARBOXYLIC ACID
YTR 830H	(2S, 3S, 5R)-3-METHYL-7-OXO-3-(1H-1,2,3-TRIAZOL-1-YLMETHYL)-4-THIA-1-AZABICYCLO[3.2.0]HEPTANE-2-CARBOXYLIC ACID 4,4-DIOXIDE
Z 321	1-[3-(2-INDANYLACETYL)-L-THIOPROLYL]PYRROLIDINE
ZAPRINAST	2-O-PROPOXYPHENYL-8-AZAPURIN-6-ONE
ZARAGOCID ACID D3	F 10863 A
ZARDAVERINE	6-(DIFLUOROMETHOXY-3-METHOXYPHENYL)-3-(2H)PYRIDAZINONE
ZD 1033	ARIMIDEX
ZD 1033	2,2′[5-(1H-1,2,4-TRIAZOL-1-YLMETHYL)-1,3-PHENYLENE]BIS(2-METHYLPROPIONONITRILE)
ZD 1033	ICI D 1033
ZD 10333	ANASTROZOLE
ZD 1542	4(Z)-6-[2S, 4S, 5R)-2-[1-METHYL-1-(2-NITRO-4-TOLYLOXY)ETHYL]-4-(3-PYRIDYL)-1,3-DIOXAN-5-YL]HEX-4-ENOIC ACID
ZD 1694	N-(5-[N-(3,4-DIHYDRO-2-METHYL-4-OXOQUINAZOLIN-6-YLMETHYL)-N-METHYLAMINO]2-THENOYL)-S-GLUTAMIC ACID
ZD 1694	RALTITREXED
ZD 1794	ICI D1694
ZD 5522	3′,5′-DIMETHYL-4′-NITROMETHYLSULPHONYL-2-(2-TOLYL)ACETANILIDE
ZENARESTAT	[3-(4-BROMO-2-FLUOROBENZYL)-7-CHLORO-2,4-DIOXO-1,2,3,4-TETRAHYDROQUINAZOLIN-1-YL]ACETATE
ZIDOVUDINE	3′AZIDO-2′,3′-DIDEOXYTHYMIDINE
ZIDOVUDINETRIPHOSPHA	AZIDOTHYMIDINE TRIPHOSPHATE
ZIFROSILONE	MDL 73745
ZILEUTON	N-(1-BENZO[b]THIEN-2-YLETHYL)-N-HYDROXYUREA
ZK 114043	(Z)-11β-(4-ACETYLPHENYL)-17β-HYDROXY-17α-(3-HYDROXY-1-PROPENYL)-13-ESTRA-4,9-DIEN-3-ONE
ZK 114863	11β-(4-ACETYLPHENYL)-17α-HYDROXY-17β-(3-HYDROXYPROPYL)-13-ESTRA-4,9-DIEN-3-ONE
ZK 138723	5-[CYCLOPENTYLIDEN-(1-IMIDAZOLE)METHYL]THIOPHENE-2-CARBONITRILE
ZK 62711	ROLIPRAM
ZK 98299	11β-[4-(DIMETHYLAMINO)PHENYL]-17α-HYDROXY-17β-(3-HYDROXYPROPYL-13α-ESTRA-4,9-DIEN-3-ONE
ZK 98299	ONAPRISTONE
ZOFENOPRILAT	SQ 29852
ZOLEDRONATE	CGP 42446
ZOLEDRONATE	2-(IMIDAZOLE-1-YL)-1-HYDROXYETHANE-1,1-BISPHOSPHONIC ACID
ZOPOLRESTAT	3,4-DIHYDRO-4-OXO-3[[5-(TRIFLUOROMETHYL)-2-BENZOTHAZOLYL]METHYL]-1-PHTHALAZINE ACETIC ACID
ZPG	N-[[[(BENZYLOXYCARBONYL)AMINO]METHYL]HYDROXYPHOSPHINYL]-L-PHENYLALANINE
ZSY 27	5-METHYL-6-[4-PYRIDINYL]-2H-1,4-THIAZINE-3[4H]ONE HCl
ZWITTERGENT 3-10	N, N-DIMETHYL-N-(3-SULPHOPROPYL)-1-DECANAMINIUM HYDROXYDE INNER SALT

EC Numbers List

EC Number	Name
1.1.1.1	ALCOHOL DEHYDROGENASE
1.1.1.2	ALCOHOL DEHYDROGENASE (NADP)
1.1.1.3	HOMOSERINE DEHYDROGENASE
1.1.1.4	(R,R)-BUTANEDIOL DEHYDROGENASE
1.1.1.5	ACETOIN DEHYDROGENASE
1.1.1.6	GLYCEROL DEHYDROGENASE
1.1.1.7	PROPANEDIOL-PHOSPHATE DEHYDROGENASE
1.1.1.8	GLYCEROL-3-PHOSPHATE DEHYDROGENASE (NAD)
1.1.1.9	D-XYLULOSE REDUCTASE
1.1.1.10	L-XYLULOSE REDUCTASE
1.1.1.11	D-ARABINITOL DEHYDROGENASE
1.1.1.14	L-IDITOL 2-DEHYDROGENASE
1.1.1.15	D-IDITOL-2-DEHYDROGENASE
1.1.1.17	MANNITOL-1-PHOSPHATE 5-DEHYDROGENASE
1.1.1.18	MYO-INOSITOL 2-DEHYDROGENASE
1.1.1.19	GLUCURONATE REDUCTASE
1.1.1.20	GLUCURONONOLATONE REDUCTASE
1.1.1.21	ALDEHYDE REDUCTASE
1.1.1.22	UDP-GLUCOSE 6-DEHYDROGENASE
1.1.1.23	HISTIDINOL DEHYDROGENASE
1.1.1.24	QUINATE DEHYDROGENASE
1.1.1.25	SHIKIMATE 5-DEHYDROGENASE
1.1.1.26	GLYOXALATE REDUCTASE
1.1.1.27	L-LACTATE DEHYDROGENASE C4
1.1.1.28	D-LACTATE DEHYDROGENASE
1.1.1.29	GLYCERATE DEHYDROGENASE
1.1.1.30	3-HYDROXYBUTYRATE DEHYDROGENASE
1.1.1.31	3-HYDROXYISOBUTYRATE DEHYDROGENASE
1.1.1.33	MEVALDATE REDUCTASE (NADPH)
1.1.1.34	HYDROXYMETHYLGLUTARYL-CoA REDUCTASE (NADPH)
1.1.1.35	3-HYDROXYACYL-CoA DEHYDROGENASE
1.1.1.36	ACETOACETYL-CoA REDUCTASE
1.1.1.37	MALATE DEHYDROGENASE
1.1.1.38	MALATE DEHYDROGENASE (OXALACETATE DECARBOXYLATING)
1.1.1.39	MALATE DEHYDROGENASE (DECARBOXYLATING)
1.1.1.40	MALATE DEHYDROGENASE (OXALACETATE-DECARBOXYLATING) (NADP)
1.1.1.41	ISOCITRATE DEHYDROGENASE (NAD)
1.1.1.42	ISOCITRATE DEHYDROGENASE (NADP)
1.1.1.43	PHOSPHOGLUCONATE 2-DEHYDROGENASE
1.1.1.44	PHOSPHOGLUCONATE DEHYDROGENASE (DECARBOXYLATING)
1.1.1.45	GULONATE DEHYDROGENASE
1.1.1.46	L-ARABINOSE 1-DEHYDROGENASE
1.1.1.47	GLUCOSE 1-DEHYDROGENASE
1.1.1.48	GALACTOSE 1-DEHYDROGENASE
1.1.1.49	GLUCOSE-6-PHOSPHATE 1-DEHYDROGENASE
1.1.1.50	3α-HYDROXYSTEROID DEHYDROGENASE (B-SPECIFIC)
1.1.1.51	3(or 17)β-HYDROXYSTEROID DEHYDROGENASE
1.1.1.53	3α(or 20β)-HYDROXYSTEROID DEHYDROGENASE
1.1.1.54	ALLYL-ALCOHOL DEHYDROGENASE
1.1.1.56	RIBITOL 2-DEHYDROGENASE
1.1.1.57	FRUCTURONATE REDUCTASE
1.1.1.58	TAGATURONATE REDUCTASE
1.1.1.60	2-HYDROXY-3-OXOPROPIONATE REDUCTASE
1.1.1.62	ESTRADIOL 17β-DEHYDROGENASE
1.1.1.63	TESTOSTERONE 17β-DEHYDROGENASE
1.1.1.64	TESTOSTERONE 17β-DEHYDROGENASE (NADP)

EC Number	Name
1.1.1.65	PYRIDOXINE 4-DEHYDROGENASE
1.1.1.67	MANNITOL 2-DEHYDROGENASE
1.1.1.69	GLUCONATE 5-DEHYDROGENASE
1.1.1.71	ALCOHOL DEHYDROGENASE (NAD(P))
1.1.1.72	GLYCEROL DEHYDROGENASE (NADP)
1.1.1.73	OCTANOL DEHYDROGENASE
1.1.1.75	(R)AMINOPROPANOL DEHYDROGENASE
1.1.1.76	LACTALDEHYDE REDUCTASE
1.1.1.78	D-LACTALDEHYDE DEHYDROGENASE
1.1.1.79	GLYOXYLATE REDUCTASE (NADP)
1.1.1.81	HYDROXYPYRUVATE REDUCTASE
1.1.1.82	MALATE DEHYDROGENASE (NADP)
1.1.1.83	D-MALATE DEHYDROGENASE (DECARBOXYLATING)
1.1.1.84	DIMETHYLMALATE DEHYDROGENASE
1.1.1.85	3-ISOPROPYLMALATE DEHYDROGENASE
1.1.1.86	KETOL-ACID REDUCTOISOMERASE
1.1.1.88	HYDROXYMETHYLGLUTARYL-CoA REDUCTASE
1.1.1.90	ARYL-ALCOHOL DEHYDROGENASE
1.1.1.91	ARYL-ALCOHOL DEHYDROGENASE (NADP)
1.1.1.93	TARTRATE DEHYDROGENASE
1.1.1.94	GLYCEROL-3-PHOSPHATE DEHYDROGENASE (NAD(P))
1.1.1.95	PHOSPHOGLYCERATE DEHYDROGENASE
1.1.1.96	DIIODOPHENYLPYRUVATE REDUCTASE
1.1.1.97	3-HYDROXYBENZYL-ALCOHOL DEHYDROGENASE
1.1.1.100	β-KETOACYL-CoA DEHYDROGENASE
1.1.1.100	3-OXOACYL-[ACYL-CARRIER-PROTEIN] REDUCTASE
1.1.1.101	ACYLGLYCERONE-PHOSPHATE REDUCTASE
1.1.1.103	L-THREONINE 3-DEHYDROGENASE
1.1.1.105	RETINOL DEHYDROGENASE
1.1.1.106	PANTOATE DEHYDROGENASE
1.1.1.107	PYRIDOXAL DEHYDROGENASE
1.1.1.108	CARNITINE 3-DEHYDROGENASE
1.1.1.112	INDANOL DEHYDROGENASE
1.1.1.115	RIBOSE 1-DEHYDROGENASE (NADP)
1.1.1.116	D-ARABINOSE 1-DEHYDROGENASE
1.1.1.119	GLUCOSE 1-DEHYDROGENASE (NADP)
1.1.1.121	ALDOSE 1-DEHYDROGENASE
1.1.1.122	D-threo-ALDOSE 1-DEHYDROGENASE
1.1.1.123	SORBOSE DEHYDROGENASE (NADP)
1.1.1.124	FRUCTOSE 5-DEHYDROGENASE (NADP)
1.1.1.126	2-DEHYDRO-3-DEOXY-3-GLUCONATE 6-DEHYDROGENASE
1.1.1.127	2-DEHYDRO-3-DEOXY-D-GLUCONATE 5-DEHYDROGENASE
1.1.1.128	IDONATE DEHYDROGENASE
1.1.1.129	THREONATE DEHYDROGENASE
1.1.1.130	3-DEHYDRO-L-GULONATE 2-DEHYDROGENASE
1.1.1.132	GDP-MANNOSE 6-DEHYDROGENASE
1.1.1.136	UDP-N-ACETYLGLUCOSAMINE DEHYDROGENASE
1.1.1.137	RIBITOL-5-PHOSPHATE 2-DEHYDROGENASE
1.1.1.138	MANNITOL 2-DEHYDROGENASE (NADP)
1.1.1.140	SORBITOL-6-PHOSPHATE 2-DEHYDROGENASE
1.1.1.141	15-HYDROXYPROSTAGLANDIN DEHYDROGENASE (NAD)
1.1.1.142	D-PINITOL DEHYDROGENASE
1.1.1.143	SEQUOYITOL DEHYDROGENASE
1.1.1.144	PERILLYL-ALCOHOL DEHYDROGENASE
1.1.1.145	3β-HYDROXY-Δ5-STEROID DEHYDROGENASE
1.1.1.146	11β-HYDROXYSTEROID DEHYDROGENASE

EC Number	Name
1.1.1.148	ESTRADIOL-17α-DEHYDROGENASE
1.1.1.149	20α-HYDROXYSTEROID DEHYDROGENASE
1.1.1.150	21-HYDROXYSTEROID DEHYDROGENASE (NAD)
1.1.1.151	21-HYDROXYSTEROID DEHYDROGENASE (NADP)
1.1.1.153	SEPIAPTERIN REDUCTASE
1.1.1.154	UREIDOGLYCOLLATE DEHYDROGENASE
1.1.1.155	HOMOISOCITRATE DEHYDROGENASE
1.1.1.156	GLYCEROL 2-DEHYDROGENASE
1.1.1.158	UDP-N-ACETYLMURAMATE DEHYDROGENASE
1.1.1.159	7α-HYDROXYSTEROID DEHYDROGENASE
1.1.1.160	CHOLESTANETETRAOL 26-DEHYDROGENASE
1.1.1.162	ERYTHRULOSE REDUCTASE
1.1.1.165	2-ALKYNOL-1-OL DEHYDROGENASE
1.1.1.168	2-DEHYDROPANTOYL-LACTONE REDUCTASE (A SPECIFIC)
1.1.1.169	2-DEHYDROPANTOATE 2-REDUCTASE
1.1.1.173	L-RHAMNOSE DEHYDROGENASE
1.1.1.176	12α-HYDROXYSTEROID DEHYDROGENASE
1.1.1.179	D-XYLOSE 1-DEHYDROGENASE
1.1.1.181	CHOLEST-5-ENE-3β,7α-DIOL 3β-DEHYDROGENASE
1.1.1.183	GERANIOL DEHYDROGENASE
1.1.1.184	CARBONYL REDUCTASE (NADPH)
1.1.1.185	L-GLYCOL DEHYDROGENASE
1.1.1.187	GDP-4-DEHYDRO-D-RHAMNOSE REDUCTASE
1.1.1.188	PROSTAGLANDIN F SYNTHASE
1.1.1.189	PROSTAGLANDIN E2 9-KETOREDUCTASE
1.1.1.189	PROSTAGLANDIN E2 9-REDUCTASE
1.1.1.190	INDOLE 3-ACETALDEHYDE REDUCTASE (NADH)
1.1.1.191	INDOLE 3-ACETALDEHYDE REDUCTASE (NADPH)
1.1.1.192	LONG-CHAIN-ALCOHOL DEHYDROGENASE
1.1.1.193	5-AMINO-6-(5PHOSPHORIBOSYLAMINO)URACIL REDUCTASE
1.1.1.195	CINNAMYL-ALCOHOL DEHYDROGENASE
1.1.1.196	15-HYDROXYPROSTAGLANDIN-D DEHYDROGENASE
1.1.1.197	15-HYDROXYPROSTAGLANDIN DEHYDROGENASE (NADP)
1.1.1.198	(+)-BORNEOL DEHYDROGENASE
1.1.1.200	ALDOSE-6-PHOSPHATE REDUCTASE (NADPH)
1.1.1.201	HYDROXYSTEROID DEHYDROGENASE (NADP)
1.1.1.202	1,3-PROPANEDIOL DEHYDROGENASE
1.1.1.203	URONATE DEHYDROGENASE
1.1.1.204	XANTHINE DEHYDROGENASE
1.1.1.205	IMP DEHYDROGENASE
1.1.1.206	TROPINE DEHYDROGENASE
1.1.1.207	(−)-MENTHOL DEHYDROGENASE
1.1.1.208	(+)-NEOMENTHOL DEHYDROGENASE
1.1.1.210	3β(or 20α)HYDROXYSTEROID DEHYDROGENASE
1.1.1.213	3α-HYDROXYSTEROID DEHYDROGENASE (A SPECIFIC)
1.1.1.216	FARNESOL DEHYDROGENASE
1.1.1.217	BENZYL-2-METHYL-HYDROXYBUTYRATE DEHYDROGENASE
1.1.1.218	MORPHINE 6-DEHYDROGENASE
1.1.1.219	DIHYDROKAEMPFEROL 4-REDUCTASE
1.1.1.222	(R)-4-HYDROXYPHENYLLACTATE DEHYDROGENASE
1.1.1.223	ISOPIPERITENOL DEHYDROGENASE
1.1.1.225	CHLORDECONE REDUCTASE
1.1.1.226	4-HYDROXYCYCLOHEXANECARBOXYLATE DEHYDROGENASE
1.1.1.229	DIETHYL-2-METHYL-3-OXOSUCCINATE DEHYDROGENASE
1.1.1.231	15-HYDROXYPROSTAGLANDIN-I DEHYDROGENASE (NADP)
1.1.1.233	N-ACYLMANNOSAMINE 1-DEHYDROGENASE

EC Number	Name
1.1.1.234	FLAVONE 4-REDUCTASE
1.1.1.236	TROPINONE REDUCTASE
1.1.1.237	HYDROXYPHENYLPYRUVATE REDUCTASE
1.1.1.239	3α(or 17β)-HYDROXYSTEROID DEHYDROGENASE
1.1.1.240	N-ACETYLHEXOSAMINE 1-DEHYDROGENASE
1.1.1.244	METHANOL DEHYDROGENASE
1.1.1.245	CYCLOHEXANOL DEHYDROGENASE
1.1.2.3	L-LACTATE DEHYDROGENASE (CYTOCHROME)
1.1.2.4	D-LACTATE DEHYDROGENASE (CYTOCHROME)
1.1.2.5	D-LACTATE DEHYDROGENASE (CYTOCHROME c-553)
1.1.3.3	MALATE OXIDASE
1.1.3.4	GLUCOSE OXIDASE
1.1.3.5	HEXOSE OXIDASE
1.1.3.6	CHOLESTEROL OXIDASE
1.1.3.7	ARYL-ALCOHOL OXIDASE
1.1.3.8	L-GULONOLACTONE OXIDASE
1.1.3.9	GALACTOSE OXIDASE
1.1.3.10	PYRANOSE OXIDASE
1.1.3.11	SORBOSE OXIDASE
1.1.3.12	PYRIDOXINE 4-OXIDASE
1.1.3.13	ALCOHOL OXIDASE
1.1.3.14	CATECHOL OXIDASE (DIMERIZING)
1.1.3.15	(S)-2-HYDROXY-ACID OXIDASE
1.1.3.16	ECDYSONE OXIDASE
1.1.3.17	CHOLINE OXIDASE
1.1.3.18	SECONDARY-ALCOHOL OXIDASE
1.1.3.19	4-HYDROXYMANDELATE OXIDASE
1.1.3.21	GLYCEROL-3-PHOSPATE OXIDASE
1.1.3.22	XANTHINE OXIDASE
1.1.3.23	THIAMIN OXIDASE
1.1.3.24	L-GALACTONOLACTONE OXIDASE
1.1.3.25	CELLOBIOSE OXIDASE
1.1.3.26	COLUMBAMINE OXIDASE
1.1.3.27	HYDROXYPHYTANATE OXIDASE
1.1.3.28	NUCLEOSIDE OXIDASE
1.1.3.29	N-ACYLHEXOSAMINE OXIDASE
1.1.3.31	METHANOL OXIDASE
1.1.4.1	VITAMIN K-EPOXIDE REDUCTASE (WARFARIN SENSITIVE)
1.1.4.2	VITAMIN K-EPOXIDE REDUCTASE (WARFARIN INSENSITIVE)
1.1.5.1	CELLOBIOSE DEHYDROGENASE (QUINONE)
1.1.99.1	CHOLINE DEHYDROGENASE
1.1.99.2	2-HYDROXYGLUTARATE DEHYDROGENASE
1.1.99.3	GLUCONATE 2-DEHYDROGENASE
1.1.99.4	DEHYDROGLUCONATE DEHYDROGENASE
1.1.99.5	GLYCEROL-3-PHOSPHATE DEHYDROGENASE
1.1.99.6	D-2-HYDROXY-ACID DEHYDROGENASE
1.1.99.6	D-2-HYDROXY-ACID DEHYDROGENASE 2
1.1.99.7	LACTATE-MALATE TRANSHYDROGENASE
1.1.99.8	ALCOHOL DEHYDROGENASE (ACCEPTOR)
1.1.99.9	PYRIDOXINE DEHYDROGENASE
1.1.99.10	GLUCOSE DEHYDROGENASE (ACCEPTOR)
1.1.99.11	FRUCTOSE 5-DEHYDROGENASE
1.1.99.12	SORBOSE DEHYDROGENASE
1.1.99.13	GLUCOSIDE 3-DEHYDROGENASE
1.1.99.14	GLYCOLATE DEHYDROGENASE
1.1.99.17	GLUCOSE DEHYDROGENASE (PYRROLOQUINOLINE-QUINONE)

EC Number	Name
1.1.99.19	URACIL DEHYDROGENASE
1.1.99.20	ALAKN-1-OL DEHYDROGENASE (ACCEPTOR)
1.1.99.22	GLYCEROL DEHYDROGENASE (ACCEPTOR)
1.1.99.25	QUINATE DEHYDROGENASE (PYRROLOQUINOLINE-QUINONE)
1.2.1.1	FORMALDEHYDE DEHYDROGENASE (GLUTATHIONE)
1.2.1.2	FORMATE DEHYDROGENASE
1.2.1.3	ALDEHYDE DEHYDROGENASE
1.2.1.4	ALDEHYDE DEHYDROGENASE (NADP)
1.2.1.5	ALDEHYDE DEHYDROGENASE (NAD(P))
1.2.1.7	BENZALDEHYDE DEHYDROGENASE (NADP)
1.2.1.8	BETAINE ALDEHYDE DEHYDROGENASE
1.2.1.8	BETAINE-ALDEHYDE DEHYDROGENASE
1.2.1.9	GLYCERALDEHYDE-3-PHOSPHATE DEHYDROGENASE (NADP)
1.2.1.10	ACETALDEHYDE DEHYDROGENASE (ACETYLATING)
1.2.1.11	ASPARTATE-SEMIALDEHYDE DEHYDROGENASE
1.2.1.12	GLYCERALDEHYDE-3-PHOSPHATE DEHYDROGENASE (PHOSPHORYLATING)
1.2.1.13	GLYCERALDEHYDE-3-PHOSPHATE DEHYDROGENASE (NADP) (PHOSPHORYLATING)
1.2.1.15	MALONATE-SEMIALDEHYDE DEHYDROGENASE
1.2.1.16	SUCCINATE-SEMIALDEHYDE DEHYDROGENASE (NAD(P))
1.2.1.17	GLYOXYLATE DEHYDROGENASE (ACYLATING)
1.2.1.18	MALONATE-SEMIALDEHYDE DEHYDROGENASE (ACYLATING)
1.2.1.19	AMINOBUTYRALDEHYDE DEHYDROGENASE
1.2.1.20	GLUTARATE-SEMIALDEHYDE DEHYDROGENASE
1.2.1.21	GLYCOLALDEHYDE DEHYDROGENASE
1.2.1.22	LACTALDEHYDE DEHYDROGENASE
1.2.1.23	2-OXOALDEHYDE DEHYDROGENASE (NAD)
1.2.1.24	SUCCINATE-SEMIALDEHYDE DEHYDROGENASE
1.2.1.25	2-OXOISOVALERATE DEHYDROGENASE (ACYLATING)
1.2.1.26	2,5-DIOXOVALERATE DEHYDROGENASE
1.2.1.26	2-OXOGLUTARATE-SEMIALDEHYDE DEHYDROGENASE
1.2.1.27	METHYLMALONATE-SEMIALDEHYDE DEHYDROGENASE (ACYLATING)
1.2.1.28	BENZALDEHYDE DEHYDROGENASE (NAD)
1.2.1.29	ARYL-ALDEHYDE DEHYDROGENASE
1.2.1.30	ARYL-ALDEHYDE DEHYDROGENASE (NADP)
1.2.1.31	L-AMINOADIPATE-SEMIALDEHYDE DEHYDROGENASE
1.2.1.32	AMINOMUCONATE-SEMIALDEHYDE DEHYDROGENASE
1.2.1.33	(R)-DEHYDROPANTOATE DEHYDROGENASE
1.2.1.36	RETINAL DEHYDROGENASE
1.2.1.38	N-ACETYL-γ-GLUTAMYL-PHOSPHATE REDUCTASE
1.2.1.39	PHENYLACETALDEHYDE DEHYDROGENASE
1.2.1.40	3α,7α,12α-TRIHYDROXYCHOLESTAN-26-AL 26-DEHYDROGENASE
1.2.1.41	GLUTAMATE-5-SEMIALDEHYDE DEHYDROGENASE
1.2.1.42	HEXADECANAL DEHYDROGENASE (ACYLATING)
1.2.1.43	FORMATE DEHYDROGENASE (NADP)
1.2.1.44	CINNAMOYL-CoA REDUCTASE
1.2.1.45	2-HYDROXY-4-CARBOXYMUCONATE-6-SEMIALDEHYDE DEHYDROGENASE
1.2.1.46	FORMALDEHYDE DEHYDROGENASE
1.2.1.47	4-TRIMETHYLAMMONIUMBUTYRALDEHYDE DEHYDROGENASE
1.2.1.48	LONG-CHAIN-ALDEHYDE DEHYDROGENASE
1.2.1.49	2-OXOALDEHYDE DEHYDROGENASE (NADP)
1.2.1.50	LONG-CHAIN-FATTY-ACYL-CoA REDUCTASE
1.2.1.51	PYRUVATE DEHYDROGENASE (NADP)
1.2.1.54	γ-GUANIDINOBUTYRATE DEHYDROGENASE
1.2.1.55	(R)-3-HYDROXYACID ESTER DEHYDROGENASE
1.2.1.55	(S)-3-HYDROXYACID ESTER DEHYDROGENASE
1.2.2.2	PYRUVATE DEHYDROGENASE (CYTOCHROME)

EC Number	Name
1.2.2.3	FORMATE DEHYDROGENASE (CYTOCHROME-C-553)
1.2.3.1	ALDEHYDE OXIDASE
1.2.3.3	PYRUVATE OXIDASE
1.2.3.4	OXALATE OXIDASE
1.2.3.7	INDOLE-3-ACETALDEHYDE OXIDASE
1.2.3.10	CARBON MONOXIDE OXIDASE
1.2.3.11	RETINAL OXIDASE
1.2.4.1	PYRUVATE DEHYDROGENASE (LIPOAMIDE)
1.2.4.2	OXOGLUTARATE DEHYDROGENASE (LIPOAMIDE)
1.2.4.4	3-METHYL-2-OXOBUTANOATE DEHYDROGENASE (LIPOAMIDE)
1.2.7.1	PYRUVATE SYNTHASE
1.2.99.2	CARBON-MONOXIDE DEHYDROGENASE
1.2.99.3	ALDEHYDE DEHYDROGENASE (PYRROLOQUINOLINE-QUINONE)
1.2.99.4	FORMALDEHYDE DISMUTASE
1.2.99.6	ALDEHYDE FERREDOXIN OXIDOREDUCTASE
1.3.1.2	DIHYDROPYRIMIDINE DEHYDROGENASE (NADP)
1.3.1.3	CORTISONE β-REDUCTASE
1.3.1.4	CORTISONE α-REDUCTASE
1.3.1.6	FUMARATE REDUCTASE (NADH)
1.3.1.8	ACYL-CoA DEHYDROGENASE (NADP)
1.3.1.9	ENOYL-[ACYL-CARRIER-PROTEIN] REDUCTASE (NADH)
1.3.1.10	ENOYL-[ACYL-CARRIER-PROTEIN] REDUCTASE (NADPH, B SPECIFIC)
1.3.1.11	COUMARATE REDUCTASE
1.3.1.12	PREPHENATE DEHYDROGENASE
1.3.1.13	PREPHENATE DEHYDROGENASE (NADP)
1.3.1.14	OROTATE REDUCTASE (NADH)
1.3.1.15	ORATE REDUCTASE (NADPH)
1.3.1.16	β-NITROACRYLATE REDUCTASE
1.3.1.17	3-METHYLENEOXINDOLE REDUCTASE A
1.3.1.17	3-METHYLENEOXINDOLE REDUCTASE B
1.3.1.19	cis-1,2-DIHYDROBENZENE-1,2-DIOL DEHYDROGENASE
1.3.1.20	trans-1,2-DIHYDROBENZENE-1,2-DIOL DEHYDROGENASE
1.3.1.21	7-DEHYDROCHOLESTEROL REDUCTASE
1.3.1.22	CHOLESTENONE 5α-REDUCTASE
1.3.1.23	CHOLESTENONE 5β-REDUCTASE
1.3.1.24	BILIVERDIN REDUCTASE
1.3.1.25	1,6-DIHYDROXYCYCLOHEXA-2,4-DIENE-1-CARBOXYLATE DEHYDROGENASE
1.3.1.26	DIHYDRODIPICOLINATE REDUCTASE
1.3.1.26	DIHYDROPICOLINATE REDUCTASE
1.3.1.27	HEXADECANAL REDUCTASE
1.3.1.29	cis-1,2-DIHYDRO-1,2-DIHYDROXYNAPHTHALENE DEHYDROGENASE
1.3.1.31	2-ENOATE REDUCTASE
1.3.1.33	PROTOCHLOROPYLI'LIDE REDUCTASE
1.3.1.34	2,4-DIENOYL-CoA REDUCTASE (NADPH)
1.3.1.35	PHOSPHATIDYLCHOLINE DESATURASE
1.3.1.37	cis-2-ENOYL-CoA REDUCTASE (NADPH)
1.3.1.38	trans-2-ENOYL-CoA REDUCTASE (NADPH)
1.3.1.40	2-HYDROXY-6-OXO-6-PHENYLHEXA-2,4-DIENOATE REDUCTASE
1.3.1.41	XANTHOMMATIN REDUCTASE
1.3.1.43	PRETYROSINE DEHYDROGENASE
1.3.1.44	trans-2-ENOYL-CoA REDUCTASE (NAD)
1.3.1.48	15-OXOPROSTAGLANDIN Δ13 REDUCTASE
1.3.2.3	GALACTONOLACTONE DEHYDROGENASE
1.3.3.1	DIHYDROOROTATE OXIDASE
1.3.3.2	LATHOSTEROL OXIDASE
1.3.3.3	COPROPORPHYRINOGEN OXIDASE

EC Number	Name
1.3.3.4	PROTOPORPHYRINOGEN OXIDASE
1.3.3.6	ACYL-CoA OXIDASE
1.3.3.7	DIHYDROURACIL OXIDASE
1.3.3.8	TETRAHYDROBERBERINE OXIDASE
1.3.4.6	AMINE OXIDASE DIAMINE
1.3.5.1	SUCCINATE DEHYDROGENASE (UBIQUINONE)
1.3.7.1	6-HYDROXYNICOTINATE REDUCTASE
1.3.99.1	SUCCINATE DEHYDROGENASE
1.3.99.2	BUTYRYL-CoA DEHYDROGENASE
1.3.99.3	ACYL-CoA DEHYDROGENASE
1.3.99.3	ACYL-CoA DEHYDROGENASE (LONG CHAIN)
1.3.99.3	ACYL-CoA DEHYDROGENASE (MEDIUM CHAIN)
1.3.99.3	ACYL-CoA DEHYDROGENASE (SHORT CHAIN)
1.3.99.4	3-OXOSTEROID Δ^1-DEHYDROGENASE
1.3.99.5	3-OXO-5α-STEROID Δ^4-DEHYDROGENASE
1.3.99.5	5α-REDUCTASE
1.3.99.5	STEROID 5α-REDUCTASE
1.3.99.5	STEROID 5β-REDUCTASE
1.3.99.6	3-OXO-5β-STEROID Δ^4-DEHYDROGENASE
1.3.99.7	GLUTARYL-CoA DEHYDROGENASE
1.3.99.8	2-FUROYL-CoA DEHYDROGENASE
1.3.99.9	β-CYCLOPIAZONATE DEHYDROGENASE
1.3.99.10	ISOVALERYL-CoA DEHYDROGENASE
1.3.99.11	DIHYDROOROTATE DEHYDROGENASE
1.3.99.12	2-METHYLACYL-CoA DEHYDROGENASE
1.3.99.13	LONG-CHAIN-ACYL-CoA DEHYDROGENASE
1.4.1.1	ALANINE DEHYDROGENASE
1.4.1.2	GLUTAMATE DEHYDROGENASE
1.4.1.3	GLUTAMATE DEHYDROGENASE (NAD(P))
1.4.1.4	GLUTAMATE DEHYDROGENASE (NADP)
1.4.1.5	L-AMINO-ACID DEHYDROGENASE
1.4.1.7	SERINE DEHYDROGENASE
1.4.1.8	VALINE DEHYDROGENASE (NADP)
1.4.1.9	LEUCINE DEHYDROGENASE
1.4.1.10	GLYCINE DEHYDROGENASE
1.4.1.11	L-ERYTHRO-3,5-DIAMINOHEXANOATE DEHYDROGENASE
1.4.1.12	2,4-DIAMINOPENTANE DEHYDROGENASE
1.4.1.13	GLUTAMATE SYNTHASE (NADPH)
1.4.1.14	GLUTAMATE SYNTHASE (NADH)
1.4.1.16	DIAMINOPIMELATE DEHYDROGENASE
1.4.1.18	LYSINE 6-DEHYDROGENASE
1.4.1.19	TRYPTOPHAN DEHYDROGENASE
1.4.1.20	PHENYLALANINE DEHYDROGENASE
1.4.2.1	GLYCINE DEHYDROGENASE (CYTOCHROME)
1.4.3.1	D-ASPARTATE OXIDASE
1.4.3.2	D-AMINO-ACID OXIDASE
1.4.3.2	L-AMINO-ACID OXIDASE
1.4.3.3	D-AMINO-ACID OXIDASE
1.4.3.4	AMINE OXIDASE
1.4.3.4	AMINE OXIDASE (FLAVIN-CONTAINING)
1.4.3.4	AMINE OXIDASE MAO
1.4.3.4	AMINE OXIDASE MAO-A
1.4.3.4	AMINE OXIDASE MAO-B
1.4.3.4	AMINE OXIDASE MAO-I
1.4.3.5	PYRIDOXAMINE-PHOSPHATE OXIDASE
1.4.3.6	AMINE OXIDASE (COPPER-CONTAINING)

EC Number	Name
1.4.3.6	AMINE OXIDASE (COPPER-CONTAINING) I
1.4.3.6	AMINE OXIDASE (COPPER-CONTAINING) II
1.4.3.6	AMINE OXIDASE DIAMINE
1.4.3.6	AMINE OXIDASE I
1.4.3.6	AMINE OXIDASE II
1.4.3.7	D-GLUTAMATE OXIDASE
1.4.3.8	ETHANOLAMINE OXIDASE
1.4.3.10	PUTRESCINE OXIDASE
1.4.3.11	L-GLUTAMATE OXIDASE
1.4.3.12	CYCLOHEXYLAMINE OXIDASE
1.4.3.13	PROTEIN-LYSINE 6-OXIDASE
1.4.3.14	L-LYSINE OXIDASE
1.4.3.15	D-GLUTAMATE(D-APSARTATE) OXIDASE
1.4.3.16	L-ASPARTATE OXIDASE
1.4.4.1	D-PROLINE REDUCTASE (DITHIOL)
1.4.4.2	GLYCINE DEHYDROGENASE (DECARBOXYLATING)
1.4.6.2	GLUTATHIONE REDUCTASE (NADPH)
1.4.7.1	GLUTAMATE SYNTHASE (FERREDOXIN)
1.4.99.1	D-AMINO-ACID DEHYDROGENASE
1.4.99.2	TAURINE DEHYDROGENASE
1.4.99.3	AMINE DEHYDROGENASE
1.4.99.4	ARALKYLAMINE DEHYDROGENASE
1.5.1.1	PYRROLINE-2-CARBOXYLATE REDUCTASE
1.5.1.2	PYRROLINE-5-CARBOXYLATE REDUCTASE
1.5.1.3	DIHYDROFOLATE REDUCTASE
1.5.1.5	METHYLELETETRAHYDROFOLATE DEHYDROGENASE (NADP)
1.5.1.6	FORMYLTETRAHYDROFOLATE DEHYDROGENASE
1.5.1.7	SACCHAROPINE DEHYDROGENASE (NAD, L-LYSINE FORMING)
1.5.1.8	SACCHAROPINE DEHYDROGENASE (NADP, L-LYSINE FORMING)
1.5.1.9	SACCHAROPINE DEHYDROGENASE (NAD, L-GLUTAMATE FORMING)
1.5.1.10	SACCHAROPINE DEHYDROGENASE (NADP, L-GLUTAMATE FORMING)
1.5.1.11	D-OCTOPINE DEHYDROGENASE
1.5.1.12	1-PYRROLINE-5-CARBOXYLATE DEHYDROGENASE
1.5.1.13	NICOTINATE DEHYDROGENASE
1.5.1.15	METHYLENETETRAHYDROFOLATE DEHYDROGENASE (NAD)
1.5.1.17	ALANOPINE DEHYDROGENASE
1.5.1.19	D-NOPALINE DEHYDROGENASE
1.5.1.20	METHYLENETETRAHYDROFOLATE REDUCTASE (NADPH)
1.5.1.21	Δ1PIPERIDINE-2-CARBOXYLATE RDUCTASE
1.5.1.22	STROMBINE DEHYDROGENASE
1.5.1.23	TAUROPINE DEHYDROGENASE
1.5.1.24	N5-(CARBOXYETHYL)ORNITHINE SYNTHASE
1.5.3.1	SARCOSINE OXIDASE
1.5.3.4	N^6-METHYL-LYSINE OXIDASE
1.5.3.5	(S)-6-HYDROXYNICOTINE OXIDASE
1.5.3.6	(R)-6-HYDROXYNICOTINE OXIDASE
1.5.3.7	L-PIPECOLATE OXIDASE
1.5.3.8	(S)-TETRAHYDROPROTOBERBERINE OXIDASE
1.5.3.9	RETICULINE OXIDASE
1.5.3.10	DIMETHYLGLYCINE OXIDASE
1.5.3.11	POLYAMINE OXIDASE
1.5.5.1	ELECTRON-TRANSFERRING-FLAVOPROTEIN DEHYDROGENASE
1.5.99.1	SARCOSINE DEHYDROGENASE
1.5.99.2	DIMETHYLGLYCINE DEHYDROGENASE
1.5.99.3	L-PIPECOLATE DEHYDROGENASE
1.5.99.4	NICOTINE DEHYDROGENASE

EC Number	Name
1.5.99.5	METHYLGLUTAMATE DEHYDROGENASE
1.5.99.6	SPERMIDINE DEHYDROGENASE
1.5.99.7	TRIMETHYLAMINE DEHYDROGENASE
1.5.99.8	PROLINE DEHYDROGENASE
1.6.1.1	NAD(P) TRANSHYDROGENASE (B-SPECIFIC)
1.6.1.2	NAD(P) TRANSHYDROGENASE
1.6.1.2	NAD(P) TRANSHYDROGENASE (AB-SPECIFIC)
1.6.2.2	CYTOCHROME-b5 REDUCTASE
1.6.2.4	NADPH-FERRIHEMOPROTEIN REDUCTASE
1.6.2.5	NADPH-CYTOCHROME-c2 REDUCTASE
1.6.2.6	LEGHEMOGLOBIN REDUCTASE
1.6.4.1	CYSTINE REDUCTASE
1.6.4.2	GLUTAHTIONE REDUCTASE (NADPH)
1.6.4.2	GLUTATHIONE REDUCTASE (NADPH)
1.6.4.4	PROTEIN-DISULFIDE REDUCTASE (NAD(P)H)
1.6.4.5	THIOREDOXIN REDUCTASE (NADPH)
1.6.4.6	CoA-GLUTATHIONE REDUCTASE (NADPH)
1.6.4.7	ASPARAGUSATE REDUCTASE (NADH)
1.6.4.8	TRYPANOTHIONE REDUCTASE
1.6.4.9	BIS-γ-GLUTAMYLCYSTEINE REDUCTASE (NADPH)
1.6.5.3	NADH DEHYDROGENASE (UBIQUINONE)
1.6.5.3	NADH DEHYDROGENASE (UBIQUINONE) 1
1.6.5.4	MONODEHYDROASCORBATE REDUCTASE (NADH)
1.6.6.1	NITRATE REDUCTASE (NADH)
1.6.6.2	NITRATE REDUCTASE (NAD(P)H)
1.6.6.3	NITRATE REDUCTASE (NADPH)
1.6.6.4	NITRITE REDUCTASE (NAD(P)H)
1.6.6.6	HYPONITRITE REDUCTASE
1.6.6.7	AZOBENZENE REDUCTASE
1.6.6.8	GMP REDUCTASE
1.6.6.9	TRIMETHYLAMINE-N-OXIDE REDUCTASE
1.6.6.11	HYDROXYLAMINE REDUCTASE (NADH)
1.6.6.13	N-HYDROXY-2-ACETAMIDOFLUORENE REDUCTASE
1.6.8.1	NAD(P)H DEHYDROGENASE (FMN)
1.6.99.1	NADPH DEHYDROGENASE
1.6.99.2	NAD(P)H DEHYDROGENASE (QUINONE)
1.6.99.3	NADH DEHYDROGENASE
1.6.99.5	NADH DEHYDROGENASE (QUINONE)
1.6.99.6	NADPH DEHYDROGENASE (QUINONE)
1.6.99.7	DIHYDROPTERIDINE REDUCTASE
1.6.99.11	AQUACOBALAMIN REDUCTASE (NADPH)
1.6.99.13	FERRIC-CHELATE REDUCTASE
1.7.2.1	NITRITE REDUCTASE (CYTOCHROME)
1.7.3.1	NITROETANE REDUCTASE
1.7.3.2	ACETYLINDOXYL OXIDASE
1.7.3.3	URATE OXIDASE
1.7.3.4	HYDROXYLAMINE OXIDASE
1.7.7.1	FERREDOXIN-NITRITE REDUCTASE
1.7.7.2	FERREDOXIN-NITRATE REDUCTASE
1.7.99.1	HYDROXYLAMINE REDUCTASE
1.7.99.3	NITRITE REDUCTASE
1.7.99.4	NITRATE REDUCTASE
1.7.99.5	5,10-METHYLENETETRAHYDROFOLATE REDUCTASE (FADH)
1.7.99.6	NITROUS OXIDE REDUCTASE
1.7.99.7	NITRIC-OXIDE REDUCTASE
1.8.1.2	SULFITE REDUCTASE (NADPH)

EC Number	Name
1.8.1.3	HYPOTAURINE DEHYDROGENASE
1.8.1.4	DIHYDROLIPOAMIDE DEHYDROGENASE
1.8.2.1	SULFITE DEHYDROGENASE
1.8.2.2	THIOSULFATE DEHYDROGENASE
1.8.3.1	SULFITE OXIDASE
1.8.3.2	THIOL OXIDASE
1.8.3.3	GLUTATHIONE OXIDASE
1.8.3.4	METHANEDIOL OXIDASE
1.8.4.2	PROTEIN-DISULFIDE REDUCTASE (GLUTATHIONE)
1.8.4.4	GLUTATHIONE-CYSTINE TRANSHYDROGENASE
1.8.4.5	METHIONINE-S-OXIDE REDUCTASE
1.8.4.6	PROTEIN-METHIONINE-S-OXIDE REDUCTASE
1.8.5.1	GLUTATHIONE DEHYDROGENASE (ASCORBATE)
1.8.7.1	SULFITE REDUCTASE (FERREDOXIN)
1.8.99.1	SULFITE REDUCTASE
1.8.99.2	ADENYLYLSULFATE REDUCTASE
1.9.3.1	CYTOCHROME-C OXIDASE
1.9.3.2	PSEUDOMONAS CYTOCHROME OXIDASE
1.9.6.1	NITRATE REDUCTASE (CYTOCHROME)
1.9.99.1	IRON-CYTOCHROME-C REDUCTASE
1.10.1.1	trans-ACENAPHTHENE-1,2-DIOL DEHYDROGENASE
1.10.2.1	L-ASCORBATE-CYTOCHROME-B5 REDUCTASE
1.10.2.2	UBIQUINOL-CYTOCHROME-C REDUCTASE
1.10.2.2	UBIQUINOL-CYTOCHROME-C2 REDUCTASE
1.10.3.1	CATECHOL OXIDASE
1.10.3.1	o-DIPHENOL OXIDASE
1.10.3.2	LACCASE
1.10.3.3	L-ASCORBATE OXIDASE
1.10.3.4	o-AMINOPHENOL OXIDASE
1.10.3.6	RIFAMYCIN-B OXIDASE
1.10.3.7	SULOCHRIN OXIDASE ((+)-BISDECHLOROGEODIN-FORMING)
1.10.3.8	SULOCHRIN OXIDASE ((–)-BISDECHLOROGEODIN-FORMING)
1.10.99.1	PLASTOQUINOL-PLASTOCYANIN REDUCTASE
1.11.1.1	NADH PEROXIDASE
1.11.1.2	NADPH PEROXIDASE
1.11.1.3	FATTY-ACID PEROXIDASE
1.11.1.5	CYTOCHROME-C PEROXIDASE
1.11.1.6	CATALASE
1.11.1.7	MYELOPEROXIDASE
1.11.1.7	PEROXIDASE
1.11.1.8	IODIDE PEROXIDASE
1.11.1.9	GLUTATHIONE PEROXIDASE
1.11.1.10	CHLORIDE PEROXIDASE
1.11.1.11	L-ASCORBATE PEROXIDASE
1.11.1.12	PHOSPHOLIPID-GLUTATHIONE GLUTATHIONE PEROXIDASE
1.11.1.13	MANGANESE PEROXIDASE
1.11.1.14	DIARYLPROPANE PEROXIDASE
1.11.7.1	PEROXIDASE
1.12.1.2	HYDROGEN DEHYDROGENASE
1.12.2.1	CYTOCHROME-C3 HYDROGENASE
1.12.99.2	COENZYME-M-7-MERCAPTOHEPTANOYLTHREONINE-PHOSPHATE-HETEROSULFIDE HYDROGENASE
1.13.11.1	CATECHOL 1,2-DIOXYGENASE
1.13.11.2	CATECHOL 2,3-DIOXYGENASE
1.13.11.3	PROTOCATECHUATE 3,4-DIOXYGENASE
1.13.11.4	GENTISATE 1,2-DIOXYGENASE
1.13.11.5	HOMOGENTISATE 1,2-DIOXYGENASE

EC Number	Name
1.13.11.6	3-HYDROXYANTHRANILATE 3,4-DIOXYGENASE
1.13.11.8	PROTOCATECHUATE 4,5-DIOXYGENASE
1.13.11.9	2,5-DIHYDROXYPYRIDINE 5,6-DIOXYGENASE
1.13.11.11	TRYPTOPHAN 2,3-DIOXYGENASE
1.13.11.12	LIPOXYGENASE
1.13.11.15	3,4-DIHYDROXYPHENYLACETATE 2,3-DIOXYGENASE
1.13.11.17	INDOLE 2,3-DIOXYGENASE
1.13.11.18	SULFUR DIOXYGENASE
1.13.11.19	CYSTEAMINE DIOXYGENASE
1.13.11.20	CYSTEINE DIOXYGENASE
1.13.11.21	β-CAROTENE 15,15′-DIOXYGENASE
1.13.11.23	2,3-DIHYDROXYINDOLE 2,3-DIOXYGENASE
1.13.11.24	QUERCETIN 2,3-DIOXYGENASE
1.13.11.25	3,4-DIHYDROXY-9,10-SECOANDROSTA-1,3,5810)TRIENE-9,17-DIONE 4,5-OXIDOREDUCTASE
1.13.11.26	peptide-TRYPTOPHAN 2,3-DIOXYGENASE
1.13.11.27	4-HYDROXYPHENYLPYRUVATE DIOXYGENASE
1.13.11.28	2,3-DIHYDROXYBENZOATE 2,3-DIOXYGENASE
1.13.11.29	STIZOLOBATE SYNTHASE
1.13.11.30	STIZOLOBINATE SYNTHASE
1.13.11.31	ARACHIDONATE 12-LIPOXYGENASE
1.13.11.32	2-NITROPROPNAE DIOXYGENASE
1.13.11.33	ARACHIDONATE 15-LIPOXYGENASE
1.13.11.33	ARACHIDONATE 15-LIPOXYGENASE 1
1.13.11.34	ARACHIDONATE 5-LIPOXYGENASE
1.13.11.40	ARACHIDONATE 8-LIPOXYGENASE
1.13.11.42	INDOLEAMINE-PYRROLE 2,3-DIOXYGENASE
1.13.12.1	ARGININE 2-MONOOXYGENASE
1.13.12.2	LYSINE 2-MONOOXYGENASE
1.13.12.3	TRYPTOPHAN 2-MONOOXYGENASE
1.13.12.4	LACTATE 2-MONOOXYGENASE
1.13.12.5	RENILLA-LUCIFERIN 2-MONOOXYGENASE
1.13.12.6	CYPRIDINA-LUCIFERIN 2-MONOOXYGENASE
1.13.12.7	PHOTINUS-LUCIFERIN 4-MONOOXYGENASE (ATP-HYDROLYSING)
1.13.12.9	PHENYLALANINE 2-MONOOXYGENASE
1.13.12.10	LYSINE 6-MONOOXYGENASE
1.13.99.1	myo-INOSITOL OXYGENASE
1.13.99.2	BENZOATE 1,2-DIOXYGENASE
1.13.99.3	TRYPTOPHAN 2′-DIOXYGENASE
1.14.11.1	γ-BUTYROBETAINE DIOXYGENASE
1.14.11.2	PROCOLLAGEN-PROLINE, 2-OXOGLUTARATE-4-DIOXYGENASE
1.14.11.3	PYRIMIDINE-DEOXYNUCLEOSIDE 2′-DIOXYGENASE
1.14.11.4	PROCOLLAGEN-LYSINE 5-DIOXYGENASE
1.14.11.6	THYMINE DIOXYGENASE
1.14.11.7	PROCOLLAGEN-PROLINE 3-DIOXYGENASE
1.14.11.8	TRIMETHYLYSINE DIOXYGENASE
1.14.11.9	NARINGENIN, 2-OXOGLUTARATE 3-DIOXYGENASE
1.14.11.11	HYOSCYAMINE (6S)-DIOXYGENASE
1.14.11.13	GIBBERELLIN 2β-DIOXIGENASE I
1.14.11.14	6β-HYDROXYHYOSCYAMINE EPOXIDASE
1.14.11.15	GIBBERELLIN 3β-DIOXYGENASE
1.14.11.16	PEPTIDE-ASPARTATE β-DIOXYGENASE
1.14.12.2	ANTHRANILATE 2,3 DIOXYGENASE
1.14.12.3	BENZENE 1,2-DIOXYGENASE
1.14.12.4	3-HYDROXY-2-METHYLPYRIDINE CARBOXYLATE DIOXYGENASE
1.14.12.7	PHTHALATE 4,5-DIOXYGENASE
1.14.12.9	4-CHLOROPHENYLACETATE 3,4-DIOXYGENASE

EC Number	Name
1.14.12.11	TOLUENE DIOXYGENASE
1.14.12.12	NAPHTHALEN 1,2-DIOXYGENASE
1.14.12.13	2-CHLOROBENZOATE 1,2-DIOXYGENASE
1.14.13.1	SALICYLATE 1-MONOOXYGENASE
1.14.13.2	4-HYDROXYBENZOATE 3-MONOOXYGENASE
1.14.13.3	4-HYDROXYPHENYLACETATE 3-MONOOXYGENASE
1.14.13.4	MELILOTATE 3-MONOOXYGENASE
1.14.13.5	IMIDAZOLEACETATE 4-MONOOXYGENASE
1.14.13.6	ORCINOL 2-MONOOXYGENASE
1.14.13.7	PHENOLL 2-MONOOXYGENASE
1.14.13.8	DIMETHYLANILINE MONOOXYGENASE (N-OXIDE FORMING)
1.14.13.8	MONOOXYGENASE FLAVINE CONTAINING
1.14.13.9	KYNURENINE 3-MONOOXYGENASE
1.14.13.10	2,6-DIHYDROXYPYRIDINE 3-MONOOXYGENASE
1.14.13.11	trans-CINNAMATE 4-MONOOXYGENASE
1.14.13.12	BENZOATE 4-MONOOXYGENASE
1.14.13.13	CALCIDOL 1-MONOOXYGENASE
1.14.13.15	CHOLESTANETRIOL 26-MONOOXYGENASE
1.14.13.16	CYCLOPENTANONE MONOOXYGENASE
1.14.13.17	CHOLESTEROL 7α-MONOOXYGENASE
1.14.13.18	4-HYDROXYPHENYLACETATE 1-MONOOXYGENASE
1.14.13.20	2,4-DICHLOROPHENOL 6-MONOOXYGENASE
1.14.13.21	FLAVONOID 3′-MONOOXYGENASE
1.14.13.22	CYCLOHEXANONE MONOOXYGENASE
1.14.13.23	3-HYDROXYBENZOATE 4-MONOOXYGENASE
1.14.13.24	3-HYDROXYBENZOATE 6-MONOOXYGENASE
1.14.13.25	METHANE MONOOXYGENASE
1.14.13.26	PHOSPHATIDYLCHOLINE 12-MONOOXYGENASE
1.14.13.27	4-AMINOBENZOATE 1-MONOOXYGENASE
1.14.13.28	3,9-DIHYDROXYPTEROCARPAN 6a-MONOOXYGENASE
1.14.13.29	4-NITROPHENOL 2-MONOOXYGENASE
1.14.13.30	LEUKOTRIENE-B4 20-MONOOXYGENASE
1.14.13.31	2-NITROPHENOL 2-MONOOXYGENASE
1.14.13.33	4-HYDROXYBENZOATE 2-MONOOXYGENASE (NAD(P)H)
1.14.13.34	LEUKOTRIENE-E4 20 MONOOXYGENASE
1.14.13.35	ANTHRANILATE 3-MONOOXYGENASE (DEAMINATING)
1.14.13.36	5-O-(4-COUMARYOYL)-D-QUINATE 3′-MONOOXYGENASE
1.14.13.37	METHYLTETRAHYDROPROTOBERBERINE 14-MONOOXYGENASE
1.14.13.38	ANHYDROTETRACYCLINE MONOOXYGENASE
1.14.13.39	NITRIC-OXIDE SYNTHASE
1.14.13.40	ANTHRANILOYL-CoA MONOOXYGENASE
1.14.13.45	CMP-N-ACETYLNEURAMINATE MONOOXYGENASE
1.14.13.46	MENTHOL MONOOXYGENASE
1.14.13.47	(–)-LIMONENE 3-MONOOXYGENASE
1.14.13.48	(–)-LIMONENE 6-MONOOXYGENASE
1.14.13.49	(–)-LIMONENE 7-MONOOXYGENASE
1.14.13.53	ISOFLAVONE 2′-HYDROXYLASE
1.14.14.1	UNSPECIFIC MONOOXYGENASE
1.14.14.3	ALKANAL MONOOXYGENASE (FMN -LINKED)
1.14.15.2	CAMPHOR 1,2-MONOOXYGENASE
1.14.15.3	ALKANE 1-MONOOXYGENASE
1.14.15.4	STEROID 11β-MONOOXYGENASE
1.14.15.5	CORTICOSTERONE 18-MONOOXYGENASE
1.14.15.6	CHOLESTEROL MONOOXYGENASE (SIDE-CHAIN-CLEAVING)
1.14.16.1	PHENYLALANINE 4-MONOOXYGENASE
1.14.16.2	TYROSINE 3-MONOOXYGENASE

EC Number	Name
1.14.16.3	ANTHRANILATE 3-MONOOXYGENASE
1.14.16.4	TRYPTOPHAN 5-MONOOXYGENASE
1.14.16.5	GLYCERYL-ETHER MONOOXYGENASE
1.14.16.6	MANDELATE 4-MONOOXYGENASE
1.14.17.1	DOPAMINE β-MONOOXYGENASE
1.14.17.3	PEPTIDYLGLYCINE MONOOXYGENASE
1.14.18.1	CATECHOL OXIDASE
1.14.18.1	CATECHOL OXIDASE ppo
1.14.18.1	MONOPHENOL MONOOXYGENASE
1.14.18.1	POLYPHENOL OXIDASE
1.14.99.1	PROSTAGLANDIN-ENDOPEROXIDE SYNTHASE
1.14.99.3	HEME OXYGENASE (DECYCLIZING)
1.14.99.4	PROGESTERONE MONOOXYGENASE
1.14.99.5	Δ9 DESATURASE
1.14.99.5	STEAROYL-CoA DESATURASE
1.14.99.6	ACYL[ACYL-CARRIER-PROTEIN] DESATURASE
1.14.99.7	SQUALENE MONOOXYGENASE
1.14.99.9	STERIOD 17α-MONOOXYGENASE
1.14.99.9	STEROID 17α-MONOOXYGENASE/17,20 LYASE
1.14.99.10	STEROID 21-MONOOXYGENASE
1.14.99.11	4-ANDROSTENE-3,17-DIONE MONOOXYGENASE
1.14.99.14	PROGESTERONE 11α-MONOOXYGENASE
1.14.99.15	4-METHOXYBENZOATE MONOOXYGENASE (O-DEMETHYLATING)
1.14.99.18	N-ACETYLNEURAMINATE MONOOXYGENASE
1.14.99.19	PLASMANYLETHANOYLAMINE DESATURASE
1.14.99.20	PHYLLOQUINONE MONOOXYGENASE (2,3-EPOXIDIZING)
1.14.99.22	ECDYSONE 20-MONOOXYGENASE
1.14.99.24	STEROID 9α-MONOOXYGENASE
1.14.99.25	LINOLEOYL-CoA DESATURASE
1.14.99.26	2-HYDROXYPYRIDINE 5-MONOOXYGENASE
1.14.99.27	JUGLONE 3-MONOOXYGENASE
1.14.99.28	LINALOOL 8-MONOOXYGENASE
1.14.99.29	DEOXYHYPUSINE MONOOXYGENASE
1.15.1.1	SUPEROXIDE DISMUTASE
1.15.1.1	SUPEROXIDE DISMUTASE (CU, ZN)
1.15.1.1	SUPEROXIDE DISMUTASE (FE)
1.15.1.1	SUPEROXIDE DISMUTASE (MN)
1.16.1.1	MERCUY(II) REDUCTASE
1.16.1.2	DIFFERIC-TRANSFERRIN REDUCTASE
1.16.1.2	DIFFERRIC-TRANSFERRIN REDUCTASE
1.16.3.1	FERROXIDASE
1.17.1.1	CDP-4-DEHYDRO-6-DEOXYGLUCOSE REDUCTASE
1.17.3.1	PTERIDINE OXIDASE
1.17.4.1	RIBONUCLEOSIDE-DIPHOSPHATE REDUCTASE
1.17.4.2	RIBONUCLEOSIDE-TRIPHOSPHATE REDUCTASE
1.18.1.1	RUBREDOXIN-NAD REDUCTASE
1.18.1.2	FERREDOXIN-NADP REDUCTASE
1.18.1.3	FERREDOXIN-NAD REDUCTASE
1.18.6.1	NITROGENASE
1.18.99.1	HYDROGENASE
1.97.1.1	CHLORATE REDUCTASE
1.97.1.2	PYROGALLOL HYDROXYLTRANSFERASE
1.97.1.3	SULFUR REDUCTASE
2.1.1.1	NICOTINAMIDE N-METHYLTRRANSFERASE
2.1.1.2	GUANIDINOACETATE N-METHYLTRRANSFERASE
2.1.1.3	THETIN-HOMOCYSTEINE S-METHYLTRRANSFERASE

EC Number	Name
2.1.1.4	ACETYLSEROTONIN O-METHYLTRANSFERASE
2.1.1.5	BETAINE-HOMOCYSTEINE S-METHYLTRANSFERASE
2.1.1.6	CATECHOL O-METHYLTRANSFERASE
2.1.1.7	NICOTINATE N-METHYLTRRANSFERASE
2.1.1.8	HISTAMINE N-METHYLTRANSFERASE
2.1.1.9	THIOL S-METHYLTRANSFERASE
2.1.1.10	HOMOCYSTEINE S-METHYLTRANSFERASE
2.1.1.11	MAGNESIUM PROTOPORPHYRIN O-METHYLTRANSFERASE
2.1.1.13	5-METHYLTETRAHYDROFOLATE-HOMOCYSTEINE S-METHYLTRANSFERASE
2.1.1.14	5-METHYLTETRAHYDROPTEROYLTRIGLUTAMATE-HOMOCYSTEINE METHYLTRANSFERASE
2.1.1.15	FATTY ACID O-METHYLTRANSFERASE
2.1.1.15	PHOSPHATIDYLETHANOLAMINE N-METHYLTRANSFERASE
2.1.1.16	METHYLENE-FATTY-ACYL-PHOSPHOLIPID SYNTHASE
2.1.1.17	PHOSPHATIDYLETHANOLAMINE N-METHYLTRANSFERASE
2.1.1.19	TRIMETHYLSULFONIUM-TETRAHYDROFOLATE N-METHYLTRANSFERASE
2.1.1.20	GLYCINE METHYLTRANSFERASE
2.1.1.21	GLUTAMATE METHYLTRANSFERASE
2.1.1.21	METHYLAMINE-GLUTAMATE METHYLTRANSFERASE
2.1.1.23	PROTEIN-ARGININE N-METHYLTRANSFERASE
2.1.1.24	PROTEIN-GLUTAMATE METHYLTRANSFERASE
2.1.1.25	PHENOL O-METHYLTRANSFERASE
2.1.1.27	TYRAMINE N-METHYLTRANSFERASE
2.1.1.28	PHENYLETHANOLAMINE N-METHYLTRANSFERASE
2.1.1.29	tRNA(CYTOSINE-5-) METHYLTRANSFERASE
2.1.1.31	tRNA(GUANINE-N^1-) METHYLTRANSFERASE
2.1.1.32	tRNA(GUANINE-N^2-) METHYLTRANSFERASE
2.1.1.33	tRNA(GUANINE-N^7-) METHYLTRANSFERASE
2.1.1.34	tRNA(GUANOSINE-2′O-) METHYLTRANSFERASE
2.1.1.35	tRNA(URACIL-5-) METHYLTRANSFERASE
2.1.1.36	tRNA(ADENINE-N^1-) METHYLTRANSFERASE
2.1.1.36	tRNA(ADENINE-N^1-) METHYLTRANSFERASE 2
2.1.1.37	DNA (CYTOSINE-5-)-METHYLTRANSFERASE
2.1.1.38	O-DEMTETHYLPUROMYCIN O-METHYLTRANSFERASE
2.1.1.39	MYO-INOSITOL 1-O-METHYLTRANSFERASE
2.1.1.40	MYO-INOSITOL 3-O-METHYLTRANSFERASE
2.1.1.41	Δ24(25)STEROL METHYLRANSFERASE
2.1.1.41	Δ24(25)STEROL METHYLTRANSFERASE
2.1.1.41	Δ(24)-STEROL C-METHYLTRANSFERASE
2.1.1.42	LUTEOLIN O-METHYLTRANSFERASE
2.1.1.43	HISTONE-LYSINE N-METHYLTRANSFERASE
2.1.1.44	DIMETHYLHISTIDINE N-METHYLTRANSFERASE
2.1.1.45	THYMIDYLATE SYNTHASE
2.1.1.46	ISOFLAVONE 4′-O-METHYLTRANSFERASE
2.1.1.47	INDOLEPYRUVATE C-METHYLTRANSFERASE
2.1.1.48	tRNA(ADENINE-N^6)-METHYLTRANSFERASE
2.1.1.49	ARYLAMINE N-METHYLTRANSFERASE
2.1.1.50	LOGANATE O-METHYLTRANSFERASE
2.1.1.53	PUTRESCINE N-METHYLTRANSFERASE
2.1.1.56	mRNA(GUANINE-N7-)-METHYLTRANSFERASE
2.1.1.57	mRNA(NUCLEOSIDE-2′O-)-METHYLTRANSFERASE
2.1.1.59	CYTOCHROME-C LYSINE N-METHYLTRANSFERASE
2.1.1.60	CALMODULIN-LYSINE N-METHYLTRANSFERASE
2.1.1.61	tRNA (5-METHYLAMINOMETHYL-2-THIOURIDYLATE) METHYLTRANSFERASE
2.1.1.63	METHYLATED-DNA-[PROTEIN]CYSTEINE S-METHYLTRANSFERASE
2.1.1.64	3-DEMETHYLUBIQUINONE-9 3-O-METHYLTRANSFERASE
2.1.1.65	LICODIONE 2-O-METHYLTRANSFERASE

EC Number	Name
2.1.1.66	rRNA(ADENOSINE-2′-O-)-METHYLTRANSFERASE
2.1.1.67	THIOPURINE S-METHYLTRANSFERASE
2.1.1.68	CAFFEATE O-METHYLTRANSFERASE
2.1.1.69	5-HYDROXYFURANOCOUMARIN 5-O-METHYLTRANSFERASE
2.1.1.70	8-HYDROXYFURANOCOUMARIN 8-O-METHYLTRANSFERASE
2.1.1.71	PHOSPHATIDYL N-METHYLETHANOLAMINE METHYLTRANSFERASE
2.1.1.73	DNA METHYLTRANSFERASE M.ALU I
2.1.1.73	DNA METHYLTRANSFERASE M.BAM HI
2.1.1.75	APIGENEIN 3-O-METHYLTRANSFERASE
2.1.1.76	QUERCETIN 3-O-METHYLTRANSFERASE
2.1.1.77	PROTEIN-D-ASPARTATE METHYLTRANSFERASE
2.1.1.78	ISOORIENTIN 3-METHYLTRANSFERASE
2.1.1.79	CYCLOPROPANE-FATTY-ACYL-PHOSPHOLIPID SYNTHASE
2.1.1.80	PROTEIN-GLUTAMATE O-METHYLTRANSFERASE
2.1.1.82	METHYLQERCITIN 7-O-METHYLTRANSFERASE
2.1.1.83	3,7-DIMETHYLQERCITIN 4′-O-METHYLTRANSFERASE
2.1.1.84	METHYLQERCETAGETIN 6-O-METHYLTRANSFERASE
2.1.1.87	PYRIDINE N-METHYLTRANSFERASE
2.1.1.88	8-HYDROXYQUERCETINE 8-O-METHYLTRANSFERASE
2.1.1.90	METHANOL-5-HYDROXYBENZIMIDAZOLYLCOBAMIDE Co-METHYLTRANSFERASE
2.1.1.91	ISOBUTYRALDOXIME O-METHYLTRANSFERASE
2.1.1.93	XANTHOTOXOL O-METHYLTRANSFERASE
2.1.1.95	TOCOPHEROL O-METHYLTRANSFERASE
2.1.1.96	THIOETHER S-METHYLTRANSFERASE
2.1.1.97	3-HYDROXYANTHRANILATE 4-C-METHYLTRANSFERASE
2.1.1.100	PROTEIN-S-ISOPRENYLCYSTEINE O-METHYLTRANSFERASE
2.1.1.101	MACROCIN O-METHYLTRANSFERASE
2.1.1.102	DEMETHYLMACROCIN O-METHYLTRANSFERASE
2.1.1.104	CAFFEOYL-CoA O-METHYLTRANSFERASE
2.1.1.106	TRYPTOPHAN 2-C-METHYLTRANSFERASE
2.1.1.107	UROPORPHYRIN-III C-METHYLTRANSFERASE
2.1.1.108	6-HYDROXYMELLEIN-O-METHYLTRANSFERASE
2.1.1.109	DEMETHYLSTERIGMATOCYSTIN 6-O-METHYLTRANSFERASE
2.1.1.112	GLUCURONOXYLAN 4-O-METHYLTRANSFERASE
2.1.2.1	GLYCINE HYDROXYMETHYLTRANSFERASE
2.1.2.2	PHOSPHORIBOSYLGLYCINAMIDE FORMYLTRANSFERASE
2.1.2.3	PHOSPHORIBOSYLAMINOIMIDAZOLECARBOXAMIDE FORMYLTRANSFERASE
2.1.2.5	GLUTAMATE FORMIMINOTRANSFERASE
2.1.2.7	D-ALANINE 2-HYDROXYMETHYLTRANSFERASE
2.1.2.8	DEOXYCYTIDYLATE HYDROXYMETHYLTRANSFERASE
2.1.2.9	METHIONYL-tRNA FORMYLTRANSFERASE
2.1.2.10	AMINOMETHYLTRANSFERASE
2.1.2.11	3-METHYL-2-OXOBUTANOATE HYDROXYMETHYLTRANSFERASE
2.1.3.1	METHYLMALONYL-CoA CARBOXYLTRANSFERASE
2.1.3.2	ASPARTATE CARBAMOYLTRANSFERASE
2.1.3.3	ORNITHINE CARBAMOYLTRANSFERASE
2.1.3.6	PUTRESCINE CARBAMOYLTRANSFERASE
2.1.3.7	3-HYDROXYMETHYLCEPHEM CARBAMOYLTRANSFERASE
2.1.3.97	MYRISTOYLTRANSFERASE
2.1.4.1	GLYCINE AMIDINOTRANSFERASE
2.1.4.2	INOSAMINE-PHOSPHATE AMIDINOTRANSFERASE
2.2.1.1	TRANSKETOLASE
2.2.1.2	TRANSALDOLASE
2.2.1.3	FORMALDEHYDE TRANSKETOLASE
2.2.7.7	DNA POLYMERASE α
2.3.1.1	AMINO-ACID N-ACETYLTRANSFERASE

EC Number	Name
2.3.1.4	GLUCOSAMINE-PHOSPHATE N-ACETYLTRANSFERASE
2.3.1.5	ARYLAMINE N-ACETYLTRANSFERASE
2.3.1.6	CHOLINE O-ACETYLTRANSFERASE
2.3.1.7	CARNITINE O-ACETYLTRANSFERASE
2.3.1.8	PHOSPHATE ACETYLTRANSFERASE
2.3.1.9	ACETYL-CoA C-ACETYLTRANSFERASE
2.3.1.12	DIHYDROLIPOAMIDE S-ACETYLTRANSFERASE
2.3.1.13	GLYCINE ACYLTRANSFERASE
2.3.1.15	GLYCEROL-3-PHOSPHATE O-ACYLTRANSFERASE
2.3.1.16	ACETYL-CoA C-ACYLTRANSFERASE
2.3.1.17	ASPARTATE N-ACETYLTRANSFERASE
2.3.1.18	GALACTOSIDE o-ACETYLTRANSFERASE
2.3.1.19	PHOSPHATE BUTYRYLTRANSFERASE
2.3.1.20	DIACYLGLYCEROL O-ACYLTRANSFERASE
2.3.1.21	CARNITINE O-PALMITOYLTRANSFERASE
2.3.1.22	2-ACYLGLYCEROL O-ACYLTRANSFERASE
2.3.1.23	1-ACYLGLYCEROPHOSPHOCHOLINE O-ACYLTRANSFERASE
2.3.1.24	SPHINGOSINE N-ACYLTRANSFERASE
2.3.1.26	STEROL O-ACYLTRANSFERASE
2.3.1.28	CHLORAMPHENICOL O-ACETYLTRANSFERASE
2.3.1.29	GLYCINE C-ACETYLTRANSFERASE
2.3.1.30	SERINE O-ACETYLTRANSFERASE
2.3.1.31	HOMOSERINE O-ACETYLTRANSFERASE
2.3.1.32	LYSINE N-ACETYLTRANSFERASE
2.3.1.34	1,3-β-GLUCAN SYNTHASE
2.3.1.35	GLUTAMATE N-ACETYLTRANSFERASE
2.3.1.36	D-AMINO-ACID N-ACETYLTRANSFERASE
2.3.1.37	5-AMINOLEVULINATE SYNTHASE
2.3.1.38	[ACYL-CARRIER-PROTEIN] S-ACETYLTRANSFERASE
2.3.1.39	[ACYL-CARRIER-PROTEIN] S-MALONYLTRANSFERASE
2.3.1.41	3-OXOACYL-[ACYL-CARRIER-PROTEIN] SYNTHASE
2.3.1.42	GLYCERONE-PHOSPHATE O-ACYLTRANSFERASE
2.3.1.43	PHOSPHATIDYLCHOLINE-STEROL O-ACYLTRANSFERASE
2.3.1.45	N-ACETYLNEURAMINATE 7-O(or 9-O) ACETYLTRANSFERASE
2.3.1.46	HOMOSERINE O-SUCCINYLTRANSFERASE
2.3.1.47	8-AMINO-7-OXONONAOATE SYNTHASE
2.3.1.48	HISTONE ACYLTANSFERASE
2.3.1.50	SERINE C-PALMITOYLTRANSFERASE
2.3.1.51	1-ACYLGLYCEROL-3-PHOSPHATE O-ACYLTRANSFERASE
2.3.1.53	PHENYLALANINE N-ACETYLTRANSEFRASE
2.3.1.54	FORMATE C-ACETYLTRANSFERASE
2.3.1.55	KANAMYCIN 6′-N-ACETYLTRANSFERASE
2.3.1.56	AROMATIC-HYDROXYLAMINE O-ACETYLTRANSFERASE
2.3.1.57	DIAMINE N-ACETYLTRANSFERASE
2.3.1.60	GENTAMICIN 3′-N-ACETYLTRANSFERASE
2.3.1.62	2-ACYLGLYCEROPHOSPHOCHOLINE O-ACYLTRANSFERASE
2.3.1.65	GLYCINE N-CHOLOYLTRANSFERASE
2.3.1.66	LEUCINE N-ACETYLTRANSFERASE
2.3.1.67	1-ALKYLGLYCEROPHOSPHOCHOLINE O-ACETYLTRANSFERASE
2.3.1.68	GLUTAMINE N-ACYLTRANSFERASE
2.3.1.69	MONOTERPENOL O-ACETYLTRANSFERASE
2.3.1.71	GLYCINE BENZOYLTRANSFERASE
2.3.1.72	INDOLEACETYLGLUCOSE-INOSITOL O-ACYLTRANSFERASE
2.3.1.73	DIACYLGLYCEROL-STEROL O-ACYLTRANSFERASE
2.3.1.74	NARINGENIN-CHALCONE SYNHTASE
2.3.1.76	RETINOL FATTY-ACYLTRANSFERASE

EC Number	Name
2.3.1.77	RETINOL O-FATTY-ACYLTRANSFERASE
2.3.1.78	HEPARAN-α-GLUCOSAMINIDINE N-ACETYLTRANSFERASE
2.3.1.82	AMINOGLYCOSIDE $N^{6\prime}$-ACETYLTRANSFERASE
2.3.1.83	PHOSPHATIDYLCHOLINE-DOLICHOL O-ACYLTRANSFERASE
2.3.1.85	FATTY ACID SYNTHASE
2.3.1.86	FATTY-ACYL-CoA SYNTHASE
2.3.1.87	ARALKYLAMINE N-ACETYLTRANSFERASE
2.3.1.88	PEPTIDE α-N-ACETYLTRANSFERASE
2.3.1.90	β-GLUCOGALLIN O-ACETYLTRANSFERASE
2.3.1.91	SINAPOYLGLUCOSE-CHOLINE O-SINAPOYLTRANSFERASE
2.3.1.92	SINAPOYLGLUCOSE-MALATE O-SINAPOYLTRANSFERASE
2.3.1.93	13-HYDROXYLUPAINE O-TIGLOYLTRANSFERASE
2.3.1.94	EROTHRONOLIDE SYNTHASE
2.3.1.96	GLYCOPROTEIN N-PALMITOYLTRANSFERASE
2.3.1.97	GLYCYLPEPTIDE N-TETRADECANOYLTRANSFERASE
2.3.1.99	QUINATE O-HYDROXYCINNAMOYLTRANSFERASE
2.3.1.100	MYELIN-PROTEOLIPID O-PALMIOYLTRANSFERASE
2.3.1.102	N^{6}-HYDROXYLYSINE O-ACETYLTRANSFERASE
2.3.1.103	SINAPOYLGLUCOSE-SINAPOYLGLUCOSE O-SINAPOYLTRANSFERASE
2.3.1.104	1-ALKENYLGLYCEROPHOSPHOCHOLINE O-ACYLTRANSFERASE
2.3.1.105	ALKYLGLYCEROPHOSPHATE 2-O-ACETYLTRANSFERASE
2.3.1.107	17-O-DEACETYLVINDOLINE O-ACETYLTRANSFERASE
2.3.1.108	TUBULIN N-ACETYLTRANSFERASE
2.3.1.110	TYRAMINE N-FERULOYLTRANSFERASE
2.3.1.111	MYCOSERATE SYNHTASE
2.3.1.114	3,4-DICHLOROANILINE N-MALONYLTRANSFERASE
2.3.1.117	2,3,4,5-TETRAHYDROPYRIDINE-2-CARBOXYLATE N-SUCCINYLTRANSFERASE
2.3.1.118	N-HYDROXYARYLAMINE O-ACETYLTRANSFERASE
2.3.1.119	ICOSANOYL-CoA SYNTHASE
2.3.1.121	1-ALKENYLGLYCEROPHOSPHOETHANOLAMINE O-ACYLTRANSFERASE
2.3.1.122	TREHALOSE O-MYCOLYLTRANSFERASE
2.3.1.123	DOLICHOL O-ACYLTRANSFERASE
2.3.1.125	1-ALKYL-2-ACETYLGLYCEROL O-ACYLTRANSFERASE
2.3.1.129	ACYL[ACYL-CARRIER-PROTEIN]-UDP-N-ACETYLGLUCOSAMINE O-ACYLTRANSFERASE
2.3.1.133	SHIKIMATE O-HYDROXYCINNAMOYLTRANSFERASE
2.3.1.135	PHOSPHATIDYLCHOLINE-RETINOL O-ACYLTRANSFERASE
2.3.1.136	POLYSIALIC-ACID O-ACETYLTRANSFERASE
2.3.1.137	CARNITINE O-OCTANOYLTRANSFERASE
2.3.1.139	ECDYSONE O-ACYLTRANSFERASE
2.3.1.140	ROSMARINATE SYNTHASE
2.3.1.141	GALACTOSYLACYLGLYCEROL O-ACYLTRANSFERASE
2.3.1.142	GLYCOPROTEIN O-FATTY-ACYLTRANSFERASE
2.3.2.2	γ-GLUTAMYLTRANSFERASE
2.3.2.4	γ-GLUTAMYLCYCLOTRANSFERASE
2.3.2.6	LEUCYLTRANSFERASE
2.3.2.7	ASPARTYLTRANSFERASE
2.3.2.8	ARGINYLTRANSFERASE
2.3.2.9	ARGARATINE γ-GLUTAMYLTRANSFERASE
2.3.2.12	PEPTIDYLTRANSFERASE
2.3.2.13	PROTEIN-GLUTAMINE γ-GLUTAMYLTRANSFERASE
2.3.2.14	D-ALANINE γ-GLUTAMYLTRANSFERASE
2.3.2.15	GLUTATHIONE γ-GLUTAMYLTRANSFERASE
2.4.1.1	GLYCEROL-3-PHOSPHATE DEHYDROGENASE
2.4.1.1	PHOSPHORYLASE
2.4.1.2	DEXTRIN DEXTRANASE
2.4.1.4	AMYLOSUCRASE

EC Number	Name
2.4.1.5	DEXTRANSUCRASE
2.4.1.7	SUCROSE PHOSPHORYLASE
2.4.1.8	MALTOSE PHOSPHORYLASE
2.4.1.10	LEVANSUCRASE
2.4.1.11	GLYCOGEN (STARCH) SYNTHASE
2.4.1.11	GLYCOGEN (STARCH) SYNTHASE D
2.4.1.12	CELLULOSE SYNTHASE (UDP FORMING)
2.4.1.13	SUCROSE SYNTHASE
2.4.1.14	SUCROSE-PHOSPHATE SYNTHASE
2.4.1.15	α,α-TREHALOSE-PHOSPHATE SYNTHASE
2.4.1.16	CHITIN SYNTHASE
2.4.1.17	UDP-GLUCURONOSYLTRANSFERASE
2.4.1.18	1,4-α-GLUCAN BRANCHING ENZYME
2.4.1.19	CYCLOMALTODEXTRIN GLUCANOTRANSFERASE
2.4.1.20	CELLOBIOSE PHOSPHORYLASE
2.4.1.21	STARCH (BACTERIAL GLYCOGEN) SYNTHASE
2.4.1.21	STARCH SYNTHASE
2.4.1.22	LACTOSE SYNTHASE
2.4.1.25	4-α-GLUCANOTRANSFERASE
2.4.1.26	DNA α-GLUCOSYLTRANSFERASE
2.4.1.30	1,3-β-OLIGIGOGLUCAN PHOSPHORYLASE
2.4.1.31	LAMINARIBOSE PHOSPHORYLASE
2.4.1.32	GLUCOMANNAN 4-β-MANNOSYLTRANSFERASE
2.4.1.34	1,3-β-GLUCAN SYNTHASE
2.4.1.35	PHENOL β-GLUCOSYLTRANSFERASE
2.4.1.37	FUCOSYLGLYCOPROTEIN 3α-GALACTOSYLTRANSFERASE
2.4.1.38	β-.-ACETYLGLUCOSAMINYL-GLYCOPEPTIDE β-1,4-GALACTOSYLTRANSFERASE
2.4.1.38	βN-ACETYLGLUCOSAMINYL-GLYCOPEPTIDE β-1,4-GALACTOSYLTRANSFERASE
2.4.1.39	STEROID N-ACETYLGLUCOSAMINYLTRANSFERASE
2.4.1.40	FUCOSYLGALACTOSE α-N-ACETYLGALACTOSAMINYLTRANSFERASE
2.4.1.41	POLYPEPTIDE N-ACETYLGALACTOSAMINYLTRANSFERASE
2.4.1.43	POLYGALACTURONATE 4-α-GALACTURONOSYLTRANSFERASE
2.4.1.45	2-HYDROXYACYLSPHINGOSINE 1β-GALACTOSYLTRANSFERASE
2.4.1.46	1,2-DIACYLGLYCEROL 3-β-GALACTOSYLTRANSFERASE
2.4.1.50	PROCOLLAGEN GALACTOSYLTRANSFERASE
2.4.1.54	UNDECAPRENL-PHOSPHATE MANNOSYLTRANSFERASE
2.4.1.62	GANGLIOSIDE GALACTOSYLTRANSFERASE
2.4.1.64	α,α-TREHALOSE PHOSPHORYLASE
2.4.1.65	GALACTOSIDE 3(4)-L-FUCOSYLTRANSFERASE
2.4.1.66	PROCOLLAGEN GLUCOSYLTRANSFERASE
2.4.1.67	GALACTINOL-RAFFINOSE GALACTOSYLTRANSFERASE
2.4.1.69	GALACTOSIDE 2-L-FUCOSYLTRANSFERASE
2.4.1.70	POLY(RIBITOL-PHOSPHATE) N-ACETYLGLUCOSAMINYLTRANSFERASE
2.4.1.71	ARYLAMINE GLUCOSYLTRANSFERASE
2.4.1.74	GLYCOSAMINOGLYCAN GALACTOSYLTRANSFERASE
2.4.1.75	UDP-GALACTURONYLTRANSFERASE
2.4.1.78	PHOSPHOPOLYPRENOL GLUCOSYLTRANSFERASE
2.4.1.79	GALACTOSYLGALACTOSYLGLUCOSYLCERAMIDE β-D-ACETYLGALACTOSAMINYLTRANSFERASE
2.4.1.80	CERAMIDE GLUCOSYLTRANSFERASE
2.4.1.81	FLAVONE 7-O-β-GLUCOSYLTRANSFERASE
2.4.1.82	GALACTINOL GALACTOSYLTRANSFERASE
2.4.1.83	DOLICHYL-PHOSPHATE β-D-MANNOSYLTRANSFERASE
2.4.1.86	GLUCOSAMINYLGALCTOSYLGLUCOSYLCERAMIDE β-GALACTOSYLTRANSFERASE
2.4.1.87	β-D-GALACTOSYL-N-ACETYLGLUCOSAMINYLGLYCOPEPTIDE α-1,3-GALACTOSYLTRANSFERASE
2.4.1.88	GLOBOSIDE α-N-ACETYLGALACTOSAMINYLTRANSFERASE
2.4.1.90	N-ACETYLLACTOSAMINE SYNTHASE

EC Number	Name
2.4.1.91	FLAVONOL 3-O-GLUCOSYLTRANSERASE
2.4.1.92	(N-ACETYLNEURAMINYL)-GALACTOSYLGLUCOSYLCERAMIDE N-ACETYLGALACTOSAMINYLTRANSFER
2.4.1.93	INULIN FRUCTOTRANSFERASE (DEPOLYMERIZING, DIFRUCTOFURANOSE-1,2′: 2′,3′-DIANHYDRIDE
2.4.1.99	SUCROSE 1^F-FRUCTOSYLTRANSFERASE
2.4.1.101	α-1,3-MANNOSYL-GLYCOPROTEIN β-1,2-N-ACETYLGLUCOSAMINYLTRANSFERASE
2.4.1.102	β-1,3-GALACTOSYL-O-GLYCOSYL-GLYCOPROTEIN β-1,6-N-ACETYLGLUCOSAMINYLTRANSFERASE
2.4.1.109	DOLICHYL-PHOSPHATE-MANNOSE-PROTEIN MANNOSYLTRANSFERASE
2.4.1.110	tRNA-QUEUOSINE β-MANNOSYLTRANSFERASE
2.4.1.112	α-1,4-GLUCAN-PROTEIN SYNTHASE (UDP-FORMING)
2.4.1.114	2-COUMARATE O-β-GLUCOSYLTRANSFERASE
2.4.1.115	ANTHOCYANIDIN 3-O-GLUCOSYLTRANSFERASE
2.4.1.116	CYANIDIN-3-RHAMNOSYLGLUCOSIDE 5-O-GLUCOSYLTRANSFERASE
2.4.1.117	DOLICHYL-PHOSPHATE β-GLUCOSYLTRANSFERASE
2.4.1.118	CYTOKININ 3-β-GLUCOSYLTRANSFERASE
2.4.1.119	DOLICHYL-DIPHOSPHOOLIGOSACCHARIDE-PROTEIN GLYCOSYLTRANSFERASE
2.4.1.120	SINAPATE 1-GLUCOSYLTRANSFERASE
2.4.1.121	INDOLE-3-ACETATE β-GLUCOSYLTRANSFERASE
2.4.1.122	GLYCOPROTEIN-N-ACETYLGALACTOSAMINE 3-β-GALACTOSYLTRANSFERASE
2.4.1.124	N-ACETYLLACTOSAMINE 3-α-GALACTOSYLTRANSFERASE
2.4.1.125	SUCROSE-1,6-GLUCAN 3(6)-α-GLUCOSYLTRANSFERASE
2.4.1.126	HYDROXYCINNAMATE 4-β-GLUCOSYLTRANSFERASE
2.4.1.127	MONOTERPENOL β-GLUCOSYLTRANSFERASE
2.4.1.129	PEPTIDOGLYCAN GLUCOSYLTRANSFERASE
2.4.1.131	GLYCOLIPID 2-α-MANNOSYLTRANSFERASE
2.4.1.132	GLYCOLIPID 3-α-MANNOSYLTRANSFERASE
2.4.1.136	GALLATE 1-β-GLUCOSYLTRANSFERASE
2.4.1.137	sn-GLYCEROL-3-PHOSPHATE 2-α-GALACTOSYLTRANSFERASE
2.4.1.138	MANNOTETRAOSE 2-α-N-ACETYLGLUCOSAMINYLTRANSFERASE
2.4.1.139	MALTOSE SYNTHASE
2.4.1.140	ALTERNASUCRASE
2.4.1.141	N-ACETYLGLUCOSAMINYLDIPHOSPHODOLICHOL N-ACETYLGLUCOSAMINYLTRANSFERASE
2.4.1.142	CHITOBIOSYLDIPHOSPHODOLICHOL α-MANNOSYLTRANSFERASE
2.4.1.143	α-1,6-MANNOSYL-GLYCOPROTEIN β-1,2-N-ACETYLGLUCOSAMINYLTRANSFERASE
2.4.1.146	β-1,3-GALACTOSYL-O-GLYCOSYL-GLYCOPROTEIN β-1,3-N-ACETYLGLUCOSAMINYLTRANSFERASE
2.4.1.147	ACETYLGALACTOSAMINYL-O-GLYCOSYL-GLYCOPROTEIN β-1,3-N-ACETYLGLUCOSAMINYLTRANSFE
2.4.1.148	ACETYLGALACTOSAMINYL-O-GLYCOSYL-GLYCOPROTEIN β-1,6-N-ACETYLGLUCOSAMINYLTRANSFE
2.4.1.149	N-ACETYLGALACTOSAMIDINE β-1,3-N-ACETYLGLUCOSAMINYLTRANSFERASE
2.4.1.150	N-ACETYLGALACTOSAMIDINE β-1,6-N-ACETYLGLUCOSAMINYLTRANSFERASE
2.4.1.151	N-ACETYLLACTOSAMINIDINE α-1,3-GALACTOSYLTRANSFERASE
2.4.1.152	GALACTOSIDE 3-FUCOSYLTRANSFERASE
2.4.1.154	GLOBOTRIOSYLCERAMIDE β-1,6-N-ACETYLGALACTOSAMINYLTRANSFERASE
2.4.1.155	N-ACETYLGLUCOSAMINYLTRANSFERASE
2.4.1.155	N-ACETYLGLUCOSAMINYLTRANSFERASE V
2.4.1.155	α-1,3(6)-MANNOSYLGLYCOPROTEIN β-1,6-N-ACETYLGLUCOSAMINYLTRANSFERASE
2.4.1.156	INDOLYLACETYL-MYO-INOSITOL GALACTOSYLTRANSFERASE
2.4.1.157	1,2-DIACYLGLYCEROL 3-GLUCOSYLTRANSFERASE
2.4.1.158	13-HYDROXYDOCOSANOATE 13-β-GLUCOSYLTRANSFERASE
2.4.1.160	PYRIDOXINE 5′-O-β-D-GLUCOSYLTRANSFERASE
2.4.1.161	OLIGOSACCHARIDE 4-α-D-GLUCOSYLTRANSFERASE
2.4.1.163	β-GALACTOSYL-N-ACETYLGLUCOSAMINYLGALACTOSYLGLUCOSYLCERAMIDE β-1,3-ACETYLGLUCOSA
2.4.1.164	GALACTOSYL-N-ACETYLGLUCOSAMINYLGALACTOSYLGLUCOSYLCERAMIDE β-1,6-ACETYLGLUCOSAMI
2.4.1.169	XYLOGLUCAN 6-XYLOSETRANSFERASE
2.4.1.172	SALICYL ALCOHOL β-D-GLUCOSYLTRANSFERASE
2.4.1.173	STEROL 3β-GLUCOSYLTRANSFERASE
2.4.1.177	CINNAMATE β-D-GLUCOSYLTRANSFERASE
2.4.1.179	LACTOSYLCERAMIDE β-1,3-GALACTOSYLTRANSFERASE

EC Number	Name
2.4.1.181	HYDROXYANTHRAQUINONE GLUCOSYLTRANSFERASE
2.4.1.182	LIPID-A-DISACCHARIDE SYNTHASE
2.4.1.184	GALACTOLIPID GALACTOSYLTRANSFERASE
2.4.1.185	FLAVANONE 7-O-β-GLUCOSYLTRANSFERASE
2.4.1.186	GLYCOGENIN GLUCOSYLTRANSFERASE
2.4.1.187	N-ACETYLGLUCOSAMINYLDIPHOSPHOUNDECAPRENOL N-ACETYL-β-D-MANNOSAMINYLTRANSFERAS
2.4.1.190	LUTEOLIN-7O-GLUCURONIDE 7-O-GLUCURONOSYLTRANSFERASE
2.4.1.191	LUTEOLIN-7O-DIGLUCURONIDE 4′-O-GLUCURONOSYLTRANSFERASE
2.4.1.192	NUATIGENIN 3-β-GLUCOSYLTRANSFERASE
2.4.1.193	SARSAPOGENIN 3-β-GLUCOSYLTRANSFERASE
2.4.1.194	4-HYDROXYBENZOATE 4O-β-D-GLUCOSYLTRANSFERASE
2.4.1.195	THIOHYDROXIMATE β-D-GLUCOSYLTRANSFERASE
2.4.1.196	NICOTINATE GLUCOSYLTRANSFERASE
2.4.1.197	HIGH-MANNOSE-OLIGOSACCHARIDE β-1,4-N-ACETYLGLUCOSAMINYLTRANSFERASE
2.4.1.198	PHOSPHATIDYLINOSITOL N-ACETYLGLUCOSAMINYLTRANSFERASE
2.4.1.200	INULIN FRUCTOTRANSFERASE (DEPOLYMERIZING, DIFRUCTOFURANOSE-1,2′: 2′,1-DIANHYDRIDE-
2.4.1.201	MANNOSYL-GLYCOPROTEIN β-1,4-N-ACETYLGLUCOSAMINYLTRANSFERASE
2.4.1.202	2,4-DIHYDROXY-7-METHOXY-2H-1,4-BENZOXAZIN-3(4H)-ONE 2D-GLUCOSYLTRANSFERASE
2.4.2.1	PURINE-NUCLEOSIDE PHOSPHORYLASE
2.4.2.2	PYRIMIDINE-NUCLEOSIDE PHOSPHORYLASE
2.4.2.3	URIDINE PHOSPHORYLASE
2.4.2.4	THYMIDINE PHOSPHORYLASE
2.4.2.6	NUCLEOSIDE DEOXYRIBOSYLTRANSFERASE
2.4.2.7	ADENINE PHOSPHORIBOSYLTRANSFERASE
2.4.2.8	HYPOXANTHINE PHOSPHORIBOSYLTRANSFERASE
2.4.2.9	URACIL PHOSPHORIBOSYLTRANSFERASE
2.4.2.10	OROTATE PHOSPHORIBOSYLTRANSFERASE
2.4.2.11	NICOTINATE PHOSPHORIBOSYLTRANSFERASE
2.4.2.12	NICOTINAMIDE PHOSPHORIBOSYLTRANSFERASE
2.4.2.14	AMIDOPHOSPHORIBOSYLTRANSFERASE
2.4.2.15	GUANOSINE PHOSPHORYLASE
2.4.2.16	URATE-RIBONUCLEOTIDE PHOSPHORYLASE
2.4.2.17	ATP PHOSPHORIBOSYLTRANSFERASE
2.4.2.18	ANTHRANILATE PHOSPHORIBOSYLTRANSFERASE
2.4.2.19	NICOTINATE-NUCLEOTIDE PYROPHOSPHORYLASE (CARBOXYLATING)
2.4.2.21	NICOTINATE-NUCLEOTIDE PHOSPHORYBOSYLTRANSFERASE
2.4.2.22	XANTHINE PHOSPHORIBOSYLTRANSFERASE
2.4.2.24	1,4-β-D-XYLAN SYNTHASE
2.4.2.25	FLAVONE APIOSYLTRANSFERASE
2.4.2.27	dTDP-DIHYDROSTREPTOSE-STREPTIDINE-6-PHOSPHATE DIHYDROSTREPTOSYLTRANSFERASE
2.4.2.28	5′-METHYLTHIOADENOSINE PHOSPHORYLASE
2.4.2.29	QUEUINE tRNA-RIBOSYLTRANSFERASE
2.4.2.30	NAD ADP-RIBOSYLTRANSFERASE
2.4.2.31	NAD(P)-ARGININE ADP-RIBOSYLTRANSFERASE
2.4.2.32	DOLICHYL-PHOSPHATE D-XYLOSYLTRANSFERASE
2.4.2.35	FLAVONOL-3O-GLYCOSIDE XYLOSYLTRANSFERASE
2.4.2.37	NAD-DINITROGEN-REDUCTASE ADP-D-RIBOSYLTRANSFERASE
2.4.4.1	PURINE-NUCLEOSIDE PHOSPHORYLASE
2.4.99.1	β-GALACTOSIDE α-2,6-SIALYLTRANSFERASE
2.4.99.2	MONOSIALOGANGLIOSIDE SIALYLTRANSFERASE
2.4.99.3	α-N-ACETYLGALACTOSAMINIDE α-2,6-SIALYLTRANSFERASE
2.4.99.3	α-N-ACETYLGALACTOSAMINIDINE α-2,6-SIALYLTRANSFERASE
2.4.99.3	α-N-ACTEYLGALACTOSAMINIDINE α-2,6-SIALYLTRANSFERASE
2.4.99.4	β-GALACTOSIDE α-2,3-SIALYLTRANSFERASE
2.4.99.6	α-2,3-SIALYLTRANSFERASE
2.4.99.8	α-N-ACETYLNEURAMINATE α-2,8-SIALYLTRANSFERASE

EC Number	Name
2.4.99.9	LACTOSYLCERAMIDE α-2,3-SIALYLTRANSFERASE
2.4.99.10	NEOLACTOTETRAOSYLCERAMIDE α-2,3-SIALYLTRANSFERASE
2.4.99.11	LACTOSYLCERAMIDE α-2,6-SIALYLTRANSFERASE
2.5.1.1	DIMETHYLALLYLtransTRANSFERASE
2.5.1.2	THIAMIN PYRIDINYLASE
2.5.1.2	THIAMIN PYRIDINYLASE I
2.5.1.2	THIAMIN PYRIDINYLASE II
2.5.1.3	THIAMIN-PHOSPHATE PYROPHOSPHORYLASE
2.5.1.4	ADENOSYLMETHIONINE CYCLOTRANSFERASE
2.5.1.5	GALACTOSE-6-SULFURYLASE
2.5.1.6	METHIONINE ADENOSYLTRANSFERASE
2.5.1.6	METHIONINE ADENOSYLTRANSFERASE I
2.5.1.6	METHIONINE ADENOSYLTRANSFERASE II
2.5.1.7	UDP-N-ACETYLGLUCOSAMINE 1-CARBOXYVINYLTRANSFERASE
2.5.1.8	tRNA ISOPENTENYLTRANSFERASE
2.5.1.9	RIBOFLAVIN SYNTHASE
2.5.1.10	GERANYLTRANSFERASE
2.5.1.11	trans-OCTAPRENYLTRANSFERASE
2.5.1.15	DIHYDROPTEROATE SYNTHASE
2.5.1.16	SPERMIDINE SYNTHASE
2.5.1.17	COB(I)ALAMIN ADENOSYLTRANSFERASE
2.5.1.18	GLUTATHIONE TRANSFERASE
2.5.1.19	3-PHOSPHOSHIKIMATE 1-CARBOXYVINYLTRANSFERASE
2.5.1.20	RUBBER cis-POLYPRENYLCISTRANSFERASE
2.5.1.21	FARNESYL-DIPHOSPHATE FARNESYLTRANSFERASE
2.5.1.22	SPERMINE SYNTHASE
2.5.1.24	DISCADENINE SYNTHASE
2.5.1.26	ALKYLGLYCERONE-PHOSPHATE SYNTHASE
2.5.1.29	FARNESYLtransTRANSFERASE
2.5.1.31	DI-trans,poly-cis-DECAPRENYLCISTRANSFERASE
2.5.1.32	GERANYLGERANYL-DIPHOSPHATE GERANYLGERANYLTRANSFERASE
2.5.1.35	ASPULVINONE DIMETHYLALLYLTRANSFERASE
2.5.1.37	LEUKOTRIENE-C4 SYNTHASE
2.5.1.39	4-HYDROXYBENZOATE NONAPRENYLTRANSFERASE
2.5.1.40	ARISTOLOCHENE SYNTHASE
2.6.1.1	ASPARTATE TRANSAMINASE
2.6.1.1	THIOSULFATE SULFURTRANSFERASE
2.6.1.2	ALANINE TRANSAMINASE
2.6.1.3	CYSTEINE TRANSAMINASE
2.6.1.4	GLYCINE TRANSAMINASE
2.6.1.5	TYROSINE TRANSAMINASE
2.6.1.6	LEUCINE TRANSAMINASE
2.6.1.7	KYNURENINE-OXOGLUTARATE AMINOTRANSFERASE
2.6.1.9	HISTIDINOL-PHOSPHATE TRANSAMINASE
2.6.1.11	ACETYLORNITHINE TRANSAMINASE
2.6.1.12	ALANINE-OXO-ACID TRANSAMINASE
2.6.1.13	ORNITHINE-OXO-ACID TRANSAMINASE
2.6.1.14	ASPARAGINE-OXO-ACID TRANSAMINASE
2.6.1.15	GLUTAMINE-PYRUVATE TRANSAMINASE
2.6.1.16	GLUCOSAMINE-FRUCTOSE-6-PHOSPHATE AMINOTRANSFERASE (ISOMERIZING)
2.6.1.16	GLUTAMINE-FRUCTOSE-6-PHOSPHATE AMINOTRANSFERASE (ISOMERIZING)
2.6.1.17	SUCCINYLAMINOPIMELATE TRANSAMINASE
2.6.1.17	SUCCINYLDIAMINOPIMELATE TRANSAMINASE
2.6.1.18	β-ALANINE-PYRUVATE TRANSAMINASE
2.6.1.19	4-AMINOBUTYRATE TRANSAMINASE
2.6.1.21	D-ALANINE TRANSAMINASE

EC Number	Name
2.6.1.23	4-HYDROXYGLUTAMATE TRANSAMINASE
2.6.1.24	DIIODOTYROSINE AMINOTRANSFERASE
2.6.1.27	TRYPTOPHAN TRANSAMINASE
2.6.1.29	DIAMINE TRANSAMINASE
2.6.1.30	PYRIDOXAMINE-PYRUVATE TRANSAMINASE
2.6.1.31	PYRIDOXAMINE-OXALOACETATE TRANSAMINASE
2.6.1.35	GLYCINE-OXALOACETATE TRANSAMINASE
2.6.1.36	L-LYSINE 6-TRANSAMINASE
2.6.1.37	(2-AMINOETHYL)PHOSPHONATE-PYRUVATE TRANSAMINASE
2.6.1.39	2-AMINOADIPATE AMINOTRANSFERASE
2.6.1.40	(R)-3-AMINO-2-METHYLPROPIONATE-PYRUVATE TRANSAMINASE
2.6.1.42	BRANCHED-CHAIN-AMINO-ACID TRANSAMINASE
2.6.1.43	AMINOLEVULINATE TRANSAMINASE
2.6.1.44	ALANINE-GLYOXALATE TRANSAMINASE
2.6.1.45	SERINE-GLYOXYLATE TRANSAMINASE
2.6.1.46	DIAMINOBUTYRATE-PYRUVATE TRANSAMINASE
2.6.1.47	ALANINE-OXOMALONATE TRANSAMINASE
2.6.1.48	5-AMINOVALERATE TRANSAMINASE
2.6.1.49	DIHYDROXYPHENYLALANINE TRANSAMINASE
2.6.1.50	SERINE-PYRUVATE TRANSAMINASE
2.6.1.52	PHOSPHOSERINE TRANSAMINASE
2.6.1.54	PYRIDOXAMINE-PHOSPHATE TRANSAMINASE
2.6.1.55	TAURINE AMINOTRANSFERASE
2.6.1.57	AROMATIC-AMINO-ACID TRANSAMINASE
2.6.1.58	PHENYLALANINE (HISTIDINE) TRANSAMINASE
2.6.1.59	dTDP-4-AMINO-4,6-DIDEOXYGALACTOSE TRANSAMINASE
2.6.1.60	AROMATIC-AMINO-ACID-GLYOXALATE AMINOTRANSFERASE
2.6.1.62	ADENOSYLMETHIONINE-8-AMINO-7-OXONONANOATE TRANSAMINASE
2.6.1.63	KYNURENINE-GLYOXALATE TRANSAMINASE
2.6.1.64	GLUTAMINE-PHENYLPYRUVATE TRANSAMINASE
2.6.1.66	VALINE-PYRUVATE TRANSAMINASE
2.6.1.67	2-AMINOHEXANOATE TRANSAMINASE
2.6.1.71	LYSINE-PYRUVATE TRANSAMINASE
2.6.1.72	D-4-HYDROXYPHENYLGLYCINE TRANSAMINASE
2.6.1.75	CYSTEINE-CONJUGATE TRANSAMINASE
2.6.99.1	dATP(dGTP)-DNA PURINETRANSFERASE
2.7.1.1	HEXOKINASE
2.7.1.2	GLUCOKINASE
2.7.1.3	KETOHEXOKINASE
2.7.1.4	FRUCTOKINASE
2.7.1.6	GALACTOKINASE
2.7.1.7	MANNOKINASE
2.7.1.11	6-PHOSPHOFRUCTOKINASE
2.7.1.12	GLUCONOKINASE
2.7.1.14	SEDOHEPTULOSEKINASE
2.7.1.18	PHOSPHORIBOKINASE
2.7.1.19	PHOSPHORIBULOKINASE
2.7.1.20	ADENOSINE KINASE
2.7.1.21	THYMIDINE KINASE
2.7.1.23	NAD KINASE
2.7.1.25	ADENYLYLSULFATE KINASE
2.7.1.26	RIBOFLAVIN KINASE
2.7.1.30	GLYCEROL KINASE
2.7.1.31	GLYCERATE KINASE
2.7.1.32	CHOLINE KINASE
2.7.1.33	PANTOTHENATE KINASE

EC Number	Name
2.7.1.34	PANTETHEINE KINASE
2.7.1.35	PYRIDOXAL KINASE
2.7.1.36	MEVALONATE KINASE
2.7.1.37	PROTEIN KINASE Cβ1
2.7.1.37	PROTEIN KINASE Cβ2
2.7.1.38	PHOSPHORYLASE KINASE
2.7.1.39	HOMOSERINE KINASE
2.7.1.40	PYRUVATE KINASE
2.7.1.40	PYRUVATE KINASE (L TYP)
2.7.1.42	RIBOFLAVIN PHOSPHOTRANSFERASE
2.7.1.43	GLUCURONOKINASE
2.7.1.48	URIDINE KINASE
2.7.1.50	HYDROXYETHYLTHIAZOLE KINASE
2.7.1.52	FUCOKINASE
2.7.1.56	1-PHOSPHOFRUCTOKINASE
2.7.1.59	N-ACETYLGLUCOSAMINE KINASE
2.7.1.60	N-ACETYLMANNOSAMINE KINASE
2.7.1.61	ACYL-PHOSPHATE PHOSPHOTRANSFERASE
2.7.1.63	POLYPHOSPHATE-GLUCOSE PHOSPHOTRANSFERASE
2.7.1.65	SCYLLO-INOSAMINE 4-KINASE
2.7.1.67	1-PHOSPHATIDYLINOSITOL 4-KINASE
2.7.1.68	1-PHOSPHATIDYLINOSITOL-4-PHOSPHATE 5-KINASE
2.7.1.68	1-PHOSPHATIDYLINOSITOL-4-PHOSPHATE 5-KINASE II
2.7.1.70	PROTAMIN KINASE
2.7.1.71	SHIKIMATE KINASE
2.7.1.72	STREPTOMYCIN KINASE
2.7.1.74	DEOXYCYTIDINE KINASE
2.7.1.76	DEOXYADENOSINE KINASE
2.7.1.77	NUCLEOSIDE PHOSPHOTRANSFERASE
2.7.1.80	PYROPHOSPHATE-SERINE PHOSPHOTRANSFERASE
2.7.1.82	ETHANOLAMINE KINASE
2.7.1.86	NADH KINASE
2.7.1.90	PYROPHOSPHATE-FRUCTOSE-6-PHOSPHATE 1-PHOSPHOTRANSFERASE
2.7.1.91	SPHINGANINE KINASE
2.7.1.94	ACYLGLYCEROL KINASE
2.7.1.95	KANAMYCIN KINASE
2.7.1.99	[PYRUVATE DEHYDROGENASE (LIPOAMIDE)] KINASE
2.7.1.100	5-METHYLTHIORIBOSE KINASE
2.7.1.102	HAMAMELOSE KINASE
2.7.1.104	PYROPHOSPHATE-PROTEIN PHOSPHOTRANSFERASE
2.7.1.105	6-PHOSPHOFRUCTO-2-KINASE
2.7.1.106	GLUCOSE-1,6-BISPHOSPHATE SYNTHASE
2.7.1.107	DIACYLGLYCEROL KINASE
2.7.1.108	DOLICHOL KINASE
2.7.1.109	[HYDROXYMETHYLGLUTARYL-CoA REDUCTASE (NADPH)] KINASE
2.7.1.110	DEPHOSPHO-[REDUCTASE KINASE] KINASE
2.7.1.112	PROTEIN TYROSINE KINASE
2.7.1.113	DEOXYGUANOSINE KINASE
2.7.1.114	AMP-THYMIDINE KINASE
2.7.1.115	[3-METHYL-2-OXOBUTANOATE DEHYDROGENASE (LIPOAMIDE)] KINASE
2.7.1.116	[ISOCITRATE DEHYDROGENASE (NADP)] KINASE
2.7.1.117	MYOSIN-LIGHT-CHAIN KINASE
2.7.1.118	ADP-THYMIDINE KINASE
2.7.1.120	CALDESMON KINASE
2.7.1.123	CAM KINASE
2.7.1.125	RHODOPSIN KINASE

EC Number	Name
2.7.1.126	β-ADRENERGIC-RECEPTOR KINASE
2.7.1.126	β2-ADRENERGIC-RECEPTOR KINASE
2.7.1.127	INOSITOL-1,4,5-TRISPHOSPHATE 3-KINASE
2.7.1.127	1-D-myo-INOSITOL-TRISPHOSPHATE 3-KINASE
2.7.1.128	[ACETYL-CoA CARBOXYLASE] KINASE
2.7.1.129	MYOSIN II HEAVY CHAIN KINASE A
2.7.1.131	LOW DENSITY LIPOPROTEIN KINASE
2.7.1.133	1-D-myo-INOSITOL-TRISPHOSPHATE 6-KINASE
2.7.1.134	1D-myo-INOSITOL-TETRAKISPHOSPHATE 1-KINASE
2.7.1.135	TAU PROTEIN KINASE
2.7.1.136	MACROLIDE KINASE
2.7.1.137	1-PHOSPHATIDYLINOSITOL 3-KINASE
2.7.1.138	CERAMIDE KINASE
2.7.1.139	1-D-myo-INOSITOL-TRISPHOSPHATE 5-KINASE
2.7.1.141	[RNA-POLYMERASE]SUBUNIT KINASE
2.7.2.1	ACETATE KINASE
2.7.2.2	CARBAMATE KINASE
2.7.2.3	PHOSPHOGLYCERATE KINASE
2.7.2.4	ASPARTATE KINASE
2.7.2.7	BUTYRATE KINASE
2.7.2.8	ACETYLGLUTAMATE KINASE
2.7.2.11	GLUTAMATE KINASE
2.7.3.1	GUANIDINOACETATE KINASE
2.7.3.2	CREATINE KINASE
2.7.3.2	CREATINE KINASE BB
2.7.3.2	CREATINE KINASE MM
2.7.3.2	PHOSPHOGLYCERATE KINASE
2.7.3.3	ARGINIIE KINASE
2.7.3.4	TAUROCYANINE KINASE
2.7.3.5	LOMBRICINE KINASE
2.7.3.6	HYPOXANTHINE-GUANINE PHOSPHORIBOSYLTRANSFERASE
2.7.3.7	OPHELINE KINASE
2.7.3.8	AMMONIA KINASE
2.7.3.9	PHOSPHOENOLPYRUVATE-PROTEIN KINASE
2.7.3.11	PROTEIN-HISTIDINE PROS-KINASE
2.7.4.1	POLYPHOSPHATE KINASE
2.7.4.2	PHOSPHOMEVALONATE KINASE
2.7.4.3	ADENYLATE KINASE
2.7.4.6	NUCLEOSIDE-DIPHOSPHATE KINASE
2.7.4.7	PHOSPHOMETHYLPYRIMIDINE KINASE
2.7.4.8	GUANYLATE KINASE
2.7.4.9	dTMP KINASE
2.7.4.14	CYTIDYLATE KINASE
2.7.4.15	THIAMIN-DIPHOSPHATE KINASE
2.7.4.16	THIAMIN-PHOSPHATE KINASE
2.7.4.19	5-METHYLDEOXYCYTIDINE-5′-PHOSPHATE KINASE
2.7.4.20	DOLICHYL-DIPHOSPHATE-POLYPHOSPHATE PHOSPHOTRANSFERASE
2.7.6.1	RIBOSE-PHOSPHATE PYROPHOSPHOKINASE
2.7.6.2	THIAMIN PYROPHOSPHOKINASE
2.7.6.3	2-AMINO-4-HYDROXY-6-HYDROXYMETHYLDIHYDROPTERIDINE PYROPHOSPHOKINASE
2.7.6.4	NUCLEOTIDE PYROPHOSPHOKINASE
2.7.6.5	GTP PYROPHOSPHOKINASE
2.7.6.5	GTP PYROPHOSPHOKINASE 1
2.7.7.1	NICOTINAMIDE-NUCLEOTIDE ADENYLYLTRANSFERASE
2.7.7.2	FMN ADENYLYLTRANSFERASE
2.7.7.3	PANTETHEINE-PHOSPHATE ADENYLYLTRANSFERASE

EC Number	Name
2.7.7.4	SULFATE ADENYLYLTRANSFERASE
2.7.7.5	SULFATE ADENYLYLTRANSFERASE (ADP)
2.7.7.6	RNA POLYMERASE
2.7.7.7	DNA POLYMERASE
2.7.7.8	POLYRIBONUCLEOTIDE NUCLEOTIDYLTRANSFERASE
2.7.7.9	UDP-GLUCOSE-1-PHOSPHATE URIDYLYLTRANSFERASE
2.7.7.10	UTP-HEXOSE-1-PHOSPHATE URIDYLYLTRANSFERASE
2.7.7.12	UDP-GLUCOSE-HEXOSE-1-PHOSPHATE URIDYLYLTRANSFERASE
2.7.7.13	MANNOSE-1-PHOSPHATE GUANYLYLTRANSFERASE
2.7.7.14	ETHANOLAMINE-PHOSPHATE CYTIDYLYLTRANSFERASE
2.7.7.15	CHOLINE-PHOSPHATE CYTIDYLYLTRANSFERASE
2.7.7.19	POLYNUCLEOTIDE ADENYLYLTRANSFERASE
2.7.7.21	tRNA CYTIDYLTRANSFERASE
2.7.7.23	UDP-N-ACETYLGLUCOSAMINE PYROPHOSPHORYLASE
2.7.7.24	GLUCOSE-1-PHOSPHATE THYMIDYLYLTRANSFERASE
2.7.7.25	tRNA ADENYLYLTRANSFERASE
2.7.7.27	GLUCOSE-1-PHOSPHATE ADENYLYLTRANSFERASE
2.7.7.28	NUCLEOSIDE-TRIPHOSPHATE-HEXOSE 1-PHOSPHATE NUCLEOTIDYLTRANSFERASE
2.7.7.29	HEXOSE-1-PHOSPHATE GUANYLYLTRANSFERASE
2.7.7.31	DNA NUCLEOTIDYLEXOTRANSFERASE
2.7.7.33	GLUCOSE-1-PHOSPHATE CYTIDYLYLTRANSFERASE
2.7.7.35	RIBOSE-5-PHOSPHATE ADENYLYLTRANSFERASE
2.7.7.37	ALDOSE-1-PHOSPHATE NUCLEOTIDYLTRANSFERASE
2.7.7.38	3-DEOXY-MANNO-OCTULOSONATE CYTIDYLYLTRANSFERASE
2.7.7.39	GLYCEROL-3-PHOSPHATE CYTIDILYLTRANSFERASE
2.7.7.41	PHOSPHATIDATE CYTIDILYLTRANSFERASE
2.7.7.42	[GLUTAMATE-AMMONIA-LIGASE] ADENYLYLTRANSFERASE
2.7.7.43	N-ACETYLNEURAMINATE CYTIDYLYLTRANSFERASE
2.7.7.48	RNA-DIRECTED RNA POLYMERASE
2.7.7.49	RNA-DIRECTED DNA POLYMERASE
2.7.7.49	RNA-DIRECTED DNA POLYMERASE 1
2.7.7.50	mRNA GUANYLYLTRANSFERASE
2.7.7.52	RNA URIDYLYLTRANSFERASE
2.7.7.53	ATP ADENYLYLTRANSFERASE
2.7.7.56	RNA NUCLEOTIDYLTRANSFERASE
2.7.8.1	ETHANOLAMINEPHOSPHOTRANSFERASE
2.7.8.2	DIACYLGLYCEROL CHOLINEPHOSPHOTRANSFERASE
2.7.8.3	CERAMIDE CHOLINEPHOSPHOTRANSFERASE
2.7.8.4	SERINE-PHOSPHOETHANOLAMINE SYNTHASE
2.7.8.5	CDP-DIACYLGLYCEROL-GLYCEROL-3-PHOSPHATE 3-PHOSPHATIDYLTRANSFERASE
2.7.8.6	UNDECAPRENYLPHOSPATE GALACTOSEPHOSPHOTRANSFERASE
2.7.8.8	CDP-DIACYLGLYCEROL-SERINE O-PHOSPHATIDYLTRANSFERASE
2.7.8.9	PHOSPHOMANNAN MANNOSEPHOSPHOTRANSFERASE
2.7.8.11	CDP-DIACYLGLYCEROL-INOSITOL 3-PHOSPHATIDYLTRANSFERASE
2.7.8.12	CDP-GLYCEROL GLYCEROPHOSPHOTRANSFERASE
2.7.8.13	PHOSPHO-N-ACETYLMURAMOYL-PENTAPEPTIDE-TRANSFERASE
2.7.8.14	CDPRIBITOL RIBITOLTRANSFERASE
2.7.8.15	UDP-N-ACETYLGLUCOSAMINE-DOLICHYL-PHOSPHATE N-ACETYLGLUCOSAMINEPHOSPHOTRANSFERA
2.7.8.16	UDP-N-ACETYLGLUCOSAMINE-LYSOSOMAL-ENZYME N-ACETYLGLUCOSAMINEPHOSPHOTRANSFERASE
2.7.8.19	UDP-GLUCOSE-GLYCOPROTEIN GLUCOSEPHOSPHOTRANSFERASE
2.7.8.20	PHOSPHATIDYLGLYCEROL-MEMBRANE-OLIGOSACCHARIDE GLYCEROPHOSPHOTRANSFERASE
2.7.8.21	MEMBRANE-OLIGOSACCHARIDE GLYCEROPHOSPHOTRANSFERASE
2.7.9.1	PYRUVATE, ORTHOPHOSPHATE DIKINASE
2.7.9.2	PYRUVATE, WATER DIKINASE
2.8.1.1	THIOSULFATE SULFURTRANSFERASE
2.8.1.2	3-MERCAPTOPYRUVATE SULFURTRANSFERASE

EC Number	Name
2.8.1.3	THIOSULFATE-THIOL SULFURTRANSFERASE
2.8.2.1	ARYL SULFOTRANSFERASE
2.8.2.1	ARYL SULFOTRANSFERASE P
2.8.2.2	ALCOHOL SULFOTRANSFERASE
2.8.2.3	AMINE SULFOTRANSFERASE
2.8.2.4	ESTRONE SULFOTRANSFERASE
2.8.2.5	CHODROTIN 4-SULFOTRANSFERASE
2.8.2.6	CHOLINE SULFOTRANSFERASE
2.8.2.8	DESULFOHEPARIN SULFOTRANSFERASE
2.8.2.9	TYROSINE-ESTER SULFOTRANSFERASE
2.8.2.9	TYROSINE-ESTER SULFOTRANSFERASE IV
2.8.2.11	GALACTOSYLCERAMIDE SULFOTRANSFERASE
2.8.2.12	HEPARITIN SULFOTRANSFERASE
2.8.2.14	BILE-SALT SULFOTRANSFERASE
2.8.2.16	THIOL SULFOTRANSFERASE
2.8.2.17	CHODROTIN 6-SULFOTRANSFERASE
2.8.2.18	CORTISOL SULFOTRANSFERASE
2.8.2.19	TRIGLUCOSYLALKYLGLYCEROL SULFOTRANSFERASE
2.8.2.20	PROTEIN TYROSINE SULFOTRANSFERASE
2.8.2.21	KERATAN SULFOTRANSFERASE
2.8.2.22	ARYLSULFATE SULFOTRANSFERASE
2.8.2.24	DESULFOGLUCINOLATE SULFOTRANSFERASE
2.8.2.25	FLAVONOL 3-SULFOTRANSFERASE
2.8.2.27	QUERCETIN-3-SULFATE 4′-SULFOTRANSFERASE
2.8.3.5	3-OXOACID CoA-TRANSFERASE
2.8.3.9	BUTYRATE-ACETOACETATE CoA-TRANSFERASE
2.8.3.13	SUCCINATE-HYDROXYMETHYLGLUTARATE CoA-TRANSFERASE
3.1.1.1	CARBOXYLESTERASE
3.1.1.1	CARBOXYLESTERASE (LONG CHAIN)
3.1.1.2	ARYLESTERASE
3.1.1.3	PREGASTRIC ESTERASE
3.1.1.3	TRIACYLGLYCEROL LIPASE
3.1.1.4	PHOSPHOLIPASE A2
3.1.1.5	LYSOPHOSPHOLIPASE
3.1.1.6	ACETYLESTERASE
3.1.1.7	ACETYLCHOLINESTERASE
3.1.1.8	CHOLINESTERASE
3.1.1.9	TROPINESTERASE
3.1.1.11	PECTINESTERASE
3.1.1.13	STEROL ESTERASE
3.1.1.13	STEROL ESTERASE (ACID)
3.1.1.14	CHLOROPHYLLASE
3.1.1.17	GLUCONOLACTONASE
3.1.1.19	URONOLACTONASE
3.1.1.20	TANNASE
3.1.1.21	RETINYL-PALMITATE ESTERASE
3.1.1.22	HYDROXYBUTYRATE-DIMER HYDROLASE
3.1.1.23	ACYLGLYCEROL LIPASE
3.1.1.24	LACTONASE
3.1.1.25	1,4-LACTONASE
3.1.1.26	GALACTOLIPASE
3.1.1.27	4-PYRIDOXOLACTONASE
3.1.1.28	ACYLCARNITINE HYDROLASE
3.1.1.29	AMINOACYL-tRNA HYDROLASE
3.1.1.31	PHOSPHOGLUCONOLACTONASE
3.1.1.32	PHOSPHOLIPASE A1

EC Number	Name
3.1.1.34	DIACYLGLYCEROL LIPASE
3.1.1.34	LIPOPROTEIN LIPASE
3.1.1.35	DIHYDROCOUMARIN HYDROLASE
3.1.1.38	TRIACETATE-LACTONASE
3.1.1.39	ACTINOMYCIN LACTONASE
3.1.1.41	CEPHALOSPORIN-C DEACETYLASE
3.1.1.42	CHLOROGENATE HYDROLASE
3.1.1.47	1-ALKYL-2-ACETYLGLYCEROPHOSPHOCHOLINE ESTERASE
3.1.1.48	FUSARININE-C ORNITHINESTERASE
3.1.1.49	SINAPINE ETSREASE
3.1.1.50	WAX-ESTER HYDROLASE
3.1.1.51	PHORBOL-DIESTER HYDROLASE
3.1.1.52	PHOSPHATIDYLINOSITOL DEACETYLASE
3.1.1.53	SIALATE -9-O-ACETYLESTERASE
3.1.1.53	SIALATE O-ACETYLESTERASE
3.1.1.59	JUVENILE HORMONE ESTERASE
3.1.2.1	ACETYL-CoA HYDROLASE
3.1.2.2	PALMITOYL-CoA HYDROLASE
3.1.2.3	SUCCINYL-CoA HYDROLASE
3.1.2.6	HYDROXYACYLGLUTATHIONE HYDROLASE
3.1.2.7	GLUTATHIONE THIOLESTERASE
3.1.2.12	S-FORMYLGLUTATHIONE HYDROLASE
3.1.2.13	S-SUCCINYLGLUTATHIONE HYDROLASE
3.1.2.14	OLEYL-[ACYL-CARRIER-PROTEIN] HYDROLASE
3.1.2.20	ACYL-CoA HYDROLASE
3.1.3.1	ALKALINE PHOSPHATASE
3.1.3.1	PHOSPHATASE ALKALINE
3.1.3.2	PHOSPHATASE ACID
3.1.3.2	PHOSPHATASE ACID PURPLE
3.1.3.3	PHOSPHOSERINE PHOSPHATASE
3.1.3.4	PHOSPHATIDATE PHOSPHATASE
3.1.3.5	5′-NUCLEOTIDASE
3.1.3.5	5′-NUCLEOTIDASE (IMP SENSITIVE)
3.1.3.6	3′-NUCLEOTIDASE
3.1.3.7	3′(2′),5′-BISPHOSPHATE NUCLEOTIDASE
3.1.3.7	PHOSPHOADENYLATE 3′-NUCLEOTIDASE
3.1.3.8	3-PHYTASE
3.1.3.9	GLUCOSE-6-PHOSPHATASE
3.1.3.10	GLUCOSE-1-PHOSPHATASE
3.1.3.11	FRUCTOSE-BISPHOSPHATASE
3.1.3.11	FRUCTOSE-BISPHOSPHATASE FERREDOXIN ACTIVATED
3.1.3.12	TREHALOSE-PHOSPHATASE
3.1.3.13	BISPHOSPHOGLYCERATE PHOSPHATASE
3.1.3.15	HISTIDINOL PHOSPHATASE
3.1.3.16	PHOSPHOPROTEIN PHOSPHATASE
3.1.3.17	PHOSPHORYLASE PHOSPHATASE
3.1.3.18	PHOSPHOGLYCOLATE PHOSPHATASE
3.1.3.19	GLYCEROL-2-PHOSPHATASE
3.1.3.20	PHOSPHOGLYCERATE PHOSPHATASE
3.1.3.21	GLYCEROL-1-PHOSPHATASE
3.1.3.22	MANNITOL-1-PHOSPHATASE
3.1.3.23	SUGAR PHOSPHATASE
3.1.3.24	SUCROSE-PHOSPHATASE
3.1.3.25	myo-INOSITOL-1(or 4)-MONOPHOSPHATASE
3.1.3.27	PHOSPHATIDYLGLYCEROPHOSPHATASE
3.1.3.30	3′-PHOSPHOADENYLSULFATE 3-PHOSPHATASE

EC Number	Name
3.1.3.31	NUCLEOTIDASE
3.1.3.32	POLYNUCLEOTIDE 3′-PHOSPHATASE
3.1.3.35	THYMIDYLATE 5′-PHOSPHATASE
3.1.3.36	PHOSPHATIDYLINOSITOL-BISPHOSPHATASE
3.1.3.37	SEDOHEPTULOSE-BISPHOSPHATASE
3.1.3.38	3-PHOSPHOGLYCERATE PHOSPHATASE
3.1.3.39	STREPTOMYCIN-6-PHOSPHATASE
3.1.3.40	GUANIDINODEOXY-SCYLLO-INOSITOL-4-PHOSPHATASE
3.1.3.41	4-NITROPHENYLPHOSPHATASE
3.1.3.42	GLYCOGEN-SYNTHASE-D PHOSPHATASE
3.1.3.43	[PYRUVATE DEHYDROGENASE (LIPOAMIDE)]-PHOSPHATASE
3.1.3.44	[ACETYL-CoA CARBOXYLASE]-PHOSPHATASE
3.1.3.45	3-DEOXY-MANNO-OCTULOSONATE-8-PHOSPHATASE
3.1.3.46	FRUCTOSE-2,6-BISPHOSPHATASE
3.1.3.47	[HYDROXYMETHYLGLUTARYL-CoA REDUCTASE (NADPH)]-PHOSPHATASE
3.1.3.48	PROTEIN TYROSINE-PHOSPHATASE
3.1.3.49	[PYRUVATE KINASE] PHOSPHATASE
3.1.3.51	DOLICHYL-PHOSPHATASE
3.1.3.52	[3-METHYL-2-OXOBUTANOATE DEHYDROGENASE (LIPOAMIDE)] PHOSPHATASE
3.1.3.53	MYOSIN-LIGHT-CHAIN PHOSPHATASE
3.1.3.56	INOSITOL-1,4,5-TRISPHOSPHATE 5-PHOSPHATASE
3.1.3.57	INOSITOL-1,4-BISPHOSPHATE 1-PHOSPHATASE
3.1.3.62	INOSITOL-1,3,4,5-TETRAKISPHOSPHATE 3-PHOSPHATASE
3.1.3.64	PHOSPHATIDYLINOSITOL 3-PHOSPHATASE
3.1.4.1	PHOSPHODIESTERASE
3.1.4.2	GLYCEROPHOSPHOCHOLINE PHOSPHODIESTERASE
3.1.4.3	PHOSPHOLIPASE C
3.1.4.4	PHOSPHOLIPASE D
3.1.4.10	1-PHOSPHATIDYLINOSITOL PHOSPHODIESTERASE
3.1.4.11	1-PHOSPHATIDYLINOSITOL-4,5-BISPHOSPHATE PHOSPHODIESTERASE
3.1.4.12	SPHINGOMYELIN PHOSPHODIESTERASE
3.1.4.12	SPHINGOMYELIN PHOSPHODIESTERASE (ACID)
3.1.4.12	SPHINGOMYELIN PHOSPHODIESTERASE (NEUTRAL)
3.1.4.13	SERINE-ETHANOLAMINEPHOSPHATE PHOSPHODIESTERASE
3.1.4.15	ADENYLYL-[GLUTAMATE-AMMONIA LIGASE] HYDROLASE
3.1.4.16	2′,3′-CYCLIC-NUCLEOTIDE 2′-PHOSPHODIESTERASE
3.1.4.17	3′,5′-CYCLIC-NUCLEOTIDE PHOSPHODIESTERASE
3.1.4.35	3′,5′-CYCLIC-GMP PHOSPHODIESTERASE
3.1.4.36	1,2-CYCLIC-INOSITOL-PHOSPHATE PHOSPHODIESTERASE
3.1.4.37	2′,3′-CYCLIC-NUCLEOTIDE 3′-PHOSPHODIESTERASE
3.1.4.38	GLYCEROPHOSPHOCHOLINE CHOLINEPHOSPHODIESTERASE (Zn^{2+})
3.1.4.39	ALKYLGLYCEROPHOSPHOETHANOLAMINE PHOSPHODIESTERASE
3.1.4.40	CMP-N-ACYLNEURAMINATE PHOSPHODIESTERASE
3.1.4.42	GLYCEROL-1,2-CYCLIC-PHOSPHATE PHOSPHODIESTERASE
3.1.4.43	GLYCEROPHOSPHOINOSITOL INOSITOLPHOSPHODIESTERASE
3.1.4.44	GLYCEROPHOSPHOINOSITOL GLYCEROPHOSPHODIESTERASE
3.1.4.45	N-ACETYLGLUCOSAMINE-1-PHOSPHODIESTER N-ACETYLGLUCOSAMINIDASE
3.1.4.48	DOILCHYLPHOSPHATE-GLUCOSE PHOSPHODIESTERASE
3.1.5.1	dGTPASE
3.1.6.1	ARYLSULFATASE
3.1.6.2	STERYL-SULFATASE
3.1.6.3	GLYCOSULFATASE
3.1.6.4	N-ACETYLGALACTOSAMINE-6-SULFATASE
3.1.6.6	CHOLINESUFATASE
3.1.6.8	CEREBROSIDE-SULFATASE
3.1.6.9	CHONDRO-4-SULFATASE

EC Number	Name
3.1.6.11	N-SULFOGLUCOSAMINE-6-SULFATASE
3.1.6.12	N-ACETYLGALACTOSAMINE-4-SULFATASE
3.1.6.13	IDURONATE 2-SULFATASE
3.1.6.14	N-ACETYLGLUCOSAMINE-6-SULFATASE
3.1.6.18	GLUCURONATE 2-SULFATASE
3.1.7.1	PRENYL-PYROPHOSPHATEASE
3.1.7.2	GUANOSINE-3′,5′-BIS(DIPHOSPHATE) 3′-PYROPHOSPHATASE
3.1.7.3	MONOTERPENYL-PYROPHOSPHATASE
3.1.8.1	ARYLDIALKYLPHOSPHATASE
3.1.11.1	EXODEOXYRIBONUCLEASE I
3.1.11.2	EXODEOXYRIBONUCLEASE III
3.1.11.3	EXODEOXYRIBONUCLEASE (LAMDA INDUCED)
3.1.11.5	EXODEOXYRIBONUCLEASE V
3.1.11.5	EXODEOXYRIBONUCLEASE Y
3.1.11.6	EXODEOXYRIBONUCLEASE VII
3.1.13.1	EXORIBONUCLEASE II
3.1.13.2	EXORIBONUCLEASE H
3.1.13.3	OLIGONUCLEOTIDASE
3.1.13.4	POLY(A)-SPECIFIC RIBONUCLEASE
3.1.14.1	YEAST RIBONUCLEASE
3.1.15.1	VENOM EXONUCLEASE
3.1.16.1	SPLEEN EXONUCLEASE
3.1.21.1	DEOXYRIBONUCLEASE I
3.1.21.2	DEOXYRIBONUCLEASE IV (PHAGE T4-INDUCED)
3.1.21.3	TYPE I SITE SPECIFIC DEOXYRIBONUCLEASE
3.1.21.4	ENDODEOXYRIBONUCLEASE Sac I
3.1.21.4	TYPE II SITE SPECIFIC DEOXYRIBONUCLEASE
3.1.21.36	ELASTASE PANCREATIC
3.1.21.37	EALASTASE LEUKOCYTE
3.1.22.1	DEOXYRIBONUCLEASE II
3.1.22.1	DNASE II
3.1.22.3	DEOXYRIBONUCLEASE V
3.1.23.6	ENDODEOXYRIBONUCLEASE BAM HI
3.1.23.6	ENDORIBONUCLEASE BAM HI
3.1.23.13	ENDODEOXYRIBONUCLEASE ECO RI
3.1.23.13	ENDORIBONUCLEASE ECO RI
3.1.23.31	ENDODEOXYRIBONUCLEASE Pst I
3.1.25.1	DEOXYRIBONUCLEASE (PYRIMIDINE DIMER)
3.1.25.2	DEOXYRIBONUCLEASE (APURINIC OR APYRIMIDIC)
3.1.26.2	RIBONUCLEASE α
3.1.26.3	RIBONUCLEASE III
3.1.26.4	CALF THYMUS RIBONUCLEASE H
3.1.26.5	RIBONUCLEASE P
3.1.26.6	RIBONUCLEASE IV
3.1.26.8	RIBONUCLEASE M5
3.1.26.9	RIBONUCLEASE (POLY-(U)-SPECIFIC)
3.1.27.1	RIBONUCLEASE T2
3.1.27.2	BACILLUS SUBTILIS RIBONUCLEASE
3.1.27.3	RIBONUCLEASE T1
3.1.27.4	RIBONUCLEASE U2
3.1.27.5	PANCREATIC RIBONUCLEASE
3.1.27.5	PANCREATIC RIBONUCLEASE A
3.1.27.5	RIBONUCLEASE PANCREATIC
3.1.27.6	ENTEROBACTER RIBONUCLEASE
3.1.27.8	RIBONUCLEASE V
3.1.30.1	ASPERGILLUS NUCLEASE S1

EC Number	Name
3.1.30.2	SERRATIA MARCESCENS NUCLEASE
3.1.31.1	MICROCOCCAL NUCLEASE
3.2.1.1	α-AMYLASE
3.2.1.2	β-AMYLASE
3.2.1.3	GLUCAN 1,4-α-GLUCOSIDASE
3.2.1.4	CELLULASE
3.2.1.6	ENDO-1,3(4)-β-GLUCANASE
3.2.1.6	ENDO 1,3(4)-β-GLUCANASE A
3.2.1.7	INULINASE
3.2.1.8	ENDO-1,4-β-XYLANASE
3.2.1.10	OLIGO-1,6-GLUCOSIDASE
3.2.1.11	DEXTRANASE
3.2.1.14	CHITINASE
3.2.1.15	POLYGALACTURONASE
3.2.1.17	LYSOZYME
3.2.1.18	EXO-α-SIALIDASE
3.2.1.18	SIALIDASE
3.2.1.20	α-GLUCOSIDASE
3.2.1.21	β-GLUCOSIDASE
3.2.1.21	β-GLUCOSIDASE (ACID)
3.2.1.22	α-GALACTOSIDASE
3.2.1.22	α-GALACTOSIDASE B
3.2.1.23	β-GALACTOSIDASE
3.2.1.24	β-GLUCOCEREBROSIDASE
3.2.1.24	α-MANNOSIDASE
3.2.1.25	β-MANNOSIDASE
3.2.1.25	exo-β-MANNOSIDASE
3.2.1.26	β-FRUCTOFURANOSIDASE
3.2.1.28	β-GLUDOSIDASE
3.2.1.28	α,α-TREHALASE
3.2.1.28	α,α-TREHALASE C
3.2.1.30	N-ACETYL-β-GLUCOSAMINIDASE
3.2.1.31	β-GLUCURONIDASE
3.2.1.32	XYLAN ENDO-1,3-XYLOSIDASE
3.2.1.33	AMYLO-1,6-GLUCOSIDASE
3.2.1.35	HYALURONOGLUCOSAMINIDASE
3.2.1.36	HYALURONOGLUCURONIDASE
3.2.1.37	XYLAN 1,4-β-XYLOSIDASE
3.2.1.38	β-D-FUCOSIDASE
3.2.1.39	GLUCAN ENDO-1,3-β-GLUCOSIDASE
3.2.1.40	α-L-RHAMNOSIDASE
3.2.1.41	α-DEXTRIN ENDO-1,6-α-GLUCOSIDASE
3.2.1.42	GDP-GLUCOSIDASE
3.2.1.44	FUCOIDANASE
3.2.1.45	GLUCOSYLCERAMIDASE
3.2.1.46	GALACTOSYLCERAMIDASE
3.2.1.47	GALACTOSYLGALACTOSYLGLUCOSYLCERAMIDASE
3.2.1.48	SUCROSE α-GLUCOSIDASE
3.2.1.49	α-N-ACETYLGALACTOSAMINIDASE
3.2.1.50	α-N-ACETYLGLUCOSAMINIDASE
3.2.1.51	α-L-FUCOSIDASE
3.2.1.52	β-N-ACETYLHEXOSAMINIDASE
3.2.1.53	β-N-ACETYLGALACTOSAMINIDASE
3.2.1.54	CYCLOMALTODEXTRINASE
3.2.1.55	α-L-ARABINOFURANOSIDASE
3.2.1.57	ISOPULLANASE

EC Number	Name
3.2.1.58	GLUCAN 1,3-β-GLUCOSIDASE
3.2.1.59	GLUCAN ENDO-1,3-α-GLUCOSIDASE
3.2.1.60	GLUCAN 1,4-α-MALTOTETRAOHYDROLASE
3.2.1.61	MYCODEXTRANASE
3.2.1.62	GLYCOSYLCERAMIDASE
3.2.1.63	1,2-α-L-FUCOSIDASE
3.2.1.65	LEVANASE
3.2.1.67	GALACTURAN 1,4-α-GALACTURONIDASE
3.2.1.68	ISOAMYLASE
3.2.1.70	GLUCAN 1,6-α-GLUCOSIDASE
3.2.1.73	LICHENINASE
3.2.1.74	GLUCAN 1,4-β-GLUCOSIDASE
3.2.1.76	L-IDURONIDASE
3.2.1.77	MANNAN 1,2-(1,3)-α-MANNOSIDASE
3.2.1.78	MANNAN ENDO-1,4-β-MANNOSIDASE
3.2.1.80	FRUCTAN β-FRUCTOSIDASE
3.2.1.81	AGARASE
3.2.1.83	κ-CARRAGEENASE
3.2.1.84	GLUCAN 1,3-α-GLUCOSIDASE
3.2.1.85	6-PHOSPHO-β-GALACTOSIDASE
3.2.1.86	6-PHOSPHO-β-GLUCOSIDASE A
3.2.1.86	6-PHOSPHO-β-GLUCOSIDASE B
3.2.1.87	CAPSULAR-POLYSACCHARIDE ENDO-1,3-α-GALACTOSIDASE
3.2.1.89	ARABINOGALACTAN ENDO-1,4-β-GALACTOSIDASE
3.2.1.91	CELLULOSE 1,4-β-CELLOBIOSIDASE
3.2.1.92	PEPTIDOGLYCAN β-N-ACETYLMURAMIDASE
3.2.1.94	GLUCAN 1,6-α-ISOMALTOSIDASE
3.2.1.95	DEXTRAN 1,6-α-ISOMALTOSIDASE
3.2.1.96	MANNOSYL-GLYCOPROTEIN ENDO-β-N-ACETYLGLUCOSAMINIDASE
3.2.1.97	GLYCOPEPTIDE α-N-ACETYLGALACTOSAMINIDASE
3.2.1.98	GLUCAN 1,4-α-MALTOHEXAOSIDASE
3.2.1.99	ARABINAN ENDO-1,5-α-L-ARABINOSIDASE
3.2.1.100	MANNAN 1,4-β-MANNOBIOSIDASE
3.2.1.101	MANNAN ENDO-1,6-β-MANNOSIDASE
3.2.1.102	BLOOD-GROUP-SUBSTANCE ENDO-1,4-β-GALACTOSIDASE
3.2.1.103	KERATAN-SULFATE ENDO-1,4-β-GALACTOSIDASE
3.2.1.104	STERYL-β-GLUCOSIDASE
3.2.1.105	STRICTOSIDINE β-GLUCOSIDASE
3.2.1.106	MANNOSYL-OLIGOSACCHARIDE GLUCOSIDASE
3.2.1.106	MANNOSYL-OLIGOSACCHERIDE GLUCOSIDASE
3.2.1.107	PROTEIN-GLUCOSYLGALACTOSYLHYDROXYLYSINE GLUCOSIDASE
3.2.1.108	LACTASE
3.2.1.109	ENDOGALACTOSAMINIDASE
3.2.1.110	MUCINAMINYLSERINE MUCINAMINIDASE
3.2.1.114	MANNOSYL-OLIGOSACCHARIDE 1,3-1,6-α-MANNOSIDASE
3.2.2.1	PURINE NUCLEOSIDASE
3.2.2.2	INOSINE NUCLEOSIDASE
3.2.2.3	URIDINE NUCLEOSIDASE
3.2.2.4	AMP NUCLEOSIDASE
3.2.2.5	NAD NUCLEOSIDASE
3.2.2.6	NAD(P) NUCLEOSIDASE
3.2.2.7	ADENOSINE NUCLEOSIDASE
3.2.2.8	RIBOSYLPYRIMIDINE NUCLEOSIDASE
3.2.2.9	ADENOSYLHOMOCYSTEINE NUCLEOSIDASE
3.2.2.10	PYRIMIDINE-5′-NUCLEOTIDE NUCLEOSIDASE
3.2.2.13	1-METHYLADENOSINE NUCLEOSIDASE

EC Number	Name
3.2.2.14	NMN NUCLEOSIDASE
3.2.2.15	DNA-DEOXYINOSINE GLYCOSIDASE
3.2.2.16	METHYLTHIOADENOSINE NUCLEOSIDASE
3.2.2.17	DEOXYRIBOPYRIMIDINE ENDONUCLEOSIDASE
3.2.2.19	ADP-RIBOSYLARGININE HYDROLASE
3.2.2.23	FORMAMIDOPYRIMIDINE-DNA GLYCOSIDASE
3.2.3.1	THIOGLUCOSIDASE
3.2.3.11	PECTIN ESTERASE
3.3.1.1	ADENOSYLHOMOCYSTEINASE
3.3.1.2	ADENOSYLMETHIONINE HYDROLASE
3.3.2.1	ISOCHORISMATASE
3.3.2.3	EPOXIDE HYDROLASE
3.3.2.4	trans-EPOXYSUCCINATE HYDROLASE
3.3.2.5	ALKENYLGLYCEROPHOSPHOETHANOLAMINE HYDROLASE
3.3.2.6	LEUKOTRIENE A4 HYDROLASE
3.4.3.12	DEOXYCYTIDYLATE DEAMINASE
3.4.11	AMINOENKEPHALINASE
3.4.11.1	LEUCYL AMINOPEPTIDASE
3.4.11.2	MEMBRANE ALANINE AMINOPEPTIDASE
3.4.11.3	CYSTINYL AMINOPEPTIDASE
3.4.11.4	TRIPEPTIDE AMINOPEPTIDASE
3.4.11.5	PROLYL AMINOPEPTIDASE
3.4.11.6	AMINOPEPTIDASE B
3.4.11.7	GLUTAMYL AMINOPEPTIDASE
3.4.11.9	XAA-PRO AMINOPEPTIDASE
3.4.11.10	BACTERIAL LEUCYL AMINOPEPTIDASE
3.4.11.11	AMINOPEPTIDASE
3.4.11.13	CLOSTRIDIUM HISTOLYTICUM AMINOPEPTIDASE
3.4.11.14	AMINOPEPTIDASE (HUMAN LIVER)
3.4.11.14	CYTOSOL ALANYL AMINOPEPTIDASE
3.4.11.15	LYSYL AMINOPEPTIDASE
3.4.11.16	XAA-TRP AMINOPEPTIDASE
3.4.11.17	TRYPTOPHAN AMINOPEPTIDASE
3.4.11.18	METHIONYL AMINOPEPTIDASE
3.4.11.18	METHIONYL AMINOPEPTIDASE 2
3.4.12.1	CARBOXYPEPTIDASE C
3.4.13.3	XAA-HIS DIPEPTIDASE
3.4.13.4	XAA-ARG DIPEPTIDASE
3.4.13.5	XAA-METHYL-HIS DIPEPTIDASE
3.4.13.6	CYSTEINYL-GLYCINE DIPEPTIDASE
3.4.13.8	PRO-XAA DIPEPTIDASE
3.4.13.9	PROLINE DIPEPTIDASE
3.4.13.9	XAA-PRO DIPEPTIDASE
3.4.13.10	β-ASPARTYLDIPEPTIDASE
3.4.13.11	DIPEPTIDASE
3.4.13.12	METHIONYL DIPEPTIDASE
3.4.13.13	HOMOCARNOSINASE
3.4.13.18	CYTOSOL NON-SPECIFIC DIPEPTIDASE
3.4.13.19	MEMBRANE DIPEPTIDASE
3.4.14.1	DIPEPTIDYL-PEPTIDASE 1
3.4.14.2	DIPEPTIDYL-PEPTIDASE 2
3.4.14.4	CARBOXYPEPTIDASE H
3.4.14.4	DIPEPTIDYL-DIPEPTIDASE 3
3.4.14.4	DIPEPTIDYL-PEPTIDASE 3
3.4.14.5	DIPEPTIDYL-PEPTIDASE 4
3.4.14.5	DIPEPTIDYL-PEPTIDASE 4β

EC Number	Name
3.4.14.5	DIPEPTIDYL-PEPTIDASE 4 CD26
3.4.14.6	TETRAPEPTIDE DIPEPTIDASE
3.4.14.7	TETRALYSINE ENDOPEPTIDASE
3.4.14.8	TRIPEPTIDYL PEPTIDASE
3.4.14.10	TRIPEPTIDYL PEPTIDASE II
3.4.15.1	PEPTIDYL-DIPEPTIDASE A
3.4.15.1	PEPTIDYL DIPEPTIDASE I
3.4.16.1	SERINE CARBOXYPEPTIDASE
3.4.16.1	SERINE-TYPE CARBOXYPEPTIDASE
3.4.16.2	PROLINE CARBOXYPEPTIDASE
3.4.16.4	D-ALANYL-D-ALANINE CARBOXYPEPTIDASE
3.4.17.1	CARBOXYPEPTIDASE A
3.4.17.2	CARBOXYPEPTIDASE B
3.4.17.3	LYSINE (ARGININE) CARBOXYPEPTIDASE
3.4.17.4	GLY-X CARBOXYPEPTIDASE
3.4.17.8	MURAMOYL-PENTAPEPTIDE CARBOXYPEPTIDASE
3.4.17.9	CARBOXYPEPTIDASE S
3.4.17.10	CARBOXYPEPTIDASE H
3.4.17.15	CARBOXYPEPTIDASE A2
3.4.18.1	LYSOSOMAL CARBOXYPEPTIDASE B
3.4.19.1	ACYLAMINOACYL-PEPTIDASE
3.4.19.3	PYROGLUTAMYL-PEPTIDASE I
3.4.19.9	γ-GLUTAMYL HYDROLASE
3.4.19.20	PEPTIDYL-GLYCINAMIDASE
3.4.21.1	CHYMOTRYPSIN
3.4.21.3	METRIDIUM PROTEINASE A
3.4.21.4	COCOONASE
3.4.21.4	TRYPSIN
3.4.21.5	THROMBIN
3.4.21.5	α-THROMBIN
3.4.21.6	COAGULATION FACTOR Xa
3.4.21.7	COLLAGENASE INTERSTITIAL
3.4.21.7	PLASMIN
3.4.21.7	PLASMIN I
3.4.21.7	PLASMIN II
3.4.21.9	ENTEROPEPTIDASE
3.4.21.10	ACROSIN
3.4.21.12	α-LYTIC ENDOPEPTIDASE
3.4.21.14	MICROBIAL SERINE PROTEINASE
3.4.21.15	CATHEPSIN L
3.4.21.18	TENEBRIO α-PROTEINASE
3.4.21.19	GLUTAMYL ENDOPEPTIDASE
3.4.21.20	CATHEPSIN G
3.4.21.21	COAGULATION FACTOR VIIa
3.4.21.21	COAGULATION FACTOR VIIa-TF
3.4.21.22	COAGULATION FACTOR IXa
3.4.21.23	VIPERA RUSSELLI PROTEINASE
3.4.21.24	RED CELL NEUTRAL ENDOPROTEINASE
3.4.21.25	CUCUMISIN
3.4.21.26	PROLYL OLIGOPEPTIDASE
3.4.21.27	COAGULATION FACTOR XIa
3.4.21.28	AGKISTRODON SERINE PROTEINASE
3.4.21.29	BOTHROPS ATROX SERINE PROTEINASE
3.4.21.30	CROTALUS ADAMANTEUS SERINE PROTEINASE
3.4.21.31	PLASMINOGEN ACTIVATOR
3.4.21.31	t-PLASMINOGEN ACTIVATOR

EC Number	Name
3.4.21.31	α-PLASMINOGEN ACTIVATOR (TISSUE TYPE)
3.4.21.32	UCA PUGILATOR COLLAGENOLYTIC PROTEINASE
3.4.21.33	ENTOMOPHTHORA COLLAGENOLYTIC PROTEINASE
3.4.21.34	KALLIKREIN PLASMA
3.4.21.35	KALLIKREIN TISSUE
3.4.21.36	ELASTASE PANCREATIC
3.4.21.37	ELASTASE LEUKOCYTE
3.4.21.37	ELASTASE NEUTROPHIL
3.4.21.38	β-CoAGULATION FACTOR XIIa
3.4.21.38	COAGULATION FACTOR XIIa
3.4.21.38	COAGULATION FACTOR XIIf
3.4.21.39	CHYMASE
3.4.21.39	CHYMASE 1
3.4.21.40	SUBMANDIBULAR PROTEINASE A
3.4.21.41	COMPLEMENT SUBCOMPONENT C1r
3.4.21.41	COMPLEMENT SUBCOMPONENT C1r (ESTEROLYTIC ACTIVITY)
3.4.21.42	COMPLEMENT SUBCOMPONENT C1s
3.4.21.43	CLASSICAL-COMPLEMENT C3/C5 CONVERTASE
3.4.21.45	COMPLEMENT FACTOR I
3.4.21.46	COMPLEMENT FACTOR D
3.4.21.47	ALTERNATIVE-COMPLEMENT-PATHWAY C3/C5 CONVERTASE
3.4.21.48	YEAST PROTEINASE B
3.4.21.49	HYPODERMIN C
3.4.21.50	LYSYL ENDOPEPTIDASE
3.4.21.50	PROTEINASE I ACHROMOBACTER
3.4.21.51	LEUKOCYTE-MEMBRANE NEUTRAL ENDOPEPTIDASE
3.4.21.52	CATHEPSIN R
3.4.21.53	ENDOPEPTIDASE La
3.4.21.56	EUPHORBAIN
3.4.21.59	TRYPTASE
3.4.21.62	SUBTILISIN
3.4.21.65	THERMOMYCOLIN
3.4.21.66	THERMITASE
3.4.21.68	t-PLASMINOGEN ACTIVATOR
3.4.21.69	PROTEIN Ca
3.4.21.73	u-PLASMINOGEN ACTIVATOR
3.4.21.74	VENOMBIN A
3.4.21.75	FURIN
3.4.21.76	MYELOBLASTIN
3.4.21.78	GRANZYME A
3.4.21.79	GRANZYME B
3.4.21.89	SIGNAL PEPTIDASE I
3.4.22.1	CATHEPSIN B
3.4.22.2	PAPAIN
3.4.22.3	FICAIN
3.4.22.4	BROMELAIN
3.4.22.6	CHYMOPAPAIN
3.4.22.7	ASCLEPAIN
3.4.22.8	CLOSTRIPAIN
3.4.22.10	STREPTOPAIN
3.4.22.12	γ-GLUTAMYL HYDROLASE
3.4.22.13	STAPHYLOCOCCAL CYSTEINE PROTEINASE
3.4.22.14	ACTINIDAIN
3.4.22.15	CATHEPSIN L
3.4.22.16	CATHEPSIN H
3.4.22.17	CALPAIN

EC Number	Name
3.4.22.21	YEAST CYSTEINE PROTEINASE E
3.4.22.22	YEAST CYSTEINE PROTEINASE D
3.4.22.23	YEAST CYSTEINE PROTEINASE F
3.4.22.25	GLYCYL ENDOPEPTIDASE
3.4.22.26	CANCER PROCOAGULANT
3.4.22.27	CATHEPSIN S
3.4.22.30	CARICAIN
3.4.22.32	STEM BROMELAIN
3.4.22.34	LEGUMAIN
3.4.22.35	HISTOLYSAIN
3.4.22.36	CASPASE-1
3.4.22.36	INTERLEUKIN CONVERTING ENZYME
3.4.22.36	INTERLEUKIN 1β CONVERTING ENZYME
3.4.22.38	CATHEPSIN K
3.4.22.38	CATHEPSIN N
3.4.23.1	PEPSIN
3.4.23.1	PEPSIN A
3.4.23.3	GASTRICSIN
3.4.23.4	CHYMOSIN
3.4.23.5	CATHEPSIN D
3.4.23.11	THYROID ASPARTIC PROTEINASE
3.4.23.12	NEPENTHES ASPARTIC PROTEINASE
3.4.23.13	LOTUS ASPARTIC PROTEINASE
3.4.23.14	SORGHUM ASPARTIC PROTEINASE
3.4.23.15	RENIN
3.4.23.15	RENIN γ
3.4.23.17	PRO-OPIOMELANOCORTIN CONVERTING ENZYME
3.4.23.20	PENICILLOPEPSIN
3.4.23.21	RHIZOPUSPEPSIN
3.4.23.22	ENDOTHIAPEPSIN
3.4.23.23	MUCOROPEPSIN
3.4.23.24	CANDIDAPEPSIN
3.4.23.25	SACCHAROPEPSIN
3.4.23.26	RHODOTORULAPEPSIN
3.4.23.27	PHYSAROPEPSIN
3.4.23.34	CATHEPSIN E
3.4.24.1	ATROLYSIN
3.4.24.2	SEPIA PROTEINASE
3.4.24.3	COLLAGENASE CLOSTRIDIUM HISTOLYTICUM
3.4.24.3	MICROBIAL COLLAGENASE
3.4.24.6	LEUCOLYSIN
3.4.24.7	INTERSTITIAL COLLAGENASE
3.4.24.8	ACHROMOBACTER IOPHAGUS COLLAGENASE
3.4.24.9	TRICHOPHYTON SCHOENLEINII COLLAGENASE
3.4.24.10	TRICHOPHYTON MENTAGROPHYTES KERATINASE
3.4.24.11	NEPRILYSIN
3.4.24.12	ENVELYSIN
3.4.24.13	IGA-SPECIFIC METALLOENDOPEPTIDASE
3.4.24.14	PROCOLLAGEN N-ENDOPEPTIDASE
3.4.24.15	THIMET OLIGOPEPTIDASE
3.4.24.16	NEUROLYSIN
3.4.24.17	STROMELYSIN 1
3.4.24.18	MEPRIN A
3.4.24.19	PROCOLLAGEN C-ENDOPEPTIDASE
3.4.24.20	PEPTIDYL-LYS METALLOENDOPEPTIDASE
3.4.24.21	ASTACIN

EC Number	Name
3.4.24.23	MATRILYSIN
3.4.24.24	GELATINASE A
3.4.24.24	GELATINASE A1
3.4.24.25	AEROMONOLYSIN
3.4.24.27	THERMOLYSIN
3.4.24.28	BACILLOLYSIN
3.4.24.29	AUREOLYSIN
3.4.24.30	COCCOLYSIN
3.4.24.34	COLLAGENASE NEUTROPHIL
3.4.24.35	GELATINASE B
3.4.24.40	SERRALYSIN
3.4.24.42	ATROLYSIN C
3.4.24.46	ADAMALYSIN II
3.4.24.55	PITRILYSIN
3.4.24.56	INSULYSIN
3.4.24.57	O-SIALOGLYCOPROTEIN ENDOPEPTIDASE
3.4.24.71	ENDOTHELIN CONVERTING ENZYME
3.4.24.71	ENDOTHELIN CONVERTING ENZYME 1
3.4.27.37	ELASTASE LEUKOCYTE
3.4.99.2	AGAVAIN
3.4.99.3	ANGIOTENSINASE
3.4.99.8	GLIOCLADIUM PROTEINASE
3.4.99.9	HURAIN
3.4.99.14	MEXICANAIN
3.4.99.16	PENICILLIUM NOTATUM EXTRACELLULAR PROTEINASE
3.4.99.18	PINGUINAIN
3.4.99.20	SCOPULARIOPSIS PROTEINASE
3.4.99.21	SOLANIN
3.4.99.29	MYXOBACTER AL-1 PROTEINASE
3.4.99.30	PROTEINASE II MYXOBACTER AL-1
3.4.99.31	TISSUE-ENDOPEPTIDASE DEGRADING COLLAGENASE-SYNTHETIC-SUBSTRATE
3.4.99.34	MYTILIDASE
3.4.99.35	PROLIPOPROTEIN-SIGNAL PEPTIDASE
3.4.99.37	REC A PEPTIDASE
3.4.99.40	PRO-GONADOLIBERIN PROTEINASE
3.4.99.46	MULTICATALYTIC ENDOPEPTIDASE COMPLEX
3.5.1.1	ASPARAGINASE
3.5.1.2	GLUTAMINASE
3.5.1.2	GLUTAMINASE (PHOSPHATE ACTIVATED)
3.5.1.3	ω-AMIDASE
3.5.1.4	AMIDASE
3.5.1.5	UREASE
3.5.1.6	β-UREIDOPROPIONASE
3.5.1.7	UREIDOSUCCINASE
3.5.1.8	FORMYLASPARTATE DEFORMYLASE II
3.5.1.9	ARYLFORMAMIDASE
3.5.1.11	PENICILLIN AMIDASE
3.5.1.12	BIOTINIDASE
3.5.1.13	ARYL-ACYLAMIDASE
3.5.1.14	AMINOACYLASE
3.5.1.15	ASPARTOACYLASE
3.5.1.16	ACETYLORNITHINE DEACETYLASE
3.5.1.17	ACYL-LYSINE DEACYLASE
3.5.1.18	SUCCINYLDIAMINOPIMELATE DESUCCINYLASE
3.5.1.19	NICOTINAMIDASE
3.5.1.21	N-ACETYL-β-ALANINE DEACYLASE

EC Number	Name
3.5.1.22	PANTOTHENASE
3.5.1.23	CERAMIDASE
3.5.1.24	CHOLYLGLYCINE HYDROLASE
3.5.1.25	N-ACETYLGLUCOSAMINE-6-PHOSPHATE DEACETYLASE
3.5.1.26	N^4(β-N-ACETYLGLUCOSAMINYL)-L-ASPARAGINASE
3.5.1.28	N-ACETYLMURAMOYL-L-ALANINE AMIDASE
3.5.1.29	2-(ACETAMIDOMETHYLENE)SUCCINATE HYDROLASE
3.5.1.30	5-AMINOPENTANAMIDASE
3.5.1.31	FORMYLMETHIONINE DEFORMYLASE
3.5.1.31	PEPTIDE DEFORMYLASE
3.5.1.32	HIPPURATE HYDROLASE
3.5.1.33	N-ACETYLGLUCOSAMINE DEACYLASE
3.5.1.36	N-METHYL-2-OXOGLUTARAMATE HYDROLASE
3.5.1.38	GLUTAMIN-(ASPARAGIN-)ASE
3.5.1.39	ALKYLAMIDASE
3.5.1.40	ACYLAGMATINE AMIDASE
3.5.1.41	CHITIN DEACYTYLASE
3.5.1.42	NICOTINAMIDE-NUCLEOTIDE AMIDASE
3.5.1.43	PEPTIDYL-GLUTAMINASE
3.5.1.44	PROTEIN-GLUTAMINE GLUTAMINASE
3.5.1.45	UREASE (ATP-HYDROLYSING)
3.5.1.46	6-AMINOHEXANOATE-DIMER HYDROLASE
3.5.1.48	ACETYLSPERMIDINE DEACETYLASE
3.5.1.49	FORMAMIDASE
3.5.1.51	4-ACETAMIDOBUTYRYL-CoA DEACETYLASE
3.5.1.52	PEPTIDE-N4-(N-ACETYL-β-GLUCOSAMINYL)ASPARAGINE AMIDASE
3.5.1.59	N-CARBAMOYLSARCOSINE AMIDASE
3.5.1.60	N-(LONG-CHAIN-ACYL)ETHANOLAMINE DEACYLASE
3.5.1.67	4-METHYLENEGLUTAMINASE
3.5.2.2	DIHYDROPYRIMIDINASE
3.5.2.3	DIHYDROOROTASE
3.5.2.4	CARBOXYMETHYLHYDANTOINASE
3.5.2.5	ALLANTOINASE
3.5.2.6	β-LACTAMASE
3.5.2.6	PENICILLINASE
3.5.2.7	IMIDAZOLONEPROPIONASE
3.5.2.9	5-OXOPROLINASE (ATP-HYDROLYSING)
3.5.2.10	CREATININASE
3.5.2.12	6-AMINOHEXANOATE-CYCLIC-DIMER HYDROLASE
3.5.3.1	ARGINASE
3.5.3.2	GLYCOCYAMINASE
3.5.3.3	CREATINASE
3.5.3.4	ALLANTOICASE
3.5.3.6	ARGININE DEIMINASE
3.5.3.7	GUANIDINOBUTYRASE
3.5.3.8	FORMIMINOGLUTAMASE
3.5.3.9	ALLANTOATE DEIMINASE
3.5.3.10	D-ARGINASE
3.5.3.11	AGMATINASE
3.5.3.12	AGMATINE DEIMINASE
3.5.3.13	FORMIMINOGLUTAMATE DEIMINASE
3.5.3.14	AMIDINOASPARTASE
3.5.3.15	PROTEIN-ARGININE DEIMINASE
3.5.3.16	METHYLGUANINASE
3.5.3.48	ACETYLSPERMIDINE DEACETYLASE
3.5.4.1	CYTOSINE DEAMINASE

EC Number	Name
3.5.4.2	ADENINE DEAMINASE
3.5.4.3	GUANINE DEAMINASE
3.5.4.3	GUANINE DEAMINASE A
3.5.4.4	ADENINE DEAMINASE
3.5.4.4	ADENOSINE DEAMINASE
3.5.4.5	CYTIDINE DEAMINASE
3.5.4.6	AMP DEAMINASE
3.5.4.7	ADP DEAMINASE
3.5.4.8	AMINOIMIDAZOLASE
3.5.4.9	METHYLENETETRAHYDROFOLATE CYCLOHYDROLASE
3.5.4.10	IMP CYCLOHYDROLASE
3.5.4.11	PTERIDIN DEAMINASE
3.5.4.12	dCMP DEAMINASE
3.5.4.13	dCTP DEAMINASE
3.5.4.14	DEOXYCYTIDINE DEAMINASE
3.5.4.16	GTP CYCLOHYDROLASE I
3.5.4.17	ADENOSINE-PHOSPHATE DEAMINASE
3.5.4.18	ATP DEAMINASE
3.5.4.21	CREATININE DEAMINASE
3.5.4.22	1-PYRROLINE-4-HYDROXY-2-CARBOXYLATE DEAMINASE
3.5.4.23	BLASTICIDIN-S-DEAMINASE
3.5.4.24	SEPIAPTERIN DEAMINASE
3.5.4.25	GTP CYCLOHYDROLASE II
3.5.5.1	NITRILASE
3.5.5.2	RICININE NITRILASE
3.5.5.3	CYANATE HYDROLASE
3.5.5.5	ARYLACETONITRILASE
3.5.99.1	RIBOFLAVINASE
3.5.99.2	THIAMINASE
3.6.1.1	INORGANIC PYROPHOSPHATASE
3.6.1.2	TRIMETASPHOSPHATASE
3.6.1.3	ADENOSINETRIPHOSPHATASE
3.6.1.5	APYRASE
3.6.1.6	NUCLEOSIDE-DIPHOSPHATASE
3.6.1.7	ACYLPHOSPHATASE
3.6.1.8	ATP PYROPHOSPHATASE
3.6.1.9	NUCLEOTIDE PYROPHOSPHATASE
3.6.1.10	ENDOPOLYPHOSPHATASE
3.6.1.11	EXOPOLYPHOSPHATASE
3.6.1.12	dCTP PYROPHOSPHATASE
3.6.1.13	ADP-RIBOSE PYROPHOSPHATASE
3.6.1.14	ADENOSINE TETRAPHOSPHATASE
3.6.1.15	NUCLEOSIDE-TRIPHOSPHATASE
3.6.1.16	CDP-GLYCEROL PYROPHOSPHATASE
3.6.1.17	BIS(5′-NUCLEOSYL)-TETRAPHOSPHATASE (ASYMMETRICAL)
3.6.1.18	FAD PYROPHOSPHATASE
3.6.1.19	NUCLEOSIDE-TRIPHOSPHATE PYROPHOSPHATASE
3.6.1.20	5′-ACYLPHOSPHOADENOSINE HYDROLASE
3.6.1.21	ADP SUGAR PYROPHOSPHATASE
3.6.1.22	NAD PYROPHOSPHATASE
3.6.1.23	dUTP PYROPHOSPHATASE
3.6.1.24	NUCLEOSIDE PHOSPHOACYLHYDROLASE
3.6.1.26	CDP-DIACYLGLYCEROLPYROPHOSPHATASE
3.6.1.27	UNDECAPRENYL-DIPHOSPHATASE
3.6.1.28	THIAMIN-TRIPHOSPHATASE
3.6.1.29	BIS(5′-ADENOSYL)-TRIPHOSPHATASE

EC Number	Name
3.6.1.30	PYROPHOSPHATASE(m7G(5′)pppN)
3.6.1.31	PHOSPHORIBOSYL-ATP PYROPHOSPHATASE
3.6.1.32	MYOSIN ATPASE
3.6.1.33	DYNEIN ATPASE
3.6.1.34	H^+-TRANSPORTING ATP SYNTHASE
3.6.1.35	H^+-TRANSPORTING ATPASE
3.6.1.35	H^+-TRANSPORTING ATPASE (V-TYPE)
3.6.1.36	H^+/K^+-EXCHANGING ATPASE
3.6.1.37	Na^+/K^+-EXCHANGING ATPASE
3.6.1.38	Ca^{2+}-TRANSPORTING ATPASE
3.6.1.39	THYMIDINE-TRIPHOSPHATASE
3.6.1.41	BIS(5′-NUCLEOSYL)-TETRAPHOSPHATASE (SYMETRICAL)
3.6.2.1	ADENYLYLSULFATASE
3.6.2.2	PHOSPHOADENYLYLSULFATASE
3.6.4.17	BIS(5′-NUCLEOSYL)-TETRAPHOSPHATASE (ASYMMETRICAL)
3.7.1.1	OXALACETASE
3.7.1.2	FUMARYLACETOACETASE
3.7.1.3	KYNURENINASE
3.7.1.4	PHLORETIN HYDROLASE
3.7.1.6	ACETYLPYRUVATE HYDROLASE
3.8.1.1	ALKYLHALIDASE
3.8.1.2	2-HALOACID DEHALOGENASE
3.8.1.3	HALOACETATE DEHALOGENASE
3.8.1.4	THYROXINE DEIODINASE
3.8.1.4	THYROXINE DEIODINASE 1
3.8.2.1	DIISOPROPYL-FLUOROPHOSPHATASE
3.9.1.1	PHOSPHOAMIDASE
3.10.1.1	N-SULFOGLUCOSAMINE SULFOHYDROLASE
3.10.1.2	CYCLAMATE SULFOHYDROLASE
3.11.1.1	PHOSPHONOACETALDEHYDE HYDROLASE
3.32.1.85	6-PHOSPHO-β-GALACTOSIDASE
4.1.1.1	PYRUVATE DECARBOXYLASE
4.1.1.3	OXALOACETATE DECARBOXYLASE
4.1.1.4	ACETOACETATE DECARBOXYLASE
4.1.1.5	ACETOLACTATE DECARBOXYLASE
4.1.1.6	ACONITATE DECARBOXYLASE
4.1.1.7	BENZOYLFORMATE DECARBOXYLASE
4.1.1.9	MALONYL-CoA DECARBOXYLASE
4.1.1.11	ASPARTATE 1-DECARBOXYLASE
4.1.1.12	ASPARTATE 4-DECARBOXYLASE
4.1.1.14	VALINE DECARBOXYLASE
4.1.1.15	GLUTAMATE DECARBOXYLASE
4.1.1.15	GLUTAMATE DECARBOXYLASE α
4.1.1.16	HYDROXYGLUTAMATE DECARBOXYLASE
4.1.1.17	ORNITHINE DECARBOXYLASE
4.1.1.18	LYSINE DECARBOXYLASE
4.1.1.19	ARGININE DECARBOXYLASE
4.1.1.20	DIAMINOPIMELATE DECARBOXYLASE
4.1.1.22	HISTIDINE DECARBOXYLASE
4.1.1.23	OROTIDINE-5′-PHOSPHATE DECARBOXYLASE
4.1.1.25	TYROSINE DECARBOXYLASE
4.1.1.28	AROMATIC-L-AMINO-ACID DECARBOXYLASE
4.1.1.29	SULFINOALANINE DECARBOXYLASE
4.1.1.31	PHOSPHOENOLPYRUVATE CARBOXYLASE
4.1.1.31	PHOSPHOENOLPYRUVATE DECARBOXYLASE
4.1.1.32	PHOSPHOENOLPYRUVATE CARBOXYKINASE (GTP)

EC Number	Name
4.1.1.33	DIPHOSPHOMEVALONATE DECARBOXYLASE
4.1.1.34	DEHYDRO-L-GULONATE DECARBOXYLASE
4.1.1.35	UDP-GLUCURONATE DECARBOXYLASE
4.1.1.36	PHOSPHOPANTOTHENOYLCYSTEINE DECARBOXYLASE
4.1.1.37	UROPORPHYRINOGEN DECARBOXYLASE
4.1.1.38	PHOSPHOENOLPYRUVATE CARBOXYKINASE (PYROPHOSPHATE)
4.1.1.39	RIBULOSE-BISPHOSPHATE CARBOXYLASE
4.1.1.40	HYDROXYPYRUVATE DECARBOXYLASE
4.1.1.41	METHYLMALONYL-CoA DECARBOXYLASE
4.1.1.45	AMINOCARBOXYMUCONATE-SEMIALDEHYDE DECARBOXYLASE
4.1.1.46	o-PYROCATECHUATE DECARBOXYLASE
4.1.1.48	INDOLE-3-GLYCEROL-PHOSPHATE SYNTHASE
4.1.1.49	PHOSPHOENOLPYRUVATE CARBOXYKINASE (ATP)
4.1.1.50	ADENOSYLMETHIONINE DECARBOXYLASE
4.1.1.52	6-METHYLSALYCILATE DECARBOXYLASE
4.1.1.53	PHENYLALANINE DECARBOXYLASE
4.1.1.55	4,5-DIHYDROOXYPHTHALATE DECARBOXYLASE
4.1.1.57	METHIONINE DECARBOXYLASE
4.1.1.58	ORSELLINATE DECARBOXYLASE
4.1.1.60	STIPITATONATE DECARBOXYLASE
4.1.1.63	PROCATECHUATE DECARBOXYLASE
4.1.1.64	2,2-DIALKYLGLYCINE DECARBOXYLASE (PYRUVATE)
4.1.1.65	PHOSPHATIDYLSERINE DECARBOXYLASE
4.1.2.4	DEOXYRIBOSE-PHOSPHATE ALDOLASE
4.1.2.5	THREONINE ALDOLASE
4.1.2.9	PHOSPHOKETOLASE
4.1.2.10	MANDELONITRILE LYASE
4.1.2.11	HYDROXYMANDELONITRILE LYASE
4.1.2.13	FRUCTOSE-BISPHOSPHATE ALDOLASE
4.1.2.14	2-DEHYDRO-3-DEOXYPHOSPHOGLUCONATE ALDOLASE
4.1.2.15	2-DEHYDRO-3-DEOXYPHOSPHOHEPTONATE ALDOLASE
4.1.2.15	3-DEOXY-D-ARABINO-HEPTULOSONATE-7-PHOSPHATE SYNTHASE
4.1.2.16	2-DEHYDRO-3-DEOXYPHOSPHOOCTONATE ALDOLASE
4.1.2.19	RHAMNULOSE-1-PHOSPHATE ALDOLASE
4.1.2.20	2-DEHYDRO-3-DEOXYGLUCARATE ALDOLASE
4.1.2.21	6-PHOSPHO-2-DEHYDRO-3-DEOXYGALACTONATE ALDOLASE
4.1.2.22	FRUCTOSE-6-PHOSPHATE PHOSPHOKETOLASE
4.1.2.24	DIMETHYLANILINE-N-OXIDE ALDOLASE
4.1.2.25	DIHYDRONEOPTERIN ALDOLASE
4.1.2.26	PHENYLSERINE ALDOLASE
4.1.2.27	SPHINGANINE-1-PHOSPHATE ALDOLASE
4.1.2.30	17α-HYDROXYPROGESTERONE ALDOLASE
4.1.2.32	TRIMETHYLAMINE-OXIDE ALDOLASE
4.1.3.1	ISOCITRATE LYASE
4.1.3.2	MALATE SYNTHASE
4.1.3.3	N-ACETYLNEURAMINATE LYASE
4.1.3.4	HYDROXYMETHYLGLUTARYL-CoA LYASE
4.1.3.5	HYDROXYMETHYLGLUTARYL-CoA SYNTHASE
4.1.3.6	CITRATE (pro-3s)-LYASE
4.1.3.7	CITRATE (si)-SYNTHASE
4.1.3.8	ATP CITRATE (pro-S)-LYASE
4.1.3.11	PROPYLMALATE SYNTHASE
4.1.3.12	2-ISOPROPYLMALATE SYNTHASE
4.1.3.15	2-HYDROXY-3-OXOADIPATE SYNTHASE
4.1.3.16	4-HYDROXY-2-OXOGLUTARATE ALDOLASE
4.1.3.17	4-HYDROXY-4-METHYL-2-OXOGLUTARATE ALDOLASE

EC Number	Name
4.1.3.18	ACETOLACTATE SYNTHASE
4.1.3.19	N-ACETYLNEURAMINATE SYNTHASE
4.1.3.20	N-ACETYLNEURAMINATE-9-PHOSPHATE SYNTHASE
4.1.3.21	HOMOCITRATE SYNTHASE
4.1.3.22	CITRAMALATE SYNTHASE
4.1.3.23	DECYLCITRATE SYNTHASE
4.1.3.24	MALYL-CoA LYASE
4.1.3.25	CITRAMALYL-CoA LYASE
4.1.3.26	3-HYDROXY-3-ISOHEXENYLGLUTARYL-CoA LYASE
4.1.3.27	ANTHRANILATE SYNTHASE
4.1.3.28	CITRATE (re)-SYNTHASE
4.1.3.29	DECYLHOMOCITRATE SYNTHASE
4.1.3.30	METHYLISOCITRATE LYASE
4.1.3.31	2-METHYLCITRATE LYASE
4.1.3.32	2,3-DIMETHYLMALATE LYASE
4.1.3.33	2-ETHYLMALATE SYNTHASE
4.1.99.1	TRYPTOPHANASE
4.1.99.2	TYROSINE PHENOL-LYASE
4.1.99.3	DEOXYRIBOPYRIMIDINE PHOTO-LYASE
4.1.99.4	1-AMINOCYCLOPROPANE-1-CARBOXYLATE DEAMINASE
4.1.99.5	OCTADECANAL DECARBOXYLASE
4.1.99.6	TRICHODIENE SYNTHASE
4.2.1.1	CARBONATE DEHYDRATASE
4.2.1.2	FUMARATE HYDRATASE
4.2.1.3	ACONITATE HYDRATASE
4.2.1.3	ACONITATE HYDRATASE m
4.2.1.6	GALACTONATE DEHYDRATASE
4.2.1.7	ALTRONATE DEHYDRATASE
4.2.1.8	MANNONATE DEHYDRATASE
4.2.1.9	DIHYDROXY ACID DEHYDRATASE
4.2.1.10	3-DEHYDROQUINATE DEHYDRATASE
4.2.1.11	PHOSPHOPYRUVATE HYDRATASE
4.2.1.12	PHOSPHOGLUCONATE DEHYDRATASE
4.2.1.13	L-SERINE DEHYDRATASE
4.2.1.14	D-SERINE DEHYDRATASE
4.2.1.16	THREONINE DEHYDRATASE
4.2.1.17	ENOYL-CoA HYDRATASE
4.2.1.17	ENOYL-CoA HYDRATASE (LONG CHAIN)
4.2.1.17	ENOYL-CoA HYDRATASE (SHORT CHAIN)
4.2.1.18	METHYLGLUTACONYL-CoA HYDRATASE
4.2.1.19	IMIDAZOLEGLYCEROL-PHOSPHATE DEHYDRATASE
4.2.1.20	TRYPTOPHAN SYNTHASE
4.2.1.22	CYSTATHIONINE β-SYNTHASE
4.2.1.24	PORPHOBILINOGEN SYNTHASE
4.2.1.26	AMINODEOXYGLUCONATE DEHYDRATASE
4.2.1.27	MALONATE-SEMIALDEHYDE DEHYDRATASE
4.2.1.28	PROPANEDIOL DEHYDRATASE
4.2.1.29	INDOLEACETALDOXIME DEHYDRATASE
4.2.1.30	GLYCEROL DEHYDRATASE
4.2.1.31	MALEATE DEHYDRATASE
4.2.1.32	L(+)-TARTRATE DEHYDRATASE
4.2.1.33	3-ISOPROPYLMALATE DEHYDRATASE
4.2.1.34	(S)-2-METHYLMALATE DEHYDRATASE
4.2.1.35	(R)-2-METHYLMALATE DEHYDRATASE
4.2.1.36	HOMOACONITATE HYDRATASE
4.2.1.38	erythro-3-HYDROXYASPARTATE DEHYDRATASE

EC Number	Name
4.2.1.39	GLUCONATE DEHYDRATASE
4.2.1.40	GLUCARATE DEHYDRATASE
4.2.1.43	2-DEHYDRO-3-DEOXY-L-ARABINONATE DEHYDRATASE
4.2.1.45	CDP-GLUCOSE 4,6-DEHYDRATASE
4.2.1.46	dTDPGLUCOSE 4,6-DEHYDRATASE
4.2.1.47	GDP-MANNOSE 4,6-DEHYDRATASE
4.2.1.49	UROCANATE HYDRATASE
4.2.1.50	PYRAZOLYLALANINE SYNTHASE
4.2.1.51	PREPHENATE DEHYDRATASE
4.2.1.52	DIHYDRODIPICOLINATE SYNTHASE
4.2.1.58	CROTONYL-[ACYL-CARRIER-PROTEIN] HYDRATASE
4.2.1.60	3-HYDROXYDECANOYL-[ACYL-CARRIER-PROTEIN] DEHYDRATASE
4.2.1.66	CYANIDE HYDRATASE
4.2.1.67	D-FUCONATE DEHYDRATASE
4.2.1.68	L-FUCONATE DEHYDRATASE
4.2.1.71	ACETYLENECARBOXYLATE HYDRATASE
4.2.1.72	ACETYLENEDICARBOXYLATE HYDRATASE
4.2.1.73	PROTAPHIN-AGLUCONE DEHYDRATASE CYCLIZING)
4.2.1.74	LONG-CHAIN-ENOYL-CoA HYDRATASE
4.2.1.75	UROPORPHYRINOGEN-III SYNTHASE
4.2.1.79	2-METHYLCITRATE SYNTHASE
4.2.1.84	NITRILE HYDRATASE
4.2.1.85	DIMETHYLMALEATE HYDRATASE
4.2.1.94	SCYTALONE DEHYDRATASE
4.2.2.1	HYALURONATE LYASE
4.2.2.2	PECTATE LYASE
4.2.2.3	ALGINATE LYASE
4.2.2.4	CHONDROITIN ABC LYASE
4.2.2.5	CHONDROITIN AC LYASE
4.2.2.7	HEPARIN LYASE
4.2.2.8	HEPARITIN-SULFATE LYASE
4.2.2.9	EXOPOLYGALACTURONATE LYASE
4.2.99.2	THREONINE SYNTHASE
4.2.99.7	ETHANOLAMINE-PHOSPHATE PHOSPHO-LYASE
4.2.99.8	O-ACETYLSERINE (THIOL)-LYASE
4.2.99.9	O-SUCCINYLHOMOSERINE (THIOL)-LYASE
4.2.99.10	O-ACETYLHOMOSERINE (THIOL)-LYASE
4.2.99.11	METHYLGLYOXAL SYNTHASE
4.2.99.13	β-(9-CYTOKININ)-ALANINE SYNTHASE
4.3.1.1	ASPARTATE AMMONIA-LYASE
4.3.1.2	METHYLASPARTATE AMMONIA-LYASE
4.3.1.3	HISTIDINE AMMONIA-LYASE
4.3.1.4	FORMIMINOTETRAHYDROFOLATE CYCLODEAMINASE
4.3.1.5	PHENYLALANINE AMMONIA-LYASE
4.3.1.7	ETHANOLAMIN AMMONIA-LYASE
4.3.1.8	HYDROXYMETHYLBILANE SYNTHASE
4.3.1.9	GLUCOSAMINATE AMMONIA-LYASE
4.3.1.10	SERINE-SULFATE AMMONIA-LYASE
4.3.1.13	CARBAMOYL-SERINE AMMONIA-LYASE
4.3.2.1	ARGININOSUCCINATE LYASE
4.3.2.2	ADENYLOSUCCINATE LYASE
4.3.2.3	UREIDOGLYCOLATE LYASE
4.3.2.5	PEPTIDYLAMIDOGLYCOLATE LYASE
4.4.1.1	CYSTATHIONINE γ-LYASE
4.4.1.2	HOMOCYSTEINE DESULFHYDRASE
4.4.1.4	ALLIIN LYASE

EC Number	Name
4.4.1.5	LACTOYLGLUTATHIONE LYASE
4.4.1.8	CYSTATHIONINE β-LYASE
4.4.1.9	L-3-CYANOALANINE SYNTHASE
4.4.1.10	CYSTEINE LYASE
4.4.1.11	METHIONINE γ-LYASE
4.4.1.12	SULFOACETALDEHYDE LYASE
4.4.1.13	CYSTEINE-CONJUGATE β-LYASE
4.4.1.14	1-AMINOCYCLOPROPANE-1-CARBOXYLATE SYNTHASE
4.5.1.1	DDT-DEHYDROCHLORINASE
4.5.1.2	3-CHLORO-D-ALANINE DEHYDROCHLORINASE
4.5.1.3	DICHLOROMETHANE DEHALOGENASE
4.6.1.1	ADENYLATE CYCLASE
4.6.1.2	GUANYLATE CYCLASE
4.6.1.3	3-DEHYDROQUINATE SYNTHASE
4.6.1.4	CHORISMATE SYNTHASE
4.99.1.1	FERROCHELATASE
4.99.1.2	ALKYLMERCURY LYASE
5.1.1.1	ALANINE RACEMASE
5.1.1.3	GLUTAMATE RACEMASE
5.1.1.4	PROLINE RACEMASE
5.1.1.5	LYSINE RACEMASE
5.1.1.7	DIAMINOPIMELATE EPIMERASE
5.1.1.8	4-HYDROXYPROLINE EPIMERASE
5.1.1.9	ARGININE RACEMASE
5.1.1.10	AMINO-ACID RACEMASE
5.1.1.11	PHENYLALANINE RACEMASE (ATP-HYDROLYSING)
5.1.1.13	ASPARTATE RACEMASE
5.1.2.1	LACTATE RACEMASE
5.1.2.2	MANDELATE RACEMASE
5.1.2.3	3-HYDROXYBUTYRYL-CoA EPIMERASE
5.1.3.1	RIBULOSE-PHOSPHATE EPIMERASE
5.1.3.2	UDP-GLUCOSE 4-EPIMERASE
5.1.3.2	UDP-GLUCOSE 4-EPIMERASE B
5.1.3.3	ALDOSE 1-EPIMERASE
5.1.3.5	UDPARABINOSE 4-EPIMERASE
5.1.3.6	UDP-GLUCURNATE 4-EPIMERASE
5.1.3.7	UDP-N-ACETYLGLUCOSAMINE 4-EPIMERASE
5.1.3.8	N-ACETYLGLUCOSAMINE 2-EPIMERASE
5.1.3.12	UDP-GLUCURONATE 5′-EPIMERASE
5.1.3.14	UDP-N-ACETYLGLUCOSAMINE 2-EPIMERASE
5.1.3.15	GLUCOSE-6-PHOSPHATE 1-EPIMERASE
5.1.3.17	HEPAROSAN-N-SULFATE-GLUCURONATE 5-EPIMERASE
5.1.3.18	GDP-MANNOSE 3,5-EPIMERASE
5.1.3.19	CHONDROITIN-GLUCURONATE 5-EPIMERASE
5.1.99.1	METHYLMALONYL-CoA EPIMERASE
5.1.99.3	ALLANTOIN RACEMASE
5.2.1.1	MALEATE ISOMERASE
5.2.1.2	MALEYLACETOACETATE ISOMERASE
5.2.1.3	RETINAL ISOMERASE
5.2.1.4	MALEYLPYRUVATE ISOMERASE
5.2.1.5	LINOLEATE ISOMERASE
5.2.1.7	RETINOL ISOMERASE
5.2.1.8	PEPTIDYLPROLYL ISOMERASE
5.3.1.1	TRIOSE-PHOSPHATE ISOMERASE
5.3.1.2	UDP-GLUCOSE 4-EPIMERASE
5.3.1.3	ARABINOSE ISOMERASE

EC Number	Name
5.3.1.4	L-ARABINOSE ISOMERASE
5.3.1.5	XYLOSE ISOMERASE
5.3.1.6	RIBOSE-5-PHOSPHATE ISOMERASE
5.3.1.7	MANNOSE ISOMERASE
5.3.1.8	MANNOSE-6-PHOSPHATE ISOMERASE
5.3.1.9	GLUCOSE-6-PHOSPHATE ISOMERASE
5.3.1.10	GLUCOSAMINE-6-PHOSPHATE ISOMERASE
5.3.1.12	GLUCURONATE ISOMERASE
5.3.1.13	ARABINOSE-5-PHOSPHATE ISOMERASE
5.3.1.14	L-RHAMNOSE ISOMERASE
5.3.1.15	D-LYXOSE KETOL ISOMERASE
5.3.1.20	RIBOSE ISOMERASE
5.3.1.21	CORTICOSTEROID SIDE-CHAIN-ISOMERASE
5.3.1.22	HYDROXYPYRUVATE ISOMERASE
5.3.2.1	PHENYLPYRUVATE TAUTOMERASE
5.3.2.2	OXALACETATE TAUTOMERASE
5.3.2.3	DOPACHROME TAUTOMERASE
5.3.3.1	STEROID Δ-ISOMERASE
5.3.3.2	ISOPENTENYL-DIPHOSPHATE Δ-ISOMERASE
5.3.3.3	VINYLACETYL-CAO Δ-ISOMERASE
5.3.3.5	CHOLESTENOL Δ-ISOMERASE
5.3.3.6	METHYLITACONATE Δ-ISOMERASE
5.3.3.8	DODECENOYL-CoA Δ-ISOMERASE
5.3.3.9	PROSTAGLANDIN-A1 Δ-ISOMERASE
5.3.3.11	ISOPIPERITENONE Δ-ISOMERASE
5.3.4.1	PROTEIN DISULFIDE ISOMERASE
5.3.99.1	HYDROPEROXIDE ISOMERASE
5.3.99.2	PROSTAGLANDIN-H2 D-ISOMERASE
5.3.99.3	PROSTAGLANDIN-H2 E-ISOMERASE
5.3.99.4	PROSTACYCLIN SYNTHASE
5.3.99.4	PROSTAGLANDIN-I SYNTHASE
5.3.99.5	THROMBOXANE A SYNTHASE
5.3.99.6	ALLENE OXIDE CYCLASE
5.4.1.1	LYSOLECITIN ACYLMUTASE
5.4.2.1	PHOSPHOGLYCERATE MUTASE
5.4.2.1	PHOSPHOGLYCERATE MUTASE (COFACTOR DEPENDENT)
5.4.2.2	PHOSPHOGLUCOMUTASE
5.4.2.3	PHOSPHOACETYLGLUCOSAMINE MUTASE
5.4.2.4	BISPHOSPHOGLYCERATE MUTASE
5.4.2.5	PHOSPHOGLUCOMUTASE (GLUCOSE COFACTOR)
5.4.2.6	β-PHOSPHOGLUCOMUTASE
5.4.2.7	PHOSPHOPENTOMUTASE
5.4.2.8	PHOSPHOMANNOMUTASE
5.4.3.2	LYSINE 2,3-AMINOMUTASE
5.4.3.3	β-LYSINE 5,6-AMINOMUTASE
5.4.3.4	D-LYSINE 5,6-AMINOMUTASE
5.4.3.5	D-ORNITHINE 4,5-AMINOMUTASE
5.4.3.6	TYROSINE 2,3-AMINOMUTASE
5.4.3.8	GLUTAMATE-1-SEMIALDEHYDE 2,1-AMINOMUTASE
5.4.99.2	METHYLMALONYL-CoA MUTASE
5.4.99.4	2-METHYLENEGLUTARATE MUTASE
5.4.99.5	CHORISMATE MUTASE
5.4.99.6	ISOCHORISMATE SYNTHASE
5.4.99.7	LANOSTEROL SYNTHASE
5.4.99.8	CYCLOARTENOL SYNTHASE
5.5.1.1	MUCONATE CYCLOISOMERASE

EC Number	Name
5.5.1.2	3-CARBOXY-cis,cis-MUCONATE CYCLOISOMERASE
5.5.1.3	TETRAHYDROXYPTERIDINE CYCLOISOMERASE
5.5.1.4	myo-INOSITOL-1-PHOSPHATE SYNTHASE
5.5.1.5	CARBOXY-cis,cis-MUCONATE CYCLASE
5.5.1.6	CHALCONE ISOMERASE
5.5.1.7	CHLOROMUCONATE CYCLOISOMERASE
5.5.1.8	GERANYL-DIPHOSPHATE CYCLASE
5.5.1.9	CYCLOEUCALENOL CYCLO-ISOMERASE
5.9.1.31	DNA TOPOISOMERASE (ATP-HYDROLYSING)
5.99.1.2	DNA TOPOISOMERASE
5.99.1.3	DNA GYRASE
5.99.1.3	DNA TOPOISOMERASE (ATP-HYDROLYSING)
6.1.1.1	TYROSINE-tRNA LIGASE
6.1.1.2	TRYPTOPHAN-tRNA LIGASE
6.1.1.3	THREONINE-tRNA LIGASE
6.1.1.4	LEUCINE-tRNA LIGASE
6.1.1.5	ISOLEUCINE-tRNA LIGASE
6.1.1.6	LYSINE-tRNA LIGASE
6.1.1.7	ALANINE-tRNA LIGASE
6.1.1.9	VALINE-tRNA LIGASE
6.1.1.10	METHIONINE-tRNA LIGASE
6.1.1.11	SERINE-tRNA LIGASE
6.1.1.12	ASPARTATE-tRNA LIGASE
6.1.1.13	D-ALANINE-POLY(PHOSPHORIBITOL)LIGASE
6.1.1.14	GLYCINE-tRNA LIGASE
6.1.1.15	PROLINE-tRNA LIGASE
6.1.1.16	CYSTEINE-tRNA LIGASE
6.1.1.17	GLUTAMATE-tRNA LIGASE
6.1.1.18	GLUTAMINE-tRNA LIGASE
6.1.1.19	ARGININE-tRNA LIGASE
6.1.1.20	PHENYLALANINE-tRNA LIGASE
6.1.1.21	HISTIDINE-tRNA LIGASE
6.1.1.22	ASPARAGINE-tRNA LIGASE
6.2.1.1	ACETATE-CoA LIGASE
6.2.1.2	BUTYRATE-CoA LIGASE
6.2.1.3	LONG-CHAIN-FATTY-ACID CoA-LIGASE
6.2.1.4	SUCCINATE-CoA LIGASE (GDP-FORMING)
6.2.1.5	SUCCINATE-CoA LIGASE (ADP-FORMING)
6.2.1.7	CHOLATE-CoA LIGASE
6.2.1.8	OXALATE-CoA LIGASE
6.2.1.9	MALATE-CoA LIGASE
6.2.1.10	ACID-CoA LIGASE (GDP-FORMING)
6.2.1.11	BIOTIN-CoA LIGASE
6.2.1.14	6-CARBOXYHEXANOATE-CoA LIGASE
6.2.1.15	ARACHIDONATE-CoA LIGASE
6.2.1.16	ACETOACETATE-CoA LIGASE
6.2.1.17	PROPIONATE-CoA LIGASE
6.2.1.21	PHENYLACETATE-CoA LIGASE
6.2.1.23	DICARBOXYLATE-CoA LIGASE
6.2.1.24	PHYTANATE-CoA LIGASE
6.2.1.121	4-COUMARATE-CoA LIGASE
6.3.1.1	ASPARTATE-AMMONIA LIGASE
6.3.1.2	GLUTAMATE-AMMONIA LIGASE
6.3.1.4	ASPARAGINE-AMMONIA LIGASE (ADP-FORMING)
6.3.1.5	NAD SYNTHASE
6.3.1.6	GLUTAMATE-ETHYLAMINE LIGASE

EC Number	Name
6.3.1.8	GLUTATHIONYLSPERMIDINE SYNTHASE
6.3.2.1	PANTOATE-β-ALANINE LIGASE
6.3.2.2	GLUTAMATE-CYSTEINE LIGASE
6.3.2.3	GLUTATHIONE SYNTHASE
6.3.2.4	D-ALANINE-D-ALANINE LIGASE
6.3.2.5	PHOSPHOPANTOTHENATE-CYSTEINE LIGASE
6.3.2.6	PHOSPHORIBOSYLAMINOIMIDAZOLE-SUCCINOCARBOXAMIDE SYNTHASE
6.3.2.8	UDP-N-ACETYLMURAMATE-ALANINE LIGASE
6.3.2.11	CARBOSINE SYNTHASE
6.3.2.12	DIHYDROFOLATE SYNTHASE
6.3.2.13	UDP-N-ACETYLMURAMOYL-D-GLUTAMATE-2,6-DIAMINOPIMELATE LIGASE
6.3.2.16	D-ALANINE-ALANYL-POLY(GLYCEROPHOSPHATE)LIGASE
6.3.2.17	FOLYLPOLYGLUTAMATE SYNTHASE
6.3.2.18	γ-GLUTAMYLHISTAMINE SYNTHASE
6.3.2.19	UBIQUITIN-PROTEIN LIGASE
6.3.2.24	TYROSINE-ARGININE LIGASE
6.3.3.2	5-FORMYLTETRAHYDROFOLATE CYCLO-LIGASE
6.3.3.3	DETHIOBIN SYNTHASE
6.3.4.1	GMP SYNTHASE
6.3.4.2	CTP SYNTHASE
6.3.4.3	FORMATE-TETRAHYDROFOLATE LIGASE
6.3.4.4	ADENYLOSUCCINATE SYNTHASE
6.3.4.5	ARGININOSUCCINATE SYNTHASE
6.3.4.6	UREA CARBOXYLASE
6.3.4.7	RIBOSE-5-PHOSPHATE-AMMONIA LIGASE
6.3.4.9	BIOTIN-[METHYLMALONYL-CoA-CARBOXYLTRANSFERASE] LIGASE
6.3.4.10	BIOTIN-[PROPIONYL-CoA-CARBOXYLASE(ATP-HYDROLYSING)] LIGASE
6.3.4.11	BIOTIN-[METHYLCROTONYL-CoA-CARBOXYLASE] LIGASE
6.3.4.13	PHOSPHORIBOSYLAMINE-GLYCINE LIGASE
6.3.4.14	BIOTIN CARBOXYLASE
6.3.4.16	CARBAMOYL-PHOSPHATE SYNTHASE (AMMONIA)
6.3.5.1	NAD SYNTHASE (GLUTAMINE-HYDROLYSING)
6.3.5.2	GMP SYNTHASE (GLUTAMINE-HYDROLYSING)
6.3.5.3	PHOSPHORIBOSYLFORMYLGLYCINAMIDINE SYNTHASE
6.3.5.4	ASPARAGINE SYNTHASE (GLUTAMINE-HYDROLYSING)
6.3.5.5	CARBAMOYL-PHOSPHATE SYNTHASE (GLUTAMINE-HYDROLYSING)
6.4.1.1	PYRUVATE CARBOXYLASE
6.4.1.2	ACETYL-CoA CARBOXYLASE
6.4.1.3	PROPIONYL-CoA CARBOXYLASE
6.4.1.4	METHYLCROTONYL-CoA CARBOXYLASE
6.4.1.5	GERANOYL-CoA CARBOXYLASE
6.5.1.1	POLYDEOXYRIBONUCLEOTIDE SYNTHASE (ATP)
6.5.1.3	POLYRIBONUCLEOTIDE SYNTHASE (ATP)